NFPA 70

National Electrical Code®

1999 Edition

This edition of NFPA 70, *National Electrical Code,* was prepared by the National Electrical Code Committee and acted on by the National Fire Protection Association, Inc., at its Annual Meeting held May 18–21, 1998, in Cincinnati, OH. It was issued by the Standards Council on July 16, 1998, with an effective date of August 5, 1998, and supersedes all previous editions.

This edition of NFPA 70 was approved as an American National Standard on August 6, 1998.

History and Development of the *National Electrical Code*

The National Fire Protection Association has acted as sponsor of the *National Electrical Code* since 1911. The original *Code* document was developed in 1897 as a result of the united efforts of various insurance, electrical, architectural, and allied interests.

In accordance with the provisions of the NFPA Regulations Governing Committee Projects, a National Electrical Code Committee Report on Proposals containing proposed amendments to the 1996 *National Electrical Code* was published by the NFPA in June 1997. This report recorded the actions of the various Code-Making Panels and the Correlating Committee of the National Electrical Code Committee on each proposal that had been made to revise the 1996 *Code*. The report was circulated to all members of the National Electrical Code Committee and was made available to other interested NFPA members and to the public for review and comment. Following the close of the public comment period, the Code-Making Panels met, acted on each comment, and reported their action to the Correlating Committee. The NFPA published the National Electrical Code Committee Report on Comments in April 1998, which recorded the actions of the Code-Making Panels and the Correlating Committee on each public comment to the National Electrical Code Committee Report on Proposals. The National Electrical Code Committee Report on Proposals and the National Electrical Code Committee Report on Comments were presented to the 1998 NFPA Annual Meeting for adoption.

NFPA has an Electrical Section that provides particular opportunity for NFPA members interested in electrical safety to become better informed and to contribute to the development of the *National Electrical Code* and other NFPA electrical standards. Each of the Code-Making Panels and the Chairman of the Correlating Committee reported their recommendations to meetings of the Electrical Section at the 1998 NFPA Annual Meeting. The Electrical Section thus had opportunity to discuss and review the report of the National Electrical Code Committee prior to the adoption of this edition of the *Code* by the Association.

This 1999 edition supersedes all other previous editions, supplements, and printings dated 1897, 1899, 1901, 1903, 1904, 1905, 1907, 1909, 1911, 1913, 1915, 1918, 1920, 1923, 1925, 1926, 1928, 1930, 1931, 1933, 1935, 1937, 1940, 1942, 1943, 1947, 1949, 1951, 1953, 1954, 1955, 1956, 1957, 1958, 1959, 1962, 1965, 1968, 1971, 1975, 1978, 1981, 1984, 1987, 1990, 1993, and 1996.

Changes in this 1999 edition of the *National Electrical Code* (as compared with the 1996 edition) are indicated by vertical lines in the margin.

The location (in the 1996 edition) of material not appearing in the 1999 edition, and not identified as a change by a vertical line, is identified by a bullet (•) in the margin. Changes in section and table numbers are not identified.

Material identified by the superscript letter "x" includes text extracted from other NFPA documents as identified in Appendix A.

This *Code* is purely advisory as far as NFPA and ANSI are concerned, but is offered for use in law and for regulatory purposes in the interest of life and property protection. Anyone noticing any errors should notify the Secretary of the National Electrical Code Committee at the NFPA Executive Office.

CONTENTS

Chapter 4 Equipment for General Use

NATIONAL ELECTRICAL CODE COMMITTEE

These lists represent the membership at the time each Committee was balloted on the text of this edition. Since that time, changes in the membership may have occurred. A key to classifications is found at the back of this document.

Technical Correlating Committee

D. Harold Ware, *Chair* [IM]
Libra Electric Co., OK
Rep. National Electrical Contractors Association

Mark W. Earley, *Secretary*
National Fire Protection Association
(nonvoting)

Jean A. O'Connor, *Recording Secretary*
National Fire Protection Association
(nonvoting)

James E. Brunssen, Bell Communications Research (Bellcore), NJ [UT]
Rep. Alliance for Telecommunications Industry Solutions
Philip H. Cox, Int'l Association of Electrical Inspectors, TX [E]
Rep. Int'l Association of Electrical Inspectors
Dale R. Deming, AM Electric Co. (T&B), MI [M]
Rep. National Electrical Manufacturers Association
William R. Drake, Marinco, CA [M]
Paul Duks, Underwriters Laboratories Inc., IL [RT]
Joseph E. Pipkin, U.S. Dept. of Labor OSHA, Washington, DC [E]
John W. Troglia, Edison Electric Institute, WI [UT]
Rep. Edison Electric Institute
John P. Widener, Jr., IBEW, Washington DC [L]
Rep. Int'l Brotherhood of Electrical Workers

Edward C. Lawry, Madison, WI [E]
(Alt. to P. H. Cox)
W. Terry Lindsay, Duncan Electric Co. Inc., TN [IM]
(Alt. to D. H. Ware)
Keku M. Mistry, Bell Communications Research (Bellcore), NJ [RT]
(Alt. to J. E. Brunssen)
James T. Pauley, Square D Co., KY [M]
(Alt. to D. R. Deming)
Robert J. Pollock, Underwriters Laboratories Inc., IL [RT]
(Alt. to P. Duks)
Craig M. Wellman, DuPont Engineering, DE [U]
(Voting Alt. to CMA Rep.)

Alternates

David P. Brown, Baltimore Gas & Electric Co., MD [UT]
(Alt. to J. W. Troglia)
Douglas A. Fisher, IBEW, Washington, DC [L]
(Alt. to J. P. Widener, Jr.)

Nonvoting

Richard G. Biermann, Biermann Electric Co. Inc., IA
(Member Emeritus)

CODE-MAKING PANEL NO. 1
Articles 90, 100, 110

Philip H. Cox, *Acting Chair* [E]
Int'l Association of Electrical Inspectors, TX
Rep. Int'l Association of Electrical Inspectors

Richard G. Biermann, Biermann Electric Co. Inc., IA [IM]
Rep. National Electrical Contractors Association
H. Landis Floyd, II, The DuPont Co., DE [U]
Rep. Institute of Electrical & Electronics Engineers, Inc.
Don B. Ivory, Idaho Electrical JATC, ID [L]
Rep. Int'l Brotherhood of Electrical Workers
John D. Minick, National Electrical Manufacturers Association, TX [M]
Rep. National Electrical Manufacturers Association
Lyman L. Parks, Canadensis, PA [UT]
Rep. Alliance for Telecommunications Industry Solutions/Bellcore
Charles L. Roach, Eastman Chemical Co., TN [U]
Rep. Chemical Manufacturers Association
Steven G. Roll, Intertek Testing Services NA, Inc., MA [RT]
John W. Troglia, Edison Electric Institute, WI [UT]
Rep. Edison Electric Institute
Robert A. Williams, Underwriters Laboratories Inc., NC [RT]
Michael P. Wissman, Northern KY Electric Service, KY [IM]
Rep. Independent Electrical Contractors

Paul Duks, Underwriters Laboratories Inc., IL [RT]
(Alt. to R. A. Williams)
James William Evans, Southport Electrical Inc., NC [IM]
(Alt. to M. P. Wissman)
Russell J. Helmick, Jr., City of Irvine, CA [E]
(Alt. to P. H. Cox)
Donald H. McCullough II, Westinghouse Savannah River Co., SC [U]
(Alt. to H. L. Floyd, II)
Deborah A. McDaniel, ICI Americas, DE [U]
(Alt. to C. L. Roach)
Larry Miller, National Electrical Manufacturers Association, VA [M]
(Alt. to J. D. Minick)
Keku M. Mistry, Bell Communications Research (Bellcore), NJ [UT]
(Alt. to L. L. Parks)
Ricky L. Oakland, IBEW, WY [L]
(Alt. to D. B. Ivory)
H. Brooke Stauffer, National Electrical Contractors Association, MD [IM]
(Alt. to R. G. Biermann)

Alternates

David P. Brown, Baltimore Gas & Electric Co., MD [UT]
(Alt. to J. W. Troglia)

CODE-MAKING PANEL NO. 2

Articles 210, 215, 220, Appendix D Examples D1 through D6

James W. Carpenter, *Chair* [E]
North Carolina Dept. of Insurance, NC
Rep. Int'l Association of Electrical Inspectors

Richard W. Becker, Engineered Electrical Systems, Inc., WA [U]
  Rep. Institute of Electrical & Electronics Engineers, Inc.
Thomas L. Harman, University of Houston - Clear Lake, TX [SE]
Billy H. McClendon, PPG Industries, Inc., LA [U]
  Rep. Chemical Manufacturers Association
Bernard Mericle, IBEW, NY [L]
  Rep. Int'l Brotherhood of Electrical Workers
Donald A. Nissen, Underwriters Laboratories Inc., IL [RT]
Gary J. O'Neil, Southern California Edison Co., CA [UT]
  Rep. Edison Electric Institute
Earl W. Roberts, REPTEC, CT [M]
  Rep. National Electrical Manufacturers Association
Albert F. Sidhom, U.S. Army Corps of Engineers, CA [U]
Michael D. Toman, MEGA Power Electrical Services, Inc., MD [IM]
  Rep. National Electrical Contractors Association
Robert G. Wilkinson, Independent Electrical Contractors of Houston, Inc., TX [IM]
  Rep. Independent Electrical Contractors

Alternates

Ernest S. Broome, City of Knoxville, TN [E]
  (Alt. to J. W. Carpenter)
Thomas J. Castor, American Electric Power Service Corp., OH [UT]
  (Alt. to G. J. O'Neil)
Ernie Howell, IEC, Rocky Mountain Chapter, CO [IM]
  (Alt. to R. G. Wilkinson)
James R. Jones, University of Alabama at Birmingham, AL [U]
  (Alt. to R. W. Becker)
Charles D. Marshall, Jr., IBEW Local 948, MI [L]
  (Alt. to B. Mericle)
James T. Pauley, Square D Co., KY [M]
  (Alt. to E. W. Roberts)
Joseph Patterson Roché, Celanese Acetate, SC [U]
  (Alt. to B. H. McClendon)
J. Morris Trimmer, University of Florida, FL [SE]
  (Alt. to T. L. Harman)

CODE-MAKING PANEL NO. 3

Articles 300, 305, 690

Raymond W. Weber, *Chair* [E]
Dept. of Commerce, WI
Rep. Int'l Association of Electrical Inspectors

Joseph J. Andrews, Westinghouse Savannah River Co., SC [U]
  Rep. Institute of Electrical & Electronics Engineers, Inc.
Charles W. Beile, Allied Tube & Conduit/Tyco, IL [M]
  Rep. National Electrical Manufacturers Association
Ward I. Bower, Sandia National Laboratories, NM [U]
  Rep. Solar Energy Industries Association (VL 690)
Michael I. Callanan, National Joint Apprentice & Training Committee, PA [L]
  Rep. Int'l Brotherhood of Electrical Workers
C. Wiley Cruise, Duke Power Co., NC [UT]
  Rep. Edison Electric Institute
Paul Duks, Underwriters Laboratories Inc., IL [RT]
Charles E. Jackson, Hoechst Celanese Chemical Group, TX [U]
  Rep. Chemical Manufacturers Association
Richard A. Pruzick, PC Electrical Construction, Inc., NJ [IM]
  Rep. National Electrical Contractors Association
Edwin K. Rodkey, Carlon Electrical Products, A Lamson & Sessions Co., OH [M]
  Rep. Society of the Plastics Industry, Inc.

Jack A. Gruber, Wheatland Tube Co., PA [M]
  (Voting Alt. to AISI Rep.)
Dennis B. Horman, PacifiCorp (Utah Power/Pacific Power), UT [UT]
  (Alt. to C. W. Cruise)
David H. Kendall, Carlon Electrical Products, A Lamson & Sessions Co., OH [M]
  (Alt. to E. K. Rodkey)
Allen F. KnicKrehm, The KnicKrehm Co., CA [IM]
  (Alt. to R. A. Pruzick)
Richard P. Owen, City of St. Paul, MN [E]
  (Alt. to R. W. Weber)
David A. Pace, Olin Corporation, AL [U]
  (Alt. to C. E. Jackson)
Ken A. Rogers, Thomas & Betts Corp., TN [M]
  (Alt. to C. W. Beile)
Melvin K. Sanders, Things Electrical Co., Inc., (TECo., Inc) IA [U]
  (Alt. to J. J. Andrews)
Robert H. Wills, Advanced Energy Systems Inc., NH [U]
  (Alt. to W. I. Bower) (VL 690)

Alternates

Joseph V. Egan, IBEW Local 456, NJ [L]
  (Alt. to M. I. Callanan)

CODE-MAKING PANEL NO. 4

Articles 225, 230

C. John Beck, *Chair* [UT]
Pacific Gas and Electric Co., CA
Rep. Edison Electric Institute

Malcolm Allison, Gould Shawmut, MA [M]
K. W. Carrick, Electrical Engineering Professional Services, Inc., MS [U]
  Rep. Institute of Electrical & Electronics Engineers, Inc.
E. T. Dickerson, University of Houston - Clear Lake, TX [IM]
  Rep. Independent Electrical Contractors

Floyd C. Ferris, New York Board of Fire Underwriters, NY [E]
Howard D. Hughes, Hughes Electric Co. Inc., AR [IM]
  Rep. National Electrical Contractors Association
James Lee Hunter, Southwire Co., GA [M]
  Rep. The Aluminum Association

William M. Lewis, Eli Lilly & Co., IN [U]
   Rep. Chemical Manufacturers Association
Robert J. Pollock, Underwriters Laboratories Inc., IL [RT]
Donald R. Strassburg, State of Washington, WA [E]
   Rep. Int'l Association of Electrical Inspectors
Mark H. Sumrall, IBEW Local 527, TX [L]
   Rep. Int'l Brotherhood of Electrical Workers
John W. Young, Siemens Energy & Automation, Inc., GA [M]
   Rep. National Electrical Manufacturers Association

Hugh D. Butler, Jr., Southwire, GA [M]
   (Alt. to J. L. Hunter)
Dennis Darling, Ayres, Lewis, Norris, & May, Inc., MI [U]
   (Alt. to K. W. Carrick)
Junior L. Owings, State of Oregon, OR [E]
   (Alt. to D. R. Strassburg)
Philip M. Piqueira, General Electric Co., CT [M]
   (Alt. to J. W. Young)
Ralph C. Prichard, Hercules Inc., DE [U]
   (Alt. to W. M. Lewis)
William J. Tipton, IBEW, OH [L]
   (Alt. to M. H. Sumrall)

Alternates

Thomas L. Adams, Commonwealth Edison, IL [UT]
   (Alt. to C. J. Beck)

CODE-MAKING PANEL NO. 5

Articles 200, 250, 280, Appendix E

J. Philip Simmons, *Acting Chair* [E]
Int'l Association of Electrical Inspectors, WA
Rep. Int'l Association of Electrical Inspectors

Jeffrey Boksiner, Bell Communications Research (Bellcore), NJ [UT]
   Rep. Alliance for Telecommunications Industry Solutions
Elio L. Checca, U.S. Dept. of Labor, VA [E]
David A. Dini, Underwriters Laboratories Inc., IL [RT]
Dan Hammel, Int'l Brotherhood of Electrical Workers, Local 704, IA [L]
   Rep. Int'l Brotherhood of Electrical Workers
Charles F. Mello, Electro-Test, Inc., OR [IM]
   Rep. Int'l Electrical Testing Association Inc.
Elliot Rappaport, Electro Technology Consultants, Inc., MI [U]
   Rep. Institute of Electrical & Electronics Engineers, Inc.
Ted G. Robertson, Robertson Electric, Inc., TX [IM]
   Rep. Independent Electrical Contractors
William F. Saffell, Phoenix, MD [IM]
   Rep. Power Tool Institute, Inc.
Melvin J. Schiff, M J Schiff & Associates, CA [SE]
   Rep. American Water Works Association (VL 250)
Deborah Ann Schouten, 3M Co., MN [U]
   Rep. Chemical Manufacturers Association
Gregory J. Steinman, Thomas & Betts Corp., TN [M]
   Rep. National Electrical Manufacturers Association
Ronald J. Toomer, Toomer Electrical Co. Inc., LA [IM]
   Rep. National Electrical Contractors Association
C. Douglas White, Houston Lighting & Power Co., TX [UT]
   Rep. Edison Electric Institute

Alternates

Paul M. Bowers, City of Iowa City, IA [E]
   (Alt. to J. P. Simmons)
Paul Dobrowsky, Eastman Kodak Co., NY [U]
   (Alt. to D. A. Schouten)
Garfield B. Gwyn, Gwyn Electrical & Plumbing Co., NC [IM]
   (Alt. to T. G. Robertson)
Ronald Lai, Framatone Connectors, NH [M]
   (Alt. to G. J. Steinman)
Dennis E. Lammert, Union Electric Co., MO [UT]
   (Alt. to C. D. White)
Chiu F. Leung, Bell Communications Research (Bellcore), NJ [UT]
   (Alt. to J. Boksiner)
Daleep C. Mohla, Union Carbide Corp., TX [U]
   (Alt. to E. Rappaport)
David Peot, Ryobi, SC [M]
   (Alt. to W. F. Saffell)
Elliot Rush, IBEW Local 231, IA [L]
   (Alt. to D. Hammel)

Nonvoting

Norman H. Davis, III, Holmes Beach, FL

CODE-MAKING PANEL NO. 6

Articles 310, 400, 402, Chapter 9 Tables 5 through 9, and Appendix B

Stephen J. Thorwegen, Jr., *Chair* [IM]
Fisk Electric Co., TX
Rep. National Electrical Contractors Association

David Brender, Copper Development Association, Inc., NY [M]
   Rep. Copper Development Association Inc.
Julian R. Burns, Burns Electrical Contractors, Inc., NC [IM]
   Rep. Independent Electrical Contractors
Kermit B. Duncan, PacifiCorp, WY [UT]
   Rep. Edison Electric Institute
William C. Ferrell, IBEW, Local 972, OH [L]
   Rep. Int'l Brotherhood of Electrical Workers
Steven Galan, Underwriters Laboratories Inc., NY [RT]
Ravindra H. Ganatra, Alcan Cable, GA [M]
   Rep. The Aluminum Association
L. Bruce McClung, Union Carbide Corp., WV [U]
   Rep. Institute of Electrical & Electronics Engineers, Inc.
Oran P. Post, City of Cuyahoga Falls, OH [E]
   Rep. Int'l Association of Electrical Inspectors

Alternates

Kenneth L. Brotherton, IBEW, Local 683, OH [L]
   (Alt. to W. C. Ferrell)
Stanley Folz, Folz Electric, Inc., IL [IM]
   (Alt. to S. J. Thorwegen, Jr.)
G. W. "Jerry" Kent, Kent Electric Systems, TX [IM]
   (Alt. to J. R. Burns)
C. David Mercier, Southwire Co., GA [M]
   (Alt. to R. H. Ganatra)
Robert E. Moore, Tampa Electric Co., FL [UT]
   (Alt. to K. B. Duncan)
Larry W. Oden, Coleman Cable Systems, Inc., IL [M]
   (Voting Alt. to NEMA Rep.)

John Stacey, City of St. Louis, MO [E]
  (Alt. to O. P. Post)
H. R. Stewart, HRS Consulting, TX [U]
  (Alt. to L. B. McClung)

Francis M. Stone, Jr., Shell Deepwater Development Systems, Inc.,
LA [U]
  (Voting Alt. to CMA Rep.)

CODE-MAKING PANEL NO. 7

Articles 320, 321, 324, 325, 326, 328, 330, 333, 334, 336, 338, 339, 340, 342, 363

Joel A. Rencsok, *Chair* [E]
City of Phoenix, AZ
Rep. Int'l Association of Electrical Inspectors

James M. Daly, BICC Cables Corp., NY [M]
  Rep. National Electrical Manufacturers Association
Robert J. Eugene, City of Spokane, WA [E]
  Rep. Int'l Conference of Building Officials
Pedro M. Garcia, Florida Power & Light Co., FL [UT]
  Rep. Edison Electric Institute
Bobby C. Gentry, Southwire Co., GA [M]
  Rep. The Aluminum Association
Robert L. Gotham, Rose City Electric Co., Inc., OR [IM]
  Rep. National Electrical Contractors Association
Thomas J. Guida, Underwriters Laboratories Inc., NY [RT]
Ralph L. Haynes, Fluor Daniel Inc., TX [U]
  Rep. Institute of Electrical & Electronics Engineers, Inc.
John E. Propst, Shell Oil Products, Co., TX [U]
  Rep. Chemical Manufacturers Association
Thomas G. Rodgers, City of Daytona Beach, FL [L]
  Rep. Int'l Brotherhood of Electrical Workers
David E. Schumacher, Schumacher Electric Co. Inc., IA [IM]
  Rep. Associated Builders and Contractors, Inc.
Robert S. Strength, Product Safety Management Inc., FL [M]
  Rep. Society of the Plastics Industry Inc.

Alternates

Harry C. Brown, IBEW Local 606, FL [L]
  (Alt. to T. G. Rodgers)
Arthur Buxbaum, San Diego Building Inspection Dept., CA [E]
  (Alt. to J. E. Rencsok)
James V. Fitzgerald, The Okonite Co., NJ [M]
  (Alt. to J. M. Daly)
Ravindra H. Ganatra, Alcan Cable, GA [M]
  (Alt. to B. C. Gentry)
M. Shan Griffith, Brown & Root, Inc., TX [U]
  (Alt. to R. L. Haynes)
Herman J. Hall, Austin, TX [M]
  (Alt. to R. S. Strength)
Bruce W. Nutt, TU Electric, TX [UT]
  (Alt. to P. M. Garcia)
Thomas H. Wood, Cecil B. Wood Inc., IL [IM]
  (Alt. to R. L. Gotham)

CODE-MAKING PANEL NO. 8

Articles 318, 331, 343, 345, 346, 347, 348, 349, 350, 351, 352, 353, 354, 356, 358, 362, 364, 365, 374, Chapter 9 Tables 1 and 4, and Appendix C

Kenneth E. Jannot, *Chair* [UT]
Detroit Edison, MI
Rep. Edison Electric Institute

Richard Berman, Underwriters Laboratories Inc., IL [RT]
John S. Corry, Corry Electric Inc., CA [IM]
  Rep. Associated Builders and Contractors, Inc.
Robert W. Cox, Zeneca Pharmaceuticals Group, DE [U]
  Rep. Chemical Manufacturers Association
George R. Dauberger, Thomas & Betts Corporation, TN [M]
  Rep. National Electrical Manufacturers Association
Charles W. Forsberg, Carlon Electrical Products, A Lamson & Sessions
Co., OH [M]
  Rep. SPI/VI
Wayne A. Lilly, City of Harrisonburg, VA [E]
  Rep. Int'l Association of Electrical Inspectors
W. Terry Lindsay, Duncan Electric Co. Inc., TN [IM]
  Rep. National Electrical Contractors Association
Richard E. Loyd, R & N Associates, AR [M]
  Rep. American Iron and Steel Institute
C. David Mercier, Southwire Co., GA [M]
  Rep. The Aluminum Association
Dennis L. Rowe, The New York Board of Fire Underwriters, NY [E]
  Rep. New York Board of Fire Underwriters
James E. Tyson, Jr., U.S. Dept. of Veterans Affairs, Washington, DC [U]
William Wusinich, IBEW Local 98, PA [L]
  Rep. Int'l Brotherhood of Electrical Workers

Alternates

Kenneth E. Christ, Solutie, Inc., MO [U]
  (Alt. to R. W. Cox)
Khimchand H. Chudasama, U.S. Dept. of Veterans Affairs, Washington,
DC [U]
  (Alt. to J. E. Tyson, Jr.)
Shari L. Duzac, Underwriters Laboratories Inc., CA [RT]
  (Alt. to R. Berman)
Jack A. Gruber, Wheatland Tube Co., PA [M]
  (Alt. to R. E. Loyd)
Wyman H. Hawley, Athens-Clarke County Government, GA [E]
  (Alt. to W. A. Lilly)
David H. Kendall, Carlon Electrical Products, A Lamson & Sessions Co.,
OH [M]
  (Alt. to C. W. Forsberg)
Danny P. Liggett, E. I. DuPont deNemours, DE [M]
  (Voting Alt. to CTI Rep.)
Gilbert McGoldrick, Square D Company, OH [M]
  (Alt. to G. R. Dauberger)
Ray R. Simpson, Int'l Brotherhood of Electrical Workers, IN [L]
  (Alt. to W. Wusinich)
Ronald J. Toomer, Toomer Electrical Co. Inc., LA [IM]
  (Alt. to W. T. Lindsay)

CODE-MAKING PANEL NO. 9

Articles 370, 373, 380, 384

Timothy M. Croushore, *Chair* [UT]
Allegheny Power Service Corp., PA
Rep. Edison Electric Institute

Joseph A. Cannatelli, Arco Chemical Co., PA [U]
Rep. Chemical Manufacturers Association
Dale R. Deming, Am Electric Co. (T&B), MI [M]
Frederic P. Hartwell, Electrical Construction & Maintenance Magazine, MA [SE]
Jeffrey H. Hidaka, Underwriters Laboratories Inc., IL [RT]
Robert Kaemmerlen, Kaemmerlen Electric Co., MO [IM]
Rep. National Electrical Contractors Association
Anthony Montuori, The New York Board of Fire Underwriters, NY [E]
Rep. Int'l Association of Electrical Inspectors
Sukanta Sengupta, FMC Corp., NJ [U]
Rep. Institute of Electrical & Electronics Engineers, Inc.
William E. Slater, RACO Inc., IN [M]
Rep. National Electrical Manufacturers Association
Paul Welnak, IBEW Local 494, WI [L]
Rep. Int'l Brotherhood of Electrical Workers

Alternates

Jeff Bernson, IBEW, IL [L]
(Alt. to P. Welnak)
Boyd H. Culp, Phillips Petroleum Co., OK [U]
(Alt. to J. A. Cannatelli)
Robert A. McCullough, Ocean County Construction Inspection Dept., NJ [E]
(Alt. to A. Montuori)
Ronald H. Reed, Square D Company, KY [M]
(Alt. to W. E. Slater)
Jerome W. Seigel, West Hartford, CT [U]
(Alt. to S. Sengupta)
Ronald L. Spees, Consumers Energy Co., MI [UT]
(Alt. to T. M. Croushore)

CODE-MAKING PANEL NO. 10

Articles 240, 780

Stanley D. Kahn, *Chair* [IM]
Heckler Electric Co., Inc., CA
Rep. National Electrical Contractors Association

Charles K. Blizard, American Electrical Testing Co., Inc., MA [IM]
Rep. Int'l Electrical Testing Association Inc.
Madeline Borthick, Mayfield Electric Co., TX [IM]
Rep. Independent Electrical Contractors
John E. Brezan, Lehigh Valley Electrical Inspection Service, PA [E]
Rep. Int'l Association of Electrical Inspectors
Robert J. Deaton, Union Carbide Corp., TX [U]
Rep. Institute of Electrical & Electronics Engineers, Inc.
Charles K. Eldridge, Indianapolis Power & Light Co., IN [UT]
Rep. Edison Electric Institute
Carl J. Fredericks, Dow Chemical U.S.A., TX [U]
Rep. Chemical Manufacturers Association
Glenn Hanson, IBEW Local 242, MN [L]
Rep. Int'l Brotherhood of Electrical Workers
Clive W. Kimblin, Cutler-Hammer, Inc., PA [M]
Rep. National Electrical Manufacturers Association
John J. Mahal, Underwriters Laboratories Inc., IL [RT]
Arden L. Munson, Hussmann Corp., MO [M]
Rep. Air Conditioning and Refrigeration Institute
George J. Ockuly, Cooper Industries, Bussmann Division, MO [M]

Alternates

David E. Chartrand, Middle Dept. Inspection Agency, Inc., NY [E]
(Alt. to J. E. Brezan)

George D. Gregory, Square D Co., IA [M]
(Alt. to C. W. Kimblin)
Roderic L. Hageman, Prit Service, Inc., IL [IM]
(Alt. to C. K. Blizard)
Charles D. Hughes, Westinghouse Savannah River Co., SC [U]
(Alt. to R. J. Deaton)
Albert H. Keisling, IBEW Local 934, TN [L]
(Alt. to G. Hanson)
Allen F. KnicKrehm, The KnicKrehm Co., CA [IM]
(Alt. to S. D. Kahn)
Kris Mantravadi, FMC Corp., NY [U]
(Alt. to C. J. Fredericks)
Vincent J. Saporita, Cooper Industries, Bussmann Division, MO [M]
(Alt. to G. J. Ockuly)
John Tolbert, Bristol Compressors, VA [M]
(Alt. to A. L. Munson)
Leslie R. Zielke, South Carolina Electric and Gas Co., SC [UT]
(Alt. to C. K. Eldridge)

Nonvoting

Rick C. Gilmour, Canadian Standards Association, ON, Canada

CODE-MAKING PANEL NO. 11

Articles 430, 440, 670, Appendix D Example D8

Thomas H. Wood, *Chair* [IM]
Cecil B. Wood Inc., IL
Rep. National Electrical Contractors Association

Rick L. Bunch, Tecumseh Products Co., MI [M]
Rep. Air Conditioning and Refrigeration Institute
Thomas E. Dye, Olin Corp., TN [U]
Rep. Chemical Manufacturers Association
William T. Fiske, ITS North America, NY [RT]
Thomas J. Garvey, State of Wisconsin, WI [E]
Rep. Int'l Association of Electrical Inspectors

Paul S. Hamer, Chevron Research & Technology Co., CA [U]
Rep. American Petroleum Institute
Michael D. Landolfi, Landolfi Electric Co. Inc., NJ [IM]
Rep. Associated Builders and Contractors, Inc.
Timothy J. Mueck, Shermco Industries, Inc., TX [IM]
Rep. Int'l Electrical Testing Association Inc.
Richard A. Rasmussen, Underwriters Laboratories Inc., NC [RT]

Vincent J. Saporita, Cooper Industries, Bussmann Division, MO [M]
Lynn F. Saunders, GM Worldwide Facilities Group., MI [U]
  Rep. Institute of Electrical & Electronics Engineers, Inc.
Charles B. Schram, Wilmette, IL [SE]
Thomas Sportini, IBEW Local 488, CT [L]
  Rep. Int'l Brotherhood of Electrical Workers
M. Edward Thomas, Alabama Power Co., AL [UT]
  Rep. Edison Electric Institute
David W. Trudeau, Factory Mutual Research, MA [I]
James R. Wright, Siemens-Furnas Controls, IL [M]
  Rep. National Electrical Manufacturers Association

Alternates

Frederick Bried, Shell Oil Products Co., TX [U]
  (Alt. to P. S. Hamer)
Joe David Cox, Eastman Chemical Co., TN [U]
  (Alt. to T. E. Dye)

Robert J. Keough, U.S. Electrical Motors, MO [M]
  (Alt. to J. R. Wright)
R. Ken Murphy, Houston Lighting & Power Co., TX [UT]
  (Alt. to M. E. Thomas)
James M. Naughton, IBEW, Local 103, MA [L]
  (Alt to T. Sportini)
George J. Ockuly, Cooper Industries, Bussmann Division, MO [M]
  (Alt. to V. J. Saporita)
Carl Radcliffe, Underwriters Laboratories Inc., NC [RT]
  (Alt. to R. A. Rasmussen)
John A. Schultz, Minnesota Board of Electricity, MN [E]
  (Alt. to T. J. Garvey)
Arthur J. Smith, III, Waldemar S. Nelson and Co., Inc., LA [U]
  (Alt. to L. F. Saunders)
Herbert P. Spiegel, Corona Industrial Electric, CA [IM]
  (Alt. to T. H. Wood)

CODE-MAKING PANEL NO. 12

Articles 426, 427, 610, 620, 625, 630, 645, 660, 665, 668, 669, 685, Appendix D Examples D9 and D10

Charles M. Trout, *Chair* [IM]
Main Electric Co. Inc., FL
Rep. National Electrical Contractors Association

Thomas M. Burke, Underwriters Laboratories Inc., CA [RT]
Andre R. Cartal, Middle Dept. Inspection Agency Inc., PA [E]
  Rep. Int'l Association of Electrical Inspectors
James F. Cook, Eagle Electric Manufacturing, NY [M]
  Rep. National Electrical Manufacturers Association
Ralph E. Droste, United Technologies/Otis Elevator Co., CT [M]
  Rep. National Elevator Industry Inc. (VL 610, 620, 630)
Robert A. Jones, Independent Electrical Contractors, TX [IM]
  Rep. Independent Electrical Contractors
William J. Kelly, Eastman Kodak Co., NY [U]
  Rep. Institute of Electrical & Electronics Engineers, Inc.
Gary M. Maurice, IBEW, Local 342, NC [L]
  Rep. Int'l Brotherhood of Electrical Workers
John H. Mortimer, Inductotherm Corp., NJ [M] (VL 665)
C. L. Pittman, Maryville, TN [M]
  Rep. The Aluminum Association (VL 427, 610, 625, 630, 645, 646, 660, 665, 668, 669, 685)
Norbert Poch, IBM Corp., MN [U]
  Rep. Information Technology Industry Council (VL 645)
Ronald L. Purvis, Georgia Power Co., GA [UT]
  Rep. Edison Electric Institute
S. L. Ralston, PPG Industries Inc., WV [U]
  Rep. Chemical Manufacturers Association
Robert H. Reuss, Material Handling, LLC, WI [M]
  Rep. Crane Manufacturers Association of America Inc. (VL 610)
T. Neil Thorla, Inland Steel Co., IN [U]
  Rep. Association of Iron & Steel Engineers (VL 610, 620, 630)
Craig B. Toepfer, Ford Motor Co., MI [U]
  Rep. Society of Automotive Engineers (VL 625)
Bernard W. Whittington, Whittington Engr. Inc., WV [SE]

Alternates

Scott Cline, McMurtrey Electric, Inc., CA [IM]
  (Alt. to C. M. Trout)
J. Kevin Curtis, Virginia Power, VA [UT]
  (Alt. to R. L. Purvis)
Gary L. Duggan, IBEW Local 270, TN [L]
  (Alt. to G. M. Maurice)
George W. Farrell, POWR-GARD Products Group, IL [M]
  (Alt. to J. F. Cook)
Robert E. Johnson, Motorola ISG, MA [U]
  (Alt. to N. Poch) (VL 645)
Mark G. Kosowski, General Motors Advanced Technology, MI [U]
  (Alt. to C. B. Toepfer) (VL 625)
John R. Kovacik, Underwriters Laboratories Inc., IL [RT]
  (Alt. to T. M. Burke)
Richard H. Laney, Siecor Corp., NC [U]
  (Voting Alt. to Insulated Cable Engineers Association Inc.)
Nick Marchitto, Otis Elevator Co., CT [M]
  (Alt. to R. E. Droste) (VL 610, 620, 630)
Robert C. Oldham, Jr., Reynolds Metals Co., VA [M]
  (Alt. to C. L. Pittman) (VL 427, 610, 625, 630, 645, 646, 660, 665, 668, 669, 685)
Merritt D. Redick, M. Redick & Associates, CA [U]
  (Alt. to W. J. Kelly)
James J. Rogers, Massachusetts St. Board of Electrical Examiners, MA [E]
  (Alt. to A. R. Cartal)
Robert C. Turner, Inductotherm Corp., NJ [M]
  (Alt. to J. H. Mortimer) (VL 665)

CODE-MAKING PANEL NO. 13

Articles 450, 455, 460, 470, 490

William T. O'Grady, *Chair* [RT]
Underwriters Laboratories Inc., NY

William A. Brunner, IBEW, Local 714, ND [L]
  Rep. Int'l Brotherhood of Electrical Workers
Hugh D. Butler, Jr., Southwire, GA [M]
  Rep. The Aluminum Association
James C. Carroll, Square D. Co., TN [M]
  Rep. National Electrical Manufacturers Association
William B. Crist, Houston Stafford Electric Co., TX [IM]
  Rep. Independent Electrical Contractors

O. L. Davis, Manzano Western, Inc., NM [IM]
  Rep. National Electrical Contractors Association
Richard P. Fogarty, Jr., Consolidated Edison Co. of N.Y., Inc., NY [UT]
  Rep. Edison Electric Institute
Walter Kroboth, Exxon Chemical Co., TX [U]
  Rep. Chemical Manufacturers Association
Hiram J. Lamb, City of Charlottesville, VA [E]
  Rep. Int'l Association of Electrical Inspectors

Milton D. Robinson, Milt Robinson Engineering Co., IN [U]
  Rep. Institute of Electrical & Electronics Engineers, Inc.
LaVerne E. Stetson, U.S. Dept. of Agriculture, NE [SE]
  Rep. American Society of Agricultural Engineers

Alternates

J. Alan Barringer, North Carolina Dept. of Insurance, NC [E]
  (Alt. to H. J. Lamb)
Jeffrey H. Hidaka, Underwriters Laboratories Inc., IL [RT]
  (Alt. to W. T. O'Grady)
Barry N. Hornberger, PECO Energy Co., PA [UT]
  (Alt. to R. P. Fogarty, Jr.)

Gene Morehart, ACME Electric Corp., NC [M]
  (Alt to J. C. Carroll)
Robert L. Simpson, Simpson Electrical Engineering Co., MO [U]
  (Alt. to M. D. Robinson)
Samuel Solomon, Alameda County Electrical JATC, CA [L]
  (Alt. to W. A. Brunner)
Lou G. Willoughby, Aluminum Company of America, OH [M]
  (Alt. to H. D. Butler, Jr.)
Ivan L. Winsett, Ronk Electrical Industries, GA [SE]
  (Alt. to L. E. Stetson)
Ralph H. Young, Eastman Chemical Co., TN [U]
  (Alt. to W. Kroboth)

CODE-MAKING PANEL NO. 14

Articles 500, 501, 502, 503, 504, 505, 510, 511, 513, 514, 515, 516

James G. Stallcup, *Acting Chair,* [SE]
GRAYBOY & Associates, TX

David N. Bishop, Chevron U.S.A. Production Co., TX [U]
  Rep. American Petroleum Institute
Edward M. Briesch, Underwriters Laboratories Inc., IL [RT]
Donald R. Cook, Shelby County Bldg. Inspections, AL [E]
  Rep. Int'l Association of Electrical Inspectors
James D. Cospolich, Waldemar S. Nelson & Co. Inc., LA [U]
  Rep. Int'l Society for Measurement and Control
Kevin P. Earley, Bay Harbour Electric, Inc., PA [IM]
  Rep. Independent Electrical Contractors
Joseph H. Kuczka, Killark Electric Mfg. Co., MO [M]
  Rep. National Electrical Manufacturers Association
Francis J. McGowan, Factory Mutual Research Corp., MA [I]
Carl R. Roselli, ANR Pipeline Co., MI [U]
  Rep. American Gas Association
Mark G. Saban, Modern Electric Co. of Illinois, IL [IM]
  Rep. National Electrical Contractors Association
Peter J. Schram, Delray Beach, FL [SE]
Harold B. Smith, Wunderlich-Malec Eng., MN [U]
  Rep. Grain Elevator & Processing Society
Lawrence C. Strachota, Mid American Energy Co., IA [UT]
  Rep. Edison Electric Institute
David Wechsler, Union Carbide Corp., WV [U]
  Rep. Chemical Manufacturers Association
James A. Weldon, IBEW Local 728, FL [L]
  Rep. Int'l Brotherhood of Electrical Workers
Donald W. Zipse, Zipse Electrical Engineering Inc., PA [U]
  Rep. Institute of Electrical & Electronics Engineers, Inc.

Alternates

Lawrence J. Adamcik, The Dow Chemical Co., MI [U]
  (Alt. to D. W. Zipse)
Alonza W. Ballard, Crouse-Hinds, NY [M]
  (Alt. to J. H. Kuczka)
Jimmy R. Bonds, Oklahoma State Dept. of Health, OK [E]
  (Alt. to D. R. Cook)
Keith E. Feigel, IBEW Local 180, CA [L]
  (Alt. to J. A. Weldon)
Mark Goodman, ARCO Products Co, CA [U]
  (Alt. to D. N. Bishop)
Paul T. Kelly, Underwriters Laboratories Inc., IL [RT]
  (Alt. to E. M. Briesch)
William G. Lawrence, Jr., Factory Mutual Research Corp., MA [I]
  (Alt. to F. J. McGowan)
Richard A. Pruzick, PC Electrical Construction, Inc., NJ [IM]
  (Alt. to M. G. Saban)
William R. Stalker, ComEd Division of Unicom, IL [UT]
  (Alt. to L. C. Strachota)
Mark C. Wirfs, R & W Engineering Inc., OR [U]
  (Alt. to H. B. Smith)

Nonvoting

John A. Bossert, Hazloc Inc., ON, Canada
Fred K. Walker, U.S. Air Force, FL

CODE-MAKING PANEL NO. 15

Articles 445, 480, 518, 520, 525, 530, 540, 695, 700, 701, 702, 705

Robert C. Duncan, *Chair* [E]
Reedy Creek Improvement District, FL
Rep. Int'l Association of Electrical Inspectors

Peter W. Amos, Consolidated Edison Co. of NY, Inc., NY [UT]
  Rep. Edison Electric Institute
Lloyd D. Bird Jr., Belco Electric, Inc., GA [IM]
  Rep. Independent Electrical Contractors
James L. Boyer, Firetrol, Inc., NC [M]
  Rep. Nat'l Electrical Manufacturers Association
Tom Dunn, Butler Amusements, CA [IM]
  Rep. Outdoor Amusement Business Association, Inc. (VL 525)
George W. Flach, Flach Consultants, LA [SE]
Michael V. Glenn, Longview Fibre Co., WA [U]
  Rep. Institute of Electrical & Electronics Engineers, Inc.
Marcelo M. Hirschler, GBH Int'l, CA [SE]
George Thomas Howard, George Thomas Howard & Associates, NV [SE]
Gordon S. Johnson, Dundee, FL [SE]
  Rep. Electrical Generating Systems Association
Jack W. Kalbfeld, Kalico Technology Inc., NY [SE] (VL 518, 520, 525, 530, 540)

Michael B. Klein, Labyrinth Electrical, Inc., MD [IM]
  Rep. Illuminating Engineering Society of North America
John R. Kovacik, Underwriters Laboratories Inc., IL [RT]
Edwin S. Kramer, Radio City Music Hall, NY [L]
  Rep. Int'l Association of Theatrical Stage Employees
Michael A. Lanni, Universal City Studios, CA [U]
  Rep. Motion Picture Association of America, Inc.
Stephen A. Mezsick, Eli Lilly & Co., IN [U]
  Rep. Chemical Manufacturers Association
Keku M. Mistry, Bell Communications Research (Bellcore), NJ [UT]
  Rep. Alliance for Telecommunications Industry Solutions (VL 445, 480, 700, 701, 702, 705)
Richard M. Shotwell, Filer, ID [IM]
  Rep. National Electrical Contractors Association
Lawrence E. Todd, Intertek Testing Services, NA, Inc. KY [RT]
Kenneth E. Vannice, NSI Corp., OR [M]
  Rep. U.S. Institute for Theatre Technology

Alternates

James T. Dollard, Jr., IBEW Local 98, PA [L]
(Voting Alt. to IBEW Rep.)
Mitchell K. Hefter, Rosco Entertainment Technology, OR [IM]
(Alt. to M. B. Klein)
James R. Iverson, Onan Corp., MN [M]
(Alt. to G. S. Johnson)
Dennis W. Marshall, T.D.B., Inc., TX [IM]
(Alt. to L. D. Bird, Jr.)
Irvin R. Miller, American Electric Power, OH [UT]
(Alt. to P. W. Amos)
Ronald H. Minter, Kaufel Tech Center, CT [M]
(Alt. to J. L. Boyer)

Michael D. Skinner, CBS Studio Center, CA [U]
(Alt. to M. A. Lanni)
Richard Sobel, Quantum Electric Corp., NY [IM]
(Alt. of R. M. Shotwell)
Steven R. Terry, Production Arts Lighting Inc., NJ [U]
(Alt. to K. E. Vannice)
Dale A. Triffo, Shell Oil Products Co., TX [U]
(Alt to. S. A. Mezsick)
Harold F. Willman, City of Lakewood, CO [E]
(Alt. to R. C. Duncan)

CODE-MAKING PANEL NO. 16

Articles 640, 650, 720, 725, 727, 760, 770, 800, 810, 820, 830, and Chapter 9
Tables 11 and 12

Dean K. Wilson, *Chair* [I]
Industrial Risk Insurers, CT
Rep. Industrial Risk Insurers

George J. Bagnall, Rural Utilities Service, Washington, DC [UT]
James E. Brunssen, Bell Communications Research (Bellcore), NJ [UT]
Rep. Alliance for Telecommunications Industry Solutions
A. William Coaker, A. W. Coaker & Associates, Inc., OH [M]
Rep. Society of the Plastics Industry Inc.
Gerald Lee Dorna, Belden Wire & Cable, IN [M]
Rep. Insulated Cable Engineers Association Inc.
S. E. Egesdal, Honeywell Inc., MN [M]
Rep. National Electrical Manufacturers Association
C. Jeff Goins, Quarry Electric, Inc., FL [IM]
Rep. Independent Electrical Contractors
Lee C. Hewitt, Underwriters Laboratories Inc., IL [RT]
Steven C. Johnson, Time Warner Cable, CO [IM]
Rep. National Cable Television Association
Ronald G. Jones, Ronald G. Jones, PE, TX [U]
Rep. Institute of Electrical & Electronics Engineers, Inc.
Harry F. Katz, South Texas Electrical JATC, TX [L]
Rep. Int'l Brotherhood of Electrical Workers
Stanley Kaufman, Lucent Technologies, GA [M]
Michael A. Lanni, Universal City Studios, CA [U]
Rep. Motion Picture Association of America, Inc. (VL 640, 720, 725, 760, 810, 820)
John Mangan, Medford City Hall, MA [E]
Rep. Int'l Association of Electrical Inspectors
Michael E. McNeil, FMC Corp/Food Ingredients Division, ME [U]
Rep. Chemical Manufacturers Association
W. D. Pirkle, Pirkle Electric Co., Inc., GA [IM]
Rep. National Electrical Contractors Association
James W. Romlein, MIS Labs, WI [M]
Rep. Building Industry Consulting Service Int'l
Kyle E. Todd, Entergy - GSU, LA [UT]
Rep. Edison Electric Institute
Inder L. Wadehra, IBM Corp., NC [U]
Rep. Information Technology Industry Council

Melvin J. Wierenga, Wierenga & Associates, MI [IM]
Rep. National Systems Contractors Association/Consumer Electronics Manufacturers Association

Alternates

Ronald P. Cantrell, IBEW Local 72, TX [L]
(Alt. to H. F. Katz)
Chrysanthos Chrysaanthou, Bell Communications Research (Bellcore), NJ [UT]
(Alt. to J. E. Brunssen)
William Hopple, Simplex Time Recorder Co., CA [M]
(Alt. to S. E. Egesdal)
Robert Jensen, Telecommunication Infrastructure Design, TX [M]
(Alt. to J. W. Romlein)
Ray A. Jones, The DuPont Company, DE [U]
(Alt. to M. E. McNeil)
William J. McCoy, GTE Telephone Operations, TX [U]
(Alt. to R. G. Jones)
Andy W. Mobley, 3M, MN [U]
(Voting Alt. to ISA Rep.)
Joseph W. Rao, RAO Electric Co., FL [IM]
(Alt. to C. J. Goins)
Jan Rowland, Houston, TX [M]
(Voting Alt. to AIOB and APOBA Rep.) (VL 640, 650, 720, 725)
Oleh Sniezko, TCI Communications, Inc., CO [IM]
(Alt. to S. C. Johnson)
Sondra K. Todd, KPL a Western Resources Co., KS [UT]
(Alt. to K. E. Todd)
William A. Zanicchi, The New York Board of Fire Underwriters, NY [E]
(Alt. to J. Mangan)

Nonvoting

Irving Mande, Edwards Systems Tech, CT
(Member Emeritus)

CODE-MAKING PANEL NO. 17

Article 517

Robert E. Bernd, *Chair* [RT]
Underwriters Laboratories Inc., IL

Stephen Benesh, Square D Co., WI [M]
Rep. National Electrical Manufacturers Association
Britton E. Berek, American Society for Healthcare Engineering, IL [U]
Rep. American Society for Healthcare Engineering
D. R. Borden, Jr., Tri-City Electric Co. Inc., FL [IM]
Rep. National Electrical Contractors Association
Thomas C. Clark, Clark Electrical Construction, Inc., TX [IM]
Rep. Associated Builders and Contractors, Inc.

James R. Duncan, Sparling, WA [U]
Rep. Institute of Electrical & Electronics Engineers, Inc.
Thomas H. Gilliam, Riverside Methodist Hospital, OH [U]
Rep. Association for the Advancement of Medical Instrumentation
James A. Meyer, Pettis Memorial VA Hospital, CA [C]
Rep. American Society of Anesthesiologists
Hugh O. Nash, Jr., Nash Lipsey Burch, LLC, TN [SE]

J. Randall Perdue, Entergy Services Inc., AR [UT]
 Rep. Edison Electric Institute
Jim M. Schmer, Boise City Building Dept., ID [E]
 Rep. Int'l Association of Electrical Inspectors
George Schuck, Jr., IBEW Local 3, NY [L]
 Rep. Int'l Brotherhood of Electrical Workers
Jeffrey L. Steplowski, U.S. Dept. of Veterans Affairs, Washington, DC [U]
Duane J. Telecky, Washoe Medical Center, NV [U]
 Rep. NFPA Health Care Section
Mike Velvikis, High Voltage Maintenance Corp., WI [IM]
 Rep. Int'l Electrical Testing Association Inc.
Walter N. Vernon, IV, Mazzetti & Associates, Inc., CA [SE]
 Rep. Consulting Engineers & Land Surveyors of CA

Matthew B. Dozier, Smith Seckman Reid, Inc., TN [U]
 (Alt. to J. R. Duncan)
Stephen D. Hewson, Underwriters Laboratories Inc., IL [RT]
 (Alt. to R. E. Bernd)
Stanley D. Kahn, Heckler Electric Co., Inc., CA [IM]
 (Alt. to D. R. Borden, Jr.)
Thomas L. Ryder, IBEW Local 692, MI [L]
 (Alt. to G. Schuck, Jr.)
Gary D. Slack, Healthcare Engineering Consultants, OH [U]
 (Alt. to B. E. Berek)
Richard H. Smith, OG&E Electric Services, OK [UT]
 (Alt. to J. R. Perdue)
James E. Tyson, Jr., U.S. Dept. of Veterans Affairs, Washington, DC [U]
 (Alt. to J. L. Steplowski)

Alternates

Steve Campolo, Leviton Mfg. Co., Inc., NY [M]
 (Alt. to S. Benesh)

CODE-MAKING PANEL NO. 18

Articles 410, 411, 600, 605

Wayne Brinkmeyer, *Chair* [IM]
Biddle Electric Corp., TX
Rep. National Electrical Contractors Association

Rudy T. Elam, Systems Engineering Services, TN [U]
 Rep. Institute of Electrical & Electronics Engineers, Inc.
Kenneth F. Kempel, Underwriters Laboratories Inc., NC [RT]
Thomas J. Lynch, IBEW Local 99, RI [L]
 Rep. Int'l Brotherhood of Electrical Workers
Bernard J. Mezger, American Lighting Association, MA [M]
 Rep. American Lighting Association (VL 410, 411)
James F. Pierce, Intertek Testing Services NA Inc., OR [RT]
Saul Rosenbaum, Leviton Mfg. Co. Inc., NY [M]
 Rep. National Electrical Manufacturers Association
Thomas E. Trainor, City of San Diego, CA [E]
 Rep. Int'l Association of Electrical Inspectors
Carl T. Wall, Alabama Power Co., AL [UT]
 Rep. Edison Electric Institute
Jack Wells, Pass & Seymour Legrand, NY [M]
David B. Wilson, Southwest Signs, TX [IM]
 Rep. Independent Electrical Contractors
Randall K. Wright, Wright Sign Co., PA [M]
 Rep. Int'l Sign Association (VL 600)

Alternates

Michael N. Ber, IEC, Houston, TX [IM]
 (Alt. to D. B. Wilson)
Mark R. Berner, PP&L, PA [UT]
 (Alt. to C. T. Wall)
Robert L. Cochran, City of Costa Mesa, CA [E]
 (Alt. to T. E. Trainor)
Howard D. Hughes, Hughes Electric Co. Inc., AR [IM]
 (Alt. to W. Brinkmeyer)
Stephen G. Kieffer, Kieffer & Co., Inc., WI [M]
 (Alt. to R. K. Wright) (VL 600)
Steven A. Larson, Lockhead Martin Energy Systems, TN [U]
 (Alt. to R. T. Elam)
Don Miletich, Cooper Lighting, IL [M]
 (Alt. to S. Rosenbaum)
Michael S. O'Boyle, Lightolier, Inc., MA [M]
 (Alt. to B. J. Mezger) (VL 410, 411)
David O. Thomas, IBEW Local 208, CT [L]
 (Alt. to T. J. Lynch)

CODE-MAKING PANEL NO. 19

Articles 545, 547, 550, 551, 552, 553, 555, 604, 675

Robert A. McCullough, *Chair* [E]
Ocean County Construction Inspection Dept., NJ
Rep. Int'l Association of Electrical Inspectors

Barry Bauman, Wisconsin Power & Light Co., WI [UT]
 Rep. American Society of Agricultural Engineers (VL 545, 547, 604, 675)
William C. Boteler, Hubbell Inc., CT [M]
 Rep. National Electrical Manufacturers Association
James W. Finch, Kampgrounds of America, Inc., MT [U] (VL 550, 551, 552, 555)
Bruce A. Hopkins, Recreation Vehicle Industry Association, VA [M] (VL 550, 551, 552)
Robert L. LaRocca, Underwriters Laboratories Inc., NY [RT]
Thomas J. LeMay, LeMay Electric, Inc., GA [IM]
 Rep. Independent Electrical Contractors
Tug Miller, California Travel Parks Association, CA [U]
 Rep. National Assn. of RV Parks & Campgrounds (VL 550, 551, 552)
John "Lou" Reid, IBEW Local 595, CA [L]
 Rep. Int'l Brotherhood of Electrical Workers
Leslie Sabin-Mercado, San Diego Gas & Electric Co., CA [UT]
 Rep. Edison Electric Institute

Dick Veenstra, Fleetwood Enterprises, Inc., CA [M]
 Rep. Manufactured Housing Institute (VL 550, 551, 552)
Paul Wahlstrom, Ahlstrom Schaeffer Electric Corp., NY [IM]
 Rep. National Electrical Contractors Association
Michael L. Zieman, RADCO, CA [RT] (VL 545, 550, 551, 552)

Alternates

Glenn H. Ankenbrand, Delmarva Power, MD [UT]
 (Alt. to L. Sabin-Mercado)
David Gorin, National Assn. of RV Parks & Campgrounds, VA [U]
 (Alt. to T. Miller) (VL 550, 551, 552)
Ronald E. Maassen, Lemberg Electric Co., Inc., WI [IM]
 (Alt. to P. Wahlstrom)
Neal Mahan, Underwriters Laboratories Inc., NC [RT]
 (Alt. to R. L. LaRocca)
Timothy P. McNeive, Thomas & Betts Corp., TN [M]
 (Alt. to W. C. Boteler)

William A. Mirand, IBEW Local 46, WA [L]
(Alt. to J. Reid)
Kent Perkins, Recreation Vehicle Industry Association, VA [M]
(Alt. to B. A. Hopkins) (VL 550, 551, 552)
John G. Sims, Champion Enterprises, MI [M]
(Alt. to D. Veenstra) (VL 550, 551, 552)

Homer Staves, Kampgrounds of America, Inc., MT [U]
(Alt. to J. W. Finch) (VL 550, 551, 552, 555)
LaVerne E. Stetson, U.S. Dept. of Agriculture, NE [U]
(Alt. to B. Bauman) (VL 545, 547, 604, 675)
Raymond F. Tucker, RADCO, CA [RT]
(Alt. to M. L. Zieman) (VL 545, 550, 551, 552)

CODE-MAKING PANEL NO. 20

Articles 422, 424, 680

Robert M. Milatovich, *Chair* [E]
Clark County, NV
Rep. Int'l Association of Electrical Inspectors

Edward S. Charkey, American Insurance Services Group, Inc., NY [I]
Robert J. Egan, IBEW Local 1, MO [L]
Rep. Int'l Brotherhood of Electrical Workers
James N. Pearse, Leviton Mfg. Co. Inc., NY [M]
Rep. National Electrical Manufacturers Association
Marcos Ramirez, Mr. Electric Service Co., Inc., NY [IM]
Rep. Independent Electrical Contractors
William F. Ryan, Jr., The New York Board of Fire Underwriters, NY [E]
Herbert P. Spiegel, Corona Industrial Electric, Inc., CA [IM]
Rep. National Electrical Contractors Association
Donald J. Talka, Underwriters Laboratories Inc., NY [RT]
Leon T. Uhl, Leon T. Uhl, PE, IL [M]
Rep. Association of Home Appliance Manufacturers (VL 422, 424)
Leonard K. VanTassel, Carrier Air Conditioning, NY [M]
Rep. Air Conditioning and Refrigeration Institute (VL 422, 424)
Lee L. West, Balboa Instruments, CA [M]
Rep. National Spa and Pool Institute (VL 680)
Robert M. Yurkanin, Electran Process Int'l Inc., NJ [U]
Rep. Institute of Electrical & Electronics Engineers, Inc.

Alternates

Xen George Anchales, San Bernardino County/Dept. of Bldg. & Safety, CA [E]
(Alt. to R. M. Milatovich)

Dennis L. Baker, Springs & Sons Electrical Contractors Inc., AZ [IM]
(Alt. to M. Ramirez)
Scott Cline, McMurtrey Electric, Inc., CA [IM]
(Alt. to H. P. Spiegel)
Neil F. LaBrake, Jr., Niagara Mohawk Power Corp., NY [UT]
(Voting Alt. to ELPG/EEI Rep.)
Earle L. McNairn, BICC Pyrotenax, ON, Canada [M]
(Alt. to J. N. Pearse)
Anthony Sardina, Carrier Corp., NY [M]
(Alt. to L. K. VanTassel) (VL 422, 424)
Stephen P. Schoemehl, IBEW Local 1, MO [L]
(Alt. to R. J. Egan)
John T. Weizeorick, Assn. of Home Appliance Manufacturers, IL [M]
(Alt. to L. T. Uhl) (VL 422, 424)
Robert E. Wisenburg, Coates Heater Co., Inc., WA [M]
(Alt. to L. L. West) (VL 680)

Nonvoting

Rick C. Gilmour, Canadian Standards Association, ON, Canada
William H. King, Jr., U.S. Consumer Product Safety Commission, MD

NOTE: Membership on a committee shall not in and of itself constitute an endorsement of the Association or any document developed by the committee on which the member serves.

Committee Scope. This Committee shall have primary responsibility for documents on minimizing the risk of electricity as a source of electric shock and as a potential ignition source of fires and explosions. It shall also be responsible for text to minimize the propagation of fire and explosions due to electrical installations.

NFPA 70

National Electrical Code®

1999 Edition

Article 90 — Introduction

90-1. Purpose.

(a) Practical Safeguarding. The purpose of this *Code* is the practical safeguarding of persons and property from hazards arising from the use of electricity.

(b) Adequacy. This *Code* contains provisions that are considered necessary for safety. Compliance therewith and proper maintenance will result in an installation that is essentially free from hazard but not necessarily efficient, convenient, or adequate for good service or future expansion of electrical use.

> FPN: Hazards often occur because of overloading of wiring systems by methods or usage not in conformity with this *Code*. This occurs because initial wiring did not provide for increases in the use of electricity. An initial adequate installation and reasonable provisions for system changes will provide for future increases in the use of electricity.

(c) Intention. This *Code* is not intended as a design specification nor an instruction manual for untrained persons.

90-2. Scope.

(a) Covered. This *Code* covers the following.

(1) Installations of electric conductors and equipment within or on public and private buildings or other structures, including mobile homes, recreational vehicles, and floating buildings; and other premises such as yards, carnival, parking, and other lots, and industrial substations.

> FPN: For additional information concerning such installations in an industrial or multibuilding complex, see the *National Electrical Safety Code*, ANSI C2-1997.

(2) Installations of conductors and equipment that connect to the supply of electricity.

(3) Installations of other outside conductors and equipment on the premises.

(4) Installations of optical fiber cables and raceways.

(5) Installations in buildings used by the electric utility, such as office buildings, warehouses, garages, machine shops, and recreational buildings, that are not an integral part of a generating plant, substation, or control center.

(b) Not Covered. This *Code* does not cover the following.

(1) Installations in ships, watercraft other than floating buildings, railway rolling stock, aircraft, or automotive vehicles other than mobile homes and recreational vehicles.

> FPN: While the scope of this *Code* indicates that the *Code* does not cover installations in ships, portions of this *Code* are incorporated by reference into Title 46, *Code of Federal Regulations*, Parts 110-113.

(2) Installations underground in mines and self-propelled mobile surface mining machinery and its attendant electrical trailing cable.

(3) Installations of railways for generation, transformation, transmission, or distribution of power used exclusively for operation of rolling stock or installations used exclusively for signaling and communications purposes.

(4) Installations of communications equipment under the exclusive control of communications utilities located outdoors or in building spaces used exclusively for such installations.

(5) Installations, including associated lighting, under the exclusive control of electric utilities for the purpose of communications, metering, generation, control, transformation, transmission, or distribution of electric energy. Such installations shall be located in buildings used exclusively by utilities for such purposes; outdoors on property owned or leased by the utility; on or along public highways, streets, roads, etc.; or outdoors on private property by established rights such as easements.

(c) Special Permission. The authority having jurisdiction for enforcing this *Code* may grant exception for the installation of conductors and equipment that are not under the exclusive control of the electric utilities and are used to connect the electric utility supply system to the service-entrance conductors of the premises served, provided such installations are outside a building or terminate immediately inside a building wall.

90-3. Code Arrangement.
This *Code* is divided into the introduction and nine chapters. Chapters 1, 2, 3, and 4 apply generally; Chapters 5, 6, and 7 apply to special occupancies, special equipment, or other special conditions. These latter chapters supplement or modify the general rules. Chapters 1 through 4 apply except as amended by Chapters 5, 6, and 7 for the particular conditions.

Chapter 8 covers communications systems and is independent of the other chapters except where they are specifically referenced therein.

Chapter 9 consists of tables.

Material identified by the superscript letter "x" includes

text extracted from other National Fire Protection Association (NFPA) documents as identified in Appendix A.

90-4. Enforcement. This *Code* is intended to be suitable for mandatory application by governmental bodies that exercise legal jurisdiction over electrical installations and for use by insurance inspectors. The authority having jurisdiction for enforcement of the *Code* will have the responsibility for making interpretations of the rules, for deciding on the approval of equipment and materials, and for granting the special permission contemplated in a number of the rules.

The authority having jurisdiction may waive specific requirements in this *Code* or permit alternate methods where it is assured that equivalent objectives can be achieved by establishing and maintaining effective safety.

This *Code* may require new products, constructions, or materials that may not yet be available at the time the *Code* is adopted. In such event, the authority having jurisdiction may permit the use of the products, constructions, or materials that comply with the most recent previous edition of this *Code* adopted by the jurisdiction.

90-5. Mandatory Rules, Permissive Rules, and Explanatory Material.

(a) Mandatory Rules. Mandatory rules of this *Code* are those that identify actions that are specifically required or prohibited and are characterized by the use of the terms *shall* or *shall not*.

(b) Permissive Rules. Permissive rules of this *Code* are those that identify actions that are allowed but not required, are normally used to describe options or alternative methods, and are characterized by the use of the terms *shall be permitted* or *shall not be required*.

(c) Explanatory Material. Explanatory material, such as references to other standards, references to related sections of this *Code,* or information related to a *Code* rule, is included in this *Code* in the form of fine print notes (FPN). Fine print notes are informational only and are not enforceable as requirements of this *Code*.

FPN: The format and language used in this *Code* follows guidelines established by NFPA and published in the *NEC Style Manual*. Copies of this manual may be obtained from NFPA.

90-6. Formal Interpretations. To promote uniformity of interpretation and application of the provisions of this *Code,* formal interpretation procedures have been established and are found in the NFPA Regulations Governing Committee Projects.

90-7. Examination of Equipment for Safety. For specific items of equipment and materials referred to in this *Code,*

examinations for safety made under standard conditions will provide a basis for approval where the record is made generally available through promulgation by organizations properly equipped and qualified for experimental testing, inspections of the run of goods at factories, and service-value determination through field inspections. This avoids the necessity for repetition of examinations by different examiners, frequently with inadequate facilities for such work, and the confusion that would result from conflicting reports as to the suitability of devices and materials examined for a given purpose.

It is the intent of this *Code* that factory-installed internal wiring or the construction of equipment need not be inspected at the time of installation of the equipment, except to detect alterations or damage, if the equipment has been listed by a qualified electrical testing laboratory that is recognized as having the facilities described above and that requires suitability for installation in accordance with this *Code*.

FPN No. 1: See requirements in Section 110-3.

FPN No. 2: *Listed* is defined in Article 100.

90-8. Wiring Planning.

(a) Future Expansion and Convenience. Plans and specifications that provide ample space in raceways, spare raceways, and additional spaces will allow for future increases in the use of electricity. Distribution centers located in readily accessible locations will provide convenience and safety of operation.

(b) Number of Circuits in Enclosures. It is elsewhere provided in this *Code* that the number of wires and circuits confined in a single enclosure be varyingly restricted. Limiting the number of circuits in a single enclosure will minimize the effects from a short circuit or ground fault in one circuit.

90-9. Metric Units of Measurement. For the purpose of this *Code*, metric units of measurement are in accordance with the modernized metric system known as the International System of Units (SI).

Values of measurement in the *Code* text will be followed by an approximate equivalent value in SI units. Tables will have a footnote for SI conversion units used in the table.

Conduit size, wire size, horsepower designation for motors, and trade sizes that do not reflect actual measurements, e.g., box sizes, will not be assigned dual designation SI units.

CHAPTER 1
General

Article 100 — Definitions

Scope. This article contains only those definitions essential to the proper application of this *Code*. It is not intended to include commonly defined general terms or commonly defined technical terms from related codes and standards. In general, only those terms that are used in two or more articles are defined in Article 100. Other definitions are included in the article in which they are used but may be referenced in Article 100.

Part A of this article contains definitions intended to apply wherever the terms are used throughout this *Code*. Part B contains definitions applicable only to the parts of articles specifically covering installations and equipment operating at over 600 volts, nominal.

A. General

Accessible (as applied to wiring methods). Capable of being removed or exposed without damaging the building structure or finish, or not permanently closed in by the structure or finish of the building.

Accessible (as applied to equipment). Admitting close approach; not guarded by locked doors, elevation, or other effective means.

Accessible, Readily (Readily Accessible). Capable of being reached quickly for operation, renewal, or inspections, without requiring those to whom ready access is requisite to climb over or remove obstacles or to resort to portable ladders, etc.

Ampacity. The current, in amperes, that a conductor can carry continuously under the conditions of use without exceeding its temperature rating.

Appliance. Utilization equipment, generally other than industrial, normally built in standardized sizes or types, that is installed or connected as a unit to perform one or more functions such as clothes washing, air conditioning, food mixing, deep frying, etc.

Approved. Acceptable to the authority having jurisdiction.

Askarel. A generic term for a group of nonflammable synthetic chlorinated hydrocarbons used as electrical insulating media. Askarels of various compositional types are used. Under arcing conditions, the gases produced, while consisting predominantly of noncombustible hydrogen chloride, can include varying amounts of combustible gases depending on the askarel type.

Attachment Plug (Plug Cap) (Plug). A device that, by insertion in a receptacle, establishes a connection between the conductors of the attached flexible cord and the conductors connected permanently to the receptacle.

Automatic. Self-acting, operating by its own mechanism when actuated by some impersonal influence, as, for example, a change in current strength, pressure, temperature, or mechanical configuration.

Bathroom. An area including a basin with one or more of the following: a toilet, a tub, or a shower.

Bonding (Bonded). The permanent joining of metallic parts to form an electrically conductive path that will ensure electrical continuity and the capacity to conduct safely any current likely to be imposed.

Bonding Jumper. A reliable conductor to ensure the required electrical conductivity between metal parts required to be electrically connected.

Bonding Jumper, Equipment. The connection between two or more portions of the equipment grounding conductor.

Bonding Jumper, Main. The connection between the grounded circuit conductor and the equipment grounding conductor at the service.

Branch Circuit. The circuit conductors between the final overcurrent device protecting the circuit and the outlet(s).

Branch Circuit, Appliance. A branch circuit that supplies energy to one or more outlets to which appliances are to be connected, and that has no permanently connected lighting fixtures that are not a part of an appliance.

Branch Circuit, General Purpose. A branch circuit that supplies a number of outlets for lighting and appliances.

Branch Circuit, Individual. A branch circuit that supplies only one utilization equipment.

Branch Circuit, Multiwire. A branch circuit that consists of two or more ungrounded conductors that have a potential difference between them, and a grounded conductor that has equal potential difference between it and each ungrounded conductor of the circuit and that is connected to the neutral or grounded conductor of the system.

Building. A structure that stands alone or that is cut off from adjoining structures by fire walls with all openings therein protected by approved fire doors.

Cabinet. An enclosure designed either for surface mounting or flush mounting and is provided with a frame, mat, or trim in which a swinging door or doors are or can be hung.

Circuit Breaker. A device designed to open and close a circuit by nonautomatic means and to open the circuit automatically on a predetermined overcurrent without damage to itself when properly applied within its rating.

FPN: The automatic opening means can be integral, direct acting with the circuit breaker, or remote from the circuit breaker.

Adjustable (as applied to circuit breakers). A qualifying term indicating that the circuit breaker can be set to trip at various values of current, time, or both, within a predetermined range.

Instantaneous Trip (as applied to circuit breakers). A qualifying term indicating that no delay is purposely introduced in the tripping action of the circuit breaker.

Inverse Time (as applied to circuit breakers). A qualifying term indicating that there is purposely introduced a delay in the tripping action of the circuit breaker, which delay decreases as the magnitude of the current increases.

Nonadjustable (as applied to circuit breakers). A qualifying term indicating that the circuit breaker does not have any adjustment to alter the value of current at which it will trip or the time required for its operation.

Setting (of circuit breakers). The value of current, time, or both, at which an adjustable circuit breaker is set to trip.

Concealed. Rendered inaccessible by the structure or finish of the building. Wires in concealed raceways are considered concealed, even though they may become accessible by withdrawing them.

Conductor.

Bare. A conductor having no covering or electrical insulation whatsoever.

Covered. A conductor encased within material of composition or thickness that is not recognized by this *Code* as electrical insulation.

Insulated. A conductor encased within material of composition and thickness that is recognized by this *Code* as electrical insulation.

Conduit Body. A separate portion of a conduit or tubing system that provides access through a removable cover(s) to the interior of the system at a junction of two or more sections of the system or at a terminal point of the system. Boxes such as FS and FD or larger cast or sheet metal boxes are not classified as conduit bodies.

Connector, Pressure (Solderless). A device that establishes a connection between two or more conductors or between one or more conductors and a terminal by means of mechanical pressure and without the use of solder.

Continuous Load. A load where the maximum current is expected to continue for 3 hours or more.

Controller. A device or group of devices that serves to govern, in some predetermined manner, the electric power delivered to the apparatus to which it is connected.

Cooking Unit, Counter-Mounted. A cooking appliance designed for mounting in or on a counter and consisting of one or more heating elements, internal wiring, and built-in or separately mountable controls.

Copper-Clad Aluminum Conductors. Conductors drawn from a copper-clad aluminum rod with the copper metallurgically bonded to an aluminum core. The copper forms a minimum of 10 percent of the cross-sectional area of a solid conductor or each strand of a stranded conductor.

Cutout Box. An enclosure designed for surface mounting that has swinging doors or covers secured directly to and telescoping with the walls of the box proper.

Dead Front. Without live parts exposed to a person on the operating side of the equipment.

Demand Factor. The ratio of the maximum demand of a system, or part of a system, to the total connected load of a system or the part of the system under consideration.

Device. A unit of an electrical system that is intended to carry but not utilize electric energy.

Disconnecting Means. A device, or group of devices, or other means by which the conductors of a circuit can be disconnected from their source of supply.

Dustproof. Constructed or protected so that dust will not interfere with its successful operation.

Dusttight. Constructed so that dust will not enter the enclosing case under specified test conditions.

Duty.

Continuous Duty. Operation at a substantially constant load for an indefinitely long time.

Intermittent Duty. Operation for alternate intervals of (1) load and no load; or (2) load and rest; or (3) load, no load, and rest.

Periodic Duty. Intermittent operation in which the load conditions are regularly recurrent.

Short-Time Duty. Operation at a substantially constant load for a short and definitely specified time.

Varying Duty. Operation at loads, and for intervals of time, both of which may be subject to wide variation.

Dwelling.

Dwelling Unit. One or more rooms for the use of one or more persons as a housekeeping unit with space for eating, living, and sleeping, and permanent provisions for cooking and sanitation.

Multifamily Dwelling. A building that contains three or more dwelling units.

One-Family Dwelling. A building that consists solely of one dwelling unit.

Two-Family Dwelling. A building that consists solely of two dwelling units.

Electric Sign. A fixed, stationary, or portable self-contained, electrically illuminated utilization equipment with words or symbols designed to convey information or attract attention.

Enclosed. Surrounded by a case, housing, fence, or walls that prevent persons from accidentally contacting energized parts.

Enclosure. The case or housing of apparatus, or the fence or walls surrounding an installation to prevent personnel from accidentally contacting energized parts, or to protect the equipment from physical damage.

FPN: See Table 430-91 for examples of enclosure types.

Energized. Electrically connected to a source of potential difference.

Equipment. A general term including material, fittings, devices, appliances, fixtures, apparatus, and the like used as a part of, or in connection with, an electrical installation.

Explosionproof Apparatus. Apparatus enclosed in a case that is capable of withstanding an explosion of a specified gas or vapor that may occur within it and of preventing the ignition of a specified gas or vapor surrounding the enclosure by sparks, flashes, or explosion of the gas or vapor within, and that operates at such an external temperature that a surrounding flammable atmosphere will not be ignited thereby.

FPN: For further information, see *Explosion-Proof and Dust-Ignition-Proof Electrical Equipment for Use in Hazardous (Classified) Locations,* ANSI/UL 1203-1994.

Exposed (as applied to live parts). Capable of being inadvertently touched or approached nearer than a safe distance by a person. It is applied to parts that are not suitably guarded, isolated, or insulated.

Exposed (as applied to wiring methods). On or attached to the surface or behind panels designed to allow access.

Externally Operable. Capable of being operated without exposing the operator to contact with live parts.

Feeder. All circuit conductors between the service equipment, the source of a separately derived system, or other power supply source and the final branch-circuit overcurrent device.

Festoon Lighting. A string of outdoor lights that is suspended between two points.

Fitting. An accessory such as a locknut, bushing, or other part of a wiring system that is intended primarily to perform a mechanical rather than an electrical function.

Garage. A building or portion of a building in which one or more self-propelled vehicles carrying volatile flammable liquid for fuel or power are kept for use, sale, storage, rental, repair, exhibition, or demonstrating purposes, and all that portion of a building that is on or below the floor or floors in which such vehicles are kept and that is not separated therefrom by suitable cutoffs.

FPN: For commercial garages, repair, and storage, see Section 511-1.

Ground. A conducting connection, whether intentional or accidental, between an electrical circuit or equipment and the earth, or to some conducting body that serves in place of the earth.

Grounded. Connected to earth or to some conducting body that serves in place of the earth.

Grounded, Effectively. Intentionally connected to earth through a ground connection or connections of sufficiently low impedance and having sufficient current-carrying capacity to prevent the buildup of voltages that may result in undue hazards to connected equipment or to persons.

Grounded Conductor. A system or circuit conductor that is intentionally grounded.

Grounding Conductor. A conductor used to connect equipment or the grounded circuit of a wiring system to a grounding electrode or electrodes.

Grounding Conductor, Equipment. The conductor used to connect the noncurrent-carrying metal parts of equipment, raceways, and other enclosures to the system grounded conductor, the grounding electrode conductor, or both, at the service equipment or at the source of a separately derived system.

Grounding Electrode Conductor. The conductor used to connect the grounding electrode to the equipment grounding conductor, to the grounded conductor, or to both, of the circuit at the service equipment or at the source of a separately derived system.

Ground-Fault Circuit Interrupter. A device intended for the protection of personnel that functions to de-energize a circuit or portion thereof within an established period of time when a current to ground exceeds some predetermined value that is less than that required to operate the overcurrent protective device of the supply circuit.

Ground-Fault Protection of Equipment. A system intended to provide protection of equipment from damaging line-to-ground fault currents by operating to cause a disconnecting means to open all ungrounded conductors of the faulted circuit. This protection is provided at current levels less than those required to protect conductors from damage through the operation of a supply circuit overcurrent device.

Guarded. Covered, shielded, fenced, enclosed, or otherwise protected by means of suitable covers, casings, barriers, rails, screens, mats, or platforms to remove the likelihood of approach or contact by persons or objects to a point of danger.

Hoistway. Any shaftway, hatchway, well hole, or other vertical opening or space in which an elevator or dumbwaiter is designed to operate.

Identified (as applied to equipment). Recognizable as suitable for the specific purpose, function, use, environment, application, etc., where described in a particular *Code* requirement.

FPN: Suitability of equipment for a specific purpose, environment, or application may be determined by a qualified testing laboratory, inspection agency, or other organization concerned with product evaluation. Such identification may include labeling or listing. (See definitions of *Labeled* and *Listed*.)

In Sight From (Within Sight From, Within Sight). Where this *Code* specifies that one equipment shall be "in sight from," "within sight from," or "within sight," etc., of another equipment, the specified equipment is to be visible and not more than 50 ft (15.24 m) distant from the other.

Interrupting Rating. The highest current at rated voltage that a device is intended to interrupt under standard test conditions.

FPN: Equipment intended to interrupt current at other than fault levels may have its interrupting rating implied in other ratings, such as horsepower or locked rotor current.

Isolated (as applied to location). Not readily accessible to persons unless special means for access are used.

Labeled. Equipment or materials to which has been attached a label, symbol, or other identifying mark of an organization that is acceptable to the authority having jurisdiction and concerned with product evaluation, that main-

tains periodic inspection of production of labeled equipment or materials, and by whose labeling the manufacturer indicates compliance with appropriate standards or performance in a specified manner.

Lighting Outlet. An outlet intended for the direct connection of a lampholder, a lighting fixture, or a pendant cord terminating in a lampholder.

Listed. Equipment, materials, or services included in a list published by an organization that is acceptable to the authority having jurisdiction and concerned with evaluation of products or services, that maintains periodic inspection of production of listed equipment or materials or periodic evaluation of services, and whose listing states that either the equipment, material, or services meets identified standards or has been tested and found suitable for a specified purpose.

FPN: The means for identifying listed equipment may vary for each organization concerned with product evaluation, some of which do not recognize equipment as listed unless it is also labeled. Use of the system employed by the listing organization allows the authority having jurisdiction to identify a listed product.

Live Parts. Electric conductors, buses, terminals, or components that are uninsulated or exposed and a shock hazard exists.

Location.

Damp Location. Partially protected locations under canopies, marquees, roofed open porches, and like locations, and interior locations subject to moderate degrees of moisture, such as some basements, some barns, and some cold-storage warehouses.

Dry Location. A location not normally subject to dampness or wetness. A location classified as dry may be temporarily subject to dampness or wetness, as in the case of a building under construction.

Wet Location. Installations underground or in concrete slabs or masonry in direct contact with the earth, and locations subject to saturation with water or other liquids, such as vehicle washing areas, and locations exposed to weather and unprotected.

Motor Control Center. An assembly of one or more enclosed sections having a common power bus and principally containing motor control units.

Multioutlet Assembly. A type of surface, flush, or free-standing raceway; designed to hold conductors and receptacles, assembled in the field or at the factory.

Nonautomatic. Action requiring personal intervention for its control. As applied to an electric controller, nonautomatic control does not necessarily imply a manual controller, but only that personal intervention is necessary.

Nonincendive Circuit. A circuit, other than field wiring, in which any arc or thermal effect produced under intended operating conditions of the equipment, is not capable, under specified test conditions, of igniting the flammable gas–, vapor–, or dust–air mixture.

FPN: For test conditions, see *Nonincendive Electrical Equipment for Use in Class I and II, Division 2 and Class III, Divisions 1 and 2 Hazardous (Classified) Locations,* ANSI/ISA-S12.12-1994.

Nonincendive Field Wiring. Wiring that enters or leaves an equipment enclosure and, under normal operating conditions of the equipment, is not capable, due to arcing or thermal effects, of igniting the flammable gas–, vapor–, or dust–air mixture. Normal operation includes opening, shorting, or grounding the field wiring.

Nonlinear Load. A load where the wave shape of the steady-state current does not follow the wave shape of the applied voltage.

FPN: Electronic equipment, electronic/electric-discharge lighting, adjustable-speed drive systems, and similar equipment may be nonlinear loads.

Outlet. A point on the wiring system at which current is taken to supply utilization equipment.

Outline Lighting. An arrangement of incandescent lamps or electric-discharge lighting to outline or call attention to certain features such as the shape of a building or the decoration of a window.

Oven, Wall-Mounted. An oven for cooking purposes and consisting of one or more heating elements, internal wiring, and built-in or separately mountable controls.

Overcurrent. Any current in excess of the rated current of equipment or the ampacity of a conductor. It may result from overload, short circuit, or ground fault.

FPN: A current in excess of rating may be accommodated by certain equipment and conductors for a given set of conditions. Therefore the rules for overcurrent protection are specific for particular situations.

Overload. Operation of equipment in excess of normal, full-load rating, or of a conductor in excess of rated ampacity that, when it persists for a sufficient length of time, would cause damage or dangerous overheating. A fault, such as a short circuit or ground fault, is not an overload.

Panelboard. A single panel or group of panel units designed for assembly in the form of a single panel; including buses, automatic overcurrent devices, and equipped with or

without switches for the control of light, heat, or power circuits; designed to be placed in a cabinet or cutout box placed in or against a wall or partition and accessible only from the front.

Plenum. A compartment or chamber to which one or more air ducts are connected and that forms part of the air distribution system.

Power Outlet. An enclosed assembly that may include receptacles, circuit breakers, fuseholders, fused switches, buses, and watt-hour meter mounting means; intended to supply and control power to mobile homes, recreational vehicles, park trailers, or boats; or to serve as a means for distributing power required to operate mobile or temporarily installed equipment.

Premises Wiring (System). That interior and exterior wiring, including power, lighting, control, and signal circuit wiring together with all of their associated hardware, fittings, and wiring devices, both permanently and temporarily installed, that extends from the service point of utility conductors or source of power such as a battery, a solar photovoltaic system, or a generator, transformer, or converter windings, to the outlet(s). Such wiring does not include wiring internal to appliances, fixtures, motors, controllers, motor control centers, and similar equipment.

Qualified Person. One familiar with the construction and operation of the equipment and the hazards involved.

Raceway. An enclosed channel of metal or nonmetallic materials designed expressly for holding wires, cables, or busbars, with additional functions as permitted in this *Code*. Raceways include, but are not limited to, rigid metal conduit, rigid nonmetallic conduit, intermediate metal conduit, liquidtight flexible conduit, flexible metallic tubing, flexible metal conduit, electrical nonmetallic tubing, electrical metallic tubing, underfloor raceways, cellular concrete floor raceways, cellular metal floor raceways, surface raceways, wireways, and busways.

Rainproof. Constructed, protected, or treated so as to prevent rain from interfering with the successful operation of the apparatus under specified test conditions.

Raintight. Constructed or protected so that exposure to a beating rain will not result in the entrance of water under specified test conditions.

Receptacle. A receptacle is a contact device installed at the outlet for the connection of an attachment plug. A single receptacle is a single contact device with no other contact device on the same yoke. A multiple receptacle is two or more contact devices on the same yoke.

Receptacle Outlet. An outlet where one or more receptacles are installed.

Remote-Control Circuit. Any electric circuit that controls any other circuit through a relay or an equivalent device.

Sealable Equipment. Equipment enclosed in a case or cabinet that is provided with a means of sealing or locking so that live parts cannot be made accessible without opening the enclosure. The equipment may or may not be operable without opening the enclosure.

Separately Derived System. A premises wiring system whose power is derived from a battery, a solar photovoltaic system, or from a generator, transformer, or converter windings, and that has no direct electrical connection, including a solidly connected grounded circuit conductor, to supply conductors originating in another system.

Service. The conductors and equipment for delivering electric energy from the serving utility to the wiring system of the premises served.

Service Cable. Service conductors made up in the form of a cable.

Service Conductors. The conductors from the service point to the service disconnecting means.

Service Drop. The overhead service conductors from the last pole or other aerial support to and including the splices, if any, connecting to the service-entrance conductors at the building or other structure.

Service-Entrance Conductors, Overhead System. The service conductors between the terminals of the service equipment and a point usually outside the building, clear of building walls, where joined by tap or splice to the service drop.

Service-Entrance Conductors, Underground System. The service conductors between the terminals of the service equipment and the point of connection to the service lateral.

> FPN: Where service equipment is located outside the building walls, there may be no service-entrance conductors, or they may be entirely outside the building.

Service Equipment. The necessary equipment, usually consisting of a circuit breaker(s) or switch(es) and fuse(s) and their accessories, connected to the load end of service conductors to a building or other structure, or an otherwise designated area, and intended to constitute the main control and cutoff of the supply.

Service Lateral. The underground service conductors between the street main, including any risers at a pole or other structure or from transformers, and the first point of connection to the service-entrance conductors in a terminal box or meter or other enclosure, inside or outside the building wall. Where there is no terminal box, meter, or other enclo-sure, the point of connection shall be considered to be the point of entrance of the service conductors into the building.

Service Point. The point of connection between the facilities of the serving utility and the premises wiring.

Show Window. Any window used or designed to be used for the display of goods or advertising material, whether it is fully or partly enclosed or entirely open at the rear and whether or not it has a platform raised higher than the street floor level.

Signaling Circuit. Any electric circuit that energizes signaling equipment.

Solar Photovoltaic System. The total components and subsystems that, in combination, convert solar energy into electrical energy suitable for connection to a utilization load.

Special Permission. The written consent of the authority having jurisdiction.

Switchboard. A large single panel, frame, or assembly of panels on which are mounted, on the face or back, or both, switches, overcurrent and other protective devices, buses, and usually instruments. Switchboards are generally accessible from the rear as well as from the front and are not intended to be installed in cabinets.

Switches.

Bypass Isolation Switch. A manually operated device used in conjunction with a transfer switch to provide a means of directly connecting load conductors to a power source, and of disconnecting the transfer switch.

General-Use Snap Switch. A form of general-use switch constructed so that it can be installed in device boxes or on box covers, or otherwise used in conjunction with wiring systems recognized by this *Code*.

General-Use Switch. A switch intended for use in general distribution and branch circuits. It is rated in amperes, and it is capable of interrupting its rated current at its rated voltage.

Isolating Switch. A switch intended for isolating an electric circuit from the source of power. It has no interrupting rating, and it is intended to be operated only after the circuit has been opened by some other means.

Motor-Circuit Switch. A switch rated in horsepower that is capable of interrupting the maximum operating overload current of a motor of the same horsepower rating as the switch at the rated voltage.

Transfer Switch. An automatic or nonautomatic device for transferring one or more load conductor connections from one power source to another.

Thermally Protected (as applied to motors). The words "Thermally Protected" appearing on the nameplate of a motor or motor-compressor indicate that the motor is provided with a thermal protector.

Thermal Protector (as applied to motors). A protective device for assembly as an integral part of a motor or motor-compressor that, when properly applied, protects the motor against dangerous overheating due to overload and failure to start.

FPN: The thermal protector may consist of one or more sensing elements integral with the motor or motor-compressor and an external control device.

Utilization Equipment. Equipment that utilizes electric energy for electronic, electromechanical, chemical, heating, lighting, or similar purposes.

Ventilated. Provided with a means to permit circulation of air sufficient to remove an excess of heat, fumes, or vapors.

Volatile Flammable Liquid. A flammable liquid having a flash point below 38°C (100°F), or a flammable liquid whose temperature is above its flash point, or a Class II combustible liquid having a vapor pressure not exceeding 40 psia (276 kPa) at 38°C (100°F) whose temperature is above its flash point.

Voltage (of a circuit). The greatest root-mean-square (rms) (effective) difference of potential between any two conductors of the circuit concerned.

FPN: Some systems, such as 3-phase 4-wire, single-phase 3-wire, and 3-wire direct current, may have various circuits of various voltages.

Voltage, Nominal. A nominal value assigned to a circuit or system for the purpose of conveniently designating its voltage class (e.g., 120/240 volts, 480Y/277 volts, 600 volts).

The actual voltage at which a circuit operates can vary from the nominal within a range that permits satisfactory operation of equipment.

FPN: See *Voltage Ratings for Electric Power Systems and Equipment (60 Hz)*, ANSI C84.1-1995.

Voltage to Ground. For grounded circuits, the voltage between the given conductor and that point or conductor of the circuit that is grounded; for ungrounded circuits, the greatest voltage between the given conductor and any other conductor of the circuit.

Watertight. Constructed so that moisture will not enter the enclosure under specified test conditions.

Weatherproof. Constructed or protected so that exposure to the weather will not interfere with successful operation.

FPN: Rainproof, raintight, or watertight equipment can fulfill the requirements for weatherproof where varying weather conditions other than wetness, such as snow, ice, dust, or temperature extremes, are not a factor.

B. Over 600 Volts, Nominal

Whereas the preceding definitions are intended to apply wherever the terms are used throughout this *Code*, the following definitions are applicable only to parts of the article specifically covering installations and equipment operating at over 600 volts, nominal.

Electronically Actuated Fuse. An overcurrent protective device that generally consists of a control module that provides current sensing, electronically derived time-current characteristics, energy to initiate tripping, and an interrupting module that interrupts current when an overcurrent occurs. Electronically actuated fuses may or may not operate in a current-limiting fashion, depending on the type of control selected.

Fuse. An overcurrent protective device with a circuit-opening fusible part that is heated and severed by the passage of overcurrent through it.

FPN: A fuse comprises all the parts that form a unit capable of performing the prescribed functions. It may or may not be the complete device necessary to connect it into an electrical circuit.

Controlled Vented Power Fuse. A fuse with provision for controlling discharge circuit interruption such that no solid material may be exhausted into the surrounding atmosphere.

FPN: The fuse is designed so that discharged gases will not ignite or damage insulation in the path of the discharge or propagate a flashover to or between grounded members or conduction members in the path of the discharge where the distance between the vent and such insulation or conduction members conforms to manufacturer's recommendations.

Expulsion Fuse Unit (Expulsion Fuse). A vented fuse unit in which the expulsion effect of gases produced by the arc and lining of the fuseholder, either alone or aided by a spring, extinguishes the arc.

Nonvented Power Fuse. A fuse without intentional provision for the escape of arc gases, liquids, or solid particles to the atmosphere during circuit interruption.

Power Fuse Unit. A vented, nonvented, or controlled vented fuse unit in which the arc is extinguished by being drawn through solid material, granular material, or liquid, either alone or aided by a spring.

Vented Power Fuse. A fuse with provision for the escape of arc gases, liquids, or solid particles to the surrounding atmosphere during circuit interruption.

Multiple Fuse. An assembly of two or more single-pole fuses.

Switching Device. A device designed to close, open, or both, one or more electric circuits.

Switching Devices.

Circuit Breaker. A switching device capable of making, carrying, and breaking currents under normal circuit conditions, and also making, carrying for a specified time, and breaking currents under specified abnormal circuit conditions, such as those of short circuit.

Cutout. An assembly of a fuse support with either a fuseholder, fuse carrier, or disconnecting blade. The fuseholder or fuse carrier may include a conducting element (fuse link), or may act as the disconnecting blade by the inclusion of a nonfusible member.

Disconnecting (or Isolating) Switch (Disconnector, Isolator). A mechanical switching device used for isolating a circuit or equipment from a source of power.

Disconnecting Means. A device, group of devices, or other means whereby the conductors of a circuit can be disconnected from their source of supply.

Interrupter Switch. A switch capable of making, carrying, and interrupting specified currents.

Oil Cutout (Oil-Filled Cutout). A cutout in which all or part of the fuse support and its fuse link or disconnecting blade is mounted in oil with complete immersion of the contacts and the fusible portion of the conducting element (fuse link) so that arc interruption by severing of the fuse link or by opening of the contacts will occur under oil.

Oil Switch. A switch having contacts that operate under oil (or askarel or other suitable liquid).

Regulator Bypass Switch. A specific device or combination of devices designed to bypass a regulator.

Article 110 — Requirements for Electrical Installations

A. General

110-2. Approval. The conductors and equipment required or permitted by this *Code* shall be acceptable only if approved.

FPN: See Examination of Equipment for Safety, Section 90-7, and Examination, Identification, Installation, and Use of Equipment, Section 110-3. See definitions of *Approved, Identified, Labeled,* and *Listed.*

110-3. Examination, Identification, Installation, and Use of Equipment.

(a) Examination. In judging equipment, considerations such as the following shall be evaluated:

(1) Suitability for installation and use in conformity with the provisions of this *Code*

FPN: Suitability of equipment use may be identified by a description marked on or provided with a product to identify the suitability of the product for a specific purpose, environment, or application. Suitability of equipment may be evidenced by listing or labeling.

(2) Mechanical strength and durability, including, for parts designed to enclose and protect other equipment, the adequacy of the protection thus provided
(3) Wire-bending and connection space
(4) Electrical insulation
(5) Heating effects under normal conditions of use and also under abnormal conditions likely to arise in service
(6) Arcing effects
(7) Classification by type, size, voltage, current capacity, and specific use
(8) Other factors that contribute to the practical safeguarding of persons using or likely to come in contact with the equipment

(b) Installation and Use. Listed or labeled equipment shall be installed and used in accordance with any instructions included in the listing or labeling.

110-4. Voltages. Throughout this *Code*, the voltage considered shall be that at which the circuit operates. The voltage rating of electrical equipment shall not be less than the nominal voltage of a circuit to which it is connected.

110-5. Conductors. Conductors normally used to carry current shall be of copper unless otherwise provided in this *Code*. Where the conductor material is not specified, the material and the sizes given in this *Code* shall apply to copper conductors. Where other materials are used, the size shall be changed accordingly.

FPN: For aluminum and copper-clad aluminum conductors, see Section 310-15.

110-6. Conductor Sizes. Conductor sizes are expressed in American Wire Gage (AWG) or in circular mils.

110-7. Insulation Integrity. Completed wiring installations shall be free from short circuits and from grounds other than as required or permitted in Article 250.

110-8. Wiring Methods. Only wiring methods recognized as suitable are included in this *Code*. The recognized methods of wiring shall be permitted to be installed in any type of building or occupancy, except as otherwise provided in this *Code*.

110-9. Interrupting Rating. Equipment intended to interrupt current at fault levels shall have an interrupting rating sufficient for the nominal circuit voltage and the current that is available at the line terminals of the equipment.

Equipment intended to interrupt current at other than fault levels shall have an interrupting rating at nominal circuit voltage sufficient for the current that must be interrupted.

110-10. Circuit Impedance and Other Characteristics. The overcurrent protective devices, the total impedance, the component short-circuit current ratings, and other characteristics of the circuit to be protected shall be selected and coordinated to permit the circuit-protective devices used to clear a fault to do so without extensive damage to the electrical components of the circuit. This fault shall be assumed to be either between two or more of the circuit conductors, or between any circuit conductor and the grounding conductor or enclosing metal raceway. Listed products applied in accordance with their listing shall be considered to meet the requirements of this section.

110-11. Deteriorating Agents. Unless identified for use in the operating environment, no conductors or equipment shall be located in damp or wet locations; where exposed to gases, fumes, vapors, liquids, or other agents that have a deteriorating effect on the conductors or equipment; or where exposed to excessive temperatures.

FPN No. 1: See Section 300-6 for protection against corrosion.

FPN No. 2: Some cleaning and lubricating compounds can cause severe deterioration of many plastic materials used for insulating and structural applications in equipment.

Equipment identified only as "dry locations," "Type 1," or "indoor use only" shall be protected against permanent damage from the weather during building construction.

110-12. Mechanical Execution of Work. Electrical equipment shall be installed in a neat and workmanlike manner.

(a) Unused Openings. Unused openings in boxes, raceways, auxiliary gutters, cabinets, equipment cases, or housings shall be effectively closed to afford protection substantially equivalent to the wall of the equipment.

(b) Subsurface Enclosures. Conductors shall be racked to provide ready and safe access in underground and subsurface enclosures, into which persons enter for installation and maintenance.

(c) Integrity of Electrical Equipment and Connections.

Internal parts of electrical equipment, including busbars, wiring terminals, insulators, and other surfaces, shall not be damaged or contaminated by foreign materials such as paint, plaster, cleaners, abrasives, or corrosive residues. There shall be no damaged parts that may adversely affect safe operation or mechanical strength of the equipment such as parts that are broken; bent; cut; or deteriorated by corrosion, chemical action, or overheating.

110-13. Mounting and Cooling of Equipment.

(a) Mounting. Electrical equipment shall be firmly secured to the surface on which it is mounted. Wooden plugs driven into holes in masonry, concrete, plaster, or similar materials shall not be used.

(b) Cooling. Electrical equipment that depends upon the natural circulation of air and convection principles for cooling of exposed surfaces shall be installed so that room airflow over such surfaces is not prevented by walls or by adjacent installed equipment. For equipment designed for floor mounting, clearance between top surfaces and adjacent surfaces shall be provided to dissipate rising warm air.

Electrical equipment provided with ventilating openings shall be installed so that walls or other obstructions do not prevent the free circulation of air through the equipment.

110-14. Electrical Connections. Because of different characteristics of dissimilar metals, devices such as pressure terminal or pressure splicing connectors and soldering lugs shall be identified for the material of the conductor and shall be properly installed and used. Conductors of dissimilar metals shall not be intermixed in a terminal or splicing connector where physical contact occurs between dissimilar conductors (such as copper and aluminum, copper and copper-clad aluminum, or aluminum and copper-clad aluminum), unless the device is identified for the purpose and conditions of use. Materials such as solder, fluxes, inhibitors, and compounds, where employed, shall be suitable for the use and shall be of a type that will not adversely affect the conductors, installation, or equipment.

FPN: Many terminations and equipment are marked with a tightening torque.

(a) Terminals. Connection of conductors to terminal parts shall ensure a thoroughly good connection without damaging the conductors and shall be made by means of pressure connectors (including set-screw type), solder lugs, or splices to flexible leads. Connection by means of wire-binding screws or studs and nuts that have upturned lugs or the equivalent shall be permitted for No. 10 or smaller conductors.

Terminals for more than one conductor and terminals used to connect aluminum shall be so identified.

(b) Splices. Conductors shall be spliced or joined with splicing devices identified for the use or by brazing, welding, or soldering with a fusible metal or alloy. Soldered splices shall first be spliced or joined so as to be mechanically and electrically secure without solder and then soldered. All splices and joints and the free ends of conductors shall be covered with an insulation equivalent to that of the conductors or with an insulating device identified for the purpose.

Wire connectors or splicing means installed on conductors for direct burial shall be listed for such use.

(c) Temperature Limitations. The temperature rating associated with the ampacity of a conductor shall be selected and coordinated so as not to exceed the lowest temperature rating of any connected termination, conductor, or device. Conductors with temperature ratings higher than specified for terminations shall be permitted to be used for ampacity adjustment, correction, or both.

(1) Termination provisions of equipment for circuits rated 100 amperes or less, or marked for Nos. 14 through 1 conductors, shall be used only for one of the following.

(a) Conductors rated 60°C (140°F), or
(b) Conductors with higher temperature ratings, provided the ampacity of such conductors is determined based on the 60°C (140°F) ampacity of the conductor size used, or
(c) Conductors with higher temperature ratings if the equipment is listed and identified for use with such conductors, or
(d) For motors marked with design letters B, C, D, or E, conductors having an insulation rating of 75°C (167°F) or higher shall be permitted to be used provided the ampacity of such conductors does not exceed the 75°C (167°F) ampacity.

(2) Termination provisions of equipment for circuits rated over 100 amperes, or marked for conductors larger than No. 1, shall be used only for

(a) Conductors rated 75°C (167°F), or
(b) Conductors with higher temperature ratings provided the ampacity of such conductors does not exceed the 75°C (167°F) ampacity of the conductor size used, or up to their ampacity if the equipment is listed and identified for use with such conductors.

(3) Separately installed pressure connectors shall be used with conductors at the ampacities not exceeding the ampacity at the listed and identified temperature rating of the connector.

FPN: With respect to Sections 110-14(c)(1), (2), and (3), equipment markings or listing information may additionally

restrict the sizing and temperature ratings of connected conductors.

110-18. Arcing Parts. Parts of electric equipment, which in ordinary operation produce arcs, sparks, flames, or molten metal, shall be enclosed or separated and isolated from all combustible material.

FPN: For hazardous (classified) locations, see Articles 500 through 517. For motors, see Section 430-14.

110-19. Light and Power from Railway Conductors. Circuits for lighting and power shall not be connected to any system that contains trolley wires with a ground return.

Exception: Car houses, power houses, or passenger and freight stations operated in connection with electric railways.

110-21. Marking. The manufacturer's name, trademark, or other descriptive marking by which the organization responsible for the product can be identified shall be placed on all electric equipment. Other markings that indicate voltage, current, wattage, or other ratings shall be provided as specified elsewhere in this *Code*. The marking shall be of sufficient durability to withstand the environment involved.

110-22. Identification of Disconnecting Means. Each disconnecting means required by this *Code* for motors and appliances, and each service, feeder, or branch circuit at the point where it originates, shall be legibly marked to indicate its purpose unless located and arranged so the purpose is evident. The marking shall be of sufficient durability to withstand the environment involved.

Where circuit breakers or fuses are applied in compliance with the series combination ratings marked on the equipment by the manufacturer, the equipment enclosure(s) shall be legibly marked in the field to indicate the equipment has been applied with a series combination rating. The marking shall be readily visible and state the following:

CAUTION — SERIES COMBINATION SYSTEM
RATED _____ AMPERES. IDENTIFIED
REPLACEMENT COMPONENTS REQUIRED.

FPN: See Section 240-83(c) for interrupting rating marking for end-use equipment.

B. 600 Volts, Nominal, or Less

110-26. Spaces About Electrical Equipment. Sufficient access and working space shall be provided and maintained about all electric equipment to permit ready and safe operation and maintenance of such equipment. Enclosures housing electrical apparatus that are controlled by lock and key shall be considered accessible to qualified persons.

(a) Working Space. Working space for equipment operating at 600 volts, nominal, or less to ground and likely to require examination, adjustment, servicing, or maintenance while energized shall comply with the dimensions of (1), (2), and (3) or as required or permitted elsewhere in this *Code.*

(1) Depth of Working Space. The depth of the working space in the direction of access to live parts shall not be less than indicated in Table 110-26(a). Distances shall be measured from the live parts if such are exposed or from the enclosure front or opening if such are enclosed.

Table 110-26(a). Working Spaces

Nominal Voltage to Ground	Minimum Clear Distance (ft)		
	Condition 1	Condition 2	Condition 3
0–150	3	3	3
151–600	3	3½	4

Notes:
1. For SI units, 1 ft = 0.3048 m.
2. Where the conditions are as follows:
Condition 1 — Exposed live parts on one side and no live or grounded parts on the other side of the working space, or exposed live parts on both sides effectively guarded by suitable wood or other insulating materials. Insulated wire or insulated busbars operating at not over 300 volts to ground shall not be considered live parts.
Condition 2 — Exposed live parts on one side and grounded parts on the other side. Concrete, brick, or tile walls shall be considered as grounded.
Condition 3 — Exposed live parts on both sides of the work space (not guarded as provided in Condition 1) with the operator between.

Exception No. 1: Working space shall not be required in back or sides of assemblies, such as dead-front switchboards or motor control centers, where there are no renewable or adjustable parts, such as fuses or switches, on the back or sides and where all connections are accessible from locations other than the back or sides. Where rear access is required to work on de-energized parts on the back of enclosed equipment, a minimum working space of 30 in. (762 mm) horizontally shall be provided.

Exception No. 2: By special permission, smaller spaces shall be permitted where all uninsulated parts are at a voltage no greater than 30 volts rms, 42 volts peak, or 60 volts dc.

Exception No. 3: In existing buildings where electrical equipment is being replaced, Condition 2 working clearance shall be permitted between dead-front switchboards, panelboards, or motor control centers located across the aisle from each other where conditions of maintenance and supervision ensure that written procedures have been adopted to prohibit equipment on both sides of the aisle from being open at the same time and qualified persons who are authorized will service the installation.

(2) Width of Working Space. The width of the working space in front of the electric equipment shall be the width of the equipment or 30 in. (762 mm), whichever is greater. In all cases, the work space shall permit at least a 90 degree opening of equipment doors or hinged panels.

(3) Height of Working Space. The work space shall be clear and extend from the grade, floor, or platform to the height required by Section 110-26(e). Within the height requirements of this section, other equipment associated with the electrical installation located above or below the electrical equipment shall be permitted to extend not more than 6 in. (153 mm) beyond the front of the electrical equipment.

(b) Clear Spaces. Working space required by this section shall not be used for storage. When normally enclosed live parts are exposed for inspection or servicing, the working space, if in a passageway or general open space, shall be suitably guarded.

(c) Access and Entrance to Working Space. At least one entrance of sufficient area shall be provided to give access to the working space about electric equipment.

For equipment rated 1200 amperes or more and over 6 ft (1.83 m) wide that contains overcurrent devices, switching devices, or control devices, there shall be one entrance not less than 24 in. (610 mm) wide and 6½ ft (1.98 m) high at each end of the working space.

Exception No. 1: Where the location permits a continuous and unobstructed way of exit travel, one means of access shall be permitted.

Exception No. 2: Where the work space required by Section 110-26(a) is doubled, only one entrance to the working space is required. It shall be located so the edge of the entrance nearest the equipment is the minimum clear distance given in Table 110-26(a) away from such equipment.

(d) Illumination. Illumination shall be provided for all working spaces about service equipment, switchboards, panelboards, or motor control centers installed indoors. Additional lighting fixtures shall not be required where the work space is illuminated by an adjacent light source. In electrical equipment rooms, the illumination shall not be controlled by automatic means only.

(e) Headroom. The minimum headroom of working spaces about service equipment, switchboards, panelboards, or motor control centers shall be 6½ ft (1.98 m). Where the electrical equipment exceeds 6½ ft (1.98 m) in height, the minimum headroom shall not be less than the height of the equipment.

Exception: Service equipment or panelboards, in existing dwelling units, that do not exceed 200 amperes.

(f) Dedicated Equipment Space. Equipment within the scope of Article 384, and motor control centers, shall be located in dedicated spaces and protected from damage as covered in (1) and (2).

Exception: Control equipment that by its very nature or because of other rules of the Code must be adjacent to or within sight of its operating machinery shall be permitted in those locations.

(1) **Indoor.** For indoor installations, the dedicated space shall comply with the following.

(a) *Dedicated Electrical Space.* The space equal to the width and depth of the equipment and extending from the floor to a height of 6 ft (1.83 m) above the equipment or to the structural ceiling, whichever is lower, shall be dedicated to the electrical installation. No piping, ducts, or equipment foreign to the electrical installation shall be located in this zone.

Exception: Equipment that is isolated from the foreign equipment by height or physical enclosures or covers that will afford adequate mechanical protection from vehicular traffic or accidental contact by unauthorized personnel or that complies with (b), shall be permitted in areas that do not have the dedicated space described in this rule.

(b) *Foreign Systems.* The space equal to the width and depth of the equipment shall be kept clear of foreign systems unless protection is provided to avoid damage from condensation, leaks, or breaks in such foreign systems. This zone shall extend from the top of the electrical equipment to the structural ceiling.

(c) *Sprinkler Protection.* Sprinkler protection shall be permitted for the dedicated space where the piping complies with this section.

(d) *Suspended Ceilings.* A dropped, suspended, or similar ceiling that does not add strength to the building structure shall not be considered a structural ceiling.

(2) **Outdoor.** Outdoor electrical equipment shall be installed in suitable enclosures and shall be protected from accidental contact by unauthorized personnel, or by vehicular traffic, or by accidental spillage or leakage from piping systems. The working clearance space shall include the zone described in Section 110-26(a). No architectural appurtenance or other equipment shall be located in this zone.

110-27. Guarding of Live Parts.

(a) **Live Parts Guarded Against Accidental Contact.** Except as elsewhere required or permitted by this *Code,* live parts of electric equipment operating at 50 volts or more shall be guarded against accidental contact by approved enclosures or by any of the following means.

(1) By location in a room, vault, or similar enclosure that is accessible only to qualified persons.
(2) By suitable permanent, substantial partitions or screens arranged so that only qualified persons will have access to the space within reach of the live parts. Any openings in such partitions or screens shall be sized and located so that persons are not likely to come into accidental contact with the live parts or to bring conducting objects into contact with them.
(3) By location on a suitable balcony, gallery, or platform elevated and arranged so as to exclude unqualified persons.
(4) By elevation of 8 ft (2.44 m) or more above the floor or other working surface.

(b) **Prevent Physical Damage.** In locations where electric equipment is likely to be exposed to physical damage, enclosures or guards shall be so arranged and of such strength as to prevent such damage.

(c) **Warning Signs.** Entrances to rooms and other guarded locations that contain exposed live parts shall be marked with conspicuous warning signs forbidding unqualified persons to enter.

FPN: For motors, see Sections 430-132 and 430-133. For over 600 volts, see Section 110-34.

C. Over 600 Volts, Nominal

110-30. General. Conductors and equipment used on circuits over 600 volts, nominal, shall comply with Part A of this article and with the following sections, which supplement or modify Part A. In no case shall the provisions of this part apply to equipment on the supply side of the service point.

110-31. Enclosure for Electrical Installations. Electrical installations in a vault, room, or closet or in an area surrounded by a wall, screen, or fence, access to which is controlled by lock and key or other approved means, shall be considered to be accessible to qualified persons only. The type of enclosure used in a given case shall be designed and constructed according to the nature and degree of the hazard(s) associated with the installation.

For installations other than equipment as described in Section 110-31(c), a wall, screen, or fence shall be used to enclose an outdoor electrical installation to deter access by persons who are not qualified. A fence shall not be less than 7 ft (2.13 m) in height or a combination of 6 ft (1.80 m) or more of fence fabric and a 1-ft (305-mm) or more extension utilizing three or more strands of barbed wire or equivalent.

FPN: See Article 450 for construction requirements for transformer vaults.

(a) Indoor Installations.

(1) In Places Accessible to Unqualified Persons. Indoor electrical installations that are open to unqualified persons shall be made with metal-enclosed equipment or shall be enclosed in a vault or in an area to which access is controlled by a lock. Metal-enclosed switchgear, unit substations, transformers, pull boxes, connection boxes, and other similar associated equipment shall be marked with appropriate caution signs. Openings in ventilated dry-type transformers or similar openings in other equipment shall be designed so that foreign objects inserted through these openings will be deflected from energized parts.

(2) In Places Accessible to Qualified Persons Only. Indoor electrical installations considered accessible only to qualified persons in accordance with this section shall comply with Sections 110-34, 110-35, and 110-36.

(b) Outdoor Installations.

(1) In Places Accessible to Unqualified Persons. Outdoor electrical installations that are open to unqualified persons shall comply with Article 225.

FPN: For clearances of conductors for system voltages over 600 volts, nominal, see *National Electrical Safety Code*, ANSI C2-1997.

(2) In Places Accessible to Qualified Persons Only. Outdoor electrical installations that have exposed live parts shall be accessible to qualified persons only in accordance with the first paragraph of this section and shall comply with Sections 110-34, 110-35, and 110-36.

(c) Enclosed Equipment Accessible to Unqualified Persons. Ventilating or similar openings in equipment shall be designed so that foreign objects inserted through these openings will be deflected from energized parts. Where exposed to physical damage from vehicular traffic, suitable guards shall be provided. Nonmetallic or metal-enclosed equipment located outdoors and accessible to the general public shall be designed so that exposed nuts or bolts cannot be readily removed, permitting access to live parts. Where nonmetallic or metal-enclosed equipment is accessible to the general public and the bottom of the enclosure is less than 8 ft (2.44 m) above the floor or grade level, the enclosure door or hinged cover shall be kept locked. Doors and covers of enclosures used solely as pull boxes, splice boxes, or junction boxes shall be locked, bolted, or screwed on. Underground box covers that weigh over 100 lb (45.4 kg) shall be considered as meeting this requirement.

110-32. Work Space About Equipment. Sufficient space shall be provided and maintained about electric equipment to permit ready and safe operation and maintenance of such equipment. Where energized parts are exposed, the minimum clear work space shall not be less than 6½ ft (1.98 m) high (measured vertically from the floor or platform), or less than 3 ft (914 mm) wide (measured parallel to the equipment). The depth shall be as required in Section 110-34(a). In all cases, the work space shall be adequate to permit at least a 90 degree opening of doors or hinged panels.

110-33. Entrance and Access to Work Space.

(a) Entrance. At least one entrance not less than 24 in. (610 mm) wide and 6½ ft (1.98 m) high shall be provided to give access to the working space about electric equipment.

(1) On switchboard and control panels exceeding 6 ft (1.83 m) in width, there shall be one entrance at each end of such boards unless the location of the switchboards and control panels permits a continuous and unobstructed way of exit travel, or unless the work space required in Section 110-34(a) is doubled.

(2) Where one entrance to the working space is permitted under the conditions described in (1), the entrance shall be located so that the edge of the entrance nearest the switchboards and control panels is the minimum clear distance given in Table 110-34(a) away from such equipment.

(3) Where bare energized parts at any voltage or insulated energized parts above 600 volts, nominal, to ground are located adjacent to such entrance, they shall be suitably guarded.

(b) Access. Permanent ladders or stairways shall be provided to give safe access to the working space around electric equipment installed on platforms, balconies, mezzanine floors, or in attic or roof rooms or spaces.

110-34. Work Space and Guarding.

(a) Working Space. Except as elsewhere required or permitted in this *Code,* the minimum clear working space in the direction of access to live parts of electrical equipment shall not be less than specified in Table 110-34(a). Distances shall be measured from the live parts, if such are exposed, or from the enclosure front or opening if such are enclosed.

Exception: Working space shall not be required in back of equipment such as dead-front switchboards or control assemblies where there are no renewable or adjustable parts (such as fuses or switches) on the back and where all connections are accessible from locations other than the back. Where rear access is required to work on de-energized parts on the back of enclosed equipment, a minimum working space of 30 in. (762 mm) horizontally shall be provided.

(b) Separation from Low-Voltage Equipment. Where switches, cutouts, or other equipment operating at 600 volts,

Table 110-34(a). Minimum Depth of Clear Working Space at Electrical Equipment

Nominal Voltage to Ground	Minimum Clear Distance (ft)		
	Condition 1	Condition 2	Condition 3
601–2500 V	3	4	5
2501–9000 V	4	5	6
9001–25,000 V	5	6	9
25,001–75 kV	6	8	10
Above 75 kV	8	10	12

Notes:
1. For SI units, 1 ft = 0.3048 m.
2. Where the conditions are as follows:

Condition 1 — Exposed live parts on one side and no live or grounded parts on the other side of the working space, or exposed live parts on both sides effectively guarded by suitable wood or other insulating materials. Insulated wire or insulated busbars operating at not over 300 volts shall not be considered live parts.

Condition 2 — Exposed live parts on one side and grounded parts on the other side. Concrete, brick, or tile walls will be considered as grounded surfaces.

Condition 3 — Exposed live parts on both sides of the work space (not guarded as provided in Condition 1) with the operator between.

nominal, or less, are installed in a room or enclosure where there are exposed live parts or exposed wiring operating at over 600 volts, nominal, the high-voltage equipment shall be effectively separated from the space occupied by the low-voltage equipment by a suitable partition, fence, or screen.

Exception: Switches or other equipment operating at 600 volts, nominal, or less, and serving only equipment within the high-voltage vault, room, or enclosure shall be permitted to be installed in the high-voltage enclosure, room, or vault if accessible to qualified persons only.

(c) Locked Rooms or Enclosures. The entrances to all buildings, rooms, or enclosures containing exposed live parts or exposed conductors operating at over 600 volts, nominal, shall be kept locked unless such entrances are under the observation of a qualified person at all times.

Where the voltage exceeds 600 volts, nominal, permanent and conspicuous warning signs shall be provided, reading as follows:

DANGER — HIGH VOLTAGE — KEEP OUT

(d) Illumination. Illumination shall be provided for all working spaces about electrical equipment. The lighting outlets shall be arranged so that persons changing lamps or making repairs on the lighting system will not be endangered by live parts or other equipment.

The points of control shall be located so that persons are not likely to come in contact with any live part or moving part of the equipment while turning on the lights.

(e) Elevation of Unguarded Live Parts. Unguarded live parts above working space shall be maintained at elevations not less than required by Table 110-34(e).

Table 110-34(e). Elevation of Unguarded Live Parts Above Working Space

Nominal Voltage Between Phases	Elevation
601–7500 V	8 ft 6 in.
7501–35,000 V	9 ft
Over 35 kV	9 ft + 0.37 in./kV above 35

Note: For SI units, 1 in. = 25.4 mm; 1 ft = 0.3048 m.

(f) Protection of Service Equipment, Metal-Enclosed Power Switchgear, and Industrial Control Assemblies. Pipes or ducts foreign to the electrical installation that require periodic maintenance or whose malfunction would endanger the operation of the electrical system shall not be located in the vicinity of the service equipment, metal-enclosed power switchgear, or industrial control assemblies. Protection shall be provided where necessary to avoid damage from condensation leaks and breaks in such foreign systems. Piping and other facilities shall not be considered foreign if provided for fire protection of the electrical installation.

110-36. Circuit Conductors. Circuit conductors shall be permitted to be installed in raceways, in cable trays, as metal-clad cable, as bare wire, cable, and busbars, or as Type MV cables, or conductors as provided in Sections 300-37, 300-39, 300-40, and 300-50. Bare live conductors shall conform with Section 490-24.

Insulators, together with their mounting and conductor attachments, where used as supports for wires, single-conductor cables, or busbars, shall be capable of safely withstanding the maximum magnetic forces that would prevail when two or more conductors of a circuit were subjected to short-circuit current.

Open runs of insulated wires and cables that have a bare lead sheath or a braided outer covering shall be supported in a manner designed to prevent physical damage to the braid or sheath. Supports for lead-covered cables shall be designed to prevent electrolysis of the sheath.

110-40. Temperature Limitations at Terminations. Conductors shall be permitted to be terminated based on the 90°C (194°F) temperature rating and ampacity as given in Tables 310-67 through 310-86, unless otherwise identified.

D. Tunnel Installations Over 600 Volts, Nominal

110-51. General.

(a) Covered. The provisions of this part shall apply to installation and use of high-voltage power distribution and utilization equipment that is portable and/or mobile, such as substations, trailers, or cars, mobile shovels, draglines, hoists, drills, dredges, compressors, pumps, conveyors, underground excavators, and the like.

(b) Other Articles. The requirements of this part shall be additional to, or amendatory of, those prescribed in Articles 100 through 490 of this *Code*. Special attention shall be paid to Article 250.

(c) Protection Against Physical Damage. Conductors and cables in tunnels shall be located above the tunnel floor and so placed or guarded to protect them from physical damage.

110-52. Overcurrent Protection. Motor-operated equipment shall be protected from overcurrent in accordance with Article 430. Transformers shall be protected from overcurrent in accordance with Article 450.

110-53. Conductors. High-voltage conductors in tunnels shall be installed in metal conduit or other metal raceway, Type MC cable, or other approved multiconductor cable. Multiconductor portable cable shall be permitted to supply mobile equipment.

110-54. Bonding and Equipment Grounding Conductors.

(a) Grounded and Bonded. All noncurrent-carrying metal parts of electric equipment and all metal raceways and cable sheaths shall be effectively grounded and bonded to all metal pipes and rails at the portal and at intervals not exceeding 1000 ft (305 m) throughout the tunnel.

(b) Equipment Grounding Conductors. An equipment grounding conductor shall be run with circuit conductors inside the metal raceway or inside the multiconductor cable jacket. The equipment grounding conductor shall be permitted to be insulated or bare.

110-55. Transformers, Switches, and Electrical Equipment. All transformers, switches, motor controllers, motors, rectifiers, and other equipment installed below ground shall be protected from physical damage by location or guarding.

110-56. Energized Parts. Bare terminals of transformers, switches, motor controllers, and other equipment shall be enclosed to prevent accidental contact with energized parts.

110-57. Ventilation System Controls. Electrical controls for the ventilation system shall be arranged so that the airflow can be reversed.

110-58. Disconnecting Means. A switching device, meeting the requirements of Article 430 or 450, shall be installed at each transformer or motor location for disconnecting the transformer or motor. The switching device shall open all ungrounded conductors of a circuit simultaneously.

110-59. Enclosures. Enclosures for use in tunnels shall be dripproof, weatherproof, or submersible as required by the environmental conditions. Switch or contactor enclosures shall not be used as junction boxes or raceways for conductors feeding through or tapping off to other switches, unless special designs are used to provide adequate space for this purpose.

CHAPTER 2
Wiring and Protection

Article 200 — Use and Identification of Grounded Conductors

200-1. Scope. This article provides requirements for the following:

(1) Identification of terminals
(2) Grounded conductors in premises wiring systems
(3) Identification of grounded conductors

> FPN: See Article 100 for definitions of *Grounded Conductor* and *Grounding Conductor*.

200-2. General. All premises wiring systems, other than circuits and systems exempted or prohibited by Sections 210-10, 215-7, 250-21, 250-22, 250-162, 503-13, 517-63, 668-11, 668-21, and Section 690-41, Exception, shall have a grounded conductor that is identified in accordance with Section 200-6.

The grounded conductor, where insulated, shall have insulation that is (1) suitable, other than color, for any ungrounded conductor of the same circuit on circuits of less than 1000 volts or impedance grounded neutral systems of 1 kV and over, or (2) rated not less than 600 volts for solidly grounded neutral systems of 1 kV and over as described in Section 250-184(a).

200-3. Connection to Grounded System. Premises wiring shall not be electrically connected to a supply system unless the latter contains, for any grounded conductor of the interior system, a corresponding conductor that is grounded. For the purpose of this section, *electrically connected* shall mean connected so as to be capable of carrying current, as distinguished from connection through electromagnetic induction.

200-6. Means of Identifying Grounded Conductors.

(a) Sizes No. 6 or Smaller. An insulated grounded conductor of No. 6 or smaller shall be identified by a continuous white or natural gray outer finish or by three continuous white stripes on other than green insulation along its entire length. Wires that have their outer covering finished to show a white or natural gray color but have colored tracer threads in the braid identifying the source of manufacture shall be considered as meeting the provisions of this section. Insulated grounded conductors shall also be permitted to be identified as follows.

(1) The grounded conductor of a mineral-insulated, metal-sheathed cable shall be identified at the time of installation by distinctive marking at its terminations.

(2) A single-conductor, sunlight-resistant, outdoor-rated cable used as a grounded conductor in photovoltaic power systems as permitted by Section 690-31 shall be identified at the time of installation by distinctive white marking at all terminations.

(3) Fixture wire shall comply with the requirements for grounded conductor identification as specified in Section 402-8.

(4) For aerial cable, the identification shall be as above, or by means of a ridge located on the exterior of the cable so as to identify it.

(b) Sizes Larger than No. 6. An insulated grounded conductor larger than No. 6 shall be identified either by a continuous white or natural gray outer finish or by three continuous white stripes on other than green insulation along its entire length or at the time of installation by a distinctive white marking at its terminations. This marking shall encircle the conductor or insulation.

(c) Flexible Cords. An insulated conductor that is intended for use as a grounded conductor, where contained within a flexible cord, shall be identified by a white or natural gray outer finish or by three continuous white stripes on other than green insulation or by methods permitted by Section 400-22.

(d) Grounded Conductors of Different Systems. Where conductors of different systems are installed in the same raceway, cable, box, auxiliary gutter, or other type of enclosure, one system grounded conductor, if required, shall have an outer covering conforming to Section 200-6(a) or 200-6(b). Each other system grounded conductor shall have an outer covering of white with a readily distinguishable different colored stripe (not green) running along the insulation, or other and different means of identification as allowed by Section 200-6(a) or (b) that will distinguish each system grounded conductor.

(e) Grounded Conductors of Multiconductor Cables. The insulated grounded conductors in a multiconductor cable shall be identified by a continuous white or natural gray outer finish or by three continuous white stripes on other than green insulation along its entire length. Multiconductor flat cable No. 4 or larger shall be permitted to employ an external ridge on the grounded conductor.

Exception No. 1: Where the conditions of maintenance and supervision ensure that only qualified persons will service the installation, grounded conductors in multiconductor cables shall be permitted to be permanently identified at their terminations at the time of installation by a distinctive white marking or other equally effective means.

Exception No. 2: The grounded conductor of a multiconductor varnished-cloth-insulated cable shall be permitted to be identified at its terminations at the time of installation by a distinctive white marking or other equally effective means.

200-7. Use of Insulation of a White or Natural Gray Color or with Three Continuous White Stripes.

(a) General. The following shall be used only for the grounded circuit conductor, unless otherwise permitted in (b) and (c):

(1) A conductor with continuous white or natural gray covering
(2) A conductor with three continuous white stripes on other than green insulation
(3) A marking at the termination of white or natural gray color

(b) Circuits of Less than 50 Volts. A conductor with white or natural gray color insulation or three continuous white stripes or having a marking of white or natural gray at the termination for circuits of less than 50 volts shall be required to be grounded only as required by Section 250-20(a).

(c) Circuits of 50 Volts or More. The use of insulation that is white or natural gray or that has three continuous white stripes for other than a grounded conductor for circuits of 50 volts or more shall be permitted only as in (1) through (3).

(1) If part of a cable assembly and where the insulation is permanently re-identified to indicate its use as an ungrounded conductor, by painting or other effective means at its termination, and at each location where the conductor is visible and accessible.

(2) Where a cable contains an insulated conductor for single-pole, 3-way, or 4-way switch loops, and the conductor with white or natural gray insulation or a marking of three continuous white stripes is used for the supply to the switch, but not as a return conductor from the switch to the switched outlet. In these applications, the conductor with white or natural gray insulation or with three continuous white stripes shall be permanently re-identified to indicate its use by painting or other effective means at its terminations and at each location where the conductor is visible and accessible.

(3) Where a flexible cord, having one conductor identified by a white or natural gray outer finish or three continuous white stripes or by any other means permitted by Section 400-22, is used for connecting an appliance or equipment permitted by Section 400-7. This shall apply to flexible cords connected to outlets whether or not the outlet is supplied by a circuit that has a grounded conductor.

200-9. Means of Identification of Terminals. The identification of terminals to which a grounded conductor is to be connected shall be substantially white in color. The identification of other terminals shall be of a readily distinguishable different color.

Exception: Where the conditions of maintenance and supervision ensure that only qualified persons will service the installations, terminals for grounded conductors shall be permitted to be permanently identified at the time of installation by a distinctive white marking or other equally effective means.

200-10. Identification of Terminals.

(a) Device Terminals. All devices, excluding lighting and appliance branch-circuit panelboards, provided with terminals for the attachment of conductors and intended for connection to more than one side of the circuit shall have terminals properly marked for identification, unless the electrical connection of the terminal intended to be connected to the grounded conductor is clearly evident.

Exception: Terminal identification shall not be required for devices that have a normal current rating of over 30 amperes, other than polarized attachment plugs and polarized receptacles for attachment plugs as required in Section 200-10(b).

(b) Receptacles, Plugs, and Connectors. Receptacles, polarized attachment plugs, and cord connectors for plugs and polarized plugs shall have the terminal intended for connection to the grounded conductor identified.

Identification shall be by a metal or metal coating that is substantially white in color or by the word *white* or the letter *W* located adjacent to the identified terminal.

If the terminal is not visible, the conductor entrance hole for the connection shall be colored white or marked with the word *white* or the letter *W*.

FPN: See Section 250-126 for identification of wiring device equipment grounding conductor terminals.

(c) Screw Shells. For devices with screw shells, the terminal for the grounded conductor shall be the one connected to the screw shell.

(d) Screw Shell Devices with Leads. For screw shell devices with attached leads, the conductor attached to the screw shell shall have a white or natural gray finish. The outer finish of the other conductor shall be of a solid color that will not be confused with the white or natural gray finish used to identify the grounded conductor.

(e) Appliances. Appliances that have a single-pole switch or a single-pole overcurrent device in the line or any line-connected screw shell lampholders, and that are to be connected by (1) a permanent wiring method or (2) field-installed attachment plugs and cords with three or more wires (including the equipment grounding conductor), shall have means to identify the terminal for the grounded circuit conductor (if any).

200-11. Polarity of Connections. No grounded conductor shall be attached to any terminal or lead so as to reverse the designated polarity.

Article 210 — Branch Circuits

A. General Provisions

210-1. Scope. This article covers branch circuits except for branch circuits that supply only motor loads, which are covered in Article 430. Provisions of this article and Article 430 apply to branch circuits with combination loads.

210-2. Other Articles for Specific-Purpose Branch Circuits. Branch circuits shall comply with this article and also with the applicable provisions of other articles of this *Code*. The provisions for branch circuits supplying equipment in the following list amend or supplement the provisions in this article and shall apply to branch circuits referred to therein.

	Article	Section
Air-conditioning and refrigerating equipment		440-6
		440-31
		440-32
Busways		364-9
Circuits and equipment operating at less than 50 volts	720	
Central heating equipment other than fixed electric space-heating equipment		422-12
Class 1, Class 2, and Class 3 remote-control, signaling, and power-limited circuits	725	
Closed-loop and programmed power distribution	780	
Cranes and hoists		610-42
Electric signs and outline lighting		600-6
Electric welders	630	
Elevators, dumbwaiters, escalators, moving walks, wheel chairlifts, and stairway chair lifts		620-61
Fire alarm systems	760	
Fixed electric heating equipment for pipelines and vessels		427-4
Fixed electric space-heating equipment		424-3

	Article	Section
Fixed outdoor electric deicing and snow-melting equipment		426-4
Information technology equipment		645-5
Infrared lamp industrial heating equipment		422-48
		424-3
Induction and dielectric heating equipment	665	
Marinas and boatyards		555-5
Mobile homes, manufactured homes, and mobile home parks	550	
Motion picture and television studios and similar locations	530	
Motors, motor circuits, and controllers	430	
Pipe organs		650-7
Recreational vehicles and recreational vehicle parks	551	
Sound-recording and similar equipment		640-6
Switchboards and panelboards		384-32
Theaters, audience areas of motion picture and television studios, and similar locations		520-41
		520-52
		520-62
X-ray equipment		660-2
		517-73

210-3. Rating. Branch circuits recognized by this article shall be rated in accordance with the maximum permitted ampere rating or setting of the overcurrent device. The rating for other than individual branch circuits shall be 15, 20, 30, 40, and 50 amperes. Where conductors of higher ampacity are used for any reason, the ampere rating or setting of the specified overcurrent device shall determine the circuit rating.

Exception: Multioutlet branch circuits greater than 50 amperes shall be permitted to supply nonlighting outlet loads on industrial premises where maintenance and supervision indicate that qualified persons will service the equipment.

210-4. Multiwire Branch Circuits.

(a) General. Branch circuits recognized by this article shall be permitted as multiwire circuits. A multiwire branch circuit shall be permitted to be considered as multiple circuits. All conductors shall originate from the same panelboard.

FPN: A 3-phase, 4-wire, wye-connected power system used to supply power to nonlinear loads may necessitate that the power system design allow for the possibility of high harmonic neutral currents.

(b) Dwelling Units. In dwelling units, a multiwire branch circuit supplying more than one device or equipment on the same yoke shall be provided with a means to disconnect simultaneously all ungrounded conductors at the panelboard where the branch circuit originated.

(c) Line-to-Neutral Loads. Multiwire branch circuits shall supply only line-to-neutral loads.

Exception No. 1: A multiwire branch circuit that supplies only one utilization equipment.

Exception No. 2: Where all ungrounded conductors of the multiwire branch circuit are opened simultaneously by the branch-circuit overcurrent device.

> FPN: See Section 300-13(b) for continuity of grounded conductor on multiwire circuits.

(d) Identification of Ungrounded Conductors. Where more than one nominal voltage system exists in a building, each ungrounded conductor of a multiwire branch circuit, where accessible, shall be identified by phase and system. This means of identification shall be permitted to be by separate color coding, marking tape, tagging, or other approved means and shall be permanently posted at each branch-circuit panelboard.

210-5. Identification for Branch Circuits.

(a) Grounded Conductor. The grounded conductor of a branch circuit shall be identified in accordance with Section 200-6.

(b) Equipment Grounding Conductor. The equipment grounding conductor shall be identified in accordance with Section 250-119.

210-6. Branch-Circuit Voltage Limitations.

(a) Occupancy Limitation. In dwelling units and guest rooms of hotels, motels, and similar occupancies, the voltage shall not exceed 120 volts, nominal, between conductors that supply the terminals of the following:

(1) Lighting fixtures
(2) Cord- and plug-connected loads 1440 volt-amperes, nominal, or less, or less than ¼ hp

(b) 120 Volts Between Conductors. Circuits not exceeding 120 volts, nominal, between conductors shall be permitted to supply the following:

(1) The terminals of lampholders applied within their voltage ratings
(2) Auxiliary equipment of electric-discharge lamps
(3) Cord- and plug-connected or permanently connected utilization equipment

(c) 277 Volts to Ground. Circuits exceeding 120 volts, nominal, between conductors and not exceeding 277 volts,

nominal, to ground shall be permitted to supply the following:

(1) Listed electric-discharge lighting fixtures
(2) Listed incandescent lighting fixtures, where supplied at 120 volts or less from the output of a stepdown autotransformer that is an integral component of the fixture and the outer shell terminal is electrically connected to a grounded conductor of the branch circuit
(3) Lighting fixtures equipped with mogul-base screw shell lampholders
(4) Lampholders, other than the screw shell type, applied within their voltage ratings
(5) Auxiliary equipment of electric-discharge lamps
(6) Cord- and plug-connected or permanently connected utilization equipment

(d) 600 Volts Between Conductors. Circuits exceeding 277 volts, nominal, to ground and not exceeding 600 volts, nominal, between conductors shall be permitted to supply the following:

(1) The auxiliary equipment of electric-discharge lamps mounted in permanently installed fixtures where the fixtures are mounted in accordance with one of the following:

  (a) Not less than a height of 22 ft (6.71 m) on poles or similar structures for the illumination of outdoor areas such as highways, roads, bridges, athletic fields, or parking lots
  (b) Not less than a height of 18 ft (5.49 m) on other structures such as tunnels

(2) Cord- and plug-connected or permanently connected utilization equipment

> FPN: See Section 410-78 for auxiliary equipment limitations.

Exception No. 1 to (b), (c), and (d): For lampholders of infrared industrial heating appliances as provided in Section 422-14.

Exception No. 2 to (b), (c), and (d): For railway properties as described in Section 110-19.

210-7. Receptacles and Cord Connectors.

(a) Grounding Type. Receptacles installed on 15- and 20-ampere branch circuits shall be of the grounding type. Grounding-type receptacles shall be installed only on circuits of the voltage class and current for which they are rated, except as provided in Tables 210-21(b)(2) and (b)(3).

Exception: Nongrounding-type receptacles installed in accordance with Section 210-7(d).

(b) To Be Grounded. Receptacles and cord connectors that have grounding contacts shall have those contacts effectively grounded.

Exception No. 1: Receptacles mounted on portable and vehicle-mounted generators in accordance with Section 250-34.

Exception No. 2: Replacement receptacles as permitted by Section 210-7(d)

(c) Methods of Grounding. The grounding contacts of receptacles and cord connectors shall be grounded by connection to the equipment grounding conductor of the circuit supplying the receptacle or cord connector.

> FPN: For installation requirements for the reduction of electrical noise, see Section 250-146(d).

The branch-circuit wiring method shall include or provide an equipment grounding conductor to which the grounding contacts of the receptacle or cord connector shall be connected.

> FPN No. 1: Section 250-118 describes acceptable grounding means.

> FPN No. 2: For extensions of existing branch circuits, see Section 250-130.

(d) Replacements. Replacement of receptacles shall comply with (1), (2), and (3) as applicable.

(1) Where a grounding means exists in the receptacle enclosure or a grounding conductor is installed in accordance with Section 250-130(c), grounding-type receptacles shall be used and shall be connected to the grounding conductor in accordance with Sections 210-7(c) or 250-130(c).

(2) Ground-fault circuit-interrupter protected receptacles shall be provided where replacements are made at receptacle outlets that are required to be so protected elsewhere in this *Code*.

(3) Where a grounding means does not exist in the receptacle enclosure, the installation shall comply with (a), (b), or (c).

(a) A nongrounding-type receptacle(s) shall be permitted to be replaced with another nongrounding-type receptacle(s).

(b) A nongrounding-type receptacle(s) shall be permitted to be replaced with a ground-fault circuit interrupter-type of receptacle(s). These receptacles shall be marked "No Equipment Ground." An equipment grounding conductor shall not be connected from the ground-fault circuit interrupter-type receptacle to any outlet supplied from the ground-fault circuit interrupter receptacle.

(c) A nongrounding-type receptacle(s) shall be permitted to be replaced with a grounding-type receptacle(s) where supplied through a ground-fault circuit interrupter. Grounding-type receptacles supplied through the ground-fault circuit interrupter shall be marked "GFCI Protected" and "No Equipment Ground." An equipment grounding conductor shall not be connected between the grounding-type receptacles.

(e) Cord- and Plug-Connected Equipment. The installation of grounding-type receptacles shall not be used as a requirement that all cord- and plug-connected equipment be of the grounded type.

> FPN: See Section 250-114 for types of cord- and plug-connected equipment to be grounded.

(f) Noninterchangeable Types. Receptacles connected to circuits that have different voltages, frequencies, or types of current (ac or dc) on the same premises shall be of such design that the attachment plugs used on these circuits are not interchangeable.

210-8. Ground-Fault Circuit-Interrupter Protection for Personnel.

> FPN: See Section 215-9 for ground-fault circuit-interrupter protection for personnel on feeders.

(a) Dwelling Units. All 125-volt, single-phase, 15- and 20-ampere receptacles installed in the locations specified below shall have ground-fault circuit-interrupter protection for personnel.

(1) Bathrooms.
(2) Garages, and also accessory buildings that have a floor located at or below grade level not intended as habitable rooms and limited to storage areas, work areas, and areas of similar use.

Exception No. 1: Receptacles that are not readily accessible.

Exception No. 2: A single receptacle or a duplex receptacle for two appliances located within dedicated space for each appliance that, in normal use, is not easily moved from one place to another, and that is cord- and plug-connected in accordance with Section 400-7(a)(6), (a)(7), or (a)(8).

Receptacles installed under the exceptions to Section 210-8(a)(2) shall not be considered as meeting the requirements of Section 210-52(g).

(3) Outdoors.

Exception: Receptacles that are not readily accessible and are supplied by a dedicated branch circuit for electric snow-melting or deicing equipment shall be permitted to be in-

stalled in accordance with the applicable provisions of Article 426.

(4) Crawl spaces. Where the crawl space is at or below grade level.
(5) Unfinished basements. For purposes of this section, unfinished basements are defined as portions or areas of the basement not intended as habitable rooms and limited to storage areas, work areas, and the like.

Exception No. 1: Receptacles that are not readily accessible.

Exception No. 2: A single receptacle or a duplex receptacle for two appliances located within dedicated space for each appliance that, in normal use, is not easily moved from one place to another, and that is cord- and plug-connected in accordance with Section 400-7(a)(6), (a)(7), or (a)(8).

Receptacles installed under the exceptions to Section 210-8(a)(5) shall not be considered as meeting the requirements of Section 210-52(g).

(6) Kitchens. Where the receptacles are installed to serve the countertop surfaces.
(7) Wet bar sinks. Where the receptacles are installed to serve the countertop surfaces and are located within 6 ft (1.83 m) of the outside edge of the wet bar sink. Receptacle outlets shall not be installed in a face-up position in the work surfaces or countertops.

(b) Other than Dwelling Units. All 125-volt, single-phase, 15- and 20-ampere receptacles installed in the locations specified below shall have ground-fault circuit-interrupter protection for personnel.

(1) Bathrooms
(2) Rooftops

Exception: Receptacles that are not readily accessible and are supplied from a dedicated branch circuit for electric snow-melting or deicing equipment shall be permitted to be installed in accordance with the applicable provisions of Article 426.

210-9. Circuits Derived from Autotransformers. Branch circuits shall not be derived from autotransformers unless the circuit supplied has a grounded conductor that is electrically connected to a grounded conductor of the system supplying the autotransformer.

Exception No. 1: An autotransformer shall be permitted without the connection to a grounded conductor where transforming from a nominal 208 volts to a nominal 240-volt supply or similarly from 240 volts to 208 volts.

Exception No. 2: In industrial occupancies, where conditions of maintenance and supervision ensure that only qualified persons will service the installation, autotransformers shall be permitted to supply nominal 600-volt loads from nominal 480-volt systems, and 480-volt loads from nominal 600-volt systems, without the connection to a similar grounded conductor.

210-10. Ungrounded Conductors Tapped from Grounded Systems. Two-wire dc circuits and ac circuits of two or more ungrounded conductors shall be permitted to be tapped from the ungrounded conductors of circuits that have a grounded neutral conductor. Switching devices in each tapped circuit shall have a pole in each ungrounded conductor. All poles of multipole switching devices shall manually switch together where such switching devices also serve as a disconnecting means as required by the following:

(1) Section 410-48 for double-pole switched lampholders
(2) Section 410-54(b) for electric-discharge lamp auxiliary equipment switching devices
(3) Section 422-31(b) for an appliance
(4) Section 424-20 for a fixed electric space-heating unit
(5) Section 426-51 for electric deicing and snow-melting equipment
(6) Section 430-85 for a motor controller
(7) Section 430-103 for a motor

210-11. Branch Circuits Required. Branch circuits for lighting and for appliances, including motor-operated appliances, shall be provided to supply the loads computed in accordance with Section 220-3. In addition, branch circuits shall be provided for specific loads not covered by Section 220-3 where required elsewhere in this *Code* and for dwelling unit loads as specified in (c).

(a) Number of Branch Circuits. The minimum number of branch circuits shall be determined from the total computed load and the size or rating of the circuits used. In all installations, the number of circuits shall be sufficient to supply the load served. In no case shall the load on any circuit exceed the maximum specified by Section 220-4.

(b) Load Evenly Proportioned Among Branch Circuits. Where the load is computed on a volt-amperes/square foot (0.093 m^2) basis, the wiring system up to and including the branch-circuit panelboard(s) shall be provided to serve not less than the calculated load. This load shall be evenly proportioned among multioutlet branch circuits within the panelboard(s). Branch-circuit overcurrent devices and circuits need only be installed to serve the connected load.

(c) Dwelling Units.

(1) Small-Appliance Branch Circuits. In addition to the number of branch circuits required by other parts of this section, two or more 20-ampere small-appliance branch circuits shall be provided for all receptacle outlets specified by Section 210-52(b).

(2) Laundry Branch Circuits. In addition to the number of branch circuits required by other parts of this section, at least one additional 20-ampere branch circuit shall be provided to supply the laundry receptacle outlet(s) required by Section 210-52(f). This circuit shall have no other outlets.

(3) Bathroom Branch Circuits. In addition to the number of branch circuits required by other parts of this section, at least one 20-ampere branch circuit shall be provided to supply the bathroom receptacle outlet(s). Such circuits shall have no other outlets.

Exception: Where the 20-ampere circuit supplies a single bathroom, outlets for other equipment within the same bathroom shall be permitted to be supplied in accordance with Section 210-23(a).

FPN: See Examples D1(a), D1(b), D2(b), and D4(a) in Appendix D.

210-12. Arc-Fault Circuit-Interrupter Protection.

(a) Definition. An *arc-fault circuit interrupter* is a device intended to provide protection from the effects of arc faults by recognizing characteristics unique to arcing and by functioning to de-energize the circuit when an arc fault is detected.

(b) Dwelling Unit Bedrooms. All branch circuits that supply 125-volt, single-phase, 15- and 20-ampere receptacle outlets installed in dwelling unit bedrooms shall be protected by an arc-fault circuit interrupter(s). This requirement shall become effective January 1, 2002.

B. Branch-Circuit Ratings

210-19. Conductors — Minimum Ampacity and Size.

(a) General. Branch-circuit conductors shall have an ampacity not less than the maximum load to be served. Where a branch circuit supplies continuous loads or any combination of continuous and noncontinuous loads, the minimum branch-circuit conductor size, before the application of any adjustment or correction factors, shall have an allowable ampacity equal to or greater than the noncontinuous load plus 125 percent of the continuous load.

Exception: Where the assembly, including the overcurrent devices protecting the branch circuit(s), is listed for operation at 100 percent of its rating, the ampacity of the branch circuit conductors shall be permitted to be not less than the sum of the continuous load plus the noncontinuous load.

FPN No. 1: See Section 310-15 for ampacity ratings of conductors.

FPN No. 2: See Part B of Article 430 for minimum rating of motor branch-circuit conductors.

FPN No. 3: See Section 310-10 for temperature limitation of conductors.

FPN No. 4: Conductors for branch circuits as defined in Article 100, sized to prevent a voltage drop exceeding 3 percent at the farthest outlet of power, heating, and lighting loads, or combinations of such loads, and where the maximum total voltage drop on both feeders and branch circuits to the farthest outlet does not exceed 5 percent, will provide reasonable efficiency of operation. See Section 215-2 for voltage drop on feeder conductors.

(b) Multioutlet Branch Circuits. Conductors of branch circuits supplying more than one receptacle for cord- and plug-connected portable loads shall have an ampacity of not less than the rating of the branch circuit.

(c) Household Ranges and Cooking Appliances. Branch-circuit conductors supplying household ranges, wall-mounted ovens, counter-mounted cooking units, and other household cooking appliances shall have an ampacity not less than the rating of the branch circuit and not less than the maximum load to be served. For ranges of 8¾ kW or more rating, the minimum branch-circuit rating shall be 40 amperes.

Exception No. 1: Tap conductors supplying electric ranges, wall-mounted electric ovens, and counter-mounted electric cooking units from a 50-ampere branch circuit shall have an ampacity of not less than 20 and shall be sufficient for the load to be served. The taps shall not be longer than necessary for servicing the appliance.

Exception No. 2: The neutral conductor of a 3-wire branch circuit supplying a household electric range, a wall-mounted oven, or a counter-mounted cooking unit shall be permitted to be smaller than the ungrounded conductors where the maximum demand of a range of 8¾ kW or more rating has been computed according to Column A of Table 220-19, but shall have an ampacity of not less than 70 percent of the branch-circuit rating and shall not be smaller than No. 10.

(d) Other Loads. Branch-circuit conductors that supply loads other than those specified in Section 210-2 and other than cooking appliances as covered above shall have an ampacity sufficient for the loads served and shall not be smaller than No. 14.

Exception No. 1: Tap conductors for such loads shall have an ampacity of not less than 15 for circuits rated less than 40 amperes and of not less than 20 for circuits rated at 40 or 50 amperes and only where these tap conductors supply any of the following loads:

(a) Individual lampholders or fixtures with taps extending not longer than 18 in. (457 mm) beyond any portion of the lampholder or fixture

(b) A fixture having tap conductors as provided in Section 410-67

(c) Individual outlets, other than receptacle outlets, with taps not over 18 in. (457 mm) long

(d) Infrared lamp industrial heating appliances

(e) Nonheating leads of deicing and snow-melting cables and mats

Exception No. 2: Fixture wires and flexible cords shall be permitted to be smaller than No. 14 as permitted by Section 240-4.

210-20. Overcurrent Protection. Branch-circuit conductors and equipment shall be protected by overcurrent protective devices that have a rating or setting that complies with (a) through (d).

(a) Continuous and Noncontinuous Loads. Where a branch circuit supplies continuous loads or any combination of continuous and noncontinuous loads, the rating of the overcurrent device shall not be less than the noncontinuous load plus 125 percent of the continuous load.

Exception: Where the assembly, including the overcurrent devices protecting the branch circuit(s), is listed for operation at 100 percent of its rating, the ampere rating of the overcurrent device shall be permitted to be not less than the sum of the continuous load plus the noncontinuous load

(b) Conductor Protection. Conductors shall be protected in accordance with Section 240-3.

Exception No. 1: Tap conductors as permitted in Section 210-19(d) shall be permitted to be protected by the branch-circuit overcurrent device.

Exception No. 2: Fixture wires and flexible cords shall be permitted to be protected in accordance with Section 240-4.

(c) Equipment. The rating or setting of the overcurrent protective device shall not exceed that specified in the applicable articles referenced in Section 240-2 for equipment.

(d) Outlet Devices. The rating or setting shall not exceed that specified in Section 210-21 for outlet devices.

210-21. Outlet Devices. Outlet devices shall have an ampere rating that is not less than the load to be served and shall comply with (a) and (b).

(a) Lampholders. Where connected to a branch circuit having a rating in excess of 20 amperes, lampholders shall be of the heavy-duty type. A heavy-duty lampholder shall have a rating of not less than 660 watts if of the admedium type and not less than 750 watts if of any other type.

(b) Receptacles.

(1) A single receptacle installed on an individual branch circuit shall have an ampere rating of not less than that of the branch circuit.

Exception No. 1: Where installed in accordance with Section 430-81(c).

Exception No. 2: A receptacle installed exclusively for the use of a cord- and plug-connected arc welder shall be permitted to have an ampere rating not less than the minimum branch-circuit conductor ampacity determined by Section 630-11(a) for arc welders.

FPN: See definition of *Receptacle* in Article 100.

(2) Where connected to a branch circuit supplying two or more receptacles or outlets, a receptacle shall not supply a total cord- and plug-connected load in excess of the maximum specified in Table 210-21(b)(2).

Table 210-21(b)(2). Maximum Cord- and Plug-Connected Load to Receptacle

Circuit Rating (Amperes)	Receptacle Rating (Amperes)	Maximum Load (Amperes)
15 or 20	15	12
20	20	16
30	30	24

(3) Where connected to a branch circuit supplying two or more receptacles or outlets, receptacle ratings shall conform to the values listed in Table 210-21(b)(3), or, where larger than 50 amperes, the receptacle rating shall not be less than the branch-circuit rating.

Exception: Receptacles for one or more cord- and plug-connected arc welders shall be permitted to have ampere ratings not less than the minimum branch-circuit conductor ampacity permitted by Sections 630-11(a) or (b) as applicable for arc welders.

Table 210-21(b)(3). Receptacle Ratings for Various Size Circuits

Circuit Rating (Amperes)	Receptacle Rating (Amperes)
15	Not over 15
20	15 or 20
30	30
40	40 or 50
50	50

(4) The ampere rating of a range receptacle shall be permitted to be based on a single range demand load as specified in Table 220-19.

210-23. Permissible Loads. In no case shall the load exceed the branch-circuit ampere rating. An individual branch circuit shall be permitted to supply any load for which it is rated. A branch circuit supplying two or more outlets or receptacles shall supply only the loads specified according to its size as specified in (a) through (d) and as summarized in Section 210-24 and Table 210-24.

(a) 15- and 20-Ampere Branch Circuits. A 15- or 20-ampere branch circuit shall be permitted to supply lighting units or other utilization equipment, or a combination of both. The rating of any one cord- and plug-connected utilization equipment shall not exceed 80 percent of the branch-circuit ampere rating. The total rating of utilization equipment fastened in place, other than lighting fixtures, shall not exceed 50 percent of the branch-circuit ampere rating where lighting units, cord- and plug-connected utilization equipment not fastened in place, or both, are also supplied.

Exception: The small appliance branch circuits, laundry branch circuits, and bathroom branch circuits required in a dwelling unit(s) by Sections 210-11(c)(1), (2), and (3) shall supply only the receptacle outlets specified in that section.

(b) 30-Ampere Branch Circuits. A 30-ampere branch circuit shall be permitted to supply fixed lighting units with heavy-duty lampholders in other than a dwelling unit(s) or utilization equipment in any occupancy. A rating of any one cord- and plug-connected utilization equipment shall not exceed 80 percent of the branch-circuit ampere rating.

(c) 40- and 50-Ampere Branch Circuits. A 40- or 50-ampere branch circuit shall be permitted to supply cooking appliances that are fastened in place in any occupancy. In other than dwelling units, such circuits shall be permitted to supply fixed lighting units with heavy-duty lampholders, infrared heating units, or other utilization equipment.

(d) Branch Circuits Larger than 50 Amperes. Branch circuits larger than 50 amperes shall supply only nonlighting outlet loads.

210-24. Branch-Circuit Requirements — Summary. The requirements for circuits that have two or more outlets or receptacles, other than the receptacle circuits of Sections 210-11(c)(1) and (2) as specifically provided for above, are summarized in Table 210-24.

210-25. Common Area Branch Circuits. Branch circuits in dwelling units shall supply only loads within that dwelling unit or loads associated only with that dwelling unit. Branch circuits required for the purpose of lighting, central alarm, signal, communications, or other needs for public or common areas of a two-family or multifamily dwelling shall not be supplied from equipment that supplies an individual dwelling unit.

C. Required Outlets

210-50. General. Receptacle outlets shall be installed as specified in Sections 210-52 through 210-63.

(a) Cord Pendants. A cord connector that is supported by a permanently installed cord pendant shall be considered a receptacle outlet.

Table 210-24. Summary of Branch-Circuit Requirements

Circuit Rating	15 A	20 A	30 A	40 A	50 A
Conductors (min. size):					
Circuit wires[1]	14	12	10	8	6
Taps	14	14	14	12	12
Fixture wires and cords — See Section 240-4					
Overcurrent Protection	**15 A**	**20 A**	**30 A**	**40 A**	**50 A**
Outlet Devices:					
Lampholders permitted	Any type	Any type	Heavy duty	Heavy duty	Heavy duty
Receptacle rating[2]	15 max. A	15 or 20 A	30 A	40 or 50 A	50 A
Maximum Load	**15 A**	**20 A**	**30 A**	**40 A**	**50 A**
Permissible load	See Section 210-23(a)	See Section 210-23(a)	See Section 210-23(b)	See Section 210-23(c)	See Section 210-23(c)

[1]These gauges are for copper conductors.

[2]For receptacle rating of cord-connected electric-discharge lighting fixtures, see Section 410-30(c).

(b) Cord Connections. A receptacle outlet shall be installed wherever flexible cords with attachment plugs are used. Where flexible cords are permitted to be permanently connected, receptacles shall be permitted to be omitted for such cords.

(c) Appliance Outlets. Appliance receptacle outlets installed in a dwelling unit for specific appliances, such as laundry equipment, shall be installed within 6 ft (1.83 m) of the intended location of the appliance.

210-52. Dwelling Unit Receptacle Outlets. Receptacle outlets required by this section shall be in addition to any receptacle that is part of a lighting fixture or appliance, located within cabinets or cupboards, or located more than 5½ ft (1.68 m) above the floor.

Permanently installed electric baseboard heaters equipped with factory-installed receptacle outlets or outlets provided as a separate assembly by the manufacturer shall be permitted as the required outlet or outlets for the wall space utilized by such permanently installed heaters. Such receptacle outlets shall not be connected to the heater circuits.

> FPN: Listed baseboard heaters include instructions that may not permit their installation below receptacle outlets.

(a) General Provisions. In every kitchen, family room, dining room, living room, parlor, library, den, sunroom, bedroom, recreation room, or similar room or area of dwelling units, receptacle outlets shall be installed in accordance with the general provisions specified in (1) through (3).

(1) Spacing. Receptacles shall be installed so that no point along the floor line in any wall space is more than 6 ft (1.83 m), measured horizontally, from an outlet in that space. Receptacle outlets shall, insofar as practicable, be spaced equal distances apart.

(2) Wall Space. As used in this section, a wall space shall include the following:

(a) Any space 2 ft (610 mm) or more in width (including space measured around corners) and unbroken along the floor line by doorways, fireplaces, and similar openings

(b) The space occupied by fixed panels in exterior walls, excluding sliding panels

(c) The space afforded by fixed room dividers such as freestanding bar-type counters or railings

(3) Floor Receptacles. Receptacle outlets in floors shall not be counted as part of the required number of receptacle outlets unless located within 18 in. (457 mm) of the wall.

(b) Small Appliances.

(1) In the kitchen, pantry, breakfast room, dining room, or similar area of a dwelling unit, the two or more 20-ampere small-appliance branch circuits required by Section 210-11(c)(1) shall serve all receptacle outlets covered by Sections 210-52(a) and (c) and receptacle outlets for refrigeration equipment.

Exception No. 1: In addition to the required receptacles specified by Section 210-52, switched receptacles supplied from a general-purpose branch circuit as defined in Section 210-70(a)(1), Exception No. 1, shall be permitted.

Exception No. 2: The receptacle outlet for refrigeration equipment shall be permitted to be supplied from an individual branch circuit rated 15 amperes or greater.

(2) The two or more small-appliance branch circuits specified in (b)(1) shall have no other outlets.

Exception No. 1: A receptacle installed solely for the electrical supply to and support of an electric clock in any of the rooms specified above.

Exception No. 2: Receptacles installed to provide power for supplemental equipment and lighting on gas-fired ranges, ovens, or counter-mounted cooking units.

(3) Receptacles installed in a kitchen to serve countertop surfaces shall be supplied by not less than two small-appliance branch circuits, either or both of which shall also be permitted to supply receptacle outlets in the same kitchen and in other rooms specified in Section 210-52(b)(1). Additional small-appliance branch circuits shall be permitted to supply receptacle outlets in the kitchen and other rooms specified in Section 210-52(b)(1). No small-appliance branch circuit shall serve more than one kitchen.

(c) Countertops. In kitchens and dining rooms of dwelling units, receptacle outlets for counter spaces shall be installed in accordance with (1) through (5).

(1) Wall Counter Spaces. A receptacle outlet shall be installed at each wall counter space that is 12 in. (305 mm) or wider. Receptacle outlets shall be installed so that no point along the wall line is more than 24 in. (610 mm), measured horizontally from a receptacle outlet in that space.

(2) Island Counter Spaces. At least one receptacle outlet shall be installed at each island counter space with a long dimension of 24 in. (610 mm) or greater and a short dimension of 12 in. (305 mm) or greater.

(3) Peninsular Counter Spaces. At least one receptacle outlet shall be installed at each peninsular counter space with a long dimension of 24 in. (610 mm) or greater and a short dimension of 12 in. (305 mm) or greater. A peninsular countertop is measured from the connecting edge.

(4) Separate Spaces. Countertop spaces separated by range tops, refrigerators, or sinks shall be considered as separate countertop spaces in applying the requirements of (1), (2), and (3).

(5) Receptacle Outlet Location. Receptacle outlets shall be located above, but not more than 18 in. (458 mm) above the countertop. Receptacle outlets shall not be installed in a face-up position in the work surfaces or countertops. Receptacle outlets rendered not readily accessible by appliances fastened in place or appliances occupying dedicated space shall not be considered as these required outlets.

Exception: To comply with the conditions as specified in (a) or (b), receptacle outlets shall be permitted to be mounted not more than 12 in. (305 mm) below the countertop. Receptacles mounted below the countertop in accordance with this exception shall not be located where the countertop extends more than 6 in. (153 mm) beyond its support base.

(a) Construction for the physically impaired
(b) On island and peninsular countertops where the countertop is flat across its entire surface (no backsplashes, dividers, etc.) and there are no means to mount a receptacle within 18 in. (458 mm) above the countertop, such as an overhead cabinet

(d) Bathrooms. In dwelling units, at least one wall receptacle outlet shall be installed in bathrooms within 36 in. (914 mm) of the outside edge of each basin. The receptacle outlet shall be located on a wall that is adjacent to the basin location. See Section 210-8(a)(1).

Receptacle outlets shall not be installed in a face-up position in the work surfaces or countertops in a bathroom basin location.

(e) Outdoor Outlets. For a one-family dwelling and each unit of a two-family dwelling that is at grade level, at least one receptacle outlet accessible at grade level and not more than 6½ ft (1.98 m) above grade shall be installed at the front and back of the dwelling. See Section 210-8(a)(3).

(f) Laundry Areas. In dwelling units, at least one receptacle outlet shall be installed for the laundry.

Exception No. 1: In a dwelling unit that is an apartment or living area in a multifamily building where laundry facilities are provided on the premises that are available to all building occupants, a laundry receptacle shall not be required.

Exception No. 2: In other than one-family dwellings where laundry facilities are not to be installed or permitted, a laundry receptacle shall not be required.

(g) Basements and Garages. For a one-family dwelling, at least one receptacle outlet, in addition to any provided for laundry equipment, shall be installed in each basement and in each attached garage, and in each detached garage with electric power. See Sections 210-8(a)(2) and (a)(5).

Where a portion of the basement is finished into a habitable room(s), the receptacle outlet required by this section shall be installed in the unfinished portion.

(h) Hallways. In dwelling units, hallways of 10 ft (3.05 m) or more in length shall have at least one receptacle outlet.

As used in this subsection, the hall length shall be considered the length along the centerline of the hall without passing through a doorway.

210-60. Guest Rooms.

(a) General. Guest rooms in hotels, motels, and similar occupancies shall have receptacle outlets installed in accordance with Section 210-52. See Section 210-8(b)(1).

(b) Receptacle Placement. In applying the provisions of Section 210-52(a), the total number of receptacle outlets shall not be less than the minimum number that would comply with the provisions of that section. These receptacle outlets shall be permitted to be located conveniently for permanent furniture layout. At least two receptacle outlets shall be readily accessible. Where receptacles are installed behind the bed, the receptacle shall be located to prevent the bed from contacting any attachment plug that may be installed, or the receptacle shall be provided with a suitable guard.

210-62. Show Windows. At least one receptacle outlet shall be installed directly above a show window for each 12 linear ft (3.66 m) or major fraction thereof of show window area measured horizontally at its maximum width.

210-63. Heating, Air-Conditioning, and Refrigeration Equipment Outlet. A 125-volt, single-phase, 15- or 20-ampere–rated receptacle outlet shall be installed at an accessible location for the servicing of heating, air-conditioning, and refrigeration equipment on rooftops and in attics and crawl spaces. The receptacle shall be located on the same level and within 25 ft (7.62 m) of the heating, air-conditioning, and refrigeration equipment. The receptacle outlet shall not be connected to the load side of the equipment disconnecting means.

Exception: Rooftop equipment on one- and two-family dwellings.

FPN: See Section 210-8 for ground-fault circuit-interrupter requirements.

210-70. Lighting Outlets Required. Lighting outlets shall be installed where specified in (a), (b), and (c).

(a) Dwelling Units. In dwelling units, lighting outlets shall be installed in accordance with (1), (2), and (3).

(1) Habitable Rooms. At least one wall switch-controlled lighting outlet shall be installed in every habitable room and bathroom.

Exception No. 1: In other than kitchens and bathrooms, one or more receptacles controlled by a wall switch shall be permitted in lieu of lighting outlets.

Exception No. 2: Lighting outlets shall be permitted to be controlled by occupancy sensors that are (1) in addition to wall switches or (2) located at a customary wall switch location and equipped with a manual override that will allow the sensor to function as a wall switch.

(2) Additional Locations. At least one wall switch-controlled lighting outlet shall be installed in hallways, stairways, attached garages, and detached garages with electric power; and to provide illumination on the exterior side of outdoor entrances or exits with grade level access. A vehicle door in a garage shall not be considered as an outdoor entrance or exit. Where lighting outlets are installed in interior stairways, there shall be a wall switch at each floor level to control the lighting outlet where the difference between floor levels is six steps or more.

Exception: In hallways, stairways, and at outdoor entrances, remote, central, or automatic control of lighting shall be permitted.

(3) Storage or Equipment Spaces. For attics, underfloor spaces, utility rooms, and basements, at least one lighting outlet containing a switch or controlled by a wall switch shall be installed where these spaces are used for storage or contain equipment requiring servicing. At least one point of control shall be at the usual point of entry to these spaces. The lighting outlet shall be provided at or near the equipment requiring servicing.

(b) Guest Rooms. At least one wall switch-controlled lighting outlet or wall switch-controlled receptacle shall be installed in guest rooms in hotels, motels, or similar occupancies.

(c) Other Locations. For attics and underfloor spaces containing equipment requiring servicing, such as heating, air-conditioning, and refrigeration equipment, at least one lighting outlet containing a switch or controlled by a wall switch shall be installed in such spaces. At least one point of control shall be at the usual point of entry to these spaces. The lighting outlet shall be provided at or near the equipment requiring servicing.

Article 215 — Feeders

215-1. Scope. This article covers the installation requirements, overcurrent protection requirements, minimum size, and ampacity of conductors for feeders supplying branch-circuit loads as computed in accordance with Article 220.

Exception: Feeders for electrolytic cells as covered in Section 668-3(c)(1) and (4).

215-2. Minimum Rating and Size.

(a) General. Feeder conductors shall have an ampacity not less than required to supply the load as computed in Parts B, C, and D of Article 220. The minimum feeder-circuit conductor size, before the application of any adjustment or correction factors, shall have an allowable ampacity equal to or greater than the noncontinuous load plus 125 percent of the continuous load.

Exception: Where the assembly, including the overcurrent devices protecting the feeder(s), is listed for operation at 100 percent of its rating, the ampacity of the feeder conductors shall be permitted to be not less than the sum of the continuous load plus the noncontinuous load.

Additional minimum sizes shall be as specified in (b), (c), and (d) under the conditions stipulated.

(b) For Specified Circuits. The ampacity of feeder conductors shall not be less than 30 amperes where the load supplied consists of any of the following number and types of circuits:

(1) Two or more 2-wire branch circuits supplied by a 2-wire feeder
(2) More than two 2-wire branch circuits supplied by a 3-wire feeder
(3) Two or more 3-wire branch circuits supplied by a 3-wire feeder
(4) Two or more 4-wire branch circuits supplied by a 3-phase, 4-wire feeder

(c) Ampacity Relative to Service-Entrance Conductors. The feeder conductor ampacity shall not be less than that of the service-entrance conductors where the feeder conductors carry the total load supplied by service-entrance conductors with an ampacity of 55 amperes or less.

(d) Individual Dwelling Unit or Mobile Home Conductors. Feeder conductors for individual dwelling units or mobile homes need not be larger than service-entrance conductors. Section 310-15(b)(6) shall be permitted to be used for conductor size.

FPN No. 1: See Examples D1 through D10 in Appendix D.

FPN No. 2: Conductors for feeders as defined in Article 100, sized to prevent a voltage drop exceeding 3 percent at the farthest outlet of power, heating, and lighting loads, or combinations of such loads, and where the maximum total voltage drop on both feeders and branch circuits to the farthest outlet does not exceed 5 percent, will provide reasonable efficiency of operation.

FPN No. 3: See Section 210-19(a), FPN No. 4, for voltage drop for branch circuits.

215-3. Overcurrent Protection. Feeders shall be protected against overcurrent in accordance with the provisions of Part A of Article 240. Where a feeder supplies continuous loads or any combination of continuous and noncontinuous loads, the rating of the overcurrent device shall not be less than the noncontinuous load plus 125 percent of the continuous load.

Exception: Where the assembly, including the overcurrent devices protecting the feeder(s), is listed for operation at 100 percent of its rating, the ampere rating of the overcurrent device shall be permitted to be not less than the sum of the continuous load plus the noncontinuous load.

215-4. Feeders with Common Neutral.

(a) Feeders with Common Neutral. Feeders containing a common neutral shall be permitted to supply two or three sets of 3-wire feeders, or two sets of 4-wire or 5-wire feeders.

(b) In Metal Raceway or Enclosure. Where installed in a metal raceway or other metal enclosure, all conductors of all feeders using a common neutral shall be enclosed within the same raceway or other enclosure as required in Section 300-20.

215-5. Diagrams of Feeders. If required by the authority having jurisdiction, a diagram showing feeder details shall be provided prior to the installation of the feeders. Such a diagram shall show the area in square feet of the building or other structure supplied by each feeder, the total connected load before applying demand factors, the demand factors used, the computed load after applying demand factors, and the size and type of conductors to be used.

215-6. Feeder Conductor Grounding Means. Where a feeder supplies branch circuits in which equipment grounding conductors are required, the feeder shall include or provide a grounding means, in accordance with the provisions of Section 250-134, to which the equipment grounding conductors of the branch circuits shall be connected.

215-7. Ungrounded Conductors Tapped from Grounded Systems. Two-wire dc circuits and ac circuits of two or more ungrounded conductors shall be permitted to be tapped from the ungrounded conductors of circuits having a grounded neutral conductor. Switching devices in each tapped circuit shall have a pole in each ungrounded conductor.

215-8. Means of Identifying Conductor with the Higher Voltage to Ground. On a 4-wire, delta-connected secondary where the midpoint of one phase winding is grounded to supply lighting and similar loads, the phase conductor having the higher voltage to ground shall be identified by an outer finish that is orange in color or by tagging or other effective means. Such identification shall be placed at each point where a connection is made if the grounded conductor is also present.

215-9. Ground-Fault Circuit-Interrupter Protection for Personnel. Feeders supplying 15- and 20-ampere receptacle branch circuits shall be permitted to be protected by a ground-fault circuit interrupter in lieu of the provisions for such interrupters as specified in Section 210-8 and Article 305.

215-10. Ground-Fault Protection of Equipment. Each feeder disconnect rated 1000 amperes or more and installed on solidly grounded wye electrical systems of more than 150 volts to ground, but not exceeding 600 volts phase-to-phase, shall be provided with ground-fault protection of equipment in accordance with the provisions of Section 230-95.

Exception No. 1: The provisions of this section shall not apply to a disconnecting means for a continuous industrial process where a nonorderly shutdown will introduce additional or increased hazards.

Exception No. 2: The provisions of this section shall not apply to fire pumps.

Exception No. 3: The provisions of this section shall not apply if ground-fault protection of equipment is provided on the supply side of the feeder.

215-11. Circuits Derived from Autotransformers. Feeders shall not be derived from autotransformers unless the system supplied has a grounded conductor that is electrically connected to a grounded conductor of the system supplying the autotransformer.

Exception No. 1: An autotransformer shall be permitted without the connection to a grounded conductor where transforming from a nominal 208 volts to a nominal 240-volt supply or similarly from 240 volts to 208 volts.

Exception No. 2: In industrial occupancies, where conditions of maintenance and supervision ensure that only qualified persons service the installation, autotransformers shall be permitted to supply nominal 600-volt loads from nominal 480-volt systems, and 480-volt loads from nominal 600-volt systems, without the connection to a similar grounded conductor.

Article 220 — Branch-Circuit, Feeder, and Service Calculations

A. General

220-1. Scope. This article provides requirements for computing branch-circuit, feeder, and service loads.

Exception: Branch-circuit and feeder calculations for electrolytic cells as covered in Section 668-3(c)(1) and (4).

220-2. Computations.

(a) Voltages. Unless other voltages are specified, for purposes of computing branch-circuit and feeder loads, nominal system voltages of 120, 120/240, 208Y/120, 240, 347, 480Y/277, 480, 600Y/347, and 600 volts shall be used.

(b) Fractions of an Ampere. Except where computations result in a fraction of an ampere 0.5 or larger, such fractions shall be permitted to be dropped.

220-3. Computation of Branch Circuit Loads. Branch-circuit loads shall be computed as shown in (a) through (c).

(a) Lighting Load for Specified Occupancies. A unit load of not less than that specified in Table 220-3(a) for occupancies specified therein shall constitute the minimum lighting load for each square foot (0.093 m^2) of floor area. The floor area for each floor shall be computed from the outside dimensions of the building, dwelling unit, or other area involved. For dwelling units, the computed floor area shall not include open porches, garages, or unused or unfinished spaces not adaptable for future use.

> FPN: The unit values herein are based on minimum load conditions and 100 percent power factor, and may not provide sufficient capacity for the installation contemplated.

(b) Other Loads — All Occupancies. In all occupancies, the minimum load for each outlet for general-use receptacles and outlets not used for general illumination shall not be less than that computed in (1) through (11), the loads shown being based on nominal branch-circuit voltages.

Exception: The loads of outlets serving switchboards and switching frames in telephone exchanges shall be waived from the computations.

(1) Specific Appliances or Loads. An outlet for a specific appliance or other load not covered in (2) through (11) shall be computed based on the ampere rating of the appliance or load served.

(2) Electric Dryers and Household Electric Cooking Appliances. Load computations shall be permitted as specified in Section 220-18 for electric dryers and Section 220-19 for electric ranges and other cooking appliances.

Table 220-3(a). General Lighting Loads by Occupancies

Type of Occupancy	Unit Load per Square Foot (Volt-Amperes)
Armories and auditoriums	1
Banks	3½[b]
Barber shops and beauty parlors	3
Churches	1
Clubs	2
Court rooms	2
Dwelling units[a]	3
Garages — commercial (storage)	½
Hospitals	2
Hotels and motels, including apartment houses without provision for cooking by tenants[a]	2
Industrial commercial (loft) buildings	2
Lodge rooms	1½
Office buildings	3½[b]
Restaurants	2
Schools	3
Stores	3
Warehouses (storage)	¼
In any of the above occupancies except one-family dwellings and individual dwelling units of two-family and multifamily dwellings:	
Assembly halls and auditoriums	1
Halls, corridors, closets, stairways	½
Storage spaces	¼

Note: For SI units, 1 ft^2 = 0.093 m^2.

[a]See Section 220-3(b)(10).

[b]In addition, a unit load of 1 volt-ampere per square foot shall be included for general-purpose receptacle outlets where the actual number of general-purpose receptacle outlets is unknown.

(3) Motor Loads. Outlets for motor loads shall be computed in accordance with the requirements in Sections 430-22 and 430-24 and Article 440.

(4) Recessed Lighting Fixtures. An outlet supplying recessed lighting fixture(s) shall be computed based on the maximum volt-ampere rating of the equipment and lamps for which the fixture(s) is rated.

(5) Heavy Duty Lampholders. Outlets for heavy-duty lampholders shall be computed at a minimum of 600 volt-amperes.

(6) Sign and Outline Lighting. Sign and outline lighting outlets shall be computed at a minimum of 1200 volt-amperes for each required branch circuit specified in Section 600-5(a).

(7) Show Windows. Show windows shall be computed in accordance with either (a) or (b).

(a) The unit load per outlet as required in other provisions of this section

(b) At 200 volt-amperes per linear foot of show window

(8) Fixed Multioutlet Assemblies. Fixed multioutlet assemblies used in other than dwelling units or the guest rooms of hotels or motels shall be computed in accordance with (a) or (b).

(a) Where appliances are unlikely to be used simultaneously, each 5 ft (1.52 m) or fraction thereof of each separate and continuous length shall be considered as one outlet of not less than 180 volt-amperes.

(b) Where appliances are likely to be used simultaneously, each 1 ft (305 mm) or fraction thereof shall be considered as an outlet of not less than 180 volt-amperes.

(9) Receptacle Outlets. Except as covered in (10), receptacle outlets shall be computed at not less than 180 volt-amperes for each single or for each multiple receptacle on one strap. A single piece of equipment consisting of a multiple receptacle comprised of four or more receptacles shall be computed at not less than 90 volt-amperes per receptacle.

This provision shall not be applicable to the receptacle outlets specified in Sections 210-11(c)(1) and (2).

(10) Dwelling Occupancies. In one-family, two-family, and multifamily dwellings and in guest rooms of hotels and motels, the outlets specified in (a), (b), and (c) are included in the general lighting load calculations of Section 220-3(a). No additional load calculations shall be required for such outlets.

(a) All general-use receptacle outlets of 20-ampere rating or less, including receptacles connected to the circuits in Section 210-11(c)(3)

(b) The receptacle outlets specified in Sections 210-52(e) and (g)

(c) The lighting outlets specified in Sections 210-70(a) and (b)

(11) Other Outlets. Other outlets not covered in (1) through (10) shall be computed based on 180 volt-amperes per outlet.

(c) Loads for Additions to Existing Installations.

(1) Dwelling Units. Loads for structural additions to an existing dwelling unit or to a previously unwired portion of an existing dwelling unit, either of which exceeds 500 ft² (46.5 m²), shall be computed in accordance with (b). Loads for new circuits or extended circuits in previously wired dwelling units shall be computed in accordance with either (a) or (b).

(2) Other than Dwelling Units. Loads for new circuits or extended circuits in other than dwelling units shall be computed in accordance with either (a) or (b).

220-4. Maximum Loads. The total load shall not exceed the rating of the branch circuit, and it shall not exceed the maximum loads specified in (a) through (c) under the conditions specified therein.

(a) Motor-Operated and Combination Loads. Where a circuit supplies only motor-operated loads, Article 430 shall apply. Where a circuit supplies only air-conditioning equipment, refrigerating equipment, or both, Article 440 shall apply. For circuits supplying loads consisting of motor-operated utilization equipment that is fastened in place and that has a motor larger than ⅛ hp in combination with other loads, the total computed load shall be based on 125 percent of the largest motor load plus the sum of the other loads.

(b) Inductive Lighting Loads. For circuits supplying lighting units that have ballasts, transformers, or autotransformers, the computed load shall be based on the total ampere ratings of such units and not on the total watts of the lamps.

(c) Range Loads. It shall be acceptable to apply demand factors for range loads in accordance with Table 220-19, including Note 4.

B. Feeders and Services

220-10. General. The computed load of a feeder or service shall not be less than the sum of the loads on the branch circuits supplied, as determined by Part A of this article, after any applicable demand factors permitted by Parts B, C, or D have been applied.

FPN: See Examples D1(a) through D10 in Appendix D. See Section 220-4(b) for the maximum load in amperes permitted for lighting units operating at less than 100 percent power factor.

220-11. General Lighting. The demand factors specified in Table 220-11 shall apply to that portion of the total branch-circuit load computed for general illumination. They shall not be applied in determining the number of branch circuits for general illumination.

220-12. Show-Window and Track Lighting.

(a) Show Windows. For show-window lighting, a load of not less than 200 volt-amperes shall be included for each linear foot (305 mm) of show window, measured horizontally along its base.

FPN: See Section 220-3(b)(7) for branch circuits supplying show windows.

(b) Track Lighting. For track lighting in other than dwelling units or guest rooms of hotels or motels, an additional load of 150 volt-amperes shall be included for every 2 ft (610 mm) of lighting track or fraction thereof.

Table 220-11. Lighting Load Demand Factors

Type of Occupancy	Portion of Lighting Load to Which Demand Factor Applies (Volt-Amperes)	Demand Factor (Percent)
Dwelling units	First 3000 or less at From 3001 to 120,000 at Remainder over 120,000 at	100 35 25
Hospitals*	First 50,000 or less at Remainder over 50,000 at	40 20
Hotels and motels, including apartment houses without provision for cooking by tenants*	First 20,000 or less at From 20,001 to 100,000 at Remainder over 100,000 at	50 40 30
Warehouses (storage)	First 12,500 or less at Remainder over 12,500 at	100 50
All others	Total volt-amperes	100

*The demand factors of this table shall not apply to the computed load of feeders or services supplying areas in hospitals, hotels, and motels where the entire lighting is likely to be used at one time, as in operating rooms, ballrooms, or dining rooms.

220-13. Receptacle Loads — Nondwelling Units. In other than dwelling units, receptacle loads computed at not more than 180 volt-amperes per outlet in accordance with Section 220-3(b)(9) and fixed multi-outlet assemblies computed in accordance with Section 220-3(b)(8) shall be permitted to be added to the lighting loads and made subject to the demand factors given in Table 220-11, or they shall be permitted to be made subject to the demand factors given in Table 220-13.

Table 220-13. Demand Factors for Nondwelling Receptacle Loads

Portion of Receptacle Load to Which Demand Factor Applies (Volt-Amperes)	Demand Factor (Percent)
First 10 kVA or less at	100
Remainder over 10 kVA at	50

220-14. Motors. Motor loads shall be computed in accordance with Sections 430-24, 430-25, and 430-26.

220-15. Fixed Electric Space Heating. Fixed electric space heating loads shall be computed at 100 percent of the total connected load; however, in no case shall a feeder or service load current rating be less than the rating of the largest branch circuit supplied.

Exception: Where reduced loading of the conductors results from units operating on duty-cycle, intermittently, or from all units not operating at one time, the authority having jurisdiction may grant permission for feeder conductors to have an ampacity less than 100 percent, provided the conductors have an ampacity for the load so determined.

220-16. Small Appliance and Laundry Loads — Dwelling Unit.

(a) Small Appliance Circuit Load. In each dwelling unit, the load shall be computed at 1500 volt-amperes for each 2-wire small-appliance branch circuit required by Section 210-11(c)(1). Where the load is subdivided through two or more feeders, the computed load for each shall include not less than 1500 volt-amperes for each 2-wire small-appliance branch circuit. These loads shall be permitted to be included with the general lighting load and subjected to the demand factors provided in Table 220-11.

Exception: The individual branch circuit permitted by Section 210-52(b)(1), Exception No. 2 shall be permitted to be excluded from the calculation required by Section 220-16

(b) Laundry Circuit Load. A load of not less than 1500 volt-amperes shall be included for each 2-wire laundry branch circuit installed as required by Section 210-11(c)(2). This load shall be permitted to be included with the general lighting load and subjected to the demand factors provided in Table 220-11.

220-17. Appliance Load — Dwelling Unit(s). It shall be permissible to apply a demand factor of 75 percent to the nameplate rating load of four or more appliances fastened in place, other than electric ranges, clothes dryers, space-heating equipment, or air-conditioning equipment, that are served by the same feeder or service in a one-family, two-family, or multifamily dwelling.

220-18. Electric Clothes Dryers — Dwelling Unit(s). The load for household electric clothes dryers in a dwelling unit(s) shall be 5000 watts (volt-amperes) or the nameplate rating, whichever is larger, for each dryer served. The use of the demand factors in Table 220-18 shall be permitted. Where two or more single-phase dryers are supplied by a 3-phase, 4-wire feeder or service, the total load shall be computed on the basis of twice the maximum number connected between any two phases.

220-19. Electric Ranges and Other Cooking Appliances — Dwelling Unit(s). The demand load for household electric ranges, wall-mounted ovens, counter-mounted cooking units, and other household cooking appliances individually rated in excess of 1¾ kW shall be permitted to be computed in accordance with Table 220-19. Where two or more single-phase ranges are supplied by a 3-phase, 4-wire feeder or service, the total load shall be computed on the basis of twice the maximum number connected between any two phases. Kilovolt-amperes (kVA) shall be considered equivalent to kilowatts (kW) for loads computed under this section.

FPN: See Example D5(a) in Appendix D.

Table 220-18. Demand Factors for Household Electric Clothes Dryers

Number of Dryers	Demand Factor (Percent)
1	100
2	100
3	100
4	100
5	80
6	70
7	65
8	60
9	55
10	50
11–13	45
14–19	40
20–24	35
25–29	32.5
30–34	30
35–39	27.5
40 and over	25

Table 220-19. Demand Loads for Household Electric Ranges, Wall-Mounted Ovens, Counter-Mounted Cooking Units, and Other Household Cooking Appliances over 1¾ kW Rating (Column A to be used in all cases except as otherwise permitted in Note 3.)

Number of Appliances	Maximum Demand (kW) (See Notes) Column A (Not over 12 kW Rating)	Demand Factor (Percent) (See Note 3) Column B (Less than 3½ kW Rating)	Column C (3½ kW to 8¾ kW Rating)
1	8	80	80
2	11	75	65
3	14	70	55
4	17	66	50
5	20	62	45
6	21	59	43
7	22	56	40
8	23	53	36
9	24	51	35
10	25	49	34
11	26	47	32
12	27	45	32
13	28	43	32
14	29	41	32
15	30	40	32

Table 220-19. (Continued)

Number of Appliances	Maximum Demand (kW) (See Notes) Column A (Not over 12 kW Rating)	Demand Factor (Percent) (See Note 3) Column B (Less than 3½ kW Rating)	Column C (3½ kW to 8¾ kW Rating)
16	31	39	28
17	32	38	28
18	33	37	28
19	34	36	28
20	35	35	28
21	36	34	26
22	37	33	26
23	38	32	26
24	39	31	26
25	40	30	26
26–30	15 kW + 1 kW for each range	30	24
31–40		30	22
41–50	25 kW + ¾ kW for each range	30	20
51–60		30	18
61 and over		30	16

Notes:

1. Over 12 kW through 27 kW ranges all of same rating. For ranges individually rated more than 12 kW but not more than 27 kW, the maximum demand in Column A shall be increased 5 percent for each additional kilowatt of rating or major fraction thereof by which the rating of individual ranges exceeds 12 kW.

2. Over 8¾ kW through 27 kW ranges of unequal ratings. For ranges individually rated more than 8¾ kW and of different ratings, but none exceeding 27 kW, an average value of rating shall be computed by adding together the ratings of all ranges to obtain the total connected load (using 12 kW for any range rated less than 12 kW) and dividing by the total number of ranges. Then the maximum demand in Column A shall be increased 5 percent for each kilowatt or major fraction thereof by which this average value exceeds 12 kW.

3. Over 1¾ kW through 8¾ kW. In lieu of the method provided in Column A, it shall be permissible to add the nameplate ratings of all household cooking appliances rated more than 1¾ kW but not more than 8¾ kW and multiply the sum by the demand factors specified in Column B or C for the given number of appliances. Where the rating of cooking appliances falls under both Column B and Column C, the demand factors for each column shall be applied to the appliances for that column, and the results added together.

4. Branch-Circuit Load. It shall be permissible to compute the branch-circuit load for one range in accordance with Table 220-19. The branch-circuit load for one wall-mounted oven or one counter-mounted cooking unit shall be the nameplate rating of the appliance. The branch-circuit load for a counter-mounted cooking unit and not more than two wall-mounted ovens, all supplied from a single branch circuit and located in the same room, shall be computed by adding the nameplate rating of the individual appliances and treating this total as equivalent to one range.

5. This table also applies to household cooking appliances rated over 1¾ kW and used in instructional programs.

FPN No. 1: See Table 220-20 for commercial cooking equipment.

FPN No. 2: See the examples in Appendix D.

220-20. Kitchen Equipment — Other Than Dwelling Unit(s). It shall be permissible to compute the load for commercial electric cooking equipment, dishwasher booster heaters, water heaters, and other kitchen equipment in accordance with Table 220-20. These demand factors shall be applied to all equipment that has either thermostatic control or intermittent use as kitchen equipment. They shall not apply to space-heating, ventilating, or air-conditioning equipment.

However, in no case shall the feeder or service demand be less than the sum of the largest two kitchen equipment loads.

Table 220-20. Demand Factors for Kitchen Equipment — Other than Dwelling Unit(s)

Number of Units of Equipment	Demand Factor (Percent)
1	100
2	100
3	90
4	80
5	70
6 and over	65

220-21. Noncoincident Loads. Where it is unlikely that two or more noncoincident loads will be in use simultaneously, it shall be permissible to use only the largest load(s) that will be used at one time, in computing the total load of a feeder or service.

220-22. Feeder or Service Neutral Load. The feeder or service neutral load shall be the maximum unbalance of the load determined by this article. The maximum unbalanced load shall be the maximum net computed load between the neutral and any one ungrounded conductor, except that the load thus obtained shall be multiplied by 140 percent for 3-wire, 2-phase or 5-wire, 2-phase systems. For a feeder or service supplying household electric ranges, wall-mounted ovens, counter-mounted cooking units, and electric dryers, the maximum unbalanced load shall be considered as 70 percent of the load on the ungrounded conductors, as determined in accordance with Table 220-19 for ranges and Table 220-18 for dryers. For 3-wire dc or single-phase ac; 4-wire, 3-phase; 3-wire, 2-phase; or 5-wire, 2-phase systems, a further demand factor of 70 percent shall be permitted for that portion of the unbalanced load in excess of 200 amperes.

There shall be no reduction of the neutral capacity for that portion of the load that consists of nonlinear loads supplied from a 4-wire, wye-connected, 3-phase system nor the grounded conductor of a 3-wire circuit consisting of two phase wires and the neutral of a 4-wire, 3-phase, wye-connected system.

FPN No. 1: See Examples D1(a), D1(b), D2(b), D4(a), and D5(a) in Appendix D.

FPN No. 2: A 3-phase, 4-wire, wye-connected power system used to supply power to nonlinear loads may necessitate that the power system design allow for the possibility of high harmonic neutral currents.

C. Optional Calculations for Computing Feeder and Service Loads

220-30. Optional Calculation — Dwelling Unit.

(a) Feeder and Service Load. For a dwelling unit having the total connected load served by a single 3-wire, 120/240-volt or 208Y/120-volt set of service-entrance or feeder conductors with an ampacity of 100 or greater, it shall be permissible to compute the feeder and service loads in accordance with this section instead of the method specified in Part B of this article. The calculated load shall be the result of adding the loads from (b) and (c). Feeder and service-entrance conductors whose demand load is determined by this optional calculation shall be permitted to have the neutral load determined by Section 220-22.

(b) General Loads. The general calculated load shall be not less than 100 percent of the first 10 kVA plus 40 percent of the remainder of the following loads:

(1) 1500 volt-amperes for each 2-wire, 20-ampere small-appliance branch circuit and each laundry branch circuit specified in Section 220-16
(2) 3 volt-amperes per square foot (0.093 m^2) for general lighting and general-use receptacles
(3) The nameplate rating of all appliances that are fastened in place, permanently connected, or located to be on a specific circuit, ranges, wall-mounted ovens, counter-mounted cooking units, clothes dryers, and water heaters
(4) The nameplate ampere or kVA rating of all motors and of all low-power-factor loads

(c) Heating and Air-Conditioning Load. Include the largest of the following six selections (load in kVA).

(1) 100 percent of the nameplate rating(s) of the air conditioning and cooling.
(2) 100 percent of the nameplate ratings of the heat pump compressors and supplemental heating unless the con-

troller prevents the compressor and supplemental heating from operating at the same time.

(3) 100 percent of the nameplate ratings of electric thermal storage and other heating systems where the usual load is expected to be continuous at the full nameplate value. Systems qualifying under this selection shall not be calculated under any other selection in (c).

(4) 65 percent of the nameplate rating(s) of the central electric space heating, including integral supplemental heating in heat pumps where the controller prevents the compressor and supplemental heating from operating at the same time.

(5) 65 percent of the nameplate rating(s) of electric space heating if less than four separately controlled units.

(6) 40 percent of the nameplate rating(s) of electric space heating if four or more separately controlled units.

220-31. Optional Calculation for Additional Loads in Existing Dwelling Unit. For an existing dwelling unit presently being served by an existing 120/240-volt or 208Y/120-volt, 3-wire service, it shall be permissible to compute load calculations as follows:

Load (kVa)	Percent of Load
First 8 kVA of load at	100
Remainder of load at	40

Load calculations shall include lighting at 3 volt-amperes/ft² (0.093 m²); 1500 volt-amperes for each 2-wire, small-appliance branch circuit and each laundry branch circuit as specified in Section 220-16; range or wall-mounted oven and counter-mounted cooking unit; and other appliances that are permanently connected or fastened in place, at nameplate rating.

If air-conditioning equipment or electric space-heating equipment is to be installed, the following formula shall be applied to determine if the existing service is of sufficient size.

Air-conditioning equipment*	100%
Central electric space heating*	100%
Less than four separately controlled space-heating units*	100%
First 8 kVA of all other loads	100%
Remainder of all other loads	40%

Other loads shall include the following:

---

*Use larger connected load of air conditioning and space heating, but not both.

(1) 1500 volt-amperes for each 20-ampere appliance circuit

(2) Lighting and portable appliances at 3 volt-amperes/ft² (0.093 m²)

(3) Household range or wall-mounted oven and counter-mounted cooking unit

(4) All other appliances fastened in place, including four or more separately controlled space-heating units, at nameplate rating

220-32. Optional Calculation — Multifamily Dwelling.

(a) Feeder or Service Load. It shall be permissible to compute the feeder or service load of a multifamily dwelling in accordance with Table 220-32 instead of Part B of this article where all the following conditions are met.

(1) No dwelling unit is supplied by more than one feeder.

(2) Each dwelling unit is equipped with electric cooking equipment.

Exception: When the computed load for multifamily dwellings without electric cooking in Part B of this article exceeds that computed under Part C for the identical load plus electric cooking (based on 8 kW per unit), the lesser of the two loads shall be permitted to be used.

(3) Each dwelling unit is equipped with either electric space heating or air conditioning or both. Feeders and service-entrance conductors whose demand load is determined by this optional calculation shall be permitted to have the neutral load determined by Section 220-22.

(b) House Loads. House loads shall be computed in accordance with Part B of this article and shall be in addition to the dwelling unit loads computed in accordance with Table 220-32.

(c) Connected Loads. The connected load to which the demand factors of Table 220-32 apply shall include the following.

(1) 1500 volt-amperes for each 2-wire, 20-ampere small-appliance branch circuit and each laundry branch circuit specified in Section 220-16.

(2) 3 volt-amperes/ft² (0.093 m²) for general lighting and general-use receptacles.

(3) The nameplate rating of all appliances that are fastened in place, permanently connected or located to be on a specific circuit, ranges, wall-mounted ovens, counter-mounted cooking units, clothes dryers, water heaters, and space heaters. If water heater elements are interlocked so that all elements cannot be used at the same time, the maximum possible load shall be considered the nameplate load.

(4) The nameplate ampere or kilovolt-ampere rating of all motors and of all low-power-factor loads.

(5) The larger of the air-conditioning load or the space-heating load.

Table 220-32. Optional Calculations — Demand Factors for Three or More Multifamily Dwelling Units

Number of Dwelling Units	Demand Factor (Percent)
3–5	45
6–7	44
8–10	43
11	42
12–13	41
14–15	40
16–17	39
18–20	38
21	37
22–23	36
24–25	35
26–27	34
28–30	33
31	32
32–33	31
34–36	30
37–38	29
39–42	28
43–45	27
46–50	26
51–55	25
56–61	24
62 and over	23

220-33. Optional Calculation — Two Dwelling Units. Where two dwelling units are supplied by a single feeder and the computed load under Part B of this article exceeds that for three identical units computed under Section 220-32, the lesser of the two loads shall be permitted to be used.

220-34. Optional Method — Schools. The calculation of a feeder or service load for schools shall be permitted in accordance with Table 220-34 in lieu of Part B of this article where equipped with electric space heating or air conditioning, or both. The connected load to which the demand factors of Table 220-34 apply shall include all of the interior and exterior lighting, power, water heating, cooking, other loads, and the larger of the air-conditioning load or space-heating load within the building or structure.

Feeders and service-entrance conductors whose demand load is determined by this optional calculation shall be permitted to have the neutral load determined by Section 220-

22. Where the building or structure load is calculated by this optional method, feeders within the building or structure shall have ampacity as permitted in Part B of this article; however, the ampacity of an individual feeder shall not be required to be larger than the ampacity for the entire building.

This section shall not apply to portable classroom buildings.

Table 220-34. Optional Method — Demand Factors for Feeders and Service-Entrance Conductors for Schools

Connected Load (VA/ft^2)	Demand Factor (Percent)
The first 3 VA/ft^2 at	100
Plus,	
Over 3 to 20 VA/ft^2 at	75
Plus,	
Remainder over 20 VA/ft^2 at	25

Note: For SI units, 1 ft^2 = 0.093 m^2.

220-35. Optional Calculations for Determining Existing Loads. The calculation of a feeder or service load for existing installations shall be permitted to use actual maximum demand to determine the existing load under the following conditions.

(1) The maximum demand data is available for a 1-year period.

Exception: If the maximum demand data for a 1-year period is not available, the calculated load shall be permitted to be based on the maximum demand (measure of average power demand over a 15-minute period) continuously recorded over a minimum 30-day period using a recording ammeter or power meter connected to the highest loaded phase of the feeder or service, based on the initial loading at the start of the recording. The recording shall reflect the maximum demand of the feeder or service by being taken when the building or space is occupied and shall include by measurement or calculation the larger of the heating or cooling equipment load, and other loads that may be periodic in nature due to seasonal or similar conditions.

(2) The maximum demand at 125 percent plus the new load does not exceed the ampacity of the feeder or rating of the service.

(3) The feeder has overcurrent protection in accordance with Section 240-3, and the service has overload protection in accordance with Section 230-90.

220-36. Optional Calculation — New Restaurants. Calculation of a service load or feeder, where the feeder serves the total load, for a new restaurant shall be permitted in accordance with Table 220-36 in lieu of Part B of this article.

The overload protection of the service-entrance conductors shall be in accordance with Sections 230-90 and 240-3.

Feeder conductors shall not be required to be of greater ampacity than the service-entrance conductors.

Service-entrance or feeder conductors whose demand load is determined by this optional calculation shall be permitted to have the neutral load determined by Section 220-22.

Table 220-36. Optional Method — Demand Factors for Service-Entrance and Feeder Conductors for New Restaurants

Total Connected Load (kVA)	All Electric Demand Factor (Percent)	Not All Electric Demand Factor (Percent)
0–250	80	100
251–280	70	90
281–325	60	80
326–375	50	70
376–800	50	65
Over 800	50	50

Note: Add all electrical loads, including both heating and cooling loads, to compute the total connected load. Select the one demand factor that applies from the table and multiply the total connected load by this single demand factor.

D. Method for Computing Farm Loads

220-40. Farm Loads — Buildings and Other Loads.

(a) Dwelling Unit. The feeder or service load of a farm dwelling unit shall be computed in accordance with the provisions for dwellings in Part B or C of this article. Where the dwelling has electric heat and the farm has electric grain-drying systems, Part C of this article shall not be used to compute the dwelling load.

(b) Other than Dwelling Unit. For each farm building or load supplied by two or more separate branch circuits, the load for feeders, service-entrance conductors, and service equipment shall be computed in accordance with demand factors not less than indicated in Table 220-40.

FPN: See Section 230-21 for overhead conductors from a pole to a building or other structure.

220-41. Farm Loads — Total. The total load of the farm for service-entrance conductors and service equipment shall be computed in accordance with the farm dwelling unit load and demand factors specified in Table 220-41. Where there is equipment in two or more farm equipment buildings or for loads having the same function, such loads shall be

Table 220-40. Method for Computing Farm Loads for Other than Dwelling Unit

Ampere Load at 240 Volts Maximum	Demand Factor (Percent)
Loads expected to operate without diversity, but not less than 125 percent full-load current of the largest motor and not less than the first 60 amperes of load	100
Next 60 amperes of all other loads	50
Remainder of other load	25

computed in accordance with Table 220-40 and shall be permitted to be combined as a single load in Table 220-41 for computing the total load.

FPN: See Section 230-21 for overhead conductors from a pole to a building or other structure.

Table 220-41. Method for Computing Total Farm Load

Individual Loads Computed in Accordance with Table 220-40	Demand Factor (Percent)
Largest load	100
Second largest load	75
Third largest load	65
Remaining loads	50

Note: To this total load, add the load of the farm dwelling unit computed in accordance with Part B or C of this article. Where the dwelling has electric heat and the farm has electric grain-drying systems, Part C of this article shall not be used to compute the dwelling load.

Article 225 — Outside Branch Circuits and Feeders

225-1. Scope. This article covers requirements for outside branch circuits and feeders run on or between buildings, structures, or poles on the premises; and electric equipment and wiring for the supply of utilization equipment that is located on or attached to the outside of buildings, structures, or poles.

FPN: For additional information on wiring over 600 volts, see *National Electrical Safety Code*, ANSI C2-1997.

225-2. Other Articles. Application of other articles, including additional requirements to specific cases of equipment and conductors, is as follows:

A. General

225-3. Calculation of Load.

(a) Branch Circuits. The load on outdoor branch circuits shall be as determined by Section 220-3.

(b) Feeders. The load on outdoor feeders shall be as determined by Part B of Article 220.

225-4. Conductor Covering.
Where within 10 ft (3.05 m) of any building or structure other than supporting poles or towers, open individual (aerial) overhead conductors shall be insulated or covered. Conductors in cables or raceways, except Type MI cable, shall be of the rubber-covered type or thermoplastic type and, in wet locations, shall comply with Section 310-8. Conductors for festoon lighting shall be of the rubber-covered or thermoplastic type.

Exception: Equipment grounding conductors and grounded circuit conductors shall be permitted to be bare or covered as specifically permitted elsewhere in this Code.

225-5. Size of Conductors.
The ampacity of outdoor branch-circuit and feeder conductors shall be in accordance with Section 310-15 based on loads as determined under Section 220-3 and Part B of Article 220.

225-6. Conductor Size and Support.

(a) Overhead Spans. Open individual conductors shall not be smaller than the following:

(1) For 600 volts, nominal, or less, No. 10 copper or No. 8 aluminum for spans up to 50 ft (15.2 m) in length and No. 8 copper or No. 6 aluminum for a longer span, unless supported by a messenger wire

(2) For over 600 volts, nominal, No. 6 copper or No. 4 aluminum where open individual conductors and No. 8 copper or No. 6 aluminum where in cable

(b) Festoon Lighting. Overhead conductors for festoon lighting shall not be smaller than No. 12 unless the conductors are supported by messenger wires. In all spans exceeding 40 ft (12.2 m), the conductors shall be supported by messenger wire. The messenger wire shall be supported by strain insulators. Conductors or messenger wires shall not be attached to any fire escape, down spout, or plumbing equipment.

225-7. Lighting Equipment Installed Outdoors.

(a) General. For the supply of lighting equipment installed outdoors, the branch circuits shall comply with Article 210 and (b) through (d).

(b) Common Neutral. The ampacity of the neutral conductor shall not be less than the maximum net computed load current between the neutral and all ungrounded conductors connected to any one phase of the circuit.

(c) 277 Volts to Ground. Circuits exceeding 120 volts, nominal, between conductors and not exceeding 277 volts, nominal, to ground shall be permitted to supply lighting fixtures for illumination of outdoor areas of industrial establishments, office buildings, schools, stores, and other commercial or public buildings where the fixtures are not less than 3 ft (914 mm) from windows, platforms, fire escapes, and the like.

(d) 600 Volts Between Conductors. Circuits exceeding 277 volts, nominal, to ground and not exceeding 600 volts, nominal, between conductors shall be permitted to supply the auxiliary equipment of electric-discharge lamps in accordance with Section 210-6(d)(1).

225-9. Overcurrent Protection.
Overcurrent protection shall be in accordance with Section 210-20 for branch circuits and Article 240 for feeders.

225-10. Wiring on Buildings. The installation of outside wiring on surfaces of buildings shall be permitted for circuits of not over 600 volts, nominal, as open wiring on insulators, as multiconductor cable, as Type MC cable, as Type MI cable, as messenger supported wiring, in rigid metal conduit, in intermediate metal conduit, in rigid nonmetallic conduit, in cable trays, as cablebus, in wireways, in auxiliary gutters, in electrical metallic tubing, in flexible metal conduit, in liquidtight flexible metal conduit, in liquidtight flexible nonmetallic conduit, and in busways. Circuits of over 600 volts, nominal, shall be installed as provided in Section 300-37. Circuits for signs and outline lighting shall be installed in accordance with Article 600.

225-11. Circuit Exits and Entrances. Where outside branch and feeder circuits leave or enter a building, the requirements of Sections 230-52 and 230-54 shall apply.

225-12. Open-Conductor Supports. Open conductors shall be supported on glass or porcelain knobs, racks, brackets, or strain insulators.

225-14. Open-Conductor Spacings.

(a) 600 Volts, Nominal, or Less. Conductors of 600 volts, nominal, or less, shall comply with the spacings provided in Table 230-51(c).

(b) Over 600 Volts, Nominal. Conductors of over 600 volts, nominal, shall comply with the spacings provided in Sections 110-36 and 490-24.

(c) Separation from Other Circuits. Open conductors shall be separated from open conductors of other circuits or systems by not less than 4 in. (102 mm).

(d) Conductors on Poles. Conductors on poles shall have a separation of not less than 1 ft (305 mm) where not placed on racks or brackets. Conductors supported on poles shall provide a horizontal climbing space not less than the following:

(1) Power conductors below communications conductors — 30 in. (762 mm)
(2) Power conductors alone or above communications conductors:
    300 volts or less — 24 in. (610 mm)
    Over 300 volts — 30 in. (762 mm)
(3) Communications conductors below power conductors — same as power conductors
(4) Communications conductors alone — no requirement

225-15. Supports Over Buildings. Supports over a building shall be in accordance with Section 230-29.

225-16. Point of Attachment to Buildings. The point of attachment to a building shall be in accordance with Section 230-26.

225-17. Means of Attachment to Buildings. The means of attachment to a building shall be in accordance with Section 230-27.

225-18. Clearance from Ground. Overhead spans of open conductors and open multiconductor cables of not over 600 volts, nominal, shall conform to the following:

10 ft (3.05 m) — above finished grade, sidewalks, or from any platform or projection from which they might be reached where the voltage does not exceed 150 volts to ground and accessible to pedestrians only

12 ft (3.66 m) — over residential property and driveways, and those commercial areas not subject to truck traffic where the voltage does not exceed 300 volts to ground

15 ft (4.57 m) — for those areas listed in the 12-ft (3.66-m) classification where the voltage exceeds 300 volts to ground

18 ft (5.49 m) — over public streets, alleys, roads, parking areas subject to truck traffic, driveways on other than residential property, and other land traversed by vehicles such as cultivated, grazing, forest, and orchard

FPN: For clearances of conductors of over 600 volts, see *National Electrical Safety Code*, ANSI C2-1997.

225-19. Clearances from Buildings for Conductors of Not Over 600 Volts, Nominal.

(a) Above Roofs. Overhead spans of open conductors and open multiconductor cables shall have a vertical clearance of not less than 8 ft (2.44 m) above the roof surface. The vertical clearance above the roof level shall be maintained for a distance not less than 3 ft (914 mm) in all directions from the edge of the roof.

Exception No. 1: The area above a roof surface subject to pedestrian or vehicular traffic shall have a vertical clearance from the roof surface in accordance with the clearance requirements of Section 225-18.

Exception No. 2: Where the voltage between conductors does not exceed 300, and the roof has a slope of 4 in. (102 mm) in 12 in. (305 mm) or greater, a reduction in clearance to 3 ft (914 mm) shall be permitted.

Exception No. 3: Where the voltage between conductors does not exceed 300, a reduction in clearance above only the overhanging portion of the roof to not less than 18 in. (457 mm) shall be permitted if (1) not more than 6 ft (1.83 m) of the conductors, 4 ft (1.22 m) horizontally, pass above the roof overhang, and (2) they are terminated at a through-the-roof raceway or approved support.

Exception No. 4: The requirement for maintaining the vertical clearance 3 ft (914 mm) from the edge of the roof shall not apply to the final conductor span where the conductors are attached to the side of a building.

(b) From Nonbuilding or Nonbridge Structures. From signs, chimneys, radio and television antennas, tanks, and other nonbuilding or nonbridge structures, clearances — vertical, diagonal, and horizontal — shall not be less than 3 ft (914 mm).

(c) Horizontal Clearances. Clearances shall not be less than 3 ft (914 mm).

(d) Final Spans. Final spans of feeders or branch circuits to a building they supply or from which they are fed shall be permitted to be attached to the building, but they shall be kept not less than 3 ft (914 mm) from windows that are designed to be opened, doors, porches, balconies, ladders, stairs, fire escapes, or similar locations. Vertical clearance of final spans above, or within 3 ft (914 mm) measured horizontally of, platforms, projections, or surfaces from which they might be reached shall be maintained in accordance with Section 225-18.

Exception: Conductors run above the top level of a window shall be permitted to be less than the 3 ft (914 mm) requirement above.

Overhead branch-circuit and feeder conductors shall not be installed beneath openings through which materials may be moved, such as openings in farm and commercial buildings, and shall not be installed where they will obstruct entrance to these building openings.

(e) Zone for Fire Ladders. Where buildings exceed three stories or 50 ft (15.2 m) in height, overhead lines shall be arranged, where practicable, so that a clear space (or zone) at least 6 ft (1.83 m) wide will be left either adjacent to the buildings or beginning not over 8 ft (2.44 m) from them to facilitate the raising of ladders when necessary for fire fighting.

> FPN: For clearance of conductors over 600 volts, see *National Electrical Safety Code*, ANSI C2-1997.

225-20. Mechanical Protection of Conductors. Mechanical protection of conductors on buildings, structures, or poles shall be as provided for services in Section 230-50.

225-21. Multiconductor Cables on Exterior Surfaces of Buildings. Supports for multiconductor cables on exterior surfaces of buildings shall be as provided in Section 230-51.

225-22. Raceways on Exterior Surfaces of Buildings. Raceways on exterior surfaces of buildings shall be raintight and arranged to drain.

Exception: Flexible metal conduit, where permitted in Section 350-5(1), shall not be required to be raintight.

225-24. Outdoor Lampholders. Where outdoor lampholders are attached as pendants, the connections to the circuit wires shall be staggered. Where such lampholders have terminals of a type that puncture the insulation and make contact with the conductors, they shall be attached only to conductors of the stranded type.

225-25. Location of Outdoor Lamps. Locations of lamps for outdoor lighting shall be below all energized conductors, transformers, or other electric utilization equipment, unless

(1) Clearances or other safeguards are provided for relamping operations, or
(2) Equipment is controlled by a disconnecting means that can be locked in the open position.

225-26. Vegetation. Vegetation such as trees shall not be used for support of overhead conductor spans.

Exception: For temporary wiring in accordance with Article 305.

B. More than One Building or Other Structure

225-30. Number of Supplies. Where more than one building or other structure is on the same property and under single management, each building or other structure served shall be supplied by one feeder or branch circuit unless permitted in (a) through (e). For the purpose of this section, a multiwire branch circuit shall be considered a single circuit.

(a) Special Conditions. Additional feeders or branch circuits shall be permitted to supply the following:

(1) Fire pumps
(2) Emergency systems
(3) Legally required standby systems
(4) Optional standby systems
(5) Parallel power production systems

(b) Special Occupancies. By special permission, additional feeders or branch circuits shall be permitted for

(1) Multiple-occupancy buildings where there is no available space for supply equipment accessible to all occupants, or
(2) A single building or other structure sufficiently large to make two or more supplies necessary.

(c) Capacity Requirements. Additional feeders or branch circuits shall be permitted where the capacity requirements are in excess of 2000 amperes at a supply voltage of 600 volts or less.

(d) Different Characteristics. Additional feeders or branch circuits shall be permitted for different voltages, frequencies, or phases, or for different uses, such as control of outside lighting from multiple locations.

(e) Documented Switching Procedures. Additional feeders or branch circuits shall be permitted to supply installations under single management where documented safe switching procedures are established and maintained for disconnection.

225-31. Disconnecting Means. Means shall be provided for disconnecting all ungrounded conductors that supply or pass through the building or structure.

225-32. Location. The disconnecting means shall be installed either inside or outside of the building or structure served or where the conductors pass through the building or structure. The disconnecting means shall be at a readily accessible location nearest the point of entrance of the conductors. For the purposes of this section, the requirements in Section 230-6 shall be permitted to be utilized.

Exception No. 1: For installations under single management, where documented safe switching procedures are established and maintained for disconnection, the disconnecting means shall be permitted to be located elsewhere on the premises.

Exception No. 2: For buildings or other structures qualifying under the provisions of Article 685, the disconnecting means shall be permitted to be located elsewhere on the premises.

Exception No. 3: For towers or poles used as lighting standards, the disconnecting means shall be permitted to be located elsewhere on the premises.

Exception No. 4: For poles or similar structures used only for support of signs installed in accordance with Article 600, the disconnecting means shall be permitted to be located elsewhere on the premises.

225-33. Maximum Number of Disconnects.

(a) General. The disconnecting means for each supply permitted by Section 225-30 shall consist of not more than six switches or six circuit breakers mounted in a single enclosure, in a group of separate enclosures, or in or on a switchboard. There shall be no more than six disconnects per supply grouped in any one location.

Exception: For the purpose of this section, disconnecting means used solely for the control circuit of the ground-fault

protection system, installed as part of the listed equipment, shall not be considered a supply disconnecting means.

(b) Single-Pole Units. Two or three single-pole switches or breakers, capable of individual operation, shall be permitted on multiwire circuits, one pole for each ungrounded conductor, as one multipole disconnect, provided they are equipped with handle ties or a master handle to disconnect all ungrounded conductors with no more than six operations of the hand.

225-34. Grouping of Disconnects.

(a) General. The two to six disconnects as permitted in Section 225-33 shall be grouped. Each disconnect shall be marked to indicate the load served.

Exception: One of the two to six disconnecting means permitted in Section 225-33, where used only for a water pump also intended to provide fire protection, shall be permitted to be located remote from the other disconnecting means.

(b) Additional Disconnecting Means. The one or more additional disconnecting means for fire pumps or for emergency, legally required standby, or optional standby system permitted by Section 225-30 shall be installed sufficiently remote from the one to six disconnecting means for normal supply to minimize the possibility of simultaneous interruption of supply.

225-35. Access to Occupants. In a multiple-occupancy building, each occupant shall have access to the occupant's supply disconnecting means.

Exception: In a multiple-occupancy building where electric supply and electrical maintenance are provided by the building management and where these are under continuous building management supervision, the supply disconnecting means supplying more than one occupancy shall be permitted to be accessible to authorized management personnel only.

225-36. Suitable for Service Equipment. The disconnecting means specified in Section 225-31 shall be suitable for use as service equipment.

Exception: For garages and outbuildings on residential property, a snap switch or a set of 3-way or 4-way snap switches shall be permitted as the disconnecting means.

225-37. Identification. Where a building or structure has any combination of feeders, branch circuits, or services passing through it or supplying it, a permanent plaque or directory shall be installed at each feeder and branch-circuit disconnect location denoting all other services, feeders, or branch circuits supplying that building or structure or passing through that building or structure and the area served by each. See Section 230-2(e).

Exception No. 1: A plaque or directory shall not be required for large capacity multibuilding industrial installations under single management, where it is ensured that disconnection can be accomplished by establishing and maintaining safe switching procedures.

Exception No. 2: This identification shall not be required for branch circuits installed from a dwelling unit to a second building or structure.

225-38. Disconnect Construction.

(a) Manually or Power Operable. The disconnecting means shall consist of either (1) a manually operable switch or a circuit breaker equipped with a handle or other suitable operating means or (2) a power-operable switch or circuit breaker, provided the switch or circuit breaker can be opened by hand in the event of a power failure.

(b) Simultaneous Opening of Poles. Each building or structure disconnecting means shall simultaneously disconnect all ungrounded supply conductors that it controls from the building or structure wiring system.

(c) Disconnection of Grounded Conductor. Where the building or structure disconnecting means does not disconnect the grounded conductor from the grounded conductors in the building or structure wiring, other means shall be provided for this purpose at the location of disconnecting means. A terminal or bus to which all grounded conductors can be attached by means of pressure connectors shall be permitted for this purpose.

In a multisection switchboard, disconnects for the grounded conductor shall be permitted to be in any section of the switchboard, provided any such switchboard section is marked.

(d) Indicating. The building or structure disconnecting means shall plainly indicate whether it is in the open or closed position.

225-39. Rating of Disconnect.
The feeder or branch-circuit disconnecting means shall have a rating of not less than the load to be carried, determined in accordance with Article 220. In no case shall the rating be lower than specified in (a), (b), (c), or (d).

(a) One-Circuit Installation. For installations to supply only limited loads of a single branch-circuit, the branch-circuit disconnecting means shall have a rating of not less than 15 amperes.

(b) Two-Circuit Installations. For installations consisting of not more than two 2-wire branch circuits, the feeder or branch-circuit disconnecting means shall have a rating of not less than 30 amperes.

(c) One-Family Dwelling. For a one-family dwelling, the feeder disconnecting means shall have a rating of not less than 100 amperes, 3-wire.

(d) All Others. For all other installations, the feeder or branch-circuit disconnecting means shall have a rating of not less than 60 amperes.

225-40. Access to Overcurrent Protective Devices.
Where a feeder overcurrent device is not readily accessible, branch-circuit overcurrent devices shall be installed on the load side, shall be mounted in a readily accessible location, and shall be of a lower ampere rating than the feeder overcurrent device.

C. Over 600 Volts

225-50. Warning Signs.
Signs with the words "WARNING — HIGH VOLTAGE — KEEP OUT" shall be posted in plain view where unauthorized persons might come in contact with live parts.

225-51. Isolating Switches.
Where oil switches or air, oil, vacuum, or sulfur hexafluoride circuit breakers constitute a building disconnecting means, an isolating switch with visible break contacts and meeting the requirements of Section 230-204(b), (c), and (d) shall be installed on the supply side of the disconnecting means and all associated equipment.

Exception: The isolating switch shall not be required where the disconnecting means is mounted on removable truck panels or metal-enclosed switchgear units that cannot be opened unless the circuit is disconnected and that, when removed from the normal operating position, automatically disconnect the circuit breaker or switch from all energized parts.

225-52. Location.
A building or structure disconnecting means shall be located in accordance with Section 225-32, or it shall be electrically operated by a similarly located remote-control device.

225-53. Type.
Each building or structure disconnect shall simultaneously disconnect all ungrounded supply conductors it controls and shall have a fault-closing rating not less than the maximum available short-circuit current available at its supply terminals.

Where fused switches or separately mounted fuses are installed, the fuse characteristics shall be permitted to contribute to the fault closing rating of the disconnecting means.

Article 230 — Services

230-1. Scope. This article covers service conductors and equipment for control and protection of services and their installation requirements.

FPN: See Figure 230-1.

General	Part A
Overhead Service-Drop Conductors	Part B
Underground Service-Lateral Conductors	Part C
Service-Entrance Conductors	Part D
Service Equipment — General	Part E
Service Equipment — Disconnecting Means	Part F
Service Equipment — Overcurrent Protection	Part G
Services Exceeding 600 Volts, Nominal	Part H

Figure 230-1   Services.

A. General

230-2. Number of Services. A building or other structure served shall be supplied by only one service unless permitted in (a) through (d). For the purpose of Section 230-40, Exception No. 2 only, underground sets of conductors, size 1/0 and larger, running to the same location and connected together at their supply end, but not connected together at their load end, shall be considered to be supplying one service.

(a) Special Conditions. Additional services shall be permitted to supply

(1) Fire pumps
(2) Emergency systems
(3) Legally required standby systems
(4) Optional standby systems
(5) Parallel power production systems

(b) Special Occupancies. By special permission, additional services shall be permitted for

(1) Multiple-occupancy buildings where there is no available space for service equipment accessible to all occupants, or
(2) A single building or other structure sufficiently large to make two or more services necessary.

(c) Capacity Requirements.

(1) Additional services shall be permitted where the capacity requirements are in excess of 2000 amperes at a supply voltage of 600 volts or less.

(2) Additional services shall be permitted where the load requirements of a single-phase installation are greater than the serving agency normally supplies through one service.

(3) Additional services shall be permitted by special permission.

(d) Different Characteristics. Additional services shall be permitted for different voltages, frequencies, or phases, or for different uses, such as for different rate schedules.

(e) Identification. Where a building or structure is supplied by more than one service, or any combination of branch circuits, feeders, and services, a permanent plaque or directory shall be installed at each service disconnect location denoting all other services, feeders, and branch circuits supplying that building or structure and the area served by each. See Section 225-37.

230-3. One Building or Other Structure Not to Be Supplied Through Another. Service conductors supplying a building or other structure shall not pass through the interior of another building or other structure.

230-6. Conductors Considered Outside the Building. Conductors shall be considered outside of a building or other structure under any of the following conditions:

(1) Where installed under not less than 2 in. (50.8 mm) of concrete beneath a building or other structure
(2) Where installed within a building or other structure in a raceway that is encased in concrete or brick not less than 2 in. (50.8 mm) thick

(3) Where installed in a transformer vault conforming to the requirements of Article 450, Part C

230-7. Other Conductors in Raceway or Cable. Conductors other than service conductors shall not be installed in the same service raceway or service cable.

Exception No. 1: Grounding conductors and bonding jumpers.

Exception No. 2: Load management control conductors having overcurrent protection.

230-8. Raceway Seal. Where a service raceway enters a building or structure from an underground distribution system, it shall be sealed in accordance with Section 300-5. Spare or unused raceways shall also be sealed. Sealants shall be identified for use with the cable insulation, shield, or other components.

230-9. Clearance from Building Openings. Service conductors installed as open conductors or multiconductor cable without an overall outer jacket shall have a clearance of not less than 3 ft (914 mm) from windows that are designed to be opened, doors, porches, balconies, ladders, stairs, fire escapes, or similar locations. Vertical clearance of final spans above, or within 3 ft (914 mm) measured horizontally of, platforms, projections, or surfaces from which they might be reached shall be maintained in accordance with Section 230-24(b).

Exception: Conductors run above the top level of a window shall be permitted to be less than the 3 ft (914 mm) requirement above.

Overhead service conductors shall not be installed beneath openings through which materials may be moved, such as openings in farm and commercial buildings, and shall not be installed where they will obstruct entrance to these building openings.

B. Overhead Service-Drop Conductors

230-21. Overhead Supply. Overhead service conductors to a building or other structure (such as a pole) on which a meter or disconnecting means is installed shall be considered as a service drop and installed accordingly.

FPN: For example, see farm loads in Part D of Article 220.

230-22. Insulation or Covering. Individual conductors shall be insulated or covered with an extruded thermoplastic or thermosetting insulating material.

Exception: The grounded conductor of a multiconductor cable shall be permitted to be bare.

230-23. Size and Rating.

(a) General. Conductors shall have sufficient ampacity to carry the current for the load as computed in accordance with Article 220 and shall have adequate mechanical strength.

(b) Minimum Size. The conductors shall not be smaller than No. 8 copper or No. 6 aluminum or copper-clad aluminum.

Exception: Conductors supplying only limited loads of a single branch circuit — such as small polyphase power, controlled water heaters, and similar loads — shall not be smaller than No. 12 hard-drawn copper or equivalent.

(c) Grounded Conductors. The grounded conductor shall not be less than the minimum size as required by Section 250-24(b).

230-24. Clearances. The vertical clearances of all service-drop conductors shall be based on conductor temperature of 60°F (15°C), no wind, with final unloaded sag in the wire, conductor, or cable.

Service-drop conductors shall not be readily accessible and shall comply with (a) through (d) for services not over 600 volts, nominal.

(a) Above Roofs. Conductors shall have a vertical clearance of not less than 8 ft (2.44 m) above the roof surface. The vertical clearance above the roof level shall be maintained for a distance of not less than 3 ft (914 mm) in all directions from the edge of the roof.

Exception No. 1: The area above a roof surface subject to pedestrian or vehicular traffic shall have a vertical clearance from the roof surface in accordance with the clearance requirements of Section 230-24(b).

Exception No. 2: Where the voltage between conductors does not exceed 300 and the roof has a slope of 4 in. (102 mm) in 12 in. (305 mm), or greater, a reduction in clearance to 3 ft (914 mm) shall be permitted.

Exception No. 3: Where the voltage between conductors does not exceed 300, a reduction in clearance above only the overhanging portion of the roof to not less than 18 in. (457 mm) shall be permitted if (1) not more than 6 ft (1.83 m) of service-drop conductors, 4 ft (1.22 m) horizontally, pass above the roof overhang, and (2) they are terminated at a through-the-roof raceway or approved support.

FPN: See Section 230-28 for mast supports.

Exception No. 4: The requirement for maintaining the vertical clearance 3 ft (914 mm) from the edge of the roof shall not apply to the final conductor span where the service drop is attached to the side of a building.

(b) Vertical Clearance from Ground. Service-drop conductors where not in excess of 600 volts, nominal, shall have the following minimum clearance from final grade:

10 ft (3.05 m) — at the electric service entrance to buildings, also at the lowest point of the drip loop of the building electric entrance, and above areas or sidewalks accessible only to pedestrians, measured from final grade or other accessible surface only for service-drop cables supported on and cabled together with a grounded bare messenger where the voltage does not exceed 150 volts to ground

12 ft (3.66 m) — over residential property and driveways, and those commercial areas not subject to truck traffic where the voltage does not exceed 300 volts to ground

15 ft (4.57 m) — for those areas listed in the 12-ft (3.66-m) classification where the voltage exceeds 300 volts to ground

18 ft (5.49 m) — over public streets, alleys, roads, parking areas subject to truck traffic, driveways on other than residential property, and other land such as cultivated, grazing, forest, and orchard

(c) Clearance from Building Openings. See Section 230-9.

(d) Clearance from Swimming Pools. See Section 680-8.

230-26. Point of Attachment. The point of attachment of the service-drop conductors to a building or other structure shall provide the minimum clearances as specified in Section 230-24. In no case shall this point of attachment be less than 10 ft (3.05 m) above finished grade.

230-27. Means of Attachment. Multiconductor cables used for service drops shall be attached to buildings or other structures by fittings identified for use with service conductors. Open conductors shall be attached to fittings identified for use with service conductors or to noncombustible, nonabsorbent insulators securely attached to the building or other structure.

230-28. Service Masts as Supports. Where a service mast is used for the support of service-drop conductors, it shall be of adequate strength or be supported by braces or guys to withstand safely the strain imposed by the service drop. Where raceway-type service masts are used, all raceway fittings shall be identified for use with service masts. Only power service-drop conductors shall be permitted to be attached to a service mast.

230-29. Supports Over Buildings. Service-drop conductors passing over a roof shall be securely supported by substantial structures. Where practicable, such supports shall be independent of the building.

C. Underground Service-Lateral Conductors

230-30. Insulation. Service-lateral conductors shall be insulated for the applied voltage.

Exception: A grounded conductor shall be permitted to be uninsulated as follows:

(a) *Bare copper used in a raceway*
(b) *Bare copper for direct burial where bare copper is judged to be suitable for the soil conditions*
(c) *Bare copper for direct burial without regard to soil conditions where part of a cable assembly identified for underground use*
(d) *Aluminum or copper-clad aluminum without individual insulation or covering where part of a cable assembly identified for underground use in a raceway or for direct burial*

230-31. Size and Rating.

(a) General. Service-lateral conductors shall have sufficient ampacity to carry the current for the load as computed in accordance with Article 220 and shall have adequate mechanical strength.

(b) Minimum Size. The conductors shall not be smaller than No. 8 copper or No. 6 aluminum or copper-clad aluminum.

Exception: Conductors supplying only limited loads of a single branch circuit — such as small polyphase power, controlled water heaters, and similar loads — shall not be smaller than No. 12 copper or No. 10 aluminum or copper-clad aluminum.

(c) Grounded Conductors. The grounded conductor shall not be less than the minimum size required by Section 250-24(b).

> FPN: Reasonable efficiency of operation can be provided when voltage drop is taken into consideration in sizing the service-lateral conductors.

230-32. Protection Against Damage. Underground service-lateral conductors shall be protected against damage in accordance with Section 300-5. Service-lateral conductors entering a building shall be installed in accordance with Section 230-6 or protected by a raceway wiring method identified in Section 230-43.

D. Service-Entrance Conductors

230-40. Number of Service-Entrance Conductor Sets.
Each service drop or lateral shall supply only one set of service-entrance conductors.

Exception No. 1: Buildings with one or more than one occupancy shall be permitted to have one set of service-entrance conductors for each class of service run to each occupancy or group of occupancies.

Exception No. 2: Where two to six service disconnecting means in separate enclosures are grouped at one location and supply separate loads from one service drop or lateral, one set of service-entrance conductors shall be permitted to supply each or several such service equipment enclosures.

Exception No. 3: A single-family dwelling unit and a separate structure shall be permitted to have one set of service-entrance conductors run to each from a single service drop or lateral.

Exception No. 4: A two-family dwelling or a multifamily dwelling shall be permitted to have one set of service-entrance conductors installed to supply the circuits covered in Section 210-25.

Exception No. 5: One set of service-entrance conductors connected to the supply side of the normal service disconnecting means shall be permitted to supply each or several systems covered by Section 230-82(3).

230-41. Insulation of Service-Entrance Conductors.
Service-entrance conductors entering or on the exterior of buildings or other structures shall be insulated.

Exception: A grounded conductor shall be permitted to be uninsulated as follows:

(a) *Bare copper used in a raceway or part of a service cable assembly*

(b) *Bare copper for direct burial where bare copper is judged to be suitable for the soil conditions*

(c) *Bare copper for direct burial without regard to soil conditions where part of a cable assembly identified for underground use*

(d) *Aluminum or copper-clad aluminum without individual insulation or covering where part of a cable assembly or identified for underground use in a raceway, or for direct burial*

230-42. Minimum Size and Rating.

(a) General. The ampacity of the service-entrance conductors before the application of any adjustment or correction factors shall not be less than either (1) or (2). Loads shall be determined in accordance with Article 220. Ampacity shall be determined from Section 310-15. The maximum

allowable current of busways shall be that value for which the busway has been listed or labeled.

(1) The sum of the noncontinuous loads plus 125 percent of continuous loads
(2) The sum of noncontinuous load plus the continuous load if the service-entrance conductors terminate in an overcurrent device where both the overcurrent device and its assembly are listed for operation at 100 percent of their rating

(b) Ungrounded Conductors. Ungrounded conductors shall have an ampacity of not less than the minimum rating of the disconnecting means specified in Section 230-79.

(c) Grounded Conductors. The grounded conductor shall not be less than the minimum size as required by Section 250-24(b).

230-43. Wiring Methods for 600 Volts, Nominal, or Less.
Service-entrance conductors shall be installed in accordance with the applicable requirements of this *Code* covering the type of wiring method used and shall be limited to the following methods:

(1) Open wiring on insulators
(2) Type IGS cable
(3) Rigid metal conduit
(4) Intermediate metal conduit
(5) Electrical metallic tubing
(6) Electrical nonmetallic tubing (ENT)
(7) Service-entrance cables
(8) Wireways
(9) Busways
(10) Auxiliary gutters
(11) Rigid nonmetallic conduit
(12) Cablebus
(13) Type MC cable
(14) Mineral-insulated, metal-sheathed cable
(15) Flexible metal conduit not over 6 ft (1.83 m) long or liquidtight flexible metal conduit not over 6 ft (1.83 m) long between raceways, or between raceway and service equipment, with equipment bonding jumper routed with the flexible metal conduit or the liquidtight flexible metal conduit according to the provisions of Section 250-102(a), (b), (c), and (e)
(16) Liquidtight flexible nonmetallic conduit

Cable tray systems shall be permitted to support cables for use as service-entrance conductors in accordance with Article 318.

230-46. Spliced Conductors.
Service-entrance conductors shall be permitted to be spliced or tapped by clamped or

bolted connections. Splices shall be made in enclosures or, if directly buried, with a listed underground splice kit. Splices of conductors shall be made in accordance with Sections 110-14, 300-5(e), 300-13, and 300-15.

230-49. Protection Against Physical Damage — Underground. Underground service-entrance conductors shall be protected against physical damage in accordance with Section 300-5.

230-50. Protection of Open Conductors and Cables Against Damage — Above Ground. Service-entrance conductors installed above ground shall be protected against physical damage as specified in (a) or (b).

(a) Service Cables. Service cables, where subject to physical damage, shall be protected by any of the following:

(1) Rigid metal conduit
(2) Intermediate metal conduit
(3) Rigid nonmetallic conduit suitable for the location
(4) Electrical metallic tubing
(5) Other approved means

(b) Other than Service Cable. Individual open conductors and cables other than service cables shall not be installed within 10 ft (3.05 m) of grade level or where exposed to physical damage.

Exception: Type MI and Type MC cable shall be permitted within 10 ft (3.05 m) of grade level where not exposed to physical damage or where protected in accordance with Section 300-5(d).

230-51. Mounting Supports. Cables or individual open service conductors shall be supported as specified in (a), (b), or (c).

(a) Service Cables. Service cables shall be supported by straps or other approved means within 12 in. (305 mm) of every service head, gooseneck, or connection to a raceway or enclosure and at intervals not exceeding 30 in. (762 mm).

(b) Other Cables. Cables that are not approved for mounting in contact with a building or other structure shall be mounted on insulating supports installed at intervals not exceeding 15 ft (4.57 m) and in a manner that will maintain a clearance of not less than 2 in. (50.8 mm) from the surface over which they pass.

(c) Individual Open Conductors. Individual open conductors shall be installed in accordance with Table 230-51(c). Where exposed to the weather, the conductors shall be mounted on insulators or on insulating supports attached to racks, brackets, or other approved means. Where not

exposed to the weather, the conductors shall be mounted on glass or porcelain knobs.

Table 230-51(c). Supports and Clearances for Individual Open Service Conductors

Maximum Volts	Maximum Distance Between Supports (ft)	Minimum Clearance (in.)	
		Between Conductors	From Surface
600	9	6	2
600	15	12	2
300	4½	3	2
600*	4½*	2½*	1*

Note: For SI units, 1 in. = 25.4 mm; 1 ft = 0.3048 m.
*Where not exposed to weather.

230-52. Individual Conductors Entering Buildings or Other Structures. Where individual open conductors enter a building or other structure, they shall enter through roof bushings or through the wall in an upward slant through individual, noncombustible, nonabsorbent insulating tubes. Drip loops shall be formed on the conductors before they enter the tubes.

230-53. Raceways to Drain. Where exposed to the weather, raceways enclosing service-entrance conductors shall be raintight and arranged to drain. Where embedded in masonry, raceways shall be arranged to drain.

Exception: As permitted in Section 350-5.

230-54. Overhead Service Locations.

(a) Raintight Service Head. Service raceways shall be equipped with a raintight service head at the point of connection to service-drop conductors.

(b) Service Cable Equipped with Raintight Service Head or Gooseneck. Service cables shall be equipped with a raintight service head.

Exception: Type SE cable shall be permitted to be formed into a gooseneck and taped with a self-sealing weather-resistant thermoplastic.

(c) Service Heads Above Service-Drop Attachment. Service heads and goosenecks in service-entrance cables shall be located above the point of attachment of the service-drop conductors to the building or other structure.

Exception: Where it is impracticable to locate the service head above the point of attachment, the service head location shall be permitted not farther than 24 in. (610 mm) from the point of attachment.

(d) Secured. Service cables shall be held securely in place.

(e) Separately Bushed Openings. Service heads shall have conductors of different potential brought out through separately bushed openings.

Exception: For jacketed multiconductor service cable without splice.

(f) Drip Loops. Drip loops shall be formed on individual conductors. To prevent the entrance of moisture, service-entrance conductors shall be connected to the service-drop conductors either (1) below the level of the service head or (2) below the level of the termination of the service-entrance cable sheath.

(g) Arranged that Water Will Not Enter Service Raceway or Equipment. Service-drop conductors and service-entrance conductors shall be arranged so that water will not enter service raceway or equipment.

230-56. Service Conductor with the Higher Voltage to Ground. On a 4-wire, delta-connected service where the midpoint of one phase winding is grounded, the service conductor having the higher phase voltage to ground shall be durably and permanently marked by an outer finish that is orange in color, or by other effective means, at each termination or junction point.

E. Service Equipment — General

230-62. Service Equipment — Enclosed or Guarded. Energized parts of service equipment shall be enclosed as specified in (a), or guarded as specified in (b).

(a) Enclosed. Energized parts shall be enclosed so that they will not be exposed to accidental contact or shall be guarded as in (b).

(b) Guarded. Energized parts that are not enclosed shall be installed on a switchboard, panelboard, or control board and guarded in accordance with Sections 110-18 and 110-27. Where energized parts are guarded as provided in Sections 110-27(a)(1) and (a)(2), a means for locking or sealing doors providing access to energized parts shall be provided.

230-66. Marking. Service equipment rated at 600 volts or less shall be marked to identify it as being suitable for use as service equipment. Individual meter socket enclosures shall not be considered service equipment.

F. Service Equipment — Disconnecting Means

230-70. General. Means shall be provided to disconnect all conductors in a building or other structure from the service-entrance conductors.

(a) Location. The service disconnecting means shall be installed at a readily accessible location either outside of a building or structure or inside nearest the point of entrance of the service conductors.

Service disconnecting means shall not be installed in bathrooms.

(b) Marking. Each service disconnect shall be permanently marked to identify it as a service disconnect.

(c) Suitable for Use. Each service disconnecting means shall be suitable for the prevailing conditions. Service equipment installed in hazardous (classified) locations shall comply with the requirements of Articles 500 through 517.

230-71. Maximum Number of Disconnects.

(a) General. The service disconnecting means for each service permitted by Section 230-2, or for each set of service-entrance conductors permitted by Section 230-40, Exception Nos. 1 or 3, shall consist of not more than six switches or six circuit breakers mounted in a single enclosure, in a group of separate enclosures, or in or on a switchboard. There shall be no more than six disconnects per service grouped in any one location. For the purpose of this section, disconnecting means used solely for power monitoring equipment or the control circuit of the ground-fault protection system, installed as part of the listed equipment, shall not be considered a service disconnecting means.

(b) Single-Pole Units. Two or three single-pole switches or breakers, capable of individual operation, shall be permitted on multiwire circuits, one pole for each ungrounded conductor, as one multipole disconnect, provided they are equipped with handle ties or a master handle to disconnect all conductors of the service with no more than six operations of the hand.

> FPN: See Section 384-16(a) for service equipment in panelboards, and see Section 430-95 for service equipment in motor control centers.

230-72. Grouping of Disconnects.

(a) General. The two to six disconnects as permitted in Section 230-71 shall be grouped. Each disconnect shall be marked to indicate the load served.

Exception: One of the two to six service disconnecting means permitted in Section 230-71, where used only for a water pump also intended to provide fire protection, shall be permitted to be located remote from the other disconnecting means.

(b) Additional Service Disconnecting Means. The one or more additional service disconnecting means for fire pumps, for legally required standby, or for optional standby services permitted by Section 230-2 shall be installed sufficiently remote from the one to six service disconnecting means for normal service to minimize the possibility of simultaneous interruption of supply.

(c) Access to Occupants. In a multiple-occupancy building, each occupant shall have access to the occupant's service disconnecting means.

Exception: In a multiple-occupancy building where electric service and electrical maintenance are provided by the building management and where these are under continuous building management supervision, the service disconnecting means supplying more than one occupancy shall be permitted to be accessible to authorized management personnel only.

230-74. Simultaneous Opening of Poles. Each service disconnect shall simultaneously disconnect all ungrounded service conductors that it controls from the premises wiring system.

230-75. Disconnection of Grounded Conductor. Where the service disconnecting means does not disconnect the grounded conductor from the premises wiring, other means shall be provided for this purpose in the service equipment. A terminal or bus to which all grounded conductors can be attached by means of pressure connectors shall be permitted for this purpose.

In a multisection switchboard, disconnects for the grounded conductor shall be permitted to be in any section of the switchboard, provided any such switchboard section is marked.

230-76. Manually or Power Operable. The service disconnecting means for ungrounded service conductors shall consist of either (1) a manually operable switch or circuit breaker equipped with a handle or other suitable operating means or (2) a power-operated switch or circuit breaker provided the switch or circuit breaker can be opened by hand in the event of a power supply failure.

230-77. Indicating. The service disconnecting means shall plainly indicate whether it is in the open or closed position.

230-79. Rating of Service Disconnecting Means. The service disconnecting means shall have a rating not less than the load to be carried, determined in accordance with Article 220. In no case shall the rating be lower than specified in (a), (b), (c), or (d).

(a) One-Circuit Installation. For installations to supply only limited loads of a single branch circuit, the service disconnecting means shall have a rating of not less than 15 amperes.

(b) Two-Circuit Installations. For installations consisting of not more than two 2-wire branch circuits, the service disconnecting means shall have a rating of not less than 30 amperes.

(c) One-Family Dwelling. For a one-family dwelling, the service disconnecting means shall have a rating of not less than 100 amperes, 3-wire.

(d) All Others. For all other installations, the service disconnecting means shall have a rating of not less than 60 amperes.

230-80. Combined Rating of Disconnects. Where the service disconnecting means consists of more than one switch or circuit breaker, as permitted by Section 230-71, the combined ratings of all the switches or circuit breakers used shall not be less than the rating required by Section 230-79.

230-81. Connection to Terminals. The service conductors shall be connected to the service disconnecting means by pressure connectors, clamps, or other approved means. Connections that depend on solder shall not be used.

230-82. Equipment Connected to the Supply Side of Service Disconnect. Only the following equipment shall be permitted to be connected to the supply side of the service disconnecting means:

(1) Cable limiters or other current-limiting devices
(2) Meters nominally rated not in excess of 600 volts, provided all metal housings and service enclosures are grounded in accordance with Article 250
(3) Instrument transformers (current and voltage), high-impedance shunts, surge-protective devices identified for use on the supply side of the service disconnect, load management devices, and surge arresters
(4) Taps used only to supply load management devices, circuits for stand-by power systems, fire pump equipment, and fire and sprinkler alarms, if provided with service equipment and installed in accordance with requirements for service-entrance conductors
(5) Solar photovoltaic systems or interconnected electric power production sources (See Articles 690 or 705 as applicable.)

(6) Control circuits for power-operable service disconnecting means, if suitable overcurrent protection and disconnecting means are provided

(7) Ground-fault protection systems where installed as part of listed equipment, if suitable overcurrent protection and disconnecting means are provided

G. Service Equipment — Overcurrent Protection

230-90. Where Required. Each ungrounded service conductor shall have overload protection.

(a) Ungrounded Conductor. Such protection shall be provided by an overcurrent device in series with each ungrounded service conductor that has a rating or setting not higher than the allowable ampacity of the conductor.

Exception No. 1: For motor-starting currents, ratings that conform with Sections 430-52, 430-62, and 430-63 shall be permitted.

Exception No. 2: Fuses and circuit breakers with a rating or setting that conform with Section 240-3(b) or (c) and Section 240-6 shall be permitted.

Exception No. 3: Two to six circuit breakers or sets of fuses shall be permitted as the overcurrent device to provide the overload protection. The sum of the ratings of the circuit breakers or fuses shall be permitted to exceed the ampacity of the service conductors, provided the calculated load in accordance with Article 220 does not exceed the ampacity of the service conductors.

Exception No. 4: Overload protection for fire pump supply conductors shall conform with Section 695-4(b)(1).

Exception No. 5: Overload protection for 120/240-volt, 3-wire, single-phase dwelling services shall be permitted in accordance with the requirements of Section 310-15(b)(6).

  FPN: See *Standard for the Installation of Centrifugal Fire Pumps*, NFPA 20-1996.

A set of fuses shall be considered all the fuses required to protect all the ungrounded conductors of a circuit. Single-pole circuit breakers, grouped in accordance with Section 230-71(b), shall be considered as one protective device.

(b) Not in Grounded Conductor. No overcurrent device shall be inserted in a grounded service conductor except a circuit breaker that simultaneously opens all conductors of the circuit.

230-91. Location. The service overcurrent device shall be an integral part of the service disconnecting means or shall be located immediately adjacent thereto.

230-92. Locked Service Overcurrent Devices. Where the service overcurrent devices are locked or sealed, or not readily accessible to the occupant, branch-circuit overcurrent devices shall be installed on the load side, shall be mounted in a readily accessible location, and shall be of lower ampere rating than the service overcurrent device.

230-93. Protection of Specific Circuits. Where necessary to prevent tampering, an automatic overcurrent device that protects service conductors supplying only a specific load, such as a water heater, shall be permitted to be locked or sealed where located so as to be accessible.

230-94. Relative Location of Overcurrent Device and Other Service Equipment. The overcurrent device shall protect all circuits and devices.

Exception No. 1: The service switch shall be permitted on the supply side.

Exception No. 2: High-impedance shunt circuits, surge arresters, surge-protective capacitors, and instrument transformers (current and voltage) shall be permitted to be connected and installed on the supply side of the service disconnecting means as permitted in Section 230-82.

Exception No. 3: Circuits for load management devices shall be permitted to be connected on the supply side of the service overcurrent device where separately provided with overcurrent protection.

Exception No. 4: Circuits used only for the operation of fire alarm, other protective signaling systems, or the supply to fire pump equipment shall be permitted to be connected on the supply side of the service overcurrent device where separately provided with overcurrent protection.

Exception No. 5: Meters nominally rated not in excess of 600 volts, provided all metal housings and service enclosures are grounded in accordance with Article 250.

Exception No. 6: Where service equipment is power operable, the control circuit shall be permitted to be connected ahead of the service equipment if suitable overcurrent protection and disconnecting means are provided.

230-95. Ground-Fault Protection of Equipment. Ground-fault protection of equipment shall be provided for solidly grounded wye electrical services of more than 150 volts to ground, but not exceeding 600 volts phase-to-phase for each service disconnect rated 1000 amperes or more.

The rating of the service disconnect shall be considered to be the rating of the largest fuse that can be installed or the highest continuous current trip setting for which the actual overcurrent device installed in a circuit breaker is rated or can be adjusted.

Definition. *Solidly grounded* means that the grounded conductor is grounded without inserting any resistor or impedance device.

Exception No. 1: The ground-fault protection provisions of this section shall not apply to a service disconnect for a continuous industrial process where a nonorderly shutdown will introduce additional or increased hazards.

Exception No. 2: The ground-fault protection provisions of this section shall not apply to fire pumps.

(a) Setting. The ground-fault protection system shall operate to cause the service disconnect to open all ungrounded conductors of the faulted circuit. The maximum setting of the ground-fault protection shall be 1200 amperes, and the maximum time delay shall be one second for ground-fault currents equal to or greater than 3000 amperes.

(b) Fuses. If a switch and fuse combination is used, the fuses employed shall be capable of interrupting any current higher than the interrupting capacity of the switch during a time when the ground-fault protective system will not cause the switch to open.

(c) Performance Testing. The ground-fault protection system shall be performance tested when first installed on site. The test shall be conducted in accordance with instructions that shall be provided with the equipment. A written record of this test shall be made and shall be available to the authority having jurisdiction.

FPN No. 1: Ground-fault protection that functions to open the service disconnect will afford no protection from faults on the line side of the protective element. It serves only to limit damage to conductors and equipment on the load side in the event of an arcing ground fault on the load side of the protective element.

FPN No. 2: This added protective equipment at the service equipment may make it necessary to review the overall wiring system for proper selective overcurrent protection coordination. Additional installations of ground-fault protective equipment may be needed on feeders and branch circuits where maximum continuity of electrical service is necessary.

FPN No. 3: Where ground-fault protection is provided for the service disconnect and interconnection is made with another supply system by a transfer device, means or devices may be needed to ensure proper ground-fault sensing by the ground-fault protection equipment.

H. Services Exceeding 600 Volts, Nominal

230-200. General. Service conductors and equipment used on circuits exceeding 600 volts, nominal, shall comply with all the applicable provisions of the preceding sections of this article and with the following sections, which supplement or modify the preceding sections. In no case shall the provisions of Part H apply to equipment on the supply side of the service point.

FPN: For clearances of conductors of over 600 volts, nominal, see *National Electrical Safety Code*, ANSI C2-1997.

230-202. Service-Entrance Conductors. Service-entrance conductors to buildings or enclosures shall be installed to conform to the following.

(a) Conductor Size. Service-entrance conductors shall not be smaller than No. 6 unless in multiconductor cable. Multiconductor cable shall not be smaller than No. 8.

(b) Wiring Methods. Service-entrance conductors shall be installed by one of the wiring methods covered in Sections 300-37 and 300-50.

230-203. Warning Signs. Signs with the words "DANGER — HIGH VOLTAGE — KEEP OUT" shall be posted in plain view where unauthorized persons might come in contact with energized parts.

230-204. Isolating Switches.

(a) Where Required. Where oil switches or air, oil, vacuum, or sulfur hexafluoride circuit breakers constitute the service disconnecting means, an isolating switch with visible break contacts shall be installed on the supply side of the disconnecting means and all associated service equipment.

Exception: An isolating switch shall not be required where the circuit breaker or switch is mounted on removable truck panels or metal-enclosed switchgear units, that

(a) Cannot be opened unless the circuit is disconnected, and
(b) Where all energized parts are automatically disconnected when the circuit breaker or switch is removed from the normal operating position.

(b) Fuses as Isolating Switch. Where fuses are of the type that can be operated as a disconnecting switch, a set of such fuses shall be permitted as the isolating switch.

(c) Accessible to Qualified Persons Only. The isolating switch shall be accessible to qualified persons only.

(d) Grounding Connection. Isolating switches shall be provided with a means for readily connecting the load side conductors to ground when disconnected from the source of supply.

A means for grounding the load side conductors shall not be required for any duplicate isolating switch installed and maintained by the electric supply company.

230-205. Disconnecting Means.

(a) Location. The service disconnecting means shall be located in accordance with Section 230-70.

(b) Type. Each service disconnect shall simultaneously disconnect all ungrounded service conductors that it controls and shall have a fault-closing rating that is not less than the maximum short-circuit current available at its supply terminals.

Where fused switches or separately mounted fuses are installed, the fuse characteristics shall be permitted to contribute to the fault-closing rating of the disconnecting means.

(c) Remote Control. For multibuilding, industrial installations under single management, the service disconnecting means shall be permitted to be located at a separate building or structure. In such cases, the service disconnecting means shall be permitted to be electrically operated by a readily accessible, remote-control device.

230-206. Overcurrent Devices as Disconnecting Means. Where the circuit breaker or alternative for it, as specified in Section 230-208 for service overcurrent devices, meets the requirements specified in Section 230-205, they shall constitute the service disconnecting means.

230-208. Protection Requirements. A short-circuit protective device shall be provided on the load side of, or as an integral part of, the service disconnect, and shall protect all ungrounded conductors that it supplies. The protective device shall be capable of detecting and interrupting all values of current, in excess of its trip setting or melting point, that can occur at its location. A fuse rated in continuous amperes not to exceed three times the ampacity of the conductor, or a circuit breaker with a trip setting of not more than six times the ampacity of the conductors, shall be considered as providing the required short-circuit protection.

FPN: See Tables 310-67 through 310-86 for ampacities of conductors rated 2001 volts and above.

Overcurrent devices shall conform to the following:

(a) Equipment Type. Equipment used to protect service-entrance conductors shall meet the requirements of Article 490, Part B.

(b) Enclosed Overcurrent Devices. The restriction to 80 percent of the rating for an enclosed overcurrent device on continuous loads shall not apply to overcurrent devices installed in services operating at over 600 volts.

230-209. Surge Arresters (Lightning Arresters). Surge arresters installed in accordance with the requirements of Article 280 shall be permitted on each ungrounded overhead service conductor.

230-210. Service Equipment — General Provisions. Service equipment, including instrument transformers, shall conform to Article 490, Part A.

230-211. Metal-Enclosed Switchgear. Metal-enclosed switchgear shall consist of a substantial metal structure and a sheet metal enclosure. Where installed over a combustible floor, suitable protection thereto shall be provided.

230-212. Services Over 15,000 Volts. Where the voltage exceeds 15,000 volts between conductors, they shall enter either metal-enclosed switchgear or a transformer vault conforming to the requirements of Sections 450-41 through 450-48.

Article 240 — Overcurrent Protection

240-1. Scope. Parts A through G of this article provide the general requirements for overcurrent protection and overcurrent protective devices not more than 600 volts, nominal. Part H covers overcurrent protection for those portions of supervised industrial installations operating at voltages of not more than 600 volts, nominal. Part I covers overcurrent protection over 600 volts, nominal.

FPN: Overcurrent protection for conductors and equipment is provided to open the circuit if the current reaches a value that will cause an excessive or dangerous temperature in conductors or conductor insulation. See also Section 110-9 for requirements for interrupting ratings and Section 110-10 for requirements for protection against fault currents.

A. General

240-2. Protection of Equipment. Equipment shall be protected against overcurrent in accordance with the article in this *Code* that covers the type of equipment as specified in the following list:

	Article
Air-conditioning and refrigerating equipment	440
Appliances	422
Audio signal processing, amplification, and reproduction equipment	640
Branch circuits	210
Busways	364
Capacitors	460
Class 1, Class 2, and Class 3 remote-control, signaling, and power-limited circuits	725
Closed-loop and programmed power distribution	780
Cranes and hoists	610
Electric signs and outline lighting	600
Electric welders	630
Electrolytic cells	668
Elevators, dumbwaiters, escalators, moving walks, wheelchair lifts, and stairway chair lifts	620
Emergency systems	700
Fire alarm systems	760
Fire pumps	695
Fixed electric heating equipment for pipelines and vessels	427

	Article
Fixed electric space-heating equipment	424
Fixed outdoor electric deicing and snow-melting equipment	426
Generators	445
Health care facilities	517
Induction and dielectric heating equipment	665
Industrial machinery	670
Lighting fixtures, lampholders, lamps, and receptacles	410
Motion picture and television studios and similar locations	530
Motors, motor circuits, and controllers	430
Phase converters	455
Pipe organs	650
Places of assembly	518
Services	230
Solar photovoltaic systems	690
Switchboards and panelboards	384
Theaters, audience areas of motion picture and television studios, and similar locations	520
Transformers and transformer vaults	450
X-ray equipment	660

240-3. Protection of Conductors. Conductors, other than flexible cords and fixture wires, shall be protected against overcurrent in accordance with their ampacities as specified in Section 310-15, unless otherwise permitted or required in (a) through (g).

(a) Power Loss Hazard. Conductor overload protection shall not be required where the interruption of the circuit would create a hazard, such as in a material handling magnet circuit or fire pump circuit. Short-circuit protection shall be provided.

> FPN: See *Standard for the Installation of Centrifugal Fire Pumps*, NFPA 20-1996.

(b) Devices Rated 800 Amperes or Less. The next higher standard overcurrent device rating (above the ampacity of the conductors being protected) shall be permitted to be used, provided all of the following conditions are met.

(1) The conductors being protected are not part of a multi-outlet branch circuit supplying receptacles for cord- and plug-connected portable loads.

(2) The ampacity of the conductors does not correspond with the standard ampere rating of a fuse or a circuit breaker without overload trip adjustments above its rating (but that shall be permitted to have other trip or rating adjustments).

(3) The next higher standard rating selected does not exceed 800 amperes.

(c) Devices Rated Over 800 Amperes. Where the overcurrent device is rated over 800 amperes, the ampacity of the conductors it protects shall be equal to or greater than the rating of the overcurrent device as defined in Section 240-6.

(d) Small Conductors. Unless specifically permitted in (e) through (g), the overcurrent protection shall not exceed 15 amperes for No. 14, 20 amperes for No. 12, and 30 amperes for No. 10 copper; or 15 amperes for No. 12 and 25 amperes for No. 10 aluminum and copper-clad aluminum after any correction factors for ambient temperature and number of conductors have been applied.

(e) Tap Conductors. Tap conductors shall be permitted to be protected against overcurrent in accordance with Sections 210-19(d), 240-21, 364-11, 364-12, and 430-53(d).

As used in this article, a *tap conductor* is defined as a conductor, other than a service conductor that has overcurrent protection ahead of its point of supply, that exceeds the value permitted for similar conductors that are protected as described elsewhere in this section.

(f) Transformer Secondary Conductors. Single-phase (other than 2-wire) and multiphase (other than delta-delta, 3-wire) transformer secondary conductors shall not be considered to be protected by the primary overcurrent protective device. Conductors supplied by the secondary side of a single-phase transformer having a 2-wire (single-voltage) secondary, or a three-phase, delta-delta connected transformer having a 3-wire (single-voltage) secondary, shall be permitted to be protected by overcurrent protection provided on the primary (supply) side of the transformer, provided this protection is in accordance with Section 450-3 and does not exceed the value determined by multiplying the secondary conductor ampacity by the secondary to primary transformer voltage ratio.

(g) Overcurrent Protection for Specific Conductor Applications. Overcurrent protection for the specific conductors shall be permitted to be provided as referenced in the following list:

	Article	Section
Air-conditioning and refrigeration equipment circuit conductors	440, Parts C, F	
Capacitor circuit conductors	460	460-8(b) and 460-25(a)–(d)
Control and instrumentation circuit conductors (Type ITC)	727	727-9
Electric welder circuit conductors	630	630-12 and 630-32

	Article	Section
Fire alarm system circuit conductors	760	760-23, 760-24, 760-41, and Chapter 9, Tables 12(a) and 12(b)
Motor-operated appliance circuit conductors	422, Part B	
Motor and motor-control circuit conductors	430, Parts C, D, E, F, H	
Phase converter supply conductors	455	455-7
Remote-control, signaling, and power-limited circuit conductors	725	725-23, 725-24, 725-41, and Chapter 9, Tables 11(a) and 11(b)
Secondary tie conductors	450	450-6

240-4. Protection of Flexible Cords and Fixture Wires. Flexible cord, including tinsel cord and extension cords, and fixture wires shall be protected against overcurrent by either (a) or (b).

(a) Ampacities. Flexible cord shall be protected by an overcurrent device in accordance with its ampacity as specified in Tables 400-5(A) and (B). Fixture wire shall be protected against overcurrent in accordance with its ampacity as specified in Table 402-5. Supplementary overcurrent protection, as in Section 240-10, shall be permitted to be an acceptable means for providing this protection.

(b) Branch Circuit Overcurrent Device. Flexible cord shall be protected where supplied by a branch circuit in accordance with one of the methods described below.

(1) Supply Cord of Listed Appliance or Portable Lamps. Where flexible cord or tinsel cord is approved for and used with a specific listed appliance or portable lamp, it shall be permitted to be supplied by a branch circuit of Article 210 in accordance with the following:

20-ampere circuits — tinsel cord or No. 18 cord and larger

30-ampere circuits — No. 16 cord and larger

40-ampere circuits — cord of 20-ampere capacity and over

50-ampere circuits — cord of 20-ampere capacity and over

(2) Fixture Wire. Fixture wire shall be permitted to be tapped to the branch circuit conductor of a branch circuit of Article 210 in accordance with the following:

20-ampere circuits — No. 18, up to 50 ft (15.2 m) of run length

20-ampere circuits — No. 16, up to 100 ft (30.5 m) of run length

20-ampere circuits — No. 14 and larger

30-ampere circuits — No. 14 and larger

40-ampere circuits — No. 12 and larger

50-ampere circuits — No. 12 and larger

(3) Extension Cord Sets. Flexible cord used in listed extension cord sets, or in extension cords made with separately listed and installed components, shall be permitted to be supplied by a branch circuit of Article 210 in accordance with the following:

20-ampere circuits — No. 16 and larger

240-6. Standard Ampere Ratings.

(a) Fuses and Fixed-Trip Circuit Breakers. The standard ampere ratings for fuses and inverse time circuit breakers shall be considered 15, 20, 25, 30, 35, 40, 45, 50, 60, 70, 80, 90, 100, 110, 125, 150, 175, 200, 225, 250, 300, 350, 400, 450, 500, 600, 700, 800, 1000, 1200, 1600, 2000, 2500, 3000, 4000, 5000, and 6000 amperes. The use of fuses and inverse time circuit breakers with nonstandard ampere ratings shall be permitted. The standard ampere rating for fuses shall be considered 1, 3, 6, 10, and 601.

(b) Adjustable-Trip Circuit Breakers. The rating of adjustable-trip circuit breakers having external means for adjusting the current setting (long-time pickup setting), not meeting the requirements of (c), shall be the maximum setting possible.

(c) Restricted Access Adjustable-Trip Circuit Breakers. A circuit breaker(s) that has restricted access to the adjusting means shall be permitted to have an ampere rating(s) that is equal to the adjusted current setting (long-time pickup setting). Restricted access shall be defined as located behind one of the following:

(1) Removable and sealable covers over the adjusting means
(2) Bolted equipment enclosure doors
(3) Locked doors accessible only to qualified personnel

240-8. Fuses or Circuit Breakers in Parallel. Fuses and circuit breakers shall be permitted to be connected in parallel where they are factory assembled in parallel and listed as a unit. Individual fuses, circuit breakers, or combinations thereof shall not otherwise be connected in parallel.

240-9. Thermal Devices. Thermal relays and other devices not designed to open short circuits shall not be used for the protection of conductors against overcurrent due to short circuits or grounds, but the use of such devices shall be permitted to protect motor branch-circuit conductors from overload if protected in accordance with Section 430-40.

240-10. Supplementary Overcurrent Protection. Where supplementary overcurrent protection is used for lighting fixtures, appliances, and other equipment or for internal circuits and components of equipment, it shall not be used as a substitute for branch-circuit overcurrent devices or in place of the branch-circuit protection specified in Article 210. Supplementary overcurrent devices shall not be required to be readily accessible.

240-11. Definition of Current-Limiting Overcurrent Protective Device. A *current-limiting overcurrent protective device* is a device that, when interrupting currents in its current-limiting range, will reduce the current flowing in the faulted circuit to a magnitude substantially less than that obtainable in the same circuit if the device were replaced with a solid conductor having comparable impedance.

240-12. Electrical System Coordination. Where an orderly shutdown is required to minimize the hazard(s) to personnel and equipment, a system of coordination based on the following two conditions shall be permitted:

(1) Coordinated short-circuit protection
(2) Overload indication based on monitoring systems or devices

For the purposes of this section, *coordination* is defined as properly localizing a fault condition to restrict outages to the equipment affected, accomplished by the choice of selective fault-protective devices.

FPN: The monitoring system may cause the condition to go to alarm, allowing corrective action or an orderly shutdown, thereby minimizing personnel hazard and equipment damage.

240-13. Ground-Fault Protection of Equipment. Ground-fault protection of equipment shall be provided in accordance with the provisions of Section 230-95 for solidly grounded wye electrical systems of more than 150 volts to ground, but not exceeding 600 volts phase-to-phase for each individual device used as a building or structure main disconnecting means rated 1000 amperes or more.

The provisions of this section shall not apply to the disconnecting means for the following:

(1) Continuous industrial processes where a nonorderly shutdown will introduce additional or increased hazards

(2) Installations where ground-fault protection is provided by other requirements for services or feeders
(3) Fire pumps installed in accordance with Article 695

B. Location

240-20. Ungrounded Conductors.

(a) Overcurrent Device Required. A fuse or an overcurrent trip unit of a circuit breaker shall be connected in series with each ungrounded conductor. A combination of a current transformer and overcurrent relay shall be considered equivalent to an overcurrent trip unit.

FPN: For motor circuits, see Parts C, D, F, and K of Article 430.

(b) Circuit Breaker as Overcurrent Device. Circuit breakers shall open all ungrounded conductors of the circuit unless otherwise permitted in (1), (2), or (3).

(1) Except where limited by Section 210-4(b), individual single-pole circuit breakers, with or without approved handle ties, shall be permitted as the protection for each ungrounded conductor of multiwire branch circuits that serve only single-phase, line-to-neutral loads.

(2) In grounded systems, individual single-pole circuit breakers with approved handle ties shall be permitted as the protection for each ungrounded conductor for line-to-line connected loads for single-phase circuits or 3-wire, direct-current circuits.

(3) For line-to-line loads in 4-wire, 3-phase systems or 5-wire, 2-phase systems having a grounded neutral and no conductor operating at a voltage greater than permitted in Section 210-6, individual single-pole circuit breakers with approved handle ties shall be permitted as the protection for each ungrounded conductor.

(c) Closed-Loop Power Distribution Systems. Listed devices that provide equivalent overcurrent protection in closed-loop power distribution systems shall be permitted as a substitute for fuses or circuit breakers.

240-21. Location in Circuit. Overcurrent protection shall be provided in each ungrounded circuit conductor and shall be located at the point where the conductors receive their supply except as specified in (a) through (g). No conductor supplied under the provisions of (a) through (g) shall supply another conductor under those provisions, except through an overcurrent protective device meeting the requirements of Section 240-3.

(a) Branch-Circuit Conductors. Branch-circuit tap conductors meeting the requirements specified in Section 210-19 shall be permitted to have overcurrent protection located as specified in that Section.

(b) Feeder Taps. Conductors shall be permitted to be tapped, without overcurrent protection at the tap, to a feeder as specified in (1) through (5).

(1) Taps Not Over 10 ft (3.05 m) Long. Where the length of the tap conductors does not exceed 10 ft (3.05 m) and the tap conductors comply with all of the following.

(a) The ampacity of the tap conductors is

  (1) Not less than the combined computed loads on the circuits supplied by the tap conductors, and

  (2) Not less than the rating of the device supplied by the tap conductors or not less than the rating of the overcurrent-protective device at the termination of the tap conductors.

(b) The tap conductors do not extend beyond the switchboard, panelboard, disconnecting means, or control devices they supply.

(c) Except at the point of connection to the feeder, the tap conductors are enclosed in a raceway, which shall extend from the tap to the enclosure of an enclosed switchboard, panelboard, or control devices, or to the back of an open switchboard.

(d) For field installations where the tap conductors leave the enclosure or vault in which the tap is made, the rating of the overcurrent device on the line side of the tap conductors shall not exceed 10 times the ampacity of the tap conductor.

FPN: For overcurrent protection requirements for lighting and appliance branch-circuit panelboards, see Sections 384-16(a) and (e).

(2) Taps Not Over 25 ft (7.62 m) Long. Where the length of the tap conductors does not exceed 25 ft (7.62 m) and the tap conductors comply with all of the following.

(a) The ampacity of the tap conductors is not less than one-third of the rating of the overcurrent device protecting the feeder conductors.

(b) The tap conductors terminate in a single circuit breaker or a single set of fuses that will limit the load to the ampacity of the tap conductors. This device shall be permitted to supply any number of additional overcurrent devices on its load side.

(c) The tap conductors are suitably protected from physical damage or are enclosed in a raceway.

(3) Taps Supplying a Transformer [Primary Plus Secondary Not Over 25 ft (7.62 m) Long]. Where the tap conductors supply a transformer and comply with all the following conditions.

(a) The conductors supplying the primary of a transformer have an ampacity at least one-third of the rating of the overcurrent device protecting the feeder conductors.

(b) The conductors supplied by the secondary of the transformer shall have an ampacity that, when multiplied by the ratio of the secondary-to-primary voltage, is at least one-third of the rating of the overcurrent device protecting the feeder conductors.

(c) The total length of one primary plus one secondary conductor, excluding any portion of the primary conductor that is protected at its ampacity, is not over 25 ft (7.62 m).

(d) The primary and secondary conductors are suitably protected from physical damage.

(e) The secondary conductors terminate in a single circuit breaker or set of fuses that will limit the load current to not more than the conductor ampacity that is permitted by Section 310-15.

(4) Taps Over 25 ft (7.62 m) Long. Where the feeder is in a high bay manufacturing building over 35 ft (10.67 m) high at walls, and the installation complies with all of the following conditions.

(a) Conditions of maintenance and supervision ensure that only qualified persons will service the systems.

(b) The tap conductors are not over 25 ft (7.62 m) long horizontally and not over 100 ft (30.5 m) total length.

(c) The ampacity of the tap conductors is not less than one-third of the rating of the overcurrent device protecting the feeder conductors.

(d) The tap conductors terminate at a single circuit breaker or a single set of fuses that will limit the load to the ampacity of the tap conductors. This single overcurrent device shall be permitted to supply any number of additional overcurrent devices on its load side.

(e) The tap conductors are suitably protected from physical damage or are enclosed in a raceway.

(f) The tap conductors are continuous from end-to-end and contain no splices.

(g) The tap conductors are sized No. 6 copper or No. 4 aluminum or larger.

(h) The tap conductors do not penetrate walls, floors, or ceilings.

(i) The tap is made no less than 30 ft (9.14 m) from the floor.

(5) Outside Taps of Unlimited Length. Where the conductors are located outdoors, except at the point of termination, and comply with all of the following conditions.

(a) The conductors are suitably protected from physical damage.

(b) The conductors terminate at a single circuit breaker or a single set of fuses that will limit the load to the ampacity of the conductors. This single overcurrent device shall be permitted to supply any number of additional overcurrent devices on its load side.

(c) The overcurrent device for the conductors is an integral part of a disconnecting means or shall be located immediately adjacent thereto.

(d) The disconnecting means for the conductors is installed at a readily accessible location either outside of a building or structure, or inside nearest the point of entrance of the conductors.

(c) Transformer Secondary Conductors. Conductors shall be permitted to be connected to a transformer secondary, without overcurrent protection at the secondary, as specified in (1) through (4).

FPN: For overcurrent protection requirements for transformers, see Section 450-3.

(1) Protection by Primary Overcurrent Device. Single-phase (other than 2-wire) and multiphase (other than delta-delta, 3-wire) transformer secondary conductors are not considered to be protected by the primary overcurrent protective device. Conductors supplied by the secondary side of a single-phase transformer having a 2-wire (single-voltage) secondary, or a three-phase, delta-delta connected transformer having a 3-wire (single-voltage) secondary, shall be permitted to be protected by overcurrent protection provided on the primary (supply) side of the transformer, provided this protection is in accordance with Section 450-3 and does not exceed the value determined by multiplying the secondary conductor ampacity by the secondary to primary transformer voltage ratio.

(2) Transformer Secondary Conductors Not Over 10 ft (3.05 m) Long. Where the length of secondary conductor does not exceed 10 ft (3.05 m) and complies with all of the following.

(a) The ampacity of the secondary conductors is

(1) Not less than the combined computed loads on the circuits supplied by the secondary conductors, and

(2) Not less than the rating of the device supplied by the secondary conductors or not less than the rating of the overcurrent-protective device at the termination of the secondary conductors.

(b) The secondary conductors do not extend beyond the switchboard, panelboard, disconnecting means, or control devices they supply.

(c) The secondary conductors are enclosed in a raceway, which shall extend from the transformer to the enclosure of an enclosed switchboard, panelboard, or control devices, or to the back of an open switchboard.

FPN: For overcurrent protection requirements for lighting and appliance branch-circuit panelboards, see Sections 384-16(a) and (e).

(3) Secondary Conductors Not Over 25 ft (7.62 m) Long. For industrial installations only, where the length of the secondary conductors does not exceed 25 ft (7.62 m) and complies with all of the following.

(a) The ampacity of the secondary conductors is not less than the secondary current rating of the transformer, and the sum of the ratings of the overcurrent devices does not exceed the ampacity of the secondary conductors.

(b) All overcurrent devices are grouped.

(c) The secondary conductors are suitably protected from physical damage.

(4) Outside Secondary Conductors. Where the conductors are located outdoors, except at the point of termination, and comply with all of the following conditions.

(a) The conductors are suitably protected from physical damage.

(b) The conductors terminate in an overcurrent device that will limit the load to the ampacity of the conductors. The overcurrent device shall be permitted to consist of not more than six circuit breakers or six sets of fuses grouped in one location.

(c) The overcurrent device for the conductors is an integral part of a disconnecting means or shall be located immediately adjacent thereto.

(d) The disconnecting means for the conductors are installed at a readily accessible location either outside of a building or structure, or inside nearest the point of entrance of the conductors.

(5) Secondary Conductors from a Feeder Tapped Transformer. Transformer secondary conductors installed in accordance with Section 240-21(b)(3) shall be permitted to have overcurrent protection as specified in that section.

(d) Service Conductors. Service-entrance conductors shall be permitted to be protected by overcurrent devices in accordance with Section 230-91.

(e) Busway Taps. Busways and busway taps shall be permitted to be protected against overcurrent in accordance with Sections 364-10 through 364-13.

(f) Motor Circuit Taps. Motor-feeder and branch-circuit conductors shall be permitted to be protected against overcurrent in accordance with Sections 430-28 and 430-53 respectively.

(g) Conductors from Generator Terminals. Conductors from generator terminals that meet the size requirement in Section 445-5 shall be permitted to be protected against overload by the generator overload protective device(s) required by Section 445-4.

240-22. Grounded Conductor. No overcurrent device shall be connected in series with any conductor that is intentionally grounded, unless one of the following two conditions are met:

(1) The overcurrent device opens all conductors of the circuit, including the grounded conductor, and is designed so that no pole can operate independently
(2) Where required by Sections 430-36 or 430-37 for motor overload protection

240-23. Change in Size of Grounded Conductor. Where a change occurs in the size of the ungrounded conductor, a similar change shall be permitted to be made in the size of the grounded conductor.

240-24. Location in or on Premises.

(a) Accessibility. Overcurrent devices shall be readily accessible unless one of the following applies.

(1) For busways, as provided in Section 364-12.
(2) For supplementary overcurrent protection, as described in Section 240-10.
(3) For overcurrent devices, as described in Sections 225-40 and 230-92.
(4) For overcurrent devices adjacent to utilization equipment that they supply, access shall be permitted to be by portable means.

(b) Occupancy. Each occupant shall have ready access to all overcurrent devices protecting the conductors supplying that occupancy. Where electric service and electrical maintenance are provided by the building management and where these are under continuous building management supervision, the service overcurrent devices and feeder overcurrent devices supplying more than one occupancy shall be permitted to be accessible to only authorized management personnel in the following:

(1) In multiple occupancy buildings
(2) For guest rooms of hotels and motels that are intended for transient occupancy

(c) Not Exposed to Physical Damage. Overcurrent devices shall be located where they will not be exposed to physical damage.

FPN: See Section 110-11, Deteriorating Agents.

(d) Not in Vicinity of Easily Ignitible Material. Overcurrent devices shall not be located in the vicinity of easily ignitible material, such as in clothes closets.

(e) Not Located in Bathrooms. In dwelling units and guest rooms of hotels and motels, overcurrent devices, other than supplementary overcurrent protection, shall not be located in bathrooms as defined in Article 100.

C. Enclosures

240-30. General.

(a) Overcurrent devices shall be protected from physical damage by one of the following:

(1) Installation in enclosures, cabinets, cutout boxes, or equipment assemblies
(2) Mounting on open-type switchboards, panelboards, or control boards that are in rooms or enclosures free from dampness and easily ignitible material, and are accessible only to qualified personnel

(b) The operating handle of a circuit breaker shall be permitted to be accessible without opening a door or cover.

240-32. Damp or Wet Locations. Enclosures for overcurrent devices in damp or wet locations shall comply with Section 373-2(a).

240-33. Vertical Position. Enclosures for overcurrent devices shall be mounted in a vertical position. Circuit breaker enclosures shall be permitted to be installed horizontally where the circuit breaker is installed in accordance with Section 240-81. Listed busway plug-in units shall be permitted to be mounted in orientations corresponding to the busway mounting position.

D. Disconnecting and Guarding

240-40. Disconnecting Means for Fuses. A disconnecting means shall be provided on the supply side of all fuses in circuits over 150 volts to ground and cartridge fuses in circuits of any voltage where accessible to other than qualified persons so that each individual circuit containing fuses can be independently disconnected from the source of power. A current-limiting device without a disconnecting means shall be permitted on the supply side of the service disconnecting means as permitted by Section 230-82. A single disconnecting means shall be permitted on the supply side of more than one set of fuses as permitted by Section 430-112, Exception, for group operation of motors and Section 424-22(c) for fixed electric space-heating equipment.

240-41. Arcing or Suddenly Moving Parts. Arcing or suddenly moving parts shall comply with (a) and (b).

(a) Location. Fuses and circuit breakers shall be located or shielded so that persons will not be burned or otherwise injured by their operation.

(b) Suddenly Moving Parts. Handles or levers of circuit breakers, and similar parts that may move suddenly in such a way that persons in the vicinity are likely to be injured by being struck by them, shall be guarded or isolated.

E. Plug Fuses, Fuseholders, and Adapters

240-50. General.

(a) Maximum Voltage. Plug fuses shall be permitted to be used in the following circuits:

(1) Circuits not exceeding 125 volts between conductors
(2) Circuits supplied by a system having a grounded neutral where the line-to-neutral voltage does not exceed 150 volts

(b) Marking. Each fuse, fuseholder, and adapter shall be marked with its ampere rating.

(c) Hexagonal Configuration. Plug fuses of 15-ampere and lower rating shall be identified by a hexagonal configuration of the window, cap, or other prominent part to distinguish them from fuses of higher ampere ratings.

(d) No Energized Parts. Plug fuses, fuseholders, and adapters shall have no exposed energized parts after fuses or fuses and adapters have been installed.

(e) Screw Shell. The screw shell of a plug-type fuseholder shall be connected to the load side of the circuit.

240-51. Edison-Base Fuses.

(a) Classification. Plug fuses of the Edison-base type shall be classified at not over 125 volts and 30 amperes and below.

(b) Replacement Only. Plug fuses of the Edison-base type shall be used only for replacements in existing installations where there is no evidence of overfusing or tampering.

240-52. Edison-Base Fuseholders. Fuseholders of the Edison-base type shall be installed only where they are made to accept Type S fuses by the use of adapters.

240-53. Type S Fuses. Type S fuses shall be of the plug type and shall comply with (a) and (b).

(a) Classification. Type S fuses shall be classified at not over 125 volts and 0 to 15 amperes, 16 to 20 amperes, and 21 to 30 amperes.

(b) Noninterchangeable. Type S fuses of an ampere classification as specified in (a) shall not be interchangeable with a lower ampere classification. They shall be designed so that they cannot be used in any fuseholder other than a Type S fuseholder or a fuseholder with a Type S adapter inserted.

240-54. Type S Fuses, Adapters, and Fuseholders.

(a) To Fit Edison-Base Fuseholders. Type S adapters shall fit Edison-base fuseholders.

(b) To Fit Type S Fuses Only. Type S fuseholders and adapters shall be designed so that either the fuseholder itself or the fuseholder with a Type S adapter inserted cannot be used for any fuse other than a Type S fuse.

(c) Nonremovable. Type S adapters shall be designed so that once inserted in a fuseholder, they cannot be removed.

(d) Nontamperable. Type S fuses, fuseholders, and adapters shall be designed so that tampering or shunting (bridging) would be difficult.

(e) Interchangeability. Dimensions of Type S fuses, fuseholders, and adapters shall be standardized to permit interchangeability regardless of the manufacturer.

F. Cartridge Fuses and Fuseholders

240-60. General.

(a) Maximum Voltage — 300-Volt Type. Cartridge fuses and fuseholders of the 300-volt type shall be permitted to be used in the following circuits:

(1) Circuits not exceeding 300 volts between conductors
(2) Single-phase line-to-neutral circuits supplied from a 3-phase, 4-wire, solidly grounded neutral source where the line-to-neutral voltage does not exceed 300 volts

(b) Noninterchangeable — 0–6000-Ampere Cartridge Fuseholders. Fuseholders shall be designed so that it will be difficult to put a fuse of any given class into a fuseholder that is designed for a current lower, or voltage higher, than that of the class to which the fuse belongs. Fuseholders for current-limiting fuses shall not permit insertion of fuses that are not current limiting.

(c) Marking. Fuses shall be plainly marked, either by printing on the fuse barrel or by a label attached to the barrel showing the following:

(1) Ampere rating
(2) Voltage rating
(3) Interrupting rating where other than 10,000 amperes
(4) Current limiting where applicable
(5) The name or trademark of the manufacturer

The interrupting rating shall not be required to be marked on fuses used for supplementary protection.

240-61. Classification. Cartridge fuses and fuseholders shall be classified according to voltage and amperage ranges. Fuses rated 600 volts, nominal or less, shall be permitted to be used for voltages at or below their ratings.

G. Circuit Breakers

240-80. Method of Operation. Circuit breakers shall be trip free and capable of being closed and opened by manual operation. Their normal method of operation by other than manual means, such as electrical or pneumatic, shall be permitted if means for manual operation is also provided.

240-81. Indicating. Circuit breakers shall clearly indicate whether they are in the open "off" or closed "on" position.

Where circuit breaker handles are operated vertically rather than rotationally or horizontally, the "up" position of the handle shall be the "on" position.

240-82. Nontamperable. A circuit breaker shall be of such design that any alteration of its trip point (calibration) or the time required for its operation will require dismantling of the device or breaking of a seal for other than intended adjustments.

240-83. Marking.

(a) Durable and Visible. Circuit breakers shall be marked with their ampere rating in a manner that will be durable and visible after installation. Such marking shall be permitted to be made visible by removal of a trim or cover.

(b) Location. Circuit breakers rated at 100 amperes or less and 600 volts or less shall have the ampere rating molded, stamped, etched, or similarly marked into their handles or escutcheon areas.

(c) Interrupting Rating. Every circuit breaker having an interrupting rating other than 5000 amperes shall have its interrupting rating shown on the circuit breaker. The interrupting rating shall not be required to be marked on circuit breakers used for supplementary protection.

(d) Used as Switches. Circuit breakers used as switches in 120-volt and 277-volt fluorescent lighting circuits shall be listed and shall be marked "SWD."

(e) Voltage Marking. Circuit breakers shall be marked with a voltage rating not less than the nominal system voltage that is indicative of their capability to interrupt fault currents between phases or phase to ground.

240-85. Applications. A circuit breaker with a straight voltage rating, such as 240V or 480V, shall be permitted to be applied in a circuit in which the nominal voltage between any two conductors does not exceed the circuit breaker's voltage rating. A two-pole circuit breaker shall not be used for protecting a 3-phase, corner-grounded delta circuit unless the circuit breaker is marked 1φ–3φ to indicate such suitability.

A circuit breaker with a slash rating, such as 120/240V or 480Y/277V, shall be permitted to be applied in a circuit where the nominal voltage of any conductor to ground does not exceed the lower of the two values of the circuit breaker's voltage rating and the nominal voltage between any two conductors does not exceed the higher value of the circuit breaker's voltage rating.

240-86. Series Ratings. Where a circuit breaker is used on a circuit having an available fault current higher than its marked interrupting rating by being connected on the load side of an acceptable overcurrent protective device having the higher rating, the following shall apply.

(a) Marking. The additional series combination interrupting rating shall be marked on the end use equipment, such as switchboards and panelboards.

(b) Motor Contribution. Series ratings shall not be used where

(1) Motors are connected on the load side of the higher-rated overcurrent device and on the line side of the lower-rated overcurrent device.
(2) The sum of the motor full-load currents exceeds 1 percent of the interrupting rating of the lower-rated circuit breaker.

H. Supervised Industrial Installations

240-90. General. Overcurrent protection in areas of supervised industrial installations shall comply with all the applicable provisions of the other sections of this article, except as provided in Part H. The provisions of Part H shall only be permitted to apply to those portions of the electrical system in the supervised industrial installation used exclusively for manufacturing or process control activities.

240-91. Definition of Supervised Industrial Installation. For the purposes of Part H, *supervised industrial in-*

stallation is defined as the industrial portions of a facility where all of the following conditions are met.

(1) Conditions of maintenance and engineering supervision ensure that only qualified persons will monitor and service the system.
(2) The premises wiring system has 2500 kVA or greater of load used in industrial process(es), manufacturing activities, or both, as calculated in accordance with Article 220.
(3) The premises has at least one service that is more than 150 volts to ground and more than 300 volts phase-to-phase.

This definition shall not apply to those installations in buildings used by the industrial facility for offices, warehouses, garages, machine shops, and recreational facilities that are not an integral part of the industrial plant, substation, or control center.

240-92. Location in Circuit. An overcurrent device shall be connected in each ungrounded circuit conductor as follows.

(a) Feeder and Branch-Circuit Conductors. Feeder and branch-circuit conductors shall be protected at the point the conductors receive their supply, as permitted in Section 240-21, or as otherwise permitted in (b) or (c).

(b) Transformer Secondary Conductors of Separately Derived Systems. Conductors shall be permitted to be connected to a transformer secondary of a separately derived system, without overcurrent protection at the connection, where the conditions of (1), (2), and (3) are met.

(1) Short-Circuit and Ground-Fault Protection. The conductors shall be protected from short-circuit and ground-fault conditions by complying with one of the following conditions.

(a) The length of the secondary conductors does not exceed 50 ft (15.24 m) and the transformer primary overcurrent device has a rating or setting that does not exceed 150 percent of the value determined by multiplying the secondary conductor ampacity by the secondary-to-primary transformer voltage ratio.
(b) The length of the secondary conductors does not exceed 75 ft (22.86 m) and the conductors are protected by a differential relay with a trip setting equal to or less than the conductor ampacity.
(c) The conductors shall be considered to be protected if the length of the secondary conductors does not exceed 75 ft (22.86 m) and if calculations, made under engineering supervision, determine that the system overcurrent devices will protect the conductors within

recognized time vs. current limits for all short-circuit and ground-fault conditions.

(2) Overload Protection. The conductors shall be protected against overload conditions by complying with one of the following.

(a) The conductors terminate in a single overcurrent device that will limit the load to the conductor ampacity.
(b) The sum of the overcurrent devices at the conductor termination limits the load to the conductor ampacity. The overcurrent devices shall consist of not more than six circuit breakers or sets of fuses, mounted in a single enclosure, in a group of separate enclosures, or in or on a switchboard. There shall be no more than six overcurrent devices grouped in any one location.
(c) Overcurrent relaying is connected [with a current transformer(s), if needed] to sense all of the secondary conductor current and limit the load to the conductor ampacity by opening upstream or downstream devices.
(d) Conductors shall be considered to be protected if calculations, made under engineering supervision, determine that the system overcurrent devices will protect the conductors from overload conditions.

(3) Physical Protection. The secondary conductors shall be suitably protected from physical damage.

(c) Outside Feeder Taps. Outside conductors shall be permitted to be tapped to a feeder or to be connected at a transformer secondary, without overcurrent protection at the tap or connection, where all the following conditions are met.

(1) The conductors are suitably protected from physical damage.
(2) The sum of the overcurrent devices at the conductor termination limits the load to the conductor ampacity. The overcurrent devices shall consist of not more than six circuit breakers or sets of fuses mounted in a single enclosure, in a group of separate enclosures, or in or on a switchboard. There shall be no more than six overcurrent devices grouped in any one location.
(3) The tap conductors are installed outdoors, except at the point of termination.
(4) The overcurrent device for the conductors is an integral part of a disconnecting means or shall be located immediately adjacent thereto.
(5) The disconnecting means for the conductors are installed at a readily accessible location either outside of a building or structure or inside nearest the point of entrance of the conductors.

I. Overcurrent Protection Over 600 Volts, Nominal

240-100. Feeders and Branch Circuits.

(a) Feeder and branch-circuit conductors shall have overcurrent protection in each ungrounded conductor located at the point where the conductor receives its supply or at a location in the circuit determined under engineering supervision. The overcurrent protection shall be permitted to be provided by one of the following.

(1) **Overcurrent Relays and Current Transformers.** Circuit breakers used for overcurrent protection of 3-phase circuits shall have a minimum of three overcurrent relays operated from three current transformers. On 3-phase, 3-wire circuits, an overcurrent relay in the residual circuit of the current transformers shall be permitted to replace one of the phase relays.

An overcurrent relay, operated from a current transformer that links all phases of a 3-phase, 3-wire circuit, shall be permitted to replace the residual relay and one of the phase-conductor current transformers. Where the neutral is not regrounded on the load side of the circuit as permitted in Section 250-184(b), the current transformer shall be permitted to link all 3-phase conductors and the grounded circuit conductor (neutral).

(2) **Fuses.** A fuse shall be connected in series with each ungrounded conductor.

(b) **Protective Devices.** The protective device(s) shall be capable of detecting and interrupting all values of current that can occur at their location in excess of their trip setting or melting point.

(c) **Conductor Protection.** The operating time of the protective device, the available short-circuit current, and the conductor used shall be coordinated to prevent damaging or dangerous temperatures in conductors or conductor insulation under short-circuit conditions.

240-101. Additional Requirements for Feeders.

(a) **Rating or Setting of Overcurrent Protective Devices.** The continuous ampere rating of a fuse shall not exceed three times the ampacity of the conductors. The long-time trip element setting of a breaker or the minimum trip setting of an electronically actuated fuse shall not exceed six times the ampacity of the conductor. For fire pumps, conductors shall be permitted to be protected for overcurrent in accordance with Section 695-4(b).

(b) **Feeder Taps.** Conductors tapped to a feeder shall be permitted to be protected by the feeder overcurrent device where that overcurrent device also protects the tap conductor.

Article 250 — Grounding

See Appendix E for a cross reference list of section numbers between the 1996 Article 250 and the 1999 Article 250.

A. General

250-1. Scope. This article covers general requirements for grounding and bonding of electrical installations, and specific requirements in (1) through (6).

(1) Systems, circuits, and equipment required, permitted, or not permitted to be grounded
(2) Circuit conductor to be grounded on grounded systems
(3) Location of grounding connections
(4) Types and sizes of grounding and bonding conductors and electrodes
(5) Methods of grounding and bonding
(6) Conditions under which guards, isolation, or insulation may be substituted for grounding

250-2. General Requirements for Grounding and Bonding. The following general requirements identify what grounding and bonding of electrical systems are required to accomplish. The prescriptive methods contained in Article 250 shall be followed to comply with the performance requirements of this section.

(a) **Grounding of Electrical Systems.** Electrical systems that are required to be grounded shall be connected to earth in a manner that will limit the voltage imposed by lightning, line surges, or unintentional contact with higher voltage lines and that will stabilize the voltage to earth during normal operation.

(b) **Grounding of Electrical Equipment.** Conductive materials enclosing electrical conductors or equipment, or forming part of such equipment, shall be connected to earth so as to limit the voltage to ground on these materials. Where the electrical system is required to be grounded, these materials shall be connected together and to the supply system grounded conductor as specified by this article. Where the electrical system is not solidly grounded, these materials shall be connected together in a manner that establishes an effective path for fault current.

(c) **Bonding of Electrically Conductive Materials and Other Equipment.** Electrically conductive materials, such as metal water piping, metal gas piping, and structural steel members, that are likely to become energized shall be bonded as specified by this article to the supply system grounded conductor or, in the case of an ungrounded electrical system, to the electrical system grounded equipment, in a manner that establishes an effective path for fault current.

(d) Performance of Fault Current Path. The fault current path shall be permanent and electrically continuous, shall be capable of safely carrying the maximum fault likely to be imposed on it, and shall have sufficiently low impedance to facilitate the operation of overcurrent devices under fault conditions.

The earth shall not be used as the sole equipment grounding conductor or fault current path.

FPN: See Figure 250-2 for information on the organization of Article 250.

Figure 250-2 Grounding.

250-4. Application of Other Articles. In other articles applying to particular cases of installation of conductors and equipment, there are requirements that are in addition to those of this article or are modifications of them.

	Article	Section
Agricultural buildings		547-8
Audio signal processing, amplification, and reproduction equipment		640-7

	Article	Section
Branch circuits		210-5
		210-6
		210-7
Cablebus		365-9
Capacitors		460-10
		460-27
Circuits and equipment operating at less than 50 volts	720	
Class 1, Class 2, and Class 3 remote-control, signaling, and power-limited circuits		725-6
Closed-loop and programmed power distribution		780-3
Communications circuits	800	
Community antenna television and radio distribution systems		820-33
		820-40
		820-41
Conductors for general wiring	310	
Cranes and hoists	610	
Electrically driven or controlled irrigation machines		675-11(c)
		675-12
		675-13
		675-14
		675-15
Electric signs and outline lighting	600	
Electrolytic cells	668	
Elevators, dumbwaiters, escalators, moving walks, wheelchair lifts, and stairway chair lifts	620	
Fire alarm systems		760-6
Fixed electric heating equipment for pipelines and vessels		427-29
		427-48
Fixed outdoor electric deicing and snow-melting equipment		426-27
Fixtures and lighting equipment		410-17
		410-18
		410-20
		410-21
		410-105(b)
Flexible cords and cables		400-22
		400-23
Floating buildings		553-8
		553-10
		553-11
Grounding-type receptacles, adapters, cord connectors, and attachment plugs		410-58
Hazardous (classified) locations	500-517	
Health care facilities	517	

	Article	Section
Induction and dielectric heating equipment	665	
Industrial machinery	670	
Information technology equipment		645-15
Intrinsically safe systems		504-50
Lighting fixtures, lampholders, lamps, and receptacles	410	
Marinas and boatyards		555-8
Mobile homes and mobile home park	550	
Motion picture and television studios and similar locations		530-20 530-66
Motors, motor circuits, and controllers	430	
Outlet, device, pull and junction boxes, conduit bodies and fittings		370-4 370-25
Over 600 volts, nominal, underground wiring methods		300-50(b)
Panelboards		384-20
Pipe organs	650	
Radio and television equipment	810	
Receptacles and cord connectors		210-7
Recreational vehicles and recreational vehicle parks	551	
Services	230	
Solar photovoltaic systems		690-41 690-42 690-43 690-45 690-47
Swimming pools, fountains, and similar installations	680	
Switchboards and panelboards		384-3(d)
Switches		380-12
Theaters, audience areas of motion picture and television studios, and similar locations		520-81
Transformers and transformer vaults		450-10
Use and identification of grounded conductors	200	
X-ray equipment	660	517-78

250-6. Objectionable Current Over Grounding Conductors.

(a) Arrangement to Prevent Objectionable Current.
The grounding of electrical systems, circuit conductors, surge arresters, and conductive noncurrent-carrying materials and equipment shall be installed and arranged in a manner that will prevent an objectionable flow of current over the grounding conductors or grounding paths.

(b) Alterations to Stop Objectionable Current.
If the use of multiple grounding connections results in an objectionable flow of current, one or more of the following alterations shall be permitted to be made, provided that the requirements of Section 250-2(d) are met.

(1) Discontinue one or more but not all of such grounding connections.
(2) Change the locations of the grounding connections.
(3) Interrupt the continuity of the conductor or conductive path interconnecting the grounding connections.
(4) Take other suitable remedial action satisfactory to the authority having jurisdiction.

(c) Temporary Currents Not Classified as Objectionable Currents.
Temporary currents resulting from accidental conditions, such as ground-fault currents, that occur only while the grounding conductors are performing their intended protective functions shall not be classified as objectionable current for the purposes specified in (a) and (b).

(d) Limitations to Permissible Alterations.
The provisions of this section shall not be considered as permitting electronic equipment from being operated on ac systems or branch circuits that are not grounded as required by this article. Currents that introduce noise or data errors in electronic equipment shall not be considered the objectionable currents addressed in this section.

(e) Isolation of Objectionable Direct-Current Ground Currents.
Where isolation of objectionable dc ground currents from cathodic protection systems is required, a listed ac coupling/dc isolating device shall be permitted in the equipment grounding path to provide an effective return path for ac ground-fault current while blocking dc current.

250-8. Connection of Grounding and Bonding Equipment.
Grounding conductors and bonding jumpers shall be connected by exothermic welding, listed pressure connectors, listed clamps, or other listed means. Connection devices or fittings that depend solely on solder shall not be used. Sheet metal screws shall not be used to connect grounding conductors to enclosures.

250-10. Protection of Ground Clamps and Fittings.
Ground clamps or other fittings shall be approved for general use without protection or shall be protected from physical damage as indicated in (1) or (2).

(1) In installations where they are not likely to be damaged
(2) Where enclosed in metal, wood, or equivalent protective covering

250-12. Clean Surfaces. Nonconductive coatings (such as paint, lacquer, and enamel) on equipment to be grounded shall be removed from threads and other contact surfaces to ensure good electrical continuity or be connected by means of fittings designed so as to make such removal unnecessary.

B. Circuit and System Grounding

250-20. Alternating-Current Circuits and Systems to Be Grounded. Alternating-current circuits and systems shall be grounded as provided for in (a), (b), (c), or (d). Other circuits and systems shall be permitted to be grounded.

> FPN: An example of a system permitted to be grounded is a corner-grounded delta transformer connection. See Section 250-26(4) for conductor to be grounded.

(a) Alternating-Current Circuits of Less than 50 Volts. Alternating-current circuits of less than 50 volts shall be grounded under any of the following conditions:

(1) Where supplied by transformers, if the transformer supply system exceeds 150 volts to ground
(2) Where supplied by transformers, if the transformer supply system is ungrounded
(3) Where installed as overhead conductors outside of buildings

(b) Alternating-Current Systems of 50 Volts to 1000 Volts. Alternating-current systems of 50 volts to 1000 volts that supply premises wiring and premises wiring systems shall be grounded under any of the following conditions:

(1) Where the system can be grounded so that the maximum voltage to ground on the ungrounded conductors does not exceed 150 volts
(2) Where the system is 3-phase, 4-wire, wye connected in which the neutral is used as a circuit conductor
(3) Where the system is 3-phase, 4-wire, delta connected in which the midpoint of one phase winding is used as a circuit conductor

(c) Alternating-Current Systems of 1 kV and Over. Alternating-current systems supplying mobile or portable equipment shall be grounded as specified in Section 250-188. Where supplying other than mobile or portable equipment, such systems shall be permitted to be grounded. Where such systems are grounded, they shall comply with the applicable provisions of this article.

(d) Separately Derived Systems. If required to be grounded as in (a) or (b), separately derived systems shall be grounded as specified in Section 250-30.

> FPN No. 1: An alternate ac power source such as an on-site generator is not a separately derived system if the neutral is solidly interconnected to a service-supplied system neutral.

> FPN No. 2: For systems that are not separately derived and are not required to be grounded as specified in Section 250-30, see Section 445-5 for minimum size of conductors that must carry fault current.

250-21. Alternating-Current Systems of 50 Volts to 1000 Volts Not Required to Be Grounded. The following ac systems of 50 volts to 1000 volts shall be permitted to be grounded but shall not be required to be grounded:

(1) Electric systems used exclusively to supply industrial electric furnaces for melting, refining, tempering, and the like
(2) Separately derived systems used exclusively for rectifiers that supply only adjustable speed industrial drives
(3) Separately derived systems supplied by transformers that have a primary voltage rating less than 1000 volts, provided that all of the following conditions are met:

(a) The system is used exclusively for control circuits.
(b) The conditions of maintenance and supervision ensure that only qualified persons will service the installation.
(c) Continuity of control power is required.
(d) Ground detectors are installed on the control system.

(4) Isolated systems as permitted or required in Articles 517 and 668

> FPN: The proper use of suitable ground detectors on ungrounded systems can provide additional protection.

(5) High-impedance grounded neutral systems as specified in Section 250-36

250-22. Circuits Not to Be Grounded. The following circuits shall not be grounded:

(1) Cranes (circuits for electric cranes operating over combustible fibers in Class III locations, as provided in Section 503-13)
(2) Health care facilities (circuits as provided in Article 517)
(3) Electrolytic cells (circuits as provided in Article 668)

250-24. Grounding Service-Supplied Alternating-Current Systems.

(a) System Grounding Connections. A premises wiring system that is supplied by an ac service that is grounded shall have at each service a grounding electrode conductor connected to the grounding electrode(s) required by Part C of this article. The grounding electrode conductor shall be connected to the grounded service conductor in accordance with (1) through (5).

(1) General. The connection shall be made at any accessible point from the load end of the service drop or service lateral to and including the terminal or bus to which the grounded service conductor is connected at the service disconnecting means.

FPN: See definition of *Service Drop* and *Service Lateral* in Article 100; see also Section 230-21 for overhead supply.

(2) Outdoor Transformer. Where the transformer supplying the service is located outside the building, at least one additional grounding connection shall be made from the grounded service conductor to a grounding electrode, either at the transformer or elsewhere outside the building.

Exception: The additional grounding connection shall not be made on high-impedance grounded neutral systems. The system shall meet the requirements of Section 250-36.

(3) Dual Fed Services. For services that are dual fed (double ended) in a common enclosure or grouped together in separate enclosures and employing a secondary tie, a single grounding electrode connection to the tie point of the grounded circuit conductors from each power source shall be permitted.

(4) Main Bonding Jumper as Wire or Busbar. Where the main bonding jumper specified in Section 250-28 is a wire or busbar, and is installed from the neutral bar or bus to the equipment grounding terminal bar or bus in the service equipment, the grounding electrode conductor shall be permitted to be connected to the equipment grounding terminal bar or bus to which the main bonding jumper is connected.

(5) Load-Side Grounding Connections. A grounding connection shall not be made to any grounded circuit conductor on the load side of the service disconnecting means except as otherwise permitted in this article.

FPN: See Section 250-30(b) for separately derived systems, Section 250-32 for connections at separate buildings or structures, and Section 250-142 for use of the grounded circuit conductor for grounding equipment.

(b) Grounded Conductor Brought to Service Equipment. Where an ac system operating at less than 1000 volts is grounded at any point, the grounded conductor(s) shall be run to each service disconnecting means, and shall be bonded to each disconnecting means enclosure. The grounded conductor(s) shall be installed in accordance with (1) through (3).

Exception: Where more than one service disconnecting means is located in an assembly listed for use as service equipment, it shall be permitted to run the grounded conductor(s) to the assembly, and the conductor(s) shall be bonded to the assembly enclosure.

(1) Routing. This conductor shall be routed with the phase conductors and shall not be smaller than the required grounding electrode conductor specified in Table 250-66, but shall not be required to be larger than the largest ungrounded service-entrance phase conductor. In addition, for service-entrance phase conductors larger than 1100 kcmil copper or 1750 kcmil aluminum, the grounded conductor shall not be smaller than 12½ percent of the area of the largest service-entrance phase conductor.

(2) Parallel Conductors. Where the service-entrance phase conductors are installed in parallel, the size of the grounded conductor shall be based on the total circular mil area of the parallel conductors as indicated in this section. Where installed in two or more raceways, the size of the grounded conductor in each raceway shall be based on the size of the ungrounded service-entrance conductor in the raceway but not smaller than No. 1/0.

FPN: See Section 310-4 for grounded conductors connected in parallel.

(3) High Impedance. The grounded conductor on a high-impedance grounded neutral system shall be grounded in accordance with Section 250-36.

(c) Grounding Electrode Conductor. A grounding electrode conductor shall be used to connect the equipment grounding conductors, the service-equipment enclosures, and, where the system is grounded, the grounded service conductor to the grounding electrode(s) required by Part C of this article.

High-impedance grounded neutral system connections shall be made as covered in Section 250-36.

FPN: See Section 250-24(a) for ac system grounding connections.

(d) Ungrounded System Grounding Connections. A premises wiring system that is supplied by an ac service that is ungrounded shall have, at each service, a grounding electrode conductor connected to the grounding electrode(s) required by Part C of this article. The grounding electrode conductor shall be connected to a metal enclosure of the service conductors at any accessible point from the load end of the service drop or service lateral to the service disconnecting means.

250-26. Conductor to Be Grounded—Alternating-Current Systems. For ac premises wiring systems, the conductor to be grounded shall be as specified in the following:

(1) Single-phase, 2-wire — one conductor
(2) Single-phase, 3-wire — the neutral conductor
(3) Multiphase systems having one wire common to all phases — the common conductor
(4) Multiphase systems requiring one grounded phase — one phase conductor

(5) Multiphase systems in which one phase is used as in (2) — the neutral conductor

250-28. Main Bonding Jumper. For a grounded system, an unspliced main bonding jumper shall be used to connect the equipment grounding conductor(s) and the service-disconnect enclosure to the grounded conductor of the system within the enclosure for each service disconnect.

Exception No. 1: Where more than one service disconnecting means is located in an assembly listed for use as service equipment, an unspliced main bonding jumper shall bond the grounded conductor(s) to the assembly enclosure.

Exception No. 2: Impedance grounded neutral systems shall be permitted to be connected as provided in Sections 250-36 and 250-186.

(a) Material. Main bonding jumpers shall be of copper or other corrosion-resistant material. A main bonding jumper shall be a wire, bus, screw, or similar suitable conductor.

(b) Construction. Where a main bonding jumper is a screw only, the screw shall be identified with a green finish that shall be visible with the screw installed.

(c) Attachment. Main bonding jumpers shall be attached in the manner specified by the applicable provisions of Section 250-8.

(d) Size. The main bonding jumper shall not be smaller than the sizes shown in Table 250-66 for grounding electrode conductors. Where the service-entrance phase conductors are larger than 1100 kcmil copper or 1750 kcmil aluminum, the bonding jumper shall have an area that is not less than 12½ percent of the area of the largest phase conductor except that where the phase conductors and the bonding jumper are of different materials (copper or aluminum), the minimum size of the bonding jumper shall be based on the assumed use of phase conductors of the same material as the bonding jumper and with an ampacity equivalent to that of the installed phase conductors.

250-30. Grounding Separately Derived Alternating-Current Systems.

(a) Grounded Systems. A separately derived ac system that is grounded shall comply with (1) through (4).

Exception: High-impedance grounded neutral system grounding connection requirements shall not be required to comply with (1) and (2) and shall be made as specified in Sections 250-36 and 250-186.

(1) Bonding Jumper. A bonding jumper in compliance with Sections 250-28(a) through (d), that is sized for the derived phase conductors, shall be used to connect the equipment grounding conductors of the separately derived system to the grounded conductor. Except as permitted by Section 250-24(a)(4), this connection shall be made at any point on the separately derived system from the source to the first system disconnecting means or overcurrent device, or it shall be made at the source of a separately derived system that has no disconnecting means or overcurrent devices. The point of connection shall be the same as the grounding electrode conductor as required in Section 250-30(a)(2).

Exception No. 1: A bonding jumper at both the source and the first disconnecting means shall be permitted where doing so does not establish a parallel path for the grounded circuit conductor. Where a grounded conductor is used in this manner, it shall not be smaller than the size specified for the bonding jumper but shall not be required to be larger than the ungrounded conductor(s). For the purposes of this exception, connection through the earth is not considered as providing a parallel path.

Exception No. 2: The size of the bonding jumper for a system that supplies a Class 1, Class 2, or Class 3 circuit, and is derived from a transformer rated not more than 1000 volt-amperes, shall not be smaller than the derived phase conductors and shall not be smaller than No. 14 copper or No. 12 aluminum.

(2) Grounding Electrode Conductor. A grounding electrode conductor, sized in accordance with Section 250-66 for the derived phase conductors, shall be used to connect the grounded conductor of the derived system to the grounding electrode as specified in (3). Except as permitted by Sections 250-24(a)(3) or (a)(4), this connection shall be made at the same point on the separately derived system where the bonding jumper is installed.

Exception: A grounding electrode conductor shall not be required for a system that supplies a Class 1, Class 2, or Class 3 circuit and is derived from a transformer rated not more than 1000 volt-amperes, provided the system grounded conductor is bonded to the transformer frame or enclosure by a jumper sized in accordance with Section 250-30(a)(1), Exception No. 2, and the transformer frame or enclosure is grounded by one of the means specified in Section 250-134.

(3) Grounding Electrode. The grounding electrode shall be as near as practicable to and preferably in the same area as the grounding electrode conductor connection to the system. The grounding electrode shall be the nearest one of the following:

(a) An effectively grounded structural metal member of the structure
(b) An effectively grounded metal water pipe within 5 ft (1.52 m) from the point of entrance into the building

Exception: In industrial and commercial buildings where conditions of maintenance and supervision ensure that only qualified persons will service the installation and the entire length of the interior metal water pipe that is being used for the grounding electrode is exposed, the connection shall be permitted at any point on the water pipe system.

(c) Other electrodes as specified in Sections 250-50 and 250-52 where the electrodes specified by (a) or (b) are not available.

Exception for (a), (b), and (c): Where a separately derived system originates in listed equipment also used as service equipment, the grounding electrode used for the service equipment shall be permitted as the grounding electrode for the separately derived system provided the grounding electrode conductor from the service equipment to the grounding electrode is of sufficient size for the separately derived system. Where the equipment ground bus internal to the service equipment is not smaller than the required grounding electrode conductor, the grounding electrode connection for the separately derived system shall be permitted to be made to the bus.

> FPN: See Section 250-104(a)(4) for bonding requirements of interior metal water piping in the area served by separately derived systems.

(4) Grounding Methods. In all other respects, grounding methods shall comply with requirements prescribed in other parts of this *Code*.

(b) Ungrounded Systems. The equipment of an ungrounded separately derived system shall be grounded as specified in (1) through (3).

(1) Grounding Electrode Conductor. A grounding electrode conductor, sized in accordance with Section 250-66 for the derived phase conductors, shall be used to connect the metal enclosures of the derived system to the grounding electrode as specified in (2). This connection shall be made at any point on the separately derived system from the source to the first system disconnecting means.

(2) Grounding Electrode. Except as permitted by Section 250-34 for portable and vehicle mounted generators, the grounding electrode shall comply with Section 250-30(a)(3).

(3) Grounding Methods. In all other respects, grounding methods shall comply with requirements prescribed in other parts of this *Code*.

250-32. Two or More Buildings or Structures Supplied from a Common Service.

(a) Grounding Electrode. Where two or more buildings or structures are supplied from a common ac service by a feeder(s) or branch circuit(s), the grounding electrode(s) required in Part C of this article at each building or structure shall be connected in the manner specified in (b) or (c). Where there are no existing grounding electrodes, the grounding electrode(s) required in Part C of this article shall be installed.

Exception: A grounding electrode at separate buildings or structures shall not be required where only one branch circuit supplies the building or structure and the branch circuit includes an equipment grounding conductor for grounding the noncurrent-carrying parts of all equipment.

(b) Grounded Systems. For a grounded system at the separate building or structure, the connection to the grounding electrode and grounding or bonding of equipment, structures, or frames required to be grounded or bonded shall comply with either (1) or (2).

(1) Equipment Grounding Conductor. An equipment grounding conductor as described in Section 250-118 shall be run with the supply conductors and connected to the building or structure disconnecting means and to the grounding electrode(s). The equipment grounding conductor shall be used for grounding or bonding of equipment, structures, or frames required to be grounded or bonded. Any installed grounded conductor shall not be connected to the equipment grounding conductor or to the grounding electrode(s).

(2) Grounded Conductor. Where (1) an equipment grounding conductor is not run with the supply to the building or structure, and (2) there are no continuous metallic paths bonded to the grounding system in both buildings or structures involved, and (3) ground-fault protection of equipment has not been installed on the common ac service, the grounded circuit conductor run with the supply to the building or structure shall be connected to the building or structure disconnecting means and to the grounding electrode(s) and shall be used for grounding or bonding of equipment, structures, or frames required to be grounded or bonded.

(c) Ungrounded Systems. The grounding electrode(s) shall be connected to the building or structure disconnecting means.

(d) Disconnecting Means Located in Separate Building or Structure on the Same Premises. Where one or more disconnecting means supply one or more additional buildings or structures under single management, and where these disconnecting means are located remote from those buildings or structures in accordance with the provisions of Section 225-32, Exception Nos. 1 and 2, all of the following conditions shall be met.

(1) The connection of the grounded circuit conductor to the grounding electrode at a separate building or structure shall not be made.

(2) An equipment grounding conductor for grounding any noncurrent-carrying equipment, interior metal piping systems, and building or structural metal frames is run with the circuit conductors to a separate building or structure and bonded to existing grounding electrode(s) required in Part C of this article, or, where there are no existing electrodes, the grounding electrode(s) required in Part C of this article shall be installed where a separate building or structure is supplied by more than one branch circuit.

(3) Bonding the equipment grounding conductor to the grounding electrode at a separate building or structure shall be made in a junction box, panelboard, or similar enclosure located immediately inside or outside the separate building or structure.

(e) Agricultural Buildings or Structures. Where livestock is housed, any portion of the equipment grounding conductor run underground to the building or structure disconnecting means shall be insulated or covered copper.

FPN: See Section 547-8 for special grounding requirements for agricultural buildings.

(f) Grounding Conductor. The size of the grounding conductor to the grounding electrode(s) shall not be less than given in Table 250-122, and shall not be required to be larger than the largest ungrounded supply conductor. The installation shall comply with Part C of this article.

250-34. Portable and Vehicle-Mounted Generators.

(a) Portable Generators. The frame of a portable generator shall not be required to be grounded and shall be permitted to serve as the grounding electrode for a system supplied by the generator under the following conditions:

(1) The generator supplies only equipment mounted on the generator or cord- and plug-connected equipment through receptacles mounted on the generator, or both, and

(2) The noncurrent-carrying metal parts of equipment and the equipment grounding conductor terminals of the receptacles are bonded to the generator frame.

(b) Vehicle-Mounted Generators. The frame of a vehicle shall be permitted to serve as the grounding electrode for a system supplied by a generator located on the vehicle under the following conditions:

(1) The frame of the generator is bonded to the vehicle frame, and

(2) The generator supplies only equipment located on the vehicle or cord- and plug-connected equipment through receptacles mounted on the vehicle, or both equipment located on the vehicle and cord- and plug-connected equipment through receptacles mounted on the vehicle or on the generator, and

(3) The noncurrent-carrying metal parts of equipment and the equipment grounding conductor terminals of the receptacles are bonded to the generator frame, and

(4) The system complies with all other provisions of this article.

(c) Grounded Conductor Bonding. A system conductor that is required to be grounded by Section 250-26 shall be bonded to the generator frame where the generator is a component of a separately derived system.

FPN: For grounding portable generators supplying fixed wiring systems, see Section 250-20(d).

250-36. High-Impedance Grounded Neutral Systems.
High-impedance grounded neutral systems in which a grounding impedance, usually a resistor, limits the ground-fault current to a low value shall be permitted for 3-phase ac systems of 480 volts to 1000 volts where all of the following conditions are met.

(1) The conditions of maintenance and supervision ensure that only qualified persons will service the installation.

(2) Continuity of power is required.

(3) Ground detectors are installed on the system.

(4) Line-to-neutral loads are not served.

High-impedance grounded neutral systems shall comply with the provisions of (a) through (f).

(a) Grounding Impedance Location. The grounding impedance shall be installed between the grounding electrode conductor and the system neutral. Where a neutral is not available, the grounding impedance shall be installed between the grounding electrode conductor and the neutral derived from a grounding transformer.

(b) Neutral Conductor. The neutral conductor from the neutral point of the transformer or generator to its connection point to the grounding impedance shall be fully insulated.

The neutral conductor shall have an ampacity of not less than the maximum current rating of the grounding impedance. In no case shall the neutral conductor be smaller than No. 8 copper or No. 6 aluminum or copper-clad aluminum.

(c) System Neutral Connection. The system neutral conductor shall not be connected to ground except through the grounding impedance.

FPN: The impedance is normally selected to limit the ground-fault current to a value slightly greater than or equal to the capacitive charging current of the system. This value of impedance will also limit transient overvoltages to safe values. For guidance, refer to criteria for limiting transient overvoltages in *Recommended Practice for Grounding of Industrial and Commercial Power Systems*, ANSI/IEEE 142-1991.

(d) Neutral Conductor Routing. The conductor connecting the neutral point of the transformer or generator to the grounding impedance shall be permitted to be installed in a separate raceway. It shall not be required to run this conductor with the phase conductors to the first system disconnecting means or overcurrent device.

(e) Equipment Bonding Jumper. The equipment bonding jumper (the connection between the equipment grounding conductors and the grounding impedance) shall be an unspliced conductor run from the first system disconnecting means or overcurrent device to the grounded side of the grounding impedance.

(f) Grounding Electrode Conductor Location. The grounding electrode conductor shall be attached at any point from the grounded side of the grounding impedance to the equipment grounding connection at the service equipment or first system disconnecting means.

C. Grounding Electrode System and Grounding Electrode Conductor

250-50. Grounding Electrode System. If available on the premises at each building or structure served, each item (a) through (d), and any made electrodes in accordance with Sections 250-52(c) and (d), shall be bonded together to form the grounding electrode system. The bonding jumper(s) shall be installed in accordance with Sections 250-64(a), (b), and (e), shall be sized in accordance with Section 250-66, and shall be connected in the manner specified in Section 250-70.

An unspliced grounding electrode conductor shall be permitted to be run to any convenient grounding electrode available in the grounding electrode system or to one or more grounding electrode(s) individually. It shall be sized for the largest grounding electrode conductor required among all the electrodes connected to it.

The grounding electrode conductor shall be unspliced or spliced by means of irreversible compression-type connectors listed for the purpose or by the exothermic welding process.

Interior metal water piping located more than 5 ft (1.52 m) from the point of entrance to the building shall not be used as a part of the grounding electrode system or as a conductor to interconnect electrodes that are part of the grounding electrode system.

Exception: In industrial and commercial buildings where conditions of maintenance and supervision ensure that only qualified persons will service the installation and the entire length of the interior metal water pipe that is being used for the conductor is exposed.

FPN: See Sections 547-8 and 547-9 for special grounding and bonding requirements for agricultural buildings.

(a) Metal Underground Water Pipe. A metal underground water pipe in direct contact with the earth for 10 ft (3.05 m) or more (including any metal well casing effectively bonded to the pipe) and electrically continuous (or made electrically continuous by bonding around insulating joints or sections or insulating pipe) to the points of connection of the grounding electrode conductor and the bonding conductors.

(1) Continuity. Continuity of the grounding path or the bonding connection to interior piping shall not rely on water meters or filtering devices and similar equipment.

(2) Supplemental Electrode Required. A metal underground water pipe shall be supplemented by an additional electrode of a type specified in Sections 250-50 or 250-52. Where the supplemental electrode is a made electrode of the rod, pipe, or plate type, it shall comply with Section 250-56. The supplemental electrode shall be permitted to be bonded to the grounding electrode conductor, the grounded service-entrance conductor, the nonflexible grounded service raceway, or any grounded service enclosure.

Exception: The supplemental electrode shall be permitted to be bonded to the interior metal water piping at any convenient point as covered in Section 250-50, Exception.

Where the supplemental electrode is a made electrode as in Section 250-52(c) or (d), that portion of the bonding jumper that is the sole connection to the supplemental grounding electrode shall not be required to be larger than No. 6 copper wire or No. 4 aluminum wire.

(b) Metal Frame of the Building or Structure. The metal frame of the building or structure, where effectively grounded.

(c) Concrete-Encased Electrode. An electrode encased by at least 2 in. (50.8 mm) of concrete, located within and near the bottom of a concrete foundation or footing that is in direct contact with the earth, consisting of at least 20 ft (6.1 m) of one or more bare or zinc galvanized or other electrically conductive coated steel reinforcing bars or rods of not less than ½-in. (12.7-mm) diameter, or consisting of at least 20 ft (6.1 m) of bare copper conductor not smaller than No. 4. Reinforcing bars shall be permitted to be bonded together by the usual steel tie wires or other effective means.

(d) Ground Ring. A ground ring encircling the building or structure, in direct contact with the earth at a depth below the earth's surface of not less than 2½ ft (762 mm), consisting of at least 20 ft (6.1 m) of bare copper conductor not smaller than No. 2.

250-52. Made and Other Electrodes. Where none of the electrodes specified in Section 250-50 is available, one or more of the electrodes specified in (b) through (d) shall be used. Where practicable, made electrodes shall be embedded below permanent moisture level. Made electrodes shall be free from nonconductive coatings such as paint or enamel. Where more than one electrode is used, each electrode of one grounding system (including that used for air terminals) shall not be less than 6 ft (1.83 m) from any other electrode of another grounding system. Two or more grounding electrodes that are effectively bonded together shall be considered a single grounding electrode system.

(a) Metal Underground Gas Piping System. A metal underground gas piping system shall not be used as a grounding electrode.

(b) Other Local Metal Underground Systems or Structures. Other local metal underground systems or structures such as piping systems and underground tanks.

(c) Rod and Pipe Electrodes. Rod and pipe electrodes shall not be less than 8 ft (2.44 m) in length, shall consist of the following materials, and shall be installed in the following manner.

(1) Electrodes of pipe or conduit shall not be smaller than ¾ in. trade size and, where of iron or steel, shall have the outer surface galvanized or otherwise metal-coated for corrosion protection.

(2) Electrodes of rods of iron or steel shall be at least ⅝ in. (15.87 mm) in diameter. Stainless steel rods less than ⅝ in. (15.87 mm) in diameter, nonferrous rods, or their equivalent shall be listed and shall not be less than ½ in. (12.7 mm) in diameter.

(3) The electrode shall be installed such that at least 8 ft (2.44 m) of length is in contact with the soil. It shall be driven to a depth of not less than 8 ft (2.44 m) except that, where rock bottom is encountered, the electrode shall be driven at an oblique angle not to exceed 45 degrees from the vertical or shall be buried in a trench that is at least 2½ ft (762 mm) deep. The upper end of the electrode shall be flush with or below ground level unless the aboveground end and the grounding electrode conductor attachment are protected against physical damage as specified in Section 250-10.

(d) Plate Electrodes. Each plate electrode shall expose not less than 2 ft^2 (0.186 m^2) of surface to exterior soil. Electrodes of iron or steel plates shall be at least ¼ in. (6.35 mm) in thickness. Electrodes of nonferrous metal shall be at least 0.06 in. (1.52 mm) in thickness. Plate electrodes shall be installed not less than 2½ ft (762 mm) below the surface of the earth.

(e) Aluminum Electrodes. Aluminum electrodes shall not be permitted.

250-54. Supplementary Grounding Electrodes. Supplementary grounding electrodes shall be permitted to be connected to the equipment grounding conductors specified in Section 250-118, but the earth shall not be used as the sole equipment grounding conductor.

250-56. Resistance of Made Electrodes. A single electrode consisting of a rod, pipe, or plate that does not have a resistance to ground of 25 ohms or less shall be augmented by one additional electrode of any of the types specified in Sections 250-50 or 250-52. Where multiple rod, pipe, or plate electrodes are installed to meet the requirements of this section, they shall not be less than 6 ft (1.83 m) apart.

FPN: The paralleling efficiency of rods longer than 8 ft (2.44 m) is improved by spacing greater than 6 ft (1.83 m).

250-58. Common Grounding Electrode. Where an ac system is connected to a grounding electrode in or at a building as specified in Sections 250-24 and 250-32, the same electrode shall be used to ground conductor enclosures and equipment in or on that building. Where separate services supply a building and are required to be connected to a grounding electrode, the same grounding electrode shall be used.

Two or more grounding electrodes that are effectively bonded together shall be considered as a single grounding electrode system in this sense.

250-60. Use of Air Terminals. Air terminal conductors and driven pipes, rods, or other made electrodes used for grounding air terminals shall not be used in lieu of the made grounding electrodes required by Section 250-52 for grounding wiring systems and equipment. This provision shall not prohibit the required bonding together of grounding electrodes of different systems.

FPN No. 1: See Section 250-106 for spacing from air terminals. See Sections 800-40(d), 810-21(j), and 820-40(d) for bonding of electrodes.

FPN No. 2: Bonding together of all separate grounding electrodes will limit potential differences between them and between their associated wiring systems.

250-62. Grounding Electrode Conductor Material. The grounding electrode conductor shall be of copper, aluminum, or copper-clad aluminum. The material selected shall be resistant to any corrosive condition existing at the installation or shall be suitably protected against corrosion. The conductor shall be solid or stranded, insulated, covered, or bare.

250-64. Grounding Electrode Conductor Installation. Grounding electrode conductors shall be installed as specified in (a) through (e).

(a) Aluminum or Copper-Clad Aluminum Conductors. Insulated or bare aluminum or copper-clad aluminum grounding conductors shall not be used where in direct contact with masonry or the earth or where subject to corrosive conditions. Where used outside, aluminum or copper-clad aluminum grounding conductors shall not be installed within 18 in. (457 mm) of the earth.

(b) Grounding Electrode Conductor. A grounding electrode conductor or its enclosure shall be securely fastened to the surface on which it is carried. A No. 4 copper or aluminum, or larger conductor shall be protected if exposed to severe physical damage. A No. 6 grounding conductor that is free from exposure to physical damage shall be permitted to be run along the surface of the building construction without metal covering or protection where it is securely fastened to the construction; otherwise, it shall be in rigid metal conduit, intermediate metal conduit, rigid nonmetallic conduit, electrical metallic tubing, or cable armor. Grounding conductors smaller than No. 6 shall be in rigid metal conduit, intermediate metal conduit, rigid nonmetallic conduit, electrical metallic tubing, or cable armor.

(c) Continuous. The grounding electrode conductor shall be installed in one continuous length without a splice or joint, unless spliced only by irreversible compression-type connectors listed for the purpose or by the exothermic welding process.

Exception: Busbars shall be permitted to have splices.

(d) Grounding Electrode Conductor Taps. Where a service consists of more than a single enclosure as permitted in Section 230-40, Exception No. 2, it shall be permitted to connect taps to the grounding electrode conductor. Each such tap conductor shall extend to the inside of each such enclosure. The grounding electrode conductor shall be sized in accordance with Section 250-66, but the tap conductors shall be permitted to be sized in accordance with the grounding electrode conductors specified in Section 250-66 for the largest conductor serving the respective enclosures. The tap conductors shall be connected to the grounding electrode conductor in such a manner that the grounding electrode conductor remains without a splice or joint.

(e) Enclosures for Grounding Electrode Conductors. Metal enclosures for grounding electrode conductors shall be electrically continuous from the point of attachment to cabinets or equipment to the grounding electrode, and shall be securely fastened to the ground clamp or fitting. Metal enclosures that are not physically continuous from cabinet or equipment to the grounding electrode shall be made electrically continuous by bonding each end to the grounding conductor. Where a raceway is used as protection for a grounding conductor, the installation shall comply with the requirements of the appropriate raceway article.

250-66. Size of Alternating-Current Grounding Electrode Conductor. The size of the grounding electrode conductor of a grounded or ungrounded ac system shall not be less than given in Table 250-66, except as permitted in (a) through (c).

Table 250-66. Grounding Electrode Conductor for Alternating-Current Systems

Size of Largest Service-Entrance Conductor or Equivalent Area for Parallel Conductors[1]		Size of Grounding Electrode Conductor	
Copper	Aluminum or Copper-Clad Aluminum	Copper	Aluminum or Copper-Clad Aluminum[2]
2 or smaller	½ or smaller	8	6
1 or 1/0	2/0 or 3/0	6	4
2/0 or 3/0	4/0 or 250 kcmil	4	2
Over 3/0 through 350 kcmil	Over 250 kcmil through 500 kcmil	2	1/0
Over 350 kcmil through 600 kcmil	Over 500 kcmil through 900 kcmil	1/0	3/0
Over 600 kcmil through 1100 kcmil	Over 900 kcmil through 1750 kcmil	2/0	4/0
Over 1100 kcmil	Over 1750 kcmil	3/0	250 kcmil

Notes:

1. Where multiple sets of service-entrance conductors are used as permitted in Section 230-40, Exception No. 2, the equivalent size of the largest service-entrance conductor shall be determined by the largest sum of the areas of the corresponding conductors of each set.

2. Where there are no service-entrance conductors, the grounding electrode conductor size shall be determined by the equivalent size of the largest service-entrance conductor required for the load to be served.

[1]This table also applies to the derived conductors of separately derived ac systems.

[2]See installation restrictions in Section 250-64(a).

FPN: See Section 250-24(b) for size of ac system conductor brought to service equipment.

(a) Connections to Made Electrodes. Where the grounding electrode conductor is connected to made electrodes as permitted in Section 250-52(c) or (d), that portion of the conductor that is the sole connection to the grounding electrode shall not be required to be larger than No. 6 copper wire or No. 4 aluminum wire.

(b) Connections to Concrete-Encased Electrodes. Where the grounding electrode conductor is connected to a concrete-encased electrode as permitted in Section 250-50(c), that portion of the conductor that is the sole connection to the grounding electrode shall not be required to be larger than No. 4 copper wire.

(c) Connections to Ground Rings. Where the grounding electrode conductor is connected to a ground ring as permitted in Section 250-50(d), that portion of the conductor that is the sole connection to the grounding electrode shall not be required to be larger than the conductor used for the ground ring.

250-68. Grounding Electrode Conductor Connection to Grounding Electrodes.

(a) Accessibility. The connection of a grounding electrode conductor to a grounding electrode shall be accessible.

Exception: An encased or buried connection to a concrete-encased, driven, or buried grounding electrode shall not be required to be accessible.

(b) Effective Grounding Path. The connection of a grounding electrode conductor shall be made in a manner that will ensure a permanent and effective grounding path. Where necessary to ensure the grounding path for a metal piping system used as a grounding electrode, effective bonding shall be provided around insulated joints and sections and around any equipment that is likely to be disconnected for repairs or replacement. Bonding conductors shall be of sufficient length to permit removal of such equipment while retaining the integrity of the bond.

250-70. Methods of Grounding Conductor Connection to Electrodes. The grounding conductor shall be connected to the grounding electrode by exothermic welding, listed lugs, listed pressure connectors, listed clamps, or other listed means. Connections depending on solder shall not be used. Ground clamps shall be listed for the materials of the grounding electrode and the grounding electrode conductor and, where used on pipe, rod, or other buried electrodes, shall also be listed for direct soil burial. Not more than one conductor shall be connected to the grounding electrode by a single clamp or fitting unless the clamp or fitting is listed for multiple conductors. One of the following methods shall be used:

(1) A listed bolted clamp of cast bronze or brass, or plain or malleable iron
(2) A pipe fitting, pipe plug, or other approved device screwed into a pipe or pipe fitting
(3) For indoor telecommunications purposes only, a listed sheet metal strap-type ground clamp having a rigid metal base that seats on the electrode and having a strap of such material and dimensions that it is not likely to stretch during or after installation
(4) An equally substantial approved means

D. Enclosure, Raceway, and Service Cable Grounding

250-80. Service Raceways and Enclosures. Metal enclosures and raceways for service conductors and equipment shall be grounded.

Exception: A metal elbow that is installed in an underground installation of rigid nonmetallic conduit and is isolated from possible contact by a minimum cover of 18 in. (457 mm) to any part of the elbow shall not be required to be grounded.

250-84. Underground Service Cable or Conduit.

(a) Underground Service Cable. The sheath or armor of a continuous underground metal-sheathed service cable system that is metallically connected to the underground system shall not be required to be grounded at the building. The sheath or armor shall be permitted to be insulated from the interior conduit or piping.

(b) Underground Service Conduit Containing Cable. An underground service conduit that contains a metal-sheathed cable bonded to the underground system shall not be required to be grounded at the building. The sheath or armor shall be permitted to be insulated from the interior conduit or piping.

250-86. Other Conductor Enclosures and Raceways. Except as permitted by Section 250-112(i), metal enclosures and raceways for other than service conductors shall be grounded.

Exception No. 1: Metal enclosures and raceways for conductors added to existing installations of open wire, knob and tube wiring, and nonmetallic-sheathed cable shall not be required to be grounded where these enclosures or wiring methods

(a) Do not provide an equipment ground;
(b) Are in runs of less than 25 ft (7.62 m);
(c) Are free from probable contact with ground, grounded metal, metal lath, or other conductive material; and
(d) Are guarded against contact by persons.

Exception No. 2: Short sections of metal enclosures or raceways used to provide support or protection of cable assemblies from physical damage shall not be required to be grounded.

Exception No. 3: A metal elbow that is installed in an underground installation of rigid nonmetallic conduit and is isolated from possible contact by a minimum cover of 18 in. (457 mm) to any part of the elbow shall not be required to be grounded.

E. Bonding

250-90. General. Bonding shall be provided where necessary to ensure electrical continuity and the capacity to conduct safely any fault current likely to be imposed.

250-92. Services.

(a) Bonding of Services. The noncurrent-carrying metal parts of equipment indicated in (1), (2), and (3) shall be effectively bonded together.

(1) The service raceways, cable trays, cablebus framework, or service cable armor or sheath except as permitted in Section 250-84.
(2) All service enclosures containing service conductors, including meter fittings, boxes, or the like, interposed in the service raceway or armor.
(3) Any metallic raceway or armor enclosing a grounding electrode conductor as specified in Section 250-64(b). Bonding shall apply at each end and to all intervening raceways, boxes, and enclosures between the service equipment and the grounding electrode.

(b) Bonding to Other Systems. An accessible means external to enclosures for connecting intersystem bonding and grounding conductors shall be provided at the service by at least one of the following means:

(1) Exposed nonflexible metallic service raceways
(2) Exposed grounding electrode conductor
(3) Approved means for the external connection of a copper or other corrosion-resistant bonding or grounding conductor to the service raceway or equipment

For the purposes of providing an accessible means for intersystem bonding, the disconnecting means at a separate building or structure as permitted in Section 250-32 and the disconnecting means at a mobile home as permitted in Section 550-23(a) shall be considered the service equipment.

FPN No. 1: A No. 6 copper conductor with one end bonded to the service raceway or equipment and with 6 in. (152 mm) or more of the other end made accessible on the outside wall is an example of the approved means covered in (b)(3).

FPN No. 2: See Sections 800-40, 810-21, and 820-40 for bonding and grounding requirements for communications circuits, radio and television equipment, and CATV circuits.

250-94. Method of Bonding at the Service. Electrical continuity at service equipment, service raceways, and service conductor enclosures shall be ensured by one of the following methods:

(1) Bonding equipment to the grounded service conductor in a manner provided in Section 250-8
(2) Connections utilizing threaded couplings or threaded bosses on enclosures where made up wrenchtight
(3) Threadless couplings and connectors where made up tight for metal raceways and metal-clad cables
(4) Other approved devices, such as bonding-type locknuts and bushings

Bonding jumpers meeting the other requirements of this article shall be used around concentric or eccentric knockouts that are punched or otherwise formed so as to impair the electrical connection to ground. Standard locknuts or bushings shall not be the sole means for the bonding required by this section.

250-96. Bonding Other Enclosures.

(a) General. Metal raceways, cable trays, cable armor, cable sheath, enclosures, frames, fittings, and other metal noncurrent-carrying parts that are to serve as grounding conductors, with or without the use of supplementary equipment grounding conductors, shall be effectively bonded where necessary to ensure electrical continuity and the capacity to conduct safely any fault current likely to be imposed on them. Any nonconductive paint, enamel, or similar coating shall be removed at threads, contact points, and contact surfaces or be connected by means of fittings designed so as to make such removal unnecessary.

(b) Isolated Grounding Circuits. Where required for the reduction of electrical noise (electromagnetic interference) on the grounding circuit, an equipment enclosure supplied by a branch circuit shall be permitted to be isolated from a raceway containing circuits supplying only that equipment by one or more listed nonmetallic raceway fittings located at the point of attachment of the raceway to the equipment enclosure. The metal raceway shall comply with provisions of this article and shall be supplemented by an internal insulated equipment grounding conductor installed in accordance with Section 250-146(d) to ground the equipment enclosure.

FPN: Use of an isolated equipment grounding conductor does not relieve the requirement for grounding the raceway system.

250-97. Bonding for Over 250 Volts. For circuits of over 250 volts to ground, the electrical continuity of metal raceways and cables with metal sheaths that contain any conductor other than service conductors shall be ensured by one or more of the methods specified for services in Sections 250-94(1) through (4).

Exception: Where oversized, concentric, or eccentric knockouts are not encountered, or where a box or enclosure with concentric or eccentric knockouts is listed for the purpose, the following methods shall be permitted:

(a) *Threadless couplings and connectors for cables with metal sheaths*

(b) *Two locknuts, on rigid metal conduit or intermediate metal conduit, one inside and one outside of boxes and cabinets*

(c) *Fittings with shoulders that seat firmly against the box or cabinet, such as electrical metallic tubing connectors, flexible metal conduit connectors, and cable connectors, with one locknut on the inside of boxes and cabinets*

(d) *Listed fittings*

250-98. Bonding Loosely Jointed Metal Raceways. Expansion fittings and telescoping sections of metal raceways shall be made electrically continuous by equipment bonding jumpers or other means.

250-100. Bonding in Hazardous (Classified) Locations. Regardless of the voltage of the electrical system, the electrical continuity of noncurrent-carrying metal parts of equipment, raceways, and other enclosures in any hazardous (classified) location as defined in Article 500 shall be ensured by any of the methods specified for services in Section 250-94 that are approved for the wiring method used.

250-102. Equipment Bonding Jumpers.

(a) Material. Equipment bonding jumpers shall be of copper or other corrosion-resistant material. A bonding jumper shall be a wire, bus, screw, or similar suitable conductor.

(b) Attachment. Equipment bonding jumpers shall be attached in the manner specified by the applicable provisions of Section 250-8 for circuits and equipment and by Section 250-70 for grounding electrodes.

(c) Size — Equipment Bonding Jumper on Supply Side of Service. The bonding jumper shall not be smaller than the sizes shown in Table 250-66 for grounding electrode conductors. Where the service-entrance phase conductors are larger than 1100 kcmil copper or 1750 kcmil aluminum, the bonding jumper shall have an area not less than 12½ percent of the area of the largest phase conductor except that, where the phase conductors and the bonding jumper are of different materials (copper or aluminum), the mini-

mum size of the bonding jumper shall be based on the assumed use of phase conductors of the same material as the bonding jumper and with an ampacity equivalent to that of the installed phase conductors. Where the service-entrance conductors are paralleled in two or more raceways or cables, the equipment bonding jumper, where routed with the raceways or cables, shall be run in parallel. The size of the bonding jumper for each raceway or cable shall be based on the size of the service-entrance conductors in each raceway or cable.

The bonding jumper for a grounding electrode conductor raceway or cable armor as covered in Section 250-64(d) shall be the same size or larger than the required enclosed grounding electrode conductor.

(d) Size — Equipment Bonding Jumper on Load Side of Service. The equipment bonding jumper on the load side of the service overcurrent devices shall be sized, as a minimum, in accordance with the sizes listed in Table 250-122, but not required to be larger than the circuit conductors supplying the equipment, and not smaller than No. 14.

A single common continuous equipment bonding jumper shall be permitted to bond two or more raceways or cables where the bonding jumper is sized in accordance with Table 250-122 for the largest overcurrent device supplying circuits therein.

(e) Installation. The equipment bonding jumper shall be permitted to be installed inside or outside of a raceway or enclosure. Where installed on the outside, the length of the equipment bonding jumper shall not exceed 6 ft (1.83 m) and shall be routed with the raceway or enclosure. Where installed inside of a raceway, the equipment bonding jumper shall comply with the requirements of Sections 250-119 and 250-148.

250-104. Bonding of Piping Systems and Exposed Structural Steel.

(a) Metal Water Piping. The interior metal water piping system shall be bonded as required in (1), (2), (3), or (4) of this section. The bonding jumper shall be installed in accordance with Section 250-64(a), (b), and (e). The points of attachment of the bonding jumper(s) shall be accessible.

(1) General. The interior metal water piping system shall be bonded to the service equipment enclosure, the grounded conductor at the service, the grounding electrode conductor where of sufficient size, or to the one or more grounding electrodes used. The bonding jumper shall be sized in accordance with Table 250-66 except as permitted in (2) and (3).

(2) Buildings of Multiple Occupancy. In buildings of multiple occupancy, where the interior metal water piping system for the individual occupancies is metallically isolated from all other occupancies by use of nonmetallic water piping, the interior metal water piping system for each occupancy shall be permitted to be bonded to the equipment grounding terminal of the panelboard or switchboard enclosure (other than service equipment) supplying that occupancy. The bonding jumper shall be sized in accordance with Table 250-122.

(3) Multiple Buildings or Structures Supplied from a Common Service. The interior metal water piping system shall be bonded to the building or structure disconnecting means enclosure where located at the building or structure, or to the equipment grounding conductor run with the supply conductors, or to the one or more grounding electrodes used. The bonding jumper shall be sized in accordance with Section 250-122 based on the rating or setting of the largest overcurrent device protecting the feeder(s) or branch circuit(s) that supply the building.

(4) Separately Derived Systems. The grounded conductor of the separately derived system shall be bonded to the nearest available point of the interior metal water piping system in the area served by the separately derived system. This connection shall be made at the same point on the separately derived system where the grounding electrode conductor is connected. The bonding jumper shall be sized in accordance with Table 250-66.

ˣ**(b) Metal Gas Piping.** Each aboveground portion of a gas piping system upstream from the equipment shutoff valve shall be electrically continuous and bonded to the grounding electrode system.

(c) Other Metal Piping. Interior metal piping that may become energized shall be bonded to the service equipment enclosure, the grounded conductor at the service, the grounding electrode conductor where of sufficient size, or to the one or more grounding electrodes used. The bonding jumper shall be sized in accordance with Table 250-122 using the rating of the circuit that may energize the piping.

The equipment grounding conductor for the circuit that may energize the piping shall be permitted to serve as the bonding means.

FPN: Bonding all piping and metal air ducts within the premises will provide additional safety.

(d) Structural Steel. Exposed interior structural steel that is interconnected to form a steel building frame and is not intentionally grounded and may become energized shall be bonded to the service equipment enclosure, the grounded conductor at the service, the grounding electrode conductor where of sufficient size, or to the one or more grounding electrodes used. The bonding jumper shall be sized in accordance with Table 250-66 and installed in accordance with

Sections 250-64(a), (b), and (e). The points of attachment of the bonding jumpers shall be accessible.

250-106. Lightning Protection Systems. The lightning protection system ground terminals shall be bonded to the building or structure grounding electrode system.

FPN No. 1: See Section 250-60 for use of air terminals. For further information, see *Standard for the Installation of Lightning Protection Systems*, NFPA 780-1997, which contains detailed information on grounding, bonding, and spacing from lightning protection systems.

FPN No. 2: Metal raceways, enclosures, frames, and other noncurrent-carrying metal parts of electric equipment installed on a building equipped with a lightning protection system may require bonding or spacing from the lightning protection conductors in accordance with *Standard for the Installation of Lightning Protection Systems*, NFPA 780-1997. Separation from lightning protection conductors is typically 6 ft (1.83 m) through air or 3 ft (0.92 m) through dense materials such as concrete, brick, or wood.

F. Equipment Grounding and Equipment Grounding Conductors

250-110. Equipment Fastened in Place or Connected by Permanent Wiring Methods (Fixed). Exposed noncurrent-carrying metal parts of fixed equipment likely to become energized shall be grounded under any of the following conditions:

(1) Where within 8 ft (2.44 m) vertically or 5 ft (1.52 m) horizontally of ground or grounded metal objects and subject to contact by persons
(2) Where located in a wet or damp location and not isolated
(3) Where in electrical contact with metal
(4) Where in a hazardous (classified) location as covered by Articles 500 through 517
(5) Where supplied by a metal-clad, metal-sheathed, metal-raceway, or other wiring method that provides an equipment ground, except as permitted by Section 250-86, Exception No. 2 for short sections of metal enclosures
(6) Where equipment operates with any terminal at over 150 volts to ground

Exception No. 1 to (1) through (6): Metal frames of electrically heated appliances, exempted by special permission, in which case the frames shall be permanently and effectively insulated from ground.

Exception No. 2 to (1) through (6): Distribution apparatus, such as transformer and capacitor cases, mounted on wooden poles, at a height exceeding 8 ft (2.44 m) above ground or grade level.

Exception No. 3 to (1) through (6): Listed equipment protected by a system of double insulation, or its equivalent, shall not be required to be grounded. Where such a system is employed, the equipment shall be distinctively marked.

250-112. Fastened in Place or Connected by Permanent Wiring Methods (Fixed) — Specific. Exposed, noncurrent-carrying metal parts of the kinds of equipment described in (a) through (k), and noncurrent-carrying metal parts of equipment and enclosures described in (l) and (m), shall be grounded regardless of voltage.

(a) Switchboard Frames and Structures. Switchboard frames and structures supporting switching equipment, except frames of 2-wire dc switchboards where effectively insulated from ground.   .

(b) Pipe Organs. Generator and motor frames in an electrically operated pipe organ, unless effectively insulated from ground and the motor driving it.

(c) Motor Frames. Motor frames, as provided by Section 430-142.

(d) Enclosures for Motor Controllers. Enclosures for motor controllers unless attached to ungrounded portable equipment.

(e) Elevators and Cranes. Electric equipment for elevators and cranes.

(f) Garages, Theaters, and Motion Picture Studios. Electric equipment in commercial garages, theaters, and motion picture studios, except pendant lampholders supplied by circuits not over 150 volts to ground.

(g) Electric Signs. Electric signs, outline lighting, and associated equipment as provided in Article 600.

(h) Motion Picture Projection Equipment. Motion picture projection equipment.

(i) Power-Limited Remote-Control, Signaling, and Fire Alarm Circuits. Equipment supplied by Class 1 power-limited circuits and Class 1, Class 2, and Class 3 remote-control and signaling circuits, and by fire alarm circuits, shall be grounded where system grounding is required by Part B of this article.

(j) Lighting Fixtures. Lighting fixtures as provided in Part E of Article 410.

(k) Skid-Mounted Equipment. Permanently mounted electrical equipment and skids shall be grounded with an equipment bonding jumper sized as required by Section 250-122.

(l) Motor-Operated Water Pumps. Motor-operated water pumps including the submersible type.

(m) Metal Well Casings. Where a submersible pump is used in a metal well casing, the well casing shall be bonded to the pump circuit equipment grounding conductor.

250-114. Equipment Connected by Cord and Plug. Under any of the conditions described in (1) through (4), exposed noncurrent-carrying metal parts of cord- and plug-connected equipment likely to become energized shall be grounded.

Exception: Listed tools, listed appliances, and listed equipment covered in (2) through (4) shall not be required to be grounded where protected by a system of double insulation or its equivalent. Double insulated equipment shall be distinctively marked.

(1) In hazardous (classified) locations (see Articles 500 through 517)
(2) Where operated at over 150 volts to ground

Exception No. 1: Motors, where guarded, shall not be required to be grounded.

Exception No. 2: Metal frames of electrically heated appliances, exempted by special permission, shall not be required to be grounded, in which case the frames shall be permanently and effectively insulated from ground.

(3) In residential occupancies

 (a) Refrigerators, freezers, and air conditioners
 (b) Clothes-washing, clothes-drying, dish-washing machines; information technology equipment; sump pumps and electrical aquarium equipment
 (c) Hand-held motor-operated tools, stationary and fixed motor-operated tools, light industrial motor-operated tools
 (d) Motor-operated appliances of the following types: hedge clippers, lawn mowers, snow blowers, and wet scrubbers
 (e) Portable handlamps

(4) In other than residential occupancies

 (a) Refrigerators, freezers, and air conditioners
 (b) Clothes-washing, clothes-drying, dish-washing machines; information technology equipment; sump pumps and electrical aquarium equipment
 (c) Hand-held motor-operated tools, stationary and fixed motor-operated tools, light industrial motor-operated tools
 (d) Motor-operated appliances of the following types: hedge clippers, lawn mowers, snow blowers, and wet scrubbers
 (e) Cord- and plug-connected appliances used in damp or wet locations or by persons standing on the

ground or on metal floors or working inside of metal tanks or boilers

(f) Tools likely to be used in wet or conductive locations
(g) Portable handlamps

Exception: Tools and portable handlamps likely to be used in wet or conductive locations shall not be required to be grounded where supplied through an isolating transformer with an ungrounded secondary of not over 50 volts.

250-116. Nonelectric Equipment. The metal parts of non-electric equipment described in this section shall be grounded.

(1) Frames and tracks of electrically operated cranes and hoists
(2) Frames of nonelectrically driven elevator cars to which electric conductors are attached
(3) Hand-operated metal shifting ropes or cables of electric elevators

> FPN: Where extensive metal in or on buildings may become energized and is subject to personal contact, adequate bonding and grounding will provide additional safety.

250-118. Types of Equipment Grounding Conductors. The equipment grounding conductor run with or enclosing the circuit conductors shall be one or more or a combination of the following.

(1) A copper or other corrosion-resistant conductor. This conductor shall be solid or stranded; insulated, covered, or bare; and in the form of a wire or a busbar of any shape.
(2) Rigid metal conduit.
(3) Intermediate metal conduit.
(4) Electrical metallic tubing.
(5) Flexible metal conduit where both the conduit and fittings are listed for grounding.
(6) Listed flexible metal conduit that is not listed for grounding, meeting all the following conditions.

   (a) The conduit is terminated in fittings listed for grounding.
   (b) The circuit conductors contained in the conduit are protected by overcurrent devices rated at 20 amperes or less.
   (c) The combined length of flexible metal conduit and flexible metallic tubing and liquidtight flexible metal conduit in the same ground return path does not exceed 6 ft (1.83 m).
   (d) The conduit is not installed for flexibility.

(7) Listed liquidight flexible metal conduit meeting all the following conditions.

   (a) The conduit is terminated in fittings listed for grounding.
   (b) For trade sizes ⅜ in. through ½ in., the circuit conductors contained in the conduit are protected by overcurrent devices rated at 20 amperes or less.
   (c) For trade sizes ¾ in. through 1¼ in., the circuit conductors contained in the conduit are protected by overcurrent devices rated not more than 60 amperes and there is no flexible metal conduit, flexible metallic tubing, or liquidtight flexible metal conduit in trade sizes ⅜ in. or ½ in. in the grounding path.
   (d) The combined length of flexible metal conduit and flexible metallic tubing and liquidtight flexible metal conduit in the same ground return path does not exceed 6 ft (1.83 m).
   (e) The conduit is not installed for flexibility.

(8) Flexible metallic tubing where the tubing is terminated in fittings listed for grounding and meeting all the following conditions.

   (a) The circuit conductors contained in the tubing are protected by overcurrent devices rated at 20 amperes or less.
   (b) The combined length of flexible metal conduit and flexible metallic tubing and liquidtight flexible metal conduit in the same ground return path does not exceed 6 ft (1.83 m).

(9) Armor of Type AC cable as provided in Section 333-21.
(10) The copper sheath of mineral-insulated, metal-sheathed cable.
(11) The metallic sheath or the combined metallic sheath and grounding conductors of Type MC cable.
(12) Cable trays as permitted in Sections 318-3(c) and 318-7.
(13) Cablebus framework as permitted in Section 365-2(a).
(14) Other electrically continuous metal raceways listed for grounding.

250-119. Identification of Equipment Grounding Conductors. Unless required elsewhere in this *Code,* equipment grounding conductors shall be permitted to be bare, covered, or insulated. Individually covered or insulated equipment grounding conductors shall have a continuous outer finish that is either green or green with one or more yellow stripes except as permitted in this section.

(a) Conductors Larger than No. 6. An insulated or covered conductor larger than No. 6 copper or aluminum shall be permitted, at the time of installation, to be permanently identified as an equipment grounding conductor at each end and at every point where the conductor is accessible. Identification shall be accomplished by one of the following:

(1) Stripping the insulation or covering from the entire exposed length
(2) Coloring the exposed insulation or covering green
(3) Marking the exposed insulation or covering with green tape or green adhesive labels

(b) Multiconductor Cable. Where the conditions of maintenance and supervision ensure that only qualified persons will service the installation, one or more insulated conductors in a multiconductor cable, at the time of installation, shall be permitted to be permanently identified as equipment grounding conductors at each end and at every point where the conductors are accessible by one of the following means:

(1) Stripping the insulation from the entire exposed length
(2) Coloring the exposed insulation green
(3) Marking the exposed insulation with green tape or green adhesive labels

(c) Flexible Cord. An uninsulated equipment grounding conductor shall be permitted, but, if individually covered, the covering shall have a continuous outer finish that is either green or green with one or more yellow stripes.

250-120. Equipment Grounding Conductor Installation. An equipment grounding conductor shall be installed as follows.

(a) Raceway, Cable Trays, Cable Armor, or Cable Sheaths. Where it consists of a raceway, cable tray, cable armor, or cable sheath or where it is a wire within a raceway or cable, it shall be installed in accordance with the applicable provisions in this *Code* using fittings for joints and terminations approved for use with the type raceway or cable used. All connections, joints, and fittings shall be made tight using suitable tools.

(b) Aluminum and Copper-Clad Aluminum Conductors. Aluminum and copper-clad aluminum conductors shall be installed in accordance with the restrictions of Section 250-64.

(c) Equipment Grounding Conductors Smaller than No. 6. Equipment grounding conductors smaller than No. 6 shall be protected from physical damage by a raceway or cable armor except where run in hollow spaces of walls or partitions, where not subject to physical damage, or where protected from physical damage.

250-122. Size of Equipment Grounding Conductors.

(a) General. Copper, aluminum, or copper-clad aluminum equipment grounding conductors of the wire type shall not be smaller than shown in Table 250-122, but shall not be required to be larger than the circuit conductors supplying the equipment. Where a raceway or a cable armor or sheath is used as the equipment grounding conductor, as provided in Sections 250-118 and 250-134(a), it shall comply with Section 250-2(d).

(b) Adjustment for Voltage Drop. Where conductors are adjusted in size to compensate for voltage drop, equipment grounding conductors, where installed, shall be adjusted proportionately according to circular mil area.

(c) Multiple Circuits. Where a single equipment grounding conductor is run with multiple circuits in the same raceway or cable, it shall be sized for the largest overcurrent device protecting conductors in the raceway or cable.

(d) Motor Circuits. Where the overcurrent device consists of an instantaneous trip circuit breaker or a motor short-circuit protector, as allowed in Section 430-52, the equipment grounding conductor size shall be permitted to be based on the rating of the motor overload protective device but not less than the size shown in Table 250-122.

(e) Flexible Cord and Fixture Wire. Equipment grounding conductors that are part of flexible cords or used with fixture wires in accordance with Section 240-4 shall be not smaller than No. 18 copper and not smaller than the circuit conductors.

(f) Conductors in Parallel. Where conductors are run in parallel in multiple raceways or cables as permitted in Section 310-4, the equipment grounding conductors, where used, shall be run in parallel in each raceway or cable. One of the following methods shall be used to ensure the equipment grounding conductors are protected.

(1) Each parallel equipment grounding conductor shall be sized on the basis of the ampere rating of the overcurrent device protecting the circuit conductors in the raceway or cable in accordance with Table 250-122.

(2) Where ground-fault protection of equipment is installed, each parallel equipment grounding conductor in a multiconductor cable shall be permitted to be sized in accordance with Table 250-122 on the basis of the trip rating of the ground-fault protection where the following conditions are met.

(1) Conditions of maintenance and supervision ensure that only qualified persons will service the installation.
(2) The ground-fault protection equipment is set to trip at not more than the ampacity of a single ungrounded conductor of one of the cables in parallel.
(3) The ground-fault protection is listed for the purpose.

Table 250-122. Minimum Size Equipment Grounding Conductors for Grounding Raceway and Equipment

Rating or Setting of Automatic Overcurrent Device in Circuit Ahead of Equipment, Conduit, etc., Not Exceeding (Amperes)	Size (AWG or kcmil)	
	Copper	Aluminum or Copper-Clad Aluminum*
15	14	12
20	12	10
30	10	8
40	10	8
60	10	8
100	8	6
200	6	4
300	4	2
400	3	1
500	2	1/0
600	1	2/0
800	1/0	3/0
1000	2/0	4/0
1200	3/0	250
1600	4/0	350
2000	250	400
2500	350	600
3000	400	600
4000	500	800
5000	700	1200
6000	800	1200

Note: Where necessary to comply with Section 250-2(d), the equipment grounding conductor shall be sized larger than this table.

*See installation restrictions in Section 250-120.

250-124. Equipment Grounding Conductor Continuity.

(a) Separable Connections. Separable connections such as those provided in drawout equipment or attachment plugs and mating connectors and receptacles shall provide for first-make, last-break of the equipment grounding conductor. First-make, last-break shall not be required where interlocked equipment, plugs, receptacles, and connectors preclude energization without grounding continuity.

(b) Switches. No automatic cutout or switch shall be placed in the equipment grounding conductor of a premises wiring system unless the opening of the cutout or switch disconnects all sources of energy.

250-126. Identification of Wiring Device Terminals.

The terminal for the connection of the equipment grounding conductor shall be identified by one of the following.

(1) A green, not readily removable terminal screw with a hexagonal head.

(2) A green, hexagonal, not readily removable terminal nut.

(3) A green pressure wire connector. If the terminal for the grounding conductor is not visible, the conductor entrance hole shall be marked with the word *green* or *ground*, the letters *G* or *GR* or the grounding symbol shown in Figure 250-126, or otherwise identified by a distinctive green color. If the terminal for the equipment grounding conductor is readily removable, the area adjacent to the terminal shall be similarly marked.

Figure 250-126   Grounding symbol.

G. Methods of Equipment Grounding

250-130. Equipment Grounding Conductor Connections. Equipment grounding conductor connections at the source of separately derived systems shall be made in accordance with Section 250-30(a)(1). Equipment grounding conductor connections at service equipment shall be made as indicated in (a) or (b). For replacement of nongrounding-type receptacles with grounding type receptacles and for branch-circuit extensions only in existing installations that do not have an equipment grounding conductor in the branch circuit, connections shall be permitted as indicated in (c).

(a) For Grounded Systems. The connection shall be made by bonding the equipment grounding conductor to the grounded service conductor and the grounding electrode conductor.

(b) For Ungrounded Systems. The connection shall be made by bonding the equipment grounding conductor to the grounding electrode conductor.

(c) Nongrounding Receptacle Replacement or Branch Circuit Extensions. The equipment grounding conductor of a grounding-type receptacle or a branch-circuit extension shall be permitted to be connected to any of the following:

(1) Any accessible point on the grounding electrode system as described in Section 250-50

(2) Any accessible point on the grounding electrode conductor

(3) The equipment grounding terminal bar within the enclosure where the branch circuit for the receptacle or branch circuit originates

(4) For grounded systems, the grounded service conductor within the service equipment enclosure

(5) For ungrounded systems, the grounding terminal bar within the service equipment enclosure

FPN: See Section 210-7(d) for the use of a ground-fault circuit-interrupting type of receptacle.

250-132. Short Sections of Raceway. Isolated sections of metal raceway or cable armor, where required to be grounded, shall be grounded in accordance with Section 250-134.

250-134. Equipment Fastened in Place or Connected by Permanent Wiring Methods (Fixed) — Grounding. Unless grounded by connection to the grounded circuit conductor as permitted by Sections 250-32, 250-140, and 250-142, noncurrent-carrying metal parts of equipment, raceways, and other enclosures, if grounded, shall be grounded by one of the following methods.

(a) Equipment Grounding Conductor Types. By any of the equipment grounding conductors permitted by Section 250-118.

(b) With Circuit Conductors. By an equipment grounding conductor contained within the same raceway, cable, or otherwise run with the circuit conductors.

Exception No. 1: As provided in Section 250-130(c), the equipment grounding conductor shall be permitted to be run separately from the circuit conductors.

Exception No. 2: For dc circuits, the equipment grounding conductor shall be permitted to be run separately from the circuit conductors.

> FPN No. 1: See Sections 250-102 and 250-168 for equipment bonding jumper requirements.
>
> FPN No. 2: See Section 400-7 for use of cords for fixed equipment.

250-136. Equipment Considered Effectively Grounded. Under the conditions specified in (a) and (b), the noncurrent-carrying metal parts of the equipment shall be considered effectively grounded.

(a) Equipment Secured to Grounded Metal Supports. Electric equipment secured to and in electrical contact with a metal rack or structure provided for its support and grounded by one of the means indicated in Section 250-134. The structural metal frame of a building shall not be used as the required equipment grounding conductor for ac equipment.

(b) Metal Car Frames. Metal car frames supported by metal hoisting cables attached to or running over metal sheaves or drums of elevator machines that are grounded by one of the methods indicated in Section 250-134.

250-138. Cord- and Plug-Connected Equipment. Noncurrent-carrying metal parts of cord- and plug-connected equipment if grounded, shall be grounded by one of the following methods.

(a) By Means of an Equipment Grounding Conductor. By means of an equipment grounding conductor run with the power supply conductors in a cable assembly or flexible cord properly terminated in a grounding-type attachment plug with one fixed grounding contact.

Exception: The grounding contacting pole of grounding-type plug-in ground-fault circuit interrupters shall be permitted to be of the movable, self-restoring type on circuits operating at not over 150 volts between any two conductors, or over 150 volts between any conductor and ground.

(b) By Means of a Separate Flexible Wire or Strap. By means of a separate flexible wire or strap, insulated or bare, protected as well as practicable against physical damage, where part of equipment.

250-140. Frames of Ranges and Clothes Dryers. This section shall apply to existing branch-circuit installations only. New branch-circuit installations shall comply with Sections 250-134 and 250-138. Frames of electric ranges, wall-mounted ovens, counter-mounted cooking units, clothes dryers, and outlet or junction boxes that are part of the circuit for these appliances shall be grounded in the manner specified by Section 250-134 or 250-138; or, except for mobile homes and recreational vehicles, shall be permitted to be grounded to the grounded circuit conductor if all of the following conditions are met.

(1) The supply circuit is 120/240-volt, single-phase, 3-wire; or 208Y/120-volt derived from a 3-phase, 4-wire wye-connected system.
(2) The grounded conductor is not smaller than No. 10 copper or No. 8 aluminum.
(3) The grounded conductor is insulated, or the grounded conductor is uninsulated and part of a Type SE service-entrance cable and the branch circuit originates at the service equipment.
(4) Grounding contacts of receptacles furnished as part of the equipment are bonded to the equipment.

250-142. Use of Grounded Circuit Conductor for Grounding Equipment.

(a) Supply-Side Equipment. A grounded circuit conductor shall be permitted to ground noncurrent-carrying metal parts of equipment, raceways, and other enclosures at any of the following locations:

(1) On the supply side or within the enclosure of the ac service-disconnecting means
(2) On the supply side or within the enclosure of the main disconnecting means for separate buildings as provided in Section 250-32(b)

(3) On the supply side or within the enclosure of the main disconnecting means or overcurrent devices of a separately derived system where permitted by Section 250-30(a)(1)

(b) Load-Side Equipment. Except as permitted in Sections 250-30(a)(1) and 250-32(b), a grounded circuit conductor shall not be used for grounding noncurrent-carrying metal parts of equipment on the load side of the service disconnecting means or on the load side of a separately derived system disconnecting means or the overcurrent devices for a separately derived system not having a main disconnecting means.

Exception No. 1: The frames of ranges, wall-mounted ovens, counter-mounted cooking units, and clothes dryers under the conditions permitted for existing installations by Section 250-140 shall be permitted to be grounded by a grounded circuit conductor.

Exception No. 2: It shall be permissible to ground meter enclosures by connection to the grounded circuit conductor on the load-side of the service disconnect if

(a) No service ground-fault protection is installed, and
(b) All meter enclosures are located near the service disconnecting means, and
(c) The size of the grounded circuit conductor is not smaller than the size specified in Table 250-122 for equipment grounding conductors.

Exception No. 3: Direct-current systems shall be permitted to be grounded on the load side of the disconnecting means or overcurrent device in accordance with Section 250-164.

250-144. Multiple Circuit Connections. Where equipment is required to be grounded, and is supplied by separate connection to more than one circuit or grounded premises wiring system, a means for grounding shall be provided for each such connection as specified in Sections 250-134 and 250-138.

250-146. Connecting Receptacle Grounding Terminal to Box. An equipment bonding jumper shall be used to connect the grounding terminal of a grounding-type receptacle to a grounded box unless grounded as in (a) through (d).

(a) Surface Mounted Box. Where the box is mounted on or at the surface, direct metal-to-metal contact between the device yoke and the box shall be permitted to ground the receptacle to the box. This provision shall not apply to cover-mounted receptacles unless the box and cover combination are listed as providing satisfactory ground continuity between the box and the receptacle.

(b) Contact Devices or Yokes. Contact devices or yokes designed and listed for the purpose shall be permitted in conjunction with the supporting screws to establish the grounding circuit between the device yoke and flush-type boxes.

(c) Floor Boxes. Floor boxes designed for and listed as providing satisfactory ground continuity between the box and the device shall be permitted.

(d) Isolated Receptacles. Where required for the reduction of electrical noise (electromagnetic interference) on the grounding circuit, a receptacle in which the grounding terminal is purposely insulated from the receptacle mounting means shall be permitted. The receptacle grounding terminal shall be grounded by an insulated equipment grounding conductor run with the circuit conductors. This grounding conductor shall be permitted to pass through one or more panelboards without connection to the panelboard grounding terminal as permitted in Section 384-20, Exception, so as to terminate within the same building or structure directly at an equipment grounding conductor terminal of the applicable derived system or service.

> FPN: Use of an isolated equipment grounding conductor does not relieve the requirement for grounding the raceway system and outlet box.

250-148. Continuity and Attachment of Equipment Grounding Conductors to Boxes. Where more than one equipment grounding conductor enters a box, all such conductors shall be spliced or joined within the box or to the box with devices suitable for the use. Connections depending solely on solder shall not be used. Splices shall be made in accordance with Section 110-14(b) except that insulation shall not be required. The arrangement of grounding connections shall be such that the disconnection or the removal of a receptacle, fixture, or other device fed from the box will not interfere with or interrupt the grounding continuity.

Exception: The equipment grounding conductor permitted in Section 250-146(d) shall not be required to be connected to the other equipment grounding conductors or to the box.

(a) Metal Boxes. A connection shall be made between the one or more equipment grounding conductors and a metal box by means of a grounding screw that shall be used for no other purpose or a listed grounding device.

(b) Nonmetallic Boxes. One or more equipment grounding conductors brought into a nonmetallic outlet box shall be arranged so that a connection can be made to any fitting or device in that box requiring grounding.

H. Direct-Current Systems

250-160. General. Direct-current systems shall comply with Part H and other sections of Article 250 not specifically intended for ac systems.

250-162. Direct-Current Circuits and Systems to Be Grounded. Direct-current circuits and systems shall be grounded as provided for in (a) and (b).

(a) Two-Wire, Direct-Current Systems. A two-wire, dc system supplying premises wiring and operating at greater than 50 volts but not greater than 300 volts, shall be grounded.

Exception No. 1: A system equipped with a ground detector and supplying only industrial equipment in limited areas shall not be required to be grounded.

Exception No. 2: A rectifier-derived dc system supplied from an ac system complying with Section 250-20 shall not be required to be grounded.

Exception No. 3: Direct-current fire alarm circuits having a maximum current of 0.030 amperes as specified in Article 760, Part C, shall not be required to be grounded.

(b) Three-Wire, Direct-Current Systems. The neutral conductor of all 3-wire, dc systems supplying premises wiring shall be grounded.

250-164. Point of Connection for Direct-Current Systems.

(a) Off-Premises Source. Direct-current systems to be grounded and supplied from an off-premises source shall have the grounding connection made at one or more supply stations. A grounding connection shall not be made at individual services or at any point on the premises wiring.

(b) On-Premises Source. Where the dc system source is located on the premises, a grounding connection shall be made at one of the following:

(1) The source
(2) The first system disconnection means or overcurrent device
(3) By other means that accomplish equivalent system protection and that utilize equipment listed and identified for the use

250-166. Size of Direct-Current Grounding Electrode Conductor. The size of the grounding electrode conductor for a dc system shall be as specified in (a) through (e).

(a) Not Smaller than the Neutral Conductor. Where the dc system consists of a 3-wire balancer set or a balancer winding with overcurrent protection as provided in Section 445-4(d), the grounding electrode conductor shall not be smaller than the neutral conductor, and not smaller than No. 8 copper or No. 6 aluminum.

(b) Not Smaller than the Largest Conductor. Where the dc system is other than as in (a), the grounding electrode conductor shall not be smaller than the largest conductor supplied by the system, and not smaller than No. 8 copper or No. 6 aluminum.

(c) Connected to Made Electrodes. Where connected to made electrodes as in Section 250-52(c) or (d), that portion of the grounding electrode conductor that is the sole connection to the grounding electrode shall not be required to be larger than No. 6 copper wire or No. 4 aluminum wire.

(d) Connected to a Concrete-Encased Electrode. Where connected to a concrete-encased electrode as in Section 250-50(c), that portion of the grounding electrode conductor that is the sole connection to the grounding electrode shall not be required to be larger than No. 4 copper wire.

(e) Connected to a Ground Ring. Where connected to a ground ring as in Section 250-50(d), that portion of the grounding electrode conductor that is the sole connection to the grounding electrode shall not be required to be larger than the conductor used for the ground ring.

250-168. Direct-Current Bonding Jumper. For dc systems, the size of the bonding jumper shall not be smaller than the system grounding conductor specified in Section 250-166.

250-169. Ungrounded Direct-Current Separately Derived Systems. Except as otherwise permitted in Section 250-34 for portable and vehicle mounted generators, an ungrounded dc separately derived system supplied from a stand-alone power source (such as an engine–generator set) shall have a grounding electrode conductor connected to an electrode that complies with Part C to provide for grounding of metal enclosures, raceways, cables, and exposed noncurrent-carrying metal parts of equipment. The grounding electrode conductor connection shall be to the metal enclosure at any point on the separately derived system from the source to the first system disconnecting means or overcurrent device, or it shall be made at the source of a separately derived system that has no disconnecting means or overcurrent devices.

The size of the grounding electrode conductor shall be in accordance with Section 250-166.

J. Instruments, Meters, and Relays

250-170. Instrument Transformer Circuits. Secondary circuits of current and potential instrument transformers shall be grounded where the primary windings are connected to circuits of 300 volts or more to ground and, where on switchboards, shall be grounded irrespective of voltage.

Exception: Circuits where the primary windings are connected to circuits of less than 1000 volts with no live parts or wiring exposed or accessible to other than qualified persons.

250-172. Instrument Transformer Cases. Cases or frames of instrument transformers shall be grounded where accessible to other than qualified persons.

Exception: Cases or frames of current transformers, the primaries of which are not over 150 volts to ground and that are used exclusively to supply current to meters.

250-174. Cases of Instruments, Meters, and Relays Operating at Less than 1000 Volts. Instruments, meters, and relays operating with windings or working parts at less than 1000 volts shall be grounded as specified in (a), (b), or (c).

(a) Not on Switchboards. Instruments, meters, and relays not located on switchboards, operating with windings or working parts at 300 volts or more to ground, and accessible to other than qualified persons, shall have the cases and other exposed metal parts grounded.

(b) On Dead-Front Switchboards. Instruments, meters, and relays (whether operated from current and potential transformers or connected directly in the circuit) on switchboards having no live parts on the front of the panels shall have the cases grounded.

(c) On Live-Front Switchboards. Instruments, meters, and relays (whether operated from current and potential transformers or connected directly in the circuit) on switchboards having exposed live parts on the front of panels shall not have their cases grounded. Mats of insulating rubber or other suitable floor insulation shall be provided for the operator where the voltage to ground exceeds 150.

250-176. Cases of Instruments, Meters, and Relays — Operating Voltage 1 kV and Over. Where instruments, meters, and relays have current-carrying parts of 1 kV and over to ground, they shall be isolated by elevation or protected by suitable barriers, grounded metal, or insulating covers or guards. Their cases shall not be grounded.

Exception: Cases of electrostatic ground detectors where the internal ground segments of the instrument are connected to the instrument case and grounded and the ground detector is isolated by elevation.

250-178. Instrument Grounding Conductor. The grounding conductor for secondary circuits of instrument transformers and for instrument cases shall not be smaller than No. 12 copper or No. 10 aluminum. Cases of instrument transformers, instruments, meters, and relays that are mounted directly on grounded metal surfaces of enclosures or grounded metal switchboard panels shall be considered to be grounded, and no additional grounding conductor shall be required.

K. Grounding of Systems and Circuits of 1 kV and Over (High Voltage)

250-180. General. Where high-voltage systems are grounded, they shall comply with all applicable provisions of the preceding sections of this article and with the following sections, which supplement and modify the preceding sections.

250-182. Derived Neutral Systems. A system neutral derived from a grounding transformer shall be permitted to be used for grounding high-voltage systems.

250-184. Solidly Grounded Neutral Systems.

(a) Neutral Conductor. The minimum insulation level for neutral conductors of solidly grounded systems shall be 600 volts.

Exception No. 1: Bare copper conductors shall be permitted to be used for the neutral of service entrances and the neutral of direct-buried portions of feeders.

Exception No. 2: Bare conductors shall be permitted for the neutral of overhead portions installed outdoors.

> FPN: See Section 225-4 for conductor covering where within 10 ft (3.05 m) of any building or other structure.

(b) Multiple Grounding. The neutral of a solidly grounded neutral system shall be permitted to be grounded at more than one point for the following:

(1) Services
(2) Direct-buried portions of feeders employing a bare copper neutral
(3) Overhead portion installed outdoors

(c) Neutral Grounding Conductor. The neutral grounding conductor shall be permitted to be a bare conductor if isolated from phase conductors and protected from physical damage.

250-186. Impedance Grounded Neutral Systems. Impedance grounded neutral systems shall comply with the provisions of (a) through (d).

(a) Location. The grounding impedance shall be inserted in the grounding conductor between the grounding electrode of the supply system and the neutral point of the supply transformer or generator.

(b) Identified and Insulated. Where the neutral conductor of an impedance grounded neutral system is used, it shall be identified, as well as fully insulated with the same insulation as the phase conductors.

(c) System Neutral Connection. The system neutral shall not be connected to ground, except through the neutral grounding impedance.

(d) Equipment Grounding Conductors. Equipment grounding conductors shall be permitted to be bare and shall be connected to the ground bus and grounding electrode conductor at the service-entrance equipment or the disconnecting means for a separately derived system and extended to the system ground.

250-188. Grounding of Systems Supplying Portable or Mobile Equipment. Systems supplying portable or mobile high-voltage equipment, other than substations installed on a temporary basis, shall comply with (a) through (f).

(a) Portable or Mobile Equipment. Portable or mobile high-voltage equipment shall be supplied from a system having its neutral grounded through an impedance. Where a delta-connected high-voltage system is used to supply portable or mobile equipment, a system neutral shall be derived.

(b) Exposed Noncurrent-Carrying Metal Parts. Exposed noncurrent-carrying metal parts of portable or mobile equipment shall be connected by an equipment grounding conductor to the point at which the system neutral impedance is grounded.

(c) Ground-Fault Current. The voltage developed between the portable or mobile equipment frame and ground by the flow of maximum ground-fault current shall not exceed 100 volts.

(d) Ground-Fault Detection and Relaying. Ground-fault detection and relaying shall be provided to automatically de-energize any high-voltage system component that has developed a ground fault. The continuity of the equipment grounding conductor shall be continuously monitored so as to de-energize automatically the high-voltage circuit to the portable or mobile equipment upon loss of continuity of the equipment grounding conductor.

(e) Isolation. The grounding electrode to which the portable or mobile equipment system neutral impedance is connected shall be isolated from and separated in the ground by at least 20 ft (6.1 m) from any other system or equipment grounding electrode, and there shall be no direct connection between the grounding electrodes, such as buried pipe, fence, etc.

(f) Trailing Cable and Couplers. High-voltage trailing cable and couplers for interconnection of portable or mobile equipment shall meet the requirements of Part C of Article 400 for cables and Section 490-55 for couplers.

250-190. Grounding of Equipment. All noncurrent-carrying metal parts of fixed, portable, and mobile equipment and associated fences, housings, enclosures, and supporting structures shall be grounded.

Exception: Where isolated from ground and located so as to prevent any person who can make contact with ground from contacting such metal parts when the equipment is energized.

Grounding conductors not an integral part of a cable assembly shall not be smaller than No. 6 copper or No. 4 aluminum.

FPN: See Section 250-110, Exception No. 2 for pole-mounted distribution apparatus.

Article 280 — Surge Arresters

A. General

280-1. Scope. This article covers general requirements, installation requirements, and connection requirements for surge arresters installed on premises wiring systems.

280-2. Definition. A *surge arrester* is a protective device for limiting surge voltages by discharging or bypassing surge current, and it also prevents continued flow of follow current while remaining capable of repeating these functions.

280-3. Number Required. Where used at a point on a circuit, a surge arrester shall be connected to each ungrounded conductor. A single installation of such surge arresters shall be permitted to protect a number of interconnected circuits, provided that no circuit is exposed to surges while disconnected from the surge arresters.

280-4. Surge Arrester Selection.

(a) Circuits of Less than 1000 Volts. The rating of the surge arrester shall be equal to or greater than the maximum continuous phase-to-ground power frequency voltage available at the point of application.

Surge arresters installed on circuits of less than 1000 volts shall be listed for the purpose.

(b) Circuits of 1 kV and Over — Silicon Carbide Types. The rating of a silicon carbide-type surge arrester shall be not less than 125 percent of the maximum continuous phase-to-ground voltage available at the point of application.

FPN No. 1: For further information on surge arresters, see *Standard for Gapped Silicon-Carbide Surge Arresters for AC Power Circuits*, ANSI/IEEE C62.1-1989; *Guide for the Application of Gapped Silicon-Carbide Surge Arresters for Alternating-Current Systems*, ANSI/IEEE C62.2-1987; *Standard for Metal-Oxide Surge Arresters for Alternating-Current Power Circuits*, ANSI/IEEE C62.11-1993; and *Guide for the Application of Metal-Oxide Surge Arresters for Alternating-Current Systems*, ANSI/IEEE C62.22-1991.

FPN No. 2: The selection of a properly rated metal oxide arrester is based on considerations of maximum continuous operating voltage and the magnitude and duration of overvoltages at the arrester location as affected by phase-to-ground faults, system grounding techniques, switching surges, and other causes. See the manufacturer's application rules for selection of the specific arrester to be used at a particular location.

B. Installation

280-11. Location. Surge arresters shall be permitted to be located indoors or outdoors. Surge arresters shall be made inaccessible to unqualified persons, unless listed for installation in accessible locations.

280-12. Routing of Surge Arrester Connections. The conductor used to connect the surge arrester to line or bus and to ground shall not be any longer than necessary and shall avoid unnecessary bends.

C. Connecting Surge Arresters

280-21. Installed at Services of Less than 1000 Volts. Line and ground connecting conductors shall not be smaller than No. 14 copper or No. 12 aluminum. The arrester grounding conductor shall be connected to one of the following:

(1) Grounded service conductor
(2) Grounding electrode conductor
(3) Grounding electrode for the service
(4) Equipment grounding terminal in the service equipment

280-22. Installed on the Load Side Services of Less than 1000 Volts. Line and ground connecting conductors shall not be smaller than No. 14 copper or No. 12 aluminum. A surge arrester shall be permitted to be connected between any two conductors—ungrounded conductor(s), grounded conductor, grounding conductor. The grounded conductor and the grounding conductor shall be interconnected only by the normal operation of the surge arrester during a surge.

280-23. Circuits of 1 kV and Over — Surge-Arrester Conductors. The conductor between the surge arrester and the line and the surge arrester and the grounding connection shall not be smaller than No. 6 copper or aluminum.

280-24. Circuits of 1 kV and Over — Interconnections. The grounding conductor of a surge arrester protecting a transformer that supplies a secondary distribution system shall be interconnected as specified in (a), (b), or (c).

(a) Metallic Interconnections. A metallic interconnection shall be made to the secondary grounded circuit conductor or the secondary circuit grounding conductor provided that, in addition to the direct grounding connection at the surge arrester, the following occurs.

(1) The grounded conductor of the secondary has elsewhere a grounding connection to a continuous metal underground water piping system. However, in urban water-pipe areas where there are at least four water-pipe connections on the neutral and not fewer than four such connections in each mile of neutral, the metallic interconnection shall be permitted to be made to the secondary neutral with omission of the direct grounding connection at the surge arrester.

(2) The grounded conductor of the secondary system is a part of a multiground neutral system of which the primary neutral has at least four ground connections in each mile of line in addition to a ground at each service.

(b) Through Spark Gap or Device. Where the surge arrester grounding conductor is not connected as in (a) or where the secondary is not grounded as in (a) but is otherwise grounded as in Sections 250-50 and 250-52, an interconnection shall be made through a spark gap or listed device as follows.

(1) For ungrounded or unigrounded primary systems, the spark gap or listed device shall have a 60-Hz breakdown voltage of at least twice the primary circuit voltage but not necessarily more than 10 kV, and there shall be at least one other ground on the grounded conductor of the secondary that is not less than 20 ft (6.1 m) distant from the surge arrester grounding electrode.

(2) For multigrounded neutral primary systems, the spark gap or listed device shall have a 60-Hz breakdown of not more than 3 kV, and there shall be at least one other ground on the grounded conductor of the secondary that is not less than 20 ft (6.1 m) distant from the surge arrester grounding electrode.

(c) By Special Permission. An interconnection of the surge arrester ground and the secondary neutral, other than as provided in (a) or (b), shall be permitted to be made only by special permission.

280-25. Grounding. Except as indicated in this article, surge arrester grounding connections shall be made as specified in Article 250. Grounding conductors shall not be run in metal enclosures unless bonded to both ends of such enclosure.

CHAPTER 3
Wiring Methods and Materials

Article 300 — Wiring Methods

A. General Requirements

300-1. Scope.

(a) All Wiring Installations. This article covers wiring methods for all wiring installations unless modified by other articles.

(b) Integral Parts of Equipment. The provisions of this article are not intended to apply to the conductors that form an integral part of equipment, such as motors, controllers, motor control centers, or factory assembled control equipment or listed utilization equipment.

300-2. Limitations.

(a) Voltage. Wiring methods specified in Chapter 3 shall be used for voltages 600 volts, nominal, or less where not specifically limited in some section of Chapter 3. They shall be permitted for voltages over 600 volts, nominal, where specifically permitted elsewhere in this *Code*.

(b) Temperature. Temperature limitation of conductors shall be in accordance with Section 310-10.

300-3. Conductors.

(a) Single Conductors. Single conductors specified in Table 310-13 shall only be installed where part of a recognized wiring method of Chapter 3.

(b) Conductors of the Same Circuit. All conductors of the same circuit and, where used, the grounded conductor and all equipment grounding conductors shall be contained within the same raceway, auxiliary gutter, cable tray, trench, cable, or cord, unless otherwise permitted in accordance with (1) through (4).

(1) Paralleled Installations. Conductors shall be permitted to be run in parallel in accordance with the provisions of Section 310-4. The requirement to run all circuit conductors within the same raceway, auxiliary gutter, cable tray, trench, cable, or cord shall apply separately to each portion of the paralleled installation, and the equipment grounding conductors shall comply with the provisions of Section 250-122. Parallel runs in cable tray shall comply with the provisions of Section 318-8(d).

Exception: Conductors installed in nonmetallic raceways run underground shall be permitted to be arranged as isolated phase installations. The raceways shall be installed in close proximity and the conductors shall comply with the provisions of Section 300-20(b).

(2) Grounding Conductors. Equipment grounding conductors shall be permitted to be installed outside a raceway or cable assembly where in accordance with the provisions of Section 250-130(c) for certain existing installations, or in accordance with Section 250-134(b), Exception No. 2, for dc circuits. Equipment bonding conductors shall be permitted to be installed on the outside of raceways in accordance with Section 250-102(e).

(3) Nonferrous Wiring Methods. Conductors in wiring methods with a nonmetallic or other nonmagnetic sheath shall, where run in different raceways, auxiliary gutters, cable trays, trenches, cables, or cords, comply with the provisions of Section 300-20(b). Conductors in single-conductor Type MI cable with a nonmagnetic sheath shall comply with the provisions of Section 330-16.

(4) Enclosures. Where an auxiliary gutter runs between a column-width panelboard and a pull box, and the pull box includes neutral terminations, the neutral conductors of circuits supplied from the panelboard shall be permitted to originate in the pull box.

(c) Conductors of Different Systems.

(1) 600 Volts, Nominal, or Less. Conductors of circuits rated 600 volts, nominal, or less, ac circuits, and dc circuits shall be permitted to occupy the same equipment wiring enclosure, cable, or raceway. All conductors shall have an insulation rating equal to at least the maximum circuit voltage applied to any conductor within the enclosure, cable, or raceway.

Exception: For solar photovoltaic systems in accordance with Section 690-4(b).

FPN: See Section 725-54(a)(1) for Class 2 and Class 3 circuit conductors.

(2) Over 600 Volts, Nominal. Conductors of circuits rated over 600 volts, nominal, shall not occupy the same equipment wiring enclosure, cable, or raceway with conductors of circuits rated 600 volts, nominal, or less unless otherwise permitted in (a) through (e).

(a) Secondary wiring to electric-discharge lamps of 1000 volts or less, if insulated for the secondary voltage involved, shall be permitted to occupy the same fixture, sign, or outline lighting enclosure as the branch-circuit conductors.

(b) Primary leads of electric-discharge lamp ballasts, insulated for the primary voltage of the ballast, where contained within the individual wiring enclosure, shall be permitted to occupy the same fixture, sign, or outline lighting enclosure as the branch-circuit conductors.

(c) Excitation, control, relay, and ammeter conductors used in connection with any individual motor or starter shall be permitted to occupy the same enclosure as the motor-circuit conductors.

(d) In motors, switchgear and control assemblies, and similar equipment, conductors of different voltage ratings shall be permitted.

(e) In manholes, if the conductors of each system are permanently and effectively separated from the conductors of the other systems and securely fastened to racks, insulators, or other approved supports, conductors of different voltage ratings shall be permitted.

Conductors having nonshielded insulation and operating at different voltage levels shall not occupy the same enclosure, cable, or raceway.

300-4. Protection Against Physical Damage. Where subject to physical damage, conductors shall be adequately protected.

(a) Cables and Raceways Through Wood Members.

(1) Bored Holes. In both exposed and concealed locations, where a cable or raceway-type wiring method is installed through bored holes in joists, rafters, or wood members, holes shall be bored so that the edge of the hole is not less than 1¼ in. (31.8 mm) from the nearest edge of the wood member. Where this distance cannot be maintained, the cable or raceway shall be protected from penetration by screws or nails by a steel plate or bushing, at least 1/16 in. (1.59 mm) thick, and of appropriate length and width installed to cover the area of the wiring.

Exception: Steel plates shall not be required to protect rigid metal conduit, intermediate metal conduit, rigid nonmetallic conduit, or electrical metallic tubing.

(2) Notches in Wood. Where there is no objection because of weakening the building structure, in both exposed and concealed locations, cables or raceways shall be permitted to be laid in notches in wood studs, joists, rafters, or other wood members where the cable or raceway at those points is protected against nails or screws by a steel plate at least 1/16 in. (1.59 mm) thick installed before the building finish is applied.

Exception: Steel plates shall not be required to protect rigid metal conduit, intermediate metal conduit, rigid nonmetallic conduit, or electrical metallic tubing.

(b) Nonmetallic-Sheathed Cables and Electrical Non-metallic Tubing Through Metal Framing Members.

(1) Nonmetallic-Sheathed Cable. In both exposed and concealed locations where nonmetallic-sheathed cables pass through either factory or field punched, cut, or drilled slots or holes in metal members, the cable shall be protected by bushings or grommets covering all metal edges and securely fastened in the opening prior to installation of the cable.

(2) Nonmetallic-Sheathed Cable and Electrical Non-metallic Tubing. Where nails or screws are likely to penetrate nonmetallic-sheathed cable or electrical nonmetallic tubing, a steel sleeve, steel plate, or steel clip not less than 1/16 in. (1.59 mm) in thickness shall be used to protect the cable or tubing.

(c) Cables Through Spaces Behind Panels Designed to Allow Access. Cables or raceway-type wiring methods, installed behind panels designed to allow access, shall be supported according to their applicable articles.

(d) Cables and Raceways Parallel to Framing Members. In both exposed and concealed locations, where a cable- or raceway-type wiring method is installed parallel to framing members, such as joists, rafters, or studs, the cable or raceway shall be installed and supported so that the nearest outside surface of the cable or raceway is not less than 1¼ in. (31.8 mm) from the nearest edge of the framing member where nails or screws are likely to penetrate. Where this distance cannot be maintained, the cable or raceway shall be protected from penetration by nails or screws by a steel plate, sleeve, or equivalent at least 1/16 in. (1.59 mm) thick.

Exception No. 1: Steel plates, sleeves, or the equivalent shall not be required to protect rigid metal conduit, intermediate metal conduit, rigid nonmetallic conduit, or electrical metallic tubing.

Exception No. 2: For concealed work in finished buildings, or finished panels for prefabricated buildings where such supporting is impracticable, it shall be permissible to fish the cables between access points.

Exception No. 3: Steel plates, sleeves, or the equivalent shall not be required to protect cables or raceways in mobile homes and recreational vehicles.

(e) Cables and Raceways Installed in Shallow Grooves. Cable- or raceway-type wiring methods installed in a groove, to be covered by wallboard, siding, paneling, carpeting, or similar finish, shall be protected by 1/16 in. (1.59 mm) thick steel plate, sleeve, or equivalent or by not less than 1¼ in. (31.8 mm) free space for the full length of the groove in which the cable or raceway is installed.

Exception: Steel plates, sleeves, or the equivalent shall not be required to protect rigid metal conduit, intermediate metal conduit, rigid nonmetallic conduit, or electrical metallic tubing.

(f) Insulated Fittings. Where raceways containing ungrounded conductors No. 4 or larger enter a cabinet, box enclosure, or raceway, the conductors shall be protected by a substantial fitting providing a smoothly rounded insulating surface, unless the conductors are separated from the fitting or raceway by substantial insulating material that is securely fastened in place.

Exception: Where threaded hubs or bosses that are an integral part of a cabinet, box enclosure, or raceway provide a smoothly rounded or flared entry for conductors.

Conduit bushings constructed wholly of insulating material shall not be used to secure a fitting or raceway. The insulating fitting or insulating material shall have a temperature rating not less than the insulation temperature rating of the installed conductors.

300-5. Underground Installations.

(a) Minimum Cover Requirements. Direct-buried cable or conduit or other raceways shall be installed to meet the minimum cover requirements of Table 300-5.

(b) Grounding. All underground installations shall be grounded and bonded in accordance with Article 250.

(c) Underground Cables Under Buildings. Underground cable installed under a building shall be in a raceway that is extended beyond the outside walls of the building.

(d) Protection from Damage. Direct-buried conductors and cables emerging from the ground shall be protected by enclosures or raceways extending from the minimum cover distance required by Section 300-5(a) below grade to a point at least 8 ft (2.44 m) above finished grade. In no case shall the protection be required to exceed 18 in. (457 mm) below finished grade.

Service laterals that are not encased in concrete and that are buried 18 in. (457 mm) or more below grade shall have their location identified by a warning ribbon that is placed in the trench at least 12 in. (305 mm) above the underground installation.

Conductors entering a building shall be protected to the point of entrance.

Where the enclosure or raceway is subject to physical damage, the conductors shall be installed in rigid metal conduit, intermediate metal conduit, Schedule 80 rigid nonmetallic conduit, or equivalent.

(e) Splices and Taps. Direct-buried conductors or cables shall be permitted to be spliced or tapped without the use of splice boxes. The splices or taps shall be made in accordance with Section 110-14(b).

(f) Backfill. Backfill that contains large rocks, paving materials, cinders, large or sharply angular substances, or corrosive material shall not be placed in an excavation where materials may damage raceways, cables, or other substructures or prevent adequate compaction of fill or contribute to corrosion of raceways, cables, or other substructures.

Where necessary to prevent physical damage to the raceway or cable, protection shall be provided in the form of granular or selected material, suitable running boards, suitable sleeves, or other approved means.

(g) Raceway Seals. Conduits or raceways through which moisture may contact energized live parts shall be sealed or plugged at either or both ends.

FPN: Presence of hazardous gases or vapors may also necessitate sealing of underground conduits or raceways entering buildings.

(h) Bushing. A bushing, or terminal fitting, with an integral bushed opening shall be used at the end of a conduit or other raceway that terminates underground where the conductors or cables emerge as a direct burial wiring method. A seal incorporating the physical protection characteristics of a bushing shall be permitted to be used in lieu of a bushing.

(i) Conductors of the Same Circuit. All conductors of the same circuit and, where used, the grounded conductor and all equipment grounding conductors shall be installed in the same raceway or shall be installed in close proximity in the same trench.

Exception No. 1: Conductors in parallel in raceways shall be permitted, but each raceway shall contain all conductors of the same circuit including grounding conductors.

Exception No. 2: Isolated phase installations shall be permitted in nonmetallic raceways in close proximity where conductors are paralleled as permitted in Section 310-4, and where the conditions of Section 300-20 are met.

(j) Ground Movement. Where direct-buried conductors, raceways, or cables are subject to movement by settlement or frost, direct-buried conductors, raceways, or cables shall be arranged to prevent damage to the enclosed conductors or to equipment connected to the raceways.

FPN: This section recognizes "S" loops in underground direct burial to raceway transitions, expansion joints in raceway risers to fixed equipment, and, generally, the provision of flexible connections to equipment subject to settlement or frost heaves.

300-6. Protection Against Corrosion. Metal raceways, cable trays, cablebus, auxiliary gutters, cable armor, boxes, cable sheathing, cabinets, elbows, couplings, fittings, supports, and support hardware shall be of materials suitable for the environment in which they are to be installed.

(a) General. Ferrous raceways, cable trays, cablebus, auxiliary gutters, cable armor, boxes, cable sheathing, cabinets, metal elbows, couplings, fittings, supports, and support hardware shall be suitably protected against corrosion inside and

Table 300-5. Minimum Cover Requirements, 0 to 600 Volts, Nominal, Burial in Inches (Cover is defined as the shortest distance in inches measured between a point on the top surface of any direct-buried conductor, cable, conduit, or other raceway and the top surface of finished grade, concrete, or similar cover.)

Location of Wiring Method or Circuit	Column 1 Direct Burial Cables or Conductors	Column 2 Rigid Metal Conduit or Intermediate Metal Conduit	Column 3 Nonmetallic Raceways Listed for Direct Burial Without Concrete Encasement or Other Approved Raceways	Column 4 Residential Branch Circuits Rated 120 Volts or Less with GFCI Protection and Maximum Overcurrent Protection of 20 Amperes	Column 5 Circuits for Control of Irrigation and Landscape Lighting Limited to Not More than 30 Volts and Installed with Type UF or in Other Identified Cable or Raceway
			Type of Wiring Method or Circuit		
All locations not specified below	24	6	18	12	6
In trench below 2-in. thick concrete or equivalent	18	6	12	6	6
Under a building	0 (in raceway only)	0	0	0 (in raceway only)	0 (in raceway only)
Under minimum of 4-in. thick concrete exterior slab with no vehicular traffic and the slab extending not less than 6 in. beyond the underground installation	18	4	4	6 (direct burial) 4 (in raceway)	6 (direct burial) 4 (in raceway)
Under streets, highways, roads, alleys, driveways, and parking lots	24	24	24	24	24
One- and two-family dwelling driveways and outdoor parking areas, and used only for dwelling-related purposes	18	18	18	12	18
In or under airport runways, including adjacent areas where trespassing prohibited	18	18	18	18	18

Notes:

1. For SI units, 1 in. = 25.4 mm.

2. Raceways approved for burial only where concrete encased shall require concrete envelope not less than 2 in. thick.

3. Lesser depths shall be permitted where cables and conductors rise for terminations or splices or where access is otherwise required.

4. Where one of the wiring method types listed in Columns 1–3 is used for one of the circuit types in Columns 4 and 5, the shallower depth of burial shall be permitted.

5. Where solid rock prevents compliance with the cover depths specified in this table, the wiring shall be installed in metal or nonmetallic raceway permitted for direct burial. The raceways shall be covered by a minimum of 2 in. of concrete extending down to rock.

outside (except threads at joints) by a coating of approved corrosion-resistant material such as zinc, cadmium, or enamel. Where protected from corrosion solely by enamel, they shall not be used outdoors or in wet locations as described in (c). Where boxes or cabinets have an approved system of organic coatings and are marked "Raintight," "Rainproof," or "Outdoor Type," they shall be permitted outdoors.

Exception: Threads at joints shall be permitted to be coated with an identified electrically conductive compound.

(b) In Concrete or in Direct Contact with the Earth. Ferrous or nonferrous metal raceways, cable armor, boxes, cable sheathing, cabinets, elbows, couplings, fittings, supports, and support hardware shall be permitted to be installed in concrete or in direct contact with the earth, or in areas subject to severe corrosive influences where made of material judged suitable for the condition, or where provided with corrosion protection approved for the condition.

(c) Indoor Wet Locations. In portions of dairies, laundries, canneries, and other indoor wet locations, and in locations where walls are frequently washed or where there are surfaces of absorbent materials, such as damp paper or wood, the entire wiring system, where installed exposed, including all boxes, fittings, conduits, and cable used therewith, shall be mounted so that there is at least a ¼-in. (6.35-mm) airspace between it and the wall or supporting surface.

Exception: Nonmetallic raceways, boxes, and fittings shall be permitted to be installed without the airspace on a concrete, masonry, tile, or similar surface.

> FPN: In general, areas where acids and alkali chemicals are handled and stored may present such corrosive conditions, particularly when wet or damp. Severe corrosive conditions may also be present in portions of meatpacking plants, tanneries, glue houses, and some stables; installations immediately adjacent to a seashore and swimming pool areas; areas where chemical deicers are used; and storage cellars or rooms for hides, casings, fertilizer, salt, and bulk chemicals.

300-7. Raceways Exposed to Different Temperatures.

(a) Sealing. Where portions of an interior raceway system are exposed to widely different temperatures, as in refrigerating or cold-storage plants, circulation of air from a warmer to a colder section through the raceway shall be prevented.

(b) Expansion Joints. Raceways shall be provided with expansion joints where necessary to compensate for thermal expansion and contraction.

> FPN: Table 347-9 provides the expansion information for polyvinyl chloride (PVC). A nominal number for steel conduit can be determined by multiplying the expansion length in this table by 0.20. The coefficient of expansion for steel electrical metallic tubing, intermediate metal conduit, and

rigid conduit is 6.50×10^{-6} (0.0000065 in. per inch of conduit for each °F in temperature change).

300-8. Installation of Conductors with Other Systems. Raceways or cable trays containing electric conductors shall not contain any pipe, tube, or equal for steam, water, air, gas, drainage, or any service other than electrical.

300-10. Electrical Continuity of Metal Raceways and Enclosures. Metal raceways, cable armor, and other metal enclosures for conductors shall be metallically joined together into a continuous electric conductor and shall be connected to all boxes, fittings, and cabinets so as to provide effective electrical continuity. Unless specifically permitted elsewhere in this *Code*, raceways and cable assemblies shall be mechanically secured to boxes, fittings, cabinets, and other enclosures.

Exception: Short sections of raceways used to provide support or protection of cable assemblies from physical damage shall not be required to be made electrically continuous.

300-11. Securing and Supporting.

(a) Secured in Place. Raceways, cable assemblies, boxes, cabinets, and fittings shall be securely fastened in place. Support wires that do not provide secure support shall not be permitted as the sole support. Support wires and associated fittings that provide secure support and that are installed in addition to the ceiling grid support wires, shall be permitted as the sole support. Where independent support wires are used, they shall be secured at both ends. Cables and raceways shall not be supported by ceiling grids.

(1) Wiring located within the cavity of a fire-rated floor–ceiling or roof–ceiling assembly shall not be secured to, or supported by, the ceiling assembly, including the ceiling support wires. An independent means of secure support shall be provided. Where independent support wires are used, they shall be distinguishable by color, tagging, or other effective means from those that are part of the fire-rated design.

Exception: The ceiling support system shall be permitted to support wiring and equipment that have been tested as part of the fire-rated assembly.

> FPN: One method of determining fire rating is testing in accordance with *Standard Methods of Tests of Fire Endurance of Building Construction and Materials*, NFPA 251-1995.

(2) Wiring located within the cavity of a nonfire-rated floor–ceiling or roof–ceiling assembly shall not be secured to, or supported by, the ceiling assembly, including the ceiling support wires. An independent means of secure support shall be provided.

Exception: The ceiling support system shall be permitted to support branch-circuit wiring and associated equipment where installed in accordance with the ceiling system manufacturer's instructions.

(b) Raceways Used as Means of Support. Raceways shall only be used as a means of support for other raceways, cables, or nonelectric equipment under the following conditions:

(1) Where the raceway or means of support is identified for the purpose; or

(2) Where the raceway contains power supply conductors for electrically controlled equipment and is used to support Class 2 circuit conductors or cables that are solely for the purpose of connection to the equipment control circuits; or

(3) Where the raceway is used to support boxes or conduit-bodies in accordance with Section 370-23 or to support fixtures in accordance with Section 410-16(f)

300-12. Mechanical Continuity — Raceways and Cables. Metal or nonmetallic raceways, cable armors, and cable sheaths shall be continuous between cabinets, boxes, fittings, or other enclosures or outlets.

Exception: Short sections of raceways used to provide support or protection of cable assemblies from physical damage shall not be required to be mechanically continuous.

300-13. Mechanical and Electrical Continuity — Conductors.

(a) General. Conductors in raceways shall be continuous between outlets, boxes, devices, etc. There shall be no splice or tap within a raceway unless permitted by Sections 300-15; 352-7; 352-29; 354-6, Exception; 362-7; 362-21; or 364-8(a).

(b) Device Removal. In multiwire branch circuits, the continuity of a grounded conductor shall not depend on device connections such as lampholders, receptacles, etc., where the removal of such devices would interrupt the continuity.

300-14. Length of Free Conductors at Outlets, Junctions, and Switch Points. At least 6 in. (152 mm) of free conductor, measured from the point in the box where it emerges from its raceway or cable sheath, shall be left at each outlet, junction, and switch point for splices or the connection of fixtures or devices. Where the opening to an outlet, junction, or switch point is less than 8 in. (203 mm) in any dimension, each conductor shall be long enough to extend at least 3 in. (76.2 mm) outside the opening.

Exception: Conductors that are not spliced or terminated at the outlet, junction, or switch point.

300-15. Boxes, Conduit Bodies, or Fittings — Where Required.

(a) Box or Conduit Body. Where the wiring method is conduit, electrical metallic tubing, Type AC cable, Type MC cable, Type MI cable, nonmetallic-sheathed cable, or other cables, a box or conduit body complying with Article 370 shall be installed at each conductor splice point, outlet, switch point, junction point, or pull point, unless otherwise permitted in (b) through (n). A box shall be installed at each outlet and switch point for concealed knob-and-tube wiring.

Fittings and connectors shall be used only with the specific wiring methods for which they are designed and listed.

(b) Equipment. An integral junction box or wiring compartment as part of approved equipment shall be permitted.

(c) Protection. Where cables enter or exit from conduit or tubing that is used to provide cable support or protection against physical damage. A fitting shall be provided on the end(s) of the conduit or tubing to protect the cable from abrasion.

(d) Type MI Cable. Where accessible fittings are used for straight-through splices in mineral-insulated metal-sheathed cable.

(e) Integral Enclosure. A wiring device with integral enclosure identified for the use, having brackets that securely fasten the device to walls or ceilings of conventional on-site frame construction, for use with nonmetallic-sheathed cable, shall be permitted in lieu of a box or conduit body.

    FPN: See Sections 336-18, Exception No. 2; 545-10; 550-10(i); and 551-47(e), Exception No. 1.

(f) Fitting. A fitting identified for the use shall be permitted in lieu of a box or conduit body where accessible after installation and not containing spliced or terminated conductors.

(g) Buried Conductors. As permitted in Section 300-5(e) for splices and taps in buried conductors and cables.

(h) Insulated Devices. As permitted in Section 336-21 for insulated devices supplied by nonmetallic-sheathed cable.

(i) Enclosures. As permitted in Section 373-8 for switches and overcurrent devices, and Section 430-10(a) for motor controllers.

(j) Fixtures. As permitted in Section 410-31 where a fixture is used as a raceway.

(k) Embedded. Where conductors are embedded as covered in Sections 424-40, 424-41(d), 426-22(b), 426-24(a), and 427-19(a).

(l) Manufactured Wiring System. Where manufactured wiring systems in accordance with Article 604 are used.

(m) Closed Loop. Where a device identified and listed as suitable for installation without a box is used with a closed-loop power distribution system.

(n) Manholes. A box or conduit body shall not be required for conductors in manholes where accessible only to qualified persons.

300-16. Raceway or Cable to Open or Concealed Wiring.

(a) Box or Fitting. A box or terminal fitting having a separately bushed hole for each conductor shall be used wherever a change is made from conduit, electrical metallic tubing, electrical nonmetallic tubing, nonmetallic-sheathed cable, Type AC cable, Type MC cable, or mineral-insulated, metal-sheathed cable and surface raceway wiring to open wiring or to concealed knob-and-tube wiring. A fitting used for this purpose shall contain no taps or splices and shall not be used at fixture outlets.

(b) Bushing. A bushing shall be permitted in lieu of a box or terminal fitting where the conductors emerge from a raceway or conduit and enter or terminate at equipment, such as open switchboards, unenclosed control equipment, or similar equipment. The bushing shall be of the insulating type for other than lead-sheathed conductors.

300-17. Number and Size of Conductors in Raceway.
The number and size of conductors in any raceway shall not be more than will permit dissipation of the heat and ready installation or withdrawal of the conductors without damage to the conductors or to their insulation.

> FPN: See the following sections of this *Code*: electrical nonmetallic tubing, Section 331-6; intermediate metal conduit, Section 345-7; rigid metal conduit, Section 346-7; rigid nonmetallic conduit, Section 347-11; electrical metallic tubing, Section 348-8; flexible metallic tubing, Section 349-12; flexible metal conduit, Section 350-12; liquidtight flexible metal conduit, Section 351-6; liquidtight nonmetallic flexible conduit, Section 351-25; surface raceways, Sections 352-4 and 352-25; underfloor raceways, Section 354-5; cellular metal floor raceways, Section 356-5; cellular concrete floor raceways, Section 358-11; wireways, Section 362-5; fixture wire, Section 402-7; theaters, Section 520-6; signs, Section 600-31(c); elevators, Section 620-33; audio signal processing, amplification, and reproduction equipment, Sections 640-23(a) and 640-24; Class 1, Class 2, and Class 3 circuits, Article 725; fire alarm circuits, Article 760, and optical fiber cables and raceways, Article 770.

300-18. Raceway Installations.

(a) Complete Runs. Raceways, other than busways or exposed raceways having hinged or removable covers, shall be installed complete between outlet, junction, or splicing points prior to the installation of conductors. Where required to facilitate the installation of utilization equipment, the raceway shall be permitted to be initially installed without a terminating connection at the equipment. Prewired raceway assemblies shall be permitted only where specifically permitted in this *Code* for the applicable wiring method.

(b) Welding. Metal raceways shall not be supported, terminated, or connected by welding to the raceway unless specifically designed to be or otherwise specifically permitted to be in this *Code*.

300-19. Supporting Conductors in Vertical Raceways.

(a) Spacing Intervals — Maximum. Conductors in vertical raceways shall be supported if the vertical rise exceeds the values in Table 300-19(a). One cable support shall be provided at the top of the vertical raceway or as close to the top as practical. Intermediate supports shall be provided as necessary to limit supported conductor lengths to not greater than those values specified in Table 300-19(a).

Exception: Steel wire armor cable shall be supported at the top of the riser with a cable support that clamps the steel wire armor. A safety device shall be permitted at the lower end of the riser to hold the cable in the event there is slippage of the cable in the wire-armored cable support. Additional

Table 300-19(a). Spacings for Conductor Supports

Size of Wire	Support of Conductors in Vertical Raceways	Conductors	
		Aluminum or Copper-Clad Aluminum	Copper
18 AWG through 8 AWG	Not greater than	100 ft	100 ft
6 AWG through 1/0 AWG	Not greater than	200 ft	100 ft
2/0 AWG through 4/0 AWG	Not greater than	180 ft	80 ft
Over 4/0 AWG through 350 kcmil	Not greater than	135 ft	60 ft
Over 350 kcmil through 500 kcmil	Not greater than	120 ft	50 ft
Over 500 kcmil through 750 kcmil	Not greater than	95 ft	40 ft
Over 750 kcmil	Not greater than	85 ft	35 ft

Note: For SI units, 1 ft = 0.3048 m.

wedge-type supports shall be permitted to relieve the strain on the equipment terminals caused by expansion of the cable under load.

(b) Support Methods. One of the following methods of support shall be used.

(1) By clamping devices constructed of or employing insulating wedges inserted in the ends of the raceways. Where clamping of insulation does not adequately support the cable, the conductor also shall be clamped.

(2) By inserting boxes at the required intervals in which insulating supports are installed and secured in a satisfactory manner to withstand the weight of the conductors attached thereto, the boxes being provided with covers.

(3) In junction boxes, by deflecting the cables not less than 90 degrees and carrying them horizontally to a distance not less than twice the diameter of the cable, the cables being carried on two or more insulating supports and additionally secured thereto by tie wires if desired. Where this method is used, cables shall be supported at intervals not greater than 20 percent of those mentioned in the preceding tabulation.

(4) By a method of equal effectiveness.

300-20. Induced Currents in Metal Enclosures or Metal Raceways.

(a) Conductors Grouped Together. Where conductors carrying alternating current are installed in metal enclosures or metal raceways, they shall be arranged so as to avoid heating the surrounding metal by induction. To accomplish this, all phase conductors and, where used, the grounded conductor and all equipment grounding conductors shall be grouped together.

Exception No. 1: Equipment grounding conductors for certain existing installations shall be permitted to be installed separate from their associated circuit conductors where run in accordance with the provisions of Section 250-130(c).

Exception No. 2: A single conductor shall be permitted to be installed in a ferromagnetic enclosure and used for skin-effect heating in accordance with the provisions of Sections 426-42 and 427-47.

(b) Individual Conductors. Where a single conductor carrying alternating current passes through metal with magnetic properties, the inductive effect shall be minimized by (1) cutting slots in the metal between the individual holes through which the individual conductors pass, or (2) passing all the conductors in the circuit through an insulating wall sufficiently large for all of the conductors of the circuit.

Exception: In the case of circuits supplying vacuum or electric-discharge lighting systems or signs, or X-ray apparatus, the currents carried by the conductors are so small that the inductive heating effect can be ignored where these conductors are placed in metal enclosures or pass through metal.

> FPN: Because aluminum is not a magnetic metal, there will be no heating due to hysteresis; however, induced currents will be present. They will not be of sufficient magnitude to require grouping of conductors or special treatment in passing conductors through aluminum wall sections.

300-21. Spread of Fire or Products of Combustion. Electrical installations in hollow spaces, vertical shafts, and ventilation or air-handling ducts shall be made so that the possible spread of fire or products of combustion will not be substantially increased. Openings around electrical penetrations through fire-resistant-rated walls, partitions, floors, or ceilings shall be firestopped using approved methods to maintain the fire resistance rating.

> FPN: Directories of electrical construction materials published by qualified testing laboratories contain many listing installation restrictions necessary to maintain the fire-resistive rating of assemblies where penetrations or openings are made. An example is the 24-in. (610-mm) minimum horizontal separation that usually applies between boxes on opposite sides of the wall. Assistance in complying with Section 300-21 can be found in these directories and product listings.

300-22. Wiring in Ducts, Plenums, and Other Air-Handling Spaces. The provisions of this section apply to the installation and uses of electric wiring and equipment in ducts, plenums, and other air-handling spaces.

> FPN: Article 424, Part F.

(a) Ducts for Dust, Loose Stock, or Vapor Removal. No wiring systems of any type shall be installed in ducts used to transport dust, loose stock, or flammable vapors. No wiring system of any type shall be installed in any duct, or shaft containing only such ducts, used for vapor removal or for ventilation of commercial-type cooking equipment.

(b) Ducts or Plenums Used for Environmental Air. Only wiring methods consisting of Type MI cable, Type MC cable employing a smooth or corrugated impervious metal sheath without an overall nonmetallic covering, electrical metallic tubing, flexible metallic tubing, intermediate metal conduit, or rigid metal conduit shall be installed in ducts or plenums specifically fabricated to transport environmental air. Flexible metal conduit and liquidtight flexible metal conduit shall be permitted, in lengths not to exceed 4 ft (1.22 m), to connect physically adjustable equipment and devices permitted to be in these ducts and plenum chambers. The connectors used with flexible metal conduit shall effectively close any openings in the connection. Equipment and devices shall be permitted within such ducts or plenum chambers only if necessary for their direct action upon, or sensing of, the contained air. Where equipment or devices are installed and illumination is necessary to facilitate maintenance and repair, enclosed gasketed-type fixtures shall be permitted.

(c) Other Space Used for Environmental Air. This section applies to space used for environmental air-handling purposes other than ducts and plenums as specified in (a) and (b). It does not include habitable rooms or areas of buildings, the prime purpose of which is not air handling.

> FPN: The space over a hung ceiling used for environmental air-handling purposes is an example of the type of other space to which this section applies.

Exception: This section shall not apply to the joist or stud spaces of dwelling units where the wiring passes through such spaces perpendicular to the long dimension of such spaces.

(1) Wiring Methods. The wiring methods for such other space shall be limited to totally enclosed, nonventilated, insulated busway having no provisions for plug-in connections, Type MI cable, Type MC cable without an overall nonmetallic covering, Type AC cable, or other factory-assembled multiconductor control or power cable that is specifically listed for the use, or listed prefabricated cable assemblies of metallic manufactured wiring systems without nonmetallic sheath. Other type cables and conductors shall be installed in electrical metallic tubing, flexible metallic tubing, intermediate metal conduit, rigid metal conduit, flexible metal conduit, or, where accessible, surface metal raceway or metal wireway with metal covers or solid bottom metal cable tray with solid metal covers.

Exception: Liquidtight flexible metal conduit shall be permitted in single lengths not exceeding 6 ft (1.83 m).

(2) Equipment. Electrical equipment with a metal enclosure, or with a nonmetallic enclosure listed for the use and having adequate fire-resistant and low-smoke-producing characteristics, and associated wiring material suitable for the ambient temperature shall be permitted to be installed in such other space unless prohibited elsewhere in this *Code*.

Exception: Integral fan systems shall be permitted where specifically identified for such use.

(d) Information Technology Equipment. Electric wiring in air-handling areas beneath raised floors for information technology equipment shall be permitted in accordance with Article 645.

300-23. Panels Designed to Allow Access. Cables, raceways, and equipment installed behind panels designed to allow access, including suspended ceiling panels, shall be arranged and secured so as to allow the removal of panels and access to the equipment.

B. Requirements for Over 600 Volts, Nominal

300-31. Covers Required. Suitable covers shall be installed on all boxes, fittings, and similar enclosures to prevent accidental contact with energized parts or physical damage to parts or insulation.

300-32. Conductors of Different Systems. See Section 300-3(c)(2).

300-34. Conductor Bending Radius. The conductor shall not be bent to a radius less than 8 times the overall diameter for nonshielded conductors or 12 times the diameter for shielded or lead-covered conductors during or after installation. For multiconductor or multiplexed single conductor cables having individually shielded conductors, the minimum bending radius is 12 times the diameter of the individually shielded conductors or 7 times the overall diameter, whichever is greater.

300-35. Protection Against Induction Heating. Metallic raceways and associated conductors shall be arranged so as to avoid heating of the raceway in accordance with the applicable provisions of Section 300-20.

300-37. Aboveground Wiring Methods. Aboveground conductors shall be installed in rigid metal conduit, in intermediate metal conduit, in electrical metallic tubing, in rigid nonmetallic conduit, in cable trays, as busways, as cablebus, in other identified raceways, or as open runs of metal-clad cable suitable for the use and purpose. In locations accessible to qualified persons only, open runs of Type MV cables, bare conductors, and bare busbars shall also be permitted. Busbars shall be permitted to be either copper or aluminum.

300-39. Braid-Covered Insulated Conductors — Open Installation. Open runs of braid-covered insulated conductors shall have a flame-retardant braid. If the conductors used do not have this protection, a flame-retardant saturant shall be applied to the braid covering after installation. This treated braid covering shall be stripped back a safe distance at conductor terminals, according to the operating voltage. This distance shall not be less than 1 in. (25.4 mm) for each kilovolt of the conductor-to-ground voltage of the circuit, where practicable.

300-40. Insulation Shielding. Metallic and semiconducting insulation shielding components of shielded cables shall be removed for a distance dependent on the circuit voltage and insulation. Stress reduction means shall be provided at all terminations of factory-applied shielding.

Metallic shielding components such as tapes, wires, or

braids, or combinations thereof, and their associated conducting or semiconducting components shall be grounded.

300-42. Moisture or Mechanical Protection for Metal-Sheathed Cables.
Where cable conductors emerge from a metal sheath and where protection against moisture or physical damage is necessary, the insulation of the conductors shall be protected by a cable sheath terminating device.

300-50. Underground Installations.

(a) **General.** Underground conductors shall be identified for the voltage and conditions under which they are installed. Direct burial cables shall comply with the provisions of Section 310-7. Underground cables shall be installed in accordance with (1) or (2), and the installation shall meet the depth requirements of Table 300-50.

(1) **Shielded Cables and Nonshielded Cables in Metal-Sheathed Cable Assemblies.** Underground cables, including nonshielded, Type MC and moisture-impervious metal sheath cables shall have those sheaths grounded through an effective grounding path meeting the requirements of Section 250-2(d). They shall be direct buried or installed in raceways identified for the use.

(2) **Other Nonshielded Cables.** Other nonshielded cables not covered in (1) shall be installed in rigid metal conduit, intermediate metal conduit, or rigid nonmetallic conduit encased in not less than 3 in. (76 mm) of concrete.

(b) **Protection from Damage.** Conductors emerging from the ground shall be enclosed in listed raceways. Raceways installed on poles shall be of rigid metal conduit, intermediate metal conduit, PVC Schedule 80, or equivalent, extending from the minimum cover depth specified in Table 300-50 to a point 8 ft (2.44 m) above finished grade. Conductors entering a building shall be protected by an approved enclosure or raceway from the minimum cover depth to the point of entrance. Where direct-buried conductors, raceways, or cables are subject to movement by settlement or frost, they shall be installed to prevent damage to the enclosed conductors or to the equipment connected to the raceways. Metallic enclosures shall be grounded.

(c) **Splices.** Direct burial cables shall be permitted to be spliced or tapped without the use of splice boxes, provided they are installed using materials suitable for the application. The taps and splices shall be watertight and protected from mechanical damage. Where cables are shielded, the shielding shall be continuous across the splice or tap.

Exception: At splices of an engineered cabling system, metallic shields of direct-buried single-conductor cables with maintained spacing between phases shall be permitted to be interrupted and overlapped. Where shields are inter-

Table 300-50. Minimum Cover Requirements (*Cover* is defined as the shortest distance in inches measured between a point on the top surface of any direct-buried conductor, cable, conduit, or other raceway and the top surface of finished grade, concrete, or similar cover.)

Circuit Voltage	Direct-Buried Cables	Rigid Nonmetallic Conduit Approved for Direct Burial*	Rigid Metal Conduit and Intermediate Metal Conduit
Over 600 V through 22 kV	30	18	6
Over 22 kV through 40 kV	36	24	6
Over 40 kV	42	30	6

Note: For SI units, 1 in. = 25.4 mm.
*Listed by a qualified testing agency as suitable for direct burial without encasement. All other nonmetallic systems shall require 2 in. (50.8 mm) of concrete or equivalent above conduit in addition to above depth.

rupted and overlapped, each shield section shall be grounded at one point.

(d) **Backfill.** Backfill containing large rocks, paving materials, cinders, large or sharply angular substances, or corrosive materials shall not be placed in an excavation where materials can damage raceways, cables, or other substructures, or prevent adequate compaction of fill, or contribute to corrosion of raceways, cables, or other substructures.

Protection in the form of granular or selected material or suitable sleeves shall be provided to prevent physical damage to the raceway or cable.

(e) **Raceway Seal.** Where a raceway enters from an underground system, the end within the building shall be sealed with an identified compound so as to prevent the entrance of moisture or gases, or it shall be so arranged to prevent moisture from contacting live parts.

Exception No. 1: Areas subject to vehicular traffic, such as thoroughfares or commercial parking areas, shall have a minimum cover of 24 in. (610 mm).

Exception No. 2: The minimum cover requirements for other than rigid metal conduit and intermediate metal conduit shall be permitted to be reduced 6 in. (152 mm) for each 2 in. (50.8 mm) of concrete or equivalent protection placed in the trench over the underground installation.

Exception No. 3: The minimum cover requirements shall not apply to conduits or other raceways that are located under a building or exterior concrete slab not less than 4 in. (102 mm) in thickness and extending not less than 6 in. (152 mm) beyond the underground installation. A warning ribbon or other effective means suitable for the conditions shall be placed above the underground installation.

Exception No. 4: Lesser depths shall be permitted where cables and conductors rise for terminations or splices or where access is otherwise required.

Exception No. 5: In airport runways, including adjacent defined areas where trespass is prohibited, cable shall be permitted to be buried not less than 18 in. (457 mm) deep and without raceways, concrete enclosement, or equivalent.

Exception No. 6: Raceways installed in solid rock shall be permitted to be buried at lesser depth where covered by 2 in. (50.8 mm) of concrete, which shall be permitted to extend to the rock surface.

Article 305 — Temporary Wiring

305-1. Scope. The provisions of this article apply to temporary electrical power and lighting wiring methods that may be of a class less than would be required for a permanent installation.

305-2. All Wiring Installations.

(a) Other Articles. Except as specifically modified in this article, all other requirements of this *Code* for permanent wiring shall apply to temporary wiring installations.

(b) Approval. Temporary wiring methods shall be acceptable only if approved based on the conditions of use and any special requirements of the temporary installation.

305-3. Time Constraints.

(a) During the Period of Construction. Temporary electrical power and lighting installations shall be permitted during the period of construction, remodeling, maintenance, repair, or demolition of buildings, structures, equipment, or similar activities.

(b) 90 Days. Temporary electrical power and lighting installations shall be permitted for a period not to exceed 90 days for Christmas decorative lighting and similar purposes.

(c) Emergencies and Tests. Temporary electrical power and lighting installations shall be permitted during emergencies and for tests, experiments, and developmental work.

(d) Removal. Temporary wiring shall be removed immediately upon completion of construction or purpose for which the wiring was installed.

305-4. General.

(a) Services. Services shall be installed in conformance with Article 230.

(b) Feeders. Feeders shall be protected as provided in Article 240. They shall originate in an approved distribution center. Conductors shall be permitted within cable assemblies, or within cords or cables of a type identified in Table 400-4 for hard usage or extra-hard usage. For the purpose of this section, Type NM and Type NMC cables shall be permitted to be used in any dwelling, building, or structure without any height limitation.

Exception: Single insulated conductors shall be permitted where installed for the purpose(s) specified in Section 305-3(c), where accessible only to qualified persons.

(c) Branch Circuits. All branch circuits shall originate in an approved power outlet or panelboard. Conductors shall be permitted within cable assemblies, or within multiconductor cord or cable of a type identified in Table 400-4 for hard usage or extra-hard usage. All conductors shall be protected as provided in Article 240. For the purposes of this section, Type NM and Type NMC cables shall be permitted to be used in any dwelling, building, or structure without any height limitation.

Branch circuits installed for the purposes specified in Sections 305-3(b) or (c) shall be permitted to be run as single insulated conductors. Where the wiring is installed in accordance with Section 305-3(b), the voltage to ground shall not exceed 150 volts, the wiring shall not be subject to physical damage, and the conductors shall be supported on insulators at intervals of not more than 10 ft (3.05 m); or, for festoon lighting, the conductors shall be arranged so that excessive strain is not transmitted to the lampholders.

(d) Receptacles. All receptacles shall be of the grounding type. Unless installed in a continuous grounded metal raceway or metal-covered cable, all branch circuits shall contain a separate equipment grounding conductor, and all receptacles shall be electrically connected to the equipment grounding conductors. Receptacles on construction sites shall not be installed on branch circuits that supply temporary lighting. Receptacles shall not be connected to the same ungrounded conductor of multiwire circuits that supply temporary lighting.

(e) Disconnecting Means. Suitable disconnecting switches or plug connectors shall be installed to permit the disconnection of all ungrounded conductors of each temporary circuit. Multiwire branch circuits shall be provided with a means to disconnect simultaneously all ungrounded conductors at the power outlet or panelboard where the branch circuit originated. Approved handle ties shall be permitted.

(f) Lamp Protection. All lamps for general illumination shall be protected from accidental contact or breakage by a suitable fixture or lampholder with a guard.

Brass shell, paper-lined sockets, or other metal-cased sockets shall not be used unless the shell is grounded.

(g) Splices. On construction sites, a box shall not be required for splices or junction connections where the circuit conductors are multiconductor cord or cable assemblies. See Sections 110-14(b) and 400-9. A box, conduit body, or terminal fitting having a separately bushed hole for each conductor shall be used wherever a change is made to a conduit or tubing system or a metal-sheathed cable system.

(h) Protection from Accidental Damage. Flexible cords and cables shall be protected from accidental damage. Sharp corners and projections shall be avoided. Where passing through doorways or other pinch points, protection shall be provided to avoid damage.

(i) Termination(s) at Devices. Flexible cords and cables entering enclosures containing devices requiring termination shall be secured to the box with fittings designed for the purpose.

(j) Support. Cable assemblies and flexible cords and cables shall be supported in place at intervals that ensure that they will be protected from physical damage. Support shall be in the form of staples, cable ties, straps, or similar type fittings installed so as not to cause damage.

305-6. Ground-Fault Protection for Personnel. Ground-fault protection for personnel for all temporary wiring installations shall be provided to comply with (a) and (b). This section shall apply only to temporary wiring installations used to supply temporary power to equipment used by personnel during construction, remodeling, maintenance, repair, or demolition of buildings, structures, equipment, or similar activities.

(a) Receptacle Outlets. All 125-volt, single-phase, 15-, 20-, and 30-ampere receptacle outlets that are not a part of the permanent wiring of the building or structure and that are in use by personnel shall have ground-fault circuit interrupter protection for personnel. If a receptacle(s) is installed or exists as part of the permanent wiring of the building or structure and is used for temporary electric power, ground-fault circuit-interrupter protection for personnel shall be provided. For the purposes of this section, cord sets or devices incorporating listed ground-fault circuit interrupter protection for personnel identified for portable use shall be permitted.

Exception No. 1: Receptacles on a 2-wire, single-phase portable or vehicle-mounted generator rated not more than 5 kW, where the circuit conductors of the generator are insulated from the generator frame and all other grounded surfaces, shall be permitted without ground-fault protection for personnel.

Exception No. 2: In industrial establishments only, where conditions of maintenance and supervision ensure that only qualified personnel are involved, an assured equipment grounding conductor program as specified in Section 305-6(b)(2) shall be permitted to be utilized for all receptacle outlets.

(b) Use of Other Outlets. Receptacles other than 125-volt, single-phase, 15-, 20-, and 30-ampere receptacles shall have protection in accordance with (1) or, the assured equipment grounding conductor program in accordance with (2).

(1) Ground-fault circuit interrupter protection for personnel.
(2) A written assured equipment grounding conductor program continuously enforced at the site by one or more designated persons to ensure that equipment grounding conductors for all cord sets, receptacles that are not a part of the permanent wiring of the building or structure, and equipment connected by cord and plug are installed and maintained in accordance with the applicable requirements of Sections 210-7(c), 250-114, 250-138, and 305-4(d).

(a) The following tests shall be performed on all cord sets, receptacles that are not part of the permanent wiring of the building or structure, and cord- and plug-connected equipment required to be grounded.

   (1) All equipment grounding conductors shall be tested for continuity and shall be electrically continuous.
   (2) Each receptacle and attachment plug shall be tested for correct attachment of the equipment grounding conductor. The equipment grounding conductor shall be connected to its proper terminal.
   (3) All required tests shall be performed

      (a) Before first use on site,
      (b) When there is evidence of damage,
      (c) Before equipment is returned to service following any repairs,
      (d) At intervals not exceeding 3 months.

(b) The tests required in (2)(a) shall be recorded and made available to the authority having jurisdiction.

305-7. Guarding. For wiring over 600 volts, nominal, suitable fencing, barriers, or other effective means shall be provided to limit access only to authorized and qualified personnel.

Article 310 — Conductors for General Wiring

310-1. Scope. This article covers general requirements for conductors and their type designations, insulations, markings, mechanical strengths, ampacity ratings, and uses. These requirements do not apply to conductors that form an integral part of equipment, such as motors, motor controllers, and similar equipment, or to conductors specifically provided for elsewhere in this *Code*.

FPN: For flexible cords and cables, see Article 400. For fixture wires, see Article 402.

310-2. Conductors.

(a) Insulated. Conductors shall be insulated.

Exception: Where covered or bare conductors are specifically permitted elsewhere in this Code.

FPN: See Section 250-184 for insulation of neutral conductors of a solidly grounded high-voltage system.

(b) Conductor Material. Conductors in this article shall be of aluminum, copper-clad aluminum, or copper unless otherwise specified.

310-3. Stranded Conductors. Where installed in raceways, conductors of size No. 8 and larger shall be stranded.

Exception: As permitted or required elsewhere in this Code.

310-4. Conductors in Parallel. Aluminum, copper-clad aluminum, or copper conductors of size No. 1/0 and larger, comprising each phase, neutral, or grounded circuit conductor, shall be permitted to be connected in parallel (electrically joined at both ends to form a single conductor).

Exception No. 1: As permitted in Section 620-12(a)(1).

Exception No. 2: Conductors in sizes smaller than No. 1/0 shall be permitted to be run in parallel to supply control power to indicating instruments, contactors, relays, solenoids, and similar control devices provided

(a) They are contained within the same raceway or cable,
(b) The ampacity of each individual conductor is sufficient to carry the entire load current shared by the parallel conductors, and
(c) The overcurrent protection is such that the ampacity of each individual conductor will not be exceeded if one or more of the parallel conductors become inadvertently disconnected.

Exception No. 3: Conductors in sizes smaller than No. 1/0 shall be permitted to be run in parallel for frequencies of 360 Hz and higher where conditions (a), (b), and (c) of Exception No. 2 are met.

Exception No. 4: Under engineering supervision, grounded neutral conductors in sizes No. 2 and larger shall be permitted to be run in parallel for existing installations.

FPN: Exception No. 4 can be utilized to alleviate overheating of neutral conductors in existing installations due to high content of triplen harmonic currents.

The paralleled conductors in each phase, neutral, or grounded circuit conductor shall

(1) Be the same length,
(2) Have the same conductor material,
(3) Be the same size in circular mil area,
(4) Have the same insulation type,
(5) Be terminated in the same manner.

Where run in separate raceways or cables, the raceways or cables shall have the same physical characteristics. Conductors of one phase, neutral, or grounded circuit conductor shall not be required to have the same physical characteristics as those of another phase, neutral, or grounded circuit conductor to achieve balance.

FPN: Differences in inductive reactance and unequal division of current can be minimized by choice of materials, methods of construction, and orientation of conductors.

Where equipment grounding conductors are used with conductors in parallel, they shall comply with the requirements of this section except that they shall be sized in accordance with Section 250-122.

Where conductors are used in parallel, space in enclosures shall be given consideration (see Articles 370 and 373).

Conductors installed in parallel shall comply with the provisions of Section 310-15(b)(2)(a).

310-5. Minimum Size of Conductors. The minimum size of conductors shall be as shown in Table 310-5.

Exception No. 1: For flexible cords as permitted by Section 400-12.

Exception No. 2: For fixture wire as permitted by Section 410-24, FPN.

Exception No. 3: For motors rated 1 hp or less as permitted by Section 430-22(c).

Exception No. 4: For cranes and hoists as permitted by Section 610-14.

Exception No. 5: For elevator control and signaling circuits as permitted by Section 620-12.

Exception No. 6: For Class 1, Class 2, and Class 3 circuits as permitted by Sections 725-27(a) and 725-51, Exception.

Exception No. 7: For fire alarm circuits as permitted by Sections 760-27(a), 760-51, Exception, and 760-71(b).

Exception No. 8: For motor-control circuits as permitted by Section 430-72.

Exception No. 9: For control and instrumentation circuits as permitted by Section 727-6.

Table 310-5. Minimum Size of Conductors

| Conductor Voltage Rating (Volts) | Minimum Conductor Size (AWG) | |
	Copper	Aluminum or Copper-Clad Aluminum
0–2000	14	12
2001–8000	8	8
8001–15,000	2	2
15,001–28,000	1	1
28,001–35,000	1/0	1/0

310-6. Shielding. Solid dielectric insulated conductors operated above 2000 volts in permanent installations shall have ozone-resistant insulation and shall be shielded. All metallic insulation shields shall be grounded through an effective grounding path meeting the requirements of Section 250-2(d). Shielding shall be for the purpose of confining the voltage stresses to the insulation.

Exception: Nonshielded insulated conductors listed by a qualified testing laboratory shall be permitted for use up to 8000 volts under the following conditions:

(a) Conductors shall have insulation resistant to electric discharge and surface tracking, or the insulated conductor(s) shall be covered with a material resistant to ozone, electric discharge, and surface tracking.

(b) Where used in wet locations, the insulated conductor(s) shall have an overall nonmetallic jacket or a continuous metallic sheath.

(c) Where operated at 5001 to 8000 volts, the insulated conductor(s) shall have a nonmetallic jacket over the insulation. The insulation shall have a specific inductive capacity not greater than 3.6, and the jacket shall have a specific inductive capacity not greater than 10 and not less than 6.

(d) Insulation and jacket thicknesses shall be in accordance with Table 310-63.

310-7. Direct Burial Conductors. Conductors used for direct burial applications shall be of a type identified for such use.
Cables rated above 2000 volts shall be shielded.

Exception: Nonshielded multiconductor cables rated 2001–5000 volts shall be permitted if the cable has an overall metallic sheath or armor.

The metallic shield, sheath, or armor shall be grounded through an effective grounding path meeting the requirements of Section 250-2(d).

FPN No. 1: See Section 300-5 for installation requirements for conductors rated 600 volts or less.

FPN No. 2: See Section 300-50 for installation requirements for conductors rated over 600 volts.

310-8. Locations.

(a) Dry Locations. Insulated conductors and cables used in dry locations shall be any of the types identified in this *Code.*

(b) Dry and Damp Locations. Insulated conductors and cables used in dry and damp locations shall be Types FEP, FEPB, MTW, PFA, RH, RHH, RHW, RHW-2, SA, THHN, THW, THW-2, THHW, THHW-2, THWN, THWN-2, TW, XHH, XHHW, XHHW-2, Z, or ZW.

(c) Wet Locations. Insulated conductors and cables used in wet locations shall be

(1) Moisture-impervious metal-sheathed;
(2) Types MTW, RHW, RHW-2, TW, THW, THW-2, THHW, THHW-2, THWN, THWN-2, XHHW, XHHW-2, ZW; or
(3) Of a type listed for use in wet locations.

(d) Locations Exposed to Direct Sunlight. Insulated conductors and cables used where exposed to direct rays of the sun shall be of a type listed or marked "sunlight resistant."

310-9. Corrosive Conditions. Conductors exposed to oils, greases, vapors, gases, fumes, liquids, or other substances having a deleterious effect on the conductor or insulation shall be of a type suitable for the application.

310-10. Temperature Limitation of Conductors. No conductor shall be used in such a manner that its operating temperature will exceed that designated for the type of insulated conductor involved. In no case shall conductors be associated together in such a way with respect to type of circuit, the wiring method employed, or the number of conductors that the limiting temperature of any conductor is exceeded.

FPN: The temperature rating of a conductor (see Tables 310-13 and 310-61) is the maximum temperature, at any location along its length, that the conductor can withstand over a prolonged time period without serious degradation. The allowable ampacity tables, the ampacity tables of Article 310 and the ampacity tables of Appendix B, the correction factors at the bottom of these tables, and the notes to the tables provide guidance for coordinating conductor sizes, types, allowable ampacities, ampacities, ambient temperatures, and number of associated conductors.

The principal determinants of operating temperature are as follows:

(1) Ambient temperature — ambient temperature may vary along the conductor length as well as from time to time.
(2) Heat generated internally in the conductor as the result of load current flow, including fundamental and harmonic currents.
(3) The rate at which generated heat dissipates into the ambient medium. Thermal insulation that covers or surrounds conductors will affect the rate of heat dissipation.
(4) Adjacent load-carrying conductors — adjacent conductors have the dual effect of raising the ambient temperature and impeding heat dissipation.

310-11. Marking.

(a) Required Information. All conductors and cables shall be marked to indicate the following information, using the applicable method described in (b):

(1) The maximum rated voltage for which the conductor was listed
(2) The proper type letter or letters for the type of wire or cable as specified elsewhere in this *Code*
(3) The manufacturer's name, trademark, or other distinctive marking by which the organization responsible for the product can be readily identified
(4) The AWG size or circular mil area
(5) Cable assemblies where the neutral conductor is smaller than the ungrounded conductors shall be so marked

(b) Method of Marking.

(1) Surface Marking. The following conductors and cables shall be durably marked on the surface. The AWG size or circular mil area shall be repeated at intervals not exceeding 24 in. (610 mm). All other markings shall be repeated at intervals not exceeding 40 in. (1.02 m).

(a) Single- and multiconductor rubber- and thermoplastic-insulated wire and cable
(b) Nonmetallic-sheathed cable
(c) Service-entrance cable
(d) Underground feeder and branch-circuit cable
(e) Tray cable
(f) Irrigation cable
(g) Power-limited tray cable
(h) Instrumentation tray cable

(2) Marker Tape. Metal-covered multiconductor cables shall employ a marker tape located within the cable and running for its complete length.

Exception No. 1: Mineral-insulated, metal-sheathed cable.
Exception No. 2: Type AC cable.

Exception No. 3: The information required in Section 310-11(a) shall be permitted to be durably marked on the outer nonmetallic covering of Type MC, Type ITC, or Type PLTC cables at intervals not exceeding 40 in. (1.02 m).

Exception No. 4: The information required in Section 310-11(a) shall be permitted to be durably marked on a nonmetallic covering under the metallic sheath of Type ITC or Type PLTC cable at intervals not exceeding 40 in. (1.02 m).

FPN: Included in the group of metal-covered cables are Type AC cable (Article 333), Type MC cable (Article 334), and lead-sheathed cable.

(3) Tag Marking. The following conductors and cables shall be marked by means of a printed tag attached to the coil, reel, or carton:

(a) Mineral-insulated, metal-sheathed cable
(b) Switchboard wires
(c) Metal-covered, single-conductor cables
(d) Conductors that have an outer surface of asbestos
(e) Type AC cable

(4) Optional Marking of Wire Size. The information required in (a)(4) shall be permitted to be marked on the surface of the individual insulated conductors for the following multiconductor cables:

(a) Type MC cable
(b) Tray cable
(c) Irrigation cable
(d) Power-limited tray cable
(e) Power-limited fire alarm cable
(f) Instrumentation tray cable

(c) Suffixes to Designate Number of Conductors. A type letter or letters used alone shall indicate a single insulated conductor. The letter suffixes shall be indicated as follows:

D — For two insulated conductors laid parallel within an outer nonmetallic covering

M — For an assembly of two or more insulated conductors twisted spirally within an outer nonmetallic covering

(d) Optional Markings. All conductors and cables contained in Chapter 3 shall be permitted to be surface marked to indicate special characteristics of the cable materials.

FPN: Examples of these markings include but are not limited to "LS" for limited smoke and markings such as "sunlight resistant."

310-12. Conductor Identification.

(a) Grounded Conductors. Insulated or covered grounded conductors shall be identified in accordance with Section 200-6.

(b) Equipment Grounding Conductors. Equipment grounding conductors shall be in accordance with Section 250-119.

(c) Ungrounded Conductors. Conductors that are intended for use as ungrounded conductors, whether used as single conductors or in multiconductor cables, shall be finished to be clearly distinguishable from grounded and grounding conductors. Ungrounded conductors shall be distinguished by colors other than white, natural gray, or green; or by a combination of color plus distinguishing marking. Distinguishing markings shall also be in a color other than white, natural gray, or green, and shall consist of a stripe or stripes or a regularly spaced series of identical marks. Distinguishing markings shall not conflict in any manner with the surface markings required by Section 310-11(b)(1).

Exception: As permitted by Section 200-7.

310-13. Conductor Constructions and Applications. Insulated conductors shall comply with the applicable provisions of one or more of the following: Tables 310-13, 310-61, 310-62, 310-63, and 310-64.

These conductors shall be permitted for use in any of the wiring methods recognized in Chapter 3 and as specified in their respective tables.

FPN: Thermoplastic insulation may stiffen at temperatures colder than minus 10°C (plus 14°F). Thermoplastic insulation may also be deformed at normal temperatures where subjected to pressure, such as at points of support. Thermoplastic insulation, where used on dc circuits in wet locations, may result in electroendosmosis between conductor and insulation.

310-14. Aluminum Conductor Material. Solid aluminum conductors No. 8, 10, and 12 shall be made of an AA-8000 series electrical grade aluminum alloy conductor material. Stranded aluminum conductors No. 8 through 1000 kcmil marked as Type XHHW, THW, THHW, THWN, THHN, service-entrance Type SE Style U and SE Style R shall be made of an AA-8000 series electrical grade aluminum alloy conductor material.

Table 310-13. Conductor Application and Insulations

Trade Name	Type Letter	Maximum Operating Temperature	Application Provisions	Insulation	Thickness of Insulation AWG or kcmil	Mils	Outer Covering[1]
Fluorinated ethylene propylene	FEP or FEPB	90°C 194°F	Dry and damp locations	Fluorinated ethylene propylene	14–10 / 8–2	20 / 30	None
		200°C 392°F	Dry locations — special applications[2]	Fluorinated ethylene propylene	14–8	14	Glass braid
					6–2	14	Asbestos or other suitable braid material
Mineral insulation (metal sheathed)	MI	90°C 194°F 250°C 482°F	Dry and wet locations For special applications[2]	Magnesium oxide	18–16[3] / 16–10 / 9–4 / 3–500	23 / 36 / 50 / 55	Copper or alloy steel
Moisture-, heat-, and oil-resistant thermoplastic	MTW	60°C 140°F 90°C 194°F	Machine tool wiring in wet locations as permitted in NFPA 79 (see Article 670) Machine tool wiring in dry locations as permitted in NFPA 79 (see Article 670)	Flame-retardant moisture-, heat-, and oil-resistant thermoplastic	22–12 / 10 / 8 / 6 / 4–2 / 1–4/0 / 213–500 / 591–1000 — (A) 30 / 30 / 45 / 60 / 60 / 80 / 95 / 110	(B) 15 / 20 / 30 / 30 / 40 / 50 / 60 / 70	(A) None (B) Nylon jacket or equivalent
Paper		85°C 185°F	For underground service conductors, or by special permission	Paper			Lead sheath

[1]Some insulations do not require an outer covering.

[2]Where design conditions require maximum conductor operating temperatures above 90°C (194°F).

[3]For signaling circuits permitting 300-volt insulation.

Table 310-13. (Continued)

Trade Name	Type Letter	Maximum Operating Temperature	Application Provisions	Insulation	Thickness of Insulation		Outer Covering[1]
					AWG or kcmil	Mils	
Perfluoroalkoxy	PFA	90°C 194°F 200°C 392°F	Dry and damp locations Dry locations — special applications[2]	Perfluoroalkoxy	14–10 8–2 1–4/0	20 30 45	None
Perfluoroalkoxy	PFAH	250°C 482°F	Dry locations only. Only for leads within apparatus or within raceways connected to apparatus (Nickel or nickel-coated copper only)	Perfluoroalkoxy	14–10 8–2 1–4/0	20 30 45	None
Thermoset Thermoset	RH RHH	75°C 167°F 90°C 194°F	Dry and damp locations Dry and damp locations	Flame-retardant thermoset	14–12[4] 10 8–2 1–4/0 213–500 501–1000 1001–2000 For 601–2000, see Table 310-62	30 45 60 80 95 110 125	Moisture-resistant, flame-retardant, non-metallic covering[1]
Moisture-resistant thermoset	RHW[5]	75°C 167°F	Dry and wet locations Where over 2000 volts insulation, shall be ozone-resistant	Flame-retardant, moisture-resistant thermoset	14–10 8–2 1–4/0 213–500 501–1000 1001–2000 For 601–2000 volts, see Table 310-62	45 60 80 95 110 125	Moisture-resistant, flame-retardant, non-metallic covering[6]
Moisture-resistant thermoset	RHW-2	90°C 194°F	Dry and wet locations	Flame-retardant moisture-resistant thermoset	14–10 8–2 1–4/0 213–500 501–1000 1001–2000 For 601–2000 volts, see Table 310-62	45 60 80 95 110 125	Moisture-resistant, flame-retardant, non-metallic covering[6]
Silicone	SA	90°C 194°F 200°C 392°F	Dry and damp locations For special application[2]	Silicone rubber	14–10 8–2 1–4/0 213–500 501–1000 1001–2000	45 60 80 95 110 125	Glass or other suitable braid material
Thermoset	SIS	90°C 194°F	Switchboard wiring only	Flame-retardant thermoset	14–10 8–2 1–4/0	30 45 95	None
Thermoplastic and fibrous outer braid	TBS	90°C 194°F	Switchboard wiring only	Thermoplastic	14–10 8 6–2 1–4/0	30 45 60 80	Flame-retardant, nonmetallic covering

[1]Some insulations do not require an outer covering.

[2]Where design conditions require maximum conductor operating temperatures above 90°C (194°F).

[4]For size Nos. 14-12, RHH insulation shall be 45 mils thickness.

[5]Listed wire types designated with the suffix ''-2,'' such as RHW-2, shall be permitted to be used at a continuous 90°C (194°F) operating temperature, wet or dry.

[6]Some rubber insulations do not require an outer covering.

(continues)

Table 310-13. (Continued)

Trade Name	Type Letter	Maximum Operating Temperature	Application Provisions	Insulation	Thickness of Insulation		Outer Covering[1]
					AWG or kcmil	Mils	
Extended polytetra-fluoroethylene	TFE	250°C 482°F	Dry locations only. Only for leads within apparatus or within raceways connected to apparatus, or as open wiring (Nickel or nickel-coated copper only)	Extruded poly-tetrafluoro-ethylene	14–10 8–2 1–4/0	20 30 45	None
Heat-resistant thermoplastic	THHN	90°C 194°F	Dry and damp locations	Flame-retardant, heat-resistant thermoplastic	14–12 10 8–6 4–2 1–4/0 250–500 501–1000	15 20 30 40 50 60 70	Nylon jacket or equivalent
Moisture- and heat-resistant thermoplastic	THHW	75°C 167°F 90°C 194°F	Wet location Dry location	Flame-retardant, moisture- and heat-resistant thermoplastic	14–10 8 6–2 1–4/0 213–500 501–1000	30 45 60 80 95 110	None
Moisture- and heat-resistant thermoplastic	THW[5]	75°C 167°F 90°C 194°F	Dry and wet locations Special applications within electric discharge lighting equipment. Limited to 1000 open-circuit volts or less (Size 14-8 only as permitted in Section 410-31)	Flame-retardant, moisture- and heat-resistant thermoplastic	14–10 8 6–2 1–4/0 213–500 501–1000 1001–2000	30 45 60 80 95 110 125	None
Moisture- and heat-resistant thermoplastic	THWN[5]	75°C 167°F	Dry and wet locations	Flame-retardant, moisture- and heat-resistant thermoplastic	14–12 10 8–6 4–2 1–4/0 250–500 501–1000	15 20 30 40 50 60 70	Nylon jacket or equivalent
Moisture-resistant thermoplastic	TW	60°C 140°F	Dry and wet locations	Flame-retardant, moisture-resistant thermoplastic	14–10 8 6–2 1–4/0 213–500 501–1000 1001–2000	30 45 60 80 95 110 125	None
Underground feeder and branch-circuit cable — single conductor (For Type UF cable employing more than one conductor, see Article 339)	UF	60°C 140°F 75°C 167°F[8]	See Article 339	Moisture-resistant Moisture- and heat-resistant	14–10 8–2 1–4/0	60[7] 80[7] 95[7]	Integral with insulation

[1]Some insulations do not require an outer covering.

[5]Listed wire types designated with the suffix ''-2,'' such as RHW-2, shall be permitted to be used at a continuous 90°C (194°F) operating temperature, wet or dry.

[7]Includes integral jacket.

[8]For ampacity limitation, see Section 339-5.

Table 310-13. (Continued)

Trade Name	Type Letter	Maximum Operating Temperature	Application Provisions	Insulation	Thickness of Insulation		Outer Covering[1]
					AWG or kcmil	Mils	
Underground service-entrance cable — single conductor (For Type USE cable employing more than one conductor, see Article 338)	USE[5]	75°C 167°F	See Article 338	Heat- and moisture-resistant	14–10 8–2 1–4/0 213–500 501–1000 1001–2000	45 60 80 95[9] 110 125	Moisture-resistant nonmetallic covering [see Section 338-1(b)]
Thermoset	XHH	90°C 194°F	Dry and damp locations	Flame-retardant thermoset	14–10 8–2 1–4/0 213–500 501–1000 1001–2000	30 45 55 65 80 95	None
Moisture-resistant thermoset	XHHW[5]	90°C 194°F 75°C 167°F	Dry and damp locations Wet locations	Flame-retardant, moisture-resistant thermoset	14–10 8–2 1–4/0 213–500 501–1000 1001–2000	30 45 55 65 80 95	None
Moisture-resistant thermoset	XHHW-2	90°C 194°F	Dry and wet locations	Flame-retardant, moisture-resistant thermoset	14–10 8–2 1–4/0 213–500 501–1000 1001–2000	30 45 55 65 80 95	None
Modified ethylene tetrafluoroethylene	Z	90°C 194°F 150°C 302°F	Dry and damp locations Dry locations — special applications[2]	Modified ethylene-tetrafluoroethylene	14–12 10 8–4 3–1 1/0–4/0	15 20 25 35 45	None
Modified ethylene tetrafluoroethylene	ZW[5]	75°C 167°F 90°C 194°F 150°C 302°F	Wet locations Dry and damp locations Dry locations — special applications[2]	Modified ethylene tetrafluoroethylene	14–10 8–2	30 45	None

[1] Some insulations do not require an outer covering.

[2] Where design conditions require maximum conductor operating temperatures above 90°C (194°F).

[5] Listed wire types designated with the suffix ''-2,'' such as RHW-2, shall be permitted to be used at a continuous 90°C (194°F) operating temperature, wet or dry.

[9] Insulation thickness shall be permitted to be 80 mils for listed Type USE conductors that have been subjected to special investigations. The nonmetallic covering over individual rubber-covered conductors of aluminum-sheathed cable and of lead-sheathed or multiconductor cable shall not be required to be flame retardant. For Type MC cable, see Section 334-20. For nonmetallic-sheathed cable, see Section 336-30. For Type UF cable, see Section 339-1.

310-15. Ampacities for Conductors Rated 0–2000 Volts.

(a) General.

(1) Tables or Engineering Supervision. Ampacities for conductors shall be permitted to be determined by tables or under engineering supervision, as provided in (b) and (c).

FPN No. 1: Ampacities provided by this section do not take voltage drop into consideration. See Section 210-19(a), FPN No. 4, for branch circuits and Section 215-2(d), FPN No. 2, for feeders.

FPN No. 2: For the allowable ampacities of Type MTW

wire, see Table 11 in the *Electrical Standard for Industrial Machinery*, NFPA 79-1997.

(2) Selection of Ampacity. Where more than one calculated or tabulated ampacity could apply for a given circuit length, the lowest value shall be used.

Exception: Where two different ampacities apply to adjacent portions of a circuit, the higher ampacity shall be permitted to be used beyond the point of transition, a distance equal to 10 ft (3.05 m) or 10 percent of the circuit length figured at the higher ampacity, whichever is less.

FPN: See Section 110-14(c) for conductor temperature limitations due to termination provisions.

(b) Tables. Ampacities for conductors rated 0 to 2000 volts shall be as specified in the Allowable Ampacity Tables 310-16 through 310-19 and Ampacity Tables 310-20 and 310-21 as modified by (1) through (7).

FPN: Tables 310-16 through 310-19 are application tables for use in determining conductor sizes on loads calculated in accordance with Article 220. Allowable ampacities result from consideration of one or more of the following:

(1) Temperature compatibility with connected equipment, especially at the connection points.
(2) Coordination with circuit and system overcurrent protection.
(3) Compliance with the requirements of product listings or certifications. See Section 110-3(b).
(4) Preservation of the safety benefits of established industry practices and standardized procedures.

(1) General. For explanation of type letters used in tables and for recognized sizes of conductors for the various conductor insulations, see Section 310-13. For installation requirements, see Sections 310-1 through 310-10 and the various articles of this *Code*. For flexible cords, see Tables 400-4, 400-5(A), and 400-5(B).

(2) Adjustment Factors.

(a) *More than Three Current-Carrying Conductors in a Raceway or Cable.* Where the number of current-carrying conductors in a raceway or cable exceeds three, or where single conductors or multiconductor cables are stacked or bundled longer than 24 in. (610 mm) without maintaining spacing and are not installed in raceways, the allowable ampacity of each conductor shall be reduced as shown in Table 310-15(b)(2)(a).

FPN: See Appendix B, Table B-310-11, for adjustment factors for more than three current-carrying conductors in a raceway or cable with load diversity.

Exception No. 1: Where conductors of different systems, as provided in Section 300-3, are installed in a common raceway or cable, the derating factors shown in Table 310-

Table 310-15(b)(2)(a). Adjustment Factors for More than Three Current-Carrying Conductors in a Raceway or Cable

Number of Current-Carrying Conductors	Percent of Values in Tables 310-16 through 310-19 as Adjusted for Ambient Temperature if Necessary
4–6	80
7–9	70
10–20	50
21–30	45
31–40	40
41 and above	35

15(b)(2)(a) shall apply to the number of power and lighting conductors only (Articles 210, 215, 220, and 230).

Exception No. 2: For conductors installed in cable trays, the provisions of Section 318-11 shall apply.

Exception No. 3: Derating factors shall not apply to conductors in nipples having a length not exceeding 24 in. (610 mm).

Exception No. 4: Derating factors shall not apply to underground conductors entering or leaving an outdoor trench if those conductors have physical protection in the form of rigid metal conduit, intermediate metal conduit, or rigid nonmetallic conduit having a length not exceeding 10 ft (3.05 m) and the number of conductors does not exceed four.

(b) *More than One Conduit, Tube, or Raceway.* Spacing between conduits, tubing, or raceways shall be maintained.

(3) Bare or Covered Conductors. Where bare or covered conductors are used with insulated conductors, their allowable ampacities shall be limited to those permitted for the adjacent insulated conductors.

(4) Neutral Conductor.
(a) A neutral conductor that carries only the unbalanced current from other conductors of the same circuit shall not be required to be counted when applying the provisions of Section 310-15(b)(2)(a).

(b) In a 3-wire circuit consisting of two phase wires and the neutral of a 4-wire, 3-phase wye-connected system, a common conductor carries approximately the same current as the line-to-neutral load currents of the other conductors and shall be counted when applying the provisions of Section 310-15(b)(2)(a).

(c) On a 4-wire, 3-phase wye circuit where the major portion of the load consists of nonlinear loads, harmonic currents are present in the neutral conductor; the neutral shall therefore be considered a current-carrying conductor.

(5) Grounding or Bonding Conductor. A grounding or bonding conductor shall not be counted when applying the provisions of Section 310-15(b)(2)(a).

(6) 120/240-Volt, 3-Wire, Single-Phase Dwelling Services and Feeders. For dwelling units, conductors, as listed in Table 310-15(b)(6) shall be permitted as 120/240-volt, 3-wire, single-phase service-entrance conductors, service lateral conductors, and feeder conductors that serve as the main power feeder to a dwelling unit and are installed in raceway or cable with or without an equipment grounding conductor. For application of this section, the main power feeder shall be the feeder(s) between the main disconnect and the lighting and appliance branch-circuit panelboard(s), and the feeder conductors to a dwelling unit shall not be required to be larger than their service-entrance conductors. The grounded conductor shall be permitted to be smaller, than the ungrounded conductors, provided the requirements of Sections 215-2, 220-22, and 230-42 are met.

Table 310-15(b)(6). Conductor Types and Sizes for 120/240-Volt, 3-Wire, Single-Phase Dwelling Services and Feeders. Conductor Types RH, RHH, RHW, RHW-2, THHN, THHW, THW, THW-2, THWN, THWN-2, XHHW, XHHW-2, SE, USE, USE-2

Conductor (AWG or kcmil)		
Copper	Aluminum or Copper-Clad Aluminum	Service or Feeder Rating (Amperes)
4	2	100
3	1	110
2	1/0	125
1	2/0	150
1/0	3/0	175
2/0	4/0	200
3/0	250	225
4/0	300	250
250	350	300
350	500	350
400	600	400

(7) Mineral-Insulated, Metal-Sheathed Cable. The temperature limitations on which the ampacities of mineral-insulated, metal-sheathed cable are based shall be determined by the insulating materials used in the end seal. Termination fittings incorporating unimpregnated, organic, insulating materials shall be limited to 90°C (194°F) operation.

(c) Engineering Supervision. Under engineering supervision, conductor ampacities shall be permitted to be calculated by means of the following general formula:

$$I = \sqrt{\frac{TC - (TA + DeltaTD)}{RDC(1 + YC)RCA}}$$

Where:

TC = Conductor temperature in degrees Celsius (°C)

TA = Ambient temperature in degrees Celsius (°C)

$Delta\ TD$ = Dielectric loss temperature rise

RDC = dc resistance of conductor at temperature TC

YC = Component ac resistance resulting from skin effect and proximity effect

RCA = Effective thermal resistance between conductor and surrounding ambient

FPN: See Appendix B for examples of formula applications.

310-60. Conductors Rated 2001 to 35,000 Volts.

(a) Definitions.

Electrical Ducts. As used in Article 310, electrical ducts shall include any of the electrical conduits recognized in Chapter 3 as suitable for use underground; and other raceways round in cross section, listed for underground use, and embedded in earth or concrete.

Thermal Resistivity. As used in this *Code, thermal resistivity* refers to the heat transfer capability through a substance by conduction. It is the reciprocal of thermal conductivity and is designated Rho and expressed in the units °C-cm/watt.

(b) Ampacities of Conductors Rated 2001 to 35,000 Volts. Ampacities for solid dielectric-insulated conductors shall be permitted to be determined by tables or under engineering supervision, as provided in (c) and (d).

(1) Selection of Ampacity. Where more than one calculated or tabulated ampacity could apply for a given circuit length, the lowest value shall be used.

Exception: Where two different ampacities apply to adjacent portions of a circuit, the higher ampacity shall be permitted to be used beyond the point of transition, a distance equal to 10 ft (3.05 m) or 10 percent of the circuit length figured at the higher ampacity, whichever is less.

FPN: See Section 110-40 for conductor temperature limitations due to termination provisions.

(c) Tables. Ampacities for conductors rated 2001 to 35,000 volts shall be as specified in the Ampacity Tables 310-67 through 310-86. Ampacities at ambient temperatures other than those shown in the tables shall be determined by the formula in (4).

FPN No. 1: For ampacities calculated in accordance with Section 310-60(b), reference IEEE *Standard Power Cable Ampacity Tables,* IEEE 835-1994 (IPCEA Pub. No. P-46-426) and the references therein for availability of all factors and constants.

Table 310-16. Allowable Ampacities of Insulated Conductors Rated 0 through 2000 Volts, 60°C through 90°C (140°F through 194°F) Not More than Three Current-Carrying Conductors in Raceway, Cable, or Earth (Directly Buried), Based on Ambient Temperature of 30°C (86°F)

Size	Temperature Rating of Conductor (See Table 310-13)						Size
	60°C (140°F)	75°C (167°F)	90°C (194°F)	60°C (140°F)	75°C (167°F)	90°C (194°F)	
AWG or kcmil	Types TW, UF	Types FEPW, RH, RHW, THHW, THW, THWN, XHHW, USE, ZW	Types TBS, SA, SIS, FEP, FEPB, MI, RHH, RHW-2, THHN, THHW, THW-2, THWN-2, USE-2, XHH, XHHW, XHHW-2, ZW-2	Types TW, UF	Types RH, RHW, THHW, THW, THWN, XHHW, USE	Types TBS, SA, SIS, THHN, THHW, THW-2, THWN-2, RHH, RHW-2, USE-2, XHH, XHHW, XHHW-2, ZW-2	AWG or kcmil
	COPPER			ALUMINUM OR COPPER-CLAD ALUMINUM			
18	—	—	14	—	—	—	—
16	—	—	18	—	—	—	—
14*	20	20	25	—	—	—	—
12*	25	25	30	20	20	25	12*
10*	30	35	40	25	30	35	10*
8	40	50	55	30	40	45	8
6	55	65	75	40	50	60	6
4	70	85	95	55	65	75	4
3	85	100	110	65	75	85	3
2	95	115	130	75	90	100	2
1	110	130	150	85	100	115	1
1/0	125	150	170	100	120	135	1/0
2/0	145	175	195	115	135	150	2/0
3/0	165	200	225	130	155	175	3/0
4/0	195	230	260	150	180	205	4/0
250	215	255	290	170	205	230	250
300	240	285	320	190	230	255	300
350	260	310	350	210	250	280	350
400	280	335	380	225	270	305	400
500	320	380	430	260	310	350	500
600	355	420	475	285	340	385	600
700	385	460	520	310	375	420	700
750	400	475	535	320	385	435	750
800	410	490	555	330	395	450	800
900	435	520	585	355	425	480	900
1000	455	545	615	375	445	500	1000
1250	495	590	665	405	485	545	1250
1500	520	625	705	435	520	585	1500
1750	545	650	735	455	545	615	1750
2000	560	665	750	470	560	630	2000

CORRECTION FACTORS

Ambient Temp. (°C)	For ambient temperatures other than 30°C (86°F), multiply the allowable ampacities shown above by the appropriate factor shown below.						Ambient Temp. (°F)
21–25	1.08	1.05	1.04	1.08	1.05	1.04	70–77
26–30	1.00	1.00	1.00	1.00	1.00	1.00	78–86
31–35	0.91	0.94	0.96	0.91	0.94	0.96	87–95
36–40	0.82	0.88	0.91	0.82	0.88	0.91	96–104
41–45	0.71	0.82	0.87	0.71	0.82	0.87	105–113
46–50	0.58	0.75	0.82	0.58	0.75	0.82	114–122
51–55	0.41	0.67	0.76	0.41	0.67	0.76	123–131
56–60	—	0.58	0.71	—	0.58	0.71	132–140
61–70	—	0.33	0.58	—	0.33	0.58	141–158
71–80	—	—	0.41	—	—	0.41	159–176

*See Section 240-3.

Table 310-17. Allowable Ampacities of Single-Insulated Conductors Rated 0 Through 2000 Volts in Free Air, Based on Ambient Air Temperature of 30°C (86°F)

Size	Temperature Rating of Conductor (See Table 310-13)						Size
	60°C (140°F)	75°C (167°F)	90°C (194°F)	60°C (140°F)	75°C (167°F)	90°C (194°F)	
AWG or kcmil	Types TW, UF	Types FEPW, RH, RHW, THHW, THW, THWN, XHHW, ZW	Types TBS, SA, SIS, FEP, FEPB, MI, RHH, RHW-2, THHN, THHW, THW-2, THWN-2, USE-2, XHH, XHHW, XHHW-2, ZW-2	Types TW, UF	Types RH, RHW, THHW, THW, THWN, XHHW	Types TBS, SA, SIS, THHN, THHW, THW-2, THWN-2, RHH, RHW-2, USE-2, XHH, XHHW, XHHW-2, ZW-2	AWG or kcmil
	COPPER			ALUMINUM OR COPPER-CLAD ALUMINUM			
18	—	—	18	—	—	—	
16	—	—	24	—	—	—	
14*	25	30	35	—	—	—	
12*	30	35	40	25	30	35	12*
10*	40	50	55	35	40	40	10*
8	60	70	80	45	55	60	8
6	80	95	105	60	75	80	6
4	105	125	140	80	100	110	4
3	120	145	165	95	115	130	3
2	140	170	190	110	135	150	2
1	165	195	220	130	155	175	1
1/0	195	230	260	150	180	205	1/0
2/0	225	265	300	175	210	235	2/0
3/0	260	310	350	200	240	275	3/0
4/0	300	360	405	235	280	315	4/0
250	340	405	455	265	315	355	250
300	375	455	505	290	350	395	300
350	420	505	570	330	395	445	350
400	455	545	615	355	425	480	400
500	515	620	700	405	485	545	500
600	575	690	780	455	540	615	600
700	630	755	855	500	595	675	700
750	655	785	885	515	620	700	750
800	680	815	920	535	645	725	800
900	730	870	985	580	700	785	900
1000	780	935	1055	625	750	845	1000
1250	890	1065	1200	710	855	960	1250
1500	980	1175	1325	795	950	1075	1500
1750	1070	1280	1445	875	1050	1185	1750
2000	1155	1385	1560	960	1150	1335	2000

CORRECTION FACTORS

Ambient Temp. (°C)	For ambient temperatures other than 30°C (86°F), multiply the allowable ampacities shown above by the appropriate factor shown below.						Ambient Temp. (°F)
21–25	1.08	1.05	1.04	1.08	1.05	1.04	70–77
26–30	1.00	1.00	1.00	1.00	1.00	1.00	78–86
31–35	0.91	0.94	0.96	0.91	0.94	0.96	87–95
36–40	0.82	0.88	0.91	0.82	0.88	0.91	96–104
41–45	0.71	0.82	0.87	0.71	0.82	0.87	105–113
46–50	0.58	0.75	0.82	0.58	0.75	0.82	114–122
51–55	0.41	0.67	0.76	0.41	0.67	0.76	123–131
56–60	—	0.58	0.71	—	0.58	0.71	132–140
61–70	—	0.33	0.58	—	0.33	0.58	141–158
71–80	—	—	0.41	—	—	0.41	159–176

*See Section 240-3.

Table 310-18. Allowable Ampacities of Three Single-Insulated Conductors, Rated 0 Through 2000 Volts, 150°C Through 250°C (302°F Through 482°F), in Raceway or Cable, Based on Ambient Air Temperature of 40°C (104°F)

Size	Temperature Rating of Conductor (See Table 310-13)				Size
	150°C (302°F)	200°C (392°F)	250°C (482°F)	150°C (302°F)	
	Type Z	Types FEP, FEPB, PFA	Types PFAH, TFE	Type Z	
AWG or kcmil	COPPER		NICKEL OR NICKEL-COATED COPPER	ALUMINUM OR COPPER-CLAD ALUMINUM	AWG or kcmil
14	34	36	39	—	14
12	43	45	54	30	12
10	55	60	73	44	10
8	76	83	93	57	8
6	96	110	117	75	6
4	120	125	148	94	4
3	143	152	166	109	3
2	160	171	191	124	2
1	186	197	215	145	1
1/0	215	229	244	169	1/0
2/0	251	260	273	198	2/0
3/0	288	297	308	227	3/0
4/0	332	346	361	260	4/0

CORRECTION FACTORS

Ambient Temp. (°C)	For ambient temperatures other than 40°C (104°F), multiply the allowable ampacities shown above by the appropriate factor shown below.				Ambient Temp. (°F)
41–50	0.95	0.97	0.98	0.95	105–122
51–60	0.90	0.94	0.95	0.90	123–140
61–70	0.85	0.90	0.93	0.85	141–158
71–80	0.80	0.87	0.90	0.80	159–176
81–90	0.74	0.83	0.87	0.74	177–194
91–100	0.67	0.79	0.85	0.67	195–212
101–120	0.52	0.71	0.79	0.52	213–248
121–140	0.30	0.61	0.72	0.30	249–284
141–160	—	0.50	0.65	—	285–320
161–180	—	0.35	0.58	—	321–356
181–200	—	—	0.49	—	357–392
201–225	—	—	0.35	—	393–437

FPN No. 2: Ampacities provided by this section do not take voltage drop into consideration. See Section 210-19(a), FPN No. 4, for branch circuits and Section 215-2(d), FPN No. 2, for feeders.

(1) Grounded Shields. Ampacities shown in Tables 310-69, 310-70, 310-81, and 310-82 are for cable with shields grounded at one point only. Where shields are grounded at more than one point, ampacities shall be adjusted to take into consideration the heating due to shield currents.

(2) Burial Depth of Underground Circuits. Where the burial depth of direct burial or electrical duct bank circuits is modified from the values shown in a figure or table, ampacities shall be permitted to be modified as indicated in (a) and (b).

(a) Where burial depths are increased in part(s) of an electrical duct run, no decrease in ampacity of the conductors is needed, provided the total length of parts of the duct run increased in depth is less than 25 percent of the total run length.

Table 310-19. Allowable Ampacities of Single-Insulated Conductors, Rated 0 Through 2000 Volts, 150°C Through 250°C (302°F Through 482°F), in Free Air, Based on Ambient Air Temperature of 40°C (104°F)

Size AWG or kcmil	Temperature Rating of Conductor (See Table 310-13)				Size AWG or kcmil
	150°C (302°F) Type Z	200°C (392°F) Types FEP, FEPB, PFA	250°C (482°F) Types PFAH, TFE	150°C (302°F) Type Z	
	COPPER	COPPER	NICKEL, OR NICKEL-COATED COPPER	ALUMINUM OR COPPER-CLAD ALUMINUM	
14	46	54	59	—	14
12	60	68	78	47	12
10	80	90	107	63	10
8	106	124	142	83	8
6	155	165	205	112	6
4	190	220	278	148	4
3	214	252	327	170	3
2	255	293	381	198	2
1	293	344	440	228	1
1/0	339	399	532	263	1/0
2/0	390	467	591	305	2/0
3/0	451	546	708	351	3/0
4/0	529	629	830	411	4/0

CORRECTION FACTORS

Ambient Temp. (°C)	For ambient temperatures other than 40°C (104°F), multiply the allowable ampacities shown above by the appropriate factor shown below.				Ambient Temp. (°F)
41–50	0.95	0.97	0.98	0.95	105–122
51–60	0.90	0.94	0.95	0.90	123–140
61–70	0.85	0.90	0.93	0.85	141–158
71–80	0.80	0.87	0.90	0.80	159–176
81–90	0.74	0.83	0.87	0.74	177–194
91–100	0.67	0.79	0.85	0.67	195–212
101–120	0.52	0.71	0.79	0.52	213–248
121–140	0.30	0.61	0.72	0.30	249–284
141–160	—	0.50	0.65	—	285–320
161–180	—	0.35	0.58	—	321–356
181–200	—	—	0.49	—	357–392
201–225	—	—	0.35	—	393–437

(b) Where burial depths are deeper than shown in a specific underground ampacity table or figure, an ampacity derating factor of 6 percent per increased foot (305 mm) of depth for all values of Rho shall be permitted.

No rating change is needed where the burial depth is decreased.

(3) Electrical Ducts in Figure 310-60. At locations where electrical ducts enter equipment enclosures from underground, spacing between such ducts, as shown in Figure 310-60, shall be permitted to be reduced without requiring the ampacity of conductors therein to be reduced.

Table 310-20. Ampacities of Two or Three Single-Insulated Conductors, Rated 0 Through 2000 Volts, Supported on a Messenger, Based on Ambient Air Temperature of 40°C (104°F)

Size	Temperature Rating of Conductor (See Table 310-13)				Size
	75°C (167°F)	90°C (194°F)	75°C (167°F)	90°C (194°F)	
AWG or kcmil	Types RH, RHW, THHW, THW, THWN, XHHW, ZW	Types THHN, THHW, THW-2, THWN-2, RHH, RWH-2, USE-2, XHHW, XHHW-2, ZW-2	Types RH, RHW, THW, THWN, THHW, XHHW	Types THHN, THHW, RHH, XHHW, RHW-2, XHHW-2, THW-2, THWN-2, USE-2, ZW-2	AWG or kcmil
	COPPER		ALUMINUM OR COPPER-CLAD ALUMINUM		
8	57	66	44	51	8
6	76	89	59	69	6
4	101	117	78	91	4
3	118	138	92	107	3
2	135	158	106	123	2
1	158	185	123	144	1
1/0	183	214	143	167	1/0
2/0	212	247	165	193	2/0
3/0	245	287	192	224	3/0
4/0	287	335	224	262	4/0
250	320	374	251	292	250
300	359	419	282	328	300
350	397	464	312	364	350
400	430	503	339	395	400
500	496	580	392	458	500
600	553	647	440	514	600
700	610	714	488	570	700
750	638	747	512	598	750
800	660	773	532	622	800
900	704	826	572	669	900
1000	748	879	612	716	1000
Ambient Temp. (°C)	For ambient temperatures other than 40°C (104°F), multiply the allowable ampacities shown above by the appropriate factor shown below.				Ambient Temp. (°F)
21–25	1.20	1.14	1.20	1.14	70–77
26–30	1.13	1.10	1.13	1.10	79–86
31–35	1.07	1.05	1.07	1.05	88–95
36–40	1.00	1.00	1.00	1.00	97–104
41–45	0.93	0.95	0.93	0.95	106–113
46–50	0.85	0.89	0.85	0.89	115–122
51–55	0.76	0.84	0.76	0.84	124–131
56–60	0.65	0.77	0.65	0.77	133–140
61–70	0.38	0.63	0.38	0.63	142–158
71–80	—	0.45	—	0.45	160–176

Table 310-21. Ampacities of Bare Covered Conductors, Based on 40°C (104°F) Ambient, 80°C (176°F) Total Conductor Temperature, 2 ft/sec (610 mm/sec) Wind Velocity

Copper Conductors			
Bare		Covered	
AWG or kcmil	Amperes	AWG or kcmil	Amperes
8	98	8	103
6	124	6	130
4	155	4	163
2	209	2	219
1/0	282	1/0	297
2/0	329	2/0	344
3/0	382	3/0	401
4/0	444	4/0	466
250	494	250	519
300	556	300	584
500	773	500	812
750	1000	750	1050
1000	1193	1000	1253
—	—	—	—
—	—	—	—
—	—	—	—

AAC Aluminum Conductors			
8	76	8	80
6	96	6	101
4	121	4	127
2	163	2	171
1/0	220	1/0	231
2/0	255	2/0	268
3/0	297	3/0	312
4/0	346	4/0	364
266.8	403	266.8	423
336.4	468	336.4	492
397.5	522	397.5	548
477.0	588	477.0	617
556.5	650	556.5	682
636.0	709	636.0	744
795.0	819	795.0	860
954.0	920	—	—
1033.5	968	1033.5	1017
1272	1103	1272	1201
1590	1267	1590	1381
2000	1454	2000	1527

Table 310-61. Conductor Application and Insulation

Trade Name	Type Letter	Maximum Operating Temperature	Application Provision	Insulation	Outer Covering
Medium voltage solid dielectric	MV-90 MV-105*	90°C 105°C	Dry or wet locations rated 2001 volts and higher	Thermoplastic or thermosetting	Jacket, sheath, or armor

*Where design conditions require maximum conductor temperatures above 90°C.

Table 310-62. Thickness of Insulation for 601 to 2000-Volt Nonshielded Types RHH and RHW, in Mils

Conductor Size (AWG or kcmil)	Column A[1]	Column B[2]
14–10	80	60
8	80	70
6–2	95	70
1–2/0	110	90
3/0–4/0	110	90
213–500	125	105
501–1000	140	120

[1]Column A insulations are limited to natural, SBR, and butyl rubbers.
[2]Column B insulations are materials such as cross-linked polyethylene, ethylene propylene rubber, and composites thereof.

(4) Ambients Not in Tables. Ampacities at ambient temperatures other than those shown in the tables shall be determined by means of the following formula:

$$I_2 = I_1 \sqrt{\frac{TC - TA_2 - Delta\ TD}{TC - TA_1 - Delta\ TD}}$$

Where:

I_1 = ampacity from tables at ambient TA_1

I_2 = ampacity at desired ambient TA_2

TC = conductor temperature in degrees Celsius (°C)

TA_1 = surrounding ambient from tables in degrees Celsius (°C)

TA_2 = desired ambient in degrees Celsius (°C)

$Delta\ TD$ = dielectric loss temperature rise

(d) Engineering Supervision. Under engineering supervision, conductor ampacities shall be permitted to be calculated by means of the following general formula:

$$I = \sqrt{\frac{TC - (TA + Delta\ TD)}{RDC\ (1 + YC)RCA}}$$

Where:

TC = conductor temperature in °C

TA = ambient temperature in °C

$Delta\ TD$ = dielectric loss temperature rise

RDC = dc resistance of conductor at temperature TC

YC = component ac resistance resulting from skin effect and proximity effect

RCA = effective thermal resistance between conductor and surrounding ambient

FPN: See Appendix B for examples of formula applications.

Table 310-63. Thickness of Insulation and Jacket for Nonshielded Solid Dielectric Insulated Conductors Rated 2001 to 8000 Volts, in Mils

Conductor Size (AWG or kcmil)	2001–5000 Volts						5001–8000 Volts 100 Percent Insulation Level Wet or Dry Locations		
	Dry Locations, Single Conductor			Wet or Dry Locations					
	Without Jacket Insulation	With Jacket		Single Conductor		Multi-conductor* Insulation	Single Conductor		Multi-conductor* Insulation
		Insulation	Jacket	Insulation	Jacket		Insulation	Jacket	
8	110	90	30	125	80	90	180	80	180
6	110	90	30	125	80	90	180	80	180
4–2	110	90	45	125	80	90	180	95	180
1–2/0	110	90	45	125	80	90	180	95	180
3/0–4/0	110	90	65	125	95	90	180	110	180
213–500	120	90	65	140	110	90	210	110	210
501–750	130	90	65	155	125	90	235	125	235
751–1000	130	90	65	155	125	90	250	140	250

*Under a common overall covering such as a jacket, sheath, or armor.

Legend

▧ Backfill (earth or concrete)

◯ Electrical duct

● Cable or cables

Note 1: Minimum burial depths to top electrical ducts or cables shall be in accordance with Section 300-50. Maximum depth to the top of electrical duct banks shall be 30 in. and maximum depth to the top of direct buried cables shall be 36 in.

Note 2: For SI units: 1 in. = 25.4 mm; 1 ft = 0.3048 m.

Figure 310-60   Cable installation dimensions for use with Tables 310-77 through 310-86.

Table 310-64. Thickness of Insulation for Shielded Solid Dielectric Insulated Conductors Rated 2001 to 35,000 Volts, in Mils

Conductor Size (AWG or kcmil)	2001–5000 Volts	5001–8000 Volts		8001–15,000 Volts		15,001–25,000 Volts		25,001–28,000 Volts		28,001–35,000 Volts	
		100 Percent Insulation Level[1]	133 Percent Insulation Level[2]	100 Percent Insulation Level[1]	133 Percent Insulation Level[2]	100 Percent Insulation Level[1]	133 Percent Insulation Level[2]	100 Percent Insulation Level[1]	133 Percent Insulation Level[2]	100 Percent Insulation Level[1]	133 Percent Insulation Level[2]
8	90	—	—	—	—	—	—	—	—	—	—
6–4	90	115	140	—	—	—	—	—	—	—	—
2	90	115	140	175	215	—	—	—	—	—	—
1	90	115	140	175	215	260	345	280	345	—	—
1/0–2000	90	115	140	175	215	260	345	280	345	345	420

[1] **100 Percent Insulation Level.** Cables in this category shall be permitted to be applied where the system is provided with relay protection such that ground faults will be cleared as rapidly as possible but, in any case, within 1 minute. While these cables are applicable to the great majority of cable installations that are on grounded systems, they shall be permitted to be used also on other systems for which the application of cables is acceptable, provided the above clearing requirements are met in completely de-energizing the faulted section.

[2] **133 Percent Insulation Level.** This insulation level corresponds to that formerly designated for grounded systems. Cables in this category shall be permitted to be applied in situations where the clearing time requirements of the 100 percent level category cannot be met, and yet there is adequate assurance that the faulted section will be de-energized in a time not exceeding 1 hour. Also, they shall be permitted to be used where additional insulation strength over the 100 percent level category is desirable.

Table 310-67. Ampacities of Insulated Single Copper Conductor Cables Triplexed in Air Based on Conductor Temperatures of 90°C (194°F) and 105°C (221°F) and Ambient Air Temperature of 40°C (104°F)

| Conductor Size (AWG or kcmil) | Temperature Rating of Conductor (See Table 310-61) | | | |
| | 2001–5000 Volts Ampacity | | 5001–35,000 Volts Ampacity | |
	90°C (194°F) Type MV-90	105°C (221°F) Type MV-105	90°C (194°F) Type MV-90	105°C (221°F) Type MV-105
8	65	74	—	—
6	90	99	100	110
4	120	130	130	140
2	160	175	170	195
1	185	205	195	225
1/0	215	240	225	255
2/0	250	275	260	295
3/0	290	320	300	340
4/0	335	375	345	390
250	375	415	380	430
350	465	515	470	525
500	580	645	580	650
750	750	835	730	820
1000	880	980	850	950

Table 310-69. Ampacities of Insulated Single Copper Conductor Isolated in Air Based on Conductor Temperatures of 90°C (194°F) and 105°C (221°F) and Ambient Air Temperature of 40°C (104°F)

| Conductor Size (AWG or kcmil) | Temperature Rating of Conductor (See Table 310-61) | | | | | |
| | 2001–5000 Volts Ampacity | | 5001–15,000 Volts Ampacity | | 15,001–35,000 Volts Ampacity | |
	90°C (194°F) Type MV-90	105°C (221°F) Type MV-105	90°C (194°F) Type MV-90	105°C (221°F) Type MV-105	90°C (194°F) Type MV-90	105°C (221°F) Type MV-105
8	83	93	—	—	—	—
6	110	120	110	125	—	—
4	145	160	150	165	—	—
2	190	215	195	215	—	—
1	225	250	225	250	225	250
1/0	260	290	260	290	260	290
2/0	300	330	300	335	300	330
3/0	345	385	345	385	345	380
4/0	400	445	400	445	395	445
250	445	495	445	495	440	490
350	550	615	550	610	545	605
500	695	775	685	765	680	755
750	900	1000	885	990	870	970
1000	1075	1200	1060	1185	1040	1160
1250	1230	1370	1210	1350	1185	1320
1500	1365	1525	1345	1500	1315	1465
1750	1495	1665	1470	1640	1430	1595
2000	1605	1790	1575	1755	1535	1710

Table 310-68. Ampacities of Insulated Single Aluminum Conductor Cables Triplexed in Air Based on Conductor Temperatures of 90°C (194°F) and 105°C (221°F) and Ambient Air Temperature of 40°C (104°F)

| Conductor Size (AWG or kcmil) | Temperature Rating of Conductor (See Table 310-61) | | | |
| | 2001–5000 Volts Ampacity | | 5001–35,000 Volts Ampacity | |
	90°C (194°F) Type MV-90	105°C (221°F) Type MV-105	90°C (194°F) Type MV-90	105°C (221°F) Type MV-105
8	50	57	—	—
6	70	77	75	84
4	90	100	100	110
2	125	135	130	150
1	145	160	150	175
1/0	170	185	175	200
2/0	195	215	200	230
3/0	225	250	230	265
4/0	265	290	270	305
250	295	325	300	335
350	365	405	370	415
500	460	510	460	515
750	600	665	590	660
1000	715	800	700	780

Table 310-70. Ampacities of Insulated Single Aluminum Conductor Isolated in Air Based on Conductor Temperatures of 90°C (194°F) and 105°C (221°F) and Ambient Air Temperature of 40°C (104°F)

| Conductor Size (AWG or kcmil) | Temperature Rating of Conductor (See Table 310-61) | | | | | |
| | 2001–5000 Volts Ampacity | | 5001–15,000 Volts Ampacity | | 15,001–35,000 Volts Ampacity | |
	90°C (194°F) Type MV-90	105°C (221°F) Type MV-105	90°C (194°F) Type MV-90	105°C (221°F) Type MV-105	90°C (194°F) Type MV-90	105°C (221°F) Type MV-105
8	64	71	—	—	—	—
6	85	95	87	97	—	—
4	115	125	115	130	—	—
2	150	165	150	170	—	—
1	175	195	175	195	175	195
1/0	200	225	200	225	200	225
2/0	230	260	235	260	230	260
3/0	270	300	270	300	270	300
4/0	310	350	310	350	310	345
250	345	385	345	385	345	380
350	430	480	430	480	430	475
500	545	605	535	600	530	590
750	710	790	700	780	685	765
1000	855	950	840	940	825	920
1250	980	1095	970	1080	950	1055
1500	1105	1230	1085	1215	1060	1180
1750	1215	1355	1195	1335	1165	1300
2000	1320	1475	1295	1445	1265	1410

Table 310-71. Ampacities of an Insulated Three-Conductor Copper Cable Isolated in Air Based on Conductor Temperatures of 90°C (194°F) and 105°C (221°F) and Ambient Air Temperature of 40°C (104°F)

Conductor Size (AWG or kcmil)	Temperature Rating of Conductor (See Table 310-61)			
	2001–5000 Volts Ampacity		5001–35,000 Volts Ampacity	
	90°C (194°F) Type MV-90	105°C (221°F) Type MV-105	90°C (194°F) Type MV-90	105°C (221°F) Type MV-105
8	59	66	—	—
6	79	88	93	105
4	105	115	120	135
2	140	154	165	185
1	160	180	185	210
1/0	185	205	215	240
2/0	215	240	245	275
3/0	250	280	285	315
4/0	285	320	325	360
250	320	355	360	400
350	395	440	435	490
500	485	545	535	600
750	615	685	670	745
1000	705	790	770	860

Table 310-72. Ampacities of Insulated Three-Conductor Aluminum Cable Isolated in Air Based on Conductor Temperatures of 90°C (194°F) and 105°C (221°F) and Ambient Air Temperature of 40°C (104°F)

Conductor Size (AWG or kcmil)	Temperature Rating of Conductor (See Table 310-61)			
	2001–5000 Volts Ampacity		5001–35,000 Volts Ampacity	
	90°C (194°F) Type MV-90	105°C (221°F) Type MV-105	90°C (194°F) Type MV-90	105°C (221°F) Type MV-105
8	46	51	—	—
6	61	68	72	80
4	81	90	95	105
2	110	120	125	145
1	125	140	145	165
1/0	145	160	170	185
2/0	170	185	190	215
3/0	195	215	220	245
4/0	225	250	255	285
250	250	280	280	315
350	310	345	345	385
500	385	430	425	475
750	495	550	540	600
1000	585	650	635	705

Table 310-73. Ampacities of an Insulated Triplexed or Three Single-Conductor Copper Cables in Isolated Conduit in Air Based on Conductor Temperatures of 90°C (194°F) and 105°C (221°F) and Ambient Air Temperature of 40°C (104°F)

Conductor Size (AWG or kcmil)	Temperature Rating of Conductor (See Table 310-61)			
	2001–5000 Volts Ampacity		5001–35,000 Volts Ampacity	
	90°C (194°F) Type MV-90	105°C (221°F) Type MV-105	90°C (194°F) Type MV-90	105°C (221°F) Type MV-105
8	55	61	—	—
6	75	84	83	93
4	97	110	110	120
2	130	145	150	165
1	155	175	170	190
1/0	180	200	195	215
2/0	205	225	225	255
3/0	240	270	260	290
4/0	280	305	295	330
250	315	355	330	365
350	385	430	395	440
500	475	530	480	535
750	600	665	585	655
1000	690	770	675	755

Table 310-74. Ampacities of an Insulated Triplexed or Three Single-Conductor Aluminum Cables in Isolated Conduit in Air Based on Conductor Temperatures of 90°C (194°F) and 105°C (221°F) and Ambient Air Temperature of 40°C (104°F)

Conductor Size (AWG or kcmil)	Temperature Rating of Conductor (See Table 310-61)			
	2001–5000 Volts Ampacity		5001–35,000 Volts Ampacity	
	90°C (194°F) Type MV-90	105°C (221°F) Type MV-105	90°C (194°F) Type MV-90	105°C (221°F) Type MV-105
8	43	48	—	—
6	58	65	65	72
4	76	85	84	94
2	100	115	115	130
1	120	135	130	150
1/0	140	155	150	170
2/0	160	175	175	200
3/0	190	210	200	225
4/0	215	240	230	260
250	250	280	255	290
350	305	340	310	350
500	380	425	385	430
750	490	545	485	540
1000	580	645	565	640

Table 310-75. Ampacities of an Insulated Three-Conductor Copper Cable in Isolated Conduit in Air Based on Conductor Temperatures of 90°C (194°F) and 105°C (221°F) and Ambient Air Temperature of 40°C (104°F)

Conductor Size (AWG or kcmil)	2001–5000 Volts Ampacity		5001–35,000 Volts Ampacity	
	90°C (194°F) Type MV-90	105°C (221°F) Type MV-105	90°C (194°F) Type MV-90	105°C (221°F) Type MV-105
8	52	58	—	—
6	69	77	83	92
4	91	100	105	120
2	125	135	145	165
1	140	155	165	185
1/0	165	185	195	215
2/0	190	210	220	245
3/0	220	245	250	280
4/0	255	285	290	320
250	280	315	315	350
350	350	390	385	430
500	425	475	470	525
750	525	585	570	635
1000	590	660	650	725

Table 310-76. Ampacities of an Insulated Three-Conductor Aluminum Cable in Isolated Conduit in Air Based on Conductor Temperatures of 90°C (194°F) and 105°C (221°F) and Ambient Air Temperature of 40°C (104°F)

Conductor Size (AWG or kcmil)	2001–5000 Volts Ampacity		5001–35,000 Volts Ampacity	
	90°C (194°F) Type MV-90	105°C (221°F) Type MV-105	90°C (194°F) Type MV-90	105°C (221°F) Type MV-105
8	41	46	—	—
6	53	59	64	71
4	71	79	84	94
2	96	105	115	125
1	110	125	130	145
1/0	130	145	150	170
2/0	150	165	170	190
3/0	170	190	195	220
4/0	200	225	225	255
250	220	245	250	280
350	275	305	305	340
500	340	380	380	425
750	430	480	470	520
1000	505	560	550	615

Table 310-77. Ampacities of Three Single-Insulated Copper Conductors in Underground Electrical Ducts (Three Conductors per Electrical Duct) Based on Ambient Earth Temperature of 20°C (68°F), Electrical Duct Arrangement per Figure 310-60, 100 Percent Load Factor, Thermal Resistance (RHO) of 90, Conductor Temperatures of 90°C (194°F) and 105°C (221°F)

Conductor Size (AWG or kcmil)	2001–5000 Volts Ampacity		5001–35,000 Volts Ampacity	
	90°C (194°F) Type MV-90	105°C (221°F) Type MV-105	90°C (194°F) Type MV-90	105°C (221°F) Type MV-105
One Circuit (See Figure 310-60, Detail 1)				
8	64	69	—	—
6	85	92	90	97
4	110	120	115	125
2	145	155	155	165
1	170	180	175	185
1/0	195	210	200	215
2/0	220	235	230	245
3/0	250	270	260	275
4/0	290	310	295	315
250	320	345	325	345
350	385	415	390	415
500	470	505	465	500
750	585	630	565	610
1000	670	720	640	690
Three Circuits (See Figure 310-60, Detail 2)				
8	56	60	—	—
6	73	79	77	83
4	95	100	99	105
2	125	130	130	135
1	140	150	145	155
1/0	160	175	165	175
2/0	185	195	185	200
3/0	210	225	210	225
4/0	235	255	240	255
250	260	280	260	280
350	315	335	310	330
500	375	405	370	395
750	460	495	440	475
1000	525	565	495	535
Six Circuits (See Figure 310-60, Detail 3)				
8	48	52	—	—
6	62	67	64	68
4	80	86	82	88
2	105	110	105	115
1	115	125	120	125
1/0	135	145	135	145
2/0	150	160	150	165
3/0	170	185	170	185
4/0	195	210	190	205
250	210	225	210	225
350	250	270	245	265
500	300	325	290	310
750	365	395	350	375
1000	410	445	390	415

Table 310-78. Ampacities of Three Single-Insulated Aluminum Conductors in Underground Electrical Ducts (Three Conductors per Electrical Duct) Based on Ambient Earth Temperature of 20°C (68°F), Electrical Duct Arrangement per Figure 310-60, 100 Percent Load Factor, Thermal Resistance (RHO) of 90, Conductor Temperatures of 90°C (194°F) and 105°C (221°F)

| Conductor Size (AWG or kcmil) | Temperature Rating of Conductor (See Table 310-61) | | | |
| | 2001–5000 Volts Ampacity | | 5001–35,000 Volts Ampacity | |
	90°C (194°F) Type MV-90	105°C (221°F) Type MV-105	90°C (194°F) Type MV-90	105°C (221°F) Type MV-105
One Circuit (See Figure 310-60, Detail 1)				
8	50	54	—	—
6	66	71	70	75
4	86	93	91	98
2	115	125	120	130
1	130	140	135	145
1/0	150	160	155	165
2/0	170	185	175	190
3/0	195	210	200	215
4/0	225	245	230	245
250	250	270	250	270
350	305	325	305	330
500	370	400	370	400
750	470	505	455	490
1000	545	590	525	565
Three Circuits (See Figure 310-60, Detail 2)				
8	44	47	—	—
6	57	61	60	65
4	74	80	77	83
2	96	105	100	105
1	110	120	110	120
1/0	125	135	125	140
2/0	145	155	145	155
3/0	160	175	165	175
4/0	185	200	185	200
250	205	220	200	220
350	245	265	245	260
500	295	320	290	315
750	370	395	355	385
1000	425	460	405	440
Six Circuits (See Figure 310-60, Detail 3)				
8	38	41	—	—
6	48	52	50	54
4	62	67	64	69
2	80	86	80	88
1	91	98	90	99
1/0	105	110	105	110
2/0	115	125	115	125
3/0	135	145	130	145
4/0	150	165	150	160
250	165	180	165	175
350	195	210	195	210
500	240	255	230	250
750	290	315	280	305
1000	335	360	320	345

Table 310-79. Ampacities of Three Insulated Copper Conductors Cabled within an Overall Covering (Three-Conductor Cable) in Underground Electrical Ducts (One Cable per Electrical Duct) Based on Ambient Earth Temperature of 20°C (68°F), Electrical Duct Arrangement per Figure 310-60, 100 Percent Load Factor, Thermal Resistance (RHO) of 90, Conductor Temperatures of 90°C (194°F) and 105°C (221°C)

| Conductor Size (AWG or kcmil) | Temperature Rating of Conductor (See Table 310-61) | | | |
| | 2001–5000 Volts Ampacity | | 5001–35,000 Volts Ampacity | |
	90°C (194°F) Type MV-90	105°C (221°F) Type MV-105	90°C (194°F) Type MV-90	105°C (221°F) Type MV-105
One Circuit (See Figure 310-60, Detail 1)				
8	59	64	—	—
6	78	84	88	95
4	100	110	115	125
2	135	145	150	160
1	155	165	170	185
1/0	175	190	195	210
2/0	200	220	220	235
3/0	230	250	250	270
4/0	265	285	285	305
250	290	315	310	335
350	355	380	375	400
500	430	460	450	485
750	530	570	545	585
1000	600	645	615	660
Three Circuits (See Figure 310-60, Detail 2)				
8	53	57	—	—
6	69	74	75	81
4	89	96	97	105
2	115	125	125	135
1	135	145	140	155
1/0	150	165	160	175
2/0	170	185	185	195
3/0	195	210	205	220
4/0	225	240	230	250
250	245	265	255	270
350	295	315	305	325
500	355	380	360	385
750	430	465	430	465
1000	485	520	485	515
Six Circuits (See Figure 310-60, Detail 3)				
8	46	50	—	—
6	60	65	63	68
4	77	83	81	87
2	98	105	105	110
1	110	120	115	125
1/0	125	135	130	145
2/0	145	155	150	160
3/0	165	175	170	180
4/0	185	200	190	200
250	200	220	205	220
350	240	270	245	275
500	290	310	290	305
750	350	375	340	365
1000	390	420	380	405

Table 310-80. Ampacities of Three Insulated Aluminum Conductors Cabled within an Overall Covering (Three-Conductor Cable) in Underground Electrical Ducts (One Cable per Electrical Duct) Based on Ambient Earth Temperature of 20°C (68°F), Electrical Duct Arrangement per Figure 310-60, 100 Percent Load Factor, Thermal Resistance (RHO) of 90, Conductor Temperatures of 90°C (194°F) and 105°C (221°F)

| | Temperature Rating of Conductor (See Table 310-61) | | | |
| | 2001–5000 Volts Ampacity | | 5001–35,000 Volts Ampacity | |
Conductor Size (AWG or kcmil)	90°C (194°F) Type MV-90	105°C (221°F) Type MV-105	90°C (194°F) Type MV-90	105°C (221°F) Type MV-105
One Circuit (See Figure 310-60, Detail 1)				
8	46	50	—	—
6	61	66	69	74
4	80	86	89	96
2	105	110	115	125
1	120	130	135	145
1/0	140	150	150	165
2/0	160	170	170	185
3/0	180	195	195	210
4/0	205	220	220	240
250	230	245	245	265
350	280	310	295	315
500	340	365	355	385
750	425	460	440	475
1000	495	535	510	545
Three Circuits (See Figure 310-60, Detail 2)				
8	41	44	—	—
6	54	58	59	64
4	70	75	75	81
2	90	97	100	105
1	105	110	110	120
1/0	120	125	125	135
2/0	135	145	140	155
3/0	155	165	160	175
4/0	175	185	180	195
250	190	205	200	215
350	230	250	240	255
500	280	300	285	305
750	345	375	350	375
1000	400	430	400	430
Six Circuits (See Figure 310-60, Detail 3)				
8	36	39	—	—
6	46	50	49	53
4	60	65	63	68
2	77	83	80	86
1	87	94	90	98
1/0	99	105	105	110
2/0	110	120	115	125
3/0	130	140	130	140
4/0	145	155	150	160
250	160	170	160	170
350	190	205	190	205
500	230	245	230	245
750	280	305	275	295
1000	320	345	315	335

Table 310-81. Ampacities of Single Insulated Copper Conductors Directly Buried in Earth Based on Ambient Earth Temperature of 20°C (68°F), Arrangement per Figure 310-60, 100 Percent Load Factor, Thermal Resistance (RHO) of 90, Conductor Temperatures of 90°C (194°F) and 105°C (221°F)

| | Temperature Rating of Conductor (See Table 310-61) | | | |
| | 2001–5000 Volts Ampacity | | 5001–35,000 Volts Ampacity | |
Conductor Size (AWG or kcmil)	90°C (194°F) Type MV-90	105°C (221°F) Type MV-105	90°C (194°F) Type MV-90	105°C (221°F) Type MV-105
One Circuit, Three Conductors (See Figure 310-60, Detail 9)				
8	110	115	—	—
6	140	150	130	140
4	180	195	170	180
2	230	250	210	225
1	260	280	240	260
1/0	295	320	275	295
2/0	335	365	310	335
3/0	385	415	355	380
4/0	435	465	405	435
250	470	510	440	475
350	570	615	535	575
500	690	745	650	700
750	845	910	805	865
1000	980	1055	930	1005
Two Circuits, Six Conductors (See Figure 310-60, Detail 10)				
8	100	110	—	—
6	130	140	120	130
4	165	180	160	170
2	215	230	195	210
1	240	260	225	240
1/0	275	295	255	275
2/0	310	335	290	315
3/0	355	380	330	355
4/0	400	430	375	405
250	435	470	410	440
350	520	560	495	530
500	630	680	600	645
750	775	835	740	795
1000	890	960	855	920

Table 310-82. Ampacities of Single Insulated Aluminum Conductors Directly Buried in Earth Based on Ambient Earth Temperature of 20°C (68°F), Arrangement per Figure 310-60, 100 Percent Load Factor, Thermal Resistance (RHO) of 90, Conductor Temperatures of 90°C (194°F) and 105°C (221°F)

Conductor Size (AWG or kcmil)	Temperature Rating of Conductor (See Table 310-61)			
	2001–5000 Volts Ampacity		5001–35,000 Volts Ampacity	
	90°C (194°F) Type MV-90	105°C (221°F) Type MV-105	90°C (194°F) Type MV-90	105°C (221°F) Type MV-105
One Circuit, Three Conductors (See Figure 310-60, Detail 9)				
8	85	90	—	—
6	110	115	100	110
4	140	150	130	140
2	180	195	165	175
1	205	220	185	200
1/0	230	250	215	230
2/0	265	285	245	260
3/0	300	320	275	295
4/0	340	365	315	340
250	370	395	345	370
350	445	480	415	450
500	540	580	510	545
750	665	720	635	680
1000	780	840	740	795
Two Circuits, Six Conductors (See Figure 310-60, Detail 10)				
8	80	85	—	—
6	100	110	95	100
4	130	140	125	130
2	165	180	155	165
1	190	200	175	190
1/0	215	230	200	215
2/0	245	260	225	245
3/0	275	295	255	275
4/0	310	335	290	315
250	340	365	320	345
350	410	440	385	415
500	495	530	470	505
750	610	655	580	625
1000	710	765	680	730

Table 310-83. Ampacities of Three Insulated Copper Conductors Cabled within an Overall Covering (Three-Conductor Cable), Directly Buried in Earth Based on Ambient Earth Temperature of 20°C (68°F), Arrangement per Figure 310-60, 100 Percent Load Factor, Thermal Resistance (RHO) of 90, Conductor Temperatures of 90°C (194°F) and 105°C (221°F)

Conductor Size (AWG or kcmil)	Temperature Rating of Conductor (See Table 310-61)			
	2001–5000 Volts Ampacity		5001–35,000 Volts Ampacity	
	90°C (194°F) Type MV-90	105°C (221°F) Type MV-105	90°C (194°F) Type MV-90	105°C (221°F) Type MV-105
One Circuit (See Figure 310-60, Detail 5)				
8	85	89	—	—
6	105	115	115	120
4	135	150	145	155
2	180	190	185	200
1	200	215	210	225
1/0	230	245	240	255
2/0	260	280	270	290
3/0	295	320	305	330
4/0	335	360	350	375
250	365	395	380	410
350	440	475	460	495
500	530	570	550	590
750	650	700	665	720
1000	730	785	750	810
Two Circuits (See Figure 310-60, Detail 10)				
8	80	84	—	—
6	100	105	105	115
4	130	140	135	145
2	165	180	170	185
1	185	200	195	210
1/0	215	230	220	235
2/0	240	260	250	270
3/0	275	295	280	305
4/0	310	335	320	345
250	340	365	350	375
350	410	440	420	450
500	490	525	500	535
750	595	640	605	650
1000	665	715	675	730

Table 310-84. Ampacities of Three Insulated Aluminum Conductors Cabled within an Overall Covering (Three-Conductor Cable), Directly Buried in Earth Based on Ambient Earth Temperature of 20°C (68°F), Arrangement per Figure 310-60, 100 Percent Load Factor, Thermal Resistance (RHO) of 90, Conductor Temperatures of 90°C (194°F) and 105°C (221°F)

Conductor Size (AWG or kcmil)	2001–5000 Volts Ampacity		5001–35,000 Volts Ampacity	
	90°C (194°F) Type MV-90	105°C (221°F) Type MV-105	90°C (194°F) Type MV-90	105°C (221°F) Type MV-105
One Circuit (See Figure 310-60, Detail 5)				
8	65	70	—	—
6	80	88	90	95
4	105	115	115	125
2	140	150	145	155
1	155	170	165	175
1/0	180	190	185	200
2/0	205	220	210	225
3/0	230	250	240	260
4/0	260	280	270	295
250	285	310	300	320
350	345	375	360	390
500	420	450	435	470
750	520	560	540	580
1000	600	650	620	665
Two Circuits (See Figure 310-60, Detail 6)				
8	60	66	—	—
6	75	83	80	95
4	100	110	105	115
2	130	140	135	145
1	145	155	150	165
1/0	165	180	170	185
2/0	190	205	195	210
3/0	215	230	220	240
4/0	245	260	250	270
250	265	285	275	295
350	320	345	330	355
500	385	413	393	423
750	480	515	485	525
1000	550	590	560	600

Table 310-85. Ampacities of Three Triplexed Single Insulated Copper Conductors Directly Buried in Earth Based on Ambient Earth Temperature of 20°C (68°F), Arrangement per Figure 310-60, 100 Percent Load Factor, Thermal Resistance (RHO) of 90, Conductor Temperatures 90°C (194°F) and 105°C (221°F)

Conductor Size (AWG or kcmil)	2001–5000 Volts Ampacity		5001–35,000 Volts Ampacity	
	90°C (194°F) Type MV-90	105°C (221°F) Type MV-105	90°C (194°F) Type MV-90	105°C (221°F) Type MV-105
One Circuit, Three Conductors (See Figure 310-60, Detail 7)				
8	90	95	—	—
6	120	130	115	120
4	150	165	150	160
2	195	205	190	205
1	225	240	215	230
1/0	255	270	245	260
2/0	290	310	275	295
3/0	330	360	315	340
4/0	375	405	360	385
250	410	445	390	410
350	490	580	470	505
500	590	635	565	605
750	725	780	685	740
1000	825	885	770	830
Two Circuits, Six Conductors (See Figure 310-60, Detail 8)				
8	85	90	—	—
6	110	115	105	115
4	140	150	140	150
2	180	195	175	190
1	205	220	200	215
1/0	235	250	225	240
2/0	265	285	255	275
3/0	300	320	290	315
4/0	340	365	325	350
250	370	395	355	380
350	445	480	425	455
500	535	575	510	545
750	650	700	615	660
1000	740	795	690	745

Table 310-86. Ampacities of Three Triplexed Single Insulated Copper Conductors Directly Buried in Earth Based on Ambient Earth Temperature of 20°C (68°F), Arrangement per Figure 310-60, 100 Percent Load Factor, Thermal Resistance (RHO) of 90, Conductor Temperatures 90°C (194°F) and 105°C (221°F)

| | Temperature Rating of Conductor (See Table 310-61) | | | |
| | 2001–5000 Volts Ampacity | | 5001–35,000 Volts Ampacity | |
Conductor Size (AWG or kcmil)	90°C (194°F) Type MV-90	105°C (221°F) Type MV-105	90°C (194°F) Type MV-90	105°C (221°F) Type MV-105
One Circuit, Three Conductors (See Figure 310-60, Detail 7)				
8	70	75	—	—
6	90	100	90	95
4	120	130	115	125
2	155	165	145	155
1	175	190	165	175
1/0	200	210	190	205
2/0	225	240	215	230
3/0	255	275	245	265
4/0	290	310	280	305
250	320	350	305	325
350	385	420	370	400
500	465	500	445	480
750	580	625	550	590
1000	670	725	635	680
Two Circuits, Six Conductors (See Figure 310-60, Detail 8)				
8	65	70	—	—
6	85	95	85	90
4	110	120	105	115
2	140	150	135	145
1	160	170	155	170
1/0	180	195	175	190
2/0	205	220	200	215
3/0	235	250	225	245
4/0	265	285	255	275
250	290	310	280	300
350	350	375	335	360
500	420	455	405	435
750	520	560	485	525
1000	600	645	565	605

Article 318 — Cable Trays

318-1. Scope. This article covers cable tray systems, including ladder, ventilated trough, ventilated channel, solid bottom, and other similar structures.

318-2. Definition.

Cable Tray System. A unit or assembly of units or sections and associated fittings forming a rigid structural system used to securely fasten or support cables and raceways.

318-3. Uses Permitted. Cable tray installations shall not be limited to industrial establishments.

(a) Wiring Methods. The following shall be permitted to be installed in cable tray systems under the conditions described in their respective articles and sections:

	Section	Article
Armored cable		333
Electrical metallic tubing		348
Electrical nonmetallic tubing		331
Fire alarm cables		760
Flexible metal conduit		350
Flexible metallic tubing		349
Instrumentation tray cable		727
Intermediate metal conduit		345
Liquidtight flexible metal conduit and liquidtight flexible nonmetallic conduit		351
Metal-clad cable		334
Mineral-insulated, metal-sheathed cable		330
Multiconductor service-entrance cable		338
Multiconductor underground feeder and branch-circuit cable		339
Multipurpose and communications cables		800
Nonmetallic-sheathed cable		336
Power and control tray cable		340
Power-limited tray cable	725-61(c) and 725-71(e)	
Optical fiber cables		770
Other factory-assembled, multiconductor control, signal, or power cables that are specifically approved for installation in cable trays		
Rigid metal conduit		346
Rigid nonmetallic conduit		347

(b) In Industrial Establishments. The wiring methods in Section 318-3(a) shall be permitted to be used in any industrial establishment under the conditions described in their respective articles. In industrial establishments only, where conditions of maintenance and supervision ensure that only qualified persons will service the installed cable tray system, any of the cables in (1) and (2) shall be permitted to be installed in ladder, ventilated trough, or ventilated channel cable trays.

(1) Single Conductors. Single conductor cables shall be permitted to be installed in accordance with the following.

(a) Single conductor cable shall be No. 1/0 or larger and shall be of a type listed and marked on the surface for use in cable trays. Where Nos. 1/0 through 4/0 single conductor cables are installed in ladder cable tray, the maximum allowable rung spacing for the ladder cable tray shall be 9 in. (229 mm). Where exposed to direct rays of the sun, cables shall be identified as being sunlight resistant.

(b) Welding cables shall comply with the provisions of Article 630, Part D.

(c) Single conductors used as equipment grounding conductors shall be insulated, covered, or bare and they shall be No. 4 or larger.

(2) Multiconductor. Multiconductor cables, Type MV (Article 326) where exposed to direct rays of the sun, shall be identified as being sunlight resistant.

(c) Equipment Grounding Conductors. Metallic cable trays shall be permitted to be used as equipment grounding conductors where continuous maintenance and supervision ensure that qualified persons will service the installed cable tray system and the cable tray complies with provisions of Section 318-7.

(d) Hazardous (Classified) Locations. Cable trays in hazardous (classified) locations shall contain only the cable types permitted in Sections 501-4, 502-4, 503-3, and 504-20.

(e) Nonmetallic Cable Tray. Nonmetallic cable tray shall be permitted in corrosive areas and in areas requiring voltage isolation.

318-4. Uses Not Permitted. Cable tray systems shall not be used in hoistways or where subject to severe physical damage. Cable tray systems shall not be used in environmental airspaces, except as permitted in Section 300-22, to support wiring methods recognized for use in such spaces.

318-5. Construction Specifications.

(a) Strength and Rigidity. Cable trays shall have suitable strength and rigidity to provide adequate support for all contained wiring.

(b) Smooth Edges. Cable trays shall not have sharp edges, burrs, or projections that may damage the insulation or jackets of the wiring.

(c) Corrosion Protection. Cable tray systems shall be corrosion resistant. If made of ferrous material, the system shall be protected from corrosion as required by Section 300-6.

(d) Side Rails. Cable trays shall have side rails or equivalent structural members.

(e) Fittings. Cable trays shall include fittings or other suitable means for changes in direction and elevation of runs.

(f) Nonmetallic Cable Tray. Nonmetallic cable trays shall be made of flame-retardant material.

318-6. Installation.

(a) Complete System. Cable trays shall be installed as a complete system. Field bends or modifications shall be made so that the electrical continuity of the cable tray system and support for the cables shall be maintained. Cable tray systems shall be permitted to have mechanically discontinuous segments between cable tray runs or between cable tray runs and equipment. The system shall provide for the support of the cables in accordance with their corresponding articles.

Where cable trays support individual conductors and where the conductors pass from one cable tray to another, or from a cable tray to raceways or to equipment where the conductors are terminated, the support distance between cable trays or between the cable tray and the equipment shall not exceed 6 ft (1.83 m). The conductors shall be secured to the cable tray(s) at the transition and they shall be protected, by guarding or by location, from physical damage.

A bonding jumper sized in accordance with Section 250-102 shall connect the two sections of cable tray, or the cable tray and the raceway or equipment. Bonding shall be in accordance with Section 250-96.

(b) Completed Before Installation. Each run of cable tray shall be completed before the installation of cables.

(c) Supports. Supports shall be provided to prevent stress on cables where they enter raceways or other enclosures from cable tray systems.

(d) Covers. In portions of runs where additional protection is required, covers or enclosures providing the required protection shall be of a material that is compatible with the cable tray.

(e) Multiconductor Cables Rated 600 Volts or Less. Multiconductor cables rated 600 volts or less shall be permitted to be installed in the same cable tray.

(f) Cables Rated Over 600 Volts. Cables rated over 600 volts and those rated 600 volts or less installed in the same cable tray shall comply with either (1) or (2).

(1) The cables rated over 600 volts are Type MC.

(2) The cables rated over 600 volts are separated from the cables rated 600 volts or less by a solid fixed barrier of a material compatible with the cable tray.

(g) Through Partitions and Walls. Cable trays shall be permitted to extend transversely through partitions and walls or vertically through platforms and floors in wet or dry locations where the installations, complete with installed cables, are made in accordance with the requirements of Section 300-21.

(h) Exposed and Accessible. Cable trays shall be exposed and accessible except as permitted by Section 318-6(g).

(i) Adequate Access. Sufficient space shall be provided and maintained about cable trays to permit adequate access for installing and maintaining the cables.

(j) Raceways, Cables, and Outlet Boxes Supported from Cable Trays. In industrial facilities where conditions of maintenance and supervision ensure only qualified persons will service the installation and where cable trays are designed to support the load, raceways, cables, and outlet boxes shall be permitted to be supported from cable trays. For raceway terminating at the tray, a listed cable tray clamp or adapter shall be used and no nearby support, such as a support within 3 ft (914 mm), shall be required.

For raceway or cable running parallel to, but under or beside, a tray, support shall be in accordance with the requirements of the appropriate raceway or cable article.

For outlet boxes located under or beside a tray, support shall be in accordance with the requirements of Article 370.

318-7. Grounding.

(a) Metallic Cable Trays. Metallic cable trays that support electrical conductors shall be grounded as required for conductor enclosures in Article 250.

(b) Steel or Aluminum Cable Tray Systems. Steel or aluminum cable tray systems shall be permitted to be used as equipment grounding conductors provided that all the following requirements are met.

(1) The cable tray sections and fittings shall be identified for grounding purposes.

(2) The minimum cross-sectional area of cable trays shall conform to the requirements in Table 318-7(b)(2).

Table 318-7(b)(2). Metal Area Requirements for Cable Trays Used as Equipment Grounding Conductors

Maximum Fuse Ampere Rating, Circuit Breaker Ampere Trip Setting, or Circuit Breaker Protective Relay Ampere Trip Setting for Ground-Fault Protection of Any Cable Circuit in the Cable Tray System	Minimum Cross-Sectional Area of Metal[a] (in.^{2})	
	Steel Cable Trays	Aluminum Cable Trays
60	0.20	0.20
100	0.40	0.20
200	0.70	0.20
400	1.00	0.40
600	1.50[b]	0.40
1000	—	0.60
1200	—	1.00
1600	—	1.50
2000	—	2.00[b]

Note: For SI units, 1 in.^{2} = 645 sq mm^2.
[a]Total cross-sectional area of both side rails for ladder or trough cable trays; or the minimum cross-sectional area of metal in channel cable trays or cable trays of one-piece construction.
[b]Steel cable trays shall not be used as equipment grounding conductors for circuits with ground-fault protection above 600 amperes. Aluminum cable trays shall not be used as equipment grounding conductors for circuits with ground-fault protection above 2000 amperes.

(3) All cable tray sections and fittings shall be legibly and durably marked to show the cross-sectional area of metal in channel cable trays, or cable trays of one-piece construction, and the total cross-sectional area of both side rails for ladder or trough cable trays.

(4) Cable tray sections, fittings, and connected raceways shall be bonded in accordance with Section 250-96 using bolted mechanical connectors or bonding jumpers sized and installed in accordance with Section 250-102.

318-8. Cable Installation.

(a) Cable Splices. Cable splices made and insulated by approved methods shall be permitted to be located within a cable tray provided they are accessible and do not project above the side rails.

(b) Fastened Securely. In other than horizontal runs, the cables shall be fastened securely to transverse members of the cable trays.

(c) Bushed Conduit and Tubing. A box shall not be required where cables or conductors are installed in bushed conduit and tubing used for support or for protection against physical damage.

(d) Connected in Parallel. Where single conductor cables comprising each phase or neutral of a circuit are connected in parallel as permitted in Section 310-4, the conductors shall be installed in groups consisting of not more than one conductor per phase or neutral to prevent current unbalance in the paralleled conductors due to inductive reactance.

Single conductors shall be securely bound in circuit groups to prevent excessive movement due to fault-current magnetic forces unless single conductors are cabled together, such as triplexed assemblies.

(e) Single Conductors. Where any of the single conductors installed in ladder or ventilated trough cable trays are Nos. 1/0 through 4/0, all single conductors shall be installed in a single layer. Conductors that are bound together to comprise each circuit group shall be permitted to be installed in other than a single layer.

318-9. Number of Multiconductor Cables, Rated 2000 Volts or Less, in Cable Trays. The number of multiconductor cables, rated 2000 volts or less, permitted in a single cable tray shall not exceed the requirements of this section. The conductor sizes herein apply to both aluminum and copper conductors.

(a) Any Mixture of Cables. Where ladder or ventilated trough cable trays contain multiconductor power or lighting cables, or any mixture of multiconductor power, lighting, control, and signal cables, the maximum number of cables shall conform to the following.

(1) Where all of the cables are No. 4/0 or larger, the sum of the diameters of all cables shall not exceed the cable tray width, and the cables shall be installed in a single layer.

(2) Where all of the cables are smaller than No. 4/0, the sum of the cross-sectional areas of all cables shall not exceed the maximum allowable cable fill area in Column 1 of Table 318-9 for the appropriate cable tray width.

(3) Where No. 4/0 or larger cables are installed in the same cable tray with cables smaller than No. 4/0, the sum of the cross-sectional areas of all cables smaller than No. 4/0 shall not exceed the maximum allowable fill area resulting from the computation in Column 2 of Table 318-9 for the appropriate cable tray width. The No. 4/0 and larger cables shall be installed in a single layer, and no other cables shall be placed on them.

(b) Multiconductor Control and/or Signal Cables Only. Where a ladder or ventilated trough cable tray, having a usable inside depth of 6 in. (152 mm) or less, contains multiconductor control and/or signal cables only, the sum of the cross-sectional areas of all cables at any cross section shall not exceed 50 percent of the interior cross-sectional area of the cable tray. A depth of 6 in. (152 mm) shall be

Table 318-9. Allowable Cable Fill Area for Multiconductor Cables in Ladder, Ventilated Trough, or Solid Bottom Cable Trays for Cables Rated 2000 Volts or Less

	Maximum Allowable Fill Area for Multiconductor Cables			
	Ladder or Ventilated Trough Cable Trays, Section 318-9(a)		Solid Bottom Cable Trays, Section 318-9(c)	
Inside Width of Cable Tray (in.)	Column 1 Applicable for Section 318-9(a)(2) Only (in.²)	Column 2[a] Applicable for Section 318-9(a)(3) Only (in.²)	Column 3 Applicable for Section 318-9(c)(2) Only (in.²)	Column 4[a] Applicable for Section 318-9(c)(3) Only (in.²)
6.0	7.0	7–(1.2 Sd)[b]	5.5	5.5–Sd[b]
9.0	10.5	10.5–(1.2 Sd)	8.0	8.0–Sd
12.0	14.0	14–(1.2 Sd)	11.0	11.0–Sd
18.0	21.0	21–(1.2 Sd)	16.5	16.5–Sd
24.0	28.0	28–(1.2 Sd)	22.0	22.0–Sd
30.0	35.0	35–(1.2 Sd)	27.5	27.5–Sd
36.0	42.0	42–(1.2 Sd)	33.0	33.0–Sd

Note: For SI units, 1 in.² = 645 mm².

[a]The maximum allowable fill areas in Columns 2 and 4 shall be computed. For example, the maximum allowable fill, in square inches, for a 6-in. (152-mm) wide cable tray in Column 2 shall be 7 minus (1.2 multiplied by Sd).

[b]The term *Sd* in Columns 2 and 4 is equal to the sum of the diameters, in inches, of all Nos. 4/0 and larger multiconductor cables in the same cable tray with smaller cables.

used to compute the allowable interior cross-sectional area of any cable tray that has a usable inside depth of more than 6 in. (152 mm).

(c) Solid Bottom Cable Trays Containing Any Mixture. Where solid bottom cable trays contain multiconductor power or lighting cables, or any mixture of multiconductor power, lighting, control, and signal cables, the maximum number of cables shall conform to the following.

(1) Where all of the cables are No. 4/0 or larger, the sum of the diameters of all cables shall not exceed 90 percent of the cable tray width, and the cables shall be installed in a single layer.

(2) Where all of the cables are smaller than No. 4/0, the sum of the cross-sectional areas of all cables shall not exceed the maximum allowable cable fill area in Column 3 of Table 318-9 for the appropriate cable tray width.

(3) Where No. 4/0 or larger cables are installed in the same cable tray with cables smaller than No. 4/0, the sum of the cross-sectional areas of all cables smaller than No. 4/0 shall not exceed the maximum allowable fill area resulting from the computation in Column 4 of Table 318-9 for the appropriate cable tray width. The No. 4/0 and larger

cables shall be installed in a single layer, and no other cables shall be placed on them.

(d) Solid Bottom Cable Tray — Multiconductor Control and/or Signal Cables Only. Where a solid bottom cable tray, having a usable inside depth of 6 in. (152 mm) or less, contains multiconductor control and/or signal cables only, the sum of the cross-sectional areas of all cables at any cross section shall not exceed 40 percent of the interior cross-sectional area of the cable tray. A depth of 6 in. (152 mm) shall be used to compute the allowable interior cross-sectional area of any cable tray that has a usable inside depth of more than 6 in. (152 mm).

(e) Ventilated Channel Cable Trays. Where ventilated channel cable trays contain multiconductor cables of any type, the following shall apply.

(1) Where only one multiconductor cable is installed, the cross-sectional area shall not exceed the value specified in Column 1 of Table 318-9(e).

(2) Where more than one multiconductor cable is installed, the sum of the cross-sectional area of all cables shall not exceed the value specified in Column 2 of Table 318-9(e).

Table 318-9(e). Allowable Cable Fill Area for Multiconductor Cables in Ventilated Channel Cable Trays for Cables Rated 2000 Volts or Less

Inside Width of Cable Tray (in.)	Maximum Allowable Fill Area for Multiconductor Cables (in.^{2})	
	Column 1 One Cable	Column 2 More than One Cable
3	2.3	1.3
4	4.5	2.5
6	7.0	3.8

Note: For SI units, 1 in.^{2} = 645 mm^2.

318-10. Number of Single Conductor Cables, Rated 2000 Volts or Less, in Cable Trays. The number of single conductor cables, rated 2000 volts or less, permitted in a single cable tray section shall not exceed the requirements of this section. The single conductors, or conductor assemblies, shall be evenly distributed across the cable tray. The conductor sizes herein apply to both aluminum and copper conductors.

(a) Ladder or Ventilated Trough Cable Trays. Where ladder or ventilated trough cable trays contain single conductor cables, the maximum number of single conductors shall conform to the following.

(1) Where all of the cables are 1000 kcmil or larger, the sum of the diameters of all single conductor cables shall not exceed the cable tray width.

(2) Where all of the cables are from 250 kcmil up to 1000 kcmil, the sum of the cross-sectional areas of all single conductor cables shall not exceed the maximum allowable cable fill area in Column 1 of Table 318-10, for the appropriate cable tray width.

(3) Where 1000 kcmil or larger single conductor cables are installed in the same cable tray with single conductor cables smaller than 1000 kcmil, the sum of the cross-sectional areas of all cables smaller than 1000 kcmil shall not exceed the maximum allowable fill area resulting from the computation in Column 2 of Table 318-10, for the appropriate cable tray width.

(4) Where any of the single conductor cables are Nos. 1/0 through 4/0, the sum of the diameters of all single conductor cables shall not exceed the cable tray width.

(b) Ventilated Channel Cable Trays. Where 3-in. (76-mm), 4-in. (102-mm), or 6-in. (152-mm) wide ventilated channel cable trays contain single conductor cables, the sum of the diameters of all single conductors shall not exceed the inside width of the channel.

Table 318-10. Allowable Cable Fill Area for Single Conductor Cables in Ladder or Ventilated Trough Cable Trays for Cables Rated 2000 Volts or Less

Inside Width of Cable Tray (in.)	Maximum Allowable Fill Area for Single Conductor Cables in Ladder or Ventilated Trough Cable Trays	
	Column 1 Applicable for Section 318-10(a)(2) Only (in.^{2})	Column 2[a] Applicable for Section 318-10(a)(3) Only (in.^{2})
6	6.5	6.5–(1.1 Sd)[b]
9	9.5	9.5–(1.1 Sd)
12	13.0	13.0–(1.1 Sd)
18	19.5	19.5–(1.1 Sd)
24	26.0	26.0–(1.1 Sd)
30	32.5	32.5–(1.1 Sd)
36	39.0	39.0–(1.1 Sd)

Note: For SI units, 1 in.^{2} = 645 mm^2.

[a]The maximum allowable fill areas in Column 2 shall be computed. For example, the maximum allowable fill, in square inches, for a 6-in. (152-mm) wide cable tray shall be 6.5 minus (1.1 multiplied by Sd).

[b]The term Sd in Column 2 is equal to the sum of the diameters, in inches, of all 1000 kcmil and larger single conductor cables in the same ladder or ventilated trough cable tray with small cables.

318-11. Ampacity of Cables, Rated 2000 Volts or Less, in Cable Trays.

(a) Multiconductor Cables. The allowable ampacity of multiconductor cables, nominally rated 2000 volts or less, installed according to the requirements of Section 318-9, shall be as given in Tables 310-16 and 310-18, subject to the provisions of (1), (2), and (3).

(1) The derating factors of Section 310-15(b)(2)(a) shall apply only to multiconductor cables with more than three current-carrying conductors. Derating shall be limited to the number of current-carrying conductors in the cable and not to the number of conductors in the cable tray.

(2) Where cable trays are continuously covered for more than 6 ft (1.83 m) with solid unventilated covers, not over 95 percent of the allowable ampacities of Tables 310-16 and 310-18 shall be permitted for multiconductor cables.

(3) Where multiconductor cables are installed in a single layer in uncovered trays, with a maintained spacing of not less than one cable diameter between cables, the ampacity shall not exceed the allowable ambient temperature-corrected ampacities of multiconductor cables, with not more than three insulated conductors rated 0 through 2000 volts in free air, in accordance with Section 310-15(c).

FPN: See Table B-310-3 in Appendix B.

(b) Single Conductor Cables. The derating factors of Section 310-15(b)(2)(a), shall not apply to the ampacity of cables in cable trays. The ampacity of single conductor cables, or single conductors cabled together (triplexed, quadruplexed, etc.), nominally rated 2000 volts or less, shall comply with the following.

(1) Where installed according to the requirements of Section 318-10, the ampacities for 600 kcmil and larger single conductor cables in uncovered cable trays shall not exceed 75 percent of the allowable ampacities in Tables 310-17 and 310-19. Where cable trays are continuously covered for more than 6 ft (1.83 m) with solid unventilated covers, the ampacities for 600 kcmil and larger cables shall not exceed 70 percent of the allowable ampacities in Tables 310-17 and 310-19.

(2) Where installed according to the requirements of Section 318-10, the ampacities for No. 1/0 through 500 kcmil single conductor cables in uncovered cable trays shall not exceed 65 percent of the allowable ampacities in Tables 310-17 and 310-19. Where cable trays are continuously covered for more than 6 ft (1.83 m) with solid unventilated covers, the ampacities for No. 1/0 through 500 kcmil cables shall not exceed 60 percent of the allowable ampacities in Tables 310-17 and 310-19.

(3) Where single conductors are installed in a single layer in uncovered cable trays, with a maintained space of not less than one cable diameter between individual conductors, the ampacity of Nos. 1/0 and larger cables shall not exceed the allowable ampacities in Tables 310-17 and 310-19.

(4) Where single conductors are installed in a triangular or square configuration in uncovered cable trays, with a maintained free air space of not less than 2.15 times one conductor diameter (2.15 × O.D.) of the largest conductor contained within the configuration and adjacent conductor configurations or cables, the ampacity of Nos. 1/0 and larger cables shall not exceed the allowable ampacities of two or three single insulated conductors rated 0 through 2000 volts supported on a messenger in accordance with Section 310-15(b).

FPN: See Table 310-20.

318-12. Number of Type MV and Type MC Cables (2001 Volts or Over) in Cable Trays. The number of cables, rated 2001 volts or over, permitted in a single cable tray shall not exceed the requirements of this section.

The sum of the diameters of single conductor and multiconductor cables shall not exceed the cable tray width, and the cables shall be installed in a single layer. Where single conductor cables are triplexed, quadruplexed, or bound together in circuit groups, the sum of the diameters of the single conductors shall not exceed the cable tray width, and these groups shall be installed in single layer arrangement.

318-13. Ampacity of Type MV and Type MC Cables (2001 Volts or Over) in Cable Trays. The ampacity of cables, rated 2001 volts, nominal, or over, installed according to Section 318-12 shall not exceed the requirements of this section.

(a) Multiconductor Cables (2001 Volts or Over). The allowable ampacity of multiconductor cables shall be as given in Tables 310-75 and 310-76, subject to the provisions of (1) and (2).

(1) Where cable trays are continuously covered for more than 6 ft (1.83 m) with solid unventilated covers, not more than 95 percent of the allowable ampacities of Tables 310-75 and 310-76 shall be permitted for multiconductor cables.

(2) Where multiconductor cables are installed in a single layer in uncovered cable trays, with maintained spacing of not less than one cable diameter between cables, the ampacity shall not exceed the allowable ampacities of Tables 310-71 and 310-72.

(b) Single Conductor Cables (2001 Volts or Over). The ampacity of single conductor cables, or single conductors cabled together (triplexed, quadruplexed, etc.), shall comply with the following.

(1) The ampacities for Nos. 1/0 and larger single conductor cables in uncovered cable trays shall not exceed 75 percent of the allowable ampacities in Tables 310-69 and 310-70. Where the cable trays are covered for more than 6 ft (1.83 m) with solid unventilated covers, the ampacities for Nos. 1/0 and larger single conductor cables shall not exceed 70 percent of the allowable ampacities in Tables 310-69 and 310-70.

(2) Where single conductor cables are installed in a single layer in uncovered cable trays, with a maintained space of not less than one cable diameter between individual conductors, the ampacity of Nos. 1/0 and larger cables shall not exceed the allowable ampacities in Tables 310-69 and 310-70.

(3) Where single conductors are installed in a triangular or square configuration in uncovered cable trays, with a maintained free air space of not less than 2.15 times the diameter (2.15 × O.D.) of the largest conductor contained within the configuration and adjacent conductor configurations or cables, the ampacity of Nos. 1/0 and larger cables shall not exceed the allowable ampacities in Tables 310-67 and 310-68.

Article 320 — Open Wiring on Insulators

320-1. Definition. Open wiring on insulators is an exposed wiring method using cleats, knobs, tubes, and flexible tubing for the protection and support of single insulated conductors run in or on buildings, and not concealed by the building structure.

320-2. Other Articles. Open wiring on insulators shall comply with this article and also with the applicable provisions of other articles in this *Code*, especially Articles 225 and 300.

320-3. Uses Permitted. Open wiring on insulators shall be permitted on systems of 600 volts, nominal, or less, only for industrial or agricultural establishments, indoors or outdoors, in wet or dry locations, where subject to corrosive vapors, and for services.

320-5. Conductors.

(a) Type. Conductors shall be of a type specified by Article 310.

(b) Ampacity. The ampacity shall comply with Section 310-15.

320-6. Conductor Supports.

(a) Conductor Sizes Smaller than No. 8. Conductors smaller than No. 8 shall be rigidly supported on noncombustible, nonabsorbent insulating materials and shall not contact any other objects.

Supports shall be installed as follows:

(1) Within 6 in. (152 mm) from a tap or splice
(2) Within 12 in. (305 mm) of a dead-end connection to a lampholder or receptacle
(3) At intervals not exceeding 4½ ft (1.37 m) and at closer intervals sufficient to provide adequate support where likely to be disturbed

(b) Conductor Sizes No. 8 and Larger. Supports for conductors No. 8 or larger installed across open spaces shall be permitted up to 15 ft (4.57 m) apart if noncombustible, nonabsorbent insulating spacers are used at least every 4½ ft (1.37 m) to maintain at least 2½ in. (64 mm) between conductors.

Where not likely to be disturbed in buildings of mill construction, No. 8 and larger conductors shall be permitted to be run across open spaces if supported from each wood cross member on approved insulators maintaining 6 in. (152 mm) between conductors.

(c) Industrial Establishments. In industrial establishments only, where conditions of maintenance and supervision ensure that only qualified persons will service the system, conductors of sizes 250 kcmil and larger shall be permitted to be run across open spaces where supported on intervals up to 30 ft (9.1 m) apart.

320-7. Mounting of Conductor Supports. Where nails are used to mount knobs, they shall not be smaller than tenpenny. Where screws are used to mount knobs, or where nails or screws are used to mount cleats, they shall be of a length sufficient to penetrate the wood to a depth equal to at least one-half the height of the knob and the full thickness of the cleat. Cushion washers shall be used with nails.

320-8. Tie Wires. No. 8 or larger conductors supported on solid knobs shall be securely tied thereto by tie wires having an insulation equivalent to that of the conductor.

320-10. Flexible Nonmetallic Tubing. In dry locations, where not exposed to severe physical damage, conductors shall be permitted to be separately enclosed in flexible nonmetallic tubing. The tubing shall be in continuous lengths not exceeding 15 ft (4.57 m) and secured to the surface by straps at intervals not exceeding 4½ ft (1.37 m).

320-11. Through Walls, Floors, Wood Cross Members, etc. Open conductors shall be separated from contact with walls, floors, wood cross members, or partitions through which they pass by tubes or bushings of noncombustible, nonabsorbent insulating material. Where the bushing is shorter than the hole, a waterproof sleeve of noninductive material shall be inserted in the hole and an insulating bushing slipped into the sleeve at each end in such a manner as to keep the conductors absolutely out of contact with the sleeve. Each conductor shall be carried through a separate tube or sleeve.

> FPN: See Section 310-10 for temperature limitation of conductors.

320-12. Clearance from Piping, Exposed Conductors, etc. Open conductors shall be separated at least 2 in. (50.8 mm) from metal raceways, piping, or other conducting material, and from any exposed lighting, power, or signaling conductor, or shall be separated therefrom by a continuous and firmly fixed nonconductor in addition to the insulation of the conductor. Where any insulating tube is used, it shall be secured at the ends. Where practicable, conductors shall pass over rather than under any piping subject to leakage or accumulations of moisture.

320-13. Entering Spaces Subject to Dampness, Wetness, or Corrosive Vapors. Conductors entering or leaving locations subject to dampness, wetness, or corrosive vapors shall have drip loops formed on them and shall then pass upward and inward from the outside of the buildings, or from the damp, wet, or corrosive location, through noncombustible, nonabsorbent insulating tubes.

> FPN: See Section 230-52 for individual conductors entering buildings or other structures.

320-14. Protection from Physical Damage. Conductors within 7 ft (2.13 m) from the floor shall be considered exposed to physical damage. Where open conductors cross ceiling joists and wall studs and are exposed to physical damage, they shall be protected by one of the following methods:

(1) Guard strips not less than 1 in. (25.4 mm) nominal in thickness and at least as high as the insulating supports, placed on each side of and close to the wiring

(2) A substantial running board at least ½ in. (12.7 mm) thick in back of the conductors with side protections. Running boards shall extend at least 1 in. (25.4 mm) outside the conductors, but not more than 2 in. (50.8 mm), and the protecting sides shall be at least 2 in. (50.8 mm) high and at least 1 in. (25.4 mm) nominal in thickness

(3) Boxing made as above and furnished with a cover kept at least 1 in. (25.4 mm) away from the conductors within. Where protecting vertical conductors on side walls, the boxing shall be closed at the top and the holes through which the conductors pass shall be bushed.

(4) Rigid metal conduit, intermediate metal conduit, rigid nonmetallic conduit, or electrical metallic tubing, in which case the rules of Articles 345, 346, 347, or 348 shall apply; or by metal piping, in which case the conductors shall be encased in continuous lengths of approved flexible tubing.

320-15. Unfinished Attics and Roof Spaces. Conductors in unfinished attics and roof spaces shall comply with (a) or (b).

(a) Accessible by Stairway or Permanent Ladder. Conductors shall be installed along the side of or through bored holes in floor joists, studs, or rafters. Where run through bored holes, conductors in the joists and in studs or rafters to a height of not less than 7 ft (2.13 m) above the floor or floor joists shall be protected by substantial running boards extending not less than 1 in. (25.4 mm) on each side of the conductors. Running boards shall be securely fastened in place. Running boards and guard strips shall not be required for conductors installed along the sides of joists, studs, or rafters.

(b) Not Accessible by Stairway or Permanent Ladder. Conductors shall be installed along the sides of or through bored holes in floor joists, studs, or rafters.

Exception: In buildings completed before the wiring is installed, attic and roof spaces that are not accessible by stairway or permanent ladder and have headroom at all points less than 3 ft (914 mm), the wiring shall be permitted to be installed on the edges of rafters or joists facing the attic or roof space.

320-16. Switches. Surface-type snap switches shall be mounted in accordance with Section 380-10(a), and boxes shall not be required. Other type switches shall be installed in accordance with Section 380-4.

Article 321 — Messenger Supported Wiring

321-1. Definition. *Messenger supported wiring* is an exposed wiring support system using a messenger wire to support insulated conductors by any one of the following:

(1) A messenger with rings and saddles for conductor support

(2) A messenger with a field-installed lashing material for conductor support

(3) Factory-assembled aerial cable

(4) Multiplex cables utilizing a bare conductor, factory assembled and twisted with one or more insulated conductors, such as duplex, triplex, or quadruplex type of construction

321-2. Other Articles. Messenger supported wiring shall comply with this article and also with the applicable provisions of other articles in this *Code*, especially Articles 225 and 300.

321-3. Uses Permitted.

(a) Cable Types. The following shall be permitted to be installed in messenger supported wiring under the conditions described in the article or section referenced for each:

	Section	Article
Metal-clad cable		334
Mineral-insulated, metal-sheathed cable		330
Multiconductor service-entrance cable		338
Multiconductor underground feeder and branch-circuit cable		339
Other factory-assembled, multiconductor control, signal, or power cables that are identified for the use		
Power and control tray cable		340
Power-limited tray cable	725-61(c) and 725-71(e)	

(b) In Industrial Establishments. In industrial establishments only, where conditions of maintenance and supervision ensure that only qualified persons will service the installed messenger supported wiring, the following shall be permitted:

(1) Any of the conductor types shown in Table 310-13 or Table 310-62

(2) MV cable

Where exposed to weather, conductors shall be listed for use in wet locations. Where exposed to direct rays of the sun, conductors or cables shall be sunlight resistant.

(c) Hazardous (Classified) Locations. Messenger supported wiring shall be permitted to be used in hazardous (classified) locations where the contained cables are permitted for such use in Sections 501-4, 502-4, 503-3, and 504-20.

321-4. Uses Not Permitted. Messenger supported wiring shall not be used in hoistways or where subject to severe physical damage.

321-5. Ampacity. The ampacity shall be determined by Section 310-15.

321-6. Messenger Support. The messenger shall be supported at dead ends and at intermediate locations so as to eliminate tension on the conductors. The conductors shall not be permitted to come into contact with the messenger supports or any structural members, walls, or pipes.

321-7. Grounding. The messenger shall be grounded as required by Sections 250-80 and 250-86 for enclosure grounding.

321-8. Conductor Splices and Taps. Conductor splices and taps made and insulated by approved methods shall be permitted in messenger supported wiring.

Article 324 — Concealed Knob-and-Tube Wiring

324-1. Definition. *Concealed knob-and-tube wiring* is a wiring method using knobs, tubes, and flexible nonmetallic tubing for the protection and support of single insulated conductors.

324-2. Other Articles. Concealed knob-and-tube wiring shall comply with this article and also with the applicable provisions of other articles in this *Code*, especially Article 300.

324-3. Uses Permitted. Concealed knob-and-tube wiring shall be permitted to be installed in the hollow spaces of walls and ceilings or in unfinished attics and roof spaces as provided in Section 324-11 only as follows:

(1) For extensions of existing installations, or

(2) Elsewhere by special permission.

324-4. Uses Not Permitted. Concealed knob-and-tube wiring shall not be used in commercial garages, theaters and similar locations, motion picture studios, hazardous (classified) locations, or in the hollow spaces of walls, ceilings, and attics where such spaces are insulated by loose, rolled, or foamed-in-place insulating material that envelops the conductors.

324-5. Conductors.

(a) Type. Conductors shall be of a type specified by Article 310.

(b) Ampacity. The ampacity shall be determined by Section 310-15.

324-6. Conductor Supports. Conductors shall be rigidly supported on noncombustible, nonabsorbent insulating materials and shall not contact any other objects. Supports shall be installed as follows:

(1) Within 6 in. (152 mm) of each side of each tap or splice
(2) At intervals not exceeding 4½ ft (1.37 m)

Where it is impracticable to provide supports, conductors shall be permitted to be fished through hollow spaces in dry locations, provided each conductor is individually enclosed in flexible nonmetallic tubing that is in continuous lengths between supports, between boxes, or between a support and a box.

324-7. Tie Wires. Where solid knobs are used, conductors shall be securely tied thereto by tie wires having insulation equivalent to that of the conductor.

324-8. Conductor Clearances.

(a) General. A clearance of not less than 3 in. (76 mm) shall be maintained between conductors and a clearance of not less than 1 in. (25.4 mm) between the conductor and the surface over which it passes.

(b) Limited Conductor Space. Where space is too limited to provide these minimum clearances, such as at meters, panelboards, outlets, and switch points, the individual conductors shall be enclosed in flexible nonmetallic tubing, which shall be continuous in length between the last support and the enclosure or terminal point.

324-9. Through Walls, Floors, Wood Cross Members, etc. Conductors shall comply with Section 320-11 where passing through holes in structural members. Where passing through wood cross members in plastered partitions, conductors shall be protected by noncombustible, nonabsorbent, insulating tubes extending not less than 3 in. (76 mm) beyond the wood member.

324-10. Clearance from Piping, Exposed Conductors, etc. Conductors shall comply with Section 320-12 for clearances from other exposed conductors, piping, etc.

324-11. Unfinished Attics and Roof Spaces. Conductors in unfinished attics and roof spaces shall comply with (a) or (b).

FPN: See Section 310-10 for temperature limitation of conductors.

(a) Accessible by Stairway or Permanent Ladder. Conductors shall be installed along the side of or through bored holes in floor joists, studs, or rafters. Where run through bored holes, conductors in the joists and in studs or rafters to a height of not less than 7 ft (2.13 m) above the floor or floor joists shall be protected by substantial running boards extending not less than 1 in. (25.4 mm) on each side of the conductors. Running boards shall be securely fastened in place. Running boards and guard strips shall not be required where conductors are installed along the sides of joists, studs, or rafters.

(b) Not Accessible by Stairway or Permanent Ladder. Conductors shall be installed along the sides of or through bored holes in floor joists, studs, or rafters.

Exception: In buildings completed before the wiring is installed, attic and roof spaces that are not accessible by stairway or permanent ladder and have headroom at all points less than 3 ft (914 mm), the wiring shall be permitted to be installed on the edges of rafters or joists facing the attic or roof space.

324-12. Splices. Splices shall be soldered unless approved splicing devices are used. In-line or strain splices shall not be used.

324-13. Boxes. Outlet boxes shall comply with Article 370.

324-14. Switches. Switches shall comply with Sections 380-4 and 380-10(b).

Article 325 — Integrated Gas Spacer Cable: Type IGS

A. General

325-1. Definition. *Type IGS cable* is a factory assembly of one or more conductors, each individually insulated and enclosed in a loose fit, nonmetallic flexible conduit as an integrated gas spacer cable rated 0 through 600 volts.

325-2. Other Articles. Type IGS cable shall comply with this article and also with the applicable provisions of other articles in this *Code*.

325-3. Uses Permitted. Type IGS cable shall be permitted for use underground, including direct burial in the earth, as service-entrance conductors, or as feeder or branch-circuit conductors.

325-4. Uses Not Permitted. Type IGS cable shall not be used as interior wiring or be exposed in contact with buildings.

B. Installation

325-11. Bending Radius. Where the coilable nonmetallic conduit and cable is bent for installation purposes or is flexed or bent during shipment or installation, the radii of bends measured to the inside of the bend shall not be less than specified in Table 325-11.

Table 325-11. Minimum Radii of Bends

Conduit Trade Size (in.)	Minimum Radii	
	in.	mm
2	24	610
3	35	889
4	45	1143

325-12. Bends. A run of Type IGS cable between pull boxes or terminations shall not contain more than the equivalent of four quarter bends (360 degrees total), including those bends located immediately at the pull box or terminations.

325-13. Fittings. Terminations and splices for Type IGS cable shall be identified as a type that is suitable for maintaining the gas pressure within the conduit. A valve and cap shall be provided for each length of the cable and conduit to check the gas pressure or to inject gas into the conduit.

325-14. Ampacity. The ampacity of Type IGS cable and conduit shall not exceed values shown in Table 325-14 for single conductor or multiconductor cable.

Table 325-14. Ampacity of Type IGS Cable

Size (kcmil)	Amperes	Size (kcmil)	Amperes
250	119	2500	376
500	168	3000	412
750	206	3250	429
1000	238	3500	445
1250	266	3750	461
1500	292	4000	476
1750	344	4250	491
2000	336	4500	505
2250	357	4750	519

C. Construction Specifications

325-20. Conductors. The conductors shall be solid aluminum rods, laid parallel, consisting of one to nineteen ½-in. (12.7-mm) diameter rods.

The minimum conductor size shall be 250 kcmil and the maximum size shall be 4750 kcmil.

325-21. Insulation. The insulation shall be dry kraft paper tapes and a pressurized sulfur hexafluoride gas (SF_6), both approved for electrical use. The nominal gas pressure shall be 20 pounds per square inch gauge (psig) (138 kPa gauge).

The thickness of the paper spacer shall be as specified in Table 325-21.

Table 325-21. Paper Spacer Thickness

Size (kcmil)	Thickness	
	in.	mm
250–1000	0.040	1.02
1250–4750	0.060	1.52

325-22. Conduit. The conduit shall be a medium density polyethylene identified as suitable for use with natural gas rated pipe in 2-in., 3-in., or 4-in. trade size. The percent fill dimensions for the conduit are shown in Table 325-22.

The size of the conduit permitted for each conductor size shall be calculated for a percent fill not to exceed those found in Table 1, Chapter 9.

Table 325-22. Conduit Dimensions

Conduit Trade Size (in.)	Outside Diameter		Inside Diameter	
	in.	mm	in.	mm
2	2.375	60	1.947	49.46
3	3.500	89	2.886	73.30
4	4.500	114	3.710	94.23

325-24. Marking. The cable shall be marked in accordance with Sections 310-11(a), 310-11(b)(1), and 310-11(d).

Article 326 — Medium Voltage Cable: Type MV

326-1. Definition. Type MV cable is a single or multiconductor solid dielectric insulated cable rated 2001 volts or higher.

326-2. Other Articles. In addition to the provisions of this article, Type MV cable shall comply with the applicable provisions of this *Code*, especially those of Articles 300, 305, 310, 318, and 490.

326-3. Uses Permitted. Type MV cables shall be permitted for use on power systems rated up to 35,000 volts, nominal, in wet or dry locations, in raceways, in cable trays as specified in Section 318-3(b)(1) or directly buried in accordance with Section 300-50, and in messenger supported wiring.

326-4. Uses Not Permitted. Type MV cable shall not be used unless identified for the use (1) where exposed to direct sunlight, and (2) in cable trays.

326-5. Construction. Type MV cables shall have copper, aluminum, or copper-clad aluminum conductors and shall be constructed in accordance with Article 310.

326-6. Ampacity. The ampacity of Type MV cable shall be in accordance with Section 310-60.

Exception: The ampacity of Type MV cable installed in cable tray shall be in accordance with Section 318-13.

326-7. Marking. Medium voltage cable shall be marked as required in Section 310-11.

Article 328 — Flat Conductor Cable: Type FCC

A. General

328-1. Scope. This article covers a field-installed wiring system for branch circuits incorporating Type FCC cable and associated accessories as defined by the article. The wiring system is designed for installation under carpet squares.

328-2. Definitions.

Type FCC Cable. *Type FCC cable* consists of three or more flat copper conductors placed edge-to-edge and separated and enclosed within an insulating assembly.

FCC System. A complete wiring system for branch circuits that is designed for installation under carpet squares. The FCC system includes Type FCC cable and associated shielding, connectors, terminators, adapters, boxes, and receptacles.

Cable Connector. A connector designed to join Type FCC cables without using a junction box.

Insulating End. An insulator designed to electrically insulate the end of a Type FCC cable.

Top Shield. A grounded metal shield covering under-carpet components of the FCC system for the purposes of providing protection against physical damage.

Bottom Shield. A protective layer that is installed between the floor and Type FCC flat conductor cable to protect the cable from physical damage and may or may not be incorporated as an integral part of the cable.

Transition Assembly. An assembly to facilitate connection of the FCC system to other wiring systems, incorporating (1) a means of electrical interconnection, and (2) a suitable box or covering for providing electrical safety and protection against physical damage.

Metal Shield Connections. Means of connection designed to electrically and mechanically connect a metal shield to another metal shield, to a receptacle housing or self-contained device, or to a transition assembly.

328-3. Other Articles. The FCC systems shall conform with the applicable provisions of Articles 210, 220, 240, 250, and 300.

328-4. Uses Permitted.

(a) Branch Circuits. Use of FCC systems shall be permitted both for general-purpose and appliance branch circuits and for individual branch circuits.

(b) Floors. Use of FCC systems shall be permitted on hard, sound, smooth, continuous floor surfaces made of concrete, ceramic, or composition flooring, wood, and similar materials.

(c) Walls. Use of FCC systems shall be permitted on wall surfaces in surface metal raceways.

(d) Damp Locations. Use of FCC systems in damp locations shall be permitted.

(e) Heated Floors. Materials used for floors heated in excess of 30°C (86°F) shall be identified as suitable for use at these temperatures.

328-5. Uses Not Permitted. FCC systems shall not be used in the following:

(1) Outdoors or in wet locations
(2) Where subject to corrosive vapors
(3) In any hazardous (classified) location
(4) In residential, school, and hospital buildings

328-6. Branch-Circuit Ratings.

(a) Voltage. Voltage between ungrounded conductors shall not exceed 300 volts. Voltage between ungrounded conductors and the grounded conductor shall not exceed 150 volts.

(b) Current. General-purpose and appliance branch circuits shall have ratings not exceeding 20 amperes. Individual branch circuits shall have ratings not exceeding 30 amperes.

B. Installation

328-10. Coverings. Floor-mounted Type FCC cable, cable connectors, and insulating ends shall be covered with carpet squares not larger than 36 in. (914 mm) square. Those carpet squares that are adhered to the floor shall be attached with release-type adhesives.

328-11. Cable Connections and Insulating Ends. All Type FCC cable connections shall use connectors identified for their use, installed such that electrical continuity, insulation, and sealing against dampness and liquid spillage are provided. All bare cable ends shall be insulated and sealed against dampness and liquid spillage using listed insulating ends.

328-12. Shields.

(a) Top Shield. A metal top shield shall be installed over all floor-mounted Type FCC cable, connectors, and insulating ends. The top shield shall completely cover all cable runs, corners, connectors, and ends.

(b) Bottom Shield. A bottom shield shall be installed beneath all Type FCC cable, connectors, and insulating ends.

328-13. Enclosure and Shield Connections. All metal shields, boxes, receptacle housings, and self-contained devices shall be electrically continuous to the equipment grounding conductor of the supplying branch circuit. All such electrical connections shall be made with connectors identified for this use. The electrical resistivity of such shield system shall not be more than that of one conductor of the Type FCC cable used in the installation.

328-14. Receptacles. All receptacles, receptacle housings, and self-contained devices used with the FCC system shall be identified for this use and shall be connected to the Type FCC cable and metal shields. Connection from any grounding conductor of the Type FCC cable shall be made to the shield system at each receptacle.

328-15. Connection to Other Systems. Power feed, grounding connection, and shield system connection between the FCC system and other wiring systems shall be accomplished in a transition assembly identified for this use.

328-16. Anchoring. All FCC system components shall be firmly anchored to the floor or wall using an adhesive or mechanical anchoring system identified for this use. Floors shall be prepared to ensure adherence of the FCC system to the floor until the carpet squares are placed.

328-17. Crossings. Crossings of more than two Type FCC cable runs shall not be permitted at any one point. Crossings of a Type FCC cable over or under a flat communications or signal cable shall be permitted. In each case, a grounded layer of metal shielding shall separate the two cables, and crossings of more than two flat cables shall not be permitted at any one point.

328-18. System Height. Any portion of an FCC system with a height above floor level exceeding 0.090 in. (2.29 mm) shall be tapered or feathered at the edges to floor level.

328-19. FCC Systems Alterations. Alterations to FCC systems shall be permitted. New cable connectors shall be used at new connection points to make alterations. It shall be permitted to leave unused cable runs and associated cable connectors in place and energized. All cable ends shall be covered with insulating ends.

328-20. Polarization of Connections. All receptacles and connections shall be constructed and installed so as to maintain proper polarization of the system.

C. Construction

328-30. Type FCC Cable. Type FCC cable shall be listed for use with the FCC system and shall consist of three, four, or five flat copper conductors, one of which shall be an equipment grounding conductor. The insulating material of the cable shall be moisture resistant and flame retardant.

328-31. Markings. Type FCC cable shall be clearly and durably marked on both sides at intervals of not more than 24 in. (610 mm) with the information required by Section 310-11(a) and with the following additional information:

(1) Material of conductors
(2) Maximum temperature rating
(3) Ampacity

328-32. Conductor Identification.

(a) Colors. Conductors shall be clearly and durably marked on both sides throughout their length as specified in Section 310-12.

(b) Order. For a 2-wire FCC system with grounding, the grounding conductor shall be central.

328-33. Corrosion Resistance. Metal components of the system shall be either corrosion resistant, coated with corrosion-resistant materials, or insulated from contact with corrosive substances.

328-34. Insulation. All insulating materials in the FCC systems shall be identified for their use.

328-35. Shields.

(a) Materials and Dimensions. All top and bottom shields shall be of designs and materials identified for their use.

Top shields shall be metal. Both metallic and nonmetallic materials shall be permitted for bottom shields.

(b) Resistivity. Metal shields shall have cross-sectional areas that provide for electrical resistivity of not more than that of one conductor of the Type FCC cable used in the installation.

(c) Metal-Shield Connectors. Metal shields shall be connected to each other and to boxes, receptacle housings, self-contained devices, and transition assemblies using metal-shield connectors.

328-36. Receptacles and Housings. Receptacle housings and self-contained devices designed either for floor mounting or for in- or on-wall mounting shall be permitted for use with the FCC system. Receptacle housings and self-contained devices shall incorporate means for facilitating entry and termination of Type FCC cable and for electrically connecting the housing or device with the metal shield. Receptacles and self-contained devices shall comply with Section 210-7. Power and communications outlets installed together in common housing shall be permitted in accordance with Section 800-52(a)(2), Exception No. 1.

328-37. Transition Assemblies. All transition assemblies shall be identified for their use. Each assembly shall incorporate means for facilitating entry of the Type FCC cable into the assembly, for connecting the Type FCC cable to grounded conductors, and for electrically connecting the assembly to the metal cable shields and to equipment grounding conductors.

Article 330 — Mineral-Insulated, Metal-Sheathed Cable: Type MI

A. General

330-1. Definition. *Type MI mineral-insulated, metal-sheathed cable* is a factory assembly of one or more conductors insulated with a highly compressed refractory mineral insulation and enclosed in a liquidtight and gastight continuous copper or alloy steel sheath.

330-2. Other Articles. Type MI cable shall comply with this article and also with the applicable provisions of other articles in this *Code*, especially Article 300.

330-3. Uses Permitted. Type MI cable shall be permitted as follows:

(1) For services, feeders, and branch circuits
(2) For power, lighting, control, and signal circuits
(3) In dry, wet, or continuously moist locations

(4) Indoors or outdoors
(5) Where exposed or concealed
(6) Embedded in plaster, concrete, fill, or other masonry, whether above or below grade
(7) In any hazardous (classified) location
(8) Where exposed to oil and gasoline
(9) Where exposed to corrosive conditions not deteriorating to its sheath
(10) In underground runs where suitably protected against physical damage and corrosive conditions

330-4. Uses Not Permitted. Type MI cable shall not be used where exposed to destructive corrosive conditions, unless protected by materials suitable for the conditions.

B. Installation

330-10. Wet Locations. Where installed in wet locations, Type MI cable shall comply with Section 300-6(c).

330-11. Through Joists, Studs, or Rafters. Type MI cable shall comply with Section 300-4 where installed through studs, joists, rafters, or similar wood members.

330-12. Supports. Type MI cable shall be supported by one of the following means:

(1) Supported securely at intervals not exceeding 6 ft (1.83 m) by straps, staples, hangers, or similar fittings designed and installed so as not to damage the cable

Exception: Where cable is fished, support shall not be required.

(2) Supported in accordance with Section 318-8(b) where installed in cable trays

330-13. Bends. Bends in Type MI cable shall be made so as not to damage the cable. The radius of the inner edge of any bend shall not be less than shown as follows:

(1) Five times the external diameter of the metallic sheath for cable not more than ¾ in. (19 mm) in external diameter
(2) Ten times the external diameter of the metallic sheath for cable greater than ¾ in. (19 mm) but not more than 1 in. (25.4 mm) in external diameter

330-14. Fittings. Fittings used for connecting Type MI cable to boxes, cabinets, or other equipment shall be identified for such use. Where single-conductor cables enter ferrous metal boxes or cabinets, the installation shall comply with Section 300-20 to prevent inductive heating.

330-15. Terminal Seals. Where Type MI cable terminates, a seal shall be provided immediately after stripping to prevent the entrance of moisture into the insulation. The conductors extending beyond the sheath shall be individually provided with an insulating material.

330-16. Single Conductors. Where single-conductor cables are used, all phase conductors and, where used, the neutral conductor shall be grouped together to minimize induced voltage on the sheath. Where single-conductor cables enter ferrous enclosures, the installation shall comply with Section 300-20 to prevent heating from induction.

C. Construction Specifications

330-20. Conductors. Type MI cable conductors shall be of solid copper or nickel-clad copper with a resistance corresponding to standard AWG sizes.

330-21. Insulation. The conductor insulation in Type MI cable shall be a highly compressed refractory mineral that provides proper spacing for all conductors.

330-22. Outer Sheath. The outer sheath shall be of a continuous construction to provide mechanical protection and moisture seal. Where made of copper, it shall provide an adequate path for equipment grounding purposes. Where made of steel, an equipment grounding conductor in accordance with Article 250 shall be provided.

Article 331 — Electrical Nonmetallic Tubing

A. General

331-1. Definition. *Electrical nonmetallic tubing* is a pliable corrugated raceway of circular cross section with integral or associated couplings, connectors, and fittings listed for the installation of electric conductors. It is composed of a material that is resistant to moisture and chemical atmospheres and is flame retardant.

A pliable raceway is a raceway that can be bent by hand with a reasonable force, but without other assistance.

Electrical nonmetallic tubing shall be made of material that does not exceed the ignitibility, flammability, smoke generation, and toxicity characteristics of rigid (nonplasticized) polyvinyl chloride.

331-2. Other Articles. Installations for electrical nonmetallic tubing shall comply with the provisions of the applicable sections of Article 300. Where equipment grounding is required by Article 250, a separate equipment grounding conductor shall be installed in the raceway.

331-3. Uses Permitted. The use of electrical nonmetallic tubing and fittings shall be permitted in the following:

(1) In any building not exceeding three floors above grade

   (a) For exposed work, where not subject to physical damage

   (b) Concealed within walls, floors, and ceilings

   FPN: See Section 336-5(a)(1) for definition of first floor.

(2) In any building exceeding three floors above grade, electrical nonmetallic tubing shall be concealed within walls, floors, and ceilings where the walls, floors, and ceilings provide a thermal barrier of material that has at least a 15-minute finish rating as identified in listings of fire-rated assemblies. The 15-minute finish rated thermal barrier shall be permitted to be used for combustible walls, floors, and ceilings.

   FPN: A finish rating is established for assemblies containing combustible (wood) supports. The finish rating is defined as the time at which the wood stud or wood joist reaches an average temperature rise of 121°C (250°F) or an individual temperature of 163°C (325°F) as measured on the plane of the wood nearest the fire. A finish rating is not intended to represent a rating for a membrane ceiling.

(3) In locations subject to severe corrosive influences as covered in Section 300-6 and where subject to chemicals for which the materials are specifically approved

(4) In concealed, dry, and damp locations not prohibited by Section 331-4

(5) Above suspended ceilings where the suspended ceilings provide a thermal barrier of material that has at least a 15-minute finish rating as identified in listings of fire-rated assemblies, except as permitted in Section 331-3(1)(a).

(6) Encased in poured concrete, or embedded in a concrete slab on grade where ENT is placed on sand or approved screenings, provided fittings identified for this purpose are used for connections

(7) For wet locations indoors as permitted in this section or in a concrete slab on or below grade, with fittings listed for the purpose

(8) ½ in. through 1 in. as listed manufactured prewired assembly

   FPN: Extreme cold may cause some types of nonmetallic conduits to become brittle and, therefore, more susceptible to damage from physical contact.

331-4. Uses Not Permitted. Electrical nonmetallic tubing shall not be used in the following:

(1) In hazardous (classified) locations, except as permitted by Section 504-20 and Section 505-15(a)(1)

(2) For the support of fixtures and other equipment

(3) Where subject to ambient temperatures in excess of 50°C (122°F) unless listed otherwise

(4) For conductors whose insulation temperature limitations would exceed those for which the tubing is listed

(5) For direct earth burial

(6) Where the voltage is over 600 volts

(7) In exposed locations, except as permitted by Sections 331-3(1), 331-3(5), and 331-3(7)

(8) In theaters and similar locations, except as provided in Articles 518 and 520

(9) Where exposed to the direct rays of the sun, unless identified as sunlight resistant

B. Installation

331-5. Size.

(a) Minimum. Tubing smaller than ½-in. electrical trade size shall not be used.

(b) Maximum. Tubing larger than 2-in. electrical trade size shall not be used.

> FPN: Metric trade numerical designations for electrical non-metallic tubing are ½ = 16, ¾ = 21, 1 = 27, 1¼ = 35, 1½ = 41, and 2 = 53.

331-6. Number of Conductors in Tubing. The number of conductors in a single tubing shall not exceed that permitted by the percentage fill in Table 1, Chapter 9.

331-7. Trimming. All cut ends of tubing shall be trimmed inside and outside to remove rough edges.

331-8. Joints. All joints between lengths of tubing and between tubing and couplings, fittings, and boxes shall be by an approved method.

331-9. Bends — How Made. Bends of electrical nonmetallic tubing shall be made so that the tubing will not be damaged and that the internal diameter of the tubing will not be effectively reduced. Bends shall be permitted to be made manually without auxiliary equipment, and the radius of the curve of the inner edge of such bends shall not be less than shown in Table 346-10.

331-10. Bends — Number in One Run. There shall not be more than the equivalent of four quarter bends (360 degrees total) between pull points, e.g., conduit bodies and boxes.

331-11. Supports. Electrical nonmetallic tubing shall be installed as a complete system as provided in Article 300 and shall be secured at intervals not exceeding 3 ft (914 mm) or, for horizontal runs, it shall be permitted to be supported by openings in framing members at intervals not exceeding 3 ft (914 mm). In addition, it shall be securely fastened in place within 3 ft (914 mm) of each outlet box, device box, junction box, cabinet, or fitting where it terminates.

Exception: Lengths not exceeding a distance of 6 ft (1.83 m) from a fixture terminal connection for tap connections to lighting fixtures shall be permitted without being secured.

331-12. Boxes and Fittings. Electrical nonmetallic tubing shall be used only with listed fittings. Boxes and fittings shall comply with the applicable provisions of Article 370.

331-13. Splices and Taps. Splices and taps shall be made only in junction boxes, outlet boxes, device boxes, or conduit bodies. See Article 370 for rules on the installation and use of boxes and conduit bodies.

331-14. Bushings. Where a tubing enters a box, fitting, or other enclosure, a bushing or adapter shall be provided to protect the wire from abrasion unless the box, fitting, or enclosure design provides equivalent protection.

> FPN: See Section 300-4(f) for the protection of conductors size No. 4 or larger.

C. Construction Specifications

331-15. General. Electrical nonmetallic tubing shall be clearly and durably marked at least every 10 ft (3.05 m) as required in the first sentence of Section 110-21. The type of material shall also be included in the marking. Tubing that has limited smoke-producing characteristics shall be permitted to be identified with the suffix LS. The type, size, and quantity of conductors used in prewired manufactured assemblies shall be identified by means of a printed tag or label attached to each end of the manufactured assembly and either the carton, coil, or reel. The enclosed conductors shall be marked in accordance with Section 310-11.

ENT, as a prewired manufactured assembly, shall be provided in continuous lengths capable of being shipped in a coil, reel, or carton without damage.

Article 333 — Armored Cable: Type AC

A. General

333-1. Definition. *Type AC cable* is a fabricated assembly of insulated conductors in a flexible metallic enclosure. See Section 333-19.

333-2. Other Articles. Type AC cable shall comply with this article and also with the applicable provisions of other articles in this *Code*, especially Article 300.

333-3. Uses Permitted. Except where otherwise specified in this *Code* and where not subject to physical damage, Type AC cable shall be permitted for branch circuits and feeders in both exposed and concealed work and in cable trays where identified for such use.

Type AC cable shall be permitted in dry locations and embedded in plaster finish on brick or other masonry, except in damp or wet locations. It shall be permissible to run or fish this cable in the air voids of masonry block or tile walls where such walls are not exposed or subject to excessive moisture or dampness.

333-4. Uses Not Permitted. Type AC cable shall not be used where prohibited elsewhere in this *Code*, including the following:

(1) In theaters and similar locations, except as provided in Article 518, Places of Assembly
(2) In motion picture studios
(3) In any hazardous (classified) location except as permitted by Sections 501-4(b), Exception, 502-4(b), Exception No. 1, and 504-20
(4) Where exposed to corrosive fumes or vapors
(5) On cranes or hoists, except as provided in Section 610-11(c)
(6) In storage battery rooms
(7) In hoistways or on elevators, except as provided in Section 620-21
(8) In commercial garages where prohibited in Article 511

B. Installation

333-7. Support. Type AC cable shall be secured by staples, cable ties, straps, hangers, or similar fittings designed and installed so as not to damage the cable at intervals not exceeding 4½ ft (1.37 m) and within 12 in. (305 mm) of every outlet box, junction box, cabinet, or fitting.

(a) Horizontal Runs. Type AC cable installed in other than vertical runs through bored or punched holes in wood or metal framing members, or through notches in wooden framing members and protected by a steel plate at least 1/16 in. (1.59 mm) thick, shall be considered secured where the support intervals do not exceed 4½ ft (1.37 m) and the armored cable is securely fastened in place by an approved means within 12 in. (505 mm) of each box, cabinet, conduit body, or other armored cable termination.

(b) Unsupported. Type AC cable shall be permitted to be unsupported where the cable is

(1) Fished between access points, where concealed in finished buildings or structures and supporting is impracticable;
(2) Not more than 2 ft (610 mm) in length at terminals where flexibility is necessary; or
(3) Not more than 6 ft (1.83 m) in length from an outlet for connections within an accessible ceiling to lighting fixtures or equipment.

(c) Cable Tray Installations. Type AC cable installed in cable trays shall comply with Section 318-8(b).

333-8. Bending Radius. All bends shall be made so that the cable will not be damaged, and the radius of the curve of the inner edge of any bend shall not be less than five times the diameter of the Type AC cable.

333-9. Boxes and Fittings. At all points where the armor of AC cable terminates, a fitting shall be provided to protect wires from abrasion, unless the design of the outlet boxes or fittings is such as to afford equivalent protection, and, in addition, an insulating bushing or its equivalent protection shall be provided between the conductors and the armor. The connector or clamp by which the Type AC cable is fastened to boxes or cabinets shall be of such design that the insulating bushing or its equivalent will be visible for inspection. Where change is made from Type AC cable to other cable or raceway wiring methods, a box, fitting, or conduit body shall be installed at junction points as required in Section 300-15.

333-10. Through or Parallel to Framing Members. Type AC cable shall comply with Section 300-4 where installed through or parallel to studs, joists, rafters, or similar wood or metal members.

333-11. Exposed Work. Exposed runs of cable shall closely follow the surface of the building finish or of running boards. Exposed runs shall also be permitted to be installed on the underside of joists where supported at each joist and located so as not to be subject to physical damage.

333-12. In Accessible Attics. Type AC cables in accessible attics or roof spaces shall be installed as specified in (a) and (b).

(a) Where Run Across the Top of Floor Joists. Where run across the top of floor joists, or within 7 ft (2.13 m) of floor or floor joists across the face of rafters or studding, in attics and roof spaces that are accessible, the cable shall be protected by substantial guard strips that are at least as high as the cable. Where this space is not accessible by permanent stairs or ladders, protection shall only be required within

6 ft (1.83 m) of the nearest edge of the scuttle hole or attic entrance.

(b) Cable Installed Parallel to Framing Members. Where the cable is installed parallel to the sides of rafters, studs, or floor joists, neither guard strips nor running boards shall be required, and the installation shall also comply with Section 300-4(d).

C. Construction Specifications

333-19. Construction. Type AC cable shall have an armor of flexible metal tape. The insulated conductors shall be in accordance with Section 333-20. Cables of the AC type shall have an internal bonding strip of copper or aluminum in intimate contact with the armor for its entire length.

333-20. Conductors. Insulated conductors shall be of a type listed in Table 310-13 or those identified for use in this cable. In addition, the conductors shall have an overall moisture-resistant and fire-retardant fibrous covering. For Type ACT, a moisture-resistant fibrous covering shall be required only on the individual conductors. The ampacity shall be determined by Section 310-15.

Armored cable installed in thermal insulation shall have conductors rated at 90°C (194°F). The ampacity of cable installed in these applications shall be that of 60°C (140°F) conductors.

333-21. Grounding. Type AC cable shall provide an adequate path for equipment grounding as required by Section 250-2(d).

333-22. Marking. The cable shall be marked in accordance with Section 310-11, except that Type AC shall have ready identification of the manufacturer by distinctive external markings on the cable sheath throughout its entire length.

Article 334 — Metal-Clad Cable: Type MC

A. General

334-1. Definition. *Type MC cable* is a factory assembly of one or more insulated circuit conductors with or without optical fiber members enclosed in an armor of interlocking metal tape, or a smooth or corrugated metallic sheath.

334-2. Other Articles. Metal-clad cable shall comply with this article and also with the applicable provisions of other articles in this *Code*, especially Article 300.

Type MC cable shall be permitted for systems in excess of 600 volts, nominal. See Section 300-2(a).

334-3. Uses Permitted. Unless specifically prohibited elsewhere in this *Code* and where not subject to physical damage, Type MC cables shall be permitted as follows:

(1) For services, feeders, and branch circuits
(2) For power, lighting, control, and signal circuits
(3) Indoors or outdoors
(4) Where exposed or concealed
(5) Direct buried where identified for such use
(6) In cable tray
(7) In any raceway
(8) As open runs of cable
(9) As aerial cable on a messenger
(10) In hazardous (classified) locations as permitted in Articles 501, 502, 503, 504, and 505
(11) In dry locations and embedded in plaster finish on brick or other masonry except in damp or wet locations
(12) In wet locations where any of the following conditions are met:

    (a) The metallic covering is impervious to moisture.
    (b) A lead sheath or moisture-impervious jacket is provided under the metal covering.
    (c) The insulated conductors under the metallic covering are listed for use in wet locations.

334-4. Uses Not Permitted. Type MC cable shall not be used where exposed to destructive corrosive conditions, such as direct burial in the earth, in concrete, or where exposed to cinder fills, strong chlorides, caustic alkalis, or vapors of chlorine or of hydrochloric acids, unless the metallic sheath is suitable for the conditions or is protected by material suitable for the conditions.

B. Installation

334-10. Installation. Type MC cable shall be installed in compliance with Articles 300, 490, 725, and Section 770-52 as applicable and in accordance with the following.

(a) Supported Cables. Type MC cable shall be supported and secured at intervals not exceeding 6 ft (1.83 m). Cables containing four or fewer conductors, sized no larger than No. 10 shall be secured within 12 in. (305 mm) of every box, cabinet, fitting, or other cable termination.

(1) Horizontal Runs. Cables installed in other than vertical runs through bored or punched holes in wood or metal framing members, or through notches in wooden framing members and protected by a steel plate at least ¹⁄₁₆ in. (1.59 mm) thick shall be considered supported and secured where such support does not exceed 6-ft (1.83-m) intervals.

(2) At Terminations. Cables containing four or fewer conductors, sized not larger than No. 10 shall be secured within 12 in. (305 mm) of every box, cabinet, fitting, or other cable termination.

(b) Unsupported Cables. Type MC cable shall not be required to be supported and secured where the cable is fished between access points, where concealed in finished buildings or structures and supporting is impracticable, or where used in lengths not more than 6 ft (1.83 m) from an outlet for connections within an accessible ceiling to lighting fixture(s) or equipment.

(c) Cable Tray. Type MC cable installed in cable tray shall comply with Article 318.

(d) Direct Buried. Direct-buried cable shall comply with Sections 300-5 or 300-50, as appropriate.

(e) Installed as Service-Entrance Cable. Type MC cable installed as service-entrance cable shall comply with Article 230.

(f) Installed Outside of Buildings or as Aerial Cable. Type MC cable installed outside of buildings or as aerial cable shall comply with Article 225 and Article 321.

(g) Through or Parallel to Joists, Studs, and Rafters. Type MC cable shall comply with Section 300-4 where installed through or parallel to joists, studs, rafters, or similar wood or metal members.

(h) In Accessible Attics. The installation of Type MC cable in accessible attics or roof spaces shall also comply with Section 333-12.

334-11. Bending Radius. All bends shall be made so that the cable will not be damaged, and the radius of the curve of the inner edge of any bend shall not be less than shown in (a) through (c).

(a) Smooth Sheath.

　(1) Ten times the external diameter of the metallic sheath for cable not more than ¾ in. (19 mm) in external diameter

　(2) Twelve times the external diameter of the metallic sheath for cable more than ¾ in. (19 mm) but not more than 1½ in. (38 mm) in external diameter

　(3) Fifteen times the external diameter of the metallic sheath for cable more than 1½ in. (38 mm) in external diameter

(b) Interlocked-Type Armor or Corrugated Sheath. Seven times the external diameter of the metallic sheath.

(c) Shielded Conductors. Twelve times the overall diameter of one of the individual conductors or seven times the overall diameter of the multiconductor cable, whichever is greater.

334-12. Fittings. Fittings used for connecting Type MC cable to boxes, cabinets, or other equipment shall be listed and identified for such use. Where single-conductor cables enter ferrous metal boxes or cabinets, the installation shall comply with Section 300-20 to prevent inductive heating.

334-13. Ampacity. The ampacity of Type MC cable shall be in accordance with Sections 310-15 or 310-60.

Exception No. 1: The ampacities for Type MC cable installed in cable tray shall be determined in accordance with Sections 318-11 and 318-13.

Exception No. 2: The ampacities of No. 18 and No. 16 conductors shall be in accordance with Table 402-5.

　FPN: See Section 310-10 for temperature limitation of conductors.

C. Construction Specifications

334-20. Conductors. The conductors shall be of copper, aluminum, or copper-clad aluminum, solid or stranded.

　The minimum conductor size shall be No. 18 copper and No. 12 aluminum or copper-clad aluminum.

334-21. Insulation. The insulated conductors shall comply with (a) or (b).

(a) 600 Volts. Insulated conductors in sizes No. 18 and No. 16 shall be of a type listed in Table 402-3, with a maximum operating temperature not less than 90°C (194°F), and as permitted by Section 725-27. Conductors larger than No. 16 shall be of a type listed in Table 310-13 or of a type identified for use in Type MC cable.

(b) Over 600 Volts. Insulated conductors shall be of a type listed in Tables 310-61 through 310-64.

334-22. Metallic Sheath. The metallic covering shall be one of the following types: smooth metallic sheath, corrugated metallic sheath, interlocking metal tape armor. The metallic sheath shall be continuous and close fitting.

　Supplemental protection of an outer covering of corrosion-resistant material shall be permitted and shall be required where such protection is needed. The sheath shall not be used as a current-carrying conductor.

　FPN: See Section 300-6 for protection against corrosion.

334-23. Grounding. Type MC cable shall provide an adequate path for equipment grounding as required by Article 250.

334-24. Marking. The cable shall be marked in accordance with Section 310-11.

Article 336 — Nonmetallic-Sheathed Cable: Types NM, NMC, and NMS

336-1. Scope. This article covers the use, installation, and construction specifications of nonmetallic-sheathed cable.

A. General

336-2. Definition. *Nonmetallic-sheathed cable* is a factory assembly of two or more insulated conductors having an outer sheath of moisture-resistant, flame-retardant, nonmetallic material.

336-3. Other Articles. Installations of nonmetallic-sheathed cable shall comply with the other applicable provisions of this *Code*, especially Articles 300 and 310.

336-4. Uses Permitted. Type NM, Type NMC, and Type NMS cables shall be permitted to be used in the following:

(1) One- and two-family dwellings
(2) Multifamily dwellings and other structures, except as prohibited in Section 336-5
(3) Cable trays, where the cables are identified for the use

> FPN: See Section 310-10 for temperature limitation of conductors.

(a) Type NM. Type NM cable shall be permitted for both exposed and concealed work in normally dry locations. It shall be permissible to install or fish Type NM cable in air voids in masonry block or tile walls where such walls are not exposed or subject to excessive moisture or dampness.

(b) Type NMC. Type NMC cable shall be permitted as follows:

(1) For both exposed and concealed work in dry, moist, damp, or corrosive locations
(2) In outside and inside walls of masonry block or tile
(3) In a shallow chase in masonry, concrete, or adobe protected against nails or screws by a steel plate at least $\frac{1}{16}$-in. (1.59-mm) thick, and covered with plaster, adobe, or similar finish

(c) Type NMS. Type NMS cable shall be permitted for both exposed and concealed work in normally dry locations. It shall be permissible to install or fish Type NMS cable in air voids in masonry block or tile walls where such walls are not exposed or subject to excessive moisture or dampness. Type NMS cable shall be used as permitted in Article 780.

336-5. Uses Not Permitted.

(a) Types NM, NMC, and NMS. Types NM, NMC, and NMS cables shall not be used in the following:

(1) In any multifamily dwelling or other structure exceeding three floors above grade

For the purpose of this article, the first floor of a building shall be that floor that has 50 percent or more of the exterior wall surface area level with or above finished grade. One additional level that is the first level and not designed for human habitation and used only for vehicle parking, storage, or similar use shall be permitted.

(2) As service-entrance cable
(3) In commercial garages having hazardous (classified) locations as provided in Section 511-3
(4) In theaters and similar locations, except as provided in Article 518, Places of Assembly
(5) In motion picture studios
(6) In storage battery rooms
(7) In hoistways
(8) Embedded in poured cement, concrete, or aggregate
(9) In any hazardous (classified) location, except as permitted by Sections 501-4(b), Exception, 502-4(b), Exception, and 504-20

(b) Types NM and NMS. Types NM and NMS cable shall not be installed in the following:

(1) Where exposed to corrosive fumes or vapors
(2) Where embedded in masonry, concrete, adobe, fill, or plaster
(3) In a shallow chase in masonry, concrete, or adobe and covered with plaster, adobe, or similar finish

B. Installation

336-6. Exposed Work — General. In exposed work, except as provided in Section 300-11(a), the cable shall be installed as specified in (a) through (d).

(a) To Follow Surface. The cable shall closely follow the surface of the building finish or of running boards.

(b) Protection from Physical Damage. The cable shall be protected from physical damage where necessary by conduit, electrical metallic tubing, Schedule 80 PVC rigid nonmetallic conduit, pipe, guard strips, listed surface metal or nonmetallic raceway, or other means. Where passing through a floor, the cable shall be enclosed in rigid metal conduit, intermediate metal conduit, electrical metallic tubing, Schedule 80 PVC rigid nonmetallic conduit, listed surface metal or nonmetallic raceway, or other metal pipe extending at least 6 in. (152 mm) above the floor.

(c) In Unfinished Basements. Where the cable is run at angles with joists in unfinished basements, it shall be permissible to secure cables not smaller than two No. 6 or three

No. 8 conductors directly to the lower edges of the joists. Smaller cables shall either be run through bored holes in joists or on running boards.

(d) In Accessible Attics. The installation of cable in accessible attics or roof spaces shall also comply with Section 333-12.

336-9. Through or Parallel to Framing Members. Types NM, NMC, or NMS cable shall comply with Section 300-4 where installed through or parallel to joists, studs, rafters, or similar wood or metal members.

336-16. Bends. Bends in cable shall be made so, and other handling shall be such, that the cable will not be damaged and the radius of the curve of the inner edge of any bend shall not be less than five times the diameter of the cable.

336-18. Supports. Nonmetallic-sheathed cable shall be secured by staples, cable ties, straps, or similar fittings designed and installed so as not to damage the cable. Cable shall be secured in place at intervals not exceeding 4½ ft (1.37 m) and within 12 in. (305 mm) from every cabinet, box, or fitting. Flat cables shall not be stapled on edge. Cables run through holes in wood or metal joists, rafters, or studs shall be considered to be supported and secured.

> FPN: See Section 370-17(c) for support where nonmetallic boxes are used.

Exception No. 1: For concealed work in finished buildings or finished panels for prefabricated buildings where such supporting is impracticable, it shall be permissible to fish cable between access points.

Exception No. 2: A wiring device identified for the use, without a separate outlet box, incorporating an integral cable clamp shall be permitted where the cable is secured in place at intervals not exceeding 4½ ft (1.37 m) and within 12 in. (305 mm) from the wiring device wall opening, and there shall be at least a 12-in. (305-mm) loop of unbroken cable or 6 in. (152 mm) of a cable end available on the interior side of the finished wall to permit replacement.

Exception No. 3: Lengths not more than 4½ ft (1.37 m) from an outlet for connection within an accessible ceiling to lighting fixture(s) or equipment.

336-20. Boxes of Insulating Material. Nonmetallic outlet boxes shall be permitted as provided in Section 370-3.

336-21. Devices of Insulating Material. Switch, outlet, and tap devices of insulating material shall be permitted to be used without boxes in exposed cable wiring and for rewiring in existing buildings where the cable is concealed and fished. Openings in such devices shall form a close fit around the outer covering of the cable and the device shall fully enclose the part of the cable from which any part of the covering has been removed.

Where connections to conductors are by binding-screw terminals, there shall be available as many terminals as conductors.

336-25. Devices with Integral Enclosures. Wiring devices with integral enclosures identified for such use shall be permitted as provided in Section 300-15(e).

336-26. Ampacity. The ampacity of Types NM, NMC, and NMS cable shall be that of 60°C (140°F) conductors and shall comply with Section 310-15.

The 90°C (194°F) rating shall be permitted to be used for ampacity derating purposes provided the final derated ampacity does not exceed that for a 60°C (140°F) rated conductor.

C. Construction Specifications

336-30. General. Nonmetallic-sheathed cable shall comply with (a) and (b).

(a) Construction. The outer cable sheath shall be a nonmetallic material.

(1) Type NM. The overall covering shall be flame retardant and moisture resistant.

(2) Type NMC. The overall covering shall be flame retardant, moisture resistant, fungus resistant, and corrosion resistant.

(3) Type NMS. Type NMS cable is a factory assembly of insulated power, communications, and signaling conductors enclosed within a common sheath of moisture-resistant, flame-retardant, nonmetallic material. The sheath shall be applied so as to separate the power conductors from the communications and signaling conductors. The signaling conductors shall be permitted to be shielded. An optional outer jacket shall be permitted.

> FPN: For composite optical cable, see Sections 770-4 and 770-52.

(b) Conductors. The insulated power conductors shall be one of the types listed in Table 310-13 that is suitable for branch-circuit wiring or one that is identified for use in these cables.

The power conductors shall be sizes No. 14 through No. 2 with copper conductors or sizes No. 12 through No. 2 with aluminum or copper-clad aluminum conductors.

The signaling conductors shall comply with Section 780-5.

In addition to the insulated conductors, the cable shall be permitted to have an insulated or bare conductor for equipment grounding purposes only. Where provided, the grounding conductor shall be sized in accordance with Article 250.

Conductors shall be rated at 90°C (194°F).

FPN: Types NM, NMC, and NMS cable identified by the markings NM-B, NMC-B, and NMS-B meet this requirement.

336-31. Marking. The cable shall be marked in accordance with Section 310-11.

Article 338 — Service-Entrance Cable: Types SE and USE

338-1. Definition. *Service-entrance cable* is a single conductor or multiconductor assembly provided with or without an overall covering, primarily used for services, and is of the following types.

(a) Type SE. Type SE, having a flame-retardant, moisture-resistant covering.

(b) Type USE. Type USE, identified for underground use, having a moisture-resistant covering, but not required to have a flame-retardant covering.

Cabled, single-conductor, Type USE constructions recognized for underground use may have a bare copper conductor cabled with the assembly. Type USE single, parallel, or cabled conductor assemblies recognized for underground use may have a bare copper concentric conductor applied. These constructions do not require an outer overall covering.

FPN: See Section 230-41, Exception, item (b) for directly buried, uninsulated service-entrance conductors.

(c) One Uninsulated Conductor. If Type SE or USE cable consists of two or more conductors, one shall be permitted to be uninsulated.

338-2. Uses Permitted as Service-Entrance Conductors. Service-entrance cable used as service-entrance conductors shall be installed as required by Article 230.

Type USE used for service laterals shall be permitted to emerge above ground outside at terminations in meter bases or other enclosures where protected in accordance with Section 300-5(d).

338-3. Uses Permitted as Branch Circuits or Feeders.

(a) Grounded Conductor Insulated. Type SE service-entrance cables shall be permitted in interior wiring systems where all of the circuit conductors of the cable are of the rubber-covered or thermoplastic type.

(b) Grounded Conductor Not Insulated. Type SE service-entrance cables without individual insulation on the grounded circuit conductor shall not be used as a branch circuit or as a feeder within a building, except a cable that has a final nonmetallic outer covering and is supplied by alternating current at not over 150 volts to ground shall be permitted as a feeder to supply only other buildings on the same premises.

Type SE service-entrance cable shall be permitted for use where the fully insulated conductors are used for circuit wiring and the uninsulated conductor is used for equipment grounding purposes.

(c) Temperature Limitations. Type SE service-entrance cable used to supply appliances shall not be subject to conductor temperatures in excess of the temperature specified for the type of insulation involved.

338-4. Installation Methods for Branch Circuits and Feeders.

(a) Interior Installations. In addition to the provisions of this article, Type SE service-entrance cable used for interior wiring shall comply with the installation requirements of Parts A and B of Article 336 and shall comply with the applicable provisions of Article 300.

FPN: See Section 310-10 for temperature limitation of conductors.

(b) Exterior Installations. In addition to the provisions of this article, service-entrance cable used for feeders or branch circuits, where installed as exterior wiring, shall be installed as required by Article 225. The cable shall be supported in accordance with Section 336-18, unless used as messenger-supported wiring as allowed by Article 321.

Type USE cable shall be installed outside in accordance with the provisions of Article 339. Where Type USE cable emerges above ground at terminations, it shall be protected in accordance with Section 300-5(d).

Multiconductor service-entrance cable shall be permitted to be installed as messenger-supported wiring in accordance with Articles 225 and 321.

338-5. Marking. Service-entrance cable shall be marked as required in Section 310-11. Cable with the neutral conductor smaller than the ungrounded conductors shall be so marked.

338-6. Bends. Bends in cable shall be made, and other handlings shall be such, that the protective coverings of the cable will not be damaged, and the radius of the curve of the inner edge of any bend shall not be less than five times the diameter of the cable.

Article 339 — Underground Feeder and Branch-Circuit Cable: Type UF

339-1. Description and Marking.

(a) Description. Underground feeder and branch-circuit cable shall be a listed Type UF cable in sizes No. 14 copper or No. 12 aluminum or copper-clad aluminum through No. 4/0. The conductors of Type UF shall be one of the moisture-resistant types listed in Table 310-13 that is suitable for branch-circuit wiring or one that is identified for such use. In addition to the insulated conductors, the cable shall be permitted to have an insulated or bare conductor for equipment grounding purposes only. The overall covering shall be flame retardant; moisture, fungus, and corrosion resistant; and suitable for direct burial in the earth.

(b) Marking. The cable shall be marked in accordance with Section 310-11.

339-2. Other Articles.
In addition to the provisions of this article, installations of underground feeder and branch-circuit cable (Type UF) shall comply with other applicable provisions of this *Code*, especially Article 300 and Section 310-13.

339-3. Use.

(a) Uses Permitted.

(1) Type UF cable shall be permitted for use underground, including direct burial in the earth, as feeder or branch-circuit cable where provided with overcurrent protection of the rated ampacity as required in Section 339-4.

(2) Where single-conductor cables are installed, all cables of the feeder circuit or branch circuit, including the neutral and equipment grounding conductor, if any, shall be run together in the same trench or raceway.

(3) For underground requirements, see Section 300-5.

(4) Type UF cable shall be permitted for interior wiring in wet, dry, or corrosive locations under the recognized wiring methods of this *Code*, and, where installed as nonmetallic-sheathed cable, the installation and conductor requirements shall comply with the provisions of Article 336 and shall be of the multiconductor type.

(5) For solar photovoltaic systems in accordance with Section 690-31.

(6) Single-conductor cables shall be permitted as the nonheating leads for heating cables as provided in Section 424-43.

Type UF cable supported by cable trays shall be of the multiconductor type.

FPN: See Section 310-10 for temperature limitation of conductors.

(b) Uses Not Permitted. Type UF cable shall not be used in the following:

(1) As service-entrance cables
(2) In commercial garages
(3) In theaters
(4) In motion picture studios
(5) In storage battery rooms
(6) In hoistways
(7) In any hazardous (classified) location
(8) Embedded in poured cement, concrete, or aggregate, except where embedded in plaster as nonheating leads as provided in Article 424
(9) Where exposed to direct rays of the sun, unless identified as sunlight resistant
(10) Where subject to physical damage

339-4. Overcurrent Protection.
Overcurrent protection shall be provided in accordance with the provisions of Section 240-3.

339-5. Ampacity.
The ampacity of Type UF cable shall be that of 60°C (140°F) conductors in accordance with Section 310-15.

Article 340 — Power and Control Tray Cable: Type TC

340-1. Definition.
Type TC power and control tray cable is a factory assembly of two or more insulated conductors, with or without associated bare or covered grounding conductors under a nonmetallic sheath, for installation in cable trays, in raceways, or where supported by a messenger wire.

340-2. Other Articles.
In addition to the provisions of this article, installations of Type TC tray cable shall comply with other applicable articles of this *Code*, especially Articles 300 and 318.

340-3. Construction.
The insulated conductors of Type TC tray cable shall be in sizes No. 18 through 1000 kcmil copper and sizes No. 12 through 1000 kcmil aluminum or copper-clad aluminum. Insulated conductors of sizes No. 14 and larger copper and sizes No. 12 and larger aluminum or copper-clad aluminum shall be one of the types listed in Tables 310-13 or 310-62 that is suitable for branch circuit and feeder circuits or one that is identified for such use. The outer sheath shall be a flame-retardant, nonmetallic material.

A metallic sheath shall not be permitted either under or over the nonmetallic sheath.

(a) Wet Locations. Where installed in wet locations, Type TC cable shall also be resistant to moisture and corrosive agents.

(b) Fire Alarm Systems. Where used for fire alarm systems, conductors shall also be in accordance with Section 760-27.

(c) Thermocouple Circuits. Conductors in Type TC cables used for thermocouple circuits in accordance with Article 725 shall also be permitted to be any of the materials used for thermocouple extension wire.

There shall be no voltage marking on a Type TC cable employing thermocouple extension wire.

(d) Class 1 Circuit Conductors. Insulated conductors of sizes No. 18 and No. 16 copper shall also be in accordance with Section 725-27.

340-4. Use Permitted. Type TC tray cable shall be permitted to be used in the following.

(1) For power, lighting, control, and signal circuits.
(2) In cable trays, or in raceways, or where supported in outdoor locations by a messenger wire.
(3) In cable trays in hazardous (classified) locations as permitted in Articles 318, 501, 502, 504, and 505 in industrial establishments where the conditions of maintenance and supervision ensure that only qualified persons will service the installation.
(4) For Class 1 circuits as permitted in Article 725.
(5) For nonpower-limited fire alarm circuits if conductors comply with the requirements of Section 760-27.
(6) In industrial establishments where the conditions of maintenance and supervision ensure that only qualified persons will service the installation, and where the cable is not subject to physical damage, Type TC tray cable that complies with the crush and impact requirements of Type MC cable and is identified for such use shall be permitted as open wiring in lengths not to exceed a total of 50 ft (15.24 m) between a cable tray and the utilization equipment or device.

The cable shall be supported and secured at intervals not exceeding 6 ft (1.83 m).

Equipment grounding for the utilization equipment shall be provided by an equipment grounding conductor within the cable.

FPN: See Section 310-10 for temperature limitation of conductors.

340-5. Uses Not Permitted. Type TC tray cable shall not be used in the following:

(1) Installed where it will be exposed to physical damage
(2) Installed as open cable on brackets or cleats
(3) Used where exposed to direct rays of the sun, unless identified as sunlight resistant
(4) Direct buried, unless identified for such use

340-6. Marking. The cable shall be marked in accordance with Section 310-11.

340-7. Ampacity. The ampacities of the conductors of Type TC tray cable shall be determined from Section 402-5 for conductors smaller than No. 14 and from Section 318-11.

340-8. Bends. Bends in Type TC cable shall be made so as not to damage the cable.

Article 342 — Nonmetallic Extensions

342-1. Definition. *Nonmetallic extensions* are an assembly of two insulated conductors within a nonmetallic jacket or an extruded thermoplastic covering. The classification includes both surface extensions, intended for mounting directly on the surface of walls or ceilings, and aerial cable containing a supporting messenger cable as an integral part of the cable assembly.

342-2. Other Articles. In addition to the provisions of this article, nonmetallic extensions shall be installed in accordance with the applicable provisions of this *Code*.

342-3. Uses Permitted. Nonmetallic extensions shall be permitted only where all of the following conditions are met:

(a) From an Existing Outlet. The extension is from an existing outlet on a 15- or 20-ampere branch circuit in conformity with the requirements of Article 210.

(b) Exposed and in a Dry Location. The extension is run exposed and in a dry location.

(c) Nonmetallic Surface Extensions. For nonmetallic surface extensions, the building is occupied for residential or office purposes and does not exceed the height limitations specified in Section 336-5(a)(1).

(c1) [Alternate to (c)]. For aerial cable, the building is occupied for industrial purposes, and the nature of the occupancy requires a highly flexible means for connecting equipment.

FPN: See Section 310-10 for temperature limitation of conductors.

342-4. Uses Not Permitted. Nonmetallic extensions shall not be used as follows.

(a) Aerial Cable. As aerial cable to substitute for one of the general wiring methods specified by this *Code*.

(b) Unfinished Areas. In unfinished basements, attics, or roof spaces.

(c) Voltage Between Conductors. Where the voltage between conductors exceeds 150 volts for nonmetallic surface extension and 300 volts for aerial cable.

(d) Corrosive Vapors. Where subject to corrosive vapors.

(e) Through a Floor or Partition. Where run through a floor or partition, or outside the room in which it originates.

342-5. Splices and Taps. Extensions shall consist of a continuous unbroken length of the assembly, without splices, and without exposed conductors between fittings. Taps shall be permitted where approved fittings completely covering the tap connections are used. Aerial cable and its tap connectors shall be provided with an approved means for polarization. Receptacle-type tap connectors shall be of the locking type.

342-6. Fittings. Each run shall terminate in a fitting that covers the end of the assembly. All fittings and devices shall be of a type identified for the use.

342-7. Installation. Nonmetallic extensions shall be installed as specified in (a) and (b).

(a) Nonmetallic Surface Extensions.
  (1) One or more extensions shall be permitted to be run in any direction from an existing outlet, but not on the floor or within 2 in. (50.8 mm) from the floor.

  (2) Nonmetallic surface extensions shall be secured in place by approved means at intervals not exceeding 8 in. (203 mm), with an allowance for 12 in. (305 mm) to the first fastening where the connection to the supplying outlet is by means of an attachment plug. There shall be at least one fastening between each two adjacent outlets supplied. An extension shall be attached to only woodwork or plaster finish, and shall not be in contact with any metal work or other conductive material other than with metal plates on receptacles.

  (3) A bend that reduces the normal spacing between the conductors shall be covered with a cap to protect the assembly from physical damage.

(b) Aerial Cable.
  (1) Aerial cable shall be supported by its messenger cable and securely attached at each end with clamps and turnbuckles. Intermediate supports shall be provided at not more than 20-ft (6.1-m) intervals. Cable tension shall be adjusted to eliminate excessive sag. The cable shall have a clearance of not less than 2 in. (50.8 mm) from steel structural members or other conductive material.

  (2) Aerial cable shall have a clearance of not less than 10 ft (3.05 m) above floor areas accessible to pedestrian traffic, and not less than 14 ft (4.27 m) above floor areas accessible to vehicular traffic.

  (3) Cable suspended over work benches, not accessible to pedestrian traffic, shall have a clearance of not less than 8 ft (2.44 m) above the floor.

  (4) Aerial cables shall be permitted as a means to support lighting fixtures where the total load on the supporting messenger cable does not exceed that for which the assembly is intended.

  (5) The supporting messenger cable, where installed in conformity with the applicable provisions of Article 250 and if properly identified as an equipment grounding conductor, shall be permitted to ground equipment. The messenger cable shall not be used as a branch-circuit conductor.

342-8. Marking. Nonmetallic extensions shall be marked in accordance with Section 110-21.

Article 343 — Nonmetallic Underground Conduit with Conductors

A. General

343-1. Description. *Nonmetallic underground conduit with conductors* is a factory assembly of conductors or cables inside a nonmetallic, smooth wall conduit with a circular cross section.

The nonmetallic conduit shall be composed of a material that is resistant to moisture and corrosive agents. It shall also be capable of being supplied on reels without damage or distortion and shall be of sufficient strength to withstand abuse, such as impact or crushing, in handling and during installation without damage to conduit or conductors.

343-2. Other Articles. Installations for nonmetallic underground conduit with conductors shall comply with the provisions of the applicable sections of Article 300. Where equipment grounding is required by Article 250, an assembly containing a separate equipment grounding conductor shall be used.

343-3. Uses Permitted. The use of listed nonmetallic underground conduit with conductors and fittings shall be permitted in the following.

(1) For direct burial underground installation. For minimum cover requirements, see Tables 300-5 and 300-50 under rigid nonmetallic conduit.
(2) Encased or embedded in concrete.
(3) In cinder fill.
(4) In underground locations subject to severe corrosive influences as covered in Section 300-6 and where subject to chemicals for which the assembly is specifically approved.

343-4. Uses Not Permitted. Nonmetallic underground conduit with conductors shall not be used in the following:

(1) In exposed locations
(2) Inside buildings

Exception: The conductor or the cable portion of the assembly, where suitable, shall be permitted to extend within the building for termination purposes in accordance with Section 300-3.

(3) In hazardous (classified) locations except as permitted by Sections 503-3(a), 504-20, 514-8, and 515-5, and in Class I, Division 2 locations as permitted in Section 501-4(b), Exception

B. Installation

343-5. Size.

(a) Minimum. Nonmetallic underground conduit with conductors smaller than ½-in. electrical trade size shall not be used.

(b) Maximum. Nonmetallic underground conduit with conductors larger than 4-in. electrical trade size shall not be used.

> FPN: Metric trade numerical designations for nonmetallic underground conduit with conductors are ½ = 16, ¾ = 21, 1 = 27, 1¼ = 35, 1½ = 41, 2 = 53, 2½ = 63, 3 = 78, 3½ = 91, and 4 = 103.

343-6. Trimming. For termination, the conduit shall be trimmed away from the conductors or cables using an approved method that will not damage the conductor or cable insulation or jacket. All ends shall be trimmed inside and out to remove rough edges.

343-7. Joints. All joints between conduit, fittings, and boxes shall be made by an approved method.

343-8. Conductor Terminations. All terminations between the conductors or cables and equipment shall be made by an approved method for that type of conductor or cable.

343-9. Bushings. Where the nonmetallic underground conduit with conductors enters a box, fitting, or other enclosure, a bushing or adapter shall be provided to protect the conductor or cable from abrasion unless the design of the box, fitting, or enclosure provides equivalent protection.

> FPN: See Section 300-4(f) for the protection of conductors size No. 4 or larger.

343-10. Bends — How Made. Bends of nonmetallic underground conduit with conductors shall be manually made so that the conduit will not be damaged and the internal diameter of the conduit will not be effectively reduced. The radius of the curve of the centerline of such bends shall not be less than shown in Table 343-10.

Table 343-10. Radius of Conduit Bends

Trade Size (in.)	Minimum Bending Radius (in.)
½	10
¾	12
1	14
1¼	18
1½	20
2	26
2½	36
3	48
4	60

Note: For SI units, in. = 25.4 mm (radius).

343-11. Bends — Number in One Run. There shall not be more than the equivalent of four quarter bends (360 degrees total) between termination points.

343-12. Splices and Taps. Splices and taps shall be made in junction boxes or other enclosures. See Article 370 for rules on the installation and use of boxes and conduit bodies.

C. Construction

343-13. General. Nonmetallic underground conduit with conductors is an assembly that is provided in continuous lengths shipped in a coil, reel, or carton.

343-14. Conductors and Cables. Conductors and cables used in nonmetallic underground conduit with conductors shall be listed, shall be suitable for use in wet locations, and shall be as follows.

(a) 600 Volts or Less. Alternating-current and direct-current circuits shall be permitted. All conductors shall have an insulation rating equal to at least the maximum nominal circuit voltage of any conductor or cable within the conduit.

(b) Over 600 Volts. Conductors or cables rated over 600 volts shall not occupy the same conduit with conductors or cables of circuits rated 600 volts or less.

343-15. Conductor Fill. The maximum number of conductors or cables in nonmetallic underground conduit with conductors shall not exceed that permitted by the percentage fill in Table 1, Chapter 9.

343-16. Marking. Nonmetallic underground conduit with conductors shall be clearly and durably marked at least every 10 ft (3.05 m) as required by Section 110-21. The type of conduit material shall also be included in the marking.

Identification of conductors or cables used in the assembly shall be provided on a tag attached to each end of the assembly or to the side of a reel. Enclosed conductors or cables shall be marked in accordance with Section 310-11.

Article 345 — Intermediate Metal Conduit

A. General

345-1. Definition. *Intermediate metal conduit* is a listed steel raceway of circular cross section with integral or associated couplings, approved for the installation of electrical conductors and used with listed fittings to provide electrical continuity.

345-2. Other Articles. Installations for intermediate metal conduit shall comply with the provisions of the applicable sections of Article 300.

345-3. Uses Permitted.

(a) All Atmospheric Conditions and Occupancies. Use of intermediate metal conduit shall be permitted under all atmospheric conditions and occupancies. Where practicable, dissimilar metals in contact anywhere in the system shall be avoided to eliminate the possibility of galvanic action. Intermediate metal conduit shall be permitted as an equipment grounding conductor.

Exception: Aluminum fittings and enclosures shall be permitted to be used with steel intermediate metal conduit.

(b) Corrosion Protection. Intermediate metal conduit, elbows, couplings, and fittings shall be permitted to be installed in concrete, in direct contact with the earth, or in areas subject to severe corrosive influences where protected by corrosion protection and judged suitable for the condition.

FPN: See Section 300-6 for protection against corrosion.

(c) Cinder Fill. Intermediate metal conduit shall be permitted to be installed in or under cinder fill where subject to permanent moisture where protected on all sides by a layer of noncinder concrete not less than 2 in. (50.8 mm) thick; where the conduit is not less than 18 in. (457 mm) under the fill; or where protected by corrosion protection and judged suitable for the condition.

FPN: See Section 300-6 for protection against corrosion.

B. Installation

345-5. Wet Locations. All supports, bolts, straps, screws, etc., shall be of corrosion-resistant materials or protected against corrosion by corrosion-resistant materials.

FPN: See Section 300-6 for protection against corrosion.

345-6. Size.

(a) Minimum. Conduit smaller than ½-in. electrical trade size shall not be used.

(b) Maximum. Conduit larger than 4-in. electrical trade size shall not be used.

FPN: Metric trade numerical designations for intermediate metal conduit are the same as those found in *Extra-heavy Duty Rigid Steel Conduits for Electrical Installations*, IEC 981-1989; namely, ½ = 16, ¾ = 21, 1 = 27, 1¼ = 35, 1½ = 41, 2 = 53, 2½ = 63, 3 = 78, 3½ = 91, and 4 = 103.

345-7. Number of Conductors in Conduit. The number of conductors in a single conduit shall not exceed that permitted by the percentage fill specified in Table 1, Chapter 9, using the conduit dimensions of Table 4, Chapter 9.

345-8. Reaming and Threading. All cut ends of conduits shall be reamed or otherwise finished to remove rough edges. Where conduit is threaded in the field, a standard cutting die with a ¾-in. taper per foot (1 in 16) shall be used.

FPN: See *Standards for Pipe Threads, General Purpose (Inch)*, ANSI/ASME B.1.20.1-1983.

345-9. Couplings and Connectors.

(a) Threadless. Threadless couplings and connectors used with conduit shall be made tight. Where buried in masonry or concrete, they shall be the concretetight type. Where installed in wet locations, they shall be the raintight type.

(b) Running Threads. Running threads shall not be used on conduit for connection at couplings.

345-10. Bends — How Made. Bends of intermediate metal conduit shall be made so that the conduit will not be damaged and so that the internal diameter of the conduit will not be effectively reduced. The radius of the curve of the inner edge of any field bend shall not be less than indicated in Table 346-10.

Exception: For field bends for conductors without lead sheath and made with a single operation (one shot) bending machine designed for the purpose, the minimum radius shall not be less than that indicated in Table 346-10, Exception.

345-11. Bends — Number in One Run. There shall not be more than the equivalent of four quarter bends (360 degrees total) between pull points, e.g., conduit bodies and boxes.

345-12. Securing and Supporting. Intermediate metal conduit shall be installed as a complete system as provided in Article 300 and shall be securely fastened in place and supported in accordance with (a) and (b).

(a) Securely Fastened. Each intermediate metal conduit shall be securely fastened within 3 ft (914 mm) of each outlet box, junction box, device box, cabinet, conduit body, or other conduit termination. Fastening shall be permitted to be increased to a distance of 5 ft (1.52 m) where structural members do not readily permit fastening within 3 ft (914 mm). Where approved, conduit shall not be required to be securely fastened within 3 ft (914 mm) of the service head for above-the-roof termination of a mast.

(b) Supports. Intermediate metal conduit shall be supported in accordance with one of the following.

(1) Conduit shall be supported at intervals not exceeding 10 ft (3.05 m).

(2) The distance between supports for straight runs of conduit shall be permitted in accordance with Table 346-12, provided the conduit is made up with threaded couplings, and such supports prevent transmission of stresses to termination where conduit is deflected between supports.

(3) Exposed vertical risers from industrial machinery shall be permitted to be supported at intervals not exceeding 20 ft (6.1 m), provided the conduit is made up with threaded couplings, is firmly supported at the top and bottom of the riser, and no other means of intermediate support is readily available.

(4) Horizontal runs of intermediate metal conduit supported by openings through framing members at intervals not exceeding 10 ft (3.05 m) and securely fastened within 3 ft (914 mm) of termination points shall be permitted.

345-13. Boxes and Fittings. See Article 370.

345-14. Splices and Taps. Splices and taps shall be made in accordance with Section 300-15. See Article 370 for rules on the installation and use of boxes and conduit bodies.

345-15. Bushings. Where a conduit enters a box, fitting, or other enclosure, a bushing shall be provided to protect the wire from abrasion unless the design of the box, fitting, or enclosure is such as to afford equivalent protection.

FPN: See Section 300-4(f) for the protection of conductors sizes No. 4 and larger at bushings.

C. Construction Specifications

345-16. General. Intermediate metal conduit shall comply with (a) through (c).

(a) Standard Lengths. The standard length of intermediate metal conduit shall be 10 ft (3.05 m), including an attached coupling, and each end shall be threaded. Longer or shorter lengths with or without coupling and threaded or unthreaded shall be permitted.

(b) Corrosion-Resistant Material. Nonferrous conduit of corrosion-resistant material shall have suitable markings.

(c) Marking. Each length shall be clearly and durably marked at least every 5 ft (1.52 m) with the letters *IMC*. Each length shall be marked as required in the first sentence of Section 110-21.

Article 346 — Rigid Metal Conduit

A. General

346-1. Definition. *Rigid metal conduit* is a listed metal raceway of circular cross section with integral or associated couplings, approved for the installation of electrical conductors and used with listed fittings to provide electrical continuity.

346-2. Other Articles. Installations for rigid metal conduit shall comply with the provisions of the applicable sections of Article 300.

346-3. Uses Permitted.

(a) All Atmospheric Conditions and Occupancies. Use of rigid metal conduit shall be permitted under all atmospheric conditions and occupancies. Where practicable, dissimilar metals in contact anywhere in the system shall be avoided to eliminate the possibility of galvanic action. Rigid metal conduit shall be permitted as an equipment grounding conductor. Ferrous raceways and fittings protected from corrosion solely by enamel shall be permitted only indoors and in occupancies not subject to severe corrosive influences.

Aluminum fittings and enclosures shall be permitted to be used with steel rigid metal conduit, and steel fittings and enclosures shall be permitted to be used with aluminum rigid metal conduit.

(b) Corrosion Protection. Rigid metal conduit, elbows, couplings, and fittings shall be permitted to be installed in concrete, in direct contact with the earth, or in areas subject to severe corrosive influences where protected by corrosion protection and judged suitable for the condition.

FPN: See Section 300-6 for protection against corrosion.

(c) Cinder Fill. Rigid metal conduit shall be permitted to be installed in or under cinder fill where subject to permanent moisture where protected on all sides by a layer of noncinder concrete not less than 2 in. (50.8 mm) thick; where the conduit is not less than 18 in. (457 mm) under the fill; or where protected by corrosion protection and judged suitable for the condition.

FPN: See Section 300-6 for protection against corrosion.

B. Installation

346-5. Wet Locations. All supports, bolts, straps, screws, etc., shall be of corrosion-resistant materials or protected against corrosion by corrosion-resistant materials.

FPN: See Section 300-6 for protection against corrosion.

346-6. Size.

(a) Minimum. Rigid metal conduit smaller than ½-in. electrical trade size shall not be used.

Exception: For enclosing the leads of motors as permitted in Section 430-145(b).

(b) Maximum. Rigid metal conduit larger than 6-in. electrical trade size shall not be used.

FPN: Metric trade numerical designations for rigid metal conduit are the same as those found in *Extra-heavy Duty Rigid Steel Conduits for Electrical Installations*, IEC 981-1989; namely, ½ = 16, ¾ = 21, 1 = 27, 1¼ = 35, 1½ = 41, 2 = 53, 2½ = 63, 3 = 78, 3½ = 91, 4 = 103, 5 = 129, and 6 = 155.

346-7. Number of Conductors in Conduit. The number of conductors in a single conduit shall not exceed that permitted by the percentage fill specified in Table 1, Chapter 9, using the conduit dimensions of Table 4, Chapter 9.

346-8. Reaming and Threading. All cut ends of conduits shall be reamed or otherwise finished to remove rough edges. Where conduit is threaded in the field, a standard cutting die with a ¾-in. taper per foot (1 in 16) shall be used.

FPN: See *Standards for Pipe Threads, General Purpose (Inch)*, ANSI/ASME B.1.20.1-1983.

346-9. Couplings and Connectors.

(a) Threadless. Threadless couplings and connectors used with conduit shall be made tight. Where buried in masonry or concrete, they shall be the concretetight type. Where installed in wet locations, they shall be the raintight type.

(b) Running Threads. Running threads shall not be used on conduit for connection at couplings.

346-10. Bends — How Made. Bends of rigid metal conduit shall be made so that the conduit will not be damaged and that the internal diameter of the conduit will not be effectively reduced. The radius of the curve of the inner edge of any field bend shall not be less than indicated in Table 346-10.

Table 346-10. Radius of Conduit Bends

Size of Conduit (in.)	Conductors Without Lead Sheath (in.)
½	4
¾	5
1	6
1¼	8
1½	10
2	12
2½	15
3	18
3½	21
4	24
5	30
6	36

Note: For SI units, 1 in. = 25.4 mm (radius).

Exception: For field bends for conductors without lead sheath and made with a single operation (one shot) bending machine designed for the purpose, the minimum radius shall not be less than that indicated in Table 346-10, Exception.

346-11. Bends — Number in One Run. There shall not be more than the equivalent of four quarter bends (360 degrees total) between pull points, e.g., conduit bodies and boxes.

Table 346-10. Exception, Radius of Conduit Bends

Size of Conduit (in.)	Radius to Center of Conduit (in.)
½	4
¾	4½
1	5¾
1¼	7¼
1½	8¼
2	9½
2½	10½
3	13
3½	15
4	16
5	24
6	30

Note: For SI units, 1 in. = 25.4 mm (radius).

346-12. Securing and Supporting. Rigid metal conduit shall be installed as a complete system as provided in Article 300 and shall be securely fastened in place and supported in accordance with (a) and (b).

(a) Securely Fastened. Each rigid metal conduit shall be securely fastened within 3 ft (914 mm) of each outlet box, junction box, device box, cabinet, conduit body, or other conduit termination. Fastening shall be permitted to be increased to a distance of 5 ft (1.52 m) where structural members do not readily permit fastening within 3 ft (914 mm). Where approved, conduit shall not be required to be securely fastened within 3 ft (914 mm) of the service head for above-the-roof termination of a mast.

(b) Supports. Rigid metal conduit shall be supported in accordance with one of the following.

(1) Conduit shall be supported at intervals not exceeding 10 ft (3.05 m).

(2) The distance between supports for straight runs of conduit shall be permitted in accordance with Table 346-12(b)(2), provided the conduit is made up with threaded couplings, and such supports prevent transmission of stresses to termination where conduit is deflected between supports.

(3) Exposed vertical risers from stationary equipment or fixtures shall be permitted to be supported at intervals not exceeding 20 ft (6.1 m), provided the conduit is made up with threaded couplings, is firmly supported at the top and bottom of the riser, and no other means of intermediate support is readily available.

(4) Horizontal runs of rigid metal conduit supported by openings through framing members at intervals not exceeding 10 ft (3.05 m) and securely fastened within 3 ft (914 mm) of termination points shall be permitted.

Table 346-12(b)(2). Supports for Rigid Metal Conduit

Conduit Size (in.)	Maximum Distance Between Rigid Metal Conduit Supports (ft)
½–¾	10
1	12
1¼–1½	14
2–2½	16
3 and larger	20

Note: For SI units, 1 ft = 0.3048 m (supports).

346-13. Boxes and Fittings. See Article 370.

346-14. Splices and Taps. Splices and taps shall be made in accordance with Section 300-15. See Article 370 for rules on the installation and use of boxes and conduit bodies.

346-15. Bushings. Where a conduit enters a box, fitting, or other enclosure, a bushing shall be provided to protect the wire from abrasion unless the design of the box, fitting, or enclosure is such as to afford equivalent protection.

FPN: See Section 300-4(f) for the protection of conductors sizes No. 4 and larger at bushings.

C. Construction Specifications

346-16. General. Rigid metal conduit shall comply with (a) through (c).

(a) Standard Lengths. The standard length of rigid metal conduit shall be 10 ft (3.05 m), including an attached coupling, and each end shall be threaded. Longer or shorter lengths with or without coupling and threaded or unthreaded shall be permitted.

(b) Corrosion-Resistant Material. Nonferrous conduit of corrosion-resistant material shall have suitable markings.

(c) Marking. Each length shall be clearly and durably identified in every 10 ft (3.05 m) as required in the first sentence of Section 110-21.

Article 347 — Rigid Nonmetallic Conduit

347-1. Description. This article shall apply to a type of conduit and fittings of suitable nonmetallic material that is resistant to moisture and chemical atmospheres. For use aboveground, it shall also be flame retardant, resistant to impact and crushing, resistant to distortion from heat under conditions likely to be encountered in service, and resistant to low temperature and sunlight effects. For use underground, the material shall be acceptably resistant to moisture and corrosive agents and shall be of sufficient strength to withstand abuse, such as by impact and crushing, in handling and during installation. Conduits listed for the purpose shall be permitted to be installed underground in continuous lengths from a reel. Where intended for direct burial, without encasement in concrete, the material shall also be capable of withstanding continued loading that is likely to be encountered after installation.

347-2. Uses Permitted. The use of listed rigid nonmetallic conduit shall be permitted under the following conditions.

> FPN: Extreme cold may cause some nonmetallic conduits to become brittle and therefore more susceptible to damage from physical contact.

(a) Concealed. In walls, floors, and ceilings.

(b) Corrosive Influences. In locations subject to severe corrosive influences as covered in Section 300-6 and where subject to chemicals for which the materials are specifically approved.

(c) Cinders. In cinder fill.

(d) Wet Locations. In portions of dairies, laundries, canneries, or other wet locations and in locations where walls are frequently washed, the entire conduit system including boxes and fittings used therewith shall be installed and equipped so as to prevent water from entering the conduit. All supports, bolts, straps, screws, etc., shall be of corrosion-resistant materials or be protected against corrosion by approved corrosion-resistant materials.

(e) Dry and Damp Locations. In dry and damp locations not prohibited by Section 347-3.

(f) Exposed. For exposed work where not subject to physical damage if identified for such use.

(g) Underground Installations. For underground installations, see Sections 300-5 and 300-50.

(h) Support of Conduit Bodies. Rigid nonmetallic conduit shall be permitted to support nonmetallic conduit bodies not larger than the largest trade size of an entering raceway. The conduit bodies shall not contain devices or support fixtures or other equipment.

347-3. Uses Not Permitted. Rigid nonmetallic conduit shall not be used in the following locations.

(a) Hazardous (Classified) Locations.

(1) In hazardous (classified) locations, except as covered in Sections 503-3(a), 504-20, 514-8, and 515-5

(2) In Class I, Division 2 locations, except as permitted in Section 501-4(b), Exception

(b) Support of Fixtures. For the support of fixtures or other equipment not described in Section 347-2(h).

(c) Physical Damage. Where subject to physical damage unless identified for such use.

(d) Ambient Temperatures. Where subject to ambient temperatures in excess of 50°C (122°F) unless listed otherwise.

(e) Insulation Temperature Limitations. For conductors whose insulation temperature limitations would exceed those for which the conduit is listed.

(f) Theaters and Similar Locations. In theaters and similar locations, except as provided in Articles 518 and 520.

347-4. Other Articles. Installation of rigid nonmetallic conduit shall comply with the applicable provisions of Article 300. Where equipment grounding is required by Article 250, a separate equipment grounding conductor shall be installed in the conduit.

Exception No. 1: As permitted in Section 250-134(b), Exception No. 2, for dc circuits and Section 250-134(b), Exception No. 1, for separately run equipment grounding conductors.

Exception No. 2: Where the grounded conductor is used to ground equipment as permitted in Section 250-142.

A. Installations

347-5. Trimming. All cut ends shall be trimmed inside and outside to remove rough edges.

347-6. Joints. All joints between lengths of conduit, and between conduit and couplings, fittings, and boxes, shall be made by an approved method.

347-8. Securing and Supporting. Rigid nonmetallic conduit shall be installed as a complete system as provided in Section 300-18 and shall be fastened so that movement from thermal expansion or contraction will be permitted. Rigid nonmetallic conduit shall be securely fastened and supported in accordance with (a) and (b).

(a) Securely Fastened. Each rigid nonmetallic conduit shall be securely fastened within 3 ft (914 mm) of each outlet box, junction box, device box, conduit body, or other conduit termination. Conduit listed for securing at other than 3 ft (914 mm) shall be permitted to be installed in accordance with the listing.

(b) Supports. Rigid nonmetallic conduit shall be supported as required in Table 347-8. Conduit listed for support at spacings other than as shown in Table 347-8 shall be permitted to be installed in accordance with the listing. Horizontal runs of rigid nonmetallic conduit supported by openings through framing members at intervals not exceeding those in Table 347-8 and securely fastened within 3 ft (914 mm) of termination points shall be permitted.

Table 347-8. Support of Rigid Nonmetallic Conduit

Conduit Size (in.)	Maximum Spacing Between Supports (ft)
½–1	3
1¼–2	5
2½–3	6
3½–5	7
6	8

Note: For SI units, 1 ft = 0.3048 m (supports).

347-9. Expansion Fittings. Expansion fittings for rigid nonmetallic conduit shall be provided to compensate for thermal expansion and contraction where the length change, in accordance with Tables 347-9(A) and (B), is expected to be ¼ in. (6.36 mm) or greater in a straight run between securely mounted items such as boxes, cabinets, elbows, or other conduit terminations.

347-10. Size.

(a) Minimum. Rigid nonmetallic conduit smaller than ½-in. electrical trade size shall not be used.

(b) Maximum. Rigid nonmetallic conduit larger than 6-in. electrical trade size shall not be used.

> FPN: Metric trade numerical designations for rigid nonmetallic conduit are ½ = 16, ¾ = 21, 1 = 27, 1¼ = 35, 1½ = 41, 2 = 53, 2½ = 63, 3 = 78, 3½ = 91, 4 = 103, 5 = 129, and 6 = 155.

347-11. Number of Conductors. The number of conductors permitted in a single conduit shall not exceed the percentage fill specified in Table 1, Chapter 9.

347-12. Bushings. Where a conduit enters a box, fitting, or other enclosure, a bushing or adapter shall be provided to protect the wire from abrasion unless the box, fitting, or enclosure design provides equivalent protection.

> FPN: See Section 300-4(f) for the protection of conductors No. 4 and larger at bushings.

347-13. Bends — How Made. Bends of rigid nonmetallic conduit shall be made so that the conduit will not be damaged and that the internal diameter of the conduit will not be effectively reduced. Field bends shall be made only with bending equipment identified for the purpose, and the radius

Table 347-9(A). Expansion Characteristics of PVC Rigid Nonmetallic Conduit Coefficient of Thermal Expansion = 3.38×10^{-5} in./in./°F

Temperature Change (°F)	Length Change of PVC Conduit (in./100 ft)	Temperature Change (°F)	Length Change of PVC Conduit (in./100 ft)
5	0.2	105	4.2
10	0.4	110	4.5
15	0.6	115	4.7
20	0.8	120	4.9
25	1.0	125	5.1
30	1.2	130	5.3
35	1.4	135	5.5
40	1.6	140	5.7
45	1.8	145	5.9
50	2.0	150	6.1
55	2.2	155	6.3
60	2.4	160	6.5
65	2.6	165	6.7
70	2.8	170	6.9
75	3.0	175	7.1
80	3.2	180	7.3
85	3.4	185	7.5
90	3.6	190	7.7
95	3.8	195	7.9
100	4.1	200	8.1

Table 347-9(B). Expansion Characteristics of Fiberglass Reinforced Conduit (Rigid Nonmetallic Conduit) Coefficient of Thermal Expansion = 1.5×10^{-5} in./in./°F

Temperature Change (°F)	Length Change of PVC Conduit (in./100 ft)	Temperature Change (°F)	Length Change of PVC Conduit (in./100 ft)
5	0.1	105	1.9
10	0.2	110	2.0
15	0.3	115	2.1
20	0.4	120	2.2
25	0.5	125	2.3
30	0.5	130	2.3
35	0.6	135	2.4
40	0.7	140	2.5
45	0.8	145	2.6
50	0.9	150	2.7
55	1.0	155	2.8
60	1.1	160	2.9
65	1.2	165	3.0
70	1.3	170	3.1
75	1.4	175	3.2
80	1.4	180	3.2
85	1.5	185	3.3
90	1.6	190	3.4
95	1.7	195	3.5
100	1.8	200	3.6

of the curve of the inner edge of such bends shall not be less than shown in Table 346-10.

347-14. Bends — Number in One Run. There shall not be more than the equivalent of four quarter bends (360 degrees total) between pull points, e.g., conduit bodies and boxes.

347-15. Boxes and Fittings. Rigid nonmetallic conduit shall be used only with listed fittings. Boxes and fittings shall comply with the applicable provisions of Article 370.

347-16. Splices and Taps. Splices and taps shall be made in accordance with Section 300-15. See Article 370 for rules on the installation and use of boxes and conduit bodies.

B. Construction Specifications

347-17. General. Rigid nonmetallic conduit shall comply with the following.

Marking. Each length of nonmetallic conduit shall be clearly and durably marked at least every 10 ft (3.05 m) as required in the first sentence of Section 110-21. The type of material shall also be included in the marking unless it is visually identifiable. For conduit recognized for use above ground, these markings shall be permanent. For conduit limited to underground use only, these markings shall be sufficiently durable to remain legible until the material is installed. Conduit shall be permitted to be surface marked to indicate special characteristics of the material.

> FPN: Examples of these optional markings include but are not limited to "LS" for limited-smoke and markings such as "sunlight resistant."

Article 348 — Electrical Metallic Tubing

A. General

348-1. Definition. *Electrical metallic tubing* is a listed metallic tubing of circular cross section approved for the installation of electrical conductors when joined together with listed fittings.

348-2. Other Articles. Installations of electrical metallic tubing shall comply with the applicable provisions of Article 300.

348-4. Uses Permitted.

(a) Exposed and Concealed. The use of listed electrical metallic tubing shall be permitted for both exposed and concealed work.

(b) Corrosion Protection. Ferrous or nonferrous electrical metallic tubing, elbows, couplings, and fittings shall be permitted to be installed in concrete, in direct contact with the earth, or in areas subject to severe corrosive influences where protected by corrosion protection and judged suitable for the condition.

FPN: See Section 300-6 for information on protection against corrosion.

348-5. Uses Not Permitted. Electrical metallic tubing shall not be used

(1) Where, during installation or afterward, it will be subject to severe physical damage.
(2) Where protected from corrosion solely by enamel.
(3) In cinder concrete or cinder fill where subject to permanent moisture unless protected on all sides by a layer of noncinder concrete at least 2 in. (50.8 mm) thick or unless the tubing is at least 18 in. (457 mm) under the fill.
(4) In any hazardous (classified) location except as permitted by Sections 502-4, 503-3, and 504-20.
(5) For the support of fixtures or other equipment except conduit bodies no larger than the largest trade size of the tubing. Where practicable, dissimilar metals in contact anywhere in the system shall be avoided to eliminate the possibility of galvanic action.

Exception: Aluminum fittings and enclosures shall be permitted to be used with steel electrical metallic tubing.

B. Installation

348-6. Wet Locations. All supports, bolts, straps, screws, etc., shall be of corrosion-resistant materials or protected against corrosion by corrosion-resistant materials.

FPN: See Section 300-6 for information on protection against corrosion.

348-7. Size.

(a) Minimum. Tubing smaller than ½-in. electrical trade size shall not be used.

Exception: For enclosing the leads of motors as permitted in Section 430-145(b).

(b) Maximum. The maximum size of tubing shall be the 4-in. electrical trade size.

FPN: Metric trade numerical designations for electrical metallic tubing are the same as those found in *Extra-heavy Duty Rigid Steel Conduits for Electrical Installations*, IEC 981-1989; namely, ½ = 16, ¾ = 21, 1 = 27, 1¼ = 35, 1½ = 41, 2 = 53, 2½ = 63, 3 = 78, 3½ = 91, and 4 = 103.

348-8. Number of Conductors in Tubing. The number of conductors permitted in a single tubing shall not exceed the percentage fill specified in Table 1, Chapter 9.

348-9. Reaming and Threading. Electrical metallic tubing shall not be threaded. Where integral couplings are utilized, such couplings shall be permitted to be factory threaded.

All cut ends of electrical metallic tubing shall be reamed or otherwise finished to remove rough edges.

348-10. Couplings and Connectors. Couplings and connectors used with tubing shall be made up tight. Where buried in masonry or concrete, they shall be concretetight type. Where installed in wet locations, they shall be of the raintight type.

348-11. Bends — How Made. Bends in the tubing shall be made so that the tubing will not be damaged and the internal diameter of the tubing will not be effectively reduced. The radius of the curve of the inner edge of any field bend shall not be less than shown in Table 346-10.

Exception: For field bends made with a bending machine designed for the purpose, the minimum radius shall not be less than indicated in Table 346-10, Exception.

348-12. Bends — Number in One Run. There shall not be more than the equivalent of four quarter bends (360 degrees total) between pull points, e.g., conduit bodies and boxes.

348-13. Supports. Electrical metallic tubing shall be installed as a complete system as provided in Article 300. Each tubing length shall be securely fastened in place at least every 10 ft (3.05 m). In addition, each tube shall be securely fastened within 3 ft (914 mm) of each outlet box, junction box, device box, cabinet, conduit body, or other tubing terminations.

Exception No. 1: Fastening of unbroken lengths shall be permitted to be increased to a distance of 5 ft (1.52 m) where structural members do not readily permit fastening within 3 ft (914 mm).

Exception No. 2: For concealed work in finished buildings or prefinished wall panels where such securing is impracticable, unbroken lengths (without coupling) of electrical metallic tubing shall be permitted to be fished.

Horizontal runs of electrical metallic tubing supported by openings through framing members at intervals not greater than 10 ft (3.05 m) and securely fastened within 3 ft (914 mm) of termination points shall be permitted.

348-14. Boxes and Fittings. Boxes and fittings shall comply with the applicable provisions of Article 370.

348-15. Splices and Taps. Splices and taps shall be made in accordance with Section 300-15. See Article 370 for rules on the installation and use of boxes and conduit bodies.

C. Construction Specifications

348-16. General. Electrical metallic tubing shall comply with (a) through (d).

(a) Cross Section. The tubing, and elbows and bends for use with the tubing, shall have a circular cross section.

(b) Finish. Tubing shall have such a finish or treatment of outer surfaces as will provide an approved durable means of readily distinguishing it, after installation, from rigid metal conduit.

(c) Connectors. Where the tubing is coupled together by threads, the connector shall be designed so as to prevent bending of the tubing at any part of the thread.

(d) Marking. Electrical metallic tubing shall be clearly and durably marked at least every 10 ft (3.05 m) as required in the first sentence of Section 110-21.

Article 349 — Flexible Metallic Tubing

A. General

349-1. Scope. The provisions of this article apply to a raceway for electric conductors that is circular in cross section, flexible, metallic, and liquidtight without a nonmetallic jacket.

349-2. Definition. Flexible metallic tubing is a listed tubing that is circular in cross section, flexible, metallic, and liquidtight without a nonmetallic jacket.

349-3. Other Articles. Installations of flexible metallic tubing shall comply with the provisions of the applicable sections of Article 300 and Section 110-21.

349-4. Uses Permitted. Flexible metallic tubing shall be permitted to be used for branch circuits

(1) In dry locations,
(2) Where concealed,
(3) In accessible locations, and
(4) For system voltages of 1000 volts maximum.

349-5. Uses Not Permitted. Flexible metallic tubing shall not be used

(1) In hoistways;
(2) In storage battery rooms;

(3) In hazardous (classified) locations unless otherwise permitted under other articles in this *Code;*
(4) Underground for direct earth burial, or embedded in poured concrete or aggregate;
(5) Where subject to physical damage; and
(6) In lengths over 6 ft (1.83 m).

B. Construction and Installation

349-10. Size.

(a) Minimum. Flexible metallic tubing smaller than ½-in. electrical trade size shall not be used.

Exception No. 1: Flexible metallic tubing of ⅜-in. trade size shall be permitted to be installed in accordance with Sections 300-22(b) and (c).

Exception No. 2: Flexible metallic tubing ⅜-in. trade size shall be permitted in lengths not in excess of 6 ft (1.83 m) as part of an approved assembly or for lighting fixtures. See Section 410-67(c).

(b) Maximum. The maximum size of flexible metallic tubing shall be the ¾-in. trade size.

> FPN: Metric trade numerical designations for flexible metallic tubing are ⅜ = 12, ½ = 16, and ¾ = 21.

349-12. Number of Conductors.

(a) Flexible Metallic Tubing — ½-in. and ¾-in. Trade Size. The number of conductors permitted in ½-in. and ¾-in. trade sizes of flexible metallic tubing shall not exceed the percentage of fill specified in Table 1, Chapter 9.

(b) Flexible Metallic Tubing — ⅜-in. Trade Size. The number of conductors permitted in ⅜-in. trade size flexible metallic tubing shall not exceed that permitted in Table 350-12.

349-16. Grounding. See Section 250-118(8) for rules on the use of flexible metallic tubing as an equipment grounding conductor.

349-17. Splices and Taps. Splices and taps shall be made in accordance with Section 300-15. See Article 370 for rules on the installation and use of boxes and conduit bodies.

349-18. Fittings. Flexible metallic tubing shall be used only with listed terminal fittings. Fittings shall effectively close any openings in the connection.

349-20. Bends.

(a) Infrequent Flexing Use. Where the flexible metallic tubing shall be infrequently flexed in service after installation, the radii of bends measured to the inside of the bend shall not be less than specified in Table 349-20(a).

Table 349-20(a). Minimum Radii for Flexing Use

Trade Size (in.)	Minimum Radii (in.)
⅜	10
½	12½
¾	17½

Note: For SI units, 1 in. = 25.4 mm (radii).

(b) Fixed Bends. Where the flexible metallic tubing is bent for installation purposes and is not flexed or bent as required by use after installation, the radii of bends measured to the inside of the bend shall not be less than specified in Table 349-20(b).

Table 349-20(b). Minimum Radii for Fixed Bends

Trade Size (in.)	Minimum Radii (in.)
⅜	3½
½	4
¾	5

Note: For SI units, 1 in. = 25.4 mm (radii).

Article 350 — Flexible Metal Conduit

A. General

350-1. Scope. This article covers the use and installation of flexible metal conduit and associated fittings.

350-2. Definition. Flexible metal conduit is a raceway of circular cross section made of helically wound, formed, interlocked metal strip.

350-3. Other Articles. Installations of flexible metal conduit shall comply with the applicable provisions of Article 300.

350-4. Uses Permitted. Flexible metal conduit shall be listed and shall be permitted to be used in exposed and concealed locations.

350-5. Uses Not Permitted. Flexible metal conduit shall not be used in the following:

(1) In wet locations unless the conductors are approved for the specific conditions and the installation is such that

liquid is not likely to enter raceways or enclosures to which the conduit is connected

(2) In hoistways, other than as permitted in Section 620-21(a)(1)

(3) In storage-battery rooms

(4) In any hazardous (classified) location other than as permitted in Sections 501-4(b) and 504-20

(5) Where exposed to materials having a deteriorating effect on the installed conductors, such as oil or gasoline

(6) Underground or embedded in poured concrete or aggregate

(7) Where subject to physical damage

B. Installation

350-10. Size.

(a) Minimum. Flexible metal conduit less than ½-in. electrical trade size shall not be used unless permitted in (1) through (5) below for ⅜-in. electrical trade size.

(1) For enclosing the leads of motors as permitted in Section 430-145(b)

(2) In lengths not in excess of 6 ft (1.83 m)
   (a) For utilizing equipment, or
   (b) As part of a listed assembly, or
   (c) For tap connections to lighting fixtures as permitted in Section 410-67(c)

(3) For manufactured wiring systems as permitted in Section 604-6(a)

(4) In hoistways, as permitted in Section 620-21(a)(1)

(5) As part of a listed assembly to connect wired fixture sections as permitted in Section 410-77(c)

(b) Maximum. Flexible metal conduit larger than 4-in. electrical trade size shall not be used.

> FPN: Metric trade numerical designations for flexible metal conduit are ⅜ = 12, ½ = 16, ¾ = 21, 1 = 27, 1¼ = 35, 1½ = 41, 2 = 53, 2½ = 63, 3 = 78, 3½ = 91, and 4 = 103.

350-12. Number of Conductors. The number of conductors permitted in a flexible metal conduit shall not exceed the percentage of fill specified in Table 1, Chapter 9, or as permitted in Table 350-12 for ⅜-in. flexible metal conduit.

350-14. Grounding. Flexible metal conduit shall be permitted as a grounding means as covered in Section 250-118. Where an equipment bonding jumper is required around flexible metal conduit, it shall be installed in accordance with Section 250-102.

Where used to connect equipment where flexibility is required, an equipment grounding conductor shall be installed.

Table 350-12. Maximum Number of Insulated Conductors in ⅜-in. Flexible Metal Conduit*

Size (AWG)	Types RFH-2, SF-2		Types TF, XHHW, AF, TW		Types TFN, THHN, THWN		Types FEP, FEPB, PF, PGF	
	Fittings Inside Conduit	Fittings Outside Conduit	Fittings Inside Conduit	Fittings Outside Conduit	Fittings Inside Conduit	Fittings Outside Conduit	Fittings Inside Conduit	Fittings Outside Conduit
18	2	3	3	5	5	8	5	8
16	1	2	3	4	4	6	4	6
14	1	2	2	3	3	4	3	4
12	—	—	1	2	2	3	2	3
10	—	—	1	1	1	1	1	2

*In addition, one covered or bare equipment grounding conductor of the same size shall be permitted.

350-16. Bends. There shall not be more than the equivalent of four quarter bends (360 degrees total) between pull points, e.g., conduit bodies and boxes. Bends in the conduit shall be made so that the conduit will not be damaged and the internal diameter of the conduit will not be effectively reduced. The radius of the curve of the inner edge of any field bend shall not be less than shown in Table 346-10.

350-18. Supports. Flexible metal conduit shall be securely fastened in place by an approved means within 12 in. (305 mm) of each box, cabinet, conduit body, or other conduit termination and shall be supported and secured at intervals not to exceed 4½ ft (1.37 m).

Exception No. 1: Where flexible metal conduit is fished.

Exception No. 2: Lengths not exceeding 3 ft (914 mm) at terminals where flexibility is required.

Exception No. 3: Lengths not exceeding 6 ft (1.83 m) from a fixture terminal connection for tap connections to light fixtures as permitted in Section 410-67(c).

Horizontal runs of flexible metal conduit supported by openings through framing members at intervals not greater than 4½ ft (1.37 m) and securely fastened within 12 in. (305 mm) of termination points shall be permitted.

350-20. Fittings. Fittings used with flexible metal conduit shall be listed. Angle connectors shall not be used for concealed raceway installations.

350-22. Trimming. All cut ends of flexible metal conduit shall be trimmed or otherwise finished to remove rough edges, except where fittings that thread into the convolutions are used.

350-24. Splices and Taps. Splices and taps shall be made in accordance with Section 300-15. See Article 370 for rules on the installation and use of boxes and conduit bodies.

Article 351 — Liquidtight Flexible Metal Conduit and Liquidtight Flexible Nonmetallic Conduit

351-1. Scope. This article covers liquidtight flexible metal conduit and liquidtight flexible nonmetallic conduit.

A. Liquidtight Flexible Metal Conduit

351-2. Definition. *Liquidtight flexible metal conduit* is a listed raceway of circular cross section having an outer liquidtight, nonmetallic, sunlight-resistant jacket over an inner flexible metal core with associated couplings, connectors, and fittings and approved for the installation of electric conductors.

351-3. Other Articles. Installations of liquidtight flexible metal conduit shall comply with the applicable provisions of Article 300 and with the specific sections of Articles 350, 501, 502, 503, and 553 referenced below.

FPN: For marking requirements, see Section 110-21.

351-4. Use.

(a) Permitted. Listed liquidtight flexible metal conduit shall be permitted to be used in exposed or concealed locations as follows:

(1) Where conditions of installation, operation, or maintenance require flexibility or protection from liquids, vapors, or solids
(2) As permitted by Sections 501-4(b), 502-4, 503-3, and 504-20 and in other hazardous (classified) locations where specifically approved, and by Section 553-7(b)
(3) For direct burial where listed and marked for the purpose

(b) Not Permitted. Liquidtight flexible metal conduit shall not be used as follows:

(1) Where subject to physical damage
(2) Where any combination of ambient and conductor tem-

perature will produce an operating temperature in excess of that for which the material is approved

351-5. Size.

(a) Minimum. Liquidtight flexible metal conduit smaller than ½-in. electrical trade size shall not be used.

Exception: Liquidtight flexible metal conduit of ⅜-in. size shall be permitted as covered in Section 350-10(a).

(b) Maximum. The maximum size of liquidtight flexible metal conduit shall be the 4-in. trade size.

> FPN: Metric trade numerical designations for liquidtight flexible metal conduit are ⅜ = 12, ½ = 16, ¾ = 21, 1 = 27, 1¼ = 35, 1½ = 41, 2 = 53, 2½ = 63, 3 = 78, 3½ = 91, and 4 = 103.

351-6. Number of Conductors.

(a) Single Conduit. The number of conductors permitted in a single conduit, ½-in. through 4-in. trade sizes, shall not exceed the percentage of fill specified in Table 1, Chapter 9.

(b) Liquidtight Flexible Metal Conduit — ⅜-in. Size. The number of conductors permitted in ⅜-in. liquidtight flexible metal conduit shall not exceed that permitted in Table 350-12.

351-7. Fittings.
Liquidtight flexible metal conduit shall be used only with listed terminal fittings. Angle connectors shall not be used for concealed raceway installations.

351-8. Supports.
Liquidtight flexible metal conduit shall be securely fastened in place by an approved means within 12 in. (305 mm) of each box, cabinet, conduit body, or other conduit termination and shall be supported and secured at intervals not to exceed 4½ ft (1.37 m).

Exception No. 1: Where liquidtight flexible metal conduit is fished.

Exception No. 2: Lengths not exceeding 3 ft (914 mm) at terminals where flexibility is necessary.

Exception No. 3: Lengths not exceeding 6 ft (1.83 m) from a fixture terminal connection for tap conductors to lighting fixtures as permitted in Section 410-67(c).

Horizontal runs of liquidtight flexible metal conduit supported by openings through framing members at intervals not greater than 4½ ft (1.37 m) and securely fastened within 12 in. (305 mm) of termination points shall be permitted.

351-9. Grounding.
Liquidtight flexible metal conduit shall be permitted as a grounding means as covered in Section 250-118. Where an equipment bonding jumper is required around liquidtight flexible metal conduit, it shall be installed in accordance with Section 250-102.

Where used to connect equipment where flexibility is required, an equipment grounding conductor shall be installed.

> FPN: See Sections 501-16(b), 502-16(b), and 503-16(b) for types of equipment grounding conductors.

351-10. Bends — Number in One Run.
There shall not be more than the equivalent of four quarter bends (360 degrees total) between pull points, e.g., conduit bodies and boxes.

351-11. Splices and Taps.
Splices and taps shall be made in accordance with Section 300-15. See Article 370 for rules on the installation and use of boxes and conduit bodies.

B. Liquidtight Flexible Nonmetallic Conduit

351-22. Definition.
Liquidtight flexible nonmetallic conduit is a listed raceway of circular cross section of various types as follows:

(1) A smooth seamless inner core and cover bonded together and having one or more reinforcement layers between the core and cover, designated as Type LFNC-A
(2) A smooth inner surface with integral reinforcement within the conduit wall, designated as Type LFNC-B
(3) A corrugated internal and external surface without integral reinforcement within the conduit wall, designated as Type LFNC-C

This conduit is flame resistant and, with fittings, is approved for the installation of electrical conductors.

351-23. Use.

(a) Permitted. Listed liquidtight flexible nonmetallic conduit shall be permitted to be used in exposed or concealed locations for the following purposes.

> FPN: Extreme cold may cause some types of nonmetallic conduits to become brittle and therefore more susceptible to damage from physical contact.

(1) Where flexibility is required for installation, operation, or maintenance.
(2) Where protection of the contained conductors is required from vapors, liquids, or solids.
(3) For outdoor locations where listed and marked as suitable for the purpose.

> FPN: For marking requirements, see Section 110-21.

(4) For direct burial where listed and marked for the purpose.

(5) Liquidtight flexible nonmetallic conduit as defined in Section 351-22(2) shall be permitted to be installed in lengths longer than 6 ft (1.83 m) where secured in accordance with Section 351-27.

(6) As a listed manufactured prewired assembly, ½-in. through 1-in. conduit, as defined in Section 351-22(2).

(b) Not Permitted. Liquidtight flexible nonmetallic conduit shall not be used in the following:

(1) Where subject to physical damage

(2) Where any combination of ambient and conductor temperatures is in excess of that for which the liquidtight flexible nonmetallic conduit is approved

(3) In lengths longer than 6 ft (1.83 m), except as permitted by Section 351-23(a)(5) or where a longer length is approved as essential for a required degree of flexibility

(4) Where voltage of the contained conductors is in excess of 600 volts, nominal

Exception: As permitted in Section 600-32(a) for electric signs over 600 volts.

351-24. Size. The electrical trade sizes of liquidtight flexible nonmetallic conduit shall be in accordance with (a) or (b):

(a) ½ in. to 4 in. inclusive

(b) ⅜ in. as permitted below

    (1) For enclosing the leads of motors as permitted in Section 430-145(b)

    (2) In lengths not exceeding 6 ft (1.83 m) as part of a listed assembly for tap connections to lighting fixtures as required in Section 410-67(c), or for utilization equipment

    (3) For electric sign conductors in accordance with Section 600-32(a)

    FPN: Metric trade numerical designations for liquidtight flexible nonmetallic conduit are ⅜ = 12, ½ = 16, ¾ = 21, 1 = 27, 1¼ = 35, 1½ = 41, 2 = 53, 2½ = 63, 3 = 78, 3½ = 91, and 4 = 103.

351-25. Number of Conductors. The number of conductors permitted in a single conduit shall be in accordance with the percentage fill specified in Table 1, Chapter 9.

351-26. Fittings. Liquidtight flexible nonmetallic conduit shall be used only with listed terminal fittings. Angle connectors shall not be used for concealed raceway installations.

351-27. Securing and Supporting. Liquidtight flexible nonmetallic conduit, as defined in Section 351-22(2), shall

be securely fastened and supported in accordance with one of the following.

(a) The conduit shall be securely fastened at intervals not exceeding 3 ft (914 mm) and within 12 in. (305 mm) on each side of every outlet box, junction box, cabinet, or fitting.

(b) Securing and supporting of the conduit shall not be required where it is fished, installed in lengths not exceeding 3 ft (914 mm) at terminals where flexibility is required, or where installed in lengths not exceeding 6 ft (1.83 m) from a fixture terminal connection for tap conductors to lighting fixtures as permitted in Section 410-67(c).

(c) Horizontal runs of liquidtight flexible nonmetallic conduit supported by openings through framing members at intervals not exceeding 3 ft (914 mm) and securely fastened within 12 in. (305 mm) of termination points shall be permitted.

351-28. Equipment Grounding. Where an equipment grounding conductor is required for the circuits installed in liquidtight flexible nonmetallic conduit, it shall be permitted to be installed on the inside or outside of the conduit. Where installed on the outside, the length of the equipment grounding conductor shall not exceed 6 ft (1.83 m) and shall be routed with the raceway or enclosure. Fittings and boxes shall be bonded or grounded in accordance with Article 250.

351-29. Splices and Taps. Splices and taps shall be made in accordance with Section 300-15. See Article 370 for rules on the installation and use of boxes and conduit bodies.

351-30. Bends — Number in One Run. There shall not be more than the equivalent of four quarter bends (360 degrees total) between pull points, e.g., conduit bodies and boxes.

Article 352 — Surface Metal Raceways and Surface Nonmetallic Raceways

A. Surface Metal Raceways

352-1. Uses.

(a) Permitted. The use of surface metal raceways shall be permitted in the following:

(1) In dry locations

(2) In Class I, Division 2 hazardous (classified) locations as permitted in Section 501-4(b), Exception

(3) Under raised floors, as permitted in Section 645-5(d)(2)

(b) Not Permitted. The use of surface metal raceways shall not be permitted in the following:

(1) Where subject to severe physical damage, unless otherwise approved

(2) Where the voltage is 300 volts or more between conductors, unless the metal has a thickness of not less than 0.040 in. (1.016 mm) nominal

(3) Where subject to corrosive vapors

(4) In hoistways

(5) Where concealed, except as permitted in Section 352-1(a)(3)

352-2. Other Articles. Surface metal raceways shall comply with the applicable provisions of Article 300.

352-3. Size of Conductors. No conductor larger than that for which the raceway is designed shall be installed in surface metal raceway.

352-4. Number of Conductors in Raceways. The number of conductors installed in any raceway shall not be greater than the number for which the raceway is designed.

The derating factors of Section 310-15(b)(2)(a) shall not apply to conductors installed in surface metal raceways where all of the following conditions are met:

(1) The cross-sectional area of the raceway exceeds 4 in.^{2} (2580 mm^2)

(2) The current-carrying conductors do not exceed 30 in number

(3) The sum of the cross-sectional areas of all contained conductors does not exceed 20 percent of the interior cross-sectional area of the surface metal raceway

352-5. Extension Through Walls and Floors. Surface metal raceways shall be permitted to pass transversely through dry walls, dry partitions, and dry floors if the length passing through is unbroken. Access to the conductors shall be maintained on both sides of the wall, partition, or floor.

352-6. Combination Raceways. Where combination surface metal raceways are used both for signaling and for lighting and power circuits, the different systems shall be run in separate compartments identified by sharply contrasting colors of the interior finish, and the same relative position of compartments shall be maintained throughout the premises.

352-7. Splices and Taps. Splices and taps shall be permitted in surface metal raceways having a removable cover that is accessible after installation. The conductors, including splices and taps, shall not fill the raceway to more than 75 percent of its area at that point. Splices and taps in surface metal raceways without removable covers shall be made only in junction boxes. All splices and taps shall be made by approved methods.

Taps of Type FC cable installed in surface metal raceway shall be made in accordance with Section 363-10.

352-8. General. Surface metal raceways shall be of such construction as will distinguish them from other raceways. Surface metal raceways and their elbows, couplings, and similar fittings shall be designed so that the sections can be electrically and mechanically coupled together and installed without subjecting the wires to abrasion.

Where covers and accessories of nonmetallic materials are used on surface metal raceways, they shall be identified for such use.

352-9. Grounding. Surface metal raceway enclosures providing a transition from other wiring methods shall have a means for connecting an equipment grounding conductor.

B. Surface Nonmetallic Raceways

352-21. Description. Part B of this article shall apply to a type of surface nonmetallic raceway and fittings of suitable nonmetallic material that is resistant to moisture and chemical atmospheres. It shall also be flame retardant, resistant to impact and crushing, resistant to distortion from heat under conditions likely to be encountered in service, and resistant to low-temperature effects. Surface nonmetallic raceways that have limited smoke-producing characteristics shall be permitted to be identified with the suffix *LS*.

352-22. Use.

(a) Permitted. The use of surface nonmetallic raceways shall be permitted in dry locations.

(b) Not Permitted. Surface nonmetallic raceways shall not be used as follows:

(1) Where concealed

(2) Where subject to severe physical damage

(3) Where the voltage is 300 volts or more between conductors, unless listed for higher voltage

(4) In hoistways

(5) In any hazardous (classified) location except Class I, Division 2 locations as permitted in Section 501-4(b), Exception

(6) Where subject to ambient temperatures exceeding those for which the nonmetallic raceway is listed

(7) For conductors whose insulation temperature limitations would exceed those for which the nonmetallic raceway is listed

352-23. Other Articles. Surface nonmetallic raceways shall comply with the applicable provisions of Article 300. Where equipment grounding is required by Article 250, a separate equipment grounding conductor shall be installed in the raceway.

352-24. Size of Conductors. No conductor larger than that for which the raceway is designed shall be installed in surface nonmetallic raceway.

352-25. Number of Conductors in Raceways. The number of conductors installed in any raceway shall not be greater than the number for which the raceway is designed.

352-26. Combination Raceways. Where combination surface nonmetallic raceways are used both for signaling and for lighting and power circuits, the different systems shall be run in separate compartments, identified by printed legend or by sharply contrasting colors of the interior finish, and the same relative position of compartments shall be maintained throughout the premises.

352-27. General. Surface nonmetallic raceways shall be of such construction as will distinguish them from other raceways. Surface nonmetallic raceways and their elbows, couplings, and similar fittings shall be designed so that the sections can be mechanically coupled together and installed without subjecting the wires to abrasion.

352-28. Extension Through Walls and Floors. Surface nonmetallic raceways shall be permitted to pass transversely through dry walls, dry partitions, and dry floors if the length passing through is unbroken. Access to the conductors shall be maintained on both sides of the wall, partition, or floor.

352-29. Splices and Taps. Splices and taps shall be permitted in surface nonmetallic raceways having a removable cover that is accessible after installation. The conductors, including splices and taps, shall not fill the raceway to more than 75 percent of its area at that point. Splices and taps in surface nonmetallic raceways without removable covers shall be made only in junction boxes. All splices and taps shall be made by approved methods.

C. Strut-Type Channel Raceway

352-40. Description. Part C of this article shall apply to strut-type channel raceways and accessories formed of metal that are resistant to moisture or protected by corrosion protection and judged suitable for the condition. These channel raceways shall be permitted to be galvanized, stainless, enameled, or PVC-coated steel, or aluminum. Covers shall be either metallic or nonmetallic.

352-41. Uses Permitted. The installation of listed strut-type channel raceways shall be permitted:

(1) Where exposed
(2) In dry locations
(3) In locations subject to corrosive vapors where protected by finishes judged suitable for the condition
(4) Where the voltage is 600 volts or less
(5) As power poles
(6) In Class I, Division 2 hazardous (classified) locations as permitted in Section 501-4(b), Exception

352-42. Uses Not Permitted.

(a) General. Strut-type channel raceways shall not be permitted to be used where concealed.

(b) Ferrous Metal. Ferrous channel raceways and fittings protected from corrosion solely by enamel shall be permitted only indoors and in occupancies not subject to severe corrosive influences.

352-43. Other Articles. Installation of strut-type channel raceways shall comply with the applicable provisions of Articles 250 and 300.

352-44. Size of Conductors. No conductor larger than that for which the raceway is listed shall be installed in strut-type channel raceways.

352-45. Number of Conductors in Raceways. The number of conductors permitted in strut-type channel raceway shall not exceed the percentage fill using Table 352-45 and applicable outside diameter (O.D.) dimensions of specific types and sizes of wire given in the Tables in Chapter 9.

The derating factors of Section 310-15(b)(2)(a) shall not apply to conductors installed in strut-type channel raceways where all of the following conditions are met:

(1) The cross-sectional area of the raceway exceeds 4 in.^{2} (2580 mm^2)
(2) The current-carrying conductors do not exceed 30 in number
(3) The sum of the cross-sectional areas of all contained conductors does not exceed 20 percent of the interior cross-sectional area of the strut-type channel raceway

Formula for wire fill: $N = \dfrac{CA}{WA}$

Where:

N = number of wires
CA = channel area in square inches
WA = wire area

352-46. Extensions Through Walls and Floors. It shall be permitted to extend unbroken lengths of strut-type channel

Table 352-45. Channel Size and Inside Diameter Area

Channel Size	Area		40% Area*		25% Area**	
	in.²	mm²	in.²	mm²	in.²	mm²
1⅝ × ¹³⁄₁₆	0.887	572	0.355	229	0.222	143
1⅝ × 1	1.151	743	0.460	297	0.288	186
1⅝ × 1⅜	1.677	1076	0.671	433	0.419	270
1⅝ × 1⅝	2.028	1308	0.811	523	0.507	327
1⅝ × 2⁷⁄₁₆	3.169	2045	1.267	817	0.792	511
1⅝ × 3¼	4.308	2780	1.723	1112	1.077	695
1½ × ¾	0.849	548	0.340	219	0.212	137
1½ × 1½	1.828	1179	0.731	472	0.457	295
1½ × 1⅞	2.301	1485	0.920	594	0.575	371
1½ × 3	3.854	2487	1.542	995	0.964	622

*Raceways with external joiners shall use a 40 percent wire fill calculation to determine the number of conductors permitted.

**Raceways with internal joiners shall use a 25 percent wire fill calculation to determine the number of conductors permitted.

raceway through walls, partitions, and floors where closure strips are removable from either side and the portion within the wall, partition, or floor remains covered.

352-47. Support of Strut-Type Channel Raceways.

(a) A surface mount strut-type channel raceway shall be secured to the mounting surface with retention straps external to the channel at intervals not exceeding 10 ft (3.05 m) and within 3 ft (914 mm) of each outlet box, cabinet, junction box, or other channel raceway termination.

(b) Suspension Mount. Strut-type channel raceways shall be permitted to be suspension mounted in air with approved appropriate methods designed for the purpose at intervals not to exceed 10 ft (3.05 m).

352-48. Splices and Taps. Splices and taps shall be permitted in raceways that are accessible after installation by having a removable cover. The conductors, including splices and taps, shall not fill the raceway to more than 75 percent of its area at that point. All splices and taps shall be made by approved methods.

352-49. General. Strut-type channel raceways shall be of a construction that distinguishes them from other raceways. Raceways and their elbows, couplings, and other fittings shall be designed so that the sections can be electrically and mechanically coupled together and installed without subjecting the wires to abrasion.

Where closure strips and accessories of nonmetallic materials are used on metallic strut-type channel raceways, they shall be listed and identified for such use.

352-50. Grounding. Strut-type channel raceway enclosures providing a transition to or from other wiring methods

shall have a means for connecting an equipment grounding conductor. Strut-type channel raceway shall be permitted as an equipment grounding conductor in accordance with Section 250-118(14). Where a snap-fit metal cover for strut-type channel raceway is used to achieve electrical continuity in accordance with the listing, this cover shall not be permitted as the means for providing electrical continuity for a receptacle mounted in the cover.

352-51. Marking. Each length shall be clearly and durably identified as required in the first sentence of Section 110-21.

Article 353 — Multioutlet Assembly

353-1. Other Articles. A multioutlet assembly shall comply with applicable provisions of Article 300.

FPN: See the definition of *multioutlet assembly* in Article 100.

353-2. Use.

(a) Permitted. The use of a multioutlet assembly shall be permitted in dry locations.

(b) Not Permitted. A multioutlet assembly shall not be installed as follows:

(1) Where concealed, except that it shall be permissible to surround the back and sides of a metal multioutlet assembly by the building finish or recess a nonmetallic multioutlet assembly in a baseboard
(2) Where subject to severe physical damage
(3) Where the voltage is 300 volts or more between conductors unless the assembly is of metal having a thickness of not less than 0.040 in. (1.02 mm)
(4) Where subject to corrosive vapors
(5) In hoistways
(6) In any hazardous (classified) locations except Class I, Division 2, locations as permitted in Section 501-4(b), Exception.

353-3. Metal Multioutlet Assembly Through Dry Partitions. It shall be permissible to extend a metal multioutlet assembly through (not run within) dry partitions, if arrangements are made for removing the cap or cover on all exposed portions and no outlet is located within the partitions.

Article 354 — Underfloor Raceways

354-1. Other Articles. Underfloor raceways shall comply with the applicable provisions of Article 300.

354-2. Use.

(a) Permitted. The installation of underfloor raceways shall be permitted beneath the surface of concrete or other flooring material or in office occupancies, where laid flush with the concrete floor and covered with linoleum or equivalent floor covering.

(b) Not Permitted. Underfloor raceways shall not be installed (1) where subject to corrosive vapors or (2) in any hazardous (classified) locations, except as permitted by Section 504-20 and in Class I, Division 2 locations as permitted in Section 501-4(b), Exception. Unless made of a material judged suitable for the condition or unless corrosion protection approved for the condition is provided, ferrous or nonferrous metal underfloor raceways, junction boxes, and fittings shall not be installed in concrete or in areas subject to severe corrosive influences.

354-3. Covering. Raceway coverings shall comply with (a) through (d).

(a) Raceways Not Over 4 in. (102 mm) Wide. Half-round and flat-top raceways not over 4 in. (102 mm) in width shall have not less than ¾ in. (19 mm) of concrete or wood above the raceway.

Exception: As permitted in (c) and (d) for flat-top raceways.

(b) Raceways Over 4 in. (102 mm) Wide but Not Over 8 in. (203 mm) Wide. Flat-top raceways over 4 in. (102 mm) but not over 8 in. (203 mm) wide with a minimum of 1 in. (25.4 mm) spacing between raceways shall be covered with concrete to a depth of not less than 1 in. (25.4 mm). Raceways spaced less than 1 in. (25.4 mm) apart shall be covered with concrete to a depth of 1½ in. (38 mm).

(c) Trench-Type Raceways Flush with Concrete. Trench-type flush raceways with removable covers shall be permitted to be laid flush with the floor surface. Such approved raceways shall be designed so that the cover plates will provide adequate mechanical protection and rigidity equivalent to junction box covers.

(d) Other Raceways Flush with Concrete. In office occupancies, approved metal flat-top raceways, if not over 4 in. (102 mm) in width, shall be permitted to be laid flush with the concrete floor surface, provided they are covered with substantial linoleum that is not less than 1/16 in. (1.59 mm) in thickness or with equivalent floor covering. Where more than one and not more than three single raceways are each installed flush with the concrete, they shall be contiguous with each other and joined to form a rigid assembly.

354-4. Size of Conductors. No conductor larger than that for which the raceway is designed shall be installed in underfloor raceways.

354-5. Maximum Number of Conductors in Raceway. The combined cross-sectional area of all conductors or cables shall not exceed 40 percent of the interior cross-sectional area of the raceway.

354-6. Splices and Taps. Splices and taps shall be made only in junction boxes.

For the purposes of this section, so-called loop wiring (continuous, unbroken conductor connecting the individual outlets) shall not be considered to be a splice or tap.

Exception: Splices and taps shall be permitted in trench-type flush raceway having a removable cover that is accessible after installation. The conductors, including splices and taps, shall not fill the raceway more than 75 percent of its area at that point.

354-7. Discontinued Outlets. When an outlet is abandoned, discontinued, or removed, the sections of circuit conductors supplying the outlet shall be removed from the raceway. No splices or reinsulated conductors, such as would be the case with abandoned outlets on loop wiring, shall be allowed in raceways.

354-8. Laid in Straight Lines. Underfloor raceways shall be laid so that a straight line from the center of one junction box to the center of the next junction box will coincide with the centerline of the raceway system. Raceways shall be firmly held in place to prevent disturbing this alignment during construction.

354-9. Markers at Ends. A suitable marker shall be installed at or near each end of each straight run of raceways to locate the last insert.

354-10. Dead Ends. Dead ends of raceways shall be closed.

354-13. Junction Boxes. Junction boxes shall be leveled to the floor grade and sealed to prevent the free entrance of water or concrete. Junction boxes used with metal raceways shall be metal and shall be electrically continuous with the raceways.

354-14. Inserts. Inserts shall be leveled and sealed to prevent the entrance of concrete. Inserts used with metal raceways shall be metal and shall be electrically continuous with the raceway. Inserts set in or on fiber raceways before the floor is laid shall be mechanically secured to the raceway. Inserts set in fiber raceways after the floor is laid shall be screwed into the raceway. When cutting through the raceway wall and setting inserts, chips and other dirt shall not be allowed to remain in the raceway, and tools shall be used that are designed so as to prevent the tool from entering the raceway and damaging conductors that may be in place.

354-15. Connections to Cabinets and Wall Outlets.
Connections between raceways and distribution centers and wall outlets shall be made by means of flexible metal conduit where not installed in concrete, rigid metal conduit, intermediate metal conduit, electrical metallic tubing, or approved fittings. Where a metallic underfloor raceway system provides for the termination of an equipment grounding conductor, rigid nonmetallic conduit, electrical nonmetallic tubing, or liquidtight flexible nonmetallic conduit where not installed in concrete, shall be permitted.

Article 356 — Cellular Metal Floor Raceways

356-1. Definitions.
For the purposes of this article, a *cellular metal floor raceway* shall be defined as the hollow spaces of cellular metal floors, together with suitable fittings, that may be approved as enclosures for electric conductors. A *cell* shall be defined as a single, enclosed tubular space in a cellular metal floor member, the axis of the cell being parallel to the axis of the metal floor member. A *header* shall be defined as a transverse raceway for electric conductors, providing access to predetermined cells of a cellular metal floor, thereby permitting the installation of electric conductors from a distribution center to the cells.

356-2. Uses Not Permitted.
Conductors shall not be installed in cellular metal floor raceways as follows:

(1) Where subject to corrosive vapor
(2) In any hazardous (classified) location except as permitted by Section 504-20, and in Class I, Division 2, locations as permitted in Section 501-4(b), Exception
(3) In commercial garages, other than for supplying ceiling outlets or extensions to the area below the floor but not above

FPN: See Section 300-8 for installation of conductors with other systems.

356-3. Other Articles.
Cellular metal floor raceways shall comply with the applicable provisions of Article 300.

A. Installation

356-4. Size of Conductors.
No conductor larger than No. 1/0 shall be installed, except by special permission.

356-5. Maximum Number of Conductors in Raceway.
The combined cross-sectional area of all conductors or cables shall not exceed 40 percent of the interior cross-sectional area of the cell or header.

356-6. Splices and Taps.
Splices and taps shall be made only in header access units or junction boxes.

For the purposes of this section, so-called loop wiring (continuous unbroken conductor connecting the individual outlets) shall not be considered to be a splice or tap.

356-7. Discontinued Outlets.
When an outlet is abandoned, discontinued, or removed, the sections of circuit conductors supplying the outlet shall be removed from the raceway. No splices or reinsulated conductors, such as would be the case with abandoned outlets on loop wiring, shall be allowed in raceways.

356-8. Markers.
A suitable number of markers shall be installed for locating cells in the future.

356-9. Junction Boxes.
Junction boxes shall be leveled to the floor grade and sealed against the free entrance of water or concrete. Junction boxes used with these raceways shall be of metal and shall be electrically continuous with the raceway.

356-10. Inserts.
Inserts shall be leveled to the floor grade and sealed against the entrance of concrete. Inserts shall be of metal and shall be electrically continuous with the raceway. In cutting through the cell wall and setting inserts, chips and other dirt shall not be allowed to remain in the raceway, and tools shall be used that are designed to prevent the tool from entering the cell and damaging the conductors.

356-11. Connection to Cabinets and Extensions from Cells.
Connections between raceways and distribution centers and wall outlets shall be made by means of flexible metal conduit where not installed in concrete, rigid metal conduit, intermediate metal conduit, electrical metallic tubing, or approved fittings. Where there are provisions for the termination of an equipment grounding conductor, nonmetallic conduit, electrical nonmetallic tubing, or liquidtight flexible nonmetallic conduit where not installed in concrete, shall be permitted.

B. Construction Specifications

356-12. General.
Cellular metal floor raceways shall be constructed so that adequate electrical and mechanical continuity of the complete system will be secured. They shall provide a complete enclosure for the conductors. The interior surfaces shall be free from burrs and sharp edges, and surfaces over which conductors are drawn shall be smooth. Suitable bushings or fittings having smooth rounded edges shall be provided where conductors pass.

Article 358 — Cellular Concrete Floor Raceways

358-1. Scope. This article covers cellular concrete floor raceways, the hollow spaces in floors constructed of precast cellular concrete slabs, together with suitable metal fittings designed to provide access to the floor cells.

358-2. Definitions. A *cell* shall be defined as a single, enclosed tubular space in a floor made of precast cellular concrete slabs, the direction of the cell being parallel to the direction of the floor member. A *header* shall be defined as transverse metal raceways for electric conductors, providing access to predetermined cells of a precast cellular concrete floor, thereby permitting the installation of electric conductors from a distribution center to the floor cells.

358-3. Other Articles. Cellular concrete floor raceways shall comply with the applicable provisions of Article 300.

358-4. Uses Not Permitted. Conductors shall not be installed in precast cellular concrete floor raceways as follows:

(1) Where subject to corrosive vapor
(2) In any hazardous (classified) locations except as permitted by Section 504-20, and in Class I, Division 2, locations as permitted in Section 501-4(b), Exception
(3) In commercial garages, other than for supplying ceiling outlets or extensions to the area below the floor but not above

  FPN: See Section 300-8 for installation of conductors with other systems.

358-5. Header. The header shall be installed in a straight line at right angles to the cells. The header shall be mechanically secured to the top of the precast cellular concrete floor. The end joints shall be closed by a metal closure fitting and sealed against the entrance of concrete. The header shall be electrically continuous throughout its entire length and shall be electrically bonded to the enclosure of the distribution center.

358-6. Connection to Cabinets and Other Enclosures. Connections from headers to cabinets and other enclosures shall be made by means of listed metal raceways and listed fittings.

358-7. Junction Boxes. Junction boxes shall be leveled to the floor grade and sealed against the free entrance of water or concrete. Junction boxes shall be of metal and shall be mechanically and electrically continuous with the header.

358-8. Markers. A suitable number of markers shall be installed for the future location of cells.

358-9. Inserts. Inserts shall be leveled and sealed against the entrance of concrete. Inserts shall be of metal and shall be fitted with grounded-type receptacles. A grounding conductor shall connect the insert receptacles to a positive ground connection provided on the header. Where cutting through the cell wall for setting inserts or other purposes (such as providing access openings between header and cells), chips and other dirt shall not be allowed to remain in the raceway, and the tool used shall be designed so as to prevent the tool from entering the cell and damaging the conductors.

358-10. Size of Conductors. No conductor larger than No. 1/0 shall be installed, except by special permission.

358-11. Maximum Number of Conductors. The combined cross-sectional area of all conductors or cables shall not exceed 40 percent of the cross-sectional area of the cell or header.

358-12. Splices and Taps. Splices and taps shall be made only in header access units or junction boxes.
  For the purposes of this section, so-called loop wiring (continuous unbroken conductor connecting the individual outlets) shall not be considered to be a splice or tap.

358-13. Discontinued Outlets. When an outlet is abandoned, discontinued, or removed, the sections of circuit conductors supplying the outlet shall be removed from the raceway. No splices or reinsulated conductors, such as would be the case of abandoned outlets on loop wiring, shall be allowed in raceways.

Article 362 — Metal Wireways and Nonmetallic Wireways

A. Metal Wireways

362-1. Definition. *Wireways* are sheet metal troughs with hinged or removable covers for housing and protecting electric wires and cable and in which conductors are laid in place after the wireway has been installed as a complete system.

362-2. Uses.

(a) Permitted. The use of wireways shall be permitted as follows:

(1) For exposed work
(2) In concealed spaces only in accordance with Section 640-24
(3) In hazardous (classified) locations as permitted by Section 501-4(b) for Class I, Division 2, locations; Section

502-4(b) for Class II, Division 2, locations; and Section 504-20 for intrinsically safe wiring

Where installed in wet locations, wireways shall be listed for the purpose.

(b) Not Permitted. The use of wireways shall not be permitted where subject to severe physical damage or corrosive vapor.

362-3. Other Articles. Installations of wireways shall comply with the applicable provisions of Article 300.

362-4. Size of Conductors. No conductor larger than that for which the wireway is designed shall be installed in any wireway.

362-5. Number of Conductors. Wireways shall not contain more than 30 current-carrying conductors at any cross section. Conductors for signaling circuits or controller conductors between a motor and its starter and used only for starting duty shall not be considered as current-carrying conductors.

The sum of cross-sectional areas of all contained conductors at any cross section of the wireway shall not exceed 20 percent of the interior cross-sectional area of the wireway.

The derating factors specified in Section 310-15(b)(2)(a) shall not be applicable to the 30 current-carrying conductors at 20 percent fill specified above.

Exception No. 1: Where the derating factors specified in Section 310-15(b)(2)(a) are applied, the number of current-carrying conductors shall not be limited, but the sum of the cross-sectional areas of all contained conductors at any cross section of the wireway shall not exceed 20 percent of the interior cross-sectional area of the wireway.

Exception No. 2: As provided in Section 520-6, the 30-conductor limitation shall not apply to theaters and similar locations.

Exception No. 3: As provided in Section 620-32, the 20 percent fill limitation shall not apply to elevators and dumbwaiters.

362-6. Deflected Insulated Conductors. Where insulated conductors are deflected within a wireway, either at the ends or where conduits, fittings, or other raceways or cables enter or leave the wireway, or where the direction of the wireway is deflected greater than 30 degrees, dimensions corresponding to Section 373-6 shall apply. Where insulated conductors No. 4 or larger enter a wireway through a raceway or cable, the distance between those raceway and cable entries shall not be less than six times the trade diameter of the larger raceway or cable connector.

Where insulated conductors No. 4 or larger enter a wireway through a raceway or cable, the distance between raceway and cable entries enclosing the same conductor shall not be less than six times the trade diameter of the larger raceway or cable connector.

362-7. Splices and Taps. Splices and taps shall be permitted within a wireway provided they are accessible. The conductors, including splices and taps, shall not fill the wireway to more than 75 percent of its area at that point.

362-8. Supports. Wireways shall be supported in accordance with the following.

(a) Horizontal Support. Wireways shall be supported where run horizontally at each end and at intervals not to exceed 5 ft (1.52 m) or for individual lengths longer than 5 ft (1.52 m) at each end or joint, unless listed for other support intervals. The distance between supports shall not exceed 10 ft (3.05 m).

(b) Vertical Support. Vertical runs of wireways shall be securely supported at intervals not exceeding 15 ft (4.57 m) and shall not have more than one joint between supports. Adjoining wireway sections shall be securely fastened together to provide a rigid joint.

362-9. Extension Through Walls. Wireways shall be permitted to pass transversely through walls if the length passing through the wall is unbroken. Access to the conductors shall be maintained on both sides of the wall.

362-10. Dead Ends. Dead ends of wireways shall be closed.

362-11. Extensions from Wireways. Extensions from wireways shall be made with cord pendants installed in accordance with Section 400-10 or any wiring method in Chapter 3 that includes a means for equipment grounding. Where a separate equipment grounding conductor is employed, connection of the equipment grounding conductors in the wiring method to the wireway shall comply with Sections 250-8 and 250-12. Where rigid nonmetallic conduit, electrical nonmetallic tubing, or liquidtight flexible nonmetallic conduit is used, connection of the equipment grounding conductor in the nonmetallic raceway to a metal wireway shall comply with Sections 250-8 and 250-12.

362-12. Marking. Wireways shall be marked so that their manufacturer's name or trademark will be visible after installation.

362-13. Grounding. Grounding shall be in accordance with the provisions of Article 250.

B. Nonmetallic Wireways

362-14. Definition. Nonmetallic wireways are flame-retardant, nonmetallic troughs with removable covers for housing and protecting electric wires and cables in which conductors are laid in place after the wireway has been installed as a complete system.

362-15. Uses Permitted. The use of listed nonmetallic wireways shall be permitted as follows:

(1) Only for exposed work, except as permitted in accordance with Section .640-24
(2) Where subject to corrosive vapors
(3) In wet locations where listed for the purpose

> FPN: Extreme cold may cause nonmetallic wireways to become brittle and, therefore, more susceptible to damage from physical contact.

362-16. Uses Not Permitted. Nonmetallic wireways shall not be used as follows:

(1) Where subject to physical damage
(2) In any hazardous (classified) location, except as permitted in Section 504-20
(3) Where exposed to sunlight unless listed and marked as suitable for the purpose
(4) Where subject to ambient temperatures other than those for which nonmetallic wireway is listed
(5) For conductors whose insulation temperature limitations would exceed those for which the nonmetallic wireway is listed

362-17. Other Articles. Installations of nonmetallic wireways shall comply with the applicable provisions of Article 300. Where equipment grounding is required by Article 250, a separate equipment grounding conductor shall be installed in the nonmetallic wireway.

Exception: Where the grounded conductor is used to ground equipment as permitted in Section 250-142.

362-18. Size of Conductors. No conductor larger than that for which the nonmetallic wireway is designed shall be installed in any nonmetallic wireway.

362-19. Number of Conductors. The sum of cross-sectional areas of all contained conductors at any cross section of the nonmetallic wireway shall not exceed 20 percent of the interior cross-sectional area of the nonmetallic wireway. Conductors for signaling circuits or controller conductors between a motor and its starter and used only for starting duty shall not be considered as current-carrying conductors.

The derating factors specified in Section 310-15(b)(2)(a) shall be applicable to the current-carrying conductors up to and including the 20 percent fill specified above.

362-20. Deflected Insulated Conductors. Where insulated conductors are deflected within a nonmetallic wireway, either at the ends or where conduits, fittings, or other raceways or cables enter or leave the nonmetallic wireway, or where the direction of the nonmetallic wireway is deflected greater than 30 degrees, dimensions corresponding to Section 373-6 shall apply.

362-21. Splices and Taps. Splices and taps shall be permitted within a nonmetallic wireway provided they are accessible. The conductors, including splices and taps, shall not fill the nonmetallic wireway to more than 75 percent of its area at that point.

362-22. Supports. Nonmetallic wireway shall be supported in accordance with (a) and (b).

(a) Horizontal Support. Nonmetallic wireways shall be supported where run horizontally at intervals not to exceed 3 ft (914 mm), and at each end or joint, unless listed for other support intervals. In no case shall the distance between supports exceed 10 ft (3.05 m).

(b) Vertical Support. Vertical runs of nonmetallic wireway shall be securely supported at intervals not exceeding 4 ft (1.22 m), unless listed for other support intervals, and shall not have more than one joint between supports. Adjoining nonmetallic wireway sections shall be securely fastened together to provide a rigid joint.

362-23. Expansion Fittings. Expansion fittings for nonmetallic wireway shall be provided to compensate for thermal expansion and contraction where the length change is expected to be 0.25 in. (6.36 mm) or greater in a straight run.

> FPN: See Table 347-9 for expansion characteristics of PVC rigid nonmetallic conduit. The expansion characteristics of PVC nonmetallic wireway are identical.

362-24. Extension Through Walls. Nonmetallic wireways shall be permitted to pass transversely through walls if the length passing through the wall is unbroken. Access to the conductors shall be maintained on both sides of the wall.

362-25. Dead Ends. Dead ends of nonmetallic wireways shall be closed using listed fittings.

362-26. Extensions from Nonmetallic Wireways. Extensions from nonmetallic wireways shall be made with cord pendants or any wiring method of Chapter 3. A separate equipment grounding conductor shall be installed in, or an equipment grounding connection shall be made to, any of the wiring methods used for the extension.

362-27. Marking. Nonmetallic wireways shall be marked so that the manufacturer's name or trademark and interior cross-sectional area in square inches shall be visible after installation. Nonmetallic wireways that have limited smoke-producing characteristics shall be permitted to be identified with the suffix *LS*.

Article 363 — Flat Cable Assemblies: Type FC

363-1. Definition. *Type FC, a flat cable assembly,* is an assembly of parallel conductors formed integrally with an insulating material web specifically designed for field installation in surface metal raceway.

363-2. Other Articles. In addition to the provisions of this article, installation of Type FC cable shall conform with the applicable provisions of Articles 210, 220, 250, 300, 310, and 352.

363-3. Uses Permitted. Flat cable assemblies shall be permitted only as branch circuits to supply suitable tap devices for lighting, small appliances, or small power loads. Flat cable assemblies shall be installed for exposed work only. Flat cable assemblies shall be installed in locations where they will not be subjected to severe physical damage.

363-4. Uses Not Permitted. Flat cable assemblies shall not be installed in the following:

(1) Where subject to corrosive vapors unless suitable for the application
(2) In hoistways
(3) In any hazardous (classified) location
(4) Outdoors or in wet or damp locations unless identified for use in wet locations

363-5. Installation. Flat cable assemblies shall be installed in the field only in surface metal raceways identified for the use. The channel portion of the surface metal raceway systems shall be installed as complete systems before the flat cable assemblies are pulled into the raceways.

363-6. Number of Conductors. The flat cable assemblies shall consist of either two, three, or four conductors.

363-7. Size of Conductors. Flat cable assemblies shall have conductors of No. 10 special stranded copper wires.

363-8. Conductor Insulation. The entire flat cable assembly shall be formed to provide a suitable insulation covering all of the conductors and using one of the materials recognized in Table 310-13 for general branch-circuit wiring.

363-9. Splices. Splices shall be made in listed junction boxes.

363-10. Taps. Taps shall be made between any phase conductor and the grounded conductor or any other phase conductor by means of devices and fittings identified for the use. Tap devices shall be rated at not less than 15 amperes, or more than 300 volts to ground, and they shall be color-coded in accordance with the requirements of Section 363-20.

363-11. Dead Ends. Each flat cable assembly dead end shall be terminated in an end-cap device identified for the use.

The dead-end fitting for the enclosing surface metal raceway shall be identified for the use.

363-12. Fixture Hangers. Fixture hangers installed with the flat cable assemblies shall be identified for the use.

363-13. Fittings. Fittings to be installed with flat cable assemblies shall be designed and installed to prevent physical damage to the cable assemblies.

363-14. Extensions. All extensions from flat cable assemblies shall be made by approved wiring methods, within the junction boxes, installed at either end of the flat cable assembly runs.

363-15. Supports. The flat cable assemblies shall be supported by means of their special design features, within the surface metal raceways.

The surface metal raceways shall be supported as required for the specific raceway to be installed.

363-16. Rating. The rating of the branch circuit shall not exceed 30 amperes.

363-17. Marking. In addition to the provisions of Section 310-11, Type FC cable shall have the temperature rating durably marked on the surface at intervals not exceeding 24 in. (610 mm).

363-18. Protective Covers. Where a flat cable assembly is installed less than 8 ft (2.44 m) above the floor or fixed working platform, it shall be protected by a metal cover identified for the use.

363-19. Identification. The grounded conductor shall be identified throughout its length by means of a distinctive and durable white or natural gray marking.

363-20. Terminal Block Identification. Terminal blocks identified for the use shall have distinctive and durable markings for color or word coding. The grounded conductor section shall have a white marking or other suitable designation. The next adjacent section of the terminal block shall have a black marking or other suitable designation. The next section shall have a red marking or other suitable designation. The final or outer section, opposite the grounded conductor section of the terminal block, shall have a blue marking or other suitable designation.

Article 364 — Busways

A. General Requirements

364-1. Scope. This article covers service-entrance, feeder, and branch-circuit busways and associated fittings.

364-2. Definition. For the purpose of this article, a *busway* is considered to be a grounded metal enclosure containing factory-mounted, bare or insulated conductors, which are usually copper or aluminum bars, rods, or tubes.

FPN: For cablebus, refer to Article 365.

364-3. Other Articles. Installations of busways shall comply with the applicable provisions of Article 300.

364-4. Use.

(a) Uses Permitted. Busways shall be permitted to be installed where they are located as follows:

(1) Located in the open and are visible, or
(2) Installed behind access panels, provided the busways are totally enclosed, of nonventilating-type construction, and installed so that the joints between sections and at fittings are accessible for maintenance purposes. Where installed behind access panels, means of access shall be provided, and the following conditions shall be met:

   (a) The space behind the access panels shall not be used for air-handling purposes, or
   (b) Where the space behind the access panels is used for environmental air, other than ducts and plenums, there shall be no provisions for plug-in connections, and the conductors shall be insulated.

(b) Uses Not Permitted. Busways shall not be installed as follows:

(1) Where subject to severe physical damage or corrosive vapors
(2) In hoistways

(3) In any hazardous (classified) location, unless specifically approved for such use

FPN: See Section 501-4(b).

(4) Outdoors or in wet or damp locations unless identified for such use

Lighting busway and trolley busway shall not be installed less than 8 ft (2.44 m) above the floor or working platform unless provided with a cover identified for the purpose.

364-5. Support. Busways shall be securely supported at intervals not exceeding 5 ft (1.52 m) unless otherwise designed and marked.

364-6. Through Walls and Floors.

(a) Walls. Unbroken lengths of busway shall be permitted to be extended through dry walls.

(b) Floors. Floor penetrations shall comply with (1) and (2).

  (1) Busways shall be permitted to be extended vertically through dry floors if totally enclosed (unventilated) where passing through and for a minimum distance of 6 ft (1.83 m) above the floor to provide adequate protection from physical damage.

  (2) In other than industrial establishments, where a vertical riser penetrates two or more dry floors, a minimum 4-in. (102-mm) high curb shall be installed around all floor openings for riser busways to prevent liquids from entering the opening. The curb shall be installed within 12 in. (304.8 mm) of the floor opening. Electrical equipment shall be located so that it will not be damaged by liquids that are retained by the curb.

FPN: See Section 300-21 for information concerning the spread of fire or products of combustion.

364-7. Dead Ends. A dead end of a busway shall be closed.

364-8. Branches from Busways. Branches from busways shall be permitted to be made by the following.

(a) Branches from busways shall be made in accordance with Articles 331, 334, 345, 346, 347, 348, 350, 351, 352, and 364. Where a nonmetallic raceway is used, connection of equipment grounding conductors in the nonmetallic raceway to the busway shall comply with Sections 250-8 and 250-12.

(b) Suitable cord and cable assemblies approved for extra-hard usage or hard usage and listed bus drop cable shall be permitted as branches from busways for the connection of portable equipment or the connection of stationary equipment to facilitate their interchange in accordance with Sections 400-7 and 400-8 and the following conditions.

(1) The cord or cable shall be attached to the building by an approved means.
(2) The length of the cord or cable from a busway plug-in device to a suitable tension take-up support device shall not exceed 6 ft (1.83 m).

Exception: In industrial establishments only, where the conditions of maintenance and supervision ensure that only qualified persons will service the installation, lengths exceeding 6 ft. (1.83 m) shall be permitted between the busway plug-in device and the tension take-up support device where the cord or cable is supported at intervals not exceeding 8 ft (2.4 m).

(3) The cord or cable shall be installed as a vertical riser from the tension take-up support device to the equipment served.
(4) Strain relief cable grips shall be provided for the cord or cable at the busway plug-in device and equipment terminations.

(c) Suitable cord and cable assemblies approved for extra-hard usage or hard usage and listed bus drop cable shall be permitted as branches from trolley-type busways for the connection of movable equipment in accordance with Sections 400-7 and 400-8.

364-9. Overcurrent Protection. Overcurrent protection shall be provided in accordance with Sections 364-10 through 364-13.

364-10. Rating of Overcurrent Protection — Feeders. A busway shall be protected against overcurrent in accordance with the allowable current rating of the busway.

Exception No. 1: The applicable provisions of Section 240-3 shall be permitted.

Exception No. 2: Where used as transformer secondary ties, the provisions of Section 450-6(a)(3) shall be permitted.

364-11. Reduction in Ampacity Size of Busway. Overcurrent protection shall be required where busways are reduced in ampacity.

Exception: For industrial establishments only, omission of overcurrent protection shall be permitted at points where busways are reduced in ampacity, provided that the length of the busway having the smaller ampacity does not exceed 50 ft (15.2 m) and has an ampacity at least equal to one-third the rating or setting of the overcurrent device next

back on the line, and provided that such busway is free from contact with combustible material.

364-12. Feeder or Branch Circuits. Where a busway is used as a feeder, devices or plug-in connections for tapping off feeder or branch circuits from the busway shall contain the overcurrent devices required for the protection of the feeder or branch circuits. The plug-in device shall consist of an externally operable circuit breaker or an externally operable fusible switch. Where such devices are mounted out of reach and contain disconnecting means, suitable means such as ropes, chains, or sticks shall be provided for operating the disconnecting means from the floor.

Exception No. 1: As permitted in Section 240-21.

Exception No. 2: For fixed or semifixed lighting fixtures, where the branch-circuit overcurrent device is part of the fixture cord plug on cord-connected fixtures.

Exception No. 3: Where fixtures without cords are plugged directly into the busway and the overcurrent device is mounted on the fixture.

364-13. Rating of Overcurrent Protection — Branch Circuits. A busway used as a branch circuit shall be protected against overcurrent in accordance with Section 210-20. Where so used, the circuit shall comply with the applicable requirements of Articles 210, 430, and 440.

364-15. Marking. Busways shall be marked with the voltage and current rating for which they are designed, and with the manufacturer's name or trademark in such manner as to be visible after installation.

B. Requirements for Over 600 Volts, Nominal

364-21. Identification. Each bus run shall be provided with a permanent nameplate on which the following information shall be provided:

(1) Rated voltage
(2) Rated continuous current; if bus is forced-cooled, both the normal forced-cooled rating and the self-cooled (not forced-cooled) rating for the same temperature rise shall be given
(3) Rated frequency
(4) Rated impulse withstand voltage
(5) Rated 60-Hz withstand voltage (dry)
(6) Rated momentary current
(7) Manufacturer's name or trademark

> FPN: See *Guide for Metal-Enclosed Bus and Calculating Losses in Isolated-Phase Bus,* ANSI C37.23-1987 (R1991), for construction and testing requirements for metal-enclosed buses.

364-22. Grounding. Metal-enclosed bus shall be grounded in accordance with Article 250.

364-23. Adjacent and Supporting Structures. Metal-enclosed busways shall be installed so that temperature rise from induced circulating currents in any adjacent metallic parts will not be hazardous to personnel or constitute a fire hazard.

364-24. Neutral. Neutral bus, where required, shall be sized to carry all neutral load current, including harmonic currents, and shall have adequate momentary and short-circuit rating consistent with system requirements.

364-25. Barriers and Seals. Bus runs that have sections located both inside and outside of buildings shall have a vapor seal at the building wall to prevent interchange of air between indoor and outdoor sections.

Exception: Vapor seals shall not be required in forced-cooled bus.

Fire barriers shall be provided where fire walls, floors, or ceilings are penetrated.

FPN: See Section 300-21 for information concerning the spread of fire or products of combustion.

364-26. Drain Facilities. Drain plugs, filter drains, or similar methods shall be provided to remove condensed moisture from low points in bus run.

364-27. Ventilated Bus Enclosures. Ventilated bus enclosures shall be installed in accordance with Article 110, Part C, and Section 490-24, unless designed so that foreign objects inserted through any opening will be deflected from energized parts.

364-28. Terminations and Connections. Where bus enclosures terminate at machines cooled by flammable gas, seal-off bushings, baffles, or other means shall be provided to prevent accumulation of flammable gas in the bus enclosures.

Flexible or expansion connections shall be provided in long, straight runs of bus to allow for temperature expansion or contraction, or where the bus run crosses building vibration insulation joints.

All conductor termination and connection hardware shall be accessible for installation, connection, and maintenance.

364-29. Switches. Switching devices or disconnecting links provided in the bus run shall have the same momentary rating as the bus. Disconnecting links shall be plainly marked to be removable only when bus is de-energized. Switching devices that are not load break shall be interlocked to prevent operation under load, and disconnecting link enclosures shall be interlocked to prevent access to energized parts.

364-30. Wiring 600 Volts or Less, Nominal. Secondary control devices and wiring that are provided as part of the metal-enclosed bus run shall be insulated by fire-retardant barriers from all primary circuit elements with the exception of short lengths of wire, such as at instrument transformer terminals.

Article 365 — Cablebus

365-1. Definition. *Cablebus* is an assembly of insulated conductors with fittings and conductor terminations in a completely enclosed, ventilated protective metal housing. Cablebus is ordinarily assembled at the point of installation from the components furnished or specified by the manufacturer in accordance with instructions for the specific job. This assembly is designed to carry fault current and to withstand the magnetic forces of such current.

365-2. Use.

(a) 600 Volts or Less. Approved cablebus shall be permitted at any voltage or current for which spaced conductors are rated and shall be installed for exposed work only. Cablebus installed outdoors or in corrosive, wet, or damp locations shall be identified for such use. Cablebus shall not be installed in hoistways or hazardous (classified) locations unless specifically approved for such use. Cablebus shall be permitted to be used for branch circuits, feeders, and services.

Cablebus framework, where bonded as required by Article 250, shall be permitted as the equipment grounding conductor for branch circuits and feeders.

(b) Over 600 Volts. Approved cablebus shall be permitted for systems in excess of 600 volts, nominal. See Section 300-37.

365-3. Conductors.

(a) Types of Conductors. The current-carrying conductors in cablebus shall have an insulation rating of 75°C (167°F) or higher of an approved type and suitable for the application in accordance with Articles 310 and 490.

(b) Ampacity of Conductors. The ampacity of conductors in cablebus shall be in accordance with Tables 310-17 and 310-19, or Tables 310-69 and 310-70 for installations over 600 volts.

(c) Size and Number of Conductors. The size and number of conductors shall be that for which the cablebus is designed, and in no case smaller than No. 1/0.

(d) Conductor Supports. The insulated conductors shall be supported on blocks or other mounting means designed for the purpose.

The individual conductors in a cablebus shall be supported at intervals not greater than 3 ft (914 mm) for horizontal runs and 1½ ft (457 mm) for vertical runs. Vertical and horizontal spacing between supported conductors shall not be less than one conductor diameter at the points of support.

365-5. Overcurrent Protection. Cablebus shall be protected against overcurrent in accordance with the allowable ampacity of the cablebus conductors in accordance with Section 240-3.

Exception: Overcurrent protection shall be permitted in accordance with Sections 240-100 and 240-101 for over 600 volts, nominal.

365-6. Support and Extension Through Walls and Floors.

(a) Support. Cablebus shall be securely supported at intervals not exceeding 12 ft (3.66 m).

Exception: Where spans longer than 12 ft (3.66 m) are required, the structure shall be specifically designed for the required span length.

(b) Transversely Routed. Cablebus shall be permitted to extend transversely through partitions or walls, other than fire walls, provided the section within the wall is continuous, protected against physical damage, and unventilated.

(c) Through Dry Floors and Platforms. Except where firestops are required, cablebus shall be permitted to extend vertically through dry floors and platforms, provided the cablebus is totally enclosed at the point where it passes through the floor or platform and for a distance of 6 ft (1.83 m) above the floor or platform.

(d) Through Floors and Platforms in Wet Locations. Except where firestops are required, cablebus shall be permitted to extend vertically through floors and platforms in wet locations where (1) there are curbs or other suitable means to prevent water flow through the floor or platform opening, and (2) where the cablebus is totally enclosed at the point where it passes through the floor or platform and for a distance of 6 ft (1.83 m) above the floor or platform.

365-7. Fittings. A cablebus system shall include approved fittings for the following:

(1) Changes in horizontal or vertical direction of the run
(2) Dead ends
(3) Terminations in or on connected apparatus or equipment or the enclosures for such equipment
(4) Additional physical protection where required, such as guards where subject to severe physical damage

365-8. Conductor Terminations. Approved terminating means shall be used for connections to cablebus conductors.

365-9. Grounding. A cablebus installation shall be grounded and bonded in accordance with Article 250, excluding Section 250-86, Exception No. 2.

365-10. Marking. Each section of cablebus shall be marked with the manufacturer's name or trade designation and the maximum diameter, number, voltage rating, and ampacity of the conductors to be installed. Markings shall be located so as to be visible after installation.

Article 370 — Outlet, Device, Pull and Junction Boxes, Conduit Bodies and Fittings

A. Scope and General

370-1. Scope. This article covers the installation and use of all boxes and conduit bodies used as outlet, junction, or pull boxes, depending on their use, and manholes and other electric enclosures intended for personnel entry. Cast, sheet metal, nonmetallic, and other boxes such as FS, FD, and larger boxes are not classified as conduit bodies. This article also includes installation requirements for fittings used to join raceways and to connect raceways and cables to boxes and conduit bodies.

370-2. Round Boxes. Round boxes shall not be used where conduits or connectors requiring the use of locknuts or bushings are to be connected to the side of the box.

370-3. Nonmetallic Boxes. Nonmetallic boxes shall be permitted only with open wiring on insulators, concealed knob-and-tube wiring, nonmetallic-sheathed cable, and nonmetallic raceways.

Exception No. 1: Where internal bonding means are provided between all entries, nonmetallic boxes shall be permitted to be used with metal raceways or metal-armored cables.

Exception No. 2: Where integral bonding means with a provision for attaching an equipment grounding jumper inside the box are provided between all threaded entries in nonmetallic boxes listed for the purpose, nonmetallic boxes shall be permitted to be used with metal raceways or metal-armored cables.

370-4. Metal Boxes. All metal boxes shall be grounded in accordance with the provisions of Article 250.

370-5. Short Radius Conduit Bodies. Conduit bodies such as capped elbows and service-entrance elbows enclosing conductors No. 6 or smaller, and that are only intended to enable the installation of the raceway and the contained conductors, shall not contain splices, taps, or devices and shall be of sufficient size to provide free space for all conductors enclosed in the conduit body.

B. Installation

370-15. Damp, Wet, or Hazardous (Classified) Locations.

(a) Damp or Wet Locations. In damp or wet locations, boxes, conduit bodies, and fittings shall be placed or equipped so as to prevent moisture from entering or accumulating within the box, conduit body, or fitting. Boxes, conduit bodies, and fittings installed in wet locations shall be listed for use in wet locations.

FPN No. 1: For boxes in floors, see Section 370-27(b).

FPN No. 2: For protection against corrosion, see Section 300-6.

(b) Hazardous (Classified) Locations. Installations in hazardous (classified) locations shall conform to Articles 500 through 517.

370-16. Number of Conductors in Outlet, Device, and Junction Boxes, and Conduit Bodies.
Boxes and conduit bodies shall be of sufficient size to provide free space for all enclosed conductors. In no case shall the volume of the box, as calculated in (a), be less than the fill calculation as calculated in (b). The minimum volume for conduit bodies shall be as calculated in (c).

The provisions of this section shall not apply to terminal housings supplied with motors. (See Section 430-12.)

Boxes and conduit bodies enclosing conductors, size No. 4 or larger, shall also comply with the provisions of Section 370-28.

(a) Box Volume Calculations. The volume of a wiring enclosure (box) shall be the total volume of the assembled sections, and, where used, the space provided by plaster rings, domed covers, extension rings, etc., that are marked with their volume in cubic inches or are made from boxes the dimensions of which are listed in Table 370-16(a).

Table 370-16(a). Metal Boxes

Box Dimension in Inches, Trade Size, or Type	Minimum Capacity (in.^{3})	Maximum Number of Conductors*						
		No. 18	No. 16	No. 14	No. 12	No. 10	No. 8	No. 6
4 × 1¼ round or octagonal	12.5	8	7	6	5	5	4	2
4 × 1½ round or octagonal	15.5	10	8	7	6	6	5	3
4 × 2⅛ round or octagonal	21.5	14	12	10	9	8	7	4
4 × 1¼ square	18.0	12	10	9	8	7	6	3
4 × 1½ square	21.0	14	12	10	9	8	7	4
4 × 2⅛ square	30.3	20	17	15	13	12	10	6
4¹¹⁄₁₆ × 1¼ square	25.5	17	14	12	11	10	8	5
4¹¹⁄₁₆ × 1½ square	29.5	19	16	14	13	11	9	5
4¹¹⁄₁₆ × 2⅛ square	42.0	28	24	21	18	16	14	8
3 × 2 × 1½ device	7.5	5	4	3	3	3	2	1
3 × 2 × 2 device	10.0	6	5	5	4	4	3	2
3 × 2 × 2¼ device	10.5	7	6	5	4	4	3	2
3 × 2 × 2½ device	12.5	8	7	6	5	5	4	2
3 × 2 × 2¾ device	14.0	9	8	7	6	5	4	2
3 × 2 × 3½ device	18.0	12	10	9	8	7	6	3
4 × 2⅛ × 1½ device	10.3	6	5	5	4	4	3	2
4 × 2⅛ × 1⅞ device	13.0	8	7	6	5	5	4	2
4 × 2⅛ × 2⅛ device	14.5	9	8	7	6	5	4	2
3¾ × 2 × 2½ masonry box/gang	14.0	9	8	7	6	5	4	2
3¾ × 2 × 3½ masonry box/gang	21.0	14	12	10	9	8	7	4
FS — Minimum internal depth 1¾ single cover/gang	13.5	9	7	6	6	5	4	2
FD — Minimum internal depth 2⅜ single cover/gang	18.0	12	10	9	8	7	6	3
FS — Minimum internal depth 1¾ multiple cover/gang	18.0	12	10	9	8	7	6	3
FD — Minimum internal depth 2⅜ multiple cover/gang	24.0	16	13	12	10	9	8	4

Note: For SI units, 1 in.^{3} = 16.4 cm^3.

*Where no volume allowances are required by Sections 370-16(b)(2) through 370-16(b)(5).

(1) Standard Boxes. The volumes of standard boxes that are not marked with a cubic inch capacity shall be as given in Table 370-16(a).

(2) Other Boxes. Boxes 100 in.³ (1640 cm³) or less, other than those described in Table 370-16(a), and nonmetallic boxes shall be durably and legibly marked by the manufacturer with their cubic inch capacity. Boxes described in Table 370-16(a) that have a larger cubic inch capacity than is designated in the table shall be permitted to have their cubic inch capacity marked as required by this section.

(b) Box Fill Calculations. The volumes in paragraphs (1) through (5), as applicable, shall be added together. No allowance shall be required for small fittings such as locknuts and bushings.

(1) Conductor Fill. Each conductor that originates outside the box and terminates or is spliced within the box shall be counted once, and each conductor that passes through the box without splice or termination shall be counted once. The conductor fill, in cubic inches, shall be computed using Table 370-16(b). A conductor, no part of which leaves the box, shall not be counted.

Exception: An equipment grounding conductor or conductors or not over four fixture wires smaller than No. 14, or both, shall be permitted to be omitted from the calculations where they enter a box from a domed fixture or similar canopy and terminate within that box.

Table 370-16(b). Volume Allowance Required per Conductor

Size of Conductor (AWG)	Free Space Within Box for Each Conductor (in.³)
18	1.50
16	1.75
14	2.00
12	2.25
10	2.50
8	3.00
6	5.00

Note: For SI units, 1 in.³ = 16.4 cm³.

(2) Clamp Fill. Where one or more internal cable clamps, whether factory or field supplied, are present in the box, a single volume allowance in accordance with Table 370-16(b) shall be made based on the largest conductor present in the box. No allowance shall be required for a cable connector with its clamping mechanism outside the box.

(3) Support Fittings Fill. Where one or more fixture studs or hickeys are present in the box, a single volume allowance in accordance with Table 370-16(b) shall be made for each type of fitting based on the largest conductor present in the box.

(4) Device or Equipment Fill. For each yoke or strap containing one or more devices or equipment, a double volume allowance in accordance with Table 370-16(b) shall be made for each yoke or strap based on the largest conductor connected to a device(s) or equipment supported by that yoke or strap.

(5) Equipment Grounding Conductor Fill. Where one or more equipment grounding conductors or equipment bonding jumpers enters a box, a single volume allowance in accordance with Table 370-16(b) shall be made based on the largest equipment grounding conductor or equipment bonding jumper present in the box. Where an additional set of equipment grounding conductors, as permitted by Section 250-146(d), is present in the box, an additional volume allowance shall be made based on the largest equipment grounding conductor in the additional set.

(c) Conduit Bodies.

(1) General. Conduit bodies enclosing No. 6 conductors or smaller, other than short radius conduit bodies as described in Section 370-5, shall have a cross-sectional area not less than twice the cross-sectional area of the largest conduit or tubing to which it is attached. The maximum number of conductors permitted shall be the maximum number permitted by Table 1 of Chapter 9 for the conduit or tubing to which it is attached.

(2) With Splices, Taps, or Devices. Only those conduit bodies that are durably and legibly marked by the manufacturer with their cubic inch capacity shall be permitted to contain splices, taps, or devices. The maximum number of conductors shall be computed in accordance with Section 370-16(b). Conduit bodies shall be supported in a rigid and secure manner.

370-17. Conductors Entering Boxes, Conduit Bodies, or Fittings. Conductors entering boxes, conduit bodies, or fittings shall be protected from abrasion and shall comply with (a) through (d).

(a) Openings to Be Closed. Openings through which conductors enter shall be adequately closed.

(b) Metal Boxes and Conduit Bodies. Where metal boxes or conduit bodies are installed with open wiring or concealed knob-and-tube wiring, conductors shall enter through insulating bushings or, in dry locations, through flexible tubing extending from the last insulating support and firmly secured to the box or conduit body. Where raceway or cable is installed with metal boxes or conduit bodies, the raceway or cable shall be secured to such boxes and conduit bodies.

(c) Nonmetallic Boxes. Nonmetallic boxes shall be suitable for the lowest temperature-rated conductor entering the box. Where nonmetallic boxes are used with open wiring or concealed knob-and-tube wiring, the conductors shall enter the box through individual holes. Where flexible tubing is used to encase the conductors, the tubing shall extend from the last insulating support to no less than ¼ in. (6.35 mm) inside the box. Where nonmetallic-sheathed cable is used, the cable assembly, including the sheath, shall extend into the box no less than ¼ in. (6.35 mm) through a nonmetallic-sheathed cable knockout opening. In all instances, all permitted wiring methods shall be secured to the boxes.

Exception: Where nonmetallic-sheathed cable or underground feeder and branch-circuit cable is used with single gang boxes no larger than a nominal size 2¼ in. × 4 in., mounted in walls or ceilings, and where the cable is fastened within 8 in. (203 mm) of the box measured along the sheath and where the sheath extends through a cable knockout no less than ¼ in. (6.35 mm), securing the cable to the box shall not be required. Multiple cable entries shall be permitted in a single cable knockout opening.

(d) Conductors No. 4 or Larger. Installation shall comply with Section 300-4(f).

370-18. Unused Openings. Unused cable or raceway openings in boxes and conduit bodies shall be effectively closed to afford protection substantially equivalent to that of the wall of the box or conduit body. Metal plugs or plates used with nonmetallic boxes or conduit bodies shall be recessed at least ¼ in. (6.35 mm) from the outer surface of the box.

370-19. Boxes Enclosing Flush Devices. Boxes used to enclose flush devices shall be of such design that the devices will be completely enclosed on back and sides, and that substantial support for the devices will be provided. Screws for supporting the box shall not be used in attachment of the device contained therein.

370-20. In Wall or Ceiling. In walls or ceilings of concrete, tile, or other noncombustible material, boxes shall be installed so that the front edge of the box will not be set back of the finished surface more than ¼ in. (6.35 mm). In walls and ceilings constructed of wood or other combustible material, boxes shall be flush with the finished surface or project therefrom.

370-21. Repairing Plaster and Drywall or Plasterboard. Plaster, drywall, or plasterboard surfaces that are broken or incomplete shall be repaired so there will be no gaps or open spaces greater than ⅛ in. (3.18 mm) at the edge of the box or fitting.

370-22. Exposed Surface Extensions. Surface extensions from a box of a concealed wiring system shall be made by mounting and mechanically securing a box or extension ring over the concealed box. Where required, equipment grounding shall be in accordance with Article 250.

Exception: A surface extension shall be permitted to be made from the cover of a concealed box where the cover is designed so it is unlikely to fall off, or be removed if its securing means becomes loose. The wiring method shall be flexible and arranged so that any required grounding continuity is independent of the connection between the box and cover.

370-23. Supports. Enclosures within the scope of this article shall be supported in accordance with one or more of the provisions in (a) through (h).

(a) Surface Mounting. An enclosure mounted on a building or other surface shall be rigidly and securely fastened in place. If the surface does not provide rigid and secure support, additional support in accordance with other provisions of this section shall be provided.

(b) Structural Mounting. An enclosure supported from a structural member of a building or from grade shall be rigidly supported either directly, or by using a metal, polymeric, or wood brace.

(1) Nails. Nails, where used as a fastening means, shall be attached by using brackets on the outside of the enclosure, or they shall pass through the interior within ¼ in. (6.35 mm) of the back or ends of the enclosure.

(2) Braces. Metal braces shall be protected against corrosion and formed from metal that is not less than 0.020 in. (508 μm) thick uncoated. Wood braces shall have a cross section not less than nominal 1 in. × 2 in. Wood braces in wet locations shall be treated for the conditions. Polymeric braces shall be identified as being suitable for the use.

(c) Mounting in Finished Surfaces. An enclosure mounted in a finished surface shall be rigidly secured thereto by clamps, anchors, or fittings identified for the application.

(d) Suspended Ceilings. An enclosure mounted to structural or supporting elements of a suspended ceiling shall be not more than 100 in.³ (1640 cm³) in size and shall be securely fastened in place in one of the following ways.

(1) Framing Members. An enclosure shall be fastened to the framing members by mechanical means such as bolts, screws, or rivets, or by the use of clips or other securing means identified for use with the type of ceiling framing member(s) and enclosure(s) employed. The framing members shall be adequately supported and securely fastened to each other and to the building structure.

(2) Support Wires. The installation shall comply with the provisions of Section 300-11(a). The enclosure shall be secured, using methods identified for the purpose, to ceiling support wires, including any additional support wires installed for that purpose. Support wires used for enclosure support shall be fastened at each end so as to be taut within the ceiling cavity.

(e) Raceway Supported Enclosure, Without Devices or Fixtures. An enclosure that does not contain a device(s) or support a fixture(s) or other equipment, and is supported by entering raceways shall not exceed 100 in.^{3} (1640 cm^3) in size. It shall have threaded entries or have hubs identified for the purpose. It shall be supported by two or more conduits threaded wrenchtight into the enclosure or hubs. Each conduit shall be secured within 3 ft (914 mm) of the enclosure, or within 18 in. (457 mm) of the enclosure if all entries are on the same side.

Exception: Rigid metal, intermediate metal, or rigid nonmetallic conduit or electrical metallic tubing shall be permitted to support a conduit body of any size, including a conduit body constructed with only one conduit entry, provided the conduit body is not larger than the largest trade size of the conduit or electrical metallic tubing.

(f) Raceway Supported Enclosures, with Devices or Fixtures. An enclosure that contains a device(s) or supports a fixture(s) or other equipment and is supported by entering raceways shall not exceed 100 in.^{3} (1640 cm^3) in size. It shall have threaded entries or have hubs identified for the purpose. It shall be supported by two or more conduits threaded wrenchtight into the enclosure or hubs. Each conduit shall be secured within 18 in. (457 mm) of the enclosure.

Exception No. 1: Rigid metal or intermediate metal conduit shall be permitted to support a conduit body of any size, including a conduit body constructed with only one conduit entry, provided the conduit bodies are not larger than the largest trade size of the conduit.

Exception No. 2: An unbroken length(s) of rigid or intermediate metal conduit shall be permitted to support a box used for fixture support, or to support a wiring enclosure within a fixture and used in lieu of a box in accordance with Section 300-15(b), where all of the following conditions are met.

(a) The conduit is securely fastened at a point so that the length of conduit beyond the last point of conduit support does not exceed 3 ft (914 mm).

(b) The unbroken conduit length before the last point of conduit support is 12 in. (305 mm) or greater, and that portion of the conduit is securely fastened at some point not less than 12 in. (305 mm) from its last point of support.

(c) Where accessible to unqualified persons, the fixture, measured to its lowest point, is at least 8 ft (2.44 m) above grade or standing area and at least 3 ft (914 mm) measured horizontally to the 8 ft (2.44 m) elevation from windows, doors, porches, fire escapes, or similar locations.

(d) A fixture supported by a single conduit does not exceed 12 in. (305 mm) in any direction from the point of conduit entry.

(e) The weight supported by any single conduit does not exceed 20 lb (9.08 kg).

(f) At the fixture end, the conduit(s) is threaded wrenchtight into the box or wiring enclosure, or into hubs identified for the purpose.

(g) Enclosures in Concrete or Masonry. An enclosure supported by embedment shall be identified as suitably protected from corrosion and securely embedded in concrete or masonry.

(h) Pendant Boxes. An enclosure supported by a pendant shall comply with (1) or (2).

(1) Flexible Cord. A box shall be supported from a multiconductor cord or cable in an approved manner that protects the conductors against strain, such as a strain-relief connector threaded into a box with a hub.

(2) Conduit. A box supporting lampholders or lighting fixtures, or wiring enclosures within fixtures used in lieu of boxes in accordance with Section 300-15(b), shall be supported by rigid or intermediate metal conduit stems. For stems longer than 18 in. (457 mm), the stems shall be connected to the wiring system with flexible fittings suitable for the location. At the fixture end, the conduit(s) shall be threaded wrenchtight into the box or wiring enclosure, or into hubs identified for the purpose.

Where supported by only a single conduit, the threaded joints shall be prevented from loosening by the use of setscrews or other effective means, or the fixture, at any point, shall be at least 8 ft (2.44 m) above grade or standing area and at least 3 ft (914 mm) measured horizontally to the 8 ft (2.44 m) elevation from windows, doors, porches, fire escapes, or similar locations. A fixture supported by a single conduit shall not exceed 12 in. (305 mm) in any horizontal direction from the point of conduit entry.

370-24. Depth of Outlet Boxes. No box shall have an internal depth of less than $\frac{1}{2}$ in. (12.7 mm). Boxes intended to enclose flush devices shall have an internal depth of not less than $\frac{15}{16}$ in. (23.8 mm).

370-25. Covers and Canopies. In completed installations, each box shall have a cover, faceplate, or fixture canopy.

(a) Nonmetallic or Metal Covers and Plates. Nonmetallic or metal covers and plates shall be permitted. Where metal covers or plates are used, they shall comply with the grounding requirements of Section 250-110.

> FPN: For additional grounding requirements, see Section 410-18(a) for metal fixture canopies, and Sections 380-12 and 410-56(d) for metal faceplates.

(b) Exposed Combustible Wall or Ceiling Finish. Where a fixture canopy or pan is used, any combustible wall or ceiling finish exposed between the edge of the canopy or pan and the outlet box shall be covered with noncombustible material.

(c) Flexible Cord Pendants. Covers of outlet boxes and conduit bodies having holes through which flexible cord pendants pass shall be provided with bushings designed for the purpose or shall have smooth, well-rounded surfaces on which the cords may bear. So-called hard rubber or composition bushings shall not be used.

370-27. Outlet Boxes.

(a) Boxes at Lighting Fixture Outlets. Boxes used at lighting fixture outlets shall be designed for the purpose. At every outlet used exclusively for lighting, the box shall be designed or installed so that a lighting fixture may be attached.

Exception: A wall-mounted fixture weighing not more than 6 lb (2.72 kg) and not exceeding 16 in. (406 mm) in any dimension shall be permitted to be supported on other boxes, provided the fixture or its supporting yoke is secured to the box with no fewer than two No. 6 or larger screws.

(b) Floor Boxes. Boxes listed specifically for this application shall be used for receptacles located in the floor.

Exception: Where the authority having jurisdiction judges them free from likely exposure to physical damage, moisture, and dirt, boxes located in elevated floors of show windows and similar locations shall be permitted to be other than those listed for floor applications. Receptacles and covers shall be listed as an assembly for this type of location.

(c) Boxes at Ceiling-Suspended (Paddle) Fan Outlets. Outlet boxes shall not be used as the sole support for ceiling-suspended (paddle) fans.

Exception: Boxes listed for the application shall be permitted as the sole means of support.

370-28. Pull and Junction Boxes. Boxes and conduit bodies used as pull or junction boxes shall comply with (a) through (d).

Exception: Terminal housings supplied with motors shall comply with the provisions of Section 430-12.

(a) Minimum Size. For raceways containing conductors of No. 4 or larger, and for cables containing conductors of No. 4 or larger, the minimum dimensions of pull or junction boxes installed in a raceway or cable run shall comply with the following.

(1) Straight Pulls. In straight pulls, the length of the box shall not be less than eight times the trade diameter of the largest raceway.

(2) Angle or U Pulls. Where splices, or where angle or U pulls are made, the distance between each raceway entry inside the box and the opposite wall of the box shall not be less than six times the trade diameter of the largest raceway in a row. This distance shall be increased for additional entries by the amount of the sum of the diameters of all other raceway entries in the same row on the same wall of the box. Each row shall be calculated individually, and the single row that provides the maximum distance shall be used.

Exception: Where a raceway or cable entry is in the wall of a box or conduit body opposite a removable cover, the distance from that wall to the cover shall be permitted to comply with the distance required for one wire per terminal in Table 373-6(a).

The distance between raceway entries enclosing the same conductor shall not be less than six times the trade diameter of the larger raceway.

When transposing cable size into raceway size in (a)(1) and (a)(2), the minimum trade size raceway required for the number and size of conductors in the cable shall be used.

(3) Boxes or conduit bodies of dimensions less than those required in (a)(1) and (a)(2) shall be permitted for installations of combinations of conductors that are less than the maximum conduit or tubing fill (of conduits or tubing being used) permitted by Table 1 of Chapter 9, provided the box or conduit body has been approved for and is permanently marked with the maximum number and maximum size of conductors permitted.

(b) Conductors in Pull or Junction Boxes. In pull boxes or junction boxes having any dimension over 6 ft (1.83 m), all conductors shall be cabled or racked up in an approved manner.

(c) Covers. All pull boxes, junction boxes, and conduit bodies shall be provided with covers compatible with the box or conduit body construction and suitable for the conditions of use. Where metal covers are used, they shall comply with the grounding requirements of Section 250-110. An extension from the cover of an exposed box shall comply with Section 370-22, Exception.

(d) Permanent Barriers. Where permanent barriers are installed in a box, each section shall be considered as a separate box.

370-29. Conduit Bodies, Junction, Pull, and Outlet Boxes to Be Accessible. Conduit bodies, junction, pull, and outlet boxes shall be installed so that the wiring contained in them can be rendered accessible without removing any part of the building or, in underground circuits, without excavating sidewalks, paving, earth, or other substance that is to be used to establish the finished grade.

Exception: Listed boxes shall be permitted where covered by gravel, light aggregate, or noncohesive granulated soil if their location is effectively identified and accessible for excavation.

C. Construction Specifications

370-40. Metal Boxes, Conduit Bodies, and Fittings.

(a) Corrosion Resistant. Metal boxes, conduit bodies, and fittings shall be corrosion resistant or shall be well-galvanized, enameled, or otherwise properly coated inside and out to prevent corrosion.

> FPN: See Section 300-6 for limitation in the use of boxes and fittings protected from corrosion solely by enamel.

(b) Thickness of Metal. Sheet steel boxes not over 100 in.^{3} (1640 cm^3) in size shall be made from steel not less than 0.0625 in. (1.59 mm) thick. The wall of a malleable iron box or conduit body and a die-cast or permanent-mold cast aluminum, brass, bronze, or zinc box or conduit body shall not be less than $3/32$ in. (2.38 mm) thick. Other cast metal boxes or conduit bodies shall have a wall thickness not less than $1/8$ in. (3.17 mm).

Exception No. 1: Listed boxes and conduit bodies shown to have equivalent strength and characteristics shall be permitted to be made of thinner or other metals.

Exception No. 2: The walls of listed short radius conduit bodies, as covered in Section 370-5, shall be permitted to be made of thinner metal.

(c) Metal Boxes Over 100 in.^{3} Metal boxes over 100 in.^{3} (1640 cm^3) in size shall be constructed so as to be of ample strength and rigidity. If of sheet steel, the metal thickness shall not be less than 0.053 in. (1.35 mm) uncoated.

(d) Grounding Provisions. A means shall be provided in each metal box for the connection of an equipment grounding conductor. The means shall be permitted to be a tapped hole or equivalent.

370-41. Covers. Metal covers shall be of the same material as the box or conduit body with which they are used, or they shall be lined with firmly attached insulating material that is not less than $1/32$ in. (0.79 mm) thick, or they shall be listed for the purpose. Metal covers shall be the same thickness as the boxes or conduit body for which they are used, or they shall be listed for the purpose. Covers of porcelain or other approved insulating materials shall be permitted if of such form and thickness as to afford the required protection and strength.

370-42. Bushings. Covers of outlet boxes and conduit bodies having holes through which flexible cord pendants may pass shall be provided with approved bushings or shall have smooth, well-rounded surfaces on which the cord may bear. Where individual conductors pass through a metal cover, a separate hole equipped with a bushing of suitable insulating material shall be provided for each conductor. Such separate holes shall be connected by a slot as required by Section 300-20.

370-43. Nonmetallic Boxes. Provisions for supports or other mounting means for nonmetallic boxes shall be outside of the box, or the box shall be constructed so as to prevent contact between the conductors in the box and the supporting screws.

370-44. Marking. All boxes and conduit bodies, covers, extension rings, plaster rings, and the like shall be durably and legibly marked with the manufacturer's name or trademark.

D. Manholes and Other Electric Enclosures Intended for Personnel Entry

370-50. General. Electric enclosures intended for personnel entry and specifically fabricated for this purpose shall be of sufficient size to provide safe work space about electric equipment with live parts that is likely to require examination, adjustment, servicing, or maintenance while energized. They shall have sufficient size to permit ready installation or withdrawal of the conductors employed without damage to the conductors or to their insulation. They shall comply with the provisions of this part.

Exception: Where electric enclosures covered by Part D of this article are part of an industrial wiring system operating under conditions of maintenance and supervision that ensure only qualified persons will monitor and supervise the system, they shall be permitted to be designed and installed in accordance with appropriate engineering practice. If required by the authority having jurisdiction, design documentation shall be provided.

370-51. Strength. Manholes, vaults, and their means of access shall be designed under qualified engineering supervision and shall withstand all loads likely to be imposed on the structures.

> FPN: See *National Electrical Safety Code*, ANSI C2-1997, for additional information on the loading that can be expected to bear on underground enclosures.

370-52. Cabling Work Space. A clear work space not less than 3 ft (914 mm) wide shall be provided where cables are located on both sides, and not less than 2½ ft (762 mm) where cables are only on one side. The vertical headroom shall not be less than 6 ft (1.83 m) unless the opening is within 1 ft (305 mm), measured horizontally, of the adjacent interior side wall of the enclosure.

Exception: A manhole containing only one or more of the following shall be permitted to have one of the horizontal work space dimensions reduced to 2 ft (608 mm) where the other horizontal clear work space is increased so the sum of the two dimensions is not less than 6 ft (1.83 m):

(a) Optical fiber cables as covered in Article 770

(b) Power-limited fire alarm circuits supplied in accordance with Section 760-41

(c) Class 2 or Class 3 remote-control and signaling circuits, or both, supplied in accordance with Section 725-41

370-53. Equipment Work Space. Where electric equipment with live parts that is likely to require examination, adjustment, servicing, or maintenance while energized is installed in a manhole, vault, or other enclosure designed for personnel access, the work space and associated requirements in Section 110-26 shall be met for installations operating at 600 volts or less. Where the installation is over 600 volts, the work space and associated requirements in Section 110-34 shall be met. A manhole access cover that weighs over 100 lb (45.4 kg) shall be considered as meeting the requirements of Section 110-34(c).

370-54. Bending Space for Conductors. Bending space for conductors operating at 600 volts or below shall be provided in accordance with the requirements of Section 370-28(a). Conductors operating over 600 volts shall be provided with bending space in accordance with Sections 370-71(a) and 370-71(b) as applicable. Where any horizontal dimension exceeds 6 ft (1.83 m), all conductors shall be cabled or racked up in an approved manner.

Exception: Where Section 370-71(b) applies, each row or column of ducts on one wall of the enclosure shall be calculated individually, and the single row or column that provides the maximum distance shall be used.

370-55. Access to Manholes.

(a) Dimensions. Rectangular access openings shall not be less than 26 in. × 22 in. (659 mm × 557 mm). Round access openings in a manhole shall not be less than 26 in. (659 mm) in diameter.

Exception: A manhole that has a fixed ladder that does not obstruct the opening, or that contains only one or more of the following shall be permitted to reduce the minimum cover diameter to 2 ft (608 mm):

(a) Optical fiber cables as covered in Article 770

(b) Power-limited fire alarm circuits supplied in accordance with Section 760-41

(c) Class 2 or Class 3 remote-control and signaling circuits, or both, supplied in accordance with Section 725-41

(b) Obstructions. Manhole openings shall be free of protrusions that could injure personnel or prevent ready egress.

(c) Location. Manhole openings for personnel shall be located where they are not directly above electric equipment or conductors in the enclosure. Where this is not practicable, either a protective barrier or a fixed ladder shall be provided.

(d) Covers. Covers shall be over 100 lb (45.4 kg) or otherwise designed to require the use of tools to open. They shall be designed or restrained so they cannot fall into the manhole or protrude sufficiently to contact electrical conductors or equipment within the manhole.

(e) Marking. Manhole covers shall have an identifying mark or logo that prominently indicates their function, such as "electric."

370-56. Access to Vaults and Tunnels.

(a) Location. Access openings for personnel shall be located where they are not directly above electric equipment or conductors in the enclosure. Other openings shall be permitted over equipment to facilitate installation, maintenance, or replacement of equipment.

(b) Locks. In addition to compliance with the requirements of Section 110-34(c), if applicable, access openings for personnel shall be arranged so that a person on the inside can exit when the access door is locked from the outside, or in the case of normally locking by padlock, the locking arrangement shall be such that the padlock can be closed on the locking system to prevent locking from the outside.

370-57. Ventilation. Where manholes, tunnels, and vaults have communicating openings into enclosed areas used by the public, ventilation to open air shall be provided wherever practicable.

370-58. Guarding. Where conductors or equipment, or both, could be contacted by objects falling or being pushed through a ventilating grating, both conductors and live parts shall be protected in accordance with the requirements of

Sections 110-27(a)(2) or 110-31(a)(1), depending on the voltage.

370-59. Fixed Ladders. Fixed ladders shall be corrosion resistant.

E. Pull and Junction Boxes for Use on Systems Over 600 Volts, Nominal

370-70. General. Where pull and junction boxes are used on systems over 600 volts, the installation shall comply with the provisions of Part E and also the following general provisions of this article:

(1) In Part A, Sections 370-2, 370-3, and 370-4
(2) In Part B, Sections 370-15; 370-17; 370-18; 370-20; 370-23(a), (b), or (f); and 370-29
(3) In Part C, Sections 370-40(a) and (c) and 370-41

370-71. Size of Pull and Junction Boxes. Pull and junction boxes shall provide adequate space and dimensions for the installation of conductors, and they shall comply with the specific requirements of this section.

Exception: Terminal housings supplied with motors shall comply with the provisions of Section 430-12.

(a) For Straight Pulls. The length of the box shall not be less than 48 times the outside diameter, over sheath, of the largest shielded or lead-covered conductor or cable entering the box. The length shall not be less than 32 times the outside diameter of the largest nonshielded conductor or cable.

(b) For Angle or U Pulls.

(1) The distance between each cable or conductor entry inside the box and the opposite wall of the box shall not be less than 36 times the outside diameter, over sheath, of the largest cable or conductor. This distance shall be increased for additional entries by the amount of the sum of the outside diameters, over sheath, of all other cables or conductor entries through the same wall of the box.

Exception No. 1: Where a conductor or cable entry is in the wall of a box opposite a removable cover, the distance from that wall to the cover shall be permitted to be not less than the bending radius for the conductors as provided in Section 300-34.

Exception No. 2: Where cables are nonshielded and not lead covered, the distance of 36 times the outside diameter shall be permitted to be reduced to 24 times the outside diameter.

(2) The distance between a cable or conductor entry and its exit from the box shall not be less than 36 times the outside diameter, over sheath, of that cable or conductor.

Exception: Where cables are nonshielded and not lead covered, the distance of 36 times the outside diameter shall be permitted to be reduced to 24 times the outside diameter.

(c) Removable Sides. One or more sides of any pull box shall be removable.

370-72. Construction and Installation Requirements.

(a) Corrosion Protection. Boxes shall be made of material inherently resistant to corrosion or shall be suitably protected, both internally and externally, by enameling, galvanizing, plating, or other means.

(b) Passing Through Partitions. Suitable bushings, shields, or fittings having smooth, rounded edges shall be provided where conductors or cables pass through partitions and at other locations where necessary.

(c) Complete Enclosure. Boxes shall provide a complete enclosure for the contained conductors or cables.

(d) Wiring Is Accessible. Boxes shall be installed so that the wiring is accessible without removing any part of the building. Working space shall be provided in accordance with Section 110-34.

(e) Suitable Covers. Boxes shall be closed by suitable covers securely fastened in place. Underground box covers that weigh over 100 lb (45.4 kg) shall be considered meeting this requirement. Covers for boxes shall be permanently marked "DANGER — HIGH VOLTAGE — KEEP OUT." The marking shall be on the outside of the box cover and shall be readily visible. Letters shall be block type and at least ½ in. (12.7 mm) in height.

(f) Suitable for Expected Handling. Boxes and their covers shall be capable of withstanding the handling to which they may likely be subjected.

Article 373 — Cabinets, Cutout Boxes, and Meter Socket Enclosures

373-1. Scope. This article covers the installation and construction specifications of cabinets, cutout boxes, and meter socket enclosures.

A. Installation

373-2. Damp, Wet, or Hazardous (Classified) Locations.

(a) Damp and Wet Locations. In damp or wet locations, surface-type enclosures within the scope of this article shall be placed or equipped so as to prevent moisture or water from entering and accumulating within the cabinet or cutout box, and shall be mounted so there is at least ¼-in. (6.35-mm) airspace between the enclosure and the wall or other supporting surface. Enclosures installed in wet locations shall be weatherproof.

Exception: Nonmetallic enclosures shall be permitted to be installed without the airspace on a concrete, masonry, tile, or similar surface.

    FPN: For protection against corrosion, see Section 300-6.

(b) Hazardous (Classified) Locations. Installations in hazardous (classified) locations shall conform to Articles 500 through 517.

373-3. Position in Wall. In walls of concrete, tile, or other noncombustible material, cabinets shall be installed so that the front edge of the cabinet will not set back of the finished surface more than ¼ in. (6.35 mm). In walls constructed of wood or other combustible material, cabinets shall be flush with the finished surface or project therefrom.

373-4. Unused Openings. Unused openings in enclosures within the scope of this article shall be effectively closed to afford protection substantially equivalent to that of the enclosures within the scope of this article. Where metal plugs or plates are used with nonmetallic cabinets or cutout boxes, they shall be recessed at least ¼ in. (6.35 mm) from the outer surface.

373-5. Cabinets, Cutout Boxes, and Meter Socket Enclosures. Conductors entering enclosures within the scope of this article shall be protected from abrasion and shall comply with (a) through (c).

(a) Openings to Be Closed. Openings through which conductors enter shall be adequately closed.

(b) Metal Cabinets, Cutout Boxes, and Meter Socket Enclosures. Where metal enclosures within the scope of this article are installed with open wiring or concealed knob-and-tube wiring, conductors shall enter through insulating bushings or, in dry locations, through flexible tubing extending from the last insulating support and firmly secured to the enclosure.

(c) Cables. Where cable is used, each cable shall be secured to the cabinet, cutout box, or meter socket enclosure.

Exception: Cables with entirely nonmetallic sheaths shall be permitted to enter the top of a surface-mounted enclosure through one or more nonflexible raceways not less than 18 in. (457 mm) or more than 10 ft (3.05 m) in length, provided all the following conditions are met.

(a) Each cable is fastened within 12 in. (305 mm), measured along the sheath, of the outer end of the raceway.

(b) The raceway extends directly above the enclosure and does not penetrate a structural ceiling.

(c) A fitting is provided on each end of the raceway to protect the cable(s) from abrasion and the fittings remain accessible after installation.

(d) The raceway is sealed or plugged at the outer end using approved means so as to prevent access to the enclosure through the raceway.

(e) The cable sheath is continuous through the raceway and extends into the enclosure beyond the fitting not less than ¼ in. (6.35 mm).

(f) The raceway is fastened at its outer end and at other points in accordance with the applicable article.

(g) Where installed as conduit or tubing, the allowable cable fill does not exceed that permitted for complete conduit or tubing systems by Table 1 of Chapter 9 of this Code, and all applicable notes thereto.

    FPN: See Table 1 in Chapter 9, including Note 9, for allowable cable fill in circular raceways. See Section 310-15(b)(2)(a) for required ampacity reductions for multiple cables installed in a common raceway.

373-6. Deflection of Conductors. Conductors at terminals or conductors entering or leaving cabinets or cutout boxes and the like shall comply with (a) through (c).

Exception: Wire-bending space in enclosures for motor controllers with provisions for one or two wires per terminal shall comply with Section 430-10(b).

(a) Width of Wiring Gutters. Conductors shall not be deflected within a cabinet or cutout box unless a gutter having a width in accordance with Table 373-6(a) is provided. Conductors in parallel in accordance with Section 310-4 shall be judged on the basis of the number of conductors in parallel.

(b) Wire-Bending Space at Terminals. Wire-bending space at each terminal shall be provided in accordance with (1) or (2).

    (1) Table 373-6(a) shall apply where the conductor does not enter or leave the enclosure through the wall opposite its terminal.

    (2) Table 373-6(b) shall apply where the conductor does enter or leave the enclosure through the wall opposite its terminal.

Exception No. 1: Where the distance between the wall and its terminal is in accordance with Table 373-6(a), a conductor shall be permitted to enter or leave an enclosure through the wall opposite its terminal provided the conductor enters or leaves the enclosure where the gutter joins an adjacent gutter that has a width that conforms to Table 373-6(b) for that conductor.

Table 373-6(a). Minimum Wire-Bending Space at Terminals and Minimum Width of Wiring Gutters in Inches

Wire Size (AWG or kcmil)	Wires per Terminal				
	1	2	3	4	5
14–10	Not specified	—	—	—	—
8–6	1½	—	—	—	—
4–3	2	—	—	—	—
2	2½	—	—	—	—
1	3	—	—	—	—
1/0–2/0	3½	5	7	—	—
3/0–4/0	4	6	8	—	—
250	4½	6	8	10	—
300–350	5	8	10	12	—
400–500	6	8	10	12	14
600–700	8	10	12	14	16
750–900	8	12	14	16	18
1000–1250	10	—	—	—	—
1500–2000	12	—	—	—	—

Notes:
1. For SI units, 1 in. = 25.4 mm.
2. Bending space at terminals shall be measured in a straight line from the end of the lug or wire connector (in the direction that the wire leaves the terminal) to the wall, barrier, or obstruction.

Exception No. 2: A conductor not larger than 350 kcmil shall be permitted to enter or leave an enclosure containing only a meter socket(s) through the wall opposite its terminal, provided the distance between the terminal and the opposite wall is not less than that specified in Table 373-6(a) and the terminal is a lay-in type where the terminal is either of the following:

(a) *Directed toward the opening in the enclosure and is within a 45 degree angle of directly facing the enclosure wall*

(b) *Directly facing the enclosure wall and offset not greater than 50 percent of the bending space specified in Table 373-6(a)*

FPN: *Offset* is the distance measured along the enclosure wall from the axis of the centerline of the terminal to a line passing through the center of the opening in the enclosure.

(c) **Conductors No. 4 or Larger.** Installation shall comply with Section 300-4(f).

373-7. Space in Enclosures. Cabinets and cutout boxes shall have sufficient space to accommodate all conductors installed in them without crowding.

373-8. Enclosures for Switches or Overcurrent Devices. Enclosures for switches or overcurrent devices shall not be used as junction boxes, auxiliary gutters, or raceways for conductors feeding through or tapping off to other switches or overcurrent devices, unless adequate space for this purpose is provided. The conductors shall not fill the wiring

Table 373-6(b). Minimum Wire-Bending Space at Terminals in Inches

Wire Size (AWG or kcmil)	Wires per Terminal						4 or More	
	1		2		3			
14–10	Not specified		—		—		—	
8	1½		—		—		—	
6	2		—		—		—	
4	3		—		—		—	
3	3		—		—		—	
2	3½		—		—		—	
1	4½		—		—		—	
1/0	5½		5½		7		—	
2/0	6		6		7½		—	
3/0	6½	(½)	6½	(½)	8		—	
4/0	7	(1)	7½	(1½)	8½	(½)	—	
250	8½	(2)	8½	(2)	9	(1)	10	
300	10	(3)	10	(2)	11	(1)	12	
350	12	(3)	12	(3)	13	(3)	14	(2)
400	13	(3)	13	(3)	14	(3)	15	(3)
500	14	(3)	14	(3)	15	(3)	16	(3)
600	15	(3)	16	(3)	18	(3)	19	(3)
700	16	(3)	18	(3)	20	(3)	22	(3)
750	17	(3)	19	(3)	22	(3)	24	(3)
800	18		20		22		24	
900	19		22		24		24	
1000	20		—		—		—	
1250	22		—		—		—	
1500	24		—		—		—	
1750	24		—		—		—	
2000	24		—		—		—	

Notes:
1. For SI units, 1 in. = 25.4 mm.
2. Bending space at terminals shall be measured in a straight line from the end of the lug or wire connector in a direction perpendicular to the enclosure wall.
3. For removable and lay-in wire terminals intended for only one wire, bending space shall be permitted to be reduced by the number of inches shown in parentheses.

space at any cross section to more than 40 percent of the cross-sectional area of the space, and the conductors, splices, and taps shall not fill the wiring space at any cross section to more than 75 percent of the cross-sectional area of that space.

373-9. Side or Back Wiring Spaces or Gutters. Cabinets and cutout boxes shall be provided with back wiring spaces, gutters, or wiring compartments as required by Section 373-11(c) and (d).

B. Construction Specifications

373-10. Material. Cabinets, cutout boxes, and meter socket enclosures shall comply with (a) through (c).

(a) **Metal Cabinets and Cutout Boxes.** Metal enclosures within the scope of this article shall be protected both inside and outside against corrosion.

FPN: For information on protection against corrosion, see Section 300-6.

(b) Strength. The design and construction of enclosures within the scope of this article shall be such as to secure ample strength and rigidity. If constructed of sheet steel, the metal thickness shall not be less than 0.053 in. (1.35 mm) uncoated.

(c) Nonmetallic Cabinets. Nonmetallic cabinets shall be listed or they shall be submitted for approval prior to installation.

373-11. Spacing. The spacing within cabinets and cutout boxes shall comply with (a) through (d).

(a) General. Spacing within cabinets and cutout boxes shall be sufficient to provide ample room for the distribution of wires and cables placed in them, and for a separation between metal parts of devices and apparatus mounted within them as follows.

(1) Base. Other than at points of support, there shall be an airspace of at least 1/16 in. (1.59 mm) between the base of the device and the wall of any metal cabinet or cutout box in which the device is mounted.

(2) Doors. There shall be an airspace of at least 1 in. (25.4 mm) between any live metal part, including live metal parts of enclosed fuses, and the door.

Exception: Where the door is lined with an approved insulating material or is of a thickness of metal not less than 0.093 in. (2.36 mm) uncoated, the airspace shall not be less than 1/2 in. (12.7 mm).

(3) Live Parts. There shall be an airspace of at least 1/2 in. (12.7 mm) between the walls, back, gutter partition, if of metal, or door of any cabinet or cutout box and the nearest exposed current-carrying part of devices mounted within the cabinet where the voltage does not exceed 250. This spacing shall be increased to at least 1 in. (25.4 mm) for voltages of 251 to 600, nominal.

Exception: Where the conditions in Section 373-11(a)(2), Exception, are met, the airspace for nominal voltages from 251 to 600 shall be permitted to be not less than 1/2 in. (12.7 mm).

(b) Switch Clearance. Cabinets and cutout boxes shall be deep enough to allow the closing of the doors when 30-ampere branch-circuit panelboard switches are in any position, when combination cutout switches are in any position, or when other single-throw switches are opened as far as their construction will permit.

(c) Wiring Space. Cabinets and cutout boxes that contain devices or apparatus connected within the cabinet or box to more than eight conductors, including those of branch circuits, meter loops, feeder circuits, power circuits, and similar circuits, but not including the supply circuit or a continuation thereof, shall have back-wiring spaces or one or more side-wiring spaces, side gutters, or wiring compartments.

(d) Wiring Space — Enclosure. Side-wiring spaces, side gutters, or side-wiring compartments of cabinets and cutout boxes shall be made tight enclosures by means of covers, barriers, or partitions extending from the bases of the devices contained in the cabinet, to the door, frame, or sides of the cabinet.

Exception: Side-wiring spaces, side gutters, and side-wiring compartments of cabinets shall not be required to be made tight enclosures where those side spaces contain only conductors that enter the cabinet directly opposite to the devices where they terminate.

Partially enclosed back-wiring spaces shall be provided with covers to complete the enclosure. Wiring spaces that are required by (c), and that are exposed when doors are open, shall be provided with covers to complete the enclosure. Where adequate space is provided for feed-through conductors and for splices as required in Section 373-8, additional barriers shall not be required.

Article 374 — Auxiliary Gutters

374-1. Use. Auxiliary gutters shall be permitted to supplement wiring spaces at meter centers, distribution centers, switchboards, and similar points of wiring systems and may enclose conductors or busbars but shall not be used to enclose switches, overcurrent devices, appliances, or other similar equipment.

374-2. Extension Beyond Equipment. An auxiliary gutter shall not extend a greater distance than 30 ft (9.14 m) beyond the equipment that it supplements.

Exception: As permitted in Section 620-35 for elevators, an auxiliary gutter shall be permitted to extend a distance greater than 30 ft (9.14 m) beyond the equipment that it supplements.

FPN: For wireways, see Article 362. For busways, see Article 364.

374-3. Supports.

(a) Sheet Metal Auxiliary Gutters. Sheet metal auxiliary gutters shall be supported throughout their entire length at intervals not exceeding 5 ft (1.52 m).

(b) Nonmetallic Auxiliary Gutters. Nonmetallic auxiliary gutters shall be supported at intervals not to exceed 3 ft

(914 mm) and at each end or joint, unless listed for other support intervals. In no case shall the distance between supports exceed 10 ft (3.05 m).

374-4. Covers. Covers shall be securely fastened to the gutter.

374-5. Number of Conductors.

(a) Sheet Metal Auxiliary Gutters. The number of conductors permitted in a sheet metal auxiliary gutter shall be in accordance with (1) through (4).

(1) Sheet metal auxiliary gutters shall not contain more than 30 current-carrying conductors at any cross section. Conductors for signaling circuits or controller conductors between a motor and its starter and used only for starting duty shall not be considered as current-carrying conductors.

(2) As provided in Section 620-35 for elevators, the 30 conductor limitation shall not apply.

(3) The sum of the cross-sectional areas of all contained conductors at any cross-section of a sheet metal auxiliary gutter shall not exceed 20 percent of the interior cross-sectional area of the sheet metal auxiliary gutter.

(4) Where the 20 percent fill specified in (3) is not exceeded and the derating factors specified in Section 310-15(b)(2)(a) is applied, the number of current-carrying conductors shall not be limited.

(b) Nonmetallic Auxiliary Gutters. The sum of cross-sectional areas of all contained conductors at any cross section of the nonmetallic auxiliary gutter shall not exceed 20 percent of the interior cross-section area of the nonmetallic auxiliary gutter.

374-6. Ampacity of Conductors.

(a) Sheet Metal Auxiliary Gutters. Where the number of current-carrying conductors contained in the sheet metal auxiliary gutter is 30 or less, the correction factors specified in Section 310-15(b)(2)(a) shall not apply. The current carried continuously in bare copper bars in sheet metal auxiliary gutters shall not exceed 1000 amperes/in.^{2} (645 mm^2) of cross section of the conductor. For aluminum bars, the current carried continuously shall not exceed 700 amperes/in.^{2} (645 mm^2) of cross section of the conductor.

(b) Nonmetallic Auxiliary Gutters. The derating factors specified in Section 310-15(b)(2)(a) shall be applicable to the current-carrying conductors in the nonmetallic auxiliary gutter.

374-7. Clearance of Bare Live Parts. Bare conductors shall be securely and rigidly supported so that the minimum

clearance between bare current-carrying metal parts of different potential mounted on the same surface will not be less than 2 in. (50.8 mm), nor less than 1 in. (25.4 mm) for parts that are held free in the air. A clearance not less than 1 in. (25.4 mm) shall be secured between bare current-carrying metal parts and any metal surface. Adequate provisions shall be made for the expansion and contraction of busbars.

374-8. Splices and Taps. Splices and taps shall comply with (a) through (d).

(a) Within Gutters. Splices or taps shall be permitted within gutters where they are accessible by means of removable covers or doors. The conductors, including splices and taps, shall not fill the gutter to more than 75 percent of its area.

(b) Bare Conductors. Taps from bare conductors shall leave the gutter opposite their terminal connections, and conductors shall not be brought in contact with uninsulated current-carrying parts of different potential.

(c) Suitably Identified. All taps shall be suitably identified at the gutter as to the circuit or equipment that they supply.

(d) Overcurrent Protection. Tap connections from conductors in auxiliary gutters shall be provided with overcurrent protection as required in Section 240-21.

374-9. Construction and Installation. Auxiliary gutters shall comply with (a) through (f).

(a) Electrical and Mechanical Continuity. Gutters shall be constructed and installed so that adequate electrical and mechanical continuity of the complete system will be secured.

(b) Substantial Construction. Gutters shall be of substantial construction and shall provide a complete enclosure for the contained conductors. All surfaces, both interior and exterior, shall be suitably protected from corrosion. Corner joints shall be made tight, and where the assembly is held together by rivets, bolts, or screws, such fasteners shall be spaced not more than 12 in. (305 mm) apart.

(c) Smooth Rounded Edges. Suitable bushings, shields, or fittings having smooth, rounded edges shall be provided where conductors pass between gutters, through partitions, around bends, between gutters and cabinets or junction boxes, and at other locations where necessary to prevent abrasion of the insulation of the conductors.

(d) Deflected Insulated Conductors. Where insulated conductors are deflected within an auxiliary gutter, either at the ends or where conduits, fittings, or other raceways or cables enter or leave the gutter, or where the direction of the gutter is deflected greater than 30 degrees, dimensions corresponding to Section 373-6 shall apply.

(e) Indoor and Outdoor Use.

(1) Sheet Metal Auxiliary Gutters. Sheet metal auxiliary gutters installed in wet locations shall be suitable for such locations.

(2) Nonmetallic Auxiliary Gutters.

(a) Nonmetallic auxiliary gutters installed outdoors shall

(1) Be listed and marked as suitable for exposure to sunlight;
(2) Be listed and marked as suitable for use in wet locations;
(3) Be listed for the maximum ambient temperature of the installation, and marked for the installed conductor insulation temperature rating; and
(4) Have expansion fittings installed where the expected length change due to expansion and contraction due to temperature change is more than 0.25 in. (6.35 mm).

(b) Nonmetallic auxiliary gutters installed indoors shall

(1) Be listed for the maximum ambient temperature of the installation and marked for the installed conductor insulation temperature rating, and
(2) Have expansion fittings installed where expected length change, due to expansion and contraction due to temperature change, is more than ¼ in. (6.35 mm).

FPN: Extreme cold may cause nonmetallic auxiliary gutter to become brittle and, therefore, more susceptible to damage from physical contact.

(f) Grounding. Grounding shall be in accordance with the provisions of Article 250.

Article 380 — Switches

A. Installation

380-1. Scope. The provisions of this article shall apply to all switches, switching devices, and circuit breakers where used as switches.

380-2. Switch Connections.

(a) Three-Way and Four-Way Switches. Three-way and four-way switches shall be wired so that all switching is done only in the ungrounded circuit conductor. Where in metal raceways or metal-armored cables, wiring between switches and outlets shall be in accordance with Section 300-20(a).

Exception: Switch loops shall not require a grounded conductor.

(b) Grounded Conductors. Switches or circuit breakers shall not disconnect the grounded conductor of a circuit.

Exception: A switch or circuit breaker shall be permitted to disconnect a grounded circuit conductor where all circuit conductors are disconnected simultaneously, or where the device is arranged so that the grounded conductor cannot be disconnected until all the ungrounded conductors of the circuit have been disconnected.

380-3. Enclosure.

(a) General. Switches and circuit breakers shall be of the externally operable type mounted in an enclosure listed for the intended use. The minimum wire-bending space at terminals and minimum gutter space provided in switch enclosures shall be as required in Section 373-6.

Exception No. 1: Pendant- and surface-type snap switches and knife switches mounted on an open-face switchboard or panelboard shall be permitted without enclosures.

Exception No. 2: Switches and circuit breakers installed in accordance with Sections 110-27(a)(1), (2), (3), or (4) shall be permitted without enclosures.

(b) Used as a Raceway. Enclosures shall not be used as junction boxes, auxiliary gutters, or raceways for conductors feeding through or tapping off to other switches or overcurrent devices, unless the enclosure complies with Section 373-8.

380-4. Wet Locations. A switch or circuit breaker in a wet location or outside of a building shall be enclosed in a weatherproof enclosure or cabinet that shall comply with Section 373-2(a). Switches shall not be installed within wet locations in tub or shower spaces unless installed as part of a listed tub or shower assembly.

380-5. Time Switches, Flashers, and Similar Devices. Time switches, flashers, and similar devices shall be of the enclosed type or shall be mounted in cabinets or boxes or equipment enclosures. Energized parts shall be barriered to prevent operator exposure when making manual adjustments or switching.

Exception: Devices mounted so they are accessible only to qualified persons shall be permitted without barriers, provided they are located within an enclosure such that any energized parts within 6 in. (152 mm) of the manual adjustment or switch are covered by suitable barriers.

380-6. Position and Connection of Switches.

(a) Single-Throw Knife Switches. Single-throw knife switches shall be placed so that gravity will not tend to close them. Single-throw knife switches, approved for use in the inverted position, shall be provided with a locking device that will ensure that the blades remain in the open position when so set.

(b) Double-Throw Knife Switches. Double-throw knife switches shall be permitted to be mounted so that the throw will be either vertical or horizontal. Where the throw is vertical, a locking device shall be provided to hold the blades in the open position when so set.

(c) Connection of Switches. Single-throw knife switches and switches with butt contacts shall be connected so that the blades are de-energized when the switch is in the open position. Single-throw knife switches, molded-case switches, switches with butt contacts, and circuit breakers used as switches shall be connected so that the terminals supplying the load are de-energized when the switch is in the open position.

Exception: The blades and terminals supplying the load of a switch shall be permitted to be energized when the switch is in the open position where the switch is connected to circuits or equipment inherently capable of providing a back-feed source of power. For such installations, a permanent sign shall be installed on the switch enclosure or immediately adjacent to open switches that reads:

WARNING — LOAD SIDE TERMINALS MAY BE ENERGIZED BY BACKFEED.

380-7. Indicating. General-use and motor-circuit switches and circuit breakers, where mounted in an enclosure as described in Section 380-3, shall clearly indicate whether they are in the open (off) or closed (on) position.

Where these switch or circuit breaker handles are operated vertically rather than rotationally or horizontally, the up position of the handle shall be the (on) position.

Exception: Vertically-operated double-throw switches shall be permitted to be in the closed (on) position with the handle in either the up or down position.

380-8. Accessibility and Grouping.

(a) Location. All switches and circuit breakers used as switches shall be located so that they may be operated from a readily accessible place. They shall be installed so that the center of the grip of the operating handle of the switch or circuit breaker, when in its highest position, will not be more than 6 ft 7 in. (2.0 m) above the floor or working platform.

Exception No. 1: On busway installations, fused switches and circuit breakers shall be permitted to be located at the same level as the busway. Suitable means shall be provided to operate the handle of the device from the floor.

Exception No. 2: Switches installed adjacent to motors, appliances, or other equipment that they supply shall be permitted to be located higher than specified in the foregoing and to be accessible by portable means.

Exception No. 3: Hookstick operable isolating switches shall be permitted at greater heights.

(b) Voltage Between Adjacent Switches. Snap switches shall not be grouped or ganged in enclosures unless they can be arranged so that the voltage between adjacent switches does not exceed 300, or unless they are installed in enclosures equipped with permanently installed barriers between adjacent switches.

380-9. Provisions for Snap Switch Faceplates.

(a) Position. Snap switches mounted in boxes shall have faceplates installed so as to completely cover the opening and seat against the finished surface.

(b) Grounding. Snap switches, including dimmer switches, shall be effectively grounded and shall provide a means to ground metal faceplates, whether or not a metal faceplate is installed. Snap switches shall be considered effectively grounded if either of the following conditions are met.

(1) The switch is mounted with metal screws to a metal box or to a nonmetallic box with integral means for grounding devices.

(2) An equipment grounding conductor or equipment bonding jumper is connected to an equipment grounding termination of the snap switch.

Exception to (b): Where no grounding means exists within the snap-switch enclosure or where the wiring method does not include or provide an equipment ground, a snap switch without a grounding connection shall be permitted for replacement purposes only. A snap switch wired under the provisions of this exception and located within reach of conducting floors or other conducting surfaces shall be provided with a faceplate of nonconducting, noncombustible material.

(c) Construction. Metal faceplates shall be of ferrous metal not less than 0.030 in. (0.762 mm) in thickness or of nonferrous metal not less than 0.040 in. (1.016 mm) in thickness. Faceplates of insulating material shall be non-combustible and not less than 0.10 in. (2.54 mm) in thickness, but they shall be permitted to be less than 0.10 in. (2.54 mm) in thickness if formed or reinforced to provide adequate mechanical strength.

380-10. Mounting of Snap Switches.

(a) Surface-Type. Snap switches used with open wiring on insulators shall be mounted on insulating material that will separate the conductors at least ½ in. (12.7 mm) from the surface wired over.

(b) Box Mounted. Flush-type snap switches mounted in boxes that are set back of the wall surface as permitted in Section 370-20 shall be installed so that the extension plaster ears are seated against the surface of the wall. Flush-type snap switches mounted in boxes that are flush with the wall surface or project therefrom shall be installed so that the mounting yoke or strap of the switch is seated against the box.

380-11. Circuit Breakers as Switches. A hand-operable circuit breaker equipped with a lever or handle, or a power-operated circuit breaker capable of being opened by hand in the event of a power failure, shall be permitted to serve as a switch if it has the required number of poles.

FPN: See the provisions contained in Sections 240-81 and 240-83.

380-12. Grounding of Enclosures. Metal enclosures for switches or circuit breakers shall be grounded as specified in Article 250. Where nonmetallic enclosures are used with metal raceways or metal-armored cables, provisions shall be made for grounding continuity.

Metal boxes for switches shall be effectively grounded. Nonmetallic boxes for switches shall be installed with a wiring method that provides or includes an equipment ground.

380-13. Knife Switches.

(a) Isolating Switches. Knife switches rated at over 1200 amperes at 250 volts or less, and at over 600 amperes at 251 to 600 volts, shall be used only as isolating switches and shall not be opened under load.

(b) To Interrupt Currents. To interrupt currents over 1200 amperes at 250 volts, nominal, or less, or over 600 amperes at 251 to 600 volts, nominal, a circuit breaker or a switch of special design listed for such purpose shall be used.

(c) General-Use Switches. Knife switches of ratings less than specified in (a) and (b) shall be considered general-use switches.

FPN: See definition of *general-use switch* in Article 100.

(d) Motor-Circuit Switches. Motor-circuit switches shall be permitted to be of the knife-switch type.

FPN: See definition of a *motor-circuit switch* in Article 100.

380-14. Rating and Use of Snap Switches. Snap switches shall be used within their ratings and as indicated in (a) through (d).

FPN No. 1: For switches on signs and outline lighting, see Section 600-6.

FPN No. 2: For switches controlling motors, see Sections 430-83, 430-109, and 430-110.

(a) Alternating Current General-Use Snap Switch. A form of general-use snap switch suitable only for use on ac circuits for controlling the following:

(1) Resistive and inductive loads, including electric-discharge lamps, not exceeding the ampere rating of the switch at the voltage involved
(2) Tungsten-filament lamp loads not exceeding the ampere rating of the switch at 120 volts
(3) Motor loads not exceeding 80 percent of the ampere rating of the switch at its rated voltage

(b) Alternating-Current or Direct-Current General-Use Snap Switch. A form of general-use snap switch suitable for use on either ac or dc circuits for controlling the following.

(1) Resistive loads not exceeding the ampere rating of the switch at the voltage applied.
(2) Inductive loads not exceeding 50 percent of the ampere rating of the switch at the applied voltage. Switches rated in horsepower are suitable for controlling motor loads within their rating at the voltage applied.
(3) Tungsten-filament lamp loads not exceeding the ampere rating of the switch at the applied voltage if T-rated.

(c) CO/ALR Snap Switches. Snap switches rated 20 amperes or less directly connected to aluminum conductors shall be listed and marked CO/ALR.

(d) Alternating-Current Specific-Use Snap Switches Rated for 347 Volts. Snap switches rated 347 volts ac shall be listed and shall be used only for controlling the following.

(1) Noninductive loads other than tungsten-filament lamps not exceeding the ampere and voltage ratings of the switch.

(2) Inductive loads not exceeding the ampere and voltage ratings of the switch. Where particular load characteristics or limitations are specified as a condition of the listing, those restrictions shall be observed regardless of the ampere rating of the load.

The ampere rating of the switch shall not be less than 15 amperes at a voltage rating of 347 volts ac. Flush-type snap switches rated 347 volts ac shall not be readily interchangeable in box mounting with switches identified in Section 380-14(a) and (b).

B. Construction Specifications

380-15. Marking. Switches shall be marked with the current and voltage and, if horsepower rated, the maximum rating for which they are designed.

380-16. 600-Volt Knife Switches. Auxiliary contacts of a renewable or quick-break type or the equivalent shall be provided on all knife switches rated 600 volts designed for use in breaking current over 200 amperes.

380-17. Fused Switches. A fused switch shall not have fuses in parallel except as permitted in Section 240-8.

380-18. Wire-Bending Space. The wire-bending space required by Section 380-3 shall meet Table 373-6(b) spacings to the enclosure wall opposite the line and load terminals.

Article 384 — Switchboards and Panelboards

A. General

384-1. Scope. This article covers the following:

(1) All switchboards, panelboards, and distribution boards installed for the control of light and power circuits
(2) Battery-charging panels supplied from light or power circuits

384-2. Other Articles. Switches, circuit breakers, and overcurrent devices used on switchboards, panelboards, and distribution boards, and their enclosures, shall comply with this article and also with the requirements of Articles 240, 250, 370, 373, 380, and other articles that apply. Switchboards and panelboards in hazardous (classified) locations shall comply with the requirements of Articles 500 through 517.

384-3. Support and Arrangement of Busbars and Conductors.

(a) Conductors and Busbars on a Switchboard or Panelboard. Conductors and busbars on a switchboard or panelboard shall comply with (1), (2), and (3) as applicable.

(1) Location. Conductors and busbars shall be located so as to be free from physical damage and shall be held firmly in place.

(2) Service Switchboards. Barriers shall be placed in all service switchboards such that no uninsulated, ungrounded service busbar or service terminal will be exposed to inadvertent contact by persons or maintenance equipment while servicing load terminations.

(3) Same Vertical Section. Other than the required interconnections and control wiring, only those conductors that are intended for termination in a vertical section of a switchboard shall be located in that section.

Exception: Conductors shall be permitted to travel horizontally through vertical sections of switchboards where such conductors are isolated from busbars by a barrier.

(b) Overheating and Inductive Effects. The arrangement of busbars and conductors shall be such as to avoid overheating due to inductive effects.

(c) Used as Service Equipment. Each switchboard or panelboard, if used as service equipment, shall be provided with a main bonding jumper sized in accordance with Section 250-28(d) or the equivalent placed within the panelboard or one of the sections of the switchboard for connecting the grounded service conductor on its supply side to the switchboard or panelboard frame. All sections of a switchboard shall be bonded together using an equipment grounding conductor sized in accordance with Table 250-122.

Exception: Switchboards and panelboards used as service equipment on high-impedance grounded-neutral systems in accordance with Section 250-36 shall not be required to be provided with a main bonding jumper.

(d) Terminals. In switchboards and panelboards, load terminals for field wiring, including grounded circuit conductor load terminals and connections to the ground bus for load equipment grounding conductors, shall be located so that it will not be necessary to reach across or beyond an uninsulated ungrounded line bus in order to make connections.

(e) High-Leg Marking. On a switchboard or panelboard supplied from a 4-wire, delta-connected system where the midpoint of one phase winding is grounded, that phase busbar or conductor having the higher voltage to ground shall be durably and permanently marked by an outer finish that is orange in color or by other effective means.

(f) Phase Arrangement. The phase arrangement on 3-phase buses shall be A, B, C from front to back, top to bottom, or left to right, as viewed from the front of the switchboard or panelboard. The B phase shall be that phase having the higher voltage to ground on 3-phase, 4-wire, delta-connected systems. Other busbar arrangements shall be permitted for additions to existing installations and shall be marked.

Exception No 1: Equipment within the same single section or multisection switchboard or panelboard as the meter on 3-phase, 4-wire, delta-connected systems shall be permitted to have the same phase configuration as the metering equipment.

(g) Minimum Wire-Bending Space. The minimum wire-bending space at terminals and minimum gutter space provided in panelboards and switchboards shall be as required in Section 373-6.

384-4. Installation.

FPN: For the dedicated space requirement, see Section 110-26(f).

B. Switchboards

384-5. Location of Switchboards. Switchboards that have any exposed live parts shall be located in permanently dry locations and then only where under competent supervision and accessible only to qualified persons. Switchboards shall be located so that the probability of damage from equipment or processes is reduced to a minimum.

384-6. Switchboards in Damp or Wet Locations. Switchboards in damp or wet locations shall be installed to comply with Section 373-2(a).

384-7. Location Relative to Easily Ignitible Material. Switchboards shall be placed so as to reduce to a minimum the probability of communicating fire to adjacent combustible materials. Where installed over a combustible floor, suitable protection thereto shall be provided.

384-8. Clearances.

(a) From Ceiling. For other than a totally enclosed switchboard, a space not less than 3 ft (914 mm) shall be provided between the top of the switchboard and any combustible ceiling, unless a noncombustible shield is provided between the switchboard and the ceiling.

(b) Around Switchboards. Clearances around switchboards shall comply with the provisions of Section 110-26.

384-9. Conductor Insulation. An insulated conductor used within a switchboard shall be listed, flame retardant, and shall be rated not less than the voltage applied to it and not less than the voltage applied to other conductors or busbars with which it may come in contact.

384-10. Clearance for Conductors Entering Bus Enclosures. Where conduits or other raceways enter a switchboard, floor-standing panelboard, or similar enclosure at the bottom, sufficient space shall be provided to permit installation of conductors in the enclosure. The wiring space shall not be less than shown in Table 384-10 where the conduit or raceways enter or leave the enclosure below the busbars, their supports, or other obstructions. The conduit or raceways, including their end fittings, shall not rise more than 3 in. (76 mm) above the bottom of the enclosure.

Table 384-10. Clearance for Conductors Entering Bus Enclosures

Conductor	Minimum Spacing Between Bottom of Enclosure and Busbars, Their Supports, or Other Obstructions	
	in.	mm
Insulated busbars, their supports, or other obstructions	8	203
Noninsulated busbars	10	254

384-12. Grounding of Instruments, Relays, Meters, and Instrument Transformers on Switchboards. Instruments, relays, meters, and instrument transformers located on switchboards shall be grounded as specified in Sections 250-170 through 250-178.

C. Panelboards

384-13. General. All panelboards shall have a rating not less than the minimum feeder capacity required for the load computed in accordance with Article 220. Panelboards shall be durably marked by the manufacturer with the voltage and the current rating and the number of phases for which they are designed and with the manufacturer's name or trademark in such a manner so as to be visible after installation, without disturbing the interior parts or wiring. All panelboard circuits and circuit modifications shall be legibly identified as to purpose or use on a circuit directory located on the face or inside of the panel doors.

FPN: See Section 110-22 for additional requirements.

384-14. Classification of Panelboards. Panelboards shall be classified for the purposes of this article as either lighting and appliance branch-circuit panelboards or power panelboards.

(a) Lighting and Appliance Branch-Circuit Panelboard. A lighting and appliance branch-circuit panelboard is one having more than 10 percent of its overcurrent devices protecting lighting and appliance branch circuits. A lighting and appliance branch circuit is a branch circuit that has a connection to the neutral of the panelboard and that has overcurrent protection of 30 amperes or less in one or more conductors.

(b) Power Panelboard. A power panelboard is one having 10 percent or fewer of its overcurrent devices protecting lighting and appliance branch circuits.

384-15. Number of Overcurrent Devices on One Panelboard. Not more than 42 overcurrent devices (other than those provided for in the mains) of a lighting and appliance branch-circuit panelboard shall be installed in any one cabinet or cutout box.

A lighting and appliance branch-circuit panelboard shall be provided with physical means to prevent the installation of more overcurrent devices than that number for which the panelboard was designed, rated, and approved.

For the purposes of this article, a 2-pole circuit breaker shall be considered two overcurrent devices; a 3-pole circuit breaker shall be considered three overcurrent devices.

384-16. Overcurrent Protection.

(a) Lighting and Appliance Branch-Circuit Panelboard Individually Protected. Each lighting and appliance branch-circuit panelboard shall be individually protected on the supply side by not more than two main circuit breakers or two sets of fuses having a combined rating not greater than that of the panelboard.

Exception No. 1: Individual protection for a lighting and appliance panelboard shall not be required if the panelboard feeder has overcurrent protection not greater than the rating of the panelboard.

Exception No. 2: For existing installations, individual protection for lighting and appliance branch-circuit panelboards shall not be required where such panelboards are used as service equipment in supplying an individual residential occupancy.

(b) Power Panelboard Protection. In addition to the requirements of Section 384-13, a power panelboard with supply conductors that include a neutral and having more than 10 percent of its overcurrent devices protecting branch circuits rated 30 amperes or less shall be protected on the supply side by an overcurrent protective device having a rating not greater than that of the panelboard.

Exception: This individual protection shall not be required for a power panelboard used as service equipment with multiple disconnecting means in accordance with Section 230-71.

(c) Snap Switches Rated at 30 Amperes or Less. Panelboards equipped with snap switches rated at 30 amperes or less shall have overcurrent protection not in excess of 200 amperes.

(d) Continuous Load. The total load on any overcurrent device located in a panelboard shall not exceed 80 percent of its rating where, in normal operation, the load will continue for three hours or more.

Exception: An assembly, including the overcurrent device, shall be permitted to be used for continuous operation at 100 percent of its rating where it is listed for this purpose.

(e) Supplied Through a Transformer. Where a panelboard is supplied through a transformer, the overcurrent protection in 384-16(a), (b), and (c) shall be located on the secondary side of the transformer.

Exception: A panelboard supplied by the secondary side of a transformer shall be considered as protected by the overcurrent protection provided on the primary side of the transformer where that protection is in accordance with Section 240-21(c)(1).

(f) Delta Breakers. A 3-phase disconnect or overcurrent device shall not be connected to the bus of any panelboard that has less than 3-phase buses. Delta breakers shall not be installed in panelboards.

(g) Back-Fed Devices. Plug-in-type overcurrent protection devices or plug-in type main lug assemblies that are back fed and used to terminate field-installed ungrounded supply conductors shall be secured in place by an additional fastener that requires other than a pull to release the device from the mounting means on the panel.

384-17. Panelboards in Damp or Wet Locations. Panelboards in damp or wet locations shall be installed to comply with Section 373-2(a).

384-18. Enclosure. Panelboards shall be mounted in cabinets, cutout boxes, or enclosures designed for the purpose and shall be dead front.

Exception: Panelboards other than of the dead-front externally operable type shall be permitted where accessible only to qualified persons.

384-19. Relative Arrangement of Switches and Fuses. In panelboards, fuses of any type shall be installed on the load side of any switches.

Exception: Fuses installed as part of service equipment in accordance with the provisions of Section 230-94 shall be permitted on the line side of the service switch.

384-20. Grounding of Panelboards. Panelboard cabinets and panelboard frames, if of metal, shall be in physical contact with each other and shall be grounded. Where the panelboard is used with nonmetallic raceway or cable or where separate grounding conductors are provided, a terminal bar for the grounding conductors shall be secured inside the cabinet. The terminal bar shall be bonded to the cabinet and panelboard frame, if of metal, otherwise it shall be connected to the grounding conductor that is run with the conductors feeding the panelboard.

Exception: Where an isolated equipment grounding conductor is provided as permitted by Section 250-146(d), the insulated equipment grounding conductor that is run with the circuit conductors shall be permitted to pass through the panelboard without being connected to the panelboard's equipment grounding terminal bar.

Grounding conductors shall not be connected to a terminal bar provided for grounded conductors (may be a neutral) unless the bar is identified for the purpose and is located where interconnection between equipment grounding conductors and grounded circuit conductors is permitted or required by Article 250.

D. Construction Specifications

384-30. Panels. The panels of switchboards shall be made of moisture-resistant, noncombustible material.

384-31. Busbars. Insulated or bare busbars shall be rigidly mounted.

384-32. Protection of Instrument Circuits. Instruments, pilot lights, potential transformers, and other switchboard devices with potential coils shall be supplied by a circuit that is protected by standard overcurrent devices rated 15 amperes or less.

Exception No. 1: Overcurrent devices rated more than 15 amperes shall be permitted where the interruption of the circuit could create a hazard. Short-circuit protection shall be provided.

Exception No. 2: For ratings of 2 amperes or less, special types of enclosed fuses shall be permitted.

384-33. Component Parts. Switches, fuses, and fuse-holders used on panelboards shall comply with the applicable requirements of Articles 240 and 380.

384-34. Knife Switches. Exposed blades of knife switches shall be de-energized when open.

FPN: See Section 380-6(c), Exception, for installation.

384-35. Wire-Bending Space in Panelboards. The enclosure for a panelboard shall have the top and bottom wire-bending space sized in accordance with Table 373-6(b) for the largest conductor entering or leaving the enclosure. Side wire-bending space shall be in accordance with Table 373-6(a) for the largest conductor to be terminated in that space.

Exception No. 1: Either the top or bottom wire-bending space shall be permitted to be sized in accordance with

Table 373-6(a) for a lighting and appliance branch-circuit panelboard rated 225 amperes or less.

Exception No. 2: Either the top or bottom wire-bending space for any panelboard shall be permitted to be sized in accordance with Table 373-6(a) where at least one side wire-bending space is sized in accordance with Table 373-6(b) for the largest conductor to be terminated in any side wire-bending space.

Exception No. 3: The top and bottom wire-bending space shall be permitted to be sized in accordance with Table 373-6(a) spacings if the panelboard is designed and constructed for wiring using only one single 90 degree bend for each conductor, including the grounded circuit conductor, and the wiring diagram shows and specifies the method of wiring that shall be used.

Exception No. 4: Either the top or the bottom wire-bending space, but not both, shall be permitted to be sized in accordance with Table 373-6(a) where there are no conductors terminated in that space.

384-36. Minimum Spacings. The distance between bare metal parts, busbars, etc., shall not be less than specified in Table 384-36.

Where close proximity does not cause excessive heating, parts of the same polarity at switches, enclosed fuses, etc., shall be permitted to be placed as close together as convenience in handling will allow.

Exception: The distance shall be permitted to be less than that specified in Table 384-36 at circuit breakers and switches and in listed components installed in switchboards and panelboards.

Table 384-36. Minimum Spacings Between Bare Metal Parts

Voltage	Opposite Polarity Where Mounted on the Same Surface (in.)	Opposite Polarity Where Held Free in Air (in.)	Live Parts to Ground* (in.)
Not over 125 volts, nominal	¾	½	½
Not over 250 volts, nominal	1¼	¾	½
Not over 600 volts, nominal	2	1	1

Note: For SI units, 1 in. = 25.4 mm.

*For spacing between live parts and doors of cabinets, see Sections 373-11(a)(1), (2), and (3).

CHAPTER 4
Equipment for General Use

Article 400 — Flexible Cords and Cables

A. General

400-1. Scope. This article covers general requirements, applications, and construction specifications for flexible cords and flexible cables.

400-2. Other Articles. Flexible cords and flexible cables shall comply with this article and with the applicable provisions of other articles of this *Code*.

400-3. Suitability. Flexible cords and cables and their associated fittings shall be suitable for the conditions of use and location.

400-4. Types. Flexible cords and flexible cables shall conform to the description in Table 400-4. Types of flexible cords and flexible cables other than those listed in the table shall be the subject of special investigation.

400-5. Ampacities for Flexible Cords and Cables. Table 400-5(A) provides the allowable ampacities and Table 400-5(B) provides the ampacities for flexible cords and cables with not more than three current-carrying conductors. These tables shall be used in conjunction with applicable end-use product standards to ensure selection of the proper size and type. If the number of current-carrying conductors exceeds three, the allowable ampacity or the ampacity of each conductor shall be reduced from the 3-conductor rating as shown in the following table.

Number of Conductors	Percent of Value in Tables 400-5(A) and 400-5(B)
4–6	80
7–9	70
10–20	50
21–30	45
31–40	40
41 and above	35

Table 400-4. Flexible Cords and Cables (See Section 400-4)

Trade Name	Type Letter	Size (AWG)	Number of Conductors	Insulation	Nominal Insulation Thickness[1]		Braid on Each Conductor	Outer Covering	Use		
					AWG	Mils					
Lamp Cord	C	18–10	2 or more	Thermset or thermo-plastic	18–16 14–10	30 45	Cotton	None	Pendant or portable	Dry locations	Not hard usage
Elevator cable	E See Note 5. See Note 9. See Note 10.	20–2	2 or more	Thermoset	20–16 14–12 12–10 8–2	20 30 45 60	Cotton	Three cotton, Outer one flame-retardant & moisture-resistant. See Note 3.	Elevator lighting and control	Nonhazardous locations		
					20–16 14–12 12–10 8–2	20 30 45 60	Flexible nylon jacket					
Elevator cable	EO See Note 5. See Note 10.	20–2	2 or more	Thermoset	20–16 14–12 12–10 8–2	20 30 45 60	Cotton	Three cotton, Outer one flame-retardant & moisture-resistant. See Note 3.	Elevator lighting and control	Nonhaz-ardous locations		
								One cotton and a neo-prene jacket. See Note 3.		Hazardous (classified) locations		

(continues)

Table 400-4. (Continued)

Trade Name	Type Letter	Size (AWG)	Number of Conductors	Insulation	Nominal Insulation Thickness[1]		Braid on Each Conductor	Outer Covering	Use		
					AWG	Mils					
Elevator cable	ET See Note 5. See Note 10.	20–2	2 or more	Thermoplastic	20–16 14–12 12–10 8–2	20 30 45 60	Rayon	Three cotton or equivalent, Outer one flame-retardant & moisture-resistant See Note 3.	Nonhazardous locations		
	ETLB See Note 5. See Note 10.						None				
	ETP See Note 5. See Note 10.						Rayon	Thermoplastic	Hazardous (classified) locations		
	ETT See Note 5. See Note 10.						None	One cotton or equivalent and a thermoplastic jacket			
Portable power cable	G	8–500 kcmil	2–6 plus grounding conductor(s)	Thermoset	8–2 1–4/0 250 kcmil– 500 kcmil	60 80 95		Oil-resistant thermoset	Portable and extra hard usage		
	G-GC	8–500 kcmil	3 plus 2 grounding conductors and 1 ground check conductor	Thermoset	8–2 1–4/0 250 kcmil– 500 kcmil	60 80 95		Oil-resistant thermoset			
Heater cord	HPD	18–12	2, 3, or 4	Thermoset	18–16 14-12	15 30	None	Cotton or rayon	Portable heaters	Dry locations	Not hard usage
Parallel heater cord	HPN See Note 6.	18–12	2 or 3	Oil-resistant thermoset	18–16 14 12	45 60 95	None	Oil-resistant thermoset	Portable	Damp locations	Not hard usage
Thermoset jacketed heater cords	HS	14–12	2, 3, or 4	Thermoset	18–16	30	None	Cotton and thermoset	Portable or portable heater	Damp	Extra hard usage
	HSJ	18–12									Hard usage
	HSO	14–12						Cotton and oil-resistant thermoset			Extra hard usage
	HSJO	18–12		Oil-resistant thermoset	14–12	45					Hard usage
	HSOO	14–12									Extra hard usage
	HSJOO	18–12									Hard usage
Twisted portable cord	PD	18–10	2 or more	Thermoset or thermoplastic	18–16 14–10	30 45	Cotton	Cotton or rayon	Pendant or portable	Dry locations	Not hard usage

Table 400-4. (Continued)

Trade Name	Type Letter	Size (AWG)	Number of Conductors	Insulation	Nominal Insulation Thickness[1]		Braid on Each Conductor	Outer Covering	Use		
					AWG	Mils					
Portable power cable	PPE	8–500 kcmil	1–6 plus optional grounding conductor(s)	Thermo-plastic elastomer	8–2 1–4/0 250 kcmil–500 kcmil	60 80 95		Oil-resistant thermo-plastic elastomer	Portable, extra hard usage		
Hard service cord	S See Note 4.	18–12	2 or more	Thermoset	18–16 14–10 8–2	30 45 60	None	Thermoset	Pendant or portable	Damp locations	Extra hard usage
Flexible stage and lighting power cable	SC	8–250 kcmil	1 or more		8–2 1–4/0 250 kcmil	60 80 95		Thermoset[2]	Portable, extra hard usage		
	SCE			Thermo-plastic elastomer				Thermo-plastic elastomer[2]			
	SCT			Thermo-plastic				Thermo-plastic[2]			
Hard service cord	SE See Note 4.	18–2	2 or more	Thermo-plastic elastomer	18–16 14–10 8–2	30 45 60	None	Thermo-plastic elastomer	Pendant or portable	Damp locations	Extra hard usage
	SEO See Note 4.							Oil-resistant thermo-plastic elastomer			
	SEOO See Note 4.			Oil-resistant thermo-plastic elastomer							
Junior hard service cord	SJ	18–10	2, 3, 4, or 5	Thermoset	18–12	30	None	Thermoset	Pendant or portable	Damp locations	Hard usage
	SJE			Thermo-plastic elastomer				Thermo-plastic elastomer			
	SJEO							Oil-resistant thermo-plastic elastomer			
	SJEOO			Oil-resistant thermo-plastic elastomer							
	SJO			Thermoset				Oil-resistant thermoset			
	SJOO			Oil-resistant thermoset							
	SJT			Thermo-plastic	10	45		Thermo-plastic			

(continues)

Table 400-4. (Continued)

Trade Name	Type Letter	Size (AWG)	Number of Conductors	Insulation	Nominal Insulation Thickness[1]		Braid on Each Conductor	Outer Covering		Use	
					AWG	Mils					
Junior hard service cord	SJTO	18–10	2, 3, 4, or 5	Thermoplastic	18–12	30	None	Oil-resistant thermoplastic	Pendant or portable	Damp locations	Hard usage
	SJTOO			Oil-resistant thermoplastic	10	45					
Hard service cord	SO See Note 4.	18–2	2 or more	Thermoset	18–16	30		Oil-resistant thermoset	Pendant or portable	Damp locations	Extra hard usage
	SOO See Note 4.			Oil-resistant thermoset	14–10 8–2	45 60					
All thermoset parallel cord	SP-1 See Note 6.	20–18	2 or 3	Thermoset	20–18	30	None	Thermoset	Pendant or portable	Damp locations	Not hard usage
	SP-2 See Note 6.	18–16			18–16	45					
	SP-3 See Note 6.	18–10			18–16 14 12 10	60 80 95 110			Refrigerators, room air conditioners, and as permitted in Section 422-16(b)		
All elastomer (thermoplastic) parallel cord	SPE-1 See Note 6.	20–18	2 or 3	Thermoplastic elastomer	20–18	30	None	Thermoplastic elastomer	Pendant or portable	Damp locations	Not hard usage
	SPE-2 See Note 6.	18–16			18–16	45					
	SPE-3 See Note 6.	18–10			18–16 14 12 10	60 80 95 110			Refrigerators, room air conditioners, and as permitted in Section 422-16(b)		
All plastic parallel cord	SPT-1 See Note 6.	20–18	2 or 3	Thermoplastic	20–18	30	None	Thermoplastic	Pendant or portable	Damp locations	Not hard usage
	SPT-2 See Note 6.	18–16			18–16	45					

Table 400-4. (Continued)

Trade Name	Type Letter	Size (AWG)	Number of Conductors	Insulation	Nominal Insulation Thickness[1] AWG	Mils	Braid on Each Conductor	Outer Covering	Use		
All plastic parallel cord	SPT-3 See Note 6.	18–10	2 or 3	Thermo-plastic	18–16 14 12 10	60 80 95 110	None	Thermo-plastic	Refrig-erators, room air condi-tioners, and as permit-ted in Section 422-16(6)	Damp locations	Not hard usage
Range, dryer cable	SRD	10–4	3 or 4	Thermoset	10–4	45	None	Thermoset	Portable	Damp locations	Ranges, dryers
	SRDE	10–4	3 or 4	Thermo-plastic elastomer			None	Thermo-plastic elastomer			
	SRDT	10–4	3 or 4	Thermo-plastic			None	Thermo-plastic			
Hard service cord	ST See Note 4.	18–2	2 or More	Thermo-plastic	18–16 14–10 8–2	30 45 60	None	Thermo-plastic	Pendant or portable	Damp locations	Extra hard usage
	STO See Note 4.							Oil-resistant thermo-plastic			
	STOO See Note 4.			Oil-resistant thermo-plastic							
Vacuum cleaner cord	SV See Note 6.	18–16	2 or 3	Thermoset	18–16	15	None	Thermoset	Pendant or portable	Damp locations	Not hard usage
	SVE See Note 6.			Thermo-plastic elastomer				Thermo-plastic elastomer			
	SVEO See Note 6.							Oil-resistant thermo-plastic elastomer			
	SVEOO See Note 6.			Oil-resistant thermo-plastic elastomer							
	SVO			Thermoset				Oil-resistant thermoset			
	SVOO			Oil-resistant thermoset				Oil-resistant thermoset			
	SVT See Note 6.			Thermo-plastic				Thermo-plastic			

(continues)

Table 400-4. (Continued)

Trade Name	Type Letter	Size (AWG)	Number of Conductors	Insulation	Nominal Insulation Thickness[1]		Braid on Each Conductor	Outer Covering	Use		
					AWG	Mils					
Vacuum cleaner cord	SVTO See Note 6.	18–16	2 or 3	Thermoplastic	18–16	15	None	Oil-resistant thermoplastic	Pendant or portable	Damp locations	Not hard usage
	SVTOO			Oil-resistant thermoplastic							
Parallel tinsel cord	TPT See Note 2.	27	2	Thermoplastic	27	30	None	Thermoplastic	Attached to an appliance	Damp locations	Not hard usage
Jacketed tinsel cord	TS See Note 2.	27	2	Thermoset	27	15	None	Thermoset	Attached to an appliance	Damp locations	Not hard usage
	TST See Note 2.			Thermoplastic				Thermoplastic			
Portable power cable	W	8–500 kcmil	1–6	Thermoset	8–2 1–4/0 250 kcmil–500 kcmil	60 80 95		Oil-resistant thermoset	Portable, extra hard usage		
Electric vehicle cable	EV	18–500 kcmil See Note 11.	2 or more plus grounding conductor(s), plus optional hybrid data, signal, communications, and optical fiber cables	Thermoset with optional nylon See Note 12.	18–16 14–10 8–2 1–4/0 250 kcmil–500 kcmil	30 (20) 45 (30) 60 (45) 80 (60) 95 (75) See Note 12.	Optional	Thermoset	Electric vehicle charging	Wet locations	Extra hard usage
	EVJ	18–12 See Note 11.			18–12	30 (20) See Note 12.					Hard usage
	EVE	18–500 kcmil See Note 11.		Thermoplastic elastomer with optional nylon See Note 12.	18–16 14–10 8–2 1–4/0 250 kcmil–500 kcmil	30 (20) 45 (30) 60 (45) 80 (60) 95 (75) See Note 12.		Thermoplastic elastomer			Extra hard usage
	EVJE	18–12 See Note 11.			18–12	30 (20) See Note 12.					Hard usage

Table 400-4. (Continued)

Trade Name	Type Letter	Size (AWG)	Number of Conductors	Insulation	Nominal Insulation Thickness[1]		Braid on Each Conductor	Outer Covering		Use	
					AWG	Mils					
Electric vehicle cable	EVT	18–500 kcmil See Note 11.	2 or more plus grounding conductor(s), plus optional hybrid data, signal, communications, and optical fiber cables	Thermoplastic with optional nylon See Note 12.	18–16 14–10 8–2 1–4/0 250 kcmil– 500 kcmil	30 (20) 45 (30) 60 (45) 80 (60) 95 (75) See Note 12.	Optional	Thermoplastic	Electric vehicle charging	Wet locations	Extra hard usage
	EVJT	18–12 See Note 11.			18–12	30 (20) See Note 12.					Hard usage

[1]See Note 8.

[2]The required outer covering on some single conductor cables may be integral with the insulation.

Notes:

1. Except for Types HPN, SP-1, SP-2, SP-3, SPE-1, SPE-2, SPE-3, SPT-1, SPT-2, SPT-3, TPT, and three-conductor parallel versions of SRD, SRDE, SRDT, individual conductors are twisted together.

2. Types TPT, TS, and TST shall be permitted in lengths not exceeding 8 ft (2.44 m) where attached directly, or by means of a special type of plug, to a portable appliance rated at 50 watts or less and of such nature that extreme flexibility of the cord is essential.

3. Rubber-filled or varnished cambric tapes shall be permitted as a substitute for the inner braids.

4. Types G, G-GC, S, SC, SCE, SCT, SE, SEO, SEOO, SO, SOO, ST, STO, STOO, PPE, and W shall be permitted for use on theater stages, in garages, and elsewhere where flexible cords are permitted by this *Code*.

5. Elevator traveling cables for operating control and signal circuits shall contain nonmetallic fillers as necessary to maintain concentricity. Cables shall have steel supporting members as required for suspension by Section 620-41. In locations subject to excessive moisture or corrosive vapors or gases, supporting members of other materials shall be permitted. Where steel supporting members are used, they shall run straight through the center of the cable assembly and shall not be cabled with the copper strands of any conductor.

In addition to conductors used for control and signaling circuits, Tyes E, EO, ET, ETLB, ETP, and ETT elevator cables shall be permitted to incorporate in the construction, one or more No. 20 telephone conductor pairs, one or more coaxial cables, or one or more optical fibers. The No. 20 conductor pairs shall be permitted to be covered with suitable shielding for telephone, audio, or higher frequency communications circuits; the coaxial cables consist of a center conductor, insulation, and shield for use in video or other radio frequency communications circuits. The optical fiber shall be suitably covered with flame-retardant thermoplastic. The insulation of the conductors shall be rubber or thermoplastic of thickness not less than specified for the other conductors of the particular type of cable. Metallic shields shall have their own protective covering. Where used, these components shall be permitted to be incorporated in any layer of the cable assembly but shall not run straight through the center.

6. The third conductor in these cables shall be used for equipment grounding purposes only. The insulation of the grounding conductor for Types SPE-1, SPE-2, SPE-3, SPT-1, SPT-2, and SPT-3 shall be permitted to be thermoset polymer.

7. The individual conductors of all cords, except those of heat-resistant cords, shall have a thermoset or thermoplastic insulation, except that the equipment grounding conductor where used shall be in accordance with Section 400-23(b).

8. Where the voltage between any two conductors exceeds 300, but does not exceed 600, flexible cord of Nos. 10 and smaller shall have thermoset or thermoplastic insulation on the individual conductors at least 45 mils in thickness, unless Type S, SE, SEO, SEOO, SO, SOO, ST, STO, or STOO cord is used.

9. Insulations and outer coverings that meet the requirements as flame retardant, limited smoke, and are so listed, shall be permitted to be designated limited smoke with the suffix *LS* after the code type designation.

10. Elevator cables in sizes No. 20 through 14 are rated 300 volts, and sizes 10 through 2 are rated 600 volts. No. 12 is rated 300 volts with a 30 mil-insulation thickness and 600 volts with a 45-mil insulation thickness.

11. Conductor size for Types EV, EVJ, EVE, EVJE, EVT, and EVJT cables apply to nonpower-limited circuits only. Conductors of power-limited (data, signal, or communications) circuits may extend beyond the stated AWG size range. All conductors shall be insulated for the same cable voltage rating.

12. Insulation thickness for Types EV, EVJ, EVE, EVJE, EVT, and EVJT cables of nylon construction is indicated in parentheses.

Table 400-5(A). Allowable Ampacity for Flexible Cords and Cables [Based on Ambient Temperature of 30°C (86°F). See Section 400-13 and Table 400-4.]

Size (AWG)	Thermoset Type TS / Thermoplastic Types TPT, TST	Thermoset Types C, E, EO, PD, S, SJ, SJO, SJOO, SO, SOO, SP-1, SP-2, SP-3, SRD, SV, SVO, SVOO / Thermoplastic Types ET, ETLB, ETP, ETT, SE, SEO, SJE, SJEO, SJT, SJTO, SJTOO, SPE-1, SPE-2, SPE-3, SPT-1, SPT-2, SPT-3, ST, SRDE, SRDT, STO, STOO, SVE, SVEO, SVT, SVTO, SVTOO		Types HPD, HPN, HS, HSJ, HSO, HSJO, HSOO, HSJOO
		A†	B†	
27*	0.5	—	—	—
20	—	5**	***	—
18	—	7	10	10
17	—	—	12	—
16	—	10	13	15
15	—	—	—	17
14	—	15	18	20
12	—	20	25	30
10	—	25	30	35
8	—	35	40	—
6	—	45	55	—
4	—	60	70	—
2	—	80	95	—

*Tinsel cord

**Elevator cables only

***7 amperes for elevator cables only; 2 amperes for other types

†The allowable currents under subheading A apply to 3-conductor cords and other multiconductor cords connected to utilization equipment so that only 3 conductors are current-carrying. The allowable currents under subheading B apply to 2-conductor cords and other multiconductor cords connected to utilization equipment so that only 2 conductors are current carrying.

Ultimate Insulation Temperature. In no case shall conductors be associated together in such a way with respect to the kind of circuit, the wiring method used, or the number of conductors such that the limiting temperature of the conductors is exceeded.

A neutral conductor that carries only the unbalanced current from other conductors of the same circuit need not be considered as a current-carrying conductor.

In a 3-wire circuit consisting of two phase wires and the neutral of a 4-wire, 3-phase wye-connected system, a common conductor carries approximately the same current as the line-to-neutral currents of the other conductors and shall be considered to be a current-carrying conductor.

On a 4-wire, 3-phase wye circuit where the major portion of the load consists of nonlinear loads, there are harmonic currents present in the neutral conductor and the neutral shall be considered to be a current-carrying conductor.

An equipment grounding conductor shall not be considered a current-carrying conductor.

Where a single conductor is used for both equipment grounding and to carry unbalanced current from other conductors, as provided for in Section 250-140 for electric ranges and electric clothes dryers, it shall not be considered as a current-carrying conductor.

Exception: For other loading conditions, adjustment factors shall be permitted to be calculated under Section 310-15(c).

FPN: See Appendix B, Table B-310-11 for adjustment factors for more than three current-carrying conductors in a raceway or cable with load diversity.

400-6. Markings.

(a) Standard Markings. Flexible cords and cables shall be marked by means of a printed tag attached to the coil reel or carton. The tag shall contain the information required in Section 310-11(a). Types S, SC, SCE, SCT, SE, SEO, SEOO, SJ, SJE, SJEO, SJEOO, SJO, SJT, SJTO, SJTOO, SO, SOO, ST, STO, and STOO flexible cords and G, G-GC, PPE, and W flexible cables shall be durably marked on the surface at intervals not exceeding 24 in. (610 mm) with the type designation, size, and number of conductors.

(b) Optional Markings. Flexible cords and cable types listed in Table 400-4 shall be permitted to be surface marked to indicate special characteristics of the cable materials.

Table 400-5(B). Ampacity of Cable Types SC, SCE, SCT, PPE, G, G-GC, and W. [Based on Ambient Temperature of 30°C (86°F). See Table 400-4.]

Size (AWG or kcmil)	Temperature Rating of Cable								
	60°C (140°F)			75°C (167°F)			90°C (194°F)		
	D[1]	E[2]	F[3]	D[1]	E[2]	F[3]	D[1]	E[2]	F[3]
8	60	55	48	70	65	57	80	74	65
6	80	72	63	95	88	77	105	99	87
4	105	96	84	125	115	101	140	130	114
3	120	113	99	145	135	118	165	152	133
2	140	128	112	170	152	133	190	174	152
1	165	150	131	195	178	156	220	202	177
1/0	195	173	151	230	207	181	260	234	205
2/0	225	199	174	265	238	208	300	271	237
3/0	260	230	201	310	275	241	350	313	274
4/0	300	265	232	360	317	277	405	361	316
250	340	296	259	405	354	310	455	402	352
300	375	330	289	445	395	346	505	449	393
350	420	363	318	505	435	381	570	495	433
400	455	392	343	545	469	410	615	535	468
500	515	448	392	620	537	470	700	613	536

[1]The ampacities under subheading D shall be permitted for single-conductor Types SC, SCE, SCT, PPE, and W cable only where the individual conductors are not installed in raceways and are not in physical contact with each other except in lengths not to exceed 24 in. (610 mm) where passing through the wall of an enclosure.

[2]The ampacities under subheading E apply to two-conductor cables and other multiconductor cables connected to utilization equipment so that only two conductors are current carrying.

[3]The ampacities under subheading F apply to three-conductor cables and other multiconductor cables connected to utilization equipment so that only three conductors are current carrying.

FPN: Examples of these markings include, but are not limited to, "LS" for limited smoke and markings such as "sunlight resistant."

400-7. Uses Permitted.

(a) Uses. Flexible cords and cables shall be used only for the following:

(1) Pendants
(2) Wiring of fixtures
(3) Connection of portable lamps, portable and mobile signs, or appliances
(4) Elevator cables
(5) Wiring of cranes and hoists
(6) Connection of stationary equipment to facilitate their frequent interchange
(7) Prevention of the transmission of noise or vibration
(8) Appliances where the fastening means and mechanical connections are specifically designed to permit ready removal for maintenance and repair, and the appliance is intended or identified for flexible cord connection
(9) Data processing cables as permitted by Section 645-5
(10) Connection of moving parts
(11) Temporary wiring as permitted in Sections 305-4(b) and 305-4(c)

(b) Attachment Plugs. Where used as permitted in subsections (a)(3), (a)(6), and (a)(8), each flexible cord shall be equipped with an attachment plug and shall be energized from a receptacle outlet.

Exception: As permitted in Section 364-8.

400-8. Uses Not Permitted. Unless specifically permitted in Section 400-7, flexible cords and cables shall not be used for the following:

(1) As a substitute for the fixed wiring of a structure
(2) Where run through holes in walls, structural ceilings suspended ceilings, dropped ceilings, or floors
(3) Where run through doorways, windows, or similar openings
(4) Where attached to building surfaces

Exception: Flexible cord and cable shall be permitted to be attached to building surfaces in accordance with the provisions of Section 364-8.

(5) Where concealed behind building walls, structural ceilings, suspended ceilings, dropped ceilings, or floors
(6) Where installed in raceways, except as otherwise permitted in this *Code*

400-9. Splices. Flexible cord shall be used only in continuous lengths without splice or tap where initially installed in applications permitted by Section 400-7(a). The repair of hard-service cord and junior hard-service cord (see Trade Name column in Table 400-4) No. 14 and larger shall be permitted if conductors are spliced in accordance with Section 110-14(b) and the completed splice retains the insulation, outer sheath properties, and usage characteristics of the cord being spliced.

400-10. Pull at Joints and Terminals. Flexible cords and cables shall be connected to devices and to fittings so that tension will not be transmitted to joints or terminals.

Exception: Listed portable single pole devices that are intended to accommodate such tension at their terminals, shall be permitted to be used with single conductor flexible cable.

FPN: Some methods of preventing pull on a cord from being transmitted to joints or terminals are knotting the cord, winding with tape, and fittings designed for the purpose.

400-11. In Show Windows and Show Cases. Flexible cords used in show windows and show cases shall be Type S, SE, SEO, SEOO, SJ, SJE, SJEO, SJEOO, SJO, SJOO, SJT, SJTO, SJTOO, SO, SOO, ST, STO, or STOO.

Exception No. 1: For the wiring of chain-supported lighting fixtures.

Exception No. 2: As supply cords for portable lamps and other merchandise being displayed or exhibited.

400-12. Minimum Size. The individual conductors of a flexible cord or cable shall not be smaller than the sizes in Table 400-4.

Exception: The size of the insulated ground-check conductor of Type G-GC cables shall be not smaller than No. 10.

400-13. Overcurrent Protection. Flexible cords not smaller than No. 18, and tinsel cords or cords having equivalent characteristics of smaller size approved for use with specific appliances, shall be considered as protected against overcurrent by the overcurrent devices described in Section 240-4.

400-14. Protection from Damage. Flexible cords and cables shall be protected by bushings or fittings where passing through holes in covers, outlet boxes, or similar enclosures.

B. Construction Specifications

400-20. Labels. Flexible cords shall be examined and tested at the factory and labeled before shipment.

400-21. Nominal Insulation Thickness. The nominal thickness of insulation for conductors of flexible cords and cables shall not be less than specified in Table 400-4.

Exception: The nominal insulation thickness for the ground-check conductors of Type G-GC cables shall not be less than 45 mils for No. 8 and not less than 30 mils for No. 10.

400-22. Grounded-Conductor Identification. One conductor of flexible cords that is intended to be used as a grounded circuit conductor shall have a continuous marker that readily distinguishes it from the other conductor or conductors. The identification shall consist of one of the methods indicated in (a) through (f).

(a) Colored Braid. A braid finished to show a white or natural gray color and the braid on the other conductor or conductors finished to show a readily distinguishable solid color or colors.

(b) Tracer in Braid. A tracer in a braid of any color contrasting with that of the braid and no tracer in the braid of the other conductor or conductors. No tracer shall be used in the braid of any conductor of a flexible cord that contains a conductor having a braid finished to show white or natural gray.

Exception: In the case of Types C and PD and cords having the braids on the individual conductors finished to show white or natural gray. In such cords, the identifying marker shall be permitted to consist of the solid white or natural gray finish on one conductor provided there is a colored tracer in the braid of each other conductor.

(c) Colored Insulation. A white or natural gray insulation on one conductor and insulation of a readily distinguishable color or colors on the other conductor or conductors for cords having no braids on the individual conductors.

For jacketed cords furnished with appliances, one conductor having its insulation colored light blue, with the other conductors having their insulation of a readily distinguishable color other than white or natural gray.

Exception: Cords that have insulation on the individual conductors integral with the jacket.

The insulation shall be permitted to be covered with an outer finish to provide the desired color.

(d) Colored Separator. A white or natural gray separator on one conductor and a separator of a readily distinguishable solid color on the other conductor or conductors of cords having insulation on the individual conductors integral with the jacket.

(e) Tinned Conductors. One conductor having the individual strands tinned and the other conductor or conductors having the individual strands untinned for cords having insulation on the individual conductors integral with the jacket.

(f) Surface Marking. One or more stripes, ridges, or grooves located on the exterior of the cord so as to identify one conductor for cords having insulation on the individual conductors integral with the jacket.

400-23. Equipment Grounding Conductor Identification. A conductor intended to be used as an equipment grounding conductor shall have a continuous identifying marker readily distinguishing it from the other conductor or conductors. Conductors having a continuous green color or a continuous green color with one or more yellow stripes shall not be used for other than equipment grounding purposes. The identifying marker shall consist of one of the methods in (a) or (b).

(a) Colored Braid. A braid finished to show a continuous green color or a continuous green color with one or more yellow stripes.

(b) Colored Insulation or Covering. For cords having no braids on the individual conductors, an insulation of a continuous green color or a continuous green color with one or more yellow stripes.

400-24. Attachment Plugs. Where a flexible cord is provided with an equipment grounding conductor and equipped with an attachment plug, the attachment plug shall comply with Sections 250-138(a) and (b).

C. Portable Cables Over 600 Volts, Nominal

400-30. Scope. This part applies to multiconductor portable cables used to connect mobile equipment and machinery.

400-31. Construction.

(a) Conductors. The conductors shall be No. 8 copper or larger and shall employ flexible stranding.

Exception: The size of the insulated ground-check conductor of Type G-GC cables shall be not smaller than No. 10.

(b) Shields. Cables operated at over 2000 volts shall be shielded. Shielding shall be for the purpose of confining the voltage stresses to the insulation.

(c) Equipment Grounding Conductor(s). An equipment grounding conductor(s) shall be provided. The total area shall not be less than that of the size of the equipment grounding conductor required in Section 250-122.

400-32. Shielding. All shields shall be grounded.

400-33. Grounding. Grounding conductors shall be connected in accordance with Part E of Article 250.

400-34. Minimum Bending Radii. The minimum bending radii for portable cables during installation and handling in service shall be adequate to prevent damage to the cable.

400-35. Fittings. Connectors used to connect lengths of cable in a run shall be of a type that lock firmly together. Provisions shall be made to prevent opening or closing these connectors while energized. Suitable means shall be used to eliminate tension at connectors and terminations.

400-36. Splices and Terminations. Portable cables shall not contain splices unless the splices are of the permanent molded, vulcanized types in accordance with Section 110-14(b). Terminations on portable cables rated over 600 volts, nominal, shall be accessible only to authorized and qualified personnel.

Article 402 — Fixture Wires

402-1. Scope. This article covers general requirements and construction specifications for fixture wires.

402-2. Other Articles. Fixture wires shall comply with this article and also with the applicable provisions of other articles of this *Code*.

FPN: For application in lighting fixtures, see Article 410.

402-3. Types. Fixture wires shall be of a type listed in Table 402-3, and they shall comply with all requirements

Table 402-3. Fixture Wires

Trade Name	Type Letter	Insulation	Thickness of Insulation			Outer Covering	Maximum Operating Temperature	Application Provisions
			AWG	Mils				
Asbestos Covered Heat-resistant Fixture Wire	AF	Impregnated asbestos or moisture-resistant insulation and impregnated asbestos		Thickness of Moisture-resistant Insulation Mils	Thickness of Asbestos Mils	None	150°C 302°F	Fixture wiring — limited to 300 volts and indoor dry locations
			18–14	—	30			
				20	10			
			12–10	—	45			
				25	20			
Heat-resistant Rubber-Covered Fixture Wire — Flexible Stranding	FFH-2	Heat-resistant rubber Cross-linked synthetic polymer	18–16 18–16	— —	30 30	Nonmetallic covering	75°C 167°F	Fixture wiring

(continues)

Table 402-3. (Continued)

Trade Name	Type Letter	Insulation	Thickness of Insulation		Outer Covering	Maximum Operating Temperature	Application Provisions
			AWG	Mils			
ECTFE — Solid or 7-Strand	HF	Ethylene chloro-trifluoro-ethylene	18–14	— 15	None	150°C 302°F	Fixture wiring
ECTFE — Flexible Stranding	HFF	Ethylene chlorotrifluoro-ethylene	18–14	— 15	None	150°C 302°F	Fixture wiring
Tape Insulated Fixture Wire — Solid or 7-Strand	KF-1	Aromatic polyimide tape	18–10	— 5.5	None	200°C 392°F	Fixture wiring — limited to 300 volts
	KF-2	Aromatic polyimide tape	18–10	— 8.4	None	200°C 392°F	Fixture wiring
Tape Insulated Fixture Wire — Flexible Stranding	KFF-1	Aromatic polyimide tape	18–10	— 5.5	None	200°C 392°F	Fixture wiring — limited to 300 volts
	KFF-2	Aromatic polyimide tape	18–10	— 8.4	None	200°C 392°F	Fixture wiring
Perfluoroalkoxy — Solid or 7-Strand (Nickel or Nickel-Coated Copper)	PAF	Perfluoroalkoxy	18–14	— 20	None	250°C 482°F	Fixture wiring (nickel or nickel-coated copper)
Perfluoroalkoxy — Flexible Stranding	PAFF	Perfluoroalkoxy	18–14	— 20	None	150°C 302°F	Fixture wiring
Fluorinated Ethylene Propylene Fixture Wire — Solid or 7-Strand	PF	Fluorinated ethylene propylene	18–14	— 20	None	200°C 392°F	Fixture wiring
Fluorinated Ethylene Propylene Fixture Wire — Flexible Stranding	PFF	Fluorinated ethylene propylene	18–14	— 20	None	150°C 302°F	Fixture wiring
Fluorinated Ethylene Propylene Fixture Wire — Solid or 7-Strand	PGF	Fluorinated ethylene propylene	18–14	— 14	Glass braid	200°C 392°F	Fixture wiring
Fluorinated Ethylene Propylene Fixture Wire — Flexible Stranding	PGFF	Fluorinated ethylene propylene	18–14	— 14	Glass braid	150°C 302°F	Fixture wiring
Extruded Polytetra-Fluoroethylene — Solid or 7-Strand (Nickel or Nickel-Coated Copper)	PTF	Extruded polytetra-fluoroethylene	18–14	— 20	None	250°C 482°F	Fixture wiring (nickel or nickel-coated copper)
Extruded Polytetra-Fluoroethylene — Flexible Stranding 26-36 AWG (Silver or Nickel-Coated Copper)	PTFF	Extruded polytetra-fluoroethylene	18–14	— 20	None	150°C 302°F	Fixture wiring (silver or nickel-coated copper)

Table 402-3. (Continued)

Trade Name	Type Letter	Insulation	Thickness of Insulation			Outer Covering	Maximum Operating Temperature	Application Provisions
			AWG		Mils			
Heat-resistant Rubber-Covered Fixture Wire — Solid or 7-Strand	RFH-1	Heat-resistant rubber	18	—	15	Nonmetallic covering	75°C 167°F	Fixture wiring — limited to 300 volts
	RFH-2	Heat-resistant rubber Cross-linked synthetic polymer	18–16	—	30	None or nonmetallic covering	75°C 167°F	Fixture wiring
Heat-resistant Cross-Linked Synthetic Polymer-Insulated Fixture Wire — Solid or Stranded	RFHH-2* RFHH-3*	Cross-linked synthetic polymer	18–16 18–16	— —	30 45	None or nonmetallic covering	90°C 194°F	Fixture wiring — multiconductor cable
Silicone Insulated Fixture Wire — Solid or 7-Strand	SF-1	Silicone rubber	18	—	15	Nonmetallic covering	200°C 392°F	Fixture wiring — limited to 300 volts
	SF-2	Silicone rubber	18–14	—	30	Nonmetallic covering	200°C 392°F	Fixture wiring
Silicone Insulated Fixture Wire — Flexible Stranding	SFF-1	Silicone rubber	18	—	15	Nonmetallic covering	150°C 302°F	Fixture wiring — limited to 300 volts
	SFF-2	Silicone rubber	18–14	—	30	Nonmetallic covering	150°C 302°F	Fixture wiring
Thermoplastic Covered Fixture Wire — Solid or 7-Strand	TF*	Thermoplastic	18–16	—	30	None	60°C 140°F	Fixture wiring
Thermoplastic Covered Fixture Wire — Flexible Stranding	TFF*	Thermoplastic	18–16	—	30	None	60°C 140°F	Fixture wiring
Heat-resistant Thermoplastic-Covered Fixture Wire — Solid or 7-Strand	TFN*	Thermoplastic	18–16	—	15	Nylon-jacketed or equivalent	90°C 194°F	Fixture wiring
Heat-resistant Thermoplastic Covered Fixture Wire — Flexible Stranded	TFFN*	Thermoplastic	18–16	—	15	Nylon-jacketed or equivalent	90°C 194°F	Fixture wiring
Cross-Linked Polyolefin Insulated Fixture Wire — Solid or 7-Strand	XF*	Cross-linked polyolefin	18–14 12–10	— —	30 45	None	150°C 302°F	Fixture wiring — limited to 300 volts
Cross-Linked Polyolefin Insulated Fixture Wire — Flexible Stranded	XFF*	Cross-linked polyolefin	18–14 12–10	— —	30 45	None	150°C 302°F	Fixture wiring — limited to 300 volts
Modified ETFE — Solid or 7-Strand	ZF	Modified ethylene tetrafluoro-ethylene	18–14	—	15	None	150°C 302°F	Fixture wiring

(continues)

Table 402-3. (Continued)

Trade Name	Type Letter	Insulation	Thickness of Insulation		Outer Covering	Maximum Operating Temperature	Application Provisions
			AWG	Mils			
Flexible Stranding	ZFF	Modified ethylene tetrafluoro-ethylene	18–14	— 15	None	150°C 302°F	Fixture wiring
High Temp. Modified ETFE — Solid or 7-Strand	ZHF	Modified ethylene tetrafluoro-ethylene	18–14	— 15	None	200°C 392°F	Fixture wiring

*Insulations and outer coverings that meet the requirements of flame retardant, limited smoke and are so listed shall be permitted to be designated limited smoke with the suffix *LS* after the code type designation.

of that table. The fixture wires listed in Table 402-3 are all suitable for service at 600 volts, nominal, unless otherwise specified.

FPN: Thermoplastic insulation may stiffen at temperatures colder than -10°C (+14°F), requiring that care be exercised during installation at such temperatures. Thermoplastic insulation may also be deformed at normal temperatures where subjected to pressure, requiring care be exercised during installation and at points of support.

402-5. Allowable Ampacities for Fixture Wires. The allowable ampacity of fixture wire shall be as specified in Table 402-5.

No conductor shall be used under such conditions that its operating temperature will exceed the temperature specified in Table 402-3 for the type of insulation involved.

FPN: See Section 310-10 for temperature limitation of conductors.

Table 402-5. Allowable Ampacity for Fixture Wires

Size (AWG)	Allowable Ampacity
18	6
16	8
14	17
12	23
10	28

402-6. Minimum Size. Fixture wires shall not be smaller than No. 18.

402-7. Number of Conductors in Conduit or Tubing. The number of fixture wires permitted in a single conduit or tubing shall not exceed the percentage fill specified in Table 1, Chapter 9.

402-8. Grounded Conductor Identification. One conductor of fixture wires that is intended to be used as a grounded conductor shall be identified by means of stripes or by the means described in Sections 400-22(a) through (e).

402-9. Marking.

(a) Required Information. All fixture wires shall be marked to indicate the information required in Section 310-11(a).

(b) Method of Marking. Thermoplastic insulated fixture wire shall be durably marked on the surface at intervals not exceeding 24 in. (610 mm). All other fixture wire shall be marked by means of a printed tag attached to the coil, reel, or carton.

(c) Optional Marking. Fixture wire types listed in Table 402-3 shall be permitted to be surface marked to indicate special characteristics of the cable materials.

FPN: Examples of these markings include, but are not limited to, "LS" for limited smoke or markings such as "sunlight resistant."

402-10. Uses Permitted. Fixture wires shall be permitted (1) for installation in lighting fixtures and in similar equipment where enclosed or protected and not subject to bending or twisting in use, or (2) for connecting lighting fixtures to the branch-circuit conductors supplying the fixtures.

402-11. Uses Not Permitted. Fixture wires shall not be used as branch-circuit conductors.

Exception: As permitted by Section 725-27 for Class 1 circuits and Section 760-27 for fire alarm circuits.

402-12. Overcurrent Protection. Overcurrent protection for fixture wires shall be as specified in Section 240-4.

Article 410 — Lighting Fixtures, Lampholders, Lamps, and Receptacles

A. General

410-1. Scope. This article covers lighting fixtures, lampholders, pendants, receptacles, incandescent filament lamps, arc lamps, electric-discharge lamps, the wiring and equipment forming part of such lamps, fixtures, and lighting installations.

> FPN: The international term for a lighting fixture is *luminaire* and is defined as a complete lighting unit consisting of a lamp or lamps together with the parts designed to distribute the light, to position and protect the lamps, and to connect the lamps to the power supply.

410-2. Application of Other Articles. Equipment for use in hazardous (classified) locations shall conform to Articles 500 through 517. Lighting systems operating at 30 volts or less shall conform to Article 411. Arc lamps used in theaters shall comply with Section 520-61, and arc lamps used in projection machines shall comply with Section 540-20. Arc lamps used on constant-current systems shall comply with the general requirements of Article 490.

410-3. Live Parts. Fixtures, lampholders, lamps, and receptacles shall have no live parts normally exposed to contact. Exposed accessible terminals in lampholders, receptacles, and switches shall not be installed in metal fixture canopies or in open bases of portable table or floor lamps.

Exception: Cleat-type lampholders and receptacles located at least 8 ft (2.44 m) above the floor shall be permitted to have exposed terminals.

B. Fixture Locations

410-4. Fixtures in Specific Locations.

(a) Wet and Damp Locations. Fixtures installed in wet or damp locations shall be installed so that water cannot enter or accumulate in wiring compartments, lampholders, or other electrical parts. All fixtures installed in wet locations shall be marked, "Suitable for Wet Locations." All fixtures installed in damp locations shall be marked, "Suitable for Wet Locations" or "Suitable for Damp Locations."

(b) Corrosive Locations. Fixtures installed in corrosive locations shall be of a type suitable for such locations.

(c) In Ducts or Hoods. Fixtures shall be permitted to be installed in commercial cooking hoods where all of the following conditions are met.

(1) The fixture shall be identified for use within commercial cooking hoods and installed so that the temperature limits of the materials used are not exceeded.

(2) The fixture shall be constructed so that all exhaust vapors, grease, oil, or cooking vapors are excluded from the lamp and wiring compartment. Diffusers shall be resistant to thermal shock.

(3) Parts of the fixture exposed within the hood shall be corrosion resistant or protected against corrosion, and the surface shall be smooth so as not to collect deposits and facilitate cleaning.

(4) Wiring methods and materials supplying the fixture(s) shall not be exposed within the cooking hood.

> FPN: See Section 110-11 for conductors and equipment exposed to deteriorating agents.

(d) Bathtub and Shower Areas. No parts of cord-connected fixtures, hanging fixtures, lighting track, pendants, or ceiling-suspended (paddle) fans shall be located within a zone measured 3 ft (914 mm) horizontally and 8 ft (2.44 m) vertically from the top of the bathtub rim or shower stall threshold. This zone is all encompassing and includes the zone directly over the tub or shower stall.

410-5. Fixtures Near Combustible Material. Fixtures shall be constructed, or installed, or equipped with shades or guards so that combustible material will not be subjected to temperatures in excess of 90°C (194°F).

410-6. Fixtures Over Combustible Material. Lampholders installed over highly combustible material shall be of the unswitched type. Unless an individual switch is provided for each fixture, lampholders shall be located at least 8 ft (2.44 m) above the floor, or shall be located or guarded so that the lamps cannot be readily removed or damaged.

410-7. Fixtures in Show Windows. Chain-supported fixtures used in a show window shall be permitted to be externally wired. No other externally wired fixtures shall be used.

410-8. Fixtures in Clothes Closets.

(a) Definition.

Storage Space. *Storage space shall be defined as a volume bounded by the sides and back closet walls and planes extending from the closet floor vertically to a height of 6 ft (1.83 m) or the highest clothes-hanging rod and parallel to the walls at a horizontal distance of 24 in. (610 mm) from the sides and back of the closet walls respectively, and continuing vertically to the closet ceiling parallel to the walls at a horizontal distance of 12 in. (305 mm) or the width of the shelf, whichever is greater.*

FPN: See Figure 410-8.

For SI units: 1 in. = 25.4 mm; 1 ft = 0.3048 m.

Figure 410-8   Closet storage space.

For a closet that permits access to both sides of a hanging rod, the storage space shall include the volume below the highest rod extending 12 in. (305 mm) on either side of the rod on a plane horizontal to the floor extending the entire length of the rod.

(b) Fixture Types Permitted. Listed fixtures of the following types shall be permitted to be installed in a closet:

(1) A surface-mounted or recessed incandescent fixture with a completely enclosed lamp
(2) A surface-mounted or recessed fluorescent fixture

(c) Fixture Types Not Permitted. Incandescent fixtures with open or partially enclosed lamps and pendant fixtures or lampholders shall not be permitted.

(d) Location. Fixtures in clothes closets shall be permitted to be installed as follows.

(1) Surface-mounted incandescent fixtures installed on the wall above the door or on the ceiling, provided there is a minimum clearance of 12 in. (305 mm) between the fixture and the nearest point of a storage space.

(2) Surface-mounted fluorescent fixtures installed on the wall above the door or on the ceiling, provided there is a minimum clearance of 6 in. (152 mm) between the fixture and the nearest point of a storage space.

(3) Recessed incandescent fixtures with a completely enclosed lamp installed in the wall or the ceiling, provided there is a minimum clearance of 6 in. (152 mm) between the fixture and the nearest point of a storage space.

(4) Recessed fluorescent fixtures installed in the wall or on the ceiling, provided there is a minimum clearance of 6 in. (152 mm) between the fixture and the nearest point of a storage space.

410-9. Space for Cove Lighting. Coves shall have adequate space and shall be located so that lamps and equipment can be properly installed and maintained.

C. Provisions at Fixture Outlet Boxes, Canopies, and Pans

410-10. Space for Conductors. Canopies and outlet boxes taken together shall provide adequate space so that fixture conductors and their connecting devices can be properly installed.

410-11. Temperature Limit of Conductors in Outlet Boxes. Fixtures shall be of such construction or installed so that the conductors in outlet boxes shall not be subjected to temperatures greater than that for which the conductors are rated.

Branch-circuit wiring, other than 2-wire or multiwire branch-circuits supplying power to fixtures connected together, shall not be passed through an outlet box that is an integral part of a fixture unless the fixture is identified for through-wiring.

FPN: See Section 410-31, Exceptions No. 2 and 3 for circuits supplying power to fixtures connected together.

410-12. Outlet Boxes to Be Covered. In a completed installation, each outlet box shall be provided with a cover unless covered by means of a fixture canopy, lampholder, receptacle, or similar device.

410-13. Covering of Combustible Material at Outlet Boxes. Any combustible wall or ceiling finish exposed between the edge of a fixture canopy or pan and an outlet box shall be covered with noncombustible material.

410-14. Connection of Electric-Discharge Lighting Fixtures.

(a) Independent of the Outlet Box. Electric-discharge lighting fixtures supported independent of the outlet box shall be connected to the branch circuit through metal raceway, nonmetallic raceway, Type MC cable, Type AC cable, Type MI cable, nonmetallic sheathed cable, or by flexible cord as permitted in Section 410-30(b) or (c).

(b) Access to Boxes. Electric-discharge lighting fixtures surface mounted over concealed outlet, pull, or junction boxes shall be installed with suitable openings in back of the fixture to provide access to the boxes.

D. Fixture Supports

410-15. Supports.

(a) General. Fixtures, lampholders, and receptacles shall be securely supported. A fixture that weighs more than 6 lb (2.72 kg) or exceeds 16 in. (406 mm) in any dimension shall not be supported by the screw shell of a lampholder.

(b) Metal Poles Supporting Lighting Fixtures. Metal poles shall be permitted to be used to support lighting fixtures and as a raceway to enclose supply conductors, provided the following conditions are met.

(1) A metal pole shall have a handhole not less than 2 in. × 4 in. (50.8 mm × 102 mm) with a raintight cover to provide access to the supply terminations within the pole or pole base.

Exception No 1: No handhole shall be required in a pole 8 ft (2.44 m) or less in height above grade where the supply wiring method continues without splice or pull point, and where the interior of the pole and any splices are accessible by removing the fixture.

Exception No. 2: No handhole shall be required in a metal pole 20 ft (6.10 m) or less in height above grade that is provided with a hinged base.

(2) Where raceway risers or cable is not installed within the pole, a threaded fitting or nipple shall be brazed or welded to the pole opposite the handhole for the supply connection.

(3) A metal pole shall be provided with a grounding terminal.

(a) A pole with a handhole shall have the grounding terminal accessible from the handhole.

(b) A pole with a hinged base shall have the grounding terminal accessible within the base.

Exception: No grounding terminal shall be required in a pole 8 ft (2.44 m) or less in height above grade where the supply wiring method continues without splice or pull, and where the interior of the pole and any splices are accessible by removing the fixture.

(4) A pole with a hinged base shall have the hinged base and pole bonded together.

(5) Metal raceways or other equipment grounding conductors shall be bonded to the pole with an equipment grounding conductor recognized by Section 250-118 and sized in accordance with Section 250-122.

(6) Conductors in vertical metal poles used as raceway shall be supported as provided in Section 300-19.

410-16. Means of Support.

(a) Outlet Boxes. Outlet boxes or fittings installed as required by Section 370-23 shall be permitted to support fixtures weighing 50 lb (22.7 kg) or less. A fixture that weighs more than 50 lb (22.7 kg) shall be supported independent of the outlet box unless the outlet box is listed for the weight to be supported.

(b) Inspection. Fixtures shall be installed so that the connections between the fixture conductors and the circuit conductors can be inspected without requiring the disconnection of any part of the wiring unless the fixtures are connected by attachment plugs and receptacles.

(c) Suspended Ceilings. Framing members of suspended ceiling systems used to support fixtures shall be securely fastened to each other and shall be securely attached to the building structure at appropriate intervals. Fixtures shall be securely fastened to the ceiling framing member by mechanical means such as bolts, screws, or rivets. Listed clips identified for use with the type of ceiling framing member(s) and fixture(s) shall also be permitted.

(d) Fixture Studs. Fixture studs that are not a part of outlet boxes, hickeys, tripods, and crowfeet shall be made of steel, malleable iron, or other material suitable for the application.

(e) Insulating Joints. Insulating joints that are not designed to be mounted with screws or bolts shall have an exterior metal casing, insulated from both screw connections.

(f) Raceway Fittings. Raceway fittings used to support a lighting fixture(s) shall be capable of supporting the weight of the complete fixture assembly and lamp(s).

(g) Busways. Fixtures shall be permitted to be connected to busways in accordance with Section 364-12.

(h) Trees. Outdoor lighting fixtures and associated equipment shall be permitted to be supported by trees.

FPN No. 1: See Section 225-26 for restrictions for support of overhead conductors.

FPN No. 2: See Section 300-5(d) for protection of conductors.

E. Grounding

410-17. General. Fixtures and lighting equipment shall be grounded as required in Article 250 and Part E of this article.

410-18. Exposed Fixture Parts.

(a) Exposed Conductive Parts. Exposed metal parts shall be grounded or insulated from ground and other conducting surfaces or inaccessible to unqualified personnel. Lamp tie wires, mounting screws, clips, and decorative bands on glass spaced at least 1½ in. (38 mm) from lamp terminals shall not be required to be grounded.

(b) Made of Insulating Material. Fixtures directly wired or attached to outlets supplied by a wiring method that does not provide a ready means for grounding shall be made of insulating material and shall have no exposed conductive parts.

410-20. Equipment Grounding Conductor Attachment. Fixtures with exposed metal parts shall be provided with a means for connecting an equipment grounding conductor for such fixtures.

410-21. Methods of Grounding. Fixtures and equipment shall be considered grounded where mechanically connected to an equipment grounding conductor as specified in Section 250-118 and sized in accordance with Section 250-122.

F. Wiring of Fixtures

410-22. Fixture Wiring — General. Wiring on or within fixtures shall be neatly arranged and shall not be exposed to physical damage. Excess wiring shall be avoided. Conductors shall be arranged so that they shall not be subjected to temperatures above those for which they are rated.

410-23. Polarization of Fixtures. Fixtures shall be wired so that the screw shells of lampholders will be connected to the same fixture or circuit conductor or terminal. The grounded conductor, where connected to a screw-shell lampholder, shall be connected to the screw shell.

410-24. Conductor Insulation. Fixtures shall be wired with conductors having insulation suitable for the environmental conditions, current, voltage, and temperature to which the conductors will be subjected.

> FPN: For ampacity of fixture wire, maximum operating temperature, voltage limitations, minimum wire size, etc., see Article 402.

410-27. Pendant Conductors for Incandescent Filament Lamps.

(a) Support. Pendant lampholders with permanently attached leads, where used for other than festoon wiring, shall be hung from separate stranded rubber-covered conductors that are soldered directly to the circuit conductors but supported independently thereof.

(b) Size. Unless part of listed decorative lighting assemblies, pendant conductors shall not be smaller than No. 14 for mogul-base or medium-base screw-shell lampholders, nor smaller than No. 18 for intermediate or candelabra-base lampholders.

(c) Twisted or Cabled. Pendant conductors longer than 3 ft (914 mm) shall be twisted together where not cabled in a listed assembly.

410-28. Protection of Conductors and Insulation.

(a) Properly Secured. Conductors shall be secured in a manner that will not tend to cut or abrade the insulation.

(b) Protection Through Metal. Conductor insulation shall be protected from abrasion where it passes through metal.

(c) Fixture Stems. Splices and taps shall not be located within fixture arms or stems.

(d) Splices and Taps. No unnecessary splices or taps shall be made within or on a fixture.

> FPN: For approved means of making connections, see Section 110-14.

(e) Stranding. Stranded conductors shall be used for wiring on fixture chains and on other movable or flexible parts.

(f) Tension. Conductors shall be arranged so that the weight of the fixture or movable parts will not put a tension on the conductors.

410-29. Cord-Connected Showcases. Individual showcases, other than fixed, shall be permitted to be connected by flexible cord to permanently installed receptacles, and groups of not more than six such showcases shall be permitted to be coupled together by flexible cord and separable locking-type connectors with one of the group connected by flexible cord to a permanently installed receptacle.

The installation shall comply with (a) through (e) of this section.

(a) Cord Requirements. Flexible cord shall be of the hard-service type, having conductors not smaller than the branch-circuit conductors, having ampacity at least equal to the branch-circuit overcurrent device, and having an equipment grounding conductor.

> FPN: See Table 250-122 for size of equipment grounding conductor.

(b) Receptacles, Connectors, and Attachment Plugs. Receptacles, connectors, and attachment plugs shall be of a listed grounding type rated 15 or 20 amperes.

(c) Support. Flexible cords shall be secured to the undersides of showcases so that (1) wiring will not be exposed to mechanical damage; (2) a separation between cases not in excess of 2 in. (50.8 mm), nor more than 12 in. (305 mm) between the first case and the supply receptacle, will be ensured; and (3) the free lead at the end of a group of showcases will have a female fitting not extending beyond the case.

(d) No Other Equipment. Equipment other than showcases shall not be electrically connected to showcases.

(e) Secondary Circuit(s). Where showcases are cord-connected, the secondary circuit(s) of each electric-discharge lighting ballast shall be limited to one showcase.

410-30. Cord-Connected Lampholders and Fixtures.

(a) Lampholders. Where a metal lampholder is attached to a flexible cord, the inlet shall be equipped with an insulating bushing that, if threaded, shall not be smaller than nominal ⅜-in. pipe size. The cord hole shall be of a size appropriate for the cord, and all burrs and fins shall be removed in order to provide a smooth bearing surface for the cord.

Bushing having holes 9/32 in. (7.14 mm) in diameter shall be permitted for use with plain pendant cord and holes 13/32 in. (10.3 mm) in diameter with reinforced cord.

(b) Adjustable Fixtures. Fixtures that require adjusting or aiming after installation shall not be required to be equipped with an attachment plug or cord connector provided the exposed cord is of the hard usage or extra-hard usage type and is not longer than that required for maximum adjustment. The cord shall not be subject to strain or physical damage.

(c) Electric-Discharge Fixtures.

(1) A listed fixture or a listed assembly shall be permitted to be cord connected if

(a) The fixture is located directly below the outlet box or busway, and

(b) The flexible cord is

  (1) Visible for its entire length outside the fixture,
  (2) Not subject to strain or physical damage, and
  (3) Terminated in a grounding-type attachment plug cap, busway plug or have a fixture assembly with a strain relief and canopy.

(2) Electric-discharge lighting fixtures provided with mogul-base, screw-shell lampholders shall be permitted to be connected to branch circuits of 50 amperes or less by cords complying with Section 240-4. Receptacles and attachment plugs shall be permitted to be of lower ampere rating than the branch circuit but not less than 125 percent of the fixture full-load current.

(3) Electric-discharge lighting fixtures equipped with a flanged surface inlet shall be permitted to be supplied by cord pendants equipped with cord connectors. Inlets and connectors shall be permitted to be of lower ampere rating than the branch circuit but not less than 125 percent of the fixture load current.

410-31. Fixtures as Raceways. Fixtures shall not be used as a raceway for circuit conductors.

Exception No. 1: Fixtures listed for use as a raceway.

Exception No. 2: Fixtures designed for end-to-end assembly to form a continuous raceway or fixtures connected together by recognized wiring methods shall be permitted to carry through conductors of a 2-wire or multiwire branch circuit supplying the fixtures.

Exception No. 3: One additional 2-wire branch circuit separately supplying one or more of the connected fixtures described in Exception No. 2 shall be permitted to be carried through the fixtures.

(FPN): See Article 100 for definition of "Multiwire Branch Circuit."

Branch-circuit conductors within 3 in. (76 mm) of a ballast within the ballast compartment shall have an insulation temperature rating not lower than 90°C (194°F), such as Types RHH, THW, THHN, THHW, FEP, FEPB, SA, and XHHW.

G. Construction of Fixtures

410-34. Combustible Shades and Enclosures. Adequate airspace shall be provided between lamps and shades or other enclosures of combustible material.

410-35. Fixture Rating.

(a) Marking. All fixtures shall be marked with the maximum lamp wattage or electrical rating, manufacturer's name, trademark, or other suitable means of identification. A fixture requiring supply wire rated higher than 60°C (140°F) shall be marked in letters not smaller than ¼ in. (6.35 mm) high, prominently displayed on the fixture and shipping carton or equivalent.

(b) Electrical Rating. The electrical rating shall include the voltage and frequency and shall indicate the current rating of the unit, including the ballast, transformer, or autotransformer.

410-36. Design and Material. Fixtures shall be constructed of metal, wood, or other material suitable for the application and shall be designed and assembled so as to secure requisite mechanical strength and rigidity. Wiring compartments, including their entrances, shall be such that conductors may be drawn in and withdrawn without physical damage.

410-37. Nonmetallic Fixtures. When fixture wiring compartments are constructed from combustible material, armored or lead-covered conductors with suitable fittings shall be used or the wiring compartment shall be lined with metal.

410-38. Mechanical Strength.

(a) Tubing for Arms. Tubing used for arms and stems where provided with cut threads shall not be less than 0.040 in. (1.016 mm) in thickness and where provided with rolled (pressed) threads shall not be less than 0.025 in. (0.635 mm) in thickness. Arms and other parts shall be fastened to prevent turning.

(b) Metal Canopies. Metal canopies supporting lampholders, shades, etc., exceeding 8 lb (3.63 kg), or incorporating attachment-plug receptacles, shall not be less than 0.020 in. (0.508 mm) in thickness. Other canopies shall not be less than 0.016 in. (0.4064 mm) if made of steel and not less than 0.020 in. (0.508 mm) if of other metals.

(c) Canopy Switches. Pull-type canopy switches shall not be inserted in the rims of metal canopies that are less than 0.025 in. (0.635 mm) in thickness unless the rims are reinforced by the turning of a bead or the equivalent. Pull-type canopy switches, whether mounted in the rims or elsewhere in sheet metal canopies, shall not be located more than 3½ in. (89 mm) from the center of the canopy. Double setscrews, double canopy rings, a screw ring, or equal method shall be used where the canopy supports a pull-type switch or pendant receptacle.

The above thickness requirements shall apply to measurements made on finished (formed) canopies.

410-39. Wiring Space. Bodies of fixtures, including portable lamps, shall provide ample space for splices and taps and for the installation of devices, if any. Splice compartments shall be of nonabsorbent, noncombustible material.

410-42. Portable Lamps.

(a) General. Portable lamps shall be wired with flexible cord recognized by Section 400-4 and an attachment plug of the polarized or grounding type. Where used with Edison-base lampholders, the grounded conductor shall be identified and attached to the screw shell and the identified blade of the attachment plug.

(b) Portable Handlamps. In addition to the provisions of Section 410-42(a), portable handlamps shall comply with the following.

(1) Metal shell, paper-lined lampholders shall not be used.
(2) Handlamps shall be equipped with a handle of molded composition or other insulating material.
(3) Handlamps shall be equipped with a substantial guard attached to the lampholder or handle.
(4) Metallic guards shall be grounded by the means of an equipment grounding conductor run with circuit conductors within the power-supply cord.
(5) Portable handlamps shall not be required to be grounded where supplied through an isolating transformer with an ungrounded secondary of not over 50 volts.

410-44. Cord Bushings. A bushing or the equivalent shall be provided where flexible cord enters the base or stem of a portable lamp. The bushing shall be of insulating material unless a jacketed type of cord is used.

410-45. Tests. All wiring shall be free from short circuits and grounds and shall be tested for these defects prior to being connected to the circuit.

410-46. Live Parts. Exposed live parts within porcelain fixtures shall be suitably recessed and located so as to make it improbable that wires will come in contact with them. There shall be a spacing of at least ½ in. (12.7 mm) between live parts and the mounting plane of the fixture.

H. Installation of Lampholders

410-47. Screw-Shell Type. Lampholders of the screw-shell type shall be installed for use as lampholders only. Where supplied by a circuit having a grounded conductor, the grounded conductor shall be connected to the screw shell.

410-48. Double-Pole Switched Lampholders. Where supplied by the ungrounded conductors of a circuit, the switching device of lampholders of the switched type shall simultaneously disconnect both conductors of the circuit.

410-49. Lampholders in Wet or Damp Locations. Lampholders installed in wet or damp locations shall be of the weatherproof type.

J. Construction of Lampholders

410-50. Insulation. The outer metal shell and the cap shall be lined with insulating material that shall prevent the shell and cap from becoming a part of the circuit. The lining shall not extend beyond the metal shell more than ⅛ in. (3.17 mm) but shall prevent any current-carrying part of the lamp base from being exposed when a lamp is in the lampholding device.

410-52. Switched Lampholders. Switched lampholders shall be of such construction that the switching mechanism interrupts the electrical connection to the center contact. The switching mechanism shall also be permitted to interrupt the electrical connection to the screw shell if the connection to the center contact is simultaneously interrupted.

K. Lamps and Auxiliary Equipment

410-53. Bases, Incandescent Lamps. An incandescent lamp for general use on lighting branch circuits shall not be equipped with a medium base if rated over 300 watts, nor with a mogul base if rated over 1500 watts. Special bases or other devices shall be used for over 1500 watts.

410-54. Electric-Discharge Lamp Auxiliary Equipment.

(a) Enclosures. Auxiliary equipment for electric-discharge lamps shall be enclosed in noncombustible cases and treated as sources of heat.

(b) Switching. Where supplied by the ungrounded conductors of a circuit, the switching device of auxiliary equipment shall simultaneously disconnect all conductors.

L. Receptacles, Cord Connectors, and Attachment Plugs (Caps)

410-56. Rating and Type.

(a) Receptacles. Receptacles installed for the attachment of portable cords shall be rated at not less than 15 amperes, 125 volts, or 15 amperes, 250 volts, and shall be of a type not suitable for use as lampholders.

(b) CO/ALR Receptacles. Receptacles rated 20 amperes or less and directly connected to aluminum conductors shall be marked CO/ALR.

(c) Isolated Ground Receptacles. Receptacles intended for the reduction of electrical noise (electromagnetic interference) as permitted in Section 250-146(d) shall be identified by an orange triangle located on the face of the receptacle. Receptacles so identified shall be used only with grounding conductors that are isolated in accordance with Section 250-146(d). Isolated ground receptacles installed in nonmetallic boxes shall be covered with a nonmetallic faceplate.

Exception: Where an isolated ground receptacle is installed in a nonmetallic box, a metal faceplate shall be permitted if the box contains a feature or accessory that permits the effective grounding of the faceplate.

(d) Faceplates. Metal faceplates shall be of ferrous metal not less than 0.030 in. (0.762 mm) in thickness or of nonferrous metal not less than 0.040 in. (1.016 mm) in thickness.

Metal faceplates shall be grounded. Faceplates of insulating material shall be noncombustible and not less than 0.10 in. (2.54 mm) in thickness but shall be permitted to be less than 0.10 in. (2.54 mm) in thickness if formed or reinforced to provide adequate mechanical strength.

(e) Position of Receptacle Faces. After installation, receptacle faces shall be flush with or project from faceplates of insulating material and shall project a minimum of 0.015 in. (0.381 mm) from metal faceplates. Faceplates shall be installed so as to completely cover the opening and seat against the mounting surface.

(f) Receptacle Mounting.

(1) Receptacles mounted in boxes that are set back of the wall surface, as permitted in Section 370-20, shall be installed so that the mounting yoke or strap of the receptacle is held rigidly at the surface of the wall.

(2) Receptacles mounted in boxes that are flush with the wall surface or project therefrom shall be installed so that the mounting yoke or strap of the receptacle is seated against the box or raised box cover.

(3) Receptacles Mounted on Covers. Receptacles mounted to and supported by a cover shall be secured by more than one screw or shall be a device assembly or box cover listed and identified for securing by a single screw.

(g) Attachment Plugs. All 15- and 20-ampere attachment plugs and connectors shall be constructed so that there are no exposed current-carrying parts except the prongs, blades, or pins. The cover for wire terminations shall be a part, which is essential for the operation of an attachment plug or connector (dead-front construction). Attachment plugs shall be installed so that their prongs, blades, or pins are not energized unless inserted into an energized receptacle. No receptacle shall be installed so as to require an energized attachment plug as its source of supply.

(h) Attachment Plug Ejector Mechanisms. Attachment plug ejector mechanisms shall not adversely affect engagement of the blades of the attachment plug with the contacts of the receptacle.

(i) Noninterchangeability. Receptacles, cord connectors, and attachment plugs shall be constructed so that receptacle or cord connectors will not accept an attachment plug with a different voltage or current rating than that for which the device is intended; however, a 20-ampere T-slot receptacle or cord connector shall be permitted to accept a 15-ampere attachment plug of the same voltage rating. Nongrounding-type receptacles and connectors shall not accept grounding-type attachment plugs.

410-57. Receptacles in Damp or Wet Locations.

(a) Damp Locations. A receptacle installed outdoors in a location protected from the weather or in other damp locations shall have an enclosure for the receptacle that is weatherproof when the receptacle is covered (attachment plug cap not inserted and receptacle covers closed).

An installation suitable for wet locations shall also be considered suitable for damp locations.

A receptacle shall be considered to be in a location protected from the weather where located under roofed open porches, canopies, marquees, and the like, and will not be subjected to a beating rain or water runoff.

(b) Wet Locations.

(1) A receptacle installed in a wet location where the product intended to be plugged into it is not attended while in use (e.g., sprinkler system controllers, landscape lighting, holiday lights, etc.) shall have an enclosure that is weatherproof with the attachment plug cap inserted or removed.

(2) A receptacle installed in a wet location where the product intended to be plugged into it will be attended while in use (e.g., portable tools, etc.) shall have an enclosure that is weatherproof when the attachment plug cap is removed.

(c) Bathtub and Shower Space. A receptacle shall not be installed within a bathtub or shower space.

(d) Protection for Floor Receptacles. Standpipes of floor receptacles shall allow floor-cleaning equipment to be operated without damage to receptacles.

(e) Flush Mounting with Faceplate. The enclosure for a receptacle installed in an outlet box flush-mounted on a wall surface shall be made weatherproof by means of a weatherproof faceplate assembly that provides a watertight connection between the plate and the wall surface.

(f) Installation. A receptacle outlet installed outdoors shall be located so that water accumulation is not likely to touch the outlet cover or plate.

410-58. Grounding-Type Receptacles, Adapters, Cord Connectors, and Attachment Plugs.

(a) Grounding Poles. Grounding-type receptacles, cord connectors, and attachment plugs shall be provided with one fixed grounding pole in addition to the circuit poles. The grounding contacting pole of grounding-type plug-in ground-fault circuit interrupters shall be permitted to be of the movable, self-restoring type on circuits operating at not over 150 volts between and to conductors nor over 150 volts between any conductor and ground.

(b) Grounding-Pole Identification. Grounding-type receptacles, adapters, cord connections, and attachment plugs shall have a means for connection of a grounding conductor to the grounding pole.

A terminal for connection to the grounding pole shall be designated by one of the following.

(1) A green-colored hexagonal-headed or shaped terminal screw or nut, not readily removable.
(2) A green-colored pressure wire connector body (a wire barrel).
(3) A similar green-colored connection device, in the case of adapters. The grounding terminal of a grounding adapter shall be a green-colored rigid ear, lug, or similar device. The grounding connection shall be designed so that it cannot make contact with current-carrying parts of the receptacle, adapter, or attachment plug. The adapter shall be polarized.
(4) If the terminal for the equipment grounding conductor is not visible, the conductor entrance hole shall be marked with the word "green" or "ground," the letters "G" or "GR" or the grounding symbol, as shown in Figure 410-58(b)(4), or otherwise identified by a distinctive green color. If the terminal for the equipment grounding conductor is readily removable, the area adjacent to the terminal shall be similarly marked.

Figure 410-58(b)(4).   Grounding symbol.

(c) Grounding Terminal Use. A grounding terminal or grounding-type device shall not be used for purposes other than grounding.

(d) Grounding-Pole Requirements. Grounding-type attachment plugs and mating cord connectors and receptacles shall be designed so that the grounding connection is made before the current-carrying connections. Grounding-type devices shall be designed so grounding poles of attachment plugs cannot be brought into contact with current-carrying parts of receptacles or cord connectors.

(e) Use. Grounding-type attachment plugs shall be used only with a cord having an equipment grounding conductor.

M. Special Provisions for Flush and Recessed Fixtures

410-64. General. Fixtures installed in recessed cavities in walls or ceilings shall comply with Sections 410-65 through 410-72.

410-65. Temperature.

(a) Combustible Material. Fixtures shall be installed so that adjacent combustible material will not be subjected to temperatures in excess of 90°C (194°F).

(b) Fire-Resistant Construction. Where a fixture is recessed in fire-resistant material in a building of fire-resistant construction, a temperature higher than 90°C (194°F) but not higher than 150°C (302°F), shall be considered acceptable if the fixture is plainly marked that it is listed for that service.

(c) Recessed Incandescent Fixtures. Incandescent fixtures shall have thermal protection and shall be identified as thermally protected.

Exception No. 1: Thermal protection shall not be required in a recessed fixture identified for use and installed in poured concrete.

Exception No. 2: Thermal protection shall not be required in a recessed fixture whose design, construction, and thermal performance characteristics are equivalent to a thermally protected fixture, and are identified as inherently protected.

410-66. Clearance and Installation.

(a) Clearance.

(1) A recessed fixture that is not identified for contact with insulation shall have all recessed parts spaced at least ½ in. (12.7 mm) from combustible materials. The points of support and the trim finishing off the opening in the ceiling or wall surface shall be permitted to be in contact with combustible materials.

(2) A recessed fixture that is identified for contact with insulation, Type IC, shall be permitted to be in contact with combustible materials at recessed parts, points of support, and portions passing through or finishing off the opening in the building structure.

(b) Installation. Thermal insulation shall not be installed above a recessed fixture or within 3 in. (76 mm) of the recessed fixture's enclosure, wiring compartment, or ballast unless it is identified for contact with insulation, Type IC.

410-67. Wiring.

(a) General. Conductors that have insulation suitable for the temperature encountered shall be used.

(b) Circuit Conductors. Branch-circuit conductors that have an insulation suitable for the temperature encountered shall be permitted to terminate in the fixture.

(c) Tap Conductors. Tap conductors of a type suitable for the temperature encountered shall be permitted to run from the fixture terminal connection to an outlet box placed at least 1 ft (305 mm) from the fixture. Such tap conductors shall be in suitable raceway or Type AC or MC cable of at least 18 in. (450 mm) but not more than 6 ft (1.83 m) in length.

N. Construction of Flush and Recessed Fixtures

410-68. Temperature. Fixtures shall be constructed so that adjacent combustible material will not be subject to temperatures in excess of 90°C (194°F).

410-70. Lamp Wattage Marking. Incandescent lamp fixtures shall be marked to indicate the maximum allowable wattage of lamps. The markings shall be permanently installed, in letters at least ¼ in. (6.35 mm) high, and shall be located where visible during relamping.

410-71. Solder Prohibited. No solder shall be used in the construction of a fixture box.

410-72. Lampholders. Lampholders of the screw-shell type shall be of porcelain or other suitable insulating materials. Where used, cements shall be of the high-heat type.

P. Special Provisions for Electric-Discharge Lighting Systems of 1000 Volts or Less

410-73. General.

(a) Open-Circuit Voltage of 1000 Volts or Less. Equipment for use with electric-discharge lighting systems and designed for an open-circuit voltage of 1000 volts or less shall be of a type intended for such service.

(b) Considered as Energized. The terminals of an electric-discharge lamp shall be considered as energized where any lamp terminal is connected to a circuit of over 300 volts.

(c) Transformers of the Oil-Filled Type. Transformers of the oil-filled type shall not be used.

(d) Additional Requirements. In addition to complying with the general requirements for lighting fixtures, such equipment shall comply with Part P of this article.

(e) Thermal Protection.

(1) The ballast of a fluorescent fixture installed indoors shall have integral thermal protection. Replacement ballasts shall also have thermal protection integral with the ballast.

(2) A simple reactance ballast in a fluorescent fixture with straight tubular lamps shall not be required to be thermally protected.

(3) A ballast in a fluorescent exit fixture shall not have thermal protection.

(4) A ballast in a fluorescent fixture that is used for egress lighting and energized only during an emergency shall not have thermal protection.

(f) High-Intensity Discharge Fixtures.

(1) Recessed high-intensity fixtures designed to be installed in wall or ceiling cavities shall have thermal protection and be identified as thermally protected.

(2) Thermal protection shall not be required in a recessed high-intensity fixture whose design, construction, and thermal performance characteristics are equivalent to a thermally protected fixture and are identified as inherently protected.

(3) Thermal protection shall not be required in a recessed high-intensity discharge fixture identified for use and installed in poured concrete.

(4) A recessed remote ballast for a high-intensity discharge fixture shall have thermal protection that is integral with the ballast and be identified as thermally protected.

410-74. Direct-Current Equipment. Fixtures installed on dc circuits shall be equipped with auxiliary equipment and resistors especially designed and for dc operation, and the fixtures shall be so marked.

410-75. Open-Circuit Voltage Exceeding 300 Volts. Equipment having an open-circuit voltage exceeding 300 volts shall not be installed in dwelling occupancies unless such equipment is designed so that there will be no exposed live parts when lamps are being inserted, are in place, or are being removed.

410-76. Fixture Mounting.

(a) Exposed Ballasts. Fixtures that have exposed ballasts or transformers shall be installed so that such ballasts or transformers will not be in contact with combustible material.

(b) Combustible Low-Density Cellulose Fiberboard. Where a surface-mounted fixture containing a ballast is to be installed on combustible low-density cellulose fiberboard, it shall be listed for this condition or shall be spaced not less than 1½ in. (38 mm) from the surface of the fiberboard. Where such fixtures are partially or wholly recessed, the provisions of Sections 410-64 through 410-72 shall apply.

FPN: Combustible low-density cellulose fiberboard includes sheets, panels, and tiles that have a density of 20 lb/ft^3 (320.36 kg/m^3) or less, and that are formed of bonded plant fiber material but does not include solid or laminated wood, nor fiberboard that has a density in excess of 20 lb/ft^3 (320.36 kg/m^3) or is a material that has been integrally treated with fire-retarding chemicals to the degree that the flame spread in any plane of the material will not exceed 25, determined in accordance with tests for surface burning characteristics of building materials. See *Test Method for Surface Burning Characteristics of Building Materials*, ANSI/ASTM E84-1997.

410-77. Equipment Not Integral with Fixture.

(a) Metal Cabinets. Auxiliary equipment, including reactors, capacitors, resistors, and similar equipment, where not installed as part of a lighting fixture assembly, shall be enclosed in accessible, permanently installed metal cabinets.

(b) Separate Mounting. Separately mounted ballasts that are intended for direct connection to a wiring system shall not be required to be separately enclosed.

(c) Wired Fixture Sections. Wired fixture sections are paired, with a ballast(s) supplying a lamp or lamps in both. For interconnection between paired units, it shall be permissible to use ⅜-in. flexible metal conduit in lengths not exceeding 25 ft (7.62 m) in conformance with Article 350. Fixture wire operating at line voltage, supplying only the ballast(s) of one of the paired fixtures, shall be permitted in the same raceway as the lamp supply wires of the paired fixtures.

410-78. Autotransformers. An autotransformer that is used to raise the voltage to more than 300 volts, as part of a ballast for supplying lighting units, shall be supplied only by a grounded system.

410-79. Switches. Snap switches shall comply with Section 380-14.

Q. Special Provisions for Electric-Discharge Lighting Systems of More than 1000 Volts

410-80. General.

(a) Open-Circuit Voltage Exceeding 1000 Volts. Equipment for use with electric-discharge lighting systems and designed for an open-circuit voltage exceeding 1000 volts shall be of a type intended for such service.

(b) Dwelling Occupancies. Equipment that has an open-circuit voltage exceeding 1000 volts shall not be installed in or on dwelling occupancies.

(c) Live Parts. The terminal of an electric-discharge lamp shall be considered as a live part where any lamp terminal is connected to a circuit of over 300 volts.

(d) Additional Requirements. In addition to complying with the general requirements for lighting fixtures, such equipment shall comply with Part Q of this article.

FPN: For signs and outline lighting, see Article 600.

410-81. Control.

(a) Disconnection. Fixtures or lamp installations shall be controlled either singly or in groups by an externally operable switch or circuit breaker that opens all ungrounded primary conductors.

(b) Within Sight or Locked Type. The switch or circuit breaker shall be located within sight from the fixtures or lamps, or it shall be permitted elsewhere if it is provided with a means for locking in the open position.

410-82. Lamp Terminals and Lampholders. Parts that must be removed for lamp replacement shall be hinged or held captive. Lamps or lampholders will be designed so that there shall be no exposed live parts when lamps are being inserted or are being removed.

410-83. Transformer Ratings. Transformers and ballasts shall have a secondary open-circuit voltage of not over 15,000 volts with an allowance on test of 1000 volts additional. The secondary-current rating shall not be more than 120 milliamperes if the open-circuit voltage is over 7500 volts, and not more than 240 milliamperes if the open-circuit voltage is 7500 volts or less.

410-84. Transformer Type. Transformers shall be enclosed and listed.

410-85. Transformers and Secondary Connections. The high-voltage windings of transformers shall not be connected in series or parallel.

410-86. Transformer Locations.

(a) Accessible. Transformers shall be accessible after installation.

(b) Secondary Conductors. Transformers shall be installed as near to the lamps as practicable to keep the secondary conductors as short as possible.

(c) Adjacent to Combustible Materials. Transformers shall be located so that adjacent combustible materials will not be subjected to temperatures in excess of 90°C (194°F).

410-87. Transformer Loading. The lamps connected to any transformer shall be of such length and characteristics so as not to cause a condition of continuous overvoltage on the transformer.

410-88. Wiring Method — Secondary Conductors. Conductors shall be installed in accordance with Section 600-32.

410-89. Lamp Supports. Lamps shall be adequately supported as required in Section 600-41.

410-90. Exposure to Damage. Lamps shall not be located where normally exposed to physical damage.

410-91. Marking. Each fixture or each secondary circuit of tubing having an open-circuit voltage of over 1000 volts shall have a clearly legible marking in letters not less than ¼ in. (6.35 mm) high reading: "Caution . . . volts." The voltage indicated shall be the rated open-circuit voltage.

410-92. Switches. Snap switches shall comply with Section 380-14.

R. Lighting Track

410-100. Definition. *Lighting track* is a manufactured assembly designed to support and energize lighting fixtures that are capable of being readily repositioned on the track. Its length may be altered by the addition or subtraction of sections of track.

410-101. Installation.

(a) Lighting Track. Lighting track shall be permanently installed and permanently connected to a branch circuit. Only lighting track fittings shall be installed on lighting track. Lighting track fittings shall not be equipped with general-purpose receptacles.

(b) Connected Load. The connected load on lighting track shall not exceed the rating of the track. Lighting track shall be supplied by a branch circuit having a rating not more than that of the track.

(c) Locations Not Permitted. Lighting track shall not be installed in the following locations:

(1) Where likely to be subjected to physical damage
(2) In wet or damp locations
(3) Where subject to corrosive vapors
(4) In storage battery rooms
(5) In hazardous (classified) locations
(6) Where concealed
(7) Where extended through walls or partitions
(8) Less than 5 ft (1.52 m) above the finished floor except where protected from physical damage or track operating at less than 30 volts rms open-circuit voltage
(9) Within the zone measured 3 ft (914 mm) horizontally and 8 ft (2.44 m) vertically from the top of the bathtub rim

(d) Support. Fittings identified for use on lighting track shall be designed specifically for the track on which they are to be installed. They shall be securely fastened to the track, maintain polarization and grounding, and shall be designed to be suspended directly from the track.

410-103. Heavy-Duty Lighting Track. Heavy-duty lighting track is lighting track identified for use exceeding 20 amperes. Each fitting attached to a heavy-duty lighting track shall have individual overcurrent protection.

410-104. Fastening. Lighting track shall be securely mounted so that each fastening will be suitable for supporting the maximum weight of fixtures that can be installed. Unless identified for supports at greater intervals, a single section 4 ft (1.22 m) or shorter in length shall have two supports, and, where installed in a continuous row, each individual section of not more than 4 ft (1.22 m) in length shall have one additional support.

410-105. Construction Requirements.

(a) Construction. The housing for the lighting track system shall be of substantial construction to maintain rigidity. The conductors shall be installed within the track housing permitting insertion of a fixture, and designed to prevent tampering and accidental contact with live parts. Components of lighting track systems of different voltages shall not be interchangeable. The track conductors shall be a minimum No. 12 or equal, and shall be copper. The track system ends shall be insulated and capped.

(b) Grounding. Lighting track shall be grounded in accordance with Article 250, and the track sections shall be securely coupled to maintain continuity of the circuitry, polarization, and grounding throughout.

Article 411 — Lighting Systems Operating at 30 Volts or Less

411-1. Scope. This article covers lighting systems operating at 30 volts or less and their associated components.

411-2. Lighting Systems Operating at 30 Volts or Less. A lighting system consisting of an isolating power supply operating at 30 volts (42.4 volts peak) or less, under any load condition, with one or more secondary circuits, each limited to 25 amperes maximum, supplying lighting fixtures and associated equipment identified for the use.

411-3. Listing Required. Lighting systems operating at 30 volts or less shall be listed for the purpose.

411-4. Locations Not Permitted. Lighting systems operating at 30 volts or less shall not be installed (1) where concealed or extended through a building wall, unless using a wiring method specified in Chapter 3, or (2) within 10 ft (3.05 m) of pools, spas, fountains, or similar locations, except as permitted by Article 680.

411-5. Secondary Circuits.

(a) Grounding. Secondary circuits shall not be grounded.

(b) Isolation. The secondary circuit shall be insulated from the branch circuit by an isolating transformer.

(c) Bare Conductors. Exposed bare conductors and current-carrying parts shall be permitted. Bare conductors shall not be installed less than 7 ft (2.2 m) above the finished floor, unless specifically listed for a lower installation height.

411-6. Branch Circuit. Lighting systems operating at 30 volts or less shall be supplied from a maximum 20-ampere branch circuit.

411-7. Hazardous (Classified) Locations. Where installed in hazardous (classified) locations, these systems shall conform with Articles 500 through 517 in addition to this article.

Article 422 — Appliances

A. General

422-1. Scope. This article covers electric appliances used in any occupancy.

422-3. Other Articles. Appliances for use in hazardous (classified) locations shall comply with Articles 500 through 517.

The requirements of Article 430 shall apply to the installation of motor-operated appliances and the requirements of Article 440 shall apply to the installation of appliances containing a hermetic refrigerant motor-compressor(s), except as specifically amended in this article.

422-4. Live Parts. Appliances shall have no live parts normally exposed to contact other than those parts functioning as open-resistance heating elements, such as the heating element of a toaster, which are necessarily exposed.

B. Installation

422-10. Branch-Circuit Rating. This section specifies the ratings of branch circuits capable of carrying appliance current without overheating under the conditions specified.

(a) Individual Circuits. The rating of an individual branch circuit shall not be less than the marked rating of the appliance or the marked rating of an appliance having combined loads as provided in Section 422-62.

The rating of an individual branch-circuit for motor-operated appliances not having a marked rating shall be in accordance with Part B of Article 430.

The branch-circuit rating for an appliance that is continuously loaded, other than a motor-operated appliance, shall not be less than 125 percent of the marked rating; or not less than 100 percent of the marked rating if the branch-circuit device and its assembly are listed for continuous loading at 100 percent of its rating.

Branch circuits for household cooking appliances shall be permitted to be in accordance with Table 220-19.

(b) Circuits Supplying Two or More Loads. For branch circuits supplying appliance and other loads, the rating shall be determined in accordance with Section 210-23.

422-11. Overcurrent Protection. Appliances shall be protected against overcurrent in accordance with (a) through (g) and Section 422-10.

(a) Branch-Circuit Overcurrent Protection. Branch circuits shall be protected in accordance with Section 240-3.

If a protective device rating is marked on an appliance, the branch-circuit overcurrent device rating shall not exceed the protective device rating marked on the appliance.

(b) Household-Type Appliance with Surface Heating Elements. A household-type appliance with surface heating elements having a maximum demand of more than 60 amperes computed in accordance with Table 220-19 shall have its power supply subdivided into two or more circuits, each of which is provided with overcurrent protection rated at not over 50 amperes.

(c) Infrared Lamp Commercial and Industrial Heating Appliances. Infrared lamp commercial and industrial heating appliances shall have overcurrent protection not exceeding 50 amperes.

(d) Open-Coil or Exposed Sheathed-Coil Types of Surface Heating Elements in Commercial-Type Heating Appliances. Open-coil or exposed sheathed-coil types of surface heating elements in commercial-type heating appliances shall be protected by overcurrent protective devices rated at not over 50 amperes.

(e) Single Nonmotor-Operated Appliance. If the branch circuit supplies a single nonmotor-operated appliance, the rating of overcurrent protection shall

(1) Not exceed that marked on the appliance;
(2) If the overcurrent protection rating is not marked and the appliance is rated 13.3 amperes or less, not exceed 20 amperes; or
(3) If the overcurrent protection rating is not marked and the appliance is rated over 13.3 amperes, not exceed 150 percent of the appliance rated current. Where 150 percent of the appliance rating does not correspond to a standard overcurrent device ampere rating, the next higher standard rating shall be permitted.

(f) Electric Heating Appliances Employing Resistance-Type Heating Elements Rated More than 48 Amperes.

(1) Electric heating appliances employing resistance-type heating elements rated more than 48 amperes, other than household appliances with surface heating elements covered by Section 422-11(b), and commercial-type heating appliances covered by Section 422-11(d), shall have the heating elements subdivided. Each subdivided load shall not exceed 48 amperes and shall be protected at not more than 60 amperes.

These supplementary overcurrent protective devices shall be (1) factory-installed within or on the heater enclosure or provided as a separate assembly by the heater manufacturer; (2) accessible; and (3) suitable for branch-circuit protection.

The main conductors supplying these overcurrent protective devices shall be considered branch-circuit conductors.

(2) Commercial kitchen and cooking appliances using sheathed-type heating elements not covered in Section 422-11(d) shall be permitted to be subdivided into circuits not exceeding 120 amperes and protected at not more than 150 amperes where one of the following is met:

(a) Elements are integral with and enclosed within a cooking surface
(b) Elements are completely contained within an enclosure identified as suitable for this use
(c) Elements are contained within an ASME-rated and stamped vessel

(3) Water heaters and steam boilers employing resistance-type immersion electric heating elements contained in an ASME-rated and stamped vessel shall be permitted to be subdivided into circuits not exceeding 120 amperes and protected at not more than 150 amperes.

(g) Motor-Operated Appliances. Motors of motor-operated appliances shall be provided with overload protection in accordance with Part C of Article 430. Hermetic refrigerant motor-compressors in air-conditioning or refrigerating equipment shall be provided with overload protection in accordance with Part F of Article 440. Where appliance overcurrent protective devices that are separate from the appliance are required, data for selection of these devices shall be marked on the appliance. The minimum marking shall be that specified in Sections 430-7 and 440-4.

422-12. Central Heating Equipment. Central heating equipment other than fixed electric space-heating equipment shall be supplied by an individual branch circuit.

Exception: Auxiliary equipment, such as a pump, valve, humidifier, or electrostatic air cleaner directly associated with the heating equipment, shall be permitted to be connected to the same branch circuit.

422-13. Storage-Type Water Heaters. A branch circuit supplying a fixed storage-type water heater that has a capacity of 120 gal (454.2 L) or less shall have a rating not less than 125 percent of the nameplate rating of the water heater.

FPN: For branch-circuit rating, see Section 422-10.

422-14. Infrared Lamp Industrial Heating Appliances. Infrared industrial heating appliance lampholders shall be permitted to be connected to any of the branch circuits in Article 210 and, in industrial occupancies, shall be permitted to be operated in series on circuits of over 150 volts to ground provided the voltage rating of the lampholders is not less than the circuit voltage.

Each section, panel, or strip carrying a number of infrared lampholders (including the internal wiring of such section, panel, or strip) shall be considered an appliance. The terminal connection block of each such assembly shall be considered an individual outlet.

422-15. Central Vacuum Outlet Assemblies.

(a) Listed central vacuum outlet assemblies shall be permitted to be connected to a branch circuit in accordance with Section 210-23(a).

(b) The ampacity of the connecting conductors shall not be less than the ampacity of the branch circuit conductors to which they are connected.

(c) An equipment grounding conductor shall be used where the central vacuum outlet assembly has accessible noncurrent-carrying metal parts.

422-16. Flexible Cords.

(a) General. Flexible cord shall be permitted (1) for the connection of appliances to facilitate their frequent interchange or to prevent the transmission of noise or vibration or (2) to facilitate the removal or disconnection of appliances that are fastened in place, where the fastening means and mechanical connections are specifically designed to permit ready removal for maintenance or repair, and the appliance is intended or identified for flexible cord connection.

(b) Specific Appliances.

(1) Electrically operated kitchen waste disposers shall be permitted to be cord- and plug-connected with a flexible cord identified as suitable for the purpose in the installation instructions of the appliance manufacturer, where all of the following conditions are met.

(a) The flexible cord shall be terminated with a grounding type attachment plug.

Exception: A listed kitchen disposer distinctly marked to identify it as protected by a system of double insulation, or its equivalent, shall not be required to be terminated with a grounding-type attachment plug.

(b) The length of the cord shall not be less than 18 in. (457 mm) and not over 36 in. (914 mm).
(c) Receptacles shall be located to avoid physical damage to the flexible cord.
(d) The receptacle shall be accessible.

(2) Built-in dishwashers and trash compactors shall be permitted to be cord- and plug-connected with a flexible cord identified as suitable for the purpose in the installation instructions of the appliance manufacturer where all of the following conditions are met.

(a) The flexible cord shall be terminated with a grounding type attachment plug.

Exception: A listed dishwasher or trash compactor distinctly marked to identify it as protected by a system of double insulation, or its equivalent, shall not be required to be terminated with a grounding-type attachment plug.

(b) The length of the cord shall be 3 ft to 4 ft (0.914 m to 1.22 m) measured from the face of the attachment plug to the plane of the rear of the appliance.
(c) Receptacles shall be located to avoid physical damage to the flexible cord.
(d) The receptacle shall be located in the space occupied by the appliance or adjacent thereto.
(e) The receptacle shall be accessible.

(3) Wall-mounted ovens and counter-mounted cooking units complete with provisions for mounting and for making electrical connections shall be permitted to be permanently connected or, only for ease in servicing or for installation, cord- and plug-connected.

A separable connector or a plug and receptacle combination in the supply line to an oven or cooking unit shall (1) not be installed as the disconnecting means required by Section 422-30 and (2) be approved for the temperature of the space in which it is located.

422-17. Protection of Combustible Material. Each electrically heated appliance that is intended by size, weight, and service to be located in a fixed position shall be placed so as to provide ample protection between the appliance and adjacent combustible material.

422-18. Support of Ceiling-Suspended (Paddle) Fans.

(a) Ceiling-Suspended (Paddle) Fans 35 lb (15.88 kg) or Less. Ceiling-suspended (paddle) fans that do not exceed

35 lb (15.88 kg) in weight, with or without accessories, shall be permitted to be supported by outlet boxes identified for such use and supported in accordance with Sections 370-23 and 370-27.

(b) Ceiling-Suspended (Paddle) Fans Exceeding 35 lb (15.88 kg). Ceiling-suspended (paddle) fans exceeding 35 lb (15.88 kg) in weight, with or without accessories, shall be supported independently of the outlet box. See Section 370-23.

Exception: Listed outlet boxes or outlet box systems that are identified for the purpose shall be permitted to support ceiling-suspended fans, with or without asccessories, that weigh 70 lb (31.76 kg) or less.

422-20. Other Installation Methods. Appliances employing methods of installation other than covered by this article shall be permitted to be used only by special permission.

C. Disconnecting Means

422-30. General. A means shall be provided to disconnect each appliance from all ungrounded conductors in accordance with the following sections of Part C. If an appliance is supplied by more than one source, the disconnecting means shall be grouped and identified.

422-31. Disconnection of Permanently Connected Appliances.

(a) Rated at Not Over 300 Volt-Amperes or ⅛ Horsepower. For permanently connected appliances rated at not over 300 volt-amperes or ⅛ hp, the branch-circuit overcurrent device shall be permitted to serve as the disconnecting means.

(b) Appliances Rated Over 300 Volt-Amperes or ⅛ Horsepower. For permanently connected appliances rated over 300 volt-amperes or ⅛ hp, the branch-circuit switch or circuit breaker shall be permitted to serve as the disconnecting means where the switch or circuit breaker is within sight from the appliance or is capable of being locked in the open position.

> FPN No. 1: For motor-driven appliances of more than ⅛ hp, see Section 422-35.

> FPN No. 2: For appliances employing unit switches, see Section 422-33.

422-32. Disconnection of Cord- and Plug-Connected Appliances.

(a) Separable Connector or an Attachment Plug and Receptacle. For cord- and plug-connected appliances, an accessible separable connector or an accessible plug and receptacle shall be permitted to serve as the disconnecting means. Where the separable connector or plug and receptacle are not accessible, cord- and plug-connected appliances shall be provided with disconnecting means in accordance with Section 422-31.

(b) Connection at the Rear Base of a Range. For cord- and plug-connected household electric ranges, an attachment plug and receptacle connection at the rear base of a range, if it is accessible from the front by removal of a drawer, shall be considered as meeting the intent of Section 422-32(a).

(c) Rating. The rating of a receptacle or of a separable connector shall not be less than the rating of any appliance connected thereto.

Exception: Demand factors authorized elsewhere in this Code shall be permitted to be applied to the rating of a receptacle or of a separable connector.

422-33. Unit Switch(es) as Disconnecting Means. A unit switch(es) with a marked off position that is a part of an appliance and disconnects all ungrounded conductors shall be permitted as the disconnecting means required by this article where other means for disconnection are provided in the following types of occupancies.

(a) Multifamily Dwellings. In multifamily dwellings, the other disconnecting means shall be within the dwelling unit, or on the same floor as the dwelling unit in which the appliance is installed, and shall be permitted to control lamps and other appliances.

(b) Two-Family Dwellings. In two-family dwellings, the other disconnecting means shall be permitted either inside or outside of the dwelling unit in which the appliance is installed. In this case, an individual switch or circuit breaker for the dwelling unit shall be permitted and shall also be permitted to control lamps and other appliances.

(c) One-Family Dwellings. In one-family dwellings, the service disconnecting means shall be permitted to be the other disconnecting means.

(d) Other Occupancies. In other occupancies, the branch-circuit switch or circuit breaker, where readily accessible for servicing of the appliance, shall be permitted as the other disconnecting means.

422-34. Switch and Circuit Breaker to Be Indicating. Switches and circuit breakers used as disconnecting means shall be of the indicating type.

422-35. Disconnecting Means for Motor-Driven Appliance. If a switch or circuit breaker serves as the disconnecting means for a permanently connected motor-driven appliance of more than ⅛ hp, it shall be located within sight from the motor controller and shall comply with Part J of Article 430.

Exception: If a motor-driven appliance of more than ⅛ hp is provided with a unit switch that complies with Section 422-33(a), (b), (c), or (d), the switch or circuit breaker serving as the other disconnecting means shall be permitted to be out of sight from the motor controller.

D. Construction

422-40. Polarity in Cord- and Plug-Connected Appliances. If the appliance is provided with a manually operated, line-connected, single-pole switch for appliance on-off operation, an Edison-base lampholder, or a 15- or 20-ampere receptacle, the attachment plug shall be of the polarized or grounding type.

A 2-wire, nonpolarized attachment plug shall be permitted to be used on a listed double-insulated shaver.

FPN: For polarity of Edison-base lampholders, see Section 410-42(a).

422-41. Cord- and Plug-Connected Appliances Subject to Immersion. Cord- and plug-connected portable free-standing hydromassage units and hand-held hair dryers shall be constructed to provide protection for personnel against electrocution when immersed while in the "on" or "off" position.

422-42. Signals for Heated Appliances. In other than dwelling-type occupancies, each electrically heated appliance or group of appliances intended to be applied to combustible material shall be provided with a signal or an integral temperature-limiting device.

422-43. Flexible Cords.

(a) Heater Cords. All cord- and plug-connected smoothing irons and electrically heated appliances that are rated at more than 50 watts and produce temperatures in excess of 121°C (250°F) on surfaces with which the cord is likely to be in contact shall be provided with one of the types of approved heater cords listed in Table 400-4.

(b) Other Heating Appliances. All other cord- and plug-connected electrically heated appliances shall be connected with one of the approved types of cord listed in Table 400-4, selected in accordance with the usage specified in that table.

422-44. Cord- and Plug-Connected Immersion Heaters. Electric heaters of the cord- and plug-connected immersion type shall be constructed and installed so that current-carrying parts are effectively insulated from electrical contact with the substance in which they are immersed.

422-45. Stands for Cord- and Plug-Connected Appliances. Each smoothing iron and other cord- and plug-connected electrically heated appliance intended to be applied to combustible material shall be equipped with an approved stand, which shall be permitted to be a separate piece of equipment or a part of the appliance.

422-46. Flatirons. Electrically heated smoothing irons shall be equipped with an identified temperature-limiting means.

422-47. Water Heater Controls. All storage or instantaneous-type water heaters other than

(a) Storage water heaters that are identified as being suitable for use with supply water temperature of 82°C (180°F) or above and a capacity of 60 kW or above, or

(b) Instantaneous-type water heaters that are identified as being suitable for such use, with a capacity of 1 gal (3.785 L) or less shall be equipped with a temperature-limiting means in addition to its control thermostat to disconnect all ungrounded conductors, and such means shall be (1) installed to sense maximum water temperature and (2) be either a trip-free, manually reset type or a type having a replacement element. Such water heaters shall be marked to require the installation of a temperature and pressure relief valve.

FPN: See *Relief Valves and Automatic Gas Shutoff Devices for Hot Water Supply Systems,* ANSI Z21.22-1986.

422-48. Infrared Lamp Industrial Heating Appliances.

(a) 300 Watts or Less. Infrared heating lamps rated at 300 watts or less shall be permitted with lampholders of the medium-base, unswitched porcelain type or other types identified as suitable for use with infrared heating lamps rated 300 watts or less.

(b) Over 300 Watts. Screw-shell lampholders shall not be used with infrared lamps rated over 300 watts, unless the lampholders are identified as being suitable for use with infrared heating lamps rated over 300 watts.

422-49. High-Pressure Spray Washers. All single-phase cord- and plug-connected high-pressure spray washing machines rated at 250 volts or less shall be provided with factory-installed ground-fault circuit-interrupter protection for personnel. The ground-fault circuit interrupter shall be an integral part of the attachment plug or shall be located in the supply cord within 12 in. (305 mm) of the attachment plug.

422-50. Cord- and Plug-Connected Pipe Heating Assemblies. Cord- and plug-connected pipe heating assemblies intended to prevent freezing of piping shall be listed.

E. Marking

422-60. Nameplate.

(a) Nameplate Marking. Each electric appliance shall be provided with a nameplate giving the identifying name and the rating in volts and amperes, or in volts and watts. If the appliance is to be used on a specific frequency or frequencies, it shall be so marked.

Where motor overload protection external to the appliance is required, the appliance shall be so marked.

FPN: See Section 422-11 for overcurrent protection requirements.

(b) To Be Visible. Marking shall be located so as to be visible or easily accessible after installation.

422-61. Marking of Heating Elements. All heating elements that are rated over one ampere, replaceable in the field, and a part of an appliance shall be legibly marked with the ratings in volts and amperes, or in volts and watts, or with the manufacturer's part number.

422-62. Appliances Consisting of Motors and Other Loads. Appliances shall be marked in accordance with (a) or (b).

(a) Marking. In addition to the marking required in Section 422-60, the marking on an appliance that consists of a motor with other load(s) or motors with or without other load(s), not provided with factory-equipped cords and attachment plugs, or where both the minimum supply circuit conductor ampacity and maximum rating of the circuit overcurrent protective device are not more than 15 amperes, shall specify the minimum supply circuit conductor ampacity and the maximum rating of the circuit overcurrent protective device.

(b) Alternate Marking Method. For appliances, other than those factory-equipped with cords and attachment plugs, complying with Section 422-60, an alternate marking method shall be permitted to specify the rating of the largest motor in volts and amperes, and the additional load(s) in volts and amperes, or volts and watts in addition to the marking required in Section 422-60. The ampere rating of a motor ⅛ hp or less or a nonmotor load 1 ampere or less shall not be required to be marked unless such loads constitute the principal load.

Article 424 — Fixed Electric Space-Heating Equipment

A. General

424-1. Scope. This article covers fixed electric equipment used for space heating. For the purpose of this article, heating equipment shall include heating cable, unit heaters, boilers, central systems, or other approved fixed electric space-heating equipment. This article shall not apply to process heating and room air conditioning.

424-2. Other Articles. All requirements of this *Code* shall apply where applicable. Fixed electric space-heating equipment for use in hazardous (classified) locations shall comply with Articles 500 through 517. Fixed electric space-heating equipment incorporating a hermetic refrigerant motor-compressor shall also comply with Article 440.

424-3. Branch Circuits.

(a) Branch-Circuit Requirements. Individual branch circuits shall be permitted to supply any size fixed electric space-heating equipment.

Branch circuits supplying two or more outlets for fixed electric space-heating equipment shall be rated 15, 20, or 30 amperes. In other than residential occupancies, fixed infrared heating equipment shall be permitted to be supplied from branch circuits rated not over 50 amperes.

(b) Branch-Circuit Sizing. The ampacity of the branch-circuit conductors and the rating or setting of overcurrent protective devices supplying fixed electric space-heating equipment consisting of resistance elements with or without a motor shall not be less than 125 percent of the total load of the motors and the heaters. The rating or setting of overcurrent protective devices shall be permitted in accordance with Section 240-3(b). A contactor, thermostat, relay, or similar device, listed for continuous operation at 100 percent of its rating, shall be permitted to supply its full-rated load as provided in Section 210-19(a), Exception.

The size of the branch-circuit conductors and overcurrent protective devices supplying fixed electric space-heating equipment, including a hermetic refrigerant motor-compressor with or without resistance units, shall be computed in accordance with Sections 440-34 and 440-35. The provisions of this section shall not apply to conductors that form an integral part of approved fixed electric space-heating equipment.

B. Installation

424-9. General. All fixed electric space-heating equipment shall be installed in an approved manner.

Permanently installed electric baseboard heaters equipped with factory-installed receptacle outlets, or outlets provided as a separate listed assembly, shall be permitted in lieu of a receptacle outlet(s) that is required by Section 210-50(b). Such receptacle outlets shall not be connected to the heater circuits.

FPN: Listed baseboard heaters include instructions that may not permit their installation below receptacle outlets.

424-10. Special Permission. Fixed electric space-heating equipment and systems installed by methods other than covered by this article shall be permitted only by special permission.

424-11. Supply Conductors. Fixed electric space-heating equipment requiring supply conductors with over 60°C insulation shall be clearly and permanently marked. This marking shall be plainly visible after installation and shall be permitted to be adjacent to the field connection box.

424-12. Locations.

(a) Exposed to Physical Damage. Where subject to physical damage, fixed electric space-heating equipment shall be protected in an approved manner.

(b) Damp or Wet Locations. Heaters and related equipment installed in damp or wet locations shall be approved for such locations and shall be constructed and installed so that water or other liquids cannot enter or accumulate in or on wired sections, electrical components, or ductwork.

FPN No. 1: See Section 110-11 for equipment exposed to deteriorating agents.

FPN No. 2: See Section 680-27 for pool deck areas.

424-13. Spacing from Combustible Materials. Fixed electric space-heating equipment shall be installed to provide the required spacing between the equipment and adjacent combustible material, unless it has been found to be acceptable where installed in direct contact with combustible material.

C. Control and Protection of Fixed Electric Space-Heating Equipment

424-19. Disconnecting Means. Means shall be provided to disconnect the heater, motor controller(s), and supplementary overcurrent protective device(s) of all fixed electric space-heating equipment from all ungrounded conductors. Where heating equipment is supplied by more than one source, the disconnecting means shall be grouped and marked.

(a) Heating Equipment with Supplementary Overcurrent Protection. The disconnecting means for fixed electric space-heating equipment with supplementary overcurrent protection shall be within sight from the supplementary overcurrent protective device(s), on the supply side of these devices, if fuses, and, in addition, shall comply with either (1) or (2).

(1) Heater Containing No Motor Rated Over ⅛ Horsepower. The above disconnecting means or unit switches complying with Section 424-19(c) shall be permitted to serve as the required disconnecting means for both the motor controller(s) and heater under either item (a) or (b).

(a) The disconnecting means provided is also within sight from the motor controller(s) and the heater.

(b) The disconnecting means provided shall be capable of being locked in the open position.

(2) Heater Containing a Motor(s) Rated Over ⅛ Horsepower. The above disconnecting means shall be permitted to serve as the required disconnecting means for both the motor controller(s) and heater by one of the means specified in items (a) through (d).

(a) Where the disconnecting means is also in sight from the motor controller(s) and the heater.

(b) Where the disconnecting means is not within sight from the heater, a separate disconnecting means shall be installed, or the disconnecting means shall be capable of being locked in the open position, or unit switches complying with Section 424-19(c) shall be permitted.

(c) Where the disconnecting means is not within sight from the motor controller location, a disconnecting means complying with Section 430-102 shall be provided.

(d) Where the motor is not in sight from the motor controller location, Section 430-102(b) shall apply.

(b) Heating Equipment Without Supplementary Overcurrent Protection.

(1) Without Motor or with Motor Not Over ⅛ Horsepower. For fixed electric space-heating equipment without a motor rated over ⅛ hp, the branch-circuit switch or circuit breaker shall be permitted to serve as the disconnecting means where the switch or circuit breaker is within sight from the heater or is capable of being locked in the open position.

(2) Over ⅛ Horsepower. For motor-driven electric space-heating equipment with a motor rated over ⅛ hp, a disconnecting means shall be located within sight from the motor controller, or shall be permitted to comply with the requirements in Section 424-19(a)(2).

(c) Unit Switch(es) as Disconnecting Means. A unit switch(es) with a marked "off" position that is part of a fixed heater and disconnects all ungrounded conductors shall be permitted as the disconnecting means required by this article where other means for disconnection are provided in the following types of occupancies.

(1) Multifamily Dwellings. In multifamily dwellings, the other disconnecting means shall be within the dwelling unit, or on the same floor as the dwelling unit in which the fixed heater is installed, and shall also be permitted to control lamps and appliances.

(2) Two-Family Dwellings. In two-family dwellings, the other disconnecting means shall be permitted either inside or outside of the dwelling unit in which the fixed heater is installed. In this case, an individual switch or circuit breaker for the dwelling unit shall be permitted and shall also be permitted to control lamps and appliances.

(3) One-Family Dwellings. In one-family dwellings, the service disconnecting means shall be permitted to be the other disconnecting means.

(4) Other Occupancies. In other occupancies, the branch-circuit switch or circuit breaker, where readily accessible for servicing of the fixed heater, shall be permitted as the other disconnecting means.

424-20. Thermostatically Controlled Switching Devices.

(a) Serving as Both Controllers and Disconnecting Means. Thermostatically controlled switching devices and combination thermostats and manually controlled switches shall be permitted to serve as both controllers and disconnecting means, provided all of the following conditions are met:

(1) Provided with a marked "off" position
(2) Directly open all ungrounded conductors when manually placed in the "off" position
(3) Designed so that the circuit cannot be energized automatically after the device has been manually placed in the "off" position
(4) Located as specified in Section 424-19

(b) Thermostats that Do Not Directly Interrupt All Ungrounded Conductors. Thermostats that do not directly interrupt all ungrounded conductors and thermostats that operate remote-control circuits shall not be required to meet the requirements of (a). These devices shall not be permitted as the disconnecting means.

424-21. Switch and Circuit Breaker to Be Indicating.
Switches and circuit breakers used as disconnecting means shall be of the indicating type.

424-22. Overcurrent Protection.

(a) Branch-Circuit Devices. Electric space-heating equipment, other than such motor-operated equipment as required by Articles 430 and 440 to have additional overcurrent protection, shall be permitted to be protected against overcurrent where supplied by one of the branch circuits in Article 210.

(b) Resistance Elements. Resistance-type heating elements in electric space-heating equipment shall be protected at not more than 60 amperes. Equipment rated more than 48 amperes and employing such elements shall have the heating elements subdivided, and each subdivided load shall not exceed 48 amperes. Where a subdivided load is less than 48 amperes, the rating of the supplementary overcurrent protective device shall comply with Section 424-3(b). A boiler employing resistance-type immersion heating elements contained in an ASME rated and stamped vessel shall be permitted to comply with Section 424-72(a).

(c) Overcurrent Protective Devices. The supplementary overcurrent protective devices for the subdivided loads specified in (b) shall be (1) factory-installed within or on the heater enclosure or supplied for use with the heater as a separate assembly by the heater manufacturer; (2) accessible, but shall not be required to be readily accessible; and (3) suitable for branch-circuit protection.

FPN: See Section 240-10.

Where cartridge fuses are used to provide this overcurrent protection, a single disconnecting means shall be permitted to be used for the several subdivided loads.

FPN No. 1: For supplementary overcurrent protection, see Section 240-10.

FPN No. 2: For disconnecting means for cartridge fuses in circuits of any voltage, see Section 240-40.

(d) Branch-Circuit Conductors. The conductors supplying the supplementary overcurrent protective devices shall be considered branch-circuit conductors.

Where the heaters are rated 50 kW or more, the conductors supplying the supplementary overcurrent protective devices specified in (c) shall be permitted to be sized at not less than 100 percent of the nameplate rating of the heater, provided all of the following conditions are met.

(1) The heater is marked with a minimum conductor size.
(2) The conductors are not smaller than the marked minimum size.
(3) A temperature-actuated device controls the cyclic operation of the equipment.

(e) Conductors for Subdivided Loads. Field-wired conductors between the heater and the supplementary overcurrent protective devices shall be sized at not less than 125 percent of the load served. The supplementary overcurrent protective devices specified in (c) shall protect these conductors in accordance with Section 240-3.

Where the heaters are rated 50 kW or more, the ampacity of field-wired conductors between the heater and the supplementary overcurrent protective devices shall be permitted to be not less than 100 percent of the load of their respective subdivided circuits, provided all of the following conditions are met.

(1) The heater is marked with a minimum conductor size.
(2) The conductors are not smaller than the marked minimum size.
(3) A temperature-activated device controls the cyclic operation of the equipment.

D. Marking of Heating Equipment

424-28. Nameplate.

(a) Marking Required. Each unit of fixed electric space-heating equipment shall be provided with a nameplate giving the identifying name and the normal rating in volts and watts, or in volts and amperes.

Electric space-heating equipment intended for use on alternating current only or direct current only shall be marked to so indicate. The marking of equipment consisting of motors over ⅛ hp and other loads shall specify the rating of the motor in volts, amperes, and frequency, and the heating load in volts and watts, or in volts and amperes.

(b) Location. This nameplate shall be located so as to be visible or easily accessible after installation.

424-29. Marking of Heating Elements. All heating elements that are replaceable in the field and are a part of an electric heater shall be legibly marked with the ratings in volts and watts, or in volts and amperes.

E. Electric Space-Heating Cables

424-34. Heating Cable Construction. Heating cables shall be furnished complete with factory-assembled nonheating leads at least 7 ft (2.13 m) in length.

424-35. Marking of Heating Cables. Each unit shall be marked with the identifying name or identification symbol, catalog number, ratings in volts and watts, or in volts and amperes.

Each unit length of heating cable shall have a permanent legible marking on each nonheating lead located within 3 in. (76 mm) of the terminal end. The lead wire shall have the following color identification to indicate the circuit voltage on which it is to be used:

120 volt, nominal — yellow

208 volt, nominal — blue

240 volt, nominal — red

277 volt, nominal — brown

480 volt, nominal — orange

424-36. Clearances of Wiring in Ceilings. Wiring located above heated ceilings shall be spaced not less than 2 in. (50.8 mm) above the heated ceiling and shall be considered as operating at an ambient temperature of 50°C (122°F). The ampacity of conductors shall be computed on the basis of the correction factors shown in the 0–2000 volt ampacity tables of Article 310. If this wiring is located above thermal insulation having a minimum thickness of 2 in. (50.8 mm), the wiring shall not require correction for temperature.

424-37. Location of Branch-Circuit and Feeder Wiring in Exterior Walls. Wiring methods shall comply with Article 300 and Section 310-10.

424-38. Area Restrictions.

(a) Shall Not Extend Beyond the Room or Area. Heating cables shall not extend beyond the room or area in which they originate.

(b) Uses Prohibited. Heating cables shall not be installed in the following:

(1) In closets
(2) Over walls
(3) Over partitions that extend to the ceiling, unless they are isolated single runs of imbedded cable
(4) Over cabinets whose clearance from the ceiling is less than the minimum horizontal dimension of the cabinet to the nearest cabinet edge that is open to the room or area

(c) In Closet Ceilings as Low-Temperature Heat Sources to Control Relative Humidity. The provisions of (b) shall not prevent the use of cable in closet ceilings as low-temperature heat sources to control relative humidity, provided they are used only in those portions of the ceiling that are unobstructed to the floor by shelves or other permanent fixtures.

424-39. Clearance from Other Objects and Openings. Heating elements of cables shall be separated at least 8 in. (203 mm) from the edge of outlet boxes and junction boxes that are to be used for mounting surface lighting fixtures.

A clearance of not less than 2 in. (50.8 mm) shall be provided from recessed fixtures and their trims, ventilating openings, and other such openings in room surfaces. Sufficient area shall be provided to ensure that no heating cable will be covered by any surface-mounted units.

424-40. Splices. Embedded cables shall be spliced only where necessary and only by approved means, and in no case shall the length of the heating cable be altered.

424-41. Installation of Heating Cables on Dry Board, in Plaster, and on Concrete Ceilings.

(a) In Walls. Cables shall not be installed in walls unless it is necessary for an isolated single run of cable to be installed down a vertical surface to reach a dropped ceiling.

(b) Adjacent Runs. Adjacent runs of cable not exceeding 2¾ watts/ft (2¾ watts/305 mm) shall not be installed less than 1½ in. (38 mm) on centers.

(c) Surfaces to Be Applied. Heating cables shall be applied only to gypsum board, plaster lath, or other fire-resistant material. With metal lath or other electrically conductive surfaces, a coat of plaster shall be applied to completely separate the metal lath or conductive surface from the cable.

FPN: See also Section 424-41(f).

(d) Splices. All heating cables, the splice between the heating cable and nonheating leads, and 3-in. (76-mm) minimum of the nonheating lead at the splice shall be embedded in plaster or dry board in the same manner as the heating cable.

(e) Ceiling Surface. The entire ceiling surface shall have a finish of thermally noninsulating sand plaster that has a nominal thickness of ½ in. (12.7 mm), or other noninsulating material identified as suitable for this use and applied according to specified thickness and directions.

(f) Secured. Cables shall be secured by means of approved stapling, tape, plaster, nonmetallic spreaders, or other approved means at either intervals not exceeding 16 in. (406 mm) or at intervals not exceeding 6 ft (1.83 m) for cables identified for such use. Staples or metal fasteners that straddle the cable shall not be used with metal lath or other electrically conductive surfaces.

(g) Dry Board Installations. In dry board installations, the entire ceiling below the heating cable shall be covered with gypsum board not exceeding ½ in. (12.7 mm) thickness. The void between the upper layer of gypsum board, plaster lath, or other fire-resistant material and the surface layer of gypsum board shall be completely filled with thermally conductive, nonshrinking plaster or other approved material or equivalent thermal conductivity.

(h) Free from Contact with Conductive Surfaces. Cables shall be kept free from contact with metal or other electrically conductive surfaces.

(i) Joists. In dry board applications, cable shall be installed parallel to the joist, leaving a clear space centered under the joist of 2½ in. (64 mm) (width) between centers of adjacent runs of cable. A surface layer of gypsum board shall be mounted so that the nails or other fasteners do not pierce the heating cable.

(j) Crossing Joists. Cables shall cross joists only at the ends of the room unless the cable is required to cross joists elsewhere in order to satisfy the manufacturer's instructions that the installer avoid placing the cable too close to ceiling penetrations and light fixtures.

424-42. Finished Ceilings. Finished ceilings shall not be covered with decorative panels or beams constructed of materials that have thermal insulating properties, such as wood, fiber, or plastic. Finished ceilings shall be permitted to be covered with paint, wallpaper, or other approved surface finishes.

424-43. Installation of Nonheating Leads of Cables.

(a) Free Nonheating Leads. Free nonheating leads of cables shall be installed in accordance with approved wiring methods from the junction box to a location within the ceiling. Such installations shall be permitted to be single conductors in approved raceways, single or multiconductor Type UF, Type NMC, Type MI, or other approved conductors.

(b) Leads in Junction Box. Not less than 6 in. (152 mm) of free nonheating lead shall be within the junction box. The marking of the leads shall be visible in the junction box.

(c) Excess Leads. Excess leads of heating cables shall not be cut but shall be secured to the underside of the ceiling and embedded in plaster or other approved material, leaving only a length sufficient to reach the junction box with not less than 6 in. (152 mm) of free lead within the box.

424-44. Installation of Cables in Concrete or Poured Masonry Floors.

(a) Watts per Linear Foot. Constant wattage heating cables shall not exceed 16½ watts/linear foot (305 mm) of cable.

(b) Spacing Between Adjacent Runs. The spacing between adjacent runs of cable shall be not less than 1 in. (25.4 mm) on centers.

(c) Secured in Place. Cables shall be secured in place by nonmetallic frames or spreaders or other approved means while the concrete or other finish is applied.

Cables shall not be installed where they bridge expansion joints unless protected from expansion and contraction.

(d) Spacings Between Heating Cable and Metal Embedded in the Floor. Spacings shall be maintained between the heating cable and metal embedded in the floor, unless the cable is a grounded metal-clad cable.

(e) Leads Protected. Leads shall be protected where they leave the floor by rigid metal conduit, intermediate metal conduit, rigid nonmetallic conduit, electrical metallic tubing, or by other approved means.

(f) Bushings or Approved Fittings. Bushings or approved fittings shall be used where the leads emerge within the floor slab.

(g) Ground-Fault Circuit-Interrupter Protection for Conductive Heated Floors of Bathrooms, Hydromassage Bathtub, Spa, and Hot Tub Locations. Ground-fault circuit-interrupter protection for personnel shall be provided for electrically heated floors in bathrooms, hydromassage bathtub, spa, and hot tub locations. This shall apply to all systems used with conductive floor coverings, whether cable, panel, or other approved types.

424-45. Inspection and Tests. Cable installations shall be made with due care to prevent damage to the cable assembly and shall be inspected and approved before cables are covered or concealed.

F. Duct Heaters

424-57. General. Part F shall apply to any heater mounted in the airstream of a forced-air system where the air-moving unit is not provided as an integral part of the equipment.

424-58. Identification. Heaters installed in an air duct shall be identified as suitable for the installation.

424-59. Airflow. Means shall be provided to ensure uniform and adequate airflow over the face of the heater in accordance with the manufacturer's instructions.

FPN: Heaters installed within 4 ft (1.22 m) of the outlet of an air-moving device, heat pump, air conditioner, elbows, baffle plates, or other obstructions in ductwork may require turning vanes, pressure plates, or other devices on the inlet side of the duct heater to ensure an even distribution of air over the face of the heater.

424-60. Elevated Inlet Temperature. Duct heaters intended for use with elevated inlet air temperature shall be identified as suitable for use at the elevated temperatures.

424-61. Installation of Duct Heaters with Heat Pumps and Air Conditioners. Heat pumps and air conditioners having duct heaters closer than 4 ft (1.22 m) to the heat pump or air conditioner shall have both the duct heater and heat pump or air conditioner identified as suitable for such installation and so marked.

424-62. Condensation. Duct heaters used with air conditioners or other air-cooling equipment that may result in condensation of moisture shall be identified as suitable for use with air conditioners.

424-63. Fan Circuit Interlock. Means shall be provided to ensure that the fan circuit is energized when any heater circuit is energized. However, time- or temperature-controlled delay in energizing the fan motor shall be permitted.

424-64. Limit Controls. Each duct heater shall be provided with an approved, integral, automatic-reset temperature-limiting control or controllers to de-energize the circuit or circuits.

In addition, an integral independent supplementary control or controllers shall be provided in each duct heater that will disconnect a sufficient number of conductors to interrupt current flow. This device shall be manually resettable or replaceable.

424-65. Location of Disconnecting Means. Duct heater controller equipment shall be either accessible with the disconnecting means installed at or within sight from the controller or as permitted by Section 424-19(a).

424-66. Installation. Duct heaters shall be installed in accordance with the manufacturer's instructions in a manner so that operation will not create a hazard to persons or property. Furthermore, duct heaters shall be located with respect to building construction and other equipment so as to permit access to the heater. Sufficient clearance shall be maintained to permit replacement of controls and heating elements and for adjusting and cleaning of controls and other parts requiring such attention. See Section 110-26.

FPN: For additional installation information, see *Standard for the Installation of Air Conditioning and Ventilating Systems*, NFPA 90A-1996, and *Standard for the Installation of Warm Air Heating and Air Conditioning Systems*, NFPA 90B-1996.

G. Resistance-Type Boilers

424-70. Scope. The provisions in Part G of this article shall apply to boilers employing resistance-type heating elements. Electrode-type boilers shall not be considered as employing resistance-type heating elements. See Part H of this article.

424-71. Identification. Resistance-type boilers shall be identified as suitable for the installation.

424-72. Overcurrent Protection.

(a) Boiler Employing Resistance-Type Immersion Heating Elements in an ASME Rated and Stamped Vessel. A boiler employing resistance-type immersion heating elements contained in an ASME rated and stamped vessel shall have the heating elements protected at not more than 150 amperes. Such a boiler rated more than 120 amperes shall have the heating elements subdivided into loads not exceeding 120 amperes.

Where a subdivided load is less than 120 amperes, the rating of the overcurrent protective device shall comply with Section 424-3(b).

(b) Boiler Employing Resistance-Type Heating Elements Rated More than 48 Amperes and Not Contained in an ASME Rated and Stamped Vessel. A boiler employing resistance-type heating elements not contained in an ASME rated and stamped vessel shall have the heating elements protected at not more than 60 amperes. Such a boiler rated more than 48 amperes shall have the heating elements subdivided into loads not exceeding 48 amperes.

Where a subdivided load is less than 48 amperes, the rating of the overcurrent protective device shall comply with Section 424-3(b).

(c) Supplementary Overcurrent Protective Devices. The supplementary overcurrent protective devices for the subdivided loads as required by Sections 424-72(a) and (b) shall be (1) factory-installed within or on the boiler enclosure or provided as a separate assembly by the boiler manufacturer; and (2) accessible, but need not be readily accessible; and (3) suitable for branch-circuit protection.

Where cartridge fuses are used to provide this overcurrent protection, a single disconnecting means shall be permitted for the several subdivided circuits. See Section 240-40.

(d) Conductors Supplying Supplementary Overcurrent Protective Devices. The conductors supplying these supplementary overcurrent protective devices shall be considered branch circuit conductors.

Where the heaters are rated 50 kW or more, the conductors supplying the overcurrent protective device specified in (c) shall be permitted to be sized at not less than 100 percent of the nameplate rating of the heater, provided all of the following conditions are met.

(1) The heater is marked with a minimum conductor size.
(2) The conductors are not smaller than the marked minimum size.
(3) A temperature- or pressure-actuated device controls the cyclic operation of the equipment.

(e) Conductors for Subdivided Loads. Field-wired conductors between the heater and the supplementary overcurrent protective devices shall be sized at not less than 125 percent of the load served. The supplementary overcurrent protective devices specified in (c) shall protect these conductors in accordance with Section 240-3.

Where the heaters are rated 50 kW or more, the ampacity of field-wired conductors between the heater and the supplementary overcurrent protective devices shall be permitted to be not less than 100 percent of the load of their respective subdivided circuits, provided all of the following conditions are met.

(1) The heater is marked with a minimum conductor size.
(2) The conductors are not smaller than the marked minimum size.
(3) A temperature-activated device controls the cyclic operation of the equipment.

424-73. Over-Temperature Limit Control. Each boiler designed so that in normal operation there is no change in state of the heat transfer medium shall be equipped with a temperature-sensitive limiting means. It shall be installed to limit maximum liquid temperature and shall directly or indirectly disconnect all ungrounded conductors to the heating elements. Such means shall be in addition to a temperature regulating system and other devices protecting the tank against excessive pressure.

424-74. Over-Pressure Limit Control. Each boiler designed so that in normal operation there is a change in state of the heat transfer medium from liquid to vapor shall be equipped with a pressure-sensitive limiting means. It shall be installed to limit maximum pressure and shall directly or indirectly disconnect all ungrounded conductors to the heating elements. Such means shall be in addition to a pressure regulating system and other devices protecting the tank against excessive pressure.

H. Electrode-Type Boilers

424-80. Scope. The provisions in Part H of this article shall apply to boilers for operation at 600 volts, nominal, or less, in which heat is generated by the passage of current between electrodes through the liquid being heated.

FPN: For over 600 volts, see Part E of Article 490.

424-81. Identification. Electrode-type boilers shall be identified as suitable for the installation.

424-82. Branch-Circuit Requirements. The size of branch-circuit conductors and overcurrent protective devices shall be calculated on the basis of 125 percent of the total load (motors not included). A contactor, relay, or other device, approved for continuous operation at 100 percent of its rating, shall be permitted to supply its full-rated load. See Section 210-19(a), Exception. The provisions of this section shall not apply to conductors that form an integral part of an approved boiler.

Where an electrode boiler is rated 50 kW or more, the conductors supplying the boiler electrode(s) shall be permitted to be sized at not less than 100 percent of the nameplate rating of the electrode boiler, provided all the following conditions are met.

(1) The electrode boiler is marked with a minimum conductor size.

(2) The conductors are not smaller than the marked minimum size.

(3) A temperature- or pressure-actuated device controls the cyclic operation of the equipment.

424-83. Over-Temperature Limit Control. Each boiler designed so that in normal operation there is no change in state of the heat transfer medium shall be equipped with a temperature-sensitive limiting means. It shall be installed to limit maximum liquid temperature and shall directly or indirectly interrupt all current flow through the electrodes. Such means shall be in addition to the temperature regulating system and other devices protecting the tank against excessive pressure.

424-84. Over-Pressure Limit Control. Each boiler designed so that in normal operation there is a change in state of the heat transfer medium from liquid to vapor shall be equipped with a pressure-sensitive limiting means. It shall be installed to limit maximum pressure and shall directly or indirectly interrupt all current flow through the electrodes. Such means shall be in addition to a pressure regulating system and other devices protecting the tank against excessive pressure.

424-85. Grounding. For those boilers designed such that fault currents do not pass through the pressure vessel, and the pressure vessel is electrically isolated from the electrodes, all exposed noncurrent-carrying metal parts, including the pressure vessel, supply, and return connecting piping, shall be grounded in accordance with Article 250.

For all other designs, the pressure vessel containing the electrodes shall be isolated and electrically insulated from ground.

424-86. Markings. All electrode-type boilers shall be marked to show the following:

(1) The manufacturer's name

(2) The normal rating in volts, amperes, and kilowatts

(3) The electrical supply required specifying frequency, number of phases, and number of wires

(4) The marking: "Electrode-Type Boiler"

(5) A warning marking: "All Power Supplies Shall Be Disconnected Before Servicing, Including Servicing the Pressure Vessel"

The nameplate shall be located so as to be visible after installation.

J. Electric Radiant Heating Panels and Heating Panel Sets

424-90. Scope. The provisions of Part J of this article shall apply to radiant heating panels and heating panel sets.

424-91. Definitions.

(a) Heating Panel. A *heating panel* is a complete assembly provided with a junction box or a length of flexible conduit for connection to a branch circuit.

(b) Heating Panel Set. A *heating panel set* is a rigid or nonrigid assembly provided with nonheating leads or a terminal junction assembly identified as being suitable for connection to a wiring system.

424-92. Markings.

(a) Markings shall be permanent and in a location that is visible prior to application of panel finish.

(b) Each unit shall be identified as suitable for the installation.

(c) Each unit shall be marked with the identifying name or identification symbol, catalog number, and rating in volts and watts, or in volts and amperes.

(d) The manufacturers of heating panels or heating panel sets shall provide marking labels that indicate that the space-heating installation incorporates heating panels or heating panel sets and instructions that the labels shall be affixed to the panelboards to identify which branch circuits supply the circuits to those space-heating installations. If the heating panels and heating panel set installations are visible and distinguishable after installation, the labels shall not be required to be provided and affixed to the panelboards.

424-93. Installation.

(a) General.

(1) Heating panels and heating panel sets shall be installed in accordance with the manufacturer's instructions.

(2) The heating portion shall not

(a) Be installed in or behind surfaces where subject to physical damage.

(b) Be run through or above walls, partitions, cupboards, or similar portions of structures that extend to the ceiling.

(c) Be run in or through thermal insulation, but shall be permitted to be in contact with the surface of thermal insulation.

(3) Edges of panels and panel sets shall be separated by not less than 8 in. (203 mm) from the edges of any outlet boxes and junction boxes that are to be used for mounting surface lighting fixtures. A clearance of not less than 2 in. (50.8 mm) shall be provided from recessed fixtures and their trims, ventilating openings, and other such openings in room surfaces, unless the heating panels and panel sets are listed and marked for lesser clearances, in which case, they shall be permitted to be installed at the marked clearances. Sufficient area shall be provided to ensure that no heating panel or heating panel set is to be covered by any surface-mounted units.

(4) After the heating panels or heating panel sets are installed and inspected, it shall be permitted to install a surface that has been identified by the manufacturer's instructions as being suitable for the installation. The surface shall be secured so that the nails or other fastenings do not pierce the heating panels or heating panel sets.

(5) Surfaces permitted by Section 424-93(a)(4) shall be permitted to be covered with paint, wallpaper, or other approved surfaces identified in the manufacturer's instructions as being suitable.

(b) Heating Panel Sets.

(1) Heating panel sets shall be permitted to be secured to the lower face of joists or mounted in between joists, headers, or nailing strips.

(2) Heating panel sets shall be installed parallel to joists or nailing strips.

(3) Nailing or stapling of heating panel sets shall be done only through the unheated portions provided for this purpose. Heating panel sets shall not be cut through or nailed through any point closer than ¼ in. (6.35 mm) to the element. Nails, staples, or other fasteners shall not be used where they penetrate current-carrying parts.

(4) Heating panel sets shall be installed as complete units unless identified as suitable for field cutting in an approved manner.

424-94. Clearances of Wiring in Ceilings. Wiring located above heated ceilings shall be spaced not less than 2 in. (50.8 mm) above the heated ceiling and shall be considered as operating at an ambient of 50°C (122°F). The ampacity shall be computed on the basis of the correction factors

given in the 0–2000 volt ampacity tables of Article 310. If this wiring is located above thermal insulations having a minimum thickness of 2 in. (50.8 mm), the wiring shall not require correction for temperature.

424-95. Location of Branch-Circuit and Feeder Wiring in Walls.

(a) Exterior Walls. Wiring methods shall comply with Article 300 and Section 310-10.

(b) Interior Walls. Any wiring behind heating panels or heating panel sets located in interior walls or partitions shall be considered as operating at an ambient temperature of 40°C (104°F), and the ampacity shall be computed on the basis of the correction factors given in the 0–2000 volt ampacity tables of Article 310.

424-96. Connection to Branch-Circuit Conductors.

(a) General. Heating panels or heating panel sets assembled together in the field to form a heating installation in one room or area shall be connected in accordance with the manufacturer's instructions.

(b) Heating Panels. Heating panels shall be connected to branch-circuit wiring by an approved wiring method.

(c) Heating Panel Sets.

(1) Heating panel sets shall be connected to branch-circuit wiring by a method identified as being suitable for the purpose.

(2) A heating panel set provided with terminal junction assembly shall be permitted to have the nonheating leads attached at the time of installation in accordance with the manufacturer's instructions.

424-97. Nonheating Leads. Excess nonheating leads of heating panels or heating panel sets shall be permitted to be cut to the required length. They shall meet the installation requirements of the wiring method employed in accordance with Section 424-96. Nonheating leads shall be an integral part of a heating panel and a heating panel set and shall not be subjected to the ampacity requirements of Section 424-3(b) for branch circuits.

424-98. Installation in Concrete or Poured Masonry.

(a) Maximum Heated Area. Heating panels or heating panel sets shall not exceed 33 watts/ft² (33 watts/ 0.093 m²) of heated area.

(b) Secured in Place and Identified as Suitable. Heating panels or heating panel sets shall be secured in place by means specified in the manufacturer's instructions and identified as suitable for the installation.

(c) Expansion Joints. Heating panels or heating panel sets shall not be installed where they bridge expansion joints unless provision is made for expansion and contraction.

(d) Spacings. Spacings shall be maintained between heating panels or heating panel sets and metal embedded in the floor. Grounded metal-clad heating panels shall be permitted to be in contact with metal embedded in the floor.

(e) Protection of Leads. Leads shall be protected where they leave the floor by rigid metal conduit, intermediate metal conduit, rigid nonmetallic conduit, electrical metallic tubing, or by other approved means.

(f) Bushings or Fittings Required. Bushings or approved fittings shall be used where the leads emerge within the floor slabs.

424-99. Installation Under Floor Covering.

(a) Identification. Heating panels or heating panel sets for installation under floor covering shall be identified as suitable for installation under floor covering.

(b) Maximum Heated Area. Heating panels or panel sets, installed under floor covering, shall not exceed 15 watts/ft^2 (15 watts/0.093 m^2) of heated area.

(c) Installation. Listed heating panels or panel sets, if installed under floor covering, shall be installed on floor surfaces that are smooth and flat in accordance with the manufacturer's instructions and shall also comply with the following.

  (1) Expansion Joints. Heating panels or heating panel sets shall not be installed where they bridge expansion joints unless protected from expansion and contraction.

  (2) Connection to Conductors. Heating panels and heating panel sets shall be connected to branch-circuit and supply wiring by wiring methods recognized in Chapter 3.

  (3) Anchoring. Heating panels and heating panel sets shall be firmly anchored to the floor using an adhesive or anchoring system identified for this use.

  (4) Coverings. After heating panels or heating panel sets are installed and inspected, they shall be permitted to be covered by a floor covering that has been identified by the manufacturer as being suitable for the installation. The covering shall be secured to the heating panel or heating panel sets with release-type adhesives or by means identified for this use.

  (5) Fault Protection. A device to open all ungrounded conductors supplying the heating panels or heating panel

sets, provided by the manufacturer, shall function when a low- or high-resistance line-to-line, line-to-grounded conductor, or line-to-ground fault occurs, such as the result of a penetration of the element or element assembly.

FPN: An integral grounding shield may be required to provide this protection.

Article 426 — Fixed Outdoor Electric Deicing and Snow-Melting Equipment

A. General

426-1. Scope. The requirements of this article shall apply to electrically energized heating systems and the installation of these systems.

(a) Embedded. Embedded in driveways, walks, steps, and other areas.

(b) Exposed. Exposed on drainage systems, bridge structures, roofs, and other structures.

426-2. Definitions. For the purpose of this article:

Heating System. A complete system consisting of components such as heating elements, fastening devices, nonheating circuit wiring, leads, temperature controllers, safety signs, junction boxes, raceways, and fittings.

Impedance Heating System. A system in which heat is generated in a pipe or rod, or combination of pipes and rods, by causing current to flow through the pipe or rod by direct connection to an ac voltage source from a dual-winding transformer. The pipe or rod shall be permitted to be embedded in the surface to be heated, or constitute the exposed components to be heated.

Resistance Heating Element. A specific separate element to generate heat that is embedded in or fastened to the surface to be heated.

FPN: Tubular heaters, strip heaters, heating cable, heating tape, and heating panels are examples of resistance heaters.

Skin-Effect Heating System. A system in which heat is generated on the inner surface of a ferromagnetic envelope embedded in or fastened to the surface to be heated.

FPN: Typically, an electrically insulated conductor is routed through and connected to the envelope at the other end. The envelope and the electrically insulated conductor are connected to an ac voltage source from a dual-winding transformer.

426-3. Application of Other Articles. All requirements of this *Code* shall apply except as specifically amended in this article. Cord- and plug-connected fixed outdoor electric deicing and snow-melting equipment intended for specific use and identified as suitable for this use shall be installed according to Article 422. Fixed outdoor electric deicing and snow-melting equipment for use in hazardous (classified) locations shall comply with Articles 500 through 516.

426-4. Branch-Circuit Sizing. The ampacity of branch-circuit conductors and the rating or setting of overcurrent protective devices supplying fixed outdoor electric deicing and snow-melting equipment shall not be less than 125 percent of the total load of the heaters. The rating or setting of overcurrent protective devices shall be permitted in accordance with Section 240-3(b).

B. Installation

426-10. General. Equipment for outdoor electric deicing and snow melting shall be identified as being suitable for the following:

(1) The chemical, thermal, and physical environment
(2) Installation in accordance with the manufacturer's drawings and instructions

426-11. Use. Electrical heating equipment shall be installed in such a manner as to be afforded protection from physical damage.

426-12. Thermal Protection. External surfaces of outdoor electric deicing and snow-melting equipment that operate at temperatures exceeding 60°C (140°F) shall be physically guarded, isolated, or thermally insulated to protect against contact by personnel in the area.

426-13. Identification. The presence of outdoor electric deicing and snow-melting equipment shall be evident by the posting of appropriate caution signs or markings where clearly visible.

426-14. Special Permission. Fixed outdoor deicing and snow-melting equipment employing methods of construction or installation other than covered by this article shall be permitted only by special permission.

C. Resistance Heating Elements

426-20. Embedded Deicing and Snow-Melting Equipment.

(a) Watt Density. Panels or units shall not exceed 120 watts/ft^2 (120 watts/0.093 m^2) of heated area.

(b) Spacing. The spacing between adjacent cable runs is dependent upon the rating of the cable, and shall be not less than 1 in. (25.4 mm) on centers.

(c) Cover. Units, panels, or cables shall be installed as follows:

(1) On a substantial asphalt or masonry base at least 2 in. (50.8 mm) thick and have at least 1½ in. (38 mm) of asphalt or masonry applied over the units, panels, or cables; or
(2) They shall be permitted to be installed over other approved bases and embedded within 3½ in. (89 mm) of masonry or asphalt but not less than 1½ in. (38 mm) from the top surface; or
(3) Equipment that has been specially investigated for other forms of installation shall be installed only in the manner for which it has been investigated.

(d) Secured. Cables, units, and panels shall be secured in place by frames or spreaders or other approved means while the masonry or asphalt finish is applied.

(e) Expansion and Contraction. Cables, units, and panels shall not be installed where they bridge expansion joints unless provision is made for expansion and contraction.

426-21. Exposed Deicing and Snow-Melting Equipment.

(a) Secured. Heating element assemblies shall be secured to the surface being heated by approved means.

(b) Over-Temperature. Where the heating element is not in direct contact with the surface being heated, the design of the heater assembly shall be such that its temperature limitations shall not be exceeded.

(c) Expansion and Contraction. Heating elements and assemblies shall not be installed where they bridge expansion joints unless provision is made for expansion and contraction.

(d) Flexural Capability. Where installed on flexible structures, the heating elements and assemblies shall have a flexural capability that is compatible with the structure.

426-22. Installation of Nonheating Leads for Embedded Equipment.

(a) Grounding Sheath or Braid. Nonheating leads having a grounding sheath or braid shall be permitted to be embedded in the masonry or asphalt in the same manner as the heating cable without additional physical protection.

(b) Raceways. All but 1 in. to 6 in. (25.4 mm to 152 mm) of nonheating leads of Type TW and other approved types not having a grounding sheath shall be enclosed in a rigid conduit, electrical metallic tubing, intermediate metal conduit, or other raceways within asphalt or masonry; and the distance from the factory splice to raceway shall not be less than 1 in. (25.4 mm) or more than 6 in. (152 mm).

(c) Bushings. Insulating bushings shall be used in the asphalt or masonry where leads enter conduit or tubing.

(d) Expansion and Contraction. Leads shall be protected in expansion joints and where they emerge from masonry or asphalt by rigid conduit, electrical metallic tubing, intermediate metal conduit, other raceways, or other approved means.

(e) Leads in Junction Boxes. Not less than 6 in. (152 mm) of free nonheating lead shall be within the junction box.

426-23. Installation of Nonheating Leads for Exposed Equipment.

(a) Nonheating Leads. Power supply nonheating leads (cold leads) for resistance elements shall be suitable for the temperature encountered. Preassembled nonheating leads on approved heaters shall be permitted to be shortened if the markings specified in Section 426-25 are retained. Not less than 6 in. (152 mm) of nonheating leads shall be provided within the junction box.

(b) Protection. Nonheating power supply leads shall be enclosed in a rigid conduit, intermediate metal conduit, electrical metallic tubing, or other approved means.

426-24. Electrical Connection.

(a) Heating Element Connections. Electrical connections, other than factory connections of heating elements to nonheating elements embedded in masonry or asphalt or on exposed surfaces, shall be made with insulated connectors identified for the use.

(b) Circuit Connections. Splices and terminations at the end of the nonheating leads, other than the heating element end, shall be installed in a box or fitting in accordance with Sections 110-14 and 300-15.

426-25. Marking. Each factory-assembled heating unit shall be legibly marked within 3 in. (76 mm) of each end of the nonheating leads with the permanent identification symbol, catalog number, and ratings in volts and watts, or in volts and amperes.

426-26. Corrosion Protection. Ferrous and nonferrous metal raceways, cable armor, cable sheaths, boxes, fittings, supports, and support hardware shall be permitted to be installed in concrete or in direct contact with the earth, or in areas subject to severe corrosive influences, where made of material suitable for the condition, or where provided with corrosion protection identified as suitable for the condition.

426-27. Grounding Braid or Sheath. Grounding means, such as copper braid, metal sheath, or other approved means, shall be provided as part of the heated section of the cable, panel, or unit.

426-28. Equipment Protection. Ground-fault protection of equipment shall be provided for fixed outdoor electric deicing and snow-melting equipment, except for equipment that employs mineral-insulated, metal-sheathed cable embedded in a noncombustible medium.

D. Impedance Heating

426-30. Personnel Protection. Exposed elements of impedance heating systems shall be physically guarded, isolated, or thermally insulated with a weatherproof jacket to protect against contact by personnel in the area.

426-31. Isolation Transformer. A dual-winding transformer with a grounded shield between the primary and secondary windings shall be used to isolate the distribution system from the heating system.

426-32. Voltage Limitations. Unless protected by a ground-fault circuit-interrupter protection for personnel, the secondary winding of the isolation transformer connected to the impedance heating elements shall not have an output voltage greater than 30 volts ac.

Where ground-fault circuit-interrupter protection for personnel is provided, the voltage shall be permitted to be greater than 30 but not more than 80 volts.

426-33. Induced Currents. All current-carrying components shall be installed in accordance with Section 300-20.

426-34. Grounding. An impedance heating system that is operating at a voltage greater than 30, but not more than 80, shall be grounded at a designated point(s).

E. Skin-Effect Heating

426-40. Conductor Ampacity. The current through the electrically insulated conductor inside the ferromagnetic envelope shall be permitted to exceed the ampacity values shown in Article 310, provided it is identified as suitable for this use.

426-41. Pull Boxes. Where pull boxes are used, they shall be accessible without excavation by location in suitable vaults or above grade. Outdoor pull boxes shall be of watertight construction.

426-42. Single Conductor in Enclosure. The provisions of Section 300-20 shall not apply to the installation of a single conductor in a ferromagnetic envelope (metal enclosure).

426-43. Corrosion Protection. Ferromagnetic envelopes, ferrous or nonferrous metal raceways, boxes, fittings, supports, and support hardware shall be permitted to be installed in concrete or in direct contact with the earth, or in areas subjected to severe corrosive influences, where made of material suitable for the condition, or where provided with corrosion protection identified as suitable for the condition. Corrosion protection shall maintain the original wall thickness of the ferromagnetic envelope.

426-44. Grounding. The ferromagnetic envelope shall be grounded at both ends; and, in addition, it shall be permitted to be grounded at intermediate points as required by its design.

The provisions of Section 250-30 shall not apply to the installation of skin-effect heating systems.

FPN: For grounding methods, see Sections 250-30(a)(4) and (b)(3).

F. Control and Protection

426-50. Disconnecting Means.

(a) Disconnection. All fixed outdoor deicing and snow-melting equipment shall be provided with a means for disconnection from all ungrounded conductors. Where readily accessible to the user of the equipment, the branch-circuit switch or circuit breaker shall be permitted to serve as the disconnecting means. Switches used as the disconnecting means shall be of the indicating type.

(b) Cord- and Plug-Connected Equipment. The factory-installed attachment plug of cord- and plug-connected equipment rated 20 amperes or less and 150 volts or less to ground shall be permitted to be the disconnecting means.

426-51. Controllers.

(a) Temperature Controller with "Off" Position. Temperature controlled switching devices that indicate an "off" position and that interrupt line current shall open all ungrounded conductors when the control device is in the "off" position. These devices shall not be permitted to serve as the disconnecting means unless provided with a positive lockout in the "off" position.

(b) Temperature Controller Without "Off" Position. Temperature controlled switching devices that do not have an "off" position shall not be required to open all ungrounded conductors and shall not be permitted to serve as the disconnecting means.

(c) Remote Temperature Controller. Remote controlled temperature-actuated devices shall not be required to meet the requirements of Section 426-51(a). These devices shall not be permitted to serve as the disconnecting means.

(d) Combined Switching Devices. Switching devices consisting of combined temperature-actuated devices and manually controlled switches that serve both as the controller and the disconnecting means shall comply with all of the following conditions:

(1) Open all ungrounded conductors when manually placed in the "off" position
(2) Be so designed that the circuit cannot be energized automatically if the device has been manually placed in the "off" position
(3) Be provided with a positive lockout in the "off" position.

426-52. Overcurrent Protection. Fixed outdoor electric deicing and snow-melting equipment shall be permitted to be protected against overcurrent where supplied by a branch circuit as specified in Section 426-4.

426-54. Cord- and Plug-Connected Deicing and Snow-Melting Equipment. Cord- and plug-connected deicing and snow-melting equipment shall be listed.

Article 427 — Fixed Electric Heating Equipment for Pipelines and Vessels

A. General

427-1. Scope. The requirements of this article shall apply to electrically energized heating systems and the installation of these systems used with pipelines or vessels or both.

427-2. Definitions. For the purpose of this article:

Impedance Heating System. A system in which heat is generated in a pipeline or vessel wall by causing current to flow through the pipeline or vessel wall by direct connection to an ac voltage source from a dual-winding transformer.

Induction Heating System. A system in which heat is generated in a pipeline or vessel wall by inducing current and hysteresis effect in the pipeline or vessel wall from an external isolated ac field source.

Integrated Heating System. A complete system consisting of components such as pipelines, vessels, heating elements, heat transfer medium, thermal insulation, moisture barrier, nonheating leads, temperature controllers, safety signs, junction boxes, raceways, and fittings.

Pipeline. A length of pipe including pumps, valves, flanges, control devices, strainers, and/or similar equipment for conveying fluids.

Resistance Heating Element. A specific separate element to generate heat that is applied to the pipeline or vessel externally or internally.

FPN: Tubular heaters, strip heaters, heating cable, heating tape, heating blankets, and immersion heaters are examples of resistance heaters.

Skin-Effect Heating System. A system in which heat is generated on the inner surface of a ferromagnetic envelope attached to a pipeline or vessel, or both.

FPN: Typically, an electrically insulated conductor is routed through and connected to the envelope at the other end. The envelope and the electrically insulated conductor are connected to an ac voltage source from a dual-winding transformer.

Vessel. A container such as a barrel, drum, or tank for holding fluids or other material.

427-3. Application of Other Articles. All requirements of this *Code* shall apply except as specifically amended in this article. Cord-connected pipe heating assemblies intended for specific use and identified as suitable for this use shall be installed according to Article 422. Fixed electric pipeline and vessel heating equipment for use in hazardous (classified) locations shall comply with Articles 500 through 516.

427-4. Branch-Circuit Sizing. The ampacity of branch-circuit conductors and the rating or setting of overcurrent protective devices that supply fixed electric heating equipment for pipelines and vessels shall be not less than 125 percent of the total load of the heaters. The rating or setting of overcurrent protective devices shall be permitted in accordance with Section 240-3(b).

B. Installation

427-10. General. Equipment for pipeline and vessel electrical heating shall be identified as being suitable for (1) the chemical, thermal, and physical environment; and (2) installation in accordance with the manufacturer's drawings and instructions.

427-11. Use. Electrical heating equipment shall be installed in such a manner as to be afforded protection from physical damage.

427-12. Thermal Protection. External surfaces of pipeline and vessel heating equipment that operate at temperatures exceeding 60°C (140°F) shall be physically guarded, isolated, or thermally insulated to protect against contact by personnel in the area.

427-13. Identification. The presence of electrically heated pipelines or vessels, or both, shall be evident by the posting of appropriate caution signs or markings at frequent intervals along the pipeline or vessel.

C. Resistance Heating Elements

427-14. Secured. Heating element assemblies shall be secured to the surface being heated by means other than the thermal insulation.

427-15. Not in Direct Contact. Where the heating element is not in direct contact with the pipeline or vessel being heated, means shall be provided to prevent overtemperature of the heating element unless the design of the heater assembly is such that its temperature limitations will not be exceeded.

427-16. Expansion and Contraction. Heating elements and assemblies shall not be installed where they bridge expansion joints unless provisions are made for expansion and contraction.

427-17. Flexural Capability. Where installed on flexible pipelines, the heating elements and assemblies shall have a flexural capability that is compatible with the pipeline.

427-18. Power Supply Leads.

(a) Nonheating Leads. Power supply nonheating leads (cold leads) for resistance elements shall be suitable for the temperature encountered. Preassembled nonheating leads on approved heaters shall be permitted to be shortened if the markings specified in Section 427-20 are retained. Not less than 6 in. (152 mm) of nonheating leads shall be provided within the junction box.

(b) Power Supply Leads Protection. Nonheating power supply leads shall be protected where they emerge from electrically heated pipeline or vessel heating units by rigid metal conduit, intermediate metal conduit, electrical metallic tubing, or other raceways identified as suitable for the application.

(c) Interconnecting Leads. Interconnecting nonheating leads connecting portions of the heating system shall be permitted to be covered by thermal insulation in the same manner as the heaters.

427-19. Electrical Connections.

(a) Nonheating Interconnections. Nonheating interconnections, where required under thermal insulation, shall be made with insulated connectors identified as suitable for this use.

(b) Circuit Connections. Splices and terminations outside the thermal insulation shall be installed in a box or fitting in accordance with Sections 110-14 and 300-15.

427-20. Marking. Each factory-assembled heating unit shall be legibly marked within 3 in. (76 mm) of each end of the nonheating leads with the permanent identification symbol, catalog number, and ratings in volts and watts, or in volts and amperes.

427-22. Equipment Protection. Ground-fault protection of equipment shall be provided for electric heat tracing and heating panels. This requirement shall not apply in industrial establishments where there is alarm indication of ground faults and

(1) Conditions of maintenance and supervision ensure that only qualified persons will service the installed systems.
(2) Continued circuit operation is necessary for safe operation of equipment or processes.

427-23. Metal Covering. Electric heating equipment shall have a grounded metal covering in accordance with (a) or (b). The metal covering shall provide an effective ground path for equipment protection.

(a) Heating Wires or Cables. Heating wires or cables shall have a grounded metal covering that surrounds the heating element and bus wires, if any, and their electrical insulation.

(b) Heating Panels. Heating panels shall have a grounded metal covering over the heating element and its electrical insulation on the side opposite the side attached to the surface to be heated.

D. Impedance Heating

427-25. Personnel Protection. All accessible external surfaces of the pipeline or vessel, or both, being heated shall be physically guarded, isolated, or thermally insulated (with a weatherproof jacket for outside installations) to protect against contact by personnel in the area.

427-26. Isolation Transformer. A dual-winding transformer with a grounded shield between the primary and secondary windings shall be used to isolate the distribution system from the heating system.

427-27. Voltage Limitations. Unless protected by ground-fault circuit-interrupter protection for personnel, the secondary winding of the isolation transformer connected to the pipeline or vessel being heated shall not have an output voltage greater than 30 volts ac.

Where ground-fault circuit-interrupter protection for personnel is provided, the voltage shall be permitted to be greater than 30 but not more than 80 volts.

427-28. Induced Currents. All current-carrying components shall be installed in accordance with Section 300-20.

427-29. Grounding. The pipeline or vessel, or both, being heated that is operating at a voltage greater than 30 but not more than 80 shall be grounded at designated points.

427-30. Secondary Conductor Sizing. The ampacity of the conductors connected to the secondary of the transformer shall be at least 100 percent of the total load of the heater.

E. Induction Heating

427-35. Scope. This part covers the installation of line frequency induction heating equipment and accessories for pipelines and vessels.

FPN: See Article 665 for other applications.

427-36. Personnel Protection. Induction coils that operate or may operate at a voltage greater than 30 volts ac shall be enclosed in a nonmetallic or split metallic enclosure, isolated or made inaccessible by location to protect personnel in the area.

427-37. Induced Current. Induction coils shall be prevented from inducing circulating currents in surrounding metallic equipment, supports, or structures by shielding, isolation, or insulation of the current paths. Stray current paths shall be bonded to prevent arcing.

F. Skin-Effect Heating

427-45. Conductor Ampacity. The ampacity of the electrically insulated conductor inside the ferromagnetic envelope shall be permitted to exceed the values given in Article 310, provided it is identified as suitable for this use.

427-46. Pull Boxes. Pull boxes for pulling the electrically insulated conductor in the ferromagnetic envelope shall be permitted to be buried under the thermal insulation, provided their locations are indicated by permanent markings on the insulation jacket surface and on drawings. For outdoor installations, pull boxes shall be of watertight construction.

427-47. Single Conductor in Enclosure. The provisions of Section 300-20 shall not apply to the installation of a single conductor in a ferromagnetic envelope (metal enclosure).

427-48. Grounding. The ferromagnetic envelope shall be grounded at both ends and, in addition, it shall be permitted to be grounded at intermediate points as required by its design. The ferromagnetic envelope shall be bonded at all joints to ensure electrical continuity.

The provisions of Section 250-30 shall not apply to the installation of skin-effect heating systems.

FPN: See Sections 250-30(a)(4) and (b)(3) for grounding methods.

G. Control and Protection

427-55. Disconnecting Means.

(a) Switch or Circuit Breaker. Means shall be provided to disconnect all fixed electric pipeline or vessel heating equipment from all ungrounded conductors. The branch-circuit switch or circuit breaker, where readily accessible to the user of the equipment, shall be permitted to serve as the disconnecting means. The disconnecting means shall be of the indicating type, and shall be provided with a positive lockout in the "off" position.

(b) Cord- and Plug-Connected Equipment. The factory-installed attachment plug of cord- and plug-connected equipment rated 20 amperes or less and 150 volts or less to ground shall be permitted to be the disconnecting means.

427-56. Controls.

(a) Temperature Control with "Off" Position. Temperature controlled switching devices that indicate an "off" position and that interrupt line current shall open all ungrounded conductors when the control device is in this "off" position. These devices shall not be permitted to serve as the disconnecting means unless provided with a positive lockout in the "off" position.

(b) Temperature Control Without "Off" Position. Temperature controlled switching devices that do not have an "off" position shall not be required to open all ungrounded conductors and shall not be permitted to serve as the disconnecting means.

(c) Remote Temperature Controller. Remote controlled temperature-actuated devices shall not be required to meet the requirements of Sections 427-56(a) and (b). These devices shall not be permitted to serve as the disconnecting means.

(d) Combined Switching Devices. Switching devices consisting of combined temperature-actuated devices and manually controlled switches that serve both as the controllers and the disconnecting means shall comply with all the following conditions:

(1) Open all ungrounded conductors when manually placed in the "off" position
(2) Be designed so that the circuit cannot be energized automatically if the device has been manually placed in the "off" position
(3) Be provided with a positive lockout in the "off" position

427-57. Overcurrent Protection. Heating equipment shall be considered as protected against overcurrent where supplied by a branch circuit as specified in Section 427-4.

Article 430 — Motors, Motor Circuits, and Controllers

A. General

430-1. Scope. This article covers motors, motor branch-circuit and feeder conductors and their protection, motor overload protection, motor control circuits, motor controllers, and motor control centers.

FPN No. 1: Installation requirements for motor control centers are covered in Section 110-26(f). Air conditioning and refrigerating equipment are covered in Article 440.

FPN No. 2: Figure 430-1 is for information only.

430-2. Adjustable-Speed Drive Systems. The incoming branch circuit or feeder to power conversion equipment included as a part of an adjustable-speed drive system shall be based on the rated input to the power conversion equipment. Where the power conversion equipment is marked to indicate that overload protection is included, additional overload protection shall not be required.

The disconnecting means shall be permitted to be in the incoming line to the conversion equipment and shall have a rating not less than 115 percent of the rated input current of the conversion unit.

FPN: Electrical resonance can result from the interaction of the nonsinusoidal currents from this type of load with power factor correction capacitors.

430-3. Part-Winding Motors. A part-winding start induction or synchronous motor is one that is arranged for starting by first energizing part of its primary (armature) winding and, subsequently, energizing the remainder of this winding in one or more steps. A standard part-winding start induction motor is arranged so that one-half of its primary winding

FPN: Figure 430-1.

430-5. Other Articles. Motors and controllers shall also comply with the applicable provisions of the following:

	Article	Section
Air-conditioning and refrigerating equipment	440	
Capacitors		460-8, 460-9
Commercial garages, aircraft hangars, gasoline dispensing and service stations, bulk storage plants, spray application, dipping, and coating processes, and inhalation anesthetizing locations	511, 513, 514, 515, 516, and 517, Part D	
Cranes and hoists	610	
Electrically driven or controlled irrigation machines	675	
Elevators, dumbwaiters, escalators, moving walks, wheelchair lifts, and stairway chair lifts	620	
Fire pumps	695	
Hazardous (classified) locations	500–503 and 505	
Industrial machinery	670	
Motion picture projectors		540-11 and 540-20
Motion picture and television studios and similar locations	530	
Resistors and reactors	470	
Theaters, audience areas of motion picture and television studios, and similar locations		520-48
Transformers and transformer vaults	450	

can be energized initially, and, subsequently, the remaining half can be energized, both halves then carrying equal current. A hermetic refrigerant compressor motor shall not be considered a standard part-winding start induction motor.

Where separate overload devices are used with a standard part-winding start induction motor, each half of the motor winding shall be individually protected in accordance with Sections 430-32 and 430-37 with a trip current one-half that specified.

Each motor-winding connection shall have branch-circuit short-circuit and ground-fault protection rated at not more than one-half that specified by Section 430-52.

Exception: A short-circuit and ground-fault protective device shall be permitted for both windings if the device will allow the motor to start. Where time-delay (dual-element) fuses are used, they shall be permitted to have a rating not exceeding 150 percent of the motor full-load current.

430-6. Ampacity and Motor Rating Determination. The size of conductors supplying equipment covered by Article 430 shall be selected from the allowable ampacity tables in accordance with Section 310-15(b) or shall be calculated in accordance with Section 310-15(c). Where flexible cord is used, the size of the conductor shall be selected in accordance with Section 400-5. The required ampacity and motor ratings shall be determined as specified in (a), (b), and (c).

(a) General Motor Applications. For general motor applications, current ratings shall be determined based on (1) and (2).

(1) Table Values. The values given in Tables 430-147, 430-148, 430-149, and 430-150, including notes, shall be used to determine the ampacity of conductors or ampere ratings of switches, branch-circuit short-circuit and ground-fault protection, instead of the actual current rating marked on the motor nameplate. Where a motor is marked in amperes, but not horsepower, the horsepower rating shall be assumed to be that corresponding to the value given in Tables 430-147, 430-148, 430-149, and 430-150, interpolated if necessary.

Exception No. 1: Multispeed motors shall be in accordance with Sections 430-22(a) and 430-52.

Exception No. 2: For equipment that employs a shaded-pole or permanent-split capacitor-type fan or blower motor that is marked with the motor type, the full load current for such motor marked on the nameplate of the equipment in which the fan or blower motor is employed shall be used instead of the horsepower rating to determine the ampacity or rating of the disconnecting means, the branch-circuit conductors, the controller, the branch-circuit short-circuit and ground-fault protection, and the separate overload protection. This marking on the equipment nameplate shall not be less than the current marked on the fan or blower motor nameplate.

Exception No. 3: For a listed motor-operated appliance that is marked with both motor horsepower and full-load current, the motor full-load current marked on the nameplate of the appliance shall be used instead of the horsepower rating on the appliance nameplate to determine the ampacity or rating of the disconnecting means, the branch-circuit conductors, the controller, the branch-circuit short-circuit and ground-fault protection, and any separate overload protection.

(2) **Nameplate Values.** Separate motor overload protection shall be based on the motor nameplate current rating.

(b) **Torque Motors.** For torque motors, the rated current shall be locked-rotor current, and this nameplate current shall be used to determine the ampacity of the branch-circuit conductors covered in Sections 430-22 and 430-24, the ampere rating of the motor overload protection, and the ampere rating of motor branch-circuit short-circuit and ground-fault protection in accordance with Section 430-52(b).

FPN: For motor controllers and disconnecting means, see Sections 430-83(d) and 430-110.

(c) **Alternating-Current Adjustable Voltage Motors.** For motors used in alternating-current, adjustable voltage, variable torque drive systems, the ampacity of conductors, or ampere ratings of switches, branch-circuit short-circuit and ground-fault protection, etc., shall be based on the maximum operating current marked on the motor or control nameplate, or both. If the maximum operating current does not appear on the nameplate, the ampacity determination shall be based

on 150 percent of the values given in Tables 430-149 and 430-150.

430-7. Marking on Motors and Multimotor Equipment.

(a) **Usual Motor Applications.** A motor shall be marked with the following information.

(1) Manufacturer's name.

(2) Rated volts and full-load amperes. For a multispeed motor, full-load amperes for each speed, except shaded-pole and permanent-split capacitor motors where amperes are required only for maximum speed.

(3) Rated frequency and number of phases, if an ac motor.

(4) Rated full-load speed.

(5) Rated temperature rise or the insulation system class and rated ambient temperature.

(6) Time rating. The time rating shall be 5, 15, 30, or 60 minutes, or continuous.

(7) Rated horsepower if ⅛ hp or more. For a multispeed motor ⅛ hp or more, rated horsepower for each speed, except shaded-pole and permanent-split capacitor motors ⅛ hp or more where rated horsepower is required only for maximum speed. Motors of arc welders are not required to be marked with the horsepower rating.

(8) Code letter or locked-rotor amperes if an alternating-current motor rated ½ hp or more. On polyphase wound-rotor motors, the code letter shall be omitted.

FPN: See Section 430-7(b).

(9) Design letter for design B, C, D, or E motors.

FPN: Motor design letter definitions are found in *Motors and Generators, Part 1, Definitions*, ANSI/NEMA MG 1-1993, and in *Standard Dictionary of Electrical and Electronic Terms*, ANSI/IEEE 100-1996.

(10) Secondary volts and full-load amperes if a wound-rotor induction motor.

(11) Field current and voltage for dc excited synchronous motors.

(12) Winding — straight shunt, stabilized shunt, compound, or series, if a dc motor. Fractional horsepower dc motors 7 in. (178 mm) or less in diameter shall not be required to be marked.

(13) A motor provided with a thermal protector complying with Sections 430-32(a)(2) or (c)(2) shall be marked "Thermally Protected." Thermally protected motors rated

100 watts or less and complying with Section 430-32(c)(2) shall be permitted to use the abbreviated marking "T.P."

(14) A motor complying with Section 430-32(c)(4) shall be marked "Impedance Protected." Impedance protected motors rated 100 watts or less and complying with Section 430-32(c)(4) shall be permitted to use the abbreviated marking "Z.P."

(b) Locked-Rotor Indicating Code Letters. Code letters marked on motor nameplates to show motor input with locked rotor shall be in accordance with Table 430-7(b).

The code letter indicating motor input with locked rotor shall be in an individual block on the nameplate, properly designated.

Table 430-7(b). Locked-Rotor Indicating Code Letters

Code Letter	Kilovolt-Amperes per Horsepower with Locked Rotor	Code Letter	Kilovolt-Amperes per Horsepower with Locked Rotor
A	0–3.14	L	9.0–9.99
B	3.15–3.54	M	10.0–11.19
C	3.55–3.99	N	11.2–12.49
D	4.0–4.49	P	12.5–13.99
E	4.5–4.99	R	14.0–15.99
F	5.0–5.59	S	16.0–17.00
G	5.6–6.29	T	18.0–19.99
H	6.3–7.09	U	29.0–22.39
J	7.1–7.99	V	22.4 and up
K	8.0–8.99		

(1) Multispeed motors shall be marked with the code letter designating the locked-rotor kilovolt-ampere (kVA) per horsepower for the highest speed at which the motor can be started.

Exception: Constant horsepower multispeed motors shall be marked with the code letter giving the highest locked-rotor kilovolt-ampere (kVA) per horsepower.

(2) Single-speed motors starting on Y connection and running on delta connections shall be marked with a code letter corresponding to the locked-rotor kilovolt-ampere (kVA) per horsepower for the Y connection.

(3) Dual-voltage motors that have a different locked-rotor kilovolt-ampere (kVA) per horsepower on the two voltages shall be marked with the code letter for the voltage giving the highest locked-rotor kilovolt-ampere (kVA) per horsepower.

(4) Motors with 60- and 50-hz ratings shall be marked with a code letter designating the locked-rotor kilovolt-ampere (kVA) per horsepower on 60 hz.

(5) Part-winding start motors shall be marked with a code letter designating the locked-rotor kilovolt-ampere (kVA) per

horsepower that is based on the locked-rotor current for the full winding of the motor.

(c) Torque Motors. Torque motors are rated for operation at standstill and shall be marked in accordance with section 430-7 (a), except that locked-rotor torque shall replace horsepower.

(d) Multimotor and Combination-Load Equipment.

(1) Multimotor and combination-load equipment shall be provided with a visible nameplate marked with the manufacturer's name, the rating in volts, frequency, number of phases, minimum supply circuit conductor ampacity, and the maximum ampere rating of the circuit short-circuit and ground-fault protective device. The conductor ampacity shall be computed in accordance with Section 430-24 and counting all of the motors and other loads that will be operated at the same time. The short-circuit and ground-fault protective device rating shall not exceed the value computed in accordance with Section 430-53. Multimotor equipment for use on two or more circuits shall be marked with the above information for each circuit.

(2) Where the equipment is not factory-wired and the individual nameplates of motors and other loads are visible after assembly of the equipment, the individual nameplates shall be permitted to serve as the required marking.

430-8. Marking on Controllers. A controller shall be marked with the manufacturer's name or identification, the voltage, the current or horsepower rating, and such other necessary data to properly indicate the motors for which it is suitable. A controller that includes motor overload protection suitable for group motor application shall be marked with the motor overload protection and the maximum branch-circuit short-circuit and ground-fault protection for such applications.

Combination controllers that employ adjustable instantaneous trip circuit breakers shall be clearly marked to indicate the ampere settings of the adjustable trip element.

Where a controller is built-in as an integral part of a motor or of a motor-generator set, individual marking of the controller shall not be required if the necessary data are on the nameplate. For controllers that are an integral part of equipment approved as a unit, the above marking shall be permitted on the equipment nameplate.

> FPN: See Section 110-10 for information on circuit impedance and other characteristics.

430-9. Terminals.

(a) Markings. Terminals of motors and controllers shall be suitably marked or colored where necessary to indicate the proper connections.

(b) Conductors. Motor controllers and terminals of control circuit devices shall be connected with copper conductors unless identified for use with a different conductor.

(c) Torque Requirements. Control circuit devices with screw-type pressure terminals used with No. 14 or smaller copper conductors shall be torqued to a minimum of 7 lb-in. (0.79 N-m) unless identified for a different torque value.

430-10. Wiring Space in Enclosures.

(a) General. Enclosures for motor controllers and disconnecting means shall not be used as junction boxes, auxiliary gutters, or raceways for conductors feeding through or tapping off to the other apparatus unless designs are employed that provide adequate space for this purpose.

> FPN: See Section 373-8 for switch and overcurrent-device enclosures.

(b) Wire-Bending Space in Enclosures. Minimum wire-bending space within the enclosures for motor controllers shall be in accordance with Table 430-10(b) where measured in a straight line from the end of the lug or wire connector (in the direction the wire leaves the terminal) to the wall or barrier. Where alternate wire termination means are substituted for that supplied by the manufacturer of the controller, they shall be of a type identified by the manufacturer for use with the controller and shall not reduce the minimum wire-bending space.

Table 430-10(b). Minimum Wire-Bending Space at the Terminals of Enclosed Motor Controllers (in.)

Size of Wire (AWG or kcmil)	Wires per Terminal*	
	1	2
14–10	Not specified	—
8–6	1½	—
4–3	2	—
2	2½	—
1	3	—
1/0	5	5
2/0	6	6
3/0–4/0	7	7
250	8	8
300	10	10
350–500	12	12
600–700	14	16
750–900	18	19

*Where provision for three or more wires per terminal exists, the minimum wire-bending space shall be in accordance with the requirements of Article 373.

430-11. Protection Against Liquids. Suitable guards or enclosures shall be provided to protect exposed current-carrying parts of motors and the insulation of motor leads where installed directly under equipment, or in other loca-

tions where dripping or spraying oil, water, or other injurious liquid may occur, unless the motor is designed for the existing conditions.

430-12. Motor Terminal Housings.

(a) Material. Where motors are provided with terminal housings, the housings shall be of metal and of substantial construction.

Exception: In other than hazardous (classified) locations, substantial, nonmetallic, nonburning housings shall be permitted provided an internal grounding means between the motor frame and the equipment grounding connection is incorporated within the housing.

(b) Dimensions and Space — Wire-to-Wire Connections. Where these terminal housings enclose wire-to-wire connections, they shall have minimum dimensions and usable volumes in accordance with Table 430-12(b).

(c) Dimensions and Space — Fixed Terminal Connections. Where these terminal housings enclose rigidly mounted motor terminals, the terminal housing shall be of sufficient size to provide minimum terminal spacings and usable volumes in accordance with Table 430-12(c)(1) and Table 430-12(c)(2).

(d) Large Wire or Factory Connections. For motors with larger ratings, greater number of leads, or larger wire sizes, or where motors are installed as a part of factory-wired equipment, without additional connection being required at the motor terminal housing during equipment installation, the terminal housing shall be of ample size to make connections, but the foregoing provisions for the volumes of terminal housings shall not be considered applicable.

(e) Equipment Grounding Connections. A means for attachment of an equipment grounding conductor termination in accordance with Section 250-8 shall be provided at motor terminal housings for wire-to-wire connections or fixed terminal connections. The means for such connections shall be permitted to be located either inside or outside the motor terminal housing.

Exception: Where a motor is installed as a part of factory-wired equipment that is required to be grounded and without additional connection being required at the motor terminal housing during equipment installation, a separate means for motor grounding at the motor terminal housing shall not be required.

430-13. Bushing. Where wires pass through an opening in an enclosure, conduit box, or barrier, a bushing shall be used to protect the conductors from the edges of openings having sharp edges. The bushing shall have smooth, well-rounded

Table 430-12(b). Terminal Housings — Wire-to-Wire Connections

Motors 11 in. in Diameter or Less

Horsepower	Cover Opening Minimum Dimension (in.)	Usable Volume Minimum (in.³)
1 and smaller[a]	1⅝	10.5
1½, 2, and 3[b]	1¾	16.8
5 and 7½	2	22.4
10 and 15	2½	36.4

Motors Over 11 in. in Diameter Alternating-Current Motors

Maximum Full Load Current for 3-Phase Motors with Maximum of 12 Leads (Amperes)	Terminal Box Cover Opening Minimum Dimension (in.)	Usable Volume Minimum (in.³)	Typical Maximum Horsepower 3-Phase	
			230 Volt	460 Volt
45	2.5	36.4	15	30
70	3.3	77	25	50
110	4.0	140	40	75
160	5.0	252	60	125
250	6.0	450	100	200
400	7.0	840	150	300
600	8.0	1540	250	500

Direct-Current Motors

Maximum Full-Load Current for Motors with Maximum of Six Leads (Amperes)	Terminal Box Minimum Dimensions (in.)	Usable Volume Minimum (in.³)
68	2.5	26
105	3.3	55
165	4.0	100
240	5.0	180
375	6.0	330
600	7.0	600
900	8.0	1100

Notes:

1. For SI units, 1 in. = 25.4 mm.

2. Auxiliary leads for such items as brakes, thermostats, space heaters, exciting fields, etc., shall be permitted to be neglected if their current-carrying area does not exceed 25 percent of the current-carrying area of the machine power leads.

[a]For motors rated 1 hp and smaller and with the terminal housing partially or wholly integral with the frame or end shield, the volume of the terminal housing shall not be less than 1.1 in.³ per wire-to-wire connection. The minimum cover opening dimension is not specified.

[b]For motors rated 1½, 2, and 3 hp and with the terminal housing partially or wholly integral with the frame or end shield, the volume of the terminal housing shall not be less than 1.4 in.³ per wire-to-wire connection. The minimum cover opening dimension is not specified.

Table 430-12(c)(1). Terminal Spacings — Fixed Terminals

	Minimum Spacing (in.)	
Nominal Volts	Between Line Terminals	Between Line Terminals and Other Uninsulated Metal Parts
240 or less	¼	¼
Over 250–600	⅜	⅜

Note: For SI units, 1 in. = 25.4 mm.

Table 430-12(c)(2). Usable Volumes — Fixed Terminals

Power-Supply Conductor Size (AWG)	Minimum Usable Volume per Power-Supply Conductor (in.³)
14	1
12 and 10	1¼
8 and 6	2¼

Note: For SI units, 1 in. = 25.4 mm.

surfaces where it may be in contact with the conductors. If used where oils, greases, or other contaminants may be present, the bushing shall be made of material not deleteriously affected.

FPN: For conductors exposed to deteriorating agents, see Section 310-9.

430-14. Location of Motors.

(a) **Ventilation and Maintenance.** Motors shall be located so that adequate ventilation is provided and so that maintenance, such as lubrication of bearings and replacing of brushes, can be readily accomplished.

(b) **Open Motors.** Open motors that have commutators or collector rings shall be located or protected so that sparks cannot reach adjacent combustible material.

Exception: Installation of these motors on wooden floors or supports shall be permitted.

430-16. Exposure to Dust Accumulations.
In locations where dust or flying material will collect on or in motors in such quantities as to seriously interfere with the ventilation or cooling of motors and thereby cause dangerous temperatures, suitable types of enclosed motors that will not overheat under the prevailing conditions shall be used.

FPN: Especially severe conditions may require the use of enclosed pipe-ventilated motors, or enclosure in separate dusttight rooms, properly ventilated from a source of clean air.

430-17. Highest Rated or Smallest Rated Motor. In determining compliance with Sections 430-24, 430-53(b), and 430-53(c), the highest rated or smallest rated motor shall be based on the rated full-load current as selected from Tables 430-147, 430-148, 430-149, and 430-150.

430-18. Nominal Voltage of Rectifier Systems. The nominal value of the ac voltage being rectified shall be used to determine the voltage of a rectifier derived system.

Exception: The nominal dc voltage of the rectifier shall be used if it exceeds the peak value of the ac voltage being rectified.

B. Motor Circuit Conductors

430-21. General. Part B specifies sizes of conductors that are capable of carrying the motor current without overheating under the conditions specified.

The provisions of Part B shall not apply to motor circuits rated over 600 volts, nominal. See Part K.

The provisions of Articles 250, 300, and 310 shall not apply to conductors that form an integral part of equipment, such as motors, motor controllers, motor control centers, or other factory-assembled control equipment.

FPN No. 1: See Sections 300-1(b) and 310-1 for similar requirements.

FPN No. 2: See Section 430-9(b) for equipment device terminal requirements.

430-22. Single Motor.

(a) General. Branch-circuit conductors that supply a single motor used in a continuous duty application shall have an ampacity of not less than 125 percent of the motor's full-load current rating as determined by Section 430-6(a)(1).

For a multispeed motor, the selection of branch-circuit conductors on the line side of the controller shall be based on the highest of the full-load current ratings shown on the motor nameplate. The selection of branch-circuit conductors between the controller and the motor shall be based on the current rating of the winding(s) that the conductors energize.

Exception No. 1: For dc motors operating from a rectified single-phase power supply, the conductors between the field wiring terminals of the rectifier and the motor shall have an ampacity of not less than the following percent of the motor full-load current rating:

(a) Where a rectifier bridge of the single-phase half-wave type is used, 190 percent

(b) Where a rectifier bridge of the single-phase full-wave type is used, 150 percent

Exception No. 2: Circuit conductors supplying power conversion equipment included as part of an adjustable-speed drive system shall have an ampacity not less than 125 percent of the rated input to the power conversion equipment

FPN: See Appendix D, Example No. D8, and Figure 430-1.

For a wye-start, delta-run connected motor, the selection of branch-circuit conductors on the line side of the controller shall be based on the motor full-load current. The selection of conductors between the controller and the motor shall be based on 58 percent of the motor full-load current.

(b) Other than Continuous Duty. Conductors for a motor used in a short-time, intermittent, periodic, or varying duty application shall have an ampacity of not less than the percentage of the motor nameplate current rating shown in Table 430-22(b), unless the authority having jurisdiction grants special permission for conductors of lower ampacity.

Table 430-22(b). Duty-Cycle Service

	Nameplate Current Rating Percentages			
Classification of Service	**5-Minute Rated Motor**	**15-Minute Rated Motor**	**30- & 60-Minute Rated Motor**	**Continuous Rated Motor**
Short-time duty operating valves, raising or lowering rolls, etc.	110	120	150	—
Intermittent duty freight and passenger elevators, tool heads, pumps, drawbridges, turntables, etc. (for arc welders, see Section 630-11)	85	85	90	140
Periodic duty rolls, ore- and coal-handling machines, etc.	85	90	95	140
Varying duty	110	120	150	200

Note: Any motor application shall be considered as continuous duty unless the nature of the apparatus it drives is such that the motor will not operate continuously with load under any condition of use.

(c) Separate Terminal Enclosure. The conductors between a stationary motor rated 1 hp or less and the separate terminal enclosure permitted in Section 430-145(b) shall be permitted to be smaller than No. 14 but not smaller than No. 18, provided they have an ampacity as specified in Section 430-22(a).

430-23. Wound-Rotor Secondary.

(a) Continuous Duty. For continuous duty, the conductors connecting the secondary of a wound-rotor ac motor to its controller shall have an ampacity not less than 125 percent of the full-load secondary current of the motor.

(b) Other than Continuous Duty. For other than continuous duty, these conductors shall have an ampacity, in percent of full-load secondary current, not less than that specified in Table 430-22(b).

(c) Resistor Separate from Controller. Where the secondary resistor is separate from the controller, the ampacity of the conductors between controller and resistor shall not be less than that shown in Table 430-23(c).

Table 430-23(c). Secondary Conductor

Resistor Duty Classification	Ampacity of Conductor in Percent of Full-Load Secondary Current
Light starting duty	35
Heavy starting duty	45
Extra-heavy starting duty	55
Light intermittent duty	65
Medium intermittent duty	75
Heavy intermittent duty	85
Continuous duty	110

430-24. Several Motors or a Motor(s) and Other Load(s). Conductors supplying several motors, or a motor(s) and other load(s), shall have an ampacity at least equal to the sum of the full-load current rating as determined by Section 430-6(a)(1) of all the motors, plus 25 percent of the highest rated motor in the group, plus the ampere rating of other loads determined in accordance with Article 220 and other applicable articles.

FPN: See Appendix D, Example No. D8.

Exception No. 1: Where one or more of the motors of the group are used for short-time, intermittent, periodic, or varying duty, the ampere rating of such motors to be used in the summation shall be determined in accordance with Section 430-22(b). For the highest rated motor, the greater of either the ampere rating from 430-22(b) or the largest continuous duty motor full-load current multiplied by 1.25 shall be used in the summation.

Exception No. 2: The ampacity of conductors supplying motor-operated fixed electric space-heating equipment shall conform with Section 424-3(b).

Exception No. 3: Where the circuitry is interlocked so as to prevent operation of selected motors or other loads at the same time, the conductor ampacity shall be permitted to be based on the summation of the currents of the motors and other loads to be operated at the same time that results in the highest total current.

430-25. Multimotor and Combination-Load Equipment. The ampacity of the conductors supplying multimotor and combination-load equipment shall not be less than the minimum circuit ampacity marked on the equipment in accordance with Section 430-7(d). Where the equipment is not factory-wired and the individual nameplates are visible in accordance with Section 430-7(d)(2), the conductor ampacity shall be determined in accordance with Section 430-24.

430-26. Feeder Demand Factor. Where reduced heating of the conductors results from motors operating on duty-cycle, intermittently, or from all motors not operating at one time, the authority having jurisdiction may grant permission for feeder conductors to have an ampacity less than specified in Section 430-24, provided the conductors have sufficient ampacity for the maximum load determined in accordance with the sizes and number of motors supplied and the character of their loads and duties.

430-27. Capacitors with Motors. Where capacitors are installed in motor circuits, conductors shall comply with Sections 460-8 and 460-9.

430-28. Feeder Taps. Feeder tap conductors shall have an ampacity not less than that required by Part B, shall terminate in a branch-circuit protective device and, in addition, shall meet one of the following requirements:

(1) Be enclosed by either an enclosed controller or by a raceway, be not more than 10 ft (3.05 m) in length, and, for field installation, be protected by an overcurrent device on the line side of the tap conductor, the rating or setting of which shall not exceed 1000 percent of the tap conductor ampacity

(2) Have an ampacity of at least one-third that of the feeder conductors, be suitably protected from physical damage or enclosed in a raceway, and be not more than 25 ft (7.62 m) in length

(3) Have the same ampacity as the feeder conductors

Exception: Feeder Taps Over 25 ft (7.62 m) Long. In high-bay manufacturing buildings [over 35 ft (10.67 m) high at walls], where conditions of maintenance and supervision ensure that only qualified persons will service the systems, conductors tapped to a feeder shall be permitted to be not over 25 ft (7.62 m) long horizontally and not over 100 ft (30.5 m) in total length where all of the following conditions are met.

(a) The ampacity of the tap conductors is not less than one-third that of the feeder conductors.

(b) The tap conductors terminate with a single circuit breaker or a single set of fuses conforming with (1) Part D, where the load-side conductors are a branch circuit or (2) Part E, where the load-side conductors are a feeder.

(c) The tap conductors are suitably protected from physical damage and are installed in raceways.

(d) *The tap conductors are continuous from end-to-end and contain no splices.*

(e) *The tap conductors shall be No. 6 copper or No. 4 aluminum or larger.*

(f) *The tap conductors shall not penetrate walls, floors, or ceilings.*

(g) *The tap shall not be made less than 30 ft (9.14 m) from the floor.*

430-29. Constant Voltage Direct-Current Motors — Power Resistors. Conductors connecting the motor controller to separately mounted power accelerating and dynamic braking resistors in the armature circuit shall have an ampacity not less than the value calculated from Table 430-29 using motor full-load current. If an armature shunt resistor is used, the power accelerating resistor conductor ampacity shall be calculated using the total of motor full-load current and armature shunt resistor current.

Armature shunt resistor conductors shall have an ampacity of not less than that calculated from Table 430-29 using rated shunt resistor current as full-load current.

Table 430-29. Conductor Rating Factors for Power Resistors

Time in Seconds		Ampacity of Conductor in Percent of Full-Load Current
On	Off	
5	75	35
10	70	45
15	75	55
15	45	65
15	30	75
15	15	85
Continuous Duty		110

C. Motor and Branch-Circuit Overload Protection

430-31. General. Part C specifies overload devices intended to protect motors, motor-control apparatus, and motor branch-circuit conductors against excessive heating due to motor overloads and failure to start.

Overload in electrical apparatus is an operating overcurrent that, when it persists for a sufficient length of time, would cause damage or dangerous overheating of the apparatus. It does not include short circuits or ground faults.

These provisions shall not be interpreted as requiring overload protection where it might introduce additional or increased hazards, as in the case of fire pumps.

FPN: For protection of fire pump supply conductors, see Section 695-6.

The provisions of Part C shall not apply to motor circuits rated over 600 volts, nominal. See Part K.

FPN: See Appendix D, Example No. D8.

430-32. Continuous-Duty Motors.

(a) **More than 1 Horsepower.** Each continuous-duty motor rated more than 1 hp shall be protected against overload by one of the following means.

(1) A separate overload device that is responsive to motor current. This device shall be selected to trip or shall be rated at no more than the following percent of the motor nameplate full-load current rating:

Motors with a marked service factor not less than 1.15	125%
Motors with a marked temperature rise not over 40°C	125%
All other motors	115%

Modification of this value shall be permitted as provided in Section 430-34. For a multispeed motor, each winding connection shall be considered separately.

Where a separate motor overload device is connected so that it does not carry the total current designated on the motor nameplate, such as for wye-delta starting, the proper percentage of nameplate current applying to the selection or setting of the overload device shall be clearly designated on the equipment, or the manufacturer's selection table shall take this into account.

FPN: Where power factor correction capacitors are installed on the load side of the motor overload device, see Section 460-9.

(2) A thermal protector integral with the motor, approved for use with the motor it protects on the basis that it will prevent dangerous overheating of the motor due to overload and failure to start. The ultimate trip current of a thermally protected motor shall not exceed the following percentage of motor full-load current given in Tables 430-148, 430-149, and 430-150:

Motor full-load current not exceeding 9 amperes	170%
Motor full-load current from 9.1 to, and including, 20 amperes	156%
Motor full-load current greater than 20 amperes	140%

If the motor current-interrupting device is separate from the motor and its control circuit is operated by a protective device integral with the motor, it shall be arranged so that the opening of the control circuit will result in interruption of current to the motor.

(3) A protective device integral with a motor that will protect the motor against damage due to failure to start shall be permitted if the motor is part of an approved assembly that does not normally subject the motor to overloads.

(4) For motors larger than 1500 hp, a protective device having embedded temperature detectors that cause current to the motor to be interrupted when the motor attains a temperature rise greater than marked on the nameplate in an ambient temperature of 40°C.

(b) One Horsepower or Less, Nonautomatically Started.

(1) Each continuous-duty motor rated at 1 hp or less that is not permanently installed, is nonautomatically started, and is within sight from the controller location shall be permitted to be protected against overload by the branch-circuit short-circuit and ground-fault protective device. This branch-circuit protective device shall not be larger than that specified in Part D of Article 430.

Exception: Any such motor shall be permitted on a nominal 120-volt branch circuit protected at not over 20 amperes.

(2) Any such motor that is not in sight from the controller location shall be protected as specified in Section 430-32(c). Any motor rated at 1 hp or less that is permanently installed shall be protected in accordance with Section 430-32(c).

(c) One Horsepower or Less, Automatically Started. Any motor of 1 hp or less that is started automatically shall be protected against overload by one of the following means.

(1) A separate overload device that is responsive to motor current. This device shall be selected to trip or shall be rated at not more than the following percentage of the motor nameplate full-load current rating:

Motors with a marked service factor not less than 1.15	125%
Motors with a marked temperature rise not over 40°C	125%
All other motors	115%

For a multispeed motor, each winding connection shall be considered separately. Modification of this value shall be permitted as provided in Section 430-34.

(2) A thermal protector integral with the motor, approved for use with the motor that it protects on the basis that it will prevent dangerous overheating of the motor due to overload and failure to start. Where the motor current-interrupting device is separate from the motor and its control circuit is operated by a protective device integral with the motor, it shall be arranged so that the opening of the control circuit will result in interruption of current to the motor.

(3) A protective device integral with a motor that will protect the motor against damage due to failure to start shall be permitted (1) if the motor is part of an approved assembly that does not subject the motor to overloads, or (2) if the assembly is also equipped with other safety controls (such as the safety combustion controls on a domestic oil burner) that protect the motor against damage due to failure to start. Where the assembly has safety controls that protect the motor, it shall be so indicated on the nameplate of the assembly where it will be visible after installation.

(4) In case the impedance of the motor windings is sufficient to prevent overheating due to failure to start, the motor shall be permitted to be protected as specified in Section 430-32(b)(1) for manually started motors if the motor is part of an approved assembly in which the motor will limit itself so that it will not be dangerously overheated.

> FPN: Many ac motors of less than 1/20 hp, such as clock motors, series motors, etc., and also some larger motors such as torque motors, come within this classification. It does not include split-phase motors having automatic switches that disconnect the starting windings.

(d) Wound-Rotor Secondaries. The secondary circuits of wound-rotor ac motors, including conductors, controllers, resistors, etc., shall be permitted to be protected against overload by the motor-overload device.

430-33. Intermittent and Similar Duty. A motor used for a condition of service that is inherently short-time, intermittent, periodic, or varying duty, as illustrated by Table 430-22(b), shall be permitted to be protected against overload by the branch-circuit short-circuit and ground-fault protective device, provided the protective device rating or setting does not exceed that specified in Table 430-152.

Any motor application shall be considered to be for continuous duty unless the nature of the apparatus it drives is such that the motor cannot operate continuously with load under any condition of use.

430-34. Selection of Overload Relay. Where the overload relay selected in accordance with Sections 430-32(a)(1) and (c)(1) is not sufficient to start the motor or to carry the load, the next higher size overload relay shall be permitted to be used, provided the trip current of the overload relay does not exceed the following percentage of motor nameplate full-load current rating:

Motors with marked service factor not less than 1.15	140%
Motors with a marked temperature rise not over 40°C	140%
All other motors	130%

If not shunted during the starting period of the motor as provided in Section 430-35, the overload device shall have sufficient time delay to permit the motor to start and accelerate its load.

FPN: A Class 20 or 30 overload relay will provide a longer motor acceleration time than a Class 10 or 20, respectively. Use of a higher class overload relay may preclude the need for selection of a higher trip current.

430-35. Shunting During Starting Period.

(a) Nonautomatically Started. For a nonautomatically started motor, the overload protection shall be permitted to be shunted or cut out of the circuit during the starting period of the motor if the device by which the overload protection is shunted or cut out cannot be left in the starting position and if fuses or inverse time circuit breakers rated or set at not over 400 percent of the full-load current of the motor are located in the circuit so as to be operative during the starting period of the motor.

(b) Automatically Started. The motor overload protection shall not be shunted or cut out during the starting period if the motor is automatically started.

Exception: The motor overload protection shall be permitted to be shunted or cut out during the starting period on an automatically started motor where

(a) The motor starting period exceeds the time delay of available motor overload protective devices, and

(b) Listed means are provided to

(1) Sense motor rotation and to automatically prevent the shunting or cutout in the event that the motor fails to start, and

(2) Limit the time of overload protection shunting or cutout to less than the locked rotor time rating of the protected motor, and

(3) Provide for shutdown and manual restart if motor running condition is not reached.

430-36. Fuses — In Which Conductor. Where fuses are used for motor overload protection, a fuse shall be inserted in each ungrounded conductor and also in the grounded conductor if the supply system is 3-wire, 3-phase ac with one conductor grounded.

430-37. Devices Other than Fuses — In Which Conductor. Where devices other than fuses are used for motor overload protection, Table 430-37 shall govern the minimum allowable number and location of overload units such as trip coils or relays.

430-38. Number of Conductors Opened by Overload Device. Motor overload devices, other than fuses or thermal protectors, shall simultaneously open a sufficient number of ungrounded conductors to interrupt current flow to the motor.

430-39. Motor Controller as Overload Protection. A motor controller shall also be permitted to serve as an over-

Table 430-37. Overload Units

Kind of Motor	Supply System	Number and Location of Overload Units, Such as Trip Coils or Relays
1-phase ac or dc	2-wire, 1-phase ac or dc ungrounded	1 in either conductor
1-phase ac or dc	2-wire, 1-phase ac or dc, one conductor grounded	1 in ungrounded conductor
1-phase ac or dc	3-wire, 1-phase ac or dc, grounded neutral	1 in either ungrounded conductor
1-phase ac	Any 3-phase	1 in ungrounded conductor
2-phase ac	3-wire, 2-phase ac, ungrounded	2, one in each phase
2-phase ac	3-wire, 2-phase ac, one conductor grounded	2 in ungrounded conductors
2-phase ac	4-wire, 2-phase ac, grounded or ungrounded	2, one per phase in ungrounded conductors
2-phase ac	5-wire, 2-phase ac, grounded neutral or ungrounded	2, one per phase in any ungrounded phase wire
3-phase ac	Any 3-phase	3, one in each phase*

Exception: An overload unit in each phase shall not be required where overload protection is provided by other approved means.

load device if the number of overload units complies with Table 430-37 and if these units are operative in both the starting and running position in the case of a dc motor, and in the running position in the case of an ac motor.

430-40. Overload Relays. Overload relays and other devices for motor overload protection that are not capable of opening short circuits shall be protected by fuses or circuit breakers with ratings or settings in accordance with Section 430-52 or by a motor short-circuit protector in accordance with Section 430-52.

Exception: Where approved for group installation and marked to indicate the maximum size of fuse or inverse time circuit breaker by which they must be protected, the overload devices shall be protected in accordance with this marking.

FPN: For instantaneous trip circuit breakers or motor short-circuit protectors, see Section 430-52.

430-42. Motors on General-Purpose Branch Circuits. Overload protection for motors used on general-purpose branch circuits as permitted in Article 210 shall be provided as specified in (a), (b), (c), or (d).

(a) Not Over 1 Horsepower. One or more motors without individual overload protection shall be permitted to be connected to a general-purpose branch circuit only where the installation complies with the limiting conditions specified in Sections 430-32(b) and (c) and Sections 430-53(a)(1) and (a)(2).

(b) Over 1 Horsepower. Motors of larger ratings than specified in Section 430-53(a) shall be permitted to be connected to general-purpose branch circuits only where each motor is protected by overload protection selected to protect the motor as specified in Section 430-32. Both the controller and the motor overload device shall be approved for group installation with the short-circuit and ground-fault protective device selected in accordance with Section 430-53.

(c) Cord- and Plug-Connected. Where a motor is connected to a branch circuit by means of an attachment plug and receptacle and individual overload protection is omitted as provided in Section 430-42(a), the rating of the attachment plug and receptacle shall not exceed 15 amperes at 125 volts or 250 volts. Where individual overload protection is required as provided in Section 430-42(b) for a motor or motor-operated appliance that is attached to the branch circuit through an attachment plug and receptacle, the overload device shall be an integral part of the motor or of the appliance. The rating of the attachment plug and receptacle shall determine the rating of the circuit to which the motor may be connected, as provided in Article 210.

(d) Time Delay. The branch-circuit short-circuit and ground- fault protective device protecting a circuit to which a motor or motor-operated appliance is connected shall have sufficient time delay to permit the motor to start and accelerate its load.

430-43. Automatic Restarting. A motor overload device that can restart a motor automatically after overload tripping shall not be installed unless approved for use with the motor it protects. A motor overload device that can restart a motor automatically after overload tripping shall not be installed if automatic restarting of the motor can result in injury to persons.

430-44. Orderly Shutdown. If immediate automatic shutdown of a motor by a motor overload protective device(s) would introduce additional or increased hazard(s) to a person(s) and continued motor operation is necessary for safe shutdown of equipment or process, a motor overload sensing device(s) conforming with the provisions of Part C of this article shall be permitted to be connected to a supervised alarm instead of causing immediate interruption of the motor circuit, so that corrective action or an orderly shutdown can be initiated.

D. Motor Branch-Circuit Short-Circuit and Ground-Fault Protection

430-51. General. Part D specifies devices intended to protect the motor branch-circuit conductors, the motor control apparatus, and the motors against overcurrent due to short circuits or grounds. These rules add to or amend the provisions of Article 240. The devices specified in Part D do not include the types of devices required by Sections 210-8, 230-95, and 305-6.

The provisions of Part D do not apply to motor circuits rated over 600 volts, nominal. See Part K.

FPN: See Appendix D, Example No. D8.

430-52. Rating or Setting for Individual Motor Circuit.

(a) General. The motor branch-circuit short-circuit and ground-fault protective device shall comply with (b) and either (c) or (d), as applicable.

(b) All Motors. The motor branch-circuit short-circuit and ground-fault protective device shall be capable of carrying the starting current of the motor.

(c) Rating or Setting.

(1) A protective device that has a rating or setting not exceeding the value calculated according to the values given in Table 430-152 shall be used.

Exception No. 1: Where the values for branch-circuit short-circuit and ground-fault protective devices determined by Table 430-152 do not correspond to the standard sizes or ratings of fuses, nonadjustable circuit breakers, thermal protective devices, or possible settings of adjustable circuit breakers, the next higher standard size, rating, or possible setting shall be permitted.

Exception No. 2: Where the rating specified in Table 430-152, as modified by Exception No. 1, is not sufficient for the starting current of the motor:

(a) The rating of a nontime-delay fuse not exceeding 600 amperes or a time-delay Class CC fuse shall be permitted to be increased but shall in no case exceed 400 percent of the full-load current.

(b) The rating of a time-delay (dual-element) fuse shall be permitted to be increased but shall in no case exceed 225 percent of the full-load current.

(c) The rating of an inverse time circuit breaker shall be permitted to be increased but shall in no case exceed

400 percent for full-load currents of 100 amperes or less or 300 percent for full-load currents greater than 100 amperes.

(d) *The rating of a fuse of 601–6000 ampere classification shall be permitted to be increased but shall in no case exceed 300 percent of the full-load current.*

FPN: See Appendix D, Example No. D8, and Figure 430-1.

(2) Where maximum branch-circuit short-circuit and ground-fault protective device ratings are shown in the manufacturer's overload relay table for use with a motor controller or are otherwise marked on the equipment, they shall not be exceeded even if higher values are allowed as shown above.

(3) An instantaneous trip circuit breaker shall be used only if adjustable and if part of a listed combination motor controller having coordinated motor overload and short-circuit and ground-fault protection in each conductor, and the setting is adjusted to no more than the value specified in Table 430-152.

FPN: For the purpose of this article, instantaneous-trip circuit breakers may include a damping means to accommodate a transient motor inrush current without nuisance tripping of the circuit breaker.

Exception No. 1: Where the setting specified in Table 430-152 is not sufficient for the starting current of the motor, the setting of an instantaneous trip circuit breaker shall be permitted to be increased but shall in no case exceed 1300 percent of the motor full-load current for other than Design E motors or Design B energy efficient motors and no more than 1700 percent of full-load motor current for Design E motors or Design B energy efficient motors. Trip settings above 800 percent for other than Design E motors or Design B energy efficient motors and above 1100 percent for Design E motors or Design B energy efficient motors shall be permitted where the need has been demonstrated by engineering evaluation. In such cases, it shall not be necessary to first apply an instantaneous-trip circuit breaker at 800 percent or 1100 percent.

FPN: For additional information on the requirements for a motor to be classified "energy efficient," see *Motors and Generators,* NEMA Standards Publication No. MG1-1993, Revision 1, Part 12.59.

Exception No. 2: Where the motor full-load current is 8 amperes or less, the setting of the instantaneous-trip circuit breaker with a continuous current rating of 15 amperes or less in a listed combination motor controller that provides coordinated motor branch-circuit overload and short-circuit and ground-fault protection shall be permitted to be increased to the value marked on the controller.

(4) For a multispeed motor, a single short-circuit and ground-fault protective device shall be permitted for two or more windings of the motor, provided the rating of the protective device does not exceed the above applicable percentage of the nameplate rating of the smallest winding protected.

Exception: For a multispeed motor, a single short-circuit and ground-fault protective device shall be permitted to be used and sized according to the full-load current of the highest current winding, where all of the following conditions are met.

(a) *Each winding is equipped with individual overload protection sized according to its full-load current.*
(b) *The branch-circuit conductors supplying each winding are sized according to the full-load current of the highest full-load current winding.*
(c) *The controller for each winding has a horsepower rating not less than that required for the winding having the highest horsepower rating.*

(5) Suitable fuses shall be permitted in lieu of devices listed in Table 430-152 for power electronic devices in a solid state motor controller system provided that the marking for replacement fuses is provided adjacent to the fuses.

(6) A listed self-protected combination controller shall be permitted in lieu of the devices specified in Table 430-152. Adjustable instantaneous-trip settings shall not exceed 1300 percent of full-load motor current for other than Design E motors or Design B energy efficient motors and not more than 1700 percent of full-load motor current for Design E motors or Design B energy efficient motors.

(7) A motor short-circuit protector shall be permitted in lieu of devices listed in Table 430-152 if the motor short-circuit protector is part of a listed combination motor controller having coordinated motor overload protection and short-circuit and ground-fault protection in each conductor and it will open the circuit at currents exceeding 1300 percent of motor full-load current for other than Design E motors or Design B energy efficient motors and 1700 percent of motor full-load motor current for Design E motors or Design B energy efficient motors.

(d) Torque Motors. Torque motor branch circuits shall be protected at the motor nameplate current rating in accordance with Section 240-3(b).

430-53. Several Motors or Loads on One Branch Circuit. Two or more motors or one or more motors and other loads shall be permitted to be connected to the same branch circuit under conditions specified in (d) and in (a), (b), or (c).

(a) Not Over 1 Horsepower. Several motors, each not exceeding 1 hp in rating, shall be permitted on a nominal 120-

volt branch circuit protected at not over 20 amperes or a branch circuit of 600 volts, nominal, or less, protected at not over 15 amperes, if all of the following conditions are met.

(1) The full-load rating of each motor does not exceed 6 amperes.
(2) The rating of the branch-circuit short-circuit and ground-fault protective device marked on any of the controllers is not exceeded.
(3) Individual overload protection conforms to Section 430-32.

(b) If Smallest Rated Motor Protected. If the branch-circuit short-circuit and ground-fault protective device is selected not to exceed that allowed by Section 430-52 for the smallest rated motor, two or more motors or one or more motors and other load(s), with each motor having individual overload protection, shall be permitted to be connected to a branch circuit where it can be determined that the branch-circuit short-circuit and ground-fault protective device will not open under the most severe normal conditions of service that might be encountered.

(c) Other Group Installations. Two or more motors of any rating or one or more motors and other load(s), with each motor having individual overload protection, shall be permitted to be connected to one branch circuit where the motor controller(s) and overload device(s) are (1) installed as a listed factory assembly and the motor branch-circuit short-circuit and ground-fault protective device is either provided as part of the assembly or is specified by a marking on the assembly, or (2) the motor branch-circuit short-circuit and ground-fault protective device, the motor controller(s), and overload device(s) are field-installed as separate assemblies listed for such use and provided with manufacturers' instructions for use with each other, and (3) all of the following conditions are complied with.

(1) Each motor overload device is listed for group installation with a specified maximum rating of fuse or inverse time circuit breaker, or both.
(2) Each motor controller is listed for group installation with a specified maximum rating of fuse or circuit breaker, or both.
(3) Each circuit breaker is one of the inverse time type and listed for group installation.
(4) The branch circuit shall be protected by fuses or inverse time circuit breakers having a rating not exceeding that specified in Section 430-52 for the highest rated motor connected to the branch circuit plus an amount equal to the sum of the full-load current ratings of all other motors and the ratings of other loads connected to the circuit. Where this calculation results in a rating less than the

ampacity of the supply conductors, it shall be permitted to increase the maximum rating of the fuses or circuit breaker to a value not exceeding that permitted by Section 240-3(b).
(5) The branch-circuit fuses or inverse time circuit breakers are not larger than allowed by Section 430-40 for the overload relay protecting the smallest rated motor of the group.

FPN: See Section 110-10 for circuit impedance and other characteristics.

(d) Single Motor Taps. For group installations described above, the conductors of any tap supplying a single motor shall not be required to have an individual branch-circuit short-circuit and ground-fault protective device, provided they comply with either of the following.

(1) No conductor to the motor shall have an ampacity less than that of the branch-circuit conductors.
(2) No conductor to the motor shall have an ampacity less than one-third that of the branch-circuit conductors, with a minimum in accordance with Section 430-22; the conductors to the motor overload device being not more than 25 ft (7.62 m) long and being protected from physical damage.

430-54. Multimotor and Combination-Load Equipment. The rating of the branch-circuit short-circuit and ground-fault protective device for multimotor and combination-load equipment shall not exceed the rating marked on the equipment in accordance with Section 430-7(d).

430-55. Combined Overcurrent Protection. Motor branch-circuit short-circuit and ground-fault protection and motor overload protection shall be permitted to be combined in a single protective device where the rating or setting of the device provides the overload protection specified in Section 430-32.

430-56. Branch-Circuit Protective Devices — In Which Conductor. Branch-circuit protective devices shall comply with the provisions of Section 240-20.

430-57. Size of Fuseholder. Where fuses are used for motor branch-circuit short-circuit and ground-fault protection, the fuseholders shall not be of a smaller size than required to accommodate the fuses specified by Table 430-152.

Exception: Where fuses having time delay appropriate for the starting characteristics of the motor are used, it shall be permitted to use fuseholders sized to fit the fuses that are used.

430-58. Rating of Circuit Breaker. A circuit breaker for motor branch-circuit short-circuit and ground-fault protection shall have a current rating in accordance with Sections 430-52 and 430-110.

E. Motor Feeder Short-Circuit and Ground-Fault Protection

430-61. General. Part E specifies protective devices intended to protect feeder conductors supplying motors against overcurrents due to short circuits or grounds.

FPN: See Appendix D, Example No. D8.

430-62. Rating or Setting — Motor Load.

(a) Specific Load. A feeder supplying a specific fixed motor load(s) and consisting of conductor sizes based on Section 430-24 shall be provided with a protective device having a rating or setting not greater than the largest rating or setting of the branch-circuit short-circuit and ground-fault protective device for any motor supplied by the feeder [based on the maximum permitted value for the specific type of a protective device shown in Table 430-152, or Section 440-22(a) for hermetic refrigerant motor-compressors], plus the sum of the full-load currents of the other motors of the group.

Where the same rating or setting of the branch-circuit short-circuit and ground-fault protective device is used on two or more of the branch circuits supplied by the feeder, one of the protective devices shall be considered the largest for the above calculations.

Exception: Where one or more instantaneous trip circuit breakers or motor short-circuit protectors are used for motor branch-circuit short-circuit and ground-fault protection as permitted in Section 430-52(c), the procedure provided above for determining the maximum rating of the feeder protective device shall apply with the following provision. For the purpose of the calculation, each instantaneous trip circuit breaker or motor short-circuit protector shall be assumed to have a rating not exceeding the maximum percentage of motor full-load current permitted by Table 430-152 for the type of feeder protective device employed.

FPN: See Appendix D, Example No. D8.

(b) Other Installations. Where feeder conductors have an ampacity greater than required by Section 430-24, the rating or setting of the feeder overcurrent protective device shall be permitted to be based on the ampacity of the feeder conductors.

430-63. Rating or Setting — Power and Light Loads. Where a feeder supplies a motor load and, in addition, a lighting or a lighting and appliance load, the feeder protective device shall be permitted to have a rating or setting that is sufficient to carry the lighting or the lighting and appliance load as determined in accordance with Articles 210 and 220 plus, for a single motor, the rating permitted by Section 430-52, and, for two or more motors, the rating permitted by Section 430-62.

F. Motor Control Circuits

430-71. General. Part F contains modifications of the general requirements and applies to the particular conditions of motor control circuits.

FPN: See Section 430-9(b) for equipment device terminal requirements.

Definition of Motor Control Circuit. The circuit of a control apparatus or system that carries the electric signals directing the performance of the controller, but does not carry the main power current.

430-72. Overcurrent Protection.

(a) General. A motor control circuit tapped from the load side of a motor branch-circuit short-circuit and ground-fault protective device(s) and functioning to control the motor(s) connected to that branch circuit shall be protected against overcurrent in accordance with Section 430-72. Such a tapped control circuit shall not be considered to be a branch circuit and shall be permitted to be protected by either a supplementary or branch-circuit overcurrent protective device(s). A motor control circuit other than such a tapped control circuit shall be protected against overcurrent in accordance with Section 725-23 or the notes to Tables 11(a) and (b) in Chapter 9, as applicable.

(b) Conductor Protection. The overcurrent protection for conductors shall be provided as specified in (1) or (2).

Exception No. 1: Where the opening of the control circuit would create a hazard as, for example, the control circuit of a fire pump motor, and the like, conductors of control circuits shall require only short-circuit and ground-fault protection and shall be permitted to be protected by the motor branch-circuit short-circuit and ground-fault protective device(s).

Exception No. 2: Conductors supplied by the secondary side of a single-phase transformer having only a two-wire (single-voltage) secondary shall be permitted to be protected by overcurrent protection provided on the primary (supply) side of the transformer, provided this protection does not exceed the value determined by multiplying the appropriate maximum rating of the overcurrent device for the secondary conductor from Table 430-72(b) by the secondary-to-primary voltage ratio. Transformer secondary conductors (other than two-wire) shall not be considered to be protected by the primary overcurrent protection.

(1) Where the motor branch-circuit short-circuit and ground-fault protective device does not provide protection in accordance with Section 430-72(b)(2), separate overcurrent protection shall be provided. The overcurrent protection shall not exceed the values specified in Column A of Table 430-72(b).

(2) Conductors shall be permitted to be protected by the motor branch-circuit short-circuit and ground-fault protective device and shall require only short-circuit and ground-fault protection. Where the conductors do not extend beyond the motor control equipment enclosure, the rating of the protective device(s) shall not exceed the value specified in Column B of Table 430-72(b). Where the conductors extend beyond the motor control equipment enclosure, the rating of the protective device(s) shall not exceed the value specified in Column C of Table 430-72(b).

(c) Control Circuit Transformer. Where a motor control circuit transformer is provided, the transformer shall be protected in accordance with (1), (2), (3), (4), or (5).

Exception: Overcurrent protection shall be omitted where the opening of the control circuit would create a hazard as, for example, the control circuit of a fire pump motor and the like.

(1) Where the transformer supplies a Class 1 power-limited circuit, Class 2, or Class 3 remote-control circuit conforming with the requirements of Article 725, the protection shall comply with Article 725.

(2) Protection shall be permitted to be provided in accordance with Section 450-3.

(3) Control circuit transformers rated less than 50 volt-amperes (VA) and that are an integral part of the motor controller and located within the motor controller enclosure shall be permitted to be protected by primary overcurrent devices, impedance limiting means, or other inherent protective means.

(4) Where the control circuit transformer rated primary current is less than 2 amperes, an overcurrent device rated or set at not more than 500 percent of the rated primary current shall be permitted in the primary circuit.

(5) Protection shall be permitted to be provided by other approved means.

430-73. Mechanical Protection of Conductor. Where damage to a motor control circuit would constitute a hazard, all conductors of such a remote motor control circuit that are outside the control device itself shall be installed in a raceway or be otherwise suitably protected from physical damage.

Where one side of the motor control circuit is grounded, the motor control circuit shall be arranged so that an accidental ground in the control circuit remote from the motor controller will (1) not start the motor, and (2) not bypass manually operated shutdown devices or automatic safety shutdown devices.

430-74. Disconnection.

(a) General. Motor control circuits shall be arranged so that they will be disconnected from all sources of supply when the disconnecting means is in the open position. The disconnecting means shall be permitted to consist of two or more separate devices, one of which disconnects the motor and the controller from the source(s) of power supply for the motor, and the other(s), the motor control circuit(s) from its power supply. Where separate devices are used, they shall be located immediately adjacent to each other.

Exception No. 1: Where more than 12 motor control circuit conductors are required to be disconnected, the disconnect-

Table 430-72(b). Maximum Rating of Overcurrent Protective Device in Amperes

| | Separate Protection Provided Column A | | Protection Provided by Motor Branch-Circuit Protective Device(s) | | | |
| | | | Conductors Within Enclosure Column B | | Conductors Extend Beyond Enclosure Column C | |
Control Circuit Conductor Size (AWG)	Copper	Aluminum or Copper-Clad Aluminum	Copper	Aluminum or Copper-Clad Aluminum	Copper	Aluminum or Copper-Clad Aluminum
18	7	—	25	—	7	—
16	10	—	40	—	10	—
14	(Note 1)	—	100	—	45	—
12	(Note 1)	(Note 1)	120	100	60	45
10	(Note 1)	(Note 1)	160	140	90	75
larger than 10	(Note 1)	(Note 1)	(Note 2)	(Note 2)	(Note 3)	(Note 3)

ing means shall be permitted to be located other than immediately adjacent to each other where all of the following conditions are complied with.

(a) Access to energized parts is limited to qualified persons in accordance with Part L of this article.

(b) A warning sign is permanently located on the outside of each equipment enclosure door or cover permitting access to the live parts in the motor control circuit(s), warning that motor control circuit disconnecting means are remotely located and specifying the location and identification of each disconnect. Where energized parts are not in an equipment enclosure as permitted by Sections 430-132 and 430-133, an additional warning sign(s) shall be located where visible to persons who may be working in the area of the energized parts.

Exception No. 2: The motor control circuit disconnecting means shall be permitted to be remote from the motor controller power supply disconnecting means where the opening of one or more motor control circuit disconnect means may result in potentially unsafe conditions for personnel or property and the conditions of items (a) and (b) of Exception No. 1 above are complied with.

(b) Control Transformer in Controller Enclosure. Where a transformer or other device is used to obtain a reduced voltage for the motor control circuit and is located in the controller enclosure, such transformer or other device shall be connected to the load side of the disconnecting means for the motor control circuit.

G. Motor Controllers

430-81. General. Part G is intended to require suitable controllers for all motors.

(a) Definition. For the definition of *Controller*, see Article 100. For the purpose of this article, a controller is any switch or device that is normally used to start and stop a motor by making and breaking the motor circuit current.

(b) Stationary Motor of ⅛ Horsepower or Less. For a stationary motor rated at ⅛ hp or less that is normally left running and is constructed so that it cannot be damaged by overload or failure to start, such as clock motors and the like, the branch-circuit protective device shall be permitted to serve as the controller.

(c) Portable Motor of ⅓ Horsepower or Less. For a portable motor rated at ⅓ hp or less, the controller shall be permitted to be an attachment plug and receptacle.

430-82. Controller Design.

(a) Starting and Stopping. Each controller shall be capable of starting and stopping the motor it controls and shall be capable of interrupting the locked-rotor current of the motor.

(b) Autotransformer. An autotransformer starter shall provide an "off" position, a running position, and at least one starting position. It shall be designed so that it cannot rest in the starting position or in any position that will render the overload device in the circuit inoperative.

(c) Rheostats. Rheostats shall be in compliance with the following.

(1) Motor-starting rheostats shall be designed so that the contact arm cannot be left on intermediate segments. The point or plate on which the arm rests when in the starting position shall have no electrical connection with the resistor.

(2) Motor-starting rheostats for dc motors operated from a constant voltage supply shall be equipped with automatic devices that will interrupt the supply before the speed of the motor has fallen to less than one-third its normal rate.

430-83. Ratings. The controller shall have a rating as specified in (a) of this section, unless otherwise permitted in (b) or (c), or as specified in (d) of this section, under the conditions specified.

(a) General.

(1) Controllers, other than inverse time circuit breakers and molded case switches, shall have horsepower ratings at the application voltage not lower than the horsepower rating of the motor. A controller for a Design E motor rated more than 2 hp shall (1) be marked as rated for use with a Design E motor, or (2) have a horsepower rating not less than 1.4 times the rating of a motor rated 3 through 100 hp, or not less than 1.3 times the rating of a motor rated over 100 hp.

(2) A branch-circuit inverse time circuit breaker rated in amperes shall be permitted as a controller for all motors, including Design E. Where this circuit breaker is also used for overload protection, it shall conform to the appropriate provisions of this article governing overload protection.

(b) Small Motors. Devices as specified in Section 430-81(b) and (c) shall be permitted as a controller.

(c) Stationary Motors of 2 Horsepower or Less. For stationary motors rated at 2 hp or less and 300 volts or less, the controller shall be permitted to be either of the following:

(1) A general-use switch having an ampere rating not less than twice the full-load current rating of the motor
(2) On ac circuits, a general-use snap switch suitable only for use on ac (not general-use ac-dc snap switches)

where the motor full-load current rating is not more than 80 percent of the ampere rating of the switch

(d) Torque Motors. For torque motors, the controller shall have a continuous-duty, full-load current rating not less than the nameplate current rating of the motor. For a motor controller rated in horsepower but not marked with the foregoing current rating, the equivalent current rating shall be determined from the horsepower rating by using Tables 430-147, 430-148, 430-149, or 430-150.

(e) Voltage Rating. A controller with a straight voltage rating, e.g., 240 volts or 480 volts, shall be permitted to be applied in a circuit in which the nominal voltage between any two conductors does not exceed the controller's voltage rating. A controller with a slash rating, e.g., 120/240 volts or 480/277 volts, shall only be applied in a circuit in which the nominal voltage to ground from any conductor does not exceed the lower of the two values of the controller's voltage rating and the nominal voltage between any two conductors does not exceed the higher value of the controller's voltage rating.

430-84. Need Not Open All Conductors. The controller shall not be required to open all conductors to the motor.

Exception: Where the controller serves also as a disconnecting means, it shall open all ungrounded conductors to the motor as provided in Section 430-111.

430-85. In Grounded Conductors. One pole of the controller shall be permitted to be placed in a permanently grounded conductor, provided the controller is designed so that the pole in the grounded conductor cannot be opened without simultaneously opening all conductors of the circuit.

430-87. Number of Motors Served by Each Controller. Each motor shall be provided with an individual controller.

Exception: For motors rated 600 volts or less, a single controller rated at not less than the equivalent horsepower, as determined in accordance with Section 430-110(c)(1), of all the motors in the group shall be permitted to serve the group under any of the following conditions:

(a) Where a number of motors drive several parts of a single machine or piece of apparatus, such as metal and woodworking machines, cranes, hoists, and similar apparatus

(b) Where a group of motors is under the protection of one overcurrent device as permitted in Section 430-53(a)

(c) Where a group of motors is located in a single room within sight from the controller location

430-88. Adjustable-Speed Motors. Adjustable-speed motors that are controlled by means of field regulation shall be equipped and connected so that they cannot be started under a weakened field.

Exception: Starting under a weakened field shall be permitted where the motor is designed for such starting.

430-89. Speed Limitation. Machines of the following types shall be provided with speed limiting devices or other speed limiting means:

(1) Separately excited dc motors
(2) Series motors
(3) Motor-generators and converters that can be driven at excessive speed from the dc end, as by a reversal of current or decrease in load

Exception: Separate speed-limiting devices or means shall not be required under either of the following conditions:

(a) Where the inherent characteristics of the machines, the system, or the load and the mechanical connection thereto are such as to safely limit the speed

(b) Where the machine is always under the manual control of a qualified operator

430-90. Combination Fuseholder and Switch as Controller. The rating of a combination fuseholder and switch used as a motor controller shall be such that the fuseholder will accommodate the size of the fuse specified in Part C of this article for motor overload protection.

Exception: Where fuses having time delay appropriate for the starting characteristics of the motor are used, fuseholders of smaller size than specified in Part C of this article shall be permitted.

430-91. Motor Controller Enclosure Types. Table 430-91 provides the basis for selecting enclosures for use in specific locations, other than hazardous (classified) locations. The enclosures are not intended to protect against conditions such as condensation, icing, corrosion, or contamination that may occur within the enclosure or enter via the conduit or unsealed openings. These internal conditions shall require special consideration by the installer and user.

H. Motor Control Centers

430-92. General. Part H covers motor control centers installed for the control of motors, lighting, and power circuits.

430-94. Overcurrent Protection. Motor control centers shall be provided with overcurrent protection in accordance with Article 240 based on the rating of the common power bus. This protection shall be provided by (1) an overcurrent

Table 430-91. Motor Controller Enclosure Selection

For Outdoor Use							
Provides a Degree of Protection Against the Following Environmental Conditions	Enclosure Type Number[1]						
	3	3R	3S	4	4X	6	6P
Incidental contact with the enclosed equipment	X	X	X	X	X	X	X
Rain, snow, and sleet	X	X	X	X	X	X	X
Sleet[2]	—	—	X	—	—	—	—
Windblown dust	X	—	X	X	X	X	X
Hosedown	—	—	—	X	X	X	X
Corrosive agents	—	—	—	—	X	—	X
Occasional temporary submersion	—	—	—	—	—	X	X
Occasional prolonged submersion	—	—	—	—	—	—	X

For Indoor Use									
Provides a Degree of Protection Against the Following Environmental Conditions	Enclosure Type Number[1]								
	1	2	4	4X	5	6	6P	12	12K	13
Incidental contact with the enclosed equipment	X	X	X	X	X	X	X	X	X	X
Falling dirt	X	X	X	X	X	X	X	X	X	X
Falling liquids and light splashing	—	X	X	X	X	X	X	X	X	X
Circulating dust, lint, fibers, and flyings	—	—	X	X	—	X	X	X	X	X
Settling airborne dust, lint, fibers, and flyings	—	—	X	X	X	X	X	X	X	X
Hosedown and splashing water	—	—	X	X	—	X	X	—	—	—
Oil and coolant seepage	—	—	—	—	—	—	—	X	X	X
Oil or coolant spraying and splashing	—	—	—	—	—	—	—	—	—	X
Corrosive agents	—	—	—	X	—	—	X	—	—	—
Occasional temporary submersion	—	—	—	—	—	X	X	—	—	—
Occasional prolonged submersion	—	—	—	—	—	—	X	—	—	—

[1]Enclosure type number shall be marked on the motor controller enclosure.

[2]Mechanism shall be operable when ice covered.

protective device located ahead of the motor control center or (2) a main overcurrent protective device located within the motor control center.

430-95. Service-Entrance Equipment. Where used as service equipment, each motor control center shall be provided with a single main disconnecting means to disconnect all ungrounded service conductors.

Exception: A second service disconnect shall be permitted to supply additional equipment.

Where a grounded conductor is provided, the motor control center shall be provided with a main bonding jumper, sized in accordance with Section 250-28(d), within one of the sections for connecting the grounded conductor, on its supply side, to the motor control center equipment ground bus.

430-96. Grounding. Multisection motor control centers shall be bonded together with an equipment grounding con-

ductor or an equivalent grounding bus sized in accordance with Table 250-122. Equipment grounding conductors shall terminate on this grounding bus or to a grounding termination point provided in a single-section motor control center.

430-97. Busbars and Conductors.

(a) Support and Arrangement. Busbars shall be protected from physical damage and be held firmly in place. Other than for required interconnections and control wiring, only those conductors that are intended for termination in a vertical section shall be located in that section.

Exception: Conductors shall be permitted to travel horizontally through vertical sections where such conductors are isolated from the busbars by a barrier.

(b) Phase Arrangement. The phase arrangement on 3-phase horizontal common power and vertical buses shall be A, B, C from front to back, top to bottom, or left to right, as viewed from the front of the motor control center. The

B phase shall be that phase having the higher voltage to ground on 3-phase, 4-wire, delta-connected systems. Other busbar arrangements shall be permitted for additions to existing installations and shall be marked.

Exception: Rear-mounted units connected to a vertical bus that is common to front-mounted units shall be permitted to have a C, B, A phase arrangement where properly identified.

(c) Minimum Wire-Bending Space. The minimum wire-bending space at the motor control center terminals and minimum gutter space shall be as required in Article 373.

(d) Spacings. Spacings between motor control center bus terminals and other bare metal parts shall not be less than specified in Table 430-97.

(e) Barriers. Barriers shall be placed in all service-entrance motor control centers to isolate service busbars and terminals from the remainder of the motor control center.

Table 430-97. Minimum Spacing Between Bare Metal Parts

	Opposite Polarity Where Mounted on the Same Surface (in.)	Opposite Polarity Where Held Free in Air (in.)	Live Parts to Ground (in.)
Not over 125 volts, nominal	¾	½	½
Not over 250 volts, nominal	1¼	¾	½
Not over 600 volts, nominal	2	1	1

Note: For SI units, 1 in. = 25.4 mm.

430-98. Marking.

(a) Motor Control Centers. Motor control centers shall be marked according to Section 110-21, and such marking shall be plainly visible after installation. Marking shall also include common power bus current rating and motor control center short-circuit rating.

(b) Motor Control Units. Motor control units in a motor control center shall comply with Section 430-8.

J. Disconnecting Means

430-101. General. Part J is intended to require disconnecting means capable of disconnecting motors and controllers from the circuit.

> FPN No. 1: See Figure 430-1.

> FPN No. 2: See Section 110-22 for identification of disconnecting means.

430-102. Location.

(a) Controller. An individual disconnecting means shall be provided for each controller and shall disconnect the controller. The disconnecting means shall be located in sight from the controller location.

Exception No. 1: For motor circuits over 600 volts, nominal, a controller disconnecting means capable of being locked in the open position shall be permitted to be out of sight of the controller, provided the controller is marked with a warning label giving the location of the disconnecting means.

Exception No. 2: A single disconnecting means shall be permitted for a group of coordinated controllers that drive several parts of a single machine or piece of apparatus. The disconnecting means and the controllers shall be located in sight from the machine or apparatus.

(b) Motor. A separate disconnecting means shall be located in sight from the motor location and the driven machinery location.

Exception: A disconnecting means, in addition to the controller disconnecting means as required in accordance with Section 430-102(a), shall not be required for the motor where the disconnecting means for the controller is individually capable of being locked in the open position.

> FPN: For information on lockout/tagout procedures, see *Standard for Electrical Safety Requirements for Employee Workplaces*, NFPA 70E-1995.

430-103. Operation.
The disconnecting means shall open all ungrounded supply conductors and shall be designed so that no pole can be operated independently. The disconnecting means shall be permitted in the same enclosure with the controller.

> FPN: See Section 430-113 for equipment receiving energy from more than one source.

430-104. To Be Indicating.
The disconnecting means shall plainly indicate whether it is in the open (off) or closed (on) position.

430-105. Grounded Conductors.
One pole of the disconnecting means shall be permitted to disconnect a permanently grounded conductor, provided the disconnecting means is designed so that the pole in the grounded conductor cannot be opened without simultaneously disconnecting all conductors of the circuit.

430-107. Readily Accessible.
One of the disconnecting means shall be readily accessible.

430-108. Every Switch.
Every disconnecting means in the motor circuit between the point of attachment to the feeder

and the point of connection to the motor shall comply with the requirements of Sections 430-109 and 430-110.

430-109. Type. The disconnecting means shall be a type specified in (a), unless otherwise permitted in (b) through (g), under the conditions specified.

(a) General.

(1) A listed motor-circuit switch rated in horsepower. For Design E motors rated greater than 2 hp, the motor circuit switch shall be either (a) marked as rated for use with Design E motors or (b) have a horsepower rating not less than 1.4 times the rating of a motor rated 3–100 hp, or not less than 1.3 times the rating of a motor rated over 100 hp.

(2) A listed molded case circuit breaker.

(3) A listed molded case switch.

(4) An instantaneous trip circuit breaker that is part of a listed combination motor controller.

(5) Listed self-protected combination controller.

(6) Listed manual motor controllers additionally marked "Suitable as Motor Disconnect" shall be permitted as a disconnecting means where installed between the final motor branch-circuit short-circuit and ground-fault protective device and the motor.

(b) Stationary Motors of ⅛ Horsepower or Less. For stationary motors of ⅛ hp or less, the branch-circuit overcurrent device shall be permitted to serve as the disconnecting means.

(c) Stationary Motors of 2 Horsepower or Less. For stationary motors rated at 2 hp or less and 300 volts or less, the disconnecting means shall be permitted to be one of the devices specified in (1), (2), or (3).

(1) A general-use switch having an ampere rating not less than twice the full-load current rating of the motor.

(2) On ac circuits, a general-use snap switch suitable only for use on ac (not general-use ac-dc snap switches) where the motor full-load current rating is not more than 80 percent of the ampere rating of the switch

(3) A listed manual motor controller having a horsepower rating not less than the rating of the motor and marked "Suitable as Motor Disconnect"

(d) Autotransformer-Type Controlled Motors. For motors of over 2 hp to and including 100 hp, the separate disconnecting means required for a motor with an autotransformer-type controller shall be permitted to be a general-use switch where all of the following provisions are met.

(1) The motor drives a generator that is provided with overload protection.

(2) The controller is capable of interrupting the locked-rotor current of the motors, is provided with a no voltage release, and is provided with running overload protection not exceeding 125 percent of the motor full-load current rating.

(3) Separate fuses or an inverse time circuit breaker rated or set at not more than 150 percent of the motor full-load current are provided in the motor branch circuit.

(e) Isolating Switches. For stationary motors rated at more than 40 hp dc or 100 hp ac, the disconnecting means shall be permitted to be a general-use or isolating switch where plainly marked "Do not operate under load."

(f) Cord- and Plug-Connected Motors. For a cord- and plug-connected motor, a horsepower-rated attachment plug and receptacle having ratings no less than the motor ratings shall be permitted to serve as the disconnecting means for other than a Design E motor, and for a Design E motor rated 2 hp or less. For a Design E motor rated more than 2 hp, an attachment plug and receptacle used as the disconnecting means shall have a horsepower rating not less than 1.4 times the motor rating. A horsepower-rated attachment plug and receptacle shall not be required for a cord- and plug-connected appliance in accordance with Section 422-32, a room air conditioner in accordance with Section 440-63, or a portable motor rated ⅓ hp or less.

(g) Torque Motors. For torque motors, the disconnecting means shall be permitted to be a general-use switch.

430-110. Ampere Rating and Interrupting Capacity.

(a) General. The disconnecting means for motor circuits rated 600 volts, nominal, or less, shall have an ampere rating of at least 115 percent of the full-load current rating of the motor.

Exception: A listed nonfused motor-circuit switch having a horsepower rating equal to or greater than the motor horsepower shall be permitted to have an ampere rating less than 115 percent of the full-load current rating of the motor.

(b) For Torque Motors. Disconnecting means for a torque motor shall have an ampere rating of at least 115 percent of the motor nameplate current.

(c) For Combination Loads. Where two or more motors are used together or where one or more motors are used in combination with other loads, such as resistance heaters, and where the combined load may be simultaneous on a single disconnecting means, the ampere and horsepower ratings of the combined load shall be determined as follows.

(1) The rating of the disconnecting means shall be determined from the sum of all currents, including resistance loads, at the full-load condition and also at the locked-rotor condition. The combined full-load current and the combined locked-rotor current so obtained shall be considered as a single motor for the purpose of this requirement as follows.

The full-load current equivalent to the horsepower rating of each motor shall be selected from Table 430-147, 430-148, 430-149, or 430-150. These full-load currents shall be added to the rating in amperes of other loads to obtain an equivalent full-load current for the combined load.

The locked-rotor current equivalent to the horsepower rating of each motor shall be selected from Table 430-151(A) or 430-151(B). The locked-rotor currents shall be added to the rating in amperes of other loads to obtain an equivalent locked-rotor current for the combined load. Where two or more motors or other loads cannot be started simultaneously, the largest sum of locked rotor currents of a motor or group of motors that can be started simultaneously and the full load currents of other concurrent loads shall be permitted to be used to determine the equivalent locked-rotor current for the simultaneous combined loads.

Exception: Where part of the concurrent load is resistance load, and where the disconnecting means is a switch rated in horsepower and amperes, the switch used shall be permitted to have a horsepower rating that is not less than the combined load of the motor(s), if the ampere rating of the switch is not less than the locked-rotor current of the motor(s) plus the resistance load.

(2) The ampere rating of the disconnecting means shall not be less than 115 percent of the sum of all currents at the full-load condition determined in accordance with Section 430-110(c)(1).

Exception: A listed nonfused motor-circuit switch having a horsepower rating equal to or greater than the equivalent horsepower of the combined loads, determined in accordance with Section 430-110(c)(1), shall be permitted to have an ampere rating less than 115 percent of the sum of all currents at the full-load condition.

(3) For small motors not covered by Table 430-147, 430-148, 430-149, or 430-150, the locked-rotor current shall be assumed to be six times the full-load current.

430-111. Switch or Circuit Breaker as Both Controller and Disconnecting Means. A switch or circuit breaker shall be permitted to be used as both the controller and disconnecting means if it complies with (a) and is one of the types specified in (b).

(a) General. The switch or circuit breaker complies with the requirements for controllers specified in Section 430-

83, opens all ungrounded conductors to the motor, and is protected by an overcurrent device in each ungrounded conductor (which shall be permitted to be the branch-circuit fuses). The overcurrent device protecting the controller shall be permitted to be part of the controller assembly or shall be permitted to be separate. An autotransformer-type controller shall be provided with a separate disconnecting means.

(b) Type. The device shall be one of the types specified in (1), (2), or (3).

(1) Air-Break Switch. An air-break switch, operable directly by applying the hand to a lever or handle.

(2) Inverse Time Circuit Breaker. An inverse time circuit breaker operable directly by applying the hand to a lever or handle. The circuit breaker shall be permitted to be both power and manually operable.

(3) Oil Switch. An oil switch used on a circuit whose rating does not exceed 600 volts or 100 amperes, or by special permission on a circuit exceeding this capacity where under expert supervision. The oil switch shall be permitted to be both power and manually operable.

430-112. Motors Served by Single Disconnecting Means. Each motor shall be provided with an individual disconnecting means.

Exception: A single disconnecting means shall be permitted to serve a group of motors under any one of the conditions of (a), (b), and (c). The single disconnecting means shall be rated in accordance with Section 430-110(c).

(a) *Where a number of motors drive several parts of a single machine or piece of apparatus, such as metal and woodworking machines, cranes, and hoists*

(b) *Where a group of motors is under the protection of one set of branch-circuit protective devices as permitted by Section 430-53(a)*

(c) *Where a group of motors is in a single room within sight from the location of the disconnecting means*

430-113. Energy from More than One Source. Motor and motor-operated equipment receiving electrical energy from more than one source shall be provided with disconnecting means from each source of electrical energy immediately adjacent to the equipment served. Each source shall be permitted to have a separate disconnecting means.

Exception No. 1: Where a motor receives electrical energy from more than one source, the disconnecting means for the main power supply to the motor shall not be required to be immediately adjacent to the motor, provided the controller disconnecting means is capable of being locked in the open position.

Exception No. 2: A separate disconnecting means shall not be required for a Class 2 remote-control circuit conforming with Article 725, rated not more than 30 volts, and that is isolated and ungrounded.

K. Over 600 Volts, Nominal

430-121. General. Part K recognizes the additional hazard due to the use of higher voltages. It adds to or amends the other provisions of this article. Other requirements for circuits and equipment operating at over 600 volts, nominal, are in Article 490.

430-122. Marking on Controllers. In addition to the marking required by Section 430-8, a controller shall be marked with the control voltage.

430-123. Conductor Enclosures Adjacent to Motors. Flexible metal conduit or liquidtight flexible metal conduit not exceeding 6 ft (1.83 m) in length shall be permitted to be employed for raceway connection to a motor terminal enclosure.

430-124. Size of Conductors. Conductors supplying motors shall have an ampacity not less than the current at which the motor overload protective device(s) is selected to trip.

430-125. Motor-Circuit Overcurrent Protection.

(a) General. Each motor circuit shall include coordinated protection to automatically interrupt overload and fault currents in the motor, the motor-circuit conductors, and the motor control apparatus.

Exception: Where a motor is vital to operation of the plant and the motor should operate to failure if necessary to prevent a greater hazard to persons, the sensing device(s) is permitted to be connected to a supervised annunciator or alarm instead of interrupting the motor circuit.

(b) Overload Protection.

(1) Each motor shall be protected against dangerous heating due to motor overloads and failure to start by a thermal protector integral with the motor or external current-sensing devices, or both.

(2) The secondary circuits of wound-rotor ac motors including conductors, controllers, and resistors rated for the application shall be considered as protected against overcurrent by the motor overload protection means.

(3) Operation of the overload interrupting device shall simultaneously disconnect all ungrounded conductors.

(4) Overload sensing devices shall not automatically reset after trip unless resetting of the overload sensing device

does not cause automatic restarting of the motor or there is no hazard to persons created by automatic restarting of the motor and its connected machinery.

(c) Fault-Current Protection.

(1) Fault-current protection shall be provided in each motor circuit by one of the following means.

(a) A circuit breaker of suitable type and rating arranged so that it can be serviced without hazard. The circuit breaker shall simultaneously disconnect all ungrounded conductors. The circuit breaker shall be permitted to sense the fault current by means of integral or external sensing elements.

(b) Fuses of a suitable type and rating placed in each ungrounded conductor. Fuses shall be used with suitable disconnecting means or they shall be of a type that can also serve as the disconnecting means. They shall be arranged so that they cannot be serviced while they are energized.

(2) Fault-current interrupting devices shall not automatically reclose the circuit.

Exception: Automatic reclosing of a circuit shall be permitted where the circuit is exposed to transient faults and where such automatic reclosing does not create a hazard to persons.

(3) Overload protection and fault-current protection shall be permitted to be provided by the same device.

430-126. Rating of Motor Control Apparatus. The ultimate trip current of overcurrent (overload) relays or other motor-protective devices used shall not exceed 115 percent of the controller's continuous current rating. Where the motor branch-circuit disconnecting means is separate from the controller, the disconnecting means current rating shall not be less than the ultimate trip setting of the overcurrent relays in the circuit.

430-127. Disconnecting Means. The controller disconnecting means shall be capable of being locked in the open position.

L. Protection of Live Parts — All Voltages

430-131. General. Part L specifies that live parts shall be protected in a manner judged adequate for the hazard involved.

430-132. Where Required. Exposed live parts of motors and controllers operating at 50 volts or more between terminals shall be guarded against accidental contact by enclosure or by location as follows:

(1) By installation in a room or enclosure that is accessible only to qualified persons
(2) By installation on a suitable balcony, gallery, or platform, elevated and arranged so as to exclude unqualified persons.
(3) By elevation 8 ft (2.44 m) or more above the floor

Exception: Live parts of motors operating at more than 50 volts between terminals shall not require additional guarding for stationary motors that have commutators, collectors, and brush rigging located inside of motor-end brackets and not conductively connected to supply circuits operating at more than 150 volts to ground.

430-133. Guards for Attendants. Where live parts of motors or controllers operating at over 150 volts to ground are guarded against accidental contact only by location as specified in Section 430-132, and where adjustment or other attendance may be necessary during the operation of the apparatus, suitable insulating mats or platforms shall be provided so that the attendant cannot readily touch live parts unless standing on the mats or platforms.

FPN: For working space, see Sections 110-26 and 110-34.

M. Grounding — All Voltages

430-141. General. Part M specifies the grounding of exposed noncurrent-carrying metal parts, likely to become energized, of motor and controller frames to prevent a voltage above ground in the event of accidental contact between energized parts and frames. Insulation, isolation, or guarding are suitable alternatives to grounding of motors under certain conditions.

430-142. Stationary Motors. The frames of stationary motors shall be grounded under any of the following conditions:

(1) Where supplied by metal-enclosed wiring
(2) Where in a wet location and not isolated or guarded
(3) If in a hazardous (classified) location as covered in Articles 500 through 517
(4) If the motor operates with any terminal at over 150 volts to ground

Where the frame of the motor is not grounded, it shall be permanently and effectively insulated from the ground.

430-143. Portable Motors. The frames of portable motors that operate at over 150 volts to ground shall be guarded or grounded.

FPN No. 1: See Section 250-114(4) for grounding of portable appliances in other than residential occupancies.

FPN No. 2: See Section 250-138(b) for color of equipment grounding conductor.

430-144. Controllers. Controller enclosures shall be grounded regardless of voltage. Controller enclosures shall have means for attachment of an equipment grounding conductor termination in accordance with Section 250-8.

Exception: Enclosures attached to ungrounded portable equipment shall not be required to be grounded.

430-145. Method of Grounding. Where required, grounding shall be done in the manner specified in Article 250.

(a) Grounding Through Terminal Housings. Where the wiring to fixed motors is metal-enclosed cable or in metal raceways, junction boxes to house motor terminals shall be provided, and the armor of the cable or the metal raceways shall be connected to them in the manner specified in Article 250.

FPN: See Section 430-12(e) for grounding connection means required at motor terminal housings.

(b) Separation of Junction Box from Motor. The junction box required by (a) shall be permitted to be separated from the motor by not more than 6 ft (1.83 m), provided the leads to the motor are Type AC cable or armored cord or are stranded leads enclosed in liquidtight flexible metal conduit, flexible metal conduit, intermediate metal conduit, rigid metal conduit, or electrical metallic tubing not smaller than $\frac{3}{8}$-in. electrical trade size, the armor or raceway being connected both to the motor and to the box.

Liquidtight flexible nonmetallic conduit and rigid nonmetallic conduit shall be permitted to enclose the leads to the motor provided the leads are stranded and the required equipment grounding conductor is connected to both the motor and to the box.

Where stranded leads are used, protected as specified above, they shall not be larger than No. 10 and shall comply with other requirements of this *Code* for conductors to be used in raceways.

(c) Grounding of Controller Mounted Devices. Instrument transformer secondaries and exposed noncurrent-carrying metal or other conductive parts or cases of instrument transformers, meters, instruments, and relays shall be grounded as specified in Sections 250-170 through 250-178.

N. Tables

Table 430-147. Full-Load Current in Amperes, Direct-Current Motors

The following values of full-load currents* are for motors running at base speed.

Horse-power	Armature Voltage Rating*					
	90 Volts	120 Volts	180 Volts	240 Volts	500 Volts	550 Volts
¼	4.0	3.1	2.0	1.6	—	—
⅓	5.2	4.1	2.6	2.0	—	—
½	6.8	5.4	3.4	2.7	—	—
¾	9.6	7.6	4.8	3.8	—	—
1	12.2	9.5	6.1	4.7	—	—
1½	—	13.2	8.3	6.6	—	—
2	—	17	10.8	8.5	—	—
3	—	25	16	12.2	—	—
5	—	40	27	20	—	—
7½	—	58	—	29	13.6	12.2
10	—	76	—	38	18	16
15	—	—	—	55	27	24
20	—	—	—	72	34	31
25	—	—	—	89	43	38
30	—	—	—	106	51	46
40	—	—	—	140	67	61
50	—	—	—	173	83	75
60	—	—	—	206	99	90
75	—	—	—	255	123	111
100	—	—	—	341	164	148
125	—	—	—	425	205	185
150	—	—	—	506	246	222
200	—	—	—	675	330	294

*These are average dc quantities.

Table 430-148. Full-Load Currents in Amperes, Single-Phase Alternating-Current Motors

The following values of full-load currents are for motors running at usual speeds and motors with normal torque characteristics. Motors built for especially low speeds or high torques may have higher full-load currents, and multispeed motors will have full-load current varying with speed, in which case the namplate current ratings shall be used.

The voltages listed are rated motor voltages. The currents listed shall be permitted for system voltage ranges of 110 to 120 and 220 to 240 volts.

Horsepower	115 Volts	200 Volts	208 Volts	230 Volts
⅙	4.4	2.5	2.4	2.2
¼	5.8	3.3	3.2	2.9
⅓	7.2	4.1	4.0	3.6
½	9.8	5.6	5.4	4.9
¾	13.8	7.9	7.6	6.9
1	16	9.2	8.8	8.0
1½	20	11.5	11.0	10
2	24	13.8	13.2	12
3	34	19.6	18.7	17
5	56	32.2	30.8	28
7½	80	46.0	44.0	40
10	100	57.5	55.0	50

Table 430-149. Full-Load Current, Two-Phase Alternating-Current Motors (4-Wire)

The following values of full-load current are for motors running at speeds usual for belted motors and motors with normal torque characteristics. Motors built for especially low speeds or high torques may require more running current, and multispeed motors will have full-load current varying with speed, in which case the nameplate current rating shall be used. Current in the common conductor of a 2-phase, 3-wire system will be 1.41 times the value given.

The voltages listed are rated motor voltages. The currents listed shall be permitted for system voltage ranges of 110 to 120, 220 to 240, 440 to 480, and 550–600 volts.

Horsepower	Induction Type Squirrel Cage and Wound Rotor (Amperes)				
	115 Volts	230 Volts	460 Volts	575 Volts	2300 Volts
½	4.0	2.0	1.0	0.8	—
¾	4.8	2.4	1.2	1.0	—
1	6.4	3.2	1.6	1.3	—
1½	9.0	4.5	2.3	1.8	—
2	11.8	5.9	3.0	2.4	—
3	—	8.3	4.2	3.3	—
5	—	13.2	6.6	5.3	—
7½	—	19	9.0	8.0	—
10	—	24	12	10	—
15	—	36	18	14	—
20	—	47	23	19	—
25	—	59	29	24	—
30	—	69	35	28	—
40	—	90	45	36	—
50	—	113	56	45	—
60	—	133	67	53	14
75	—	166	83	66	18
100	—	218	109	87	23
125	—	270	135	108	28
150	—	312	156	125	32
200	—	416	208	167	43

Table 430-150. Full-Load Current Three-Phase Alternating-Current Motors
The following values of full-load currents are typical for motors running at speeds usual for belted motors and motors with normal torque characteristics.

Motors built for low speeds (1200 rpm or less) or high torques may require more running current, and multispeed motors will have full-load current varying with speed. In these cases, the nameplate current rating shall be used.

The voltages listed are rated motor voltages. The currents listed shall be permitted for system voltage ranges of 110 to 120, 220 to 240, 440 to 480, and 550 to 600 volts.

Horsepower	Induction Type Squirrel Cage and Wound Rotor (Amperes)							Synchronous-Type Unity Power Factor* (Amperes)			
	115 Volts	200 Volts	208 Volts	230 Volts	460 Volts	575 Volts	2300 Volts	230 Volts	460 Volts	575 Volts	2300 Volts
½	4.4	2.5	2.4	2.2	1.1	0.9	—	—	—	—	—
¾	6.4	3.7	3.5	3.2	1.6	1.3	—	—	—	—	—
1	8.4	4.8	4.6	4.2	2.1	1.7	—	—	—	—	—
1½	12.0	6.9	6.6	6.0	3.0	2.4	—	—	—	—	—
2	13.6	7.8	7.5	6.8	3.4	2.7	—	—	—	—	—
3	—	11.0	10.6	9.6	4.8	3.9	—	—	—	—	—
5	—	17.5	16.7	15.2	7.6	6.1	—	—	—	—	—
7½	—	25.3	24.2	22	11	9	—	—	—	—	—
10	—	32.2	30.8	28	14	11	—	—	—	—	—
15	—	48.3	46.2	42	21	17	—	—	—	—	—
20	—	62.1	59.4	54	27	22	—	—	—	—	—
25	—	78.2	74.8	68	34	27	—	53	26	21	—
30	—	92	88	80	40	32	—	63	32	26	—
40	—	120	114	104	52	41	—	83	41	33	—
50	—	150	143	130	65	52	—	104	52	42	—
60	—	177	169	154	77	62	16	123	61	49	12
75	—	221	211	192	96	77	20	155	78	62	15
100	—	285	273	248	124	99	26	202	101	81	20
125	—	359	343	312	156	125	31	253	126	101	25
150	—	414	396	360	180	144	37	302	151	121	30
200	—	552	528	480	240	192	49	400	201	161	40
250	—	—	—	—	302	242	60	—	—	—	—
300	—	—	—	—	361	289	72	—	—	—	—
350	—	—	—	—	414	336	83	—	—	—	—
400	—	—	—	—	477	382	95	—	—	—	—
450	—	—	—	—	515	412	103	—	—	—	—
500	—	—	—	—	590	472	118	—	—	—	—

*For 90 and 80 percent power factor, the figures shall be multiplied by 1.1 and 1.25, respectively.

Table 430-151(A). Conversion Table of Single-Phase Locked-Rotor Currents for Selection of Disconnecting Means and Controllers as Determined from Horsepower and Voltage Rating

For use only with Sections 430-110, 440-12, 440-41, and 455-8(c).

Rated Horsepower	Maximum Locked-Rotor Current in Amperes, Single Phase		
	115 Volts	208 Volts	230 Volts
½	58.8	32.5	29.4
¾	82.8	45.8	41.4
1	96	53	48
1½	120	66	60
2	144	80	72
3	204	113	102
5	336	186	168
7½	480	265	240
10	600	332	300

Table 430-151(B). Conversion Table of Polyphase Design B, C, D, and E Maximum Locked-Rotor Currents for Selection of Disconnecting Means and Controllers as Determined from Horsepower and Voltage Rating and Design Letter

For use only with Sections 430-110, 440-12,* 440-41,* and 455-8(c).

Rated Horsepower	Maximum Motor Locked-Rotor Current in Amperes Two- and Three-Phase, Design B, C, D and E											
	115 Volts		200 Volts		208 Volts		230 Volts		460 Volts		575 Volts	
	B, C, D	E	B, C, D	E	B, C, D	E	B, C, D	E	B, C, D	E	B, C, D	E
½	40	40	23	23	22.1	22.1	20	20	10	10	8	8
¾	50	50	28.8	28.8	27.6	27.6	25	25	12.5	12.5	10	10
1	60	60	34.5	34.5	33	33	30	30	15	15	12	12
1½	80	80	46	46	44	44	40	40	20	20	16	16
2	100	100	57.5	57.5	55	55	50	50	25	25	20	20
3	—	—	73.6	84	71	81	64	73	32	36.5	25.6	29.2
5	—	—	105.8	140	102	135	92	122	46	61	36.8	48.8
7½	—	—	146	210	140	202	127	183	63.5	91.5	50.8	73.2
10	—	—	186.3	259	179	249	162	225	81	113	64.8	90
15	—	—	267	388	257	373	232	337	116	169	93	135
20	—	—	334	516	321	497	290	449	145	225	116	180
25	—	—	420	646	404	621	365	562	183	281	146	225
30	—	—	500	775	481	745	435	674	218	337	174	270
40	—	—	667	948	641	911	580	824	290	412	232	330
50	—	—	834	1185	802	1139	725	1030	363	515	290	412
60	—	—	1001	1421	962	1367	870	1236	435	618	348	494
75	—	—	1248	1777	1200	1708	1085	1545	543	773	434	618
100	—	—	1668	2154	1603	2071	1450	1873	725	937	580	749
125	—	—	2087	2692	2007	2589	1815	2341	908	1171	726	936
150	—	—	2496	3230	2400	3106	2170	2809	1085	1405	868	1124
200	—	—	3335	4307	3207	4141	2900	3745	1450	1873	1160	1498
250	—	—	—	—	—	—	—	—	1825	2344	1460	1875
300	—	—	—	—	—	—	—	—	2200	2809	1760	2247
350	—	—	—	—	—	—	—	—	2550	3277	2040	2622
400	—	—	—	—	—	—	—	—	2900	3745	2320	2996
450	—	—	—	—	—	—	—	—	3250	4214	2600	3371
500	—	—	—	—	—	—	—	—	3625	4682	2900	3746

| *In determining compliance with Sections 440-12 and 440-41, the values in the B, C, D columns shall be used.

Table 430-152. Maximum Rating or Setting of Motor Branch-Circuit Short-Circuit and Ground-Fault Protective Devices

	Percentage of Full-Load Current			
Type of Motor	Nontime Delay Fuse[1]	Dual Element (Time-Delay) Fuse[1]	Instan-taneous Trip Breaker	Inverse Time Breaker[2]
Single-phase motors	300	175	800	250
AC polyphase motors other than wound-rotor Squirrel cage—				
Other than Design E	300	175	800	250
Design E	300	175	1100	250
Synchronous[3]	300	175	800	250
Wound rotor	150	150	800	150
Direct current (constant voltage)	150	150	250	150

Note: For certain exceptions to the values specified, see Sections 430-52 through 430-54.

[1]The values in the Nontime Delay Fuse column apply to Time-Delay Class CC fuses.

[2]The values given in the last column also cover the ratings of nonadjustable inverse time types of circuit breakers that may be modified as in Section 430-52.

[3]Synchronous motors of the low-torque, low-speed type (usually 450 rpm or lower), such as are used to drive reciprocating compressors, pumps, etc., that start unloaded, do not require a fuse rating or circuit-breaker setting in excess of 200 percent of full-load current.

Article 440 — Air-Conditioning and Refrigerating Equipment

A. General

440-1. Scope. The provisions of this article apply to electric motor-driven air-conditioning and refrigerating equipment, and to the branch circuits and controllers for such equipment. It provides for the special considerations necessary for circuits supplying hermetic refrigerant motor-compressors and for any air-conditioning or refrigerating equipment that is supplied from a branch circuit that supplies a hermetic refrigerant motor-compressor.

440-2. Definitions.

Branch-Circuit Selection Current. *Branch-circuit selection current* is the value in amperes to be used instead of the rated-load current in determining the ratings of motor branch-circuit conductors, disconnecting means, controllers, and branch-circuit short-circuit and ground-fault protective devices wherever the running overload protective device permits a sustained current greater than the specified percentage of the rated-load current. The value of branch-circuit selection current will always be equal to or greater than the marked rated-load current.

Hermetic Refrigerant Motor-Compressor. A combination consisting of a compressor and motor, both of which are enclosed in the same housing, with no external shaft or shaft seals, the motor operating in the refrigerant.

Rated-Load Current. The rated-load current for a hermetic refrigerant motor-compressor is the current resulting when the motor-compressor is operated at the rated load, rated voltage, and rated frequency of the equipment it serves.

440-3. Other Articles.

(a) Article 430. These provisions are in addition to, or amendatory of, the provisions of Article 430 and other articles in this *Code*, which apply except as modified in this article.

(b) Articles 422, 424, or 430. The rules of Articles 422, 424, or 430, as applicable, shall apply to air-conditioning and refrigerating equipment that does not incorporate a hermetic refrigerant motor-compressor. This equipment includes devices that employ refrigeration compressors driven by conventional motors, furnaces with air-conditioning evaporator coils installed, fan-coil units, remote forced air-cooled condensers, remote commercial refrigerators, etc.

(c) Article 422. Devices such as room air conditioners, household refrigerators and freezers, drinking water coolers, and beverage dispensers shall be considered appliances, and the provisions of Article 422 shall also apply.

(d) Other Applicable Articles. Hermetic refrigerant motor-compressors, circuits, controllers, and equipment shall also comply with the applicable provisions of the following:

	Article	Section
Capacitors		460-9
Commercial garages, aircraft hangars, gasoline dispensing and service stations, bulk storage plants, spray application, dipping, and coating processes, and inhalation anesthetizing locations	511, 513, 514, 515, 516, and 517, Part D	
Hazardous (classified) locations	500–503 and 505	
Motion picture and television studios and similar locations	530	
Resistors and reactors	470	

440-4. Marking on Hermetic Refrigerant Motor-Compressors and Equipment.

(a) Hermetic Refrigerant Motor-Compressor Nameplate. A hermetic refrigerant motor-compressor shall be provided with a nameplate that shall indicate the manufacturer's name, trademark, or symbol; identifying designation; phase; voltage; and frequency. The rated-load current in amperes of the motor-compressor shall be marked by the equipment manufacturer on either or both the motor-compressor nameplate and the nameplate of the equipment in which the motor-compressor is used. The locked-rotor current of each single-phase motor-compressor having a rated-load current of more than 9 amperes at 115 volts, or more than 4.5 amperes at 230 volts, and each polyphase motor-compressor shall be marked on the motor-compressor nameplate. Where a thermal protector complying with Sections 440-52(a)(2) and (b)(2) is used, the motor-compressor nameplate or the equipment nameplate shall be marked with the words "thermally protected." Where a protective system complying with Sections 440-52(a)(4) and (b)(4) is used and is furnished with the equipment, the equipment nameplate shall be marked with the words, "thermally protected system." Where a protective system complying with Sections 440-52(a)(4) and (b)(4) is specified, the equipment nameplate shall be appropriately marked.

(b) Multimotor and Combination-Load Equipment. Multimotor and combination-load equipment shall be provided with a visible nameplate marked with the maker's name, the rating in volts, frequency and number of phases, minimum supply circuit conductor ampacity, and the maximum rating of the branch-circuit short-circuit and ground-fault protective device. The ampacity shall be calculated by using Part D and counting all the motors and other loads that will be operated at the same time. The branch-circuit short-circuit and ground-fault protective device rating shall not exceed the value calculated by using Part C. Multimotor or combination-load equipment for use on two or more circuits shall be marked with the above information for each circuit.

Exception No. 1: Multimotor and combination-load equipment that is suitable under the provisions of this article for connection to a single 15- or 20-ampere, 120-volt, or a 15-ampere, 208- or 240-volt, single-phase branch circuit shall be permitted to be marked as a single load.

Exception No. 2: The minimum supply circuit conductor ampacity and the maximum rating of the branch-circuit short-circuit and ground-fault protective device shall not be required to be marked on a room air conditioner conforming with Section 440-62(a).

(c) Branch-Circuit Selection Current. A hermetic refrigerant motor-compressor, or equipment containing such a compressor, having a protection system that is approved for use with the motor-compressor that it protects and that permits continuous current in excess of the specified percentage of nameplate rated-load current given in Section 440-52(b)(2) or (b)(4), shall also be marked with a branch-circuit selection current that complies with Section 440-52(b)(2) or (b)(4). This marking shall be provided by the equipment manufacturer and shall be on the nameplate(s) where the rated-load current(s) appears.

440-5. Marking on Controllers. A controller shall be marked with the manufacturer's name, trademark, or symbol; identifying designation; voltage; phase; full-load and locked-rotor current (or horsepower) rating; and such other data as may be needed to properly indicate the motor-compressor for which it is suitable.

440-6. Ampacity and Rating. The size of conductors for equipment covered by this article shall be selected from Tables 310-16 through 310-19 or calculated in accordance with Section 310-15 as applicable. The required ampacity of conductors and rating of equipment shall be determined as follows.

(a) Hermetic Refrigerant Motor-Compressor. For a hermetic refrigerant motor-compressor, the rated-load current marked on the nameplate of the equipment in which the motor-compressor is employed shall be used in determining the rating or ampacity of the disconnecting means, the branch-circuit conductors, the controller, the branch-circuit short-circuit and ground-fault protection, and the separate motor overload protection. Where no rated-load current is shown on the equipment nameplate, the rated-load current shown on the compressor nameplate shall be used. For disconnecting means and controllers, see also Sections 440-12 and 440-41.

Exception No. 1: Where so marked, the branch-circuit selection current shall be used instead of the rated-load current to determine the rating or ampacity of the disconnecting means, the branch-circuit conductors, the controller, and the branch-circuit short-circuit and ground-fault protection.

Exception No. 2: For cord- and plug-connected equipment, the nameplate marking shall be used in accordance with Section 440-22(b), Exception No. 2.

(b) Multimotor Equipment. For multimotor equipment employing a shaded-pole or permanent split-capacitor-type fan or blower motor, the full-load current for such motor marked on the nameplate of the equipment in which the fan or blower motor is employed shall be used instead of the horsepower rating to determine the ampacity or rating of the disconnecting means, the branch-circuit conductors, the controller, the branch-circuit short-circuit and ground-fault protection, and the separate overload protection. This marking on the equipment nameplate shall not be less than the current marked on the fan or blower motor nameplate.

440-7. Highest Rated (Largest) Motor. In determining compliance with this article and with Sections 430-24, 430-53(b) and (c), and 430-62(a), the highest rated (largest) motor shall be considered to be the motor that has the highest rated-load current. Where two or more motors have the same highest rated-load current, only one of them shall be considered as the highest rated (largest) motor. For other than hermetic refrigerant motor-compressors, and fan or blower motors as covered in Section 440-6(b), the full-load current used to determine the highest rated motor shall be the equivalent value corresponding to the motor horsepower rating selected from Tables 430-148, 430-149, or 430-150.

Exception: Where so marked, the branch-circuit selection current shall be used instead of the rated-load current in determining the highest rated (largest) motor-compressor.

440-8. Single Machine. An air-conditioning or refrigerating system shall be considered to be a single machine under the provisions of Section 430-87, Exception, and Section 430-112, Exception. The motors shall be permitted to be located remotely from each other.

B. Disconnecting Means

440-11. General. The provisions of Part B are intended to require disconnecting means capable of disconnecting air-conditioning and refrigerating equipment including motor-compressors and controllers from the circuit conductors. See Figure 430-1.

440-12. Rating and Interrupting Capacity.

(a) Hermetic Refrigerant Motor-Compressor. A disconnecting means serving a hermetic refrigerant motor-compressor shall be selected on the basis of the nameplate rated-load current or branch-circuit selection current, whichever is greater, and locked-rotor current, respectively, of the motor-compressor as follows.

(1) The ampere rating shall be at least 115 percent of the nameplate rated-load current or branch-circuit selection current, whichever is greater.

(2) To determine the equivalent horsepower in complying with the requirements of Section 430-109, the horsepower rating shall be selected from Tables 430-148, 430-149, or 430-150 corresponding to the rated-load current or branch-circuit selection current, whichever is greater, and also the horsepower rating from Tables 430-151(A) or 430-151(B) corresponding to the locked-rotor current. In case

the nameplate rated-load current or branch-circuit selection current and locked-rotor current do not correspond to the currents shown in Tables 430-148, 430-149, 430-150, 430-151(A), or 430-151(B), the horsepower rating corresponding to the next higher value shall be selected. In case different horsepower ratings are obtained when applying these tables, a horsepower rating at least equal to the larger of the values obtained shall be selected.

(b) Combination Loads. Where one or more hermetic refrigerant motor-compressors are used together or are used in combination with other motors or loads, and where the combined load may be simultaneous on a single disconnecting means, the rating for the combined load shall be determined as follows.

(1) The horsepower rating of the disconnecting means shall be determined from the sum of all currents, including resistance loads, at the rated-load condition and also at the locked-rotor condition. The combined rated-load current and the combined locked-rotor current so obtained shall be considered as a single motor for the purpose of this requirement as follows.

(a) The full-load current equivalent to the horsepower rating of each motor, other than a hermetic refrigerant motor-compressor, and fan or blower motors as covered in Section 440-6(b) shall be selected from Tables 430-148, 430-149, or 430-150. These full-load currents shall be added to the motor-compressor rated-load current(s) or branch-circuit selection current(s), whichever is greater, and to the rating in amperes of other loads to obtain an equivalent full-load current for the combined load.

(b) The locked-rotor current equivalent to the horsepower rating of each motor, other than a hermetic refrigerant motor-compressor, shall be selected from Table 430-151(A) or 430-151(B), and, for fan and blower motors of the shaded-pole or permanent split-capacitor type marked with the locked-rotor current, the marked value shall be used. The locked-rotor currents shall be added to the motor-compressor locked-rotor current(s) and to the rating in amperes of other loads to obtain an equivalent locked-rotor current for the combined load. Where two or more motors or other loads such as resistance heaters, or both, cannot be started simultaneously, appropriate combinations of locked-rotor and rated-load current or branch-circuit selection current, whichever is greater, shall be an acceptable means of determining the equivalent locked-rotor current for the simultaneous combined load.

Exception: Where part of the concurrent load is a resistance load and the disconnecting means is a switch rated in horsepower and amperes, the switch used shall be permitted to have a horsepower rating not less than the combined load to the motor-compressor(s) and other motor(s) at the locked-rotor condition, if the ampere rating of the switch is not less than this locked-rotor load plus the resistance load.

(2) The ampere rating of the disconnecting means shall be at least 115 percent of the sum of all currents at the rated-load condition determined in accordance with Section 440-12(b)(1).

(c) Small Motor-Compressors. For small motor-compressors not having the locked-rotor current marked on the nameplate, or for small motors not covered by Tables 430-147, 430-148, 430-149, or 430-150, the locked-rotor current shall be assumed to be six times the rated-load current. See Section 440-3(a).

(d) Every Switch. Every disconnecting means in the refrigerant motor-compressor circuit between the point of attachment to the feeder and the point of connection to the refrigerant motor-compressor shall comply with the requirements of Section 440-12.

(e) Disconnecting Means Rated in Excess of 100 Horsepower. Where the rated-load or locked-rotor current as determined above would indicate a disconnecting means rated in excess of 100 hp, the provisions of Section 430-109(e) shall apply.

440-13. Cord-Connected Equipment. For cord-connected equipment such as room air conditioners, household refrigerators and freezers, drinking water coolers, and beverage dispensers, a separable connector or an attachment plug and receptacle shall be permitted to serve as the disconnecting means. See also Section 440-63.

440-14. Location. Disconnecting means shall be located within sight from and readily accessible from the air-conditioning or refrigerating equipment. The disconnecting means shall be permitted to be installed on or within the air-conditioning or refrigerating equipment.

Exception No. 1: Where the disconnecting means provided in accordance with Section 430-102(a) is capable of being locked in the open position, and the refrigerating or air conditioning equipment is essential to an industrial process in a facility where the conditions of maintenance and the supervision ensure that only qualified persons will service the equipment, a disconnecting means within sight from the equipment shall not be required.

Exception No. 2: Where an attachment plug and receptacle serve as the disconnecting means in accordance with Section 440-13, their location shall be accessible, but shall not be required to be readily accessible.

FPN: See Parts G and J of Article 430 for additional requirements.

C. Branch-Circuit Short-Circuit and Ground-Fault Protection

440-21. General. The provisions of Part C specify devices intended to protect the branch-circuit conductors, control apparatus, and motors in circuits supplying hermetic refrigerant motor-compressors against overcurrent due to short circuits and grounds. They are in addition to or amendatory of the provisions of Article 240.

440-22. Application and Selection.

(a) Rating or Setting for Individual Motor-Compressor. The motor-compressor branch-circuit short-circuit and ground-fault protective device shall be capable of carrying the starting current of the motor. A protective device having a rating or setting not exceeding 175 percent of the motor-compressor rated-load current or branch-circuit selection current, whichever is greater, shall be permitted, provided that, where the protection specified is not sufficient for the starting current of the motor, the rating or setting shall be permitted to be increased, but shall not exceed 225 percent of the motor rated-load current or branch-circuit selection current, whichever is greater.

Exception: The rating of the branch-circuit short-circuit and ground-fault protective device shall not be required to be less than 15 amperes.

(b) Rating or Setting for Equipment. The equipment branch-circuit short-circuit and ground-fault protective device shall be capable of carrying the starting current of the equipment. Where the hermetic refrigerant motor-compressor is the only load on the circuit, the protection shall conform with Section 440-22(a). Where the equipment incorporates more than one hermetic refrigerant motor-compressor or a hermetic refrigerant motor-compressor and other motors or other loads, the equipment short-circuit and ground-fault protection shall conform with Section 430-53 and the following.

(1) Where a hermetic refrigerant motor-compressor is the largest load connected to the circuit, the rating or setting of the branch-circuit short-circuit and ground-fault protective device shall not exceed the value specified in Section 440-22(a) for the largest motor-compressor plus the sum of the rated-load current or branch-circuit selection current, whichever is greater, of the other motor-compressor(s) and the ratings of the other loads supplied.

(2) Where a hermetic refrigerant motor-compressor is not the largest load connected to the circuit, the rating or setting of the branch-circuit short-circuit and ground-fault protective device shall not exceed a value equal to the sum of the rated-load current or branch-circuit selection current, whichever is greater, rating(s) for the motor-compressor(s) plus the value specified in Section 430-53(c)(4) where other motor loads are supplied, or the value specified in Section 240-3 where only nonmotor loads are supplied in addition to the motor-compressor(s).

Exception No. 1: Equipment that will start and operate on a 15- or 20-ampere 120-volt, or 15-ampere 208- or 240-volt single-phase branch circuit, shall be permitted to be protected by the 15- or 20-ampere overcurrent device protecting the branch circuit, but if the maximum branch-circuit short-circuit and ground-fault protective device rating marked on the equipment is less than these values, the circuit protective device shall not exceed the value marked on the equipment nameplate.

Exception No. 2: The nameplate marking of cord- and plug-connected equipment rated not greater than 250 volts, single-phase, such as household refrigerators and freezers, drinking water coolers, and beverage dispensers, shall be used in determining the branch-circuit requirements, and each unit shall be considered as a single motor unless the nameplate is marked otherwise.

(c) Protective Device Rating Not to Exceed the Manufacturer's Values. Where maximum protective device ratings shown on a manufacturer's heater table for use with a motor controller are less than the rating or setting selected in accordance with Sections 440-22(a) and (b), the protective device rating shall not exceed the manufacturer's values marked on the equipment.

D. Branch-Circuit Conductors

440-31. General. The provisions of Part D and Article 310 specify ampacities of conductors required to carry the motor current without overheating under the conditions specified, except as modified in Section 440-6(a), Exception No. 1.

The provisions of these articles shall not apply to integral conductors of motors, motor controllers and the like, or to conductors that form an integral part of approved equipment.

FPN: See Sections 300-1(b) and 310-1 for similar requirements.

440-32. Single Motor-Compressor. Branch-circuit conductors supplying a single motor-compressor shall have an ampacity not less than 125 percent of either the motor-compressor rated-load current or the branch-circuit selection current, whichever is greater.

440-33. Motor-Compressor(s) With or Without Additional Motor Loads. Conductors supplying one or more motor-compressor(s) with or without an additional load(s) shall have an ampacity not less than the sum of the rated-load or branch-circuit selection current ratings, whichever is larger, of all the motor-compressors plus the full-load

currents of the other motors, plus 25 percent of the highest motor or motor-compressor rating in the group.

Exception No. 1: Where the circuitry is interlocked so as to prevent the starting and running of a second motor-compressor or group of motor-compressors, the conductor size shall be determined from the largest motor-compressor or group of motor-compressors that is to be operated at a given time.

Exception No. 2: The branch circuit conductors for room air conditioners shall be in accordance with Part G of Article 440.

440-34. Combination Load. Conductors supplying a motor-compressor load in addition to a lighting or appliance load as computed from Article 220 and other applicable articles shall have an ampacity sufficient for the lighting or appliance load plus the required ampacity for the motor-compressor load determined in accordance with Section 440-33, or, for a single motor-compressor, in accordance with Section 440-32.

Exception: Where the circuitry is interlocked so as to prevent simultaneous operation of the motor-compressor(s) and all other loads connected, the conductor size shall be determined from the largest size required for the motor-compressor(s) and other loads to be operated at a given time.

440-35. Multimotor and Combination-Load Equipment. The ampacity of the conductors supplying multimotor and combination-load equipment shall not be less than the minimum circuit ampacity marked on the equipment in accordance with Section 440-4(b).

E. Controllers for Motor-Compressors

440-41. Rating.

(a) Motor-Compressor Controller. A motor-compressor controller shall have both a continuous-duty full-load current rating and a locked-rotor current rating, not less than the nameplate rated-load current or branch-circuit selection current, whichever is greater, and locked-rotor current, respectively of the compressor (see Sections 440-6 and 440-7). In case the motor controller is rated in horsepower, but is without one or both of the foregoing current ratings, equivalent currents shall be determined from the ratings as follows. Use Tables 430-148, 430-149, or 430-150 to determine the equivalent full-load current rating. Use Tables 430-151(A) or 430-151(B) to determine the equivalent locked-rotor current ratings.

(b) Controller Serving More than One Load. A controller, serving more than one motor-compressor or a motor-compressor and other loads, shall have a continuous-duty full-load current rating and a locked-rotor current rating not less than the combined load as determined in accordance with Section 440-12(b).

F. Motor-Compressor and Branch-Circuit Overload Protection

440-51. General. The provisions of Part F specify devices intended to protect the motor-compressor, the motor-control apparatus, and the branch-circuit conductors against excessive heating due to motor overload and failure to start. See Sections 240-3(g) and 430-31.

440-52. Application and Selection.

(a) Protection of Motor-Compressor. Each motor-compressor shall be protected against overload and failure to start by one of the following means.

(1) A separate overload relay that is responsive to motor-compressor current. This device shall be selected to trip at not more than 140 percent of the motor-compressor rated-load current.

(2) A thermal protector integral with the motor-compressor, approved for use with the motor-compressor that it protects on the basis that it will prevent dangerous overheating of the motor-compressor due to overload and failure to start. If the current-interrupting device is separate from the motor-compressor and its control circuit is operated by a protective device integral with the motor-compressor, it shall be arranged so that the opening of the control circuit will result in interruption of current to the motor-compressor.

(3) A fuse or inverse time circuit breaker responsive to motor current, which shall also be permitted to serve as the branch-circuit short-circuit and ground-fault protective device. This device shall be rated at not more than 125 percent of the motor-compressor rated-load current. It shall have sufficient time delay to permit the motor-compressor to start and accelerate its load. The equipment or the motor-compressor shall be marked with this maximum branch-circuit fuse or inverse time circuit breaker rating.

(4) A protective system, furnished or specified and approved for use with the motor-compressor that it protects on the basis that it will prevent dangerous overheating of the motor-compressor due to overload and failure to start. If the current-interrupting device is separate from the motor-compressor and its control circuit is operated by a protective device that is not integral with the current-interrupting device, it shall be arranged so that the opening of the control circuit will result in interruption of current to the motor-compressor.

(b) Protection of Motor-Compressor Control Apparatus and Branch-Circuit Conductors. The motor-compressor controller(s), the disconnecting means, and the branch-circuit conductors shall be protected against overcurrent due to motor overload and failure to start by one of the following means, which shall be permitted to be the same device or system protecting the motor-compressor in accordance with Section 440-52(a).

Exception: Overload protection of motor-compressors and equipment on 15- and 20-ampere, single-phase, branch circuits shall be permitted to be in accordance with Sections 440-54 and 440-55.

(1) An overload relay selected in accordance with Section 440-52(a)(1)
(2) A thermal protector applied in accordance with Section 440-52(a)(2) and that will not permit a continuous current in excess of 156 percent of the marked rated-load current or branch-circuit selection current
(3) A fuse or inverse time circuit breaker selected in accordance with Section 440-52(a)(3)
(4) A protective system in accordance with Section 440-52(a)(4) and that will not permit a continuous current in excess of 156 percent of the marked rated-load current or branch-circuit selection current

440-53. Overload Relays. Overload relays and other devices for motor overload protection that are not capable of opening short circuits shall be protected by fuses or inverse time circuit breakers with ratings or settings in accordance with Part C unless approved for group installation or for part-winding motors and marked to indicate the maximum size of fuse or inverse time circuit breaker by which they shall be protected.

Exception: The fuse or inverse time circuit breaker size marking shall be permitted on the nameplate of approved equipment in which the overload relay or other overload device is used.

440-54. Motor-Compressors and Equipment on 15- or 20-Ampere Branch Circuits — Not Cord- and Attachment Plug-Connected. Overload protection for motor-compressors and equipment used on 15- or 20-ampere 120-volt, or 15-ampere 208- or 240-volt single-phase branch circuits as permitted in Article 210 shall be permitted as indicated in (a) and (b).

(a) Overload Protection. The motor-compressor shall be provided with overload protection selected as specified in Section 440-52(a). Both the controller and motor overload protective device shall be approved for installation with the short-circuit and ground-fault protective device for the branch circuit to which the equipment is connected.

(b) Time Delay. The short-circuit and ground-fault protective device protecting the branch circuit shall have sufficient time delay to permit the motor-compressor and other motors to start and accelerate their loads.

440-55. Cord- and Attachment Plug-Connected Motor-Compressors and Equipment on 15- or 20-Ampere Branch Circuits. Overload protection for motor-compressors and equipment that are cord- and attachment plug-connected and used on 15- or 20-ampere 120-volt, or 15-ampere 208- or 240-volt single-phase branch circuits as permitted in Article 210 shall be permitted as indicated in (a), (b), and (c).

(a) Overload Protection. The motor-compressor shall be provided with overload protection as specified in Section 440-52(a). Both the controller and the motor overload protective device shall be approved for installation with the short-circuit and ground-fault protective device for the branch circuit to which the equipment is connected.

(b) Attachment Plug and Receptacle Rating. The rating of the attachment plug and receptacle shall not exceed 20 amperes at 125 volts or 15 amperes at 250 volts.

(c) Time Delay. The short-circuit and ground-fault protective device protecting the branch circuit shall have sufficient time delay to permit the motor-compressor and other motors to start and accelerate their loads.

G. Provisions for Room Air Conditioners

440-60. General. The provisions of Part G shall apply to electrically energized room air conditioners that control temperature and humidity. For the purpose of Part G, a room air conditioner (with or without provisions for heating) shall be considered as an ac appliance of the air-cooled window, console, or in-wall type that is installed in the conditioned room and that incorporates a hermetic refrigerant motor-compressor(s). The provisions of Part G cover equipment rated not over 250 volts, single phase, and such equipment shall be permitted to be cord- and attachment plug-connected.

A room air conditioner that is rated three phase or rated over 250 volts shall be directly connected to a wiring method recognized in Chapter 3, and provisions of Part G shall not apply.

440-61. Grounding. Room air conditioners shall be grounded in accordance with Sections 250-110, 250-112, and 250-114.

440-62. Branch-Circuit Requirements.

(a) Room Air Conditioner as a Single Motor Unit. A room air conditioner shall be considered as a single motor unit in determining its branch-circuit requirements where all the following conditions are met.

(1) It is cord- and attachment plug-connected.
(2) Its rating is not more than 40 amperes and 250 volts, single phase.
(3) Total rated-load current is shown on the room air-conditioner nameplate rather than individual motor currents.
(4) The rating of the branch-circuit short-circuit and ground-fault protective device does not exceed the ampacity of the branch-circuit conductors or the rating of the receptacle, whichever is less.

(b) Where No Other Loads Are Supplied. The total marked rating of a cord- and attachment plug-connected room air conditioner shall not exceed 80 percent of the rating of a branch circuit where no other loads are supplied.

(c) Where Lighting Units or Other Appliances Are Also Supplied. The total marked rating of a cord- and attachment plug-connected room air conditioner shall not exceed 50 percent of the rating of a branch circuit where lighting units or other appliances are also supplied.

440-63. Disconnecting Means. An attachment plug and receptacle shall be permitted to serve as the disconnecting means for a single-phase room air conditioner rated 250 volts or less if (1) the manual controls on the room air conditioner are readily accessible and located within 6 ft (1.83 m) of the floor or (2) an approved manually operable switch is installed in a readily accessible location within sight from the room air conditioner.

440-64. Supply Cords. Where a flexible cord is used to supply a room air conditioner, the length of such cord shall not exceed 10 ft (3.05 m) for a nominal, 120-volt rating or 6 ft (1.83 m) for a nominal, 208- or 240-volt rating.

Article 445 — Generators

445-1. General. Generators and their associated wiring and equipment shall also comply with the applicable provisions of Articles 695, 700, 701, 702, and 705.

445-2. Location. Generators shall be of a type suitable for the locations in which they are installed. They shall also meet the requirements for motors in Section 430-14. Generators installed in hazardous (classified) locations as described in Articles 500 through 503 and Article 505, or in other locations as described in Articles 510 through 517, and in Articles 518, 520, 525, 530, 665, and 695 shall also comply with the applicable provisions of those articles.

445-3. Marking. Each generator shall be provided with a nameplate giving the manufacturer's name, the rated frequency, power factor, number of phases if of alternating current, the rating in kilowatts or kilovolt amperes, the normal volts and amperes corresponding to the rating, rated revolutions per minute, insulation system class and rated ambient temperature or rated temperature rise, and time rating.

445-4. Overcurrent Protection.

(a) Constant-Voltage Generators. Constant-voltage generators, except ac generator exciters, shall be protected from overloads by inherent design, circuit breakers, fuses, or other acceptable overcurrent protective means suitable for the conditions of use.

(b) Two-Wire Generators. Two-wire, dc generators shall be permitted to have overcurrent protection in one conductor only if the overcurrent device is actuated by the entire current generated other than the current in the shunt field. The overcurrent device shall not open the shunt field.

(c) 65 Volts or Less. Generators operating at 65 volts or less and driven by individual motors shall be considered as protected by the overcurrent device protecting the motor if these devices will operate when the generators are delivering not more than 150 percent of their full-load rated current.

(d) Balancer Sets. Two-wire, dc generators used in conjunction with balancer sets to obtain neutrals for 3-wire systems shall be equipped with overcurrent devices that will disconnect the 3-wire system in case of excessive unbalancing of voltages or currents.

(e) Three-Wire, Direct-Current Generators. Three-wire, dc generators, whether compound or shunt wound, shall be equipped with overcurrent devices, one in each armature lead, and connected so as to be actuated by the entire current from the armature. Such overcurrent devices shall consist either of a double-pole, double-coil circuit breaker, or of a 4-pole circuit breaker connected in the main and equalizer leads and tripped by two overcurrent devices, one in each armature lead. Such protective devices shall be interlocked so that no one pole can be opened without simultaneously disconnecting both leads of the armature from the system.

Exception to (a) through (e): Where deemed by the authority having jurisdiction, a generator is vital to the operation of an electrical system and the generator should operate to failure to prevent a greater hazard to persons. The overload sensing device(s) shall be permitted to be connected to an annunciator or alarm supervised by authorized personnel instead of interrupting the generator circuit.

445-5. Ampacity of Conductors. The ampacity of the conductors from the generator terminals to the first overcurrent device shall not be less than 115 percent of the nameplate current rating of the generator. It shall be permitted to size the neutral conductors in accordance with Section 220-22.

Conductors that must carry ground-fault currents shall not be smaller than required by Section 250-24(b). Neutral conductors of dc generators that must carry ground-fault currents shall not be smaller than the minimum required size of the largest conductor.

Exception: Where the design and operation of the generator prevent overloading, the ampacity of the conductors shall not be less than 100 percent of the nameplate current rating of the generator.

445-6. Protection of Live Parts. Live parts of generators operated at more than 50 volts to ground shall not be exposed to accidental contact where accessible to unqualified persons.

445-7. Guards for Attendants. Where necessary for the safety of attendants, the requirements of Section 430-133 shall apply.

445-8. Bushings. Where wires pass through an opening in an enclosure, conduit box, or barrier, a bushing shall be used to protect the conductors from the edges of an opening having sharp edges. The bushing shall have smooth, well-rounded surfaces where it may be in contact with the conductors. If used where oils, grease, or other contaminants may be present, the bushing shall be made of a material not deleteriously affected.

445-9. Generator Terminal Housings. Generator terminal housings shall comply with Section 430-12.

445-10. Disconnecting Means Required for Generators. Generators shall be equipped with a disconnect by means of which the generator and all protective devices and control apparatus are able to be disconnected entirely from the circuits supplied by the generator except where

(1) The driving means for the generator can be readily shut down; and
(2) The generator is not arranged to operate in parallel with another generator or other source of voltage.

Article 450 — Transformers and Transformer Vaults (Including Secondary Ties)

450-1. Scope. This article covers the installation of all transformers.

Exception No. 1: Current transformers.

Exception No. 2: Dry-type transformers that constitute a component part of other apparatus and comply with the requirements for such apparatus.

Exception No. 3: Transformers that are an integral part of an X-ray, high-frequency, or electrostatic-coating apparatus.

Exception No. 4: Transformers used with Class 2 and Class 3 circuits that comply with Article 725.

Exception No. 5: Transformers for sign and outline lighting that comply with Article 600.

Exception No. 6: Transformers for electric-discharge lighting that comply with Article 410.

Exception No. 7: Transformers used for power-limited fire alarm circuits that comply with Part C of Article 760.

Exception No. 8: Transformers used for research, development, or testing, where effective arrangements are provided to safeguard persons from contacting energized parts.

This article covers the installation of transformers dedicated to supplying power to a fire pump installation as modified by Article 695.

This article also covers the installation of transformers in hazardous (classified) locations as modified by Articles 501 through 504.

A. General Provisions

450-2. Definitions. For the purpose of this article:

Transformer. The word *transformer* is intended to mean an individual transformer, single- or polyphase, identified by a single nameplate, unless otherwise indicated in this article.

450-3. Overcurrent Protection. Overcurrent protection of transformers shall comply with (a), (b), or (c). As used in this section, the word *transformer* shall mean a transformer or polyphase bank of two or more single-phase transformers operating as a unit.

FPN No. 1: See Sections 240-3, 240-21, 240-100, and 240-101 for overcurrent protection of conductors.

FPN No. 2: Nonlinear loads can increase heat in a transformer without operating its overcurrent protective device.

(a) Transformers Over 600 Volts, Nominal. Overcurrent protection shall be provided in accordance with Table 450-3(a).

Table 450-3(a). Maximum Rating or Setting of Overcurrent Protection for Transformers Over 600 Volts (as a Percentage of Transformer-Related Current)

| Location Limitations | Transformer Rated Impedance | Primary Protection Over 600 Volts | | Secondary Protection (see Note 2) | | |
| | | | | Over 600 Volts | | 600 Volts or Below |
		Circuit Breaker (see Note 4)	Fuse Rating	Circuit Breaker (see Note 4)	Fuse Rating	Circuit Breaker or Fuse Rating
Any location	Not more than 6%	600% (see Note 1)	300% (see Note 1)	300% (see Note 1)	250% (see Note 1)	125% (see Note 1)
	More than 6% and not more than 10%	400% (see Note 1)	300% (see Note 1)	250% (see Note 1)	225% (see Note 1)	125% (see Note 1)
Supervised locations only (see Note 3)	Any	300% (see Note 1)	250% (see Note 1)	Not required	Not required	Not required
	Not more than 6%	600%	300%	300% (see Note 5)	250% (see Note 5)	250% (see Note 5)
	More than 6% and not more than 10%	400%	300%	250% (see Note 5)	225% (see Note 5)	250% (see Note 5)

Notes:

1. Where the required fuse rating or circuit breaker setting does not correspond to a standard rating or setting, the next higher standard rating or setting shall be permitted.

2. Where secondary overcurrent protection is required, the secondary overcurrent device shall be permitted to consist of not more than six circuit breakers or six sets of fuses grouped in one location. Where multiple overcurrent devices are utilized, the total of all the device ratings shall not exceed the allowed value of a single overcurrent device. If both breakers and fuses are utilized as the overcurrent device, the total of the device ratings shall not exceed that allowed for fuses.

3. A supervised location where conditions of maintenance and supervision ensure that only qualified persons will monitor and service the transformer installation.

4. Electronically actuated fuses that may be set to open at a specific current shall be set in accordance with settings for circuit breakers.

5. A transformer equipped with a coordinated thermal overload protection by the manufacturer shall be permitted to have separate secondary protection omitted.

(b) Transformers 600 Volts, Nominal, or Less. Overcurrent protection shall be provided in accordance with Table 450-3(b).

Exception: Where the transformer is installed as a motor-control circuit transformer in accordance with Sections 430-72(c)(1) through (5).

(c) Voltage Transformers. Voltage transformers installed indoors or enclosed shall be protected with primary fuses.

> FPN: For protection of instrument circuits including voltage transformers, see Section 384-32.

450-4. Autotransformers 600 Volts, Nominal, or Less.

(a) Overcurrent Protection. Each autotransformer 600 volts, nominal, or less, shall be protected by an individual overcurrent device installed in series with each ungrounded input conductor. Such overcurrent device shall be rated or set at not more than 125 percent of the rated full-load input current of the autotransformer. Where this calculation does not correspond to a standard rating of a fuse or nonadjustable circuit breaker and the rated input current is 9 amperes or more, the next higher standard rating described in Section 240-6 shall be permitted. An overcurrent device shall not be installed in series with the shunt winding (the winding common to both the input and the output circuits) of the autotransformer between Points A and B as shown in Figure 450-4.

Figure 450-4   Autotransformer.

Exception: Where the rated input current of the autotransformer is less than 9 amperes, an overcurrent device rated or set at not more than 167 percent of the input current shall be permitted.

(b) Transformer Field-Connected as an Autotransformer. A transformer field-connected as an autotransformer shall be identified for use at elevated voltage.

> FPN: For information on permitted uses of autotransformers, see Section 210-9 and Section 215-11.

450-5. Grounding Autotransformers.
Grounding autotransformers covered in this section are zigzag or T-connected transformers connected to 3-phase, 3-wire ungrounded systems for the purpose of creating a 3-phase, 4-wire distribution system or to provide a neutral reference for grounding purposes. Such transformers shall have a continuous per phase current rating and a continuous neutral current rating.

Table 450-3(b). Maximum Rating or Setting of Overcurrent Protection for Transformers 600 Volts and Less (as a Percentage of Transformer-Rated Current)

Protection Method	Primary Protection			Secondary Protection (see Note 2)	
	Currents of 9 Amperes or More	Currents Less than 9 Amperes	Currents Less than 2 Amperes	Currents of 9 Amperes or More	Currents Less than 9 Amperes
Primary only protection	125% (see Note 1)	167%	300%	Not required	Not required
Primary and secondary protection	250% (see Note 3)	250% (see Note 3)	250% (see Note 3)	125% (see Note 1)	167%

Notes:

1. Where 125 percent of this current does not correspond to a standard rating of a fuse or nonadjustable circuit breaker, the next higher standard rating described in Section 240-6 shall be permitted.

2. Where secondary overcurrent protection is required, the secondary overcurrent device shall be permitted to consist of not more than six circuit breakers or six sets of fuses grouped in one location. Where multiple overcurrent devices are utilized, the total of all the device ratings shall not exceed that allowed value of a single overcurrent device. If both breakers and fuses are utilized as the overcurrent device, the total of the device ratings shall not exceed that allowed for fuses.

3. A transformer equipped with coordinated thermal overload protection by the manufacturer and arranged to interrupt the primary current, shall be permitted to have primary overcurrent protection rated or set at a current value that is not more than six times the rated current of the transformer for transformers having not more than 6 percent impedance, and not more than four times the rated current of the transformer for transformers having more than 6 percent but not more than 10 percent impedance.

FPN: The phase current in a grounding autotransformer is one-third the neutral current.

(a) Three-Phase, 4-Wire System. A grounding autotransformer used to create a 3-phase, 4-wire distribution system from a 3-phase, 3-wire ungrounded system shall conform to the following.

(1) Connections. The transformer shall be directly connected to the ungrounded phase conductors and shall not be switched or provided with overcurrent protection that is independent of the main switch and common-trip overcurrent protection for the 3-phase, 4-wire system.

(2) Overcurrent Protection. An overcurrent sensing device shall be provided that will cause the main switch or common-trip overcurrent protection referred to in (a)(1) to open if the load on the autotransformer reaches or exceeds 125 percent of its continuous current per phase or neutral rating. Delayed tripping for temporary overcurrents sensed at the autotransformer overcurrent device shall be permitted for the purpose of allowing proper operation of branch or feeder protective devices on the 4-wire system.

(3) Transformer Fault Sensing. A fault-sensing system that will cause the opening of a main switch or common-trip overcurrent device for the 3-phase, 4-wire system shall be provided to guard against single-phasing or internal faults.

FPN: This can be accomplished by the use of two subtractive-connected donut-type current transformers installed to sense and signal when an unbalance occurs in the line current to the autotransformer of 50 percent or more of rated current.

(4) Rating. The autotransformer shall have a continuous neutral-current rating that is sufficient to handle the maximum possible neutral unbalanced load current of the 4-wire system.

(b) Ground Reference for Fault Protection Devices. A grounding autotransformer used to make available a specified magnitude of ground-fault current for operation of a ground-responsive protective device on a 3-phase, 3-wire ungrounded system shall conform to the following requirements.

(1) Rating. The autotransformer shall have a continuous neutral-current rating sufficient for the specified ground-fault current.

(2) Overcurrent Protection. An overcurrent protective device of adequate short-circuit rating that will open simultaneously all ungrounded conductors when it operates shall be applied in the grounding autotransformer branch circuit and shall be rated or set at a current not exceeding 125 percent of the autotransformer continuous per phase-current rating or 42 percent of the continuous-current rating of any series connected devices in the autotransformer neutral connection. Delayed tripping for temporary overcurrents to permit the proper operation of ground-responsive tripping devices on the main system shall be permitted, but shall not exceed values that would be more than the short-time current rating of the grounding autotransformer or any series connected devices in the neutral connection thereto.

(c) Ground Reference for Damping Transitory Overvoltages. A grounding autotransformer used to limit transitory overvoltages shall be of suitable rating and connected in accordance with Section 450-5(a)(1).

450-6. Secondary Ties. A secondary tie is a circuit operating at 600 volts, nominal, or less, between phases that connects two power sources or power supply points, such as the secondaries of two transformers. The tie shall be permitted to consist of one or more conductors per phase.

As used in this section, the word *transformer* means a transformer or a bank of transformers operating as a unit.

(a) Tie Circuits. Tie circuits shall be provided with overcurrent protection at each end as required in Article 240.

Under the conditions described in (a)(1) and (a)(2), the overcurrent protection shall be permitted to be in accordance with (a)(3).

(1) Loads at Transformer Supply Points Only. Where all loads are connected at the transformer supply points at each end of the tie and overcurrent protection is not provided in accordance with Article 240, the rated ampacity of the tie shall not be less than 67 percent of the rated secondary current of the largest transformer connected to the secondary tie system.

(2) Loads Connected Between Transformer Supply Points. Where load is connected to the tie at any point between transformer supply points and overcurrent protection is not provided in accordance with Article 240, the rated ampacity of the tie shall not be less than 100 percent of the rated secondary current of the largest transformer connected to the secondary tie system.

Exception: As otherwise provided in Section 450-6(a)(4).

(3) Tie Circuit Protection. Under the conditions described in (a)(1) and (a)(2), both ends of each tie conductor shall be equipped with a protective device that will open at a predetermined temperature of the tie conductor under short-circuit conditions. This protection shall consist of one of the following: (1) a fusible link cable connector, terminal, or lug, commonly known as a limiter, each being of a size corresponding with that of the conductor and of construction and characteristics according to the operating voltage and the type of insulation on the tie conductors or (2) automatic circuit breakers actuated by devices having comparable current-time characteristics.

(4) Interconnection of Phase Conductors Between Transformer Supply Points. Where the tie consists of more than one conductor per phase, the conductors of each phase shall comply with one of the following provisions.

(a) *Interconnected.* The conductors shall be interconnected in order to establish a load supply point, and the protection specified in (a)(3) shall be provided in each tie conductor at this point.

(b) *Not Interconnected.* The loads shall be connected to one or more individual conductors of a paralleled conductor tie without interconnecting the conductors of each phase and without the protection specified in (a)(3) at load connection points. Where this is done, the tie conductors of each phase shall have a combined capacity of not less than 133 percent of the rated secondary current of the largest transformer connected to the secondary tie system, the total load of such taps shall not exceed the rated secondary current of the largest transformer, and the loads shall be equally divided on each phase and on the individual conductors of each phase as far as practicable.

(5) Tie Circuit Control. Where the operating voltage exceeds 150 volts to ground, secondary ties provided with limiters shall have a switch at each end that, when open, will de-energize the associated tie conductors and limiters. The current rating of the switch shall not be less than the rated current of the conductors connected to the switch. It shall be capable of opening its rated current, and it shall be constructed so that it will not open under the magnetic forces resulting from short-circuit current.

(b) Overcurrent Protection for Secondary Connections. Where secondary ties are used, an overcurrent device rated or set at not more than 250 percent of the rated secondary current of the transformers shall be provided in the secondary connections of each transformer. In addition, an automatic circuit breaker actuated by a reverse-current relay set to open the circuit at not more than the rated secondary current of the transformer shall be provided in the secondary connection of each transformer.

450-7. Parallel Operation. Transformers shall be permitted to be operated in parallel and switched as a unit provided the overcurrent protection for each transformer meets the requirements of Section 450-3(a) for primary and secondary protective devices over 600 volts or Section 450-3(b) for primary and secondary protective devices 600 volts or less.

450-8. Guarding. Transformers shall be guarded as specified in (a) through (d).

(a) Mechanical Protection. Appropriate provisions shall be made to minimize the possibility of damage to transform-

ers from external causes where the transformers are exposed to physical damage.

(b) Case or Enclosure. Dry-type transformers shall be provided with a noncombustible moisture-resistant case or enclosure that will provide protection against the accidental insertion of foreign objects.

(c) Exposed Energized Parts. Switches or other equipment operating at 600 volts, nominal, or less, and serving only equipment within a transformer enclosure shall be permitted to be installed in the transformer enclosure if accessible to qualified persons only. All energized parts shall be guarded in accordance with Sections 110-27 and 110-34.

(d) Voltage Warning. The operating voltage of exposed live parts of transformer installations shall be indicated by signs or visible markings on the equipment or structures.

450-9. Ventilation. The ventilation shall be adequate to dispose of the transformer full-load losses without creating a temperature rise that is in excess of the transformer rating.

FPN No. 1: See *General Requirements for Liquid-Immersed Distribution, Power, and Regulating Transformers*, ANSI/IEEE C57.12.00-1993, and *General Requirements for Dry-Type Distribution and Power Transformers*, ANSI/IEEE C57.12.01-1989.

FPN No. 2: Additional losses may occur in some transformers where nonsinusoidal currents are present, resulting in increased heat in the transformer above its rating. See *Recommended Practice for Establishing Transformer Capability When Supplying Nonsinusoidal Load Currents*, ANSI/IEEE C57.110-1993, where transformers are utilized with nonlinear loads.

Transformers with ventilating openings shall be installed so that the ventilating openings are not blocked by walls or other obstructions. The required clearances shall be clearly marked on the transformer.

450-10. Grounding. Exposed noncurrent-carrying metal parts of transformer installations, including fences, guards, etc., shall be grounded where required under the conditions and in the manner specified for electric equipment and other exposed metal parts in Article 250.

450-11. Marking. Each transformer shall be provided with a nameplate giving the name of the manufacturer, rated kilovolt-amperes, frequency, primary and secondary voltage, impedance of transformers 25 kVA and larger, required clearances for transformers with ventilating openings, and the amount and kind of insulating liquid where used. In addition, the nameplate of each dry-type transformer shall include the temperature class for the insulation system.

450-12. Terminal Wiring Space. The minimum wire-bending space at fixed, 600-volt and below terminals of transformer line and load connections shall be as required in Section 373-6. Wiring space for pigtail connections shall conform to Table 370-16(b).

450-13. Accessibility. All transformers and transformer vaults shall be readily accessible to qualified personnel for inspection and maintenance, or meet the requirements of (a) or (b).

(a) Open Installations. Dry-type transformers 600 volts, nominal, or less, located in the open on walls, columns, or structures, shall not be required to be readily accessible.

(b) Hollow Space Installations. Dry-type transformers 600 volts, nominal, or less and not exceeding 50 kVA shall be permitted in hollow spaces of buildings not permanently closed in by structure, and provided they meet the ventilation requirements of Section 450-9 and separation from combustible materials requirements of Section 450-21(a). Transformers so installed shall not be required to be readily accessible.

B. Specific Provisions Applicable to Different Types of Transformers

450-21. Dry-Type Transformers Installed Indoors.

(a) Not Over 112½ kVA. Dry-type transformers installed indoors and rated 112½ kVA or less shall have a separation of at least 12 in. (305 mm) from combustible material unless separated from the combustible material by a fire-resistant, heat-insulated barrier.

Exception: This rule shall not apply to transformers rated for 600 volts, nominal, or less, completely enclosed, with or without ventilating openings.

(b) Over 112½ kVA. Individual dry-type transformers of more than 112½ kVA rating shall be installed in a transformer room of fire-resistant construction. Unless specified otherwise in this article, the term *fire resistant* means a construction having a minimum fire rating of 1 hour.

Exception No. 1: Transformers with Class 155 or higher insulation systems and separated from combustible material by a fire-resistant, heat-insulating barrier or by not less than 6 ft (1.83 m) horizontally and 12 ft (3.66 m) vertically.

Exception No. 2: Transformers with Class 155 or higher insulation systems and completely enclosed except for ventilating openings.

FPN: See *Method for Fire Tests of Building Construction and Materials*, ANSI/ASTM E119-1995, and *Standard Methods of Tests of Fire Endurance of Building Construction and Materials*, NFPA 251-1995.

(c) Over 35,000 Volts. Dry-type transformers rated over 35,000 volts shall be installed in a vault complying with Part C of this article.

450-22. Dry-Type Transformers Installed Outdoors. Dry-type transformers installed outdoors shall have a weatherproof enclosure.

Transformers exceeding 112½ kVA shall not be located within 12 in. (305 mm) of combustible materials of buildings unless the transformer has Class 155 insulation systems or higher and is completely enclosed except for ventilating openings.

450-23. Less-Flammable Liquid-Insulated Transformers. Transformers insulated with listed less-flammable liquids that have a fire point of not less than 300°C shall be permitted to be installed in accordance with (a) or (b).

(a) Indoor Installations. In accordance with one of the following.

(1) In Type I or Type II buildings, in areas where all of the following requirements are met:

   (a) The transformer is rated 35,000 volts or less.
   (b) No combustible materials are stored.
   (c) A liquid confinement area is provided.
   (d) The installation complies with all restrictions provided for in the listing of the liquid.

(2) With an automatic fire extinguishing system and a liquid confinement area, provided the transformer is rated 35,000 volts or less.

(3) In accordance with Section 450-26.

(b) Outdoor Installations. Less-flammable liquid-filled transformers shall be permitted to be installed outdoors attached to, adjacent to, or on the roof of buildings, where installed in accordance with (1) or (2).

(1) For Type I and Type II buildings, the installation shall comply with all restrictions provided for in the listing of the liquid.

FPN: Installations adjacent to combustible material, fire escapes, or door and window openings may require additional safeguards such as those listed in Section 450-27.

(2) In accordance with Section 450-27.

FPN No. 1: As used in this section, *Type I and Type II buildings* refers to Type I and Type II building construction as defined in *Standard on Types of Building Construction*, NFPA 220-1995. *Combustible materials* refers to those materials not classified as noncombustible or limited-combustible as defined in *Standard on Types of Building Construction*, NFPA 220-1995.

FPN No. 2: See definition of *Listed* in Article 100.

450-24. Nonflammable Fluid-Insulated Transformers. Transformers insulated with a dielectric fluid identified as nonflammable shall be permitted to be installed indoors or outdoors. Such transformers installed indoors and rated over 35,000 volts shall be installed in a vault. Such transformers installed indoors shall be furnished with a liquid confinement area and a pressure-relief vent. The transformers shall be furnished with a means for absorbing any gases generated by arcing inside the tank, or the pressure-relief vent shall be connected to a chimney or flue that will carry such gases to an environmentally safe area.

FPN: Safety may be increased if fire hazard analyses are performed for such transformer installations.

For the purposes of this section, a nonflammable dielectric fluid is one that does not have a flash point or fire point, and is not flammable in air.

450-25. Askarel-Insulated Transformers Installed Indoors. Askarel-insulated transformers installed indoors and rated over 25 kVA shall be furnished with a pressure-relief vent. Where installed in a poorly ventilated place, they shall be furnished with a means for absorbing any gases generated by arcing inside the case, or the pressure-relief vent shall be connected to a chimney or flue that will carry such gases outside the building. Askarel-insulated transformers rated over 35,000 volts shall be installed in a vault.

450-26. Oil-Insulated Transformers Installed Indoors. Oil-insulated transformers installed indoors shall be installed in a vault constructed as specified in Part C of this article.

Exception No. 1: Where the total capacity does not exceed 112½ kVA, the vault specified in Part C of this article shall be permitted to be constructed of reinforced concrete that is not less than 4 in. (102 mm) thick.

Exception No. 2: Where the nominal voltage does not exceed 600, a vault shall not be required if suitable arrangements are made to prevent a transformer oil fire from igniting other materials, and the total capacity in one location does not exceed 10 kVA in a section of the building classified as combustible, or 75 kVA where the surrounding structure is classified as fire-resistant construction.

Exception No. 3: Electric furnace transformers that have a total rating not exceeding 75 kVA shall be permitted to be installed without a vault in a building or room of fire-resistant construction, provided suitable arrangements are made to prevent a transformer oil fire from spreading to other combustible material.

Exception No. 4: A transformer that has a total rating not exceeding 75 kVA and a supply voltage of 600 volts or less that is an integral part of charged particle accelerating equipment shall be permitted to be installed without a vault in a building or room of noncombustible or fire-resistant construction, provided suitable arrangements are made to prevent a transformer oil fire from spreading to other combustible material.

Exception No. 5: Transformers shall be permitted to be installed in a detached building that does not comply with Part C of this article if neither the building nor its contents present a fire hazard to any other building or property, and if the building is used only in supplying electric service and the interior is accessible only to qualified persons.

Exception No. 6: Oil-insulated transformers shall be permitted to be used without a vault in portable and mobile surface mining equipment (such as electric excavators) if each of the following conditions is met.

(a) Provision is made for draining leaking fluid to the ground.

(b) Safe egress is provided for personnel.

(c) A minimum ¼-in. (6.35-mm) steel barrier is provided for personnel protection.

450-27. Oil-Insulated Transformers Installed Outdoors. Combustible material, combustible buildings, and parts of buildings, fire escapes, and door and window openings shall be safeguarded from fires originating in oil-insulated transformers installed on roofs, attached to or adjacent to a building or combustible material.

Space separations, fire-resistant barriers, automatic water spray systems, and enclosures that confine the oil of a ruptured transformer tank are recognized safeguards. One or more of these safeguards shall be applied according to the degree of hazard involved in cases where the transformer installation presents a fire hazard.

Oil enclosures shall be permitted to consist of fire-resistant dikes, curbed areas or basins, or trenches filled with coarse, crushed stone. Oil enclosures shall be provided with trapped drains where the exposure and the quantity of oil involved are such that removal of oil is important.

FPN: For additional information on transformers installed on poles or structures or underground, see *National Electrical Safety Code*, ANSI C2-1997.

450-28. Modification of Transformers. When modifications are made to a transformer in an existing installation that change the type of the transformer with respect to Part B of this article, such transformer shall be marked to show the type of insulating liquid installed, and the modified transformer installation shall comply with the applicable requirements for that type of transformer.

C. Transformer Vaults

450-41. Location. Vaults shall be located where they can be ventilated to the outside air without using flues or ducts wherever such an arrangement is practicable.

450-42. Walls, Roofs, and Floors. The walls and roofs of vaults shall be constructed of materials that have adequate structural strength for the conditions with a minimum fire resistance of 3 hours. The floors of vaults in contact with the earth shall be of concrete that is not less than 4 in. (102 mm) thick, but where the vault is constructed with a vacant space or other stories below it, the floor shall have adequate structural strength for the load imposed thereon and a minimum fire resistance of 3 hours. For the purposes of this section, studs and wallboard construction shall not be acceptable.

Exception: Where transformers are protected with automatic sprinkler, water spray, carbon dioxide, or halon, construction of 1-hour rating shall be permitted.

FPN No. 1: For additional information, see *Method for Fire Tests of Building Construction and Materials*, ANSI/ASTM E119-1995, and *Standard Methods of Tests of Fire Endurance of Building Construction and Materials*, NFPA 251-1995.

FPN No. 2: A typical 3-hour construction is 6 in. (152-mm) thick reinforced concrete.

450-43. Doorways. Vault doorways shall be protected as follows.

(a) Type of Door. Each doorway leading into a vault from the building interior shall be provided with a tight-fitting door that has a minimum fire rating of 3 hours. The authority having jurisdiction shall be permitted to require such a door for an exterior wall opening where conditions warrant.

Exception: Where transformers are protected with automatic sprinkler, water spray, carbon dioxide, or halon, construction of 1-hour rating shall be permitted.

FPN: For additional information, see *Standard for Fire Doors and Fire Windows*, NFPA 80-1995.

(b) Sills. A door sill or curb that is of sufficient height to confine the oil from the largest transformer within the vault shall be provided, and in no case shall the height be less than 4 in. (102 mm).

(c) Locks. Doors shall be equipped with locks, and doors shall be kept locked, access being allowed only to qualified persons. Personnel doors shall swing out and be equipped with panic bars, pressure plates, or other devices that are normally latched but open under simple pressure.

450-45. Ventilation Openings. Where required by Section 450-9, openings for ventilation shall be provided in accordance with (a) through (f).

(a) Location. Ventilation openings shall be located as far away as possible from doors, windows, fire escapes, and combustible material.

(b) Arrangement. A vault ventilated by natural circulation of air shall be permitted to have roughly half of the total area of openings required for ventilation in one or more openings near the floor and the remainder in one or more openings in the roof or in the sidewalls near the roof, or all of the area required for ventilation shall be permitted in one or more openings in or near the roof.

(c) Size. For a vault ventilated by natural circulation of air to an outdoor area, the combined net area of all ventilating openings, after deducting the area occupied by screens, gratings, or louvers, shall not be less than 3 in.^{2} (1936 mm^2) per kVA of transformer capacity in service, and in no case shall the net area be less than 1 ft^2 (0.093 m^2) for any capacity under 50 kVA.

(d) Covering. Ventilation openings shall be covered with durable gratings, screens, or louvers, according to the treatment required in order to avoid unsafe conditions.

(e) Dampers. All ventilation openings to the indoors shall be provided with automatic closing fire dampers that operate in response to a vault fire. Such dampers shall possess a standard fire rating of not less than 1½ hours.

FPN: See *Standard for Fire Dampers*, ANSI/UL 555-1995.

(f) Ducts. Ventilating ducts shall be constructed of fire-resistant material.

450-46. Drainage. Where practicable, vaults containing more than 100 kVA transformer capacity shall be provided with a drain or other means that will carry off any accumulation of oil or water in the vault unless local conditions make this impracticable. The floor shall be pitched to the drain where provided.

450-47. Water Pipes and Accessories. Any pipe or duct system foreign to the electrical installation shall not enter or pass through a transformer vault. Piping or other facilities provided for vault fire protection, or for transformer cooling, shall not be considered foreign to the electrical installation.

450-48. Storage in Vaults. Materials shall not be stored in transformer vaults.

Article 455 — Phase Converters

A. General

455-1. Scope. This article covers the installation and use of phase converters.

455-2. Definitions.

Manufactured Phase. The manufactured or derived phase originates at the phase converter and is not solidly connected to either of the single-phase input conductors.

Phase Converter. An electrical device that converts single-phase power to 3-phase electrical power.

FPN: Phase converters have characteristics that modify the starting torque and locked-rotor current of motors served, and consideration is required in selecting a phase converter for a specific load.

Rotary-Phase Converter. A device that consists of a rotary transformer and capacitor panel(s) that permits the operation of 3-phase loads from a single-phase supply.

Static-Phase Converter. A device without rotating parts, sized for a given 3-phase load to permit operation from a single-phase supply.

455-3. Other Articles. All applicable requirements of this *Code* shall apply to phase converters except as amended by this article.

455-4. Marking. Each phase converter shall be provided with a permanent nameplate indicating the following:

(1) Manufacturer's name
(2) Rated input and output voltages
(3) Frequency
(4) Rated single-phase input full-load amperes
(5) Rated minimum and maximum single load in kilovolt-amperes (kVA) or horsepower
(6) Maximum total load in kilovolt-ampere (kVA) or horsepower
(7) For a rotary-phase converter, 3-phase amperes at full load

455-5. Equipment Grounding Connection. A means for attachment of an equipment grounding conductor termination in accordance with Section 250-8 shall be provided.

455-6. Conductors.

(a) Ampacity. The ampacity of the single-phase supply conductors shall be determined by (1) or (2).

FPN: Single-phase conductors sized to prevent a voltage drop not exceeding 3 percent from the source of supply to the phase converter may help ensure proper starting and operation of motor loads.

(1) Variable Loads. Where the loads to be supplied are variable, the conductor ampacity shall not be less than 125 percent of the phase converter nameplate single-phase input full-load amperes.

(2) Fixed Loads. Where the phase converter supplies specific fixed loads, and the conductor ampacity is less than 125 percent of the phase converter nameplate single-phase input full-load amperes, the conductors shall have an ampacity not less than 250 percent of the sum of the full-load, 3-phase current rating of the motors and other loads served where the input and output voltages of the phase converter are identical. Where the input and output voltages of the phase converter are different, the current as determined by this section shall be multiplied by the ratio of output to input voltage.

(b) Manufactured Phase Marking. The manufactured phase conductors shall be identified in all accessible locations with a distinctive marking.

455-7. Overcurrent Protection. The single-phase supply conductors and phase converter shall be protected from overcurrent by (a) or (b). Where the required fuse rating or circuit breaker setting does not correspond to a standard rating or setting, the next higher standard rating or setting shall be permitted.

(a) Variable Loads. Where the loads to be supplied are variable, overcurrent protection shall be set at not more than 125 percent of the phase converter nameplate single-phase input full-load amperes.

(b) Fixed Loads. Where the phase converter supplies specific fixed loads, and the conductors are sized in accordance with Section 455-6(a)(2), the conductors shall be protected in accordance with their ampacity. The overcurrent protection determined from this section shall not exceed 125 percent of the phase converter nameplate single-phase input amperes.

455-8. Disconnecting Means. Means shall be provided to disconnect simultaneously all ungrounded single-phase supply conductors to the phase converter.

(a) Location. The disconnecting means shall be readily accessible and located in sight from the phase converter.

(b) Type. The disconnecting means shall be a switch rated in horsepower, a circuit breaker, or a molded-case switch. Where only nonmotor loads are served, an ampere-rated switch shall be permitted.

(c) Rating. The ampere rating of the disconnecting means shall not be less than 115 percent of the rated maximum single-phase input full-load amperes or, for specific fixed loads, shall be permitted to be selected from (1) or (2).

(1) Current Rated Disconnect. The disconnecting means shall be a circuit breaker or molded-case switch with an ampere rating not less than 250 percent of the sum of the following:

(a) Full-load, 3-phase current ratings of the motors

(b) Other loads served

(2) Horsepower Rated Disconnect. The disconnecting means shall be a switch with a horsepower rating. The equivalent locked rotor current of the horsepower rating of the switch shall not be less than 200 percent of the sum of the following:

(a) Nonmotor loads

(b) The 3-phase, locked-rotor current of the largest motor as determined from Table 430-151(B) and

(c) The full-load current of all other 3-phase motors operating at the same time

(d) Voltage Ratios. The calculations in (c) shall apply directly where the input and output voltages of the phase converter are identical. Where the input and output voltages of the phase converter are different, the current shall be multiplied by the ratio of the output to input voltage.

455-9. Connection of Single-Phase Loads. Where single-phase loads are connected on the load side of a phase converter, they shall not be connected to the manufactured phase.

455-10. Terminal Housings. A terminal housing shall be provided on a phase converter, and the terminal housing shall be in accordance with the provisions of Section 430-12.

B. Specific Provisions Applicable to Different Types of Phase Converters

455-20. Disconnecting Means. The single-phase disconnecting means for the input of a static phase converter shall be permitted to serve as the disconnecting means for the phase converter and a single load if the load is within sight of the disconnecting means.

455-21. Start-Up. Power to the utilization equipment shall not be supplied until the rotary-phase converter has been started.

455-22. Power Interruption. Utilization equipment supplied by a rotary-phase converter shall be controlled in such a manner that power to the equipment will be disconnected in the event of a power interruption.

> FPN: Magnetic motor starters, magnetic contactors, and similar devices, with manual or time delay restarting for the load, will provide restarting after power interruption.

455-23. Capacitors. Capacitors that are not an integral part of the rotary-phase conversion system but are installed for a motor load shall be connected to the line side of that motor overload protective device.

Article 460 — Capacitors

460-1. Scope. This article covers the installation of capacitors on electric circuits.

Surge capacitors or capacitors included as a component part of other apparatus and conforming with the requirements of such apparatus are excluded from these requirements.

This article also covers the installation of capacitors in hazardous (classified) locations as modified by Articles 501 through 503.

460-2. Enclosing and Guarding.

(a) Containing More than 3 gal (11.36 L) of Flammable Liquid. Capacitors containing more than 3 gal (11.36 L) of flammable liquid shall be enclosed in vaults or outdoor fenced enclosures complying with Article 110, Part C. This limit shall apply to any single unit in an installation of capacitors.

(b) Accidental Contact. Where capacitors are accessible to unauthorized and unqualified persons, they shall be enclosed, located, or guarded so that persons cannot come into accidental contact or bring conducting materials into accidental contact with exposed energized parts, terminals, or buses associated with them. However, no additional guarding is required for enclosures accessible only to authorized and qualified persons.

A. 600 Volts, Nominal, and Under

460-6. Discharge of Stored Energy. Capacitors shall be provided with a means of discharging stored energy.

(a) Time of Discharge. The residual voltage of a capacitor shall be reduced to 50 volts, nominal, or less, within 1 minute after the capacitor is disconnected from the source of supply.

(b) Means of Discharge. The discharge circuit shall be either permanently connected to the terminals of the capacitor or capacitor bank, or provided with automatic means of connecting it to the terminals of the capacitor bank on removal of voltage from the line. Manual means of switching or connecting the discharge circuit shall not be used.

460-8. Conductors.

(a) Ampacity. The ampacity of capacitor circuit conductors shall not be less than 135 percent of the rated current of the capacitor. The ampacity of conductors that connect a capacitor to the terminals of a motor or to motor circuit conductors shall not be less than one-third the ampacity of the motor circuit conductors and in no case less than 135 percent of the rated current of the capacitor.

(b) Overcurrent Protection.

(1) An overcurrent device shall be provided in each ungrounded conductor for each capacitor bank.

Exception: A separate overcurrent device shall not be required for a capacitor connected on the load side of a motor overload protective device.

(2) The rating or setting of the overcurrent device shall be as low as practicable.

(c) Disconnecting Means.

(1) A disconnecting means shall be provided in each ungrounded conductor for each capacitor bank.

Exception: A separate disconnecting means shall not be required where a capacitor is connected on the load side of a motor controller.

(2) The disconnecting means shall open all ungrounded conductors simultaneously.

(3) The disconnecting means shall be permitted to disconnect the capacitor from the line as a regular operating procedure.

(4) The rating of the disconnecting means shall not be less than 135 percent of the rated current of the capacitor.

460-9. Rating or Setting of Motor Overload Device. Where a motor installation includes a capacitor connected on the load side of the motor overload device, the rating or setting of the motor overload device shall be based on the improved power factor of the motor circuit.

The effect of the capacitor shall be disregarded in determining the motor circuit conductor rating in accordance with Section 430-22.

460-10. Grounding. Capacitor cases shall be grounded in accordance with Article 250.

Exception: Capacitor cases shall not be grounded where the capacitor units are supported on a structure designed to operate at other than ground potential.

460-12. Marking. Each capacitor shall be provided with a nameplate giving the name of the manufacturer, rated voltage, frequency, kilovar or amperes, number of phases, and, if filled with a combustible liquid, the amount of liquid in gallons. Where filled with a nonflammable liquid, the nameplate shall so state. The nameplate shall also indicate if a capacitor has a discharge device inside the case.

B. Over 600 Volts, Nominal

460-24. Switching.

(a) Load Current. Group-operated switches shall be used for capacitor switching and shall be capable of the following:

(1) Carrying continuously not less than 135 percent of the rated current of the capacitor installation
(2) Interrupting the maximum continuous load current of each capacitor, capacitor bank, or capacitor installation that will be switched as a unit
(3) Withstanding the maximum inrush current, including contributions from adjacent capacitor installations
(4) Carrying currents due to faults on capacitor side of switch

(b) Isolation.

(1) A means shall be installed to isolate from all sources of voltage each capacitor, capacitor bank, or capacitor installation that will be removed from service as a unit.

(2) The isolating means shall provide a visible gap in the electrical circuit adequate for the operating voltage.

(3) Isolating or disconnecting switches (with no interrupting rating) shall be interlocked with the load-interrupting device or shall be provided with prominently displayed caution signs in accordance with Section 490-22 to prevent switching load current.

(c) Additional Requirements for Series Capacitors. The proper switching sequence shall be ensured by use of one of the following:

(1) Mechanically sequenced isolating and bypass switches
(2) Interlocks
(3) Switching procedure prominently displayed at the switching location

460-25. Overcurrent Protection.

(a) Provided to Detect and Interrupt Fault Current. A means shall be provided to detect and interrupt fault current likely to cause dangerous pressure within an individual capacitor.

(b) Single-Phase or Multiphase Devices. Single-phase or multiphase devices shall be permitted for this purpose.

(c) Protected Individually or in Groups. Capacitors shall be permitted to be protected individually or in groups.

(d) Protective Devices Rated or Adjusted. Protective devices for capacitors or capacitor equipment shall be rated or adjusted to operate within the limits of the safe zone for individual capacitors. If the protective devices are rated or adjusted to operate within the limits for Zone 1 or Zone 2, the capacitors shall be enclosed or isolated.

In no event shall the rating or adjustment of the protective devices exceed the maximum limit of Zone 2.

FPN: For definitions of *Safe Zone*, *Zone 1*, and *Zone 2*, see *Shunt Power Capacitors*, ANSI/IEEE 18-1992.

460-26. Identification. Each capacitor shall be provided with a permanent nameplate giving the manufacturer's name, rated voltage, frequency, kilovar or amperes, number of phases, and the amount of liquid in gallons identified as flammable, if such is the case.

460-27. Grounding. Capacitor neutrals and cases, if grounded, shall be grounded in accordance with Article 250.

Exception: Where the capacitor units are supported on a structure that is designed to operate at other than ground potential.

460-28. Means for Discharge.

(a) Means to Reduce the Residual Voltage. A means shall be provided to reduce the residual voltage of a capacitor to 50 volts or less within 5 minutes after the capacitor is disconnected from the source of supply.

(b) Connection to Terminals. A discharge circuit shall be either permanently connected to the terminals of the capacitor or provided with automatic means of connecting it to the terminals of the capacitor bank after disconnection of the capacitor from the source of supply. The windings of motors, or transformers, or of other equipment directly connected to capacitors without a switch or overcurrent device interposed shall meet the requirements of (a).

Article 470 — Resistors and Reactors (For Rheostats, See Section 430-82.)

A. 600 Volts, Nominal, and Under

470-1. Scope. This article covers the installation of separate resistors and reactors on electric circuits.

Exception: Resistors and reactors that are component parts of other apparatus.

This article also covers the installation of resistors and reactors in hazardous (classified) locations as modified by Articles 501 through 504.

470-2. Location. Resistors and reactors shall not be placed where exposed to physical damage.

470-3. Space Separation. A thermal barrier shall be required if the space between the resistors and reactors and any combustible material is less than 12 in. (305 mm).

470-4. Conductor Insulation. Insulated conductors used for connections between resistance elements and controllers shall be suitable for an operating temperature of not less than 90°C (194°F).

Exception: Other conductor insulations shall be permitted for motor starting service.

B. Over 600 Volts, Nominal

470-18. General.

(a) Protected Against Physical Damage. Resistors and reactors shall be protected against physical damage.

(b) Isolated by Enclosure or Elevation. Resistors and reactors shall be isolated by enclosure or elevation to protect personnel from accidental contact with energized parts.

(c) Combustible Materials. Resistors and reactors shall not be installed in close enough proximity to combustible materials to constitute a fire hazard and shall have a clearance of not less than 1 ft (305 mm) from combustible materials.

(d) Clearances. Clearances from resistors and reactors to grounded surfaces shall be adequate for the voltage involved.

FPN: See Article 490.

(e) Temperature Rise from Induced Circulating Currents. Metallic enclosures of reactors and adjacent metal parts shall be installed so that the temperature rise from induced circulating currents will not be hazardous to personnel or constitute a fire hazard.

470-19. Grounding. Resistor and reactor cases or enclosures shall be grounded in accordance with Article 250.

Exception: Resistor or reactor cases or enclosures supported on a structure designed to operate at other than ground potential shall not be grounded.

470-20. Oil-Filled Reactors. Installation of oil-filled reactors, in addition to the above requirements, shall comply with applicable requirements of Article 450.

Article 480 — Storage Batteries

480-1. Scope. The provisions of this article shall apply to all stationary installations of storage batteries.

480-2. Definitions.

Nominal Battery Voltage. The voltage computed on the basis of 2 volts per cell for the lead-acid type and 1.2 volts per cell for the alkali type.

Sealed Cell or Battery. A sealed cell or battery is one that has no provision for the addition of water or electrolyte or for external measurement of electrolyte specific gravity. The individual cells shall be permitted to contain a venting arrangement as described in Section 480-9(b).

Storage Battery. A battery comprised of one or more rechargeable cells of the lead-acid, nickel-cadmium, or other rechargeable electrochemical types.

480-3. Wiring and Equipment Supplied from Batteries. Wiring and equipment supplied from storage batteries shall be subject to the requirements of this *Code* applying to wiring and equipment operating at the same voltage.

480-4. Grounding. The requirements of Article 250 shall apply.

480-5. Insulation of Batteries Not Over 250 Volts. This section shall apply to storage batteries having cells connected so as to operate at a nominal battery voltage of not over 250 volts.

(a) Vented Lead-Acid Batteries. Cells and multicompartment batteries with covers sealed to containers of nonconductive, heat-resistant material shall not require additional insulating support.

(b) Vented Alkaline-Type Batteries. Cells with covers sealed to jars of nonconductive, heat-resistant material shall require no additional insulation support. Cells in jars of conductive material shall be installed in trays of nonconductive material with not more than 20 cells (24 volts, nominal) in the series circuit in any one tray.

(c) Rubber Jars. Cells in rubber or composition containers shall require no additional insulating support where the total nominal voltage of all cells in series does not exceed 150 volts. Where the total voltage exceeds 150 volts, batteries shall be sectionalized into groups of 150 volts or less, and each group shall have the individual cells installed in trays or on racks.

(d) Sealed Cells or Batteries. Sealed cells and multicompartment sealed batteries constructed of nonconductive, heat-resistant material shall not require additional insulating support. Batteries constructed of a conducting container shall have insulating support if a voltage is present between the container and ground.

480-6. Insulation of Batteries of Over 250 Volts. The provisions of Section 480-5 shall apply to storage batteries having the cells connected so as to operate at a nominal voltage exceeding 250 volts, and, in addition, the provisions of this section shall also apply to such batteries. Cells shall be installed in groups having a total nominal voltage of not over 250 volts. Insulation, which can be air, shall be provided between groups and shall have a minimum separation between live battery parts of opposite polarity of 2 in. (50.8 mm) for battery voltages not exceeding 600 volts.

480-7. Racks and Trays. Racks and trays shall comply with (a) and (b).

(a) Racks. Racks, as required in this article, are rigid frames designed to support cells or trays. They shall be substantial and made of the following:

(1) Metal, treated so as to be resistant to deteriorating action by the electrolyte and provided with nonconducting members directly supporting the cells or with continuous insulating material other than paint on conducting members, or

(2) Other construction such as fiberglass or other suitable nonconductive materials

(b) Trays. Trays are frames, such as crates or shallow boxes usually of wood or other nonconductive material, constructed or treated so as to be resistant to deteriorating action by the electrolyte.

480-8. Battery Locations. Battery locations shall conform to (a), (b), and (c).

(a) Ventilation. Provisions shall be made for sufficient diffusion and ventilation of the gases from the battery to prevent the accumulation of an explosive mixture.

(b) Live Parts. Guarding of live parts shall comply with Section 110-27.

(c) Working Space. Working space about the battery systems shall comply with Section 110-26. Working clearance shall be measured from the edge of the battery rack.

480-9. Vents.

(a) Vented Cells. Each vented cell shall be equipped with a flame arrester that is designed to prevent destruction of the cell due to ignition of gases within the cell by an external spark or flame under normal operating conditions.

(b) Sealed Cells. Sealed battery or cells shall be equipped with a pressure-release vent to prevent excessive accumulation of gas pressure, or the battery or cell shall be designed to prevent scatter of cell parts in event of a cell explosion.

Article 490 — Equipment, Over 600 Volts, Nominal

A. General

490-1. Scope. This article covers the general requirements for equipment operating at more than 600 volts, nominal.

FPN: See *Standard for Electrical Safety Requirements for Employee Workplaces*, NFPA 70E-1995, for electrical safety requirements for employee workplaces.

490-2. Definition. For the purposes of this article, *high voltage* shall be defined as more than 600 volts, nominal.

490-3. Oil-Filled Equipment. Installation of electrical equipment, other than transformers covered in Article 450, containing more than 10 gal (37.85 L) of flammable oil per unit shall meet the requirements of Parts B and C of Article 450.

B. Equipment — Specific Provisions

490-21. Circuit-Interrupting Devices.

(a) Circuit Breakers.

(1) Circuit breakers installed indoors shall be mounted either in metal-enclosed units or fire-resistant cell-mounted units, or they shall be permitted to be open mounted in locations accessible to qualified persons only.

(2) Circuit breakers used to control oil-filled transformers shall be located either outside the transformer vault or be capable of operation from outside the vault.

(3) Oil circuit breakers shall be arranged or located so that adjacent readily combustible structures or materials are safeguarded in an approved manner.

(4) Circuit breakers shall have the following equipment or operating characteristics.

(a) An accessible mechanical or other approved means for manual tripping, independent of control power.

(b) Be release free (trip free).

(c) If capable of being opened or closed manually while energized, the main contacts shall operate independently of the speed of the manual operation.

(d) A mechanical position indicator at the circuit breaker to show the open or closed position of the main contacts.

(e) A means of indicating the open and closed position of the breaker at the point(s) from which they may be operated.

(f) A permanent and legible nameplate showing manufac-turer's name or trademark, manufacturer's type or identification number, continuous current rating, interrupting rating in millivolt-amperes (mVA) or amperes, and maximum voltage rating. Modification of a circuit breaker affecting its rating(s) shall be accompanied by an appropriate change of nameplate information.

(5) The continuous current rating of a circuit breaker shall not be less than the maximum continuous current through the circuit breaker.

(6) The interrupting rating of a circuit breaker shall not be less than the maximum fault current the circuit breaker will be required to interrupt, including contributions from all connected sources of energy.

(7) The closing rating of a circuit breaker shall not be less than the maximum asymmetrical fault current into which the circuit breaker can be closed.

(8) The momentary rating of a circuit breaker shall not be less than the maximum asymmetrical fault current at the point of installation.

(9) The rated maximum voltage of a circuit breaker shall not be less than the maximum circuit voltage.

(b) Power Fuses and Fuseholders.

(1) Use. Where fuses are used to protect conductors and equipment, a fuse shall be placed in each ungrounded conductor. Two power fuses shall be permitted to be used in parallel to protect the same load, if both fuses have identical ratings, and both fuses are installed in an identified common mounting with electrical connections that will divide the current equally. Power fuses of the vented type shall not be used indoors, underground, or in metal enclosures unless identified for the use.

(2) Interrupting Rating. The interrupting rating of power fuses shall not be less than the maximum fault current the fuse will be required to interrupt, including contributions from all connected sources of energy.

(3) Voltage Rating. The maximum voltage rating of power fuses shall not be less than the maximum circuit voltage. Fuses having a minimum recommended operating voltage shall not be applied below this voltage.

(4) Identification of Fuse Mountings and Fuse Units. Fuse mountings and fuse units shall have permanent and legible nameplates showing the manufacturer's type or designation, continuous current rating, interrupting current rating, and maximum voltage rating.

(5) Fuses. Fuses that expel flame in opening the circuit shall be designed or arranged so that they will function properly without hazard to persons or property.

(6) Fuseholders. Fuseholders shall be designed or installed so that they will be de-energized while replacing a fuse.

Exception: Fuses and fuseholders designed to permit fuse replacement by qualified persons using equipment designed for the purpose without de-energizing the fuseholder shall be permitted.

(7) High-Voltage Fuses. Metal-enclosed switchgear and substations that utilize high-voltage fuses shall be provided with a gang-operated disconnecting switch. Isolation of the fuses from the circuit shall be provided by either connecting a switch between the source and the fuses or providing roll-out switch and fuse-type construction. The switch shall be of the load-interrupter type, unless mechanically or electrically interlocked with a load-interrupting device arranged to reduce the load to the interrupting capability of the switch.

Exception: More than one switch shall be permitted as the disconnecting means for one set of fuses where the switches are installed to provide connection to more than one set of supply conductors. The switches shall be mechanically or electrically interlocked to permit access to the fuses only when all switches are open. A conspicuous sign shall be placed at the fuses reading

> WARNING — FUSES MAY BE ENERGIZED
> FROM MORE THAN ONE SOURCE.

(c) Distribution Cutouts and Fuse Links — Expulsion Type.

(1) Installation. Cutouts shall be located so that they may be readily and safely operated and re-fused, and so that the exhaust of the fuses will not endanger persons. Distribution cutouts shall not be used indoors, underground, or in metal enclosures.

(2) Operation. Where fused cutouts are not suitable to interrupt the circuit manually while carrying full load, an approved means shall be installed to interrupt the entire load. Unless the fused cutouts are interlocked with the switch to prevent opening of the cutouts under load, a conspicuous sign shall be placed at such cutouts reading:

> WARNING — DO NOT OPEN UNDER LOAD.

(3) Interrupting Rating. The interrupting rating of distribution cutouts shall not be less than the maximum fault current the cutout will be required to interrupt, including contributions from all connected sources of energy.

(4) Voltage Rating. The maximum voltage rating of cutouts shall not be less than the maximum circuit voltage.

(5) Identification. Distribution cutouts shall have on their body, door, or fuse tube a permanent and legible nameplate or identification showing the manufacturer's type or designation, continuous current rating, maximum voltage rating, and interrupting rating.

(6) Fuse Links. Fuse links shall have a permanent and legible identification showing continuous current rating and type.

(7) Structure Mounted Outdoors. The height of cutouts mounted outdoors on structures shall provide safe clearance between lowest energized parts (open or closed position) and standing surfaces, in accordance with Section 110-34(e).

(d) Oil-Filled Cutouts.

(1) Continuous Current Rating. The continuous current rating of oil-filled cutouts shall not be less than the maximum continuous current through the cutout.

(2) Interrupting Rating. The interrupting rating of oil-filled cutouts shall not be less than the maximum fault current the oil-filled cutout will be required to interrupt, including contributions from all connected sources of energy.

(3) Voltage Rating. The maximum voltage rating of oil-filled cutouts shall not be less than the maximum circuit voltage.

(4) Fault Closing Rating. Oil-filled cutouts shall have a fault closing rating not less than the maximum asymmetrical fault current that can occur at the cutout location, unless suitable interlocks or operating procedures preclude the possibility of closing into a fault.

(5) Identification. Oil-filled cutouts shall have a permanent and legible nameplate showing the rated continuous current, rated maximum voltage, and rated interrupting current.

(6) Fuse Links. Fuse links shall have a permanent and legible identification showing the rated continuous current.

(7) Location. Cutouts shall be located so that they will be readily and safely accessible for re-fusing, with the top of the cutout not over 5 ft (1.52 m) above the floor or platform.

(8) Enclosure. Suitable barriers or enclosures shall be provided to prevent contact with nonshielded cables or energized parts of oil-filled cutouts.

(e) Load Interrupters. Load-interrupter switches shall be permitted if suitable fuses or circuit breakers are used in conjunction with these devices to interrupt fault currents. Where these devices are used in combination, they shall be coordinated electrically so that they will safely withstand the effects of closing, carrying, or interrupting all possible currents up to the assigned maximum short-circuit rating.

Where more than one switch is installed with interconnected load terminals to provide for alternate connection to different supply conductors, each switch shall be provided with a conspicuous sign reading

WARNING — SWITCH MAY BE ENERGIZED
BY BACKFEED.

(1) Continuous Current Rating. The continuous current rating of interrupter switches shall equal or exceed the maximum continuous current at the point of installation.

(2) Voltage Rating. The maximum voltage rating of interrupter switches shall equal or exceed the maximum circuit voltage.

(3) Identification. Interrupter switches shall have a permanent and legible nameplate including the following information: manufacturer's type or designation, continuous current rating, interrupting current rating, fault closing rating, maximum voltage rating.

(4) Switching of Conductors. The switching mechanism shall be arranged to be operated from a location where the operator is not exposed to energized parts and shall be arranged to open all ungrounded conductors of the circuit simultaneously with one operation. Switches shall be arranged to be locked in the open position. Metal-enclosed switches shall be operable from outside the enclosure.

(5) Stored Energy for Opening. The stored-energy operator shall be permitted to be left in the uncharged position after the switch has been closed if a single movement of the operating handle charges the operator and opens the switch.

(6) Supply Terminals. The supply terminals of fused interrupter switches shall be installed at the top of the switch enclosure or, if the terminals are located elsewhere, the equipment shall have barriers installed so as to prevent persons from accidentally contacting energized parts or dropping tools or fuses into energized parts.

490-22. Isolating Means. Means shall be provided to completely isolate an item of equipment. The use of isolating switches shall not be required where there are other ways of de-energizing the equipment for inspection and repairs, such as drawout-type metal-enclosed switchgear units and removable truck panels.

Isolating switches not interlocked with an approved circuit-interrupting device shall be provided with a sign warning against opening them under load.

A fuseholder and fuse, designed for the purpose, shall be permitted as an isolating switch.

490-23. Voltage Regulators. Proper switching sequence for regulators shall be ensured by use of one of the following:

(1) Mechanically sequenced regulator bypass switch(es)
(2) Mechanical interlocks
(3) Switching procedure prominently displayed at the switching location

490-24. Minimum Space Separation. In field-fabricated installations, the minimum air separation between bare live conductors and between such conductors and adjacent grounded surfaces shall not be less than the values given in Table 490-24. These values shall not apply to interior portions or exterior terminals of equipment designed, manufactured, and tested in accordance with accepted national standards.

C. Equipment — Metal-Enclosed Power Switchgear and Industrial Control Assemblies

490-30. General. This part covers assemblies of metal-enclosed power switchgear and industrial control, including but not limited to switches, interrupting devices and their control, metering, protection and regulating equipment, where an integral part of the assembly, with associated interconnections and supporting structures. This part also includes metal-enclosed power switchgear assemblies that form a part of unit substations, power centers, or similar equipment.

490-31. Arrangement of Devices in Assemblies. Arrangement of devices in assemblies shall be such that individual components can safely perform their intended function without adversely affecting the safe operation of other components in the assembly.

490-32. Guarding of High-Voltage Energized Parts Within a Compartment. Where access for other than visual inspection is required to a compartment that contains energized high-voltage parts, barriers shall be provided to prevent accidental contact by persons, tools, or other equipment with energized parts. Exposed live parts shall only be permitted in compartments accessible to qualified persons. Fuses and fuseholders designed to enable future replacement without de-energizing the fuse holder shall only be permitted for use by qualified persons.

490-33. Guarding of Low Voltage Energized Parts Within a Compartment. Energized bare parts mounted on doors shall be guarded where the door must be opened for maintenance of equipment or removal of draw-out equipment.

490-34. Clearance for Cable Conductors Entering Enclosure. The unobstructed space opposite terminals or opposite raceways or cables entering a switchgear or control assembly shall be adequate for the type of conductor and method of termination.

Table 490-24. Minimum Clearance of Live Parts*

Nominal Voltage Rating (kV)	Impulse Withstand, B.I.L (kV)		Minimum Clearance of Live Parts (in.)			
			Phase-to-Phase		Phase-to-Ground	
	Indoors	Outdoors	Indoors	Outdoors	Indoors	Outdoors
2.4–4.16	60	95	4.5	7	3.0	6
7.2	75	95	5.5	7	4.0	6
13.8	95	110	7.5	12	5.0	7
14.4	110	110	9.0	12	6.5	7
23	125	150	10.5	15	7.5	10
34.5	150	150	12.5	15	9.5	10
	200	200	18.0	18	13.0	13
46	—	200	—	18	—	13
	—	250	—	21	—	17
69	—	250	—	21	—	17
	—	350	—	31	—	25
115	—	550	—	53	—	42
138	—	550	—	53	—	42
	—	650	—	63	—	50
161	—	650	—	63	—	50
	—	750	—	72	—	58
230	—	750	—	72	—	58
	—	900	—	89	—	71
	—	1050	—	105	—	83

Note: For SI units, 1 in. = 25.4 mm.

*The values given are the minimum clearance for rigid parts and bare conductors under favorable service conditions. They shall be increased for conductor movement or under unfavorable service conditions or wherever space limitations permit. The selection of the associated impulse withstand voltage for a particular system voltage is determined by the characteristics of the surge protective equipment.

490-35. Accessibility of Energized Parts.

(a) Doors that would provide unqualified persons access to high-voltage energized parts shall be locked.

(b) Low-voltage control equipment, relays, motors, and the like shall not be installed in compartments with exposed high-voltage energized parts or high-voltage wiring unless the access means is interlocked with the high-voltage switch or disconnecting means to prevent the access means from being opened or removed unless the high-voltage switch or disconnecting means is in the isolating position.

Exception: High-voltage instrument or control transformers and space heaters shall be permitted to be installed in the high-voltage compartment without access restrictions beyond those that apply to the high-voltage compartment generally.

490-36. Grounding. Frames of switchgear and control assemblies shall be grounded.

490-37. Grounding of Devices. Devices with metal cases or frames, or both, such as instruments, relays, meters, and instrument and control transformers, located in or on switchgear or control, shall have the frame or case grounded.

490-38. Door Stops and Cover Plates. External hinged doors or covers shall be provided with stops to hold them in the open position. Cover plates intended to be removed for inspection of energized parts or wiring shall be equipped with lifting handles and shall not exceed 12 ft^2(1.11 m^2) in area or 60 lb (27.22 kg) in weight, unless they are hinged and bolted or locked.

490-39. Gas Discharge from Interrupting Devices. Gas discharged during operating of interrupting devices shall be directed so as not to endanger personnel.

490-40. Inspection Windows. Windows intended for inspection of disconnecting switches or other devices shall be of suitable transparent material.

490-41. Location of Devices.

(a) Control and instrument transfer switch handles or push buttons other than those covered in (b), shall be in a readily accessible location at an elevation of not over 78 in. (1.98 m).

Exception: Operating handles requiring more than 50 lb (22.68 kg) of force shall be located no higher than 66 in. (1.68 m) in either the open or closed position.

(b) Operating handles for infrequently operated devices, such as drawout fuses, fused potential or control transformers and their primary disconnects, and bus transfer switches, shall be permitted to be located where they are safely operable and serviceable from a portable platform.

Exception: Operating handles requiring more than 50 lb (22.68 kg) of force shall be located no higher than 66 in. (1.68 m) in either the open or closed position.

490-42. Interlocks — Interrupter Switches. Interrupter switches equipped with stored energy mechanisms shall have mechanical interlocks to prevent access to the switch compartment unless the stored energy mechanism is in the discharged or blocked position.

490-43. Stored Energy for Opening. The stored energy operator shall be permitted to be left in the uncharged position after the switch has been closed if a single movement of the operating handle charges the operator and opens the switch.

490-44. Fused Interrupter Switches.

(a) Supply Terminals. The supply terminals of fused interrupter switches shall be installed at the top of the switch enclosure or, if the terminals are located elsewhere, the equipment shall have barriers installed so as to prevent persons from accidentally contacting energized parts or dropping tools or fuses into energized parts.

(b) Backfeed. Where fuses can be energized by backfeed, a sign shall be placed on the enclosure door reading

WARNING — FUSES MAY BE ENERGIZED
BY BACKFEED.

(c) Switching Mechanism. The switching mechanism shall be arranged to be operated from a location outside the enclosure where the operator is not exposed to energized parts and shall be arranged to open all ungrounded conductors of the circuit simultaneously with one operation. Switches shall be capable of being locked in the open position.

490-45. Interlocks — Circuit Breakers.

(a) Circuit breakers equipped with stored energy mechanisms shall be designed to prevent the release of the stored energy unless the mechanism has been fully charged.

(b) Mechanical interlocks shall be provided in the housing to prevent the complete withdrawal of the circuit breaker from the housing when the stored energy mechanism is in the fully charged position, unless a suitable device is provided to block the closing function of the circuit breaker before complete withdrawal.

D. Mobile and Portable Equipment

490-51. General.

(a) Covered. The provisions of this part shall apply to installations and use of high-voltage power distribution and utilization equipment that is portable or mobile, or both, such as substations and switch houses mounted on skids, trailers, or cars; mobile shovels; draglines; cranes; hoists; drills; dredges; compressors; pumps; conveyors; underground excavators; and the like.

(b) Other Requirements. The requirements of this part shall be additional to, or amendatory of, those prescribed in Articles 100 through 725 of this *Code*. Special attention shall be paid to Article 250.

(c) Protection. Adequate enclosures or guarding, or both, shall be provided to protect portable and mobile equipment from physical damage.

(d) Disconnecting Means. Disconnecting means shall be installed for mobile and portable high-voltage equipment according to the requirements of Part H of Article 230 and shall disconnect all ungrounded conductors.

490-52. Overcurrent Protection. Motors driving single or multiple dc generators supplying a system operating on a cyclic load basis do not require overload protection, provided that the thermal rating of the ac drive motor cannot be exceeded under any operating condition. The branch-circuit protective device(s) shall provide short-circuit and locked-rotor protection, and shall be permitted to be external to the equipment.

490-53. Enclosures. All energized switching and control parts shall be enclosed in effectively grounded metal cabinets or enclosures. These cabinets or enclosures shall be marked "DANGER — HIGH VOLTAGE — KEEP OUT" and shall be locked so that only authorized and qualified persons can enter. Circuit breakers and protective equipment shall have the operating means projecting through the metal cabinet or enclosure so these units can be reset without opening locked doors. With doors closed, reasonable safe access for normal operation of these units shall be provided.

490-54. Collector Rings. The collector ring assemblies on revolving-type machines (shovels, draglines, etc.) shall be guarded to prevent accidental contact with energized parts by personnel on or off the machine.

490-55. Power Cable Connections to Mobile Machines. A metallic enclosure shall be provided on the mobile machine for enclosing the terminals of the power cable. The enclosure shall include provisions for a solid connection for the ground wire(s) terminal to effectively ground the machine frame. Ungrounded conductors shall be attached to insulators or be terminated in approved high-voltage cable couplers (which include ground wire connectors) of proper voltage and ampere rating. The method of cable termination used shall prevent any strain or pull on the cable from stressing the electrical connections. The enclosure shall have provision for locking so only authorized and qualified persons may open it and shall be marked

DANGER — HIGH VOLTAGE — KEEP OUT.

490-56. High-Voltage Portable Cable for Main Power Supply. Flexible high-voltage cable supplying power to portable or mobile equipment shall comply with Article 250 and Article 400, Part C.

E. Electrode-Type Boilers

490-70. General. The provisions of this part shall apply to boilers operating over 600 volts, nominal, in which heat is generated by the passage of current between electrodes through the liquid being heated.

490-71. Electric Supply System. Electrode-type boilers shall be supplied only from a 3-phase, 4-wire solidly grounded wye system, or from isolating transformers arranged to provide such a system. Control circuit voltages shall not exceed 150 volts, shall be supplied from a grounded system, and shall have the controls in the ungrounded conductor.

490-72. Branch-Circuit Requirements.

(a) Rating. Each boiler shall be supplied from an individual branch circuit rated not less than 100 percent of the total load.

(b) Common-Trip Fault Interrupting Device. The circuit shall be protected by a 3-phase, common-trip fault- interrupting device, which shall be permitted to automatically reclose the circuit upon removal of an overload condition but shall not reclose after a fault condition.

(c) Phase-Fault Protection. Phase-fault protection shall be provided in each phase, consisting of a separate phase-overcurrent relay connected to a separate current transformer in the phase.

(d) Ground Current Detection. Means shall be provided for detection of the sum of the neutral and ground currents and shall trip the circuit-interrupting device if the sum of those currents exceeds the greater of 5 amperes or 7½ percent of the boiler full-load current for 10 seconds or exceeds an instantaneous value of 25 percent of the boiler full-load current.

(e) Grounded Neutral Conductor. The grounded neutral conductor shall be as follows:

(1) Connected to the pressure vessel containing the electrodes
(2) Insulated for not less than 600 volts
(3) Have not less than the ampacity of the largest ungrounded branch-circuit conductor
(4) Installed with the ungrounded conductors in the same raceway, cable, or cable tray, or where installed as open conductors, in close proximity to the ungrounded conductors
(5) Not be used for any other circuit

490-73. Pressure and Temperature Limit Control. Each boiler shall be equipped with a means to limit the maximum temperature or pressure, or both, by directly or indirectly interrupting all current flow through the electrodes. Such means shall be in addition to the temperature or pressure, or both, regulating systems and pressure relief or safety valves.

490-74. Grounding. All exposed noncurrent-carrying metal parts of the boiler and associated exposed grounded structures or equipment shall be bonded to the pressure vessel or to the neutral conductor to which the vessel is connected in accordance with Section 250-102, except the ampacity of the bonding jumper shall not be less than the ampacity of the neutral conductor.

CHAPTER 5
Special Occupancies

Article 500 — Hazardous (Classified) Locations, Classes I, II, and III, Divisions 1 and 2

500-1. Scope — Articles 500 through 504. Articles 500 through 504 cover the requirements for electrical and electronic equipment and wiring for all voltages in Class I, Divisions 1 and 2; Class II, Divisions 1 and 2; and Class III, Divisions 1 and 2 locations where fire or explosion hazards may exist due to flammable gases or vapors, flammable liquids, combustible dust, or ignitible fibers or flyings.

> FPN: For the requirements for electrical and electronic equipment and wiring for all voltages in Class I, Zone 0, Zone 1, and Zone 2 hazardous (classified) locations where fire or explosion hazards may exist due to flammable gases or vapors, or flammable liquids, refer to Article 505.

500-2. Other Articles. Except as modified in Articles 500 through 504, all other applicable rules contained in this Code shall apply to electrical equipment and wiring installed in hazardous (classified) locations.

500-3. General.

(a) Classifications of Locations. Locations shall be classified depending on the properties of the flammable vapors, liquids, or gases, or combustible dusts or fibers that may be present and the likelihood that a flammable or combustible concentration or quantity is present. Where pyrophoric materials are the only materials used or handled, these locations shall not be classified.

Each room, section, or area shall be considered individually in determining its classification.

> FPN: Through the exercise of ingenuity in the layout of electrical installations for hazardous (classified) locations, it is frequently possible to locate much of the equipment in less hazardous or in nonhazardous locations and, thus, to reduce the amount of special equipment required.

(b) Documentation. All areas designated as hazardous (classified) locations shall be properly documented. This documentation shall be available to those authorized to design, install, inspect, maintain, or operate electrical equipment at the location.

(c) Reference Standards.

> FPN No. 1: It is important that the authority having jurisdiction be familiar with recorded industrial experience as well as with the standards of the National Fire Protection Association, the American Petroleum Institute, and the Instrument Society of America that may be of use in the classification of various locations, the determination of adequate ventilation, and the protection against static electricity and lightning hazards.

> FPN No. 2: For further information on the classification of locations, see *Flammable and Combustible Liquids Code*, NFPA 30-1996; *Standard for Drycleaning Plants*, NFPA 32-1996; *Standard for Spray Application Using Flammable or Combustible Materials*, NFPA 33-1995; *Standard for Dipping and Coating Processes Using Flammable or Combustible Liquids*, NFPA 34-1995; *Standard for the Manufacture of Organic Coatings*, NFPA 35-1995; *Standard for Solvent Extraction Plants*, NFPA 36-1997; *Standard on Fire Protection for Laboratories Using Chemicals*, NFPA 45-1996; *Standard for Gaseous Hydrogen Systems at Consumer Sites*, NFPA 50A-1994; *Standard for Liquefied Hydrogen Systems at Consumer Sites*, NFPA 50B-1994; *Liquefied Petroleum Gas Code*, NFPA 58-1998; *Standard for the Storage and Handling of Liquefied Petroleum Gases at Utility Gas Plants*, NFPA 59-1998; *Recommended Practice for the Classification of Flammable Liquids, Gases, or Vapors and of Hazardous (Classified) Locations for Electrical Installations in Chemical Process Areas*, NFPA 497-1997; *Recommended Practice for the Classification of Combustible Dusts and of Hazardous (Classified) Locations for Electrical Installations in Chemical Process Areas*, NFPA 499-1997; *Standard for Fire Protection in Wastewater Treatment and Collection Facilities*, NFPA 820-1995; *Recommended Practice for Classification of Locations for Electrical Installations at Petroleum Facilities Classified as Class I, Division 1 and Division 2*, ANSI/API RP500-1997; *Area Classification In Hazardous (Classified) Dust Locations*, ANSI/ISA-S12.10-1988.

> FPN No. 3: For further information on protection against static electricity and lightning hazards in hazardous (classified) locations, see *Recommended Practice on Static Electricity*, NFPA 77-1993; *Standard for the Installation of Lightning Protection Systems*, NFPA 780-1997; and *Protection Against Ignitions Arising Out of Static Lightning and Stray Currents*, API RP 2003-1991.

> FPN No. 4: For further information on ventilation, see *Flammable and Combustible Liquids Code*, NFPA 30-1996; and *Recommended Practice for Classification of Locations for Electrical Installations at Petroleum Facilities Classified as Class I, Division 1 and Division 2*, API RP 500-1997, Section 4.6.

> FPN No. 5: For further information on electrical systems for hazardous (classified) locations on offshore oil and gas producing platforms, see *Recommended Practice for Design and Installation of Electrical Systems for Offshore Production Platforms*, ANSI/API RP 14F-1991.

(d) Threaded Conduit. All threaded conduit referred to herein shall be threaded with an NPT standard conduit cutting die that provides ¾-in. taper per foot. Such conduit shall be made wrenchtight to prevent sparking when fault current flows through the conduit system, and ensure the explosionproof or dust-ignitionproof integrity of the conduit system where applicable.

For equipment provided with metric threaded entries, such entries shall be identified as being metric, or listed adapters to permit connection to conduit or NPT-threaded fittings shall be provided with the equipment. Adapters shall be used for connection to conduit or NPT-threaded fittings. Listed cable fittings that have metric threads shall be permitted to be used.

FPN: Threading specifications for metric threaded entries are located in *Metric Screw Threads,* ISO 965/1:1980 and *Metric Screw Threads,* ISO 965/3:1980.

(e) Fiber Optic Cable Assembly. Where a fiber optic cable assembly contains conductors that are capable of carrying current, the fiber optic cable assembly shall be installed in accordance with the requirements of Articles 500, 501, 502, or 503, as applicable.

500-4. Protection Techniques. The following shall be acceptable protection techniques for electrical and electronic equipment in hazardous (classified) locations.

(a) Explosionproof Apparatus. This protection technique shall be permitted for equipment in those Class I, Division 1 and 2 locations for which it is approved.

FPN: Explosionproof apparatus is defined in Article 100. For further information, see *Explosionproof and Dust-Ignitionproof Electrical Equipment for Use in Hazardous (Classified) Locations,* ANSI/UL 1203-1994.

(b) Dust Ignitionproof. This protection technique shall be permitted for equipment in those Class II, Division 1 and 2 locations for which it is approved.

FPN: Dust-ignitionproof equipment is defined in Article 100. For further information, see *Explosionproof and Dust-Ignitionproof Electrical Equipment for Use in Hazardous (Classified) Locations,* ANSI/UL 1203-1994.

(c) Dusttight. This protection technique shall be permitted for equipment in those Class II, Division 2 and Class III locations for which it is approved.

FPN No. 1: Dusttight enclosures are constructed so that dust will not enter the enclosing cases under specified test conditions.

FPN No. 2: For further information, see *Nonincendive Electrical Equipment for Use in Class I and II, Division 2, and Class III, Divisions 1 and 2 Hazardous (Classified) Locations,* ANSI/ISA S12.12-1994, and *Electrical Equipment for Use in Class I and II, Division 2 and Class III Hazardous (Classified) Locations,* UL 1604-1994.

FPN No. 3: For further information on test conditions for equipment other than rotating equipment, see *Enclosures for Electrical Equipment (1000 volts Maximum),* ANSI/NEMA 250-1991.

(d) Purged and Pressurized. This protection technique shall be permitted for equipment in any hazardous (classified) location for which it is approved.

FPN No. 1: In some cases, hazards may be reduced or hazardous (classified) locations limited or eliminated by adequate positive-pressure ventilation from a source of clean air

in conjunction with effective safeguards against ventilation failure.

FPN No. 2: For further information, see *Standard for Purged and Pressurized Enclosures for Electrical Equipment,* NFPA 496-1998.

(e) Intrinsically Safe Systems. Intrinsically safe apparatus and wiring shall be permitted in any hazardous (classified) location for which it is approved, and the provisions of Articles 501 through 503 and 510 through 516 shall not be considered applicable to such installations, except as required by Article 504.

Installation of intrinsically safe apparatus and wiring shall be in accordance with the requirements of Article 504.

FPN: For further information, see *Intrinsically Safe Apparatus and Associated Apparatus for Use in Class I, II, and III, Division 1, Hazardous Locations,* ANSI/UL 913-1997.

(f) Nonincendive. A protective technique where, under normal operating conditions, any arcing or thermal effects are not capable of igniting the flammable gas, vapor, or dust-in-air mixture. This protection technique shall be permitted for equipment in those Class I, Division 2; Class II, Division 2; and Class III locations for which it is approved.

(1) Nonincendive Circuit. A circuit in which any arc or thermal effect produced, under intended operating conditions of the equipment or due to opening, shorting, or grounding of field wiring, is not capable, under specified test conditions, of igniting the flammable gas–, vapor–, or dust–air mixture.

FPN No. 1: For test conditions, see *Nonincendive Electrical Equipment for Use in Class I and II, Division 2, and Class III, Division 1 and 2 Hazardous (Classified) Locations,* ANSI/ISA S12.12-1994.

FPN No. 2: Nonincendive circuit is defined in Article 100. For further information, see *Nonincendive Electrical Equipment for Use in Class I and II, Division 2, and Class III, Division 1 and 2 Hazardous (Classified) Locations,* ANSI/ISA S12.12-1994.

(2) Nonincendive Equipment. Equipment having electrical/electronic circuitry that is incapable, under normal operating conditions, of causing ignition of a specified flammable gas–, vapor–, or dust–air mixture due to arcing or thermal means.

FPN: For further information, see *Nonincendive Electrical Equipment for Use in Class I and II, Division 2, and Class III, Divisions 1 and 2 Hazardous (Classified) Locations,* ANSI/ISA-S12.12-1994.

(a) Nonincendive Component. A component having contacts for making or breaking an incendive circuit and the contacting mechanism shall be constructed so that the component is incapable of igniting the specified flammable gas– or vapor–air mixture. The housing of a nonincendive

component is not intended to exclude the flammable atmosphere or contain an explosion.

This protection technique shall be permitted for current-interrupting contacts in those Class I, Division 2 locations for which the component is approved.

FPN: For further information, see *Electrical Equipment for Use in Class I and II, Division 2, and Class III Hazardous (Classified) Locations,* UL 1604-1994.

(g) Oil Immersion. This protection technique shall be permitted for current-interrupting contacts in Class I, Division 2 locations as described in Section 501-6(b)(1)(b).

FPN: See Sections 501-3(b)(1), Exception (a); 501-5(a)(1), Exception (b), 501-6(b)(1); 501-14(b)(1), Exception (a); 502-14(a)(2), Exception; and 502-14(a)(3), Exception. For further information, see *Industrial Control Equipment for Use in Hazardous (Classified) Locations,* ANSI/UL 698-1995.

(h) Hermetically Sealed. A hermetically sealed device shall be sealed against the entrance of an external atmosphere and the seal shall be made by fusion, e.g., soldering, brazing, welding, or the fusion of glass to metal.

This protection technique shall be permitted for current-interrupting contacts in Class I, Division 2 locations.

FPN: See Sections 501-3(b)(1), Exception (b); 501-5(a)(1), Exception (a); 501-6(b)(1)(a); and 501-14(b)(1), Exception (b). For further information, see *Nonincendive Electrical Equipment for Use in Class I and II, Division 2, and Class III, Division 1 and 2 Hazardous (Classified) Locations,* ANSI/ISA S12.12-1994.

(i) Other Protection Techniques. Other protection techniques used in equipment listed for use in hazardous (classified) locations.

500-5. Special Precaution. Articles 500 through 504 require equipment construction and installation that will ensure safe performance under conditions of proper use and maintenance.

FPN No. 1: It is important that inspection authorities and users exercise more than ordinary care with regard to installation and maintenance.

FPN No. 2: Low ambient conditions require special consideration. Explosionproof or dust-ignitionproof equipment may not be suitable for use at temperatures lower than −25°C (−13°F) unless they are identified for low-temperature service. However, at low ambient temperatures, flammable concentrations of vapors may not exist in a location classified Class I, Division 1 at normal ambient temperature.

For purposes of testing, approval, and area classification, various air mixtures (not oxygen-enriched) shall be grouped in accordance with Sections 500-5(a) and 500-5(b).

Exception: Equipment approved for a specific gas, vapor, or dust.

FPN: This grouping is based on the characteristics of the materials. Facilities are available for testing and identifying equipment for use in the various atmospheric groups.

(a) Class I Group Classifications. Class I groups shall be as follows.

^x**(1) Group A.** Acetylene.

^x**(2) Group B.** Flammable gas, flammable liquid–produced vapor, or combustible liquid–produced vapor mixed with air that may burn or explode, having either a maximum experimental safe gap (MESG) value less than or equal to 0.45 mm or a minimum igniting current ratio (MIC ratio) less than or equal to 0.40.

FPN: A typical Class I, Group B material is hydrogen.

Exception No. 1: Group D equipment shall be permitted to be used for atmospheres containing butadiene provided all conduit runs into explosionproof equipment are provided with explosionproof seals installed within 18 in. (457 mm) of the enclosure.

Exception No. 2: Group C equipment shall be permitted to be used for atmospheres containing allyl glycidyl ether, n-butyl glycidyl ether, ethylene oxide, propylene oxide, and acrolein provided all conduit runs into explosionproof equipment are provided with explosionproof seals installed within 18 in. (457 mm) of the enclosure.

^x**(3) Group C.** Flammable gas, flammable liquid–produced vapor, or combustible liquid–produced vapor mixed with air that may burn or explode, having either a maximum experimental safe gap (MESG) value greater than 0.45 mm and less than or equal to 0.75 mm, or a minimum igniting current ratio (MIC ratio) greater than 0.40 and less than or equal to 0.80.

FPN: A typical Class I, Group C material is ethylene.

^x**(4) Group D.** Flammable gas, flammable liquid–produced vapor, or combustible liquid–produced vapor mixed with air that may burn or explode, having either a maximum experimental safe gap (MESG) value greater than 0.75 mm or a minimum igniting current ratio (MIC ratio) greater than 0.80.

FPN: A typical Class I, Group D material is propane.

Exception: For atmospheres containing ammonia, the authority having jurisdiction for enforcement of this Code shall be permitted to reclassify the location to a less hazardous location or a nonhazardous location.

FPN No. 1: For additional information on the properties and group classification of Class I materials, see *Recommended Practice for the Classification of Flammable Liquids, Gases, or Vapors and of Hazardous (Classified) Locations for Electrical Installations in Chemical Process Areas,* NFPA 497-

1997, and *Guide to Fire Hazard Properties of Flammable Liquids, Gases, and Volatile Solids,* NFPA 325-1994.

FPN No. 2: The explosion characteristics of air mixtures of gases or vapors vary with the specific material involved. For Class I locations, Groups A, B, C, and D, the classification involves determinations of maximum explosion pressure and maximum safe clearance between parts of a clamped joint in an enclosure. It is necessary, therefore, that equipment be approved not only for class but also for the specific group of the gas or vapor that will be present.

FPN No. 3: Certain chemical atmospheres may have characteristics that require safeguards beyond those required for any of the above groups. Carbon disulfide is one of these chemicals because of its low ignition temperature [100°C (212°F)] and the small joint clearance permitted to arrest its flame.

FPN No. 4: For classification of areas involving ammonia atmosphere, see *Safety Code for Mechanical Refrigeration,* ANSI/ASHRAE 15-1994, and *Safety Requirements for the Storage and Handling of Anhydrous Ammonia,* ANSI/CGA G2.1-1989.

(b) Class II Group Classifications. Class II groups shall be as follows.

ˣ**(1) Group E.** Atmospheres containing combustible metal dusts, including aluminum, magnesium, and their commercial alloys, or other combustible dusts whose particle size, abrasiveness, and conductivity present similar hazards in the use of electrical equipment.

FPN: Certain metal dusts may have characteristics that require safeguards beyond those required for atmospheres containing the dusts of aluminum, magnesium, and their commercial alloys. For example, zirconium, thorium, and uranium dusts have extremely low ignition temperatures [as low as 20°C (68°F)] and minimum ignition energies lower than any material classified in any of the Class I or Class II Groups.

ˣ**(2) Group F.** Atmospheres containing combustible carbonaceous dusts that have more than 8 percent total entrapped volatiles (see *Standard Test Method for Volatile Material in the Analysis Sample for Coal and Coke,* ASTM D3175-89, for coal and coke dusts) or that have been sensitized by other materials so that they present an explosion hazard. Coal, carbon black, charcoal, and coke dusts are examples of carbonaceous dusts.

(3) Group G. Atmospheres containing combustible dusts not included in Group E or F, including flour, grain, wood, plastic, and chemicals.

FPN No. 1: For additional information on group classification of Class II materials, see *Recommended Practice for the Classification of Combustible Dusts and of Hazardous (Classified) Locations for Electrical Installations in Chemical Process Areas,* NFPA 499-1997.

FPN No. 2: The explosion characteristics of air mixtures of dust vary with the materials involved. For Class II locations, Groups E, F, and G, the classification involves the tightness of the joints of assembly and shaft openings to prevent the entrance of dust in the dust-ignitionproof enclosure, the blanketing effect of layers of dust on the equipment that may cause overheating, and the ignition temperature of the dust. It is necessary, therefore, that equipment be approved not only for the class, but also for the specific group of dust that will be present.

FPN No. 3: Certain dusts may require additional precautions due to chemical phenomena that can result in the generation of ignitible gases. See *National Electrical Safety Code,* ANSI C2-1997, Section 127A, Coal Handling Areas.

(c) Approval for Class and Properties. Equipment, regardless of the classification of the location in which it is installed, that depends on a single compression seal, diaphragm, or tube to prevent flammable or combustible fluids from entering the equipment, shall be approved for a Class I, Division 2 location.

Exception: Equipment installed in a Class I, Division 1 location shall be suitable for the Division 1 location.

FPN: See Section 501-5(f)(3) for additional requirements.

Equipment shall be approved not only for the class of location but also for the explosive, combustible, or ignitible properties of the specific gas, vapor, dust, fiber, or flyings that will be present. In addition, Class I equipment shall not have any exposed surface that operates at a temperature in excess of the ignition temperature of the specific gas or vapor. Class II equipment shall not have an external temperature higher than that specified in Section 500-5(f). Class III equipment shall not exceed the maximum surface temperatures specified in Section 503-1.

Equipment that has been approved for a Division 1 location shall be permitted in a Division 2 location of the same class and group.

Where specifically permitted in Articles 501 through 503, general-purpose equipment or equipment in general-purpose enclosures shall be permitted to be installed in Division 2 locations if the equipment does not constitute a source of ignition under normal operating conditions.

Unless otherwise specified, normal operating conditions for motors shall be assumed to be rated full-load steady conditions.

Where flammable gases or combustible dusts are or may be present at the same time, the simultaneous presence of both shall be considered when determining the safe operating temperature of the electrical equipment.

FPN: The characteristics of various atmospheric mixtures of gases, vapors, and dusts depend on the specific material involved.

(d) Marking. Approved equipment shall be marked to show the class, group, and operating temperature or temperature range referenced to a 40°C ambient.

Exception No. 1: Equipment of the non-heat-producing type, such as junction boxes, conduit, and fittings, and equipment of the heat-producing type having a maximum temperature not more than 100°C (212°F) shall not be required to have a marked operating temperature or temperature range.

Exception No. 2: Fixed lighting fixtures marked for use in Class I, Division 2 or Class II, Division 2 locations only shall not be required to be marked to indicate the group.

Exception No. 3: Fixed general-purpose equipment in Class I locations, other than fixed lighting fixtures, that is acceptable for use in Class I, Division 2 locations shall not be required to be marked with the class, group, division, or operating temperature.

Exception No. 4: Fixed dusttight equipment other than fixed lighting fixtures that are acceptable for use in Class II, Division 2 and Class III locations shall not be required to be marked with the class, group, division, or operating temperature.

Exception No. 5: Electric equipment suitable for ambient temperatures exceeding 40°C (104°F) shall be marked with both the maximum ambient temperature and the operating temperature or temperature range at that ambient temperature.

FPN: Equipment not marked to indicate a division, or marked "Division 1" or "Div. 1," is suitable for both Division 1 and 2 locations. Equipment marked "Division 2" or "Div. 2" is suitable for Division 2 locations only.

The temperature range, if provided, shall be indicated in identification numbers, as shown in Table 500-5(d).

Identification numbers marked on equipment nameplates shall be in accordance with Table 500-5(d).

Table 500-5(d). Identification Numbers

Maximum Temperature		Identification Number
°C	°F	
450	842	T1
300	572	T2
280	536	T2A
260	500	T2B
230	446	T2C
215	419	T2D
200	392	T3
180	356	T3A
165	329	T3B
160	320	T3C
135	275	T4
120	248	T4A
100	212	T5
85	185	T6

Equipment that is approved for Class I and Class II shall be marked with the maximum safe operating temperature, as determined by simultaneous exposure to the combinations of Class I and Class II conditions.

FPN: Since there is no consistent relationship between explosion properties and ignition temperature, the two are independent requirements.

(e) Class I Temperature. The temperature marking specified in (d) shall not exceed the ignition temperature of the specific gas or vapor to be encountered.

FPN: For information regarding ignition temperatures of gases and vapors, see *Recommended Practice for the Classification of Combustible Dusts and of Hazardous (Classified) Locations for Electrical Installations in Chemical Process Areas,* NFPA 499-1997, and *Guide to Fire Hazard Properties of Flammable Liquids, Gases, and Volatile Solids,* NFPA 325-1994.

(f) Class II Temperature. The temperature marking specified in (d) shall be less than the ignition temperature of the specific dust to be encountered. For organic dusts that may dehydrate or carbonize, the temperature marking shall not exceed the lower of either the ignition temperature or 165°C (329°F).

FPN: See *Recommended Practice for the Classification of Combustible Dusts and of Hazardous (Classified) Locations for Electrical Installations in Chemical Process Areas,* NFPA 499-1997, for minimum ignition temperatures of specific dusts.

The ignition temperature for which equipment was approved prior to this requirement shall be assumed to be as shown in Table 500-5(f).

Table 500-5(f).

Class II Group	Equipment that Is Not Subject to Overloading		Equipment (Such as Motors or Power Transformers) that May Be Overloaded			
			Normal Operation		Abnormal Operation	
	°C	°F	°C	°F	°C	°F
E	200	392	200	392	200	392
F	200	392	150	302	200	392
G	165	329	120	248	165	329

500-6. Specific Occupancies. Articles 510 through 517 cover garages, aircraft hangars, gasoline dispensing and service stations, bulk storage plants, spray application, dipping and coating processes, and health care facilities.

500-7. Class I Locations. Class I locations are those in which flammable gases or vapors are or may be present in the air in quantities sufficient to produce explosive or ignitible mixtures. Class I locations shall include those specified in (a) and (b).

(a) Class I, Division 1. A Class I, Division 1 location is a location

(1) In which ignitible concentrations of flammable gases or vapors can exist under normal operating conditions, or
(2) In which ignitible concentrations of such gases or vapors may exist frequently because of repair or maintenance operations or because of leakage, or
(3) In which breakdown or faulty operation of equipment or processes might release ignitible concentrations of flammable gases or vapors, and might also cause simultaneous failure of electrical equipment in such a way as to directly cause the electrical equipment to become a source of ignition.

FPN No. 1: This classification usually includes the following locations:

(1) Where volatile flammable liquids or liquefied flammable gases are transferred from one container to another
(2) Interiors of spray booths and areas in the vicinity of spraying and painting operations where volatile flammable solvents are used
(3) Locations containing open tanks or vats of volatile flammable liquids
(4) Drying rooms or compartments for the evaporation of flammable solvents
(5) Locations containing fat and oil extraction equipment using volatile flammable solvents
(6) Portions of cleaning and dyeing plants where flammable liquids are used
(7) Gas generator rooms and other portions of gas manufacturing plants where flammable gas may escape
(8) Inadequately ventilated pump rooms for flammable gas or for volatile flammable liquids
(9) The interiors of refrigerators and freezers in which volatile flammable materials are stored in open, lightly stoppered, or easily ruptured containers
(10) All other locations where ignitible concentrations of flammable vapors or gases are likely to occur in the course of normal operations

FPN No. 2: In some Division 1 locations, ignitible concentrations of flammable gases or vapors may be present continuously or for long periods of time. Examples include the following:

(1) The inside of inadequately vented enclosures containing instruments normally venting flammable gases or vapors to the interior of the enclosure
(2) The inside of vented tanks containing volatile flammable liquids
(3) The area between the inner and outer roof sections of a floating roof tank containing volatile flammable fluids

(4) Inadequately ventilated areas within spraying or coating operations using volatile flammable fluids
(5) The interior of an exhaust duct that is used to vent ignitible concentrations of gases or vapors

Experience has demonstrated the prudence of avoiding the installation of instrumentation or other electric equipment in these particular areas altogether or where it cannot be avoided because it is essential to the process and other locations are not feasible [see Section 500-3(a), FPN], using electric equipment or instrumentation approved for the specific application or consisting of intrinsically safe systems as described in Article 504.

(b) Class I, Division 2. A Class I, Division 2 location is a location

(1) In which volatile flammable liquids or flammable gases are handled, processed, or used, but in which the liquids, vapors, or gases will normally be confined within closed containers or closed systems from which they can escape only in case of accidental rupture or breakdown of such containers or systems, or in case of abnormal operation of equipment, or
(2) In which ignitible concentrations of gases or vapors are normally prevented by positive mechanical ventilation, and which might become hazardous through failure or abnormal operation of the ventilating equipment, or
(3) That is adjacent to a Class I, Division 1 location, and to which ignitible concentrations of gases or vapors might occasionally be communicated unless such communication is prevented by adequate positive-pressure ventilation from a source of clean air, and effective safeguards against ventilation failure are provided.

FPN No. 1: This classification usually includes locations where volatile flammable liquids or flammable gases or vapors are used but that, in the judgment of the authority having jurisdiction, would become hazardous only in case of an accident or of some unusual operating condition. The quantity of flammable material that might escape in case of accident, the adequacy of ventilating equipment, the total area involved, and the record of the industry or business with respect to explosions or fires are all factors that merit consideration in determining the classification and extent of each location.

FPN No. 2: Piping without valves, checks, meters, and similar devices would not ordinarily introduce a hazardous condition even though used for flammable liquids or gases. Depending on factors such as the quantity and size of the containers and ventilation, locations used for the storage of flammable liquids or liquefied or compressed gases in sealed containers may either be considered hazardous (classified) or unclassified locations. See *Flammable and Combustible Liquids Code,* NFPA 30-1996, and *Liquefied Petroleum Gas Code,* NFPA 58-1998.

500-8. Class II Locations. Class II locations are those that are hazardous because of the presence of combustible dust. Class II locations shall include those specified in (a) and (b).

(a) Class II, Division 1. A Class II, Division 1 location is a location

(1) In which combustible dust is in the air under normal operating conditions in quantities sufficient to produce explosive or ignitible mixtures, or

(2) Where mechanical failure or abnormal operation of machinery or equipment might cause such explosive or ignitible mixtures to be produced, and might also provide a source of ignition through simultaneous failure of electric equipment, operation of protection devices, or from other causes, or

(3) In which combustible dusts of an electrically conductive nature may be present in hazardous quantities.

FPN: Combustible dusts that are electrically nonconductive include dusts produced in the handling and processing of grain and grain products, pulverized sugar and cocoa, dried egg and milk powders, pulverized spices, starch and pastes, potato and wood-flour, oil meal from beans and seed, dried hay, and other organic materials that may produce combustible dusts when processed or handled. Only Group E dusts are considered to be electrically conductive for classification purposes. Dusts containing magnesium or aluminum are particularly hazardous, and the use of extreme precaution will be necessary to avoid ignition and explosion.

(b) Class II, Division 2. A Class II, Division 2 location is a location

(1) Where combustible dust is not normally in the air in quantities sufficient to produce explosive or ignitible mixtures, and dust accumulations are normally insufficient to interfere with the normal operation of electrical equipment or other apparatus, but combustible dust may be in suspension in the air as a result of infrequent malfunctioning of handling or processing equipment and

(2) Where combustible dust accumulations on, in, or in the vicinity of the electrical equipment may be sufficient to interfere with the safe dissipation of heat from electrical equipment or may be ignitible by abnormal operation or failure of electrical equipment.

FPN No. 1: The quantity of combustible dust that may be present and the adequacy of dust removal systems are factors that merit consideration in determining the classification and may result in an unclassified area.

FPN No. 2: Where products such as seed are handled in a manner that produces low quantities of dust, the amount of dust deposited may not warrant classification.

500-9. Class III Locations. Class III locations are those that are hazardous because of the presence of easily ignitible fibers or flyings, but in which such fibers or flyings are not likely to be in suspension in the air in quantities sufficient to produce ignitible mixtures. Class III locations shall include those specified in (a) and (b).

(a) Class III, Division 1. A Class III, Division 1 location is a location in which easily ignitible fibers or materials producing combustible flyings are handled, manufactured, or used.

FPN No. 1: Such locations usually include some parts of rayon, cotton, and other textile mills; combustible fiber manufacturing and processing plants; cotton gins and cotton-seed mills; flax-processing plants; clothing manufacturing plants; woodworking plants; and establishments and industries involving similar hazardous processes or conditions.

FPN No. 2: Easily ignitible fibers and flyings include rayon, cotton (including cotton linters and cotton waste), sisal or henequen, istle, jute, hemp, tow, cocoa fiber, oakum, baled waste kapok, Spanish moss, excelsior, and other materials of similar nature.

(b) Class III, Division 2. A Class III, Division 2 location is a location in which easily ignitible fibers are stored or handled other than in the process of manufacture.

Article 501 — Class I Locations

501-1. General. The general rules of this *Code* shall apply to the electric wiring and equipment in locations classified as Class I in Section 500-7.

Equipment listed and marked in accordance with Section 505-10 for use in Class I, Zone 0, 1, or 2 locations shall be permitted in Class I, Division 2 locations for the same gas and with a suitable temperature rating.

Exception: As modified by this article.

501-2. Transformers and Capacitors.

(a) Class I, Division 1. In Class I, Division 1 locations, transformers and capacitors shall comply with the following.

(1) **Containing Liquid that Will Burn.** Transformers and capacitors containing a liquid that will burn shall be installed only in approved vaults that comply with Sections 450-41 through 450-48, and, in addition,

(a) There shall be no door or other communicating opening between the vault and the Division 1 location, and

(b) Ample ventilation shall be provided for the continuous removal of flammable gases or vapors, and

(c) Vent openings or ducts shall lead to a safe location outside of buildings, and

(d) Vent ducts and openings shall be of sufficient area to relieve explosion pressures within the vault, and all portions of vent ducts within the buildings shall be of reinforced concrete construction.

(2) Not Containing Liquid that Will Burn. Transformers and capacitors that do not contain a liquid that will burn shall be installed in vaults complying with (a)(1) or be approved for Class I locations.

(b) Class I, Division 2. In Class I, Division 2 locations, transformers and capacitors shall comply with Sections 450-21 through 450-27.

501-3. Meters, Instruments, and Relays.

(a) Class I, Division 1. In Class I, Division 1 locations, meters, instruments, and relays, including kilowatt-hour meters, instrument transformers, resistors, rectifiers, and thermionic tubes, shall be provided with enclosures approved for Class I, Division 1 locations.

Enclosures approved for Class I, Division 1 locations include explosionproof enclosures and purged and pressurized enclosures.

FPN: See *Standard for Purged and Pressurized Enclosures for Electrical Equipment*, NFPA 496-1998.

(b) Class I, Division 2. In Class I, Division 2 locations, meters, instruments, and relays shall comply with the following.

(1) Contacts. Switches, circuit breakers, and make-and-break contacts of pushbuttons, relays, alarm bells, and horns shall have enclosures approved for Class I, Division 1 locations in accordance with Section 501-3(a).

Exception: General-purpose enclosures shall be permitted, if current-interrupting contacts are

(a) Immersed in oil, or
(b) Enclosed within a chamber that is hermetically sealed against the entrance of gases or vapors, or
(c) In nonincendive circuits, or
(d) Part of a listed nonincendive component.

(2) Resistors and Similar Equipment. Resistors, resistance devices, thermionic tubes, rectifiers, and similar equipment that are used in or in connection with meters, instruments, and relays shall comply with Section 501-3(a).

Exception: General-purpose–type enclosures shall be permitted if such equipment is without make-and-break or sliding contacts [other than as provided in (b)(1)] and if the maximum operating temperature of any exposed surface will not exceed 80 percent of the ignition temperature in degrees Celsius of the gas or vapor involved or has been tested and found incapable of igniting the gas or vapor. This exception shall not apply to thermionic tubes.

(3) Without Make-or-Break Contacts. Transformer windings, impedance coils, solenoids, and other windings that do not incorporate sliding or make-or-break contacts

shall be provided with enclosures. General-purpose–type enclosures shall be permitted.

(4) General-Purpose Assemblies. Where an assembly is made up of components for which general-purpose enclosures are acceptable as provided in (b)(1), (b)(2), and (b)(3), a single general-purpose enclosure shall be acceptable for the assembly. Where such an assembly includes any of the equipment described in (b)(2), the maximum obtainable surface temperature of any component of the assembly shall be clearly and permanently indicated on the outside of the enclosure. Alternatively, approved equipment shall be permitted to be marked to indicate the temperature range for which it is suitable, using the identification numbers of Table 500-5(d).

(5) Fuses. Where general-purpose enclosures are permitted in (b)(1), (b)(2), (b)(3), and (b)(4), fuses for overcurrent protection of instrument circuits not subject to overloading in normal use shall be permitted to be mounted in general-purpose enclosures if each such fuse is preceded by a switch complying with (b)(1).

(6) Connections. To facilitate replacements, process control instruments shall be permitted to be connected through flexible cord, attachment plug, and receptacle, provided the following:

(1) A switch complying with (b)(1) is provided so that the attachment plug is not depended on to interrupt current; and
(2) The current does not exceed 3 amperes at 120 volts, nominal; and
(3) The power-supply cord does not exceed 3 ft (914 mm), is of a type approved for extra-hard usage or for hard usage if protected by location, and is supplied through an attachment plug and receptacle of the locking and grounding type; and
(4) Only necessary receptacles are provided; and
(5) The receptacle carries a label warning against unplugging under load.

501-4. Wiring Methods. Wiring methods shall comply with (a) and (b).

(a) Class I, Division 1.

(1) Fixed Wiring. In Class I, Division 1 locations, threaded rigid metal conduit, threaded steel intermediate metal conduit, or Type MI cable with termination fittings approved for the location shall be the wiring method employed. All boxes, fittings, and joints shall be threaded for connection to conduit or cable terminations and shall be explosionproof. Threaded joints shall be made up with at least five threads fully engaged. Type MI cable shall be installed and supported in a manner to avoid tensile stress at the termination fittings.

Exception No. 1: Rigid nonmetallic conduit complying with Article 347 shall be permitted where encased in a concrete envelope a minimum of 2 in. (50.8 mm) thick and provided with not less than 24 in. (610 mm) of cover measured from the top of the conduit to grade. The concrete encasement shall be permitted to be omitted where subject to the provisions of Section 511-4, Exception; 514-8, Exception No. 2; and Section 515-5(a).

Where rigid nonmetallic conduit is used, threaded rigid metal conduit or threaded steel intermediate metal conduit shall be used for the last 24 in. (610 mm) of the underground run to emergence or to the point of connection to the aboveground raceway. An equipment grounding conductor shall be included to provide for electrical continuity of the raceway system and for grounding of noncurrent-carrying metal parts.

Exception No. 2: In industrial establishments with restricted public access, where the conditions of maintenance and supervision ensure that only qualified persons will service the installation, Type MC cable, listed for use in Class I, Division 1 locations, with a gas/vaportight continuous corrugated aluminum sheath, an overall jacket of suitable polymeric material, separate grounding conductors in accordance with Section 250-122, and provided with termination fittings listed for the application shall be permitted.

FPN: See Sections 334-3 and 334-4 for restrictions on use of Type MC cable.

Exception No. 3: In industrial establishments with restricted public access, where the conditions of maintenance and supervision ensure that only qualified persons will service the installation, Type ITC cable, listed for use in Class I, Division 1 locations, with a gas/vaportight continuous corrugated aluminum sheath, an overall jacket of suitable polymeric material and provided with termination fittings listed for the application shall be permitted.

(2) Flexible Connections. Where necessary to employ flexible connections, as at motor terminals, flexible fittings listed for Class I locations shall be used.

Exception: Flexible cord installed in accordance with the provisions of Section 501-11 shall be permitted.

(b) Class I, Division 2. In Class I, Division 2 locations, threaded rigid metal conduit, threaded steel intermediate metal conduit, enclosed gasketed busways, enclosed gasketed wireways, or Type PLTC cable in accordance with the provisions of Article 725, or Type ITC cable in cable trays, in raceways, supported by messenger wire, or directly buried where the cable is listed for this use; Type MI, MC, MV,

or TC cable with approved termination fittings shall be the wiring method employed. Type ITC, PLTC, MI, MC, MV, or TC cable shall be permitted to be installed in cable tray systems and shall be installed in a manner to avoid tensile stress at the termination fittings. Boxes, fittings, and joints shall not be required to be explosionproof except as required by Sections 501-3(b)(1), 501-6(b)(1), and 501-14(b)(1). Where provision must be made for limited flexibility, as at motor terminals, flexible metal fittings, flexible metal conduit with approved fittings, liquidtight flexible metal conduit with approved fittings, liquidtight flexible nonmetallic conduit with approved fittings, or flexible cord approved for extra-hard usage and provided with approved bushed fittings shall be used. An additional conductor for grounding shall be included in the flexible cord.

FPN: See Section 501-16(b) for grounding requirements where flexible conduit is used.

Exception: Nonincendive field wiring shall be permitted using any of the methods suitable for wiring in ordinary locations.

501-5. Sealing and Drainage. Seals in conduit and cable systems shall comply with (a) through (f). Sealing compound shall be of a type approved for the conditions and use. Sealing compound shall be used in Type MI cable termination fittings to exclude moisture and other fluids from the cable insulation.

FPN No. 1: Seals are provided in conduit and cable systems to minimize the passage of gases and vapors and prevent the passage of flames from one portion of the electrical installation to another through the conduit. Such communication through Type MI cable is inherently prevented by construction of the cable. Unless specifically designed and tested for the purpose, conduit and cable seals are not intended to prevent the passage of liquids, gases, or vapors at a continuous pressure differential across the seal. Even at differences in pressure across the seal equivalent to a few inches of water, there may be a slow passage of gas or vapor through a seal, and through conductors passing through the seal. See Section 501-5(e)(2). Temperature extremes and highly corrosive liquids and vapors can affect the ability of seals to perform their intended function. See Section 501-5(c)(2).

FPN No. 2: Gas or vapor leakage and propagation of flames may occur through the interstices between the strands of standard stranded conductors larger than No. 2. Special conductor constructions, e.g., compacted strands or sealing of the individual strands, are means of reducing leakage and preventing the propagation of flames.

(a) Conduit Seals, Class I, Division 1. In Class I, Division 1 locations, conduit seals shall be located as follows.

(1) In each conduit entry into an explosionproof enclosure where either (a) the enclosure contains apparatus, such as switches, circuit breakers, fuses, relays, or resistors, that may produce arcs, sparks, or high temperatures that are considered to be an ignition source in normal operation, or (b) the entry is 2-in. size or larger and the enclosure contains terminals, splices, or taps. For the purposes of this section high temperatures shall be considered to be any temperatures exceeding 80 percent of the autoignition temperature in degrees Celsius of the gas or vapor involved.

Exception to (a)(1)(a): Conduit entering an enclosure where such switches, circuit breakers, fuses, relays, or resistors are

(a) *Enclosed within a chamber hermetically sealed against the entrance of gases or vapors, or*

(b) *Immersed in oil in accordance with Section 501-6(b)(1)(b), or*

(c) *Enclosed within a factory-sealed explosionproof chamber located within the enclosure, approved for the location, and marked "factory sealed" or equivalent.*

Factory-sealed enclosures shall not be considered to serve as a seal for another adjacent explosionproof enclosure that is required to have a conduit seal.

Conduit seals shall be installed within 18 in. (457 mm) from the enclosure. Only explosionproof unions, couplings, reducers, elbows, capped elbows, and conduit bodies similar to L, T, and Cross types that are not larger than the trade size of the conduit shall be permitted between the sealing fitting and the explosionproof enclosure.

(2) In each conduit entry into a pressurized enclosure where the conduit is not pressurized as part of the protection system. Conduit seals shall be installed within 18 in. (457 mm) from the pressurized enclosure.

FPN No. 1: Installing the seal as close as possible to the enclosure will reduce problems with purging the dead airspace in the pressurized conduit.

FPN No. 2: For further information, see *Standard for Purged and Pressurized Enclosures for Electrical Equipment*, NFPA 496-1998.

(3) Where two or more explosionproof enclosures for which conduit seals are required under (a)(1) are connected by nipples or by runs of conduit not more than 36 in. (914 mm) long, a single conduit seal in each such nipple connection or run of conduit shall be considered sufficient if located not more than 18 in. (457 mm) from either enclosure.

(4) In each conduit run leaving a Class I, Division 1 location. The sealing fitting shall be permitted on either side of the boundary of such location within 10 ft (3.05 m) of the boundary, and shall be designed and installed so to minimize the amount of gas or vapor within the Division 1 portion of the conduit from being communicated to the conduit beyond the seal. Except for approved explosionproof

reducers at the conduit seal, there shall be no union, coupling, box, or fitting between the conduit seal and the point at which the conduit leaves the Division 1 location.

Exception: Metal conduit containing no unions, couplings, boxes, or fittings that passes completely through a Class I, Division 1 location with no fittings less than 12 in. (305 mm) beyond each boundary shall not require a conduit seal if the termination points of the unbroken conduit are in unclassified locations.

(b) Conduit Seals, Class I, Division 2. In Class I, Division 2 locations, conduit seals shall be located as follows.

(1) For connections to enclosures that are required to be explosionproof, a conduit seal shall be provided in accordance with (a)(1)(a) and (a)(3). All portions of the conduit run or nipple between the seal and such enclosure shall comply with Section 501-4(a).

(2) In each conduit run passing from a Class I, Division 2 location into an unclassified location. The sealing fitting shall be permitted on either side of the boundary of such location within 10 ft (3.05 m) of the boundary, and shall be designed and installed so to minimize the amount of gas or vapor within the Division 2 portion of the conduit from being communicated to the conduit beyond the seal. Rigid metal conduit or threaded steel intermediate metal conduit shall be used between the sealing fitting and the point at which the conduit leaves the Division 2 location, and a threaded connection shall be used at the sealing fitting. Except for approved explosionproof reducers at the conduit seal, there shall be no union, coupling, box, or fitting between the conduit seal and the point at which the conduit leaves the Division 2 location.

Exception No. 1: Metal conduit containing no unions, couplings, boxes or fittings that passes completely through a Class I, Division 2 location with no fittings less than 12 in. (305 mm) beyond each boundary shall not be required to be sealed if the termination points of the unbroken conduit are in unclassified locations.

Exception No. 2: Conduit systems terminating at an unclassified location where a wiring method transition is made to cable tray, cablebus, ventilated busway, Type MI cable, or open wiring shall not be required to be sealed where passing from the Class I, Division 2 location into the unclassified location. The unclassified location shall be outdoors or, if the conduit system is all in one room, it shall be permitted to be indoors. The conduits shall not terminate at an enclosure containing an ignition source in normal operation.

Exception No. 3: Conduit systems passing from an enclosure or room that is unclassified as a result of pressurization into a Class I, Division 2 location shall not require a seal at the boundary.

FPN: For further information, refer to *Standard for Purged and Pressurized Enclosures for Electrical Equipment,* NFPA 496-1998.

Exception No. 4: Segments of aboveground conduit systems shall not be required to be sealed where passing from a Class I, Division 2 location into an unclassified location if the following conditions are met:

(a) *No part of the conduit system segment passes through a Class I, Division 1 location where the conduit contains unions, couplings, boxes, or fittings within 12 in. (305 mm) of the Class I, Division 1 location; and*

(b) *The conduit system segment is located entirely in outdoor locations; and*

(c) *The conduit system segment is not directly connected to canned pumps, process or service connections for flow, pressure, or analysis measurement, etc., that depend on a single compression seal, diaphragm, or tube to prevent flammable or combustible fluids from entering the conduit system; and*

(d) *The conduit system segment contains only threaded metal conduit, unions, couplings, conduit bodies, and fittings in the unclassified location; and*

(e) *The conduit system segment is sealed at its entry to each enclosure or fitting housing terminals, splices, or taps in Class I, Division 2 locations.*

(c) Class I, Divisions 1 and 2. Where required, seals in Class I, Division 1 and 2 locations shall comply with the following.

(1) Fittings. Enclosures for connections or equipment shall be provided with an approved integral means for sealing, or sealing fittings approved for Class I locations shall be used. Sealing fittings shall be accessible.

(2) Compound. Sealing compound shall be approved and shall provide a seal against passage of gas or vapors through the seal fitting, shall not be affected by the surrounding atmosphere or liquids, and shall not have a melting point of less than 93°C (200°F).

(3) Thickness of Compounds. In a completed seal, the minimum thickness of the sealing compound shall not be less than the trade size of the sealing fitting and, in no case, less than ⅝ in. (16 mm).

Exception: Listed cable sealing fittings shall not be required to have a minimum thickness equal to the trade size of the fitting.

(4) Splices and Taps. Splices and taps shall not be made in fittings intended only for sealing with compound, nor shall other fittings in which splices or taps are made be filled with compound.

(5) Assemblies. In an assembly where equipment that may produce arcs, sparks, or high temperatures is located in a compartment separate from the compartment containing splices or taps, and an integral seal is provided where conductors pass from one compartment to the other, the entire assembly shall be approved for Class I locations. Seals in conduit connections to the compartment containing splices or taps shall be provided in Class I, Division 1 locations where required by (a)(1)(b).

(6) Conductor Fill. The cross-sectional area of the conductors permitted in a seal shall not exceed 25 percent of the cross-sectional area of a rigid metal conduit of the same trade size unless it is specifically approved for a higher percentage of fill.

(d) Cable Seals, Class I, Division 1. In Class I, Division 1 locations, cable seals shall be located as follows.

(1) Cable shall be sealed at all terminations. The sealing fitting shall comply with (c). Multiconductor Type MC cables with a gas/vaportight continuous corrugated aluminum sheath and an overall jacket of suitable polymeric material shall be sealed with an approved fitting after removing the jacket and any other covering so that the sealing compound will surround each individual insulated conductor in such a manner as to minimize the passage of gases and vapors.

Exception: Shielded cables and twisted pair cables shall not require the removal of the shielding material or separation of the twisted pairs provided the termination is by an approved means to minimize the entrance of gases or vapors and prevent propagation of flame into the cable core.

(2) Cables in conduit with a gas/vaportight continuous sheath capable of transmitting gases or vapors through the cable core shall be sealed in the Division 1 location after removing the jacket and any other coverings so that the sealing compound will surround each individual insulated conductor and the outer jacket.

Exception: Multiconductor cables with a gas/vaportight continuous sheath capable of transmitting gases or vapors through the cable core shall be permitted to be considered as a single conductor by sealing the cable in the conduit within 18 in. (457 mm) of the enclosure and the cable end within the enclosure by an approved means to minimize the entrance of gases or vapors and prevent the propagation of flame into the cable core, or by other approved methods. For shielded cables and twisted pair cables, it shall not be required to remove the shielding material or separate the twisted pair.

(3) Each multiconductor cable in conduit shall be considered as a single conductor if the cable is incapable of transmitting gases or vapors through the cable core. These cables shall be sealed in accordance with (a).

(e) Cable Seals, Class I, Division 2. In Class I, Division 2 locations, cable seals shall be located as follows.

(1) Cables entering enclosures that are required to be approved for Class I locations shall be sealed at the point of entrance. The sealing fitting shall comply with (b)(1). Multiconductor cables with a gas/vaportight continuous sheath capable of transmitting gases or vapors through the cable core shall be sealed in an approved fitting in the Division 2 location after removing the jacket and any other coverings so that the sealing compound will surround each individual insulated conductor in such a manner as to minimize the passage of gases and vapors. Multiconductor cables in conduit shall be sealed as described in (d).

Exception No. 1: Cables passing from an enclosure or room that is unclassified as a result of Type Z pressurization into a Class I, Division 2 location shall not require a seal at the boundary.

Exception No. 2: Shielded cables and twisted pair cables shall not require the removal of the shielding material or separation of the twisted pairs provided the termination is by an approved means to minimize the entrance of gases or vapors and prevent propagation of flame into the cable core.

(2) Cables with a gas/vaportight continuous sheath and that will not transmit gases or vapors through the cable core in excess of the quantity permitted for seal fittings shall not be required to be sealed except as required in (e)(1). The minimum length of such cable run shall not be less than that length that limits gas or vapor flow through the cable core to the rate permitted for seal fittings [0.007 ft^3/hour (198 cm^3/hour) of air at a pressure of 6 in. of water (1493 pascals)].

FPN No. 1: See *Outlet Boxes and Fittings for Use in Hazardous (Classified) Locations,* ANSI/UL 886-1994.

FPN No. 2: The cable core does not include the interstices of the conductor strands.

(3) Cables with a gas/vaportight continuous sheath capable of transmitting gases or vapors through the cable core shall not be required to be sealed except as required in (e)(1), unless the cable is attached to process equipment or devices that may cause a pressure in excess of 6 in. of water (1493 pascals) to be exerted at a cable end, in which case a seal, barrier, or other means shall be provided to prevent migration of flammables into an unclassified area.

Exception: Cables with an unbroken gas/vaportight continuous sheath shall be permitted to pass through a Class I, Division 2 location without seals.

(4) Cables that do not have gas/vaportight continuous sheath shall be sealed at the boundary of the Division 2 and unclassified location in such a manner as to minimize the passage of gases or vapors into an unclassified location.

FPN: The sheath mentioned in (d) and (e) may be either metal or a nonmetallic material.

(f) Drainage.

(1) Control Equipment. Where there is a probability that liquid or other condensed vapor may be trapped within enclosures for control equipment or at any point in the raceway system, approved means shall be provided to prevent accumulation or to permit periodic draining of such liquid or condensed vapor.

(2) Motors and Generators. Where the authority having jurisdiction judges that there is a probability that liquid or condensed vapor may accumulate within motors or generators, joints and conduit systems shall be arranged to minimize entrance of liquid. If means to prevent accumulation or to permit periodic draining are judged necessary, such means shall be provided at the time of manufacture and shall be considered an integral part of the machine.

(3) Canned Pumps, Process or Service Connections, etc. For canned pumps, process or service connections for flow, pressure, or analysis measurement, etc., that depend on a single compression seal, diaphragm, or tube to prevent flammable or combustible fluids from entering the electrical raceway or cable system capable of transmitting fluids, an additional approved seal, barrier, or other means shall be provided to prevent the flammable or combustible fluid from entering the raceway or cable system capable of transmitting fluids beyond the additional devices or means, if the primary seal fails.

The additional approved seal or barrier and the interconnecting enclosure shall meet the temperature and pressure conditions to which they will be subjected upon failure of the primary seal, unless other approved means are provided to accomplish the purpose above.

Drains, vents, or other devices shall be provided so that primary seal leakage will be obvious.

FPN: See also the fine print notes to Section 501-5.

501-6. Switches, Circuit Breakers, Motor Controllers, and Fuses.

(a) Class I, Division 1. In Class I, Division 1 locations, switches, circuit breakers, motor controllers, and fuses, including pushbuttons, relays, and similar devices, shall be provided with enclosures, and the enclosure in each case, together with the enclosed apparatus, shall be approved as a complete assembly for use in Class I locations.

(b) Class I, Division 2. Switches, circuit breakers, motor controllers, and fuses in Class I, Division 2 locations shall comply with the following:

(1) Type Required. Circuit breakers, motor controllers, and switches intended to interrupt current in the normal performance of the function for which they are installed shall be provided with enclosures approved for Class I, Division 1 locations in accordance with Section 501-3(a), unless general-purpose enclosures are provided and

(a) The interruption of current occurs within a chamber hermetically sealed against the entrance of gases and vapors, or

(b) The current make-and-break contacts are oil-immersed and of the general-purpose type having a 2-in. (50.8-mm) minimum immersion for power contacts and a 1-in. (25.4-mm) minimum immersion for control contacts, or

(c) The interruption of current occurs within a factory-sealed explosionproof chamber approved for the location, or

(d) The device is a solid state, switching control without contacts, where the surface temperature does not exceed 80 percent of the ignition temperature in degrees Celsius of the gas or vapor involved.

(2) Isolating Switches. Fused or unfused disconnect and isolating switches for transformers or capacitor banks that are not intended to interrupt current in the normal performance of the function for which they are installed shall be permitted to be installed in general-purpose enclosures.

(3) Fuses. For the protection of motors, appliances, and lamps, other than as provided in (b)(4), standard plug or cartridge fuses shall be permitted, provided they are placed within enclosures approved for the location; or fuses shall be permitted if they are within general-purpose enclosures, and if they are of a type in which the operating element is immersed in oil or other approved liquid, or the operating element is enclosed within a chamber hermetically sealed against the entrance of gases and vapors, or the fuse is a nonindicating, filled, current-limiting type.

(4) Fuses Internal to Lighting Fixtures. Approved cartridge fuses shall be permitted as supplementary protection within lighting fixtures.

501-7. Control Transformers and Resistors. Transformers, impedance coils, and resistors used as, or in conjunction with, control equipment for motors, generators, and appliances shall comply with (a) and (b).

(a) Class I, Division 1. In Class I, Division 1 locations, transformers, impedance coils, and resistors, together with any switching mechanism associated with them, shall be provided with enclosures approved for Class I, Division 1 locations in accordance with Section 501-3(a).

(b) Class I, Division 2. In Class I, Division 2 locations, control transformers and resistors shall comply with the following.

(1) Switching Mechanisms. Switching mechanisms used in conjunction with transformers, impedance coils, and resistors shall comply with Section 501-6(b).

(2) Coils and Windings. Enclosures for windings of transformers, solenoids, or impedance coils shall be permitted to be of the general-purpose type.

(3) Resistors. Resistors shall be provided with enclosures; and the assembly shall be approved for Class I locations, unless resistance is nonvariable and maximum operating temperature, in degrees Celsius, will not exceed 80 percent of the ignition temperature of the gas or vapor involved, or has been tested and found incapable of igniting the gas or vapor.

501-8. Motors and Generators.

(a) Class I, Division 1. In Class I, Division 1 locations, motors, generators, and other rotating electric machinery shall be as follows:

(1) Approved for Class I, Division 1 locations; or
(2) Of the totally enclosed type supplied with positive-pressure ventilation from a source of clean air with discharge to a safe area, so arranged to prevent energizing of the machine until ventilation has been established and the enclosure has been purged with at least 10 volumes of air, and also arranged to automatically de-energize the equipment when the air supply fails; or
(3) Of the totally enclosed inert gas-filled type supplied with a suitable reliable source of inert gas for pressuring the enclosure, with devices provided to ensure a positive pressure in the enclosure and arranged to automatically de-energize the equipment when the gas supply fails; or
(4) Of a type designed to be submerged in a liquid that is flammable only when vaporized and mixed with air, or in a gas or vapor at a pressure greater than atmospheric and that is flammable only when mixed with air; and the machine is arranged so to prevent energizing it until it has been purged with the liquid or gas to exclude air, and also arranged to automatically de-energize the equipment when the supply of liquid or gas or vapor fails or the pressure is reduced to atmospheric.

Totally enclosed motors of Types (2) or (3) shall have no external surface with an operating temperature in degrees Celsius in excess of 80 percent of the ignition temperature of the gas or vapor involved. Appropriate devices shall be provided to detect and automatically de-energize the motor or provide an adequate alarm if there is any increase in temperature of the motor beyond designed limits. Auxiliary equipment shall be of a type approved for the location in which it is installed.

FPN: See ASTM *Test Procedure,* D 2155-69.

(b) Class I, Division 2. In Class I, Division 2 locations, motors, generators, and other rotating electric machinery in which are employed sliding contacts, centrifugal or other types of switching mechanism (including motor overcurrent, overloading, and overtemperature devices), or integral resistance devices, either while starting or while running, shall be approved for Class I, Division 1 locations, unless such sliding contacts, switching mechanisms, and resistance devices are provided with enclosures approved for Class I, Division 2 locations in accordance with Section 501-3(b). The exposed surface of space heaters used to prevent condensation of moisture during shutdown periods shall not exceed 80 percent of the ignition temperature in degrees Celsius of the gas or vapor involved when operated at rated voltage, and the maximum surface temperature [based on a 40°C (104°F) ambient] shall be permanently marked on a visible nameplate mounted on the motor. Otherwise, space heaters shall be approved for Class I, Division 2 locations.

In Class I, Division 2 locations, the installation of open or nonexplosionproof enclosed motors, such as squirrel-cage induction motors without brushes, switching mechanisms, or similar arc-producing devices that are not identified for use in a Class I, Division 2 location, shall be permitted.

FPN No. 1: It is important to consider the temperature of internal and external surfaces that may be exposed to the flammable atmosphere.

FPN No. 2: It is important to consider the risk of ignition due to currents arcing across discontinuities and overheating of parts in multisection enclosures of large motors and generators. Such motors and generators may need equipotential bonding jumpers across joints in the enclosure and from enclosure to ground. Where the presence of ignitible gases or vapors is suspected, clean-air purging may be needed immediately prior to and during start-up periods.

501-9. Lighting Fixtures. Lighting fixtures shall comply with (a) or (b).

(a) Class I, Division 1. In Class I, Division 1 locations, lighting fixtures shall comply with the following.

(1) Approved Fixtures. Each fixture shall be approved as a complete assembly for the Class I, Division 1 location and shall be clearly marked to indicate the maximum wattage of lamps for which it is approved. Fixtures intended for portable use shall be specifically approved as a complete assembly for that use.

(2) Physical Damage. Each fixture shall be protected against physical damage by a suitable guard or by location.

(3) Pendant Fixtures. Pendant fixtures shall be suspended by and supplied through threaded rigid metal conduit stems or threaded steel intermediate conduit stems, and threaded joints shall be provided with set-screws or other effective means to prevent loosening. For stems longer than 12 in. (305 mm), permanent and effective bracing against lateral displacement shall be provided at a level not more than 12 in. (305 mm) above the lower end of the stem, or flexibility in the form of a fitting or flexible connector approved for the Class I, Division 1 location shall be provided not more than 12 in. (305 mm) from the point of attachment to the supporting box or fitting.

(4) Supports. Boxes, box assemblies, or fittings used for the support of lighting fixtures shall be approved for Class I locations.

(b) Class I, Division 2. In Class I, Division 2 locations, lighting fixtures shall comply with the following.

(1) Portable Lighting Equipment. Portable lighting equipment shall comply with (a)(1).

Exception: Where portable lighting equipment are mounted on movable stands and are connected by flexible cords, as covered in Section 501-11, they shall be permitted, where mounted in any position, if they conform to Section 501-9(b)(2).

(2) Fixed Lighting. Lighting fixtures for fixed lighting shall be protected from physical damage by suitable guards or by location. Where there is danger that falling sparks or hot metal from lamps or fixtures might ignite localized concentrations of flammable vapors or gases, suitable enclosures or other effective protective means shall be provided. Where lamps are of a size or type that may, under normal operating conditions, reach surface temperatures exceeding 80 percent of the ignition temperature in degrees Celsius of the gas or vapor involved, fixtures shall comply with (a)(1) or shall be of a type that has been tested in order to determine the marked operating temperature or temperature range.

(3) Pendant Fixtures. Pendant fixtures shall be suspended by threaded rigid metal conduit stems, threaded steel intermediate metal conduit stems, or by other approved means. For rigid stems longer than 12 in. (305 mm), permanent and effective bracing against lateral displacement shall be provided at a level not more than 12 in. (305 mm) above the lower end of the stem, or flexibility in the form of an approved fitting or flexible connector shall be provided not more than 12 in. (305 mm) from the point of attachment to the supporting box or fitting.

(4) Switches. Switches that are a part of an assembled fixture or of an individual lampholder shall comply with Section 501-6(b)(1).

(5) Starting Equipment. Starting and control equipment for electric-discharge lamps shall comply with Section 501-7(b).

Exception: A thermal protector potted into a thermally protected fluorescent lamp ballast if the lighting fixture is approved for locations of this class and division.

501-10. Utilization Equipment.

(a) Class I, Division 1. In Class I, Division 1 locations, all utilization equipment shall be approved for Class I, Division 1 locations.

(b) Class I, Division 2. In Class I, Division 2 locations, all utilization equipment shall comply with the following.

(1) Heaters. Electrically heated utilization equipment shall conform with either item (a) or (b).

(a) The heater shall not exceed 80 percent of the ignition temperature in degrees Celsius of the gas or vapor involved on any surface that is exposed to the gas or vapor when continuously energized at the maximum rated ambient temperature. If a temperature controller is not provided, these conditions shall apply when the heater is operated at 120 percent of rated voltage.

Exception No. 1: For motor-mounted anticondensation space heaters, see Section 501-8(b).

Exception No. 2: A current-limiting device is applied to the circuit serving the heater that will limit the current in the heater to a value less than that required to raise the heater surface temperature to 80 percent of the ignition temperature.

(b) The heater shall be approved for Class I, Division 1 locations.

Exception: Electrical resistance heat tracing approved for Class I, Division 2 locations.

(2) Motors. Motors of motor-driven utilization equipment shall comply with Section 501-8(b).

(3) Switches, Circuit Breakers, and Fuses. Switches, circuit breakers, and fuses shall comply with Section 501-6(b).

501-11. Flexible Cords, Class I, Divisions 1 and 2. A
flexible cord shall be permitted for connection between portable lighting equipment or other portable utilization equipment and the fixed portion of their supply circuit. Flexible cord shall also be permitted for that portion of the circuit where the fixed wiring methods of Section 501-4(a) cannot provide the necessary degree of movement for fixed and mobile electrical utilization equipment, in an industrial establishment where conditions of maintenance and engineering supervision ensure that only qualified persons will install and service the installation, and the flexible cord is protected by location or by a suitable guard from damage. The length of the flexible cord shall be continuous. Where flexible cords are used, the cords shall be as follows:

(1) Of a type approved for extra-hard usage;
(2) Contain, in addition to the conductors of the circuit, a grounding conductor complying with Section 400-23;
(3) Connected to terminals or to supply conductors in an approved manner;
(4) Be supported by clamps or by other suitable means in such a manner that there will be no tension on the terminal connections; and
(5) Be provided with suitable seals where the flexible cord enters boxes, fittings, or enclosures of the explosion-proof type.

Exception: As provided in Sections 501-3(b)(6) and 501-4(b).

Electric submersible pumps with means for removal without entering the wet-pit shall be considered portable utilization equipment. The extension of the flexible cord within a suitable raceway between the wet-pit and the power source shall be permitted.

Electric mixers intended for travel into and out of open-type mixing tanks or vats shall be considered portable utilization equipment.

FPN: See Section 501-13 for flexible cords exposed to liquids having a deleterious effect on the conductor insulation.

501-12. Receptacles and Attachment Plugs, Class I, Divisions 1 and 2. Receptacles and attachment plugs shall be of the type providing for connection to the grounding conductor of a flexible cord and shall be approved for the location.

Exception: As provided in Section 501-3(b)(6).

501-13. Conductor Insulation, Class I, Divisions 1 and 2.
Where condensed vapors or liquids may collect on, or come in contact with, the insulation on conductors, such insulation shall be of a type approved for use under such conditions; or the insulation shall be protected by a sheath of lead or by other approved means.

501-14. Signaling, Alarm, Remote-Control, and Communications Systems.

(a) Class I, Division 1. In Class I, Division 1 locations, all apparatus and equipment of signaling, alarm, remote-control, and communications systems, regardless of voltage, shall be approved for Class I, Division 1 locations, and all wiring shall comply with Sections 501-4(a) and 501-5(a) and (c).

(b) Class I, Division 2. In Class I, Division 2 locations, signaling, alarm, remote-control, and communications systems shall comply with the following.

(1) Contacts. Switches, circuit breakers, and make-and-break contacts of pushbuttons, relays, alarm bells, and horns shall have enclosures approved for Class I, Division 1 locations in accordance with Section 501-3(a).

Exception: General-purpose enclosures shall be permitted if current-interrupting contacts are one of the following:

(a) Immersed in oil, or
(b) Enclosed within a chamber hermetically sealed against the entrance of gases or vapors, or
(c) In nonincendive circuits, or
(d) Part of a listed nonincendive component

(2) Resistors and Similar Equipment. Resistors, resistance devices, thermionic tubes, rectifiers, and similar equipment shall comply with Section 501-3(b)(2).

(3) Protectors. Enclosures shall be provided for lightning protective devices and for fuses. Such enclosures shall be permitted to be of the general-purpose type.

(4) Wiring and Sealing. All wiring shall comply with Sections 501-4(b) and 501-5(b) and (c).

501-15. Live Parts, Class I, Divisions 1 and 2. There shall be no exposed live parts.

501-16. Grounding, Class I, Divisions 1 and 2. Wiring and equipment in Class I, Division 1 and 2 locations shall be grounded as specified in Article 250 and with the following additional requirements.

(a) Bonding. The locknut-bushing and double-locknut types of contacts shall not be depended on for bonding purposes, but bonding jumpers with proper fittings or other approved means of bonding shall be used. Such means of bonding shall apply to all intervening raceways, fittings, boxes, enclosures, etc., between Class I locations and the point of grounding for service equipment or point of grounding of a separately derived system.

Exception: The specific bonding means shall only be required to the nearest point where the grounded circuit conductor and the grounding electrode are connected together on the line side of the building or structure disconnecting means as specified in Sections 250-32(a), (b), and (c), provided the branch-circuit overcurrent protection is located on the load side of the disconnecting means.

FPN: See Section 250-100 for additional bonding requirements in hazardous (classified) locations.

(b) Types of Equipment Grounding Conductors. Where flexible metal conduit or liquidtight flexible metal conduit is used as permitted in Section 501-4(b) and is to be relied on to complete a sole equipment grounding path, it shall be installed with internal or external bonding jumpers in parallel with each conduit and complying with Section 250-102.

Exception: In Class I, Division 2 locations, the bonding jumper shall be permitted to be deleted where all the following conditions are met.

(a) Listed liquidtight flexible metal conduit 6 ft (1.83 m) or less in length, with fittings listed for grounding, is used.
(b) Overcurrent protection in the circuit is limited to 10 amperes or less.
(c) The load is not a power utilization load.

501-17. Surge Protection.

(a) Class I, Division 1. Surge arresters, including their installation and connection, shall comply with Article 280. The surge arresters and capacitors shall be installed in enclosures approved for Class I, Division 1 locations. Surge-protective capacitors shall be of a type designed for specific duty.

(b) Class I, Division 2. Surge arresters shall be nonarcing, such as metal-oxide varistor (MOV), sealed type, and surge-protective capacitors shall be of a type designed for specific duty. Installation and connection shall comply with Article 280.

Enclosures shall be permitted to be of the general-purpose type. Surge protection of types other than described above shall be installed in enclosures approved for Class I, Division 1 locations.

501-18. Multiwire Branch Circuits. In a Class I, Division 1 location, a multiwire branch circuit shall not be permitted.

Exception: Where the disconnect device(s) for the circuit opens all ungrounded conductors of the multiwire circuit simultaneously.

Article 502 — Class II Locations

502-1. General. The general rules of this *Code* shall apply to the electric wiring and equipment in locations classified as Class II locations in Section 500-8.

Exception: As modified by this article.

Dust-ignitionproof, as used in this article, shall mean enclosed in a manner that will exclude dusts and, where installed and protected in accordance with this *Code*, will not permit arcs, sparks, or heat otherwise generated or liberated inside of the enclosure to cause ignition of exterior accumulations or atmospheric suspensions of a specified dust on or in the vicinity of the enclosure.

FPN: For further information on dust-ignitionproof enclosures, see Type 9 enclosure in *Enclosures for Electrical Equipment*, ANSI/NEMA 250-1991, and *Explosionproof and Dust-Ignitionproof Electrical Equipment for Hazardous (Classified) Locations*, ANSI/UL 1203-1994.

Equipment installed in Class II locations shall be able to function at full rating without developing surface temperatures high enough to cause excessive dehydration or gradual carbonization of any organic dust deposits that may occur.

FPN: Dust that is carbonized or excessively dry is highly susceptible to spontaneous ignition.

Equipment and wiring of the type defined in Article 100 as explosionproof shall not be required and shall not be acceptable in Class II locations unless approved for such locations.

Where Class II, Group E dusts are present in hazardous quantities, there are only Division 1 locations.

502-2. Transformers and Capacitors.

(a) Class II, Division 1. In Class II, Division 1 locations, transformers and capacitors shall comply with the following.

(1) Containing Liquid that Will Burn. Transformers and capacitors containing a liquid that will burn shall be installed only in approved vaults complying with Sections 450-41 through 450-48, and, in addition, the following shall apply.

(a) Doors or other openings communicating with the Division 1 location shall have self-closing fire doors on both sides of the wall, and the doors shall be carefully fitted and provided with suitable seals (such as weather stripping) to minimize the entrance of dust into the vault.
(b) Vent openings and ducts shall communicate only with the outside air.
(c) Suitable pressure-relief openings communicating with the outside air shall be provided.

(2) Not Containing Liquid that Will Burn. Transformers and capacitors that do not contain a liquid that will burn shall be installed in vaults complying with Sections 450-41 through 450-48 or be approved as a complete assembly, including terminal connections for Class II locations.

(3) Metal Dusts. No transformer or capacitor shall be installed in a location where dust from magnesium, aluminum, aluminum bronze powders, or other metals of similarly hazardous characteristics may be present.

(b) Class II, Division 2. In Class II, Division 2 locations, transformers and capacitors shall comply with the following.

(1) Containing Liquid that Will Burn. Transformers and capacitors containing a liquid that will burn shall be installed in vaults that comply with Sections 450-41 through 450-48.

(2) Containing Askarel. Transformers containing askarel and rated in excess of 25 kVA shall be as follows:

(a) Provided with pressure-relief vents
(b) Provided with a means for absorbing any gases generated by arcing inside the case, or the pressure-relief vents shall be connected to a chimney or flue that will carry such gases outside the building
(c) Have an airspace of not less than 6 in. (152 mm) between the transformer cases and any adjacent combustible material

(3) Dry-Type Transformers. Dry-type transformers shall be installed in vaults or shall have their windings and terminal connections enclosed in tight metal housings without ventilating or other openings and shall operate at not over 600 volts, nominal.

502-4. Wiring Methods. Wiring methods shall comply with (a) and (b).

(a) Class II, Division 1. In Class II, Division 1 locations, threaded rigid metal conduit, threaded steel intermediate metal conduit, or Type MI cable with termination fittings approved for the location shall be the wiring method employed. Type MI cable shall be installed and supported in a manner to avoid tensile stress at the termination fittings.

Exception: In industrial establishments with limited public access, where the conditions of maintenance and supervision ensure that only qualified persons will service the installation, Type MC cable, listed for use in Class II, Division 1 locations, with a gas/vaportight continuous corrugated aluminum sheath, an overall jacket of suitable polymeric material, separate grounding conductors in accordance with Section 250-122, and provided with termination fittings listed for the application shall be permitted.

(1) Fittings and Boxes. Fittings and boxes shall be provided with threaded bosses for connection to conduit or cable terminations, shall have close-fitting covers, and shall have no openings (such as holes for attachment screws) through which dust might enter or through which sparks or burning material might escape. Fittings and boxes in which taps, joints, or terminal connections are made, or that are used in locations where dusts are of a combustible, electrically conductive nature, shall be approved for Class II locations.

(2) Flexible Connections. Where necessary to employ flexible connections, dusttight flexible connectors, liquidtight flexible metal conduit with approved fittings, liquidtight flexible nonmetallic conduit with approved fittings, or flexible cord approved for extra-hard usage and provided with bushed fittings shall be used. Where flexible cords are used, they shall comply with Section 502-12. Where flexible connections are subject to oil or other corrosive conditions, the insulation of the conductors shall be of a type approved for the condition or shall be protected by means of a suitable sheath.

FPN: See Section 502-16(b) for grounding requirements where flexible conduit is used.

(b) Class II, Division 2. In Class II, Division 2 locations, rigid metal conduit, intermediate metal conduit, electrical metallic tubing, dusttight wireways, Type MC or MI cable with approved termination fittings, Type PLTC in cable trays, Type ITC in cable trays, or Type MC, MI, or TC cable installed in ladder, ventilated trough, or ventilated channel cable trays in a single layer, with a space not less than the larger cable diameter between the two adjacent cables, shall be the wiring method employed.

Exception No. 1: Nonincendive field wiring shall be permitted using any of the methods suitable for wiring in ordinary locations.

Exception No. 2: Type MC cable listed for use in Class II, Division 1 locations shall be permitted to be installed without the above required spacings.

(1) Wireways, Fittings, and Boxes. Wireways, fittings, and boxes in which taps, joints, or terminal connections are made shall be designed to minimize the entrance of dust and (1) shall be provided with telescoping or close-fitting covers or other effective means to prevent the escape of sparks or burning material and (2) shall have no openings (such as holes for attachment screws) through which, after installation, sparks or burning material might escape or through which adjacent combustible material might be ignited.

(2) Flexible Connections. Where flexible connections are necessary, (a)(2) shall apply.

502-5. Sealing, Class II, Divisions 1 and 2. Where a raceway provides communication between an enclosure that is required to be dust-ignitionproof and one that is not, suitable means shall be provided to prevent the entrance of dust into the dust-ignitionproof enclosure through the raceway. One of the following means shall be permitted:

(1) A permanent and effective seal,
(2) A horizontal raceway not less than 10 ft (3.05 m) long, or

(3) A vertical raceway not less than 5 ft (1.52 m) long and extending downward from the dust-ignitionproof enclosure

Where a raceway provides communication between an enclosure that is required to be dust-ignitionproof and an enclosure in an unclassified location, seals shall not be required.

Sealing fittings shall be accessible.

Seals shall not be required to be explosionproof.

FPN: Electrical sealing putty is a method of sealing.

502-6. Switches, Circuit Breakers, Motor Controllers, and Fuses.

(a) Class II, Division 1. In Class II, Division 1 locations, switches, circuit breakers, motor controllers, and fuses shall comply with the following:

(1) Type Required. Switches, circuit breakers, motor controllers, and fuses, including pushbuttons, relays, and similar devices that are intended to interrupt current during normal operation or that are installed where combustible dusts of an electrically conductive nature may be present, shall be provided with approved dust-ignitionproof enclosures.

(2) Isolating Switches. Disconnecting and isolating switches containing no fuses and not intended to interrupt current and not installed where dusts may be of an electrically conductive nature shall be provided with tight metal enclosures that shall be designed to minimize the entrance of dust and that shall (1) be equipped with telescoping or close-fitting covers or with other effective means to prevent the escape of sparks or burning material and (2) have no openings (such as holes for attachment screws) through which, after installation, sparks or burning material might escape or through which exterior accumulations of dust or adjacent combustible material might be ignited.

(3) Metal Dusts. In locations where dust from magnesium, aluminum, aluminum bronze powders, or other metals of similarly hazardous characteristics may be present, fuses, switches, motor controllers, and circuit breakers shall have enclosures specifically approved for such locations.

(b) Class II, Division 2. In Class II, Division 2 locations, enclosures for fuses, switches, circuit breakers, and motor controllers, including pushbuttons, relays, and similar devices, shall be dusttight.

502-7. Control Transformers and Resistors.

(a) Class II, Division 1. In Class II, Division 1 locations, control transformers, solenoids, impedance coils, resistors, and any overcurrent devices or switching mechanisms associated with them shall have dust-ignitionproof enclosures approved for Class II locations. No control transformer, impedance coil, or resistor shall be installed in a location where dust from magnesium, aluminum, aluminum bronze powders, or other metals of similarly hazardous characteristics may be present unless provided with an enclosure approved for the specific location.

(b) Class II, Division 2. In Class II, Division 2 locations, transformers and resistors shall comply with the following.

(1) Switching Mechanisms. Switching mechanisms (including overcurrent devices) associated with control transformers, solenoids, impedance coils, and resistors shall be provided with dusttight enclosures.

(2) Coils and Windings. Where not located in the same enclosure with switching mechanisms, control transformers, solenoids, and impedance coils shall be provided with tight metal housings without ventilating openings.

(3) Resistors. Resistors and resistance devices shall have dust-ignitionproof enclosures approved for Class II locations.

Exception: Where the maximum normal operating temperature of the resistor will not exceed 120°C (248°F), nonadjustable resistors or resistors that are part of an automatically timed starting sequence shall be permitted to have enclosures complying with (b)(2).

502-8. Motors and Generators.

(a) Class II, Division 1. In Class II, Division 1 locations, motors, generators, and other rotating electrical machinery shall be

(1) Approved for Class II, Division 1 locations, or
(2) Totally enclosed pipe-ventilated, meeting temperature limitations in Section 502-1.

(b) Class II, Division 2. In Class II, Division 2 locations, motors, generators, and other rotating electrical equipment shall be totally enclosed nonventilated, totally enclosed pipe-ventilated, totally enclosed water–air cooled, totally enclosed fan-cooled or dust-ignitionproof for which maximum full-load external temperature shall be in accordance with Section 500-5(f) for normal operation when operating in free air (not dust blanketed) and shall have no external openings.

Exception: If the authority having jurisdiction believes accumulations of nonconductive, nonabrasive dust will be moderate and if machines can be easily reached for routine cleaning and maintenance, the following shall be permitted to be installed:

(a) Standard open-type machines without sliding contacts, centrifugal or other types of switching mechanism (in-

cluding motor overcurrent, overloading, and overtemperature devices), or integral resistance devices
(b) Standard open-type machines with such contacts, switching mechanisms, or resistance devices enclosed within dusttight housings without ventilating or other openings
(c) Self-cleaning textile motors of the squirrel-cage type

502-9. Ventilating Piping. Ventilating pipes for motors, generators, or other rotating electric machinery, or for enclosures for electric equipment, shall be of metal not less than 0.021 in. (533 μm) in thickness, or of equally substantial noncombustible material, and shall comply with the following:

(1) Lead directly to a source of clean air outside of buildings,
(2) Be screened at the outer ends to prevent the entrance of small animals or birds, and
(3) Be protected against physical damage and against rusting or other corrosive influences

Ventilating pipes shall also comply with (a) and (b).

(a) Class II, Division 1. In Class II, Division 1 locations, ventilating pipes, including their connections to motors or to the dust-ignitionproof enclosures for other equipment, shall be dusttight throughout their length. For metal pipes, seams and joints shall comply with one of the following:

(1) Be riveted and soldered,
(2) Be bolted and soldered,
(3) Be welded, or
(4) Be rendered dusttight by some other equally effective means

(b) Class II, Division 2. In Class II, Division 2 locations, ventilating pipes and their connections shall be sufficiently tight to prevent the entrance of appreciable quantities of dust into the ventilated equipment or enclosure and to prevent the escape of sparks, flame, or burning material that might ignite dust accumulations or combustible material in the vicinity. For metal pipes, lock seams and riveted or welded joints shall be permitted; and tight-fitting slip joints shall be permitted where some flexibility is necessary, as at connections to motors.

502-10. Utilization Equipment.

(a) Class II, Division 1. In Class II, Division 1 locations, all utilization equipment shall be approved for Class II locations. Where dust from magnesium, aluminum, aluminum bronze powders, or other metals of similarly hazardous characteristics may be present, such equipment shall be approved for the specific location.

(b) Class II, Division 2. In Class II, Division 2 locations, all utilization equipment shall comply with the following.

(1) Heaters. Electrically heated utilization equipment shall be approved for Class II locations.

Exception: Metal-enclosed radiant heating panel equipment shall be dusttight and marked in accordance with Section 500-5(d).

(2) Motors. Motors of motor-driven utilization equipment shall comply with Section 502-8(b).

(3) Switches, Circuit Breakers, and Fuses. Enclosures for switches, circuit breakers, and fuses shall be dusttight.

(4) Transformers, Solenoids, Impedance Coils, and Resistors. Transformers, solenoids, impedance coils, and resistors shall comply with Section 502-7(b).

502-11. Lighting Fixtures. Lighting fixtures shall comply with (a) and (b).

(a) Class II, Division 1. In Class II, Division 1 locations, lighting fixtures for fixed and portable lighting shall comply with the following.

(1) Approved Fixtures. Each fixture shall be approved for Class II locations and shall be clearly marked to indicate the maximum wattage of the lamp for which it is approved. In locations where dust from magnesium, aluminum, aluminum bronze powders, or other metals of similarly hazardous characteristics may be present, fixtures for fixed or portable lighting and all auxiliary equipment shall be approved for the specific location.

(2) Physical Damage. Each fixture shall be protected against physical damage by a suitable guard or by location.

(3) Pendant Fixtures. Pendant fixtures shall be suspended by threaded rigid metal conduit stems, threaded steel intermediate metal conduit stems, by chains with approved fittings, or by other approved means. For rigid stems longer than 12 in. (305 mm), permanent and effective bracing against lateral displacement shall be provided at a level not more than 12 in. (305 mm) above the lower end of the stem, or flexibility in the form of a fitting or a flexible connector approved for the location shall be provided not more than 12 in. (305 mm) from the point of attachment to the supporting box or fitting. Threaded joints shall be provided with set-screws or other effective means to prevent loosening. Where wiring between an outlet box or fitting and a pendant fixture is not enclosed in conduit, flexible cord approved for hard usage shall be used, and suitable seals shall be provided where the cord enters the fixture and the outlet box or fitting. Flexible cord shall not serve as the supporting means for a fixture.

(4) Supports. Boxes, box assemblies, or fittings used for the support of lighting fixtures shall be approved for Class II locations.

(b) Class II, Division 2. In Class II, Division 2 locations, lighting fixtures shall comply with the following.

(1) Portable Lighting Equipment. Portable lighting equipment shall be approved for Class II locations. They shall be clearly marked to indicate the maximum wattage of lamps for which they are approved.

(2) Fixed Lighting. Lighting fixtures for fixed lighting, where not of a type approved for Class II locations, shall provide enclosures for lamps and lampholders that shall be designed to minimize the deposit of dust on lamps and to prevent the escape of sparks, burning material, or hot metal. Each fixture shall be clearly marked to indicate the maximum wattage of the lamp that shall be permitted without exceeding an exposed surface temperature in accordance with Section 500-5(f) under normal conditions of use.

(3) Physical Damage. Lighting fixtures for fixed lighting shall be protected from physical damage by suitable guards or by location.

(4) Pendant Fixtures. Pendant fixtures shall be suspended by threaded rigid metal conduit stems, threaded steel intermediate metal conduit stems, by chains with approved fittings, or by other approved means. For rigid stems longer than 12 in. (305 mm), permanent and effective bracing against lateral displacement shall be provided at a level not more than 12 in. (305 mm) above the lower end of the stem, or flexibility in the form of an approved fitting or a flexible connector shall be provided not more than 12 in. (305 mm) from the point of attachment to the supporting box or fitting. Where wiring between an outlet box or fitting and a pendant fixture is not enclosed in conduit, flexible cord approved for hard usage shall be used. Flexible cord shall not serve as the supporting means for a fixture.

(5) Electric-Discharge Lamps. Starting and control equipment for electric-discharge lamps shall comply with the requirements of Section 502-7(b).

502-12. Flexible Cords — Class II, Divisions 1 and 2. Flexible cords used in Class II locations shall comply with the following:

(1) Be of a type approved for extra-hard usage,
(2) Contain, in addition to the conductors of the circuit, a grounding conductor complying with Section 400-23,
(3) Be connected to terminals or to supply conductors in an approved manner,
(4) Be supported by clamps or by other suitable means in such a manner that there will be no tension on the terminal connections, and

(5) Be provided with suitable seals to prevent the entrance of dust where the flexible cord enters boxes or fittings that are required to be dust-ignitionproof

502-13. Receptacles and Attachment Plugs.

(a) Class II, Division 1. In Class II, Division 1 locations, receptacles and attachment plugs shall be of the type providing for connection to the grounding conductor of the flexible cord and shall be approved for Class II locations.

(b) Class II, Division 2. In Class II, Division 2 locations, receptacles and attachment plugs shall be of the type that provide for connection to the grounding conductor of the flexible cord and shall be designed so that connection to the supply circuit cannot be made or broken while live parts are exposed.

502-14. Signaling, Alarm, Remote-Control, and Communications Systems; and Meters, Instruments, and Relays.

FPN: See Article 800 for rules governing the installation of communications circuits.

(a) Class II, Division 1. In Class II, Division 1 locations, signaling, alarm, remote-control, and communications systems; and meters, instruments, and relays shall comply with the following:

(1) Wiring Methods. The wiring method shall comply with Section 502-4(a).

(2) Contacts. Switches, circuit breakers, relays, contactors, fuses and current-breaking contacts for bells, horns, howlers, sirens, and other devices in which sparks or arcs may be produced shall be provided with enclosures approved for a Class II location.

Exception: Where current-breaking contacts are immersed in oil or where the interruption of current occurs within a chamber sealed against the entrance of dust, enclosures shall be permitted to be of the general-purpose type.

(3) Resistors and Similar Equipment. Resistors, transformers, choke coils, rectifiers, thermionic tubes, and other heat-generating equipment shall be provided with enclosures approved for Class II locations.

Exception: Where resistors or similar equipment are immersed in oil or enclosed in a chamber sealed against the entrance of dust, enclosures shall be permitted to be of the general-purpose type.

(4) Rotating Machinery. Motors, generators, and other rotating electric machinery shall comply with Section 502-8(a).

(5) Combustible, Electrically Conductive Dusts. Where dusts are of a combustible, electrically conductive nature, all wiring and equipment shall be approved for Class II locations.

(6) Metal Dusts. Where dust from magnesium, aluminum, aluminum bronze powders, or other metals of similarly hazardous characteristics may be present, all apparatus and equipment shall be approved for the specific conditions.

(b) Class II, Division 2. In Class II, Division 2 locations, signaling, alarm, remote-control, and communications systems; and meters, instruments, and relays shall comply with the following.

(1) Contacts. Enclosures shall comply with (a)(2), or contacts shall have tight metal enclosures designed to minimize the entrance of dust and shall have telescoping or tight-fitting covers and no openings through which, after installation, sparks or burning material might escape.

Exception: In nonincendive circuits, enclosures shall be permitted to be of the general-purpose type.

(2) Transformers and Similar Equipment. The windings and terminal connections of transformers, choke coils, and similar equipment shall be provided with tight metal enclosures without ventilating openings.

(3) Resistors and Similar Equipment. Resistors, resistance devices, thermionic tubes, rectifiers, and similar equipment shall comply with (a)(3).

Exception: Enclosures for thermionic tubes, nonadjustable resistors, or rectifiers for which maximum operating temperature will not exceed 120°C (248°F) shall be permitted to be of the general-purpose type.

(4) Rotating Machinery. Motors, generators, and other rotating electric machinery shall comply with Section 502-8(b).

(5) Wiring Methods. The wiring method shall comply with Section 502-4(b).

502-15. Live Parts, Class II, Divisions 1 and 2. Live parts shall not be exposed.

502-16. Grounding, Class II, Divisions 1 and 2. Wiring and equipment in Class II, Divisions 1 and 2 locations shall be grounded as specified in Article 250 and with the following additional requirements.

(a) Bonding. The locknut-bushing and double-locknut types of contact shall not be depended on for bonding purposes, but bonding jumpers with proper fittings or other approved means of bonding shall be used. Such means of bonding shall apply to all intervening raceways, fittings, boxes, enclosures, etc., between Class II locations and the point of grounding for service equipment or point of grounding of a separately derived system.

Exception: The specific bonding means shall only be required to the nearest point where the grounded circuit conductor and the grounding electrode conductor are connected together on the line side of the building or structure disconnecting means as specified in Sections 250-32(a), (b), and (c), if the branch-circuit overcurrent protection is located on the load side of the disconnecting means.

> FPN: See Section 250-100 for additional bonding requirements in hazardous (classified) locations.

(b) Types of Equipment Grounding Conductors. Where flexible conduit is used as permitted in Section 502-4, it shall be installed with internal or external bonding jumpers in parallel with each conduit and complying with Section 250-102.

Exception: In Class II, Division 2 locations, the bonding jumper shall be permitted to be deleted where all the following conditions are met.

(a) Listed liquidtight flexible metal conduit 6 ft (1.83 m) or less in length, with fittings listed for grounding, is used.
(b) Overcurrent protection in the circuit is limited to 10 amperes or less.
(c) The load is not a power utilization load.

502-17. Surge Protection — Class II, Divisions 1 and 2. Surge arresters, including their installation and connection, shall comply with Article 280. In addition, surge arresters, if installed in a Class II, Division 1 location, shall be in suitable enclosures.

Surge-protective capacitors shall be of a type designed for specific duty.

502-18. Multiwire Branch Circuits. In a Class II, Division 1 location, a multiwire branch circuit shall not be permitted.

Exception: Where the disconnect device(s) for the circuit opens all ungrounded conductors of the multiwire circuit simultaneously.

Article 503 — Class III Locations

503-1. General. The general rules of this *Code* shall apply to electric wiring and equipment in locations classified as Class III locations in Section 500-9.

Exception: As modified by this article.

Equipment installed in Class III locations shall be able to function at full rating without developing surface temperatures high enough to cause excessive dehydration or gradual carbonization of accumulated fibers or flyings. Organic material that is carbonized or excessively dry is highly susceptible to spontaneous ignition. The maximum surface temperatures under operating conditions shall not exceed 165°C (329°F) for equipment that is not subject to overloading, and 120°C (248°F) for equipment (such as motors or power transformers) that may be overloaded.

> FPN: For electric trucks, see *Fire Safety Standard for Powered Industrial Trucks Including Type Designations, Areas of Use, Conversions, Maintenance, and Operation,* NFPA 505-1996.

503-2. Transformers and Capacitors — Class III, Divisions 1 and 2. Transformers and capacitors shall comply with Section 502-2(b).

503-3. Wiring Methods. Wiring methods shall comply with (a) and (b).

(a) Class III, Division 1. In Class III, Division 1 locations, the wiring method shall be rigid metal conduit, rigid nonmetallic conduit, intermediate metal conduit, electrical metallic tubing, dusttight wireways, or Type MC or MI cable with approved termination fittings.

(1) Boxes and Fittings. All boxes and fittings shall be dusttight.

(2) Flexible Connections. Where necessary to employ flexible connections, dusttight flexible connectors, liquidtight flexible metal conduit with approved fittings, liquidtight flexible nonmetallic conduit with approved fittings, or flexible cord in conformance with Section 503-10 shall be used.

> FPN: See Section 503-16(b) for grounding requirements where flexible conduit is used.

(b) Class III, Division 2. In Class III, Division 2 locations, the wiring method shall comply with (a).

Exception: In sections, compartments, or areas used solely for storage and containing no machinery, open wiring on insulators shall be permitted where installed in accordance with Article 320, but only on condition that protection as required by Section 320-14 be provided where conductors are not run in roof spaces and are well out of reach of sources of physical damage.

503-4. Switches, Circuit Breakers, Motor Controllers, and Fuses — Class III, Divisions 1 and 2. Switches, circuit breakers, motor controllers, and fuses, including push-buttons, relays, and similar devices, shall be provided with dusttight enclosures.

503-5. Control Transformers and Resistors — Class III, Divisions 1 and 2. Transformers, impedance coils, and resistors used as or in conjunction with control equipment for motors, generators, and appliances shall be provided with dusttight enclosures complying with the temperature limitations in Section 503-1.

503-6. Motors and Generators — Class III, Divisions 1 and 2. In Class III, Divisions 1 and 2 locations, motors, generators, and other rotating machinery shall be totally enclosed nonventilated, totally enclosed pipe ventilated, or totally enclosed fan cooled.

Exception: In locations where, in the judgment of the authority having jurisdiction, only moderate accumulations of lint or flyings will be likely to collect on, in, or in the vicinity of a rotating electric machine and where such machine is readily accessible for routine cleaning and maintenance, one of the following shall be permitted:

(a) Self-cleaning textile motors of the squirrel-cage type;
(b) Standard open-type machines without sliding contacts, centrifugal or other types of switching mechanism, including motor overload devices; or
(c) Standard open-type machines having such contacts, switching mechanisms, or resistance devices enclosed within tight housings without ventilating or other openings

503-7. Ventilating Piping — Class III, Divisions 1 and 2. Ventilating pipes for motors, generators, or other rotating electric machinery, or for enclosures for electric equipment, shall be of metal not less than 0.021 in. (533 μm) in thickness, or of equally substantial noncombustible material, and shall comply with the following:

(1) Lead directly to a source of clean air outside of buildings,
(2) Be screened at the outer ends to prevent the entrance of small animals or birds, and
(3) Be protected against physical damage and against rusting or other corrosive influences

Ventilating pipes shall be sufficiently tight, including their connections, to prevent the entrance of appreciable quantities of fibers or flyings into the ventilated equipment or enclosure and to prevent the escape of sparks, flame, or burning material that might ignite accumulations of fibers or flyings or combustible material in the vicinity. For metal pipes, lock seams and riveted or welded joints shall be permitted; and tight-fitting slip joints shall be permitted where some flexibility is necessary, as at connections to motors.

503-8. Utilization Equipment — Class III, Divisions 1 and 2.

(a) Heaters. Electrically heated utilization equipment shall be approved for Class III locations.

(b) Motors. Motors of motor-driven utilization equipment shall comply with Section 503-6.

(c) Switches, Circuit Breakers, Motor Controllers, and Fuses. Switches, circuit breakers, motor controllers, and fuses shall comply with Section 503-4.

503-9. Lighting Fixtures — Class III, Divisions 1 and 2.

(a) Fixed Lighting. Lighting fixtures for fixed lighting shall provide enclosures for lamps and lampholders that are designed to minimize entrance of fibers and flyings and to prevent the escape of sparks, burning material, or hot metal. Each fixture shall be clearly marked to show the maximum wattage of the lamps that shall be permitted without exceeding an exposed surface temperature of 165°C (329°F) under normal conditions of use.

(b) Physical Damage. A fixture that may be exposed to physical damage shall be protected by a suitable guard.

(c) Pendant Fixtures. Pendant fixtures shall be suspended by stems of threaded rigid metal conduit, threaded intermediate metal conduit, threaded metal tubing of equivalent thickness, or by chains with approved fittings. For stems longer than 12 in. (305 mm), permanent and effective bracing against lateral displacement shall be provided at a level not more than 12 in. (305 mm) above the lower end of the stem, or flexibility in the form of an approved fitting or a flexible connector shall be provided not more than 12 in. (305 mm) from the point of attachment to the supporting box or fitting.

(d) Portable Lighting Equipment. Portable lighting equipment shall be equipped with handles and protected with substantial guards. Lampholders shall be of the unswitched type with no provision for receiving attachment plugs. There shall be no exposed current-carrying metal parts, and all exposed noncurrent-carrying metal parts shall be grounded. In all other respects, portable lighting equipment shall comply with (a).

503-10. Flexible Cords — Class III, Divisions 1 and 2. Flexible cords shall comply with the following:

(1) Be of a type approved for extra-hard usage;
(2) Contain, in addition to the conductors of the circuit, a grounding conductor complying with Section 400-23;
(3) Be connected to terminals or to supply conductors in an approved manner;
(4) Be supported by clamps or other suitable means in such a manner that there will be no tension on the terminal connections; and
(5) Be provided with suitable means to prevent the entrance of fibers or flyings where the cord enters boxes or fittings

503-11. Receptacles and Attachment Plugs — Class III, Divisions 1 and 2. Receptacles and attachment plugs shall be of the grounding type and shall be designed so to minimize the accumulation or the entry of fibers or flyings, and shall prevent the escape of sparks or molten particles.

Exception: In locations where, in the judgment of the authority having jurisdiction, only moderate accumulations of lint or flyings will be likely to collect in the vicinity of a receptacle, and where such receptacle is readily accessible for routine cleaning, general-purpose grounding-type receptacles mounted so as to minimize the entry of fibers or flyings shall be permitted.

503-12. Signaling, Alarm, Remote-Control, and Local Loudspeaker Intercommunications Systems — Class III, Divisions 1 and 2. Signaling, alarm, remote-control, and local loudspeaker intercommunications systems shall comply with the requirements of Article 503 regarding wiring methods, switches, transformers, resistors, motors, lighting fixtures, and related components.

503-13. Electric Cranes, Hoists, and Similar Equipment — Class III, Divisions 1 and 2. Where installed for operation over combustible fibers or accumulations of flyings, traveling cranes and hoists for material handling, traveling cleaners for textile machinery, and similar equipment shall comply with (a) through (d).

(a) Power Supply. Power supply to contact conductors shall be isolated from all other systems and shall be equipped with an acceptable ground detector that will give an alarm and automatically de-energize the contact conductors in case of a fault to ground or will give a visual and audible alarm as long as power is supplied to the contact conductors and the ground fault remains.

(b) Contact Conductors. Contact conductors shall be located or guarded so as to be inaccessible to other than authorized persons and shall be protected against accidental contact with foreign objects.

(c) Current Collectors. Current collectors shall be arranged or guarded so as to confine normal sparking and prevent escape of sparks or hot particles. To reduce sparking, two or more separate surfaces of contact shall be provided for each contact conductor. Reliable means shall be provided to keep contact conductors and current collectors free of accumulations of lint or flyings.

(d) Control Equipment. Control equipment shall comply with Sections 503-4 and 503-5.

503-14. Storage Battery Charging Equipment — Class III, Divisions 1 and 2. Storage battery charging equipment shall be located in separate rooms built or lined with substantial noncombustible materials. The rooms shall be constructed to prevent ignitible amounts of flyings or lint and shall be well ventilated.

503-15. Live Parts — Class III, Divisions 1 and 2. Live parts shall not be exposed.

Exception: As provided in Section 503-13.

503-16. Grounding — Class III, Divisions 1 and 2. Wiring and equipment in Class III, Divisions 1 and 2 locations shall be grounded as specified in Article 250 and with the following additional requirements.

(a) Bonding. The locknut-bushing and double-locknut types of contacts shall not be depended on for bonding purposes, but bonding jumpers with proper fittings or other approved means of bonding shall be used. Such means of bonding shall apply to all intervening raceways, fittings, boxes, enclosures, etc., between Class III locations and the point of grounding for service equipment or point of grounding of a separately derived system.

Exception: The specific bonding means shall only be required to the nearest point where the grounded circuit conductor and the grounding electrode conductor are connected together on the line side of the building or structure disconnecting means as specified in Sections 250-32(a), (b), and (c), if the branch-circuit overcurrent protection is located on the load side of the disconnecting means.

FPN: See Section 250-100 for additional bonding requirements in hazardous (classified) locations.

(b) Types of Equipment Grounding Conductors. Where flexible conduit is used as permitted in Section 503-3, it shall be installed with internal or external bonding jumpers in parallel with each conduit and complying with Section 250-102.

Exception: In Class III, Division 1 and 2 locations, the bonding jumper shall be permitted to be deleted where all the following conditions are met.

(a) Listed liquidtight flexible metal 6 ft (1.83 m) or less in length, with fittings listed for grounding, is used;
(b) Overcurrent protection in the circuit is limited to 10 amperes or less; and
(c) The load is not a power utilization load.

Article 504 — Intrinsically Safe Systems

504-1. Scope. This article covers the installation of intrinsically safe (I.S.) apparatus, wiring, and systems for Class I, II, and III locations.

> FPN: For further information, see *Wiring Practices for Hazardous (Classified) Locations Instrumentation Part 1: Intrinsic Safety,* ANSI/ISA RP 12.6-1995.

504-2. Definitions.

Associated Apparatus. Apparatus in which the circuits are not necessarily intrinsically safe themselves, but that affect the energy in the intrinsically safe circuits and are relied on to maintain intrinsic safety. Associated apparatus may be either of the following:

(1) Electrical apparatus that has an alternative-type protection for use in the appropriate hazardous (classified) location, or

(2) Electrical apparatus not so protected that shall not be used within a hazardous (classified) location

> FPN No. 1: Associated apparatus has identified intrinsically safe connections for intrinsically safe apparatus and also may have connections for nonintrinsically safe apparatus.

> FPN No. 2: An example of associated apparatus is an intrinsic safety barrier, which is a network designed to limit the energy (voltage and current) available to the protected circuit in the hazardous (classified) location, under specified fault conditions.

Control Drawing. A drawing or other document provided by the manufacturer of the intrinsically safe or associated apparatus that details the allowed interconnections between the intrinsically safe and associated apparatus.

Different Intrinsically Safe Circuits. Different intrinsically safe circuits are intrinsically safe circuits in which the possible interconnections have not been evaluated and approved as intrinsically safe.

Intrinsically Safe Apparatus. Apparatus in which all the circuits are intrinsically safe.

Intrinsically Safe Circuit. A circuit in which any spark or thermal effect is incapable of causing ignition of a mixture of flammable or combustible material in air under prescribed test conditions.

> FPN: Test conditions are described in *Standard for Safety, Intrinsically Safe Apparatus and Associated Apparatus for Use in Class I, II, and III, Division 1, Hazardous (Classified) Locations,* ANSI/UL 913-1997.

Intrinsically Safe System. An assembly of interconnected intrinsically safe apparatus, associated apparatus, and interconnecting cables in that those parts of the system that may be used in hazardous (classified) locations are intrinsically safe circuits.

> FPN: An intrinsically safe system may include more than one intrinsically safe circuit.

Simple Apparatus. A device that will neither generate nor store more than 1.2 volts, 0.1 ampere, 25 milliwatts, or 20 microjoules.

> FPN: Examples are switches, thermocouples, light-emitting diodes (LEDs), connectors, and resistance temperature devices (RTDs).

504-3. Application of Other Articles. Except as modified by this article, all applicable articles of this *Code* shall apply.

504-4. Equipment Approval. All intrinsically safe apparatus and associated apparatus shall be approved.

Exception: Simple apparatus, as described on the control drawing, shall not be required to be approved.

504-10. Equipment Installation.

(a) Control Drawing. Intrinsically safe apparatus, associated apparatus, and other equipment shall be installed in accordance with the control drawing(s).

Exception: A simple apparatus that does not interconnect intrinsically safe circuits.

> FPN: The control drawing identification is marked on the apparatus.

(b) Location. Intrinsically safe apparatus shall be permitted to be installed in any hazardous (classified) locations for which it has been approved. General-purpose enclosures shall be permitted for intrinsically safe apparatus.

Associated apparatus shall be permitted to be installed in any hazardous (classified) location for which it has been approved, or if protected by other means permitted by Articles 501 through 503 and 505.

504-20. Wiring Methods. Intrinsically safe apparatus and wiring shall be permitted to be installed using any of the wiring methods suitable for unclassified locations, including Chapter 7 and Chapter 8. Sealing shall be as provided in Section 504-70, and separation shall be as provided in Section 504-30.

504-30. Separation of Intrinsically Safe Conductors.

(a) From Nonintrinsically Safe Circuit Conductors.

(1) Open Wiring. Conductors and cables of intrinsically safe circuits not in raceways or cable trays shall be separated at least 1.97 in. (50 mm) and secured from conductors and cables of any nonintrinsically safe circuits.

Exception: Where either (1) all of the intrinsically safe circuit conductors are in Type MI or MC cables or (2) all of the nonintrinsically safe circuit conductors are in raceways or Type MI or MC cables where the sheathing or cladding is capable of carrying fault current to ground.

(2) In Raceways, Cable Trays, and Cables. Conductors of intrinsically safe circuits shall not be placed in any raceway, cable tray, or cable with conductors of any nonintrinsically safe circuit.

Exception No. 1: Where conductors of intrinsically safe circuits are separated from conductors of nonintrinsically safe circuits by a distance of at least 1.97 in. (50 mm) and secured, or by a grounded metal partition or an approved insulating partition.

FPN: No. 20 gauge sheet metal partitions 0.0359 in. (912 μm) or thicker are generally considered acceptable.

Exception No. 2: Where either (1) all of the intrinsically safe circuit conductors or (2) all of the nonintrinsically safe circuit conductors are in grounded metal-sheathed or metal-clad cables where the sheathing or cladding is capable of carrying fault current to ground.

FPN: Cables meeting the requirements of Articles 330 and 334 are typical of those considered acceptable.

(3) Within Enclosures.

(a) Conductors of intrinsically safe circuits shall be separated at least 1.97 in. (50 mm) from conductors of any nonintrinsically safe circuits, or as specified in Section 504-30(a)(2).

(b) All conductors shall be secured so that any conductor that might come loose from a terminal cannot come in contact with another terminal.

FPN No. 1: The use of separate wiring compartments for the intrinsically safe and nonintrinsically safe terminals is the preferred method of complying with this requirement.

FPN No. 2: Physical barriers such as grounded metal partitions or approved insulating partitions or approved restricted access wiring ducts separated from other such ducts by at least ¾ in. (19 mm) can be used to help ensure the required separation of the wiring.

(b) From Different Intrinsically Safe Circuit Conductors. Different intrinsically safe circuits shall be in separate cables or shall be separated from each other by one of the following means.

(1) The conductors of each circuit are within a grounded metal shield.
(2) The conductors of each circuit have insulation with a minimum thickness of 0.01 in. (254 μm).

Exception: Unless otherwise approved.

504-50. Grounding.

(a) Intrinsically Safe Apparatus, Associated Apparatus, and Raceways. Intrinsically safe apparatus, associated apparatus, cable shields, enclosures, and raceways, if of metal, shall be grounded.

FPN: Supplementary bonding to the grounding electrode may be needed for some associated apparatus, e.g., zener diode barriers, if specified in the control drawing. See *Wiring Practices for Hazardous (Classified) Locations Instrumentation Part 1: Intrinsic Safety,* ANSI/ISA RP 12.6-1995.

(b) Connection to Grounding Electrodes. Where connection to a grounding electrode is required, the grounding electrode shall be as specified in Sections 250-50(a), (b), (c), and (d) and shall comply with Section 250-30(c). Section 250-52 shall not be used if electrodes specified in Section 250-50 are available.

(c) Shields. Where shielded conductors or cables are used, shields shall be grounded.

Exception: Where a shield is part of an intrinsically safe circuit.

504-60. Bonding.

(a) Hazardous Locations. In hazardous (classified) locations, intrinsically safe apparatus shall be bonded in the hazardous (classified) location in accordance with Section 250-100.

(b) Nonhazardous Locations. In nonhazardous locations, where metal raceways are used for intrinsically safe system wiring in hazardous locations, associated apparatus shall be bonded in accordance with Sections 501-16(a), 502-16(a), 503-16(a), or 505-25, as applicable.

504-70. Sealing. Conduits and cables that are required to be sealed by Sections 501-5 and 502-5 shall be sealed to minimize the passage of gases, vapors, or dusts. Such seals shall not be required to be explosionproof or flameproof.

Exception: Seals shall not be required for enclosures that contain only intrinsically safe apparatus, except as required by Section 501-5(f)(3).

504-80. Identification. Labels required by this section shall be suitable for the environment where they are installed with consideration given to exposure to chemicals and sunlight.

(a) Terminals. Intrinsically safe circuits shall be identified at terminal and junction locations in a manner that will prevent unintentional interference with the circuits during testing and servicing.

(b) Wiring. Raceways, cable trays, and open wiring for intrinsically safe system wiring shall be identified with permanently affixed labels with the wording "Intrinsic Safety Wiring" or equivalent. The labels shall be located so as to be visible after installation and placed so that they may be readily traced through the entire length of the installation. Spacing between labels shall not be more than 25 ft (7.62 m).

Exception: Circuits run underground shall be permitted to be identified where they become accessible after emergence from the ground.

> FPN No. 1: Wiring methods permitted in unclassified locations may be used for intrinsically safe systems in hazardous (classified) locations. Without labels to identify the application of the wiring, enforcement authorities cannot determine that an installation is in compliance with the *Code*.

> FPN No. 2: In unclassified locations, the identification is necessary to ensure that nonintrinsically safe wire will not be inadvertently added to existing raceways at a later date.

(c) Color Coding. Color coding shall be permitted to identify intrinsically safe conductors where they are colored light blue and where no other conductors colored light blue are used. Likewise, color coding shall be permitted to identify raceways, cable trays, and junction boxes where they are colored light blue and contain only intrinsically safe wiring.

Article 505 — Class I, Zone 0, 1, and 2 Locations

505-1. Scope. This article covers the requirements for the zone classification system as an alternative to the division classification system covered in Article 500 for electrical and electronic equipment and wiring for all voltages in Class I, Zone 0, Zone 1, and Zone 2 hazardous (classified) locations where fire or explosion hazards may exist due to flammable gases, vapors, or liquids.

> FPN: For the requirements for electrical and electronic equipment and wiring for all voltages in Class I, Division 1 or Division 2; Class II, Division 1 or Division 2; and Class III, Division 1 or Division 2 hazardous (classified) locations where fire or explosion hazards may exist due to flammable gases or vapors, flammable liquids, or combustible dusts or fibers, refer to Articles 500 through 504.

505-2. Other Articles. All other applicable rules contained in this *Code* shall apply to electrical equipment and wiring installed in hazardous (classified) locations.

Exception: As modified by Article 504 and this article.

505-3. Location and General Requirements.

(a) Classification of Locations. Locations shall be classified depending on the properties of the flammable vapors, liquids, or gases that may be present and the likelihood that a flammable or combustible concentration or quantity is present. Where pyrophoric materials are the only materials used or handled, these locations shall not be classified.

Each room, section, or area shall be considered individually in determining its classification.

> FPN No. 1: See Section 505-6 for restrictions on area classification.

> FPN No. 2: Through the exercise of ingenuity in the layout of electrical installations for hazardous (classified) locations, it is frequently possible to locate much of the equipment in less hazardous or in nonhazardous locations and, thus, to reduce the amount of special equipment required.

(b) Threading. All threaded conduit referred to herein shall be threaded with an NPT standard conduit cutting die that provides ¾-in. taper per foot. Such conduit shall be made wrenchtight to prevent sparking when fault current flows through the conduit system, and to ensure the explosionproof or flameproof integrity of the conduit system where applicable.

Equipment provided with threaded entries for field wiring connections shall be installed in accordance with (1) or (2).

(1) Equipment Provided with Threaded Entries for NPT Threaded Conduit or Fittings. For equipment provided with threaded entries for NPT threaded conduit or fittings, listed conduit, conduit fittings, or cable fittings shall be used.

(2) Equipment Provided with Threaded Entries for Metric Threaded Conduit or Fittings. For equipment with metric threaded entries, such entries shall be identified as being metric, or listed adapters to permit connection to conduit or NPT-threaded fittings shall be provided with the equipment. Adapters shall be used for connection to conduit or NPT-threaded fittings. Listed cable fittings that have metric threads shall be permitted to be used.

> FPN: Threading specifications for metric threaded entries are located in *Metric Screw Threads*, ISO 965/1:1980, and *Metric Screw Threads*, ISO 965/3:1980.

505-4. Protection Techniques. The following shall be acceptable protection techniques for electrical and electronic equipment in hazardous (classified) locations.

> FPN: For additional information, see *Electrical Apparatus for Use in Class I, Zone 0, 1 Hazardous (Classified) Locations General Requirements*, ISA S12.0.01-1997; *Electrical Equipment for Use in Class I, Zone 0, 1, and 2 Hazardous (Classified) Locations*, ANSI/UL 2279, 1997; and *Electrical Apparatus for Explosive Gas Atmospheres - Part 0: General*

Requirements, IEC 79-0-1983, Amendment No. 1 (1987), and Amendment No. 2 (1991).

(a) Flameproof "d." This protection technique shall be permitted for equipment in those Class I, Zone 1 locations for which it is approved.

FPN No. 1: *Flameproof* is a type of protection of electrical equipment in which the enclosure will withstand an internal explosion of a flammable mixture that has penetrated into the interior, without suffering damage and without causing ignition, through any joints or structural openings in the enclosure, of an external explosive atmosphere consisting of one or more of the gases or vapors for which it is designed.

FPN No. 2: For further information, see *Electrical Apparatus for Use In Class I, Zone 1 and 2 Hazardous (Classified) Locations, Type of Protection — Flameproof "d,"* ISA S12.22.01-1996; *Electrical Apparatus for Explosive Gas Atmospheres, Part 1 — Construction and Verification Test of Flameproof Enclosures of Electrical Apparatus,* IEC 79-1-1990 and Amendment No. 1 (1993).

(b) Purged and Pressurized. This protection technique shall be permitted for equipment in those Class I, Zone 1 or Zone 2 locations for which it is approved.

FPN No. 1: In some cases, hazards may be reduced or hazardous (classified) locations limited or eliminated by adequate positive-pressure ventilation from a source of clean air in conjunction with effective safeguards against ventilation failure.

FPN No. 2: For further information, see *Standard for Purged and Pressurized Enclosures for Electrical Equipment,* NFPA 496-1998.

FPN No. 3: Pressurized "p" is a type of protection of electrical equipment that uses the technique of guarding against the ingress of the external atmosphere, which may be explosive, into an enclosure by maintaining a protective gas therein at a pressure above that of the external atmosphere. For further information, see *Electrical Apparatus for Explosive Gas Atmospheres - Part 2: Electrical Apparatus, Type of Protection "p,"* IEC 79-2-1983; and *Electrical Apparatus for Explosive Gas Atmospheres - Part 13: Construction and Use of Rooms or Buildings Protected by Pressurization,* IEC 79-13-1982.

(c) Intrinsic Safety. This protection technique shall be permitted for equipment in those Class I, Zone 0 or Zone 1 locations for which it is approved.

FPN No. 1: *Intrinsic safety* is designated type of protection "ia" by IEC 79-11 for use in Zone 0 locations. *Intrinsic safety* is designated type of protection "ib" by IEC 79-11 for use in Zone 1 locations.

FPN No. 2: For further information, see *Intrinsically Safe Apparatus and Associated Apparatus for Use in Class I, II, and III, Hazardous Locations,* ANSI/UL 913-1997; *Electrical Apparatus for Explosive Gas Atmospheres — Part 11: Intrinsic Safety "i,"* IEC 79-11-1991; and *Electrical Apparatus for Explosive Gas Atmospheres — Part 3: Spark-test Apparatus for Intrinsically-safe Circuits,* IEC 79-3-1990.

FPN No. 3: Intrinsically safe associated apparatus, designated by [ia] or [ib], is connected to intrinsically safe equipment ("ia" or "ib" respectively), but is located outside the hazardous (classified) location unless also protected by another type of protection (such as flameproof).

(d) Type of Protection "n." This protection technique shall be permitted for equipment in those Class I, Zone 2 locations for which it is approved. Type of protection "n" is further subdivided into nA, nC, and nR.

FPN No. 1: See Table 505-10(b) 1 for the descriptions of subdivisions for type of protection "n."

FPN No. 2: *Type "n" protection* is a type of protection applied to electrical equipment such that, in normal operation, the electrical equipment is not capable of igniting a surrounding explosive gas atmosphere and a fault capable of causing ignition is not likely to occur.

FPN No. 3: For further information, see *Electrical Apparatus for Explosive Gas Atmospheres, Part 15 — Electrical Apparatus with Type of Protection "n,"* IEC 79-15-1987.

(e) Oil Immersion "o." This protection technique shall be permitted for equipment in those Class I, Zone 1 locations for which it is approved.

FPN No. 1: *Oil immersion* is a type of protection in which the electrical equipment or parts of the electrical equipment are immersed in a protective liquid in such a way that an explosive atmosphere that may be above the liquid or outside the enclosure cannot be ignited.

FPN No. 2: For further information, see *Electrical Apparatus for Use in Class I, Zone 1 Hazardous (Classified) Locations, Type of Protection — Oil-Immersion "o,"* ISA S12.26.01 — 1996; and *Electrical Apparatus for Explosive Gas Atmospheres, Part 6 — Oil-Immersion "o,"* IEC 79-6-1995.

(f) Increased Safety "e." This protection technique shall be permitted for equipment in those Class I, Zone 1 locations for which it is approved.

FPN No. 1: *Increased safety* is a type of protection applied to electrical equipment that does not produce arcs or sparks in normal service and under specified abnormal conditions, in which additional measures are applied so as to give increased security against the possibility of excessive temperatures and of the occurrence of arcs and sparks.

FPN No. 2: For further information, see *Electrical Apparatus for Use in Class I, Zone 1 Hazardous (Classified) Locations, Type of Protection — Increased Safety "e,"* ISA S12.16.01-1996; and *Electrical Apparatus for Explosive Gas Atmospheres — Part 7: Increased Safety "e,"* IEC 79-7-1990, Amendment No. 1 (1991), and Amendment No. 2 (1993).

(g) Encapsulation "m." This protection technique shall be permitted for equipment in those Class I, Zone 1 locations for which it is approved.

FPN No. 1: *Encapsulation* is a type of protection in which the parts that could ignite an explosive atmosphere by either sparking or heating are enclosed in a compound in such a way that this explosive atmosphere cannot be ignited.

FPN No. 2: For further information, see *Electrical Apparatus for Use in Class I, Zone 1 Hazardous (Classified) Locations Type of Protection — Encapsulation "m,"* ISA S12.23.01-1996, and *Electrical Apparatus for Explosive Gas Atmospheres — Part 18: Encapsulation "m,"* IEC 79-18-1992.

(h) Powder Filling "q." This protection technique shall be permitted for equipment in those Class I, Zone 1 locations for which it is approved.

FPN No. 1: *Powder filling* is a type of protection in which the parts capable of igniting an explosive atmosphere are fixed in position and completely surrounded by filling material (glass or quartz powder) to prevent the ignition of an external explosive atmosphere.

FPN No. 2: For further information, see *Electrical Apparatus for Use in Class I, Zone 1 Hazardous (Classified) Locations Type of Protection — Powder Filling "q,"* ISA S12.25.01-1996, and *Electrical Apparatus for Explosive Gas Atmospheres — Part 5: Powder Filling, Type of Protection "q,"* IEC 79-5-1967.

505-5. Reference Standards.

FPN No. 1: It is important that the authority having jurisdiction be familiar with recorded industrial experience as well as with standards of the National Fire Protection Association, the American Petroleum Institute, and the International Society for Measurement and Control (ISA), that may be of use in the classification of various locations, the determination of adequate ventilation, and the protection against static electricity and lightning hazards.

FPN No. 2: For further information on the classification of locations, see *Electrical Apparatus for Explosive Gas Atmospheres, Classification of Hazardous Areas,* IEC 79-10-1995; *Classification of Locations for Electrical Installations at Petroleum Facilities Classified as Class I, Zone 0, Zone 1, or Zone 2,* API RP 505-1996; *Electrical Apparatus for Explosive Gas Atmospheres, Classifications of Hazardous (Classified) Locations,* ISA S12.24.01-1997; and *Model Code of Safe Practice in the Petroleum Industry, Part 15: Area Classification Code for Petroleum Installations,* IP 15, The Institute of Petroleum, London.

FPN No. 3: For further information on protection against static electricity and lightning hazards in hazardous (classified) locations, see *Recommended Practice on Static Electricity,* NFPA 77-1993; *Standard for the Installation of Lightning Protection Systems,* NFPA 780-1997; and *Protection Against Ignitions Arising Out of Static Lightning and Stray Currents,* API RP 2003-1991.

FPN No. 4: For further information on ventilation, see *Flammable and Combustible Liquids Code,* NFPA 30-1996, and *Recommended Practice for Classification of Locations for Electrical Installations at Petroleum Facilities Classified as Class I, Division 1 or Division 2,* API RP 500-1997, Section 6.3.

FPN No. 5: For further information on electrical systems for hazardous (classified) locations on offshore oil and gas producing platforms, see *Design and Installation of Electrical Systems for Offshore Production Platforms,* ANSI/API RP 14F-1991.

FPN No. 6: For further information on the installation of electrical equipment in hazardous (classified) locations in general, see *Electrical Apparatus for Explosive Gas Atmospheres — Part 14: Electrical Installations in Explosive Gas Atmospheres (Other than Mines),* IEC 79-14-1996, and *Electrical Apparatus for Explosive Gas Atmospheres — Part 16: Artificial Ventilation for the Protection of Analyzer(s) Houses,* IEC 79-16-1990.

505-6. Special Precaution. Article 505 requires equipment construction and installation that will ensure safe performance under conditions of proper use and maintenance.

FPN No. 1: It is important that inspection authorities and users exercise more than ordinary care with regard to the installation and maintenance of electrical equipment in hazardous (classified) locations.

FPN No. 2: Low ambient conditions require special consideration. Electrical equipment depending on the protection techniques described by Section 505-4(a) may not be suitable for use at temperatures lower than $-20°C$ ($-13°F$) unless they are approved for use at lower temperatures. However, at low ambient temperatures, flammable concentrations of vapors may not exist in a location classified Class I, Zones 0, 1, or 2 at normal ambient temperature.

(a) Supervision of Work. Classification of areas and selection of equipment and wiring methods shall be under the supervision of a qualified Registered Professional Engineer.

(b) Dual Classification. In instances of areas within the same facility classified separately, Class I, Zone 2 locations shall be permitted to abut, but not overlap, Class I, Division 2 locations. Class I, Zone 0 or Zone 1 locations shall not abut Class I, Division 1 or Division 2 locations.

(c) Reclassification Permitted. A Class I, Division 1 or Division 2 location shall be permitted to be reclassified as a Class I, Zone 0, Zone 1, or Zone 2 location provided all of the space that is classified because of a single flammable gas or vapor source is reclassified under the requirements of this article.

505-7. Grouping and Classification. For purposes of testing, approval, and area classification, various air mixtures (not oxygen enriched) shall be grouped as required in (a), (b), and (c).

FPN: Group I is intended for use in describing atmospheres that contain firedamp (a mixture of gases, composed mostly of methane, found underground, usually in mines). This *Code* does not apply to installations underground in mines. See Section 90-2(b).

Group II shall be subdivided into IIC, IIB, and IIA, as noted in (a), (b), and (c), according to the nature of the gas or vapor, for protection techniques "d," "ia," "ib," "[ia]," and "[ib]," and, where applicable, "n" and "o."

FPN No. 1: The gas and vapor subdivision as described above is based on the maximum experimental safe gap (MESG), minimum igniting current (MIC), or both. Test equipment for determining the MESG is described in *Construction and Verification Tests of Flameproof Enclosures of Electrical Apparatus*, IEC 79-1A-1975, Amendment No. 1 (1993) and UL *Technical Report No. 58* (1993). The test equipment for determining MIC is described in *Spark-Test Apparatus for Intrinsically-Safe Circuits*, IEC 79-3-1990. The classification of gases or vapors according to their maximum experimental safe gaps and minimum igniting currents is described in *Classification of Mixtures of Gases or Vapours with Air According to Their Maximum Experimental Safe Gaps and Minimum Igniting Currents*, IEC 79-12-1978.

FPN No. 2: Verification of electrical equipment utilizing protection techniques "e," "m," "p," and "q," due to design technique, does not require tests involving MESG or MIC. Therefore, Group II is not required to be subdivided for these protection techniques.

FPN No. 3: It is necessary that the meanings of the different equipment markings and Group II classifications be carefully observed to avoid confusion with Class I, Divisions 1 and 2, Groups A, B, C, and D.

ˣ**(a) Group IIC.** Atmospheres containing acetylene, hydrogen, or flammable gas, flammable liquid-produced vapor, or combustible liquid-produced vapor mixed with air that may burn or explode, having either a maximum experimental safe gap (MESG) value less than or equal to 0.50 mm or minimum igniting current ratio (MIC ratio) less than or equal to 0.45.

FPN: Group IIC is equivalent to a combination of Class I, Group A, and Class I, Group B, as described in Sections 500-5(a)(1) and (a)(2).

ˣ**(b) Group IIB.** Atmospheres containing acetaldehyde, ethylene, or flammable gas, flammable liquid-produced vapor, or combustible liquid-produced vapor mixed with air that may burn or explode, having either maximum experimental safe gap (MESG) values greater than 0.50 mm and less than or equal to 0.90 mm or minimum igniting current ratio (MIC ratio) greater than 0.45 and less than or equal to 0.80.

FPN: Group IIB is equivalent to Class I, Group C, as described in Section 500-5(a)(3).

ˣ**(c) Group IIA.** Atmospheres containing acetone, ammonia, ethyl alcohol, gasoline, methane, propane, or flammable gas, flammable liquid-produced vapor, or combustible liquid-produced vapor mixed with air that may burn or explode, having either a maximum experiment safe gap (MESG) value greater than 0.90 mm or minimum igniting current ratio (MIC ratio) greater than 0.80.

FPN: Group IIA is equivalent to Class I, Group D as described in Section 500-5(a)(4).

(d) Other. Equipment shall be permitted to be listed for a specific gas or vapor, specific mixtures of gases or vapors, or any specific combination of gases or vapors.

FPN: One common example is equipment marked for "IIB + H_2."

505-8. Class I Temperature. The temperature marking specified in Section 505-10(b) shall not exceed the ignition temperature of the specific gas or vapor to be encountered.

FPN: For information regarding ignition temperatures of gases and vapors, see *Recommended Practice for the Classification of Flammable Liquids, Gases, or Vapors and of Hazardous (Classified) Locations for Electrical Installations in Chemical Process Areas*, NFPA 497-1997; *Guide to Fire Hazard Properties of Flammable Liquids, Gases, and Volatile Solids*, NFPA 325-1994; and *Electrical Apparatus for Explosive Gas Atmospheres, Data for Flammable Gases and Vapours, Relating to the Use of Electrical Apparatus*, IEC 79-20-1996.

505-9. Zone Classification. The classification into zones shall be in accordance with the following.

(a) Class I, Zone 0. A Class I, Zone 0 location is a location

(1) In which ignitible concentrations of flammable gases or vapors are present continuously, or
(2) In which ignitible concentrations of flammable gases or vapors are present for long periods of time.

FPN No. 1: As a guide in determining when flammable gases or vapors are present continuously or for long periods of time, refer to *Recommended Practice for Classification of Locations for Electrical Installations of Petroleum Facilities Classified as Class I, Zone 0, Zone 1 or Zone 2*, API RP 505-1996; *Electrical Apparatus for Explosive Gas Atmospheres, Classifications of Hazardous Areas*, IEC 79-10-1995; and *Area Classification Code for Petroleum Installations, Model Code, Part 15*, Institute of Petroleum; and *Electrical Apparatus for Explosive Gas Atmospheres, Classifications of Hazardous (Classified) Locations*, ISA S12.24.01-1997.

FPN No. 2: This classification includes locations inside vented tanks or vessels that contain volatile flammable liquids; inside inadequately vented spraying or coating enclosures, where volatile flammable solvents are used; between the inner and outer roof sections of a floating roof tank containing volatile flammable liquids; inside open vessels, tanks and pits containing volatile flammable liquids; the interior of an exhaust duct that is used to vent ignitible concentrations of gases or vapors; and inside inadequately ventilated enclosures that contain normally venting instruments utilizing or analyzing flammable fluids and venting to the inside of the enclosures.

FPN No. 3: It is not good practice to install electrical equipment in Zone 0 locations except when the equipment is essential to the process or when other locations are not feasible. [See Section 505-3(a) FPN No. 2.] If it is necessary to install electrical systems in a Zone 0 location, it is good practice to install intrinsically safe systems as described by Article 504.

(b) Class I, Zone 1. A Class I, Zone 1 location is a location

(1) In which ignitible concentrations of flammable gases or vapors are likely to exist under normal operating conditions; or

(2) In which ignitible concentrations of flammable gases or vapors may exist frequently because of repair or maintenance operations or because of leakage; or

(3) In which equipment is operated or processes are carried on, of such a nature that equipment breakdown or faulty operations could result in the release of ignitible concentrations of flammable gases or vapors and also cause simultaneous failure of electrical equipment in a mode to cause the electrical equipment to become a source of ignition; or

(4) That is adjacent to a Class I, Zone 0 location from which ignitible concentrations of vapors could be communicated, unless communication is prevented by adequate positive pressure ventilation from a source of clean air and effective safeguards against ventilation failure are provided.

FPN No. 1: Normal operations is considered the situation when plant equipment is operating within its design parameters. Minor releases of flammable material may be part of normal operations. Minor releases include the releases from mechanical packings on pumps. Failures that involve repair or shutdown (such as the breakdown of pump seats and flange gaskets, and spillage caused by accidents) are not considered normal operation.

FPN No. 2: This classification usually includes locations where volatile flammable liquids or liquefied flammable gases are transferred from one container to another. In areas in the vicinity of spraying and painting operations where flammable solvents are used; adequately ventilated drying rooms or compartments for evaporation of flammable solvents; adequately ventilated locations containing fat and oil extraction equipment using volatile flammable solvents; portions of cleaning and dyeing plants where volatile flammable liquids are used; adequately ventilated gas generator rooms and other portions of gas manufacturing plants where flammable gas may escape; inadequately ventilated pump rooms for flammable gas or for volatile flammable liquids; the interiors of refrigerators and freezers in which volatile flammable materials are stored in the open, lightly stoppered, or easily ruptured containers; and other locations where ignitible concentrations of flammable vapors or gases are likely to occur in the course of normal operation, but not classified Zone 0.

(c) Class I, Zone 2. A Class I, Zone 2 location is a location

(1) In which ignitible concentrations of flammable gases or vapors are not likely to occur in normal operation and if they do occur will exist only for a short period; or

(2) In which volatile flammable liquids, flammable gases, or flammable vapors are handled, processed, or used, but in which the liquids, gases, or vapors normally are confined within closed containers of closed systems from which they can escape, only as a result of accidental rupture or breakdown of the containers or system, or as a result of the abnormal operation of the equipment with which the liquids or gases are handled, processed, or used; or

(3) In which ignitible concentrations of flammable gases or vapors normally are prevented by positive mechanical ventilation, but which may become hazardous as a result of failure or abnormal operation of the ventilation equipment; or

(4) That is adjacent to a Class I, Zone 1 location, from which ignitible concentrations of flammable gases or vapors could be communicated, unless such communication is prevented by adequate positive-pressure ventilation from a source of clean air, and effective safeguards against ventilation failure are provided.

FPN: The Zone 2 classification usually includes locations where volatile flammable liquids or flammable gases or vapors are used, but which would become hazardous only in case of an accident or of some unusual operating condition.

505-10. Listing, Marking, and Documentation.

(a) Listing. Equipment that is listed for a Zone 0 location shall be permitted in a Zone 1 or Zone 2 location of the same gas or vapor. Equipment that is listed for a Zone 1 location shall be permitted in a Zone 2 location of the same gas or vapor.

(b) Marking. Equipment shall be marked in accordance with (1) or (2).

   (1) Division Equipment. Equipment approved for Class I, Division 1 or Class I, Division 2 shall, in addition to being marked in accordance with Section 500-5(d), be permitted to be marked with the following:

(a) Class I, Zone 1 or Class I, Zone 2 (as applicable), and
(b) Applicable gas classification group(s) in accordance with Table 505-10(b)(2), and
(c) Temperature classification in accordance with Section 505-10(b)(3)

Figure 505-10(b)(1).

Table 505-10(b)(1). Types of Protection Designation

Designation	Technique	Zone*
d	Flameproof enclosure	1
e	Increased safety	1
ia	Intrinsic safety	0
ib	Intrinsic safety	1
[ia]	Intrinsically safe associated apparatus	Nonhazardous
[ib]	Intrinsically safe associated apparatus	Nonhazardous
m	Encapsulation	1
nA	Nonsparking equipment	2
nC	Sparking equipment in which the contacts are suitably protected other than by restricted breathing enclosure	2
nR	Restricted breathing enclosure	2
o	Oil immersion	1
p	Purged and pressurized	1 or 2
q	Powder filled	1

*Does not address use where a combination of techniques is used.

(2) Zone Equipment. Equipment meeting one or more of the protection techniques described in Section 505-4 shall be marked with the following in the order shown:

(a) Class
(b) Zone
(c) Symbol "AEx"
(d) Protection technique(s) in accordance with Table 505-10(b)(2)
(e) Applicable gas classification group(s) in accordance with Table 505-10(b)(2)
(f) Temperature classification in accordance with Section 505-10(b)(3)

Exception: Intrinsically safe associated apparatus shall be required to be marked only with (c), (d), and (e).

FPN No. 1: An example of such a required marking is "Class I, Zone 0, AEx ia IIC T6."

Electrical equipment of types of protection "e," "m," "p," or "q," shall be marked Group II. Electrical equipment of types of protection "d," "ia," "ib," "[ia]," or "[ib]" shall be marked Group IIA, or IIB, or IIC, or for a specific gas or vapor. Electrical equipment of types of protection "n" shall be marked Group II unless it contains enclosed-break devices, nonincendive components, or energy-limited equipment or circuits, in which case it shall be marked Group IIA, IIB, or IIC, or a specific gas or vapor. Electrical equipment of other types of protection shall be marked Group II unless the type of protection utilized by the equipment requires

that it shall be marked Group IIA, IIB, or IIC, or a specific gas or vapor.

FPN No. 2: An explanation of the marking that is required follows.

Table 505-10(b)(2). Gas Classification Groups

Gas Group	Comment
IIC	See Section 505-7(a)
7IIB	See Section 505-7(b)
IIA	See Section 505-7(c)

(3) Temperature Classifications. Approved equipment shall be marked to show the operating temperature or temperature range referenced to a 40°C (104°F) ambient. The temperature range, if provided, shall be indicated in identification numbers, as shown in Table 505-10(b)(3).

Table 505-10(b)(3). Classification of Maximum Surface Temperature for Group II Electrical Equipment

Temperature Class	Maximum Surface Temperature (°C)
T1	≤450
T2	≤300
T3	≤200
T4	≤135
T5	≤100
T6	≤85

Electrical equipment designed for use in the ambient temperature range between $-20°C$ and $+40°C$ shall require no additional temperature marking.

Electrical equipment that is designed for use in a range of ambient temperatures other than $-20°C$ and $+40°C$ is considered to be special; and the ambient temperature range shall then be marked on the equipment, including either the symbol "Ta" or "Tamb" together with the special range of ambient temperatures. As an example, such a marking might be "$-30°C$ Ta $+40°C$."

Electrical equipment suitable for ambient temperatures exceeding 40°C (104°F) shall be marked with both the maximum ambient temperature and the operating temperature or temperature range at that ambient temperature.

Exception No. 1: Equipment of the nonheat-producing type, such as conduit fittings, and equipment of the heat-producing type having a maximum temperature of not more than 100°C (212°F) shall not be required to have a marked operating temperature or temperature range.

Exception No. 2: Equipment approved for Class I, Division 1 or Division 2 locations as permitted by Sections 505-20(b) and (c) shall be permitted to be marked in accordance with Section 500-5(d) and Table 500-5(d).

(c) Documentation for Industrial Occupancies. All areas in industrial occupancies designated as hazardous (classified) locations shall be properly documented. This documentation shall be available to those authorized to design, install, inspect, maintain, or operate electrical equipment at the location.

> FPN: For examples of area classification drawings, see *Electrical Apparatus for Explosive Gas Atmospheres, Classification of Hazardous Areas,* IEC 79-10-1995; *Classification of Locations for Electrical Installations at Petroleum Facilities Classified as Class I, Zone 0, Zone 1, or Zone 2,* API RP 505-1997; *Electrical Apparatus for Explosive Gas Atmospheres, Classifications of Hazardous (Classified) Locations,* ISA S12.24.01-1997; *Recommended Practice for Classification of Locations for Electrical Installations at Petroleum Facilities Classified as Class I, Division I or Division 2,* API RP 500-1997, *Model Code of Safe Practice in the Petroleum Industry, Part 15: Area Classification Code for Petroleum Installations,* IP 15, The Institute of Petroleum, London.

505-15. Wiring Methods.

(a) Zone 0. In Class I, Zone 0 locations, only the following wiring methods shall be permitted.

(1) Intrinsically safe wiring in accordance with Article 504.

> FPN: Article 504 only includes protection technique "ia."

(2) Seals shall be provided within 10 ft (3.05 m) of where a conduit leaves a Zone 0 location. There shall be no unions, couplings, boxes, or fittings, except reducers at the seal, in the conduit run between the seal and the point at which the conduit leaves the location.

Exception: A rigid unbroken conduit that passes completely through the Zone 0 location with no fittings less than 12 in. (305 mm) beyond each boundary, shall not be required to be sealed, if the termination points of the unbroken conduit are in unclassified locations.

(3) Seals shall be provided on cables at the first point of termination after entry into the Zone 0 location.

(4) Seals shall not be required to be explosionproof or flameproof.

(b) Zone 1. In Class I, Zone 1 locations, all wiring methods permitted for Class I, Division 1 locations shall be permitted.

Where Class I, Division 1 wiring methods are used, sealing and drainage shall be provided in accordance with Sections 501-5(a), (c), (d), and (f), except where the term "Division 1" is used "Zone 1" shall be substituted.

An explosionproof seal, constructed in accordance with Section 501-5(c), shall be provided for each conduit entering an enclosure having type of protection "e" or "d," except where the type of protection "d" enclosure is marked to indicate that a seal is not required.

Wiring methods shall maintain the integrity of protection techniques.

> FPN No. 1: For example, equipment with type of protection "e" requires that conduit seals or cable fittings incorporate suitable methods to maintain the "ingress protection" (minimum IP54) of the enclosure; and, for conduit, serve to maintain the explosionproof integrity of the conduit system.

> FPN No. 2: Different electrical enclosures provide different degrees of "ingress protection." The measures applied to enclosures of electrical apparatus include

> (1) The protection of persons against contact with or approach to live parts and against contact with moving parts (other than smooth rotating shafts and the like) inside the enclosure;
> (2) The protection of the apparatus inside the enclosure against ingress of solid foreign bodies; and
> (3) The protection of the apparatus inside the enclosure against harmful ingress of water.

(c) Zone 2. In Class I, Zone 2 locations, all wiring methods permitted for Class I, Division 2 locations shall be permitted. Sealing and drainage shall be provided in accordance with Sections 501-5(b), (c), (e), and (f), except where the term "Division 2" is used, "Zone 2" shall be substituted and where the term "Division 1" is used, "Zone 1" shall be substituted.

Wiring methods shall maintain the integrity of protection techniques.

(d) Solid Obstacles. Flameproof equipment with flanged joints shall not be installed such that the flange openings are closer than the distances shown in Table 505-15 to any solid obstacle that is not a part of the equipment (such as steelworks, walls, weather guards, mounting brackets, pipes, or other electrical equipment) unless the equipment is listed for a smaller distance of separation.

Table 505-15. Minimum Distance of Obstructions from Flameproof "d" Flange Openings

Gas Group	Minimum Distance	
	in.	mm
IIC	$1^{37}/_{64}$	40
IIB	$1^{3}/_{16}$	30
IIA	$^{25}/_{64}$	10

505-20. Equipment.

(a) Zone 0. In Class I, Zone 0 locations, only equipment specifically listed and marked as suitable for the location shall be permitted.

Exception: Intrinsically safe equipment listed for use in Class I, Division 1 locations for the same gas, or as permitted by Section 505-7(d), and with a suitable temperature rating shall be permitted.

(b) Zone 1. In Class I, Zone 1 locations, only equipment specifically listed and marked as suitable for the location shall be permitted.

Exception: Equipment approved for use in Class I, Division 1 or listed for use in Class I, Zone 0 locations for the same gas, or as permitted by Section 505-7(d) and with a suitable temperature rating shall be permitted.

(c) Zone 2. In Class I, Zone 2 locations, only equipment specifically listed and marked as suitable for the location shall be permitted.

Exception No. 1: Equipment listed for use in Class I, Zone 0 or Zone 1 locations for the same gas, or as permitted by Section 505-7(d), and with a suitable temperature rating, shall be permitted.

Exception No. 2: Equipment approved for use in Class I, Division 1 or Division 2 locations for the same gas, or as permitted by Section 505-7(d), and with a suitable temperature rating shall be permitted.

Exception No. 3: In Class I, Zone 2 locations, the installation of open or nonexplosionproof or nonflameproof enclosed motors, such as squirrel-cage induction motors without brushes, switching mechanisms, or similar arc-producing devices that are not identified for use in a Class I, Zone 2 location shall be permitted.

FPN No. 1: It is important to consider the temperature of internal and external surfaces that may be exposed to the flammable atmosphere.

FPN No. 2: It is important to consider the risk of ignition due to currents arcing across discontinuities and overheating of parts in multisection enclosures of large motors and generators. Such motors and generators may need equipotential bonding jumpers across joints in the enclosure and from enclosure to ground. Where the presence of ignitible gases or vapors is suspected, clean air purging may be needed immediately prior to and during start-up periods.

(d) Manufacturer's Instructions. Electrical equipment installed in hazardous (classified) locations shall be installed in accordance with the instructions (if any) provided by the manufacturer.

505-21. Increased Safety "e" Motors and Generators. In Class I, Zone 1 locations, Increased Safety "e" motors and generators of all voltage ratings shall be listed for Class I, Zone 1 locations, and shall comply with the following.

(1) Motors shall be marked with the current ratio, I_A/I_N, and time, t_E;
(2) Motors shall have controllers marked with the model or identification number, output rating (horsepower or kilowatt), full-load amperes, starting current ratio (I_A/I_N), and time (t_E) of the motors that they are intended to protect; the controller marking shall also include the specific overload protection type (and setting, if applicable) that is listed with the motor or generator;
(3) Connections shall be made with the specific terminals listed with the motor or generator;
(4) Terminal housings shall be permitted to be of substantial, nonmetallic, nonburning material provided an internal grounding means between the motor frame and the equipment grounding connection is incorporated within the housing;
(5) The provisions of Part C of Article 430 shall apply regardless of the voltage rating of the motor;
(6) The motors shall be protected against overload by a separate overload device that is responsive to motor current. This device shall be selected to trip or shall be rated in accordance with the listing of the motor and its overload protection;
(7) Sections 430-34 and 430-44 shall not apply to such motors; and
(8) The motor overload protection shall not be shunted or cut out during the starting period.

505-25. Grounding and Bonding. Grounding and bonding shall comply with Article 250 and Section 501-16.

Article 510 — Hazardous (Classified) Locations — Specific

510-1. Scope. Articles 511 through 517 cover occupancies or parts of occupancies that are or may be hazardous because of atmospheric concentrations of flammable liquids, gases, or vapors, or because of deposits or accumulations of materials that may be readily ignitible.

510-2. General. The general rules of this *Code* and the provisions of Articles 500 through 504, shall apply to electric wiring and equipment in occupancies within the scope of Articles 511 through 517, except as such rules are modified in Articles 511 through 517. Where unusual conditions exist in a specific occupancy, the authority having jurisdiction shall judge with respect to the application of specific rules.

Article 511 — Commercial Garages, Repair and Storage

511-1. Scope. These occupancies shall include locations used for service and repair operations in connection with self-propelled vehicles (including, but not limited to, passenger automobiles, buses, trucks, and tractors) in which volatile flammable liquids are used for fuel or power.

511-2. Locations. Areas in which flammable fuel is transferred to vehicle fuel tanks shall conform to Article 514. Parking garages used for parking or storage and where no repair work is done except exchange of parts and routine maintenance requiring no use of electrical equipment, open flame, welding, or the use of volatile flammable liquids are not classified.

> FPN: For further information, see *Standard for Parking Structures,* NFPA 88A-1995, and *Standard for Repair Garages,* NFPA 88B-1997.

511-3. Class I Locations. Classification under Article 500.

(a) Up to a Level of 18 in. (457 mm) Above the Floor. For each floor, the entire area up to a level of 18 in. (457 mm) above the floor shall be considered to be a Class I, Division 2 location.

Exception: Where the enforcing agency determines that there is mechanical ventilation providing a minimum of four air changes per hour.

(b) Any Pit or Depression Below Floor Level. Any pit or depression below floor level shall be considered to be a Class I, Division 1 location, which shall extend up to said floor level, except that any pit or depression in which six air changes per hour are exhausted at the floor level of the pit shall be permitted to be judged by the enforcing agency to be a Class I, Division 2 location.

Exception: Lubrication and service rooms without dispensing shall be classified in accordance with Table 514-2.

(c) Areas Adjacent to Defined Locations or with Positive-Pressure Ventilation. Areas adjacent to defined locations in which flammable vapors are not likely to be released, such as stock rooms, switchboard rooms, and other similar locations, shall not be classified where mechanically ventilated at a rate of four or more air changes per hour or where effectively cut off by walls or partitions.

(d) Adjacent Areas by Special Permission. Adjacent areas that by reason of ventilation, air pressure differentials, or physical spacing are such that, in the opinion of the authority enforcing this *Code,* no ignition hazard exists, shall be unclassified.

(e) Fuel-Dispensing Units. Where fuel-dispensing units (other than liquid petroleum gas, which is prohibited) are located within buildings, the requirements of Article 514 shall govern.

Where mechanical ventilation is provided in the dispensing area, the controls shall be interlocked so that the dispenser cannot operate without ventilation as prescribed in Section 500-7(b).

(f) Portable Lighting Equipment. Portable lighting equipment shall be equipped with handle, lampholder, hook, and substantial guard attached to the lampholder or handle. All exterior surfaces that might come in contact with battery terminals, wiring terminals, or other objects shall be of nonconducting material or shall be effectively protected with insulation. Lampholders shall be of an unswitched type and shall not provide means for plug-in of attachment plugs. The outer shell shall be of molded composition or other suitable material. Unless the lamp and its cord are supported or arranged in such a manner that they cannot be used in the locations classified in Section 511-3, they shall be of a type approved for Class I, Division 1 locations.

511-4. Wiring and Equipment in Class I Locations. Within Class I locations as defined in Section 511-3, wiring and equipment shall conform to applicable provisions of Article 501. Raceways embedded in a masonry wall or buried beneath a floor shall be considered to be within the Class I location above the floor if any connections or extensions lead into or through such areas.

Exception: Rigid nonmetallic conduit that complies with Article 347 shall be permitted where buried under not less than 2 ft (610 mm) of cover. Where rigid nonmetallic conduit is used, threaded rigid metal conduit or threaded steel intermediate metal conduit shall be used for the last 2 ft (610 mm) of the underground run to emergence or to the point of connection to the aboveground raceway and an equipment grounding conductor shall be included to provide electrical continuity of the raceway system and for grounding of noncurrent-carrying metal parts.

511-5. Sealing. Approved seals conforming to the requirements of Section 501-5 shall be provided, and Section 501-5(b)(2) shall apply to horizontal as well as vertical boundaries of the defined Class I locations.

511-6. Wiring in Spaces Above Class I Locations.

(a) Fixed Wiring Above Class I Locations. All fixed wiring above Class I locations shall be in metal raceways, rigid nonmetallic conduit, electrical nonmetallic tubing, flexible metal conduit, liquidtight flexible metal conduit, or liquidtight flexible nonmetallic conduit or shall be Type MC, MI, manufactured wiring systems, or PLTC cable in accordance with Article 725, or Type TC cable. Cellular metal floor raceways or cellular concrete floor raceways shall be permitted to be used only for supplying ceiling outlets or extensions to the area below the floor, but such raceways shall have no connections leading into or through any Class I location above the floor.

(b) Pendants. For pendants, flexible cord suitable for the type of service and approved for hard usage shall be used.

(c) Grounded and Grounding Conductors. Where a circuit supplies portables or pendants and includes a grounded conductor as provided in Article 200, receptacles, attachment plugs, connectors, and similar devices shall be of the grounding type, and the grounded conductor of the flexible cord shall be connected to the screw shell of any lampholder or to the grounded terminal of any utilization equipment supplied. Approved means shall be provided for maintaining continuity of the grounding conductor between the fixed wiring system and the noncurrent-carrying metal portions of pendant fixtures, portable lamps, and portable utilization equipment.

(d) Attachment Plug Receptacles. Attachment plug receptacles in a fixed position shall be located above the level of any defined Class I location or shall be approved for the location.

511-7. Equipment Above Class I Locations.

(a) Arcing Equipment. Equipment that is less than 12 ft (3.66 m) above the floor level and that may produce arcs, sparks, or particles of hot metal, such as cutouts, switches, charging panels, generators, motors, or other equipment (excluding receptacles, lamps, and lampholders) having make-and-break or sliding contacts, shall be of the totally enclosed type or constructed so as to prevent the escape of sparks or hot metal particles.

(b) Fixed Lighting. Lamps and lampholders for fixed lighting that is located over lanes through which vehicles are commonly driven or that may otherwise be exposed to physical damage shall be located not less than 12 ft (3.66 m) above floor level, unless of the totally enclosed type or constructed so as to prevent escape of sparks or hot metal particles.

511-8. Battery Charging Equipment.
Battery chargers and their control equipment, and batteries being charged, shall not be located within locations classified in Section 511-3.

511-9. Electric Vehicle Charging.

(a) General. All electrical equipment and wiring shall be installed in accordance with Article 625, except as noted in Section 511-9(b) and (c). Flexible cords shall be of a type approved for extra-hard usage.

(b) Connector Location. No connector shall be located within a Class I location as defined in Section 511-3.

(c) Plug Connections to Vehicles. Where plugs are provided for direct connection to vehicles, the point of connection shall not be within a Class I location as defined in Section 511-3, and, where the cord is suspended from overhead, it shall be arranged so that the lowest point of sag is at least 6 in. (152 mm) above the floor. Where an automatic arrangement is provided to pull both cord and plug beyond the range of physical damage, no additional connector shall be required in the cable or at the outlet.

511-10. Ground-Fault Circuit-Interrupter Protection for Personnel.
All 125-volt, single-phase, 15- and 20-ampere receptacles installed in areas where electrical diagnostic equipment, electrical hand tools, or portable lighting equipment are to be used shall have ground-fault circuit-interrupter protection for personnel.

511-16. Grounding.
All metal raceways, the metal armor or metallic sheath on cables, and all noncurrent-carrying metal parts of fixed or portable electrical equipment, regardless of voltage, shall be grounded as provided in Article 250. Grounding in Class I locations shall comply with Section 501-16.

Article 513 — Aircraft Hangars

513-1. Scope.
This article shall apply to buildings or structures inside any part of which aircraft containing Class I (flammable) liquids or Class II (combustible) liquids whose temperatures are above their flash points are housed or stored and in which aircraft might undergo service, repairs, or alterations. It shall not apply to locations used exclusively for aircraft that have never contained fuel or unfueled aircraft.

> FPN No. 1: For definitions of aircraft hangar and unfueled aircraft, see *Standard on Aircraft Hangars*, NFPA 409-1995.

> FPN No. 2: For further information on fuel classification, see *Flammable and Combustible Liquids Code*, NFPA 30-1996, and *Guide to Fire Hazard Properties of Flammable Liquids, Gases, and Volatile Solids*, NFPA 325-1994.

513-2. Definitions.
For the purpose of this article, the following definitions shall apply.

Mobile Equipment. Equipment with electric components suitable to be moved only with mechanical aids or which are provided with wheels for movements by person(s) or powered devices.

Portable Equipment. Equipment with electric components suitable to be moved by a single person without mechanical aids.

513-3. Classification of Locations.

(a) Below Floor Level. Any pit or depression below the level of the hangar floor shall be classified as a Class I, Division 1 location that shall extend up to said floor level.

(b) Areas Not Cut Off or Ventilated. The entire area of the hangar, including any adjacent and communicating areas not suitably cut off from the hangar, shall be classified as a Class I, Division 2 location up to a level 18 in. (457 mm) above the floor.

(c) Vicinity of Aircraft. The area within 5 ft (1.52 m) horizontally from aircraft power plants or aircraft fuel tanks shall be classified as a Class I, Division 2 location that shall extend upward from the floor to a level 5 ft (1.52 m) above the upper surface of wings and of engine enclosures.

(d) Areas Suitably Cut Off and Ventilated. Adjacent areas in which flammable liquids or vapors are not likely to be released, such as stock rooms, electrical control rooms, and other similar locations, shall not be classified where adequately ventilated and where effectively cut off from the hangar itself by walls or partitions.

513-4. Wiring and Equipment in Class I Locations.

All wiring and equipment that is or may be installed or operated within any of the Class I locations defined in Section 513-3 shall comply with the applicable provisions of Article 501. All wiring installed in or under the hangar floor shall comply with the requirements for Class I, Division 1 locations. Where such wiring is located in vaults, pits, or ducts, adequate drainage shall be provided.

Attachment plugs and receptacles in Class I locations shall be approved for Class I locations or shall be designed so that they cannot be energized while the connections are being made or broken.

513-5. Wiring Not Within Class I Locations.

(a) Fixed Wiring. All fixed wiring in a hangar, but not within a Class I location as defined in Section 513-3, shall be installed in metal raceways or shall be Type MI, TC, or MC cable.

Exception: Wiring in unclassified locations, as defined in Section 513-3(d), shall be of a type recognized in Chapter 3.

(b) Pendants. For pendants, flexible cord suitable for the type of service and approved for hard usage shall be used.

Each such cord shall include a separate equipment grounding conductor.

(c) Portable Equipment. For portable utilization equipment and lamps, flexible cord suitable for the type of service and approved for extra-hard usage shall be used. Each such cord shall include a separate equipment grounding conductor.

(d) Grounded and Grounding Conductors. Where a circuit supplies portables or pendants and includes a grounded conductor as provided in Article 200, receptacles, attachment plugs, connectors, and similar devices shall be of the grounding type, and the grounded conductor of the flexible cord shall be connected to the screw shell of any lampholder or to the grounded terminal of any utilization equipment supplied. Approved means shall be provided for maintaining continuity of the grounding conductor between the fixed wiring system and the noncurrent-carrying metal portions of pendant fixtures, portable lamps, and portable utilization equipment.

513-6. Equipment Not Within Class I Locations.

(a) Arcing Equipment. In locations other than those described in Section 513-3, equipment that is less than 10 ft (3.05 m) above wings and engine enclosures of aircraft and that may produce arcs, sparks, or particles of hot metal, such as lamps and lampholders for fixed lighting, cutouts, switches, receptacles, charging panels, generators, motors, or other equipment having make-and-break or sliding contacts, shall be of the totally enclosed type or constructed so as to prevent the escape of sparks or hot metal particles.

Exception: Equipment in areas described in Section 513-3(d) shall be permitted to be of the general-purpose type.

(b) Lampholders. Lampholders of metal-shell, fiber-lined types shall not be used for fixed incandescent lighting.

(c) Portable Lighting Equipment. Portable lighting equipment that are used within a hangar shall be approved for the location in which they are used.

(d) Portable Equipment. Portable utilization equipment that is or may be used within a hangar shall be of a type suitable for use in Class I, Division 2 locations.

513-7. Stanchions, Rostrums, and Docks.

(a) In Class I Location. Electric wiring, outlets, and equipment (including lamps) on or attached to stanchions, rostrums, or docks that are located or likely to be located in a Class I location, as defined in Section 513-3(c), shall comply with the requirements for Class I, Division 2 locations.

(b) Not in Class I Location. Where stanchions, rostrums, or docks are not located or likely to be located in a Class I location, as defined in Section 513-3(c), wiring and equipment shall comply with Sections 513-5 and 513-6, except that such wiring and equipment not more than 18 in. (457 mm) above the floor in any position shall comply with Section 513-7(a). Receptacles and attachment plugs shall be of a locking type that will not readily disconnect.

(c) Mobile Type. Mobile stanchions with electric equipment complying with (b) above shall carry at least one permanently affixed warning sign to read

WARNING — KEEP 5 FT CLEAR OF AIRCRAFT
ENGINES AND FUEL TANK AREAS.

513-8. Sealing. Approved seals shall be provided in accordance with Section 501-5. Sealing requirements specified in Sections 501-5(a)(4) and (b)(2) shall apply to horizontal as well as to vertical boundaries of the defined Class I locations. Raceways embedded in a concrete floor or buried beneath a floor shall be considered to be within the Class I location above the floor.

513-9. Aircraft Electrical Systems. Aircraft electrical systems shall be de-energized when the aircraft is stored in a hangar and, whenever possible, while the aircraft is undergoing maintenance.

513-10. Aircraft Batteries — Charging and Equipment. Aircraft batteries shall not be charged where installed in an aircraft located inside or partially inside a hangar.

Battery chargers and their control equipment shall not be located or operated within any of the Class I locations defined in Section 513-3 and shall preferably be located in a separate building or in an area such as defined in Section 513-3(d). Mobile chargers shall carry at least one permanently affixed warning sign to read "WARNING — KEEP 5 FT CLEAR OF AIRCRAFT ENGINES AND FUEL TANK AREAS." Tables, racks, trays, and wiring shall not be located within a Class I location and, in addition, shall comply with Article 480.

513-11. External Power Sources for Energizing Aircraft.

(a) Not Less than 18 in. (457 mm) Above Floor. Aircraft energizers shall be designed and mounted so that all electric equipment and fixed wiring will be at least 18 in. (457 mm) above floor level and shall not be operated in a Class I location as defined in Section 513-3(c).

(b) Marking for Mobile Units. Mobile energizers shall carry at least one permanently affixed warning sign to read

WARNING — KEEP 5 FT CLEAR OF AIRCRAFT
ENGINES AND FUEL TANK AREAS.

(c) Cords. Flexible cords for aircraft energizers and ground support equipment shall be approved for the type of service

and extra-hard usage and shall include an equipment grounding conductor.

513-12. Mobile Servicing Equipment with Electric Components.

(a) General. Mobile servicing equipment (such as vacuum cleaners, air compressors, air movers, etc.) having electric wiring and equipment not suitable for Class I, Division 2 locations shall be designed and mounted so that all such fixed wiring and equipment will be at least 18 in. (457 mm) above the floor. Such mobile equipment shall not be operated within the Class I location defined in Section 513-3(c) and shall carry at least one permanently affixed warning sign to read

WARNING — KEEP 5 FT CLEAR OF AIRCRAFT
ENGINES AND FUEL TANK AREAS.

(b) Cords and Connectors. Flexible cords for mobile equipment shall be suitable for the type of service and approved for extra-hard usage and shall include an equipment grounding conductor. Attachment plugs and receptacles shall be approved for the location in which they are installed and shall provide for connection of the equipment grounding conductor.

(c) Restricted Use. Equipment that is not identified as suitable for Class I, Division 2 locations shall not be operated in locations where maintenance operations likely to release flammable liquids or vapors are in progress.

513-16. Grounding. All metal raceways, the metal armor or metallic sheath on cables, and all noncurrent-carrying metal parts of fixed or portable electrical equipment, regardless of voltage, shall be grounded as provided in Article 250. Grounding in Class I locations shall comply with Section 501-16.

Article 514 — Gasoline Dispensing and Service Stations

514-1. Definition. A *gasoline dispensing and service station* is a location where gasoline or other volatile flammable liquids or liquefied flammable gases are transferred to the fuel tanks (including auxiliary fuel tanks) of self-propelled vehicles or approved containers.

Other areas used as lubritoriums, service rooms, repair rooms, offices, salesrooms, compressor rooms, and similar locations shall comply with Articles 510 and 511 with respect to electric wiring and equipment.

Where the authority having jurisdiction can satisfactorily determine that flammable liquids having a flash point below 38°C (100°F), such as gasoline, will not be handled, such location shall not be required to be classified.

FPN No. 1: For further information regarding safeguards for gasoline dispensing and service stations, see *Automotive and Marine Service Station Code,* NFPA 30A-1996.

FPN No. 2: For information on classified areas pertaining to LP-Gas systems other than residential or commercial, see *Standard for the Storage and Handling of Liquefied Petroleum Gases,* NFPA 58-1995, and *Standard for the Storage and Handling of Liquefied Petroleum Gases at Utility Gas Plants,* NFPA 59-1995.

FPN No. 3: See Section 555-10 for gasoline dispensing stations in marinas and boatyards.

ˣ514-2. Class I Locations. Table 514-2 shall be applied where Class I liquids are stored, handled, or dispensed and

ˣTable 514-2. Class I Locations — Service Stations

Location	Class I, Group D Division	Extent of Classified Location
Underground Tank		
Fill Opening	1	Any pit, box, or space below grade level, any part of which is within the Division 1 or 2 classified location.
	2	Up to 18 in. above grade level within a horizontal radius of 10 ft from a loose fill connection and within a horizontal radius of 5 ft from a tight fill connection.
Vent — Discharging Upward	1	Within 3 ft of open end of vent, extending in all directions.
	2	Space between 3 ft and 5 ft of open end of vent, extending in all directions.
Dispensing Device[1,4] (except overhead type)[2]		
Pits	1	Any pit, box, or space below grade level, any part of which is within the Division 1 or 2 classified location.
Dispenser		FPN: Space classification inside the dispenser enclosure is covered in *Power Operated Dispensing Devices for Petroleum Products,* ANSI/UL 87-1995.
	2	Within 18 in. horizontally in all directions extending to grade from the dispenser enclosure or that portion of the dispenser enclosure containing liquid handling components.
		FPN: Space classification inside the dispenser enclosure is covered in *Power Operated Dispensing Devices for Petroleum Products,* ANSI/UL 87-1995.
Outdoor	2	Up to 18 in. above grade level within 20 ft horizontally of any edge of enclosure.
Indoor with Mechanical Ventilation	2	Up to 18 in. above grade or floor level within 20 ft horizontally of any edge of enclosure.
with Gravity Ventilation	2	Up to 18 in. above grade or floor level within 25 ft horizontally of any edge of enclosure.
Dispensing Device[4]		
Overhead Type[2]	1	The space within the dispenser enclosure, and all electrical equipment integral with the dispensing hose or nozzle.
	2	A space extending 18 in. horizontally in all directions beyond the enclosure and extending to grade.
	2	Up to 18 in. above grade level within 20 ft horizontally measured from a point vertically below the edge of any dispenser enclosure.
Remote Pump — Outdoor	1	Any pit, box, or space below grade level if any part is within a horizontal distance of 10 ft from any edge of pump.
	2	Within 3 ft of any edge of pump, extending in all directions. Also up to 18 in. above grade level within 10 ft horizontally from any edge of pump.

(continues)

Table 514-2. Continued

Location	Class I, Group D Division	Extent of Classified Location
Remote Pump — Indoor	1	Entire space within any pit.
	2	Within 5 ft of any edge of pump, extending in all directions. Also up to 3 ft above grade level within 25 ft horizontally from any edge of pump.
Lubrication or Service Room — with Dispensing	1	Any pit within any unventilated space.
	2	Any pit with ventilation.
	2	Space up to 18 in. above floor or grade level and 3 ft horizontally from a lubrication pit.
Dispenser for Class I Liquids	2	Within 3 ft of any fill or dispensing point, extending in all directions.
Lubrication or Service Room — Without Dispensing	2	Entire area within any pit used for lubrication or similar services where Class I liquids may be released.
	2	Area up to 18 in. above any such pit, and extending a distance of 3 ft horizontally from any edge of the pit.
	2	Entire unventilated area within any pit, below grade area, or subfloor area.
	2	Area up to 18 in. above any such unventilated pit, below grade work area, or subfloor work area and extending a distance of 3 ft horizontally from the edge of any such pit, belowgrade work area, or subfloor work area.
	Nonclassified	Any pit, belowgrade work area, or subfloor work area that is provided with exhaust ventilation at a rate of not less than 1 cfm/ft^2 (0.3 m^3/minute/m^2) of floor area at all times that the building is occupied or when vehicles are parked in or over this area and where exhaust air is taken from a point within 12 in. (0.3 m) of the floor of the pit, belowgrade work area, or subfloor work area.
Special Enclosure Inside Building[3]	1	Entire enclosure.
Sales, Storage, and Rest Rooms	Nonclassified	If there is any opening to these rooms within the extent of a Division 1 location, the entire room shall be classified as Division 1.
Vapor Processing Systems Pits	1	Any pit, box, or space below grade level, any part of which is within a Division 1 or 2 classified location or that houses any equipment used to transport or process vapors.
Vapor Processing Equipment Located Within Protective Enclosures FPN: See *Automotive and Marine Service Station Code,* NFPA 30A-1996, Section 4-5.7	2	Within any protective enclosure housing vapor processing equipment.
Vapor Processing Equipment Not Within Protective Enclosures (excluding piping and combustion devices)	2	The space within 18 in. in all directions of equipment containing flammable vapor or liquid extending to grade level. Up to 18 in. above grade level within 10 ft horizontally of the vapor processing equipment.
Equipment Enclosures	1	Any space within the enclosure where vapor or liquid is present under normal operating conditions.
Vacuum-Assist Blowers	2	The space within 18 in. in all directions extending to grade level. Up to 18 in. above grade level within 10 ft horizontally.

Note: For SI units, 1 in. = 2.5 cm; 1 ft = 0.3048 m.

[1] Refer to Figure 514-2 for an illustration of classified location around dispensing devices.

[2] Ceiling mounted hose reel.

[3] FPN: See *Automotive and Marine Service Station Code,* NFPA 30A-1996, Section 2-2.

[4] FPN: Area classification inside the dispenser enclosure is covered in *Power-Operated Dispensing Devices for Petroleum Products,* ANSI/UL 87-1995.

shall be used to delineate and classify service stations. A Class I location shall not extend beyond an unpierced wall, roof, or other solid partition.

FPN: For information on area classification where liquefied petroleum gases are dispensed, see *Standard for the Storage and Handling of Liquefied Petroleum Gases*, NFPA 58-1995.

ˣFigure 514-2   Classified locations adjacent to dispensers as detailed in Table 514-2.

514-3. Wiring and Equipment Within Class I Locations. All electrical equipment and wiring within Class I locations defined in Section 514-2 shall comply with the applicable provisions of Article 501.

Exception: As permitted in Section 514-8.

FPN. For special requirements for conductor insulation, see Section 501-13.

514-4. Wiring and Equipment Above Class I Locations. Wiring and equipment above the Class I locations defined in Section 514-2 shall comply with Sections 511-6 and 511-7.

514-5. Circuit Disconnects.

(a) General. Each circuit leading to or through dispensing equipment, including equipment for remote pumping sys-

tems, shall be provided with a clearly identified and readily accessible switch or other acceptable means, located remote from the dispensing devices, to disconnect simultaneously from the source of supply, all conductors of the circuit, including the grounded conductor, if any.

Single-pole breakers utilizing handle ties shall not be permitted.

ˣ(b) Attended Self-Service Stations. Emergency controls as specified in Section 514-5(a) shall be installed at a location acceptable to the authority having jurisdiction, but controls shall not be more than 100 ft (30 m) from dispensers.

ˣ(c) Unattended Self-Service Stations. Emergency controls as specified in Section 514-5(a) shall be installed at a location acceptable to the authority having jurisdiction, but the controls shall be more than 20 ft (7 m) but less than 100 ft (30 m) from the dispensers. Additional emergency controls shall be installed on each group of dispensers or the outdoor equipment used to control the dispensers. Emergency controls shall shut off all power to all dispensing equipment at the station. Controls shall be manually reset only in a manner approved by the authority having jurisdiction.

FPN: For additional information, see 9-4.5 and 9-5.3 of *Automotive and Marine Service Station Code*, NFPA 30A-1996.

514-6. Provisions for Maintenance and Service of Dispensing Equipment. Each dispensing device shall be provided with a means to remove all external voltage sources, including feedback, during periods of maintenance and service of the dispensing equipment.

514-7. Sealing.

(a) At Dispenser. An approved seal shall be provided in each conduit run entering or leaving a dispenser or any cavities or enclosures in direct communication therewith. The sealing fitting shall be the first fitting after the conduit emerges from the earth or concrete.

(b) At Boundary. Additional seals shall be provided in accordance with Section 501-5. Sections 501-5(a)(4) and (b)(2) shall apply to horizontal as well as to vertical boundaries of the defined Class I locations.

514-8. Underground Wiring. Underground wiring shall be installed in threaded rigid metal conduit or threaded steel intermediate metal conduit. Any portion of electrical wiring or equipment that is below the surface of a Class I, Division 1 or Division 2 location (as defined in Table 514-2) shall be considered to be in a Class I, Division 1 location, which shall extend at least to the point of emergence above grade. Refer to Table 300-5.

Exception No. 1: Type MI cable shall be permitted where it is installed in accordance with Article 330.

Exception No. 2: Rigid nonmetallic conduit complying with Article 347 shall be permitted where buried under not less than 2 ft (610 mm) of cover. Where rigid nonmetallic conduit is used, threaded rigid metal conduit or threaded steel intermediate metal conduit shall be used for the last 2 ft (610 mm) of the underground run to emergence or to the point of connection to the aboveground raceway, and an equipment grounding conductor shall be included to provide electrical continuity of the raceway system and for grounding of noncurrent-carrying metal parts.

514-16. Grounding. All metal raceways, the metal armor or metallic sheath on cables, and all noncurrent-carrying metal parts of fixed or portable electrical equipment, regardless of voltage, shall be grounded as provided in Article 250. Grounding in Class I locations shall comply with Section 501-16.

Article 515 — Bulk Storage Plants

515-1. Definition. A *bulk storage plant* is that portion of a property where flammable liquids are received by tank vessel, pipelines, tank car, or tank vehicle, and are stored or blended in bulk for the purpose of distributing such liquids by tank vessel, pipeline, tank car, tank vehicle, portable tank, or container.

> FPN: For further information, see *Flammable and Combustible Liquids Code,* NFPA 30-1996.

ˣ515-2. Class I Locations. Table 515-2 shall be applied where Class I liquids are stored, handled, or dispensed and shall be used to delineate and classify bulk storage plants. The Class I location shall not extend beyond a floor, wall, roof, or other solid partition that has no communicating openings.

ˣTable 515-2. Class I Locations — Bulk Plants

Location	Class I, Division	Extent of Classified Location
Indoor equipment installed in accordance with *Flammable and Combustible Liquids Code,* NFPA 30-1996, Section 5-3.4.5, where flammable vapor-air mixtures may exist under normal operation	1	Space within 5 ft of any edge of such equipment, extending in all directions.
	2	Space between 5 ft and 8 ft of any edge of such equipment, extending in all directions. Also, space up to 3 ft above floor or grade level within 5 ft to 25 ft horizontally from any edge of such equipment.[1]
Outdoor equipment of the type covered in *Flammable and Combustible Liquids Code,* NFPA 30-1996, Section 5-3.4.5, where flammable vapor-air mixtures may exist under normal operation	1	Space within 3 ft of any edge of such equipment, extending in all directions.
	2	Space between 3 ft and 8 ft of any edge of such equipment, extending in all directions. Also, space up to 3 ft above floor or grade level within 3 ft to 10 ft horizontally from any edge of such equipment.
Tank — Aboveground[2]	1	Space inside dike where dike height is greater than the distance from the tank to the dike for more than 50 percent of the tank circumference.
Shell, Ends, or Roof and Dike Space	2	Within 10 ft from shell, ends, or roof of tank. Space inside dikes to level of top of dike.
Vent	1	Within 5 ft of open end of vent, extending in all directions.
	2	Space between 5 ft and 10 ft from open end of vent, extending in all directions.
Floating Roof	1	Space above the roof and within the shell.
Underground Tank Fill Opening	1	Any pit, box, or space below grade level, if any part is within a Division 1 or 2 classified location.
	2	Up to 18 in. above grade level within a horizontal radius of 10 ft from a loose fill connection, and within a horizontal radius of 5 ft from a tight fill connection.

ˣTable 515-2. Continued

Location	Class I, Division	Extent of Classified Location
Vent — Discharging Upward	1	Within 3 ft of open end of vent, extending in all directions.
	2	Space between 3 ft and 5 ft of open end of vent, extending in all directions.
Drum and Container Filling Outdoors, or Indoors with Adequate Ventilation	1	Within 3 ft of vent and fill openings, extending in all directions.
	2	Space between 3 ft and 5 ft from vent or fill opening, extending in all directions. Also, up to 18 in. above floor or grade level within a horizontal radius of 10 ft from vent or fill openings.
Pumps, Bleeders, Withdrawal Fittings, Meters, and Similar Devices Indoors	2	Within 5 ft of any edge of such devices, extending in all directions. Also up to 3 ft above floor or grade level within 25 ft horizontally from any edge of such devices.
Outdoors	2	Within 3 ft of any edge of such devices, extending in all directions. Also up to 18 in. above grade level within 10 ft horizontally from any edge of such devices.
Pits Without Mechanical Ventilation	1	Entire space within pit if any part is within a Division 1 or 2 classified location.
With Adequate Mechanical Ventilation	2	Entire space within pit if any part is within a Division 1 or 2 classified location.
Containing Valves, Fittings, or Piping, and Not Within a Division 1 or 2 Classified Location	2	Entire pit.
Drainage Ditches, Separators, Impounding Basins Outdoor	2	Space up to 18 in. above ditch, separator, or basin. Also up to 18 in. above grade within 15 ft horizontally from any edge.
Indoor		Same as pits.
Tank Vehicle and Tank Car[3] Loading Through Open Dome	1	Within 3 ft of edge of dome, extending in all directions.
	2	Space between 3 ft and 15 ft from edge of dome, extending in all directions.
Loading through Bottom Connections with Atmospheric Venting	1	Within 3 ft of point of venting to atmosphere, extending in all directions.
	2	Space between 3 ft and 15 ft from point of venting to atmosphere, extending in all directions. Also up to 18 in. above grade within a horizontal radius of 10 ft from point of loading connection.
Office and Rest Rooms	Unclassified	If there is any opening to these rooms within the extent of an indoor classified location, the room shall be classified the same as if the wall, curb, or partition did not exist.
Loading through Closed Dome with Atmospheric Venting	1	Within 3 ft of open end of vent, extending in all directions.
	2	Space between 3 ft and 15 ft from open end of vent, extending in all directions. Also within 3 ft of edge of dome, extending in all directions.
Loading through Closed Dome with Vapor Control	2	Within 3 ft of point of connection of both fill and vapor lines, extending in all directions.

(continues)

ˣTable 515-2. Continued

Location	Class I, Division	Extent of Classified Location
Bottom Loading with Vapor Control Any Bottom Unloading	2	Within 3 ft of point of connections, extending in all directions. Also up to 18 in. above grade within a horizontal radius of 10 ft from point of connections.
Storage and Repair Garage for Tank Vehicles	1	All pits or spaces below floor level.
	2	Space up to 18 in. above floor or grade level for entire storage or repair garage.
Garages for other than Tank Vehicles	Unclassified	If there is any opening to these rooms within the extent of an outdoor classified location, the entire room shall be classified the same as the space classification at the point of the opening.
Outdoor Drum Storage	Unclassified	
Indoor Warehousing Where There Is No Flammable Liquid Transfer	Unclassified	If there is any opening to these rooms within the extent of an indoor classified location, the room shall be classified the same as if the wall, curb, or partition did not exist.
Piers and Wharves		See Figure 515-2.

Note: For SI units, 1 in. = 25.4 mm; 1 ft = 0.3048 m.

[1]The release of Class I liquids may generate vapors to the extent that the entire building, and possibly a zone surrounding it, should be considered a Class I, Division 2 location.

[2]For Tanks — Underground, see Section 514-2.

[3]When classifying extent of space, consideration shall be given to the fact that tank cars or tank vehicles may be spotted at varying points. Therefore, the extremities of the loading or unloading positions shall be used.

FPN: The area classifications listed in Table 515-2 are based on the premise that the installation meets the applicable requirements of *Flammable and Combustible Liquids Code,* NFPA 30-1996, Chapter 5, in all respects. Should this not be the case, the authority having jurisdiction has the authority to classify the extent of the classified space.

515-3. Wiring and Equipment Within Class I Locations. All electric wiring and equipment within the Class I locations defined in Section 515-2 shall comply with the applicable provisions of Article 501.

Exception: As permitted in Section 515-5.

515-4. Wiring and Equipment Above Class I Locations. All fixed wiring above Class I locations shall be in metal raceways or PVC Schedule 80 rigid nonmetallic conduit, or equivalent, or be Type MI, TC, or MC cable. Fixed equipment that may produce arcs, sparks, or particles of hot metal, such as lamps and lampholders for fixed lighting, cutouts, switches, receptacles, motors, or other equipment having make-and-break or sliding contacts, shall be of the totally enclosed type or be constructed so as to prevent the escape of sparks or hot metal particles. Portable lamps or other utilization equipment and their flexible cords shall comply with the provisions of Article 501 for the class of location above which they are connected or used.

515-5. Underground Wiring.

(a) Wiring Method. Underground wiring shall be installed in threaded rigid metal conduit or threaded steel intermediate metal conduit or, where buried under not less than 2 ft (610 mm) of cover, shall be permitted in rigid nonmetallic conduit or an approved cable. Where rigid nonmetallic conduit is used, threaded rigid metal conduit or threaded steel intermediate metal conduit shall be used for the last 2 ft (610 mm) of the conduit run to emergence or to the point of connection to the aboveground raceway. Where cable is used, it shall be enclosed in threaded rigid metal conduit or threaded steel intermediate metal conduit from the point of lowest buried cable level to the point of connection to the aboveground raceway.

(b) Insulation. Conductor insulation shall comply with Section 501-13.

(c) Nonmetallic Wiring. Where rigid nonmetallic conduit or cable with a nonmetallic sheath is used, an equipment grounding conductor shall be included to provide for electrical continuity of the raceway system and for grounding of noncurrent-carrying metal parts.

Deck

Open sump in deck for draining lines and hoses

50 ft

25 ft

25 ft

Operating envelope and stored position of loading arms or hoses

50 ft

25 ft

2 ft

Approach

Pier

Shore

Water level

▨ Division 1   ▨ Division 2   ☐ Unclassified

Note 1: The "source of vapor" shall be the operating envelope and stored position of the outboard flange connection of the loading arm (or hose).

Note 2: The berth area adjacent to tanker and barge cargo tanks is to be Division 2 to the following extent:
  a. 25 ft (7.6 m) horizontally in all directions on the pier side from that portion of the hull containing cargo tanks
  b. From the water level to 25 ft (7.6 m) above the cargo tanks at their highest position

Note 3: Additional locations may have to be classified as required by the presence of other sources of flammable liquids on the berth, or by Coast Guard or other regulations.

For SI units: 1 ft = 0.3048 m.

ˣFigure 515-2 Marine terminal handling flammable liquids.

515-6. Sealing. Sealing requirements in Sections 501-5(a)(4) and (b)(2) shall apply to horizontal as well as to vertical boundaries of the defined Class I locations. Buried raceways under defined Class I locations shall be considered to be within a Class I, Division 1 location.

515-7. Gasoline Dispensing. Where gasoline or other volatile flammable liquids or liquefied flammable gases are dispensed at bulk stations, the applicable provisions of Article 514 shall apply.

515-16. Grounding. All metal raceways, the metal armor or metallic sheath on cables, and all noncurrent-carrying metal parts of fixed or portable electrical equipment, regardless of voltage, shall be grounded as provided in Article 250. Grounding in Class I locations shall comply with Section 501-16.

FPN: For information on grounding for static protection, see 5-6.3.4 and 5-6.3.5 of *Flammable and Combustible Liquids Code*, NFPA 30-1996.

Article 516 — Spray Application, Dipping, and Coating Processes

516-1. Scope. This article covers the regular or frequent application of flammable liquids, combustible liquids, and combustible powders by spray operations and the application of flammable liquids, or combustible liquids at temperatures above their flashpoint, by dipping, coating, or other means.

FPN: For further information regarding safeguards for these processes, such as fire protection, posting of warning signs, and maintenance, see *Standard for Spray Application Using Flammable and Combustible Materials*, NFPA 33-1995, and *Standard for Dipping and Coating Processes Using Flam-*

mable or Combustible Liquids, NFPA 34-1995. For additional information regarding ventilation, see *Standard for Exhaust Systems for Air Conveying of Materials*, NFPA 91-1995.

516-2. Classification of Locations. Classification is based on dangerous quantities of flammable vapors, combustible mists, residues, dusts, or deposits.

(a) Class I or Class II, Division 1 Locations. The following spaces shall be considered Class I or Class II, Division 1 locations, as applicable.

ˣ**(1)** The interiors of spray booths and rooms except as specifically provided in Section 516-3(d).

ˣ**(2)** The interior of exhaust ducts.

ˣ**(3)** Any area in the direct path of spray operations.

ˣ**(4)** For dipping and coating operations, all space within a 5-ft (1.52-m) radial distance from the vapor sources extending from these surfaces to the floor. The vapor source shall be the liquid exposed in the process and the drainboard, and any dipped or coated object from which it is possible to measure vapor concentrations exceeding 25 percent of the lower flammable limit at a distance of 1 ft (305 mm), in any direction, from the object.

ˣ**(5)** Pits within 25 ft (7.62 m) horizontally of the vapor source. If pits extend beyond 25 ft (7.62 m) of the vapor source, a vapor stop shall be provided, or the entire pit shall be classified as Class I, Division 1.

ˣ**(6)** The interior of any enclosed coating or dipping process.

(b) Class I or Class II, Division 2 Locations. The following spaces shall be considered Class I or Class II, Division 2 as applicable.

ˣ**(1)** For open spraying, all space outside of but within 20 ft (6.10 m) horizontally and 10 ft (3.05 m) vertically of the Class I, Division 1 location as defined in Section 516-2(a), and not separated from it by partitions. See Figure 516-2(b)(1).

ˣ**(2)** If spray application operations are conducted within a closed-top, open-face, or open-front booth or room, any electrical wiring or utilization equipment located outside of the booth or room but within the boundaries designated as Division 2 in Figure 516-2(b)(2) shall be suitable for Class I, Division 2 or Class II, Division 2 locations, whichever is applicable.

The Class I, Division 2 or Class II, Division 2 locations shown in Figure 516-2(b)(2) shall extend from the edges of the open face or open front of the booth or room in accordance with the following.

(a) If the exhaust ventilation system is interlocked with the spray application equipment, then the Division 2 location shall extend 5 ft (1525 mm) horizontally and 3 ft (915 mm) vertically from the open face or open front of the booth or room, as shown in Figure 516-2(b)(2) top.

(b) If the exhaust ventilation system is not interlocked with the spray application equipment, then the Division 2 location shall extend 10 ft (3050 mm) horizontally and 3 ft (915 mm) vertically from the open face or open front of the booth or room, as shown in Figure 516-2(b)(2) bottom.

For the purposes of this subsection, *interlocked* shall mean that the spray application equipment cannot be operated unless the exhaust ventilation system is operating and functioning properly and spray application is automatically stopped if the exhaust ventilation system fails.

ˣ**(3)** For spraying operations conducted within an open top spray booth, the space 3 ft (914 mm) vertically above the booth and within 3 ft (914 mm) of other booth openings shall be considered Class I or Class II, Division 2.

ˣ**(4)** For spraying operations confined to an enclosed spray booth or room, the space within 3 ft (914 mm) in all

ˣ**Figure 516-2(b)(1)   Class I or Class II, Division 2 location adjacent to an unenclosed spray operation.**

^x**Figure 516-2(b)(2)   Class I or Class II, Division 2 locations adjacent to a closed top, open face, or open front spray booth or room.**

directions from any openings shall be considered Class I or Class II, Division 2 as shown in Figure 516-2(b)(4).

^x**(5)** For dip tanks and drain boards, the 3-ft (914-mm) space surrounding the Class I, Division 1 location as defined in Section 516-2(a)(4) and as shown in Figure 516-2(b)(5).

^x**(6)** For dip tanks and drain boards, the space 3 ft (914 mm) above the floor and extending 20 ft (6.1 m) horizontally in all directions from the Class I, Division 1 location.

Exception: This space shall not be required to be considered a hazardous (classified) location where the vapor source area is 5 ft² (0.46 m²) or less, and where the contents of the open tank, trough, or container do not exceed 5 gal (18.9 L). In addition, the vapor concentration during operation and shutdown periods shall not exceed 25 percent of the lower flammable limit outside the Class I location specified in Section 516-2(a)(4).

^x**(c) Enclosed Coating and Dipping Operations.** The space adjacent to an enclosed dipping or coating process or apparatus shall be considered unclassified.

For SI units: 1 ft = 0.3048 m.

^x**Figure 516-2(b)(4)   Class I (or Class II), Division 2 locations adjacent to an enclosed spray booth or spray room.**

Exception: The space within 3 ft (914 mm) in all directions from any opening in the enclosure shall be classified as Class I, Division 2.

(d) Adjacent Locations. Adjacent locations that are cut off from the defined Class I or Class II locations by tight partitions without communicating openings, and within which flammable vapors or combustible powders are not likely to be released, shall be unclassified.

(e) Unclassified Locations. Locations using drying, curing, or fusion apparatus and provided with positive mechanical ventilation adequate to prevent accumulation of flammable concentrations of vapors, and provided with effective interlocks to de-energize all electrical equipment (other than equipment approved for Class I locations) in case the ventilating equipment is inoperative, shall be permitted to be unclassified where the authority having jurisdiction so judges.

    FPN: For further information regarding safeguards, see *Standard for Ovens and Furnaces*, NFPA 86-1995.

516-3. Wiring and Equipment in Class I Locations.

(a) Wiring and Equipment — Vapors. All electric wiring and equipment within the Class I location (containing vapor ˣFigure 516-2(b)(5)   Electrical area classification for open processes without vapor containment or ventilation.

only — not residues) defined in Section 516-2 shall comply with the applicable provisions of Article 501.

ˣ**(b) Wiring and Equipment — Vapors and Residues.** Unless specifically listed for locations containing deposits of dangerous quantities of flammable or combustible vapors, mists, residues, dusts, or deposits (as applicable), there shall be no electrical equipment in any spray area as herein defined whereon deposits of combustible residue may readily accumulate, except wiring in rigid metal conduit, intermediate metal conduit, Type MI cable, or in metal boxes or fittings containing no taps, splices, or terminal connections.

(c) Illumination. Illumination of readily ignitible areas through panels of glass or other transparent or translucent material shall be permitted only if it complies with the following.

(1) Fixed lighting units are used as the source of illumination;
(2) The panel effectively isolates the Class I location from the area in which the lighting unit is located;
(3) The lighting unit is approved for its specific location;
(4) The panel is of a material or is protected so that breakage will be unlikely; and
(5) The arrangement is such that normal accumulations of hazardous residue on the surface of the panel will not be raised to a dangerous temperature by radiation or conduction from the source of illumination.

ˣ**(d) Portable Equipment.** Portable electric lamps or other utilization equipment shall not be used in a spray area during spray operations.

Exception No. 1: Where portable electric lamps are required for operations in spaces not readily illuminated by fixed lighting within the spraying area, they shall be of the type approved for Class I, Division 1 locations where readily ignitible residues may be present.

Exception No. 2: Where portable electric drying apparatus are used in automobile refinishing spray booths and the following requirements are met.

(a) The apparatus and its electrical connections are not located within the spray enclosure during spray operations.
(b) Electrical equipment within 18 in. (45.7 cm) of the floor is approved for Class I, Division 2 locations.
(c) All metallic parts of the drying apparatus are electrically bonded and grounded.
(d) Interlocks are provided to prevent the operation of spray equipment while drying apparatus is within the spray enclosure, to allow for a 3-minute purge of the enclosure before energizing the drying apparatus, and to shut off drying apparatus on failure of ventilation system.

(e) Electrostatic Equipment. Electrostatic spraying or detearing equipment shall be installed and used only as provided in Section 516-4.

FPN: For further information, see *Standard for Spray Application Using Flammable and Combustible Materials,* NFPA 33-1995.

ˣ**516-4. Fixed Electrostatic Equipment.** This section shall apply to any equipment using electrostatically charged elements for the atomization, charging, and/or precipitation of hazardous materials for coatings on articles or for other similar purposes in which the charging or atomizing device is attached to a mechanical support or manipulator. This shall include robotic devices. This section shall not apply to devices that are held or manipulated by hand. Where robotic programming procedures involve manual manipulation of the robot arm while spraying with the high voltage on, the provisions of Section 516-5 shall apply. The installation of electrostatic spraying equipment shall comply with Sections 516-4(a) through (j). Spray equipment shall be listed or approved.

All automatic electrostatic equipment systems shall comply with Sections 516-4(a) through (i).

(a) Power and Control Equipment. Transformers, high-voltage supplies, control apparatus, and all other electric portions of the equipment shall be installed outside of the Class I location as defined in Section 516-2 or be of a type approved for the location.

Exception: High-voltage grids, electrodes, electrostatic atomizing heads, and their connections shall be permitted within the Class I location.

(b) Electrostatic Equipment. Electrodes and electrostatic atomizing heads shall be adequately supported in permanent locations and shall be effectively insulated from ground. Electrodes and electrostatic atomizing heads that are permanently attached to their bases, supports, reciprocators, or robots shall be deemed to comply with this section.

(c) High-Voltage Leads. High-voltage leads shall be properly insulated and protected from mechanical damage or exposure to destructive chemicals. Any exposed element at high voltage shall be effectively and permanently supported on suitable insulators and shall be effectively guarded against accidental contact or grounding.

(d) Support of Goods. Goods being coated using this process shall be supported on conveyors or hangers. The conveyors or hangers shall be arranged to (1) ensure that the parts being coated are electrically connected to ground with a resistance of 1 megohm or less and (2) to prevent parts from swinging.

(e) Automatic Controls. Electrostatic apparatus shall be equipped with automatic means that will rapidly de-energize the high-voltage elements under any of the following conditions:

(1) Stoppage of ventilating fans or failure of ventilating equipment from any cause
(2) Stoppage of the conveyor carrying goods through the high-voltage field unless stoppage is required by the spray process

(3) Occurrence of excessive current leakage at any point in the high-voltage system
(4) De-energizing the primary voltage input to the power supply

(f) Grounding. All electrically conductive objects in the spray area, except those objects required by the process to be at high voltage, shall be adequately grounded. This requirement shall apply to paint containers, wash cans, guards, hose connectors, brackets, and any other electrically conductive objects or devices in the area.

(g) Isolation. Safeguards such as adequate booths, fencing, railings, interlocks, or other means shall be placed about the equipment or incorporated therein so that they, either by their location or character, or both, ensure that a safe separation of the process is maintained.

(h) Signs. Signs shall be conspicuously posted to

(1) Designate the process zone as dangerous with regard to fire and accident,
(2) Identify the grounding requirements for all electrically conductive objects in the spray area, and
(3) Restrict access to qualified personnel only.

(i) Insulators. All insulators shall be kept clean and dry.

(j) Other Than Nonincendive Equipment. Spray equipment that cannot be classified as nonincendive shall comply with (1) and (2).

(1) Conveyors or hangers shall be arranged so to maintain a safe distance of at least twice the sparking distance between goods being painted and electrodes, electrostatic atomizing heads, or charged conductors. Warnings defining this safe distance shall be posted.

(2) The equipment shall provide an automatic means of rapidly de-energizing the high-voltage elements in the event the distance between the goods being painted and the electrodes or electrostatic atomizing heads falls below that specified in (1).

ˣ516-5. Electrostatic Hand-Spraying Equipment. This section shall apply to any equipment using electrostatically charged elements for the atomization, charging, and/or precipitation of materials for coatings on articles, or for other similar purposes in which the atomizing device is hand held or manipulated during the spraying operation. Electrostatic hand-spraying equipment and devices used in connection with paint-spraying operations shall be of approved types and shall comply with Sections 516-5(a) through (e).

(a) General. The high-voltage circuits shall be designed so as not to produce a spark of sufficient intensity to ignite the most readily ignitible of those vapor–air mixtures likely to be encountered, nor result in appreciable shock hazard upon coming in contact with a grounded object under all normal operating conditions. The electrostatically charged exposed elements of the hand gun shall be capable of being energized only by an actuator that also controls the coating material supply.

(b) Power Equipment. Transformers, power packs, control apparatus, and all other electric portions of the equipment shall be located outside of the Class I location or be approved for the location.

Exception: The hand gun itself and its connections to the power supply shall be permitted within the Class I location.

(c) Handle. The handle of the spraying gun shall be electrically connected to ground by a metallic connection and be constructed so that the operator in normal operating position is in intimate electrical contact with the grounded handle to prevent buildup of a static charge on the operator's body. Signs indicating the necessity for grounding other persons entering the spray area shall be conspicuously posted.

(d) Electrostatic Equipment. All electrically conductive objects in the spraying area shall be adequately grounded. This requirement shall apply to paint containers, wash cans, and any other electrically conductive objects or devices in the area. The equipment shall carry a prominent, permanently installed warning regarding the necessity for this grounding feature.

(e) Support of Objects. Objects being painted shall be maintained in metallic contact with the conveyor or other grounded support. Hooks shall be regularly cleaned to ensure adequate grounding of 1 megohm or less. Areas of contact shall be sharp points or knife edges where possible. Points of support of the object shall be concealed from random spray where feasible; and, where the objects being sprayed are supported from a conveyor, the point of attachment to the conveyor shall be located so as to not collect spray material during normal operation.

ˣ516-6. Powder Coating. This section shall apply to processes in which combustible dry powders are applied. The hazards associated with combustible dusts are present in such a process to a degree, depending on the chemical composition of the material, particle size, shape, and distribution.

> FPN: The hazards associated with combustible dusts are inherent in this process. Generally speaking, the hazard rating of the powders employed depends on the chemical composition of the material, particle size, shape, and distribution.

(a) Electric Equipment and Sources of Ignition. Electric equipment and other sources of ignition shall comply with the requirements of Article 502. Portable electric lamps and other utilization equipment shall not be used within a Class II location during operation of the finishing processes. Where such lamps or utilization equipment are used during cleaning or repairing operations, they shall be of a type approved for Class II, Division 1 locations, and all exposed metal parts shall be effectively grounded.

Exception: Where portable electric lamps are required for operations in spaces not readily illuminated by fixed lighting within the spraying area, they shall be of the type approved for Class II, Division 1 locations where readily ignitible residues may be present.

(b) Fixed Electrostatic Spraying Equipment. The provisions of Section 516-4 and Section 516-6(a) shall apply to fixed electrostatic spraying equipment.

(c) Electrostatic Hand-Spraying Equipment. The provisions of Section 516-5 and Section 516-6(a) shall apply to electrostatic hand-spraying equipment.

(d) Electrostatic Fluidized Beds. Electrostatic fluidized beds and associated equipment shall be of approved types. The high-voltage circuits shall be designed so that any discharge produced when the charging electrodes of the bed are approached or contacted by a grounded object shall not be of sufficient intensity to ignite any powder–air mixture likely to be encountered nor to result in an appreciable shock hazard.

(1) Transformers, power packs, control apparatus, and all other electric portions of the equipment shall be located outside the powder-coating area or shall otherwise comply with the requirements of Section 516-6(a).

Exception: The charging electrodes and their connections to the power supply shall be permitted within the powder-coating area.

(2) All electrically conductive objects within the powder-coating area shall be adequately grounded. The powder-coating equipment shall carry a prominent, permanently installed warning regarding the necessity for grounding these objects.

(3) Objects being coated shall be maintained in electrical contact (less than 1 megohm) with the conveyor or other support in order to ensure proper grounding. Hangers shall be regularly cleaned to ensure effective electrical contact. Areas of electrical contact shall be sharp points or knife edges where possible.

(4) The electric equipment and compressed air supplies shall be interlocked with a ventilation system so that the equipment cannot be operated unless the ventilating fans are in operation.

516-7. Wiring and Equipment Above Class I and II Locations.

(a) Wiring. All fixed wiring above the Class I and II locations shall be in metal raceways, rigid nonmetallic conduit, or electrical nonmetallic tubing, or shall be Type MI, TC, or MC cable. Cellular metal floor raceways shall be permitted only for supplying ceiling outlets or extensions to the area below the floor of a Class I or II location, but such raceways shall have no connections leading into or through the Class I or II location above the floor unless suitable seals are provided.

(b) Equipment. Equipment that may produce arcs, sparks, or particles of hot metal, such as lamps and lampholders for fixed lighting, cutouts, switches, receptacles, motors, or other equipment having make-and-break or sliding contacts, where installed above a Class I or II location or above a location where freshly finished goods are handled, shall be of the totally enclosed type or be constructed so as to prevent the escape of sparks or hot metal particles.

516-16. Grounding.
All metal raceways, the metal armor or metallic sheath on cables, and all noncurrent-carrying metal parts of fixed or portable electrical equipment, regardless of voltage, shall be grounded as provided in Article 250. Grounding in Class I locations shall comply with Section 501-16.

Article 517 — Health Care Facilities

A. General

517-1. Scope.
The provisions of this article shall apply to electrical construction and installation criteria in health care facilities.

> FPN No. 1: This article is not intended to apply to veterinary facilities.
>
> FPN No. 2: For information concerning performance, maintenance, and testing criteria, refer to the appropriate health care facilities documents.

517-2. General.
The requirements in Parts B and C apply not only to single-function buildings, but are also intended to be individually applied to their respective forms of occupancy within a multifunction building (e.g., a doctor's examining room located within a limited care facility would be required to meet the provisions of Section 517-10).

517-3. Definitions.

Alternate Power Source. One or more generator sets, or battery systems where permitted, intended to provide power during the interruption of the normal electrical services or the public utility electrical service intended to provide power during interruption of service normally provided by the generating facilities on the premises.

Ambulatory Health Care Center. A building or part thereof used to provide services or treatment to four or more patients at the same time and meeting either (1) or (2).

(1) Those facilities that provide, on an outpatient basis, treatment for patients that would render them incapable of taking action for self-preservation under emergency conditions without assistance from others, such as hemodialysis units or freestanding emergency medical units.

(2) Those facilities that provide, on an outpatient basis, surgical treatment requiring general anesthesia.

Anesthetizing Location. Any area of a health care facility that has been designated to be used for the administration of any flammable or nonflammable inhalation anesthetic agent in the course of examination or treatment, including the use of such agents for relative analgesia.

Critical Branch. A subsystem of the emergency system consisting of feeders and branch circuits supplying energy to task illumination, special power circuits, and selected receptacles serving areas and functions related to patient care, and which are connected to alternate power sources by one or more transfer switches during interruption of the normal power source.

Electrical Life-Support Equipment. Electrically powered equipment whose continuous operation is necessary to maintain a patient's life.

Emergency System. A system of feeders and branch circuits meeting the requirements of Article 700, except as amended by Article 517, and intended to supply alternate power to a limited number of prescribed functions vital to the protection of life and patient safety, with automatic restoration of electrical power within 10 seconds of power interruption.

Equipment System. A system of feeders and branch circuits arranged for delayed, automatic, or manual connection to the alternate power source and that serves primarily 3-phase power equipment.

Essential Electrical System. A system comprised of alternate sources of power and all connected distribution systems and ancillary equipment, designed to ensure continuity of electrical power to designated areas and functions of a health care facility during disruption of normal power sources, and also designed to minimize disruption within the internal wiring system.

Exposed Conductive Surfaces. Those surfaces that are capable of carrying electric current and that are unprotected, unenclosed, or unguarded, permitting personal contact. Paint, anodizing, and similar coatings are not considered suitable insulation, unless they are listed for such use.

Flammable Anesthetics. Gases or vapors, such as fluroxene, cyclopropane, divinyl ether, ethyl chloride, ethyl ether, and ethylene, which may form flammable or explosive mixtures with air, oxygen, or reducing gases such as nitrous oxide.

Flammable Anesthetizing Location. Any area of the facility that has been designated to be used for the administration of any flammable inhalation anesthetic agents in the normal course of examination or treatment.

Hazard Current. For a given set of connections in an isolated power system, the total current that would flow through a low impedance if it were connected between either isolated conductor and ground.

Fault Hazard Current. The hazard current of a given isolated system with all devices connected except the line isolation monitor.

Monitor Hazard Current. The hazard current of the line isolation monitor alone.

Total Hazard Current. The hazard current of a given isolated system with all devices, including the line isolation monitor, connected.

Health Care Facilities. Buildings or portions of buildings that contain, but are not limited to, occupancies such as hospitals; nursing homes; limited care; supervisory care; clinics; medical and dental offices; and ambulatory care, whether permanent or movable.

Hospital. A building or part thereof used for the medical, psychiatric, obstetrical, or surgical care, on a 24-hour basis, of four or more inpatients. *Hospital*, wherever used in this *Code*, shall include general hospitals, mental hospitals, tuberculosis hospitals, children's hospitals, and any such facilities providing inpatient care.

Isolated Power System. A system comprising an isolating transformer or its equivalent, a line isolation monitor, and its ungrounded circuit conductors.

Isolation Transformer. A transformer of the multiple-winding type, with the primary and secondary windings physically separated, which inductively couples its secondary winding to the grounded feeder systems that energize its primary winding.

Life Safety Branch. A subsystem of the emergency system consisting of feeders and branch circuits, meeting the requirements of Article 700 and intended to provide adequate power needs to ensure safety to patients and personnel, and which are automatically connected to alternate power sources during interruption of the normal power source.

Limited Care Facility. A building or part thereof used on a 24-hour basis for the housing of four or more persons who are incapable of self-preservation because of age, physical limitation due to accident or illness, or mental limitations, such as mental retardation/developmental disability, mental illness, or chemical dependency.

Line Isolation Monitor. A test instrument designed to continually check the balanced and unbalanced impedance from each line of an isolated circuit to ground and equipped with a built-in test circuit to exercise the alarm without adding to the leakage current hazard.

Nursing Home. A building or part thereof used for the lodging, boarding, and nursing care, on a 24-hour basis, of four or more persons who, because of mental or physical incapacity, may be unable to provide for their own needs and safety without the assistance of another person. *Nursing home*, wherever used in this *Code*, shall include nursing and convalescent homes, skilled nursing facilities, intermediate care facilities, and infirmaries of homes for the aged.

Nurses' Stations. Areas intended to provide a center of nursing activity for a group of nurses serving bed patients, where the patient calls are received, nurses are dispatched, nurses' notes written, inpatient charts prepared, and medications prepared for distribution to patients. Where such activities are carried on in more than one location within a nursing unit, all such separate areas are considered a part of the nurses' station.

Patient Bed Location. The location of an inpatient sleeping bed; or the bed or procedure table used in a critical patient care area.

Patient Care Area. Any portion of a health care facility wherein patients are intended to be examined or treated. Areas of a health care facility in which patient care is administered are classified as general care areas or critical care areas, either of which may be classified as a wet location. The governing body of the facility designates these areas in accordance with the type of patient care anticipated and with the following definitions of the area classification.

FPN: Business offices, corridors, lounges, day rooms, dining rooms, or similar areas typically are not classified as patient care areas.

(a) General Care Areas. General care areas are patient bedrooms, examining rooms, treatment rooms, clinics, and similar areas in which it is intended that the patient shall come in contact with ordinary appliances such as a nurse call system, electrical beds, examining lamps, telephone, and entertainment devices. In such areas, it may also be intended that patients be connected to electromedical devices (such as heating pads, electrocardiographs, drainage pumps, monitors, otoscopes, ophthalmoscopes, intravenous lines, etc.).

(b) Critical Care Areas. Critical care areas are those special care units, intensive care units, coronary care units, angiography laboratories, cardiac catheterization laboratories, delivery rooms, operating rooms, and similar areas in which patients are intended to be subjected to invasive procedures and connected to line-operated, electromedical devices.

(c) Wet Locations. Wet locations are those patient care areas that are normally subject to wet conditions while patients are present. These include standing fluids on the floor or drenching of the work area, either of which condition is intimate to the patient or staff. Routine housekeeping procedures and incidental spillage of liquids do not define a wet location.

Patient Equipment Grounding Point. A jack or terminal bus that serves as the collection point for redundant grounding of electric appliances serving a patient vicinity or for grounding other items in order to eliminate electromagnetic interference problems.

Patient Vicinity. In an area in which patients are normally cared for, the *patient vicinity* is the space with surfaces likely to be contacted by the patient or an attendant who can touch the patient. Typically in a patient room, this encloses a space within the room not less than 6 ft (1.83 m) beyond the perimeter of the bed in its nominal location, and extending vertically not less than 7½ ft (2.29 m) above the floor.

Psychiatric Hospital. A building used exclusively for the psychiatric care, on a 24-hour basis, of four or more inpatients.

Reference Grounding Point. The ground bus of the panelboard or isolated power system panel supplying the patient care area.

Selected Receptacles. A minimum number of electric receptacles to accommodate appliances ordinarily required for local tasks or likely to be used in patient care emergencies.

Task Illumination. Provision for the minimum lighting required to carry out necessary tasks in the described areas, including safe access to supplies and equipment, and access to exits.

Therapeutic High-Frequency Diathermy Equipment. Therapeutic high-frequency diathermy equipment is therapeutic induction and dielectric heating equipment.

X-Ray Installations (Long-Time Rating). A rating based on an operating interval of 5 minutes or longer.

X-Ray Installations (Mobile). X-ray equipment mounted on a permanent base with wheels, casters, or a combination of both to facilitate moving the equipment while completely assembled.

X-Ray Installations (Momentary Rating). A rating based on an operating interval that does not exceed 5 seconds.

X-Ray Installations (Portable). X-ray equipment designed to be hand carried.

X-Ray Installations (Transportable). X-ray equipment to be installed in a vehicle or that may be readily disassembled for transport in a vehicle.

B. Wiring and Protection

517-10. Applicability.

(a) Part B shall apply to patient care areas of all health care facilities.

(b) Part B shall not apply to the following:

(1) Business offices, corridors, waiting rooms, and the like in clinics, medical and dental offices, and outpatient facilities

(2) Areas of nursing homes and limited care facilities wired in accordance with Chapters 1 through 4 of this *Code* where these areas are used exclusively as patient sleeping rooms

FPN: See *Life Safety Code®*, NFPA *101®*-1997.

517-11. General Installation — Construction Criteria. It is the purpose of this article to specify the installation criteria and wiring methods that will minimize electrical hazards by the maintenance of adequately low-potential differences only between exposed conductive surfaces that are likely to become energized and could be contacted by a patient.

FPN: In a health care facility, it is difficult to prevent the occurrence of a conductive or capacitive path from the patient's body to some grounded object, because that path may be established accidentally or through instrumentation directly connected to the patient. Other electrically conductive surfaces that may make an additional contact with the patient, or instruments that may be connected to the patient, then become possible sources of electric currents that can traverse the patient's body. The hazard is increased as more apparatus is associated with the patient, and, therefore, more intensive precautions are needed. Control of electric shock hazard requires the limitation of electric current that might flow in an electric circuit involving the patient's body by raising the resistance of the conductive circuit that includes the patient, or by insulating exposed surfaces that might become energized, in addition to reducing the potential difference that can appear between exposed conductive surfaces in the patient vicinity, or by combinations of these methods. A special problem is presented by the patient with an externalized direct conductive path to the heart muscle. The patient may be electrocuted at current levels so low that additional protection in the design of appliances, insulation of the catheter, and control of medical practice are required.

517-12. Wiring Methods. Except as modified in this article, wiring methods shall comply with the applicable requirements of Chapters 1 through 4 of this *Code*.

517-13. Grounding of Receptacles and Fixed Electric Equipment.

(a) Patient Care Area. In an area used for patient care, the grounding terminals of all receptacles and all noncurrent-carrying conductive surfaces of fixed electric equipment likely to become energized that are subject to personal contact, operating at over 100 volts, shall be grounded by an insulated copper conductor. The grounding conductor shall be sized in accordance with Table 250-122 and installed in metal raceways with the branch-circuit conductors supplying these receptacles or fixed equipment.

Exception No. 1: Metal raceways shall not be required where listed Types MI, MC, or AC cables are used, provided the outer metal armor or sheath of the cable is identified as an acceptable grounding return path.

Exception No. 2: Metal faceplates shall be permitted to be grounded by means of a metal mounting screw(s) securing the faceplate to a grounded outlet box or grounded wiring device.

Exception No. 3: Light fixtures more than 7½ ft (2.2 m) above the floor and switches located outside of the patient vicinity shall not be required to be grounded by an insulated grounding conductor.

(b) Methods. In addition to the requirements of Section 517-13(a), all branch circuits serving patient care areas shall be provided with a ground path for fault current by installation in a metal raceway system or cable assembly. The metal raceway system, or cable armor or sheath assembly, shall itself qualify as an equipment grounding return path in accordance with Section 250-118. Type MC cable and Type MI cable shall have an outer metal armor or sheath that is identified as an acceptable grounding return path.

517-14. Panelboard Bonding. The equipment grounding terminal buses of the normal and essential branch-circuit panelboards serving the same individual patient vicinity shall be bonded together with an insulated continuous copper conductor not smaller than No. 10. Where more than two panels serve the same location, this conductor shall be continuous from panel to panel, but shall be permitted to be broken in order to terminate on the ground bus in each panel.

517-16. Receptacles with Insulated Grounding Terminals. Receptacles with insulated grounding terminals, as

permitted in Section 250-146(d), shall be identified; such identification shall be visible after installation.

FPN: Caution is important in specifying such a system with receptacles having insulated grounding terminals, since the grounding impedance is controlled only by the grounding conductors and does not benefit functionally from any parallel grounding paths.

517-17. Ground-Fault Protection.

(a) Feeders. Where ground-fault protection is provided for operation of the service disconnecting means or feeder disconnecting means as specified by Sections 230-95 or 215-10, an additional step of ground-fault protection shall be provided in the next level of feeder disconnecting means downstream toward the load. Such protection shall consist of overcurrent devices and current transformers or other equivalent protective equipment that shall cause the feeder disconnecting means to open.

The additional levels of ground-fault protection shall not be installed

(1) On the load side of an essential electrical system transfer switch, or
(2) Between the on-site generating unit(s) described in Section 517-35(b) and the essential electrical system transfer switch(es), or
(3) On electrical systems that are not solidly grounded wye systems with greater than 150 volts to ground, but not exceeding 600 volts phase-to-phase.

(b) Selectivity. Ground-fault protection for operation of the service and feeder disconnecting means shall be fully selective such that the feeder device and not the service device shall open on ground faults on the load side of the feeder device. A six-cycle minimum separation between the service and feeder ground-fault tripping bands shall be provided. Operating time of the disconnecting devices shall be considered in selecting the time spread between these two bands to achieve 100 percent selectivity.

FPN: See Section 230-95, fine print note, for transfer of alternate source where ground-fault protection is applied.

(c) Testing. When equipment ground-fault protection is first installed, each level shall be performance tested to ensure compliance with Section 517-17(b).

517-18. General Care Areas.

(a) Patient Bed Location. Each patient bed location shall be supplied by at least two branch circuits, one from the emergency system and one from the normal system. All branch circuits from the normal system shall originate in the same panelboard.

Exception No. 1: Branch circuits serving only special-purpose outlets or receptacles, such as portable X-ray outlets, shall not be required to be served from the same distribution panel or panels.

Exception No. 2: Requirements of Section 517-18(a) shall not apply to patient bed locations in clinics, medical and dental offices, and outpatient facilities; psychiatric, substance abuse, and rehabilitation hospitals; sleeping rooms of nursing homes and limited care facilities meeting the requirements of Section 517-10(b)(2).

Exception No. 3: A general care patient bed location served from two separate transfer switches on the emergency system shall not be required to have circuits from the normal system.

(b) Patient Bed Location Receptacles. Each patient bed location shall be provided with a minimum of four receptacles. They shall be permitted to be of the single or duplex types or a combination of both. All receptacles, whether four or more, shall be listed "hospital grade" and so identified. Each receptacle shall be grounded by means of an insulated copper conductor sized in accordance with Table 250-122.

Exception No. 1: Requirements of Section 517-18(b) shall not apply to psychiatric, substance abuse, and rehabilitation hospitals meeting the requirements of Section 517-10(b)(2).

Exception No. 2: Psychiatric security rooms shall not be required to have receptacle outlets installed in the room.

FPN: It is not intended that there be a total, immediate replacement of existing non-hospital grade receptacles. It is intended, however, that non-hospital grade receptacles be replaced with hospital grade receptacles upon modification of use, renovation, or as existing receptacles need replacement.

(c) Pediatric Locations. Fifteen- and 20-ampere, 125-volt receptacles intended to supply patient care areas of pediatric wards, rooms, or areas shall be listed tamper resistant or shall employ a listed tamper resistant cover.

517-19. Critical Care Areas.

(a) Patient Bed Location Branch Circuits. Each patient bed location shall be supplied by at least two branch circuits, one or more from the emergency system and one or more circuits from the normal system. At least one branch circuit from the emergency system shall supply an outlet(s) only at that bed location. All branch circuits from the normal system shall be from a single panelboard. Emergency system receptacles shall be identified and shall also indicate the panelboard and circuit number supplying them.

Exception No. 1: Branch circuits serving only special-purpose receptacles or equipment in critical care areas shall be permitted to be served by other panelboards.

Exception No. 2: Critical care locations served from two separate transfer switches on the emergency system shall not be required to have circuits from the normal system.

(b) Patient Bed Location Receptacles.

(1) Each patient bed location shall be provided with a minimum of six receptacles, at least one of which shall be connected to:

(a) The normal system branch circuit required in Section 517-19(a), or

(b) An emergency system branch circuit supplied by a different transfer switch than the other receptacles at the same location.

(2) The above receptacles shall be permitted to be of the single or duplex types, or a combination of both. All receptacles, whether six or more, shall be listed "hospital grade" and so identified. Each receptacle shall be grounded to the reference grounding point by means of an insulated copper equipment grounding conductor.

(c) Patient Vicinity Grounding and Bonding (Optional). A patient vicinity shall be permitted to have a patient equipment grounding point. The patient equipment grounding point, where supplied, shall be permitted to contain one or more jacks listed for the purpose. An equipment bonding jumper, not smaller than No. 10, shall be used to connect the grounding terminal of all grounding-type receptacles to the patient equipment grounding point. The bonding conductor shall be permitted to be arranged centrically or looped as convenient.

FPN: Where there is no patient equipment grounding point, it is important that the distance between the reference grounding point and the patient vicinity be as short as possible to minimize any potential differences.

(d) Panelboard Grounding. Where a grounded electrical distribution system is used, and metal feeder raceway or Type MC or MI cable is installed, grounding of a panelboard or switchboard shall be ensured by one of the following means at each termination or junction point of the raceway or Type MC or MI cable.

(1) A grounding bushing and a continuous copper bonding jumper, sized in accordance with Section 250-122, with the bonding jumper connected to the junction enclosure or the ground bus of the panel

(2) Connection of feeder raceways or Type MC or MI cable to threaded hubs or bosses on terminating enclosures

(3) Other approved devices such as bonding-type locknuts or bushings

(e) Additional Protective Techniques in Critical Care Areas (Optional). Isolated power systems shall be permitted to be used for critical care areas, and, if used, the isolated power system equipment shall be listed for the purpose and the system designed and installed so that it meets the provisions of and is in accordance with Section 517-160.

Exception: The audible and visual indicators of the line isolation monitor shall be permitted to be located at the nursing station for the area being served.

(f) Isolated Power System Grounding. Where an isolated ungrounded power source is used and limits the first-fault current to a low magnitude, the grounding conductor associated with the secondary circuit shall be permitted to be run outside of the enclosure of the power conductors in the same circuit.

> FPN: Although it is permitted to run the grounding conductor outside of the conduit, it is safer to run it with the power conductors to provide better protection in case of a second ground fault.

(g) Special-Purpose Receptacle Grounding. The equipment grounding conductor for special-purpose receptacles, such as the operation of mobile X-ray equipment, shall be extended to the reference grounding points of branch circuits for all locations likely to be served from such receptacles. Where such a circuit is served from an isolated ungrounded system, the grounding conductor shall not be required to be run with the power conductors; however, the equipment grounding terminal of the special-purpose receptacle shall be connected to the reference grounding point.

517-20. Wet Locations.

(a) All receptacles and fixed equipment within the area of the wet location shall have ground-fault circuit-interrupter protection for personnel if interruption of power under fault conditions can be tolerated, or be served by an isolated power system if such interruption cannot be tolerated.

Exception: Branch circuits supplying only listed, fixed, therapeutic and diagnostic equipment shall be permitted to be supplied from a normal grounded service, single- or 3-phase system, provided that

(a) Wiring for grounded and isolated circuits does not occupy the same raceway, and
(b) All conductive surfaces of the equipment are grounded.

(b) Where an isolated power system is utilized, the equipment shall be listed for the purpose and installed so that it meets the provisions of and is in accordance with Section 517-160.

> FPN: For requirements for installation of therapeutic pools and tubs, see Part F of Article 680.

517-21. Ground-Fault Circuit-Interrupter Protection for Personnel. Ground-fault circuit-interrupter protection

for personnel shall not be required for receptacles installed in those critical care areas where the toilet and basin are installed within the patient room.

C. Essential Electrical System

517-25. Scope. The essential electrical system for these facilities shall comprise a system capable of supplying a limited amount of lighting and power service, which is considered essential for life safety and orderly cessation of procedures during the time normal electrical service is interrupted for any reason. This includes clinics, medical and dental offices, outpatient facilities, nursing homes, limited care facilities, hospitals, and other health care facilities serving patients.

> FPN: For information as to the need for an essential electrical system, see *Standard for Health Care Facilities,* NFPA 99-1996.

517-30. Essential Electrical Systems for Hospitals.

(a) Applicability. The requirements of Part C, Sections 517-30 through 517-35, shall apply to hospitals where an essential electrical system is required.

> FPN No. 1: For performance, maintenance, and testing requirements of essential electrical systems in hospitals, see *Standard for Health Care Facilities,* NFPA 99-1996. For installation of centrifugal fire pumps, see *Standard for the Installation of Centrifugal Fire Pumps,* NFPA 20-1996.

> FPN No. 2: For additional information, see *Standard for Health Care Facilities,* NFPA 99-1996.

(b) General.

[x](1) Essential electrical systems for hospitals shall be comprised of two separate systems capable of supplying a limited amount of lighting and power service, which is considered essential for life safety and effective hospital operation during the time the normal electrical service is interrupted for any reason. These two systems shall be the emergency system and the equipment system.

[x](2) The emergency system shall be limited to circuits essential to life safety and critical patient care. These are designated the life safety branch and the critical branch.

[x](3) The equipment system shall supply major electrical equipment necessary for patient care and basic hospital operation.

[x](4) The number of transfer switches to be used shall be based on reliability, design, and load considerations. Each

branch of the essential electrical system shall be served by one or more transfer switches as shown in Figures 517-30(a) and 517-30(b). One transfer switch shall be permitted to serve one or more branches or systems in a facility with a maximum demand on the essential electrical system of 150 kVA as shown in Figure 517-30(c).

FPN: See *Standard for Health Care Facilities,* NFPA 99-1996: 3-4.3.2, Transfer Switch Operation Type I; 3-4.2.1.4, Automatic Transfer Switch Features; and 3-4.2.1.6, Nonautomatic Transfer Device Features.

(5) Other Loads. Loads served by the generating equipment not specifically named in Sections 517-32, 517-33,

Figure 517-30(a)   Small electrical system — hospitals.

Figure 517-30(c)   Small electrical system — hospitals (single transfer switch).

Figure 517-30(b)   Typical large electrical system — hospitals.

and 517-34 shall be served by their own transfer switches such that these loads

(a) Shall not be transferred if the transfer will overload the generating equipment, and

(b) Shall be automatically shed upon generating equipment overloading.

(6) Hospital power sources and alternate power sources shall be permitted to serve the essential electrical systems of contiguous or same site facilities.

(c) Wiring Requirements.

(1) Separation from Other Circuits. The life safety branch and critical branch of the emergency system shall be kept entirely independent of all other wiring and equipment and shall not enter the same raceways, boxes, or cabinets with each other or other wiring.

Wiring of the life safety branch and the critical branch shall be permitted to occupy the same raceways, boxes, or cabinets of other circuits not part of the branch where such wiring is as follows:

(1) In transfer equipment enclosures, or

(2) In exit or emergency lighting fixtures supplied from two sources, or

(3) In a common junction box attached to exit or emergency lighting fixtures supplied from two sources, or

(4) For two or more emergency circuits supplied from the same branch

The wiring of the equipment system shall be permitted to occupy the same raceways, boxes, or cabinets of other circuits that are not part of the emergency system.

(2) Isolated Power Systems. Where isolated power systems are installed in any of the areas in Sections 517-33(a)(1) and (a)(2), each system shall be supplied by an individual circuit serving no other load.

(3) Mechanical Protection of the Emergency System. The wiring of the emergency system of a hospital shall be mechanically protected by installation in nonflexible metal raceways, or shall be wired with Type MI cable. Where installed as branch circuit conductors serving patient care areas, the installation shall comply with the requirements of Section 517-13.

Exception No. 1: Flexible power cords of appliances, or other utilization equipment, connected to the emergency system shall not be required to be enclosed in raceways.

Exception No. 2: Secondary circuits of transformer-powered communications or signaling systems shall not be required to be enclosed in raceways unless otherwise specified by Chapters 7 or 8.

Exception No. 3: Schedule 80 rigid nonmetallic conduit shall be permitted if the branch circuits do not serve patient care areas and it is not prohibited elsewhere in this Code.

Exception No. 4: Where encased in not less than 2 in. (50.8 mm) of concrete, Schedule 40 rigid nonmetallic conduit or electrical nonmetallic tubing shall be permitted if the branch ciruits do not serve patient care areas.

Exception No. 5: Flexible metal raceways and cable assemblies shall be permitted to be used in listed prefabricated medical headwalls, listed office furnishings, or where necessary for flexible connection to equipment.

FPN: See Section 517-13(b) for additional grounding requirements in patient care areas.

(d) Capacity of Systems. The essential electrical system shall have adequate capacity to meet the demand for the operation of all functions and equipment to be served by each system and branch.

Feeders shall be sized in accordance with Articles 215 and 220. The generator set(s) shall have sufficient capacity and proper rating to meet the demand produced by the load of the essential electrical system(s) at any one time.

Demand calculations for sizing of the generator set(s) shall be based on the following:

(1) Prudent demand factors and historical data, or

(2) Connected load, or

(3) Feeder calculation procedures described in Article 220, or

(4) Any combination of the above

[X]517-31. Emergency System. Those functions of patient care depending on lighting or appliances that are connected to the emergency system shall be divided into two mandatory branches: the life safety branch and the critical branch, described in Sections 517-32 and 517-33.

The branches of the emergency system shall be installed and connected to the alternate power source so that all functions specified herein for the emergency system shall be automatically restored to operation within 10 seconds after interruption of the normal source.

[X]517-32. Life Safety Branch. No function other than those listed in (a) through (f) shall be connected to the life safety branch. The life safety branch of the emergency system shall supply power for the following lighting, receptacles, and equipment.

(a) Illumination of Means of Egress. Illumination of means of egress, such as lighting required for corridors, passageways, stairways, and landings at exit doors, and all necessary ways of approach to exits. Switching arrangements to transfer patient corridor lighting in hospitals from general illumination circuits to night illumination circuits shall be permitted provided only one of two circuits can be selected and both circuits cannot be extinguished at the same time.

> FPN: See *Life Safety Code, NFPA 101*-1997, Sections 5-8 and 5-9.

(b) Exit Signs. Exit signs and exit directional signs.

> FPN: See *Life Safety Code, NFPA 101*-1997, Section 5-10.

(c) Alarm and Alerting Systems. Alarm and alerting systems including the following:

(1) Fire alarms

> FPN: See *Life Safety Code, NFPA 101*-1997, Sections 7-6 and 12-3.4.

(2) Alarms required for systems used for the piping of nonflammable medical gases

> FPN: See *Standard for Health Care Facilities, NFPA 99*-1996, 12-3.4.1.

(d) Communications Systems. Hospital communications systems, where used for issuing instructions during emergency conditions.

(e) Generator Set Location. Task illumination battery charger for emergency battery-powered lighting unit(s) and selected receptacles at the generator set location.

(f) Elevators. Elevator cab lighting, control, communications, and signal systems.

ˣ517-33. Critical Branch.

ˣ(a) Task Illumination and Selected Receptacles. The critical branch of the emergency system shall supply power for task illumination, fixed equipment, selected receptacles, and special power circuits serving the following areas and functions related to patient care.

(1) Critical care areas that utilize anesthetizing gases — task illumination, selected receptacles, and fixed equipment
(2) The isolated power systems in special environments
(3) Patient care areas — task illumination and selected receptacles in the following:

   (a) Infant nurseries
   (b) Medication preparation areas
   (c) Pharmacy dispensing areas
   (d) Selected acute nursing areas
   (e) Psychiatric bed areas (omit receptacles)
   (f) Ward treatment rooms
   (g) Nurses' stations (unless adequately lighted by corridor luminaires)

(4) Additional specialized patient care task illumination and receptacles, where needed
(5) Nurse call systems
(6) Blood, bone, and tissue banks
(7) Telephone equipment rooms and closets
(8) Task illumination, selected receptacles, and selected power circuits for the following:

   (a) General care beds (at least one duplex receptacle per patient bedroom)
   (b) Angiographic labs
   (c) Cardiac catheterization labs
   (d) Coronary care units
   (e) Hemodialysis rooms or areas
   (f) Emergency room treatment areas (selected)
   (g) Human physiology labs
   (h) Intensive care units
   (i) Postoperative recovery rooms (selected)

(9) Additional task illumination, receptacles, and selected power circuits needed for effective hospital operation. Single-phase fractional horsepower motors shall be permitted to be connected to the critical branch.

ˣ(b) Subdivision of the Critical Branch. It shall be permitted to subdivide the critical branch into two or more branches.

> FPN: It is important to analyze the consequences of supplying an area with only critical care branch power when failure occurs between the area and the transfer switch. Some proportion of normal and critical power, or critical power from separate transfer switches, may be appropriate.

ˣ(c) Receptacle Identification. The receptacles or the faceplates for receptacles supplied by the critical branch shall have a distinctive color or marking so as to be readily recognizable.

ˣ517-34. Equipment System Connection to Alternate Power Source. The equipment system shall be installed and connected to the alternate power source, such that the equipment described in Section 517-34(a) is automatically restored to operation at appropriate time-lag intervals following the energizing of the emergency system. Its arrangement shall also provide for the subsequent connection of equipment described in Section 517-34(b).

ˣ(a) Equipment for Delayed Automatic Connection. The following equipment shall be arranged for delayed automatic connection to the alternate power source.

(1) Central suction systems serving medical and surgical functions, including controls. Such suction systems shall be permitted on the critical branch.

(2) Sump pumps and other equipment required to operate for the safety of major apparatus, including associated control systems and alarms.

(3) Compressed air systems serving medical and surgical functions, including controls. Such air systems shall be permitted on the critical branch.

(4) Smoke control and stair pressurization systems, or both.

(5) Kitchen hood supply or exhaust systems, or both, if required to operate during a fire in or under the hood.

Exception: Sequential delayed automatic connection to the alternate power source to prevent overloading the generator shall be permitted where engineering studies indicate it is necessary.

ˣ**(b) Equipment for Delayed Automatic or Manual Connection.** The following equipment shall be arranged for either delayed automatic or manual connection to the alternate power source:

(1) Heating equipment to provide heating for operating, delivery, labor, recovery, intensive care, coronary care, nurseries, infection/isolation rooms, emergency treatment spaces, and general patient rooms

Exception: Heating of general patient rooms and infection/isolation rooms during disruption of the normal source shall not be required under any of the following conditions:

(a) The outside design temperature is higher than +20°F (–6.7°C), or

(b) The outside design temperature is lower than +20°F (–6.7°C) and where a selected room(s) is provided for the needs of all confined patients, then only such room(s) need be heated, or

(c) The facility is served by a dual source of normal power.

FPN No. 1: The design temperature is based on the 97½ percent design value as shown in Chapter 24 of the ASHRAE *Handbook of Fundamentals* (1997).

FPN No. 2: For a description of a dual source of normal power, see Section 517-35(c), FPN.

(2) An elevator(s) selected to provide service to patient, surgical, obstetrical, and ground floors during interruption of normal power

In instances where interruption of normal power would result in other elevators stopping between floors, throw-over facilities shall be provided to allow the temporary operation of any elevator for the release of patients or other persons who may be confined between floors.

(3) Supply, return, and exhaust ventilating systems for surgical and obstetrical delivery suites, intensive care,

coronary care, nurseries, infection/isolation rooms, emergency treatment spaces, and exhaust fans for laboratory fume hoods, nuclear medicine areas where radioactive material is used, ethylene oxide evacuation, and anesthesia evacuation

(4) Hyperbaric facilities

(5) Hypobaric facilities

(6) Automatically operated doors

(7) Minimal electrically heated autoclaving equipment shall be permitted to be arranged for either automatic or manual connection to the alternate source

(8) Controls for equipment listed in Section 517-34

(9) Other selected equipment shall be permitted to be served by the equipment system.

517-35. Sources of Power.

ˣ**(a) Two Independent Sources of Power.** Essential electrical systems shall have a minimum of two independent sources of power: a normal source generally supplying the entire electrical system and one or more alternate sources for use when the normal source is interrupted.

(b) Alternate Source of Power. The alternate source of power shall be one of the following:

(1) Generator(s) driven by some form of prime mover(s) and located on the premises

(2) Another generating unit(s) where the normal source consists of a generating unit(s) located on the premises

(3) An external utility service when the normal source consists of a generating unit(s) located on the premises

(c) Location of Essential Electrical System Components. Careful consideration shall be given to the location of the spaces housing the components of the essential electrical system to minimize interruptions caused by natural forces common to the area (e.g., storms, floods, earthquakes, or hazards created by adjoining structures or activities). Consideration shall also be given to the possible interruption of normal electrical services resulting from similar causes as well as possible disruption of normal electrical service due to internal wiring and equipment failures.

FPN: Facilities in which the normal source of power is supplied by two or more separate central station-fed services experience greater than normal electrical service reliability than those with only a single feed. Such a dual source of normal power consists of two or more electrical services fed from separate generator sets or a utility distribution network that has multiple power input sources and is arranged to provide mechanical and electrical separation so that a fault between the facility and the generating sources will not likely cause an interruption of more than one of the facility service feeders.

517-40. Essential Electrical Systems for Nursing Homes and Limited Care Facilities.

ˣ**(a) Applicability.** The requirements of Part C, Sections 517-40(c) through 517-44, shall apply to nursing homes and limited care facilities.

Exception: The requirements of Part C, Section 517-40(c) through 517-44, shall not apply to freestanding buildings used as nursing homes and limited care facilities, provided that:

(a) Admitting and discharge policies are maintained that preclude the provision of care for any patient or resident who may need to be sustained by electrical life-support equipment.

(b) No surgical treatment requiring general anesthesia is offered.

(c) An automatic battery-operated system(s) or equipment is provided that shall be effective for at least 1½ hours and is otherwise in accordance with Section 700-12 and that shall be capable of supplying lighting for exit lights, exit corridors, stairways, nursing stations, medical preparation areas, boiler rooms, and communications areas. This system shall also supply power to operate all alarm systems.

    FPN: See *Life Safety Code, NFPA 101-1997.*

(b) Inpatient Hospital Care Facilities. Nursing homes and limited care facilities that provide inpatient hospital care shall comply with the requirements of Part C, Sections 517-30 through 517-35.

(c) Facilities Contiguous or Located on the Same Site with Hospitals. Nursing homes and limited care facilities that are contiguous or located on the same site with a hospital shall be permitted to have their essential electrical systems supplied by that of the hospital.

    FPN: For performance, maintenance, and testing requirements of essential electrical systems in nursing homes and limited care facilities, see *Standard for Health Care Facilities, NFPA 99-1996.*

517-41. Essential Electrical Systems.

ˣ**(a) General.** Essential electrical systems for nursing homes and limited care facilities shall be comprised of two separate branches capable of supplying a limited amount of lighting and power service, which is considered essential for the protection of life safety and effective operation of the institution during the time normal electrical service is interrupted for any reason. These two separate branches shall be the life safety branch and the critical branch.

ˣ**(b) Transfer Switches.** The number of transfer switches to be used shall be based on reliability, design, and load considerations. Each branch of the essential electrical system

shall be served by one or more transfer switches as shown in Figures 517-41(a) and 517-41(b). One transfer switch shall be permitted to serve one or more branches or systems in a facility with a maximum demand on the essential electrical system of 150 kVa as shown in Figure 517-41(c).

    FPN: See *Standard for Health Care Facilities,* NFPA 99-1996, 3-5.3.2, Transfer Switch Operation Type II; 3-4.2.1.4, Automatic Transfer Switch Features; and 3-4.2.1.6, Nonautomatic Transfer Device Features.

(c) Capacity of System. The essential electrical system shall have adequate capacity to meet the demand for the operation of all functions and equipment to be served by each branch at one time.

(d) Separation from Other Circuits. The life safety branch shall be kept entirely independent of all other wiring and equipment and shall not enter the same raceways, boxes, or cabinets with other wiring except as follows:

(1) In transfer switches
(2) In exit or emergency lighting fixtures supplied from two sources, or
(3) In a common junction box attached to exit or emergency lighting fixtures supplied from two sources

The wiring of the critical branch shall be permitted to occupy the same raceways, boxes, or cabinets of other circuits that are not part of the life safety branch.

Figure 517-41(a)   Small electrical system — nursing homes and limited care facilities.

Figure 517-41(b)   Typical large electrical system — nursing homes and limited care facilities.

Figure 517-41(c)   Small electrical system — nursing homes and limited care facilities (single transfer switch).

ˣ517-42. Automatic Connection to Life Safety Branch. The life safety branch shall be installed and connected to the alternate source of power so that all functions specified herein shall be automatically restored to operation within 10 seconds after the interruption of the normal source. No functions other than those listed in (a) through (g) shall be connected to the life safety branch. The life safety branch shall supply power for the following lighting, receptacles, and equipment.

FPN: The life safety branch is called the emergency system in *Standard for Health Care Facilities,* NFPA 99-1996.

(a) Illumination of Means of Egress. Illumination of means of egress as is necessary for corridors, passageways, stairways, landings, and exit doors and all ways of approach to exits. Switching arrangement to transfer patient corridor lighting from general illumination circuits shall be permitted providing only one of two circuits can be selected and both circuits cannot be extinguished at the same time.

FPN: See *Life Safety Code,* NFPA *101*-1997, Sections 5-8 and 5-9.

(b) Exit Signs. Exit signs and exit directional signs.

FPN: See *Life Safety Code*, NFPA *101*-1997, Section 5-10.

(c) Alarm and Alerting Systems. Alarm and alerting systems, including the following:

(1) Fire alarms

FPN: See *Life Safety Code,* NFPA *101*-1997, Sections 7-6 and 12-3.4.

(2) Alarms required for systems used for the piping of non-flammable medical gases

> FPN: See *Standard for Health Care Facilities,* NFPA 99-1996, Section 16-3.4.1.

(d) Communications Systems. Communications systems, where used for issuing instructions during emergency conditions.

(e) Dining and Recreation Areas. Sufficient lighting in dining and recreation areas to provide illumination to exit ways.

(f) Generator Set Location. Task illumination and selected receptacles in the generator set location.

(g) Elevators. Elevator cab lighting, control, communications, and signal systems.

ˣ517-43. Connection to Critical Branch. The critical branch shall be installed and connected to the alternate power source so that the equipment listed in Section 517-43(a) shall be automatically restored to operation at appropriate time-lag intervals following the restoration of the life safety branch to operation. Its arrangement shall also provide for the additional connection of equipment listed in Section 517-43(b) by either delayed automatic or manual operation.

(a) Delayed Automatic Connection. The following equipment shall be connected to the critical branch and shall be arranged for delayed automatic connection to the alternate power source.

(1) Patient care areas — task illumination and selected receptacles in the following:

    (a) Medication preparation areas
    (b) Pharmacy dispensing areas
    (c) Nurses' stations (unless adequately lighted by corridor luminaires)

(2) Sump pumps and other equipment required to operate for the safety of major apparatus and associated control systems and alarms
(3) Smoke control and stair pressurization systems
(4) Kitchen hood supply and/or exhaust systems, if required to operate during a fire in or under the hood

(b) Delayed Automatic or Manual Connection. The following equipment shall be connected to the critical branch and shall be arranged for either delayed automatic or manual connection to the alternate power source.

  (1) Heating equipment to provide heating for patient rooms.

Exception: Heating of general patient rooms during disruption of the normal source shall not be required under any of the following conditions.

(a) *The outside design temperature is higher than +20°F (−6.7°C), or*
(b) *The outside design temperature is lower than +20°F (−6.7°C) and where a selected room(s) is provided for the needs of all confined patients, then only such room(s) need be heated.*
(c) *The facility is served by a dual source of normal power as described in Section 517-44(c), FPN.*

> FPN: The outside design temperature is based on the 97½ percent design values as shown in Chapter 24 of the ASHRAE *Handbook of Fundamentals* (1997).

(2) Elevator Service. In instances where disruption of power would result in elevators stopping between floors, throw-over facilities shall be provided to allow the temporary operation of any elevator for the release of passengers. For elevator cab lighting, control, and signal system requirements, see Section 517-42(g).

(3) Additional illumination, receptacles, and equipment shall be permitted to be connected only to the critical branch.

517-44. Sources of Power.

ˣ(a) Two Independent Sources of Power. Essential electrical systems shall have a minimum of two independent sources of power: a normal source generally supplying the entire electrical system and one or more alternate sources for use when the normal source is interrupted.

ˣ(b) Alternate Source of Power. The alternate source of power shall be a generator(s) driven by some form of prime mover(s) and located on the premises.

Exception No. 1: Where the normal source consists of generating units on the premises, the alternate source shall be either another generator set or an external utility service.

Exception No. 2: Nursing homes or limited care facilities meeting the requirements of Section 517-40(a), Exception, shall be permitted to use a battery system or self-contained battery integral with the equipment.

(c) Location of Essential Electrical System Components. Careful consideration shall be given to the location of the spaces housing the components of the essential electrical system to minimize interruptions caused by natural forces common to the area (e.g., storms, floods, earthquakes, or hazards created by adjoining structures or activities). Consideration shall also be given to the possible interruption of normal electrical services resulting from similar causes as well as possible disruption of normal electrical service due to internal wiring and equipment failures.

FPN: Facilities in which the normal source of power is supplied by two or more separate central station-fed services experience greater than normal electrical service reliability than those with only a single feed. Such a dual source of normal power consists of two or more electrical services fed from separate generator sets or a utility distribution network that has multiple power input sources and is arranged to provide mechanical and electrical separation so that a fault between the facility and the generating sources will not likely cause an interruption of more than one of the facility service feeders.

517-45. Essential Electrical Systems for Ambulatory Health Care Centers.

(a) **Applicability.** The requirements of this section shall apply to those health care facilities described in Section 517-45.

ˣ(b) **Connections.** The essential electrical system shall supply power for the following:

(1) Task illumination that is related to the safety of life and that is necessary for the safe cessation of procedures in progress
(2) All anesthesia and resuscitative equipment used in areas where inhalation anesthetics are administered to patients, including alarm and alerting devices

> FPN: See *Standard for Health Care Facilities,* NFPA 99-1996, 13-3.4.1.

(3) All electrical life-support equipment in areas where procedures are performed that require such equipment for the support of the patient's life

(c) **Alternate Source of Power.**

ˣ(1) **Power Source.** The alternate source of power for the system shall be specifically designed for this purpose and shall be either a generator, battery system, or self-contained battery integral with the equipment. Where critical care areas are present in the facility, the essential electrical system shall be as required in Sections 517-30 through 517-35.

ˣ(2) **System Capacity.** The alternate source of power shall be separate and independent of the normal source and shall have a capacity to sustain its connected loads for a minimum of 1½ hours after loss of the normal source.

ˣ(3) **System Operation.** The system shall be arranged so that, in the event of a failure of the normal power source, the alternate source of power shall be automatically connected to the load within 10 seconds.

> FPN: See *Standard for Health Care Facilities,* NFPA 99-1996, 3-6.3.2, Transfer Switch Operation for Type III with Generator Sets, and 3-6.3.3, Transfer Switch Operation for Type III with Battery Systems.

517-50. Essential Electrical Systems for Clinics, Medical and Dental Offices, and Other Health Care Facilities Not Covered in Sections 517-30, 517-40, and 517-45.

(a) **Applicability.** The requirements of this section shall apply to those health care facilities described in Section 517-50.

ˣ(b) **Connections.** The essential electrical system shall supply power for the following:

(1) Task illumination that is related to the safety of life and that is necessary for the safe cessation of procedures in progress
(2) All anesthesia and resuscitative equipment used in areas where inhalation anesthetics are administered to patients, including alarm and alerting devices

> FPN: See *Standard for Health Care Facilities*, NFPA 99-1996, 14-3.4 and 15-3.4.1.

(c) **Alternate Source of Power.**

ˣ(1) **Power Source.** The alternate source of power for the system shall be specifically designed for this purpose and shall be either a generator, battery system, or self-contained battery integral with the equipment. Where electrical life-support equipment is required, the essential electrical system shall be as required in Sections 517-30 through 517-35.

ˣ(2) **System Capacity.** The alternate source of power shall be separate and independent of the normal source and shall have a capacity to sustain its connected loads for a minimum of 1½ hours after loss of the normal source.

ˣ(3) **System Operation.** The system shall be arranged so that, in the event of a failure of the normal power source, the alternate source of power shall be automatically connected to the load within 10 seconds.

> FPN: See *Standard for Health Care Facilities,* NFPA 99-1996, 3-6.3.2, Transfer Switch Operation for Type III with Generator Sets, and 3-6.3.3, Transfer Switch Operation for Type III with Battery Systems.

D. Inhalation Anesthetizing Locations

> FPN: For further information regarding safeguards for anesthetizing locations, see *Standard for Health Care Facilities* NFPA 99-1996.

517-60. Anesthetizing Location Classification.

> FPN: If either of the following anesthetizing locations is designated a wet location, refer to Section 517-20.

(a) **Hazardous (Classified) Location.**

ˣ(1) In a location where flammable anesthetics are employed, the entire area shall be considered to be a Class I,

Division 1 location that shall extend upward to a level 5 ft (1.52 m) above the floor. The remaining volume up to the structural ceiling is considered to be above a hazardous (classified) location.

(2) Any room or location in which flammable anesthetics or volatile flammable disinfecting agents are stored shall be considered to be a Class I, Division 1 location from floor to ceiling.

(b) Other-than-Hazardous (Classified) Location. Any inhalation anesthetizing location designated for the exclusive use of nonflammable anesthetizing agents shall be considered to be an other-than-hazardous (classified) location.

517-61. Wiring and Equipment.

(a) Within Hazardous (Classified) Anesthetizing Locations.

ˣ**(1)** Except as permitted in Section 517-160, each power circuit within, or partially within, a flammable anesthetizing location as referred to in Section 517-60 shall be isolated from any distribution system by the use of an isolated power system.

(2) Isolated power system equipment shall be listed for the purpose and the system designed and installed so that it meets the provisions and is in accordance with Part G.

ˣ**(3)** In hazardous (classified) locations referred to in Section 517-60, all fixed wiring and equipment, and all portable equipment, including lamps and other utilization equipment, operating at more than 10 volts between conductors shall comply with the requirements of Sections 501-1 through 501-15 and Sections 501-16(a) and (b) for Class I, Division 1 locations. All such equipment shall be specifically approved for the hazardous atmospheres involved.

(4) Where a box, fitting, or enclosure is partially, but not entirely, within a hazardous (classified) location(s), the hazardous (classified) location(s) shall be considered to be extended to include the entire box, fitting, or enclosure.

(5) Receptacles and attachment plugs in a hazardous (classified) location(s) shall be listed for use in Class I, Group C hazardous (classified) locations and shall have provision for the connection of a grounding conductor.

(6) Flexible cords used in hazardous (classified) locations for connection to portable utilization equipment, including lamps operating at more than 8 volts between conductors, shall be of a type approved for extra-hard usage in accordance with Table 400-4 and shall include an additional conductor for grounding.

(7) A storage device for the flexible cord shall be provided and shall not subject the cord to bending at a radius of less than 3 in. (76 mm).

(b) Above Hazardous (Classified) Anesthetizing Locations.

(1) Wiring above a hazardous (classified) location referred to in Section 517-60 shall be installed in rigid metal conduit, electrical metallic tubing, intermediate metal conduit, Type MI cable, or Type MC cable that employs a continuous, gas/vaportight metal sheath.

(2) Installed equipment that may produce arcs, sparks, or particles of hot metal, such as lamps and lampholders for fixed lighting, cutouts, switches, generators, motors, or other equipment having make-and-break or sliding contacts, shall be of the totally enclosed type or be constructed so as to prevent escape of sparks or hot metal particles.

Exception: Wall-mounted receptacles installed above the hazardous (classified) location in flammable anesthetizing locations shall not be required to be totally enclosed or have openings guarded or screened to prevent dispersion of particles.

(3) Surgical and other lighting fixtures shall conform to Section 501-9(b).

Exception No. 1: The surface temperature limitations set forth in Section 501-9(b)(2) shall not apply.

Exception No. 2: Integral or pendant switches that are located above and cannot be lowered into the hazardous (classified) location(s) shall not be required to be explosionproof.

(4) Approved seals shall be provided in conformance with Section 501-5, and Section 501-5(a)(4) shall apply to horizontal as well as to vertical boundaries of the defined hazardous (classified) locations.

(5) Receptacles and attachment plugs located above hazardous (classified) anesthetizing locations shall be listed for hospital use for services of prescribed voltage, frequency, rating, and number of conductors with provision for the connection of the grounding conductor. This requirement shall apply to attachment plugs and receptacles of the 2-pole, 3-wire grounding type for single-phase, 120-volt, nominal, ac service.

(6) Plugs and receptacles rated 250-volts, for connection of 50-ampere, and 60-ampere ac medical equipment for use above hazardous (classified) locations shall be arranged so that the 60-ampere receptacle will accept either the 50-ampere or the 60-ampere plug. Fifty-ampere receptacles shall be designed so as not to accept the 60-ampere attachment plug. The plugs shall be of the 2-pole, 3-wire design with a third contact connecting to the insulated (green or green with yellow stripe) equipment grounding conductor of the electrical system.

(c) Other-than-Hazardous (Classified) Anesthetizing Locations.

(1) Wiring serving other-than-hazardous (classified) locations, as defined in Section 517-60, shall be installed in a metal raceway system or cable assembly. The metal raceway system, or cable armor or sheath assembly, shall qualify as an equipment grounding return path in accordance with Section 250-118. Type MC and Type MI cable shall have an outer metal armor or sheath that is identified as an acceptable grounding return path.

Exception: Pendant receptacle constructions that employ at least Type SJO or equivalent flexible cords suspended not less than 6 ft (1.83 m) from the floor shall not be required to be installed in a metal raceway or cable assembly.

(2) Receptacles and attachment plugs installed and used in other-than-hazardous (classified) locations shall be listed for hospital use for services of prescribed voltage, frequency, rating, and number of conductors with provision for connection of the grounding conductor. This requirement shall apply to 2-pole, 3-wire grounding type for single-phase, 120-, 208-, or 240-volt, nominal, ac service.

(3) Plugs and receptacles rated 250 volts, for connection of 50-ampere, and 60-ampere ac medical equipment for use in other-than-hazardous (classified) locations shall be arranged so that the 60-ampere receptacle will accept either the 50-ampere or the 60-ampere plug. Fifty-ampere receptacles shall be designed so as not to accept the 60-ampere attachment plug. The plugs shall be of the 2-pole, 3-wire design with a third contact connecting to the insulated (green or green with yellow stripe) equipment grounding conductor of the electrical system.

517-62. Grounding. In any anesthetizing area, all metal raceways and metal-sheathed cables, and all noncurrent-carrying conductive portions of fixed electric equipment, shall be grounded. Grounding in Class I locations shall comply with Section 501-16.

Exception: Equipment operating at not more than 10 volts between conductors shall not be required to be grounded.

517-63. Grounded Power Systems in Anesthetizing Locations.

(a) Battery-Powered Emergency Lighting Units. One or more battery-powered emergency lighting units shall be provided in accordance with Section 700-12(e).

(b) Branch-Circuit Wiring. Branch circuits supplying only listed, fixed, therapeutic and diagnostic equipment, permanently installed above the hazardous (classified) location and in other-than-hazardous (classified) locations, shall be

permitted to be supplied from a normal grounded service, single- or three-phase system, provided the following:

(1) Wiring for grounded and isolated circuits does not occupy the same raceway or cable,
(2) All conductive surfaces of the equipment are grounded,
(3) Equipment (except enclosed X-ray tubes and the leads to the tubes) are located at least 8 ft (2.44 m) above the floor or outside the anesthetizing location, and
(4) Switches for the grounded branch circuit are located outside the hazardous (classified) location.

Exception: Sections 517-63(b)(3) and (b)(4) shall not apply in other-than-hazardous (classified) locations.

(c) Fixed Lighting Branch Circuits. Branch circuits supplying only fixed lighting shall be permitted to be supplied by a normal grounded service, provided the following:

(1) Such fixtures are located at least 8 ft (2.44 m) above the floor,
(2) All conductive surfaces of fixtures are grounded,
(3) Wiring for circuits supplying power to fixtures does not occupy the same raceway or cable for circuits supplying isolated power, and
(4) Switches are wall-mounted and located above hazardous (classified) locations.

Exception: Sections 517-63(c)(1) and (c)(4) shall not apply in other-than-hazardous (classified) locations.

(d) Remote-Control Stations. Wall-mounted remote-control stations for remote-control switches operating at 24 volts or less shall be permitted to be installed in any anesthetizing location.

(e) Location of Isolated Power Systems. An isolated power center listed for the purpose and its grounded primary feeder shall be permitted to be located in an anesthetizing location, provided it is installed above a hazardous (classified) location or in an other-than-hazardous (classified) location.

(f) Circuits in Anesthetizing Locations. Except as permitted above, each power circuit within, or partially within, a flammable anesthetizing location as referred to in Section 517-60 shall be isolated from any distribution system supplying other-than-anesthetizing locations.

517-64. Low-Voltage Equipment and Instruments.

(a) Equipment Requirements. Low-voltage equipment that is frequently in contact with the bodies of persons or has exposed current-carrying elements shall be as follows:

(1) Operate on an electrical potential of 10 volts or less, or

(2) Approved as intrinsically safe or double-insulated equipment, or

(3) Moisture resistant

(b) Power Supplies. Power shall be supplied to low-voltage equipment from the following:

(1) An individual portable isolating transformer (autotransformers shall not be used) connected to an isolated power circuit receptacle by means of an appropriate cord and attachment plug, or

(2) A common low-voltage isolating transformer installed in an other-than-hazardous (classified) location, or

(3) Individual dry-cell batteries, or

(4) Common batteries made up of storage cells located in an other-than-hazardous (classified) location

(c) Isolated Circuits. Isolating-type transformers for supplying low-voltage circuits shall have the following:

(1) Approved means for insulating the secondary circuit from the primary circuit, and

(2) Have the core and case grounded

(d) Controls. Resistance or impedance devices shall be permitted to control low-voltage equipment but shall not be used to limit the maximum available voltage to the equipment.

(e) Battery-Powered Appliances. Battery-powered appliances shall not be capable of being charged while in operation unless their charging circuitry incorporates an integral isolating-type transformer.

(f) Receptacles or Attachment Plugs. Any receptacle or attachment plug used on low-voltage circuits shall be of a type that does not permit interchangeable connection with circuits of higher voltage.

> FPN: Any interruption of the circuit, even circuits as low as 10 volts, either by any switch, or loose or defective connections anywhere in the circuit, may produce a spark that is sufficient to ignite flammable anesthetic agents. See 7-5.1.2.3 of *Standard for Health Care Facilities,* NFPA 99-1996.

E. X-Ray Installations

Nothing in this part shall be construed as specifying safeguards against the useful beam or stray X-ray radiation.

> FPN No. 1: Radiation safety and performance requirements of several classes of X-ray equipment are regulated under Public Law 90-602 and are enforced by the Department of Health and Human Services.

> FPN No. 2: In addition, information on radiation protection by the National Council on Radiation Protection and Measurements is published as *Reports of the National Council on Radiation Protection and Measurement.* These reports are obtainable from NCRP Publications, P.O. Box 30175, Washington, DC 20014.

517-71. Connection to Supply Circuit.

(a) Fixed and Stationary Equipment. Fixed and stationary X-ray equipment shall be connected to the power supply by means of a wiring method that meets the general requirements of this *Code.*

Exception: Equipment properly supplied by a branch circuit rated at not over 30 amperes shall be permitted to be supplied through a suitable attachment plug and hard-service cable or cord.

(b) Portable, Mobile, and Transportable Equipment. Individual branch circuits shall not be required for portable, mobile, and transportable medical X-ray equipment requiring a capacity of not over 60 amperes.

(c) Over 600-Volt Supply. Circuits and equipment operated on a supply circuit of over 600 volts shall comply with Article 490.

517-72. Disconnecting Means.

(a) Capacity. A disconnecting means of adequate capacity for at least 50 percent of the input required for the momentary rating or 100 percent of the input required for the long-time rating of the X-ray equipment, whichever is greater, shall be provided in the supply circuit.

(b) Location. The disconnecting means shall be operable from a location readily accessible from the X-ray control.

(c) Portable Equipment. For equipment connected to a 120-volt branch circuit of 30 amperes or less, a grounding-type attachment plug and receptacle of proper rating shall be permitted to serve as a disconnecting means.

517-73. Rating of Supply Conductors and Overcurrent Protection.

(a) Diagnostic Equipment.

(1) The ampacity of supply branch-circuit conductors and the current rating of overcurrent protective devices shall not be less than 50 percent of the momentary rating or 100 percent of the long-time rating, whichever is greater.

(2) The ampacity of supply feeders and the current rating of overcurrent protective devices supplying two or more branch circuits supplying X-ray units shall not be less than 50 percent of the momentary demand rating of the largest unit plus 25 percent of the momentary demand rating of the

next largest unit plus 10 percent of the momentary demand rating of each additional unit. Where simultaneous biplane examinations are undertaken with the X-ray units, the supply conductors and overcurrent protective devices shall be 100 percent of the momentary demand rating of each X-ray unit.

> FPN: The minimum conductor size for branch and feeder circuits is also governed by voltage regulation requirements. For a specific installation, the manufacturer usually specifies minimum distribution transformer and conductor sizes, rating of disconnecting means, and overcurrent protection.

(b) Therapeutic Equipment. The ampacity of conductors and rating of overcurrent protective devices shall not be less than 100 percent of the current rating of medical X-ray therapy equipment.

> FPN: The ampacity of the branch-circuit conductors and the ratings of disconnecting means and overcurrent protection for X-ray equipment are usually designated by the manufacturer for the specific installation.

517-74. Control Circuit Conductors.

(a) Number of Conductors in Raceway. The number of control circuit conductors installed in a raceway shall be determined in accordance with Section 300-17.

(b) Minimum Size of Conductors. Size No. 18 or No. 16 fixture wires as specified in Section 725-27 and flexible cords shall be permitted for the control and operating circuits of X-ray and auxiliary equipment where protected by not larger than 20-ampere overcurrent devices.

517-75. Equipment Installations.
All equipment for new X-ray installations and all used or reconditioned X-ray equipment moved to and reinstalled at a new location shall be of an approved type.

517-76. Transformers and Capacitors.
Transformers and capacitors that are part of X-ray equipment shall not be required to comply with Articles 450 and 460.

Capacitors shall be mounted within enclosures of insulating material or grounded metal.

517-77. Installation of High-Tension X-ray Cables.
Cables with grounded shields connecting X-ray tubes and image intensifiers shall be permitted to be installed in cable trays or cable troughs along with X-ray equipment control and power supply conductors without the need for barriers to separate the wiring.

517-78. Guarding and Grounding.

(a) High-Voltage Parts. All high-voltage parts, including X-ray tubes, shall be mounted within grounded enclosures. Air, oil, gas, or other suitable insulating media shall be used to insulate the high-voltage from the grounded enclosure. The connection from the high-voltage equipment to X-ray tubes and other high-voltage components shall be made with high-voltage shielded cables.

(b) Low-Voltage Cables. Low-voltage cables connecting to oil-filled units that are not completely sealed, such as transformers, condensers, oil coolers, and high-voltage switches, shall have insulation of the oil-resistant type.

(c) Noncurrent-Carrying Metal Parts. Noncurrent-carrying metal parts of X-ray and associated equipment (controls, tables, X-ray tube supports, transformer tanks, shielded cables, X-ray tube heads, etc.) shall be grounded in the manner specified in Article 250, as modified by Sections 517-13(a) and (b).

F. Communications, Signaling Systems, Data Systems, Fire Alarm Systems, and Systems Less than 120 Volts, Nominal

517-80. Patient Care Areas.
Equivalent insulation and isolation to that required for the electrical distribution systems in patient care areas shall be provided for communications, signaling systems, data system circuits, fire alarm systems, and systems less than 120 volts, nominal.

> FPN: An acceptable alternate means of providing isolation for patient/nurse call systems is by the use of nonelectrified signaling, communications, or control devices held by the patient or within reach of the patient.

517-81. Other-than-Patient-Care Areas.
In other-than-patient-care areas, installations shall be in accordance with the appropriate provisions of Articles 640, 725, 760, and 800.

517-82. Signal Transmission Between Appliances.

(a) General. Permanently installed signal cabling from an appliance in a patient location to remote appliances shall employ a signal transmission system that prevents hazardous grounding interconnection of the appliances.

> FPN: See Section 517-13(b).

(b) Common Signal Grounding Wire. Common signal grounding wires (i.e., the chassis ground for single-ended transmission) shall be permitted to be used between appliances all located within the patient vicinity, provided the appliances are served from the same reference grounding point.

G. Isolated Power Systems

517-160. Isolated Power Systems.

(a) Installations.

(1) Each isolated power circuit shall be controlled by a switch that has a disconnecting pole in each isolated circuit conductor to simultaneously disconnect all power. Such isolation shall be accomplished by means of one or more transformers having no electrical connection between primary and secondary windings, by means of motor generator sets, or by means of suitably isolated batteries.

ˣ(2) Circuits supplying primaries of isolating transformers shall operate at not more than 600 volts between conductors and shall be provided with proper overcurrent protection. The secondary voltage of such transformers shall not exceed 600 volts between conductors of each circuit. All circuits supplied from such secondaries shall be ungrounded and shall have an approved overcurrent device of proper ratings in each conductor. Circuits supplied directly from batteries or from motor generator sets shall be ungrounded, and shall be protected against overcurrent in the same manner as transformer-fed secondary circuits. If an electrostatic shield is present, it shall be connected to the reference grounding point.

(3) The isolating transformers, motor generator sets, batteries and battery chargers, and associated primary or secondary overcurrent devices shall not be installed in hazardous (classified) locations. The isolated secondary circuit wiring extending into a hazardous anesthetizing location shall be installed in accordance with Section 501-4.

ˣ(4) Isolation Transformers. An isolation transformer shall not serve more than one operating room except as covered in (a) and (b).

For purposes of this section, anesthetic induction rooms are considered part of the operating room or rooms served by the induction rooms.

(a) *Induction Rooms.* Where an induction room serves more than one operating room, the isolated circuits of the induction room shall be permitted to be supplied from the isolation transformer of any one of the operating rooms served by that induction room.

(b) *Higher Voltages.* Isolation transformers shall be permitted to serve single receptacles in several patient areas where

(1) The receptacles are reserved for supplying power to equipment requiring 150 volts or higher, such as portable X-ray units, and

(2) The receptacles and mating plugs are not interchangeable with the receptacles on the local isolated power system.

(5) The isolated circuit conductors shall be identified as follows:

Isolated Conductor No. 1 — Orange

Isolated Conductor No. 2 — Brown

For 3-phase systems, the third conductor shall be identified as yellow. Where isolated circuit conductors supply 125-volt, single-phase, 15- and 20-ampere receptacles, the orange conductor(s) shall be connected to the terminal(s) on the receptacles that are identified in accordance with Section 200-10(b) for connection to the grounded circuit conductor.

(6) Wire-pulling compounds that increase the dielectric constant shall not be used on the secondary conductors of the isolated power supply.

FPN No. 1: It is desirable to limit the size of the isolation transformer to 10 kVA or less and to use conductor insulation with low leakage to meet impedance requirements.

FPN No. 2: Minimizing the length of branch-circuit conductors and using conductor insulations with a dielectric constant less than 3.5 and insulation resistance constant greater than 6100 megohm-meters (20,000 megohm-ft) at 16°C (60°F) reduces leakage from line to ground reducing the hazard current.

(b) Line Isolation Monitor.

(1) In addition to the usual control and overcurrent protective devices, each isolated power system shall be provided with a continually operating line isolation monitor that indicates total hazard current. The monitor shall be designed so that a green signal lamp, conspicuously visible to persons in each area served by the isolated power system, remains lighted when the system is adequately isolated from ground. An adjacent red signal lamp and an audible warning signal (remote if desired) shall be energized when the total hazard current (consisting of possible resistive and capacitive leakage currents) from either isolated conductor to ground reaches a threshold value of 5 mA under nominal line voltage conditions. The line monitor shall not alarm for a fault hazard of less than 3.7 mA or for a total hazard current of less than 5 mA.

Exception: A system shall be permitted to be designed to operate at a lower threshold value of total hazard current. A line isolation monitor for such a system shall be permitted to be approved with the provision that the fault hazard current shall be permitted to be reduced but not to less than 35 percent of the corresponding threshold value of the total hazard current, and the monitor hazard current is to be correspondingly reduced to not more than 50 percent of the alarm threshold value of the total hazard current.

(2) The line isolation monitor shall be designed to have sufficient internal impedance such that, when properly connected to the isolated system, the maximum internal current that can flow through the line isolation monitor, when any point of the isolated system is grounded, shall be 1 mA.

Exception: The line isolation monitor shall be permitted to be of the low-impedance type such that the current through the line isolation monitor, when any point of the isolated system is grounded, will not exceed twice the alarm threshold value for a period not exceeding 5 milliseconds.

FPN: Reduction of the monitor hazard current, provided this reduction results in an increased "not alarm" threshold value for the fault hazard current, will increase circuit capacity.

(3) An ammeter calibrated in the total hazard current of the system (contribution of the fault hazard current plus monitor hazard current) shall be mounted in a plainly visible place on the line isolation monitor with the "alarm on" zone at approximately the center of the scale.

Exception: The line isolation monitor shall be permitted to be a composite unit, with a sensing section cabled to a separate display panel section on which the alarm or test functions are located.

FPN: It is desirable to locate the ammeter so that it is conspicuously visible to persons in the anesthetizing location.

Article 518 — Places of Assembly

518-1. Scope. This article covers all buildings or portions of buildings or structures designed or intended for the assembly of 100 or more persons.

518-2. General Classifications.

(a) Examples. Places of assembly shall include, but not be limited to, the following:

Armories	Courtrooms
Assembly halls	Dance halls
Auditoriums	Dining facilities
Auditoriums within	Exhibition halls
Business establishments	Gymnasiums
Mercantile establishments	Mortuary chapels
Other occupancies	Multipurpose rooms
Schools	Museums
Bowling lanes	Places of awaiting transportation
Church chapels	Pool rooms
Club rooms	Restaurants
Conference rooms	Skating rinks

(b) Multiple Occupancies. Occupancy of any room or space for assembly purposes by less than 100 persons in a building of other occupancy, and incidental to such other occupancy, shall be classified as part of the other occupancy and subject to the provisions applicable thereto.

(c) Theatrical Areas. Where any such building structure, or portion thereof, contains a projection booth or stage platform or area for the presentation of theatrical or musical productions, either fixed or portable, the wiring for that area, including associated audience seating areas, and all equipment that is used in the referenced area, and portable equipment and wiring for use in the production that will not be connected to permanently installed wiring, shall comply with Article 520.

FPN: For methods of determining population capacity, see local building code or, in its absence, the *Life Safety Code*, NFPA *101*-1997.

518-3. Other Articles.

(a) Hazardous (Classified) Areas. Electrical installations in hazardous (classified) areas located in places of assembly shall comply with Article 500.

(b) Temporary Wiring. In exhibition halls used for display booths, as in trade shows, the temporary wiring shall be installed in accordance with Article 305. Flexible cables and cords approved for hard or extra-hard usage shall be permitted to be laid on floors where protected from contact by the general public. The ground-fault circuit-interrupter requirements of Section 305-6 shall not apply.

Exception: Where conditions of supervision and maintenance ensure that only qualified persons will service the installation, flexible cords or cables identified in Table 400-4 for hard usage or extra-hard usage shall be permitted in cable trays used only for temporary wiring. All cords or cables shall be installed in a single layer. A permanent sign shall be attached to the cable tray at intervals not to exceed 25 ft (6.1 m). The sign shall read

CABLE TRAY FOR TEMPORARY WIRING ONLY

(c) Emergency Systems. Control of emergency systems shall comply with Article 700.

518-4. Wiring Methods.

(a) General. The fixed wiring methods shall be metal raceways, flexible metal raceways, nonmetallic raceways encased in not less than 2 in. (50.8 mm) of concrete, Type MI, MC, or AC cable containing an insulated equipment grounding conductor sized in accordance with Table 250-122.

Exception: Fixed wiring methods shall be as provided in

(a) Audio signal processing, amplification, and reproduction equipment — Article 640

(b) Communications circuits — Article 800

(c) Class 2 and Class 3 remote-control and signaling circuits — Article 725

(d) Fire alarm circuits — Article 760

(b) Nonrated Construction. Nonmetallic-sheathed cable, Type AC cable, electrical nonmetallic tubing, and rigid nonmetallic conduit shall be permitted to be installed in those buildings or portions thereof that are not required to be of fire-rated construction by the applicable building code.

> FPN: Fire-rated construction is the fire-resistive classification used in building codes.

(c) Spaces with Finish Rating. Electrical nonmetallic tubing and rigid nonmetallic conduit shall be permitted to be installed in restaurants, conference and meeting rooms in hotels or motels, dining facilities and church chapels where

(1) The electrical nonmetallic tubing or rigid nonmetallic conduit is installed concealed within walls, floors, and ceilings where the walls, floors, and ceilings provide a thermal barrier of material that has at least a 15-minute finish rating as identified in listings of fire-rated assemblies.

(2) The electrical nonmetallic tubing or rigid nonmetallic conduit is installed above suspended ceilings where the suspended ceilings provide a thermal barrier of material that has at least a 15-minute finish rating as identified in listings of fire-rated assemblies.

Electrical nonmetallic tubing and rigid nonmetallic conduit are not recognized for use in other space used for environmental air in accordance with Section 300-22(c).

> FPN: A finish rating is established for assemblies containing combustible (wood) supports. The finish rating is defined as the time at which the wood stud or wood joist reaches an average temperature rise of 121°C (250°F) or an individual temperature rise of 163°C (325°F) as measured on the plane of the wood nearest the fire. A finish rating is not intended to represent a rating for a membrane ceiling.

518-5. Supply. Portable switchboards and portable power distribution equipment shall be supplied only from listed power outlets of sufficient voltage and ampere rating. Such power outlets shall be protected by overcurrent devices. Such overcurrent devices and power outlets shall not be accessible to the general public. Provisions for connection of an equipment grounding conductor shall be provided. The neutral of feeders supplying solid-state, 3-phase, 4-wire dimmer systems shall be considered a current-carrying conductor.

Article 520 — Theaters, Audience Areas of Motion Picture and Television Studios, and Similar Locations

A. General

520-1. Scope. This article covers all buildings or that part of a building or structure designed or used for presentation, dramatic, musical, motion picture projection, or similar purposes and to specific audience seating areas within motion picture or television studios.

520-2. Definitions.

Border Light. A permanently installed overhead strip light.

Breakout Assembly. An adapter used to connect a multipole connector containing two or more branch-circuits to multiple individual branch circuit connectors.

Bundled. Cables or conductors that are physically tied, wrapped, taped, or otherwise periodically bound together.

Connector Strip. A metal wireway containing pendant or flush receptacles.

Drop Box. A box containing pendant- or flush-mounted receptacles attached to a multiconductor cable via strain relief, or a multipole connector.

Footlight. A border light installed on or in the stage.

Grouped. Cables or conductors positioned adjacent to one another but not in continuous contact with each other.

Portable Equipment. Equipment fed with portable cords or cables intended to be moved from one place to another.

Portable Power Distribution Unit. A power distribution box containing receptacles and overcurrent devices.

Proscenium. The wall and arch that separates the stage from the auditorium (house).

Stand Lamp (Work Light). A portable stand that contains a general-purpose lighting fixture or lamp holder with guard for the purpose of providing general illumination on the stage or in the auditorium.

Strip Light. A lighting fixture with multiple lamps arranged in a row.

Two-Fer. An adapter cable containing one male plug and two female cord connectors used to connect two loads to one branch circuit.

520-3. Motion Picture Projectors. Motion picture equipment and its installation and use shall comply with Article 540.

520-4. Audio Signal Processing, Amplification, and Reproduction Equipment. Audio signal processing, amplification, and reproduction equipment and its installation shall comply with Article 640.

520-5. Wiring Methods.

(a) General. The fixed wiring method shall be metal raceways, nonmetallic raceways encased in at least 2 in. (50.8 mm) of concrete, Type MI cable, or Type MC cable.

Exception: Fixed wiring methods shall be as provided in Article 640 for audio signal processing, amplification and reproduction equipment, in Article 800 for communication circuits, in Article 725 for Class 2 and Class 3 remote-control and signaling circuits, and in Article 760 for fire alarm circuits.

(b) Portable Equipment. The wiring for portable switchboards, stage set lighting, stage effects, and other wiring not fixed as to location shall be permitted with approved flexible cords and cables as provided elsewhere in Article 520. Fastening such cables and cords by uninsulated staples or nailing shall not be permitted.

(c) Nonrated Construction. Nonmetallic-sheathed cable, Type AC cable, electrical nonmetallic tubing, and rigid nonmetallic conduit shall be permitted to be installed in those buildings or portions thereof that are not required to be of fire-rated construction by the applicable building code.

520-6. Number of Conductors in Raceway. The number of conductors permitted in any metal conduit, rigid nonmetallic conduit as permitted in this article, or electrical metallic tubing for border or stage pocket circuits or for remote-control conductors shall not exceed the percentage fill shown in Table 1 of Chapter 9. Where contained within an auxiliary gutter or a wireway, the sum of the cross-sectional areas of all contained conductors at any cross section shall not exceed 20 percent of the interior cross-sectional area of the auxiliary gutter or wireway. The 30-conductor limitation of Sections 362-5 and 374-5 shall not apply.

520-7. Enclosing and Guarding Live Parts. Live parts shall be enclosed or guarded to prevent accidental contact by persons and objects. All switches shall be of the externally operable type. Dimmers, including rheostats, shall be placed in cases or cabinets that enclose all live parts.

520-8. Emergency Systems. Control of emergency systems shall comply with Article 700.

520-9. Branch Circuits. A branch circuit of any size supplying one or more receptacles shall be permitted to supply stage set lighting. The voltage rating of the receptacles shall not be less than the circuit voltage. Receptacle ampere ratings and branch-circuit conductor ampacity shall not be less than the branch-circuit overcurrent device ampere rating. Table 210-21(b)(2) shall not apply.

520-10. Portable Equipment. Portable stage and studio lighting equipment and portable power distribution equipment shall be permitted for temporary use outdoors provided the equipment is supervised by qualified personnel while energized and barriered from the general public.

B. Fixed Stage Switchboards

520-21. Dead Front. Stage switchboards shall be of the dead-front type and shall comply with Part D of Article 384 unless approved based on suitability as a stage switchboard as determined by a qualified testing laboratory and recognized test standards and principles.

520-22. Guarding Back of Switchboard. Stage switchboards having exposed live parts on the back of such boards shall be enclosed by the building walls, wire mesh grills, or by other approved methods. The entrance to this enclosure shall be by means of a self-closing door.

520-23. Control and Overcurrent Protection of Receptacle Circuits. Means shall be provided at a stage-lighting switchboard to which load circuits are connected for overcurrent protection of stage-lighting branch circuits, including branch circuits supplying stage and auditorium receptacles used for cord- and plug-connected stage equipment. Where the stage switchboard contains dimmers to control nonstage lighting, the locating of the overcurrent protective devices for these branch circuits at the stage switchboard shall be permitted.

520-24. Metal Hood. A stage switchboard that is not completely enclosed dead-front and dead-rear or recessed into a wall shall be provided with a metal hood extending the full length of the board to protect all equipment on the board from falling objects.

520-25. Dimmers. Dimmers shall comply with (a) through (d).

(a) Disconnection and Overcurrent Protection. Where dimmers are installed in ungrounded conductors, each dimmer shall have overcurrent protection not greater than

125 percent of the dimmer rating, and shall be disconnected from all ungrounded conductors when the master or individual switch or circuit breaker supplying such dimmer is in the open position.

(b) Resistance- or Reactor-Type Dimmers. Resistance- or series reactor-type dimmers shall be permitted to be placed in either the grounded or the ungrounded conductor of the circuit. Where designed to open either the supply circuit to the dimmer or the circuit controlled by it, the dimmer shall then comply with Section 380-1. Resistance- or reactor-type dimmers placed in the grounded neutral conductor of the circuit shall not open the circuit.

(c) Autotransformer-Type Dimmers. The circuit supplying an autotransformer-type dimmer shall not exceed 150 volts between conductors. The grounded conductor shall be common to the input and output circuits.

> FPN: See Section 210-9 for circuits derived from autotransformers.

(d) Solid-State-Type Dimmers. The circuit supplying a solid-state dimmer shall not exceed 150 volts between conductors unless the dimmer is listed specifically for higher voltage operation. Where a grounded conductor supplies a dimmer, it shall be common to the input and output circuits. Dimmer chassis shall be connected to the equipment grounding conductor.

520-26. Type of Switchboard. A stage switchboard shall be either one or a combination of the following types.

(a) Manual. Dimmers and switches are operated by handles mechanically linked to the control devices.

(b) Remotely Controlled. Devices are operated electrically from a pilot-type control console or panel. Pilot control panels shall either be part of the switchboard or shall be permitted to be at another location.

(c) Intermediate. A stage switchboard with circuit interconnections is a secondary switchboard (patch panel) or panelboard remote to the primary stage switchboard. It shall contain overcurrent protection. Where the required branch-circuit overcurrent protection is provided in the dimmer panel, it shall be permitted to be omitted from the intermediate switchboard.

520-27. Stage Switchboard Feeders.

(a) Type of Feeder. Feeders supplying stage switchboards shall be one of the following.

(1) Single Feeder. A single feeder disconnected by a single disconnect device.

(2) Multiple Feeders to Intermediate Stage Switchboard (Patch Panel). Multiple feeders of unlimited quantity shall be permitted, provided that all multiple feeders are part of a single system. Where combined, neutral conductors in a given raceway shall be of sufficient ampacity to carry the maximum unbalanced current supplied by multiple feeder conductors in the same raceway, but need not be greater than the ampacity of the neutral supplying the primary stage switchboard. Parallel neutral conductors shall comply with Section 310-4.

(3) Separate Feeders to Single Primary Stage Switchboard (Dimmer Bank). Installations with separate feeders to a single primary stage switchboard shall have a disconnecting means for each feeder. The primary stage switchboard shall have a permanent and obvious label stating the number and location of disconnecting means. If the disconnecting means are located in more than one distribution switchboard, the primary stage switchboard shall be provided with barriers to correspond with these multiple locations.

(b) Neutral. The neutral of feeders supplying solid-state, 3-phase, 4-wire dimming systems shall be considered a current-carrying conductor.

(c) Supply Capacity. For the purposes of computing supply capacity to switchboards, it shall be permissible to consider the maximum load that the switchboard is intended to control in a given installation, provided that

(1) All feeders supplying the switchboard shall be protected by an overcurrent device with a rating not greater than the ampacity of the feeder.
(2) The opening of the overcurrent device shall not affect the proper operation of the egress or emergency lighting systems.

> FPN: For computation of stage switchboard feeder loads, see Section 220-10.

C. Fixed Stage Equipment Other than Switchboards

520-41. Circuit Loads.

(a) Circuits Rated 20 Amperes or Less. Footlights, border lights, and proscenium side lights shall be arranged so that no branch circuit supplying such equipment will carry a load exceeding 20 amperes.

(b) Circuits Rated Greater than 20 Amperes. Where heavy-duty lampholders only are used, such circuits shall be permitted to comply with Article 210 for circuits supplying heavy-duty lampholders.

520-42. Conductor Insulation. Foot, border, proscenium, or portable strip light fixtures and connector strips shall be wired with conductors that have insulation suitable for the temperature at which the conductors will be operated, but not less than 125°C (257°F). The ampacity of the 125°C (257°F) conductors shall be that of 60°C (140°F) conductors. All drops from connector strips shall be 90°C (194°F) wire sized to the ampacity of 60°C (140°F) cords and cables with no more than 6 in. (152 mm) of conductor extending into the connector strip. Section 310-15(b)(2)(a) shall not apply.

FPN: See Table 310-13 for conductor types.

520-43. Footlights.

(a) Metal Trough Construction. Where metal trough construction is employed for footlights, the trough containing the circuit conductors shall be made of sheet metal not lighter than No. 20 MSG treated to prevent oxidation. Lampholder terminals shall be kept at least ½ in. (12.7 mm) from the metal of the trough. The circuit conductors shall be soldered to the lampholder terminals.

(b) Other-than-Metal Trough Construction. Where the metal trough construction specified in Section 520-43(a) is not used, footlights shall consist of individual outlets with lampholders wired with rigid metal conduit, intermediate metal conduit, or flexible metal conduit, Type MC cable, or mineral-insulated, metal-sheathed cable. The circuit conductors shall be soldered to the lampholder terminals.

(c) Disappearing Footlights. Disappearing footlights shall be arranged so that the current supply will be automatically disconnected when the footlights are replaced in the storage recesses designed for them.

520-44. Borders and Proscenium Sidelights.

(a) General. Borders and proscenium sidelights shall be as follows:

(1) Constructed as specified in Section 520-43,
(2) Suitably stayed and supported; and
(3) Designed so that the flanges of the reflectors or other adequate guards will protect the lamps from mechanical damage and from accidental contact with scenery or other combustible material.

(b) Cords and Cables for Border Lights.

(1) General. Cords and cables for supply to border lights shall be listed for extra-hard usage. The cords and cables shall be suitably supported. Such cords and cables shall be employed only where flexible conductors are neces-

sary. Ampacity of the conductors shall be as provided in Section 400-5.

Table 520-44. Ampacity of Listed Extra-Hard Usage Cords and Cables with Temperature Ratings of 75°C (167°F) and 90°C (194°F)*
[Based on Ambient Temperature of 30°C (86°F)]

Size (AWG)	Temperature Rating of Cords and Cables		Maximum Rating of Overcurrent Device
	75°C (167°F)	90°C (194°F)	
14	24	28	15
12	32	35	20
10	41	47	25
8	57	65	35
6	77	87	45
4	101	114	60
2	133	152	80

*Ampacity shown is the ampacity for multiconductor cords and cables where only three copper conductors are current-carrying. If the number of current-carrying conductors in a cord or cable exceeds three and the load diversity factor is a minimum of 50 percent, the ampacity of each conductor shall be reduced as shown in the following table.

Number of Conductors	Percent of Ampacity
4–6	80
7–24	70
25–42	60
43 and above	50

Note: Ultimate insulation temperature. In no case shall conductors be associated together in such a way with respect to the kind of circuit, the wiring method used, or the number of conductors such that the temperature limit of the conductors will be exceeded.

A neutral conductor that carries only the unbalanced current from other conductors of the same circuit need not be considered as a current-carrying conductor.

In a 3-wire circuit consisting of two phase wires and the neutral of a 4-wire, 3-phase, wye-connected system, a common conductor carries approximately the same current as the line-to-neutral currents of the other conductors and shall be considered to be a current-carrying conductor.

On a 4-wire, 3-phase, wye circuit where the major portion of the load consists of nonlinear loads such as electric-discharge lighting, electronic computer/data processing, or similar equipment, there are harmonic currents present in the neutral conductor, and the neutral shall be considered to be a current-carrying conductor.

(2) Cords and Cables Not in Contact with Heat-Producing Equipment. Listed multiconductor extra-hard usage-type cords and cables not in direct contact with equipment containing heat-producing elements shall be permitted to have their ampacity determined by Table 520-44. Maximum load current in any conductor with an ampacity determined by Table 520-44 shall not exceed the values in Table 520-44.

520-45. Receptacles. Receptacles for electrical equipment or fixtures on stages shall be rated in amperes. Conductors supplying receptacles shall be in accordance with Articles 310 and 400.

520-46. Connector Strips, Drop Boxes, Floor Pockets, and Other Outlet Enclosures. Receptacles for the connection of portable stage-lighting equipment shall be pendant or mounted in suitable pockets or enclosures and shall comply with Section 520-45. Supply cables for connector strips and drop boxes shall be as specified in Section 520-44(b).

520-47. Backstage Lamps (Bare Bulbs). Lamps (bare bulbs) installed in backstage and ancillary areas where they can come in contact with scenery shall be located and guarded so as to be free from physical damage and shall provide an air space of not less than 2 in. (50.8 mm) between such lamps and any combustible material.

Exception: Decorative lamps installed in scenery shall not be considered to be backstage lamps for the purpose of this section.

520-48. Curtain Machines. Curtain machines shall be listed.

520-49. Smoke Ventilator Control. Where stage smoke ventilators are released by an electrical device, the circuit operating the device shall be normally closed and shall be controlled by at least two externally operable switches, one switch being placed at a readily accessible location on stage and the other where designated by the authority having jurisdiction. The device shall be designed for the full voltage of the circuit to which it is connected, no resistance being inserted. The device shall be located in the loft above the scenery and shall be enclosed in a suitable metal box having a tight, self-closing door.

D. Portable Switchboards on Stage

520-50. Road Show Connection Panel (A Type of Patch Panel). A panel designed to allow for road show connection of portable stage switchboards to fixed lighting outlets by means of permanently installed supplementary circuits. The panel, supplementary circuits, and outlets shall comply with (a) through (d).

(a) Load Circuits. Circuits shall terminate in grounding-type polarized inlets of current and voltage rating that match the fixed-load receptacle.

(b) Circuit Transfer. Circuits that are transferred between fixed and portable switchboards shall have all circuit conductors transferred simultaneously.

(c) Overcurrent Protection. The supply devices of these supplementary circuits shall be protected by branch-circuit overcurrent protective devices. The individual supplementary circuit, within the road show connection panel and theater, shall be protected by branch-circuit overcurrent protective devices of suitable ampacity installed within the road show connection panel.

(d) Enclosure. Panel construction shall be in accordance with Article 384.

520-51. Supply. Portable switchboards shall be supplied only from power outlets of sufficient voltage and ampere rating. Such power outlets shall include only externally operable, enclosed fused switches or circuit breakers mounted on stage or at the permanent switchboard in locations readily accessible from the stage floor. Provisions for connection of an equipment grounding conductor shall be provided. The neutral of feeders supplying solid-state, 3-phase, 4-wire dimmer systems shall be considered a current-carrying conductor.

520-52. Overcurrent Protection. Circuits from portable switchboards directly supplying equipment containing incandescent lamps of not over 300 watts shall be protected by overcurrent protective devices having a rating or setting of not over 20 amperes. Circuits for lampholders over 300 watts shall be permitted where overcurrent protection complies with Article 210.

520-53. Construction and Feeders. Portable switchboards and feeders for use on stages shall comply with (a) through (p).

(a) Enclosure. Portable switchboards shall be placed within an enclosure of substantial construction, which shall be permitted to be arranged so that the enclosure is open during operation. Enclosures of wood shall be completely lined with sheet metal of not less than No. 24 MSG and shall be well galvanized, enameled, or otherwise properly coated to prevent corrosion or be of a corrosion-resistant material.

(b) Energized Parts. There shall not be exposed energized parts within the enclosure.

(c) Switches and Circuit Breakers. All switches and circuit breakers shall be of the externally operable, enclosed type.

(d) Circuit Protection. Overcurrent devices shall be provided in each ungrounded conductor of every circuit supplied through the switchboard. Enclosures shall be provided for all overcurrent devices in addition to the switchboard enclosure.

(e) Dimmers. The terminals of dimmers shall be provided with enclosures, and dimmer faceplates shall be arranged so that accidental contact cannot be readily made with the faceplate contacts.

(f) Interior Conductors.

(1) Type. All conductors other than busbars within the switchboard enclosure shall be stranded. Conductors shall be approved for an operating temperature at least equal to the approved operating temperature of the dimming devices used in the switchboard and in no case less than the following:

(1) Resistance-type dimmers — 200°C (392°F); or
(2) Reactor-type, autotransformer, and solid-state dimmers — 125°C (257°F).

All control wiring shall comply with Article 725.

(2) Protection. Each conductor shall have an ampacity not less than the rating of the circuit breaker, switch, or fuse that it supplies. Circuit interrupting and bus bracing shall be in accordance with Sections 110-9 and 110-10. The short-circuit current rating shall be marked on the switchboard.

Conductors shall be enclosed in metal wireways or shall be securely fastened in position and shall be bushed where they pass through metal.

(g) Pilot Light. A pilot light shall be provided within the enclosure and shall be connected to the circuit supplying the board so that the opening of the master switch will not cut off the supply to the lamp. This lamp shall be on an individual branch circuit having overcurrent protection rated or set at not over 15 amperes.

(h) Supply Conductors.

(1) General. The supply to a portable switchboard shall be by means of listed extra-hard usage cords or cables. The supply cords or cable shall terminate within the switchboard enclosure, in an externally operable fused master switch or circuit breaker, or in a connector assembly identified for the purpose. The supply cords or cable (and connector assembly) shall have sufficient ampacity to carry the total load connected to the switchboard and shall be protected by overcurrent devices.

(2) Single-Conductor Cables. Single-conductor portable supply cable sets shall not be smaller than No. 2 conductors. The equipment grounding conductor shall not be smaller than No. 6 conductor. Single conductor grounded neutral cables for a supply shall be sized as per Section 520-53(o)(2). Where single conductors are paralleled for increased ampacity, the paralleled conductors shall be of the same length and size. Single conductor supply cables shall be grouped together but not bundled. The equipment grounding conductor shall be permitted to be of a different type, provided it meets the other requirements of this section,

and it shall be permitted to be reduced in size as permitted by Section 250-122. Grounded (neutral) and equipment grounding conductors shall be identified in accordance with Sections 200-6, 250-134(b), and 310-12. Grounded conductors shall be permitted to be identified by marking at least the first 6 in. (152.4 mm) from both ends of each length of conductor with white or natural gray. Equipment grounding conductors shall be permitted to be identified by marking at least the first 6 in. (152.4 mm) from both ends of each length of conductor with green or green with yellow stripes. Where more than one nominal voltage exists within the same premises, each ungrounded conductor shall be identified by system.

(3) Supply Conductors Not Over 10 ft (3.05 m) Long. Where supply conductors do not exceed 10 ft (3.05 m) in length between supply and switchboard or supply and a subsequent overcurrent device, the supply conductors shall be permitted to be reduced in size where all of the following conditions are met.

(a) The ampacity of the supply conductors shall be at least one-quarter of the ampacity of the supply overcurrent protection device.
(b) The supply conductors shall terminate in a single overcurrent protection device that will limit the load to the ampacity of the supply conductors. This single overcurrent device shall be permitted to supply additional overcurrent devices on its load side.
(c) The supply conductors shall not penetrate walls, floors, or ceilings or be run through doors or traffic areas. The supply conductors shall be adequately protected from physical damage.
(d) The supply conductors shall be suitably terminated in an approved manner.
(e) Conductors shall be continuous without splices or connectors.
(f) Conductors shall not be bundled.
(g) Conductors shall be supported above the floor in an approved manner.

(4) Supply Conductors Not Over 20 ft (6.1 m) Long. Where supply conductors do not exceed 20 ft (6.1 m) in length between supply and switchboard or supply and a subsequent overcurrent protection device, the supply conductors shall be permitted to be reduced in size where all of the following conditions are met.

(a) The ampacity of the supply conductors shall be at least one-half of the ampacity of the supply overcurrent protection device.
(b) The supply conductors shall terminate in a single overcurrent protection device that will limit the load to the

ampacity of the supply conductors. This single overcurrent device shall be permitted to supply additional overcurrent devices on its load side.

(c) The supply conductors shall not penetrate walls, floors, or ceilings or be run through doors or traffic areas. The supply conductors shall be adequately protected from physical damage.

(d) The supply conductors shall be suitably terminated in an approved manner.

(e) The supply conductors shall be supported in an approved manner at least 7 ft (2.13 m) above the floor except at terminations.

(f) The supply conductors shall not be bundled.

(g) Tap conductors shall be in unbroken lengths.

(5) Supply Conductors Not Reduced in Ampacity. Supply conductors not reduced in ampacity shall be permitted to pass through holes in walls specifically designed for the purpose. If penetration is through the fire-resistant–rated wall, it shall be in accordance with Section 300-21.

(i) Cable Arrangement. Cables shall be protected by bushings where they pass through enclosures and shall be arranged so that tension on the cable will not be transmitted to the connections. Where power conductors pass through metal, the requirements of Section 300-20 shall apply.

(j) Number of Supply Interconnections. Where connectors are used in a supply conductor, there shall be a maximum number of three interconnections (mated connector pairs) where the total length from supply to switchboard does not exceed 100 ft (30.5 m). In cases where the total length from supply to switchboard exceeds 100 ft (30.5 m), one additional interconnection shall be permitted for each additional 100 ft (30.5 m) of supply conductor.

(k) Single-Pole Separable Connectors. Where single-pole portable cable connectors are used, they shall be listed and of the locking type. Sections 400-10 and 410-56 shall not apply to listed single-pole separable connectors and single-conductor cable assemblies utilizing listed single-pole separable connectors. Where paralleled sets of current-carrying, single-pole separable connectors are provided as input devices, they shall be prominently labeled with a warning indicating the presence of internal parallel connections. The use of single-pole separable connectors shall comply with at least one of the following conditions.

(1) Connection and disconnection of connectors are only possible where the supply connectors are interlocked to the source and it is not possible to connect or disconnect connectors when the supply is energized.

(2) Line connectors are of the listed sequential-interlocking type so that load connectors shall be connected in the following sequence:

(a) Equipment grounding conductor connection

(b) Grounded circuit conductor connection, if provided

(c) Ungrounded conductor connection, and that disconnection shall be in the reverse order

(3) A caution notice shall be provided adjacent to the line connectors indicating that plug connection shall be in the following order:

(a) Equipment grounding conductor connectors

(b) Grounded circuit conductor connectors, if provided

(c) Ungrounded conductor connectors, and that disconnection shall be in the reverse order

(l) Protection of Supply Conductors and Connectors. All supply conductors and connectors shall be protected against physical damage by an approved means. This protection shall not be required to be raceways.

(m) Flanged Surface Inlets. Flanged surface inlets (recessed plugs) that are used to accept the power shall be rated in amperes.

(n) Terminals. Terminals to which stage cables are connected shall be located so as to permit convenient access to the terminals.

(o) Neutral.

(1) Neutral Terminal. In portable switchboard equipment designed for use with 3-phase, 4-wire with ground supply, the supply neutral terminal, its associated busbar, or equivalent wiring, or both, shall have an ampacity equal to at least twice the ampacity of the largest ungrounded supply terminal.

Exception: Where portable switchboard equipment is specifically constructed and identified to be internally converted in the field, in an approved manner, from use with a balanced 3-phase, 4-wire with ground supply to a balanced single-phase, 3-wire with ground supply, the supply neutral terminal and its associated busbar, equivalent wiring, or both, shall have an ampacity equal to at least that of the largest ungrounded single-phase supply terminal.

(2) Supply Neutral. The power supply conductors for portable switchboards shall be sized considering the neutral as a current-carrying conductor. Where single-conductor feeder cables, not installed in raceways, are used on multiphase circuits, the grounded neutral conductor shall have an ampacity of at least 130 percent of the ungrounded circuit conductors feeding the portable switchboard.

(p) Qualified Personnel. The routing of portable supply conductors, the making and breaking of supply connectors and other supply connections, and the energization and de-energization of supply services shall be performed by qualified personnel, and portable switchboards shall be so marked, indicating this requirement in a permanent and conspicuous manner.

Exception: A portable switchboard shall be permitted to be connected to a permanently installed supply receptacle by other than qualified personnel, provided that the supply receptacle is protected for its rated ampacity by an overcurrent device of not greater than 150 amperes, and where the receptacle, interconnection, and switchboard further

(a) Employ listed multipole connectors suitable for the purpose for every supply interconnection, and

(b) Prevent access to all supply connections by the general public, and

(c) Employ listed extra-hard usage multiconductor cords or cables with an ampacity suitable for the type of load and not less than the ampere rating of the connectors.

E. Portable Stage Equipment Other than Switchboards

520-61. Arc Lamp Fixtures. Arc lamp fixtures, including enclosed arc lamp fixtures and associated ballasts, shall be listed. Interconnecting cord sets and interconnecting cords and cables shall be extra-hard usage type and listed.

520-62. Portable Power Distribution Units. Portable power distribution units shall comply with (a) through (e).

(a) Enclosure. The construction shall be such that no current-carrying part will be exposed.

(b) Receptacles and Overcurrent Protection. Receptacles shall comply with Section 520-45 and shall have branch-circuit overcurrent protection in the box. Fuses and circuit breakers shall be protected against physical damage. Cords or cables supplying pendant receptacles shall be listed for extra-hard usage.

(c) Busbars and Terminals. Busbars shall have an ampacity equal to the sum of the ampere ratings of all the circuits connected to the busbar. Lugs shall be provided for the connection of the master cable.

(d) Flanged Surface Inlets. Flanged surface inlets (recessed plugs) that are used to accept the power shall be rated in amperes.

(e) Cable Arrangement. Cables shall be adequately protected where they pass through enclosures and be arranged so that tension on the cable will not be transmitted to the terminations.

520-63. Bracket Fixture Wiring.

(a) Bracket Wiring. Brackets for use on scenery shall be wired internally, and the fixture stem shall be carried through to the back of the scenery where a bushing shall be placed on the end of the stem. Externally wired brackets or other fixtures shall be permitted where wired with cords designed for hard usage that extend through scenery and without joint or splice in canopy of fixture back and terminate in an approved-type stage connector located, where practical, within 18 in. (457 mm) of the fixture.

(b) Mounting. Fixtures shall be securely fastened in place.

520-64. Portable Strips. Portable strips shall be constructed in accordance with the requirements for border lights and proscenium side lights in Section 520-44(a). The supply cable shall be protected by bushings where it passes through metal and shall be arranged so that tension on the cable will not be transmitted to the connections.

FPN No. 1: See Section 520-42 for wiring of portable strips.

FPN No. 2: See Section 520-68(a)(3) for insulation types required on single conductors.

520-65. Festoons. Joints in festoon wiring shall be staggered. Lamps enclosed in lanterns or similar devices of combustible material shall be equipped with guards.

520-66. Special Effects. Electrical devices used for simulating lightning, waterfalls, and the like shall be constructed and located so that flames, sparks, or hot particles cannot come in contact with combustible material.

520-67. Multipole Branch-Circuit Cable Connectors. Multipole branch-circuit cable connectors, male and female, for flexible conductors shall be constructed so that tension on the cord or cable will not be transmitted to the connections. The female half shall be attached to the load end of the power supply cord or cable. The connector shall be rated in amperes and designed so that differently rated devices cannot be connected together. Alternating-current multipole connectors shall be polarized and comply with Sections 410-56(g) and 410-58.

FPN: See Section 400-10 for pull at terminals.

520-68. Conductors for Portables.

(a) Conductor Type.

(1) General. Flexible conductors, including cable extensions, used to supply portable stage equipment shall be listed extra-hard usage cords or cables.

(2) Stand Lamps. Reinforced cord shall be permitted to supply stand lamps where the cord is not subject to severe physical damage and is protected by an overcurrent device rated at not over 20 amperes.

(3) High-Temperature Applications. A special assembly of conductors in sleeving not longer than 3.3 ft (1 m) shall be permitted to be employed in lieu of flexible cord if the individual wires are stranded and rated not less than 125°C (257°F) and the outer sleeve is glass fiber with a wall thickness of at least 0.025 in. (0.635 mm).

Portable stage equipment requiring flexible supply conductors with a higher temperature rating where one end is permanently attached to the equipment shall be permitted to employ alternate, suitable conductors as determined by a qualified testing laboratory and recognized test standards.

(4) Breakouts. Listed, hard usage (junior hard service) cords shall be permitted in breakout assemblies where all of the following conditions are met.

(a) The cords are utilized to connect between a single multipole connector containing two or more branch circuits and multiple two-pole, 3-wire connectors.

(b) The longest cord in the breakout assembly does not exceed 20 ft (6.1 m).

(c) The breakout assembly is protected from physical damage by attachment over its entire length to a pipe, truss, tower, scaffold, or other substantial support structure.

(d) All branch circuits feeding the breakout assembly are protected by overcurrent devices rated at not over 20 amperes.

(b) Conductor Ampacity. The ampacity of conductors shall be as given in Section 400-5, except multiconductor, listed, extra-hard usage portable cords that are not in direct contact with equipment containing heat-producing elements shall be permitted to have their ampacity determined by Table 520-44. Maximum load current in any conductor with an ampacity determined by Table 520-44 shall not exceed the values in Table 520-44.

Exception: Where alternate conductors are allowed in Section 520-68(a)(3), their ampacity shall be as given in the appropriate table in this Code for the types of conductors employed.

520-69. Adapters. Adapters, two-fers, and other single and multiple circuit outlet devices shall comply with (a), (b), and (c).

(a) No Reduction in Current Rating. Each receptacle and its corresponding cable shall have the same current and voltage rating as the plug supplying it. It shall not be utilized in a stage circuit with a greater current rating.

(b) Connectors. All connectors shall be wired in accordance with Section 520-67.

(c) Conductor Type. Conductors for adapters and two-fers shall be listed, extra-hard usage or listed, hard usage (junior hard service) cord. Hard usage (junior hard service) cord shall be restricted in overall length to 3.3 ft (1 m).

F. Dressing Rooms

520-71. Pendant Lampholders. Pendant lampholders shall not be installed in dressing rooms.

520-72. Lamp Guards. All exposed incandescent lamps in dressing rooms, where less than 8 ft (2.44 m) from the floor, shall be equipped with open-end guards riveted to the outlet box cover or otherwise sealed or locked in place.

520-73. Switches Required. All lights and any receptacles adjacent to the mirror(s) and above the dressing table counter(s) installed in dressing rooms shall be controlled by wall switches installed in the dressing room(s). Each switch controlling receptacles adjacent to the mirror(s) and above the dressing table counter(s) shall be provided with a pilot light located outside the dressing room, adjacent to the door to indicate when the receptacles are energized. Other outlets installed in the dressing room shall not be required to be switched.

G. Grounding

520-81. Grounding. All metal raceways and metal-sheathed cables shall be grounded. The metal frames and enclosures of all equipment, including border lights and portable lighting fixtures, shall be grounded. Grounding, where used, shall be in accordance with Article 250.

Article 525 — Carnivals, Circuses, Fairs, and Similar Events

A. General Requirements

525-1. Scope. This article covers the installation of portable wiring and equipment for carnivals, circuses, exhibitions, fairs, traveling attractions, and similar functions, including wiring in or on all structures.

525-3. Other Articles.

(a) Permanent Structures. Articles 518 and 520 shall apply to wiring in permanent structures.

(b) Portable Wiring and Equipment. Wherever the requirements of other articles of this *Code* and Article 525 differ, the requirements of Article 525 shall apply to the portable wiring and equipment.

(c) Audio Signal Processing, Amplification, and Reproduction Equipment. Article 640 shall apply to the wiring and installation of audio signal processing, amplification, and reproduction equipment.

525-6. Protection of Electrical Equipment. Electrical equipment and wiring methods in or on rides, concessions, or other units shall be provided with mechanical protection where such equipment or wiring methods are subject to physical damage.

B. Installation

525-10. Power Sources.

(a) Separately Derived Systems.

(1) **Generators.** Generators shall comply with the requirements of Article 445.

(2) **Transformers.** Transformers shall comply with applicable requirements of Sections 240-3(a), (b)(3), and (c); Section 250-30; and Article 450.

(b) Services. Services shall be installed in accordance with applicable requirements of Article 230 and, in addition, shall comply with the following.

(1) **Guarding.** Service equipment shall not be installed in a location that is accessible to unqualified persons, unless the equipment is lockable.

(2) **Mounting and Location.** Service equipment shall be mounted on a solid backing and be installed so as to be protected from the weather, unless of weatherproof construction.

525-12. Overhead Conductor Clearances.

(a) Vertical Clearances. Conductors shall have a vertical clearance to ground in accordance with Section 225-18. These clearances shall apply only to wiring installed outside of tents and concessions.

(b) Clearance to Rides and Attractions. Amusement rides and amusement attractions shall be maintained not less than 15 ft (4.57 m) in any direction from overhead conductors operating at 600 volts or less, except for the conductors supplying the amusement ride or attraction. Amusement rides or attractions shall not be located under or within 15 ft (4.57 m) horizontally of conductors operating in excess of 600 volts.

525-13. Wiring Methods.

(a) Type. Unless otherwise provided for in this article, wiring methods shall comply with the applicable requirements of Chapters 1 through 4 of this *Code*. Where flexible cords or cables are used and are not subject to physical damage, they shall be permitted to be listed for hard usage. When used outdoors, flexible cords and cables shall also be listed for wet locations and shall be sunlight resistant.

(b) Single Conductor. Single conductor cable shall be permitted only in sizes No. 2 or larger.

(c) Open Conductors. Open conductors are prohibited except as part of a listed assembly or festoon lighting installed in accordance with Article 225.

(d) Splices. Flexible cords or cables shall be continuous without splice or tap between boxes or fittings. Cord connectors shall not be laid on the ground unless listed for wet locations. Connectors and cable connections shall not be placed in audience traffic paths or within areas accessible to the public unless guarded.

(e) Support. Wiring for an amusement ride, attraction, tent, or similar structure shall not be supported by any other ride or structure unless specifically designed for the purpose.

(f) Protection. Flexible cords or cables run on the ground, where accessible to the public, shall be covered with approved nonconductive mats. Cables and mats shall be arranged so as not to present a tripping hazard.

(g) Inside Tents and Concessions. Electrical wiring for temporary lighting, where installed inside of tents and concessions, shall be securely installed, and where subject to physical damage, shall be provided with mechanical protection. All temporary lamps for general illumination shall be protected from accidental breakage by a suitable fixture or lampholder with a guard.

525-14. Boxes and Fittings. A box or fitting shall be installed at each connection point, outlet, switchpoint, or junction point.

525-15. Portable Distribution or Termination Boxes. Portable distribution or termination boxes shall comply with (a) through (d).

(a) Construction. Boxes shall be designed so that no live parts are exposed to accidental contact. Where installed outdoors the box shall be of weatherproof construction and mounted so that the bottom of the enclosure is not less than 6 in. (152 mm) above the ground.

(b) Busbars and Terminals. Busbars shall have an ampere rating not less than the overcurrent device supplying the feeder supplying the box. Where conductors terminate directly on busbars, busbar connectors shall be provided.

(c) Receptacles and Overcurrent Protection. Receptacles shall have overcurrent protection installed within the box. The overcurrent protection shall not exceed the ampere rating of the receptacle, except as permitted in Article 430 for motor loads.

(d) Single-Pole Connectors. Where single-pole connectors are used, they shall comply with Section 530-22.

525-16. Overcurrent Protection. Overcurrent protection of equipment and conductors shall be provided in accordance with Article 240.

525-17. Motors. Motors and associated equipment shall be installed in accordance with Article 430.

525-18. Ground-Fault Circuit-Interrupter Protection for Personnel.

(a) General-Use 15- and 20-Ampere, 125-Volt Receptacles. All 125-volt, single-phase, 15- and 20-ampere receptacle outlets that are in use by personnel shall have listed ground-fault circuit-interrupter protection for personnel. The ground-fault circuit interrupter shall be permitted to be an integral part of the attachment plug or located in the power-supply cord, within 12 in. (305 mm) of the attachment plug. For the purposes of this section, listed cord sets incorporating ground-fault circuit-interrupter protection for personnel shall be permitted. Egress lighting shall not be connected to the load side terminals of a ground-fault circuit-interrupter receptacle.

(b) Appliance Receptacles. Receptacles supplying items, such as cooking and refrigeration equipment, which are incompatible with ground-fault circuit-interrupter devices shall not be required to have ground-fault circuit-interrupter protection.

(c) Other Receptacles. Other receptacle outlets not covered in (a) or (b) shall be permitted to have ground-fault circuit-interrupter protection for personnel, or a written procedure shall be continuously enforced at the site by one or more designated persons to ensure the safety of equipment grounding conductors for all cord sets and receptacles, as described in Section 305-6(b)(2).

C. Grounding and Bonding

525-20. General. All system and equipment grounding shall be in accordance with Article 250.

525-21. Equipment. The following equipment connected to the same source shall be bonded:

(1) Metal raceways and metal sheathed cable
(2) Metal enclosures of electric equipment
(3) Metal frames and metal parts of rides, concessions, trailers, trucks, or other equipment that contain or support electrical equipment

525-22. Equipment Grounding Conductor. All equipment requiring grounding shall be grounded by an equipment grounding conductor of a type and size recognized by Section 250-118 and installed in accordance with Article 250. The equipment grounding conductor shall be bonded to the system grounded conductor at the service disconnecting means, or in the case of a separately derived system such as a generator, at the generator or first disconnecting means supplied by the generator. The grounded circuit conductor shall not be connected to the equipment grounding conductor on the load side of the service disconnecting means or on the load side of a separately derived system disconnecting means.

D. Disconnecting Means

525-30. Type and Location. Each ride and concession shall be provided with a fused disconnect switch or circuit breaker located within sight and within 6 ft (1.83 m) of the operator's station. The disconnecting means shall be readily accessible to the operator, including when the ride is in operation. Where accessible to unqualified persons, the enclosure for the switch or circuit breaker shall be of the lockable type. A shunt trip device that opens the fused disconnect or circuit breaker when a switch located in the ride operator's console is closed shall be a permissible method of opening the circuit.

E. Attractions Utilizing Pools, Fountains, and Similar Installations with Contained Volumes of Water

525-40. Wiring and Equipment. This equipment shall be installed to comply with the applicable requirements of Article 680.

Article 530 — Motion Picture and Television Studios and Similar Locations

A. General

530-1. Scope. The requirements of this article shall apply to television studios and motion picture studios using either film or electronic cameras, except as provided in Section 520-1, and exchanges, factories, laboratories, stages, or a portion of the building in which film or tape more than

⅞ in. (22 mm) in width is exposed, developed, printed, cut, edited, rewound, repaired, or stored.

FPN: For methods of protecting against cellulose nitrate film hazards, see *Standard for the Storage and Handling of Cellulose Nitrate Motion Picture Film,* NFPA 40-1997.

530-2. Definitions.

Alternating-Current Power Distribution Box (Alternating-Current Plugging Box, Scatter Box). An ac distribution center or box that contains one or more grounding-type, polarized receptacles, that may contain overcurrent protection devices.

Bull Switch. An externally-operated wall-mounted safety switch, which may or may not contain overcurrent protection, that is designed for the connection of portable cables and cords.

Location (Shooting Location). A place outside a motion picture studio where a production or part of it is filmed or recorded.

Location Board (Deuce Board). Portable equipment containing a lighting contactor or contactors and overcurrent protection designed for remote control of stage lighting.

Motion Picture Studio (Lot). A building or group of buildings and other structures designed, constructed, or permanently altered for use by the entertainment industry for the purpose of motion picture or television production.

Portable Equipment. Equipment intended to be moved from one place to another.

Plugging Box. A dc device consisting of one or more 2-pole, 2-wire, nonpolarized, nongrounding-type receptacles intended to be used on dc circuits only.

Single-Pole Separable Connector. A device that is installed at the ends of portable, flexible, single-conductor cable that is used to establish connection or disconnection between two cables or one cable and a single-pole, panel-mounted separable connector.

Spider (Cable Splicing Block). A device that contains busbars that are insulated from each other for the purpose of splicing or distributing power to portable cables and cords that are terminated with single-pole busbar connectors.

Stage Effect (Special Effect). An electrical or electro-mechanical piece of equipment used to simulate a distinctive visual or audible effect such as wind machines, lightning simulators, sunset projectors, and the like.

Stage Property. An article or object used as a visual element in a motion picture or television production, except painted backgrounds (scenery) and costumes.

Stage Set. A specific area set up with temporary scenery and properties designed and arranged for a particular scene in a motion picture or television production.

Stand Lamp (Work Light). A portable stand that contains a general-purpose lighting fixture or lampholder with guard for the purpose of providing general illumination in the studio or stage.

Television Studio or Motion Picture Stage (Sound Stage). A building or portion of a building usually insulated from the outside noise and natural light for use by the entertainment industry for the purpose of motion picture, television, or commercial production.

530-6. Portable Equipment.
Portable stage and studio lighting equipment and portable power distribution equipment shall be permitted for temporary use outdoors if the equipment is supervised by qualified personnel while energized and barriered from the general public.

B. Stage or Set

530-11. Permanent Wiring.
The permanent wiring shall be Type MC cable, Type MI cable, or in approved raceways.

Exception: Communications circuits; audio signal processing, amplification, and reproduction circuits; Class 1, Class 2, and Class 3 remote-control or signaling circuits and power-limited fire alarm circuits shall be permitted to be wired in accordance with Articles 800, 640, 725, and 760.

530-12. Portable Wiring.

(a) Stage Set Wiring. The wiring for stage set lighting and other supply wiring not fixed as to location shall be done with listed hard usage flexible cords and cables. Where subject to physical damage, such wiring shall be listed extra-hard usage flexible cords and cables. Splices or taps in cables shall be permitted if the total connected load does not exceed the maximum ampacity of the cable.

(b) Stage Effects and Electrical Equipment Used as Stage Properties. The wiring for stage effects and electrical equipment used as stage properties shall be permitted to be wired with single- or multi-conductor listed flexible cords or cables if the conductors are protected from physical damage and secured to the scenery by approved cable ties or by insulated staples. Splices or taps shall be permitted where such are made with listed devices and the circuit is protected at not more than 20 amperes.

(c) Other Electrical Equipment. Cords and cables other than extra-hard usage, where supplied as a part of a listed assembly, shall be permitted.

530-13. Stage Lighting and Effects Control. Switches used for studio stage set lighting and effects (on the stages and lots and on location) shall be of the externally operable type. Where contactors are used as the disconnecting means for fuses, an individual externally operable switch, such as a tumbler switch, for the control of each contactor shall be located at a distance of not more than 6 ft (1.83 m) from the contactor, in addition to remote-control switches. A single externally operable switch shall be permitted to simultaneously disconnect all the contactors on any one location board, where located at a distance of not more than 6 ft (1.83 m) from the location board.

530-14. Plugging Boxes. Each receptacle of dc plugging boxes shall be rated at not less than 30 amperes.

530-15. Enclosing and Guarding Live Parts.

(a) Live Parts. Live parts shall be enclosed or guarded to prevent accidental contact by persons and objects.

(b) Switches. All switches shall be of the externally operable type.

(c) Rheostats. Rheostats shall be placed in approved cases or cabinets that enclose all live parts, having only the operating handles exposed.

(d) Current-Carrying Parts. Current-carrying parts of bull switches, location boards, spiders, and plugging boxes shall be enclosed, guarded, or located so that persons cannot accidentally come into contact with them or bring conductive material into contact with them.

530-16. Portable Lamps. Portable lamps and work lights shall be equipped with flexible cords, composition or metal-sheathed porcelain sockets, and substantial guards.

Exception: Portable lamps used as properties in a motion picture set or television stage set, on a studio stage or lot, or on location shall not be considered to be portable lamps for the purpose of this section.

530-17. Portable Arc Lamp Fixtures.

(a) Portable Carbon Arc Lamps. Portable carbon arc lamps shall be substantially constructed. The arc shall be provided with an enclosure designed to retain sparks and carbons and to prevent persons or materials from coming into contact with the arc or bare live parts. The enclosures shall be ventilated. All switches shall be of the externally operable type.

(b) Portable Noncarbon Arc Electric-Discharge Lamp Fixtures. Portable noncarbon arc lamp fixtures, including enclosed arc lamp fixtures, and associated ballasts shall be

listed. Interconnecting cord sets, and interconnecting cords and cables, shall be extra-hard usage type and listed.

530-18. Overcurrent Protection.

General. Automatic overcurrent protective devices (circuit breakers or fuses) for motion picture studio stage set lighting and the stage cables for such stage set lighting shall be as given in (a) through (g). The maximum ampacity allowed on a given conductor, cable, or cord size shall be as given in the applicable tables of Articles 310 and 400.

(a) Stage Cables. Stage cables for stage set lighting shall be protected by means of overcurrent devices set at not more than 400 percent of the ampacity given in the applicable tables of Articles 310 and 400.

(b) Feeders. In buildings used primarily for motion picture production, the feeders from the substations to the stages shall be protected by means of overcurrent devices (generally located in the substation) having a suitable ampere rating. The overcurrent devices shall be permitted to be multipole or single-pole gang operated. No pole shall be required in the neutral conductor. The overcurrent device setting for each feeder shall not exceed 400 percent of the ampacity of the feeder, as given in the applicable tables of Article 310.

(c) Cable Protection. Cables shall be protected by bushings where they pass through enclosures and shall be arranged so that tension on the cable will not be transmitted to the connections. Where power conductors pass through metal, the requirements of Section 300-20 shall apply.

Portable feeder cables shall be permitted to temporarily penetrate fire-rated walls, floors, or ceilings provided that

(1) The opening be of noncombustible material;
(2) When in use, the penetration is sealed with a temporary seal of a listed firestop material; and
(3) When not in use, the opening shall be capped with a material of equivalent fire rating.

(d) Location Boards. Overcurrent protection (fuses or circuit breakers) shall be provided at the location boards. Fuses in the location boards shall have an ampere rating of not over 400 percent of the ampacity of the cables between the location boards and the plugging boxes.

(e) Plugging Boxes. Cables and cords supplied through plugging boxes shall be of copper. Cables and cords smaller than No. 8 shall be attached to the plugging box by means of a plug containing two cartridge fuses or a 2-pole circuit breaker. The rating of the fuses or the setting of the circuit breaker shall not be over 400 percent of the rated ampacity of the cables or cords as given in the applicable tables of Articles 310 and 400. Plugging boxes shall not be permitted on ac systems.

(f) Alternating-Current Power Distribution Boxes. Alternating-current power distribution boxes used on sound stages and shooting locations shall contain connection receptacles of a polarized, grounding type.

(g) Lighting. Work lights, stand lamps, and fixtures, rated 1000 watts or less and connected to dc plugging boxes, shall be by means of plugs containing two cartridge fuses not larger than 20 amperes, or they shall be permitted to be connected to special outlets on circuits protected by fuses or circuit breakers rated at not over 20 amperes. Plug fuses shall not be used unless they are on the load side of the fuse or circuit breakers on the location boards.

530-19. Sizing of Feeder Conductors for Television Studio Sets.

(a) General. It shall be permissible to apply the demand factors listed in Table 530-19(a) to that portion of the maximum possible connected load for studio or stage set lighting for all permanently installed feeders between substations and stages and to all permanently installed feeders between the main stage switchboard and stage distribution centers or location boards.

(b) Portable Feeders. A demand factor of 50 percent of maximum possible connected load shall be permitted for all portable feeders.

Table 530-19(a). Demand Factors for Stage Set Lighting

Portion of Stage Set Lighting Load to Which Demand Factor Applied (volt-amperes)	Feeder Demand Factor
First 50,000 or less at	100%
From 50,001 to 100,000 at	75%
From 100,001 to 200,000 at	60%
Remaining over 200,000 at	50%

530-20. Grounding. Type MC cable, Type MI cable, metal raceways, and all noncurrent-carrying metal parts of appliances, devices, and equipment shall be grounded as specified in Article 250. This shall not apply to pendant and portable lamps, to stage lighting and stage sound equipment, or to other portable and special stage equipment operating at not over 150 volts dc to ground.

530-21. Plugs and Receptacles.

(a) Rating. Plugs and receptacles shall be rated in amperes. The voltage rating of the plugs and receptacles shall not be less than the circuit voltage. Plug and receptacle ampere ratings for ac circuits shall not be less than the feeder or branch-circuit overcurrent device ampere rating. Table 210-21(b)(2) shall not apply.

(b) Interchangeability. Plugs and receptacles used in portable professional motion picture and television equipment shall be permitted to be interchangeable for ac or dc use on the same premises provided they are listed for ac/dc use and marked in a suitable manner to identify the system to which they are connected.

530-22. Single-Pole Separable Connectors.

(a) General. Where ac single-pole portable cable connectors are used, they shall be listed and of the locking type. Section 400-10 and 410-56 shall not apply to listed single-pole separable connections and single-conductor cable assemblies utilizing listed single-pole separable connectors. Where paralleled sets of current-carrying single-pole separable connectors are provided as input devices, they shall be prominently labeled with a warning indicating the presence of internal parallel connections. The use of single-pole separable connectors shall comply with at least one of the following conditions.

(1) Connection and disconnection of connectors are only possible where the supply connectors are interlocked to the source and it is not possible to connect or disconnect connectors when the supply is energized.

(2) Line connectors are of the listed sequential-interlocking type so that load connectors shall be connected in the following sequence:

(a) Equipment grounding conductor connection
(b) Grounded circuit conductor connection, if provided
(c) Ungrounded conductor connection, and that disconnection shall be in the reverse order

(3) A caution notice shall be provided adjacent to the line connectors indicating that plug connection shall be in the following order:

(a) Equipment grounding conductor connectors
(b) Grounded circuit-conductor connectors, if provided
(c) Ungrounded conductor connectors, and that disconnection shall be in the reverse order

(b) Interchangeability. Single-pole separable connectors used in portable professional motion picture and television equipment shall be permitted to be interchangeable for ac or dc use or for different current ratings on the same premises provided they are listed for ac/dc use and marked in a suitable manner to identify the system to which they are connected.

530-23. Branch Circuits. A branch circuit of any size supplying one or more receptacles shall be permitted to supply stage set lighting loads.

C. Dressing Rooms

530-31. Dressing Rooms. Fixed wiring in dressing rooms shall be installed in accordance with the wiring methods covered in Chapter 3. Wiring for portable dressing rooms shall be approved.

D. Viewing, Cutting, and Patching Tables

530-41. Lamps at Tables. Only composition or metal-sheathed, porcelain, keyless lampholders equipped with suitable means to guard lamps from physical damage and from film and film scrap shall be used at patching, viewing, and cutting tables.

E. Cellulose Nitrate Film Storage Vaults

530-51. Lamps in Cellulose Nitrate Film Storage Vaults. Lamps in cellulose nitrate film storage vaults shall be installed in rigid fixtures of the glass-enclosed and gasketed type. Lamps shall be controlled by a switch having a pole in each ungrounded conductor. This switch shall be located outside of the vault and provided with a pilot light to indicate whether the switch is on or off. This switch shall disconnect from all sources of supply all ungrounded conductors terminating in any outlet in the vault.

530-52. Motors and Other Equipment in Cellulose Nitrate Film Storage Vaults. Except as permitted in Section 530-51, no receptacles, outlets, electric motors, heaters, portable lights, or other portable electric equipment shall be located in cellulose nitrate film storage vaults.

F. Substations

530-61. Substations. Wiring and equipment of over 600 volts, nominal, shall comply with Article 490.

530-62. Portable Substations. Wiring and equipment in portable substations shall conform to the sections applying to installations in permanently fixed substations, but, due to the limited space available, the working spaces shall be permitted to be reduced, provided that the equipment shall be arranged so that the operator can work safely and so that other persons in the vicinity cannot accidentally come into contact with current-carrying parts or bring conducting objects into contact with them while they are energized.

530-63. Overcurrent Protection of Direct-Current Generators. Three-wire generators shall have overcurrent protection in accordance with Section 445-4(e).

530-64. Direct-Current Switchboards.

(a) Switchboards of not over 250 volts dc between conductors, where located in substations or switchboard rooms accessible to qualified persons only, shall not be required to be dead-front.

(b) Frames of dc circuit breakers installed on switchboards shall not be required to be grounded.

G. Separately Derived Systems with 60 Volts to Ground

530-70. General. Use of a separately derived 120-volt, single-phase, 3-wire system with 60 volts on each of two ungrounded conductors to a grounded neutral conductor shall be permitted for the purpose of reducing objectionable noise in audio/video production or other similar sensitive electronic equipment locations provided that its use is restricted to electronic equipment only and that all of the requirements in Sections 530-71 through 530-73 are met.

530-71. Wiring Methods.

(a) Panelboards and Overcurrent Protection. Use of standard single-phase panelboards and distribution equipment with a higher voltage rating shall be permitted. The system shall be clearly marked on the face of the panel or on the inside of the panel doors. Common-trip, two-pole circuit breakers that are identified for operation at the system voltage shall be provided for both ungrounded conductors in all feeders and branch circuits.

(b) Junction Boxes. All junction box covers shall be clearly marked to indicate the distribution panel and the system voltage.

(c) Color Coding. All feeders and branch-circuit conductors installed under this section shall be identified as to system at all splices and terminations by color, marking, tagging, or equally effective means. The means of identification shall be posted at each branch-circuit panelboard and at the disconnecting means for the building.

(d) Voltage Drop. The voltage drop on any branch circuit shall not exceed 1.5 percent. The combined voltage drop of feeder and branch-circuit conductors shall not exceed 2.5 percent.

530-72. Grounding.

(a) General. The system shall be grounded as provided in Section 250-30 as a separately derived single-phase, 3-wire system.

(b) Grounding Conductors Required. Permanently wired utilization equipment and receptacles shall be grounded by means of an equipment grounding conductor run with the circuit conductors to an equipment grounding bus prominently marked "Technical Equipment Ground" in the originating branch-circuit panelboard. The grounding bus shall be connected to the grounded conductor on the line side of the separately derived system's disconnecting means. The grounding conductor shall not be smaller than that specified in Table 250-122 and run with the feeder conductors. The technical equipment grounding bus need not be bonded to the panelboard enclosure. Other grounding methods authorized elsewhere in this *Code* shall be permitted where the impedance of the grounding return path does not exceed the impedance of equipment grounding conductors sized and installed in accordance with Part G of this article.

FPN No. 1: See Section 250-122 for equipment grounding conductor sizing requirements where circuit conductors are adjusted in size to compensate for voltage drop.

FPN No. 2: These requirements limit the impedance of the ground-fault path where only 60 volts applies to a fault condition instead of the usual 120 volts.

530-73. Receptacles.

(a) **General.** Where receptacles are used as a means of connecting equipment, the following conditions shall be met:

(1) All 15- and 20-ampere receptacle outlets shall be ground-fault circuit-interrupter protected.
(2) All outlet strips, adapters, receptacle covers, and faceplates shall be marked as follows:

> WARNING — TECHNICAL POWER
> Do not connect to lighting equipment
> For electronic equipment use only
> 60/120 volt 1ø ac
> GFCI protected

(3) A 125-volt, single-phase, 15- or 20-ampere rated receptacle having one of its current-carrying poles connected to a grounded circuit conductor shall be located within 6 ft (1.83 m) of all permanently installed 15- or 20-ampere rated 60/120-volt technical power-system receptacles.
(4) All 125-volt receptacles used for 60/120-volt technical power shall be uniquely configured and identified for use with this class of system. All 125-volt, single-phase, 15- or 20-ampere rated receptacle outlets and attachment plugs that are identified for use with grounded circuit conductors shall be permitted in machine rooms, control rooms, equipment rooms, equipment racks, and other similar locations that are restricted to use by qualified personnel.

(b) **Isolated Ground Receptacles.** Isolated ground receptacles shall be permitted as described in Section 250-146(d), however, the branch-circuit equipment grounding conductor shall be terminated as required in Section 530-72(b).

Article 540 — Motion Picture Projectors

A. General

540-1. Scope. The provisions of this article apply to motion picture projection rooms, motion picture projectors, and associated equipment of the professional and nonprofessional types using incandescent, carbon arc, xenon, or other light source equipment that develops hazardous gases, dust, or radiation.

FPN: For further information, see *Standard for the Storage and Handling of Cellulose Nitrate Motion Picture Film,* NFPA 40-1997.

B. Definitions

540-2. Professional Projector. A type of projector using 35- or 70-mm film that has a minimum width of 1⅜ inch (35 mm) and has on each edge 5.4 perforations per inch, or a type using carbon arc, xenon, or other light source equipment that develops hazardous gases, dust, or radiation.

540-3. Nonprofessional Projector. Nonprofessional projectors are those types other than described in Section 540-2.

C. Equipment and Projectors of the Professional Type

540-10. Motion Picture Projection Room Required. Every professional-type projector shall be located within a projection room. Every projection room shall be of permanent construction, approved for the type of building in which the projection room is located. All projection ports, spotlight ports, viewing ports, and similar openings shall be provided with glass or other approved material so as to completely close the opening. Such rooms shall not be considered as hazardous (classified) locations as defined in Article 500.

FPN: For further information on protecting openings in projection rooms handling cellulose nitrate motion picture film, see *Life Safety Code,* NFPA *101*-1997.

540-11. Location of Associated Electrical Equipment.

(a) **Motor Generator Sets, Transformers, Rectifiers, Rheostats, and Similar Equipment.** Motor generator sets, transformers, rectifiers, rheostats, and similar equipment for the supply or control of current to projection or spotlight equipment shall, where nitrate film is used, be located in a separate room. Where placed in the projection room, they shall be located or guarded so that arcs or sparks cannot come in contact with film, and the commutator end or ends of motor generator sets shall comply with one of the conditions in (1) through (6).

(1) Types. Be of the totally enclosed, enclosed fan-cooled, or enclosed pipe-ventilated type.

(2) Separate Rooms or Housings. Be enclosed in separate rooms or housings built of noncombustible material constructed so as to exclude flyings or lint, and properly ventilated from a source of clean air.

(3) Solid Metal Covers. Have the brush or sliding-contact end of motor-generator enclosed by solid metal covers.

(4) Tight Metal Housings. Have brushes or sliding contacts enclosed in substantial, tight metal housings.

(5) Upper and Lower Half Enclosures. Have the upper half of the brush or sliding-contact end of the motor-generator enclosed by a wire screen or perforated metal and the lower half enclosed by solid metal covers.

(6) Wire Screens or Perforated Metal. Have wire screens or perforated metal placed at the commutator of brush ends. No dimension of any opening in the wire screen or perforated metal shall exceed 0.05 in. (1.27 mm), regardless of the shape of the opening and of the material used.

(b) Switches, Overcurrent Devices, or Other Equipment. Switches, overcurrent devices, or other equipment not normally required or used for projectors, sound reproduction, flood or other special effect lamps, or other equipment shall not be installed in projection rooms.

Exception No. 1: In projection rooms approved for use only with cellulose acetate (safety) film, the installation of appurtenant electrical equipment used in conjunction with the operation of the projection equipment and the control of lights, curtains, and audio equipment, etc., shall be permitted. In such projection rooms, a sign reading "Safety Film Only Permitted in This Room" shall be posted on the outside of each projection room door and within the projection room itself in a conspicuous location.

Exception No. 2: Remote-control switches for the control of auditorium lights or switches for the control of motors operating curtains and masking of the motion picture screen shall be permitted to be installed in projection rooms.

(c) Emergency Systems. Control of emergency systems shall comply with Article 700.

540-12. Work Space. Each motion picture projector, floodlight, spotlight, or similar equipment shall have clear working space not less than 30 in. (762 mm) wide on each side and at the rear thereof.

Exception: One such space shall be permitted between adjacent pieces of equipment.

540-13. Conductor Size. Conductors supplying outlets for arc and xenon projectors of the professional type shall not be smaller than No. 8 and shall be of sufficient size for the projector employed. Conductors for incandescent-type projectors shall conform to normal wiring standards as provided in Section 210-24.

540-14. Conductors on Lamps and Hot Equipment. Insulated conductors having a rated operating temperature of not less than 200°C (392°F) shall be used on all lamps or other equipment where the ambient temperature at the conductors as installed will exceed 50°C (122°F).

540-15. Flexible Cords. Cords approved for hard usage, as provided in Table 400-4, shall be used on portable equipment.

540-20. Approval. Projectors and enclosures for arc, xenon and incandescent lamps and rectifiers, transformers, rheostats, and similar equipment shall be listed.

540-21. Marking. Projectors and other equipment shall be marked with the manufacturer's name or trademark and with the voltage and current for which they are designed in accordance with Section 110-21.

D. Nonprofessional Projectors

540-31. Motion Picture Projection Room Not Required. Projectors of the nonprofessional or miniature type, where employing cellulose acetate (safety) film, shall be permitted to be operated without a projection room.

540-32. Approval. Projection equipment shall be listed.

E. Audio Signal Processing, Amplification, and Reproduction Equipment

540-50. Audio Signal Processing, Amplification, and Reproduction Equipment. Audio signal processing, amplification, and reproduction equipment shall be installed as provided in Article 640.

Article 545 — Manufactured Buildings

A. General

545-1. Scope. This article covers requirements for a manufactured building and building components as herein defined.

545-2. Other Articles. Wherever the requirements of other articles of this *Code* and Article 545 differ, the requirements of Article 545 shall apply.

545-3. Definitions.

Building Component. Any subsystem, subassembly, or other system designed for use in or integral with or as part of a structure, which can include structural, electrical, mechanical, plumbing, and fire protection systems, and other systems affecting health and safety.

Building System. Plans, specifications, and documentation for a system of manufactured building or for a type or a system of building components, which can include structural, electrical, mechanical, plumbing, and fire protection systems, and other systems affecting health and safety, and including such variations thereof as are specifically permitted by regulation, and which variations are submitted as part of the building system or amendment thereto.

Closed Construction. Any building, building component, assembly, or system manufactured in such a manner that all concealed parts of processes of manufacture cannot be inspected before installation at the building site without disassembly, damage, or destruction.

Manufactured Building. Any building that is of closed construction and is made or assembled in manufacturing facilities on or off the building site for installation, or assembly and installation on the building site, other than manufactured homes, mobile homes, park trailers, or recreational vehicles.

545-4. Wiring Methods.

(a) Methods Permitted. All raceway and cable wiring methods included in this *Code* and such other wiring systems specifically intended and listed for use in manufactured buildings shall be permitted with listed fittings and with fittings listed and identified for manufactured buildings.

(b) Securing Cables. In closed construction, cables shall be permitted to be secured only at cabinets, boxes, or fittings where No. 10 or smaller conductors are used and protection against physical damage is provided.

545-5. Service-Entrance Conductors. Service-entrance conductors shall meet the requirements of Article 230. Provisions shall be made to route the service-entrance conductors from the service equipment to the point of attachment of the service drop or service lateral.

545-6. Installation of Service-Entrance Conductors. Service-entrance conductors shall be installed after erection at the building site.

Exception: Where point of attachment is known prior to manufacture.

545-7. Service Equipment. Service equipment shall be installed in accordance with Section 230-70.

545-8. Protection of Conductors and Equipment. Protection shall be provided for exposed conductors and equipment during processes of manufacturing, packaging, in transit, and erection at the building site.

545-9. Boxes.

(a) Other Dimensions. Boxes of dimensions other than those required in Table 370-16(a) shall be permitted to be installed where tested, identified, and listed to applicable standards.

(b) Not Over 100 in.³ (1640 cm³). Any box not over 100 in.³ (1640 cm³) in size, intended for mounting in closed construction, shall be affixed with anchors or clamps so as to provide a rigid and secure installation.

545-10. Receptacle or Switch with Integral Enclosure. A receptacle or switch with integral enclosure and mounting means, where tested, identified, and listed to applicable standards, shall be permitted to be installed.

545-11. Bonding and Grounding. Prewired panels and building components shall provide for the bonding, or bonding and grounding, of all exposed metals likely to become energized, in accordance with Article 250, Parts E, F, and G.

545-12. Grounding Electrode Conductor. Provisions shall be made to route a grounding electrode conductor from the service equipment to the point of attachment to the grounding electrode.

545-13. Component Interconnections. Fittings and connectors that are intended to be concealed at the time of on-site assembly, where tested, identified, and listed to applicable standards, shall be permitted for on-site interconnection of modules or other building components. Such fittings and connectors shall be equal to the wiring method employed in insulation, temperature rise, and fault-current withstand and shall be capable of enduring the vibration and minor relative motions occurring in the components of manufactured building.

Article 547 — Agricultural Buildings

547-1. Scope. The provisions of this article shall apply to the following agricultural buildings or that part of a building or adjacent areas of similar or like nature as specified in (a) and (b).

(a) Excessive Dust and Dust with Water. Agricultural buildings where excessive dust and dust with water may accumulate, including all areas of poultry, livestock, and fish confinement systems, where litter dust or feed dust, including mineral feed particles, may accumulate.

(b) Corrosive Atmosphere. Agricultural buildings where a corrosive atmosphere exists. Such buildings include areas:

(1) Where poultry and animal excrement may cause corrosive vapors

(2) Where corrosive particles may combine with water

(3) Where the area is damp and wet by reason of periodic washing for cleaning and sanitizing with water and cleansing agents

(4) Where similar conditions exist

547-2. Other Articles. For agricultural buildings not having conditions as specified in Section 547-1, the electrical installations shall be made in accordance with the applicable articles in this *Code*.

547-3. Surface Temperatures. Electrical equipment or devices installed in accordance with the provisions of this article shall be installed in a manner such that they will function at full rating without developing surface temperatures in excess of the specified normal safe operating range of the equipment or device.

547-4. Wiring Methods.

(a) Wiring Systems. Types UF, NMC, copper SE cables, rigid nonmetallic conduit, liquidtight flexible nonmetallic conduit, or other cables or raceways suitable for the location, with approved termination fittings, shall be the wiring methods employed. Article 320 and Article 502 wiring methods shall be permitted for areas described in Section 547-1(a).

> FPN: See Sections 300-7 and 347-9 for installation of raceway systems exposed to widely different temperatures.

(b) Mounting. All cables shall be secured within 8 in. (203 mm) of each cabinet, box, or fitting. The ¼-in. (6.35-mm) airspace required for nonmetallic boxes, fittings, conduit, and cables in Section 300-6(c) shall not be required in buildings covered by this article.

(c) Boxes and Fittings. All boxes and fittings shall comply with Section 547-5.

(d) Flexible Connections. Where necessary to employ flexible connections, dusttight flexible connectors, liquidtight flexible conduit, or flexible cord listed and identified for hard usage shall be used. All connectors and fittings used shall be listed and identified for the purpose.

(e) Physical Protection. All electrical wiring and equipment subject to physical damage shall be protected.

(f) Separate Equipment Grounding Conductor. Non-current-carrying metal parts of equipment, raceways, and other enclosures, where required to be grounded, shall be grounded by a copper equipment grounding conductor installed between the equipment and the building disconnecting means. If installed underground, the equipment grounding conductor shall be insulated or covered.

547-5. Switches, Circuit Breakers, Controllers, and Fuses. Switches, circuit breakers, controllers, and fuses, including pushbuttons, relays, and similar devices, shall be provided with enclosures as specified in (a) and (b).

(a) Excessive Dust and Dust with Water. Buildings described in Section 547-1(a) shall utilize dustproof and weatherproof enclosures.

(b) Corrosive Atmosphere. Buildings described in Section 547-1(b) shall utilize enclosures suitable for the conditions encountered in the application.

> FPN No. 1: See Table 430-91 for appropriate enclosure type designations.

> FPN No. 2: Cast aluminum and magnetic steels may corrode in agricultural environments.

547-6. Motors. Motors and other rotating electrical machinery shall be totally enclosed or designed so as to minimize the entrance of dust, moisture, or corrosive particles.

547-7. Lighting Fixtures. Lighting fixtures shall comply with the following.

(a) Minimize the Entrance of Dust. Lighting fixtures shall be installed to minimize the entrance of dust, foreign matter, moisture, and corrosive material.

(b) Exposed to Physical Damage. Any lighting fixture that may be exposed to physical damage shall be protected by a suitable guard.

(c) Exposed to Water. A fixture that may be exposed to water from condensation, building cleansing water, or solution shall be watertight.

547-8. Service Equipment, Separately Derived Systems, Feeders, Disconnecting Means, and Grounding. Where one or more agricultural buildings are supplied from a distribution point, the disconnecting means and grounding of services and feeders shall comply with (a), (b), or (c).

(a) Disconnecting Means and Overcurrent Protection at a Building(s). Where the disconnecting means and overcurrent protection are located at the load end of the service conductors, service equipment grounding shall meet the requirements of Section 250-24. A disconnecting means shall also be installed at the distribution point when two or more buildings are supplied from that distribution point.

(b) Disconnecting Means and Overcurrent Protection at Distribution Point. Where the disconnecting means and overcurrent protection are located at the distribution point, feeders to buildings shall meet the requirements of Section 250-32 and Article 225, Part B.

(c) Disconnecting Means Without Overcurrent Protection at the Distribution Point. Where the disconnecting means without overcurrent protection is located at the distribution point and a disconnecting means and overcurrent protection is located at the building(s), the grounded circuit conductor connection to the grounding electrode shall not be permitted at the building disconnecting means at each building and all of the following conditions shall be met.

(1) All buildings and premises wiring are under single management.
(2) Disconnecting means suitable for use as service equipment is provided at the distribution point.
(3) An equipment grounding conductor is run with the supply conductors and is of the same size as the largest supply conductor, if of the same material, or is adjusted in size in accordance with the equivalent size columns of Table 250-122 if of different materials.
(4) The equipment grounding conductor is bonded to the grounded circuit conductor at the distribution point or at the source of a separately derived system.
(5) A grounding electrode system is provided and connected to the equipment grounding conductor at the building disconnecting means.

Distribution Point. A centrally located electrical supply structure from which services or feeders to agricultural buildings and other buildings, including the associated farm dwelling, are normally supplied.

547-9. Bonding and Equipotential Plane.

(a) Definition of Equipotential Plane. An area accessible to livestock where a wire mesh or other conductive elements are embedded in concrete, are bonded to all metal structures and fixed nonelectrical metal equipment that may become energized and are connected to the electrical grounding system to prevent a difference in voltage from developing within the plane. For this section, livestock does not include poultry.

(b) General. Wire mesh or other conductive elements shall be installed in the concrete floor of livestock confinement areas and be bonded to the building grounding electrode system to provide an equipotential plane that may have voltage gradient ramps at entrances and exits that are traversed daily by the same livestock. The bonding conductor

shall be copper, insulated, covered or bare, and not smaller than No. 8. The means of bonding to wire mesh or conductive elements shall be by pressure connectors or clamps of brass, copper, copper alloy, or an equally substantial approved means.

Exception No. 1: An equipotential plane shall not be required where there is no electric service to the building nor metal equipment accessible to livestock that is likely to become energized.

Exception No. 2: Slatted floors that are supported by structures that are a part of an equipotential plane shall not be required to be bonded.

> FPN No. 1: A natural voltage gradient exists at the edge of an equipotential plane. Typically, voltage gradients exceeding 1 volt per foot at the edge of the equipotential plane will require a voltage gradient ramp.
>
> FPN No. 2: Methods to establish equipotential planes and voltage gradient ramps are described in *Equipotential Planes in Animal Containment Areas*, American Society of Agricultural Engineers (ASAE) EP473-1997.
>
> FPN No. 3: Low grounding electrode system resistances may reduce potential differences in livestock facilities.

(c) Receptacles. All 125-volt, single-phase, 15- and 20-ampere general purpose receptacles in areas having an equipotential plane shall have ground-fault circuit-interrupter protection for personnel.

Article 550 — Mobile Homes, Manufactured Homes, and Mobile Home Parks

A. General

550-1. Scope. The provisions of this article cover the electrical conductors and equipment installed within or on mobile homes, the conductors that connect mobile homes to a supply of electricity, and the installation of electrical wiring, fixtures, equipment, and appurtenances related to electrical installations within a mobile home park up to the mobile home service-entrance conductors or, if none, the mobile home service equipment.

550-2. Definitions.

Appliance, Fixed. An appliance that is fastened or otherwise secured at a specific location.

Appliance, Portable. An appliance that is actually moved or can easily be moved from one place to another in normal use.

> FPN: For the purpose of this article, the following major appliances, other than built-in, are considered portable if cord connected: refrigerators, range equipment, clothes washers,

dishwashers without booster heaters, or other similar appliances.

Appliance, Stationary. An appliance that is not easily moved from one place to another in normal use.

Distribution Panelboard. See definition of *panelboard* in Article 100.

Feeder Assembly. The overhead or under-chassis feeder conductors, including the grounding conductor, together with the necessary fittings and equipment or a power-supply cord listed for mobile home use, designed for the purpose of delivering energy from the source of electrical supply to the distribution panelboard within the mobile home.

Laundry Area. An area containing or designed to contain a laundry tray, clothes washer, or a clothes dryer.

Manufactured Home. A factory-assembled structure or structures that bears a label identifying it as a manufactured home that is transportable in one or more sections, that is built on a permanent chassis and designed to be used as a dwelling with or without a permanent foundation where connected to the required utilities, and that includes the plumbing, heating, air-conditioning, and electrical systems contained therein.

> FPN No. 1: See the applicable building code for definition of the term *permanent foundation.*
>
> FPN No. 2: See Part 3280, *Manufactured Home Construction and Safety Standards,* of the federal Department of Housing and Urban Development for additional information on the definition.

For the purpose of this *Code* and unless otherwise indicated, the term *mobile home* includes manufactured homes.

Mobile Home. A factory-assembled structure or structures transportable in one or more sections that is built on a permanent chassis and designed to be used as a dwelling without a permanent foundation where connected to the required utilities, and includes the plumbing, heating, air-conditioning, and electric systems contained therein.

For the purpose of this *Code* and unless otherwise indicated, the term *mobile home* includes manufactured homes.

Mobile Home Accessory Building or Structure. Any awning, cabana, ramada, storage cabinet, carport, fence, windbreak, or porch established for the use of the occupant of the mobile home on a mobile home lot.

Mobile Home Lot. A designated portion of a mobile home park designed for the accommodation of one mobile home and its accessory buildings or structures for the exclusive use of its occupants.

Mobile Home Park. A contiguous parcel of land that is used for the accommodation of occupied mobile homes.

Mobile Home Service Equipment. The equipment containing the disconnecting means, overcurrent protective devices, and receptacles or other means for connecting a mobile home feeder assembly.

Park Electrical Wiring Systems. All of the electrical wiring, fixtures, equipment, and appurtenances related to electrical installations within a mobile home park, including the mobile home service equipment.

550-3. Other Articles. Wherever the requirements of other articles of this *Code* and Article 550 differ, the requirements of Article 550 shall apply.

550-4. General Requirements.

(a) Mobile Home Not Intended as a Dwelling Unit. A mobile home not intended as a dwelling unit, for example, those equipped for sleeping purposes only, contractor's on-site offices, construction job dormitories, mobile studio dressing rooms, banks, clinics, mobile stores, or intended for the display or demonstration of merchandise or machinery, shall not be required to meet the provisions of this article pertaining to the number or capacity of circuits required. It shall, however, meet all other applicable requirements of this article if provided with an electrical installation intended to be energized from a 120-volt or 120/240-volt ac power supply system. Where different voltage is required by either design or available power supply system, adjustment shall be made in accordance with other articles and sections for the voltage used.

(b) In Other than Mobile Home Parks. Mobile homes installed in other than mobile home parks shall comply with the provisions of this article.

(c) Connection to Wiring System. The provisions of this article apply to mobile homes intended for connection to a wiring system rated 120/240 volts, nominal, 3-wire ac, with grounded neutral.

(d) Listed or Labeled. All electrical materials, devices, appliances, fittings, and other equipment shall be listed or labeled by a qualified testing agency and shall be connected in an approved manner when installed.

B. Mobile Homes

550-5. Power Supply.

(a) Feeder. The power supply to the mobile home shall be a feeder assembly consisting of not more than one listed 50-ampere mobile home power-supply cord with an integrally molded or securely attached plug cap, or a permanently installed feeder.

Exception No. 1: A mobile home that is factory equipped with gas or oil-fired central heating equipment and cooking appliances shall be permitted to be provided with a listed mobile home power-supply cord rated 40 amperes.

Exception No. 2: Manufactured homes constructed in accordance with Section 550-23(b).

(b) Power-Supply Cord. If the mobile home has a power-supply cord, it shall be permanently attached to the distribution panelboard or to a junction box permanently connected to the distribution panelboard, with the free end terminating in an attachment plug cap.

Cords with adapters and pigtail ends, extension cords, and similar items shall not be attached to, or shipped with, a mobile home.

A suitable clamp or the equivalent shall be provided at the distribution panelboard knockout to afford strain relief for the cord to prevent strain from being transmitted to the terminals when the power-supply cord is handled in its intended manner.

The cord shall be a listed type with four conductors, one of which shall be identified by a continuous green color or a continuous green color with one or more yellow stripes for use as the grounding conductor.

(c) Attachment Plug Cap. The attachment plug cap shall be a 3-pole, 4-wire, grounding type, rated 50 amperes, 125/250 volts with a configuration as shown in Figure 550-5(c) and intended for use with the 50-ampere, 125/250-volt receptacle configuration shown in Figure 550-5(c). It shall be listed, by itself or as part of a power-supply cord assembly, for the purpose, and shall be molded to or installed on the flexible cord so that it is secured tightly to the cord at the point where the cord enters the attachment plug cap. If a right-angle cap is used, the configuration shall be oriented so that the grounding member is farthest from the cord.

FPN: Complete details of the 50-ampere plug and receptacle configuration can be found in the National Electrical Manufacturers Association *Standard for Dimensions of Attachment Plugs and Receptacles*, ANSI/NEMA WD6-1989, Figure 14-50.

Receptacle   Cap

125/250-V, 50-A, 3-pole, 4-wire, grounding type

Figure 550-5(c)   50-ampere, 125/250-volt receptacle and attachment plug cap configurations, 3-pole, 4-wire, grounding-types, used for mobile home supply cords and mobile home parks.

(d) Overall Length of a Power-Supply Cord. The overall length of a power-supply cord, measured from the end of the cord, including bared leads, to the face of the attachment plug cap shall not be less than 21 ft (6.4 m) and shall not exceed 36½ ft (11.13 m). The length of the cord from the face of the attachment plug cap to the point where the cord enters the mobile home shall not be less than 20 ft (6.1 m).

(e) Marking. The power-supply cord shall bear the following marking:

FOR USE WITH MOBILE HOMES — 40 AMPERES.

or

FOR USE WITH MOBILE HOMES — 50 AMPERES.

(f) Point of Entrance. The point of entrance of the feeder assembly to the mobile home shall be in the exterior wall, floor, or roof.

(g) Protected. Where the cord passes through walls or floors, it shall be protected by means of conduits and bushings or equivalent. The cord shall be permitted to be installed within the mobile home walls, provided a continuous raceway having a maximum size of 1¼ in. (31.8 mm) is installed from the branch-circuit panelboard to the underside of the mobile home floor.

(h) Protection Against Corrosion and Mechanical Damage. Permanent provisions shall be made for the protection of the attachment plug cap of the power-supply cord and any connector cord assembly or receptacle against corrosion and mechanical damage if such devices are in an exterior location while the mobile home is in transit.

(i) Mast Weatherhead or Raceway. Where the calculated load exceeds 50 amperes or where a permanent feeder is used, the supply shall be by means of the following:

(1) One mast weatherhead installation, installed in accordance with Article 230, containing four continuous, insulated, color-coded feeder conductors, one of which shall be an equipment grounding conductor; or
(2) A metal raceway or rigid nonmetallic conduit from the disconnecting means in the mobile home to the underside of the mobile home, with provisions for the attachment to a suitable junction box or fitting to the raceway on the underside of the mobile home [with or without conductors as in Section 550-5(i)(1)].

550-6. Disconnecting Means and Branch-Circuit Protective Equipment. The branch-circuit equipment shall be permitted to be combined with the disconnecting means as a single assembly. Such a combination shall be permitted to be designated as a distribution panelboard. If a fused distribution panelboard is used, the maximum fuse size for the mains shall be plainly marked with lettering at least ¼ in. (6.4 mm) high and visible when fuses are changed.

Where plug fuses and fuseholders are used, they shall be tamper-resistant Type S, enclosed in dead-front fuse panelboards. Electrical distribution panelboards containing circuit breakers shall also be dead-front type.

> FPN: See Section 110-22 concerning identification of each disconnecting means and each service, feeder, or branch circuit at the point where it originated and the type marking needed.

(a) Disconnecting Means. A single disconnecting means shall be provided in each mobile home consisting of a circuit breaker, or a switch and fuses and its accessories installed in a readily accessible location near the point of entrance of the supply cord or conductors into the mobile home. The main circuit breakers or fuses shall be plainly marked "Main." This equipment shall contain a solderless type of grounding connector or bar for the purposes of grounding, with sufficient terminals for all grounding conductors. The neutral bar termination of the grounded circuit conductors shall be insulated in accordance with Section 550-11(a). The disconnecting equipment shall have a rating suitable for the connected load. The distribution equipment, either circuit breaker or fused type, shall be located a minimum of 24 in. (610 mm) from the bottom of such equipment to the floor level of the mobile home.

> FPN: See Section 550-15(b) for information on disconnecting means for branch circuits designed to energize heating or air-conditioning equipment, or both, located outside the mobile home, other than room air conditioners.

A distribution panelboard shall be rated not less than 50 amperes and employ a 2-pole circuit breaker rated 40 amperes for a 40-ampere supply cord, or 50 amperes for a 50-ampere supply cord. A distribution panelboard employing a disconnect switch and fuses shall be rated 60 amperes and shall employ a single 2-pole, 60-ampere fuseholder with 40- or 50-ampere main fuses for 40- or 50-ampere supply cords, respectively. The outside of the distribution panelboard shall be plainly marked with the fuse size.

The distribution panelboard shall be located in an accessible location but shall not be located in a bathroom or a clothes closet. A clear working space at least 30 in. (762 mm) wide and 30 in. (762 mm) in front of the distribution panelboard shall be provided. This space shall extend from the floor to the top of the distribution panelboard.

(b) Branch-Circuit Protective Equipment. Branch-circuit distribution equipment shall be installed in each mobile home and shall include overcurrent protection for each branch circuit consisting of either circuit breakers or fuses.

The branch-circuit overcurrent devices shall be rated as follows:

(1) Not more than the circuit conductors; and

(2) Not more than 150 percent of the rating of a single appliance rated 13.3 amperes or more that is supplied by an individual branch circuit; but

(3) Not more than the overcurrent protection size and of the type marked on the air conditioner or other motor-operated appliance.

A 15-ampere multiple receptacle shall be permitted where connected to a 20-ampere laundry circuit.

(c) Two-Pole Circuit Breakers. Where circuit breakers are provided for branch-circuit protection, 240-volt circuits shall be protected by a 2-pole common or companion trip, or handle-tied paired circuit breakers.

(d) Electrical Nameplates. A metal nameplate on the outside adjacent to the feeder assembly entrance shall read:

> THIS CONNECTION FOR 120/240-VOLT,
> 3-POLE, 4-WIRE, 60 HERTZ, _____ AMPERE
> SUPPLY

The correct ampere rating shall be marked in the blank space.

550-7. Branch Circuits. The number of branch circuits required shall be determined in accordance with (a) through (c).

(a) Lighting. Based on 3 volt-amperes/ft² (32.26 VA/m²) times outside dimensions of the mobile home (coupler excluded) divided by 120 volts to determine the number of 15- or 20-ampere lighting area circuits, e.g.,

$$\frac{3 \times \text{length} \times \text{width}}{120 \times 15 \text{ (or 20)}} = \text{No. of 15- (or 20-) ampere circuits}$$

The lighting circuits shall be permitted to serve built-in gas ovens with electric service only for lights, clocks or timers, or listed cord-connected garbage disposal units.

(b) Small Appliances. Small appliance branch circuits shall be installed in accordance with Section 210-52(b).

(c) General Appliances. (Including furnace, water heater, range, and central or room air conditioner, etc.) There shall be one or more circuits of adequate rating in accordance with the following:

> FPN No. 1: For the laundry branch circuit, see Section 210-11(c)(2).

> FPN No. 2: For central air conditioning, see Article 440.

(1) Ampere rating of fixed appliances not over 50 percent of circuit rating if lighting outlets (receptacles, other than kitchen, dining area, and laundry, considered as lighting outlets) are on the same circuit.

(2) For fixed appliances on a circuit without lighting outlets, the sum of rated amperes shall not exceed the branch-circuit rating. Motor loads or other continuous duty loads shall not exceed 80 percent of the branch-circuit rating.

(3) The rating of a single cord- and plug-connected appliance on a circuit having no other outlets shall not exceed 80 percent of the circuit rating.

(4) The rating of a range branch circuit shall be based on the range demand as specified for ranges in Section 550-13(b)(5).

550-8. Receptacle Outlets.

(a) Grounding-Type Receptacle Outlets. All receptacle outlets (1) shall be of grounding type; (2) shall be installed according to Section 210-7; and (3), except where supplying specific appliances, receptacles shall be 15- or 20-ampere, 125-volt, either single or duplex, and shall accept parallel-blade attachment plugs.

(b) Ground-Fault Circuit Interrupters. All 120-volt, single-phase, 15- and 20-ampere receptacle outlets installed outdoors and in bathrooms, including receptacles in light fixtures, shall have ground-fault circuit-interrupter protection for personnel. Ground-fault circuit-interrupter protection for personnel shall be provided for receptacle outlets located within 6 ft (1.83 m) of any lavatory or sink.

Exception: Receptacles installed for appliances in dedicated spaces, such as for dishwashers, disposals, refrigerators, freezers, and laundry equipment.

No receptacle shall be required in the area occupied by a toilet, shower, tub, or any combination thereof. If a receptacle is installed in such an area, it shall have ground-fault circuit-interrupter protection for personnel.

Feeders supplying branch circuits shall be permitted to be protected by a ground-fault circuit-interrupter in lieu of the provision for such interrupters specified herein.

(c) Cord-Connected Fixed Appliance. A grounding-type receptacle outlet shall be provided for each cord-connected fixed appliance installed.

(d) Required Receptacle Outlets. Receptacle outlets shall be provided in all rooms other than the bath, closet, and hall areas, and shall be installed so that no point along the floor line is more than 6 ft (1.83 m) measured horizontally from an outlet in that space. Countertops shall have receptacles located every 6 ft (1.83 m). The contiguous measurement of countertop and floor line shall be permitted where measured from the required receptacle in rooms requiring small appliance circuits. Receptacle outlets on small appliance circuits

shall not be included in determining the spacing for receptacle outlets of other circuits.

Exception No. 1: Where the measured distance is interrupted by an interior doorway, sink, refrigerator, range, oven, or cooktop, an additional receptacle outlet shall be provided where the interrupted space is at least 2 ft (610 mm) wide at the floor line and at least 12 in. (305 mm) wide at the countertop.

Exception No. 2: Receptacles rendered not readily accessible by stationary appliances shall not be considered as the required outlets.

Exception No. 3: The distance along a floor line occupied by a door opened fully against that space shall not be required to be included in establishing the horizontal measurement if the door swing is limited to 90 degrees nominal by that wall space.

Exception No. 4: Receptacle requirements for bar-type counters and for fixed room dividers shall be permitted to be provided by a receptacle outlet in the wall at the nearest point where the counter or room divider attaches to the wall provided the following:

(a) The divider does not exceed 8 ft (2.44 m) in length; and
(b) The divider does not exceed 4 ft (1.22 m) in height; and
(c) The divider is attached to a wall at one end only.

(e) Outdoor Receptacle Outlets. At least one receptacle outlet shall be installed outdoors. A receptacle outlet located in a compartment accessible from the outside of the mobile home shall be considered an outdoor receptacle. Outdoor receptacle outlets shall be protected as required in Section 550-8(b).

(f) Receptacle Outlets Not Permitted.

(1) Shower or Bathtub Space. Receptacle outlets shall not be installed in or within reach [30 in. (762 mm)] of a shower or bathtub space.

(2) Face-Up Position. A receptacle shall not be installed in a face-up position in any countertop.

(g) Pipe Heating Cable Outlet. Where a pipe heating cable outlet is installed, the outlet shall be as follows:

(1) Located within 2 ft (610 mm) of the cold water inlet.
(2) Connected to an interior branch circuit, other than a small appliance branch circuit. It shall be permitted to utilize a bathroom receptacle circuit for this purpose.
(3) On a circuit where all of the outlets are on the load side of the ground-fault circuit-interrupter protection for personnel.
(4) Mounted on the underside of the mobile home and shall not be considered to be the outdoor receptacle outlet required in Section 550-8(e).

550-9. Fixtures and Appliances.

(a) Fasten Appliances in Transit. Means shall be provided to securely fasten appliances when the mobile home is in transit. (See Section 550-11 for provisions on grounding.)

(b) Accessibility. Every appliance shall be accessible for inspection, service, repair, or replacement without removal of permanent construction.

(c) Pendants. Pendant-type fixtures or pendant cords shall be listed and identified for the interconnection of building components.

(d) Bathtub and Shower Fixtures. Where a lighting fixture is installed over a bathtub or in a shower stall, it shall be of the enclosed and gasketed type listed for wet locations.

(e) Location of Switches. The switch for shower lighting fixtures and exhaust fans located over a tub or in a shower stall shall be located outside the tub or shower space.

550-10. Wiring Methods and Materials. Except as specifically limited in this section, the wiring methods and materials included in this *Code* shall be used in mobile homes.

(a) Nonmetallic Boxes. Nonmetallic boxes shall be permitted only with nonmetallic cable or nonmetallic raceways.

(b) Nonmetallic Cable Protection. Nonmetallic cable located 15 in. (381 mm) or less above the floor, if exposed, shall be protected from physical damage by covering boards, guard strips, or raceways. Cable likely to be damaged by stowage shall be so protected in all cases.

(c) Metal-Covered and Nonmetallic Cable Protection. Metal-covered and nonmetallic cables shall be permitted to pass through the centers of the wide side of 2 in. by 4 in. studs. However, they shall be protected where they pass through 2 in. by 2 in. studs or at other studs or frames where the cable or armor would be less than 1¼ in. (31.8 mm) from the inside or outside surface of the studs where the wall covering materials are in contact with the studs. Steel plates on each side of the cable, or a tube, with not less than No. 16 MSG wall thickness shall be required to protect the cable. These plates or tubes shall be securely held in place.

(d) Metal Faceplates. Where metal faceplates are used, they shall be effectively grounded.

(e) Installation Requirements. If a range, clothes dryer, or similar appliance is connected by metal-covered cable or flexible metal conduit, a length of not less than 3 ft (914 mm) of free cable or conduit shall be provided to permit moving the appliance. The cable or flexible metal conduit shall be secured to the wall. Type NM or Type SE cable shall not be used to connect a range or dryer. This shall not prohibit the use of Type NM or Type SE cable between the branch-circuit overcurrent-protective device and a junction box or range or dryer receptacle.

(f) Raceways. Where rigid metal conduit or intermediate metal conduit is terminated at an enclosure with a locknut and bushing connection, two locknuts shall be provided, one inside and one outside of the enclosure. Rigid nonmetallic conduit, or electrical nonmetallic tubing, or surface raceway shall be permitted. All cut ends of conduit and tubing shall be reamed or otherwise finished to remove rough edges.

(g) Switches. Switches shall be rated as follows.

(1) For lighting circuits, switches shall be rated not less than 10 amperes, 120 to 125 volts, and in no case less than the connected load.

(2) For motors or other loads, switches shall have ampere or horsepower ratings, or both, adequate for loads controlled. (An ac general-use snap switch shall be permitted to control a motor 2 hp or less with full-load current not over 80 percent of the switch ampere rating.)

(h) Under-Chassis Wiring (Exposed to Weather).

(1) Where outdoor or under-chassis line-voltage (120 volts, nominal, or higher) wiring is exposed to moisture or physical damage, it shall be protected by rigid metal conduit or intermediate metal conduit. The conductors shall be suitable for wet locations.

Exception: Electrical metallic tubing or rigid nonmetallic conduit shall be permitted where closely routed against frames and equipment enclosures.

(2) The cables or conductors shall be Type NMC, Type TW, or equivalent.

(i) Boxes, Fittings, and Cabinets. Boxes, fittings, and cabinets shall be securely fastened in place and shall be supported from a structural member of the home, either directly or by using a substantial brace.

Exception: Snap-in-type boxes. Boxes provided with special wall or ceiling brackets and wiring devices with integral enclosures that securely fasten to walls or ceilings and are identified for the use shall be permitted without support from a structural member or brace. The testing and approval shall include the wall and ceiling construction systems for which the boxes and devices are intended to be used.

(j) Appliance Terminal Connections. Appliances having branch-circuit terminal connections that operate at temperatures higher than 60°C (140°F) shall have circuit conductors as described in (1) or (2).

(1) Branch-circuit conductors having an insulation suitable for the temperature encountered shall be permitted to be run directly to the appliance.

(2) Conductors having an insulation suitable for the temperature encountered shall be run from the appliance terminal connection to a readily accessible outlet box placed at least 1 ft (305 mm) from the appliance. These conductors shall be in a suitable raceway that shall extend for at least 4 ft (1.22 m).

(k) Component Interconnections. Fittings and connectors that are intended to be concealed at the time of assembly shall be listed and identified for the interconnection of building components. Such fittings and connectors shall be equal to the wiring method employed in insulation, temperature rise, and fault-current withstanding and shall be capable of enduring the vibration and shock occurring in mobile home transportation.

550-11. Grounding. Grounding of both electrical and non-electrical metal parts in a mobile home shall be through connection to a grounding bus in the mobile home distribution panelboard. The grounding bus shall be grounded through the green-colored insulated conductor in the supply cord or the feeder wiring to the service ground in the service-entrance equipment located adjacent to the mobile home location. Neither the frame of the mobile home nor the frame of any appliance shall be connected to the grounded circuit conductor (neutral) in the mobile home.

(a) Insulated Neutral.

(1) The grounded circuit conductor (neutral) shall be insulated from the grounding conductors and from equipment enclosures and other grounded parts. The grounded (neutral) circuit terminals in the distribution panelboard and in ranges, clothes dryers, counter-mounted cooking units, and wall-mounted ovens shall be insulated from the equipment enclosure. Bonding screws, straps, or buses in the distribution panelboard or in appliances shall be removed and discarded.

(2) Connections of ranges and clothes dryers with 120/240-volt, 3-wire ratings shall be made with 4-conductor cord and 3-pole, 4-wire, grounding-type plugs or by Type AC cable, Type MC cable, or conductors enclosed in flexible metal conduit.

(b) Equipment Grounding Means.

(1) The green-colored insulated grounding wire in the supply cord or permanent feeder wiring shall be connected to the grounding bus in the distribution panelboard or disconnecting means.

(2) In the electrical system, all exposed metal parts, enclosures, frames, lamp fixture canopies, etc., shall be effec-tively bonded to the grounding terminal or enclosure of the distribution panelboard.

(3) Cord-connected appliances, such as washing machines, clothes dryers, refrigerators, and the electrical system of gas ranges, etc., shall be grounded by means of a cord with grounding conductor and grounding-type attachment plug.

(c) Bonding of Noncurrent-Carrying Metal Parts.

(1) All exposed noncurrent-carrying metal parts that may become energized shall be effectively bonded to the grounding terminal or enclosure of the distribution panelboard. A bonding conductor shall be connected between the distribution panelboard and accessible terminal on the chassis.

(2) Grounding terminals shall be of the solderless type and listed as pressure-terminal connectors recognized for the wire size used. The bonding conductor shall be solid or stranded, insulated or bare, and shall be No. 8 copper minimum, or equivalent. The bonding conductor shall be routed so as not to be exposed to physical damage.

(3) Metallic gas, water, and waste pipes and metallic air-circulating ducts shall be considered bonded if they are connected to the terminal on the chassis [see Section 550-11(c)(1)] by clamps, solderless connectors, or by suitable grounding-type straps.

(4) Any metallic roof and exterior covering shall be considered bonded if (1) the metal panels overlap one another and are securely attached to the wood or metal frame parts by metallic fasteners and (2) if the lower panel of the metallic exterior covering is secured by metallic fasteners at a cross member of the chassis by two metal straps per mobile home unit or section at opposite ends.

The bonding strap material shall be a minimum of 4 in. (102 mm) in width of material equivalent to the skin or a material of equal or better electrical conductivity. The straps shall be fastened with paint-penetrating fittings such as screws and starwashers or equivalent.

550-12. Testing.

(a) Dielectric Strength Test. The wiring of each mobile home shall be subjected to a 1-minute, 900-volt, dielectric strength test (with all switches closed) between live parts (including neutral) and the mobile home ground. Alternatively, the test shall be permitted to be performed at 1080 volts for 1 second. This test shall be performed after branch circuits are complete and after fixtures or appliances are installed.

Exception: Listed fixtures or appliances shall not be required to withstand the dielectric strength test.

(b) Continuity and Operational Tests and Polarity Checks. Each mobile home shall be subjected to the following:

(1) An electrical continuity test to ensure that all exposed electrically conductive parts are properly bonded;
(2) An electrical operational test to demonstrate that all equipment, except water heaters and electric furnaces, is connected and in working order; and
(3) Electrical polarity checks of permanently wired equipment and receptacle outlets to determine that connections have been properly made.

550-13. Calculations. The following method shall be employed in computing the supply-cord and distribution-panelboard load for each feeder assembly for each mobile home in lieu of the procedure shown in Article 220 and shall be based on a 3-wire, 120/240-volt supply with 120-volt loads balanced between the two legs of the 3-wire system.

(a) Lighting and Small Appliance Load.

Lighting Volt-Amperes: Length times width of mobile home floor (outside dimensions) times 3 volt-amperes/ft^2, e.g.,

$$\text{Length} \times \text{width} \times 3 = \text{lighting volt-amperes}$$

Small Appliance Volt-Amperes: Number of circuits times 1500 volt-amperes for each 20-ampere appliance receptacle circuit (see definition of *Appliance, Portable* with note) including 1500 volt-amperes for laundry circuit, e.g.,

$$\text{No. of circuits} \times 1500 = \text{small appliance volt-amperes.}$$

Total: Lighting volt-amperes plus small appliance volt-amperes = total volt-amperes

First 3000 total volt-amperes at 100 percent plus remainder at 35 percent = volt-amperes to be divided by 240 volts to obtain current (amperes) per leg.

(b) Total Load for Determining Power Supply. Total load for determining power supply is the sum of the following.

(1) Lighting and small appliance load as calculated in Section 550-13(a).
(2) Nameplate amperes for motors and heater loads (exhaust fans, air conditioners, electric, gas, or oil heating). Omit smaller of the heating and cooling loads, except include blower motor if used as air-conditioner evaporator motor. Where an air conditioner is not installed and a 40-ampere power-supply cord is provided, allow 15 amperes per leg for air conditioning.
(3) Twenty-five percent of current of largest motor in (2).

(4) Total of nameplate amperes for waste disposer, dishwasher, water heater, clothes dryer, wall mounted oven, cooking units. Where the number of these appliances exceeds three, use 75 percent of total.
(5) Derive amperes for freestanding range (as distinguished from separate ovens and cooking units) by dividing the following values by 240 volts.

Nameplate Rating (watts)	Use (volt-amperes)
0–10,000	80 percent of rating
Over 10,000–12,500	8,000
Over 12,500–13,500	8,400
Over 13,500–14,500	8,800
Over 14,500–15,500	9,200
Over 15,500–16,500	9,600
Over 16,500–17,500	10,000

(6) If outlets or circuits are provided for other than factory-installed appliances, include the anticipated load.

FPN: Refer to Appendix D, Example D11, for an illustration of the application of this calculation.

(c) Optional Method of Calculation for Lighting and Appliance Load. For mobile homes, the optional method for calculating lighting and appliance load shown in Section 220-30 shall be permitted.

550-14. Interconnection of Multiple Section Mobile Home Units. Approved and listed fixed-type wiring methods shall be used to join portions of a circuit that must be electrically joined that are located in adjacent sections of mobile homes after the home is installed on its support foundation. The circuit's junction shall be accessible for disassembly when the home is prepared for relocation.

550-15. Outdoor Outlets, Fixtures, Air-Cooling Equipment, etc.

(a) Listed for Outdoor Use. Outdoor fixtures and equipment shall be listed for outdoor use. Outdoor receptacle or convenience outlets shall be of a gasketed-cover type for use in wet locations.

(b) Outside Heating Equipment, Air-Conditioning Equipment, or Both. A mobile home provided with a branch circuit designed to energize outside heating equipment or air-conditioning equipment, or both, located outside the mobile home, other than room air conditioners, shall have such branch-circuit conductors terminate in a listed outlet box, or disconnecting means, located on the outside of the mobile home. A label shall be permanently affixed adjacent to the outlet box and shall contain the following information:

> THIS CONNECTION IS FOR HEATING AND/OR AIR-CONDITIONING EQUIPMENT. THE BRANCH CIRCUIT IS RATED AT NOT MORE THAN _____ AMPERES, AT _____ VOLTS, 60-HERTZ, _____ CONDUCTOR AMPACITY. A DISCONNECTING MEANS SHALL BE LOCATED WITHIN SIGHT OF THE EQUIPMENT.

The correct voltage and ampere rating shall be given. The tag shall be not less than 0.020-in. (508-μm) thick etched brass, stainless steel, anodized or alclad aluminum, or equivalent. The tag shall not be less than 3 in. (76 mm) by 1¾ in. (44.5 mm) minimum size.

C. Services and Feeders

550-21. Distribution System. The mobile home park secondary electrical distribution system to mobile home lots shall be single-phase, 120/240 volts, nominal. For the purpose of Part C, where the park service exceeds 240 volts, nominal, transformers and secondary distribution panelboards shall be treated as services.

550-22. Minimum Allowable Demand Factors. Park electrical wiring systems shall be calculated (at 120/240 volts) on the larger of (1) 16,000 volt-amperes for each mobile home lot or (2) the load calculated in accordance with Section 550-13 for the largest typical mobile home that each lot will accept. It shall be permissible to compute the feeder or service load in accordance with Table 550-22. No demand factor shall be allowed for any other load, except as provided in this *Code.*

Service and feeder conductors to a mobile home in compliance with Section 310-15(b)(6) shall be permitted.

Table 550-22. Demand Factors for Feeders and Service-Entrance Conductors

Number of Mobile Homes	Demand Factor (percent)
1	100
2	55
3	44
4	39
5	33
6	29
7–9	28
10–12	27
13–15	26
16–21	25
22–40	24
41–60	23
61 and over	22

550-23. Service Equipment.

(a) Mobile Home Service Equipment. The mobile home service equipment shall be located adjacent to the mobile home and not mounted in or on the mobile home. The service equipment shall be located in sight from and not more than 30 ft (9.14 m) from the exterior wall of the mobile home it serves. The service equipment shall be permitted to be located elsewhere on the premises, provided that a disconnecting means suitable for service equipment is located in sight from and not more than 30 ft (9.14 m) from the exterior wall of the mobile home it serves. Grounding at the disconnecting means shall be in accordance with Section 250-32.

(b) Manufactured Home Service Equipment. The manufactured home service equipment shall be permitted to be installed in or on a manufactured home, provided that all of the following conditions are met.

(1) The manufactured home is secured to a permanent foundation that complies with applicable building codes.
(2) The service equipment is installed in a manner acceptable to the authority having jurisdiction.
(3) The installation of the service equipment complies with Article 230.
(4) Means are provided for the connection of a grounding electrode conductor to the service equipment and routing it outside the structure.

(c) Rating. Mobile home service equipment shall be rated at not less than 100 amperes at 120/240 volts, and provisions shall be made for connecting a mobile home feeder assembly by a permanent wiring method. Power outlets used as mobile home service equipment shall also be permitted to contain receptacles rated up to 50 amperes with appropriate overcurrent protection. Fifty-ampere receptacles shall conform to the configuration shown in Figure 550-5(c).

> FPN: Complete details of the 50-ampere plug and receptacle configuration can be found in *National Electrical Manufacturers Association Standard for Wiring Devices — Dimensional Requirements,* ANSI/NEMA WD 6-1988, Figure 14-50.

(d) Additional Outside Electrical Equipment. Mobile home service equipment shall also contain a means for connecting a mobile home accessory building or structure, or additional electrical equipment located outside a mobile home by a fixed wiring method.

(e) Additional Receptacles. Additional receptacles shall be permitted for connection of electrical equipment located outside the mobile home, and all such 125-volt, single-phase, 15- and 20-ampere receptacles shall be protected by a listed ground-fault circuit interrupter.

(f) Mounting Height. Outdoor mobile home disconnecting means shall be installed so the bottom of the enclosure containing the disconnecting means is not less than 2 ft (610 mm) above finished grade or working platform. The disconnecting means shall be installed so that the center of the grip of the operating handle, when in the highest position, will not be more than 6 ft 7 in. (2.0 m) above the finished grade or working platform.

(g) Marking. Where a 125/250-volt receptacle is used in mobile home service equipment, the service equipment shall be marked as follows.

> TURN DISCONNECTING SWITCH OR CIRCUIT BREAKER OFF BEFORE INSERTING OR REMOVING PLUG. PLUG MUST BE FULLY INSERTED OR REMOVED.

The marking shall be located on the service equipment adjacent to the receptacle outlet.

550-24. Feeder.

(a) Feeder Conductors. Mobile home feeder conductors shall consist of either a listed cord, factory-installed in accordance with Section 550-5(b), or a permanently installed feeder consisting of four, insulated, color-coded conductors that shall be identified by the factory or field marking of the conductors in compliance with Section 310-12. Equipment grounding conductors shall not be identified by stripping the insulation.

Exception: Where a mobile home feeder is installed between service equipment and a mobile home disconnecting means as covered in Section 550-23(a), it shall be permitted to omit the equipment grounding conductor where the grounded circuit conductor is grounded at the disconnecting means as required in Section 250-32(b).

(b) Adequate Feeder Capacity. Mobile home lot feeder circuit conductors shall have adequate capacity for the loads supplied and shall be rated at not less than 100 amperes at 120/240 volts.

Article 551 — Recreational Vehicles and Recreational Vehicle Parks

A. General

551-1. Scope. The provisions of this article cover the electrical conductors and equipment installed within or on recreational vehicles, the conductors that connect recreational vehicles to a supply of electricity, and the installation of equipment and devices related to electrical installations within a recreational vehicle park.

551-2. Definitions. (See Article 100 for additional definitions.)

Air-Conditioning or Comfort-Cooling Equipment. All of that equipment intended or installed for the purpose of processing the treatment of air so as to control simultaneously its temperature, humidity, cleanliness, and distribution to meet the requirements of the conditioned space.

Appliance, Fixed. An appliance that is fastened or otherwise secured at a specific location.

Appliance, Portable. An appliance that is actually moved or can easily be moved from one place to another in normal use.

> FPN: For the purpose of this article, the following major appliances, other than built-in, are considered portable if cord connected: refrigerators, range equipment, clothes washers, dishwashers without booster heaters, or other similar appliances.

Appliance, Stationary. An appliance that is not easily moved from one place to another in normal use.

Camping Trailer. A vehicular portable unit mounted on wheels and constructed with collapsible partial side walls that fold for towing by another vehicle and unfold at the campsite to provide temporary living quarters for recreational, camping, or travel use. (See Recreational Vehicle.)

Converter. A device that changes electrical energy from one form to another, as from alternating current to direct current.

Dead Front (as applied to switches, circuit breakers, switchboards, and distribution panelboards). Designed, constructed, and installed so that no current-carrying parts are normally exposed on the front.

Disconnecting Means. The necessary equipment usually consisting of a circuit breaker or switch and fuses, and their accessories, located near the point of entrance of supply conductors in a recreational vehicle and intended to constitute the means of cutoff for the supply to that recreational vehicle.

Distribution Panelboard. A single panel or group of panel units designed for assembly in the form of a single panel; including buses, and with or without switches and/or automatic overcurrent-protective devices for the control of light, heat, or power circuits of small individual as well as aggregate capacity; designed to be placed in a cabinet or cutout box placed in or against a wall or partition and accessible only from the front.

Frame. Chassis rail and any welded addition thereto of metal thickness of 16 MSG or greater.

Low Voltage. An electromotive force rated 24 volts, nominal, or less, supplied from a transformer, converter, or battery.

Motor Home. A vehicular unit designed to provide temporary living quarters for recreational, camping, or travel use built on or permanently attached to a self-propelled motor vehicle chassis or on a chassis cab or van that is an integral part of the completed vehicle. (See Recreational Vehicle.)

Power-Supply Assembly. The conductors, including ungrounded, grounded, and equipment grounding conductors, the connectors, attachment plug caps, and all other fittings, grommets, or devices installed for the purpose of delivering energy from the source of electrical supply to the distribution panel within the recreational vehicle.

Recreational Vehicle. A vehicular-type unit primarily designed as temporary living quarters for recreational, camping, or travel use, which either has its own motive power or is mounted on or drawn by another vehicle. The basic entities are travel trailer, camping trailer, truck camper, and motor home.

Recreational Vehicle Park. A plot of land upon which two or more recreational vehicle sites are located, established, or maintained for occupancy by recreational vehicles of the general public as temporary living quarters for recreation or vacation purposes.

Recreational Vehicle Site. A plot of ground within a recreational vehicle park intended for the accommodation of either a recreational vehicle, tent, or other individual camping unit on a temporary basis.

Recreational Vehicle Site Feeder Circuit Conductors. The conductors from the park service equipment to the recreational vehicle site supply equipment.

Recreational Vehicle Site Supply Equipment. The necessary equipment, usually a power outlet, consisting of a circuit breaker or switch and fuse and their accessories, located near the point of entrance of supply conductors to a recreational vehicle site and intended to constitute the disconnecting means for the supply to that site.

Recreational Vehicle Stand. That area of a recreational vehicle site intended for the placement of a recreational vehicle.

Transformer. A device that, when used, will raise or lower the voltage of alternating current of the original source.

Travel Trailer. A vehicular unit, mounted on wheels, designed to provide temporary living quarters for recreational, camping, or travel use, of such size or weight as not to require special highway movement permits when towed by a motorized vehicle, and of gross trailer area less than 320 ft^2 (29.7 m^2). (See Recreational Vehicle.)

Truck Camper. A portable unit constructed to provide temporary living quarters for recreational, travel, or camping use, consisting of a roof, floor, and sides, designed to be loaded onto and unloaded from the bed of a pick-up truck. (See Recreational Vehicle.)

551-3. Other Articles. Wherever the requirements of other articles of this *Code* and Article 551 differ, the requirements of Article 551 shall apply.

551-4. General Requirements.

(a) Not Covered. A recreational vehicle not used for the purposes as defined in Section 551-2 shall not be required to meet the provisions of Part A pertaining to the number or capacity of circuits required. It shall, however, meet all other applicable requirements of this article if the recreational vehicle is provided with an electrical installation intended to be energized from a 120- or 120/240-volt, nominal, ac power-supply system.

(b) Systems. This article covers battery and other low-voltage power systems (24 volts or less), combination electrical systems, generator installations, and 120- or 120/240-volt, nominal, systems.

B. Low-Voltage Systems

551-10. Low-Voltage Systems.

(a) Low-Voltage Circuits. Low-voltage circuits furnished and installed by the recreational vehicle manufacturer, other than automotive vehicle circuits or extensions thereof, are subject to this *Code*. Circuits supplying lights subject to federal or state regulations shall comply with applicable government regulations and this *Code*.

(b) Low-Voltage Wiring.

(1) Copper conductors shall be used for low-voltage circuits.

Exception: Metal chassis or frame shall be permitted as the return path to the source of supply.

(2) Conductors shall conform to the requirements for Type GXL, HDT, SGT, SGR, or Type SXL or shall have insulation in accordance with Table 310-13 or the equivalent. Conductor sizes No. 6 through No. 18 or SAE shall be listed.

FPN: See SAE Standard J1128-1995 for Types GXL, HDT, and SXL and SAE Standard J1127-1995 for Types SGT and SGR.

(3) Single-wire, low-voltage conductors shall be of the stranded type.

(4) All insulated low-voltage conductors shall be surface marked at intervals not greater than 4 ft (1.22 m) as follows.

(a) Listed conductors shall be marked as required by the listing agency.

(b) SAE conductors shall be marked with the name or logo of the manufacturer, specification designation, and wire gauge.

(c) Other conductors shall be marked with the name or logo of the manufacturer, temperature rating, wire gauge, conductor material, and insulation thickness.

(5) Conductors shall have a minimum insulation rating of 90°C (194°F) for interior installations and 125°C (257°F) for all engine compartment wiring or any under-chassis installations where conductors are located less than 18 in. (457.2 mm) from any component of an internal combustion engine exhaust system.

(c) Low-Voltage Wiring Methods.

(1) Conductors shall be protected against physical damage and shall be secured. Where insulated conductors are clamped to the structure, the conductor insulation shall be supplemented by an additional wrap or layer of equivalent material, except that jacketed cables shall not be required to be so protected. Wiring shall be routed away from sharp edges, moving parts, or heat sources.

(2) Conductors shall be spliced or joined with splicing devices that provide a secure connection or by brazing, welding, or soldering with a fusible metal or alloy. Soldered splices shall first be spliced or joined so as to be mechanically and electrically secure without solder and then soldered. All splices, joints, and free ends of conductors shall be covered with an insulation equivalent to that on the conductors.

(3) Battery and dc circuits shall be physically separated by at least a ½-in. (12.7-mm) gap or other approved means from circuits of a different power source. Acceptable methods shall be by clamping, routing, or equivalent means that ensure permanent total separation. Where circuits of different power sources cross, the external jacket of the nonmetallic-sheathed cables shall be deemed adequate separation.

(4) Ground connections to the chassis or frame shall be made in an accessible location and shall be mechanically secure. Ground connections shall be by means of copper conductors and copper or copper-alloy terminals of the solderless type identified for the size of wire used. The surface on which ground terminals make contact shall be cleaned and be free from oxide or paint or shall be electrically connected through the use of a cadmium, tin, or zinc-plated internal/external-toothed lockwasher or locking terminals. Ground terminal attaching screws, rivets or bolts, nuts, and lockwashers shall be cadmium, tin, or zinc-plated except

rivets shall be permitted to be unanodized aluminum where attaching to aluminum structures.

(5) The chassis-grounding terminal of the battery shall be bonded to the vehicle chassis with a minimum No. 8 copper conductor. In the event the power lead from the battery exceeds No. 8, then the bonding conductor shall be of an equal size.

(d) Battery Installations. Storage batteries subject to the provisions of this *Code* shall be securely attached to the vehicle and installed in an area vaportight to the interior and ventilated directly to the exterior of the vehicle. Where batteries are installed in a compartment, the compartment shall be ventilated with openings having a minimum area of 1.7 in.^{2} (1100 mm^2) at both the top and at the bottom. Where compartment doors are equipped for ventilation, the openings shall be within 2 in. (50.8 mm) of the top and bottom. Batteries shall not be installed in a compartment containing spark- or flame-producing equipment, except that they shall be permitted to be installed in the engine generator compartment if the only charging source is from the engine generator.

(e) Overcurrent Protection.

(1) Low-voltage circuit wiring shall be protected by overcurrent protective devices rated not in excess of the ampacity of copper conductors, as follows.

Table 551-10(e)(1). Low-Voltage Overcurrent Protection

Wire Size (AWG)	Ampacity	Wire Type
18	6	Stranded only
16	8	Stranded only
14	15	Stranded or solid
12	20	Stranded or solid
10	30	Stranded or solid

(2) Circuit breakers or fuses shall be of an approved type, including automotive types. Fuseholders shall be clearly marked with maximum fuse size, and both circuit breakers and fuses shall be protected against shorting and physical damage by a cover or equivalent means.

FPN: For further information, see *Standard for Electric Fuses (Cartridge Type)*, ANSI/SAE J554-1987; *Standard for Blade Type Electric Fuses*, SAE J1284-1988; and *Standard for Automotive Glass Tube Fuses*, UL 275-1993.

(3) Higher current-consuming, dc appliances, such as pumps, compressors, heater blowers, and similar motor-driven appliances, shall be installed in accordance with the manufacturer's instructions.

Motors that are controlled by automatic switching or by latching-type manual switches shall be protected in accordance with Section 430-32(c).

(4) The overcurrent protective device shall be installed in an accessible location on the vehicle within 18 in. (457 mm) of the point where the power supply connects to the vehicle circuits. If located outside the recreational vehicle, the device shall be protected against weather and physical damage.

Exception: External low-voltage supply shall be permitted to have the overcurrent protective device within 18 in. (457 mm) after entering the vehicle or after leaving a metal raceway.

(f) Switches. Switches shall have a dc rating not less than the connected load.

(g) Lighting Fixtures. All low-voltage interior lighting fixtures rated more than 4 watts, employing lamps rated more than 1.2 watts, shall be listed.

(h) Cigarette Lighter Receptacles. Twelve-volt receptacles that will accept and energize cigarette lighters shall be installed in a noncombustible outlet box, or the assembly shall be identified by the manufacturer of the product as thermally protected.

C. Combination Electrical Systems

551-20. Combination Electrical Systems.

(a) General. Vehicle wiring suitable for connection to a battery or dc supply source shall be permitted to be connected to a 120-volt source, provided the entire wiring system and equipment are rated and installed in full conformity with Parts A, C, D, E, and F requirements of this article covering 120-volt electrical systems. Circuits fed from ac transformers shall not supply dc appliances.

(b) Voltage Converters (120-Volt Alternating Current to Low-Voltage Direct Current). The 120-volt ac side of the voltage converter shall be wired in full conformity with Parts A, C, D, E, and F requirements of this article for 120-volt electrical systems.

Exception: Converters supplied as an integral part of a listed appliance shall not be subject to the above.

All converters and transformers shall be listed for use in recreation vehicles and designed or equipped to provide over-temperature protection. To determine the converter rating, the following formula shall be applied to the total connected load, including average battery charging rate, of all 12-volt equipment:

The first 20 amperes of load at 100 percent, plus

The second 20 amperes of load at 50 percent, plus

All load above 40 amperes at 25 percent

Exception: A low-voltage appliance that is controlled by a momentary switch (normally open) that has no means for holding in the closed position shall not be considered as a connected load when determining the required converter rating. Momentarily energized appliances shall be limited to those used to prepare the vehicle for occupancy or travel.

(c) Bonding Voltage Converter Enclosures. The noncurrent-carrying metal enclosure of the voltage converter shall be bonded to the frame of the vehicle with a minimum No. 8 copper conductor. The grounding conductor for the battery and the metal enclosure shall be permitted to be the same conductor.

(d) Dual-Voltage Fixtures or Appliances. Fixtures or appliances having both 120-volt and low-voltage connections shall be listed for dual voltage.

(e) Autotransformers. Autotransformers shall not be used.

(f) Receptacles and Plug Caps. Where a recreational vehicle is equipped with a 120-volt or 120/240-volt ac system, a low-voltage system, or both, receptacles and plug caps of the low-voltage system shall differ in configuration from those of the 120- or 120/240-volt system. Where a vehicle equipped with a battery or other low-voltage system has an external connection for low-voltage power, the connector shall have a configuration that will not accept 120-volt power.

D. Other Power Sources

551-30. Generator Installations.

(a) Mounting. Generators shall be mounted in such a manner as to be effectively bonded to the recreational vehicle chassis.

(b) Generator Protection. Equipment shall be installed to ensure that the current-carrying conductors from the engine generator and from an outside source are not connected to a vehicle circuit at the same time.

Receptacles used as disconnecting means shall be accessible (as applied to wiring methods) and capable of interrupting their rated current without hazard to the operator.

(c) Installation of Storage Batteries and Generators. Storage batteries and internal-combustion-driven generator units (subject to the provisions of this *Code*) shall be secured in place to avoid displacement from vibration and road shock.

(d) Ventilation of Generator Compartments. Compartments accommodating internal-combustion-driven generator units shall be provided with ventilation in accordance with instructions provided by the manufacturer of the generator unit.

> FPN: For generator compartment construction requirements, see *Standard on Recreational Vehicles*, NFPA 501C-1996.

(e) Supply Conductors. The supply conductors from the engine generator to the first termination on the vehicle shall be of the stranded type and be installed in listed flexible conduit or listed liquidtight flexible conduit. The point of first termination shall be in a

(1) Panelboard,
(2) Junction box with a blank cover,
(3) Junction box with a receptacle,
(4) Enclosed transfer switch, or
(5) Receptacle assembly listed in conjunction with the generator.

The panelboard or junction box with a receptacle shall be installed within the vehicle's interior and within 18 in. (457 mm) of the compartment wall but not inside the compartment. If the generator is below the floor level and not in a compartment, the panelboard or junction box with receptacle shall be installed within the vehicle interior within 18 in. (457 mm) of the point of entry into the vehicle. A junction box with a blank cover shall be mounted on the compartment wall and shall be permitted inside or outside the compartment. A receptacle assembly listed in conjunction with the generator shall be mounted in accordance with its listing. If the generator is below floor level and not in a compartment, the junction box with blank cover shall be mounted either to any part of the generator supporting structure (but not to the generator) or to the vehicle floor within 18 in. (457 mm) of any point directly above the generator on either the inside or outside of the floor surface. Overcurrent protection in accordance with Section 240-3 shall be provided for supply conductors as an integral part of a listed generator or shall be located within 18 in. (457 mm) of their point of entry into the vehicle.

551-31. Multiple Supply Source.

(a) Multiple Supply Sources. Where a multiple supply system consisting of an alternate power source and a power-supply cord is installed, the feeder from the alternate power source shall be protected by an overcurrent-protective device. Installation shall be in accordance with Sections 551-30(a) and (b) and 551-40.

(b) Calculation of Loads. Calculation of loads shall be in accordance with Section 551-42.

(c) Multiple Supply Sources Capacity. The multiple supply sources shall not be required to be of the same capacity.

(d) Alternate Power Sources Exceeding 30 Amperes. If an alternate power source exceeds 30 amperes, 120 volts, nominal, it shall be permissible to wire it as a 120-volt, nominal, system or a 120/240-volt, nominal, system, provided an overcurrent-protective device of the proper rating is installed in the feeder.

(e) Power-Supply Assembly Not Less than 30 Amperes. The external power-supply assembly shall be permitted to be less than the calculated load but not less than 30 amperes and shall have overcurrent protection not greater than the capacity of the external power-supply assembly.

551-32. Other Sources. Other sources of ac power, such as inverters or motor generators, shall be listed for use in recreational vehicles and shall be installed in accordance with the terms of the listing. Other sources of ac power shall be wired in full conformity with the requirements in Parts A, C, D, E, and F of this article covering 120-volt electrical systems.

551-33. Alternate Source Restriction. Transfer equipment, if not integral with the listed power source, shall be installed to ensure that the current-carrying conductors from other sources of ac power and from an outside source are not connected to the vehicle circuit at the same time.

E. Nominal 120-Volt or 120/240-Volt Systems

551-40. 120-Volt or 120/240-Volt, Nominal, Systems.

(a) General Requirements. The electrical equipment and material of recreational vehicles indicated for connection to a wiring system rated 120 volts, nominal, 2-wire with ground, or a wiring system rated 120/240 volts, nominal, 3-wire with ground, shall be listed and installed in accordance with the requirements of Parts A, C, D, E, and F of this article.

(b) Materials and Equipment. Electrical materials, devices, appliances, fittings, and other equipment installed, intended for use in, or attached to the recreational vehicle shall be listed. All products shall be used only in the manner in which they have been tested and found suitable for the intended use.

(c) Ground-Fault Circuit-Interrupter Protection. The internal wiring of a recreational vehicle having only one 15- or 20-ampere branch circuit as permitted in Sections 551-42(a) and (b) shall have ground-fault circuit-interrupter protection for personnel. The ground-fault circuit interrupter shall be installed at the point where the power supply assembly terminates within the recreational vehicle. Where a separable cord set is not employed, the ground-fault circuit interrupter shall be permitted to be an integral part of the attachment plug of the power supply assembly. The ground-fault circuit interrupter shall provide protection also under the conditions of an open grounded circuit conductor, interchanged circuit conductors, or both.

551-41. Receptacle Outlets Required.

(a) Spacing. Receptacle outlets shall be installed at wall spaces 2 ft (610 mm) wide or more so that no point along the floor line is more than 6 ft (1.83 m), measured horizontally, from an outlet in that space.

Exception No. 1: Bath and hall areas.

Exception No. 2: Wall spaces occupied by kitchen cabinets, wardrobe cabinets, built-in furniture, behind doors that may open fully against a wall surface, or similar facilities.

(b) Location. Receptacle outlets shall be installed as follows:

(1) Adjacent to countertops in the kitchen [at least one on each side of the sink if countertops are on each side and are 12 in. (305 mm) or over in width]
(2) Adjacent to the refrigerator and gas range space, except where a gas-fired refrigerator or cooking appliance, requiring no external electrical connection, is factory installed
(3) Adjacent to countertop spaces of 12 in. (305 mm) or more in width that cannot be reached from a receptacle required in Section 551-41(b)(1) by a cord of 6 ft (1.83 m) without crossing a traffic area, cooking appliance, or sink

(c) Ground-Fault Circuit-Interrupter Protection. Where provided, each 125-volt, single-phase, 15- or 20-ampere receptacle outlet shall have ground-fault circuit-interrupter protection for personnel in the following locations:

(1) Adjacent to a bathroom lavatory
(2) Where the receptacles are installed to serve the countertop surfaces, and are within 6 ft (1.83 m) of any lavatory or sink

Exception No. 1: Receptacles installed for appliances in dedicated spaces, such as for dishwashers, disposals, refrigerators, freezers, and laundry equipment.

Exception No. 2: Single receptacles for interior connections of expandable room sections.

Exception No. 3: De-energized receptacles that are within 6 ft (18.3 m) of any sink or lavatory due to the retraction of the expandable room section.

(3) In the area occupied by a toilet, shower, tub, or any combination thereof
(4) On the exterior of the vehicle

Exception: Receptacles that are located inside of an access panel that is installed on the exterior of the vehicle to supply power for an installed appliance shall not be required to have ground-fault circuit-interrupter protection.

The receptacle outlet shall be permitted in a listed lighting fixture. A receptacle outlet shall not be installed in a tub or combination tub–shower compartment.

(d) Face-Up Position. A receptacle shall not be installed in a face-up position in any countertop or similar horizontal surfaces within the living area.

551-42. Branch Circuits Required. Each recreational vehicle containing a 120-volt electrical system shall contain one of the following.

(a) One 15-Ampere Circuit. One 15-ampere circuit to supply lights, receptacle outlets, and fixed appliances. Such recreational vehicles shall be equipped with one 15-ampere switch and fuse or one 15-ampere circuit breaker.

(b) One 20-Ampere Circuit. One 20-ampere circuit to supply lights, receptacle outlets, and fixed appliances. Such recreational vehicles shall be equipped with one 20-ampere switch and fuse or one 20-ampere circuit breaker.

(c) Two to Five 15- or 20-Ampere Circuits. A maximum of five 15- or 20-ampere circuits to supply lights, receptacle outlets, and fixed appliances shall be permitted. Such recreational vehicles shall be equipped with a distribution panelboard rated at 120 volts maximum with a 30-ampere rated main power supply assembly. Not more than two 120-volt thermostatically controlled appliances (i.e., air conditioner and water heater) shall be installed in such systems unless appliance isolation switching, energy management systems, or similar methods are used.

Exception: Additional 15- or 20-ampere circuits shall be permitted where a listed energy management system rated at 30-ampere maximum is employed within the system.

FPN: See Section 210-23(a) for permissible loads. See Section 551-45(c) for main disconnect and overcurrent protection requirements.

(d) More than Five Circuits Without a Listed Energy Management System. A 50-ampere, 120/240-volt power-supply assembly shall be used where six or more circuits are employed. The load distribution shall ensure a reasonable current balance between phases.

551-43. Branch-Circuit Protection.

(a) Rating. The branch-circuit overcurrent devices shall be rated as follows:

(1) Not more than the circuit conductors, and
(2) Not more than 150 percent of the rating of a single appliance rated 13.3 amperes or more and supplied by an individual branch circuit, but
(3) Not more than the overcurrent protection size marked on an air conditioner or other motor-operated appliances

(b) Protection for Smaller Conductors. A 20-ampere fuse or circuit breaker shall be permitted for protection for fixture leads, cords, or small appliances, and No. 14 tap conductors, not over 6 ft (1.83 m) long for recessed lighting fixtures.

(c) Fifteen-Ampere Receptacle Considered Protected by 20 Amperes. If more than one receptacle or load is on a branch circuit, a 15-ampere receptacle shall be permitted to be protected by a 20-ampere fuse or circuit breaker.

551-44. Power-Supply Assembly.

(a) Fifteen-Ampere Main Power-Supply Assembly. Recreational vehicles wired in accordance with Section 551-42(a) shall use a listed 15-ampere, or larger, main power-supply assembly.

(b) Twenty-Ampere Main Power-Supply Assembly. Recreational vehicles wired in accordance with Section 551-42(b) shall use a listed 20-ampere, or larger, main power-supply assembly.

(c) Thirty-Ampere Main Power-Supply Assembly. Recreational vehicles wired in accordance with Section 551-42(c) shall use a listed 30-ampere, or larger, main power-supply assembly.

(d) Fifty-Ampere Power-Supply Assembly. Recreational vehicles wired in accordance with Section 551-42(d) shall use a listed 50-ampere, 120/240-volt main power-supply assembly.

551-45. Distribution Panelboard.

(a) Listed and Appropriately Rated. A listed and appropriately rated distribution panelboard or other equipment specifically listed for the purpose shall be used. The grounded conductor termination bar shall be insulated from the enclosure as provided in Section 551-54(c). An equipment grounding terminal bar shall be attached inside the metal enclosure of the panelboard.

(b) Location. The distribution panelboard shall be installed in a readily accessible location. Working clearance for the panelboard shall be not less than 24 in. (610 mm) wide and 30 in. (762 mm) deep.

Exception No. 1: Where the panelboard cover is exposed to the inside aisle space, then one of the working clearance dimensions shall be permitted to be reduced to a minimum of 22 in. (559 mm). A panelboard is considered exposed where the panelboard cover is within 2 in. (50.8 mm) of the aisle's finished surface.

Exception No. 2: Compartment doors used for access to a generator shall be permitted to be equipped with a locking system.

(c) Dead-Front Type. The distribution panelboard shall be of the dead-front type and shall consist of one or more circuit breakers or Type S fuseholders. A main disconnecting means shall be provided where fuses are used or where more than two circuit breakers are employed. A main overcurrent protective device not exceeding the power-supply assembly rating shall be provided where more than two branch circuits are employed.

551-46. Means for Connecting to Power Supply.

(a) Assembly. The power-supply assembly or assemblies shall be factory supplied or factory installed and be of one of the types specified herein.

(1) Separable. Where a separable power-supply assembly consisting of a cord with a female connector and molded attachment plug cap is provided, the vehicle shall be equipped with a permanently mounted, flanged surface inlet (male, recessed-type motor-base attachment plug) wired directly to the distribution panelboard by an approved wiring method. The attachment plug cap shall be of a listed type.

(2) Permanently Connected. Each power-supply assembly shall be connected directly to the terminals of the distribution panelboard or conductors within a junction box and provided with means to prevent strain from being transmitted to the terminals. The ampacity of the conductors between each junction box and the terminals of each distribution panelboard shall be at least equal to the ampacity of the power-supply cord. The supply end of the assembly shall be equipped with an attachment plug of the type described in Section 551-46(c). Where the cord passes through the walls or floors, it shall be protected by means of conduit and bushings or equivalent. The cord assembly shall have permanent provisions for protection against corrosion and mechanical damage while the vehicle is in transit.

(b) Cord. The cord exposed usable length shall be measured from the point of entrance to the recreational vehicle or the face of the flanged surface inlet (motor-base attachment plug) to the face of the attachment plug at the supply end.

The cord exposed usable length, measured to the point of entry on the vehicle exterior, shall be a minimum of 23 ft (7.0 m) where the point of entrance is at the side of the vehicle, or shall be a minimum 28 ft (8.5 m) where the point of entrance is at the rear of the vehicle.

Where the cord entrance into the vehicle is more than 3 ft (0.9 m) above the ground, the minimum cord lengths above shall be increased by the vertical distance of the cord entrance heights above 3 ft (0.9 m).

FPN: See Section 551-46(e).

(c) Attachment Plugs.

(1) Recreational vehicles having only one 15-ampere branch circuit as permitted by Section 551-42(a) shall have an attachment plug that shall be 2-pole, 3-wire, grounding type, rated 15 amperes, 125 volts, conforming to the configuration shown in Figure 551-46(c).

FPN: Complete details of this configuration can be found in National Electrical Manufacturers Association's *Standard for Dimensions of Attachment Plugs and Receptacle,* ANSI/ NEMA WD 6-1989, Figure 5-15.

(2) Recreational vehicles having only one 20-ampere branch circuit as permitted in Section 551-42(b) shall have an attachment plug that shall be 2-pole, 3-wire, grounding type, rated 20 amperes, 125 volts, conforming to the configuration shown in Figure 551-46(c).

FPN: Complete details of this configuration can be found in National Electrical Manufacturers Association's *Standard for Dimensions of Attachment Plugs and Receptacles,* ANSI/ NEMA WD 6-1989, Figure 5-20.

(3) Recreational vehicles wired in accordance with Section 551-42(c) shall have an attachment plug that shall be 2-pole, 3-wire, grounding type, rated 30 amperes, 125 volts, conforming to the configuration shown in Figure 551-46(c) intended for use with units rated at 30 amperes, 125 volts.

FPN: Complete details of this configuration can be found in National Electrical Manufacturers Association's *Standard for Dimensions of Attachment Plugs and Receptacles,* ANSI/ NEMA WD 6-1989, Figure TT.

(4) Recreational vehicles having a power-supply assembly rated 50 amperes as permitted by Section 551-42(d) shall have a 3-pole, 4-wire, grounding-type attachment plug rated 50 amperes, 125/250 volts, conforming to the configuration shown in Figure 551-46(c).

FPN: Complete details of this configuration can be found in National Electrical Manufacturers Association's *Standard for Dimensions of Attachment Plugs and Receptacles,* ANSI/ NEMA WD 6-1989, Figure 14-50.

Receptacles   **Caps**

20-A, 125-V, 2-pole, 3-wire, grounding type

125-V, 20-A, 2-pole, 3-wire, grounding type

125-V, 15-A, 2-pole, 3-wire, grounding type

30-A, 125-V, 2-pole, 3-wire, grounding type

50-A, 125/250-V, 3-pole, 4-wire, grounding type

Figure 551-46(c)   Configurations for grounding-type receptacles and attachment plug caps used for recreational vehicle supply cords and recreational vehicle lots.

(d) Labeling at Electrical Entrance. Each recreational vehicle shall have permanently affixed to the exterior skin, at or near the point of entrance of the power-supply cord(s), a label 3 in. (76 mm) × 1¾ in. (44.5 mm) minimum size, made of etched, metal-stamped, or embossed brass, stainless steel, or anodized or alclad aluminum not less than 0.020 in. (508 μm) thick, or other suitable material [e.g., 0.005-in. (127-μm) thick plastic laminate] that reads, as appropriate, either

THIS CONNECTION IS FOR 110–125-VOLT AC, 60 HZ, _____ AMPERE SUPPLY.

or

THIS CONNECTION IS FOR 120/240-VOLT AC, 3-POLE, 4-WIRE, 60 HZ, _____ AMPERE SUPPLY.

The correct ampere rating shall be marked in the blank space.

(e) Location. The point of entrance of a power-supply assembly shall be located within 15 ft (4.57 m) of the rear, on the left (road) side or at the rear, left of the longitudinal center of the vehicle, within 18 in. (457 mm) of the outside wall.

Exception No. 1: A recreational vehicle equipped with only a listed flexible drain system or a side-vent drain system shall be permitted to have the electrical point of entrance located on either side, provided the drain(s) for the plumbing system is (are) located on the same side.

Exception No. 2: A recreational vehicle shall be permitted to have the electrical point of entrance located more than 15 ft (4.57 m) from the rear. Where this occurs, the distance beyond the 15-ft (4.57-m) dimension shall be added to the cord's minimum length as specified in Section 551-46(b).

551-47. Wiring Methods.

(a) Wiring Systems. Cables and raceways installed in accordance with Articles 330 through 352 shall be permitted in accordance with their applicable article, except as otherwise specified in this article. An equipment grounding means shall be provided in accordance with Section 250-118.

(b) Conduit and Tubing. Where rigid metal conduit or intermediate metal conduit is terminated at an enclosure with a locknut and bushing connection, two locknuts shall be provided, one inside and one outside of the enclosure. All cut ends of conduit and tubing shall be reamed or otherwise finished to remove rough edges.

(c) Nonmetallic Boxes. Nonmetallic boxes shall be acceptable only with nonmetallic-sheathed cable or nonmetallic raceways.

(d) Boxes. In walls and ceilings constructed of wood or other combustible material, boxes and fittings shall be flush with the finished surface or project therefrom.

(e) Mounting. Wall and ceiling boxes shall be mounted in accordance with Article 370.

Exception No. 1: Snap-in-type boxes or boxes provided with special wall or ceiling brackets that securely fasten boxes in walls or ceilings shall be permitted.

Exception No. 2: A wooden plate providing a 1½-in. (38-mm) minimum width backing around the box and of a thickness of ½ in. (12.7 mm) or greater (actual) attached directly to the wall panel shall be considered as approved means for mounting outlet boxes.

(f) Sheath Armor. The sheath of nonmetallic-sheathed cable, metal-clad cable, and Type AC cable shall be continuous between outlet boxes and other enclosures.

(g) Protected. Metal-clad, Type AC, or nonmetallic-sheathed cables and electrical nonmetallic tubing shall be permitted to pass through the centers of the wide side of 2 in. by 4 in. wood studs. However, they shall be protected where they pass through 2 in. by 2 in. wood studs or at other wood studs or frames where the cable or tubing would be less than 1¼ in. (31.8 mm) from the inside or outside surface. Steel plates on each side of the cable or tubing, or a steel tube, with not less than No. 16 MSG wall thickness, shall be installed to protect the cable or tubing. These plates or tubes shall be securely held in place. Where nonmetallic-sheathed cables pass through punched, cut, or drilled slots or holes in metal members, the cable shall be protected by bushings or grommets securely fastened in the opening prior to installation of the cable.

(h) Bends. No bend shall have a radius of less than five times the cable diameter.

(i) Cable Supports. Where connected with cable connectors or clamps, cables shall be supported within 12 in. (305 mm) of outlet boxes, distribution panelboards, and splice boxes on appliances. Supports shall be provided every 4½ ft (1.37 m) at other places.

(j) Nonmetallic Box Without Cable Clamps. Nonmetallic-sheathed cables shall be supported within 8 in. (203 mm) of a nonmetallic outlet box without cable clamps. Where wiring devices with integral enclosures are employed with a loop of extra cable to permit future replacement of the device, the cable loop shall be considered as an integral portion of the device.

(k) Physical Damage. Where subject to physical damage, exposed nonmetallic cable shall be protected by covering boards, guard strips, raceways, or other means.

(l) Metal Faceplates. Metal faceplates shall be of ferrous metal not less than 0.030 in. (762 µm) in thickness or of nonferrous metal not less than 0.040 in. (1.016 mm) in thickness. Nonmetallic faceplates shall be listed.

(m) Metal Faceplates Effectively Grounded. Where metal faceplates are used, they shall be effectively grounded.

(n) Moisture or Physical Damage. Where outdoor or under-chassis wiring is 120 volts, nominal, or over and is exposed to moisture or physical damage, the wiring shall be protected by rigid metal conduit, intermediate metal conduit, or by electrical metallic tubing or rigid nonmetallic conduit that is closely routed against frames and equipment enclosures or other raceway or cable identified for the application.

(o) Component Interconnections. Fittings and connectors that are intended to be concealed at the time of assembly shall be listed and identified for the interconnection of building components. Such fittings and connectors shall be equal to the wiring method employed in insulation, temperature rise, and fault-current withstanding, and shall be capable of enduring the vibration and shock occurring in recreational vehicles.

(p) Method of Connecting Expandable Units.

(1) That portion of a branch circuit that is installed in an expandable unit shall be permitted to be connected to the portion of the branch circuit in the main body of the vehicle

by means of a flexible cord or attachment plug and cord listed for hard usage. The cord and its connections shall conform to all provisions of Article 400 and shall be considered as a permitted use under Section 400-7. Where the attachment plug and cord are located within the vehicle's interior, use of plastic thermoset or elastomer parallel cord Type SPT-3, SP-3, or SPE shall be permitted.

(2) If the receptacle provided for connection of the cord to the main circuit is located on the outside of the vehicle, it shall be protected with a ground-fault circuit interrupter for personnel and be listed for wet locations. A cord located on the outside of a vehicle shall be identified for outdoor use.

(3) Unless removable or stored within the vehicle interior, the cord assembly shall have permanent provisions for protection against corrosion and mechanical damage while the vehicle is in transit.

(4) If an attachment plug and cord is used, it shall be installed so as not to permit exposed live attachment plug pins.

(q) **Prewiring for Air-Conditioning Installation.** Prewiring installed for the purpose of facilitating future air-conditioning installation shall conform to the following and other applicable portions of this article. The circuit shall serve no other purpose.

(1) An overcurrent protective device with a rating compatible with the circuit conductors shall be installed in the distribution panelboard and wiring connections completed.

(2) The load end of the circuit shall terminate in a junction box with a blank cover or a device listed for the purpose. Where a junction box with a blank cover is used, the free ends of the conductors shall be adequately capped or taped.

(3) A label conforming to Section 551-46(d) shall be placed on or adjacent to the junction box and shall read

> AIR-CONDITIONING CIRCUIT. THIS
> CONNECTION IS FOR AIR CONDITIONERS
> RATED 110–125-VOLT AC, 60 HZ, _____
> AMPERES MAXIMUM. DO NOT EXCEED
> CIRCUIT RATING.

An ampere rating, not to exceed 80 percent of the circuit rating, shall be legibly marked in the blank space.

(r) **Prewiring for Generator Installation.** Prewiring installed for the purpose of facilitating future generator installation shall conform to the following and other applicable portions of this article.

(1) Circuit conductors shall be appropriately sized in relation to the anticipated load and shall be protected by an overcurrent device in accordance with their ampacities.

(2) Where junction boxes are utilized at either of the circuit originating or terminus points, free ends of the conductors shall be adequately capped or taped.

(3) Where devices such as receptacle outlet, transfer switch, etc., are installed, the installation shall be complete, including circuit conductor connections. All devices shall be listed and appropriately rated.

(4) A label conforming to Section 551-46(d) shall be placed on the cover of each junction box containing incomplete circuitry and shall read, as appropriate, either

> GENERATOR CIRCUIT. THIS CONNECTION
> IS FOR GENERATORS RATED 110–125-VOLT
> AC, 60 HZ, _____ AMPERES MAXIMUM.

or

> GENERATOR CIRCUIT. THIS CONNECTION
> IS FOR GENERATORS RATED 120/240-VOLT
> AC, 60 HZ, _____ AMPERES MAXIMUM.

The correct ampere rating shall be legibly marked in the blank space.

551-48. Conductors and Boxes. The maximum number of conductors permitted in boxes shall be in accordance with Section 370-16.

551-49. Grounded Conductors. The identification of grounded conductors shall be in accordance with Section 200-6.

551-50. Connection of Terminals and Splices. Conductor splices and connections at terminals shall be in accordance with Section 110-14.

551-51. Switches. Switches shall be rated as follows.

(a) **Lighting Circuits.** For lighting circuits, switches shall be rated not less than 10 amperes, 120–125 volts and in no case less than the connected load.

(b) **Motors or Other Loads.** For motors or other loads, switches shall have ampere or horsepower ratings, or both, adequate for loads controlled. (An ac general-use snap switch shall be permitted to control a motor 2 hp or less with full-load current not over 80 percent of the switch ampere rating.)

551-52. Receptacles. All receptacle outlets shall be of the grounding type and installed in accordance with Sections 210-7 and 210-21.

551-53. Lighting Fixtures.

(a) **General.** Any combustible wall or ceiling finish exposed between the edge of a fixture canopy, or pan and the outlet box, shall be covered with noncombustible material or a material identified for the purpose.

(b) Shower Fixtures. If a lighting fixture is provided over a bathtub or in a shower stall, it shall be of the enclosed and gasketed type and listed for the type of installation, and it shall be ground-fault circuit-interrupter protected.

The switch for shower lighting fixtures and exhaust fans, located over a tub or in a shower stall, shall be located outside the tub or shower space.

(c) Outdoor Outlets, Fixtures, Air-Cooling Equipment, etc. Outdoor fixtures and other equipment shall be listed for outdoor use.

551-54. Grounding. (See also Section 551-56 on bonding of noncurrent-carrying metal parts.)

(a) Power-Supply Grounding. The grounding conductor in the supply cord or feeder shall be connected to the grounding bus or other approved grounding means in the distribution panelboard.

(b) Distribution Panelboard. The distribution panelboard shall have a grounding bus with sufficient terminals for all grounding conductors or other approved grounding means.

(c) Insulated Neutral.

(1) The grounded circuit conductor (neutral) shall be insulated from the equipment grounding conductors and from equipment enclosures and other grounded parts. The grounded (neutral) circuit terminals in the distribution panelboard and in ranges, clothes dryers, counter-mounted cooking units, and wall-mounted ovens shall be insulated from the equipment enclosure. Bonding screws, straps, or buses in the distribution panelboard or in appliances shall be removed and discarded.

(2) Connection of electric ranges and electric clothes dryers utilizing a grounded (neutral) conductor, if cord-connected, shall be made with 4-conductor cord and 3-pole, 4-wire, grounding-type plug caps and receptacles.

551-55. Interior Equipment Grounding.

(a) Exposed Metal Parts. In the electrical system, all exposed metal parts, enclosures, frames, lighting fixture canopies, etc., shall be effectively bonded to the grounding terminals or enclosure of the distribution panelboard.

(b) Equipment Grounding Conductors. Bare wires, green-colored wires, or green wires with a yellow stripe(s) shall be used for equipment grounding conductors only.

(c) Grounding of Electrical Equipment. Where grounding of electrical equipment is specified, it shall be permitted as follows.

(1) Connection of metal raceway (conduit or electrical metallic tubing), the sheath of Type MC and Type MI cable where the sheath is identified for grounding, or the armor of Type AC cable to metal enclosures.

(2) A connection between the one or more equipment grounding conductors and a metal box by means of a grounding screw, which shall be used for no other purpose, or a listed grounding device.

(3) The equipment grounding conductor in nonmetallic-sheathed cable shall be permitted to be secured under a screw threaded into the fixture canopy other than a mounting screw or cover screw, or attached to a listed grounding means (plate) in a nonmetallic outlet box for fixture mounting. (Grounding means shall also be permitted for fixture attachment screws.)

(d) Grounding Connection in Nonmetallic Box. A connection between the one or more grounding conductors brought into a nonmetallic outlet box shall be so arranged that a connection can be made to any fitting or device in that box that requires grounding.

(e) Grounding Continuity. Where more than one equipment grounding conductor of a branch circuit enters a box, all such conductors shall be in good electrical contact with each other, and the arrangement shall be such that the disconnection or removal of a receptacle, fixture, or other device fed from the box will not interfere with or interrupt the grounding continuity.

(f) Cord-Connected Appliances. Cord-connected appliances, such as washing machines, clothes dryers, refrigerators, and the electrical system of gas ranges, etc., shall be grounded by means of an approved cord with equipment grounding conductor and grounding-type attachment plug.

551-56. Bonding of Noncurrent-Carrying Metal Parts.

(a) Required Bonding. All exposed noncurrent-carrying metal parts that may become energized shall be effectively bonded to the grounding terminal or enclosure of the distribution panelboard.

(b) Bonding Chassis. A bonding conductor shall be connected between any distribution panelboard and an accessible terminal on the chassis. Aluminum or copper-clad aluminum conductors shall not be used for bonding if such conductors or their terminals are exposed to corrosive elements.

Exception: Any recreational vehicle that employs a unitized metal chassis-frame construction to which the distribution panelboard is securely fastened with a bolt(s) and nut(s) or by welding or riveting shall be considered to be bonded.

(c) Bonding Conductor Requirements. Grounding terminals shall be of the solderless type and listed as pressure terminal connectors recognized for the wire size used. The bonding conductor shall be solid or stranded, insulated or bare, and shall be No. 8 copper minimum, or equal.

(d) Metallic Roof and Exterior Bonding. The metal roof and exterior covering shall be considered bonded where

(1) The metal panels overlap one another and are securely attached to the wood or metal frame parts by metal fasteners, and
(2) The lower panel of the metal exterior covering is secured by metal fasteners at each cross member of the chassis, or the lower panel is bonded to the chassis by a metal strap.

(e) Gas, Water, and Waste Pipe Bonding. The gas, water, and waste pipes shall be considered grounded if they are bonded to the chassis.

(f) Furnace and Metal Air Duct Bonding. Furnace and metal circulating air ducts shall be bonded.

551-57. Appliance Accessibility and Fastening. Every appliance shall be accessible for inspection, service, repair, and replacement without removal of permanent construction. Means shall be provided to securely fasten appliances in place when the recreational vehicle is in transit.

F. Factory Tests

551-60. Factory Tests (Electrical). Each recreational vehicle shall be subjected to the following tests.

(a) Circuits of 120 Volts or 120/240 Volts. Each recreational vehicle designed with a 120-volt or a 120/240-volt electrical system shall withstand the applied potential without electrical breakdown of a 1-minute, 900-volt dielectric strength test, or a 1-second, 1080-volt dielectric strength test, with all switches closed, between ungrounded and grounded conductors and the recreational vehicle ground. During the test, all switches and other controls shall be in the "on" position. Fixtures and permanently installed appliances shall not be required to withstand this test. The test shall be performed after branch circuits are complete prior to energizing the system and again after all outer coverings and cabinetry have been secured.

Each recreational vehicle shall be subjected to the following:

(1) A continuity test to ensure that all metal parts are properly bonded; and
(2) Operational tests to demonstrate that all equipment is properly connected and in working order; and
(3) Polarity checks to determine that connections have been properly made.

(b) Low-Voltage Circuits. An operational test of all low-voltage circuits shall be conducted to demonstrate that all equipment is connected and in electrical working order. This test shall be performed in the final stages of production after all outer coverings and cabinetry have been secured.

G. Recreational Vehicle Parks

551-71. Type Receptacles Provided. Every recreational vehicle site with electrical supply shall be equipped with at least one 20-ampere, 125-volt receptacle. A minimum of 5 percent of all recreational vehicle sites, with electrical supply, shall each be equipped with a 50-ampere, 125/250-volt receptacle conforming to the configuration as identified in Figure 551-46(c). These electrical supplies shall be permitted to include additional receptacles that have configurations in accordance with Section 551-81. A minimum of 70 percent of all recreational vehicle sites with electrical supply shall each be equipped with a 30-ampere, 125-volt receptacle conforming to Figure 551-46(c). This supply shall be permitted to include additional receptacle configurations conforming to Section 551-81. The remainder of all recreational vehicle sites with electrical supply shall be equipped with one or more of the receptacle configurations conforming to Section 551-81. Dedicated tent sites with a 15- or 20-ampere electrical supply shall be permitted to be excluded when determining the percentage of recreational vehicle sites with 30- or 50-ampere receptacles.

Additional receptacles shall be permitted for the connection of electrical equipment outside the recreational vehicle within the recreational vehicle park.

All 125-volt, single-phase, 15- and 20-ampere receptacles shall have listed ground-fault circuit-interrupter protection for personnel.

551-72. Distribution System. The recreational vehicle park secondary electrical distribution system to 50-ampere recreational vehicle sites shall be derived from a single-phase, 120/240-volt, 3-wire system. Other recreational vehicle sites with 125-volt, 20- and 30-ampere receptacles may be derived from any grounded distribution system that supplies 120-volt single-phase power. The neutral conductors shall not be reduced in size below the size of the ungrounded conductors for the site distribution. The neutral conductors shall be permitted to be reduced in size below the minimum required size of the ungrounded conductors for 240-volt, line-to-line, permanently connected loads only.

551-73. Calculated Load.

(a) Basis of Calculations. Electrical service and feeders shall be calculated on the basis of not less than 9600 volt-amperes per site equipped with 50-ampere, 120/240-volt supply facilities; 3600 volt-amperes per site equipped with both 20-ampere and 30-ampere supply facilities; 2400 volt-amperes per site equipped with only 20-ampere supply facilities; and 600 volt-amperes per site equipped with only 20-ampere supply facilities that are dedicated to tent sites. The demand factors set forth in Table 551-73 shall be the minimum allowable demand factors that shall be permitted in calculating load for service and feeders. Where the electrical supply for a recreational vehicle site has more than one receptacle, the calculated load shall only be computed for the highest rated receptacle.

Table 551-73. Demand Factors for Site Feeders and Service-Entrance Conductors for Park Sites

Number of Recreational Vehicle Sites	Demand Factor (percent)
1	100
2	90
3	80
4	75
5	65
6	60
7–9	55
10–12	50
13–15	48
16–18	47
19–21	45
22–24	43
25–35	42
36 plus	41

(b) Transformers and Secondary Distribution Panelboards. For the purpose of this *Code*, where the park service exceeds 240 volts, transformers and secondary distribution panelboards shall be treated as services.

(c) Demand Factors. The demand factor for a given number of sites shall apply to all sites indicated. For example, 20 sites calculated at 45 percent of 3600 volt-amperes results in a permissible demand of 1620 volt-amperes per site or a total of 32,400 volt-amperes for 20 sites.

FPN: These demand factors may be inadequate in areas of extreme hot or cold temperature with loaded circuits for heating or air conditioning.

(d) Feeder-Circuit Capacity. Recreational vehicle site feeder-circuit conductors shall have adequate ampacity for the loads supplied and shall be rated at not less than 30 amperes. The grounded conductors shall have the same ampacity as the ungrounded conductors.

FPN: Due to the long circuit lengths typical in most recreational vehicle parks, feeder conductor sizes found in the ampacity tables of Article 310 may be inadequate to maintain

the voltage regulation suggested in the fine print note to Section 210-19. Total circuit voltage drop is a sum of the voltage drops of each serial circuit segment, where the load for each segment is calculated using the load that segment sees and the demand factors of Section 551-73(a).

Loads for other amenities such as, but not limited to, service buildings, recreational buildings, and swimming pools shall be sized separately and then be added to the value calculated for the recreational vehicle sites where they are all supplied by one service.

551-74. Overcurrent Protection. Overcurrent protection shall be provided in accordance with Article 240.

551-75. Grounding. All electrical equipment and installations in recreational vehicle parks shall be grounded as required by Article 250.

551-76. Grounding — Recreational Vehicle Site Supply Equipment.

(a) Exposed Noncurrent-Carrying Metal Parts. Exposed noncurrent-carrying metal parts of fixed equipment, metal boxes, cabinets, and fittings that are not electrically connected to grounded equipment, shall be grounded by a continuous equipment grounding conductor run with the circuit conductors from the service equipment or from the transformer of a secondary distribution system. Equipment grounding conductors shall be sized in accordance with Section 250-122 and shall be permitted to be spliced by listed means.

The arrangement of equipment grounding connnections shall be such that the disconnection or removal of a receptacle or other device will not interfere with, or interrupt, the grounding continuity.

(b) Secondary Distribution System. Each secondary distribution system shall be grounded at the transformer.

(c) Neutral Conductor Not to Be Used as an Equipment Ground. The neutral conductor shall not be used as an equipment ground for recreational vehicles or equipment within the recreational vehicle park.

(d) No Connection on the Load Side. No connection to a grounding electrode shall be made to the neutral conductor on the load side of the service disconnecting means or transformer distribution panelboard.

551-77. Recreational Vehicle Site Supply Equipment.

(a) Location. Where provided, the recreational vehicle site electrical supply equipment shall be located on the left (road) side of the parked vehicle, on a line that is 9 ft (2.74 m), ± 1 ft (0.3 m), from the longitudinal centerline of the stand and shall be located at any point on this line from the rear of the stand to 15 ft (4.57 m) forward of the rear of the stand.

For pull-through sites, the electrical supply equipment shall be permitted to be located at any point along the line from 16 ft (4.88 m) forward of the rear of the stand to 32 ft (9.75 m) forward of the rear of the stand.

(b) Disconnecting Means. A disconnecting switch or circuit breaker shall be provided in the site supply equipment for disconnecting the power supply to the recreational vehicle.

(c) Access. All site supply equipment shall be accessible by an unobstructed entrance or passageway not less than 2 ft (610 mm) wide and 6½ ft (1.98 m) high.

(d) Mounting Height. Site supply equipment shall be located not less than 2 ft (610 mm) nor more than 6½ ft (1.98 m) above the ground.

(e) Working Space. Sufficient space shall be provided and maintained about all electrical equipment to permit ready and safe operation, in accordance with Section 110-16.

(f) Marking. Where the site supply equipment contains a 125/250-volt receptacle, the equipment shall be marked as follows: "Turn disconnecting switch or circuit breaker off before inserting or removing plug. Plug must be fully inserted or removed." The marking shall be located on the equipment adjacent to the receptacle outlet.

551-78. Protection of Outdoor Equipment.

(a) Wet Locations. All switches, circuit breakers, receptacles, control equipment, and metering devices located in wet locations or outside of a building shall be rainproof equipment.

(b) Meters. If secondary meters are installed, meter sockets without meters installed shall be blanked off with an approved blanking plate.

551-79. Clearance for Overhead Conductors. Open conductors of not over 600 volts, nominal, shall have a vertical clearance of not less than 18 ft (5.49 m) and a horizontal clearance of not less than 3 ft (914 mm) in all areas subject to recreational vehicle movement. In all other areas, clearances shall conform to Sections 225-18 and 225-19.

FPN: For clearances of conductors over 600 volts, nominal, see *National Electrical Safety Code*, ANSI C2-1997.

551-80. Underground Service, Feeder, Branch-Circuit, and Recreational Vehicle Site Feeder-Circuit Conductors.

(a) General. All direct-burial conductors, including the equipment grounding conductor if of aluminum, shall be insulated and identified for the use. All conductors shall be continuous from equipment to equipment. All splices and taps shall be made in approved junction boxes or by use of material listed and identified for the purpose.

(b) Protection Against Physical Damage. Direct-buried conductors and cables entering or leaving a trench shall be protected by rigid metal conduit, intermediate metal conduit, electrical metallic tubing with supplementary corrosion protection, rigid nonmetallic conduit, liquidtight flexible nonmetallic conduit, liquidtight flexible metal conduit, or other approved raceways or enclosures. Where subject to physical damage, the conductors or cables shall be protected by rigid metal conduit, intermediate metal conduit, or Schedule 80 rigid nonmetallic conduit. All such protection shall extend at least 18 in. (457 mm) into the trench from finished grade.

FPN: See Section 300-5 and Article 339 for conductors or Type UF cable used underground or in direct burial in earth.

551-81. Receptacles. A receptacle to supply electric power to a recreational vehicle shall be one of the configurations shown in Figure 551-46(c) in the following ratings.

(a) 50-Ampere. 125/250-volt, 50-ampere, 3-pole, 4-wire, grounding type for 120/240-volt systems

(b) 30-Ampere. 125-volt, 30-ampere, 2-pole, 3-wire, grounding type for 120-volt systems

(c) 20-Ampere. 125-volt, 20-ampere, 2-pole, 3-wire, grounding type for 120-volt systems

FPN: Complete details of these configurations can be found in National Electrical Manufacturers Association's *Standard for Dimensions of Attachment Plugs and Receptacles*, ANSI/NEMA WD 6-1989, Figures 14-50, TT, and 5-20.

Article 552 — Park Trailers

A. General

552-1. Scope. The provisions of this article cover the electrical conductors and equipment installed within or on park trailers not covered fully under Articles 550 and 551.

552-2. Definitions. (See Articles 100, 550, and 551 for additional definitions.)

Park Trailer. A unit that meets the following criteria:

(1) Built on a single chassis mounted on wheels, and
(2) Having a gross trailer area not exceeding 400 ft^2 (37.2 m^2) in the set-up mode.

552-3. Other Articles. Wherever the requirements of other articles of this *Code* and Article 552 differ, the requirements of Article 552 shall apply.

552-4. General Requirements. A park trailer as specified in Section 552-2 is intended for seasonal use. It is not intended as a permanent dwelling unit or for commercial uses such as banks, clinics, offices, or similar.

B. Low-Voltage Systems

552-10. Low-Voltage Systems.

(a) Low-Voltage Circuits. Low-voltage circuits furnished and installed by the park trailer manufacturer, other than those related to braking, are subject to this *Code*. Circuits supplying lights subject to federal or state regulations shall comply with applicable government regulations and this *Code*.

(b) Low-Voltage Wiring.

(1) Copper conductors shall be used for low-voltage circuits.

Exception: A metal chassis or frame shall be permitted as the return path to the source of supply.

(2) Conductors shall conform to the requirements for Type GXL, HDT, SGT, SGR, or Type SXL or shall have insulation in accordance with Table 310-13 or the equivalent. Conductor sizes No. 6 through 18 or SAE shall be listed.

> FPN: See SAE Standard J1128-1995 for Types GXL, HDT, and SXL and SAE Standard J1127-1995 for Types SGT and SGR.

(3) Single-wire, low-voltage conductors shall be of the stranded type.

(4) All insulated low-voltage conductors shall be surface marked at intervals not greater than 4 ft (1.22 m) as follows.

(a) Listed conductors shall be marked as required by the listing agency.

(b) SAE conductors shall be marked with the name or logo of the manufacturer, specification designation, and wire gauge.

(c) Other conductors shall be marked with the name or logo of the manufacturer, temperature rating, wire gauge, conductor material, and insulation thickness.

(c) Low-Voltage Wiring Methods.

(1) Conductors shall be protected against physical damage and shall be secured. Where insulated conductors are clamped to the structure, the conductor insulation shall be supplemented by an additional wrap or layer of equivalent material, except that jacketed cables shall not be required to be so protected. Wiring shall be routed away from sharp edges, moving parts, or heat sources.

(2) Conductors shall be spliced or joined with splicing devices that provide a secure connection or by brazing, welding, or soldering with a fusible metal or alloy. Soldered splices shall first be spliced or joined to be mechanically and electrically secure without solder and then soldered. All splices, joints, and free ends of conductors shall be covered with an insulation equivalent to that on the conductors.

(3) Battery and other low-voltage circuits shall be physically separated by at least a ½-in. (12.7-mm) gap or other approved means from circuits of a different power source. Acceptable methods shall be by clamping, routing, or equivalent means that ensure permanent total separation. Where circuits of different power sources cross, the external jacket of the nonmetallic-sheathed cables shall be deemed adequate separation.

(4) Ground connections to the chassis or frame shall be made in an accessible location and shall be mechanically secure. Ground connections shall be by means of copper conductors and copper or copper-alloy terminals of the solderless type identified for the size of wire used. The surface on which ground terminals make contact shall be cleaned and be free from oxide or paint or shall be electrically connected through the use of a cadmium, tin, or zinc-plated internal/external-toothed lockwasher or locking terminals. Ground terminal attaching screws, rivets or bolts, nuts, and lockwashers shall be cadmium, tin, or zinc-plated except rivets shall be permitted to be unanodized aluminum where attaching to aluminum structures.

(5) The chassis-grounding terminal of the battery shall be bonded to the unit chassis with a minimum No. 8 copper conductor. In the event the power lead from the battery exceeds No. 8, then the bonding conductor shall be of an equal size.

(d) Battery Installations. Storage batteries subject to the provisions of this *Code* shall be securely attached to the unit and installed in an area vaportight to the interior and ventilated directly to the exterior of the unit. Where batteries are installed in a compartment, the compartment shall be ventilated with openings having a minimum area of 1.7 in.^{2} (1100 mm^2) at both the top and at the bottom. Where compartment doors are equipped for ventilation, the openings shall be within 2 in. (50.8 mm) of the top and bottom. Batteries shall not be installed in a compartment containing spark- or flame-producing equipment.

(e) Overcurrent Protection.

(1) Low-voltage circuit wiring shall be protected by overcurrent protective devices rated not in excess of the ampacity of copper conductors, as follows.

Table 552-10(e)(1). Low-Voltage Overcurrent Protection

Wire Size (AWG)	Ampacity	Wire Type
18	6	Stranded only
16	8	Stranded only
14	15	Stranded or solid
12	20	Stranded or solid
10	30	Stranded or solid

(2) Circuit breakers or fuses shall be of an approved type, including automotive types. Fuseholders shall be clearly marked with maximum fuse size and shall be protected against shorting and physical damage by a cover or equivalent means.

> FPN: For further information, see *Standard for Electric Fuses (Cartridge Type), ANSI/SAE J554-1987; Standard for Blade Type Electric Fuses, SAE J1284-1988; and Standard for Automotive Glass Tube Fuses, UL 275-1993.*

(3) Higher current-consuming, dc appliances such as pumps, compressors, heater blowers, and similar motor-driven appliances shall be installed in accordance with the manufacturer's instructions.

Motors that are controlled by automatic switching or by latching-type manual switches shall be protected in accordance with Section 430-32(c).

(4) The overcurrent protective device shall be installed in an accessible location on the unit within 18 in. (457 mm) of the point where the power supply connects to the unit circuits. If located outside the park trailer, the device shall be protected against weather and physical damage.

Exception: External low-voltage supply shall be permitted to have the overcurrent protective device within 18 in. (457 mm) after entering the unit or after leaving a metal raceway.

(f) Switches. Switches shall have a dc rating not less than the connected load.

(g) Lighting Fixtures. All low-voltage interior lighting fixtures rated more than 4 watts, employing lamps rated more than 1.2 watts, shall be listed.

C. Combination Electrical Systems

552-20. Combination Electrical Systems.

(a) General. Unit wiring suitable for connection to a battery or other low-voltage supply source shall be permitted to be connected to a 120-volt source, provided that the entire wiring system and equipment are rated and installed in full conformity with Parts A, C, D, and E requirements of this article covering 120-volt electrical systems. Circuits fed from ac transformers shall not supply dc appliances.

(b) Voltage Converters (120-Volt Alternating Current to Low-Voltage Direct Current). The 120-volt ac side of the voltage converter shall be wired in full conformity with Parts A, C, D, and E requirements of this article for 120-volt electrical systems.

Exception: Converters supplied as an integral part of a listed appliance shall not be subject to the above.

All converters and transformers shall be listed for use in recreation units and designed or equipped to provide over-temperature protection. To determine the converter rating, the following formula shall be applied to the total connected load, including average battery charging rate, of all 12-volt equipment:

The first 20 amperes of load at 100 percent; plus

The second 20 amperes of load at 50 percent; plus

All load above 40 amperes at 25 percent

Exception: A low-voltage appliance that is controlled by a momentary switch (normally open) that has no means for holding in the closed position shall not be considered as a connected load when determining the required converter rating. Momentarily energized appliances shall be limited to those used to prepare the unit for occupancy or travel.

(c) Bonding Voltage Converter Enclosures. The non-current-carrying metal enclosure of the voltage converter shall be bonded to the frame of the unit with a No. 8 copper conductor minimum. The grounding conductor for the battery and the metal enclosure shall be permitted to be the same conductor.

(d) Dual-Voltage Fixtures or Appliances. Fixtures or appliances having both 120-volt and low-voltage connections shall be listed for dual voltage.

(e) Autotransformers. Autotransformers shall not be used.

(f) Receptacles and Plug Caps. Where a park trailer is equipped with a 120-volt or 120/240-volt ac system, a low-voltage system, or both, receptacles and plug caps of the low-voltage system shall differ in configuration from those of the 120-volt or 120/240-volt system. Where a unit equipped with a battery or dc system has an external connection for low-voltage power, the connector shall have a configuration that will not accept 120-volt power.

D. Nominal 120-Volt or 120/240-Volt Systems

552-40. 120-Volt or 120/240-Volt, Nominal, Systems.

(a) General Requirements. The electrical equipment and material of park trailers indicated for connection to a wiring system rated 120 volts, nominal, 2-wire with ground, or a wiring system rated 120/240 volts, nominal, 3-wire with ground, shall be listed and installed in accordance with the requirements of Parts A, C, D, and E of this article.

(b) Materials and Equipment. Electrical materials, devices, appliances, fittings, and other equipment installed, intended for use in, or attached to the park trailer shall be listed. All products shall be used only in the manner in which they have been tested and found suitable for the intended use.

552-41. Receptacle Outlets Required.

(a) Spacing. Receptacle outlets shall be installed at wall spaces 2 ft (610 mm) wide or more so that no point along the floor line is more than 6 ft (1.83 m), measured horizontally, from an outlet in that space.

Exception No. 1: Bath and hall areas.

Exception No. 2: Wall spaces occupied by kitchen cabinets, wardrobe cabinets, built-in furniture, behind doors that may open fully against a wall surface, or similar facilities.

(b) Location. Receptacle outlets shall be installed as follows:

(1) Adjacent to countertops in the kitchen [at least one on each side of the sink if countertops are on each side and are 12 in. (305 mm) or over in width]
(2) Adjacent to the refrigerator and gas range space, except where a gas-fired refrigerator or cooking appliance, requiring no external electrical connection, is factory-installed
(3) Adjacent to countertop spaces of 12 in. (305 mm) or more in width that cannot be reached from a receptacle required in Section 552-41(b)(1) by a cord of 6 ft 1.83 m) without crossing a traffic area, cooking appliance, or sink

(c) Ground-Fault Circuit-Interrupter Protection. Where provided, each 125-volt, single-phase, 15- or 20-ampere receptacle outlet shall have ground-fault circuit-interrupter protection for personnel in the following locations:

(1) Adjacent to a bathroom lavatory
(2) Within 6 ft (1.83 m) of any lavatory or sink

Exception: Receptacles installed for appliances in dedicated spaces, such as for dishwashers, disposals, refrigerators, freezers, and laundry equipment.

(3) In the area occupied by a toilet, shower, tub, or any combination thereof
(4) On the exterior of the unit

Exception: Receptacles that are located inside of an access panel that is installed on the exterior of the unit to supply power for an installed appliance shall not be required to have ground-fault circuit-interrupter protection.

The receptacle outlet shall be permitted in a listed lighting fixture. A receptacle outlet shall not be installed in a tub or combination tub–shower compartment.

(d) Pipe Heating Cable Outlet. Where a pipe heating cable outlet is installed, the outlet shall be as follows:

(1) Located within 2 ft (610 mm) of the cold water inlet
(2) Connected to an interior branch circuit, other than a small appliance branch circuit
(3) On a circuit where all of the outlets are on the load side of the ground-fault circuit-interrupter protection for personnel
(4) Mounted on the underside of the park trailer and shall not be considered to be the outdoor receptacle outlet required in Section 552-41(e)

(e) Outdoor Receptacle Outlets. At least one receptacle outlet shall be installed outdoors. A receptacle outlet located in a compartment accessible from the outside of the park trailer shall be considered an outdoor receptacle. Outdoor receptacle outlets shall be protected as required in Section 552-41(c)(4).

(f) Receptacle Outlets Not Permitted.

(1) Shower or Bathtub Space. Receptacle outlets shall not be installed in or within reach [30 in. (762 mm)] of a shower or bathtub space.

(2) Face-Up Position. A receptacle shall not be installed in a face-up position in any countertop.

552-43. Power Supply.

(a) Feeder. The power supply to the park trailer shall be a feeder assembly consisting of not more than one listed 30-ampere or 50-ampere park trailer power-supply cord with an integrally molded or securely attached cap, or a permanently installed feeder.

(b) Power-Supply Cord. If the park trailer has a power-supply cord, it shall be permanently attached to the distribution panelboard or to a junction box permanently connected to the distribution panelboard, with the free end terminating in a molded-on attachment plug cap.

Cords with adapters and pigtail ends, extension cords,

and similar items shall not be attached to, or shipped with, a park trailer.

A suitable clamp or the equivalent shall be provided at the distribution panelboard knockout to afford strain relief for the cord to prevent strain from being transmitted to the terminals when the power-supply cord is handled in its intended manner.

The cord shall be a listed type with 3-wire, 120-volt or 4-wire, 120/240-volt conductors, one of which shall be identified by a continuous green color or a continuous green color with one or more yellow stripes for use as the grounding conductor.

(c) Mast Weatherhead or Raceway. Where the calculated load exceeds 50 amperes or where a permanent feeder is used, the supply shall be by means of one of the following:

(1) One mast weatherhead installation, installed in accordance with Article 230, containing four continuous, insulated, color-coded feeder conductors, one of which shall be an equipment grounding conductor; or

(2) A metal raceway, rigid nonmetallic conduit, or liquid-tight flexible nonmetallic conduit from the disconnecting means in the park trailer to the underside of the park trailer, with provisions for the attachment to a suitable junction box or fitting to the raceway on the underside of the park trailer [with or without conductors as in Section 550-5(i)(1)]

552-44. Cord.

(a) Permanently Connected. Each power-supply assembly shall be factory supplied or factory installed and connected directly to the terminals of the distribution panelboard or conductors within a junction box and provided with means to prevent strain from being transmitted to the terminals. The ampacity of the conductors between each junction box and the terminals of each distribution panelboard shall be at least equal to the ampacity of the power-supply cord. The supply end of the assembly shall be equipped with an attachment plug of the type described in Section 552-44(c). Where the cord passes through the walls or floors, it shall be protected by means of conduit and bushings or equivalent. The cord assembly shall have permanent provisions for protection against corrosion and mechanical damage while the unit is in transit.

(b) Cord Length. The cord exposed usable length shall be measured from the point of entrance to the park trailer or the face of the flanged surface inlet (motor-base attachment plug) to the face of the attachment plug at the supply end.

The cord exposed usable length, measured to the point of entry on the unit exterior, shall be a minimum of 23 ft (7.0 m) where the point of entrance is at the side of the unit, or shall be a minimum 28 ft (8.5 m) where the point of

entrance is at the rear of the unit. The maximum length shall not exceed 36½ ft (11.13 m).

Where the cord entrance into the unit is more than 3 ft (0.9 m) above the ground, the minimum cord lengths above shall be increased by the vertical distance of the cord entrance heights above 3 ft (0.9 m).

(c) Attachment Plugs.

(1) Park trailers wired in accordance with Section 552-46(a) shall have an attachment plug that shall be 2-pole, 3-wire, grounding-type, rated 30 amperes, 125 volts, conforming to the configuration shown in Figure 552-44(c) intended for use with units rated at 30 amperes, 125 volts.

FPN: Complete details of this configuration can be found in National Electrical Manufacturers Association's *Standard for Dimensions of Attachment Plugs and Receptacles*, ANSI/ NEMA WD 6-1989, Figure TT.

(2) Park trailers having a power-supply assembly rated 50 amperes as permitted by Section 552-43(b) shall have a 3-pole, 4-wire, grounding-type attachment plug rated 50 amperes, 125/250 volts, conforming to the configuration shown in Figure 552-44(c).

FPN: Complete details of this configuration can be found in National Electrical Manufacturers Association *Standard for Dimensions of Attachment Plugs and Receptacles,* ANSI/ NEMA WD 6-1989, Figure 14-50.

30-A, 125-V, 2-pole, 3-wire, grounding type

50-A, 125/250-V, 3-pole, 4-wire, grounding type

Figure 552-44(c).

(d) Labeling at Electrical Entrance. Each park trailer shall have permanently affixed to the exterior skin, at or near the point of entrance of the power-supply assembly, a label 3 in. × 1¾ in. (76 mm × 44.5 mm) minimum size, made of etched, metal-stamped, or embossed brass, stainless steel, or anodized or alclad aluminum not less than 0.020-in. (0.508-mm) thick, or other suitable material [e.g., 0.005-in. (0.127-mm) thick plastic laminate], that reads, as appropriate, either

THIS CONNECTION IS FOR 110–125-VOLT AC,
60 HZ, 30 AMPERE SUPPLY.

or

THIS CONNECTION IS FOR 120/240-VOLT AC, 3-POLE,
4-WIRE, 60 HZ, _____ AMPERE SUPPLY.

The correct ampere rating shall be marked in the blank space.

(e) Location. The point of entrance of a power-supply assembly shall be located within 15 ft (4.57 m) of the rear, on the left (road) side or at the rear, left of the longitudinal center of the unit, within 18 in. (457 mm) of the outside wall.

Exception: A park trailer shall be permitted to have the electrical point of entrance located more than 15 ft (4.57 m) from the rear. Where this occurs, the distance beyond the 15-ft (4.57-m) dimension shall be added to the cord's minimum length as specified in Section 551-46(b).

552-45. Distribution Panelboard.

(a) Listed and Appropriately Rated. A listed and appropriately rated distribution panelboard or other equipment specifically listed for the purpose shall be used. The grounded conductor termination bar shall be insulated from the enclosure as provided in Section 552-55(c). An equipment grounding terminal bar shall be attached inside the metal enclosure of the panelboard.

(b) Location. The distribution panelboard shall be installed in a readily accessible location. Working clearance for the panelboard shall be not less than 24 in. (610 mm) wide and 30 in. (762 mm) deep.

Exception: Where the panelboard cover is exposed to the inside aisle space, one of the working clearance dimensions shall be permitted to be reduced to a minimum of 22 in. (559 mm). A panelboard shall be considered exposed where the panelboard cover is within 2 in. (50.8 mm) of the aisle's finished surface.

(c) Dead-Front Type. The distribution panelboard shall be of the dead-front type. A main disconnecting means shall be provided where fuses are used or where more than two circuit breakers are employed. A main overcurrent protective device not exceeding the power-supply assembly rating shall be provided where more than two branch circuits are employed.

552-46. Branch Circuits. Branch circuits shall be determined in accordance with the following.

(a) Two to Five 15-or 20-Ampere Circuits. Two to five 15- or 20-ampere circuits to supply lights, receptacle outlets,

and fixed appliances shall be permitted. Such park trailers shall be equipped with a distribution panelboard rated at 120 volts maximum with a 30-ampere rated main power supply assembly. Not more than two 120-volt thermostatically controlled appliances (e.g., air conditioner and water heater) shall be installed in such systems unless appliance isolation switching, energy management systems, or similar methods are used.

Exception: Additional 15- or 20-ampere circuits shall be permitted where a listed energy management system rated at 30 amperes maximum is employed within the system.

(b) More than Five Circuits. Where more than five circuits are needed, they shall be determined in accordance with the following.

(1) Lighting. Based on 3 volt-amperes/ft^2 (32.26 volt-ampere/m^2) multiplied by the outside dimensions of the park trailer (coupler excluded) divided by 120 volts to determine the number of 15- or 20-ampere lighting area circuits, e.g.,

$$\frac{3 \times \text{length} \times \text{width}}{120 \times 15 \text{ (or 20)}} = \text{No. of 15- (or 20-) ampere circuits}$$

The lighting circuits shall be permitted to serve built-in gas ovens with electric service only for lights, clocks or timers, or listed cord-connected garbage disposal units.

(2) Small Appliances. Small appliance branch circuits shall be installed in accordance with Section 210-11(c)(1).

(3) General Appliances. (including furnace, water heater, range, and central or room air conditioner, etc.). There shall be one or more circuits of adequate rating in accordance with the following.

FPN No. 1: For the laundry branch circuit, see Section 210-11(c)(2).

FPN No. 2: For central air conditioning, see Article 440.

(a) Ampere rating of fixed appliances not over 50 percent of circuit rating if lighting outlets (receptacles, other than kitchen, dining area, and laundry, considered as lighting outlets) are on the same circuit.

(b) For fixed appliances on a circuit without lighting outlets, the sum of rated amperes shall not exceed the branch-circuit rating. Motor loads or other continuous duty loads shall not exceed 80 percent of the branch-circuit rating.

(c) The rating of a single cord- and plug-connected appliance on a circuit having no other outlets shall not exceed 80 percent of the circuit rating.

(d) The rating of a range branch circuit shall be based on the range demand as specified for ranges in Section 552-47(b)(5).

552-47. Calculations. The following method shall be employed in computing the supply-cord and distribution-panelboard load for each feeder assembly for each park trailer in lieu of the procedure shown in Article 220 and shall be based on a 3-wire, 120/240-volt supply with 120-volt loads balanced between the two phases of the 3-wire system.

(a) Lighting and Small Appliance Load.

Lighting Volt-Amperes: Length times width of park trailer floor (outside dimensions) times 3 volt-amperes/ft^2, e.g.,

$$\text{Length} \times \text{width} \times 3 = \text{lighting volt-amperes}$$

Small Appliance Volt-Amperes: Number of circuits times 1500 volt-amperes for each 20-ampere appliance receptacle circuit (see definition of *Appliance, Portable* with note) including 1500 volt-amperes for laundry circuit, e.g.,

$$\text{No. of circuits} \times 1500 = \text{small appliance volt-amperes}$$

Total: Lighting volt-amperes plus small appliance volt-amperes = total volt-amperes

First 3000 total volt-amperes at 100 percent plus remainder at 35 percent = volt-amperes to be divided by 240 volts to obtain current (amperes) per leg

(b) Total Load for Determining Power Supply. Total load for determining power supply is the sum of the following:

(1) Lighting and small appliance load as calculated in Section 552-47(a).
(2) Nameplate amperes for motors and heater loads (exhaust fans, air conditioners, electric, gas, or oil heating). Omit smaller of the heating and cooling loads, except include blower motor if used as air-conditioner evaporator motor. Where an air conditioner is not installed and a 50-ampere power-supply cord is provided, allow 15 amperes per phase for air conditioning.
(3) Twenty-five percent of current of largest motor in (2).
(4) Total of nameplate amperes for disposal, dishwasher, water heater, clothes dryer, wall-mounted oven, cooking units. Where the number of these appliances exceeds three, use 75 percent of total.
(5) Derive amperes for freestanding range (as distinguished from separate ovens and cooking units) by dividing the following values by 240 volts.

Nameplate Rating (watts)	Use (volt-amperes)
0–10,000	80 percent of rating
Over 10,000–12,500	8,000
Over 12,500–13,500	8,400
Over 13,500–14,500	8,800
Over 14,500–15,500	9,200
Over 15,500–16,500	9,600
Over 16,500–17,500	10,000

(6) If outlets or circuits are provided for other than factory-installed appliances, include the anticipated load.

> FPN: Refer to Appendix D, Example D12, for an illustration of the application of this calculation.

(c) Optional Method of Calculation for Lighting and Appliance Load. For park trailers, the optional method for calculating lighting and appliance load shown in Section 220-30 shall be permitted.

552-48. Wiring Methods.

(a) Wiring Systems. Cables and raceways installed in accordance with Articles 330 through 352 shall be permitted in accordance with their applicable article, except as otherwise specified in this article. An equipment grounding means shall be provided in accordance with Section 250-118.

(b) Conduit and Tubing. Where rigid metal conduit or intermediate metal conduit is terminated at an enclosure with a locknut and bushing connection, two locknuts shall be provided, one inside and one outside of the enclosure. All cut ends of conduit and tubing shall be reamed or otherwise finished to remove rough edges.

(c) Nonmetallic Boxes. Nonmetallic boxes shall be acceptable only with nonmetallic-sheathed cable or nonmetallic raceways.

(d) Boxes. In walls and ceilings constructed of wood or other combustible material, boxes and fittings shall be flush with the finished surface or project therefrom.

(e) Mounting. Wall and ceiling boxes shall be mounted in accordance with Article 370.

Exception No. 1: Snap-in-type boxes or boxes provided with special wall or ceiling brackets that securely fasten boxes in walls or ceilings shall be permitted.

Exception No. 2: A wooden plate providing a 1½-in. (38-mm) minimum width backing around the box and of a thickness of ½ in. (12.7 mm) or greater (actual) attached directly to the wall panel shall be considered as approved means for mounting outlet boxes.

(f) Sheath Armor. The sheath of nonmetallic-sheathed cable, metal-clad cable, and Type AC cable shall be continuous between outlet boxes and other enclosures.

(g) Protected. Metal-clad, Type AC, or nonmetallic-sheathed cables and electrical nonmetallic tubing shall be permitted to pass through the centers of the wide side of 2 in. by 4 in. wood studs. However, they shall be protected where they pass through 2 in. by 2 in. wood studs or at other wood studs or frames where the cable or tubing would be less than 1¼ in. (31.8 mm) from the inside or outside surface. Steel plates on each side of the cable or tubing, or a steel tube, with not less than No. 16 MSG wall thickness, shall be installed to protect the cable or tubing. These plates or tubes shall be securely held in place. Where nonmetallic-sheathed cables pass through punched, cut, or drilled slots or holes in metal members, the cable shall be protected by bushings or grommets securely fastened in the opening prior to installation of the cable.

(h) Cable Supports. Where connected with cable connectors or clamps, cables shall be supported within 12 in. (305 mm) of outlet boxes, distribution panelboards, and splice boxes on appliances. Supports shall be provided every 4½ ft (1.37 m) at other places.

(i) Nonmetallic Box Without Cable Clamps. Nonmetallic-sheathed cables shall be supported within 8 in. (203 mm) of a nonmetallic outlet box without cable clamps.

Exception: Where wiring devices with integral enclosures are employed with a loop of extra cable to permit future replacement of the device, the cable loop shall be considered as an integral portion of the device.

(j) Physical Damage. Where subject to physical damage, exposed nonmetallic cable shall be protected by covering boards, guard strips, raceways, or other means.

(k) Metal Faceplates. Metal faceplates shall be of ferrous metal not less than 0.030 in. (0.762 mm) in thickness or of nonferrous metal not less than 0.040 in. (1.016 mm) in thickness. Nonmetallic faceplates shall be listed.

(l) Metal Faceplates Effectively Grounded. Where metal faceplates are used, they shall be effectively grounded.

(m) Moisture or Physical Damage. Where outdoor or under-chassis wiring is 120 volts, nominal, or over and is exposed to moisture or physical damage, the wiring shall be protected by rigid metal conduit, intermediate metal conduit, or by electrical metallic tubing or rigid nonmetallic conduit that is closely routed against frames and equipment enclosures or other raceway or cable identified for the application.

(n) Component Interconnections. Fittings and connectors that are intended to be concealed at the time of assembly shall be listed and identified for the interconnection of building components. Such fittings and connectors shall be equal to the wiring method employed in insulation, temperature rise, and fault-current withstanding, and shall be capable of enduring the vibration and shock occurring in park trailers.

(o) Method of Connecting Expandable Units.

(1) That portion of a branch circuit that is installed in an expandable unit shall be permitted to be connected to the branch circuit in the main body of the vehicle by means of a flexible cord or attachment plug and cord listed for hard usage. The cord and its connections shall conform to all provisions of Article 400 and shall be considered as a permitted use under Section 400-7.

(2) If the receptacle provided for connection of the cord to the main circuit is located on the outside of the unit, it shall be protected with a ground-fault circuit interrupter for personnel and be listed for wet locations. A cord located on the outside of a unit shall be identified for outdoor use.

(3) Unless removable or stored within the unit interior, the cord assembly shall have permanent provisions for protection against corrosion and mechanical damage while the unit is in transit.

(4) If an attachment plug and cord is used it shall be installed so as not to permit exposed live attachment plug pins.

(p) Prewiring for Air-Conditioning Installation. Prewiring installed for the purpose of facilitating future air-conditioning installation shall conform to the following and other applicable portions of this article. The circuit shall serve no other purpose.

(1) An overcurrent protective device with a rating compatible with the circuit conductors shall be installed in the distribution panelboard and wiring connections completed.

(2) The load end of the circuit shall terminate in a junction box with a blank cover or a device listed for the purpose. Where a junction box with a blank cover is used, the free ends of the conductors shall be adequately capped or taped.

(3) A label conforming to Section 552-44(b) shall be placed on or adjacent to the junction box and shall read

> AIR-CONDITIONING CIRCUIT. THIS
> CONNECTION IS FOR AIR CONDITIONERS
> RATED 110–125-VOLT AC, 60 HZ, _____
> AMPERES MAXIMUM. DO NOT EXCEED
> CIRCUIT RATING.

An ampere rating, not to exceed 80 percent of the circuit rating, shall be legibly marked in the blank space.

552-49. Conductors and Boxes.

(a) Maximum Number of Conductors. The maximum number of conductors permitted in boxes shall be in accordance with Section 370-16.

(b) Free Conductor at Each Box. At least 6 in. (152 mm) of free conductor shall be left at each box except where conductors are intended to loop without joints.

552-50. Grounded Conductors. The identification of grounded conductors shall be in accordance with Section 200-6.

552-51. Connection of Terminals and Splices. Conductor splices and connections at terminals shall be in accordance with Section 110-14.

552-52. Switches. Switches shall be rated as follows.

(a) Lighting Circuits. For lighting circuits, switches shall be rated not less than 10 amperes, 120/125 volts and in no case less than the connected load.

(b) Motors or Other Loads. For motors or other loads, switches shall have ampere or horsepower ratings, or both, adequate for loads controlled. (An ac general-use snap switch shall be permitted to control a motor 2 hp or less with full-load current not over 80 percent of the switch ampere rating.)

552-53. Receptacles. All receptacle outlets shall be of the grounding type and installed in accordance with Sections 210-7 and 210-21.

552-54. Lighting Fixtures.

(a) General. Any combustible wall or ceiling finish exposed between the edge of a fixture canopy, or pan and the outlet box, shall be covered with noncombustible material or a material identified for the purpose.

(b) Shower Fixtures. If a lighting fixture is provided over a bathtub or in a shower stall, it shall be of the enclosed and gasketed type and listed for the type of installation, and it shall be ground-fault circuit-interrupter protected.

The switch for shower lighting fixtures and exhaust fans, located over a tub or in a shower stall, shall be located outside the tub or shower space.

(c) Outdoor Outlets, Fixtures, Air-Cooling Equipment, etc. Outdoor fixtures and other equipment shall be listed for outdoor use.

552-55. Grounding. (See also Section 552-57 on bonding of noncurrent-carrying metal parts.)

(a) Power-Supply Grounding. The grounding conductor in the supply cord or feeder shall be connected to the grounding bus or other approved grounding means in the distribution panelboard.

(b) Distribution Panelboard. The distribution panelboard shall have a grounding bus with sufficient terminals for all grounding conductors or other approved grounding means.

(c) Insulated Neutral.

(1) The grounded circuit conductor (neutral) shall be insulated from the equipment grounding conductors and from equipment enclosures and other grounded parts. The grounded (neutral) circuit terminals in the distribution panelboard and in ranges, clothes dryers, counter-mounted cooking units, and wall-mounted ovens shall be insulated from the equipment enclosure. Bonding screws, straps, or buses in the distribution panelboard or in appliances shall be removed and discarded.

(2) Connection of electric ranges and electric clothes dryers utilizing a grounded (neutral) conductor, if cord-connected, shall be made with 4-conductor cord and 3-pole, 4-wire, grounding-type plug caps and receptacles.

552-56. Interior Equipment Grounding.

(a) Exposed Metal Parts. In the electrical system, all exposed metal parts, enclosures, frames, lighting fixture canopies, etc., shall be effectively bonded to the grounding terminals or enclosure of the distribution panelboard.

(b) Equipment Grounding Conductors. Bare wires, green-colored wires, or green wires with a yellow stripe(s) shall be used for equipment grounding conductors only.

(c) Grounding of Electrical Equipment. Where grounding of electrical equipment is specified, it shall be permitted as follows.

(1) Connection of metal raceway (conduit or electrical metallic tubing), the sheath of Type MC and Type MI cable where the sheath is identified for grounding, or the armor of Type AC cable to metal enclosures.
(2) A connection between the one or more equipment grounding conductors and a metal box by means of a grounding screw, which shall be used for no other purpose, or a listed grounding device.
(3) The equipment grounding conductor in nonmetallic-sheathed cable shall be permitted to be secured under a screw threaded into the fixture canopy other than a mounting screw or cover screw, or attached to a listed grounding means (plate) in a nonmetallic outlet box for fixture mounting (grounding means shall also be permitted for fixture attachment screws).

(d) Grounding Connection in Nonmetallic Box. A connection between the one or more grounding conductors brought into a nonmetallic outlet box shall be arranged so that a connection can be made to any fitting or device in that box that requires grounding.

(e) Grounding Continuity. Where more than one equipment grounding conductor of a branch circuit enters a box, all such conductors shall be in good electrical contact with each other, and the arrangement shall be such that the disconnection or removal of a receptacle, fixture, or other device fed from the box will not interfere with or interrupt the grounding continuity.

(f) Cord-Connected Appliances. Cord-connected appliances, such as washing machines, clothes dryers, refrigerators, and the electrical system of gas ranges, etc., shall be grounded by means of an approved cord with equipment grounding conductor and grounding-type attachment plug.

552-57. Bonding of Noncurrent-Carrying Metal Parts.

(a) Required Bonding. All exposed noncurrent-carrying metal parts that may become energized shall be effectively bonded to the grounding terminal or enclosure of the distribution panelboard.

(b) Bonding Chassis. A bonding conductor shall be connected between any distribution panelboard and an accessible terminal on the chassis. Aluminum or copper-clad aluminum conductors shall not be used for bonding if such conductors or their terminals are exposed to corrosive elements.

Exception: Any park trailer that employs a unitized metal chassis-frame construction to which the distribution panelboard is securely fastened with a bolt(s) and nut(s) or by welding or riveting shall be considered to be bonded.

(c) Bonding Conductor Requirements. Grounding terminals shall be of the solderless type and listed as pressure terminal connectors recognized for the wire size used. The bonding conductor shall be solid or stranded, insulated or bare, and shall be No. 8 copper minimum, or equivalent.

(d) Metallic Roof and Exterior Bonding. The metal roof and exterior covering shall be considered bonded where

(1) The metal panels overlap one another and are securely attached to the wood or metal frame parts by metal fasteners, and
(2) The lower panel of the metal exterior covering is secured by metal fasteners at each cross member of the chassis, or the lower panel is bonded to the chassis by a metal strap.

(e) Gas, Water, and Waste Pipe Bonding. The gas, water, and waste pipes shall be considered grounded if they are bonded to the chassis.

(f) Furnace and Metal Air Duct Bonding. Furnace and metal circulating air ducts shall be bonded.

552-58. Appliance Accessibility and Fastening. Every appliance shall be accessible for inspection, service, repair, and replacement without removal of permanent construction. Means shall be provided to securely fasten appliances in place when the park trailer is in transit.

552-59. Outdoor Outlets, Fixtures, Air-Cooling Equipment, etc.

(a) Listed for Outdoor Use. Outdoor fixtures and equipment shall be listed for outdoor use. Outdoor receptacle or convenience outlets shall be of a gasketed-cover type for use in wet locations.

(b) Outside Heating Equipment, Air-Conditioning Equipment, or Both. A park trailer provided with a branch circuit designed to energize outside heating equipment or air-conditioning equipment, or both, located outside the park trailer, other than room air conditioners, shall have such branch-circuit conductors terminate in a listed outlet box or disconnecting means located on the outside of the park trailer. A label shall be permanently affixed adjacent to the outlet box and shall contain the following information:

> THIS CONNECTION IS FOR HEATING AND/OR AIR-CONDITIONING EQUIPMENT. THE BRANCH CIRCUIT IS RATED AT NOT MORE THAN _____ AMPERES, AT _____ VOLTS, 60-Hz, _____ CONDUCTOR AMPACITY. A DISCONNECTING MEANS SHALL BE LOCATED WITHIN SIGHT OF THE EQUIPMENT.

The correct voltage and ampere rating shall be given. The tag shall not be less than 0.020-in. (508-μm) thick etched brass, stainless steel, anodized or alclad aluminum, or equivalent. The tag shall not be less than 3 in. $\times$ 1¾ in. (76 mm $\times$ 44.5 mm) minimum size.

E. Factory Tests

552-60. Factory Tests (Electrical). Each park trailer shall be subjected to the following tests.

(a) Circuits of 120 Volts or 120/240 Volts. Each park trailer designed with a 120-volt or a 120/240-volt electrical system shall withstand the applied potential without electrical breakdown of a 1-minute, 900-volt dielectric strength test, or a 1-second, 1080-volt dielectric strength test, with all switches closed, between ungrounded and grounded conductors and the park trailer ground. During the test, all switches and other controls shall be in the on position. Fixtures and permanently installed appliances shall not be required to withstand this test.

Each park trailer shall be subjected to the following.

(1) A continuity test to ensure that all metal parts are properly bonded;
(2) Operational tests to demonstrate that all equipment is properly connected and in working order;

(3) Polarity checks to determine that connections have been properly made; and

(4) Receptacles requiring GFCI protection shall be tested for correct function by the use of a GFCI testing device.

(b) Low-Voltage Circuits. Low-voltage circuit conductors in each park trailer shall withstand the applied potential without electrical breakdown of a 1-minute, 500-volt or a 1-second, 600-volt dielectric strength test. The potential shall be applied between ungrounded and grounded conductors.

The test shall be permitted on running light circuits before the lights are installed, provided the unit's outer covering and interior cabinetry have been secured. The braking circuit shall be permitted to be tested before being connected to the brakes, provided the wiring has been completely secured.

Article 553 — Floating Buildings

A. General

553-1. Scope. This article covers wiring, services, feeders, and grounding for floating buildings.

553-2. Definition.

Floating Building. A building unit as defined in Article 100 that floats on water, is moored in a permanent location, and has a premises wiring system served through connection by permanent wiring to an electricity supply system not located on the premises.

553-3. Application of Other Articles. Wiring for floating buildings shall comply with the applicable provisions of other articles of this *Code*, except as modified by this article.

B. Services and Feeders

553-4. Location of Service Equipment. The service equipment for a floating building shall be located adjacent to, but not in or on, the building.

553-5. Service Conductors. One set of service conductors shall be permitted to serve more than one set of service equipment.

553-6. Feeder Conductors. Each floating building shall be supplied by a single set of feeder conductors from its service equipment.

Exception: Where the floating building has multiple occupancy, each occupant shall be permitted to be supplied by a single set of feeder conductors extended from the occupant's service equipment to the occupant's panelboard.

553-7. Installation of Services and Feeders.

(a) Flexibility. Flexibility of the wiring system shall be maintained between floating buildings and the supply conductors. All wiring shall be installed so that motion of the water surface and changes in the water level will not result in unsafe conditions.

(b) Wiring Methods. Liquidtight flexible metal conduit or liquidtight flexible nonmetallic conduit with approved fittings shall be permitted for feeders and where flexible connections are required for services. Extra-hard usage portable power cable listed for both wet locations and sunlight resistance shall be permitted for a feeder to a floating building where flexibility is required. Other raceways, suitable for the location, shall be permitted to be installed where flexibility is not required.

FPN: See Sections 555-1 and 555-6.

C. Grounding

553-8. General Requirements. Grounding of both electrical and nonelectrical parts in a floating building shall be through connection to a grounding bus in the building panelboard. The grounding bus shall be grounded through a green-colored insulated equipment grounding conductor run with the feeder conductors and connected to a grounding terminal in the service equipment. The grounding terminal in the service equipment shall be grounded by connection through an insulated grounding electrode conductor to a grounding electrode on shore.

553-9. Insulated Neutral. The grounded circuit conductor (neutral) shall be an insulated conductor identified in conformance with Section 200-6. The neutral conductor shall be connected to the equipment grounding terminal in the service equipment, and, except for that connection, it shall be insulated from the equipment grounding conductors, equipment enclosures, and all other grounded parts. The neutral circuit terminals in the panelboard and in ranges, clothes dryers, counter-mounted cooking units, and the like shall be insulated from the enclosures.

553-10. Equipment Grounding.

(a) Electrical Systems. All enclosures and exposed metal parts of electrical systems shall be bonded to the grounding bus.

(b) Cord-Connected Appliances. Where required to be grounded, cord-connected appliances shall be grounded by means of an equipment grounding conductor in the cord and a grounding-type attachment plug.

553-11. Bonding of Noncurrent-Carrying Metal Parts. All metal parts in contact with the water, all metal piping, and all noncurrent-carrying metal parts that may become energized, shall be bonded to the grounding bus in the panelboard.

Article 555 — Marinas and Boatyards

555-1. Scope. This article covers the installation of wiring and equipment in the areas comprising fixed or floating piers, wharfs, docks, and other areas in marinas, boatyards, boat basins, boathouses, and similar occupancies that are used, or intended for use, for the purpose of repair, berthing, launching, storage, or fueling of small craft and the moorage of floating buildings.

555-2. Application of Other Articles. Wiring and equipment for marinas and boatyards shall comply with this article and also with the applicable provisions of other articles of this *Code*.

> FPN: For disconnection of auxiliary power from boats, see *Fire Protection Standard for Pleasure and Commercial Motor Craft*, NFPA 302-1994.

555-3. Receptacles. Where shore power is supplied, those accommodations for boats 20 ft (6.1 m) or less in length shall be equipped with shore-power receptacles of a locking and grounding type rated at not less than 20 amperes.

Where shore power is supplied to accommodations for boats longer than 20 ft (6.1 m) in length, shore-power receptacles of a locking and grounding type rated at 30 amperes or more shall be provided.

Fifteen- and 20-ampere, single-phase, 125-volt receptacles other than those supplying shore power to boats located at piers, wharfs, and other locations shall be protected by ground-fault circuit interrupters.

> FPN No. 1: For various configurations and ratings of locking- and grounding-type receptacles and caps, see National Electrical Manufacturers Association's *Standard for Dimensions of Attachment Plugs and Receptacles*, ANSI/NEMA 18WD 6-1989.

> FPN No. 2: For locking- and grounding-type receptacles for auxiliary power to boats, see *Fire Protection Standard for Marinas and Boatyards*, NFPA 303-1995.

> FPN No. 3: In locating receptacles, consideration should be given to the maximum tide level and wave action. See *Fire Protection Standard for Marinas and Boatyards*, NFPA 303-1995, for establishment of datum plane.

555-4. Disconnecting Means. A readily accessible disconnecting means shall be provided by which each boat can be isolated from its supply circuit. The disconnecting means shall consist of a circuit breaker or switch, or both, and shall be located within sight of the shore power connection and is intended to constitute the means of cutoff of the supply to the boat.

555-5. Branch Circuits. Each single receptacle that supplies shore power to boats shall be supplied from a power outlet or panelboard by an individual branch circuit of the voltage class and rating corresponding to the rating of the receptacle.

> FPN: Supplying receptacles at voltages other than the voltages marked on the receptacle may cause overheating or malfunctioning of connected equipment, for example, supplying single-phase, 120/240-volt, 3-wire loads from a 208Y/120-volt, 3-wire source.

555-6. Feeders and Services. The load for each feeder and/or service circuit supplying receptacles that supply shore power for boats shall be calculated as follows. These calculations may be modified as indicated in sections (a) and (b).

Number of Receptacles	Sum of the Rating of the Receptacles (percent)
1–4	100
5–8	90
9–14	80
15–30	70
31–40	60
41–50	50
51–70	40
71–plus	30

(a) Where shore power accommodations provide two receptacles specifically for an individual boat slip and these receptacles have different voltages (for example, one 30 ampere, 125 volt and one 50 ampere, 125/250 volt), only the receptacle with the larger kilowatt demand shall be required to be calculated.

(b) If the facility being installed includes individual kilowatt-hour submeters for each slip, and is being calculated using the criteria listed in Section 555-6, the total demand amperes may be multiplied by 0.9 to achieve the final demand amperes.

> FPN: These demand factors may be inadequate in areas of extreme hot or cold temperatures with loaded circuits for heating, air-conditioning, or refrigerating equipment.

555-7. Wiring Methods. The wiring method shall be of a type identified for use in wet locations. Extra-hard usage portable power cable listed for both wet locations and sunlight resistance shall be permitted for a feeder where flexibility is required.

FPN: For further information on wiring methods for various locations and for establishment of datum plane, see *Fire Protection Standard for Marinas and Boatyards*, NFPA 303-1995.

555-8. Grounding.

(a) Equipment to Be Grounded. The following items shall be connected to an equipment grounding conductor run with the circuit conductors in a raceway or cable:

(1) Boxes, cabinets, and all other metal enclosures
(2) Metal frames of utilization equipment
(3) Grounding terminals of grounding-type receptacles

(b) Type of Equipment Grounding Conductor. The equipment grounding conductor shall be an insulated copper conductor with a continuous outer finish that is either green or green with one or more yellow stripes.

Exception: The equipment grounding conductor of Type MI cable shall be permitted to be identified at terminations.

(c) Size of Equipment Grounding Conductor. The insulated copper equipment grounding conductor shall be sized in accordance with Section 250-122 but not smaller than No. 12.

(d) Branch-Circuit Equipment Grounding Conductor. The insulated equipment grounding conductor for branch circuits shall terminate at a grounding terminal in a remote panelboard or the grounding terminal in the main service equipment.

(e) Feeder Equipment Grounding Conductors. Where a feeder supplies a remote panelboard, an insulated equipment grounding conductor shall extend from a grounding terminal in the service equipment to a grounding terminal in the remote panelboard.

555-9. Wiring Over and Under Navigable Water. Wiring over and under navigable water shall be subject to approval by the authority having jurisdiction.

555-10. Gasoline Dispensing Stations — Hazardous (Classified) Locations. Electrical equipment and wiring located in gasoline dispensing stations shall comply with Article 514.

FPN: For further information, see *Automotive and Marine Service Station Code*, NFPA 30A-1996, and *Fire Protection Standard for Marinas and Boatyards*, NFPA 303-1995.

555-11. Location of Service Equipment. The service equipment for floating docks or marinas shall be located adjacent to, but not on or in, the floating structure.

CHAPTER 6
Special Equipment

Article 600 — Electric Signs and Outline Lighting

A. General

600-1. Scope. This article covers the installation of conductors and equipment for electric signs and outline lighting as defined in Article 100.

> FPN: As defined in Article 100, electric signs and outline lighting includes all products and installations utilizing neon tubing, such as signs, decorative elements, skeleton tubing, or art forms.

600-2. Definitions.

Electric-Discharge Lighting. Systems of illumination utilizing fluorescent lamps, high intensity discharge (HID) lamps, or neon tubing.

Neon Tubing. Electric-discharge tubing manufactured into shapes that form letters, parts of letters, skeleton tubing, outline lighting, other decorative elements, or art forms, and filled with various inert gases.

Sign Body. A portion of a sign that may provide protection from the weather, but is not an electrical enclosure.

Skeleton Tubing. Neon tubing that is itself the sign or outline lighting and not attached to an enclosure or sign body.

600-3. Listing. Electric signs and outline lighting — fixed, mobile, or portable — shall be listed and installed in conformance with that listing, unless otherwise approved by special permission.

(a) Field installed skeleton tubing shall not be required to be listed where installed in conformance with this *Code*.

(b) Outline lighting shall not be required to be listed as a system when it consists of listed lighting fixtures wired in accordance with Chapter 3.

600-4. Markings.

(a) Signs and Outline Lighting Systems. Signs and outline lighting systems shall be marked with the manufacturer's name, trademark, or other means of identification; and, input voltage and current rating.

(b) Signs and outline lighting systems with incandescent lamp holders shall be marked to indicate the maximum allowable wattage of lamps. The markings shall be permanently installed, in letters at least ¼ in. (6.35 mm) high, and shall be located where visible during relamping.

600-5. Branch Circuits.

(a) Required Branch Circuit. Each commercial building and each commercial occupancy accessible to pedestrians shall be provided with at least one outlet in an accessible location at each entrance to each tenant space for sign or outline lighting system use. The outlet(s) shall be supplied by a branch circuit rated at least 20 amperes that supplies no other load. Service hallways or corridors shall not be considered accessible to pedestrians.

(b) Rating.

(1) Branch circuits that supply signs and outline lighting systems containing incandescent and fluorescent forms of illumination shall be rated not to exceed 20 amperes.

(2) Branch circuits that supply neon tubing installations shall not be rated in excess of 30 amperes.

(3) Computed Load. The load for the required branch circuit shall be computed at a minimum of 1200 volt-amperes.

(c) Wiring Methods.

(1) Supply. The wiring method used to supply signs and outline lighting systems shall terminate within a sign, an outline lighting system enclosure, a suitable box, or a conduit body.

(2) Enclosures as Pull Boxes. Signs and transformer enclosures shall be permitted to be used as pull or junction boxes for conductors supplying other adjacent signs, outline lighting systems, or floodlights that are part of a sign, and shall be permitted to contain both branch and secondary circuit conductors.

(3) Metal poles used to support signs shall be permitted to enclose supply conductors, provided the poles and conductors are installed in accordance with Section 410-15(b).

600-6. Disconnects. Each sign and outline lighting system, or feeder circuit or branch circuit supplying a sign or outline lighting system, shall be controlled by an externally operable switch or circuit breaker that will open all ungrounded conductors. Signs and outline lighting systems located within fountains shall have the disconnect located in accordance with Section 680-12.

Exception No. 1: A disconnecting means shall not be required for an exit directional sign located within a building.

Exception No. 2: A disconnecting means shall not be required for cord-connected signs with an attachment plug.

(a) Location. The disconnecting means shall be within sight of the sign or outline lighting system that it controls. Where the disconnecting means is out of the line of sight from any section that may be energized, the disconnecting means shall be capable of being locked in the open position.

Signs or outline lighting systems operated by electronic or electromechanical controllers located external to the sign or outline lighting system shall be permitted to have a disconnecting means located within sight of the controller or in the same enclosure with the controller. The disconnecting means shall disconnect the sign or outline lighting system and the controller from all ungrounded supply conductors. It shall be designed so that no pole can be operated independently and it shall be capable of being locked in the open position.

(b) Control Switch Rating. Switches, flashers, and similar devices controlling transformers and electronic power supplies shall be rated for controlling inductive loads or have a current rating not less than twice the current rating of the transformer.

FPN: See Section 380-14 for rating of snap switches.

600-7. Grounding. Signs and metal equipment of outline lighting systems shall be grounded. Listed flexible metal conduit or listed liquidtight flexible metal conduit that encloses the secondary wiring of a transformer or power supply for use with electric discharge tubing shall be permitted as a bonding means in lengths not exceeding 100 ft (30.5 m). Small metal parts not exceeding 2 in. (50.8 mm) in any dimension, not likely to be energized, and spaced at least ¾ in. (19 mm) from neon tubing shall not require bonding. Where listed nonmetallic conduit is used to enclose the secondary wiring of a transformer or power supply and a bonding conductor is required, the bonding conductor shall be installed separate and remote from the nonmetallic conduit and be spaced at least 1½ in. (38 mm) from the conduit when the circuit is operated at 100 Hz or 1¾ in. (44.45 mm) when the circuit is operated at over 100 Hz. Bonding conductors shall be copper and not smaller than No. 14. Metal parts of a building shall not be permitted as a grounded or equipment grounding conductor.

FPN: Refer to Section 600-32(j) for additional restrictions on length of high-voltage secondary conductors.

600-8. Enclosures. Live parts other than lamps and neon tubing shall be enclosed.

Exception: A transformer or electronic power supply provided with an integral enclosure, including a primary and secondary circuit splice enclosure, shall not be required to be provided with an additional enclosure.

(a) Strength. Enclosures shall have ample structural strength and rigidity.

(b) Material. Sign and outline lighting system enclosures shall be constructed of metal or shall be listed.

(c) Minimum Thickness of Enclosure Metal. Sheet copper or aluminum shall be at least 0.020 in. (508 μm) thick. Sheet steel shall be at least 0.016 in. (406 μm) (No. 28 MSG) thick.

(d) Protection of Metal. Metal parts of equipment shall be protected from corrosion.

600-9. Location.

(a) Vehicles. Sign or outline lighting system equipment shall be at least 14 ft (4.4 m) above areas accessible to vehicles unless protected from physical damage.

(b) Pedestrians. Neon tubing accessible to pedestrians, other than dry-location portable signs, shall be protected from physical damage.

(c) Adjacent to Combustible Materials. Signs and outline lighting systems shall be installed so that adjacent combustible materials shall not be subjected to temperatures in excess of 90°C (194°F).

The spacing between wood or other combustible materials and an incandescent or HID lamp or lampholder shall not be less than 2 in. (50.8 mm).

(d) Wet Location. Signs and outline lighting system equipment for wet location use, other than listed watertight type, shall be weatherproof and have drain holes, as necessary, in accordance with the following.

(1) Drain holes shall not be larger than ½ in. (12.7 mm) or smaller than ¼ in. (6.35 mm).
(2) Every low point or isolated section of the equipment shall have at least one drain hole.
(3) Drain holes shall be positioned such that there will be no external obstructions.

600-10. Portable or Mobile Signs.

(a) Support. Portable or mobile signs shall be adequately supported and readily movable without the use of tools.

(b) Attachment Plug. An attachment plug shall be provided for each portable or mobile sign.

(c) Wet or Damp Location. Portable or mobile signs in wet or damp locations shall meet all of the following.

(1) Cords. All cords shall be junior hard service or hard service types as designated in Table 400-4, and have an equipment grounding conductor.

(2) Ground-Fault Circuit Interrupter. Portable or mobile signs shall be provided with factory-installed ground-fault circuit-interrupter protection for personnel. The ground-fault circuit interrupter shall be an integral part of the attachment plug or shall be located in the power-supply cord within 12 in. (305 mm) of the attachment plug.

(d) Dry Location. Portable or mobile signs in dry locations shall meet the following.

(1) Cords shall be SP-2, SPE-2, SPT-2, or heavier, as designated in Table 400-4.
(2) The cord shall not exceed 15 ft (4.57 m) in length.

600-21. Ballasts, Transformers, and Electronic Power Supplies.

(a) Accessibility. Ballasts, transformers, and electronic power supplies shall be located where accessible and shall be securely fastened in place.

(b) Location. Ballasts, transformers, and electronic power supplies shall be installed as near to the lamps or neon tubing as practicable to keep the secondary conductors as short as possible.

(c) Wet Location. Ballasts, transformers, and electronic power supplies used in wet locations shall be of the weather-proof type or be of the outdoor type and protected from the weather by placement in a sign body or separate enclosure.

(d) Working Space. A working space at least 3 ft (914 mm) high, 3 ft (914 mm) wide, by 3 ft (914 mm) deep shall be provided at each ballast, transformer, and electronic power supply or its enclosure where not installed in a sign.

(e) Attic Locations. Ballasts, transformers, and electronic power supplies shall be permitted to be located in attics and soffits, provided there is an access door at least 3 ft (914 mm) by 2 ft (610 mm) and a passageway of at least 3 ft (914 mm) high by 2 ft (610 mm) wide with a suitable permanent walkway at least 12 in. (305 mm) wide extending from the point of entry to each component.

(f) Suspended Ceilings. Ballasts, transformers, and electronic power supplies shall be permitted to be located above suspended ceilings, provided their enclosures are securely fastened in place and not dependent on the suspended ceiling grid for support.

600-22. Ballasts.

(a) Type. Ballasts shall be identified for the use and shall be listed.

(b) Thermal Protection. Ballasts shall be thermally protected.

600-23. Transformers and Electronic Power Supplies.

(a) Type. Transformers and electronic power supplies shall be identified for the use and shall be listed.

(b) Secondary-Circuit Ground-Fault Protection. Transformers and electronic power supplies other than the following shall have secondary-circuit ground-fault protection:

(1) Transformers with isolated secondaries and with a maximum open circuit voltage of 7500 volts or less
(2) Transformers with integral porcelain or glass secondary housing for the neon tubing and requiring no field wiring of the secondary circuit

(c) Voltage. Secondary-circuit voltage shall not exceed 15,000 volts, nominal, under any load condition. The voltage to ground of any output terminals of the secondary circuit shall not exceed 7500 volts, under any load conditions.

(d) Rating. Transformers and electronic power supplies shall have a secondary-circuit current rating of not more than 300 mA.

(e) Secondary Connections. Secondary circuit outputs shall not be connected in parallel or in series.

(f) Marking. A transformer or power supply must be marked to indicate that it has secondary fault protection.

B. Field-Installed Skeleton Tubing

600-30. Applicability. Part B of this article shall apply only to field-installed skeleton tubing. These requirements are in addition to the requirements of Part A.

600-31. Neon Secondary-Circuit Conductors, 1000 Volts or Less, Nominal.

(a) Wiring Method. Conductors shall be installed using any wiring method included in Chapter 3 suitable for the conditions.

(b) Insulation and Size. Conductors shall be insulated, listed for the purpose, and not smaller than No. 18.

(c) Number of Conductors in Raceway. The number of conductors in a raceway shall be in accordance with Table 1 of Chapter 9.

(d) Installation. Conductors shall be installed so they are not subject to physical damage.

(e) Protection of Leads. Bushings shall be used to protect wires passing through an opening in metal.

600-32. Neon Secondary Circuit Conductors, Over 1000 Volts, Nominal.

(a) Wiring Method. Conductors shall be installed on insulators, in rigid metal conduit, intermediate metal conduit, rigid nonmetallic conduit, liquidtight flexible nonmetallic conduit, flexible metal conduit, liquidtight flexible metal conduit, electrical metallic tubing, metal enclosures, or other equipment listed for the purpose. Wiring methods shall be installed in accordance with the requirements of Chapter 3. Only one conductor shall be installed per length of conduit or tubing. Nonmetallic conduit or flexible nonmetallic conduit, when operated at 100 Hz or less, shall be spaced at least 1½ in. (38 mm) from grounded or bonded parts; and, when operated at over 100 Hz, shall be spaced at least 1¾ in. (44.45 mm) from grounded or bonded parts. Metal parts of a building shall not be used as a grounded or equipment grounding conductor.

(b) Insulation and Size. Conductors shall be insulated, listed for the purpose, rated for the voltage, not smaller than No. 18, and have a minimum temperature rating of 105°C (221°F).

(c) Installation. Conductors shall be installed so they are not subject to physical damage.

(d) Bends in Conductors. Sharp bends in insulated conductors shall be avoided.

(e) Spacing. Conductors shall be separated from each other and from all objects other than insulators or neon tubing by a spacing of not less than 1½ in. (38 mm). GTO cable installed in metal conduit or tubing requires no spacing between the cable insulation and the conduit or tubing.

(f) Insulators and Bushings. Insulators and bushings for conductors shall be listed for the purpose.

(g) Conductors in Raceways.

(1) Damp or Wet Locations. In damp or wet locations, the insulation on all conductors shall extend not less than 4 in. (102 mm) beyond the metal conduit or tubing.

(2) Dry Locations. In dry locations, the insulation on all conductors shall extend not less than 2½ in. (64 mm) beyond the metal conduit or tubing.

(h) Between Neon Tubing and Grounded Midpoint. Conductors shall be permitted to run from the ends of neon tubing to the grounded midpoint of transformers or electronic power supplies listed for the purpose and provided with terminals at the midpoint. Where such connections are made to the grounded midpoint, the connections between the high-voltage terminals and the line ends of the neon tubing shall be as short as possible.

(i) Dwelling Occupancies. Equipment having an open circuit voltage exceeding 1000 volts shall not be installed in or on dwelling occupancies.

(j) Length of High-Voltage Cable. Not more than 20 ft (7 m) of high-voltage cable shall be permitted in metal conduit or tubing from a high-voltage terminal of a transformer/ power supply to the first neon tube. Not more than 50 ft (15.2 m) of high-voltage cable shall be permitted in nonmetallic conduit from a high-voltage terminal of a transformer/power supply to the first neon tube.

600-41. Neon Tubing.

(a) Design. The length and design of the tubing shall not cause a continuous overcurrent beyond the design loading of the transformer or electronic power supply.

(b) Support. Tubing shall be supported by listed tube supports.

(c) Spacing. A spacing of not less than ¼ in. (6.5 mm) shall be maintained between the tubing and the nearest surface, other than its support.

600-42. Electrode Connections.

(a) Accessibility. Terminals of the electrode shall not be accessible to unqualified persons.

(b) Electrode Connections. Connections shall be made by use of a connection device, twisting the wires together, or use of an electrode receptacle. Connections shall be electrically and mechanically secure and shall be in an enclosure listed for the purpose.

(c) Support. The neon tubing and conductor shall be supported not more than 6 in. (152 mm) from the electrode connection.

(d) Receptacles. Electrode receptacles shall be listed for the purpose.

(e) Bushings. Where electrodes penetrate an enclosure, bushings listed for the purpose shall be used, unless receptacles are provided.

(f) Wet Locations. A listed cap shall be used to close the opening between neon tubing and a receptacle where the receptacle penetrates a building. Where a bushing or neon tubing penetrates a building, the opening between neon tubing and the bushing shall be sealed.

(g) Electrode Enclosures. Electrode enclosures shall be listed for the purpose.

Article 604 — Manufactured Wiring Systems

604-1. Scope. The provisions of this article apply to field-installed wiring using off-site manufactured subassemblies for branch circuits, remote-control circuits, signaling circuits, and communications circuits in accessible areas.

604-2. Definition.

Manufactured Wiring System. A system containing component parts that are assembled in the process of manufacture and cannot be inspected at the building site without damage or destruction to the assembly.

604-3. Other Articles. Except as modified by the requirements of this article, all other applicable articles of this *Code* shall apply.

604-4. Uses Permitted. The manufactured wiring systems shall be permitted in accessible and dry locations and in plenums and spaces used for environmental air, where listed for this application and installed in accordance with Section 300-22.

Exception No. 1: In concealed spaces, one end of tapped cable shall be permitted to extend into hollow walls for direct termination at switch and outlet points.

Exception No. 2: For use in outdoor locations where listed for the purpose.

604-5. Uses Not Permitted. Where conductors or cables are limited by the provisions in Articles 333 and 334.

604-6. Construction.

(a) Cable or Conduit Types.

(1) Cable shall be listed armored cable or metal-clad cable containing nominal 600-volt No. 10 or 12 copper-insulated conductors with a bare or insulated copper equipment grounding conductor equivalent in size to the ungrounded conductor.

(2) Conduit shall be listed flexible metal conduit or listed liquidtight flexible conduit containing nominal 600-volt No. 10 or 12 copper-insulated conductors with a bare or insulated copper equipment grounding conductor equivalent in size to the ungrounded conductor.

Exception No. 1 to (1) and (2): A fixture tap, maximum 6 ft (1.83 m) long, intended for connection to a single fixture shall be permitted to contain conductors smaller than No. 12 but not smaller than No. 18.

Exception No. 2 to (1) and (2): Conductors smaller than No. 12 shall be permitted for remote-control, signaling, or communications circuits. The assembly shall be listed for the purpose.

Exception No. 3 to (1) and (2): Flexible cord suitable for hard usage, with minimum No. 12 conductors, shall be permitted as part of a listed factory-made assembly not exceeding 6 ft (1.83 m) in length when making a transition between components of a manufactured wiring system and utilization equipment, not permanently secured to the building structure. The cord shall be visible for its entire length and shall not be subject to strain or physical damage.

(3) Each section shall be marked to identify the type of cable or conduit.

(b) Receptacles and Connectors. Receptacles and connectors shall be of the locking type, uniquely polarized and identified for the purpose, and shall be part of a listed assembly for the appropriate system.

(c) Other Component Parts. Other component parts shall be listed for the appropriate system.

604-7. Unused Outlets. All unused outlets shall be capped to effectively close the connector openings.

Article 605 — Office Furnishings (Consisting of Lighting Accessories and Wired Partitions)

605-1. Scope. This article covers electrical equipment, lighting accessories, and wiring systems used to connect, or contained within, or installed on relocatable wired partitions.

605-2. General. Wiring systems shall be identified as suitable for providing power for lighting accessories and appliances in wired partitions. These partitions shall not extend from floor to ceiling.

Exception: Where permitted by the authority having jurisdiction, these relocatable wired partitions shall be permitted to extend to the ceiling but shall not penetrate the ceiling.

(a) Use. These assemblies shall be installed and used only as provided for by this article.

(b) Other Articles. Except as modified by the requirements of this article, all other articles of this *Code* shall apply.

(c) Hazardous (Classified) Locations. Where used in hazardous (classified) locations, these assemblies shall conform with Articles 500 through 517 in addition to this article.

605-3. Wireways. All conductors and connections shall be contained within wiring channels of metal or other material identified as suitable for the conditions of use. Wiring channels shall be free of projections or other conditions that may damage conductor insulation.

605-4. Partition Interconnections. The electrical connection between partitions shall be a flexible assembly identified for use with wired partitions or shall be permitted to be installed using flexible cord provided all the following conditions are met.

(1) The cord is extra-hard usage type.
(2) The partitions are mechanically contiguous.
(3) The cord is not longer than necessary for maximum positioning of the partitions but is in no case to exceed 2 ft (610 mm).
(4) The cord is terminated at an attachment plug and cord connector with strain relief.

605-5. Lighting Accessories. Lighting equipment listed and identified for use with wired partitions shall comply with all of the following.

(a) Support. A means for secure attachment or support shall be provided.

(b) Connection. Where cord and plug connection is provided, the cord length shall be suitable for the intended application but shall not exceed 9 ft (2.74 m) in length. The cord shall not be smaller than No. 18, shall contain an equipment grounding conductor, and shall be of the hard usage type. Connection by other means shall be identified as suitable for the condition of use.

(c) Receptacle Outlet. Convenience receptacles shall not be permitted in lighting accessories.

605-6. Fixed-Type Partitions. Wired partitions that are fixed (secured to building surfaces) shall be permanently connected to the building electrical system by one of the wiring methods of Chapter 3.

605-7. Freestanding-Type Partitions. Partitions of the freestanding type (not fixed) shall be permitted to be permanently connected to the building electrical system by one of the wiring methods of Chapter 3.

605-8. Freestanding-Type Partitions, Cord and Plug Connected. Individual partitions of the freestanding type, or groups of individual partitions that are electrically connected, mechanically contiguous, and do not exceed 30 ft (9.14 m) when assembled, shall be permitted to be connected to the building electrical system by a single flexible cord and plug, provided all of the following conditions are met.

(a) Flexible Power-Supply Cord. The flexible power-supply cord shall be extra-hard usage type with No. 12 or larger conductors with an insulated equipment grounding conductor and not exceeding 2 ft (610 mm) in length.

(b) Receptacle Supplying Power. The receptacle(s) supplying power shall be on a separate circuit serving only panels and no other loads and shall be located not more than 12 in. (305 mm) from the partition that is connected to it.

(c) Receptacle Outlets, Maximum. Individual partitions or groups of interconnected individual partitions shall not contain more than thirteen 15-ampere, 125-volt receptacle outlets.

(d) Multiwire Circuits, Not Permitted. Individual partitions or groups of interconnected individual partitions shall not contain multiwire circuits.

> FPN: See Section 210-4 for circuits supplying partitions in Sections 605-6 and 605-7.

Article 610 — Cranes and Hoists

A. General

610-1. Scope. This article covers the installation of electrical equipment and wiring used in connection with cranes, monorail hoists, hoists, and all runways.

> FPN: For further information, see *Safety Code for Cranes, Derricks, Hoists, Jacks, and Slings,* ANSI B-30.

610-2. Special Requirements for Particular Locations.

(a) Hazardous (Classified) Locations. All equipment that operates in a hazardous (classified) location shall conform to Article 500.

(1) Equipment used in locations that are hazardous because of the presence of flammable gases or vapors shall conform to Article 501.

(2) Equipment used in locations that are hazardous because of combustible dust shall conform to Article 502.

(3) Equipment used in locations that are hazardous because of the presence of easily ignitible fibers or flyings shall conform to Article 503.

(b) Combustible Materials. Where a crane, hoist, or monorail hoist operates over readily combustible material, the resistors shall be located as permitted in the following:

(1) In a well-ventilated cabinet composed of noncombustible material constructed so that it will not emit flames or molten metal

(2) In a cage or cab constructed of noncombustible material that encloses the sides of the cage or cab from the floor to a point at least 6 in. (152 mm) above the top of the resistors

(c) Electrolytic Cell Lines. See Section 668-32.

B. Wiring

610-11. Wiring Method. Conductors shall be enclosed in raceways or be Type AC cable with insulated grounding conductor, Type MC cable, or Type MI cable unless otherwise permitted in (a) through (e).

(a) Contact conductors are not required to be enclosed in raceways.

(b) Short lengths of open conductors at resistors, collectors, and other equipment are not required to be enclosed in raceways.

(c) Where flexible connections are necessary to motors and similar equipment, flexible stranded conductors shall be installed in flexible metal conduit, liquidtight flexible metal conduit, liquidtight flexible nonmetallic conduit, multiconductor cable, or an approved nonmetallic enclosure.

(d) Where multiconductor cable is used with a suspended pushbutton station, the station shall be supported in some satisfactory manner that protects the electric conductors against strain.

(e) Where flexibility is required for power or control to moving parts, a cord suitable for the purpose shall be permitted provided

(1) Suitable strain relief and protection from physical damage is provided, and
(2) In Class 1, Division 2 locations, cord shall be approved for extra-hard usage.

610-12. Raceway or Cable Terminal Fittings. Conductors leaving raceways or cables shall comply with one of the following.

(a) Separately Bushed Hole. A box or terminal fitting that has a separately bushed hole for each conductor shall be used wherever a change is made from a raceway or cable to open wiring. A fitting used for this purpose shall not contain taps or splices and shall not be used at fixture outlets.

(b) Bushing in Lieu of a Box. A bushing shall be permitted to be used in lieu of a box at the end of a rigid metal conduit, intermediate metal conduit, or electrical metallic tubing where the raceway terminates at unenclosed controls or similar equipment including contact conductors, collectors, resistors, brakes, power-circuit limit switches, and dc split-frame motors.

610-13. Types of Conductors. Conductors shall comply with Table 310-13 unless otherwise permitted in (a) through (d).

(a) A conductor(s) exposed to external heat or connected to resistors shall have a flame-resistant outer covering or be covered with flame-resistant tape individually or as a group.

(b) Contact conductors along runways, crane bridges, and monorails shall be permitted to be bare, and shall be copper, aluminum, steel, or other alloys or combinations thereof in the form of hard drawn wire, tees, angles, tee rails, or other stiff shapes.

(c) Where flexibility is required, flexible cord or cable shall be permitted to be used and, where necessary, cable reels or take-up devices shall be used.

(d) Conductors for Class 1, Class 2, and Class 3 remote-control, signaling, and power-limited circuits, installed in accordance with Article 725, shall be permitted.

610-14. Rating and Size of Conductors.

(a) Ampacity. The allowable ampacities of conductors shall be as shown in Table 610-14(a).

FPN: For the ampacities of conductors between controllers and resistors, see Section 430-23.

(b) Secondary Resistor Conductors. Where the secondary resistor is separate from the controller, the minimum size of the conductors between controller and resistor shall be calculated by multiplying the motor secondary current by the appropriate factor from Table 610-14(b) and selecting a wire from Table 610-14(a).

(c) Minimum Size. Conductors external to motors and controls shall not be smaller than No. 16 unless otherwise permitted in (1) and (2).

(1) No. 18 wire in multiconductor cord shall be permitted for control circuits at not over 7 amperes.

(2) Wires not smaller than No. 20 shall be permitted for electronic circuits.

Table 610-14(a). Ampacities of Insulated Copper Conductors Used with Short-Time Rated Crane and Hoist Motors. Based on Ambient Temperature of 30°C (86°F). Up to Four Conductors in Raceway or Cable[1] Up to 3 ac[2] or 4 dc[1] Conductors in Raceway or Cable

Maximum Operating Temperature	75°C (167°F)		90°C (194°F)		125°C (257°F)		Maximum Operating Temperature
	Types MTW, RH, RHW, THW, THWN, XHHW, USE, ZW		Types TA, TBS, SA, SIS, PFA, FEP, FEPB, RHH, THHN, XHHW, Z, ZW		Types FEP, FEPB, PFA, PFAH, SA, TFE, Z, ZW		
Size (AWG or kcmil)	60 Min	30 Min	60 Min	30 Min	60 Min	30 Min	Size (AWG or kcmil)
16	10	12	—	—	—	—	16
14	25	26	31	32	38	40	14
12	30	33	36	40	45	50	12
10	40	43	49	52	60	65	10
8	55	60	63	69	73	80	8
6	76	86	83	94	101	119	6
5	85	95	95	106	115	134	5
4	100	117	111	130	133	157	4
3	120	141	131	153	153	183	3
2	137	160	148	173	178	214	2
1	143	175	158	192	210	253	1
1/0	190	233	211	259	253	304	1/0
2/0	222	267	245	294	303	369	2/0
3/0	280	341	305	372	370	452	3/0
4/0	300	369	319	399	451	555	4/0
250	364	420	400	461	510	635	250
300	455	582	497	636	587	737	300
350	486	646	542	716	663	837	350
400	538	688	593	760	742	941	400
450	600	765	660	836	818	1042	450
500	660	847	726	914	896	1143	500

AMPACITY CORRECTION FACTORS

Ambient Temperature (°C)	For ambient temperatures other than 30°C (86°F), multiply the ampacities shown above by the appropriate factor shown below.						Ambient Temperature (°F)
21–25	1.05	1.05	1.04	1.04	1.02	1.02	70–77
26–30	1.00	1.00	1.00	1.00	1.00	1.00	79–86
31–35	0.94	0.94	0.96	0.96	0.97	0.97	88–95
36–40	0.88	0.88	0.91	0.91	0.95	0.95	97–104
41–45	0.82	0.82	0.87	0.87	0.92	0.92	106–113
46–50	0.75	0.75	0.82	0.82	0.89	0.89	115–122
51–55	0.67	0.67	0.76	0.76	0.86	0.86	124–131
56–60	0.58	0.58	0.71	0.71	0.83	0.83	133–140
61–70	0.33	0.33	0.58	0.58	0.76	0.76	142–158
71–80	—	—	0.41	0.41	0.69	0.69	160–176
81–90	—	—	—	—	0.61	0.61	177–194
91–100	—	—	—	—	0.51	0.51	195–212
101–120	—	—	—	—	0.40	0.40	213–248

Note: Other insulations shown in Table 310-13 and approved for the temperature and location shall be permitted to be substituted for those shown in Table 610-14(a). The allowable ampacities of conductors used with 15-minute motors shall be the 30-minute ratings increased by 12 percent.

[1]For 5 to 8 simultaneously energized power conductors in raceway or cable, the ampacity of each power conductor shall be reduced to a value of 80 percent of that shown in the table.

[2]For 4 to 6 simultaneously energized 125°C (257°F) ac power conductors in raceway or cable, the ampacity of each power conductor shall be reduced to a value of 80 percent of that shown in the table.

Table 610-14(b). Secondary Conductor Rating Factors

Time in Seconds		Ampacity of Wire in Percent of Full-Load Secondary Current
On	Off	
5	75	35
10	70	45
15	75	55
15	45	65
15	30	75
15	15	85
Continuous Duty		110

(d) Contact Conductors. Contact wires shall have an ampacity not less than that required by Table 610-14(a) for 75°C (167°F) wire, and in no case shall they be smaller than the following:

Distance Between End Strain Insulators or Clamp-Type Intermediate Supports (ft)	Size of Wire (AWG)
Less than 30	6
30–60	4
Over 60	2

Note: For SI units, 1 ft = 0.3048 m.

(e) Calculation of Motor Load.

(1) For one motor, use 100 percent of motor nameplate full-load ampere rating.

(2) For multiple motors on a single crane or hoist, the minimum ampacity of the power supply conductors shall be the nameplate full-load ampere rating of the largest motor or group of motors for any single crane motion, plus 50 percent of the nameplate full-load ampere rating of the next largest motor or group of motors, using that column of Table 610-14(a) that applies to the longest time-rated motor.

(3) For multiple cranes or hoists, or both, supplied by a common conductor system, compute the motor minimum ampacity for each crane as defined in Section 610-14(e), add them together, and multiply the sum by the appropriate demand factor from Table 610-14(e).

(f) Other Loads. Additional loads, such as heating, lighting, and air conditioning, shall be provided for by application of the appropriate sections of this *Code*.

(g) Nameplate. Each crane, monorail, or hoist shall be provided with a visible nameplate marked with the manufacture's name, the rating in volts, frequency, number of phases, and circuit amperes as calculated in Sections 610-14(e) and (f).

Table 610-14(e). Demand Factors

Number of Cranes or Hoists	Demand Factor
2	0.95
3	0.91
4	0.87
5	0.84
6	0.81
7	0.78

610-15. Common Return. Where a crane or hoist is operated by more than one motor, a common-return conductor of proper ampacity shall be permitted.

C. Contact Conductors

610-21. Installation of Contact Conductors. Contact conductors shall comply with (a) through (h).

(a) Locating or Guarding Contact Conductors. Runway contact conductors shall be guarded, and bridge contact conductors shall be located or guarded in a manner that persons cannot inadvertently touch energized current-carrying parts.

(b) Contact Wires. Wires that are used as contact conductors shall be secured at the ends by means of approved strain insulators and shall be mounted on approved insulators so that the extreme limit of displacement of the wire will not bring the latter within less than 1½ in. (38 mm) from the surface wired over.

(c) Supports Along Runways. Main contact conductors carried along runways shall be supported on insulating supports placed at intervals not exceeding 20 ft (6.1 m) unless otherwise permitted in (f).

Such conductors shall be separated not less than 6 in. (152 mm), other than for monorail hoists where a spacing of not less than 3 in. (76 mm) shall be permitted. Where necessary, intervals between insulating supports shall be permitted to be increased up to 40 ft (12.2 m), the separation between conductors being increased proportionately.

(d) Supports on Bridges. Bridge wire contact conductors shall be kept at least 2½ in. (64 mm) apart, and, where the span exceeds 80 ft (24.4 m), insulating saddles shall be placed at intervals not exceeding 50 ft (15.2 m).

(e) Supports for Rigid Conductors. Conductors along runways and crane bridges, that are of the rigid type specified in Section 610-13(c) and not contained within an approved enclosed assembly, shall be carried on insulating supports spaced at intervals of not more than 80 times the vertical dimension of the conductor, but in no case greater than 15 ft

(4.57 m), and spaced apart sufficiently to give a clear electrical separation of conductors or adjacent collectors of not less than 1 in. (25.4 mm).

(f) Track as Circuit Conductor. Monorail, tram rail, or crane runway tracks shall be permitted as a conductor of current for one phase of a 3-phase, ac system furnishing power to the carrier, crane, or trolley, provided all of the following conditions are met.

(1) The conductors supplying the other two phases of the power supply are insulated.
(2) The power for all phases is obtained from an insulating transformer.
(3) The voltage does not exceed 300 volts.
(4) The rail serving as a conductor is effectively grounded at the transformer and also shall be permitted to be grounded by the fittings used for the suspension or attachment of the rail to a building or structure.

(g) Electrical Continuity of Contact Conductors. All sections of contact conductors shall be mechanically joined to provide a continuous electrical connection.

(h) Not to Supply Other Equipment. Contact conductors shall not be used as feeders for any equipment other than the crane(s) or hoist(s) that they are primarily designed to serve.

610-22. Collectors. Collectors shall be designed so as to reduce to a minimum sparking between them and the contact conductor; and, where operated in rooms used for the storage of easily ignitible combustible fibers and materials, they shall comply with Section 503-13.

D. Disconnecting Means

610-31. Runway Conductor Disconnecting Means. A disconnecting means that has a continuous ampere rating not less than that computed in Sections 610-14(e) and (f) shall be provided between the runway contact conductors and the power supply. Such disconnecting means shall consist of a motor-circuit switch, circuit breaker, or molded case switch. This disconnecting means shall be as follows:

(1) Readily accessible and operable from the ground or floor level
(2) Arranged to be locked in the open position
(3) Open all ungrounded conductors simultaneously
(4) Placed within view of the runway contact conductors

610-32. Disconnecting Means for Cranes and Monorail Hoists. A motor-circuit switch or circuit breaker arranged to be locked in the open position shall be provided in the leads from the runway contact conductors or other power supply on all cranes and monorail hoists. A motor-circuit switch, circuit breaker, or molded case switch arranged to be locked in the open position shall be provided.

Where a monorail hoist or hand-propelled crane bridge installation meets all of the following, the disconnecting means shall be permitted to be omitted.

(1) The unit is controlled from the ground or floor level.
(2) The unit is within view of the power supply disconnecting means.
(3) No fixed work platform has been provided for servicing the unit.

Where the disconnecting means is not readily accessible from the crane or monorail hoist operating station, means shall be provided at the operating station to open the power circuit to all motors of the crane or monorail hoist.

610-33. Rating of Disconnecting Means. The continuous ampere rating of the switch or circuit breaker required by Section 610-32 shall not be less than 50 percent of the combined short-time ampere rating of the motors, nor less than 75 percent of the sum of the short-time ampere rating of the motors required for any single motion.

E. Overcurrent Protection

610-41. Feeders, Runway Conductors.

(a) Single Feeder. The runway supply conductors and main contact conductors of a crane or monorail shall be protected by an overcurrent device(s) that shall not be greater than the largest rating or setting of any branch-circuit protective device plus the sum of the nameplate ratings of all the other loads with application of the demand factors from Table 610-14(e).

(b) More than One Feeder Circuit. Where more than one feeder circuit is installed to supply runway conductors, each feeder circuit shall be sized and protected in compliance with Section 610-41(a).

610-42. Branch-Circuit Short-Circuit and Ground-Fault Protection. Branch circuits shall be protected as follows.

(a) Fuse or Circuit Breaker Rating. Crane, hoist, and monorail hoist motor branch circuits shall be protected by fuses or inverse-time circuit breakers that have a rating in accordance with Table 430-152. Taps to control circuits shall be permitted to be taken from the load side of a branch-circuit protective device, provided each tap and piece of equipment is properly protected.

Where two or more motors operate a single motion, the sum of their nameplate current ratings shall be considered as a single motor current in the above calculations.

Two or more motors shall be permitted to be connected to the same branch circuit if no tap conductor to an individual motor has an ampacity less than one-third that of the branch circuit and if each motor is protected from overload according to Section 610-43.

(b) Taps to Brake Coils. Taps to brake coils do not require separate overcurrent protection.

610-43. Overload Protection.

(a) Motor and Branch-Circuit Overload Protection. Each motor, motor controller, and branch-circuit conductor shall be protected from overload by one of the following means.

(1) A single motor shall be considered as protected where the branch-circuit overcurrent device meets the rating requirements of Section 610-42.
(2) Overload relay elements in each ungrounded circuit conductor, with all relay elements protected from short circuit by the branch-circuit protection.
(3) Thermal sensing devices, sensitive to motor temperature or to temperature and current, that are thermally in contact with the motor winding(s). A hoist or trolley is considered to be protected if the sensing device is connected in the hoist's upper limit switch circuit so as to prevent further hoisting during an overload condition of either motor.

(b) Manually Controlled Motor. If the motor is manually controlled, with spring return controls, the overload protective device shall not be required to protect the motor against stalled rotor conditions.

(c) Multimotor. Where two or more motors drive a single trolley, truck, or bridge and are controlled as a unit and protected by a single set of overload devices with a rating equal to the sum of their rated full-load currents. A hoist or trolley shall be considered to be protected if the sensing device is connected in the hoist's upper limit switch circuit so as to prevent further hoisting during an overtemperature condition of either motor.

(d) Hoists and Monorail Hoists. Hoists and monorail hoists and their trolleys that are not used as part of an overhead traveling crane shall not require individual motor overload protection, provided the largest motor does not exceed 7½ hp and all motors are under manual control of the operator.

F. Control

610-51. Separate Controllers. Each motor shall be provided with an individual controller unless otherwise permitted in (a) or (b).

(a) Where two or more motors drive a single hoist, carriage, truck, or bridge, they shall be permitted to be controlled by a single controller.

(b) One controller shall be permitted to be switched between motors, provided the following:

(1) The controller has a horsepower rating that is not lower than the horsepower rating of the largest motor.
(2) Only one motor is operated at one time.

610-53. Overcurrent Protection. Conductors of control circuits shall be protected against overcurrent. Control circuits shall be considered as protected by overcurrent devices that are rated or set at not more than 300 percent of the ampacity of the control conductors, unless otherwise permitted in (a) or (b).

(a) Taps to control transformers shall be considered as protected where the secondary circuit is protected by a device rated or set at not more than 200 percent of the rated secondary current of the transformer and not more than 200 percent of the ampacity of the control circuit conductors.

(b) Where the opening of the control circuit would create a hazard, as for example, the control circuit of a hot metal crane, the control circuit conductors shall be considered as being properly protected by the branch-circuit overcurrent devices.

610-55. Limit Switch. A limit switch or other device shall be provided to prevent the load block from passing the safe upper limit of travel of all hoisting mechanisms.

610-57. Clearance. The dimension of the working space in the direction of access to live parts that are likely to require examination, adjustment, servicing, or maintenance while energized shall be a minimum of 2½ ft (762 mm). Where controls are enclosed in cabinets, the door(s) shall either open at least 90 degrees or be removable.

G. Grounding

610-61. Grounding. All exposed noncurrent-carrying metal parts of cranes, monorail hoists, hoists, and accessories, including pendant controls, shall be metallically joined together into a continuous electrical conductor so that the entire crane or hoist will be grounded in accordance with Article 250. Moving parts, other than removable accessories or attachments, that have metal-to-metal bearing surfaces shall be considered to be electrically connected to each other through the bearing surfaces for grounding purposes. The trolley frame and bridge frame shall be considered as electrically grounded through the bridge and trolley wheels and its respective tracks unless local conditions, such as paint or other insulating material, prevent reliable metal-to-metal contact. In this case, a separate bonding conductor shall be provided.

Article 620 — Elevators, Dumbwaiters, Escalators, Moving Walks, Wheelchair Lifts, and Stairway Chair Lifts

A. General

620-1. Scope. This article covers the installation of electrical equipment and wiring used in connection with elevators, dumbwaiters, escalators, moving walks, wheelchair lifts, and stairway chair lifts.

> FPN No. 1: For further information, see *Safety Code for Elevators and Escalators,* ASME/ANSI A17.1-1996.

> FPN No. 2: For further information, see *Elevator and Escalator Electrical Equipment Certification Standard,* ASME/ANSI A17.5-1996 (CSA B44.1-1996).

620-2. Definitions.

Control System. The overall system governing the starting, stopping, direction of motion, acceleration, speed, and retardation of the moving member.

Controller, Motion. The electric device(s) for that part of the control system that governs the acceleration, speed, retardation, and stopping of the moving member.

Controller, Motor. The operative units of the control system comprised of the starter device(s) and power conversion equipment used to drive an electric motor, or the pumping unit used to power hydraulic control equipment.

Controller, Operation. The electric device(s) for that part of the control system that initiates the starting, stopping, and direction of motion in response to a signal from an operating device.

Operating Device. The car switch, push buttons, key or toggle switch(s), or other devices used to activate the operation controller.

Signal Equipment. Includes audible and visual equipment such as chimes, gongs, lights, and displays that convey information to the user.

> FPN No. 1: The motor controller, motion controller, and operation controller may be located in a single enclosure or a combination of enclosures.

> FPN No. 2: Figure 620-2 is for information only.

Figure 620-2 Control system.

620-3. Voltage Limitations. The supply voltage shall not exceed 300 volts between conductors unless otherwise permitted in (a) through (c).

(a) Power Circuits. Branch circuits to door operator controllers and door motors and branch circuits and feeders to motor controllers, driving machine motors, machine brakes, and motor-generator sets shall not have a circuit voltage in excess of 600 volts. Internal voltages of power conversion and functionally associated equipment, including the interconnecting wiring, shall be permitted to have higher voltages provided that all such equipment and wiring shall be listed for the higher voltages. Where the voltage exceeds 600 volts, warning labels or signs that read "DANGER — HIGH VOLTAGE" shall be attached to the equipment and shall be plainly visible.

(b) Lighting Circuits. Lighting circuits shall comply with the requirements of Article 410.

(c) Heating and Air-Conditioning Circuits. Branch circuits for heating and air-conditioning equipment located on the elevator car shall not have a circuit voltage in excess of 600 volts.

620-4. Live Parts Enclosed. All live parts of electrical apparatus in the hoistways, at the landings, in or on the cars of elevators and dumbwaiters, in the wellways or the landings of escalators or moving walks, or in the runways and machinery spaces of wheelchair lifts and stairway chair lifts shall be enclosed to protect against accidental contact.

FPN: See Section 110-27 for guarding of live parts (600 volts, nominal, or less).

620-5. Working Clearances. Working space shall be provided about controllers, disconnecting means, and other electrical equipment. The minimum working space shall not be less than that specified in Section 110-26(a).

Where conditions of maintenance and supervision ensure that only qualified persons will examine, adjust, service, and maintain the equipment, the clearance requirements of Section 110-26(a) shall be waived as permitted in (a) through (d).

(a) Flexible Connections to Equipment. Electrical equipment in (1) through (4) is provided with flexible leads to all external connections so that it can be repositioned to meet the clear working space requirements of Section 110-26(a).

(1) Controllers and disconnecting means for dumbwaiters, escalators, moving walks, wheelchair lifts, and stairway chair lifts installed in the same space with the driving machine

(2) Controllers and disconnecting means for elevators installed in the hoistway or on the car
(3) Controllers for door operators
(4) Other electrical equipment installed in the hoistway or on the car

(b) Guards. Live parts of the electrical equipment are suitably guarded, isolated, or insulated, and the equipment can be examined, adjusted, serviced, or maintained while energized without removal of this protection.

FPN: See definition of *Exposed* in Article 100.

(c) Examination, Adjusting, and Servicing. Electrical equipment is not required to be examined, adjusted, serviced, or maintained while energized.

(d) Low Voltage. Uninsulated parts are at a voltage not greater than 30 volts rms, 42 volts peak, or 60 volts dc.

B. Conductors

620-11. Insulation of Conductors. The insulation of conductors shall comply with (a) through (d).

FPN: One method of determining that conductors are flame retardant is by testing the conductors to the VW-1 (Vertical-Wire) Flame Test in the *Reference Standard for Electrical Wires, Cables, and Flexible Cords,* ANSI/UL 1581-1991.

(a) Hoistway Door Interlock Wiring. The conductors to the hoistway door interlocks from the hoistway riser shall be flame retardant and suitable for a temperature of not less than 200°C (392°F). Conductors shall be Type SF or equivalent.

(b) Traveling Cables. Traveling cables used as flexible connections between the elevator or dumbwaiter car or counterweight and the raceway shall be of the types of elevator cable listed in Table 400-4 or other approved types.

(c) Other Wiring. All conductors in raceways shall have flame-retardant insulation.

Conductors shall be Type MTW, TF, TFF, TFN, TFFN, THHN, THW, THWN, TW, XHHW, hoistway cable, or any other conductor with insulation designated as flame retardant. Shielded conductors shall be permitted, if such conductors are insulated for the maximum nominal circuit voltage applied to any conductor within the cable or raceway system.

(d) Insulation. All conductors shall have an insulation voltage rating equal to at least the maximum nominal circuit voltage applied to any conductor within the enclosure, cable, or raceway. Insulations and outer coverings that are designated with the suffix "/LS" and are so listed shall be permitted.

620-12. Minimum Size of Conductors. The minimum size of conductors, other than conductors that form an integral part of control equipment, shall be as follows.

(a) Traveling Cables.

(1) For lighting circuits: No. 14 copper. No. 20 copper or larger conductors shall be permitted in parallel, provided the ampacity is equivalent to at least that of No. 14 copper.

(2) For other circuits, No. 20 copper.

(b) Other Wiring. No. 24 copper. Smaller size listed conductors shall be permitted.

620-13. Feeder and Branch-Circuit Conductors. Conductors shall have an ampacity in accordance with (a) through (d). With generator field control, the conductor ampacity shall be based on the nameplate current rating of the driving motor of the motor-generator set that supplies power to the elevator motor.

FPN No. 1: The heating of conductors depends on root-mean-square current values, which, with generator field control, are reflected by the nameplate current rating of the motor-generator driving motor rather than by the rating of the elevator motor, which represents actual but short-time and intermittent full-load current values.

FPN No. 2: See Figure 620-13.

To supply	
Common feeder	620-13
Feeder demand factor	620-14
Motor feeder short-circuit and ground-fault protection	620-61(c)
Disconnecting means	620-51
Motor branch-circuit short-circuit and ground-fault protection	620-61(d)
Branch-circuit conductors	620-13
External transformer (optional)	Article 450
Motor controller	620-13, 620-15
Motor control circuits	620-61(a)
Motor overload protection	620-61(b)
Motor conductors	620-13
Motor	Article 430, Part A
Thermal protection	Article 430, Part C

Figure 620-13.

(a) Conductors Supplying Single Motor. Conductors supplying a single motor shall have an ampacity not less than the percentage of motor nameplate current determined from Sections 430-22(a) and (b).

FPN: Elevator motor currents, or those of similar functions, may exceed the nameplate value, but since they are inherently intermittent duty, and the heating of the motor and conductors is dependent on the root-mean-square (rms) current value, conductors are sized for duty cycle service as shown in Table 430-22(b).

(b) Conductors Supplying a Single Motor Controller. Conductors supplying a single motor controller shall have an ampacity not less than the motor controller nameplate current rating, plus all other connected loads.

FPN: Motor controller nameplate current rating may be derived based on the rms value of the motor current using an intermittent duty cycle and other control system loads, if applicable.

(c) Conductors Supplying a Single Power Transformer. Conductors supplying a single power transformer shall have an ampacity not less than the nameplate current rating of the power transformer plus all other connected loads.

FPN No. 1: The nameplate current rating of a power transformer supplying a motor controller reflects the nameplate current rating of the motor controller at line voltage (transformer primary).

FPN No. 2: See Appendix D, Example No. D10.

(d) Conductors Supplying More than One Motor, Motor Controller, or Power Transformer. Conductors supplying more than one motor, motor controller, or power transformer shall have an ampacity not less than the sum of the nameplate current ratings of the equipment plus all other connected loads. The ampere ratings of motors to be used in the summation shall be determined from Table 430-22(b), and Sections 430-24 and 430-24, Exception No. 1.

FPN: See Appendix D, Examples No. D9 and D10.

620-14. Feeder Demand Factor. Feeder conductors of less ampacity than required by Section 620-13 shall be permitted subject to the requirements of Table 620-14.

FPN: Demand factors are based on 50 percent duty cycle (i.e., half time on and half time off).

620-15. Motor Controller Rating. The motor controller rating shall comply with Section 430-83. The rating shall be permitted to be less than the nominal rating of the elevator motor, when the controller inherently limits the available power to the motor and is marked as power limited.

FPN: For controller markings, see Section 430-8.

C. Wiring

620-21. Wiring Methods. Conductors and optical fibers located in hoistways, in escalator and moving walk well-

Table 620-14. Feeder Demand Factors for Elevators

Number of Elevators on a Single Feeder	Demand Factor
1	1.00
2	0.95
3	0.90
4	0.85
5	0.82
6	0.79
7	0.77
8	0.75
9	0.73
10 or more	0.72

ways, in wheelchair lifts, stairway chair lift runways, and machinery spaces, in or on cars, and in machine and control rooms, not including the traveling cables connecting the car or counterweight and hoistway wiring, shall be installed in rigid metal conduit, intermediate metal conduit, electrical metallic tubing, rigid nonmetallic conduit, or wireways, or shall be Type MC, MI, or AC cable unless otherwise permitted in (a) through (c).

(a) Elevators.

(1) Hoistways.

(a) Flexible metal conduit, liquidtight flexible metal conduit, or liquidtight flexible nonmetallic conduit shall be permitted in hoistways between risers and limit switches, interlocks, operating buttons, and similar devices.

(b) Cables used in Class 2 power-limited circuits shall be permitted to be installed between risers and signal equipment and operating devices provided the cables are supported and protected from physical damage and are of a jacketed and flame-retardant type.

(2) Cars.

(a) Flexible metal conduit, liquidtight flexible metal conduit, or liquidtight flexible nonmetallic conduit of ³⁄₈-in. nominal trade size or larger, not exceeding 6 ft (1.83 m) in length shall be permitted on cars where located so as to be free from oil and if securely fastened in place.

Exception: Liquidtight flexible nonmetallic conduit of ³⁄₈-in. nominal trade size or larger, as defined by Section 351-22(2), shall be permitted in lengths in excess of 6 ft (1.83 m).

(b) Hard-service cords and junior hard-service cords that conform to the requirements of Article 400 (Table 400-4) shall be permitted as flexible connections between the fixed wiring on the car and devices on the car doors or gates. Hard-service cords only shall be permitted as flexible connections for the top-of-car operating device or the car-top work light. Devices or fixtures shall be grounded by means of an equipment grounding conductor run with the circuit

conductors. Cables with smaller conductors and other types and thicknesses of insulation and jackets shall be permitted as flexible connections between the fixed wiring on the car and devices on the car doors or gates, if listed for this use.

(c) Flexible cords and cables that are components of listed equipment and used in circuits operating at 30 volts rms or less or 42 volts dc or less shall be permitted in lengths not to exceed 6 ft (1.83 m) provided the cords and cables are supported and protected from physical damage and are of a jacketed and flame-retardant type.

(d) Flexible metal conduit, liquidtight flexible metal conduit, liquidtight flexible nonmetallic conduit or flexible cords and cables, or conductors grouped together and taped or corded that are part of listed equipment, a driving machine, or a driving machine brake shall be permitted on the car assembly, in lengths not to exceed 6 ft (1.83 m) without being installed in a raceway and where located to be protected from physical damage and are of a flame-retardant type.

(3) Machine Room and Machinery Spaces.

(a) Flexible metal conduit, liquidtight flexible metal conduit, or liquidtight flexible nonmetallic conduit of ³⁄₈-in. nominal trade size or larger, not exceeding 6 ft (1.83 m) in length, shall be permitted between control panels and machine motors, machine brakes, motor-generator sets, disconnecting means, and pumping unit motors and valves.

Exception: Liquidtight flexible nonmetallic conduit, as defined in Section 351-22(2), shall be permitted to be installed in lengths in excess of 6 ft (1.83 m).

(b) Where motor-generators, machine motors, or pumping unit motors and valves are located adjacent to or underneath control equipment and are provided with extra-length terminal leads not exceeding 6 ft (1.83 m) in length, such leads shall be permitted to be extended to connect directly to controller terminal studs without regard to the carrying-capacity requirements of Articles 430 and 445. Auxiliary gutters shall be permitted in machine and control rooms between controllers, starters, and similar apparatus.

(c) Flexible cords and cables that are components of listed equipment and used in circuits operating at 30 volts rms or less or 42 volts dc or less shall be permitted in lengths not to exceed 6 ft (1.83 m) provided the cords and cables are supported and protected from physical damage and are of a jacketed and flame-retardant type.

(d) On existing or listed equipment, conductors shall also be permitted to be grouped together and taped or corded without being installed in a raceway. Such cable groups shall be supported at intervals not over 3 ft (914 mm) and located so as to be protected from physical damage.

(4) Counterweight. Flexible metal conduit, liquidtight flexible metal conduit, liquidtight flexible nonmetallic conduit or flexible cords and cables, or conductors grouped together and taped or corded that are part of listed equipment, a driving machine, or a driving machine brake shall be permitted on the counterweight assembly, in lengths not to exceed 6 ft (1.83 m) without being installed in a raceway and where located to be protected from physical damage and are of a flame-retardant type.

(b) Escalators.

   (1) Flexible metal conduit, liquidtight flexible metal conduit, or liquidtight flexible nonmetallic conduit shall be permitted in escalator and moving walk wellways. Flexible metal conduit or liquidtight flexible conduit of ⅜-in. nominal trade size shall be permitted in lengths not in excess of 6 ft (1.83 m).

Exception: ⅜-in. nominal trade size or larger liquidtight flexible nonmetallic conduit, as defined in Section 351-22(2), shall be permitted to be installed in lengths in excess of 6 ft (1.83 m).

   (2) Cables used in Class 2 power-limited circuits shall be permitted to be installed within escalators and moving walkways provided the cables are supported and protected from physical damage and are of a jacketed and flame-retardant type.

   (3) Hard-service cords that conform to the requirements of Article 400 (Table 400-4) shall be permitted as flexible connections on escalators and moving walk control panels and disconnecting means where the entire control panel and disconnecting means are arranged for removal from machine spaces as permitted in Section 620-5.

(c) Wheelchair Lifts and Stairway Chair Lift Raceways.

   (1) Flexible metal conduit or liquidtight flexible metal conduit shall be permitted in wheelchair lifts and stairway chair lift runways and machinery spaces. Flexible metal conduit or liquidtight flexible conduit of ⅜-in. nominal trade size shall be permitted in lengths not in excess of 6 ft (1.83 m).

Exception: ⅜-in. nominal trade size or larger liquidtight flexible nonmetallic conduit, as defined in Section 351-22(2), shall be permitted to be installed in lengths in excess of 6 ft (1.83 m).

   (2) Cables used in Class 2 power-limited circuits shall be permitted to be installed within wheelchair lifts and stairway chair lift runways and machinery spaces provided the cables are supported and protected from physical damage and are of a jacketed and flame-retardant type.

620-22. Branch Circuits for Car Lighting, Receptacle(s), Ventilation, Heating, and Air Conditioning.

(a) Car Light Source. A separate branch circuit shall supply the car lights, receptacle(s), auxiliary lighting power source, and ventilation on each elevator car. The overcurrent device protecting the branch circuit shall be located in the elevator machine room/machinery space.

(b) Air-Conditioning and Heating Source. A dedicated branch circuit shall supply the air-conditioning and heating units on each elevator car. The overcurrent device protecting the branch circuit shall be located in the elevator machine room/machinery space.

620-23. Branch Circuit for Machine Room/Machinery Space Lighting and Receptacle(s).

(a) A separate branch circuit shall supply the machine room/machinery space lighting and receptacle(s).
   Required lighting shall not be connected to the load side of a ground-fault circuit interrupter.

(b) The machine room lighting switch shall be located at the point of entry to such machine rooms/machinery spaces.

(c) At least one 125-volt, single-phase, duplex receptacle shall be provided in each machine room and machinery space.

   FPN: See *Safety Code for Elevators and Escalators,* ANSI/ASME A17.1-1996, for illumination levels.

620-24. Branch Circuit for Hoistway Pit Lighting and Receptacle(s).

(a) A separate branch circuit shall supply the hoistway pit lighting and receptacle(s).
   Required lighting shall not be connected to the load side of a ground-fault circuit interrupter.

(b) The lighting switch shall be located so as to be readily accessible from the pit access door.

(c) At least one 125-volt, single-phase, duplex receptacle shall be provided in the hoistway pit.

   FPN: See *Safety Code for Elevators and Escalators*, ANSI/ASME A17.1-1996, for illumination levels.

D. Installation of Conductors

620-32. Metal Wireways and Nonmetallic Wireways. The sum of the cross-sectional area of the individual conductors in a wireway shall not be more than 50 percent of the interior cross-sectional area of the wireway.
   Vertical runs of wireways shall be securely supported at intervals not exceeding 15 ft (4.57 m) and shall have not more than one joint between supports. Adjoining wireway sections shall be securely fastened together to provide a rigid joint.

620-33. Number of Conductors in Raceways. The sum of the cross-sectional area of the individual conductors in raceways shall not exceed 40 percent of the interior cross-sectional area of the raceway, except as permitted in Section 620-32 for wireways.

620-34. Supports. Supports for cables or raceways in a hoistway or in an escalator or moving walk wellway or wheelchair lift and stairway chair lift runway shall be securely fastened to the guide rail; escalator or moving walk truss; or to the hoistway, wellway, or runway construction.

620-35. Auxiliary Gutters. Auxiliary gutters shall not be subject to the restrictions of Section 374-2 as to length or of Section 374-5 as to number of conductors.

620-36. Different Systems in One Raceway or Traveling Cable. Optical fiber cables and conductors for operating devices, operation and motion control, power, signaling, fire alarm, lighting, heating, and air-conditioning circuits of 600 volts or less shall be permitted to be run in the same traveling cable or raceway system if all conductors are insulated for the maximum voltage applied to any conductor within the cables or raceway system and if all live parts of the equipment are insulated from ground for this maximum voltage. Such a traveling cable or raceway shall also be permitted to include shielded conductors and/or one or more coaxial cables, if such conductors are insulated for the maximum voltage applied to any conductor within the cable or raceway system. Conductors shall be permitted to be covered with suitable shielding for telephone, audio, video, or higher frequency communications circuits.

620-37. Wiring in Hoistways and Machine Rooms.

(a) Uses Permitted. Only such electric wiring, raceways, and cables used directly in connection with the elevator or dumbwaiter, including wiring for signals; for communication with the car; for lighting, heating, air conditioning, and ventilating the elevator car; for fire detecting systems; for pit sump pumps; and for heating, lighting, and ventilating the hoistway, shall be permitted inside the hoistway and the machine room.

(b) Lightning Protection. Bonding of elevator rails (car and/or counterweight) to a lightning protection system grounding down conductor(s), shall be permitted. The lightning protection system grounding down conductor(s) shall not be located within the hoistway. Elevator rails or other hoistway equipment shall not be used as the grounding down conductor for lightning protection systems.

> FPN: See Section 250-106 for bonding requirements. For further information, see *Standard for the Installation of Lightning Protection Systems,* NFPA 780-1997.

(c) Main Feeders. Main feeders for supplying power to elevators and dumbwaiters shall be installed outside the hoistway unless as follows.

(1) By special permission, feeders for elevators shall be permitted within an existing hoistway if no conductors are spliced within the hoistway.
(2) Feeders shall be permitted inside the hoistway for elevators with driving machine motors located in the hoistway or on the car or counterweight.

620-38. Electrical Equipment in Garages and Similar Occupancies. Electrical equipment and wiring used for elevators, dumbwaiters, escalators, moving walks, and wheelchair lifts and stairway chair lifts in garages shall comply with the requirements of Article 511.

> FPN: Garages used for parking or storage and where no repair work is done in accordance with Section 511-2 are not classified.

E. Traveling Cables

620-41. Suspension of Traveling Cables. Traveling cables shall be suspended at the car and hoistways' ends, or counterweight end where applicable, so as to reduce the strain on the individual copper conductors to a minimum.

Traveling cables shall be supported by one of the following means:

(1) By its steel supporting member(s)
(2) By looping the cables around supports for unsupported lengths less than 100 ft (30.5 m)
(3) By suspending from the supports by a means that automatically tightens around the cable when tension is increased for unsupported lengths up to 200 ft (61 m)

> FPN: Unsupported length for the hoistway suspension means is that length of cable as measured from the point of suspension in the hoistway to the bottom of the loop, with the elevator car located at the bottom landing. Unsupported length for the car suspension means is that length of cable as measured from the point of suspension on the car to the bottom of the loop, with the elevator car located at the top landing.

620-42. Hazardous (Classified) Locations. In hazardous (classified) locations, traveling cables shall be of a type approved for hazardous (classified) locations and shall comply with Sections 501-11, 502-12, or 503-10, as applicable.

620-43. Location of and Protection for Cables. Traveling cable supports shall be located so as to reduce to a minimum the possibility of damage due to the cables coming in contact with the hoistway construction or equipment in the hoistway.

Where necessary, suitable guards shall be provided to protect the cables against damage.

620-44. Installation of Traveling Cables. Traveling cable shall be permitted to be run without the use of a raceway for a distance not exceeding 6 ft (1.83 m) in length as measured from the first point of support on the elevator car or hoistway wall, or counterweight where applicable, provided the conductors are grouped together and taped or corded, or in the original sheath.

Traveling cables shall be permitted to be continued to elevator controller enclosures and to elevator car and machine room connections, as fixed wiring, provided they are suitably supported and protected from physical damage.

F. Disconnecting Means and Control

620-51. Disconnecting Means. A single means for disconnecting all ungrounded main power supply conductors for each unit shall be provided and be designed so that no pole can be operated independently. Where multiple driving machines are connected to a single elevator, escalator, moving walk, or pumping unit, there shall be one disconnecting means to disconnect the motor(s) and control valve operating magnets.

The disconnecting means for the main power supply conductors shall not disconnect the branch circuit required in Sections 620-22, 620-23, and 620-24.

(a) Type. The disconnecting means shall be an enclosed externally operable fused motor circuit switch or circuit breaker capable of being locked in the open position. The disconnecting means shall be a listed device.

> FPN: For additional information, see *Safety Code for Elevators and Escalators,* ASME/ANSI A17.1-1996.

(b) Operation. No provision shall be made to open or close this disconnecting means from any other part of the premises. If sprinklers are installed in hoistways, machine rooms, or machinery spaces, the disconnecting means shall be permitted to automatically open the power supply to the affected elevator(s) prior to the application of water. No provision shall be made to automatically close this disconnecting means. Power shall only be restored by manual means.

> FPN: To reduce hazards associated with water on live elevator electrical equipment.

(c) Location. The disconnecting means shall be located where it is readily accessible to qualified persons.

(1) On Elevators Without Generator Field Control. On elevators without generator field control, the disconnecting means shall be located within sight of the motor controller. Driving machines or motion and operation controllers not within sight of the disconnecting means shall be provided with a manually operated switch installed in the control circuit to prevent starting. The manually operated switch(s) shall be installed adjacent to this equipment.

Where the driving machine is located in a remote machinery space, a single means for disconnecting all ungrounded main power supply conductors shall be provided and be capable of being locked in the open position.

(2) On Elevators with Generator Field Control. On elevators with generator field control, the disconnecting means shall be located within sight of the motor controller for the driving motor of the motor-generator set. Driving machines, motor-generator sets, or motion and operation controllers not within sight of the disconnecting means shall be provided with a manually operated switch installed in the control circuit to prevent starting. The manually operated switch(s) shall be installed adjacent to this equipment.

Where the driving machine or the motor-generator set is located in a remote machinery space, a single means for disconnecting all ungrounded main power supply conductors, shall be provided and be capable of being locked in the open position.

(3) On Escalators and Moving Walks. On escalators and moving walks, the disconnecting means shall be installed in the space where the controller is located.

(4) On Wheelchair Lifts and Stairway Chair Lifts. On wheelchair lifts and stairway chair lifts, the disconnecting means shall be located within sight of the motor controller.

(d) Identification and Signs. Where there is more than one driving machine in a machine room, the disconnecting means shall be numbered to correspond to the identifying number of the driving machine that they control.

The disconnecting means shall be provided with a sign to identify the location of the supply side overcurrent protective device.

620-52. Power from More than One Source.

(a) Single-car and Multicar Installations. On single-car and multicar installations, equipment receiving electrical power from more than one source shall be provided with a disconnecting means for each source of electrical power. The disconnecting means shall be within sight of the equipment served.

(b) Warning Sign for Multiple Disconnecting Means. Where multiple disconnecting means are used and parts of the controllers remain energized from a source other than the one disconnected, a warning sign shall be mounted on or next to the disconnecting means. The sign shall be clearly legible and shall read

WARNING — PARTS OF THE CONTROLLER ARE
NOT DE-ENERGIZED BY THIS SWITCH.

(c) Interconnection Multicar Controllers. Where interconnections between controllers are necessary for the operation of the system on multicar installations that remain energized from a source other than the one disconnected, a warning sign in accordance with Section 620-52(b) shall be mounted on or next to the disconnecting means.

620-53. Car Light, Receptacle(s), and Ventilation Disconnecting Means. Elevators shall have a single means for disconnecting all ungrounded car light, receptacle(s), and ventilation power-supply conductors for that elevator car.

The disconnecting means shall be capable of being locked in the open position and shall be located in the machine room for that elevator car. Where there is no machine room, the disconnecting means shall be located in the machinery space for that elevator car.

Where there is equipment for more than one elevator car in the machine room, the disconnecting means shall be numbered to correspond to the identifying number of the elevator car whose light source they control.

The disconnecting means shall be provided with a sign to identify the location of the supply side overcurrent protective device.

620-54. Heating and Air-Conditioning Disconnecting Means. Elevators shall have a single means for disconnecting all ungrounded car heating and air-conditioning power-supply conductors for that elevator car.

The disconnecting means shall be capable of being locked in the open position and shall be located in the machine room for that elevator car. Where there is no machine room, the disconnecting means shall be located in the machinery space for that elevator car.

Where there is equipment for more than one elevator car in the machine room, the disconnecting means shall be numbered to correspond to the identifying number of the elevator car whose heating and air conditioning source they control.

The disconnecting means shall be provided with a sign to identify the location of the supply side overcurrent protective device.

G. Overcurrent Protection

620-61. Overcurrent Protection. Overcurrent protection shall be provided as follows.

(a) Operating Devices, Control, and Signaling Circuits. Operating devices, control and signaling circuits shall be protected against overcurrent in accordance with the requirements of Sections 725-23 and 725-24.

Class 2 power-limited circuits shall be protected against overcurrent in accordance with the requirements of Chapter 9, Notes to Tables 11(a) and 11(b).

(b) Overload Protection for Motors.

(1) Duty on elevator and dumbwaiter driving machine motors and driving motors of motor-generators used with generator field control shall be rated as intermittent. Such motors shall be protected against overload in accordance with Section 430-33.

(2) Duty on escalator and moving walk driving machine motors shall be rated as continuous. Such motors shall be protected against overload in accordance with Section 430-32.

(3) Escalator and moving walk driving machine motors and driving motors of motor-generator sets shall be protected against running overload as provided in Table 430-37.

(4) Duty on wheelchair lift and stairway chair lift driving machine motors shall be rated as intermittent. Such motors shall be protected against overload in accordance with Section 430-33.

FPN: For further information, see Section 430-44 for orderly shutdown.

(c) Motor Feeder Short-Circuit and Ground-Fault Protection. Motor feeder short-circuit and ground-fault protection shall be as required in Article 430, Part E.

(d) Motor Branch-Circuit Short-Circuit and Ground-Fault Protection. Motor branch-circuit short-circuit and ground-fault protection shall be as required in Article 430, Part D.

620-62. Selective Coordination. Where more than one driving machine disconnecting means is supplied by a single feeder, the overcurrent protective devices in each disconnecting means shall be selectively coordinated with any other supply side overcurrent protective devices.

H. Machine Room

620-71. Guarding Equipment. Elevator, dumbwaiter, escalator, and moving walk driving machines; motor-generator sets; motor controllers; and disconnecting means shall be installed in a room or enclosure set aside for that purpose unless otherwise permitted in (a) or (b). The room or enclosure shall be secured against unauthorized access.

(a) Motor Controllers. Motor controllers shall be permitted outside the spaces herein specified, provided they are in enclosures with doors or removable panels that are capable of being locked in the closed position and the disconnecting means is located adjacent to or is an integral part of the motor controller. Motor controller enclosures for escalator or moving walks shall be permitted in the balustrade on the side located away from the moving steps or moving treadway. If the disconnecting means is an integral part of the motor controller, it shall be operable without opening the enclosure.

(b) Driving Machines. Elevators with driving machines located on the car, counterweight, or in the hoistway, and driving machines for dumbwaiters, wheelchair lifts, and stairway lifts shall be permitted outside the spaces herein specified.

J. Grounding

620-81. Metal Raceways Attached to Cars. Metal raceways, Type MC cable, Type MI cable, or Type AC cable attached to elevator cars shall be bonded to grounded metal parts of the car that they contact.

620-82. Electric Elevators. For electric elevators, the frames of all motors, elevator machines, controllers, and the metal enclosures for all electrical equipment in or on the car or in the hoistway shall be grounded in accordance with Article 250.

620-83. Nonelectric Elevators. For elevators other than electric having any electric conductors attached to the car, the metal frame of the car, where normally accessible to persons, shall be grounded in accordance with Article 250.

620-84. Escalators, Moving Walks, Wheelchair Lifts, and Stairway Chair Lifts. Escalators, moving walks, wheelchair lifts, and stairway chair lifts shall comply with Article 250.

620-85. Ground-Fault Circuit-Interrupter Protection for Personnel. Each 125-volt, single-phase, 15- and 20-ampere receptacle installed in pits, on elevator car tops, and in escalator and moving walk wellways shall be of the ground-fault circuit-interrupter type.

All 125-volt, single-phase, 15- and 20-ampere receptacles installed in machine rooms and machinery spaces shall have ground-fault circuit-interrupter protection for personnel.

A single receptacle supplying a permanently installed sump pump shall not require ground-fault circuit-interrupter protection.

K. Emergency and Standby Power Systems

620-91. Emergency and Standby Power Systems. An elevator(s) shall be permitted to be powered by an emergency or standby power system.

FPN: See ASME/ANSI A17.1-1996, Rule 211.2, and CAN/CSA-B44-1994, Clause 3.12.13, for additional information.

(a) Regenerative Power. For elevator systems that regenerate power back into the power source, which is unable to absorb the regenerative power under overhauling elevator load conditions, a means shall be provided to absorb this power.

(b) Other Building Loads. Other building loads, such as power and lighting shall be permitted as the energy absorption means required in (a) provided that such loads are automatically connected to the emergency or standby power system operating the elevators and are large enough to absorb the elevator regenerative power.

(c) Disconnecting Means. The disconnecting means required by Section 620-51 shall disconnect the elevator from both the emergency or standby power system and the normal power system.

Where an additional power source is connected to the load side of the disconnecting means, the disconnecting means required in Section 620-51 shall be provided with an auxiliary contact that is positively opened mechanically and the opening shall not be solely dependent on springs. This contact shall cause the additional power source to be disconnected from its load when the disconnecting means is in the open position.

Article 625 — Electric Vehicle Charging System

A. General

625-1. Scope. The provisions of this article cover the electrical conductors and equipment external to an electric vehicle that connect an electric vehicle to a supply of electricity by conductive or inductive means, and the installation of equipment and devices related to electric vehicle charging.

FPN: For industrial trucks, see *Fire Safety Standard for Powered Industrial Trucks Including Type Designations, Areas of Use, Conversions, Maintenance, and Operation,* NFPA 505-1996.

625-2. Definitions.

Electric Vehicle. An automotive-type vehicle for highway use, such as passenger automobiles, buses, trucks, vans, and the like, primarily powered by an electric motor that draws current from a rechargeable storage battery, fuel cell, photovoltaic array, or other source of electric current. For the purpose of this article, electric motorcycles and similar type vehicles and off-road self-propelled electric vehicles, such as industrial trucks, hoists, lifts, transports, golf carts, airline ground support equipment, tractors, boats, and the like, are not included.

Electric Vehicle Connector. A device that by insertion into an electric vehicle inlet, establishes an electrical connection to the electric vehicle for the purpose of charging and information exchange. This is part of the electric vehicle coupler.

Electric Vehicle Coupler. A mating electric vehicle inlet and electric vehicle connector set.

Electric Vehicle Inlet. The device on the electric vehicle into which the electric vehicle connector is inserted for charging and information exchange. This is part of the electric vehicle coupler. For the purposes of this *Code*, the electric vehicle inlet is considered to be part of the electric vehicle and not part of the electric vehicle supply equipment.

Electric Vehicle Nonvented Storage Battery. A hermetically-sealed battery comprised of one or more rechargeable electrochemical cells that has no provision for release of excessive gas pressure, or the addition of water or electrolyte, or for external measurements of electrolyte specific gravity.

Electric Vehicle Supply Equipment. The conductors, including the ungrounded, grounded, and equipment grounding conductors and the electric vehicle connectors, attachment plugs, and all other fittings, devices, power outlets, or apparatuses installed specifically for the purpose of delivering energy from the premises wiring to the electric vehicle.

Personnel Protection System. A system of personnel protection devices and constructional features that when used together provide protection against electric shock of personnel.

625-3. Other Articles. Wherever the requirements of other articles of this *Code* and Article 625 differ, the requirements of Article 625 shall apply.

625-4. Voltages. Unless other voltages are specified, the nominal ac system voltages of 120, 120/240, 208Y/120, 240, 480Y/277, 480, 600Y/347, and 600 volts shall be used to supply equipment covered by this article.

625-5. Listed or Labeled. All electrical materials, devices, fittings, and associated equipment shall be listed or labeled.

B. Wiring Methods

625-9. Electric Vehicle Coupler. The electric vehicle coupler shall comply with (a) through (f).

(a) Polarization. The electric vehicle coupler shall be polarized unless part of a system identified and listed as suitable for the purpose.

(b) Noninterchangeability. The electric vehicle coupler shall have a configuration that is noninterchangeable with wiring devices in other electrical systems. Nongrounding-type electric vehicle couplers shall not be interchangeable with grounding-type electric vehicle couplers.

(c) Construction and Installation. The electric vehicle coupler shall be constructed and installed so as to guard against inadvertent contact by persons with parts made live from the electric vehicle supply equipment or the electric vehicle battery.

(d) Unintentional Disconnection. The electric vehicle coupler shall be provided with a positive means to prevent unintentional disconnection.

(e) Grounding Pole. The electric vehicle coupler shall be provided with a grounding pole, unless part of a system identified and listed as suitable for the purpose in accordance with Article 250.

(f) Grounding Pole Requirements. If a grounding pole is provided, the electric vehicle coupler shall be designed so that the grounding pole connection is the first to make and the last to break contact.

C. Equipment Construction

625-13. Electric Vehicle Supply Equipment. Electric vehicle supply equipment rated at 125 volt, single phase, 15 or 20 amperes or part of a system identified and listed as suitable for the purpose and meeting the requirements of Sections 625-18, 625-19, and 625-29 shall be permitted to be cord and plug connected. All other electric vehicle supply equipment shall be permanently connected and fastened in place. This equipment shall have no exposed live parts.

625-14. Rating. Electric vehicle supply equipment shall have sufficient rating to supply the load served. For the purposes of this article, electric vehicle charging loads shall be considered to be continuous loads.

625-15. Markings. The electric vehicle supply equipment shall comply with (a) through (c).

(a) General. All electric vehicle supply equipment shall be marked by the manufacturer

FOR USE WITH ELECTRIC VEHICLES

(b) Ventilation Not Required. Where marking is required by Section 625-29(c), the electric vehicle supply equipment shall be clearly marked by the manufacturer

VENTILATION NOT REQUIRED

The marking shall be located so as to be clearly visible after installation.

(c) Ventilation Required. Where marking is required by Section 625-29(d), the electric vehicle supply equipment shall be clearly marked by the manufacturer "Ventilation Required." The marking shall be located so as to be clearly visible after installation.

625-16. Means of Coupling. The means of coupling to the electric vehicle shall be either conductive or inductive. Attachment plugs, electric vehicle connectors, and electric vehicle inlets shall be listed or labeled for the purpose.

625-17. Cable. The electric vehicle supply equipment cable shall be Type EV, EVJ, EVE, EVJE, EVT, or EVJT flexible cable as specified in Article 400 and Table 400-4. Ampacities shall be as specified in Table 400-5(A) for No. 10 and smaller and Table 400-5(B) for No. 8 and larger. The overall length of the cable shall not exceed 25 ft (7.63 m). Other cable types and assemblies listed as being suitable for the purpose, including optional hybrid communications, signal, and optical fiber cables, shall be permitted.

625-18. Interlock. Electric vehicle supply equipment shall be provided with an interlock that de-energizes the electric vehicle connector and its cable whenever the electric connector is uncoupled from the electric vehicle. An interlock shall not be required for portable cord- and plug-connected electric vehicle supply equipment intended for connection to receptacle outlets rated 125 volts, single phase, 15 and 20 amperes.

625-19. Automatic De-energization of Cable. The electric vehicle supply equipment or the cable-connector combination of the equipment shall be provided with an automatic means to de-energize the cable conductors and electric vehicle connector upon exposure to strain that could result in either cable rupture or separation of the cable from the electric connector and exposure of live parts. Automatic means to de-energize the cable conductors and electric vehicle connector shall not be required for portable cord- and plug-connected electric vehicle supply equipment intended for connection to receptacle outlets rated at 125 volts, single phase, 15 and 20 amperes.

D. Control and Protection

625-21. Overcurrent Protection. Overcurrent protection for feeders and branch circuits supplying electric vehicle supply equipment shall be sized for continuous duty and shall have a rating of not less than 125 percent of the maximum load of the electric vehicle supply equipment. Where noncontinuous loads are supplied from the same feeder or branch circuit, the overcurrent device shall have a rating of not less than the sum of the noncontinuous loads plus 125 percent of the continuous loads.

625-22. Personnel Protection System. The electric vehicle supply equipment shall have a listed system of protection against electric shock of personnel. The personnel protection system shall be composed of listed personnel protection devices and constructional features. Where cord- and plug-connected electric vehicle supply equipment is used, the interrupting device of a listed personnel protection system shall be provided and shall be an integral part of the attachment plug or shall be located in the power supply cable not more than 12 in. (305 mm) from the attachment plug.

625-23. Disconnecting Means. For electric vehicle supply equipment rated more than 60 amperes or more than 150 volts to ground, the disconnecting means shall be provided and installed in a readily accessible location. The disconnecting means shall be capable of being locked in the open position.

625-25. Loss of Primary Source. Means shall be provided such that upon loss of voltage from the utility or other electric system(s), energy cannot be backfed through the electric vehicle supply equipment to the premises wiring system. The electric vehicle shall not be permitted to serve as a standby power supply.

E. Electric Vehicle Supply Equipment Locations

625-28. Hazardous (Classified) Locations. Where electric vehicle supply equipment or wiring is installed in a hazardous (classified) location, the requirements of Articles 500 through 516 shall apply.

625-29. Indoor Sites. Indoor sites shall include, but not be limited to, integral, attached, and detached residential garages; enclosed and underground parking structures; repair and nonrepair commercial garages; and agricultural buildings.

(a) Location. The electric vehicle supply equipment shall be located to permit direct connection to the electric vehicle.

(b) Height. Unless specifically listed for the purpose and location, the coupling means of the electric vehicle supply equipment shall be stored or located at a height of not less than 18 in. (457 mm) and not more than 4 ft (1.22 m) above the floor level.

(c) Ventilation Not Required. Where electric vehicle non-vented storage batteries are used or where the electric vehicle supply equipment is listed or labeled as suitable for charging electric vehicles indoors without ventilation and marked in accordance with Section 625-15(b), mechanical ventilation shall not be required.

(d) Ventilation Required. Where the electric vehicle supply equipment is listed or labeled as suitable for charging electric vehicles that require ventilation for indoor charging and marked in accordance with Section 625-15(c), mechanical ventilation, such as a fan, shall be provided. The ventilation shall include both supply and exhaust equipment and shall be permanently installed and located to intake from, and vent directly to, the outdoors. Positive pressure ventilation systems shall only be permitted in buildings or areas that have been specifically designed and approved for that application. Mechanical ventilation requirements shall be determined by one of the following methods.

(1) For supply voltages and currents specified in Table 625-29(d), the minimum ventilation requirements in cubic feet per minute (cfm) shall be as specified in Table 625-29(d) for each of the total number of electric vehicles that can be charged at one time. This table allows for sufficient ventilation for any configuration of electric vehicle supply equipment and electric vehicle charging space.

(2) For supply voltages and currents other than specified in Table 625-29(d), the minimum ventilation requirements in cubic feet per minute (cfm) shall be calculated by means of the following general formulas as applicable.

(a) Single phase:

$$\text{Ventilation}_{\text{single phase}} = \frac{(\text{volts})(\text{amperes})}{48.7}$$

(b) Three phase:

$$\text{Ventilation}_{\text{three phase}} = \frac{1.732\,(\text{volts})(\text{amperes})}{48.7}$$

(3) For an electric vehicle supply equipment ventilation system designed by a person qualified to perform such calculations as an integral part of a building's total ventilation system, the minimum ventilation requirements shall be permitted to be determined per calculations specified in the engineering study.

The supply circuit to the mechanical ventilation equipment shall be electrically interlocked with the electric vehicle supply equipment and shall remain energized during the entire electric vehicle charging cycle. Electric vehicle supply equipment shall be marked in accordance with Section 625-15. Electric vehicle supply equipment receptacles rated at 125 volts, single phase, 15 and 20 amperes shall be marked in accordance with Section 625-15(c) and shall be switched and the mechanical ventilation system shall be electrically interlocked through the switch supply power to the receptacle.

For receptacles at each electric vehicle charging space identified "For Use with Electric Vehicles," and rated 125 volts, single phase, 15 and 20 amperes, such receptacles shall be switched, and mechanical ventilation shall be electrically

Table 625-29(d). Minimum Ventilation Required in Cubic Feet per Minute (cfm) for Each of the Total Number of Electric Vehicles that Can Be Charged at One Time

Branch-Circuit Ampere Rating	Branch-Circuit Voltage						
	Single Phase			3 Phase			
	120 V	208 V	240 V or 120/240 V	208 V or 208 Y/120 V	240 V	480 V or 480 Y/277 V	600 V or 600 Y/347 V
15	37	64	74	—	—	—	—
20	49	85	99	148	171	342	427
30	74	128	148	222	256	512	641
40	99	171	197	296	342	683	854
50	123	214	246	370	427	854	1066
60	148	256	296	444	512	1025	1281
100	246	427	493	740	854	1708	2135
150	—	—	—	1110	1281	2562	3203
200	—	—	—	1480	1708	3416	4270
250	—	—	—	1850	2135	4270	5338
300	—	—	—	2221	2562	5125	6406
350	—	—	—	2591	2989	5979	7473
400	—	—	—	2961	3416	6832	8541

interlocked through the switch supply power to the receptacle.

625-30. Outdoor Sites. Outdoor sites shall include, but not be limited to, residential carports and driveways, curbside, open parking structures, parking lots, and commercial charging facilities.

(a) **Location.** The electric vehicle supply equipment shall be located to permit direct connection to the electric vehicle.

(b) **Height.** Unless specifically listed for the purpose and location, the coupling means of electric vehicle supply equipment shall be stored or located at a height of not less than 24 in. (610 mm) and not more than 4 ft (1.22 m) above the parking surface.

Article 630 — Electric Welders

A. General

630-1. Scope. This article covers electric arc welding, resistance welding apparatus, and other similar welding equipment that is connected to an electric supply system.

B. Arc Welders

630-11. Ampacity of Supply Conductors. The ampacity of conductors for arc welders shall be as follows.

(a) Individual Welders. The ampacity of the supply conductors shall not be less than the $I_{1\text{eff}}$ value on the rating plate. Alternatively, if the $I_{1\text{eff}}$ is not given, the ampacity of the supply conductors shall not be less than the current value determined by multiplying the rated primary current in amperes given on the welder rating plate and the following factor based on the duty cycle of the welder.

| | Multiplier for Arc Welders | |
| | Nonmotor | Motor |
Duty Cycle	Generator	Generator
100	1.00	1.00
90	0.95	0.96
80	0.89	0.91
70	0.84	0.86
60	0.78	0.81
50	0.71	0.75
40	0.63	0.69
30	0.55	0.62
20 or less	0.45	0.55

(b) Group of Welders. The ampacity of conductors that supply a group of welders shall be permitted to be less than the sum of the currents, as determined in accordance with (a), of the welders supplied. The conductor rating shall be determined in each case according to the welder loading based on the use to be made of each welder and the allowance permissible in the event that all the welders supplied by the conductors will not be in use at the same time. The load value used for each welder shall take into account both the magnitude and the duration of the load while the welder is in use.

> FPN: Conductor ratings based on 100 percent of the current, as determined in accordance with (a), of the two largest welders, 85 percent for the third largest welder, 70 percent for the fourth largest welder, and 60 percent for all remaining welders, can be assumed to provide an ample margin of safety under high-production conditions with respect to the maximum permissible temperature of the conductors. Percentage values lower than those given are permissible in cases where the work is such that a high-operating duty cycle for individual welders is impossible.

630-12. Overcurrent Protection. Overcurrent protection for arc welders shall be as provided in (a) and (b). Where the values as determined by this section do not correspond with the standard ampere ratings provided in Section 240-6 or the rating or setting specified results in unnecessary opening of the overcurrent device, the next higher standard rating or setting shall be permitted.

(a) For Welders. Each welder shall have overcurrent protection rated or set at not more than 200 percent of $I_{1\text{max}}$.

Alternatively, if the $I_{1\text{max}}$ is not given, the overcurrent protection shall be rated or set at not more than 200 percent of the rated primary current of the welder.

An overcurrent device shall not be required for a welder that has supply conductors protected by an overcurrent device rated or set at not more than 200 percent of $I_{1\text{max}}$ or the rated primary current of the welder.

If the supply conductors for a welder are protected by an overcurrent device rated or set at not more than 200 percent of I_{max} or rated primary current of the welder, a separate overcurrent device shall not be required.

(b) For Conductors. Conductors that supply one or more welders shall be protected by an overcurrent device rated or set at not more than 200 percent of the conductor rating.

> FPN: $I_{1\text{max}}$ is the maximum value of the rated supply current at maximum rated output.
>
> $I_{1\text{eff}}$ is the maximum value of the effective supply current, calculated from the rated supply current (I_1), the corresponding duty cycle (duty factor) (X), and the supply current at no-load (I_0) by the following formula.
>
> $$I_{1\text{eff}} = \sqrt{I_1^2 X + I_0^2(1 - X)}$$

630-13. Disconnecting Means. A disconnecting means shall be provided in the supply circuit for each arc welder that is not equipped with a disconnect mounted as an integral part of the welder.

The disconnecting means shall be a switch or circuit breaker, and its rating shall not be less than that necessary to accommodate overcurrent protection as specified under Section 630-12.

630-14. Marking. A rating plate shall be provided for arc welders giving the following information:

(1) Name of manufacturer
(2) Frequency
(3) Number of phases
(4) Primary voltage
(5) $I_{1\text{max}}$ and $I_{1\text{eff}}$, or rated primary current
(6) Maximum open-circuit voltage
(7) Rated secondary current and
(8) Basis of rating, such as the duty cycle

C. Resistance Welders

630-31. Ampacity of Supply Conductors. The ampacity of the supply conductors for resistance welders necessary to limit the voltage drop to a value permissible for the satisfactory performance of the welder is usually greater than that required to prevent overheating as described in (a) and (b).

(a) Individual Welders. The rated ampacity for conductors for individual welders shall comply with the following.

(1) The ampacity of the supply conductors for a welder that may be operated at different times at different values of primary current or duty cycle shall not be less than 70 percent of the rated primary current for seam and automatically fed welders, and 50 percent of the rated primary current for manually operated nonautomatic welders.

(2) The ampacity of the supply conductors for a welder wired for a specific operation for which the actual primary current and duty cycle are known and remain unchanged shall not be less than the product of the actual primary current and the multiplier given below for the duty cycle at which the welder will be operated.

Duty Cycle (percent)	Multiplier
50	0.71
40	0.63
30	0.55
25	0.50
20	0.45
15	0.39
10	0.32
7.5	0.27
5 or less	0.22

(b) Groups of Welders. The ampacity of conductors that supply two or more welders shall not be less than the sum of the value obtained in accordance with (a) for the largest welder supplied and 60 percent of the values obtained for all the other welders supplied.

> FPN: **Explanation of Terms**
>
> (1) The *rated primary current* is the rated kilovolt-amperes (kVA) multiplied by 1000 and divided by the rated primary voltage, using values given on the nameplate.
> (2) The *actual primary current* is the current drawn from the supply circuit during each welder operation at the particular heat tap and control setting used.
> (3) The *duty cycle* is the percentage of the time during which the welder is loaded. For instance, a spot welder supplied by a 60-Hz system (216,000 cycles per hour) making four hundred 15-cycle welds per hour would have a duty cycle of 2.8 percent (400 multiplied by 15, divided by 216,000, multiplied by 100). A seam welder operating 2 cycles "on" and 2 cycles "off" would have a duty cycle of 50 percent.

630-32. Overcurrent Protection. Overcurrent protection for resistance welders shall be as provided in (a) and (b). Where the values as determined by this section do not correspond with the standard ampere ratings provided in Section 240-6 or the rating or setting specified results in unnecessary opening of the overcurrent device, the next higher standard rating or setting shall be permitted.

(a) For Welders. Each welder shall have an overcurrent device rated or set at not more than 300 percent of the rated primary current of the welder. If the supply conductors for a welder are protected by an overcurrent device rated or set at not more than 200 percent of the rated primary current of the welder, a separate overcurrent device shall not be required.

(b) For Conductors. Conductors that supply one or more welders shall be protected by an overcurrent device rated or set at not more than 300 percent of the conductor rating.

630-33. Disconnecting Means. A switch or circuit breaker shall be provided by which each resistance welder and its control equipment can be disconnected from the supply circuit. The ampere rating of this disconnecting means shall not be less than the supply conductor ampacity determined in accordance with Section 630-31. The supply circuit switch shall be permitted as the welder disconnecting means where the circuit supplies only one welder.

630-34. Marking. A nameplate shall be provided for each resistance welder giving the following information:

(1) Name of manufacturer
(2) Frequency
(3) Primary voltage
(4) Rated kilovolt-amperes (kVA) at 50 percent duty cycle
(5) Maximum and minimum open-circuit secondary voltage
(6) Short-circuit secondary current at maximum secondary voltage, and
(7) Specified throat and gap setting

D. Welding Cable

630-41. Conductors. Insulation of conductors intended for use in the secondary circuit of electric welders shall be flame retardant.

630-42. Installation. Cables shall be permitted to be installed in a dedicated cable tray as provided in (a), (b), and (c).

(a) Cable Support. The cable tray shall provide support at not greater than 6-in. (152-mm) intervals.

(b) Spread of Fire and Products of Combustion. The installation shall comply with Section 300-21.

(c) Signs. A permanent sign shall be attached to the cable tray at intervals not greater than 20 ft (6.1 m). The sign shall read

CABLE TRAY FOR WELDING CABLES ONLY

Article 640 — Audio Signal Processing, Amplification, and Reproduction Equipment

A. General

640-1. Scope. This article covers equipment and wiring for audio signal generation, recording, processing, amplification and reproduction; distribution of sound; public address; speech input systems; temporary audio system installations; and electronic organs or other electronic musical instruments. This also includes audio systems subject to Article 517, Part F, and Articles 518, 520, 525, and 530.

> FPN No. 1: Examples of permanently installed distributed audio system locations include, but are not limited to, restaurant, hotel, business office, commercial and retail sales environments, churches, and schools. Both portable and permanently installed equipment locations include, but are not limited to, residences, auditoriums, theaters, stadiums, and movie and television studios. Temporary installations include, but are not limited to, auditoriums, theaters, stadiums (which use both temporary and permanently installed systems), and outdoor events such as fairs, festivals, circuses, public events, and concerts.

> FPN No. 2: Fire and burglary alarm signaling devices are specifically not encompassed by this article.

640-2. Definitions. For purposes of this article, the following definitions apply.

Audio Amplifier or Pre-Amplifier. Electronic equipment that increases the current or voltage, or both, potential of an audio signal intended for use by another piece of audio equipment. *Amplifier* is the term used to denote an audio amplifier within this article.

Audio Autotransformer. A transformer with a single winding and multiple taps intended for use with an amplifier loudspeaker signal output.

Audio Signal Processing Equipment. Electrically operated equipment that produces or processes, or both, electronic signals that, when appropriately amplified and reproduced by a loudspeaker, produce an acoustic signal within the range of normal human hearing (typically 20-20 kHz). Within this article, the terms *equipment* and *audio equipment* are assumed to be equivalent to audio signal processing equipment.

> FPN: This equipment includes, but is not limited to, loudspeakers; headphones; pre-amplifiers; microphones and their power supplies; mixers; MIDI (musical instrument digital interface) equipment or other digital control systems; equalizers, compressors, and other audio signal processing equipment; audio media recording and playback equipment including turntables, tape decks and disk players (audio and multimedia), synthesizers, tone generators, and electronic organs. Electronic organs and synthesizers may have integral or separate amplification and loudspeakers. With the exception of amplifier outputs, virtually all such equipment is used to process signals (utilizing analog or digital techniques) that have nonhazardous levels of voltage or current potential.

Audio System. Within this article, the term *audio system* means the totality of all equipment and interconnecting wiring used to fabricate a fully functional audio signal processing, amplification, and reproduction system.

Audio Transformer. A transformer with two or more electrically isolated windings and multiple taps intended for use with an amplifier loudspeaker signal output.

Equipment Rack. A framework for the support enclosure, or both, of equipment. May be portable or stationary. See ANSI/EIA/310-D-1992, *Cabinets, Racks, Panels and Associated Equipment*.

Loudspeaker. Equipment that converts an ac electric signal into an acoustic signal. The term *speaker* is commonly used to mean loudspeaker.

Maximum Output Power. The maximum output power delivered by an amplifier into its rated load as determined under specified test conditions. This may exceed the manufacturer's rated output power for the same amplifier.

Mixer. Equipment used to combine and level match a multiplicity of electronic signals, such as from microphones, electronic instruments, and recorded audio.

Mixer–Amplifier. Equipment that combines the functions of a mixer and amplifier within a single enclosure.

Portable Equipment. Equipment fed with portable cords or cables intended to be moved from one place to another.

Powered Loudspeaker. Equipment that consists of a loudspeaker and amplifier within the same enclosure. Other signal processing may also be included.

Rated Load Impedance. The amplifier manufacturer's stated or marked speaker impedance into which an amplifier will deliver its rated output power. 2Ω, 4Ω, and 8Ω are typical ratings.

Rated Output Power. The amplifier manufacturer's stated or marked output power capability into its rated load.

Rated Output Voltage. For audio amplifiers of the constant-voltage type, this is the nominal output voltage when the amplifier is delivering full rated power. Rated output voltage is used for determining approximate acoustic output in distributed speaker systems that typically employ impedance matching transformers. Typical ratings are 25 volts, 70.7 volts, and 100 volts.

Technical Power System. An electrical distribution system with grounding in accordance with Section 250-146(d), where the equipment grounding conductor is isolated from the premises grounded conductor except at a single grounded termination point within a branch circuit panelboard, the originating (main breaker) branch-circuit panelboard, or at the premises grounding electrode.

Temporary Equipment. Portable wiring and equipment intended for use with events of a transient or temporary nature where all equipment is presumed to be removed at the conclusion of the event.

640-3. Locations and Other Articles. Circuits and equipment shall comply with (a) through (k), as applicable.

(a) Spread of Fire or Products of Combustion. See Section 300-21.

(b) Ducts, Plenums, and Other Air-Handling Spaces. See Section 300-22, where installed in ducts or plenums or other space used for environmental air.

> FPN: *Standard for the Installation of Air Conditioning and Ventilation Systems,* NFPA 90A-1996, 2-3.10.1(a), Exception No. 3, permits loudspeakers, loudspeaker assemblies, and their accessories listed in accordance with *Fire Test for Heat and Visible Smoke Release for Discrete Products and Their Accessories Installed in Air-Handling Spaces,* UL 2043-1996, to be installed in other spaces used for environmental air (ceiling cavity plenums).

(c) Cable Trays. Cable trays shall be used in accordance with Article 318.

> FPN: See Section 725-61(c) for the use of Class 2, Class 3, and Type PLTC cable in cable trays.

(d) Hazardous (Classified) Locations. Equipment used in hazardous (classified) locations shall comply with the applicable requirements of Chapter 5.

(e) Places of Assembly. Equipment used in places of assembly shall comply with Article 518.

(f) Theaters, Audience Areas of Motion Picture and Television Studios, and Similar Locations. Equipment used in theaters, audience areas of motion picture and television studios, and similar locations shall comply with Article 520.

(g) Carnivals, Circuses, Fairs, and Similar Events. Equipment used in carnivals, circuses, fairs, and similar events shall comply with Article 525.

(h) Motion Picture and Television Studios. Equipment used in motion picture and television studios shall comply with Article 530.

(i) Swimming Pools, Fountains, and Similar Locations. Audio equipment used in or near swimming pools, fountains, and similar locations shall comply with Article 680.

(j) Combination Systems. Where the authority having jurisdiction permits audio systems for paging or music, or both, to be combined with fire alarm systems, the wiring shall comply with Article 760.

> FPN: For installation requirements for such combination systems, refer to *National Fire Alarm Code®,* NFPA 72-1996, and *Life Safety Code®,* NFPA *101®,* 1997.

(k) Antennas. Equipment used in audio systems that contain an audio or video tuner and an antenna input shall comply with Article 810. Wiring other than antenna wiring that connects such equipment to other audio equipment shall comply with this article.

640-4. Protection of Electrical Equipment. Amplifiers, loudspeakers, and other equipment shall be so located or protected so as to guard against environmental exposure or physical damage, such as might result in fire, shock, or personal hazard.

640-5. Access to Electrical Equipment Behind Panels Designed to Allow Access. Access to equipment shall not be denied by an accumulation of wires and cables that prevents removal of panels, including suspended ceiling panels.

640-6. Mechanical Execution of Work. Equipment and cabling shall be installed in a neat and workmanlike manner. Cables for installed systems shall be supported by the building structure in such a manner that the cable will not be damaged by normal building use.

> FPN: One way to determine accepted industry practice is to refer to nationally recognized standards such as the following: *Commercial Building Telecommunications Wiring Standard,* ANSI/EIA/TIA 568-A-1995; *Commercial Building Standard for Telecommunications Pathways and Spaces,* ANSI/EIA/TIA 569-A-1997; and *Residential and Light Commercial Telecommunications Wiring Standard,* ANSI/EIA/TIA 570-1991, or other ANSI-approved installation standards.

640-7. Grounding.

(a) General. Wireways and auxiliary gutters shall be grounded and bonded in accordance with the requirements of Article 250. Where the wireway or auxiliary gutter does not contain power-supply wires, the equipment grounding conductor shall not be required to be larger than No. 14 copper or its equivalent. Where the wireway or auxiliary gutter contains power-supply wires, the equipment grounding conductor shall not be smaller than specified in Section 250-122.

(b) Separately Derived Systems with 60 Volts to Ground. Grounding of separately derived systems with 60 volts to ground shall be in accordance with Section 530-72.

(c) Isolated Ground Receptacles. Isolated grounding-type receptacles shall be permitted as described in Section 250-146(d), and for the implementation of other technical power systems in compliance with Article 250. For separately derived systems with 60 volts to ground, the branch-circuit equipment grounding conductor shall be terminated as required in Section 530-72(b).

> FPN: See Section 410-56(c) for grounding-type receptacles and required identification.

640-8. Grouping of Conductors. Insulated conductors of different systems grouped or bundled so as to be in close physical contact with each other in the same raceway or other enclosure, or in portable cords or cables, shall comply with Section 300-3(c)(1).

640-9. Wiring Methods.

(a) Wiring to and Between Audio Equipment.

(1) Wiring and equipment from source of power to and between devices connected to the premises wiring systems shall comply with the requirements of Chapters 1 through 4, except as modified by this article.

(2) Separately derived systems shall comply with the applicable articles of this *Code*, except as modified by this article. Separately derived systems with 60 volts to ground shall be permitted for use in audio system installations as specified in Article 530, Part G.

(3) All wiring not connected to the premises wiring system or to a wiring system separately derived from the premises wiring system shall comply with Article 725.

(b) Auxiliary Power Supply Wiring. Equipment that has a separate input for an auxiliary power supply shall be wired in compliance with Article 725. Battery installation shall be in accordance with Article 480.

> FPN No. 1: This section does not apply to the use of uninterruptable power supply (ups) equipment, or other sources of supply, that are intended to act as a direct replacement for the primary circuit power source and are connected to the primary circuit input.

> FPN No. 2: Refer to *National Fire Alarm Code,* NFPA 72-1996, where equipment is used for a fire alarm system.

(c) Output Wiring and Listing of Amplifiers. Amplifiers with output circuits carrying audio program signals shall be permitted to employ Class 1, Class 2, or Class 3 wiring where the amplifier is listed and marked for use with the specific class of wiring method. Such listing shall ensure the energy output is equivalent to the shock and fire risk of the same class as stated in Article 725. Overcurrent protection shall be provided and shall be permitted to be inherent to the amplifier.

Audio circuits wired using Class 1 wiring methods shall not occupy the same raceway or enclosure with other than audio circuits wired using Class 1 wiring methods.

Audio circuits wired using Class 2 wiring methods shall not occupy the same raceway or enclosure with other than audio circuits wired using Class 2 wiring methods.

Audio circuits wired using Class 3 wiring methods shall not occupy the same raceway or enclosure with other than audio circuits wired using Class 3 wiring methods.

> FPN No. 1: *Amplifiers for Fire Protective Signaling Systems,* ANSI/UL 1711-1994, contains requirements for the listing of amplifiers used for fire alarm systems in compliance with *National Fire Alarm Code,* NFPA 72-1996.

> FPN No. 2: Examples of requirements for listing amplifiers used in residential, commercial, and professional use are found in *Commercial Audio Equipment,* ANSI/UL 813-1996, *Professional Video and Audio Equipment,* ANSI/UL 1419-1997, *Audio-Video Products and Accessories,* ANSI/UL 1492-1996, or *Audio/Video and Musical Instrument Apparatus for Household, Commercial, and Similar Use,* ANSI/UL 6500-1996.

(d) Use of Audio Transformers and Autotransformers. Audio transformers and autotransformers shall only be used for audio signals in a manner so as not to exceed the manufacturer's stated input or output voltage, impedance, or power limitations. The input or output wires of an audio transformer or autotransformer shall be allowed to connect directly to the amplifier or loudspeaker terminals. No electrical terminal or lead shall be required to be grounded or bonded.

640-10. Audio Systems Near Bodies of Water. Audio systems near bodies of water, either natural or artificial, shall be subject to the following restrictions.

Exception: This section does not include audio systems intended for use on boats, yachts, or other forms of land or water transportation used near bodies of water, whether or not supplied by branch-circuit power.

> FPN: See Section 680-23 for installation of underwater audio equipment.

(a) Equipment Supplied by Branch-Circuit Power. Audio system equipment supplied by branch-circuit power shall not be placed laterally within 5 ft (1.52 m) of the inside wall of a pool, spa, hot tub, or fountain, nor within 5 ft (1.52 m) of the prevailing or tidal high water mark. The equipment shall be provided with branch-circuit power protected by a ground-fault circuit interrupter where required by other articles.

(b) Equipment Not Supplied by Branch-Circuit Power. Audio system equipment powered by a listed Class 2 power supply or by the output of an amplifier listed as permitting the use of Class 2 wiring shall only be restricted in placement by the manufacturer's recommendations.

> FPN: Placement of the power supply or amplifier, if supplied by branch-circuit power, is still subject to Section 640-10(a).

B. Permanent Audio System Installations

640-21. Use of Flexible Cords and Cables.

(a) Between Equipment and Branch-Circuit Power. Power supply cords for audio equipment shall be suitable for the use and shall be permitted to be used where the interchange, maintenance, or repair of such equipment is facilitated through the use of a power supply cord.

(b) Between Loudspeakers and Amplifiers, or Between Loudspeakers. Cables used to connect loudspeakers to each other or to an amplifier shall comply with Article 725. Other listed cable types and assemblies, including optional hybrid communications, signal, and optical fiber cables, shall be permitted.

(c) Between Equipment. Cables used for the distribution of audio signals between equipment shall comply with Article 725. Other listed cable types and assemblies, including optional hybrid communications, signal, and optical fiber cables, shall be permitted. Other cable types and assemblies specified by the equipment manufacturer as acceptable for the use shall be permitted in accordance with Section 110-3(b).

(d) Between Equipment and Power Supplies Other than Branch-Circuit Power.

(1) Storage batteries shall be installed and wired in accordance with the requirements of this *Code* for the voltage and power delivered.

(2) Transformers, transformer rectifiers, and other ac or dc power supplies shall be installed and wired in accordance with the requirements of this *Code* for the voltage and power delivered.

> FPN: For some equipment, sources such as in (1) and (2) will serve as the only source of power. These could, in turn, be supplied with intermittent or continuous branch-circuit power.

(e) Between Equipment Racks and Premises Wiring System. Flexible cords and cables shall be permitted for the electrical connection of permanently installed equipment racks to the premises wiring system to facilitate access to equipment or for the purpose of isolating the technical ground of the rack from the premises ground. Connection shall be made using either approved plugs and receptacles or by direct connection within an approved enclosure. Flexible cords and cables shall not be subjected to physical manipulation or abuse while the rack is in use.

640-22. Wiring of Equipment Racks. Equipment racks shall be fabricated of metal and grounded. Bonding shall not be required if the rack is connected to a technical power ground.

Equipment racks shall be wired in a neat and workmanlike manner. Wires, cables, structural components, or other equipment shall not be placed in such a manner as to prevent reasonable access to equipment power switches and resettable or replaceable circuit overcurrent protection devices.

Supply cords or cables, if used, shall terminate within the equipment rack enclosure in an identified connector assembly. The supply cords or cable (and connector assembly, if used) shall have sufficient ampacity to carry the total load connected to the equipment rack and shall be protected by overcurrent devices.

640-23. Conduit or Tubing.

(a) Number of Conductors. The number of conductors permitted in a single conduit or tubing shall not exceed the percentage fill specified in Table 1, Chapter 9.

(b) Nonmetallic Conduit or Tubing and Insulating Bushings. The use of nonmetallic conduit or tubing and insulating bushings shall be permitted where a technical ground system is employed and shall comply with applicable articles.

640-24. Wireways, Gutters, and Auxiliary Gutters. The use of metallic and nonmetallic wireways, gutters, and auxiliary gutters shall be permitted for use with audio signal conductors and shall comply with applicable articles with respect to permitted locations, construction, and fill.

640-25. Loudspeaker Installation in Fire Resistance-Rated Partitions, Walls, and Ceilings. Loudspeakers installed in a fire resistance-rated partition, wall, or ceiling shall be listed for the purpose or installed in an enclosure or recess that maintains the fire resistance rating.

> FPN: Fire-rated construction is the fire-resistive classification used in building codes. One method of determining fire rating is testing in accordance with *Standard Methods of Tests of Fire Endurance of Building Construction and Materials,* NFPA 251-1995.

C. Portable and Temporary Audio System Installations

640-41. Multipole Branch-Circuit Cable Connectors. Multipole branch-circuit cable connectors, male and female, for power supply cords and cables shall be constructed so that tension on the cord or cable will not be transmitted to the connections. The female half shall be attached to the load end of the power supply cord or cable. The connector shall be rated in amperes and designed so that differently rated devices cannot be connected together. Alternating-current multipole connectors shall be polarized and comply with Sections 410-56(g) and 410-58. Alternating-current or direct-current multipole connectors utilized for connection between loudspeakers and amplifiers, or between loudspeakers, shall not be compatible with nonlocking 15- or 20-ampere rated connectors intended for branch-circuit power, nor with connectors rated 250 volts or greater of either the locking or nonlocking type. Signal cabling not intended for such loudspeaker and amplifier interconnection shall not be permitted to be compatible with multipole branch-circuit cable connectors of any accepted configuration.

FPN: See Section 400-10 for pull at terminals.

640-42. Use of Flexible Cords and Cables.

(a) Between Equipment and Branch-Circuit Power. Power supply cords for audio equipment shall be listed and shall be permitted to be used where the interchange, maintenance, or repair of such equipment is facilitated through the use of a power supply cord.

(b) Between Loudspeakers and Amplifiers, or Between Loudspeakers. Flexible cords and cables used to connect loudspeakers to each other or to an amplifier shall comply with Article 400 and Article 725, respectively. Cords and cables listed for portable use, either hard or extra-hard usage as defined by Article 400, shall also be permitted. Other listed cable types and assemblies, including optional hybrid communications, signal, and optical fiber cables, shall be permitted.

(c) Between Equipment and/or Between Equipment Racks. Flexible cords and cables used for the distribution of audio signals between equipment shall comply with Article 400 and Article 725, respectively. Cords and cables listed for portable use, either hard or extra-hard service as defined by Article 400, shall also be permitted. Other listed cable types and assemblies, including optional hybrid communications, signal, and optical fiber cables, shall be permitted.

(d) Between Equipment, Equipment Racks, and Power Supplies Other than Branch-Circuit Power.

(1) Storage batteries shall be installed and wired in accordance with the requirements of this *Code* for the voltage and power delivered.

(2) Transformers, transformer rectifiers, and other ac or dc power supplies shall be installed and wired in accordance with the requirements of this *Code* for the voltage and power delivered.

(3) Generators shall be installed in accordance with Article 445.

(e) Between Equipment Racks and Branch-Circuit Power. The supply to a portable equipment rack shall be by means of listed extra-hard usage cords or cables, as defined in Article 400. For outdoor portable or temporary use, the cords or cables shall be further listed as being suitable for wet locations and sunlight resistant.

(1) Where equipment racks include audio and lighting and/or power equipment, Articles 520 and 525 shall apply as appropriate.

(2) The usage and construction of cable extensions, adapters, and breakout assemblies shall be in accordance with Article 520 or 525, as appropriate.

640-43. Wiring of Equipment Racks. Equipment racks fabricated of metal shall be grounded. Nonmetallic racks with covers (if provided) removed shall not allow access to Class 1, Class 3, or primary circuit power without the removal of covers over terminals or the use of tools.

Equipment racks shall be wired in a neat and workmanlike manner. Wires, cables, structural components, or other equipment shall not be placed in such a manner as to prevent reasonable access to equipment power switches and resettable or replaceable circuit overcurrent protection devices.

Wiring that exits the equipment rack for connection to other equipment or to a power supply shall be relieved of strain or otherwise suitably terminated such that a pull on the flexible cord or cable shall not increase the risk of damage to the cable or connected equipment such as to cause an unreasonable risk of fire or electric shock.

640-44. Environmental Protection of Equipment. Temporary outdoor, unsheltered placement or use of portable equipment not listed for the purpose shall be permitted only where appropriate protection of such equipment from adverse weather conditions is provided to prevent risk of fire or electrical shock. Where the system is intended to remain operable during adverse weather, arrangements shall be made for maintaining operation and ventilation of heat dissipating equipment.

640-45. Protection of Wiring. Where accessible to the public, flexible cords and cables laid or run on the ground or on the floor shall be covered with approved nonconductive mats. Cables and mats shall be arranged so as not to present a tripping hazard.

640-46. Equipment Access. Equipment likely to present a risk of fire, electrical shock, or physical injury to the public shall be protected by barriers or supervised by qualified personnel so as to prevent public access.

Article 645 — Information Technology Equipment

645-1. Scope. This article covers equipment, power-supply wiring, equipment interconnecting wiring, and grounding of information technology equipment and systems, including terminal units, in an information technology equipment room.

> FPN: For further information, see *Standard for the Protection of Electronic Computer/Data Processing Equipment,* NFPA 75-1995.

645-2. Special Requirements for Information Technology Equipment Room. This article applies, provided all the following conditions are met.

(a) Disconnecting means complying with Section 645-10 are provided.

(b) A separate heating/ventilating/air-conditioning (HVAC) system is provided that is dedicated for information technology equipment use and is separated from other areas of occupancy. Any HVAC system that serves other occupancies shall be permitted to also serve the information technology equipment room if fire/smoke dampers are provided at the point of penetration of the room boundary. Such dampers shall operate on activation of smoke detectors and also by operation of the disconnecting means required by Section 645-10.

> FPN: For further information, see *Standard for the Protection of Electronic Computer/Data Processing Equipment,* NFPA 75-1995.

(c) Listed information technology equipment is installed.

> FPN: For further information, see *Standard for the Protection of Electronic Computer/Data Processing Equipment,* NFPA 75-1995.

(d) Occupied only by those personnel needed for the maintenance and functional operation of the installed information technology equipment.

> FPN: For further information, see *Standard for the Protection of Electronic Computer/Data Processing Equipment,* NFPA 75-1995.

(e) The room is separated from other occupancies by fire-resistant-rated walls, floors, and ceilings with protected openings.

(f) The building construction, rooms, or areas and occupancy comply with the applicable building code.

645-5. Supply Circuits and Interconnecting Cables.

(a) Branch-Circuit Conductors. The branch-circuit conductors supplying one or more units of a data processing system shall have an ampacity not less than 125 percent of the total connected load.

(b) Connecting Cables. The data processing system shall be permitted to be connected to a branch circuit by any of the following means listed for the purpose.

(1) Computer/data processing cable and attachment plug cap.
(2) Flexible cord and an attachment plug cap.
(3) Cord set assembly. Where run on the surface of the floor, they shall be protected against physical damage.

(c) Interconnecting Cables. Separate data processing units shall be permitted to be interconnected by means of cables and cable assemblies listed for the purpose. Where run on the surface of the floor, they shall be protected against physical damage.

(d) Under Raised Floors. Power cables, communications cables, connecting cables, interconnecting cables, and receptacles associated with the information technology equipment shall be permitted under a raised floor, provided the following.

(1) The raised floor is of suitable construction and the area under the floor is accessible.

> FPN: See *Standard for the Protection of Electronic Computer/Data Processing Equipment,* NFPA 75-1995.

(2) The branch-circuit supply conductors to receptacles or field-wired equipment are in rigid metal conduit, rigid nonmetallic conduit, intermediate metal conduit, electrical metallic tubing, metal wireway, surface metal raceway with metal cover, flexible metal conduit, liquidtight flexible metal or nonmetallic conduit, Type MI cable, Type MC cable, or Type AC cable. These supply conductors shall be installed in accordance with the requirements of Section 300-11.

(3) Ventilation in the underfloor area is used for the information technology equipment room only.

(4) Openings in raised floors for cables protect cables against abrasions and minimize the entrance of debris beneath the floor.

(5) Cables, other than those covered in (2) and those complying with (a), (b), and (c) below, shall be listed as Type DP cable having adequate fire-resistant characteristics suitable for use under raised floors of an information technology equipment room.

(a) Interconnecting cables enclosed in a raceway.

(b) Interconnecting cables listed with equipment manufactured prior to July 1, 1994, being installed with that equipment.

(c) Cable type designations Type TC (Article 340); Types CL2, CL3, and PLTC (Article 725); Types NPLF and FPL (Article 760); Types OFC and OFN (Article 770); Types CM and MP (Article 800); Type CATV (Article 820). These designations shall be permitted to have an additional letter P or R or G. Green insulated single conductor cables, No. 4 and larger, marked "for use in cable trays" or "for CT use" shall be permitted for equipment grounding.

FPN: One method of defining fire resistance is by establishing that the cables do not spread fire to the top of the tray in the "Vertical Tray Flame Test" referenced in the *Standard for Electrical Wires, Cables, and Flexible Cords*, ANSI/UL 1581-1991. Another method of defining fire resistance is for the damage (char length) not to exceed 4 ft 11 in. (1.5 m) when performing the CSA "Vertical Flame Test — Cables in Cable Trays," as described in *Test Methods for Electrical Wires and Cables*, CSA C22.2 No. 0.3-M-1985.

(e) **Securing in Place.** Power cables; communications cables; connecting cables; interconnecting cables; and associated boxes, connectors, plugs, and receptacles that are listed as part of, or for, information technology equipment shall not be required to be secured in place.

645-6. Cables Not in Information Technology Equipment Room. Cables extending beyond the information technology equipment room shall be subject to the applicable requirements of this *Code*.

FPN: For signaling circuits, refer to Article 725; for fiber optic circuits, refer to Article 770; and for communications circuits, refer to Article 800. For fire alarm systems, refer to Article 760.

645-7. Penetrations. Penetrations of the fire-resistant room boundary shall be in accordance with Section 300-21.

645-10. Disconnecting Means. A means shall be provided to disconnect power to all electronic equipment in the information technology equipment room. There shall also be a similar means to disconnect the power to all dedicated HVAC systems serving the room and cause all required fire/ smoke dampers to close. The control for these disconnecting means shall be grouped and identified and shall be readily accessible at the principal exit doors. A single means to control both the electronic equipment and HVAC systems shall be permitted.

Exception: Installations qualifying under the provisions of Article 685.

645-11. Uninterruptible Power Supplies (UPS). Unless otherwise permitted in (a) or (b), UPS systems installed within the information technology room, and their supply and output circuits, shall comply with Section 645-10. The disconnecting means shall also disconnect the battery from its load.

(a) Installations qualifying under the provisions of Article 685.

(b) A disconnecting means complying with Section 645-10 shall not be required for power sources capable of supplying 750 volt-amperes or less derived either from UPS equipment or from battery circuits integral to electronic equipment, provided all other requirements of Section 645-11 are met.

645-15. Grounding. All exposed noncurrent-carrying metal parts of an information technology system shall be grounded in accordance with Article 250 or shall be double insulated. Power systems derived within listed information technology equipment that supply information technology systems through receptacles or cable assemblies supplied as part of this equipment shall not be considered separately derived for the purpose of applying Section 250-20(d). Where signal reference structures are installed, they shall be bonded to the equipment grounding system provided for the information technology equipment.

FPN No. 1: The bonding and grounding requirements in the product standards governing this listed equipment ensure that it complies with Article 250.

FPN No. 2: Where isolated grounding-type receptacles are used, see Sections 250-146(d) and 410-56(c).

645-16. Marking. Each unit of an information technology system supplied by a branch circuit shall be provided with a manufacturer's nameplate, which shall also include the input power requirements for voltage, frequency, and maximum rated load in amperes.

Article 650 — Pipe Organs

650-1. Scope. This article covers those electrical circuits and parts of electrically operated pipe organs that are employed for the control of the sounding apparatus and keyboards.

650-2. Other Articles. Electronic organs shall comply with the appropriate provisions of Article 640.

650-3. Source of Energy. The source of power shall be a transformer-type rectifier, the dc potential of which shall not exceed 30 volts dc.

650-4. Grounding. The rectifier shall be grounded according to the provisions in Section 250-112(b).

650-5. Conductors. Conductors shall comply with (a) through (d).

(a) Size. Not less than No. 28 for electronic signal circuits and not less than No. 26 for electromagnetic valve supply and the like. A main common-return conductor in the electromagnetic supply shall not be less than No. 14.

(b) Insulation. Conductors shall have thermoplastic or thermosetting insulation.

(c) Conductors to Be Cabled. Except for the common-return conductor and conductors inside the organ proper, the organ sections and the organ console conductors shall be cabled. The common-return conductors shall be permitted under an additional covering enclosing both cable and return conductor, or shall be permitted as a separate conductor and shall be permitted to be in contact with the cable.

(d) Cable Covering. Each cable shall be provided with an outer covering, either overall or on each of any subassemblies of grouped conductors. Tape shall be permitted in place of a covering. Where not installed in metal raceway, the covering shall be flame retardant or the cable or each cable subassembly shall be covered with a closely wound fireproof tape.

650-6. Installation of Conductors. Cables shall be securely fastened in place and shall be permitted to be attached directly to the organ structure without insulating supports. Cables shall not be placed in contact with other conductors.

650-7. Overcurrent Protection. Circuits shall be arranged so that all conductors shall be protected from overcurrent by an overcurrent device rated at not more than 6 amperes.

Exception: The main supply conductors and the common-return conductors.

Article 660 — X-Ray Equipment

A. General

660-1. Scope. This article covers all X-ray equipment operating at any frequency or voltage for industrial or other nonmedical or nondental use.

> FPN: See Article 517, Part E, for X-ray installations in health care facilities.

Nothing in this article shall be construed as specifying safeguards against the useful beam or stray X-ray radiation.

> FPN No. 1: Radiation safety and performance requirements of several classes of X-ray equipment are regulated under Public Law 90-602 and are enforced by the Department of Health and Human Services.

> FPN No. 2: In addition, information on radiation protection by the National Council on Radiation Protection and Measurements is published as *Reports of the National Council on Radiation Protection and Measurement.* These reports can be obtained from NCRP Publications, 7910 Woodmont Ave., Suite 1016, Bethesda, MD 20814.

660-2. Definitions.

Long-Time Rating. A rating based on an operating interval of 5 minutes or longer.

Mobile. X-ray equipment mounted on a permanent base with wheels and/or casters for moving while completely assembled.

Momentary Rating. A rating based on an operating interval that does not exceed 5 seconds.

Portable. X-ray equipment designed to be hand-carried.

Transportable. X-ray equipment to be installed in a vehicle or that may be readily disassembled for transport in a vehicle.

660-3. Hazardous (Classified) Locations. Unless approved for the location, X-ray and related equipment shall not be installed or operated in hazardous (classified) locations.

> FPN: See Article 517, Part D.

660-4. Connection to Supply Circuit.

(a) Fixed and Stationary Equipment. Fixed and stationary X-ray equipment shall be connected to the power supply by means of a wiring method meeting the general requirements of this *Code.* Equipment properly supplied by a branch circuit rated at not over 30 amperes shall be permitted to be supplied through a suitable attachment plug cap and hard-service cable or cord.

(b) Portable, Mobile, and Transportable Equipment. Individual branch circuits shall not be required for portable, mobile, and transportable X-ray equipment requiring a capacity of not over 60 amperes. Portable and mobile types of X-ray equipment of any capacity shall be supplied through a suitable hard-service cable or cord. Transportable X-ray equipment of any capacity shall be permitted to be connected to its power supply by suitable connections and hard-service cable or cord.

(c) Over 600 Volts, Nominal. Circuits and equipment operated at more than 600 volts, nominal, shall comply with Article 490.

660-5. Disconnecting Means. A disconnecting means of adequate capacity for at least 50 percent of the input required for the momentary rating or 100 percent of the input required for the long-time rating of the X-ray equipment, whichever is greater, shall be provided in the supply circuit. The disconnecting means shall be operable from a location readily accessible from the X-ray control. For equipment connected to a 120-volt, nominal, branch circuit of 30 amperes or less, a grounding-type attachment plug cap and receptacle of proper rating shall be permitted to serve as a disconnecting means.

660-6. Rating of Supply Conductors and Overcurrent Protection.

(a) Branch-Circuit Conductors. The ampacity of supply branch-circuit conductors and the overcurrent protective devices shall not be less than 50 percent of the momentary rating or 100 percent of the long-time rating, whichever is greater.

(b) Feeder Conductors. The rated ampacity of conductors and overcurrent devices of a feeder for two or more branch circuits supplying X-ray units shall not be less than 100 percent of the momentary demand rating [as determined by (a)] of the two largest X-ray apparatus plus 20 percent of the momentary ratings of other X-ray apparatus.

> FPN: The minimum conductor size for branch and feeder circuits is also governed by voltage regulation requirements. For a specific installation, the manufacturer usually specifies minimum distribution transformer and conductor sizes, rating of disconnect means, and overcurrent protection.

660-7. Wiring Terminals. X-ray equipment not provided with a permanently attached cord or cord set shall be provided with suitable wiring terminals or leads for the connection of power-supply conductors of the size required by the rating of the branch circuit for the equipment.

660-8. Number of Conductors in Raceway. The number of control circuit conductors installed in a raceway shall be determined in accordance with Section 300-17.

660-9. Minimum Size of Conductors. Size No. 18 or 16 fixture wires, as specified in Section 725-27, and flexible cords shall be permitted for the control and operating circuits of X-ray and auxiliary equipment where protected by not larger than 20-ampere overcurrent devices.

660-10. Equipment Installations. All equipment for new X-ray installations and all used or reconditioned X-ray equipment moved to and reinstalled at a new location shall be of an approved type.

B. Control

660-20. Fixed and Stationary Equipment.

(a) Separate Control Device. A separate control device, in addition to the disconnecting means, shall be incorporated in the X-ray control supply or in the primary circuit to the high-voltage transformer. This device shall be a part of the X-ray equipment but shall be permitted in a separate enclosure immediately adjacent to the X-ray control unit.

(b) Protective Device. A protective device, which shall be permitted to be incorporated into the separate control device, shall be provided to control the load resulting from failures in the high-voltage circuit.

660-21. Portable and Mobile Equipment. Portable and mobile equipment shall comply with Section 660-20, but the manually controlled device shall be located in or on the equipment.

660-23. Industrial and Commercial Laboratory Equipment.

(a) Radiographic and Fluoroscopic Types. All radiographic- and fluoroscopic-type equipment shall be effectively enclosed or shall have interlocks that de-energize the equipment automatically to prevent ready access to live current-carrying parts.

(b) Diffraction and Irradiation Types. Diffraction- and irradiation-type equipment or installations not effectively enclosed or provided with interlocks to prevent access to live current-carrying parts during operation shall be provided with a positive means to indicate when they are energized. The indicator shall be a pilot light, readable meter deflection, or equivalent means.

660-24. Independent Control. Where more than one piece of equipment is operated from the same high-voltage circuit, each piece or each group of equipment as a unit shall be provided with a high-voltage switch or equivalent disconnecting means. This disconnecting means shall be constructed, enclosed, or located so as to avoid contact by persons with its live parts.

C. Transformers and Capacitors

660-35. General. Transformers and capacitors that are part of an X-ray equipment shall not be required to comply with Articles 450 and 460.

660-36. Capacitors. Capacitors shall be mounted within enclosures of insulating material or grounded metal.

D. Guarding and Grounding

660-47. General.

(a) High-Voltage Parts. All high-voltage parts, including X-ray tubes, shall be mounted within grounded enclosures. Air, oil, gas, or other suitable insulating media shall be used to insulate the high voltage from the grounded enclosure. The connection from the high-voltage equipment to X-ray tubes and other high-voltage components shall be made with high-voltage shielded cables.

(b) Low-Voltage Cables. Low-voltage cables connecting to oil-filled units that are not completely sealed, such as transformers, condensers, oil coolers, and high-voltage switches, shall have insulation of the oil-resistant type.

660-48. Grounding. Noncurrent-carrying metal parts of X-ray and associated equipment (controls, tables, X-ray tube supports, transformer tanks, shielded cables, X-ray tube heads, etc.) shall be grounded in the manner specified in Article 250. Portable and mobile equipment shall be provided with an approved grounding-type attachment plug cap.

Exception: Battery-operated equipment.

Article 665 — Induction and Dielectric Heating

A. General

665-1. Scope. This article covers the construction and installation of induction and dielectric heating equipment and accessories for industrial and scientific applications, but not for medical or dental applications, appliances, or line frequency pipelines and vessels heating.

FPN No. 1: See Article 422 for appliances.

FPN No. 2: See Article 427, Part E, for line frequency pipelines and vessels heating.

665-2. Definitions.

Dielectric Heating. Dielectric heating is the heating of a nominally insulating material due to its own dielectric losses when the material is placed in a varying electric field.

Heating Equipment. The term *heating equipment* as used in this article includes any equipment used for heating purposes whose heat is generated by induction or dielectric methods.

Induction Heating. The heating of a nominally conductive material due to its own I^2R losses when the material is placed in a varying electromagnetic field.

665-3. Other Articles. Wiring from the source of power to the heating equipment shall comply with Chapters 1 through 4. Circuits and equipment operated at more than 600 volts, nominal, shall comply with Article 490.

665-4. Hazardous (Classified) Locations. Induction and dielectric heating equipment shall not be installed in hazardous (classified) locations as defined in Article 500 unless the equipment and wiring are designed and approved for the hazardous (classified) locations.

B. Guarding, Grounding, and Labeling

665-20. Enclosures. The converting apparatus (including the dc line) and high-frequency electric circuits (excluding the output circuits and remote-control circuits) shall be completely contained within an enclosure or enclosures of noncombustible material.

665-21. Panel Controls. All panel controls shall be of dead-front construction.

665-22. Access to Internal Equipment. Doors or detachable panels shall be employed for internal access. Where doors are used giving access to voltages from 500 to 1000 volts ac or dc, either door locks shall be provided or interlocking shall be installed. Where doors are used giving access to voltages of over 1000 volts ac or dc, either mechanical lockouts with a disconnecting means to prevent access until voltage is removed from the cubicle, or both door interlocking and mechanical door locks, shall be provided. Detachable panels not normally used for access to such parts shall be fastened in a manner that will make them inconvenient to remove.

665-23. Warning Labels or Signs. Warning labels or signs that read "DANGER — HIGH VOLTAGE — KEEP OUT" shall be attached to the equipment and shall be plainly visible where unauthorized persons might come in contact with energized parts, even when doors are open or when panels are removed from compartments containing over 250 volts ac or dc.

665-24. Capacitors. Where capacitors in excess of 0.1 microfarad are used in dc circuits, either as rectifier filter components or suppressers, etc., having circuit voltages of over 240 volts to ground, bleeder resistors or grounding switches shall be used as grounding devices. The time of discharge shall be in accordance with Section 460-6(a).

Where capacitors are individually switched out of a circuit, a bleeder resistor or automatic switch shall be used as a discharge means.

Where auxiliary rectifiers are used with filter capacitors in the output for bias supplies, tube keyers, etc., bleeder resistors shall be used, even though the dc voltage may not exceed 240 volts.

665-25. Work Applicator Shielding. Protective cages or adequate shielding shall be used to guard work applicators other than induction heating coils. Induction heating coils shall be permitted to be protected by insulation or refractory materials, or both. Interlock switches shall be used on all hinged access doors, sliding panels, or other easy means of access to the applicator. All interlock switches shall be connected in such a manner as to remove all power from the applicator when any one of the access doors or panels is open. Interlocks on access doors or panels shall not be required if the applicator is an induction heating coil at dc ground potential or operating at less than 150 volts ac.

665-26. Grounding and Bonding. Grounding or inter-unit bonding, or both, shall be used wherever required for circuit operation, for limiting to a safe value radio frequency potentials between all exposed noncurrent-carrying parts of the equipment and earth ground, between all equipment parts and surrounding objects, and between such objects and earth ground. Such grounding and bonding shall be installed in accordance with Article 250.

665-27. Marking. Each heating equipment shall be provided with a nameplate giving the manufacturer's name and model identification and the following input data: line volts, frequency, number of phases, maximum current, full-load kilovolt-amperes (kVA), and full-load power factor.

665-28. Control Enclosures. Direct-current or low-frequency ac shall be permitted in the control portion of the heating equipment. This shall be limited to not over 150 volts. Solid or stranded wire No. 18 or larger shall be used. A step-down transformer with proper overcurrent protection shall be permitted in the control enclosure to obtain an ac voltage of less than 150 volts. The higher-voltage terminals shall be guarded to prevent accidental contact. 60-Hz components shall be permitted to control high frequency where

properly rated by the induction heating equipment manufacturer. Electronic circuits utilizing solid-state devices and tubes shall be permitted printed circuits or wires smaller than No. 18.

C. Motor-Generator Equipment

665-40. General. Motor-generator equipment shall include all rotating equipment designed to operate from an ac or dc motor or by mechanical drive from a prime mover, producing an alternating current of any frequency for induction or dielectric heating, or both.

665-41. Ampacity of Supply Conductors. The ampacity of supply conductors shall be determined in accordance with Article 430.

665-42. Overcurrent Protection. Overcurrent protection shall be provided as specified in Article 430 for the electric supply circuit.

665-43. Disconnecting Means. The disconnecting means shall be provided as specified in Article 430.

A readily accessible disconnecting means shall be provided by which each heating equipment can be isolated from its supply circuit. The ampere rating of this disconnecting means shall not be less than the nameplate current rating of the equipment. The supply circuit disconnecting means shall be permitted as a heating equipment disconnecting means where the circuit supplies only one equipment.

665-44. Output Circuit. The output circuit shall include all output components external to the generator, including contactors, transformers, busbars, and other conductors, and shall comply with (a) and (b).

(a) Generator Output. The output circuit shall be isolated from ground unless otherwise permitted as follows:

(1) Where the capacitive coupling inherent in the generator causes the generator terminals to have voltages from terminal to ground that are equal or
(2) Where a vacuum or controlled atmosphere is used with a coil in a tank or chamber, the center point of the coil shall be grounded to maintain an equal potential between each terminal and ground.

Where rated at over 500 volts, the output circuit shall incorporate a dc ground protector unit. The dc impressed on the output circuit shall not exceed 30 volts and shall not exceed a current capability of 5 mA.

An isolating transformer for matching the load and the source shall be permitted in the output circuit if the output secondary is not at dc ground potential.

(b) Component Interconnections. The various components required for a complete induction heating equipment installation shall be connected by properly protected multiconductor cable, busbar, or coaxial cable. Cables shall be installed in nonferrous raceways. Busbars shall be protected, where required, by nonferrous enclosures.

665-47. Remote Control.

(a) Selector Switch. Where remote controls are used for applying power, a selector switch shall be provided and interlocked to provide power from only one control point at a time.

(b) Foot Switches. Switches operated by foot pressure shall be provided with a shield over the contact button to avoid accidental closing of a switch.

D. Equipment Other than Motor-Generators

665-60. General. Equipment other than motor-generators shall consist of all static multipliers and oscillator-type units utilizing vacuum tubes or solid-state devices, or both. The equipment shall be capable of converting ac or dc to an ac frequency suitable for induction or dielectric heating, or both.

665-61. Ampacity of Supply Conductors. The ampacity of supply conductors shall be determined in accordance with (a) and (b).

(a) Nameplate Rating. The ampacity of the circuit conductors shall not be less than the nameplate current rating of the equipment.

(b) Two or More. The ampacity of conductors supplying two or more equipments shall not be less than the sum of the nameplate current ratings on all equipments.

If simultaneous operation of two or more equipments supplied from the same feeder is not possible, the ampacity of the feeder shall not be less than the sum of the nameplate ratings for the largest group of machines capable of simultaneous operation, plus 100 percent of the stand-by currents of the remaining machines supplied.

665-62. Overcurrent Protection. Overcurrent protection shall be provided as specified in Article 240 for the equipment as a whole. This overcurrent protection shall be provided separately or as a part of the equipment.

665-63. Disconnecting Means. A readily accessible disconnecting means shall be provided by which each heating equipment can be isolated from its supply circuit. The disconnecting means shall be located within sight from the controller or be capable of being locked in the open position.

The rating of this disconnecting means shall not be less than the nameplate rating of the equipment. The supply circuit disconnecting means shall be permitted for disconnecting the heating equipment where the circuit supplies only one equipment.

665-64. Output Circuit. The output circuit shall include all output components external to the converting device, including contactors, transformers, busbars, and other conductors, and shall comply with (a) and (b).

(a) Converter Output. The output circuit shall be isolated from ground unless a dc voltage can exist at the terminals from an internal component failure; then the output circuit (direct or coupled) shall be at dc ground potential.

(b) Converter and Applicator Connection. Where the connections between the converter and the work applicator exceed 2 ft (610 mm) in length, the connections shall be enclosed or guarded with nonferrous, noncombustible material.

665-66. Line Frequency in Converter Equipment Output. Commercial frequencies of 25- to 60-Hz ac output shall be permitted to be coupled for control purposes, but shall be limited to not over 150 volts during periods of circuit operation.

665-67. Keying. Where high-speed keying circuits dependent on the effect of "oscillator blocking" are employed, the peak radio frequency output voltage during the blocked portion of the cycle shall not exceed 100 volts in units employing radio frequency converters.

665-68. Remote Control.

(a) Selector Switch. Where remote controls are used for applying power, a selector switch shall be provided and interlocked to provide power from only one control point at a time.

(b) Foot Switches. Switches operated by foot pressure shall be provided with a shield over the contact button to avoid accidental closing of the switch.

Article 668 — Electrolytic Cells

668-1. Scope. The provisions of this article apply to the installation of the electrical components and accessory equipment of electrolytic cells, electrolytic cell lines, and process power supply for the production of aluminum, cadmium, chlorine, copper, fluorine, hydrogen peroxide, magnesium, sodium, sodium chlorate, and zinc.

Not covered by this article are cells used as a source of

electric energy and for electroplating processes and cells used for the production of hydrogen.

> FPN No. 1: In general, any cell line or group of cell lines operated as a unit for the production of a particular metal, gas, or chemical compound may differ from any other cell lines producing the same product because of variations in the particular raw materials used, output capacity, use of proprietary methods or process practices, or other modifying factors to the extent that detailed *Code* requirements become overly restrictive and do not accomplish the stated purpose of this *Code*.

> FPN No. 2: For further information, see *Standard for Electrical Safety Practices in Electrolytic Cell Line Working Zones*, IEEE 463-1993.

668-2. Definitions.

Cell Line. An assembly of electrically interconnected electrolytic cells supplied by a source of direct-current power.

Cell Line Attachments and Auxiliary Equipment. As applied to this article, cell line attachments and auxiliary equipment include, but are not limited to, auxiliary tanks; process piping; duct work; structural supports; exposed cell line conductors; conduits and other raceways; pumps, positioning equipment, and cell cutout or bypass electrical devices. Auxiliary equipment includes tools, welding machines, crucibles, and other portable equipment used for operation and maintenance within the electrolytic cell line working zone.

In the cell line working zone, auxiliary equipment includes the exposed conductive surfaces of ungrounded cranes and crane-mounted cell-servicing equipment.

Electrically Connected. A connection capable of carrying current as distinguished from connection through electromagnetic induction.

Electrolytic Cell. A tank or vat in which electrochemical reactions are caused by applying electrical energy for the purpose of refining or producing usable materials.

Electrolytic Cell Line Working Zone. The cell line working zone is the space envelope wherein operation or maintenance is normally performed on or in the vicinity of exposed energized surfaces of electrolytic cell lines or their attachments.

668-3. Other Articles.

(a) Lighting, Ventilating, Material Handling. Chapters 1 through 4 shall apply to services, feeders, branch circuits, and apparatus for supplying lighting, ventilating, material handling, and the like, that are outside the electrolytic cell line working zone.

(b) Systems Not Electrically Connected. Those elements of a cell line power-supply system that are not electrically connected to the cell supply system, such as the primary winding of a two-winding transformer, the motor of a motor-generator set, feeders, branch circuits, disconnecting means, motor controllers, and overload protective equipment shall be required to comply with all applicable provisions of this *Code*.

(c) Electrolytic Cell Lines. Electrolytic cell lines shall comply with the provisions of Chapters 1, 2, 3, and 4 except as amended in (1), (2), (3), or (4).

(1) The electrolytic cell line conductors shall not be required to comply with the provisions of Articles 110, 210, 215, 220, and 225. See Section 668-11.

(2) Overcurrent protection of electrolytic cell dc process power circuits shall not be required to comply with the requirements of Article 240.

(3) Equipment located or used within the electrolytic cell line working zone or associated with the cell line dc power circuits shall not be required to comply with the provisions of Article 250.

(4) The electrolytic cells, cell line attachments, and the wiring of auxiliary equipments and devices within the cell line working zone shall not be required to comply with the provisions of Articles 110, 210, 215, 220, and 225. See Section 668-30.

> FPN: See Section 668-15 for equipment, apparatus, and structural component grounding.

668-10. Cell Line Working Zone.

(a) Area Covered. The space envelope of the cell line working zone shall encompass any space

(1) Within 96 in. (2.44 m) above energized surfaces of electrolytic cell lines or their energized attachments.
(2) Below energized surfaces of electrolytic cell lines or their energized attachments, provided the headroom in the space beneath is less than 96 in. (2.44 m).
(3) Within 42 in. (1.07 m) horizontally from energized surfaces of electrolytic cell lines or their energized attachments or from the space envelope described in Sections 668-10(a)(1) or (a)(2).

(b) Area Not Covered. The cell line working zone shall not be required to extend through or beyond walls, floors, roofs, partitions, barriers, or the like.

668-11. Direct-Current Cell Line Process Power Supply.

(a) Not Grounded. The dc cell line process power-supply conductors shall not be required to be grounded.

(b) Metal Enclosures Grounded. All metal enclosures of dc cell line process power-supply apparatus operating at a power-supply potential between terminals of over 50 volts shall be grounded as follows:

(1) Through protective relaying equipment, or
(2) By a minimum 2/0 copper grounding conductor or a conductor of equal or greater conductance

(c) Grounding Requirements. The grounding connections required by Section 668-11(b) shall be installed in accordance with Sections 250-8, 250-10, 250-12, 250-68, and 250-70.

668-12. Cell Line Conductors.

(a) Insulation and Material. Cell line conductors shall be either bare, covered, or insulated and of copper, aluminum, copper-clad aluminum, steel, or other suitable material.

(b) Size. Cell line conductors shall be of such cross-sectional area that the temperature rise under maximum load conditions and at maximum ambient shall not exceed the safe operating temperature of the conductor insulation or the material of the conductor supports.

(c) Connections. Cell line conductors shall be joined by bolted, welded, clamped, or compression connectors.

668-13. Disconnecting Means.

(a) More than One Process Power Supply. Where more than one dc cell line process power supply serves the same cell line, a disconnecting means shall be provided on the cell line circuit side of each power supply to disconnect it from the cell line circuit.

(b) Removable Links or Conductors. Removable links or removable conductors shall be permitted to be used as the disconnecting means.

668-14. Shunting Means.

(a) Partial or Total Shunting. Partial or total shunting of cell line circuit current around one or more cells shall be permitted.

(b) Shunting One or More Cells. The conductors, switches, or combination of conductors and switches used for shunting one or more cells shall comply with the applicable requirements of Section 668-12.

668-15. Grounding. For equipment, apparatus, and structural components that are required to be grounded by provisions of Article 668, the provisions of Article 250 shall apply, except a water pipe electrode shall not be required to be used. Any electrode or combination of electrodes described in Sections 250-50 and 250-52 shall be permitted.

668-20. Portable Electrical Equipment.

(a) Portable Electrical Equipment Not to Be Grounded. The frames and enclosures of portable electrical equipment used within the cell line working zone shall not be grounded.

Exception No. 1: Where the cell line voltage does not exceed 200 volts dc, these frames and enclosures shall be permitted to be grounded.

Exception No. 2: These frames and enclosures shall be permitted to be grounded where guarded.

(b) Isolating Transformers. Electrically powered, hand-held, cord-connected portable equipment with ungrounded frames or enclosures used within the cell line working zone shall be connected to receptacle circuits that have only ungrounded conductors such as a branch circuit supplied by an isolating transformer with an ungrounded secondary.

(c) Marking. Ungrounded portable electrical equipment shall be distinctively marked and shall employ plugs and receptacles of a configuration that prevents connection of this equipment to grounding receptacles and that prevents inadvertent interchange of ungrounded and grounded portable electrical equipments.

668-21. Power Supply Circuits and Receptacles for Portable Electrical Equipment.

(a) Isolated Circuits. Circuits supplying power to ungrounded receptacles for hand-held, cord-connected equipments shall be electrically isolated from any distribution system supplying areas other than the cell line working zone and shall be ungrounded. Power for these circuits shall be supplied through isolating transformers. Primaries of such transformers shall operate at not more than 600 volts between conductors and shall be provided with proper overcurrent protection. The secondary voltage of such transformers shall not exceed 300 volts between conductors, and all circuits supplied from such secondaries shall be ungrounded and shall have an approved overcurrent device of proper rating in each conductor.

(b) Noninterchangeability. Receptacles and their mating plugs for ungrounded equipment shall not have provision for a grounding conductor and shall be of a configuration that prevents their use for equipment required to be grounded.

(c) Marking. Receptacles on circuits supplied by an isolating transformer with an ungrounded secondary shall be a distinctive configuration, distinctively marked, and shall not be used in any other location in the plant.

668-30. Fixed and Portable Electrical Equipment.

(a) Electrical Equipment Not Required to Be Grounded. Alternating-current systems supplying fixed and portable electrical equipments within the cell line working zone shall not be required to be grounded.

(b) Exposed Conductive Surfaces Not Required to Be Grounded. Exposed conductive surfaces, such as electrical equipment housings, cabinets, boxes, motors, raceways, and the like, that are within the cell line working zone shall not be required to be grounded.

(c) Wiring Methods. Auxiliary electrical equipment such as motors, transducers, sensors, control devices, and alarms, mounted on an electrolytic cell or other energized surface, shall be connected to premises wiring systems by any of the following means.

(1) Multiconductor hard usage cord.
(2) Wire or cable in suitable raceways or metal or nonmetallic cable trays. If metal conduit, cable tray, armored cable, or similar metallic systems are used, they shall be installed with insulating breaks such that they will not cause a potentially hazardous electrical condition.

(d) Circuit Protection. Circuit protection shall not be required for control and instrumentation that are totally within the cell line working zone.

(e) Bonding. Bonding of fixed electrical equipment to the energized conductive surfaces of the cell line, its attachments, or auxiliaries shall be permitted. Where fixed electrical equipment is mounted on an energized conductive surface, it shall be bonded to that surface.

668-31. Auxiliary Nonelectric Connections.
Auxiliary nonelectric connections, such as air hoses, water hoses, and the like, to an electrolytic cell, its attachments, or auxiliary equipments shall not have continuous conductive reinforcing wire, armor, braids, and the like. Hoses shall be of a nonconductive material.

668-32. Cranes and Hoists.

(a) Conductive Surfaces to Be Insulated from Ground. The conductive surfaces of cranes and hoists that enter the cell line working zone shall not be required to be grounded. The portion of an overhead crane or hoist that contacts an energized electrolytic cell or energized attachments shall be insulated from ground.

(b) Hazardous Electrical Conditions. Remote crane or hoist controls that may introduce hazardous electrical condi-

tions into the cell line working zone shall employ one or more of the following systems:

(1) Isolated and ungrounded control circuit in accordance with Section 668-21(a)
(2) Nonconductive rope operator
(3) Pendant pushbutton with nonconductive supporting means and having nonconductive surfaces or ungrounded exposed conductive surfaces
(4) Radio

668-40. Enclosures.
General-purpose electrical equipment enclosures shall be permitted where a natural draft ventilation system prevents the accumulation of gases.

Article 669 — Electroplating

669-1. Scope.
The provisions of this article apply to the installation of the electrical components and accessory equipment that supply the power and controls for electroplating, anodizing, electropolishing, and electrostripping. For purposes of this article, the term *electroplating* shall be used to identify any or all of these processes.

669-2. Other Articles.
Except as modified by this article, wiring and equipment used for electroplating processes shall comply with the applicable requirements of Chapters 1 through 4.

669-3. General.
Equipment for use in electroplating processes shall be identified for such service.

669-5. Branch-Circuit Conductors.
Branch-circuit conductors supplying one or more units of equipment shall have an ampacity of not less than 125 percent of the total connected load. The ampacities for busbars shall be in accordance with Section 374-6.

669-6. Wiring Methods.
Conductors connecting the electrolyte tank equipment to the conversion equipment shall be as follows.

(a) Systems Not Exceeding 50 Volts Direct Current. Insulated conductors shall be permitted to be run without insulated support provided they are protected from physical damage. Bare copper or aluminum conductors shall be permitted where supported on insulators.

(b) Systems Exceeding 50 Volts Direct Current. Insulated conductors shall be permitted to be run on insulated supports, provided they are protected from physical damage. Bare copper or aluminum conductors shall be permitted where supported on insulators and guarded against accidental contact up to the point of termination in accordance with Section 110-27.

669-7. Warning Signs. Warning signs shall be posted to indicate the presence of bare conductors.

669-8. Disconnecting Means.

(a) More than One Power Supply. Where more than one power supply serves the same dc system, a disconnecting means shall be provided on the dc side of each power supply.

(b) Removable Links or Conductors. Removable links or removable conductors shall be permitted to be used as the disconnecting means.

669-9. Overcurrent Protection. Direct-current conductors shall be protected from overcurrent by one or more of the following:

(1) Fuses or circuit breakers,
(2) A current-sensing device that operates a disconnecting means, or
(3) Other approved means

Article 670 — Industrial Machinery

670-1. Scope. This article covers the definition of, the nameplate data for, and the size and overcurrent protection of supply conductors to industrial machinery.

> FPN: For further information, see *Electrical Standard for Industrial Machinery*, NFPA 79-1997.

670-2. Definitions.

Industrial Machinery (Machine). A power-driven machine (or a group of machines working together in a coordinated manner), not portable by hand while working, that is used to process material by cutting; forming; pressure; electrical, thermal, or optical techniques; lamination; or a combination of these processes. It can include associated equipment used to transfer material or tooling (including fixtures), assemble/disassemble, inspect or test, or package. [The associated electrical equipment including the logic controller(s) and associated software or logic together with the machine actuators and sensors are considered as part of the industrial machine.]

Industrial Manufacturing System. A systematic array of one or more industrial machines not portable by hand and that includes any associated material handling, manipulating, gauging, measuring, or inspection equipment.

670-3. Machine Nameplate Data.

(a) Permanent Nameplate. A permanent nameplate that lists supply voltage, phase, frequency, full-load current, the maximum ampere rating of the short-circuit and ground-fault protective device, ampere rating of largest motor or load, short-circuit interrupting capacity of the machine overcurrent-protective device, if furnished, and diagram number shall be attached to the control equipment enclosure or machine where plainly visible after installation.

The full-load current shown on the nameplate shall not be less than the sum of the full-load currents required for all motors and other equipment that may be in operation at the same time under normal conditions of use. Where unusual type loads, duty cycles, etc., require oversized conductors, the required capacity shall be included in the marked "full-load current." Where more than one incoming supply circuit is to be provided, the nameplate shall state the above information for each circuit.

(b) Overcurrent Protection. Where overcurrent protection is provided in accordance with Section 670-4(b), the machine shall be marked "overcurrent protection provided at machine supply terminals."

670-4. Supply Conductors and Overcurrent Protection.

(a) Size. The size of the supply conductor shall be such as to have an ampacity not less than 125 percent of the full-load current rating of all resistance heating loads plus 125 percent of the full-load current rating of highest rated motor plus the sum of the full-load current ratings of all other connected motors and apparatus that may be in operation at the same time.

> FPN: See the 0–2000-volt ampacity tables of Article 310 for ampacity of conductors rated 600 volts and below.

(b) Overcurrent Protection. A machine shall be considered as an individual unit and therefore shall be provided with a disconnecting means. The disconnecting means shall be permitted to be supplied by branch circuits protected by either fuses or circuit breakers. The disconnecting means shall not be required to incorporate overcurrent protection. Where furnished as part of the machine, overcurrent protection shall consist of a single circuit breaker or set of fuses, the machine shall bear the marking required in Section 670-3, and the supply conductors shall be considered either as feeders or taps as covered by Section 240-21.

The rating or setting of the overcurrent protective device for the circuit supplying the machine shall not be greater than the sum of the largest rating or setting of the branch-circuit short-circuit and ground-fault protective device provided with the machine, plus 125 percent of the full-load current rating of all resistance heating loads, plus the sum of the full-load currents of all other motors and apparatus that may be in operation at the same time.

Exception: Where one or more instantaneous trip circuit breakers or motor short-circuit protectors are used for motor branch-circuit short-circuit and ground-fault protection as permitted by Section 430-52(c), the procedure specified above for determining the maximum rating of the protective device for the circuit supplying the machine shall apply with the following provision. For the purpose of the calculation, each instantaneous trip circuit breaker or motor short-circuit protector shall be assumed to have a rating not exceeding the maximum percentage of motor full-load current permitted by Table 430-152 for the type of machine supply circuit protective device employed.

Where no branch-circuit short-circuit and ground-fault protective device is provided with the machine, the rating or setting of the overcurrent protective device shall be based on Sections 430-52 and 430-53, as applicable.

670-5. Clearance. Where the conditions of maintenance and supervision ensure that only qualified persons will service the installation, the dimensions of the working space in the direction of access to live parts operating at not over 150 volts line-to-line or line-to-ground that are likely to require examination, adjustment, servicing, or maintenance while energized shall be a minimum of 2½ ft (762 mm). Where controls are enclosed in cabinets, the door(s) shall open at least 90 degrees or be removable.

Exception: Where the enclosure requires a tool to open, and where only diagnostic and troubleshooting testing is involved on live parts, the clearances shall be permitted to be less than 2½ ft (762 mm).

Article 675 — Electrically Driven or Controlled Irrigation Machines

A. General

675-1. Scope. The provisions of this article apply to electrically driven or controlled irrigation machines, and to the branch circuits and controllers for such equipment.

675-2. Definitions.

Center Pivot Irrigation Machines. A multimotored irrigation machine that revolves around a central pivot and employs alignment switches or similar devices to control individual motors.

Collector Rings. An assembly of slip rings for transferring electrical energy from a stationary to a rotating member.

Irrigation Machines. An electrically driven or controlled machine, with one or more motors, not hand portable,

and used primarily to transport and distribute water for agricultural purposes.

675-3. Other Articles. These provisions are in addition to, or amendatory of, the provisions of Article 430 and other articles in this *Code* that apply except as modified in this article.

675-4. Irrigation Cable.

(a) Construction. The cable used to interconnect enclosures on the structure of an irrigation machine shall be an assembly of stranded, insulated conductors with nonhygroscopic and nonwicking filler in a core of moisture- and flame-resistant nonmetallic material overlaid with a metallic covering and jacketed with a moisture-, corrosion-, and sunlight-resistant nonmetallic material.

The conductor insulation shall be of a type listed in Table 310-13 for an operating temperature of 75°C (167°F) and for use in wet locations. The core insulating material thickness shall not be less than 30 mils (762 μm), and the metallic overlay thickness shall not be less than 8 mils (203 μm). The jacketing material thickness shall not be less than 50 mils (1.27 mm).

A composite of power, control, and grounding conductors in the cable shall be permitted.

(b) Alternate Wiring Methods. Other cables listed for the purpose.

(c) Supports. Irrigation cable shall be secured by straps, hangers, or similar fittings identified for the purpose and installed as not to damage the cable. Cable shall be supported at intervals not exceeding 4 ft (1.22 m).

(d) Fittings. Fittings shall be used at all points where irrigation cable terminates. The fittings shall be designed for use with the cable and shall be suitable for the conditions of service.

675-5. More than Three Conductors in a Raceway or Cable. The signal and control conductors of a raceway or cable shall not be counted for the purpose of derating the conductors as required in Section 310-15(b)(2)(a).

675-6. Marking on Main Control Panel. The main control panel shall be provided with a nameplate that shall give the following information:

(1) The manufacturer's name, the rated voltage, the phase, and the frequency
(2) The current rating of the machine, and
(3) The rating of the main disconnecting means and size of overcurrent protection required

675-7. Equivalent Current Ratings. Where intermittent duty is not involved, the provisions of Article 430 shall be used for determining ratings for controllers, disconnecting means, conductors, and the like. Where irrigation machines have inherent intermittent duty, the following determinations of equivalent current ratings shall be used.

(a) Continuous-Current Rating. The equivalent continuous-current rating for the selection of branch-circuit conductors and overcurrent protection shall be equal to 125 percent of the motor nameplate full-load current rating of the largest motor plus a quantity equal to the sum of each of the motor nameplate full-load current ratings of all remaining motors on the circuit multiplied by the maximum percent duty cycle at which they can continuously operate.

(b) Locked-Rotor Current. The equivalent locked-rotor current rating shall be equal to the numerical sum of the locked-rotor current of the two largest motors plus 100 percent of the sum of the motor nameplate full-load current ratings of all the remaining motors on the circuit.

675-8. Disconnecting Means.

(a) Main Controller. A controller that is used to start and stop the complete machine shall meet all of the following requirements:

(1) An equivalent continuous current rating not less than specified in Sections 675-7(a) or 675-22(a)
(2) A horsepower rating not less than the value from Tables 430-151(A) and (B) based on the equivalent locked-rotor current specified in Sections 675-7(b) or 675-22(b)

(b) Main Disconnecting Means. The main disconnecting means for the machine shall provide overcurrent protection and shall be at the point of connection of electrical power to the machine or shall be visible and not more than 50 ft (15.2 m) from the machine and shall be readily accessible and capable of being locked in the open position. This disconnecting means shall have a horsepower and current rating not less than required for the main controller.

Exception: Circuit breakers without marked horsepower ratings shall be permitted in accordance with Section 430-109.

(c) Disconnecting Means for Individual Motors and Controllers. A disconnecting means shall be provided to simultaneously disconnect all ungrounded conductors for each motor and controller and shall be located as required by Article 430, Part J. The disconnecting means shall not be required to be readily accessible.

675-9. Branch-Circuit Conductors. The branch-circuit conductors shall have an ampacity not less than specified in Sections 675-7(a) or 675-22(a).

675-10. Several Motors on One Branch Circuit.

(a) Protection Required. Several motors, each not exceeding 2-hp rating, shall be permitted to be used on an irrigation machine circuit protected at not more than 30 amperes at 600 volts, nominal, or less, provided all of the following conditions are met.

(1) The full-load rating of any motor in the circuit shall not exceed 6 amperes.
(2) Each motor in the circuit shall have individual overload protection in accordance with Section 430-32.
(3) Taps to individual motors shall not be smaller than No. 14 copper and not more than 25 ft (7.62 m) in length.

(b) Individual Protection Not Required. Individual branch-circuit short-circuit protection for motors and motor controllers shall not be required where the requirements of Section 675-10(a) are met.

675-11. Collector Rings.

(a) Transmitting Current for Power Purposes. Collector rings shall have a current rating not less than 125 percent of the full-load current of the largest device served plus the full-load current of all other devices served, or as determined from Sections 675-7(a) or 675-22(a).

(b) Control and Signal Purposes. Collector rings for control and signal purposes shall have a current rating not less than 125 percent of the full-load current of the largest device served plus the full-load current of all other devices served.

(c) Grounding. The collector ring used for grounding shall have a current rating of not less than that sized in accordance with Section 675-11(a).

(d) Protection. Collector rings shall be protected from the expected environment and from accidental contact by means of a suitable enclosure.

675-12. Grounding. The following equipment shall be grounded:

(1) All electrical equipment on the irrigation machine
(2) All electrical equipment associated with the irrigation machine
(3) Metal junction boxes and enclosures, and
(4) Control panels or control equipment that supply or control electrical equipment to the irrigation machine

Exception: Grounding shall not be required on machines where all of the following provisions are met.

(a) *The machine is electrically controlled but not electrically driven.*

(b) *The control voltage is 30 volts or less.*

(c) *The control or signal circuits are current limited as specified in Chapter 9, Tables 11(a) and 11(b).*

675-13. Methods of Grounding. Machines that require grounding shall have a noncurrent-carrying equipment grounding conductor provided as an integral part of each cord, cable, or raceway. This grounding conductor shall be sized not less than the largest supply conductor in each cord, cable, or raceway. Feeder circuits supplying power to irrigation machines shall have an equipment grounding conductor sized according to Table 250-122.

675-14. Bonding. Where electrical grounding is required on an irrigation machine, the metallic structure of the machine, metallic conduit, or metallic sheath of cable shall be bonded to the grounding conductor. Metal-to-metal contact with a part that is bonded to the grounding conductor and the noncurrent-carrying parts of the machine shall be considered as an acceptable bonding path.

675-15. Lightning Protection. If an irrigation machine has a stationary point, a grounding electrode system in accordance with Article 250, Part C, shall be connected to the machine at the stationary point for lightning protection.

675-16. Energy from More than One Source. Equipment within an enclosure receiving electrical energy from more than one source shall not be required to have a disconnecting means for the additional source, provided that its voltage is 30 volts or less and it meets the requirements of Part C of Article 725.

675-17. Connectors. External plugs and connectors on the equipment shall be of the weatherproof type.

Unless provided solely for the connection of circuits meeting the requirements of Part C of Article 725, external plugs and connectors shall be constructed as specified in Section 250-124(a).

B. Center Pivot Irrigation Machines

675-21. General. The provisions of Part B are intended to cover additional special requirements that are peculiar to center pivot irrigation machines. See Section 675-2 for definition *of Center Pivot Irrigation Machines.*

675-22. Equivalent Current Ratings. In order to establish ratings of controllers, disconnecting means, conductors, and

the like, for the inherent intermittent duty of center pivot irrigation machines, the following determination shall be used.

(a) Continuous-Current Rating. The equivalent continuous-current rating for the selection of branch-circuit conductors and branch-circuit devices shall be equal to 125 percent of the motor nameplate full-load current rating of the largest motor plus 60 percent of the sum of the motor nameplate full-load current ratings of all remaining motors on the circuit.

(b) Locked-Rotor Current. The equivalent locked-rotor current rating shall be equal to the numerical sum of two times the locked-rotor current of the largest motor plus 80 percent of the sum of the motor nameplate full-load current ratings of all the remaining motors on the circuit.

Article 680 — Swimming Pools, Fountains, and Similar Installations

A. General

680-1. Scope. The provisions of this article apply to the construction and installation of electrical wiring for and equipment in or adjacent to all swimming, wading, therapeutic, and decorative pools, fountains, hot tubs, spas, and hydromassage bathtubs, whether permanently installed or storable, and to metallic auxiliary equipment, such as pumps, filters, and similar equipment.

680-2. Approval of Equipment. All electrical equipment installed in the water, walls, or decks of pools, fountains, and similar installations shall comply with the provisions of this article.

680-3. Other articles.

(a) Except as modified by this section, wiring and equipment in or adjacent pools and fountains shall comply with the applicable requirements of Chapters 1 through 4.

> FPN: See Section 370-23 for junction boxes and Section 347-3 for rigid nonmetallic conduit.

(b) The installation and wiring of audio equipment adjacent to pools and fountains shall comply with the applicable requirements of Article 640. Underwater loudspeakers shall be installed in accordance with Section 680-23.

680-4. Definitions.

Cord- and Plug-Connected Lighting Assembly. A lighting assembly consisting of a lighting fixture intended for installation in the wall of a spa, hot tub, or storable pool, and a cord- and plug-connected transformer.

Dry-Niche Lighting Fixture. A lighting fixture intended for installation in the wall of a pool or fountain in a niche that is sealed against the entry of pool water.

Forming Shell. A structure designed to support a wet-niche lighting fixture assembly and intended for mounting in a pool or fountain structure.

Fountain. As used in this article, the term includes fountains, ornamental pools, display pools, and reflection pools. It does not include drinking fountains.

Hydromassage Bathtub. A permanently installed bathtub equipped with a recirculating piping system, pump, and associated equipment. It is designed so it can accept, circulate, and discharge water upon each use.

No-Niche Lighting Fixture. A lighting fixture intended for installation above or below the water without a niche.

Packaged Spa or Hot Tub Equipment Assembly. A factory-fabricated unit consisting of water-circulating, heating, and control equipment mounted on a common base, intended to operate a spa or hot tub. Equipment may include pumps, air blowers, heaters, lights, controls, sanitizer generators, etc.

Packaged Therapeutic Tub or Hydrotherapeutic Tank Equipment Assembly. A factory-fabricated unit consisting of water-circulating, heating, and control equipment mounted on a common base, intended to operate a therapeutic tub or hydrotherapeutic tank. Equipment may include pumps, air blowers, heaters, lights, controls, sanitizer generators, etc.

Permanently Installed Decorative Fountains and Reflection Pools. Those that are constructed in the ground, on the ground, or in a building in such a manner that the fountain cannot be readily disassembled for storage, whether or not served by electrical circuits of any nature. These units are primarily constructed for their aesthetic value and are not intended for swimming or wading.

Permanently Installed Swimming, Wading, and Therapeutic Pools. Those that are constructed in the ground or partially in the ground, and all others capable of holding water in a depth greater than 42 in. (1.07 m), and all pools installed inside of a building, regardless of water depth, whether or not served by electrical circuits of any nature.

Pool. As used in this article, the term includes swimming, wading, and permanently installed therapeutic pools.

Pool Cover, Electrically Operated. Motor-driven equipment designed to cover and uncover the water surface of a pool by means of a flexible sheet or rigid frame.

Self-Contained Spa or Hot Tub. Factory-fabricated unit consisting of a spa or hot tub vessel with all water-circulating, heating, and control equipment integral to the unit. Equipment may include pumps, air blowers, heaters, lights, controls, sanitizer generators, etc.

Self-Contained Therapeutic Tubs or Hydrotherapeutic Tanks. A factory-fabricated unit consisting of a therapeutic tub or hydrotherapeutic tank with all water-circulating, heating, and control equipment integral to the unit. Equipment may include pumps, air blowers, heaters, light controls, sanitizer generators, etc.

Spa or Hot Tub. A hydromassage pool, or tub for recreational or therapeutic use, not located in health care facilities, designed for immersion of users, and usually having a filter, heater, and motor-driven blower. It may be installed indoors or outdoors, on the ground or supporting structure, or in the ground or supporting structure. Generally, a spa or hot tub is not designed or intended to have its contents drained or discharged after each use.

Storable Swimming or Wading Pool. Those that are constructed on or above the ground and are capable of holding water to a maximum depth of 42 in. (1.07 m), or a pool with nonmetallic, molded polymeric walls or inflatable fabric walls regardless of dimension.

Wet-Niche Lighting Fixture. A lighting fixture intended for installation in a forming shell mounted in a pool or fountain structure where the fixture will be completely surrounded by water.

680-5. Transformers and Ground-Fault Circuit Interrupters.

(a) Transformers. Transformers used for the supply of underwater fixtures, together with the transformer enclosure, shall be identified for the purpose. The transformer shall be an isolated winding type that has a grounded metal barrier between the primary and secondary windings.

(b) Ground-Fault Circuit Interrupters. Ground-fault circuit interrupters shall be self-contained units, circuit-breaker types, receptacle types, or other approved types.

(c) Wiring. Conductors on the load side of a ground-fault circuit interrupter or of a transformer, used to comply with the provisions of Section 680-20(a)(1), shall not occupy raceway, boxes, or enclosures containing other conductors unless the other conductors are protected by ground-fault circuit interrupters or are grounding conductors. Supply conductors to a feed-through type ground-fault circuit interrupter shall be permitted in the same enclosure.

Ground-fault circuit interrupters shall be permitted in a panelboard that contains circuits protected by other than ground-fault circuit interrupters.

680-6. Receptacles, Lighting Fixtures, Lighting Outlets, Switching Devices, and Ceiling-Suspended (Paddle) Fans.

(a) Receptacles.

(1) A receptacle(s) that provides power for a water-pump motor(s) for, or other loads directly related to the circulation and sanitation system, a permanently installed pool or fountain, as permitted in Section 680-7, shall be permitted between 5 ft and 10 ft (1.52 m and 3.05 m) from the inside walls of the pool or fountain, and, where so located, shall be single and of the locking and grounding types and shall be protected by a ground-fault circuit interrupter(s).

Other receptacles on the property shall be located at least 10 ft (3.05 m) from the inside walls of a pool or fountain.

(2) Where a permanently installed pool is installed at a dwelling unit(s), at least one 125-volt 15- or 20-ampere receptacle on a general-purpose branch circuit shall be located a minimum of 10 ft (3.05 m) from and not more than 20 ft (6.08 m) from the inside wall of the pool. This receptacle shall be located not more than 6 ft 6 in. (1.98 m) above the floor, platform, or grade level serving the pool.

(3) All 125-volt receptacles located within 20 ft (6.08 m) of the inside walls of a pool or fountain shall be protected by a ground-fault circuit interrupter.

FPN: In determining the above dimensions, the distance to be measured is the shortest path the supply cord of an appliance connected to the receptacle would follow without piercing a floor, wall, ceiling, doorway with hinged or sliding door, window opening, or other effective permanent barrier.

(b) Lighting Fixtures, Lighting Outlets, and Ceiling-Suspended (Paddle) Fans.

(1) In outdoor pool areas, lighting fixtures, lighting outlets, and ceiling-suspended (paddle) fans shall not be installed over the pool or over the area extending 5 ft (1.52 m) horizontally from the inside walls of a pool unless no part of the lighting fixture or ceiling-suspended (paddle) fan is less than 12 ft (3.66 m) above the maximum water level.

(2) Existing lighting fixtures and lighting outlets located less than 5 ft (1.52 m) measured horizontally from the inside walls of a pool shall be at least 5 ft (1.52 m) above the surface of the maximum water level, shall be rigidly attached to the existing structure, and shall be protected by a ground-fault circuit interrupter.

(3) In indoor pool areas, the limitations of Section 680-6(b)(1) shall not apply if all of the following conditions are complied with

(a) Fixtures are of a totally enclosed type,
(b) A ground-fault circuit interrupter is installed in the branch circuit supplying the fixture(s) or ceiling-suspended (paddle) fans, and
(c) The distance from the bottom of the fixture or ceiling-suspended (paddle) fan to the maximum water level is not less than 7 ft 6 in. (2.29 m).

(4) Lighting fixtures and lighting outlets installed in the area extending between 5 ft (1.52 m) and 10 ft (3.05 m) horizontally from the inside walls of a pool shall be protected by a ground-fault circuit interrupter unless installed 5 ft (1.52 m) above the maximum water level and rigidly attached to the structure adjacent to or enclosing the pool.

(5) Cord-connected lighting fixtures shall meet the same specifications as other cord- and plug-connected equipment as set forth in Section 680-7 where installed within 16 ft (4.88 m) of any point on the water surface, measured radially.

(c) Switching Devices. Switching devices on the property shall be located at least 5 ft (1.52 m) horizontally from the inside walls of a pool unless separated from the pool by a solid fence, wall, or other permanent barrier.

(d) Motors in Other than Dwelling Units. Wiring supplying pool pump motors rated 15 and 20 amperes, 125 volt or 240 volt, single phase, whether by receptacle or direct connection, shall be provided with ground-fault circuit-interrupter protection for personnel.

680-7. Cord- and Plug-Connected Equipment. Fixed or stationary equipment rated 20 amperes or less, other than an underwater lighting fixture for a permanently installed pool, shall be permitted to be connected with a flexible cord to facilitate the removal or disconnection for maintenance or repair. For other than storable pools, the flexible cord shall not exceed 3 ft (914 mm) in length and shall have a copper equipment grounding conductor not smaller than No. 12 with a grounding-type attachment plug.

FPN: See Section 680-25(e) for connection with flexible cords.

680-8. Overhead Conductor Clearances. The following parts of pools shall not be placed under existing service-drop conductors or any other open overhead wiring; nor shall such wiring be installed above the following:

(1) Pools and the area extending 10 ft (3.05 m) horizontally from the inside of the walls of the pool,
(2) Diving structure, or
(3) Observation stands, towers, or platforms unless the installations provide the clearances in Table 680-8

Table 680-8. Clearances

	Insulated Supply or Service Drop Cables, 0–750 Volts to Ground, Supported on and Cabled Together with an Effectively Grounded Bare Messenger or Effectively Grounded Neutral Conductor	All Other Supply or Service-Drop Conductors Voltage to Ground	
		0–15 kV	Greater than 15–50 kV
A Clearance in any direction to the water level, edge of water surface, base of diving platform, or permanently anchored raft	22 ft (6.7 m)	25 ft (7.62 m)	27 ft (8.23 m)
B Clearance in any direction to the diving platform or tower	14 ft (4.27 m)	17 ft (5.2 m)	18 ft (5.49 m)
C Horizontal limit of clearance measured from inside wall of the pool	This limit shall extend to the outer edge of the structures listed in (1) and (2) but not less than 10 ft (3.05 m).		

Figure 680-8.

Community antenna system coaxial cables complying with Article 820 and the supporting messengers shall be permitted at a height of not less than 10 ft (3.05 m) above swimming and wading pools, diving structures, and observation stands, towers, or platforms.

FPN: See Sections 225-18 and 225-19 for clearances for conductors not covered by this section.

680-9. Electric Pool Water Heaters. All electric pool water heaters shall have the heating elements subdivided into loads not exceeding 48 amperes and protected at not more than 60 amperes.

The ampacity of the branch-circuit conductors and the rating or setting of overcurrent protective devices shall not be less than 125 percent of the total load of the nameplate rating.

680-10. Underground Wiring Location. Underground wiring shall not be permitted under the pool or within the area extending 5 ft (1.52 m) horizontally from the inside wall of the pool unless this wiring is necessary to supply pool equipment permitted by this article. Where space limitations prevent wiring from being routed 5 ft (1.52 m) or more from the pool, such wiring shall be permitted where installed in rigid metal conduit, intermediate metal conduit, or a nonmetallic raceway system. All metal conduit shall be corrosion resistant and suitable for the location. The minimum burial depth shall be as follows.

Wiring Method	Minimum Burial (in.)
Rigid metal conduit	6
Intermediate metal conduit	6
Nonmetallic raceways listed for direct burial without concrete encasement	18
Other approved raceways*	18

Note: For SI units, 1 in. = 25.4 mm.

*Raceways approved for burial only where concrete encased shall require a concrete envelope not less than 2 in. (50.8 mm) thick.

680-11. Equipment Rooms and Pits. Electric equipment shall not be installed in rooms or pits that do not have adequate drainage to prevent water accumulation during normal operation or filter maintenance.

680-12. Disconnecting Means. A disconnecting means shall be provided and be accessible, located within sight from all pools, spas, and hot tub equipment, and shall be located at least 5 ft (1.52 m) from the inside walls of the pool, spa, or hot tub.

B. Permanently Installed Pools

680-20. Underwater Lighting Fixtures. Paragraphs (a) through (d) apply to all lighting fixtures installed below the normal water level of the pool.

(a) General.

(1) The design of an underwater lighting fixture supplied from a branch circuit either directly or by way of a transformer meeting the requirements of Section 680-5(a) shall be such that, where the fixture is properly installed without a ground-fault circuit interrupter, there is no shock hazard with any likely combination of fault conditions during normal use (not relamping).

In addition, a ground-fault circuit interrupter shall be installed in the branch circuit supplying fixtures operating at more than 15 volts, so that there is no shock hazard during relamping. The installation of the ground-fault circuit interrupter shall be such that there is no shock hazard with any likely fault-condition combination that involves a person in a conductive path from any ungrounded part of the branch circuit or the fixture to ground.

Compliance with this requirement shall be obtained by the use of a listed underwater lighting fixture and by installation of a listed ground-fault circuit interrupter in the branch circuit.

(2) No lighting fixtures shall be installed for operation on supply circuits over 150 volts between conductors.

(3) Lighting fixtures mounted in walls shall be installed with the top of the fixture lens at least 18 in. (457 mm) below the normal water level of the pool, unless the lighting fixture is listed and identified for use at a depth of not less than 4 in. (102 mm) below the normal water level of the pool.

(4) A lighting fixture facing upward shall have the lens adequately guarded to prevent contact by any person.

(5) Fixtures that depend on submersion for safe operation shall be inherently protected against the hazards of overheating when not submerged.

(b) Wet-Niche Fixtures.

(1) Forming shells shall be installed for the mounting of all wet-niche underwater fixtures and shall be equipped with provisions for conduit entries.

Conduit shall extend from the forming shell to a suitable junction box or other enclosure located as provided in Section 680-21. Conduit shall be rigid metal, intermediate metal, liquidtight flexible nonmetallic, or rigid nonmetallic.

Metal conduit shall be of brass or other approved corrosion-resistant metal.

Where a nonmetallic conduit is used, a No. 8 insulated copper conductor shall be installed in this conduit with provisions for terminating in the forming shell, junction box or transformer enclosure, or ground-fault circuit-interrupter enclosure unless a listed low-voltage lighting system is used, not requiring grounding. The termination of the No. 8 conductor in the forming shell shall be covered with, or encapsulated in, a listed potting compound to protect such connection from the possible deteriorating effect of pool water. Metal parts of the fixture and forming shell in contact with the pool water shall be of brass or other approved corrosion-resistant metal.

(2) The end of the flexible-cord jacket and the flexible-cord conductor terminations within a fixture shall be covered with, or encapsulated in, a suitable potting compound to prevent the entry of water into the fixture through the cord or its conductors. In addition, the grounding connection within a fixture shall be similarly treated to protect such connection from the deteriorating effect of pool water in the event of water entry into the fixture.

(3) The fixture shall be bonded to and secured to the forming shell by a positive locking device that ensures a low-resistance contact and requires a tool to remove the fixture from the forming shell. Bonding is not required for fixtures listed for the application, having no noncurrent-carrying metal parts.

(c) Dry-Niche Fixtures.
A dry-niche lighting fixture shall be provided with a provision for drainage of water and a means for accommodating one equipment grounding conductor for each conduit entry.

Approved rigid metal conduit, intermediate metal conduit, liquidtight flexible nonmetallic conduit, or rigid nonmetallic conduit shall be installed from the fixture to the service equipment or panelboard. Where installed on buildings, electrical metallic tubing shall be permitted to be used to protect conductors. Where installed within buildings, electrical nonmetallic tubing or electrical metallic tubing shall be permitted to be used to protect conductors. A junction box shall not be required but, if used, shall not be required to be elevated or located as specified in Section 680-21(a)(4), if the fixture is specifically identified for the purpose.

(d) No-Niche Fixtures.
A no-niche fixture shall be as follows:

(1) Listed for the purpose
(2) Installed in accordance with the requirements of Section 680-20(b)

Where connection to a forming shell is specified, the connection shall be to the mounting bracket.

680-21. Junction Boxes and Enclosures for Transformers or Ground-Fault Circuit Interrupters.

(a) Junction Boxes. A junction box connected to a conduit that extends directly to a forming shell or mounting bracket of a no-niche fixture shall be as follows.

(1) Listed and labeled for the purpose.

(2) Equipped with threaded entries or hubs or a nonmetallic hub listed for the purpose.

(3) Of copper, brass, suitable plastic, or other approved corrosion-resistant material.

(4) Provided with electrical continuity between every connected metal conduit and the grounding terminals by means of copper, brass, or other approved corrosion-resistant metal that is integral with the box.

(5) Located not less than 4 in. (102 mm), measured from the inside of the bottom of the box, above the ground level, or pool deck, or not less than 8 in. (203 mm) above the maximum pool water level, whichever provides the greater elevation, and located not less than 4 ft (1.22 m) from the inside wall of the pool, unless separated from the pool by a solid fence, wall, or other permanent barrier. If used on a lighting system operating at 15 volts or less, a flush deck box shall be permitted provided

(a) An approved potting compound is used to fill the box to prevent the entrance of moisture, and

(b) The flush deck box is located not less than 4 ft (1.22 m) from the inside wall of the pool.

(b) Other Enclosures. An enclosure for a transformer, ground-fault circuit interrupter, or a similar device connected to a conduit that extends directly to a forming shell or mounting bracket of a no-niche fixture shall be as follows:

(1) Listed and labeled for the purpose

(2) Equipped with threaded entries or hubs or a nonmetallic hub listed for the purpose

(3) Provided with an approved seal, such as duct seal at the conduit connection, that prevents circulation of air between the conduit and the enclosures

(4) Provided with electrical continuity between every connected metal conduit and the grounding terminals by means of copper, brass, or other approved corrosion-resistant metal that is integral with the enclosures

(5) Located not less than 4 in. (102 mm), measured from the inside bottom of the enclosure, above the ground level or pool deck, or not less than 8 in. (203 mm) above the maximum pool water level, whichever provides the greater elevation, and located not less than 4 ft (1.22 m) from the inside wall of the pool, unless separated from the pool by a solid fence, wall, or other permanent barrier

(c) Protection. Junction boxes and enclosures mounted above the grade of the finished walkway around the pool shall not be located in the walkway unless afforded additional protection, such as by location under diving boards, adjacent to fixed structures, and the like.

(d) Grounding Terminals. Junction boxes, transformer enclosures, and ground-fault circuit-interrupter enclosures connected to a conduit that extends directly to a forming shell or mounting bracket of a no-niche fixture shall be provided with a number of grounding terminals that shall be at least one more than the number of conduit entries.

(e) Strain Relief. The termination of a flexible cord of an underwater lighting fixture within a junction box, transformer enclosure, ground-fault circuit interrupter, or other enclosure shall be provided with a strain relief.

680-22. Bonding. It shall not be the intent of this section to require that the No. 8 or larger solid copper bonding conductor be extended or attached to any remote panelboard, service equipment, or any electrode, but only that it shall be employed to eliminate voltage gradients in the pool area as prescribed.

(a) Bonded Parts. The following parts shall be bonded together.

(1) All metallic parts of the pool structure, including the reinforcing metal of the pool shell, coping stones, and deck. The usual steel tie wires shall be considered suitable for bonding the reinforcing steel together, and welding or special clamping shall not be required. These tie wires shall be made tight. Where reinforcing steel is effectively insulated by a listed encapsulating nonconductive compound, at the time of manufacture and installation, it shall not be required to be bonded.

(2) All forming shells and mounting brackets of a no-niche fixture unless a listed low-voltage lighting system is used, not requiring bonding.

(3) All metal fittings within or attached to the pool structure. Isolated parts that are not over 4 in. (102 mm) in any dimension and do not penetrate into the pool structure more than 1 in. (25.4 mm) shall not require bonding.

(4) Metal parts of electrical equipment associated with the pool water circulating system, including pump motors and metal parts of equipment associated with pool covers, including electric motors. Metal parts of listed equipment incorporating an approved system of double insulation and

providing a means for grounding internal nonaccessible, noncurrent-carrying metal parts shall not be bonded.

(5) Metal-sheathed cables and raceways, metal piping, and all fixed metal parts that are within 5 ft (1.52 m) horizontally of the inside walls of the pool, and within 12 ft (3.66 m) above the maximum water level of the pool, or any observation stands, towers, or platforms, or from any diving structures, and that are not separated from the pool by a permanent barrier.

(b) Common Bonding Grid. The parts specified in (a) shall be connected to a common bonding grid with a solid copper conductor, insulated, covered, or bare, not smaller than No. 8. Connection shall be made by exothermic welding or by pressure connectors or clamps that are labeled as being suitable for the purpose and are of the following material: stainless steel, brass, copper, or copper alloy. The common bonding grid shall be permitted to be any of the following:

(1) The structural reinforcing steel of a concrete pool where the reinforcing rods are bonded together by the usual steel tie wires or the equivalent
(2) The wall of a bolted or welded metal pool
(3) A solid copper conductor, insulated, covered, or bare, not smaller than No. 8
(4) Rigid metal conduit or intermediate metal conduit of brass or other identified corrosion-resistant metal conduit

Structural reinforcing steel or the walls of bolted or welded metal pool structures shall be permitted as a common bonding grid for nonelectrical parts where connections can be made in accordance with Section 250-8.

(c) Pool Water Heaters. For pool water heaters rated at more than 50 amperes that have specific instructions regarding bonding and grounding, only those parts designated to be bonded shall be bonded, and only those parts designated to be grounded shall be grounded.

680-23. Underwater Audio Equipment. All underwater audio equipment shall be identified for the purpose.

(a) Speakers. Each speaker shall be mounted in an approved metal forming shell, the front of which is enclosed by a captive metal screen, or equivalent, that is bonded to and secured to the forming shell by a positive locking device that ensures a low-resistance contact and requires a tool to open for installation or servicing of the speaker. The forming shell shall be installed in a recess in the wall or floor of the pool.

(b) Wiring Methods. Rigid metal conduit or intermediate metal conduit of brass or other identified corrosion-resistant

metal or rigid nonmetallic conduit shall extend from the forming shell to a suitable junction box or other enclosure as provided in Section 680-21. Where rigid nonmetallic conduit is used, a No. 8 insulated copper conductor shall be installed in this conduit with provisions for terminating in the forming shell and the junction box. The termination of the No. 8 conductor in the forming shell shall be covered with, or encapsulated in, a suitable potting compound to protect such connection from the possible deteriorating effect of pool water.

(c) Forming Shell and Metal Screen. The forming shell and metal screen shall be of brass or other approved corrosion-resistant metal.

680-24. Grounding. The following equipment shall be grounded:

(1) Wet-niche and no-niche underwater lighting fixtures, other than those low-voltage systems listed for the application without a grounding conductor
(2) Dry-niche underwater lighting fixtures
(3) All electrical equipment located within 5 ft (1.52 m) of the inside wall of the pool
(4) All electrical equipment associated with the recirculating system of the pool
(5) Junction boxes
(6) Transformer enclosures
(7) Ground-fault circuit interrupters
(8) Panelboards that are not part of the service equipment and that supply any electrical equipment associated with the pool

680-25. Methods of Grounding.

(a) General. The following provisions shall apply to the grounding of underwater lighting fixtures, junction boxes, metal transformer enclosures, panelboards, motors, and other electrical enclosures and equipment.

(b) Pool Lighting Fixtures and Related Equipment.

(1) Wet-niche, dry-niche, or no-niche lighting fixtures shall be connected to an equipment grounding conductor sized in accordance with Table 250-122 but not smaller than No. 12.

Exception: An equipment grounding conductor between the wiring chamber of the secondary winding of a transformer and a junction box shall be sized in accordance with the overcurrent device in this circuit.

(2) The equipment grounding conductor shall be an insulated copper conductor and shall be installed with the circuit conductors in rigid metal conduit, intermediate metal conduit, liquidtight flexible nonmetallic conduit, or rigid non-metallic conduit.

(3) Where installed on buildings, electrical metallic tubing shall be permitted to be used to protect conductors. Where installed within buildings, electrical nonmetallic tubing or electrical metallic tubing shall be permitted to be used to protect conductors.

FPN: For requirements of electrical nonmetallic tubing, see Article 331.

Exception: Where connecting to transformers for pool lights, liquidtight flexible metal conduit or liquidtight flexible nonmetallic conduit shall be permitted to be used when installed in accordance with Article 351 and does not exceed 6 ft (1.83 m) for any one length or 10 ft (3.05 m) of total length used.

(4) The junction box, transformer enclosure, or other enclosure in the supply circuit to a wet-niche or no-niche lighting fixture and the field-wiring chamber of a dry-niche lighting fixture shall be grounded to the equipment grounding terminal of the panelboard. This terminal shall be directly connected to the panelboard enclosure. The equipment grounding conductor shall be installed without joint or splice except as permitted in (a) and (b).

(a) Where more than one underwater lighting fixture is supplied by the same branch circuit, the equipment grounding conductor, installed between the junction boxes, transformer enclosures, or other enclosures in the supply circuit to wet-niche fixtures, or between the field-wiring compartments of dry-niche fixtures, shall be permitted to be terminated on grounding terminals.

(b) Where the underwater lighting fixture is supplied from a transformer, ground-fault circuit interrupter, clock-operated switch, or a manual snap switch that is located between the panelboard and a junction box connected to the conduit that extends directly to the underwater lighting fixture, the equipment grounding conductor shall be permitted to terminate on grounding terminals on the transformer, ground-fault circuit interrupter, clock-operated switch enclosure, or an outlet box used to enclose a snap switch.

(5) Wet-niche or no-niche lighting fixtures that are supplied by a flexible cord or cable shall have all exposed noncurrent-carrying metal parts grounded by an insulated copper equipment grounding conductor that is an integral part of the cord or cable. This grounding conductor shall be connected to a grounding terminal in the supply junction box, transformer enclosure, or other enclosure. The grounding conductor shall not be smaller than the supply conductors and not smaller than No. 16.

(c) Motors. Pool-associated motors shall be connected to an equipment grounding conductor sized in accordance with Table 250-122 but not smaller than No. 12. It shall be an insulated copper conductor and shall be installed with the circuit conductors in rigid metal conduit, intermediate metal conduit, rigid nonmetallic conduit, or Type MC cable listed for the application. Where installed on or within buildings, electrical metallic tubing shall be permitted to be used to protect the conductors.

Where necessary to employ flexible connections at or adjacent to the motor, liquidtight flexible metal or nonmetallic conduit with approved fittings shall be permitted.

In the interior of a one-family dwelling or in the interior of another building or structure associated with a one-family dwelling, any of the wiring methods recognized in Chapter 3 of this *Code* that contain a copper equipment grounding conductor that is insulated or covered by the outer sheath of the wiring method and is not smaller than No. 12 shall be permitted to be used for the connection of pool-associated motors. Flexible cord shall be permitted in accordance with Section 680-7.

(d) Panelboards. A panelboard and, where installed, a disconnecting means, that are not part of the service equipment or source of a separately derived system, shall have an equipment grounding conductor installed between its grounding terminal and the grounding terminal of the applicable service equipment or source of a separately derived system. This conductor shall be sized in accordance with Table 250-122 but not smaller than No. 12. On separately derived systems, this conductor shall be sized in accordance with Table 250-66 but not smaller than No. 8. It shall be an insulated conductor and shall be installed with the feeder conductors in rigid metal conduit, intermediate metal conduit, liquidtight flexible nonmetallic conduit, or rigid nonmetallic conduit. Electrical metallic tubing shall be permitted to be used to protect conductors where installed on or within the building in accordance with Article 348. Electrical nonmetallic tubing shall be permitted to be used to enclose the conductors where installed within the building in accordance with Article 331. The equipment grounding conductor shall be connected to an equipment grounding terminal of the panelboard and, where installed, to the enclosure for a disconnecting means.

(1) Where the equipment grounding conductor between an existing remote panelboard and the service equipment is connected by means of a flexible metal conduit or an approved cable assembly with an insulated or covered equipment grounding conductor, the conduits listed above shall not be required.

(2) A panelboard at a separate building shall be permitted to supply swimming pool equipment if the feeder meets the requirements for grounding in Section 250-32. Where installed, an equipment grounding conductor shall be an insulated conductor.

(e) **Cord-Connected Equipment.** Where fixed or stationary equipment is connected with a flexible cord to facilitate removal or disconnection for maintenance, repair, or storage as provided in Section 680-7, the equipment grounding conductors shall be connected to a fixed metal part of the assembly. The removable part shall be mounted on or bonded to the fixed metal part.

(f) **Other Equipment.** Other electrical equipment shall be grounded in accordance with Article 250 and connected by wiring methods of Chapter 3.

680-26. Electrically Operated Pool Covers.

(a) **Motors and Controllers.** The electric motors, controllers, and wiring shall be located at least 5 ft (1.52 m) from the inside wall of the pool unless separated from the pool by a wall, cover, or other permanent barrier. Electric motors installed below grade level shall be of the totally enclosed type.

> FPN No. 1: For cabinets installed in damp and wet locations, see Section 373-2(a).
>
> FPN No. 2: For switches or circuit breakers installed in wet locations, see Section 380-4.
>
> FPN No. 3: For protection against liquids, see Section 430-11.

(b) **Wiring Methods.** The electric motor and controller shall be connected to a circuit protected by a ground-fault circuit interrupter.

680-27. Deck Area Heating. The provisions of this section shall apply to all pool deck areas, including a covered pool, where electrically operated comfort heating units are installed within 20 ft (6.1 m) of the inside wall of the pool.

(a) **Unit Heaters.** Unit heaters shall be rigidly mounted to the structure and shall be of the totally enclosed or guarded types. Unit heaters shall not be mounted over the pool or within the area extending 5 ft (1.52 m) horizontally from the inside walls of a pool.

(b) **Permanently Wired Radiant Heaters.** Radiant electric heaters shall be suitably guarded and securely fastened to their mounting device(s). Heaters shall not be installed over a pool or within the area extending 5 ft (1.52 m) horizontally from the inside walls of the pool and shall be

mounted at least 12 ft (3.66 m) vertically above the pool deck unless otherwise approved.

(c) **Radiant Heating Cables Not Permitted.** Radiant heating cables embedded in or below the deck shall not be permitted.

680-28. Double Insulated Pool Pumps. A permanently installed pool shall be permitted to be provided with listed cord- and plug-connected pool pumps incorporating an approved system of double insulation that provides a means for grounding only the internal and nonaccessible, noncurrent-carrying metal parts of the pump.

C. Storable Pools

680-30. Pumps. A cord-connected pool filter pump shall incorporate an approved system of double insulation or its equivalent and shall be provided with means for grounding only the internal and nonaccessible noncurrent-carrying metal parts of the appliance.

The means for grounding shall be an equipment grounding conductor run with the power-supply conductors in the flexible cord that is properly terminated in a grounding-type attachment plug having a fixed grounding contact member.

680-31. Ground-Fault Circuit Interrupters Required. All electrical equipment, including power-supply cords, used with storable pools shall be protected by ground-fault circuit interrupters.

> FPN: Where flexible cords are used, see Section 400-4.

680-32. Lighting Fixtures.

(a) **15 Volts or Less.** A lighting fixture installed in or on the wall of a storable pool shall be part of a cord- and plug-connected lighting assembly. This assembly shall be as follows:

(1) Have no exposed metal parts
(2) Have a fixture lamp that operates at 15 volts or less
(3) Have an impact-resistant polymeric lens, fixture body, and transformer enclosure
(4) Have a transformer meeting the requirements of Section 680-5(a) with a primary rating not over 150 volts, and
(5) Be listed as an assembly for the purpose

(b) **Not Over 150 Volts.** A lighting assembly without a transformer, and with the fixture lamp(s) operating at not over 150 volts, shall be permitted to be cord- and plug-connected where the assembly complies with all of the following.

(1) It has no exposed metal parts.

(2) It has an impact-resistant polymeric lens and fixture body.

(3) A ground-fault circuit interrupter with open neutral protection is provided as an integral part of the assembly.

(4) The fixture lamp is permanently connected to the ground-fault circuit interrupter with open-neutral protection.

(5) It complies with the requirements of Section 680-20(a).

(6) It is listed as an assembly for the purpose.

D. Spas and Hot Tubs

680-38. Emergency Switch for Spas and Hot Tubs. A clearly labeled emergency shutoff or control switch for the purpose of stopping the motor(s) that provide power to the recirculation system and jet system shall be installed readily accessible to the users and at least 5 ft (1.52 m) away, adjacent to, and within sight of the spa or hot tub. This requirement shall not apply to single-family dwellings.

680-40. Outdoor Installations. A spa or hot tub installed outdoors shall comply with the provisions of Parts A and B of this article except as permitted in (a) and (b).

(a) Flexible Connections. Listed packaged units utilizing a factory-installed remote panelboard shall be permitted to be connected with not more than 6 ft (1.828 m) of liquidtight flexible conduit or be cord and plug connected with a cord not longer than 15 ft (4.57 m) if protected by a ground-fault circuit interrupter.

(b) Bonding. Bonding by metal-to-metal mounting on a common frame or base shall be permitted. The metal bands or hoops used to secure wooden staves shall not be required to be bonded as required in Section 680-22.

680-41. Indoor Installations. A spa or hot tub installed indoors shall comply with the provisions of Parts A and B except as modified by this section, and shall be connected by the wiring methods of Chapter 3.

Listed spa and hot tub packaged units rated 20 amperes or less shall be permitted to be cord and plug connected to facilitate the removal or disconnection of the unit for maintenance and repair.

(a) Receptacles. At least one 125-volt, 15- or 20-ampere receptacle on a general-purpose branch circuit shall be located a minimum of 5 ft (1.52 m) from and not more than 10 ft (3.05 m) from the inside wall of the spa or hot tub.

(1) Receptacles on the property shall be located at least 5 ft (1.52 m) measured horizontally from the inside walls of the spa or hot tub.

(2) Receptacles of 125 volts located within 10 ft (3.05 m) of the inside walls of a spa or hot tub shall be protected by a ground-fault circuit interrupter.

FPN: In determining the above dimensions, the distance to be measured is the shortest path the supply cord of an appliance connected to the receptacle would follow without piercing a floor, wall, or ceiling of a building or other effective permanent barrier.

(3) Receptacles that provide power for a spa or hot tub shall be ground-fault circuit-interrupter protected.

(b) Mounting Height of Lighting Fixtures, Lighting Outlets, and Ceiling-Suspended (Paddle) Fans.

(1) Lighting fixtures, lighting outlets, and ceiling-suspended (paddle) fans located over the spa or hot tub or within 5 ft (1.52 m) from the inside walls of the spa or hot tub shall be a minimum of 7 ft 6 in. (2.29 m) above the maximum water level and shall be protected by a ground-fault circuit interrupter.

Lighting fixtures, lighting outlets, and ceiling-suspended (paddle) fans that are located 12 ft (3.66 m) or more above the maximum water level shall not require a ground-fault circuit interrupter for protection.

(2) Lighting fixtures meeting the requirements of (a) or (b) and protected by a ground-fault circuit interrupter shall be permitted to be installed less than 7 ft 6 in. (2.29 m) over a spa or hot tub.

(a) Recessed fixtures with a glass or plastic lens and non-metallic or electrically isolated metal trim, suitable for use in damp locations.

(b) Surface-mounted fixtures with a glass or plastic globe and a nonmetallic body or a metallic body isolated from contact. Such fixtures shall be suitable for use in damp locations.

(c) Wall Switches. Switches shall be located at least 5 ft (1.52 m), measured horizontally, from the inside walls of the spa or hot tub.

(d) Bonding. The following parts shall be bonded together.

(1) All metal fittings within or attached to the spa or hot tub structure.

(2) Metal parts of electrical equipment associated with the spa or hot tub water circulating system, including pump motors.

(3) Metal conduit and metal piping within 5 ft (1.52 m) of the inside walls of the spa or hot tub and that are not separated from the spa or hot tub by a permanent barrier.

(4) All metal surfaces that are within 5 ft (1.52 m) of the inside walls of the spa or hot tub and not separated from the spa or hot tub area by a permanent barrier.

Exception: Small conductive surfaces not likely to become energized, such as air and water jets and drain fittings, where not connected to metallic piping, towel bars, mirror frames, and similar nonelectrical equipment, shall not be required to be bonded.

(5) Electrical devices and controls not associated with the spas or hot tubs shall be located a minimum of 5 ft (1.52 m) away from such units or be bonded to the spa or hot tub system.

(e) Methods of Bonding. All metal parts associated with the spa or hot tub shall be bonded by any of the following methods:

(1) The interconnection of threaded metal piping and fittings
(2) Metal-to-metal mounting on a common frame or base
(3) The provisions of a copper bonding jumper, insulated, covered, or bare, not smaller than No. 8 solid

(f) Grounding. The following equipment shall be grounded:

(1) All electric equipment located within 5 ft (1.52 m) of the inside wall of the spa or hot tub
(2) All electric equipment associated with the circulating system of the spa or hot tub

(g) Methods of Grounding.

(1) All electrical equipment shall be grounded in accordance with Article 250 and be connected by the wiring methods of Chapter 3.

(2) Where equipment is connected with a flexible cord, the equipment grounding conductor shall be connected to a fixed metal part of the assembly.

(h) Electric Water Heaters. All electric spa or hot tub water heaters shall be listed and shall have the heating elements subdivided into loads not exceeding 48 amperes and protected at not more than 60 amperes.

The ampacity of the branch-circuit conductors, and the rating or setting of overcurrent protective devices, shall not be less than 125 percent of the total load of the nameplate rating.

(i) Underwater Audio Equipment. Underwater audio equipment shall comply with the provisions of Parts B or C of this article.

680-42. Protection. An outlet that supplies a self-contained spa or hot tub, or a packaged spa or hot tub equipment assembly, shall be protected by a ground-fault circuit interrupter.

Exception: Where a listed self-contained unit or listed packaged equipment assembly is marked to indicate that integral ground-fault circuit-interrupter protection is provided for all electrical parts within the unit or assembly (pumps, air blowers, heaters, lights, controls, sanitizer generators, wiring, etc.).

FPN: See Section 680-4 for definitions of *self-contained spa or hot tub,* and for *packaged spa or hot tub equipment assembly.*

E. Fountains

680-50. General. The provisions of Part E shall apply to all permanently installed fountains as defined in Section 680-4. Self-contained, portable fountains not larger than 5 ft (1.52 m) in any dimension are not covered by Part E. Fountains that have water common to a pool shall comply with the pool requirements of this article.

680-51. Lighting Fixtures, Submersible Pumps, and Other Submersible Equipment.

(a) Ground-Fault Circuit Interrupter. A ground-fault circuit interrupter shall be installed in the branch circuit supplying fountain equipment unless the equipment is listed for operation at 15 volts or less and is supplied by a transformer that complies with Section 680-5(a).

(b) Operating Voltage. No lighting fixtures shall be installed for operation on supply circuits over 150 volts between conductors. Submersible pumps and other submersible equipment shall operate at 300 volts or less between conductors.

(c) Lighting Fixture Lenses. Lighting fixtures shall be installed with the top of the fixture lens below the normal water level of the fountain unless approved for above-water locations. A lighting fixture facing upward shall have the lens adequately guarded to prevent contact by any person.

(d) Overheating Protection. Electrical equipment that depends on submersion for safe operation shall be protected against overheating by a low-water cutoff or other approved means when not submerged.

(e) Wiring. Equipment shall be equipped with provisions for threaded conduit entries or be provided with a suitable flexible cord. The maximum length of exposed cord in the fountain shall be limited to 10 ft (3.05 m). Cords extending beyond the fountain perimeter shall be enclosed in approved wiring enclosures. Metal parts of equipment in contact with water shall be of brass or other approved corrosion-resistant metal.

(f) Servicing. All equipment shall be removable from the water for relamping or normal maintenance. Fixtures shall not be permanently imbedded into the fountain structure so that the water level must be reduced or the fountain drained for relamping, maintenance, or inspection.

(g) Stability. Equipment shall be inherently stable or be securely fastened in place.

680-52. Junction Boxes and Other Enclosures.

(a) General. Junction boxes and other enclosures used for other than underwater installation shall comply with Sections 680-21(a), (b), (c), (d), and (e).

(b) Underwater Junction Boxes and Other Underwater Enclosures. Junction boxes and other underwater enclosures shall be submersible and

(1) Be equipped with provisions for threaded conduit entries or compression glands or seals for cord entry;
(2) Be of copper, brass, or other approved corrosion-resistant material;
(3) Be filled with an approved potting compound to prevent the entry of moisture; and
(4) Be firmly attached to the supports or directly to the fountain surface and bonded as required.

Where the junction box is supported only by the conduit, the conduit shall be of copper, brass, or other approved corrosion-resistant metal. Where the box is fed by nonmetallic conduit, it shall have additional supports and fasteners of copper, brass, or other approved corrosion-resistant material.

FPN: See Section 370-23 for support of enclosures.

680-53. Bonding. All metal piping systems associated with the fountain shall be bonded to the equipment grounding conductor of the branch circuit supplying the fountain.

FPN: See Section 250-122 for sizing of these conductors.

680-54. Grounding. The following equipment shall be grounded:

(1) All electrical equipment located within the fountain or within 5 ft (1.52 m) of the inside wall of the fountain
(2) All electrical equipment associated with the recirculating system of the fountain
(3) Panelboards that are not part of the service equipment and that supply any electrical equipment associated with the fountain

680-55. Methods of Grounding.

(a) Applied Provisions. The provisions of Section 680-25 shall apply, excluding paragraph (e).

(b) Supplied by a Flexible Cord. Electrical equipment that is supplied by a flexible cord shall have all exposed non-current-carrying metal parts grounded by an insulated copper equipment grounding conductor that is an integral part of this cord. This grounding conductor shall be connected to a grounding terminal in the supply junction box, transformer enclosure, or other enclosure.

680-56. Cord- and Plug-Connected Equipment.

(a) Ground-Fault Circuit Interrupter. All electrical equipment, including power-supply cords, shall be protected by ground-fault circuit interrupters.

(b) Cord Type. Flexible cord immersed in or exposed to water shall be of the hard-service type as designated in Table 400-4 and shall be marked water resistant.

(c) Sealing. The end of the flexible cord jacket and the flexible cord conductor termination within equipment shall be covered with, or encapsulated in, a suitable potting compound to prevent the entry of water into the equipment through the cord or its conductors. In addition, the ground connection within equipment shall be similarly treated to protect such connections from the deteriorating effect of water that may enter into the equipment.

(d) Terminations. Connections with flexible cord shall be permanent, except that grounding-type attachment plugs and receptacles shall be permitted to facilitate removal or disconnection for maintenance, repair, or storage of fixed or stationary equipment not located in any water-containing part of a fountain.

680-57. Signs.

(a) General. Includes only fixed, stationary electrically illuminated utilization equipment with words or symbols designed to convey information or attract attention.

(b) Ground-Fault Circuit-Interrupter Protection for Personnel. All circuits supplying the sign shall have ground-fault circuit-interrupter protection for personnel.

(c) Location. Any sign installed inside a fountain shall be at least 5 ft (1.52 m) inside the fountain measured from the outside edges of the fountain.

(d) Disconnect. Shall comply with Section 600-6.

(e) Bonding. Shall comply with Section 600-7.

(f) Grounding. Any equipment associated with the sign shall be grounded as per Article 250.

F. Pools and Tubs for Therapeutic Use

680-60. General. The provisions of Part F shall apply to pools and tubs for therapeutic use in health care facilities, gymnasiums, athletic training rooms, and similar areas. See Section 517-3 for definition of *health care facilities*. Portable therapeutic appliances shall comply with Article 422.

680-61. Permanently Installed Therapeutic Pools. Therapeutic pools that are constructed in the ground, on the ground, or in a building in such a manner that the pool cannot be readily disassembled shall comply with Parts A and B of this article.

Exception: The limitations of Sections 680-6(b)(1) and (2) shall not apply where all lighting fixtures are of the totally enclosed type.

680-62. Therapeutic Tubs (Hydrotherapeutic Tanks). Therapeutic tubs, used for the submersion and treatment of patients, that are not easily moved from one place to another in normal use or that are fastened or otherwise secured at a specific location, including associated piping systems, shall conform to this part.

(a) Protection. The outlet(s) that supplies the following shall be protected by a ground-fault circuit interrupter:

(1) A self-contained therapeutic tub or hydrotherapeutic tank, or
(2) A packaged therapeutic tub or hydrotherapeutic tank, or
(3) A field-assembled therapeutic tub or hydrotherapeutic tank with a heater load of 50 amperes or less

A listed self-contained unit or listed packaged equipment assembly marked to indicate that integral ground-fault circuit-interrupter protection is provided for all electrical parts within the unit or assembly (pumps, air blowers, heaters, lights, controls, sanitizer generators, wiring, etc.) shall not require that the outlet supply be protected by a ground-fault circuit interrupter.

A field-assembled therapeutic tub or hydrotherapeutic tank rated greater than 250 volts or rated 3 phase shall not require the supply to be protected by a ground-fault circuit interrupter.

(b) Ground-Fault Circuit Interrupter. A ground-fault circuit interrupter shall protect all therapeutic equipment.

Exception: Portable therapeutic appliances shall comply with Section 250-114.

(c) Bonding. The following parts shall be bonded together.

(1) All metal fittings within or attached to the tub structure.

(2) Metal parts of electrical equipment associated with the tub water circulating system, including pump motors.

(3) Metal-sheathed cables and raceways and metal piping that are within 5 ft (1.52 m) of the inside walls of the tub and not separated from the tub by a permanent barrier.

(4) All metal surfaces that are within 5 ft (1.52 m) of the inside walls of the tub and not separated from the tub area by a permanent barrier.

(5) Electrical devices and controls not associated with the therapeutic tubs shall be located a minimum of 5 ft (1.52 m) away from such units or be bonded to the therapeutic tub system.

(d) Methods of Bonding. All metal parts associated with the tub shall be bonded by any of the following methods:

(1) The interconnection of threaded metal piping and fittings,
(2) Metal-to-metal mounting on a common frame or base,
(3) Connections by suitable metal clamps, or
(4) By the provisions of a solid copper bonding jumper, insulated, covered, or bare, not smaller than No. 8

(e) Grounding. The following equipment shall be grounded:

(1) All electrical equipment located within 5 ft (1.52 m) of the inside wall of the tub
(2) All electrical equipment associated with the circulating system of the tub

(f) Methods of Grounding.

(1) All electrical equipment shall be grounded in accordance with Article 250 and connected by wiring methods of Chapter 3.

(2) Where equipment is connected with a flexible cord, the equipment grounding conductor shall be connected to a fixed metal part of the assembly.

(g) Receptacles. All receptacles within 5 ft (1.52 m) of a therapeutic tub shall be protected by a ground-fault circuit interrupter.

(h) Lighting Fixtures. All lighting fixtures used in therapeutic tub areas shall be of the totally enclosed type.

G. Hydromassage Bathtubs

680-70. Protection. Hydromassage bathtubs and their associated electrical components shall be protected by a ground-fault circuit interrupter. All 125-volt, single-phase receptacles within 5 ft (1.52 m) measured horizontally of the inside walls of a hydromassage tub shall be protected by a ground-fault circuit interrupter(s).

680-71. Other Electrical Equipment. Lighting fixtures, switches, receptacles, and other electrical equipment located in the same room, and not directly associated with a hydromassage bathtub, shall be installed in accordance with the requirements of Chapters 1 through 4 in this *Code* covering the installation of that equipment in bathrooms.

680-72. Accessibility. Hydromassage bathtub electrical equipment shall be accessible without damaging the building structure or building finish.

680-73. Bonding. All metal piping systems, metal parts of electrical equipment, and pump motors associated with the hydromassage tub shall be bonded together using a copper bonding jumper, insulated, covered, or bare, not smaller than No. 8 solid.

Metal parts of listed equipment incorporating an approved system of double insulation and providing a means for grounding internal nonaccessible, noncurrent-carrying metal parts shall not be bonded.

Article 685 — Integrated Electrical Systems

A. General

685-1. Scope. This article covers integrated electrical systems, other than unit equipment, in which orderly shutdown is necessary to ensure safe operation. An *integrated electrical system* as used in this article is a unitized segment of an industrial wiring system where all of the following conditions are met:

(1) An orderly shutdown is required to minimize personnel hazard and equipment damage,
(2) The conditions of maintenance and supervision ensure that qualified persons will service the system, and
(3) Effective safeguards, acceptable to the authority having jurisdiction, are established and maintained

685-2. Application of Other Articles. In the following other articles applying to particular cases of installation of conductors and equipment, there are orderly shutdown requirements that are in addition to those of this article or are modifications of them.

	Section
More than one building or other structure	225, Part B
Ground-fault protection of equipment	230-95, Exception No. 1
Protection of conductors	240-3
Electrical system coordination	240-12
Ground-fault protection of equipment	240-13(1)
Grounding ac systems of 50 to 1000 volts	250-21(c)
Equipment protection	427-22
Orderly shutdown	430-44
Disconnection	430-74, Exception Nos. 1 and 2
Disconnecting means in sight from controller	430-102(a), Exception No. 2
Energy from more than one source	430-113, Exception Nos. 1 and 2
Disconnecting means	645-10, Exception
Point of connection	705-12(a)

B. Orderly Shutdown

685-10. Location of Overcurrent Devices in or on Premises. Location of overcurrent devices that are critical to integrated electrical systems shall be permitted to be accessible, with mounting heights permitted to ensure security from operation by nonqualified personnel.

685-12. Direct-Current System Grounding. Two-wire dc circuits shall be permitted to be ungrounded.

685-14. Ungrounded Control Circuits. Where operational continuity is required, control circuits of 150 volts, or less, from separately derived systems shall be permitted to be ungrounded.

Article 690 — Solar Photovoltaic Systems

A. General

690-1. Scope. The provisions of this article apply to solar photovoltaic electrical energy systems including the array circuit(s), inverter(s), and controller(s) for such systems. [See Figures 690-1(a) and (b).] Solar photovoltaic systems covered by this article may be interactive with other electrical power production sources or stand alone, with or without electrical energy storage such as batteries. These systems may have ac or dc output for utilization.

690-2. Definitions.

Alternating-Current Module (Alternating-Current Photovoltaic Module). A complete, environmentally protected unit consisting of solar cells, optics, inverter, and other components, exclusive of tracker, designed to generate ac power when exposed to sunlight.

Array. A mechanically integrated assembly of modules or panels with a support structure and foundation, tracker, and other components, as required, to form a direct-current power-producing unit.

Blocking Diode. A diode used to block reverse flow of current into a photovoltaic source circuit.

Note 1: These diagrams are intended to be a means of identification for photovoltaic system components, circuits, and connections.
Note 2: Disconnecting means required by Article 690, Part C are not shown.
Note 3: System grounding and equipment grounding are not shown. See Article 690, Part E.

Figure 690-1(a) Identification of solar photovoltaic system components.

Note 1: These diagrams are intended to be a means of identification for photovoltaic system components, circuits, and connections.
Note 2: Disconnecting means and overcurrent protection required by Article 690 are not shown.
Note 3: System grounding and equipment grounding are not shown. See Article 690, Part E.
Note 4: Custom designs occur in each configuration and some components are optional.

Figure 690-1(b) Identification of solar photovoltaic system components in common system configurations.

Charge Controller. Equipment that controls dc voltage or dc current, or both, used to charge a battery.

Electrical Production and Distribution Network. A power production, distribution, and utilization system, such as a utility system and connected loads, that is external to and not controlled by the photovoltaic power system.

Hybrid System. A system comprised of multiple power sources. These power sources may include photovoltaic, wind, micro-hydro generators, engine-driven generators, and others, but do not include electrical production and distribution network systems. Energy storage systems, such as batteries, do not constitute a power source for the purpose of this definition.

Interactive System. A solar photovoltaic system that operates in parallel with and may deliver power to an electrical production and distribution network. For the purpose of this definition, an energy storage subsystem of a solar photovoltaic system, such as a battery, is not another electrical production source.

Inverter. Equipment that is used to change voltage level or waveform, or both, of electrical energy. Commonly, an inverter [also known as a power conditioning unit (PCU) or power conversion system (PCS)] is a device that changes dc input to an ac output. Inverters may also function as battery chargers that use alternating current from another source and convert it into direct current for charging batteries.

Inverter Input Circuit. Conductors between the inverter and the battery in stand-alone systems or the conductors between the inverter and the photovoltaic output circuits for electrical production and distribution network.

Inverter Output Circuit. Conductors between the inverter and an ac load center for stand-alone systems or the conductors between the inverter and the service equipment or another electric power production source, such as a utility, for electrical production and distribution network.

Module. A complete, environmentally protected unit consisting of solar cells, optics, and other components, exclusive of tracker, designed to generate dc power when exposed to sunlight.

Panel. A collection of modules mechanically fastened together, wired, and designed to provide a field-installable unit.

Photovoltaic Output Circuit. Circuit conductors between the photovoltaic source circuit(s) and the inverter or dc utilization equipment.

Photovoltaic Power Source. An array or aggregate of arrays that generates dc power at system voltage and current.

Photovoltaic Source Circuit. Circuits between modules and from modules to the common connection point(s) of the dc system.

Solar Cell. The basic photovoltaic device that generates electricity when exposed to light.

Solar Photovoltaic System. The total components and subsystems that, in combination, convert solar energy into electrical energy suitable for connection to a utilization load.

Stand-Alone System. A solar photovoltaic system that supplies power independently of an electrical production and distribution network.

System Voltage. The dc voltage of any photovoltaic source circuit or output circuit. For 3-wire or multiwire installations, including 2-wire circuits connected to 3-wire systems, the system voltage shall be the highest voltage between any two conductors.

690-3. Other Articles. Wherever the requirements of other articles of this *Code* and Article 690 differ, the requirements of Article 690 shall apply.

690-4. Installation.

(a) Solar Photovoltaic System. A solar photovoltaic system shall be permitted to supply a building or other structure in addition to any service(s) of another electricity supply system(s).

(b) Conductors of Different Systems. Photovoltaic source circuits and photovoltaic output circuits shall not be contained in the same raceway, cable tray, cable, outlet box, junction box, or similar fitting as feeders or branch circuits of other systems, unless the conductors of the different systems are separated by a partition or are connected together.

(c) Module Connection Arrangement. The connections to a module or panel shall be arranged so that removal of a module or panel from a photovoltaic source circuit does not interrupt a grounded conductor to another photovoltaic

source circuit. Sets of modules interconnected as systems rated at 50 volts or less, with or without blocking diodes, and having a single overcurrent device shall be considered as a single-source circuit. Supplementary overcurrent devices used for the exclusive protection of the photovoltaic modules are not considered as overcurrent devices for the purpose of this section.

(d) Equipment. Inverters or motor generators shall be identified for use in solar photovoltaic systems.

690-5. Ground-Fault Protection. Roof-mounted dc photovoltaic arrays located on dwellings shall be provided with dc ground-fault protection to reduce fire hazards.

(a) Ground-Fault Detection and Interruption. The ground-fault protection device or system shall be capable of detecting a ground fault, interrupting the flow of fault current, and providing an indication of the fault.

(b) Disconnection of Conductors. The ungrounded conductors of the faulted source circuit shall be automatically disconnected. If the grounded conductors of the faulted source circuit are disconnected to comply with the requirements of Section 690-5(a), all conductors of the faulted source circuit shall be opened automatically and simultaneously. Interrupting the ground-fault current path to open the grounded conductor of the array or faulted sections of the array shall be permitted.

(c) Labels and Markings. Labels and markings shall be applied near the ground-fault indicator at a visible location stating that if a ground fault is indicated, the normally grounded conductors may be energized and ungrounded.

690-6. Alternating-Current Modules.

(a) Photovoltaic Source Circuits. The requirements of Article 690 pertaining to photovoltaic source circuits shall not apply to ac modules. The photovolaic source circuit, conductors, and inverters shall be considered as internal wiring of an ac module.

(b) Inverter Output Circuit. The output of an ac module shall be considered an inverter output circuit.

(c) Disconnecting Means. A single disconnecting means, in accordance with Sections 690-15 and 690-17, shall be permitted for the combined ac output of one or more ac modules. Additionally, each ac module in a multiple ac-module system shall be provided with a connector, bolted, or terminal-type disconnecting means.

(d) Ground-Fault Detection. Alternating-current-module systems shall be permitted to use a single detection device to detect only ac ground faults and to disable the array by removing ac power to the ac module(s).

(e) Overcurrent Protection. The output circuits of ac modules shall be permitted to have overcurrent protection and conductor sizing in accordance with Section 240-4(b)(2).

B. Circuit Requirements

690-7. Maximum Voltage.

(a) Maximum System Voltage. In a dc photovoltaic source circuit or output circuit, the maximum system voltage for that circuit shall be computed as the sum of the rated open-circuit voltage of the series-connected photovoltaic modules corrected for the lowest expected ambient temperature. For crystalline and multi-crystalline silicon modules, the rated open-circuit voltage shall be multiplied by the correction factor provided in Table 690-7. This voltage shall be used to determine the voltage rating of cables, disconnects, overcurrent devices, and other equipment. Where the lowest expected ambient temperature is below −40°C (−40°F), or where other than crystalline or multi-crystalline silicon photovoltaic modules are used, the system voltage adjustment shall be made in accordance with the manufacturer's instructions.

Table 690-7. Voltage Correction Factors for Crystalline and Multi-Crystalline Silicon Modules

Ambient Temperature (°C)	Correction factors for ambient temperatures below 25°C (77°F), multiply the rated open-circuit voltage by the appropriate correction factor shown below.	Ambient Temperature (°F)
25 to 10	1.06	77 to 50
9 to 0	1.10	49 to 32
−1 to −10	1.13	31 to 14
−11 to −20	1.17	13 to −4
−21 to −40	1.25	−5 to −40

(b) Direct-Current Utilization Circuits. The voltage of dc utilization circuits shall conform with Section 210-6.

(c) Photovoltaic Source and Output Circuits. In one- and two-family dwellings, photovoltaic source circuits and photovoltaic output circuits that do not include lampholders, fixtures, or receptacles shall be permitted to have a maximum system voltage up to 600 volts. Other installations with a maximum system voltage over 600 volts shall comply with Article 690, Part I.

(d) Circuits Over 150 Volts to Ground. In one- and two-family dwellings, live parts in photovoltaic source circuits and photovoltaic output circuits over 150 volts to ground shall not be accessible to other than qualified persons while energized.

FPN: See Section 110-27 for guarding of live parts, and Section 210-6 for voltage to ground and between conductors.

690-8. Circuit Sizing and Current.

(a) Computation of Maximum Circuit Current. The maximum current for the specific circuit shall be computed as follows.

(1) Photovoltaic Source Circuit Currents. The maximum current shall be the sum of parallel module rated short-circuit currents multiplied by 125 percent.

(2) Photovoltaic Output Circuit Currents. The maximum current shall be the sum of parallel source circuit maximum currents as calculated in (1).

(3) Inverter Output Circuit Current. The maximum current shall be the inverter continuous output current rating.

(4) Stand-Alone Inverter Input Circuit Current. The maximum current shall be the stand-alone continuous inverter input current rating when the inverter is producing rated power at the lowest input voltage.

(b) Ampacity and Overcurrent Device Ratings. Photovoltaic system currents shall be considered continuous. The circuit conductors and overcurrent devices shall be sized to carry not less than 125 percent of the maximum currents as computed in (a). The rating or setting of overcurrent devices shall be permitted in accordance with Sections 240-3(b) and (c).

Exception: Circuits containing an assembly together with its overcurrent device(s) that is listed for continuous operation at 100 percent of its rating shall be permitted to be utilized at 100 percent of its rating.

(c) Systems with Multiple Direct-Current Voltages. For a photovoltaic power source that has multiple output circuit voltages and employs a common-return conductor, the ampacity of the common-return conductor shall not be less than the sum of the ampere ratings of the overcurrent devices of the individual output circuits.

690-9. Overcurrent Protection.

(a) Circuits and Equipment. Photovoltaic source circuit, photovoltaic output circuit, inverter output circuit, and storage battery circuit conductors and equipment shall be protected in accordance with the requirements of Article 240. Circuits connected to more than one electrical source shall have overcurrent devices located so as to provide overcurrent protection from all sources.

Exception: An overcurrent device shall not be required for circuit conductors sized in accordance with Section 690-8(b) and located where

(a) *There are no external sources such as parallel-connected source circuits, batteries, or backfeed from inverters, or*
(b) *The short-circuit currents from all sources do not exceed the ampacity of the conductors.*

> FPN: Possible backfeed of current from any source of supply, including a supply through an inverter into the photovoltaic output circuit and photovoltaic source circuits, is a consideration in determining whether adequate overcurrent protection from all sources is provided for conductors and modules.

(b) Power Transformers. Overcurrent protection for a transformer with a source(s) on each side shall be provided in accordance with Section 450-3 by considering first one side of the transformer, then the other side of the transformer, as the primary.

Exception: A power transformer with a current rating on the side connected toward the photovoltaic power source not less than the short-circuit output current rating of the inverter shall be permitted without overcurrent protection from that source.

(c) Photovoltaic Source Circuits. Branch-circuit or supplementary-type overcurrent devices shall be permitted to provide overcurrent protection in photovoltaic source circuits. The overcurrent devices shall be accessible, but shall not be required to be readily accessible.

(d) Direct-Current Rating. Overcurrent devices, either fuses or circuit breakers, used in any dc portion of a photovoltaic power system shall be listed for use in dc circuits and shall have the appropriate voltage, current, and interrupt ratings.

690-10. Stand-Alone Systems. The premises wiring system shall be adequate to meet the requirements of this *Code* for a similar installation connected to a service. The wiring on the supply side of the building or structure disconnecting means shall comply with this *Code* except as modified by (a), (b), and (c).

(a) Inverter Output. The ac inverter output from a stand-alone system shall be permitted to supply ac power to the building or structure disconnecting means at current levels below the rating of that disconnecting means.

(b) Sizing and Protection. The circuit conductors between the inverter output and the building or structure dis-

connecting means shall be sized based on the output rating of the inverter. These conductors shall be protected from overcurrents in accordance with Article 240. The overcurrent protection shall be located at the output of the inverter.

(c) Single 120-Volt Supply. The inverter output of a stand-alone solar photovoltaic system shall be permitted to supply 120 volts to single-phase, 3-wire 120/240-volt service equipment or distribution panels where there are no 240-volt outlets and where there are no multiwire branch circuits. In all installations, the rating of the overcurrent device connected to the output of the inverter shall be less than the rating of the neutral bus in the service equipment. This equipment shall be marked

> WARNING — SINGLE 120-VOLT SUPPLY —
> DO NOT CONNECT MULTIWIRE BRANCH CIRCUITS!

C. Disconnecting Means

690-13. All Conductors. Means shall be provided to disconnect all current-carrying conductors of a photovoltaic power source from all other conductors in a building or other structure. Where a circuit grounding connection is not designed to be automatically interrupted as part of the ground-fault protection system required by Section 690-5, a switch or circuit breaker used as a disconnecting means shall not have a pole in the grounded conductor.

> FPN: The grounded conductor may have a bolted or terminal disconnecting means to allow maintenance or troubleshooting by qualified personnel.

690-14. Additional Provisions. The provisions of Article 230, Part F, as modified by (a) and (b), shall apply to the photovoltaic power source disconnecting means.

(a) Disconnecting Means. The disconnecting means shall not be required to be suitable as service equipment and shall be rated in accordance with Section 690-17.

(b) Equipment. Equipment such as photovoltaic source circuit isolating switches, overcurrent devices, and blocking diodes shall be permitted on the photovoltaic side of the photovoltaic disconnecting means.

690-15. Disconnection of Photovoltaic Equipment. Means shall be provided to disconnect equipment, such as inverters, batteries, charge controllers, and the like, from all ungrounded conductors of all sources. If the equipment is energized from more than one source, the disconnecting means shall be grouped and identified.

A single disconnecting means in accordance with Section 690-17 shall be permitted for the combined ac output of one or more inverters or ac modules in an interactive system.

690-16. Fuses. Disconnecting means shall be provided to disconnect a fuse from all sources of supply if the fuse is energized from both directions and is accessible to other than qualified persons. Such a fuse in a photovoltaic source circuit shall be capable of being disconnected independently of fuses in other photovoltaic source circuits.

690-17. Switch or Circuit Breaker. The disconnecting means for ungrounded conductors shall consist of a manually operable switch(es) or circuit breaker(s)

(1) Located where readily accessible,
(2) Externally operable without exposing the operator to contact with live parts,
(3) Plainly indicating whether in the open or closed position, and
(4) Shall have an interrupting rating sufficient for the nominal circuit voltage and the current that is available at the line terminals of the equipment.

Where all terminals of the disconnecting means may be energized in the open position, a warning sign shall be mounted on or adjacent to the disconnecting means. The sign shall be clearly legible and shall read substantially:

WARNING — ELECTRIC SHOCK HAZARD —
DO NOT TOUCH TERMINALS —
TERMINALS ON BOTH THE LINE AND LOAD SIDES
MAY BE ENERGIZED IN THE OPEN POSITION.

Exception No. 1: A disconnecting means located on the dc side shall be permitted to have an interrupting rating less than the current-carrying rating where the system is designed so that the dc switch cannot be opened under load.

Exception No. 2: A connector shall be permitted to be used as an ac or dc disconnecting means provided that it complies with the requirements of Section 690-33 and is listed and identified for the use.

690-18. Installation and Service of an Array. Open circuiting, short circuiting, or opaque covering shall be used to disable an array or portions of an array for installation and service.

> FPN: Photovoltaic modules are energized while exposed to light. Installation, replacement, or servicing of array components while a module(s) is irradiated may expose persons to electric shock.

D. Wiring Methods

690-31. Methods Permitted.

(a) Wiring Systems. All raceway and cable wiring methods included in this *Code* and other wiring systems and fittings specifically intended and identified for use on photovoltaic arrays shall be permitted. Where wiring devices with integral enclosures are used, sufficient length of cable shall be provided to facilitate replacement.

(b) Single Conductor Cable. Types SE, UF, and USE single-conductor cable shall be permitted in photovoltaic source circuits where installed in the same manner as a Type UF multiconductor cable in accordance with Article 339. Where exposed to direct rays of the sun, Type UF cable identified as sunlight-resistant or Type USE cable shall be used.

(c) Flexible Cords and Cables. Flexible cords and cables, where used to connect the moving parts of tracking PV modules, shall comply with Article 400 and shall be of a type identified as a hard service cord or portable power cable; shall be suitable for extra-hard usage, listed for outdoor use, water resistant, and sunlight resistant. Allowable ampacities shall be in accordance with Section 400-5. For ambient temperatures exceeding 30°C (86°F), the ampacities shall be derated by the appropriate factors given in Table 690-31(c).

(d) Small Conductor Cables. Single-conductor cables listed for outdoor use that are sunlight resistant and moisture resistant with sizes No. 16 and No. 18 shall be permitted for module interconnections where such cables meet the ampacity requirements of Section 690-8. Section 310-15 shall be used to determine the cable ampacity and temperature derating factors.

690-32. Component Interconnections. Fittings and connectors that are intended to be concealed at the time of on-site assembly, where listed for such use, shall be permitted for on-site interconnection of modules or other array components. Such fittings and connectors shall be equal to the wiring method employed in insulation, temperature rise, and fault-current withstand, and shall be capable of resisting the effects of the environment in which they are used.

690-33. Connectors. The connectors permitted by Article 690 shall comply with (a) through (e).

(a) Configuration. The connectors shall be polarized and shall have a configuration that is noninterchangeable with receptacles in other electrical systems on the premises.

(b) Guarding. The connectors shall be constructed and installed so as to guard against inadvertent contact with live parts by persons.

(c) Type. The connectors shall be of the latching or locking type.

(d) Grounding Member. The grounding member shall be the first to make and the last to break contact with the mating connector.

Table 690-31(c). Correction Factors

Ambient Temperature (°C)	Temperature Rating of Conductor				Ambient Temperature (°F)
	60°C (140°F)	75°C (167°F)	90°C (194°F)	105°C (221°F)	
30	1.00	1.00	1.00	1.00	86
31–35	0.91	0.94	0.96	0.97	87–95
36–40	0.82	0.88	0.91	0.93	96–104
41–45	0.71	0.82	0.87	0.89	105–113
46–50	0.58	0.75	0.82	0.86	114–122
51–55	0.41	0.67	0.76	0.82	123–131
56–60	—	0.58	0.71	0.77	132–140
61–70	—	0.33	0.58	0.68	141–149
71–80	—	—	0.41	0.58	150–158

(e) Interruption of Circuit. The connectors shall be capable of interrupting the circuit current without hazard to the operator.

690-34. Access to Boxes. Junction, pull, and outlet boxes located behind modules or panels shall be installed so that the wiring contained in them can be rendered accessible directly or by displacement of a module(s) or panel(s) secured by removable fasteners and connected by a flexible wiring system.

E. Grounding

690-41. System Grounding. For a photovoltaic power source, one conductor of a 2-wire system rated over 50 volts and a neutral conductor of a 3-wire system shall be solidly grounded.

Exception: Other methods that accomplish equivalent system protection and that utilize equipment listed and identified for the use shall be permitted.

FPN: See Section 250-2(a).

690-42. Point of System Grounding Connection. The dc circuit grounding connection shall be made at any single point on the photovoltaic output circuit.

FPN: Locating the grounding connection point as close as practicable to the photovoltaic source will better protect the system from voltage surges due to lightning.

690-43. Equipment Grounding. Exposed noncurrent-carrying metal parts of module frames, equipment, and conductor enclosures shall be grounded regardless of voltage.

690-45. Size of Equipment Grounding Conductor. The equipment grounding conductor shall be not smaller than the required size of the circuit conductors in systems where the available photovoltaic power source short-circuit current is less than twice the current rating of the overcurrent device. In other systems, the equipment grounding conductor shall be sized in accordance with Section 250-122.

690-47. Grounding Electrode System. A grounding electrode system shall be provided in accordance with Sections 250-50 through 250-60.

F. Marking

690-51. Modules. Modules shall be marked with identification of terminals or leads as to polarity, maximum overcurrent device rating for module protection, and with rated

(1) Open-circuit voltage
(2) Operating voltage
(3) Maximum permissible system voltage
(4) Operating current
(5) Short-circuit current, and
(6) Maximum power

690-52. Alternating-Current Photovoltaic Modules. Alternating-current modules shall be marked with identification of terminals or leads, and with identification of the rated

(1) Nominal operating ac voltage
(2) Nominal operating ac frequency
(3) Maximum ac power
(4) Maximum ac current, and
(5) Maximum overcurrent device rating for ac module protection

690-53. Photovoltaic Power Source. A marking, specifying the photovoltaic power source rated as follows shall be provided by the installer at the site at an accessible location at the disconnecting means for the photovoltaic power source:

(1) Operating current
(2) Operating voltage
(3) Maximum system voltage, and
(4) Short-circuit current

FPN: Reflecting systems used for irradiance enhancement may result in increased levels of output current and power.

690-54. Interactive System Point of Interconnection. All interactive system(s) points of interconnection with other sources shall be marked at an accessible location at the disconnecting means as a power source with the maximum ac output operating current and the operating ac voltage.

G. Connection to Other Sources

690-60. Identified Interactive Equipment. Only inverters and ac modules listed and identified as interactive shall be permitted in interactive systems.

690-61. Loss of Interactive System Power. An inverter or an ac module in an interactive solar photovoltaic system shall automatically de-energize its output to the connected electrical production and distribution network upon loss of voltage in that system and shall remain in that state until the electrical production and distribution network voltage has been restored.

A normally interactive solar photovoltaic system shall be permitted to operate as a stand-alone system to supply loads that have been disconnected from electrical production and distribution network sources.

690-62. Ampacity of Neutral Conductor. If a single-phase, 2-wire inverter output is connected to the neutral and one ungrounded conductor (only) of a 3-wire system or of a 3-phase, 4-wire, wye-connected system, the maximum load connected between the neutral and any one ungrounded conductor plus the inverter output rating shall not exceed the ampacity of the neutral conductor.

690-63. Unbalanced Interconnections.

(a) Single Phase. Single-phase inverters for photovoltaic systems and ac modules in interactive solar photovoltaic systems shall not be connected to 3-phase power systems unless the interconnected system is designed so that significant unbalanced voltages cannot result.

(b) Three Phase. Three-phase inverters and 3-phase ac modules in interactive systems shall have all phases automatically de-energized upon loss of, or unbalanced, voltage in one or more phases unless the interconnected system is designed so that significant unbalanced voltages will not result.

690-64. Point of Connection. The output of a photovoltaic power source shall be connected as specified in (a) or (b).

(a) Supply Side. A photovoltaic power source shall be permitted to be connected to the supply side of the service disconnecting means as permitted in Section 230-82(5).

(b) Load Side. A photovoltaic power source shall be permitted to be connected to the load side of the service disconnecting means of the other source(s) at any distribution equipment on the premises provided that all of the following conditions are met.

(1) Each source interconnection shall be made at a dedicated circuit breaker or fusible disconnecting means.

(2) The sum of the ampere ratings of overcurrent devices in circuits supplying power to a busbar or conductor shall not exceed the rating of the busbar or conductor.

Exception: For a dwelling unit, the sum of the ampere ratings of the overcurrent devices shall not exceed 120 percent of the rating of the busbar or conductor.

(3) The interconnection point shall be on the line side of all ground-fault protection equipment.

Exception: Connection shall be permitted to be made to the load side of ground-fault protection, provided that there is ground-fault protection for equipment from all ground-fault current sources.

(4) Equipment containing overcurrent devices in circuits supplying power to a busbar or conductor shall be marked to indicate the presence of all sources.

Exception: Equipment with power supplied from a single point of connection.

(5) Equipment such as circuit breakers, if backfed, shall be identified for such operation.

H. Storage Batteries

690-71. Installation.

(a) General. Storage batteries in a solar photovoltaic system shall be installed in accordance with the provisions of Article 480. The interconnected battery cells shall be considered grounded where the photovoltaic power source is installed in accordance with Section 690-41, Exception.

(b) Dwellings.

(1) Storage batteries for dwellings shall have the cells connected so as to operate at less than 50 volts.

Exception: Where live parts are not accessible during routine battery maintenance, a battery system voltage in accordance with Section 690-7 shall be permitted.

(2) Live parts of battery systems for dwellings shall be guarded to prevent accidental contact by persons or objects, regardless of voltage or battery type.

> FPN: Batteries in solar photovoltaic systems are subject to extensive charge-discharge cycles and typically require frequent maintenance, such as checking electrolyte and cleaning connections.

(c) Current Limiting. A listed, current-limiting, overcurrent device shall be installed in each circuit adjacent to the batteries where the available short-circuit current from a battery or battery bank exceeds the interrupting or withstand ratings of other equipment in that circuit. The installation of current-limiting fuses shall comply with Section 690-16.

690-72. Charge Control. Equipment shall be provided to control the charging process of the battery. Charge control shall not be required where the design of the photovoltaic source circuit is matched to the voltage rating and charge current requirements of the interconnected battery cells, and the maximum charging current multiplied by 1 hour is less than 3 percent of the rated battery capacity expressed in ampere-hours or as recommended by the battery manufacturer. All adjusting means for control of the charging process shall be accessible only to qualified persons.

> FPN: Certain battery types such as valve-regulated lead acid or nickel cadmium can experience thermal failure when overcharged.

690-74. Battery Interconnections. Flexible cables, as identified in Article 400, in sizes No. 2/0 and larger shall be permitted within the battery enclosure from battery terminals to nearby junction box where they shall be connected to an approved wiring method. Flexible battery cables shall also be permitted between batteries and cells within the battery enclosure. Such cables shall be listed for hard service use and identified as moisture resistant.

I. Systems Over 600 Volts

690-80. General. Solar photovoltaic systems with a maximum system voltage over 600 volts dc shall comply with Article 490 and other requirements applicable to installations rated over 600 volts.

690-85. Definitions. For the purposes of Part I of this article, the voltages used to determine cable and equipment ratings shall be defined as follows.

Battery Circuits. In battery circuits, the voltage shall be the highest voltage experienced under charging conditions.

Photovoltaic Circuits. In dc photovoltaic source circuits and photovoltaic output circuits, the voltage shall be the maximum system voltage.

Article 695 — Fire Pumps

695-1. Scope.

(a) Covered. This article covers the installation of the following:

(1) Electric power sources and interconnecting circuits
(2) Switching and control equipment dedicated to fire pump drivers

(b) Not Covered. This article does not cover the following:

(1) The performance, maintenance, and acceptance testing of the fire pump system, and the internal wiring of the components of the system
(2) Pressure maintenance (jockey or makeup) pumps

> FPN: See *Standard for the Installation of Centrifugal Fire Pumps*, NFPA 20-1996, for further information.

695-2. Application of Other Articles. Except as required or permitted by this article, the installation of electrical conductors and equipment for fire pumps shall comply with all applicable requirements in Chapters 1 through 4 of this *Code*.

695-3. Power Source(s) for Electric Motor-Driven Fire Pumps. Electric motor-driven fire pumps shall have a reliable source of power.

ˣ(a) Individual Sources. Where reliable, and where capable of carrying indefinitely the sum of the locked-rotor current of the fire pump motor(s) and the pressure maintenance pump motor(s) and the full-load current of the associated fire pump accessory equipment when connected to this power supply, the power source for an electric motor-driven fire pump shall be one or more of the following.

(1) Electric Utility Service Connection. A fire pump shall be permitted to be supplied by a separate service, or by a tap located ahead of and not within the same cabinet, enclosure, or vertical switchboard section as the service disconnecting means. The connection shall be located and arranged so as to minimize the possibility of damage by fire from within the premises and from exposing hazards. A tap ahead of the service disconnecting means shall comply with Section 230-82(4). The service equipment shall comply with the labeling requirements in Section 230-2 and the location requirements in Section 230-72(b).

(2) On-Site Power Production Facility. A fire pump shall be permitted to be supplied by an on-site power production facility. The source facility shall be located and protected to minimize the possibility of damage by fire.

ˣ(b) Multiple Sources. Where reliable power cannot be obtained from a source described in Section 695-3(a), power shall be supplied from an approved combination of two or more of either of such sources or from an approved combination of one or more of such sources in combination with an on-site standby generator complying with (1) and (3) or two or more feeders complying with (2) or (3) or from two or more sets of feeders complying with (2) and (3).

(1) Generator Capacity. An on-site generator(s) used to comply with this section shall be of sufficient capacity to allow normal starting and running of the motor(s) driving the fire pump(s) while supplying all other simultaneously operated load. Automatic shedding of one or more optional standby loads in order to comply with this capacity requirement shall be permitted. A tap ahead of the on-site generator disconnecting means shall not be required.

(2) Feeder Sources. This section applies to multi-building campus-style complexes with fire pumps at one or more buildings. Where sources in Section 695-3(a) are not practicable, and with the approval of the authority having jurisdiction, two or more feeder sources shall be permitted as one power source or as more than one power source where such feeders are connected to or derived from separate utility services. The connection(s), overcurrent protective device(s), and disconnecting means for such feeders shall meet the requirements of Section 695-4(b).

(3) Arrangement. The power sources shall be arranged so that a fire at one source will not cause an interruption at the other source.

ˣ695-4. Continuity of Power. Circuits that supply electric motor-driven fire pumps shall be supervised from inadvertent disconnection as covered in (a) or (b).

(a) Direct Connection. The supply conductors shall directly connect the power source to either a listed fire pump controller or listed combination fire pump controller and power transfer switch.

(b) Supervised Connection. A single disconnecting means and associated overcurrent protective device(s) shall be permitted to be installed between a remote power source and one of the following:

a. A listed fire pump controller
b. A listed fire pump power transfer switch, or
c. A listed combination fire pump controller and power transfer switch

For systems installed under the provisions of Section 695-3(b)(2) only, such additional disconnecting means and associated overcurrent protective device(s) shall be permitted as required to comply with other provisions of this *Code*. For on-site generators installed under the provisions of Section 695-3(b)(1), the overcurrent devices shall not be required. All disconnecting means and overcurrent protective devices that are unique to the fire pump loads shall comply with the following.

(1) Overcurrent Device Selection. The overcurrent protective device(s) shall be selected or set to carry indefinitely the sum of the locked-rotor current of the fire pump motor(s) and the pressure maintenance pump motor(s) and the full-load current of the associated fire pump accessory equipment when connected to this power supply.

(2) Disconnecting Means. The disconnecting means shall be

a. Identified as suitable for use as service equipment,
b. Lockable in the closed position, and
c. Located sufficiently remote from other building or other fire pump source disconnecting means that inadvertent contemporaneous operation would be unlikely.

(3) Disconnect Marking. The disconnecting means shall be marked "Fire Pump Disconnecting Means." The letters shall be at least 1 in. (25.4 mm) in height, and they shall be visible without opening enclosure doors or covers.

(4) Controller Marking. A placard shall be placed adjacent to the fire pump controller stating the location of this disconnecting means and the location of the key (if the disconnecting means is locked).

(5) Supervision. The disconnecting means shall be supervised in the closed position by one of the following methods:

(a) Central station, proprietary, or remote station signal device
(b) Local signaling service that will cause the sounding of an audible signal at a constantly attended point
(c) Locking the disconnecting means in the closed position

(d) Sealing of disconnecting means and approved weekly recorded inspections when the disconnecting means are located within fenced enclosures or in buildings under the control of the owner

695-5. Transformers. Where the service or system voltage is different from the utilization voltage of the fire pump motor, transformer(s) protected by disconnecting means and overcurrent protective devices shall be permitted to be installed between the system supply and the fire pump controller in accordance with (a) and (b), or (c). Only transformers covered in (c) shall be permitted to supply loads not directly associated with the fire pump system.

(a) Size. Where a transformer supplies an electric motor-driven fire pump, it shall be rated at a minimum of 125 percent of the sum of the fire pump motor(s) and pressure maintenance pump(s) motor loads, and 100 percent of the associated fire pump accessory equipment supplied by the transformer.

(b) Overcurrent Protection. The primary overcurrent protective device(s) shall be selected or set to carry indefinitely the sum of the locked-rotor current of the fire pump motor(s) and the pressure maintenance pump motor(s) and the full-load current of the associated fire pump accessory equipment when connected to this power supply. Secondary overcurrent protection shall not be permitted.

(c) Feeder Source. Where a feeder source is provided in accordance with Section 695-3(b)(2), transformers supplying the fire pump system shall be permitted to supply other loads. All other loads shall be calculated in accordance with Article 220, including demand factors as applicable.

(1) **Size.** Transformers shall be rated at a minimum of 125 percent of the sum of the fire pump motor(s) and pressure maintenance pump(s) motor loads, and 100 percent of the remaining load supplied by the transformer.

(2) **Overcurrent Protection.** The transformer size, the feeder size, and the overcurrent protective device(s) shall be coordinated such that overcurrent protection is provided for the transformer in accordance with Section 450-3 and for the feeder in accordance with Section 215-3, and such that the overcurrent protective device(s) is selected or set to carry indefinitely the sum of the locked-rotor current of the fire pump motor(s), the pressure maintenance pump motor(s), the full-load current of the associated fire pump accessory equipment, and 100 percent of the remaining loads supplied by the transformer.

695-6. Power Wiring. Power circuits and wiring methods shall comply with the requirements in (a) through (g), and as permitted in Section 230-90(a), Exception No. 4; Section 230-94, Exception No. 4; Section 230-95, Exception No. 2; Section 240-13(3); Section 230-208; Section 240-3(a); and Section 430-31.

(a) Service Conductors. Supply conductors shall be physically routed outside a building(s) and shall be installed as service entrance conductors in accordance with Article 230. Where supply conductors cannot be physically routed outside buildings, they shall be permitted to be routed through buildings where installed in accordance with Condition No. 1 or Condition No. 2 of Section 230-6. Where a fire pump is wired under the provisions of Section 695-3(b)(2), this requirement shall apply to all supply conductors on the load side of the service disconnecting means that constitute the normal source of supply to that fire pump.

Exception: Where there are multiple sources of supply with means for automatic connection from one source to the other, the requirement shall only apply to those conductors on the load side of that point of automatic connection between sources.

(b) Circuit Conductors. Fire pump supply conductors on the load side of the final disconnecting means and overcurrent device(s) permitted by Section 695-4(b) shall be kept entirely independent of all other wiring. They shall only supply loads that are directly associated with the fire pump system, and they shall be protected to resist potential damage by fire, structural failure, or operational accident. They shall be permitted to be routed through a building(s) encased in 2 in. (50.8 mm) of concrete or within enclosed construction dedicated to the fire pump circuit(s) and having a minimum of a 1-hour fire resistance rating, or they shall be permitted to be within listed electrical circuit protective systems with a minimum of 1-hour fire resistance. The installation shall comply with any restrictions provided in the listing of the electrical circuit protective system used.

Exception: The supply conductors located in the electrical equipment room where they originate and in the fire pump room shall not be required to have the minimum 1-hour fire separation or fire resistance rating, unless otherwise required by Section 700-9(d) of this Code.

(c) Conductor Size.

(1) Conductors supplying a fire pump motor(s), pressure maintenance pumps, and associated fire pump accessory equipment shall have a minimum rating of 125 percent of the sum of the fire pump motor(s) and pressure maintenance motor(s) full-load current(s), and 100 percent of the associated fire pump accessory equipment.

(2) Conductors supplying only a fire pump motor(s) shall have a minimum rating of 125 percent of the fire pump motor(s) full-load current(s).

(d) Overload Protection. Power circuits shall not have automatic protection against overloads. Except as provided in Section 695-5(c)(2), branch-circuit and feeder conductors shall be protected against short circuit only. Where a tap is made to supply a fire pump, and the tap wiring is run in accordance with Section 230-6, the applicable distance and size restrictions in Section 240-21 shall not apply.

Exception: Conductors between storage batteries and the engine shall not require overcurrent protection or disconnecting means.

(e) Pump Wiring. All wiring from the controllers to the pump motors shall be in rigid metal conduit, intermediate metal conduit, liquidtight flexible metal conduit, or liquidtight flexible nonmetallic conduit Type LFNC-B, or Type MI cable.

(f) Junction Points. Where wire connectors are used in the fire pump circuit, the connectors shall be listed. A fire pump controller or fire pump power transfer switch, where provided, shall not be used as a junction box to supply other equipment including a pressure maintenance (jockey) pump(s). A fire pump controller and fire pump power transfer switch, where provided, shall not serve any load other than the fire pump for which it is intended.

(g) Mechanical Protection. All wiring from engine controllers and batteries shall be protected against mechanical injury, and shall be installed in accordance with the controller and engine manufacturer's instructions.

ˣ695-7. Voltage Drop. The voltage at the controller line terminals shall not drop more than 15 percent below normal (controller-rated voltage) under motor starting conditions. The voltage at the motor terminals shall not drop more than 5 percent below the voltage rating of the motor when the motor is operating at 115 percent of the full-load current rating of the motor.

Exception: This limitation shall not apply for emergency run mechanical starting.

ˣ695-10. Listed Equipment. Diesel engine fire pump controllers, electric fire pump controllers, electric motors, fire pump power transfer switches, foam pump controllers, and limited service controllers shall be listed for fire pump service.

695-12. Equipment Location.

(a) Controllers and Transfer Switches. Electric motor-driven fire pump controllers and power transfer switches shall be located as close as practicable to the motors that they control and shall be within sight of the motors.

(b) Engine-Drive Controllers. Engine-drive fire pump controllers shall be located as close as is practical to the engines that they control and shall be within sight of the engines.

(c) Storage Batteries. Storage batteries for diesel engine drives shall be rack supported above the floor, secured against displacement, and located where they will not be subjected to excessive temperature, vibration, mechanical injury, or flooding with water.

(d) Energized Equipment. All energized equipment parts shall be located at least 12 in. (305 mm) above the floor level.

(e) Protection Against Pump Water. Fire pump controllers and power transfer switches shall be located or protected so that they will not be damaged by water escaping from pumps or pump connections.

(f) Mounting. All fire pump control equipment shall be mounted in a substantial manner on noncombustible supporting structures.

695-14. Control Wiring.

ˣ(a) Control Circuit Failures. External control circuits shall be arranged so that failure of any external circuit (open or short circuit) shall not prevent the operation of a pump(s) from all other internal or external means. Breakage, disconnecting, shorting of the wires, or loss of power to these circuits may cause continuous running of the fire pump, but shall not prevent the controller(s) from starting the fire pump(s) due to causes other than these external control circuits.

ˣ(b) Sensor Functioning. No undervoltage, phase-loss, frequency-sensitive, or other sensor(s) shall be installed that automatically or manually prohibit actuation of the motor contactor.

ˣ(c) Remote Device(s). No remote device(s) shall be installed that will prevent automatic operation of the transfer switch.

ˣ(d) Engine-Drive Control Wiring. All wiring between the controller and the diesel engine shall be stranded and sized to continuously carry the charging or control currents as required by the controller manufacturer. Such wiring shall be protected against mechanical injury. Controller manufacturer's specifications for distance and wire size shall be followed.

(e) Electric Fire Pump Control Wiring Methods. All electric motor-driven fire pump control wiring shall be in rigid metal conduit, intermediate metal conduit, liquidtight flexible metal conduit, or Type MI cable.

CHAPTER 7
Special Conditions

Article 700 — Emergency Systems

A. General

700-1. Scope. The provisions of this article apply to the electrical safety of the installation, operation, and maintenance of emergency systems consisting of circuits and equipment intended to supply, distribute, and control electricity for illumination or power, or both, to required facilities when the normal electrical supply or system is interrupted.

Emergency systems are those systems legally required and classed as emergency by municipal, state, federal, or other codes, or by any governmental agency having jurisdiction. These systems are intended to automatically supply illumination or power, or both, to designated areas and equipment in the event of failure of the normal supply or in the event of accident to elements of a system intended to supply, distribute, and control power and illumination essential for safety to human life.

FPN No. 1: For further information regarding wiring and installation of emergency systems in health care facilities, see Article 517.

FPN No. 2: For further information regarding performance and maintenance of emergency systems in health care facilities, see *Standard for Health Care Facilities,* NFPA 99-1996.

FPN No. 3: Emergency systems are generally installed in places of assembly where artificial illumination is required for safe exiting and for panic control in buildings subject to occupancy by large numbers of persons, such as hotels, theaters, sports arenas, health care facilities, and similar institutions. Emergency systems may also provide power for such functions as ventilation where essential to maintain life, fire detection and alarm systems, elevators, fire pumps, public safety communications systems, industrial processes where current interruption would produce serious life safety or health hazards, and similar functions.

FPN No. 4: For specification of locations where emergency lighting is considered essential to life safety, see *Life Safety Code,*® NFPA *101*®-1997.

FPN No. 5: For further information regarding performance of emergency and standby power systems, see *Standard for Emergency and Standby Power Systems,* NFPA 110-1996.

700-2. Application of Other Articles. Except as modified by this article, all applicable articles of this *Code* shall apply.

700-3. Equipment Approval. All equipment shall be approved for use on emergency systems.

700-4. Tests and Maintenance.

(a) Conduct or Witness Test. The authority having jurisdiction shall conduct or witness a test of the complete system upon installation and periodically afterward.

(b) Tested Periodically. Systems shall be tested periodically on a schedule acceptable to the authority having jurisdiction to ensure the systems are maintained in proper operating condition.

(c) Battery Systems Maintenance. Where battery systems or unit equipments are involved, including batteries used for starting, control, or ignition in auxiliary engines, the authority having jurisdiction shall require periodic maintenance.

(d) Written Record. A written record shall be kept of such tests and maintenance.

(e) Testing Under Load. Means for testing all emergency lighting and power systems during maximum anticipated load conditions shall be provided.

700-5. Capacity.

(a) Capacity and Rating. An emergency system shall have adequate capacity and rating for all loads to be operated simultaneously. The emergency system equipment shall be suitable for the maximum available fault current at its terminals.

(b) Selective Load Pickup, Load Shedding, and Peak Load Shaving. The alternate power source shall be permitted to supply emergency, legally required standby and optional standby system loads where automatic selective load pickup and load shedding is provided as needed to ensure adequate power to (1) the emergency circuits, (2) the legally required standby circuits, and (3) the optional standby circuits, in that order of priority. The alternate power source shall be permitted to be used for peak load shaving, provided the above conditions are met.

Peak load-shaving operation shall be permitted for satisfying the test requirement of Section 700-4(b), provided all other conditions of Section 700-4 are met.

A portable or temporary alternate source shall be available whenever the emergency generator is out of service for major maintenance or repair.

700-6. Transfer Equipment.

(a) Transfer equipment, including automatic transfer switches, shall be automatic, identified for emergency use, and approved by the authority having jurisdiction. Transfer equipment shall be designed and installed to prevent the inadvertent interconnection of normal and emergency sources of supply in any operation of the transfer equipment.

(b) Means shall be permitted to bypass and isolate the transfer equipment. Where bypass isolation switches are used, inadvertent parallel operation shall be avoided.

(c) Automatic transfer switches shall be electrically operated and mechanically held.

(d) Transfer equipment shall supply only emergency loads.

700-7. Signals. Audible and visual signal devices shall be provided, where practicable, for the following purposes.

(a) Derangement. To indicate derangement of the emergency source.

(b) Carrying Load. To indicate that the battery is carrying load.

(c) Not Functioning. To indicate that the battery charger is not functioning.

(d) Ground Fault. To indicate a ground fault in solidly grounded wye emergency systems of more than 150 volts to ground and circuit-protective devices rated 1000 amperes or more. The sensor for the ground-fault signal devices shall be located at, or ahead of, the main system disconnecting means for the emergency source, and the maximum setting of the signal devices shall be for a ground-fault current of 1200 amperes. Instructions on the course of action to be taken in event of indicated ground fault shall be located at or near the sensor location.

FPN: For signals for generator sets, see *Standard for Emergency and Standby Power Systems,* NFPA 110-1996.

700-8. Signs.

(a) Emergency Sources. A sign shall be placed at the service entrance equipment indicating type and location of on-site emergency power sources.

Exception: A sign shall not be required for individual unit equipment as specified in Section 700-12(e).

(b) Grounding. Where the grounded circuit conductor connected to the emergency source is connected to a grounding electrode conductor at a location remote from the emergency source, there shall be a sign at the grounding location that shall identify all emergency and normal sources connected at that location.

B. Circuit Wiring

700-9. Wiring, Emergency System.

(a) Identification. All boxes and enclosures (including transfer switches, generators, and power panels) for emergency circuits shall be permanently marked so they will be readily identified as a component of an emergency circuit or system.

(b) Wiring. Unless otherwise permitted in (1) through (4), wiring from an emergency source or emergency source distribution overcurrent protection to emergency loads shall be kept entirely independent of all other wiring and equipment. Wiring of two or more emergency circuits supplied from the same source shall be permitted in the same raceway, cable, box, or cabinet.

(1) The normal power source wiring shall be permitted to be located in transfer equipment enclosures.

(2) In exit or emergency lighting fixtures, wiring supplied from two sources shall be permitted.

(3) In a common junction box, attached to exit or emergency lighting fixtures, wiring supplied from two sources shall be permitted.

(4) The wiring within a common junction box attached to unit equipment, containing only the branch circuit supplying the unit equipment and the emergency circuit supplied by the unit equipment shall be permitted.

(c) Wiring Design and Location. Emergency wiring circuits shall be designed and located to minimize the hazards that might cause failure due to flooding, fire, icing, vandalism, and other adverse conditions.

(d) Fire Protection. Emergency systems shall meet the following additional requirements in assembly occupancies greater than 1000 persons or in buildings above 75 ft (23 m) in height with any of the following occupancy classes: assembly, educational, residential, detention and correctional, business, and mercantile.

(1) Feeder-circuit wiring shall meet one of the following conditions:

(a) Be installed within buildings that are fully protected by an approved automatic fire suppression system

(b) Be a listed electrical circuit protective system with a minimum 1-hour fire rating

(c) Be protected by a listed thermal barrier system for electrical system components

(d) Be protected by a fire-rated assembly having a minimum fire rating of 1 hour

(e) Be embedded in a minimum of 2 in. (50.8 mm) of concrete

(f) Be a cable listed to maintain circuit integrity for a minimum of 1 hour when installed in accordance with the listing requirements

(2) Equipment for feeder circuits (including transfer switches, transformers, panelboards, etc.) shall be located either in spaces fully protected by approved automatic fire suppression systems (sprinklers, carbon dioxide systems, etc.) or in spaces with a 1-hour fire resistance rating.

FPN: For the definition of *occupancy class*, see Section 4-1 of *Life Safety Code,* NFPA *101*-1997.

C. Sources of Power

700-12. General Requirements. Current supply shall be such that, in the event of failure of the normal supply to, or within, the building or group of buildings concerned, emergency lighting, emergency power, or both shall be available within the time required for the application but not to exceed 10 seconds. The supply system for emergency purposes, in addition to the normal services to the building and meeting the general requirements of this section, shall be one or more of the types of systems described in (a) through (d). Unit equipment in accordance with (e) shall satisfy the applicable requirements of this article.

In selecting an emergency source of power, consideration shall be given to the occupancy and the type of service to be rendered, whether of minimum duration, as for evacuation of a theater, or longer duration, as for supplying emergency power and lighting due to an indefinite period of current failure from trouble either inside or outside the building.

Equipment shall be designed and located to minimize the hazards that might cause complete failure due to flooding, fires, icing, and vandalism.

Equipment for sources of power as described in (a) through (d) where located within assembly occupancies greater than 1000 persons or in buildings above 75 ft (23 m) in height with any of the following occupancy classes — assembly, educational, residential, detention and correctional, business, and mercantile — shall be installed either in spaces fully protected by approved automatic fire suppression systems (sprinklers, carbon dioxide systems, etc.), or in spaces with a 1-hour fire rating.

FPN No. 1: For the definition of *occupancy class*, see Section 4-1 of *Life Safety Code,* NFPA *101*-1997.

FPN No. 2: Assignment of degree of reliability of the recognized emergency supply system depends on the careful evaluation of the variables at each particular installation.

(a) Storage Battery. Storage batteries used as a source of power for emergency systems shall be of suitable rating and capacity to supply and maintain the total load for a period of 1½ hours minimum, without the voltage applied to the load falling below 87½ percent of normal.

Batteries, whether of the acid or alkali type, shall be designed and constructed to meet the requirements of emergency service and shall be compatible with the charger for that particular installation.

For a sealed battery, the container shall not be required to be transparent. However, for the lead acid battery that requires water additions, transparent or translucent jars shall be furnished. Automotive-type batteries shall not be used.

An automatic battery charging means shall be provided.

(b) Generator Set.

(1) A generator set driven by a prime mover acceptable to the authority having jurisdiction and sized in accordance with Section 700-5. Means shall be provided for automatically starting the prime mover on failure of the normal service and for automatic transfer and operation of all required electrical circuits. A time-delay feature permitting a 15-minute setting shall be provided to avoid retransfer in case of short-time reestablishment of the normal source.

(2) Where internal combustion engines are used as the prime mover, an on-site fuel supply shall be provided with an on-premise fuel supply sufficient for not less than 2 hours full-demand operation of the system. Where power is needed for the operation of the fuel transfer pumps to deliver fuel to a generator set day tank, this pump shall be connected to the emergency power system.

(3) Prime movers shall not be solely dependent on a public utility gas system for their fuel supply or municipal water supply for their cooling systems. Means shall be provided for automatically transferring from one fuel supply to another where dual fuel supplies are used.

Exception: Where acceptable to the authority having jurisdiction, the use of other than on-site fuels shall be permitted where there is a low probability of a simultaneous failure of both the off-site fuel delivery system and power from the outside electrical utility company.

(4) Where a storage battery is used for control or signal power, or as the means of starting the prime mover, it shall be suitable for the purpose and shall be equipped with an automatic charging means independent of the generator set. Where the battery charger is required for the operation of the generator set, it shall be connected to the emergency system. Where power is required for the operation of dampers used to ventilate the generator set, the dampers shall be connected to the emergency system.

(5) Generator sets that require more than 10 seconds to develop power shall be acceptable, provided an auxiliary power supply will energize the emergency system until the generator can pick up the load.

(c) Uninterruptible Power Supplies. Uninterruptible power supplies used to provide power for emergency systems shall comply with the applicable provisions of Section 700-12(a) and (b).

(d) Separate Service. Where acceptable to the authority having jurisdiction as suitable for use as an emergency source, a second service shall be permitted. This service shall be in accordance with Article 230, with separate service drop or lateral, widely separated electrically and physically from the normal service to minimize the possibility of simultaneous interruption of supply.

(e) Unit Equipment. Individual unit equipment for emergency illumination shall consist of the following:

(1) A rechargeable battery
(2) A battery charging means
(3) Provisions for one or more lamps mounted on the equipment, or shall be permitted to have terminals for remote lamps, or both, and
(4) A relaying device arranged to energize the lamps automatically upon failure of the supply to the unit equipment

The batteries shall be of suitable rating and capacity to supply and maintain at not less than 87½ percent of the nominal battery voltage for the total lamp load associated with the unit for a period of at least 1½ hours, or the unit equipment shall supply and maintain not less than 60 percent of the initial emergency illumination for a period of at least 1½ hours. Storage batteries, whether of the acid or alkali type, shall be designed and constructed to meet the requirements of emergency service

Unit equipment shall be permanently fixed in place (i.e., not portable) and shall have all wiring to each unit installed in accordance with the requirements of any of the wiring methods in Chapter 3. Flexible cord and plug connection shall be permitted, provided that the cord does not exceed 3 ft (914 mm) in length. The branch circuit feeding the unit equipment shall be the same branch circuit as that serving the normal lighting in the area and connected ahead of any local switches. The branch circuit that feeds unit equipment shall be clearly identified at the distribution panel. Emergency illumination fixtures that obtain power from a unit equipment and are not part of the unit equipment shall be wired to the unit equipment as required by Section 700-9 and by one of the wiring methods of Chapter 3.

Exception: In a separate and uninterrupted area supplied by a minimum of three normal lighting circuits, a separate branch circuit for unit equipment shall be permitted if it originates from the same panelboard as that of the normal lighting circuits and is provided with a lock-on feature.

D. Emergency System Circuits for Lighting and Power

700-15. Loads on Emergency Branch Circuits. No appliances and no lamps, other than those specified as required for emergency use, shall be supplied by emergency lighting circuits.

700-16. Emergency Illumination. Emergency illumination shall include all required means of egress lighting, illuminated exit signs, and all other lights specified as necessary to provide required illumination.

Emergency lighting systems shall be designed and installed so that the failure of any individual lighting element, such as the burning out of a light bulb, cannot leave in total darkness any space that requires emergency illumination.

Where high-intensity discharge lighting such as high- and low-pressure sodium, mercury vapor, and metal halide is used as the sole source of normal illumination, the emergency lighting system shall be required to operate until normal illumination has been restored.

Exception: Alternative means that ensure emergency lighting illumination level is maintained shall be permitted.

700-17. Circuits for Emergency Lighting. Branch circuits that supply emergency lighting shall be installed to provide service from a source complying with Section 700-12 when the normal supply for lighting is interrupted. Such installations shall provide either one of the following: (1) an emergency lighting supply, independent of the general lighting supply, with provisions for automatically transferring the emergency lights upon the event of failure of the general lighting system supply, or (2) two or more separate and complete systems with independent power supply, each system providing sufficient current for emergency lighting purposes. Unless both systems are used for regular lighting purposes and are both kept lighted, means shall be provided for automatically energizing either system upon failure of the other. Either or both systems shall be permitted to be a part of the general lighting system of the protected occupancy if circuits supplying lights for emergency illumination are installed in accordance with other sections of this article.

700-18. Circuits for Emergency Power. For branch circuits that supply equipment classed as emergency, there shall be an emergency supply source to which the load will be transferred automatically upon the failure of the normal supply.

E. Control — Emergency Lighting Circuits

700-20. Switch Requirements. The switch or switches installed in emergency lighting circuits shall be arranged so that only authorized persons will have control of emergency lighting.

Exception No. 1: Where two or more single-throw switches are connected in parallel to control a single circuit, at least

one of these switches shall be accessible only to authorized persons.

Exception No. 2: Additional switches that act only to put emergency lights into operation but not disconnect them shall be permissible.

Switches connected in series or 3- and 4-way switches shall not be used.

700-21. Switch Location. All manual switches for controlling emergency circuits shall be in locations convenient to authorized persons responsible for their actuation. In places of assembly, such as theaters, a switch for controlling emergency lighting systems shall be located in the lobby or at a place conveniently accessible thereto.

In no case shall a control switch for emergency lighting in a theater, or motion-picture theater or place of assembly, be placed in a motion-picture projection booth or on a stage or platform.

Exception: Where multiple switches are provided, one such switch shall be permitted in such locations where arranged so that it can energize the circuit only, but it cannot de-energize the circuit.

700-22. Exterior Lights. Those lights on the exterior of a building that are not required for illumination when there is sufficient daylight shall be permitted to be controlled by an automatic light-actuated device.

F. Overcurrent Protection

700-25. Accessibility. The branch-circuit overcurrent devices in emergency circuits shall be accessible to authorized persons only.

> FPN: Fuses and circuit breakers for emergency circuit overcurrent protection, where coordinated to ensure selective clearing of fault currents, increase overall reliability of the system.

700-26. Ground-Fault Protection of Equipment. The alternate source for emergency systems shall not be required to have ground-fault protection of equipment with automatic disconnecting means. Ground-fault indication of the emergency source shall be provided per Section 700-7(d).

Article 701 — Legally Required Standby

A. General

701-1. Scope. The provisions of this article apply to the electrical safety of the installation, operation, and maintenance of legally required standby systems consisting of circuits and equipment intended to supply, distribute, and control electricity to required facilities for illumination or power, or both, when the normal electrical supply or system is interrupted.

The systems covered by this article consist only of those that are permanently installed in their entirety, including the power source.

> FPN No. 1: For additional information, see *Standard for Health Care Facilities,* NFPA 99-1996.

> FPN No. 2: For further information regarding performance of emergency and standby power systems, see *Standard for Emergency and Standby Power Systems,* NFPA 110-1996.

> FPN No. 3: For further information, see *Recommended Practice for Emergency and Standby Power Systems for Industrial and Commercial Applications,* ANSI/IEEE 446-1995.

701-2. Legally Required Standby Systems. Legally required standby systems are those systems required and so classed as legally required standby by municipal, state, federal, or other codes or by any governmental agency having jurisdiction. These systems are intended to automatically supply power to selected loads (other than those classed as emergency systems) in the event of failure of the normal source.

> FPN: Legally required standby systems are typically installed to serve loads, such as heating and refrigeration systems, communications systems, ventilation and smoke removal systems, sewerage disposal, lighting systems, and industrial processes, that, when stopped during any interruption of the normal electrical supply, could create hazards or hamper rescue or fire-fighting operations.

701-3. Application of Other Articles. Except as modified by this article, all applicable articles of this *Code* shall apply.

701-4. Equipment Approval. All equipment shall be approved for the intended use.

701-5. Tests and Maintenance for Legally Required Standby Systems.

(a) Conduct or Witness Test. The authority having jurisdiction shall conduct or witness a test of the complete system upon installation.

(b) Tested Periodically. Systems shall be tested periodically on a schedule and in a manner acceptable to the authority having jurisdiction to ensure the systems are maintained in proper operating condition.

(c) Battery Systems Maintenance. Where batteries are used for control, starting, or ignition of prime movers, the authority having jurisdiction shall require periodic maintenance.

(d) Written Record. A written record shall be kept on such tests and maintenance.

(e) Testing Under Load. Means for testing legally required standby systems under load shall be provided.

701-6. Capacity and Rating. A legally required standby system shall have adequate capacity and rating for the supply of all equipment intended to be operated at one time. Legally required standby system equipment shall be suitable for the maximum available fault current at its terminals.

The alternate power source shall be permitted to supply legally required standby and optional standby system loads where automatic selective load pickup and load shedding is provided as needed to ensure adequate power to the legally required standby circuits.

701-7. Transfer Equipment.

(a) Transfer equipment, including automatic transfer switches, shall be automatic and identified for standby use and approved by the authority having jurisdiction. Transfer equipment shall be designed and installed to prevent the inadvertent interconnection of normal and alternate sources of supply in any operation of the transfer equipment.

(b) Means to bypass and isolate the transfer switch equipment shall be permitted. Where bypass isolation switches are used, inadvertent parallel operation shall be avoided.

(c) Automatic transfer switches shall be electrically operated and mechanically held.

701-8. Signals. Audible and visual signal devices shall be provided, where practicable, for the following purposes.

(a) Derangement. To indicate derangement of the standby source.

(b) Carrying Load. To indicate that the standby source is carrying load.

(c) Not Functioning. To indicate that the battery charger is not functioning.

FPN: For signals for generator sets, see *Standard for Emergency and Standby Power Systems,* NFPA 110-1996.

701-9. Signs.

(a) Mandated Standby. A sign shall be placed at the service entrance indicating type and location of on-site legally required standby power sources.

Exception: A sign shall not be required for individual unit equipment as specified in Section 701-11(f).

(b) Grounding. Where the grounded circuit conductor connected to the emergency source is connected to a grounding electrode conductor at a location remote from the emergency source, there shall be a sign at the grounding location that shall identify all emergency and normal sources connected at that location.

B. Circuit Wiring

701-10. Wiring Legally Required Standby Systems. The legally required standby system wiring shall be permitted to occupy the same raceways, cables, boxes, and cabinets with other general wiring.

C. Sources of Power

701-11. Legally Required Standby Systems. Current supply shall be such that, in the event of failure of the normal supply to, or within, the building or group of buildings concerned, legally required standby power will be available within the time required for the application but not to exceed 60 seconds. The supply system for legally required standby purposes, in addition to the normal services to the building, shall be permitted to comprise one or more of the types of systems described in (a) through (e). Unit equipment in accordance with (f) shall satisfy the applicable requirements of this article.

In selecting a legally required standby source of power, consideration shall be given to the type of service to be rendered, whether of short-time duration or long duration.

Consideration shall be given to the location or design, or both, of all equipment to minimize the hazards that might cause complete failure due to floods, fires, icing, and vandalism.

FPN: Assignment of degree of reliability of the recognized legally required standby supply system depends on the careful evaluation of the variables at each particular installation.

(a) Storage Battery. A storage battery shall be of suitable rating and capacity to supply and maintain at not less than 87½ percent of system voltage the total load of the circuits supplying legally required standby power for a period of at least 1½ hours.

Batteries, whether of the acid or alkali type, shall be designed and constructed to meet the service requirements of emergency service and shall be compatible with the charger for that particular installation.

For a sealed battery, the container shall not be required to be transparent. However, for the lead acid battery that requires water additions, transparent or translucent jars shall be furnished. Automotive-type batteries shall not be used.

An automatic battery charging means shall be provided.

(b) Generator Set.

(1) A generator set driven by a prime mover acceptable to the authority having jurisdiction and sized in accordance with Section 701-6. Means shall be provided for automatically starting the prime mover upon failure of the normal service and for automatic transfer and operation of all required electrical circuits. A time-delay feature permitting a 15-minute setting shall be provided to avoid retransfer in case of short-time re-establishment of the normal source.

(2) Where internal combustion engines are used as the prime mover, an on-site fuel supply shall be provided with an on-premise fuel supply sufficient for not less than 2 hours full-demand operation of the system.

(3) Prime movers shall not be solely dependent on a public utility gas system for their fuel supply or municipal water supply for their cooling systems. Means shall be provided for automatically transferring one fuel supply to another where dual fuel supplies are used.

Exception: Where acceptable to the authority having jurisdiction, the use of other than on-site fuels shall be permitted where there is a low probability of a simultaneous failure of both the off-site fuel delivery system and power from the outside electrical utility company.

(4) Where a storage battery is used for control or signal power, or as the means of starting the prime mover, it shall be suitable for the purpose and shall be equipped with an automatic charging means independent of the generator set.

(c) Uninterruptible Power Supplies. Uninterruptible power supplies used to provide power for legally required standby systems shall comply with the applicable provisions of Section 701-11(a) and (b).

(d) Separate Service. Where acceptable to the authority having jurisdiction, a second service shall be permitted. This service shall be in accordance with Article 230, with separate service drop or lateral widely separated electrically and physically from the normal service to minimize the possibility of simultaneous interruption of supply.

(e) Connection Ahead of Service Disconnecting Means. Where acceptable to the authority having jurisdiction, connections ahead of, but not within, the main service disconnecting means shall be permitted. The legally required standby service shall be sufficiently separated from the normal main service disconnecting means to prevent simultaneous interruption of supply through an occurrence within the building or groups of buildings served.

FPN: See Section 230-82 for equipment permitted on the supply side of a service disconnecting means.

(f) Unit Equipment. Individual unit equipment for legally required standby illumination shall consist of the following:

(1) A rechargeable battery
(2) A battery charging means
(3) Provisions for one or more lamps mounted on the equipment and shall be permitted to have terminals for remote lamps, and
(4) A relaying device arranged to energize the lamps automatically upon failure of the supply to the unit equipment

The batteries shall be of suitable rating and capacity to supply and maintain at not less than 87½ percent of the nominal battery voltage for the total lamp load associated with the unit for a period of at least 1½ hours, or the unit equipment shall supply and maintain not less than 60 percent of the initial legally required standby illumination for a period of at least 1½ hours. Storage batteries, whether of the acid or alkali type, shall be designed and constructed to meet the requirements of emergency service.

Unit equipment shall be permanently fixed in place (i.e., not portable) and shall have all wiring to each unit installed in accordance with the requirements of any of the wiring methods in Chapter 3. Flexible cord and plug connection shall be permitted, provided that the cord does not exceed 3 ft (914 mm) in length. The branch circuit feeding the unit equipment shall be the same branch circuit as that serving the normal lighting in the area and connected ahead of any local switches. Legally required standby illumination fixtures that obtain power from a unit equipment and are not part of the unit equipment shall be wired to the unit equipment by one of the wiring methods of Chapter 3.

Exception: In a separate and uninterrupted area supplied by a minimum of three normal lighting circuits, a separate branch circuit for unit equipment shall be permitted if it originates from the same panelboard as that of the normal lighting circuits and is provided with a lock-on feature.

D. Overcurrent Protection

701-15. Accessibility. The branch-circuit overcurrent devices in legally required standby circuits shall be accessible to authorized persons only.

701-17. Ground-Fault Protection of Equipment. The alternate source for legally required standby systems shall not be required to have ground-fault protection of equipment.

Article 702 — Optional Standby Systems

A. General

702-1. Scope. The provisions of this article apply to the installation and operation of optional standby systems.

The systems covered by this article consist only of those that are permanently installed in their entirety, including prime movers.

702-2. Optional Standby Systems. Optional standby systems are intended to protect public or private facilities or property where life safety does not depend on the performance of the system. Optional standby systems are intended to supply on-site generated power to selected loads either automatically or manually.

FPN: Optional standby systems are typically installed to provide an alternate source of electric power for such facilities as industrial and commercial buildings, farms, and residences, and to serve loads such as heating and refrigeration systems, data processing and communications systems, and industrial processes that, when stopped during any power outage, could cause discomfort, serious interruption of the process, damage to the product or process, or the like.

702-3. Application of Other Articles. Except as modified by this article, all applicable articles of this *Code* shall apply.

702-4. Equipment Approval. All equipment shall be approved for the intended use.

702-5. Capacity and Rating. An optional standby system shall have adequate capacity and rating for the supply of all equipment intended to be operated at one time. Optional standby system equipment shall be suitable for the maximum available fault current at its terminals. The user of the optional standby system shall be permitted to select the load connected to the system.

702-6. Transfer Equipment. Transfer equipment shall be suitable for the intended use and designed and installed so as to prevent the inadvertent interconnection of normal and alternate sources of supply in any operation of the transfer equipment.

Transfer equipment, located on the load side of branch-circuit protection, shall be permitted to contain supplementary overcurrent protection having an interrupting rating sufficient for the available fault current that the generator can deliver.

702-7. Signals. Audible and visual signal devices shall be provided, where practicable, for the following purposes.

(1) Derangement. To indicate derangement of the optional standby source.

(2) Carrying Load. To indicate that the optional standby source is carrying load.

702-8. Signs.

(a) Standby. A sign shall be placed at the service-entrance equipment that indicates the type and location of on-site optional standby power sources. A sign shall not be required for individual unit equipment for standby illumination.

(b) Grounding. Where the grounded circuit conductor connected to the emergency source is connected to a grounding electrode conductor at a location remote from the emergency source, there shall be a sign at the grounding location that shall identify all emergency and normal sources connected at that location.

B. Circuit Wiring

702-9. Wiring Optional Standby Systems. The optional standby system wiring shall be permitted to occupy the same raceways, cables, boxes, and cabinets with other general wiring.

Article 705 — Interconnected Electric Power Production Sources

705-1. Scope. This article covers installation of one or more electric power production sources operating in parallel with a primary source(s) of electricity.

FPN: Examples of the types of primary sources are a utility supply, on-site electric power source(s), or other sources.

705-2. Definition. For purposes of this article, the following definition applies.

Interactive System. An electric power production system that is operating in parallel with and capable of delivering energy to an electric primary source supply system.

705-3. Other Articles. Interconnected electric power production sources shall comply with this article and also the applicable requirements of the following articles:

	Article
Generators	445
Emergency systems	700
Legally required standby systems	701
Optional standby systems	702

Exception: Installation of solar photovoltaic systems operated as interconnected power sources shall be in accordance with Article 690.

705-10. Directory. A permanent plaque or directory, denoting all electrical power sources on or in the premises, shall be installed at each service equipment location and at locations of all electric power production sources capable of being interconnected.

Exception: Installations with large numbers of power production sources shall be permitted to be designated by groups.

705-12. Point of Connection. The outputs of electric power production systems shall be interconnected at the premises service disconnecting means.

(a) The outputs shall be permitted to be interconnected at a point or points elsewhere on the premises where the system qualifies as an integrated electric system and incorporates protective equipment in accordance with all applicable sections of Article 685.

(b) The outputs shall be permitted to be interconnected at a point or points elsewhere on the premises where all of the following conditions are met.

(1) The aggregate of nonutility sources of electricity has a capacity in excess of 100 kW, or the service is above 1000 volts;
(2) The conditions of maintenance and supervision ensure that qualified persons will service and operate the system; and
(3) Safeguards and protective equipment are established and maintained.

705-14. Output Characteristics. The output of a generator or other electric power production source operating in parallel with an electric supply system shall be compatible with the voltage, wave shape, and frequency of the system to which it is connected.

> FPN: The term *compatible* does not necessarily mean matching the primary source wave shape.

705-16. Interrupting and Short-Circuit Current Rating. Consideration shall be given to the contribution of fault currents from all interconnected power sources for the interrupting and short-circuit current ratings of equipment on interactive systems.

705-20. Disconnecting Means, Sources. Means shall be provided to disconnect all ungrounded conductors of an electric power production source(s) from all other conductors. See Article 230.

705-21. Disconnecting Means, Equipment. Means shall be provided to disconnect equipment, such as inverters or transformers, associated with a power production source, from all ungrounded conductors of all sources of supply. Equipment intended to be operated and maintained as an integral part of a power production source exceeding 1000 volts shall not be required to have a disconnecting means.

705-22. Disconnect Device. The disconnecting means for ungrounded conductors shall consist of a manually or power operable switch(es) or circuit breaker(s)

(1) Located where accessible;
(2) Externally operable without exposing the operator to contact with live parts and, if power operable, of a type that can be opened by hand in the event of a power supply failure;
(3) Plainly indicating whether in the open or closed position; and
(4) Having ratings not less than the load to be carried and the fault current to be interrupted.

For disconnect equipment energized from both sides, a marking shall be provided to indicate that all contacts of the disconnect equipment may be energized.

> FPN No. 1: In parallel generation systems, some equipment, including knife blade switches and fuses, are likely to be energized from both directions. See Section 240-40.

> FPN No. 2: Interconnection to an off-premises primary source could require a visibly verifiable disconnecting device.

705-30. Overcurrent Protection. Conductors shall be protected in accordance with Article 240. Equipment overcurrent protection shall be in accordance with the articles referenced in Article 240. Equipment and conductors connected to more than one electrical source shall have a sufficient number of overcurrent devices located so as to provide protection from all sources.

(a) Generators shall be protected in accordance with Section 445-4.

(b) Solar photovoltaic systems shall be protected in accordance with Article 690.

(c) Overcurrent protection for a transformer with a source(s) on each side shall be provided in accordance with Section 450-3 by considering first one side of the transformer, then the other side of the transformer, as the primary.

705-32. Ground-Fault Protection. Where ground-fault protection is used, the output of an interactive system shall be connected to the supply side of the ground-fault protection.

Exception: Connection shall be permitted to be made to the load side of ground-fault protection provided that there is ground-fault protection for equipment from all ground-fault current sources.

705-40. Loss of Primary Source. Upon loss of primary source, an electric power production source shall be automatically disconnected from all ungrounded conductors of the primary source and shall not be reconnected until the primary source is restored.

> FPN No. 1: Risks to personnel and equipment associated with the primary source could occur if an interactive electric power production source can operate as an island. Special detection methods can be required to determine that a primary source supply system outage has occurred, and whether there should be automatic disconnection. When the primary source supply system is restored, special detection methods can be required to limit exposure of power production sources to out-of-phase reconnection.

> FPN No. 2: Induction-generating equipment on systems with significant capacitance can become self-excited upon loss of primary source and experience severe over-voltage as a result.

705-42. Unbalanced Interconnections. A 3-phase electric power production source shall be automatically disconnected from all ungrounded conductors of the interconnected systems when one of the phases of that source opens. This requirement shall not be applicable to an electric power production source providing power for an emergency or legally required standby system.

705-43. Synchronous Generators. Synchronous generators in a parallel system shall be provided with the necessary equipment to establish and maintain a synchronous condition.

705-50. Grounding. Interconnected electric power production sources shall be grounded in accordance with Article 250.

Exception: For direct-current systems connected through an inverter directly to a grounded service, other methods that accomplish equivalent system protection and that utilize equipment listed and identified for the use shall be permitted.

Article 720 — Circuits and Equipment Operating at Less than 50 Volts

720-1. Scope. This article covers installations operating at less than 50 volts, direct current or alternating current.

Exception: As covered in Articles 411, 551, 650, 669, 690, 725, and 760.

720-2. Hazardous (Classified) Locations. Installations coming within the scope of this article and installed in hazardous (classified) locations shall also comply with the appropriate provisions of Articles 500 through 517.

720-4. Conductors. Conductors shall not be smaller than No. 12 copper or equivalent. Conductors for appliance branch circuits supplying more than one appliance or appli-

ance receptacle shall not be smaller than No. 10 copper or equivalent.

720-5. Lampholders. Standard lampholders that have a rating of not less than 660 watts shall be used.

720-6. Receptacle Rating. Receptacles shall have a rating of not less than 15 amperes.

720-7. Receptacles Required. Receptacles of not less than 20-ampere rating shall be provided in kitchens, laundries, and other locations where portable appliances are likely to be used.

720-8. Overcurrent Protection. Overcurrent protection shall comply with Article 240.

720-9. Batteries. Installations of storage batteries shall comply with Article 480.

720-10. Grounding. Grounding shall be as provided in Article 250.

720-11. Mechanical Execution of Work. Circuits operating at less than 50 volts shall be installed in a neat and workmanlike manner. Cables shall be supported by the building structure in such a manner that the cable will not be damaged by normal building use.

Article 725 — Class 1, Class 2, and Class 3 Remote-Control, Signaling, and Power-Limited Circuits

A. General

725-1. Scope. This article covers remote-control, signaling, and power-limited circuits that are not an integral part of a device or appliance.

> FPN: The circuits described herein are characterized by usage and electrical power limitations that differentiate them from electric light and power circuits; therefore, alternative requirements to those of Chapters 1 through 4 are given with regard to minimum wire sizes, derating factors, overcurrent protection, insulation requirements, and wiring methods and materials.

725-2. Definitions. For purposes of this article, the following definitions apply.

Class 1 Circuit. The portion of the wiring system between the load side of the overcurrent device or power-limited supply and the connected equipment. The voltage and power limitations of the source are in accordance with Section 725-21.

Class 2 Circuit. The portion of the wiring system between the load side of a Class 2 power source and the connected equipment. Due to its power limitations, a Class 2 circuit considers safety from a fire initiation standpoint and provides acceptable protection from electric shock.

Class 3 Circuit. The portion of the wiring system between the load side of a Class 3 power source and the connected equipment. Due to its power limitations, a Class 3 circuit considers safety from a fire initiation standpoint. Since higher levels of voltage and current than Class 2 are permitted, additional safeguards are specified to provide protection from an electric shock hazard that could be encountered.

725-3. Locations and Other Articles. Circuits and equipment shall comply with (a) through (e). Only those sections of Article 300 referenced in this article shall apply to Class 1, Class 2, and Class 3 circuits.

(a) Spread of Fire or Products of Combustion. Section 300-21.

(b) Ducts, Plenums, and Other Air-Handling Spaces. Section 300-22, where installed in ducts or plenums or other space used for environmental air.

Exception to (b): As permitted in Section 725-61(a).

(c) Hazardous (Classified) Locations. Articles 500 through 516 and Article 517, Part D, where installed in hazardous (classified) locations.

(d) Cable Trays. Article 318, where installed in cable tray.

(e) Motor Control Circuits. Article 430, Part F, where tapped from the load side of the motor branch-circuit protective device(s) as specified in Section 430-72(a).

725-5. Access to Electrical Equipment Behind Panels Designed to Allow Access. Access to equipment shall not be denied by an accumulation of conductors and cables that prevents removal of panels, including suspended ceiling panels.

725-6. Class 1, Class 2, and Class 3 Circuit Grounding. Class 1, Class 2, and Class 3 circuits and equipment shall be grounded in accordance with Article 250.

725-7. Mechanical Execution of Work. Class 1, Class 2, and Class 3 circuits shall be installed in a neat and workmanlike manner. Cables shall be supported by the building structure in such a manner that the cable will not be damaged by normal building use.

FPN: One way to determine accepted industry practice is to refer to nationally recognized standards such as *Commercial*

Building Telecommunications Wiring Standard, ANSI/EIA/TIA 568-1991; *Commercial Building Standard for Telecommunications Pathways and Spaces,* ANSI/EIA/TIA 569-1990; and *Residential and Light Commercial Telecommunications Wiring Standard,* ANSI/EIA/TIA 570-1991.

725-8. Safety-Control Equipment.

(a) Remote-Control Circuits. Remote-control circuits for safety-control equipment shall be classified as Class 1 if the failure of the equipment to operate introduces a direct fire or life hazard. Room thermostats, water temperature regulating devices, and similar controls used in conjunction with electrically controlled household heating and air conditioning shall not be considered safety-control equipment.

(b) Physical Protection. Where damage to remote-control circuits of safety control equipment would introduce a hazard, as covered in Section 725-8(a), all conductors of such remote-control circuits shall be installed in rigid metal conduit, intermediate metal conduit, rigid nonmetallic conduit, electrical metallic tubing, Type MI cable, Type MC cable, or be otherwise suitably protected from physical damage.

725-15. Class 1, Class 2, and Class 3 Circuit Requirements. A remote-control, signaling, or power-limited circuit shall comply with the following parts of this article:

(1) Class 1 Circuits, Parts A and B
(2) Class 2 and Class 3 Circuits, Parts A and C

B. Class 1 Circuits

725-21. Class 1 Circuit Classifications and Power Source Requirements. Class 1 circuits shall be classified as either Class 1 power-limited circuits where they comply with the power limitations of (a), or as Class 1 remote-control and signaling circuits where they are used for remote control or signaling purposes and comply with the power limitations of (b).

(a) Class 1 Power-Limited Circuits. These circuits shall be supplied from a source that has a rated output of not more than 30 volts and 1000 volt-amperes.

(1) Class 1 Transformers. Transformers used to supply power-limited Class 1 circuits shall comply with Article 450.

(2) Other Class 1 Power Sources. Power sources other than transformers shall be protected by overcurrent devices rated at not more than 167 percent of the volt-ampere rating of the source divided by the rated voltage. The overcurrent devices shall not be interchangeable with overcurrent devices of higher ratings. The overcurrent device shall be permitted to be an integral part of the power supply.

To comply with the 1000 volt-ampere limitation of (a),

the maximum output (VA_{max}) of power sources other than transformers shall be limited to 2500 volt-amperes, and the product of the maximum current (I_{max}) and maximum voltage (V_{max}) shall not exceed 10,000 volt-amperes. These ratings shall be determined with any overcurrent-protective device bypassed.

VA_{max} is the maximum volt-ampere output after one minute of operation regardless of load and with overcurrent protection bypassed, if used. Current-limiting impedance shall not be bypassed when determining VA_{max}.

I_{max} is the maximum output current under any noncapacitive load, including short circuit, and with overcurrent protection bypassed, if used. Current-limiting impedance should not be bypassed when determining I_{max}. Where a current-limiting impedance, listed for the purpose, or as part of a listed product, is used in combination with a stored energy source, e.g., storage battery, to limit the output current, I_{max} limits apply after 5 seconds.

V_{max} is the maximum output voltage regardless of load with rated input applied.

(b) Class 1 Remote-Control and Signaling Circuits. These circuits shall not exceed 600 volts. The power output of the source shall not be required to be limited.

725-23. Class 1 Circuit Overcurrent Protection. Overcurrent protection for conductors No. 14 and larger shall be provided in accordance with the conductor ampacity, without applying the derating factors of Section 310-15 to the ampacity calculation. Overcurrent protection shall not exceed 7 amperes for No. 18 conductors and 10 amperes for No. 16.

Exception: Where other articles of this Code permit or require other overcurrent protection.

 FPN: For example, see Section 430-72 for motors, Section 610-53 for cranes and hoists, and Sections 517-74(b) and 660-9 for X-ray equipment.

725-24. Class 1 Circuit Overcurrent Device Location. Overcurrent devices shall be located at the point where the conductor to be protected receives its supply.

Exception No. 1: Where the overcurrent device protecting the larger conductor also protects the smaller conductor.

Exception No. 2: Transformer secondary conductors. Class 1 circuit conductors supplied by the secondary of a single-phase transformer having only a 2-wire (single-voltage) secondary shall be permitted to be protected by overcurrent protection provided on the primary (supply) side of the transformer, provided this protection is in accordance with Section 450-3 and does not exceed the value determined by multiplying the secondary conductor ampacity by the secondary-to-primary transformer voltage ratio. Transformer secondary conductors other than 2 wire shall not be considered to be protected by the primary overcurrent protection.

Exception No. 3: Electronic power source output conductors. Class 1 circuit conductors supplied by the output of a single-phase, listed electronic power source, other than a transformer, having only a 2-wire (single-voltage) output for connection to Class 1 circuits shall be permitted to be protected by overcurrent protection provided on the input side of the electronic power source, provided this protection does not exceed the value determined by multiplying the Class 1 circuit conductor ampacity by the output-to-input voltage ratio. Electronic power source outputs, other than 2 wire (single voltage), connected to Class 1 circuits shall not be considered to be protected by overcurrent protection on the input of the electronic power source.

 FPN: A single-phase, listed electronic power supply whose output supplies a 2-wire (single-voltage) circuit is an example of a Class 1 power source meeting the requirements of Section 725-23.

Exception No. 4: Class 1 circuit conductors No. 14 and larger that are tapped from the load side of the overcurrent-protective device(s) of a controlled light and power circuit shall require only short-circuit and ground-fault protection and shall be permitted to be protected by the branch-circuit overcurrent-protective device(s) where the rating of the protective device(s) is not more than 300 percent of the ampacity of the Class 1 circuit conductor.

725-25. Class 1 Circuit Wiring Methods. Installations of Class 1 circuits shall be in accordance with the appropriate articles in Chapter 3.

Exception No. 1: As provided in Sections 725-26 through 725-28.

Exception No. 2: Where other articles of this Code permit or require other methods.

725-26. Conductors of Different Circuits in Same Cable, Enclosure, or Raceway.

(a) Two or More Class 1 Circuits. Class 1 circuits shall be permitted to occupy the same cable, enclosure, or raceway without regard to whether the individual circuits are alternating current or direct current, provided all conductors are insulated for the maximum voltage of any conductor in the cable, enclosure, or raceway.

(b) Class 1 Circuits with Power Supply Circuits. Class 1 circuits and power supply circuits shall be permitted to occupy the same cable, enclosure, or raceway only where the equipment powered is functionally associated.

Exception No. 1: Where installed in factory- or field-assembled control centers.

Exception No. 2: Underground conductors in a manhole where one of the following conditions is met.

(a) *The power-supply or Class 1 circuit conductors are in a metal-enclosed cable or Type UF cable.*

(b) *The conductors are permanently separated from the power-supply conductors by a continuous firmly fixed nonconductor, such as flexible tubing, in addition to the insulation on the wire.*

(c) *The conductors are permanently and effectively separated from the power supply conductors and securely fastened to racks, insulators, or other approved supports.*

725-27. Class 1 Circuit Conductors.

(a) Sizes and Use. Conductors of sizes No. 18 and 16 shall be permitted to be used, provided they supply loads that do not exceed the ampacities given in Section 402-5 and are installed in a raceway, an approved enclosure, or a listed cable. Conductors larger than No. 16 shall not supply loads greater than the ampacities given in Section 310-15. Flexible cords shall comply with Article 400.

(b) Insulation. Insulation on conductors shall be suitable for 600 volts. Conductors larger than No. 16 shall comply with Article 310. Conductors in sizes No. 18 and 16 shall be Type FFH-2, KF-2, KFF-2, PAF, PAFF, PF, PFF, PGF, PGFF, PTF, PTFF, RFH-2, RFHH-2, RFHH-3, SF-2, SFF-2, TF, TFF, TFFN, TFN, ZF, or ZFF. Conductors with other types and thicknesses of insulation shall be permitted if listed for Class 1 circuit use.

725-28. Number of Conductors in Cable Trays and Raceway, and Derating.

(a) Class 1 Circuit Conductors. Where only Class 1 circuit conductors are in a raceway, the number of conductors shall be determined in accordance with Section 300-17. The derating factors given in Section 310-15(b)(2)(a) shall apply only if such conductors carry continuous loads in excess of 10 percent of the ampacity of each conductor.

(b) Power-Supply Conductors and Class 1 Circuit Conductors. Where power-supply conductors and Class 1 circuit conductors are permitted in a raceway in accordance with Section 725-26, the number of conductors shall be determined in accordance with Section 300-17. The derating factors given in Section 310-15(b)(2)(a) shall apply as follows:

(1) To all conductors where the Class 1 circuit conductors carry continuous loads in excess of 10 percent of the ampacity of each conductor and where the total number of conductors is more than three

(2) To the power-supply conductors only, where the Class 1 circuit conductors do not carry continuous loads in excess of 10 percent of the ampacity of each conductor and where the number of power-supply conductors is more than three

(c) Class 1 Circuit Conductors in Cable Trays. Where Class 1 circuit conductors are installed in cable trays, they shall comply with the provisions of Sections 318-9 through 318-11.

725-29. Circuits Extending Beyond One Building. Class 1 circuits that extend aerially beyond one building shall also meet the requirements of Article 225.

C. Class 2 and Class 3 Circuits

725-41. Power Sources for Class 2 and Class 3 Circuits.

(a) Power Source. The power source for a Class 2 or a Class 3 circuit shall be as specified in (1), (2), (3), or (4):

> FPN No. 1: Figure 725-41 illustrates the relationships between Class 2 or Class 3 power sources, their supply, and the Class 2 or Class 3 circuits.

> FPN No. 2: Tables 11(a) and 11(b) in Chapter 9 provide the requirements for listed Class 2 and Class 3 power sources.

(1) A listed Class 2 or Class 3 transformer
(2) A listed Class 2 or Class 3 power supply
(3) Other listed equipment marked to identify the Class 2 or Class 3 power source

> FPN: Examples of other listed equipment are as follows:

> (1) A circuit card listed for use as a Class 2 or Class 3 power source where used as part of a listed assembly
> (2) A current-limiting impedance, listed for the purpose, or part of a listed product, used in conjunction with a nonpower-limited transformer or a stored energy source, e.g., storage battery, to limit the output current
> (3) A thermocouple

Exception: Thermocouples shall not require listing as a Class 2 power source.

(4) Listed information technology (computer) equipment limited power circuits

> FPN: One way to determine applicable requirements for listing of information technology (computer) equipment is to refer to *Standard for Safety of Information Technology Equipment, Including Electrical Business Equipment,* UL 1950-1995. Typically such circuits are used to interconnect information technology equipment for the purpose of exchanging information (data).

(5) A dry cell battery shall be considered an inherently limited Class 2 power source, provided the voltage is 30 volts or less and the capacity is equal to or less than that available from series connected No. 6 carbon zinc cells.

Figure 725-41   Class 2 and Class 3 circuits.

(b) Interconnection of Power Sources. Class 2 or Class 3 power sources shall not have the output connections paralleled or otherwise interconnected unless listed for such interconnection.

725-51. Wiring Methods on Supply Side of the Class 2 or Class 3 Power Source. Conductors and equipment on the supply side of the power source shall be installed in accordance with the appropriate requirements of Chapters 1 through 4. Transformers or other devices supplied from electric light or power circuits shall be protected by an overcurrent device rated not over 20 amperes.

Exception: The input leads of a transformer or other power source supplying Class 2 and Class 3 circuits shall be permitted to be smaller than No. 14, but not smaller than No. 18 if they are not over 12 in. (305 mm) long and if they have insulation that complies with Section 725-27(b).

725-52. Wiring Methods and Materials on Load Side of the Class 2 or Class 3 Power Source. Conductors on the load side of the power source shall be insulated at not less than the requirements of Section 725-71 and shall be installed in accordance with Sections 725-54 and 725-61.

725-54. Installation of Conductors and Equipment.

(a) Separation from Electric Light, Power, Class 1, Non-power-Limited Fire Alarm Circuit Conductors, and Medium Power Network-Powered Broadband Communications Cables.

(1) In Cables, Compartments, Cable Trays, Enclosures, Manholes, Outlet Boxes, Device Boxes, and Raceways. Cables and conductors of Class 2 and Class 3 circuits shall not be placed in any cable, cable tray, compartment, enclosure, manhole, outlet box, device box, raceway, or similar fitting with conductors of electric light, power, Class 1, nonpower-limited fire alarm circuits, and medium power network-powered broadband communications cables.

Exception No. 1: Where the conductors of the electric light, power, Class 1, nonpower-limited fire alarm, and medium power network-powered broadband communications circuits are separated by a barrier from the Class 2 and Class 3 circuits. In enclosures, Class 2 or Class 3 circuits shall be permitted to be installed in a raceway within the enclosure to separate them from Class 1, electric light, power, nonpower-limited fire alarm, and medium power network-powered broadband communications circuits.

Exception No. 2: Conductors in compartments, enclosures, device boxes, outlet boxes, or similar fittings, where electric light, power, Class 1, nonpower-limited fire alarm, and medium power network-powered broadband communications circuit conductors are introduced solely to connect to the equipment connected to Class 2 or Class 3 circuits to which the other conductors are connected, and

(a) The electric light, power, Class 1, nonpower-limited fire alarm, and medium power network-powered broadband communications circuit conductors are routed to maintain a minimum of 0.25 in. (6.35 mm) separation from the conductors and cables of Class 2 and Class 3 circuits, or

(b) The circuit conductors operate at 150 volts or less to ground and also comply with one of the following:

(1) The Class 2 and Class 3 circuits are installed using Type CL3, CL3R, or CL3P or permitted substitute cables, provided these Class 3 cable conductors extending beyond the jacket are separated by a minimum of 0.25 in. (6.35 mm) or by a nonconductive sleeve or nonconductive barrier from all other conductors, or

(2) The Class 2 and Class 3 circuit conductors are installed as a Class 1 circuit in accordance with Section 725-21.

Exception No. 3: Conductors entering compartments, enclosures, device boxes, outlet boxes, or similar fittings, where electric light, power, Class 1, nonpower-limited fire alarm, and medium power network-powered communications circuit conductors are introduced solely to connect the equipment connected to Class 2 or Class 3 circuits to which the other conductors in the enclosure are connected. If the conductors must enter an enclosure that is provided with a single opening, they shall be permitted to enter through a single fitting (such as a tee) provided the conductors are separated from the conductors of the other circuits by a continuous and firmly fixed nonconductor, such as flexible tubing.

Exception No. 4: Underground conductors in a manhole where one of the following conditions is met.

(a) The electric light, power, Class 1, nonpower-limited fire alarm, and medium power network-powered broadband

communications circuit conductors are in a metal-enclosed cable or Type UF cable.

(b) The conductors are permanently and effectively separated from the conductors of the other circuits by a continuous and firmly fixed nonconductor, such as flexible tubing, in addition to the insulation or covering on the wire.

(c) The conductors are permanently and effectively separated from conductors of the other circuits and securely fastened to racks, insulators, or other approved supports.

Exception No. 5: As permitted by Section 780-6(a) and installed in accordance with Article 780.

Exception No. 6: In cable trays, where the conductors of the electric light, power, Class 1, and nonpower-limited fire alarm circuits are separated by a solid fixed barrier of a material compatible with the cable tray, or where the Class 2 and Class 3 circuits are installed in Type MC cable.

(2) In Hoistways. Class 2 or Class 3 circuit conductors shall be installed in rigid metal conduit, rigid nonmetallic conduit, intermediate metal conduit, or electrical metallic tubing in hoistways.

Exception: As provided for in Section 620-21 for elevators and similar equipment.

(3) Other Applications. Conductors of Class 2 and Class 3 circuits shall be separated by at least 2 in. (50.8 mm) from conductors of any electric light, power, Class 1, nonpower-limited fire alarm, or medium power network-powered broadband communications circuits.

Exception No. 1: Where either (1) all of the electric light, power, Class 1, nonpower-limited fire alarm, and medium power network-powered broadband communications circuit conductor or (2) all of the Class 2 and Class 3 circuit conductors are in raceway or in metal-sheathed, metal-clad, nonmetallic-sheathed, or Type UF cables.

Exception No. 2: Where all of the electric light, power, Class 1, nonpower-limited fire alarm, and medium power network-powered broadband communications circuit conductors are permanently separated from all of the Class 2 and Class 3 circuit conductors by a continuous and firmly fixed nonconductor, such as porcelain tubes or flexible tubing, in addition to the insulation on the conductors.

(b) Conductors of Different Circuits in Same Cable, Enclosure, or Raceway.

(1) Two or More Class 2 Circuits. Conductors of two or more Class 2 circuits shall be permitted within the same cable, enclosure, or raceway.

(2) Two or More Class 3 Circuits. Conductors of two or more Class 3 circuits shall be permitted within the same cable, enclosure, or raceway.

(3) Class 2 Circuits with Class 3 Circuits. Conductors of one or more Class 2 circuits shall be permitted within the same cable, enclosure, or raceway with conductors of Class 3 circuits, provided that the insulation of the Class 2 circuit conductors in the cable, enclosure, or raceway is at least that required for Class 3 circuits.

(4) Class 2 and Class 3 Circuits with Communications Circuits. Class 2 and Class 3 circuit conductors shall be permitted in the same cable with communications circuits, in which case the Class 2 and Class 3 circuits shall be classified as communications circuits and shall meet the requirements of Article 800. The cables shall be listed as communications cables or multipurpose cables.

Exception: Cables constructed of individually listed Class 2, Class 3, and communications cables under a common jacket shall not be required to be classified as communications cables. The fire-resistance rating of the composite cable shall be determined by the performance of the composite cable.

(5) Class 2 or Class 3 Cables with Other Circuit Cables. Jacketed cables of Class 2 or Class 3 circuits shall be permitted in the same enclosure or raceway with jacketed cables of any of the following:

(a) Power-limited fire alarm systems in compliance with Article 760
(b) Nonconductive and conductive optical fiber cables in compliance with Article 770
(c) Communications circuits in compliance with Article 800
(d) Community antenna television and radio distribution systems in compliance with Article 820
(e) Low power network-powered broadband communications in compliance with Article 830

(c) Circuit Conductors Extending Beyond One Building. Where Class 2 or Class 3 circuit conductors extend beyond one building and are run so as to be subject to accidental contact with electric light or power conductors operating at over 300 volts to ground, or are exposed to lightning on interbuilding circuits on the same premises, the requirements of the following shall also apply:

(1) Sections 800-10, 800-12, 800-13, 800-30, 800-31, 800-32, 800-33, and 800-40 for other than coaxial conductors
(2) Sections 820-10, 820-33, and 820-40 for coaxial conductors

(d) Support of Conductors. Class 2 or Class 3 circuit conductors shall not be strapped, taped, or attached by any means to the exterior of any conduit or other raceway as a means of support.

Exception: Except as permitted by Section 300-11(b)(2).

725-61. Applications of Listed Class 2, Class 3, and PLTC Cables. Class 2, Class 3, and PLTC cables shall comply with (a) through (g).

(a) Plenum. Cables installed in ducts, plenums, and other spaces used for environmental air shall be Type CL2P or CL3P.

Exception: Listed wires and cables installed in compliance with Section 300-22.

(b) Riser. Cables installed in vertical runs and penetrating more than one floor, or cables installed in vertical runs in a shaft, shall be Type CL2R or CL3R. Floor penetrations requiring Type CL2R or CL3R shall contain only cables suitable for riser or plenum use.

Exception No. 1: Other cables as covered in Table 725-61 and other listed wiring methods as covered in Chapter 3, where installed in metal raceways or located in a fireproof shaft having firestops at each floor.

Exception No. 2: Types CL2, CL3, CL2X, and CL3X cable in one- and two-family dwellings.

FPN: See Section 300-21 for firestop requirements for floor penetrations.

(c) Cable Trays. Cables installed in cable trays outdoors shall be Type PLTC. Cables installed in cable trays indoors shall be Types PLTC, CL3P, CL3R, CL3, CL2P, CL2R, and CL2.

FPN: See Section 800-52(d) for cables permitted in cable trays.

(d) Hazardous (Classified) Locations. Cables installed in hazardous (classified) locations shall be Type PLTC. Where the use of Type PLTC cable is permitted in Sections 501-4(b), 502-4(b), and 504-20, the cable shall be installed in cable trays; in raceways; supported by messenger wire, or otherwise adequately supported and mechanically protected by angles, struts, channels, other mechanical means; or directly buried where the cable is listed for this use.

Exception No. 1: For Class 2 circuits as permitted by Section 501-4(b), Exception.

Exception No. 2: Conductors in Type PLTC cables used for Class 2 thermocouple circuits shall be permitted to be any of the materials used for thermocouple extension wire.

Exception No. 3: In industrial establishments where the conditions of maintenance and supervision ensure that only qualified persons will service the installation, and where the cable is not subject to physical damage, Type PLTC cable that complies with the crush and impact requirements of Type MC cable and is identified for such use shall be permitted as open wiring between cable tray and utilization equipment in lengths not to exceed 50 ft (15.24 m), where the

cable is supported and protected against physical damage using mechanical protection such as dedicated struts, angles, or channels. The cable shall be supported and secured at intervals not exceeding 6 ft (1.83 m).

(e) Other Wiring Within Buildings. Cables installed in building locations other than the locations covered in (a) through (d) shall be Type CL2 or CL3.

Exception No. 1: Type CL2X or CL3X where installed in a raceway, or other wiring methods as covered in Chapter 3.

Exception No. 2: In nonconcealed spaces where the exposed length of cable does not exceed 10 ft (3.05 m).

Exception No. 3: Listed Type CL2X, Class 2 cables less than 0.25 in. (6.4 mm) in diameter and listed Type CL3X, Class 3 cables less than 0.25 in. (6.4 mm) in diameter installed in one- or two-family dwellings.

Exception No. 4: Listed Type CL2X, Class 2 cables less than 0.25 in. (6.4 mm) in diameter and listed Type CL3X, Class 3 cables less than 0.25 in. (6.4 mm) installed in nonconcealed spaces in multifamily dwellings.

Exception No. 5: Type CMUC undercarpet communications wires and cables installed under carpet.

(f) Cross-Connect Arrays. Type CL2 or CL3 conductors or cable shall be used.

(g) Class 2 and Class 3 Cable Uses and Permitted Substitutions. The uses and permitted substitutions for Class 2 and Class 3 cables listed in Table 725-61 shall be considered suitable for the purpose and shall be permitted.

FPN: For information on Types CMP, CMR, CMG, CM, and CMX cables, see Section 800-50.

Type CM—Communications wires and cables
Type CL2 and CL3—Class 2 and Class 3 remote-control, signaling, and power-limited cables
Type PLTC—Power-limited tray cable

A → B Cable A shall be permitted to be used in place of cable B.

Figure 725-61 Cable substitution hierarchy.

Table 725-61. Cable Uses and Permitted Substitutions

Cable Type	Use	References	Permitted Substitutions
CL3P	Class 3 plenum cable	725-61(a)	CMP
CL2P	Class 2 plenum cable	725-61(a)	CMP, CL3P
CL3R	Class 3 riser cable	725-61(b)	CMP, CL3P, CMR
CL2R	Class 2 riser cable	725-61(b)	CMP, CL3P, CL2P, CMR, CL3R
PLTC	Power-limited tray cable	725-61(c) and (d)	
CL3	Class 3 cable	725-61(b), (e), and (f)	CMP, CL3P, CMR, CL3R, CMG, CM, PLTC
CL2	Class 2 cable	725-61(b), (e), and (f)	CMP, CL3P, CL2P, CMR, CL3R, CL2R, CMG, CM, PLTC, CL3
CL3X	Class 3 cable, limited use	725-61(b) and (e)	CMP, CL3P, CMR, CL3R, CMG, CM, PLTC, CL3, CMX
CL2X	Class 2 cable, limited use	725-61(b) and (e)	CMP, CL3P, CL2P, CMR, CL3R, CL2R, CMG, CM, PLTC, CL3, CL2, CMX, CL3X

725-71. Listing and Marking of Class 2, Class 3, and Type PLTC Cables. Class 2, Class 3, and Type PLTC cables installed as wiring within buildings shall be listed as being resistant to the spread of fire and other criteria in accordance with (a) through (g) and shall be marked in accordance with (h).

(a) Types CL2P and CL3P. Types CL2P and CL3P plenum cables shall be listed as being suitable for use in ducts, plenums, and other space used for environmental air and shall also be listed as having adequate fire-resistant and low smoke-producing characteristics.

> FPN: One method of defining low smoke-producing cable is by establishing an acceptable value of the smoke produced when tested in accordance with *Standard Method of Test for Fire and Smoke Characteristics of Wires and Cables,* NFPA 262-1994, to a maximum peak optical density of 0.5 and a maximum average optical density of 0.15. Similarly, one method of defining fire-resistant cables is by establishing a maximum allowable flame travel distance of 5 ft (1.52 m) when tested in accordance with the same test.

(b) Types CL2R and CL3R. Types CL2R and CL3R riser cables shall be listed as being suitable for use in a vertical run in a shaft or from floor to floor and shall also be listed as having fire-resistant characteristics capable of preventing the carrying of fire from floor to floor.

> FPN: One method of defining fire-resistant characteristics capable of preventing the carrying of fire from floor to floor is that the cables pass the requirements of *Test for Flame Propagation Height of Electrical and Optical-Fiber Cable Installed Vertically in Shafts,* ANSI/UL 1666-1997.

(c) Types CL2 and CL3. Types CL2 and CL3 cables shall be listed as being suitable for general-purpose use, with the exception of risers, ducts, plenums, and other space used for environmental air, and shall also be listed as being resistant to the spread of fire.

> FPN: One method of defining resistant to the spread of fire is that the cables do not spread fire to the top of the tray in the vertical tray flame test in *Reference Standard for Electrical Wires, Cables and Flexible Cords,* ANSI/UL 1581-1991.
>
> Another method of defining resistant to the spread of fire is for the damage (char length) not to exceed 4 ft 11 in. (1.5 m) when performing the CSA vertical flame test for cables in cable trays, as described in *Test Methods for Electrical Wires and Cables,* CSA C22.2 No. 0.3-M-1985.

(d) Types CL2X and CL3X. Types CL2X and CL3X limited-use cables shall be listed as being suitable for use in dwellings and for use in raceway and shall also be listed as being flame retardant.

> FPN: One method of determining that cable is resistant to flame spread is by testing the cable to the VW-1 (vertical-wire) flame test in the *Reference Standard for Electrical Wires, Cables and Flexible Cords,* ANSI/UL 1581-1991.

(e) Type PLTC. Type PLTC nonmetallic-sheathed, power-limited tray cable shall be listed as being suitable for cable trays and shall consist of a factory assembly of two or more insulated conductors under a nonmetallic jacket, marked in accordance with Section 310-11 and Table 725-71. The insulated conductors shall be Nos. 22 through 12. The conductor material shall be copper (solid or stranded). Insulation on conductors shall be suitable for 300 volts. The cable core shall be either (1) two or more parallel conductors, (2) one or more group assemblies of twisted or parallel conductors, or (3) a combination thereof. A metallic shield or a metallized foil shield with drain wire(s) shall be permitted to be applied either over the cable core, over groups of conductors, or both. The cable shall be listed as being resistant to the spread of fire. The outer jacket shall be a sunlight- and moisture-resistant nonmetallic material.

Exception No. 1: Where a smooth metallic sheath, welded and corrugated metallic sheath, or interlocking tape armor is applied over the nonmetallic jacket, an overall nonmetallic jacket shall not be required. On metallic-sheathed cable without an overall nonmetallic jacket, the information required in Section 310-11 shall be located on the nonmetallic jacket under the sheath.

Exception No. 2: Conductors in PLTC cables used for Class 2 thermocouple circuits shall be permitted to be any of the materials used for thermocouple extension wire.

> FPN: One method of defining resistant to the spread of fire is that the cables do not spread fire to the top of the tray in the vertical tray flame test in the *Reference Standard for Electrical Wires, Cables and Flexible Cords,* ANSI/UL 1581-1991.
>
> Another method of defining resistant to the spread of fire is for the damage (char length) not to exceed 4 ft 11 in. (1.5 m) when performing the CSA vertical flame test for cables in cable trays, as described in T*est Methods for Electrical Wires and Cables,* CSA C22.2 No. 0.3-M-1985.

(f) Class 3 Cable Voltage Rating. Class 3 cables shall have a voltage rating of not less than 300 volts.

(g) Class 3 Single Conductors. Class 3 single conductors used as other wiring within buildings shall not be smaller than No. 18 and shall be Type CL3. Conductor types described in Section 725-27(b) that are also listed as Type CL3 shall be permitted.

> FPN: One method of defining resistant to the spread of fire is that the cables do not spread fire to the top of the tray in the "Vertical Tray Flame Test" in *Reference Standard for Electrical Wires, Cables and Flexible Cords,* ANSI/UL 1581-1991.
>
> Another method of defining resistant to the spread of fire is for the damage (char length) not to exceed 4 ft 11 in. (1.5 m) when performing the CSA "Vertical Flame Test — Cables in Cable Trays" as described in *Test Methods for Electrical Wires and Cables,* CSA C22.2 No. 0.3-M-1985.

(h) Marking. Cables shall be marked in accordance with Table 725-71. Voltage ratings shall not be marked on the cables.

> FPN: Voltage markings on cables may be misinterpreted to suggest that the cables may be suitable for Class 1 electric light and power applications.

Exception: Voltage markings shall be permitted where the cable has multiple listings and a voltage marking is required for one or more of the listings.

Table 725-71. Cable Markings

Cable Marking	Type	Listing References
CL3P	Class 3 plenum cable	725-71(a), (f), and (h)
CL2P	Class 2 plenum cable	725-71(a) and (h)
CL3R	Class 3 riser cable	725-71(b), (f), and (h)
CL2R	Class 2 riser cable	725-71(b) and (h)
PLTC	Power-limited tray cable	725-71(e) and (h)
CL3	Class 3 cable	725-71(c), (f), and (h)
CL2	Class 2 cable	725-71(c), (f), and (h)
CL3X	Class 3 cable, limited use	725-71(d), (f), and (h)
CL2X	Class 2 cable, limited use	725-71(d), (f), and (h)

> FPN: Class 2 and Class 3 cable types are listed in descending order of fire resistance rating, and Class 3 cables are listed above Class 2 cables, because Class 3 cables can substitute for Class 2 cables.

Article 727 — Instrumentation Tray Cable: Type ITC

727-1. Scope. This article covers the use, installation, and construction specifications of instrumentation tray cable for application to instrumentation and control circuits operating at 150 volts or less and 5 amperes or less.

727-2. Definition. *Type ITC instrumentation tray cable* is a factory assembly of two or more insulated conductors, with or without a grounding conductor(s), enclosed in a nonmetallic sheath.

727-3. Other Articles. In addition to the provisions of this article, installation of Type ITC cable shall comply with other applicable articles of this *Code,* such as Articles 240, 250, 300, and 318.

727-4. Uses Permitted. Type ITC cable shall be permitted to be used in industrial establishments where the conditions of maintenance and supervision ensure that only qualified persons will service the installation

(1) In cable trays
(2) In raceways
(3) In hazardous locations as permitted in Articles 501, 502, 503, 504, and 505
(4) As open wiring where equipped with a smooth metallic sheath, continuous corrugated metallic sheath, or interlocking tape armor applied over the nonmetallic sheath in accordance with Section 727-6

The cable shall be supported and secured at intervals not exceeding 6 ft (1.83 m).

of Article 300 referenced in this article shall apply to fire alarm systems.

(a) Spread of Fire or Products of Combustion. See Section 300-21.

(b) Ducts, Plenums, and Other Air-Handling Spaces. Section 300-22, where installed in ducts or plenums or other spaces used for environmental air.

Exception to (b): As permitted in Sections 760-30(b)(1) and (2) and Section 760-61(a).

(c) Hazardous (Classified) Locations. Articles 500 through 516 and Article 517, Part D, where installed in hazardous (classified) locations.

(d) Corrosive, Damp, or Wet Locations. Sections 110-11, 300-6, and 310-9 where installed in corrosive, damp, or wet locations.

(e) Building Control Circuits. Article 725 where building control circuits (e.g., elevator capture, fan shutdown) are associated with the fire alarm system.

(f) Optical Fiber Cables. Where optical fiber cables are utilized for fire alarm circuits, the cables shall be installed in accordance with Article 770.

760-5. Access to Electrical Equipment Behind Panels Designed to Allow Access. Access to equipment shall not be denied by an accumulation of conductors and cables that prevents removal of panels, including suspended ceiling panels.

760-6. Fire Alarm Circuit and Equipment Grounding. Fire alarm circuits and equipment shall be grounded in accordance with Article 250.

760-7. Fire Alarm Circuits Extending Beyond One Building. Power-limited fire alarm circuits that extend beyond one building and run outdoors shall either meet the installation requirements of Parts B, C, and D of Article 800, or shall meet the installation requirements of Article 225. Nonpower-limited fire alarm circuits that extend beyond one building and run outdoors shall meet the requirements of Article 225.

760-8. Mechanical Execution of Work. Fire alarm circuits shall be installed in a neat and workmanlike manner. Cables shall be supported by the building structure in such a manner that the cable will not be damaged by normal building use.

> FPN: One way to determine accepted industry practice is to refer to nationally recognized standards such as *Commercial Building Telecommunications Wiring Standard*, ANSI/EIA/TIA 568-A-1995; *Commercial Building Standard for Telecommunications Pathways and Spaces*, ANSI/EIA/TIA 569-

1990; and *Residential and Light Commercial Telecommunications Wiring Standard*, ANSI/EIA/TIA 570-1991.

760-10. Fire Alarm Circuit Identification. Fire alarm circuits shall be identified at terminal and junction locations, in a manner that will prevent unintentional interference with the signaling circuit during testing and servicing.

760-15. Fire Alarm Circuit Requirements. Fire alarm circuits shall comply with the following parts of this article.

(a) Nonpower-Limited Fire Alarm (NPLFA) Circuits. See Parts A and B.

(b) Power-Limited Fire Alarm (PLFA) Circuits. See Parts A and C.

B. Nonpower-Limited Fire Alarm (NPLFA) Circuits

760-21. NPLFA Circuit Power Source Requirements. The power source of nonpower-limited fire alarm circuits shall comply with Chapters 1 through 4, and the output voltage shall not be more than 600 volts, nominal.

760-23. NPLFA Circuit Overcurrent Protection. Overcurrent protection for conductors No. 14 and larger shall be provided in accordance with the conductor ampacity without applying the derating factors of Section 310-15 to the ampacity calculation. Overcurrent protection shall not exceed 7 amperes for No. 18 conductors and 10 amperes for No. 16 conductors.

Exception: Where other articles of this Code permit or require other overcurrent protection.

760-24. NPLFA Circuit Overcurrent Device Location. Overcurrent devices shall be located at the point where the conductor to be protected receives its supply.

Exception No. 1: Where the overcurrent device protecting the larger conductor also protects the smaller conductor.

Exception No. 2: Transformer secondary conductors. Nonpower-limited fire alarm circuit conductors supplied by the secondary of a single-phase transformer that has only a 2-wire (single-voltage) secondary shall be permitted to be protected by overcurrent protection provided by the primary (supply) side of the transformer, provided the protection is in accordance with Section 450-3 and does not exceed the value determined by multiplying the secondary conductor ampacity by the secondary-to-primary transformer voltage ratio. Transformer secondary conductors other than 2-wire shall not be considered to be protected by the primary overcurrent protection.

Exception No. 3: Electronic power source output conductors. Nonpower-limited circuit conductors supplied by the output of a single-phase, listed electronic power source, other than a transformer, having only a 2 wire (single-voltage) output for connection to nonpower-limited circuits shall be permitted to be protected by overcurrent protection provided on the input side of the electronic power source, provided this protection does not exceed the value determined by multiplying the nonpower-limited circuit conductor ampacity by the output-to-input voltage ratio. Electronic power source outputs, other than 2 wire (single voltage), connected to nonpower-limited circuits shall not be considered to be protected by overcurrent protection on the input of the electronic power source.

FPN: A single-phase, listed electronic power supply whose output supplies a 2-wire (single-voltage) circuit is an example of a nonpower-limited power source that meets the requirements of Section 760-21.

760-25. NPLFA Circuit Wiring Methods. Installation of nonpower-limited fire alarm circuits shall be in accordance with Sections 110-3(b), 300-11(a), 300-15, 300-17, and other appropriate articles of Chapter 3.

Exception No. 1: As provided in Sections 760-26 through 760-30.

Exception No. 2: Where other articles of this Code require other methods.

760-26. Conductors of Different Circuits in Same Cable, Enclosure, or Raceway.

(a) Class 1 with NPLFA Circuits. Class 1 and nonpower-limited fire alarm circuits shall be permitted to occupy the same cable, enclosure, or raceway without regard to whether the individual circuits are alternating current or direct current, provided all conductors are insulated for the maximum voltage of any conductor in the enclosure or raceway.

(b) Fire Alarm with Power-Supply Circuits. Power-supply and fire alarm circuit conductors shall be permitted in the same cable, enclosure, or raceway only where connected to the same equipment.

760-27. NPLFA Circuit Conductors.

(a) Sizes and Use. Only copper conductors shall be permitted to be used for fire alarm systems. No. 18 conductors and No. 16 conductors shall be permitted to be used, provided they supply loads that do not exceed the ampacities given in Table 402-5 and are installed in a raceway, an approved enclosure, or a listed cable. Conductors larger than No. 16 shall not supply loads greater than the ampacities given in Section 310-15, as applicable.

(b) Insulation. Insulation on conductors shall be suitable for 600 volts. Conductors larger than No. 16 shall comply with Article 310. Conductors in sizes No. 18 and 16 shall

be Type KF-2, KFF-2, PAFF, PTFF, PF, PFF, PGF, PGFF, RFH-2, RFHH-2, RFHH-3, SF-2, SFF-2, TF, TFF, TFN, TFFN, ZF, or ZFF. Conductors with other types and thickness of insulation shall be permitted if listed for nonpower-limited fire alarm circuit use.

FPN: For application provisions, see Table 402-3.

(c) Conductor Materials. Conductors shall be solid or stranded copper.

Exception to (b) and (c): Wire Types PAF and PTF shall be permitted only for high-temperature applications between 90°C (194°F) and 250°C (482°F).

760-28. Number of Conductors in Cable Trays and Raceways, and Derating.

(a) NPLFA Circuits and Class 1 Circuits. Where only nonpower-limited fire alarm circuit and Class 1 circuit conductors are in a raceway, the number of conductors shall be determined in accordance with Section 300-17. The derating factors given in Section 310-15(b)(2)(a) shall apply if such conductors carry continuous load in excess of 10 percent of the ampacity of each conductor.

(b) Power-Supply Conductors and Fire Alarm Circuit Conductors. Where power-supply conductors and fire alarm circuit conductors are permitted in a raceway in accordance with Section 760-26, the number of conductors shall be determined in accordance with Section 300-17. The derating factors given in Section 310-15(b)(2)(a), shall apply as follows:

(1) To all conductors where the fire alarm circuit conductors carry continuous loads in excess of 10 percent of the ampacity of each conductor and where the total number of conductors is more than three

(2) To the power-supply conductors only, where the fire alarm circuit conductors do not carry continuous loads in excess of 10 percent of the ampacity of each conductor and where the number of power-supply conductors is more than three

(c) Cable Trays. Where fire alarm circuit conductors are installed in cable trays, they shall comply with Sections 318-9 through 318-11.

760-30. Multiconductor NPLFA Cables. Multiconductor nonpower-limited fire alarm cables that meet the requirements of Section 760-31 shall be permitted to be used on fire alarm circuits operating at 150 volts or less and shall be installed in accordance with (a) and (b).

(a) NPLFA Wiring Method. Multiconductor nonpower-limited fire alarm circuit cables shall be installed as follows.

(1) In raceway or exposed on surface of ceiling and sidewalls or fished in concealed spaces. Cable splices or

terminations shall be made in listed fittings, boxes, enclosures, fire alarm devices, or utilization equipment. Where installed exposed, cables shall be adequately supported and installed in such a way that maximum protection against physical damage is afforded by building construction such as baseboards, door frames, ledges, etc. Where located within 7 ft (2.13 m) of the floor, cables shall be securely fastened in an approved manner at intervals of not more than 18 in. (457 mm).

(2) In metal raceway or rigid nonmetallic conduit where passing through a floor or wall to a height of 7 ft (2.13 m) above the floor unless adequate protection can be afforded by building construction such as detailed in (1), or unless an equivalent solid guard is provided.

(3) In rigid metal conduit, rigid nonmetallic conduit, intermediate metal conduit, or electrical metallic tubing where installed in hoistways.

Exception: As provided for in Section 620-21 for elevators and similar equipment.

(b) Applications of Listed NPLFA Cables. The use of nonpower-limited fire alarm circuit cables shall comply with (1) through (4).

(1) Ducts and Plenums. Multiconductor nonpower-limited fire alarm circuit cables, Types NPLFP, NPLFR, and NPLF, shall not be installed exposed in ducts or plenums. See Section 300-22(b).

(2) Other Spaces Used for Environmental Air. Cables installed in other spaces used for environmental air shall be Type NPLFP.

Exception No. 1: Types NPLFR and NPLF cables installed in compliance with Section 300-22(c).

Exception No. 2: Other wiring methods in accordance with Section 300-22(c) and conductors in compliance with Section 760-27(c).

(3) Riser. Cables installed in vertical runs and penetrating more than one floor or cables installed in vertical runs in a shaft shall be Type NPLFR. Floor penetrations requiring Type NPLFR shall contain only cables suitable for riser or plenum use.

Exception No. 1: Type NPLF or other cables specified in Chapter 3 that are in compliance with Section 760-27(c) and encased in metal raceway.

Exception No. 2: Type NPLF cables located in a fireproof shaft having firestops at each floor.

FPN: See Section 300-21 for firestop requirements for floor penetrations.

(4) Other Wiring Within Buildings. Cables installed in building locations other than the locations covered in Section 760-30(b)(1), (2), and (3) shall be Type NPLF.

Exception No. 1: Chapter 3 wiring methods with conductors in compliance with Section 760-27(c).

Exception No. 2: Type NPLFP or Type NPLFR cables shall be permitted.

760-31. Listing and Marking of NPLFA Cables. Nonpower-limited fire alarm cables installed as wiring within buildings shall be listed in accordance with (a) and (b) and as being resistant to the spread of fire in accordance with (c) through (f), and shall be marked in accordance with (g).

(a) NPLFA Conductor Materials. Conductors shall be No. 18 or larger solid or stranded copper.

(b) Insulated Conductors. Insulated conductors shall be suitable for 600 volts. Insulated conductors No. 14 and larger shall be one of the types listed in Table 310-13 or one that is identified for this use. Insulated conductors No. 18 and No. 16 shall be in accordance with Section 760-27.

(c) Type NPLFP. Type NPLFP nonpower-limited fire alarm cable for use in other space used for environmental air shall be listed as being suitable for use in other space used for environmental air as described in Section 300-22(c) and shall also be listed as having adequate fire-resistant and low smoke-producing characteristics.

FPN: One method of defining low smoke-producing cable is by establishing an acceptable value of the smoke produced when tested in accordance with the *Standard Method of Test for Fire and Smoke Characteristics of Wires and Cables*, NFPA 262-1994, to a maximum peak optical density of 0.5 and a maximum average optical density of 0.15. Similarly, one method of defining fire-resistant cables is by establishing a maximum allowable flame travel distance of 5 ft (1.52 m) when tested in accordance with the same test.

(d) Type NPLFR. Type NPLFR nonpower-limited fire alarm riser cable shall be listed as being suitable for use in a vertical run in a shaft or from floor to floor and shall also be listed as having fire-resistant characteristics capable of preventing the carrying of fire from floor to floor.

FPN: One method of defining fire-resistant characteristics capable of preventing the carrying of fire from floor to floor is that the cables pass the *Test for Flame Propagation Height of Electrical and Optical-Fiber Cables Installed Vertically in Shafts*, ANSI/UL 1666-1997.

(e) Type NPLF. Type NPLF nonpower-limited fire alarm cable shall be listed as being suitable for general-purpose fire alarm use, with the exception of risers, ducts, plenums, and other space used for environmental air, and shall also be listed as being resistant to the spread of fire.

FPN: One method of defining resistant to the spread of fire is that the cables do not spread fire to the top of the tray in the vertical-tray flame test in the *Reference Standard for Electrical Wires, Cables and Flexible Cords,* ANSI/UL 1581-1991.

Another method of defining resistant to the spread of fire is for the damage (char length) not to exceed 4 ft 11 in. (1.5 m) when performing the CSA vertical flame test — cables in cable trays, as described in *Test Methods for Electrical Wires and Cables,* CSA C22.2 No. 0.3-M-1985.

(f) Fire Alarm Circuit Integrity (CI) Cable. Cables suitable for use in fire alarm systems to ensure survivability of critical circuits during a specified time under fire conditions shall be listed as circuit integrity (CI) cable. Cables identified in Sections 760-31(c), (d), and (e) meeting the requirements for circuit integrity shall have the additional classification using the suffix "CI" (for example, NPLFP-CI, NPLFR-CI and NPLF-CI).

FPN No. 1: This cable may be used for fire alarm circuits to comply with the survivability requirements of *National Fire Alarm Code,*® NFPA 72-1996, subsections 3-2.4, 3-4.4, 3-12.4, and 3-12.4.3, that the cable maintain its electrical function during fire conditions for a defined period of time.

FPN No. 2: One method of defining circuit integrity (CI) cable is by establishing a minimum 2-hour fire resistance rating for the cable when tested in accordance with the *Standard for Tests of Fire Resistive Cables,* UL 2196-1995.

(g) NPLFA Cable Markings. Multiconductor nonpower-limited fire alarm cables shall be marked in accordance with Table 760-31(g). Nonpower-limited fire alarm circuit cables shall be permitted to be marked with a maximum usage voltage rating of 150 volts. Cables that are listed for circuit integrity shall be identified with the suffix "CI" as defined in (f).

Table 760-31(g). NPLFA Cable Markings

Cable Marking	Type	Reference
NPLFP	Nonpower-limited fire alarm circuit cable for use in other space used for environmental air	760-31(c) and (g)
NPLFR	Nonpower-limited fire alarm circuit riser cable	760-31(d) and (g)
NPLF	Nonpower-limited fire alarm circuit cable	760-31(e) and (g)

Note: Cables identified in (c), (d), and (e) meeting the requirements for circuit integrity shall have the additional classification using the suffix "CI" (for example, NPLFP-CI, NPLFR-CI, and NPLF-CI).

FPN: Cable types are listed in descending order of fire-resistance rating.

C. Power-Limited Fire Alarm (PLFA) Circuits

760-41. Power Sources for PLFA Circuits. The power source for a power-limited fire alarm circuit shall be as specified in (a), (b), or (c).

FPN: Tables 12(a) and 12(b) in Chapter 9 provide the listing requirements for power-limited fire alarm circuit sources.

(a) Transformers. A listed PLFA or Class 3 transformer.

(b) Power Supplies. A listed PLFA or Class 3 power supply.

(c) Listed Equipment. Listed equipment marked to identify the PLFA power source.

FPN: Examples of listed equipment are a fire alarm control panel with integral power source; a circuit card listed for use as a PLFA source, where used as part of a listed assembly; a current-limiting impedance, listed for the purpose or part of a listed product, used in conjunction with a nonpower-limited transformer or a stored energy source, e.g., storage battery, to limit the output current.

760-42. Circuit Marking. The equipment shall be durably marked where plainly visible to indicate each circuit that is a power-limited fire alarm circuit.

FPN: See Section 760-52(a), Exception No. 3, where a power-limited circuit is to be reclassified as a nonpower-limited circuit.

760-51. Wiring Methods on Supply Side of the PLFA Power Source. Conductors and equipment on the supply side of the power source shall be installed in accordance with the appropriate requirements of Part B and Chapters 1 through 4. Transformers or other devices supplied from power-supply conductors shall be protected by an overcurrent device rated not over 20 amperes.

Exception: The input leads of a transformer or other power source supplying power-limited fire alarm circuits shall be permitted to be smaller than No. 14, but not smaller than No. 18, if they are not over 12 in. (305 mm) long and if they have insulation that complies with Section 760-27(b).

760-52. Wiring Methods and Materials on Load Side of the PLFA Power Source. Fire alarm circuits on the load side of the power source shall be permitted to be installed using wiring methods and materials in accordance with either (a) or (b).

(a) NPLFA Wiring Methods and Materials. Installation shall be in accordance with Section 760-25, and conductors shall be solid or stranded copper.

Exception No. 1: The derating factors given in Section 310-15(b)(2)(a) shall not apply.

Exception No. 2: Conductors and multiconductor cables described in and installed in accordance with Sections 760-27 and 760-30 shall be permitted.

Exception No. 3: Power-limited circuits shall be permitted to be reclassified and installed as nonpower-limited circuits if the power-limited fire alarm circuit markings required by Section 760-42 are eliminated and the entire circuit is installed using the wiring methods and materials in accordance with Part B, Nonpower-Limited Fire Alarm Circuits.

FPN: Power-limited circuits reclassified and installed as nonpower-limited circuits are no longer power-limited circuits, regardless of the continued connection to a power-limited source.

(b) PLFA Wiring Methods and Materials. Power-limited fire alarm conductors and cables described in Section 760-71 shall be installed as detailed in (1), (2), or (3) of this section. Devices shall be installed in accordance with Sections 110-3(b), 300-11(a), and 300-15.

(1) In raceway or exposed on the surface of ceiling and sidewalls or fished in concealed spaces. Cable splices or terminations shall be made in listed fittings, boxes, enclosures, fire alarm devices, or utilization equipment. Where installed exposed, cables shall be adequately supported and installed in such a way that maximum protection against physical damage is afforded by building construction such as baseboards, door frames, ledges, etc. Where located within 7 ft (2.13 m) of the floor, cables shall be securely fastened in an approved manner at intervals of not more than 18 in. (457 mm).

(2) In metal raceways or rigid nonmetallic conduit where passing through a floor or wall to a height of 7 ft (2.13 m) above the floor, unless adequate protection can be afforded by building construction such as detailed in (1), or unless an equivalent solid guard is provided.

(3) In rigid metal conduit, rigid nonmetallic conduit, intermediate metal conduit, or electrical metallic tubing where installed in hoistways.

Exception No. 1: As provided for in Section 620-21 for elevators and similar equipment.

Exception No. 2: Other wiring methods and materials installed in accordance with the requirements of Section 760-3 shall be permitted to extend or replace the conductors and cables described in Section 760-71 and permitted by Section 760-52(b).

760-54. Installation of Conductors and Equipment.

(a) Separation from Electric Light, Power, Class 1, NPLFA, and Medium Power Network-Powered Broadband Communications Circuit Conductors.

(1) In Cables, Compartments, Enclosures, Outlet Boxes, or Raceways. Power-limited circuit cables and conductors shall not be placed in any cable, cable tray, compartment, enclosure, outlet box, raceway, or similar fitting with conductors of electric light, power, Class 1, nonpower-limited fire alarm circuit conductors, or medium power network-powered broadband communications circuits.

Exception No. 1: Where the conductors of the electric light, power, Class 1, nonpower-limited fire alarm, and medium power network-powered broadband communications circuit conductors are separated by a barrier from the power-limited fire alarm circuits. In enclosures, power-limited fire alarm circuits shall be permitted to be installed in a raceway within the enclosure to separate them from Class 1, electric light, power, nonpower-limited fire alarm, and medium power network-powered broadband communications circuits.

Exception No. 2: Conductors in compartments, enclosures, device boxes, outlet boxes, or similar fittings, where electric light, power, Class 1, nonpower-limited fire alarm, or medium power network-powered broadband communications circuit conductors are introduced solely to connect to the equipment connected to power-limited circuits to which the other conductors are connected, and

(a) The electric light, power, Class 1, nonpower-limited fire alarm, and medium power network-powered broadband communications circuit conductors are routed to maintain a minimum of 0.25 in. (6.35 mm) separation from the conductors and cables of power-limited fire alarm circuits, or
(b) The circuit conductors operate at 150 volts or less to ground and also comply with one of the following.

(1) The fire alarm power-limited circuits are installed using Types FPL, FPLR, FPLP or permitted substitute cables, provided these power-limited cable conductors extending beyond the jacket are separated by a minimum of 0.25 in. (6.35 mm) or by a nonconductive sleeve or nonconductive barrier from all other conductors, or
(2) The fire alarm power-limited circuit conductors are installed as nonpower-limited fire alarm circuits in accordance with Section 760-25.

Exception No. 3: Conductors entering compartments, enclosures, device boxes, outlet boxes, or similar fittings, where electric light, power, Class 1, nonpower-limited fire alarm, or medium power network-powered broadband communications circuit conductors are introduced solely to connect to the equipment connected to power-limited fire alarm circuits or to other circuits controlled by the fire alarm system to which the other conductors in the enclosure are connected. If the conductors must enter an enclosure that is provided with a single opening, they shall be permitted to enter through a single fitting (such as a tee) provided the conductors are separated from the conductors of the other circuits by a continuous and firmly fixed nonconductor, such as flexible tubing.

(2) In Hoistways. Power-limited fire alarm circuit conductors shall be installed in rigid metal conduit, intermediate metal conduit, or electrical metallic tubing in hoistways.

Exception: As provided for in Section 620-21 for elevators and similar equipment.

(3) Other Applications. Power-limited fire alarm circuit conductors shall be separated at least 2 in. (50.8 mm) from conductors of any electric light, power, Class 1, nonpower limited fire alarm, or medium power network-powered broadband communications circuits.

Exception No. 1: Where either (1) all of the electric light, power, Class 1, nonpower-limited fire alarm, and medium power network-powered broadband communications circuit conductors or (2) all of the power-limited fire alarm circuit conductors are in raceway or in metal-sheathed, metal-clad, nonmetallic-sheathed, or Type UF cables.

Exception No. 2: Where all of the electric light, power, Class 1, nonpower-limited fire alarm, and medium power network-powered broadband communications circuit conductors are permanently separated from all of the power-limited fire alarm circuit conductors by a continuous and firmly fixed nonconductor, such as porcelain tubes or flexible tubing in addition to the insulation on the wire.

(b) Conductors of Different PLFA Circuits, Class 2, Class 3, and Communications Circuits in Same Cable, Enclosure, or Raceway.

(1) Two or More PLFA Circuits. Cable and conductors of two or more power-limited fire alarm circuits, communications circuits, or Class 3 circuits shall be permitted in the same cable, enclosure, or raceway.

(2) Class 2 Circuits with PLFA Circuits. Conductors of one or more Class 2 circuits shall be permitted within the same cable, enclosure, or raceway with conductors of power-limited fire alarm circuits, provided that the insulation of the Class 2 circuit conductors in the cable, enclosure, or raceway is at least that required by the power-limited fire alarm circuits.

(3) Low Power Network-Powered Broadband Communications Cables and PLFA Cables. Low power network-powered broadband communications circuits shall be permitted in the same enclosure or raceway with PLFA cables.

(c) Support of Conductors. Power-limited fire alarm circuit conductors shall not be strapped, taped, or attached by any means to the exterior of any conduit or other raceway as a means of support.

(d) Conductor Size. Conductors of No. 26 shall only be permitted where spliced with a connector listed as suitable for No. 26 to No. 24 or larger conductors that are terminated on equipment or where the No. 26 conductors are terminated on equipment listed as suitable for No. 26 conductors. Single conductors shall not be smaller than No. 18.

760-55. Current-Carrying Continuous Line-Type Fire Detectors.

(a) Application. Listed continuous line-type fire detectors, including insulated copper tubing of pneumatically operated detectors, employed for both detection and carrying signaling currents shall be permitted to be used in power-limited circuits.

(b) Installation. Continuous line-type fire detectors shall be installed in accordance with Sections 760-42 through 760-52 and Section 760-54.

760-61. Applications of Listed PLFA Cables. Power-limited fire alarm cables shall comply with (a) through (c) or, where cable substitution is being made, with (d).

(a) Plenum. Cables installed in ducts, plenums, and other spaces used for environmental air shall be Type FPLP.

Exception: Types FPLP, FPLR, and FPL cable installed in compliance with Section 300-22.

(b) Riser. Cables installed in vertical runs and penetrating more than one floor or cables installed in vertical runs in a shaft shall be Type FPLR. Floor penetrations requiring Type FPLR shall contain only cables suitable for riser or plenum use.

Exception No. 1: Where the cables are installed in metal raceway or are located in a fireproof shaft having firestops at each floor.

Exception No. 2: Type FPL in one- and two-family dwellings.

FPN: See Section 300-21 for firestop requirements for floor penetrations.

(c) Other Wiring Within Buildings. Cables installed in building locations other than the locations covered in (a) and (b) shall be Type FPL.

Exception No. 1: Where the cables are enclosed in raceway.

Exception No. 2: Cables specified in Chapter 3 that meet the requirements of Sections 760-71(a) and (b) and are installed in nonconcealed spaces where the exposed length of cable does not exceed 10 ft (3.05 m).

Exception No. 3: A portable fire alarm system provided to protect a stage or set when not in use shall be permitted to use wiring methods in accordance with Section 530-12.

(d) Cable Uses and Permitted Substitutions. The uses and permitted substitutions for power-limited fire alarm circuit cables listed in Table 760-61 shall be considered suitable for the purpose and shall be permitted.

FPN: For information on multipurpose cables (Types MPP, MPR, MPG, MP) and communications cables (Types CMP, CMR, CMG, CM), see Section 800-50.

Type CM—Communications wires and cables
Type FPL—Power-limited fire alarm cables
Type MP—Multipurpose cables (coaxial cables only)

[A]→[B] Cable A shall be permitted to be used in place of cable B.
No. 26 minimum

Figure 760-61 Cable substitution hierarchy.

760-71. Listing and Marking of PLFA Cables and Insulated Continuous Line-Type Fire Detectors. Type FPL cables installed as wiring within buildings shall be listed as being resistant to the spread of fire and other criteria in accordance with (a) through (h) and shall be marked in accordance with (i). Insulated continuous line-type fire detectors shall be listed in accordance with (j).

(a) Conductor Materials. Conductors shall be solid or stranded copper.

(b) Conductor Size. The size of conductors in a multiconductor cable shall not be smaller than No. 26. Single conductors shall not be smaller than No. 18.

(c) Ratings. The cable shall have a voltage rating of not less than 300 volts.

(d) Type FPLP. Type FPLP power-limited fire alarm plenum cable shall be listed as being suitable for use in ducts, plenums, and other space used for environmental air and shall also be listed as having adequate fire-resistant and low smoke-producing characteristics.

FPN: One method of defining low smoke-producing cable is by establishing an acceptable value of the smoke produced when tested in accordance with the *Standard Method of Test for Fire and Smoke Characteristics of Wires and Cables,* NFPA 262-1994, to a maximum peak optical density of 0.5 and a maximum average optical density of 0.15. Similarly, one method of defining fire-resistant cables is by establishing maximum allowable flame travel distance of 5 ft (1.52 m) when tested in accordance with the same test.

(e) Type FPLR. Type FPLR power-limited fire alarm riser cable shall be listed as being suitable for use in a vertical run in a shaft or from floor to floor and shall also be listed as having fire-resistant characteristics capable of preventing the carrying of fire from floor to floor.

FPN: One method of defining fire-resistant characteristics capable of preventing the carrying of fire from floor to floor is that the cables pass the requirements of the *Standard Test for Flame Propagation Height of Electrical and Optical-Fiber Cable Installed Vertically in Shafts,* ANSI/UL 1666-1997.

(f) Type FPL. Type FPL power-limited fire alarm cable shall be listed as being suitable for general-purpose fire alarm use, with the exception of risers, ducts, plenums, and other spaces used for environmental air and shall also be listed as being resistant to the spread of fire.

FPN: One method of defining resistant to the spread of fire is that the cables do not spread fire to the top of the tray in the vertical-tray flame test in the *Reference Standard for Electrical Wires, Cables and Flexible Cords,* ANSI/UL 1581-1991. Another method of defining resistant to the

Table 760-61. Cable Uses and Permitted Substitutions

Cable Type	Use	References	Permitted Substitutions	
			Multiconductor	Coaxial
FPLP	Power-limited fire alarm plenum cable	760-61(a)	CMP	MPP
FPLR	Power-limited fire alarm riser cable	760-61(b)	CMP, FPLP, CMR	MPP, MPR
FPL	Power-limited fire alarm cable	760-61(c)	CMP, FPLP, CMR, FPLR, CMG, CM	MPP, MPR, MPG, MP

spread of fire is for the damage (char length) not to exceed 4 ft 11 in. (1.5 m) when performing the CSA vertical flame test — cables in cable trays, as described in *Test Methods for Electrical Wires and Cables*, CSA C22.2 No. 0.3-M-1985.

(g) Fire Alarm Circuit Integrity (CI) Cable. Cables suitable for use in fire alarm systems to ensure survivability of critical circuits during a specified time under fire conditions shall be listed as circuit integrity (CI) cable. Cables identified in Sections 760-71(d), (e), and (f) meeting the requirements for circuit integrity shall have the additional classification using the suffix "CI" (for example, FPLP-CI, FPLR-CI and FPL-CI).

> FPN No. 1: This cable is used for fire alarm circuits as one method of complying with the survivability requirements of the *National Fire Alarm Code,* NFPA 72-1996, 3-2.4, 3-4.4, 3-12.4, and 3-12.4.3, that the cable maintain its electrical function during fire conditions for a defined period of time.

> FPN No. 2: One method of defining circuit integrity (CI) cable is by establishing a minimum 2-hour fire resistance rating for the cable when tested in accordance with the *Standard for Tests of Fire Resistive Cables*, UL 2196-1995.

(h) Coaxial Cables. Coaxial cables shall be permitted to use 30 percent conductivity copper-covered steel center conductor wire and shall be listed as Type FPLP, FPLR, or FPL cable.

(i) Cable Marking. The cable shall be marked in accordance with Table 760-71(i). The voltage rating shall not be marked on the cable. Cables that are listed for circuit integrity shall be identified with the suffix CI as defined in Section 760-71(g).

> FPN: Voltage ratings on cables may be misinterpreted to suggest that the cables may be suitable for Class 1, electric light, and power applications.

Exception: Voltage markings shall be permitted where the cable has multiple listings and voltage marking is required for one or more of the listings.

Table 760-71(i). Cable Markings

Cable Marking	Type	Listing References
FPLP	Power-limited fire alarm plenum cable	760-71(d) and (i)
FPLR	Power-limited fire alarm riser cable	760-71(e) and (i)
FPL	Power-limited fire alarm cable	760-71(f) and (i)

Note: Cables identified in (d), (e), and (f) meeting the requirements for circuit integrity shall have the additional classification using the suffix "CI" (for example, FPLP-CI, FPLR-CI, and FPL-CI).

> FPN: Cable types are listed in descending order of fire-resistance rating.

(j) Insulated Continuous Line-Type Fire Detectors. Insulated continuous line-type fire detectors shall be rated in accordance with (c), listed as being resistant to the spread of fire in accordance with (d) through (f), marked in accordance with (i), and the jacket compound shall have a high degree of abrasion resistance.

Article 770 — Optical Fiber Cables and Raceways

A. General

770-1. Scope. The provisions of this article apply to the installation of optical fiber cables and raceways. This article does not cover the construction of optical fiber cables and raceways.

770-2. Definitions.

Exposed. The circuit is in such a position that, in case of failure of supports and insulation, contact with another circuit may result.

> FPN: See Article 100 for two other definitions of *Exposed*.

Optical Fiber Raceway. A raceway designed for enclosing and routing nonconductive optical fiber cables.

Point of Entrance. The point at which the wire or cable emerges from an external wall, from a concrete floor slab, or from a rigid metal conduit or an intermediate metal conduit grounded to an electrode in accordance with Section 800-40(b).

770-3. Locations and Other Articles. Circuits and equipment shall comply with (a) and (b). Only those sections of Article 300 referenced in this article shall apply to optical fiber cables and raceways.

(a) Spread of Fire or Products of Combustion. See Section 300-21.

(b) Ducts, Plenums, and Other Air-Handling Spaces. Section 300-22, where installed in ducts or plenums or other space used for environmental air.

Exception to (b): As permitted in Section 770-53(a).

770-4. Optical Fiber Cables. Optical fiber cables transmit light for control, signaling, and communications through an optical fiber.

770-5. Types. Optical fiber cables can be grouped into three types.

(a) Nonconductive. These cables contain no metallic members and no other electrically conductive materials.

(b) Conductive. These cables contain noncurrent-carrying conductive members such as metallic strength members, metallic vapor barriers, and metallic armor or sheath.

(c) Composite. These cables contain optical fibers and current-carrying electrical conductors, and shall be permitted to contain noncurrent-carrying conductive members such as metallic strength members and metallic vapor barriers. Composite optical fiber cables shall be classified as electrical cables in accordance with the type of electrical conductors.

770-6. Raceways for Optical Fiber Cables. The raceway shall be of a type permitted in Chapter 3 and installed in accordance with Chapter 3.

Exception: Listed nonmetallic optical fiber raceway identified as general purpose, riser, or plenum optical fiber raceway in accordance with Section 770-51 and installed in accordance with Sections 331-7 through 331-14, where the requirements applicable to electrical nonmetallic tubing shall apply. Unlisted underground or outside plant construction plastic innerduct shall be terminated at the point of entrance.

> FPN: For information on listing requirements for optical fiber raceways, see *Standard for Optical Fiber Raceways,* UL 2024.

Where optical fiber cables are installed within the raceway without current-carrying conductors, the raceway fill tables of Chapter 3 and Chapter 9 shall not apply.

Where nonconductive optical fiber cables are installed with electric conductors in a raceway, the raceway fill tables of Chapter 3 and Chapter 9 shall apply.

770-7. Access to Electrical Equipment Behind Panels Designed to Allow Access. Access to equipment shall not be denied by an accumulation of cables that prevents removal of panels, including suspended ceiling panels.

770-8. Mechanical Execution of Work. Optical fiber cables shall be installed in a neat and workmanlike manner. Cables shall be supported by the building structure in such a manner that the cable will not be damaged by normal building use.

> FPN: One way to determine accepted industry practice is to refer to nationally recognized standards such as *Commercial Building Telecommunications Wiring Standard,* ANSI/EIA/ TIA 568A-1995; *Commercial Building Standard for Telecommunications Pathways and Spaces,* ANSI/EIA/TIA 569-A-1997; and *Residential and Light Commercial Telecommunications Wiring Standard,* ANSI/EIA/TIA 570-1991, or other ANSI-approved installation standards.

B. Protection

770-33. Grounding of Entrance Cables. Where exposed to contact with electric light or power conductors, the noncurrent-carrying metallic members of optical fiber cables entering buildings shall be grounded as close to the point of entrance as practicable or shall be interrupted as close to the point of entrance as practicable by an insulating joint or equivalent device.

C. Cables Within Buildings

770-49. Fire Resistance of Optical Fiber Cables. Optical fiber cables installed as wiring within buildings shall be listed as being resistant to the spread of fire in accordance with Sections 770-50 and 770-51.

770-50. Listing, Marking, and Installation of Optical Fiber Cables. Optical fiber cables in a building shall be listed as being suitable for the purpose, and cables shall be marked in accordance with Table 770-50.

Exception No. 1: Optical fiber cables shall not be required to be listed and marked where the length of the cable within the building, measured from its point of entrance, does not exceed 50 ft (15.2 m) and the cable enters the building from the outside and is terminated in an enclosure.

> FPN: Splice cases or terminal boxes, both metallic and plastic types, are typically used as enclosures for splicing or terminating optical fiber cables.

Exception No. 2: Conductive optical fiber cable shall not be required to be listed and marked where the cable enters

Table 770-50. Cable Markings

Cable Marking	Type	Reference
OFNP	Nonconductive optical fiber plenum cable	770-51(a) and 770-53(a)
OFCP	Conductive optical fiber plenum cable	770-51(a) and 770-53(a)
OFNR	Nonconductive optical fiber riser cable	770-51(b) and 770-53(b)
OFCR	Conductive optical fiber riser cable	770-51(b) and 770-53(b)
OFNG	Nonconductive optical fiber general-purpose cable	770-51(c) and 770-53(c)
OFCG	Conductive optical fiber general-purpose cable	770-51(c) and 770-53(c)
OFN	Nonconductive optical fiber general-purpose cable	770-51(d) and 770-53(c)
OFC	Conductive optical fiber general-purpose cable	770-51(d) and 770-53(c)

the building from the outside and is run in rigid metal conduit or intermediate metal conduit and such conduits are grounded to an electrode in accordance with Section 800-40(b).

Exception No. 3: Nonconductive optical fiber cables shall not be required to be listed and marked where the cable enters the building from the outside and is run in raceway installed in compliance with Chapter 3.

> FPN No. 1: Cable types are listed in descending order of fire resistance rating. Within each fire resistance rating, nonconductive cable is listed first, since it may substitute for the conductive cable.

> FPN No. 2: See the referenced sections for requirements and permitted uses.

770-51. Listing Requirements for Optical Fiber Cables and Raceways. Optical fiber cables shall be listed in accordance with (a) through (d), and optical fiber raceways shall be listed in accordance with (e) through (g).

(a) Types OFNP and OFCP. Types OFNP and OFCP nonconductive and conductive optical fiber plenum cables shall be listed as being suitable for use in ducts, plenums, and other space used for environmental air and shall also be listed as having adequate fire-resistant and low smoke-producing characteristics.

> FPN: One method of defining low smoke-producing cables is by establishing an acceptable value of the smoke produced when tested in accordance with the *Standard Method of Test for Fire and Smoke Characteristics of Wires and Cables,* NFPA 262-1994, to a maximum peak optical density of 0.5 and a maximum average optical density of 0.15. Similarly, one method of defining fire-resistant cables is by defining maximum allowable flame travel distance of 5 ft (1.52 m) when tested in accordance with the same test.

(b) Types OFNR and OFCR. Types OFNR and OFCR nonconductive and conductive optical fiber riser cables shall be listed as being suitable for use in a vertical run in a shaft or from floor to floor and shall also be listed as having fire-resistant characteristics capable of preventing the carrying of fire from floor to floor.

> FPN: One method of defining fire-resistant characteristics capable of preventing the carrying of fire from floor to floor is that the cables pass the requirements of the *Standard Test for Flame Propagation Height of Electrical and Optical-Fiber Cable Installed Vertically in Shafts,* ANSI/UL 1666-1997.

(c) Types OFNG and OFCG. Types OFNG and OFCG nonconductive and conductive general-purpose optical fiber cables shall be listed as being suitable for general-purpose use, with the exception of risers and plenums, and shall also be listed as being resistant to the spread of fire.

> FPN: One method of defining resistance to the spread of fire is for the damage (char length) not to exceed 4 ft 11 in.

(1.5 m) when performing the vertical flame test — cables in cable trays, as described in *Test Methods for Electrical Wires and Cables,* CSA C22.2 No. 0.3-M 1985.

(d) Types OFN and OFC. Types OFN and OFC nonconductive and conductive optical fiber cables shall be listed as being suitable for general-purpose use, with the exception of risers, plenums, and other space used for environmental air, and shall also be listed as being resistant to the spread of fire.

> FPN: One method of defining resistant to the spread of fire is that the cables do not spread fire to the top of the tray in the vertical-tray flame test in the *Reference Standard for Electrical Wires, Cables and Flexible Cords,* ANSI/UL 1581-1991.
>
> Another method of defining resistant to the spread of fire is for the damage (char length) not to exceed 4 ft 11 in. (1.5 m) when performing the vertical flame test — cables in cable trays, as described in *Test Methods for Electrical Wires and Cables,* CSA C22.2 No. 0.3-M-1985.

(e) Plenum Optical Fiber Raceway. Plenum optical fiber raceways shall be listed as having adequate fire-resistant and low smoke-producing characteristics.

(f) Riser Optical Fiber Raceway. Riser optical fiber raceways shall be listed as having fire-resistant characteristics capable of preventing the carrying of fire from floor to floor.

(g) General-Purpose Optical Fiber Cable Raceway. General-purpose optical fiber cable raceway shall be listed as being resistant to the spread of fire.

770-52. Installation of Optical Fibers and Electrical Conductors.

(a) With Conductors for Electric Light, Power, Class 1, Nonpower-Limited Fire Alarm, or Medium Power Network-Powered Broadband Communications Circuits. Optical fibers shall be permitted within the same composite cable for electric light, power, Class 1, nonpower-limited fire alarm, or medium power network-powered broadband communications circuits operating at 600 volts or less only where the functions of the optical fibers and the electrical conductors are associated.

Nonconductive optical fiber cables shall be permitted to occupy the same cable tray or raceway with conductors for electric light, power, Class 1, nonpower-limited fire alarm, or medium power network-powered broadband communications circuits operating at 600 volts or less. Conductive optical fiber cables shall not be permitted to occupy the same cable tray or raceway with conductors for electric light, power, Class 1, nonpower-limited fire alarm, or medium power network-powered broadband communications circuits.

Composite optical fiber cables containing only current-carrying conductors for electric light, power, Class 1 circuits rated 600 volts or less shall be permitted to occupy the same cabinet, cable tray, outlet box, panel, raceway, or other termination enclosure with conductors for electric light, power, or Class 1 circuits operating at 600 volts or less.

Nonconductive optical fiber cables shall not be permitted to occupy the same cabinet, outlet box, panel, or similar enclosure housing the electrical terminations of an electric light, power, Class 1, nonpower-limited fire alarm, or medium power network-powered broadband communications circuit.

Exception No. 1: Occupancy of the same cabinet, outlet box, panel, or similar enclosure shall be permitted where nonconductive optical fiber cable is functionally associated with the electric light, power, Class 1, nonpower-limited fire alarm, or medium power network-powered broadband communications circuit.

Exception No. 2: Occupancy of the same cabinet, outlet box, panel, or similar enclosure shall be permitted where nonconductive optical fiber cables are installed in factory- or field-assembled control centers.

Exception No. 3: In industrial establishments only, where conditions of maintenance and supervision ensure that only qualified persons will service the installation, nonconductive optical fiber cables shall be permitted with circuits exceeding 600 volts.

Exception No. 4: In industrial establishments only, where conditions of maintenance and supervision ensure that only qualified persons will service the installation, composite optical fiber cables shall be permitted to contain current-carrying conductors operating over 600 volts.

Installations in raceway shall comply with Section 300-17.

(b) With Other Conductors. Optical fibers shall be permitted in the same cable, and conductive and nonconductive optical fiber cables shall be permitted in the same cable tray, enclosure, or raceway with conductors of any of the following:

(1) Class 2 and Class 3 remote-control, signaling, and power-limited circuits in compliance with Article 725
(2) Power-limited fire alarm systems in compliance with Article 760
(3) Communications circuits in compliance with Article 800
(4) Community antenna television and radio distribution systems in compliance with Article 820
(5) Low power network-powered broadband communications circuits in compliance with Article 830

(c) Grounding. Noncurrent-carrying conductive members of optical fiber cables shall be grounded in accordance with Article 250.

770-53. Applications of Listed Optical Fiber Cables and Raceways. Nonconductive and conductive optical fiber cables shall comply with (a) through (f) as applicable.

(a) Plenum. Cables installed in ducts, plenums, and other spaces used for environmental air shall be Type OFNP or OFCP.

Also, listed plenum optical fiber raceways shall be permitted to be installed in ducts and plenums as described in Section 300-22(b) and in other space used for environmental air as described in Section 300-22(c). Only Type OFNP cable shall be permitted to be installed in these raceways.

Exception: Types OFNR, OFCR, OFNG, OFN, OFCG, and OFC cables installed in compliance with Section 300-22.

(b) Riser. Cables installed in vertical runs and penetrating more than one floor or cables installed in vertical runs in a shaft shall be Types OFNR or OFCR. Floor penetrations requiring Types OFNR or OFCR shall contain only cables suitable for riser or plenum use.

Also, listed riser optical fiber raceways shall be permitted to be installed in vertical runs in a shaft or from floor to floor. Only Types OFNR and OFNP cables shall be permitted to be installed in these raceways.

Exception No. 1: Where Types OFNG, OFN, OFCG, and OFC cables are encased in metal raceway or are located in a fireproof shaft having firestops at each floor.

Exception No. 2: Type OFNG, OFN, OFCG, or OFC cable in one- and two-family dwellings.

FPN: See Section 300-21 for firestop requirements for floor penetrations.

(c) Other Wiring Within Buildings. Cables installed in building locations other than the locations covered in (a) and (b) shall be Type OFNG, OFN, OFCG, or OFC. Such cables shall be permitted to be installed in listed general-purpose optical fiber raceways.

(d) Hazardous (Classified) Locations. Cables installed in hazardous (classified) locations shall be any type indicated in Table 770-53.

(e) Cable Trays. Optical fiber cables of the types listed in Table 770-50 shall be permitted to be installed in cable trays.

FPN: It is not the intent to require that these optical fiber cables be listed specifically for use in cable trays.

(f) Cable Substitutions. The substitutions for optical fiber cables listed in Table 770-53 shall be permitted.

Table 770-53. Cable Substitutions

Cable Type	Permitted Substitutions
OFNP	None
OFCP	OFNP
OFNR	OFNP
OFCR	OFNP, OFCP, OFNR
OFNG, OFN	OFNP, OFNR
OFCG, OFC	OFNP, OFCP, OFNR, OFCR, OFNG, OFN

Figure 770-53   Cable substitution hierarchy.

Article 780 — Closed-Loop and Programmed Power Distribution

780-1. Scope. The provisions of this article apply to premise power distribution systems jointly controlled by a signaling between the energy controlling equipment and utilization equipment.

780-2. General.

(a) Other Articles. Except as modified by the requirements of this article, all other applicable articles of this *Code* shall apply.

(b) Component Parts. All equipment and conductors shall be listed and identified.

780-3. Control. The control equipment and all power switching devices operated by the control equipment shall be listed and identified. The system shall operate as follows.

(a) Characteristic Electrical Identification Required. Outlets of a closed-loop power distribution system shall not be energized unless the utilization equipment first exhibits a characteristic electrical identification.

(b) Conditions for De-energization. Outlets shall be de-energized when any of the following conditions occur.

(1) A nominal-operation acknowledgement signal is not being received from the utilization equipment connected to the outlet in a closed-loop power distribution system.
(2) A ground-fault condition exists.
(3) An overcurrent condition exists.

(c) Additional Conditions for De-energization When an Alternate Source of Power is Used. In addition to the requirements in (b), outlets shall be de-energized when any of the following conditions occur.

(1) The grounded conductor is not properly grounded.
(2) Any ungrounded conductor is not at nominal voltage.

(d) Controller Malfunction. In the event of a controller malfunction, all associated outlets shall be de-energized.

780-5. Power Limitation in Signaling Circuits. For signaling circuits not exceeding 24 volts, the current required shall not exceed 1 ampere where protected by an overcurrent device or an inherently limited power source.

780-6. Cables and Conductors.

(a) Hybrid Cable. Listed hybrid cable consisting of power, communications, and signaling conductors shall be permitted under a common jacket. The jacket shall be applied so as to separate the power conductors from the communications and signaling conductors. An optional outer jacket shall be permitted to be applied. The individual conductors of a hybrid cable shall conform to the *Code* provisions applicable to their current, voltage, and insulation rating. The signaling conductors shall not be smaller than No. 24 copper.

(b) Cables and Conductors in the Same Cabinet, Panel, or Box. The power, communications, and signaling conductors of listed hybrid cable are permitted to occupy the same cabinet, panel, or outlet box (or similar enclosure housing the electrical terminations of electric light or power circuits) only if connectors specifically listed for hybrid cable are employed.

780-7. Noninterchangeability. Receptacles, cord connectors, and attachment plugs used on closed-loop power distribution systems shall be constructed so that they are not interchangeable with other receptacles, cord connectors, and attachment plugs.

CHAPTER 8
Communications Systems

Article 800 — Communications Circuits

A. General

800-1. Scope. This article covers telephone, telegraph (except radio), outside wiring for fire alarm and burglar alarm, and similar central station systems; and telephone systems not connected to a central station system but using similar types of equipment, methods of installation, and maintenance.

> FPN No. 1: For further information for fire alarm, guard tour, sprinkler waterflow, and sprinkler supervisory systems, see Article 760.

> FPN No. 2: For installation requirements of optical fiber cables, see Article 770.

> FPN No. 3: For installation requirements for network-powered broadband communications circuits, see Article 830.

800-2. Definitions. See Article 100. For purposes of this article, the following additional definitions apply.

Block. A square or portion of a city, town, or village enclosed by streets and including the alleys so enclosed, but not any street.

Cable. A factory assembly of two or more conductors having an overall covering.

Cable Sheath. A covering over the conductor assembly that may include one or more metallic members, strength members, or jackets.

Exposed. A circuit that is in such a position that, in case of failure of supports and insulation, contact with another circuit may result.

> FPN: See Article 100 for two other definitions of *Exposed*.

Point of Entrance. The point of entrance within a building is the point at which the wire or cable emerges from an external wall, from a concrete floor slab, or from a rigid metal conduit or an intermediate metal conduit grounded to an electrode in accordance with Section 800-40(b).

Premises. The land and buildings of a user located on the user side of the utility-user network point of demarcation.

Wire. A factory assembly of one or more insulated conductors without an overall covering.

800-3. Hybrid Power and Communications Cables. The provisions of Section 780-6 shall apply for listed hybrid power and communications cables in closed-loop and programmed power distribution.

> FPN: See Section 800-51(i) for hybrid power and communications cable in other applications.

800-4. Equipment. Equipment intended to be electrically connected to a telecommunications network shall be listed for the purpose. Installation of equipment shall also comply with Section 110-3(b).

> FPN: One way to determine applicable requirements is to refer to the *Standard for Safety of Information Technology Equipment, Including Electrical Business Equipment*, UL 1950-1993, third edition; the *Standard for Safety, Telephone Equipment*, UL 1459-1995, third edition; or the *Standard for Safety, Communications Circuit Accessories*, UL 1863-1995, second edition. For information on listing requirements for communications raceways, see UL 2024-1995, *Standard for Optical Fiber Raceways*.

Exception: This listing requirement shall not apply to test equipment that is intended for temporary connection to a telecommunications network by qualified persons during the course of installation, maintenance, or repair of telecommunications equipment or systems.

800-5. Access to Electrical Equipment Behind Panels Designed to Allow Access. Access to equipment shall not be denied by an accumulation of wires and cables that prevents removal of panels, including suspended ceiling panels.

800-6. Mechanical Execution of Work. Communications circuits and equipment shall be installed in a neat and workmanlike manner. Cables shall be supported by the building structure in such a manner that the cable will not be damaged by normal building use.

> FPN: One way to determine accepted industry practice is to refer to nationally recognized standards such as *Commercial Building Telecommunications Cabling Standard*, ANSI/EIA/TIA 568-A-1995; *Commercial Building Standard for Telecommunications Pathways and Spaces*, ANSI/TIA/EIA 569-A-1997; and *Residential and Light Commercial Telecommunications Wiring Standard*, ANSI/EIA/TIA 570-1991, or other ANSI-approved installation standards.

800-7. Hazardous (Classified) Locations. Communications circuits and equipment installed in a location that is classified in accordance with Article 500 shall comply with the applicable requirements of Chapter 5.

B. Conductors Outside and Entering Buildings

800-10. Overhead Communications Wires and Cables. Overhead communications wires and cables entering buildings shall comply with (a) through (c).

(a) On Poles and In-Span. Where communications wires and cables and electric light or power conductors are supported by the same pole or run parallel to each other in-span, the following conditions shall be met.

(1) Relative Location. Where practicable, the communications wires and cables shall be located below the electric light or power conductors.

(2) Attachment to Crossarms. Communications wires and cables shall not be attached to a crossarm that carries electric light or power conductors.

(3) Climbing Space. The climbing space through communications wires and cables shall comply with the requirements of Section 225-14(d).

(4) Clearance. Supply service drops of 0–750 volts running above and parallel to communications service drops shall have a minimum separation of 12 in. (30.48 cm) at any point in the span, including the point of and at their attachment to the building, provided the nongrounded conductors are insulated and that a clearance of not less than 40 in. (1.02 m) is maintained between the two services at the pole.

(b) Above Roofs. Communications wires and cables shall have a vertical clearance of not less than 8 ft (2.44 m) from all points of roofs above which they pass.

Exception No. 1: Auxiliary buildings, such as garages and the like.

Exception No. 2: A reduction in clearance above only the overhanging portion of the roof to not less than 18 in. (457 mm) shall be permitted if (1) not more than 4 ft (1.22 mm) of communications service-drop conductors pass above the roof overhang and (2) they are terminated at a through- or above-the-roof raceway or approved support.

Exception No. 3: Where the roof has a slope of not less than 4 in. (102 mm) in 12 in. (305 mm), a reduction in clearance to not less than 3 ft (914 mm) shall be permitted.

800-11. Underground Circuits Entering Buildings. Underground communications wires and cables entering buildings shall comply with (a) through (c).

(a) With Electric Light or Power Conductors. Underground communications wires and cables in a raceway, handhole, or manhole containing electric light, power, Class 1, or nonpower-limited fire alarm circuit conductors shall be in a section separated from such conductors by means of brick, concrete, or tile partitions or by means of a suitable barrier.

(b) Underground Block Distribution. Where the entire street circuit is run underground and the circuit within the block is placed so as to be free from likelihood of accidental contact with electric light or power circuits of over 300 volts to ground, the insulation requirements of Sections 800-12(a) and (c) shall not apply, insulating supports shall not be required for the conductors, and bushings shall not be required where the conductors enter the building.

(c) Point of Entry. The point of entry for communications wiring and cables shall be within 20 ft (7.0 m) of the electrical service entry point.

Exception: Where it is impracticable to install the communications service in this manner, a separate grounding electrode, installed in compliance with Section 800-40(b)(3) and bonded in accordance with Sections 800-40(c) and (d).

　FPN: Under certain conditions, the length of the bonding conductor has a direct relationship to the difference in potential between the communications and the power circuits.

800-12. Circuits Requiring Primary Protectors. Circuits that require primary protectors as provided in Section 800-30 shall comply with the following.

(a) Insulation, Wires, and Cables. Communications wires and cables without a metallic shield, running from the last outdoor support to the primary protector, shall be listed as being suitable for the purpose and shall have current-carrying capacity as specified in Section 800-30(a)(1)(b) or 800-30(a)(1)(c).

(b) On Buildings. Communications wires and cables in accordance with Section 800-12(a) shall be separated at least 4 in. (102 mm) from electric light or power conductors not in a raceway or cable, or be permanently separated from conductors of the other system by a continuous and firmly fixed nonconductor in addition to the insulation on the wires, such as porcelain tubes or flexible tubing. Communications wires and cables in accordance with Section 800-12(a) exposed to accidental contact with electric light and power conductors operating at over 300 volts to ground and attached to buildings shall be separated from woodwork by being supported on glass, porcelain, or other insulating material.

Exception: Separation from woodwork shall not be required where fuses are omitted as provided for in Section 800-30(a)(1), or where conductors are used to extend circuits to a building from a cable having a grounded metal sheath.

(c) Entering Buildings. Where a primary protector is installed inside the building, the communications wires and cables shall enter the building either through a noncombustible, nonabsorbent insulating bushing or through a metal raceway. The insulating bushing shall not be required where the entering communications wires and cables (1) are in metal-sheathed cable, (2) pass through masonry, (3) meet the requirements of Section 800-12(a) and fuses are omitted as provided in Section 800-30(a)(1), or (4) meet the requirements of Section 800-12(a) and are used to extend circuits to a building from a cable having a grounded metallic sheath. Raceways or bushings shall slope upward from the outside or, where this cannot be done, drip loops shall be formed in the communications wires and cables immediately before they enter the building.

Raceways shall be equipped with an approved service head. More than one communications wire and cable shall be permitted to enter through a single raceway or bushing. Conduits or other metal raceways located ahead of the primary protector shall be grounded.

800-13. Lightning Conductors. Where practicable, a separation of at least 6 ft (1.83 m) shall be maintained between communications wires and cables on buildings and lightning conductors.

C. Protection

800-30. Protective Devices.

(a) Application. A listed primary protector shall be provided on each circuit run partly or entirely in aerial wire or aerial cable not confined within a block. Also, a listed primary protector shall be provided on each circuit, aerial or underground, located within the block containing the building served so as to be exposed to accidental contact with electric light or power conductors operating at over 300 volts to ground. In addition, where there exists a lightning exposure, each interbuilding circuit on a premises shall be protected by a listed primary protector at each end of the interbuilding circuit. Installation of primary protectors shall also comply with Section 110-3(b).

FPN No. 1: On a circuit not exposed to accidental contact with power conductors, providing a listed primary protector in accordance with this article will help protect against other hazards, such as lightning and above-normal voltages induced by fault currents on power circuits in proximity to the communications circuit.

FPN No. 2: Interbuilding circuits are considered to have a lightning exposure unless one or more of the following conditions exist.

(1) Circuits in large metropolitan areas where buildings are close together and sufficiently high to intercept lightning.
(2) Interbuilding cable runs of 140 ft (42.7 m) or less, directly buried or in underground conduit, where a continuous metallic cable shield or a continuous metallic conduit containing the cable is bonded to each building grounding electrode system.
(3) Areas having an average of five or fewer thunderstorm days per year and earth resistivity of less than 100 ohm-meters. Such areas are found along the Pacific coast.

(1) Fuseless Primary Protectors. Fuseless-type primary protectors shall be permitted under any of the following conditions:

(a) Where conductors enter a building through a cable with grounded metallic sheath member(s) and if the conductors in the cable safely fuse on all currents greater than the current-carrying capacity of the primary protector and of the primary protector grounding conductor
(b) Where insulated conductors in accordance with Section 800-12(a) are used to extend circuits to a building from a cable with an effectively grounded metallic sheath member(s) and if the conductors in the cable or cable stub, or the connections between the insulated conductors and the exposed plant, safely fuse on all currents greater than the current-carrying capacity of the primary protector, or the associated insulated conductors and of the primary protector grounding conductor
(c) Where insulated conductors in accordance with Section 800-12(a) or (b) are used to extend circuits to a building from other than a cable with a metallic sheath member(s) if (1) the primary protector is listed for this purpose, and (2) the connections of the insulated conductors to the exposed plant or the conductors of the exposed plant safely fuse on all currents greater than the current-carrying capacity of the primary protector, or the associated insulated conductors and of the primary protector grounding conductor
(d) Where insulated conductors in accordance with Section 800-12(a) are used to extend circuits aerially to a building from an unexposed buried or underground circuit
(e) Where insulated conductors in accordance with Section 800-12(a) are used to extend circuits to a building from cable with an effectively grounded metallic sheath member(s) and if (1) the combination of the primary protector and insulated conductors is listed for this purpose, and (2) the insulated conductors safely fuse on all currents greater than the current-carrying capacity of the primary protector and of the primary protector grounding conductor

(2) Fused Primary Protectors. Where the requirements listed under Sections 800-30(a)(1)(a) through (e) are not met, fused-type primary protectors shall be used. Fused-type primary protectors shall consist of an arrester connected between each line conductor and ground, a fuse in series with each line conductor, and an appropriate mounting arrangement. Primary protector terminals shall be marked to indicate line, instrument, and ground, as applicable.

(b) Location. The primary protector shall be located in, on, or immediately adjacent to the structure or building served and as close as practicable to the point at which the exposed conductors enter or attach.

For purposes of this section, the point at which the exposed conductors enter shall be considered to be the point of emergence through an exterior wall, a concrete floor slab, or from a rigid metal conduit or an intermediate metal conduit grounded to an electrode in accordance with Section 800-40(b).

For purposes of this section, primary protectors located at mobile home service equipment located in sight from and not more than 30 ft (9.14 m) from the exterior wall of the mobile home it serves, or at a mobile home disconnecting means grounded in accordance with Section 250-32 and located in sight from and not more than 30 ft (9.14 m) from the exterior wall of the mobile home it serves, shall be considered to meet the requirements of this section.

FPN: Selecting a primary protector location to achieve the shortest practicable primary protector grounding conductor will help limit potential differences between communications circuits and other metallic systems.

(c) Hazardous (Classified) Locations. The primary protector shall not be located in any hazardous (classified) location as defined in Article 500, nor in the vicinity of easily ignitible material.

Exception: As permitted in Sections 501-14, 502-14, and 503-12.

800-31. Primary Protector Requirements. The primary protector shall consist of an arrester connected between each line conductor and ground in an appropriate mounting. Primary protector terminals shall be marked to indicate line and ground as applicable.

FPN: One way to determine applicable requirements for a listed primary protector is to refer to the *Standard for Protectors for Paired Conductor Communications Circuits,* ANSI/UL 497-1995.

800-32. Secondary Protector Requirements. Where a secondary protector is installed in series with the indoor communications wire and cable between the primary protector and the equipment, it shall be listed for the purpose. The secondary protector shall provide means to safely limit currents to less than the current-carrying capacity of listed indoor communications wire and cable, listed telephone set line cords, and listed communications terminal equipment having ports for external wire line communications circuits. Any overvoltage protection, arresters, or grounding connection shall be connected on the equipment terminals side of the secondary protector current-limiting means.

FPN No. 1: One way to determine applicable requirements for a listed secondary protector is to refer to the *Standard for Secondary Protectors for Communications Circuits,* UL 497A-1996.

FPN No. 2: Secondary protectors on exposed circuits are not intended for use without primary protectors.

800-33. Cable Grounding. The metallic sheath of communications cables entering buildings shall be grounded as close as practicable to the point of entrance or shall be interrupted as close to the point of entrance as practicable by an insulating joint or equivalent device.

For purposes of this section, the point of entrance shall be considered to be at the point of emergence through an exterior wall, a concrete floor slab, or from a rigid metal conduit or an intermediate metal conduit grounded to an electrode in accordance with Section 800-40(b).

D. Grounding Methods

800-40. Cable and Primary Protector Grounding. The metallic member(s) of the cable sheath, where required to be grounded by Section 800-33, and primary protectors shall be grounded as specified in (a) through (d).

(a) Grounding Conductor.

(1) Insulation. The grounding conductor shall be insulated and shall be listed as suitable for the purpose.

(2) Material. The grounding conductor shall be copper or other corrosion-resistant conductive material, stranded or solid.

(3) Size. The grounding conductor shall not be smaller than No. 14.

(4) Run in Straight Line. The grounding conductor shall be run to the grounding electrode in as straight a line as practicable.

(5) Physical Damage. Where necessary, the grounding conductor shall be guarded from physical damage. Where the grounding conductor is run in a metal raceway, both ends of the raceway shall be bonded to the grounding conductor or the same terminal or electrode to which the grounding conductor is connected.

(b) Electrode. The grounding conductor shall be connected as follows.

(1) To the nearest accessible location on the following:

(a) The building or structure grounding electrode system as covered in Section 250-50
(b) The grounded interior metal water piping system as covered in Section 250-104(a)
(c) The power service accessible means external to enclosures as covered in Section 250-92(b)
(d) The metallic power service raceway
(e) The service equipment enclosure
(f) The grounding electrode conductor or the grounding electrode conductor metal enclosure; or

(g) To the grounding conductor or the grounding electrode of a building or structure disconnecting means that is grounded to an electrode as covered in Section 250-32.

For purposes of this section, the mobile home service equipment or the mobile home disconnecting means, as described in Section 800-30(b), shall be considered accessible.

(2) If the building or structure served has no grounding means, as described in (b)(1), to any one of the individual electrodes described in Section 250-50; or

(3) If the building or structure served has no grounding means, as described in (b)(1) or (b)(2), to an effectively grounded metal structure or to a ground rod or pipe not less than 5 ft (1.52 m) in length and ½ in. (12.7 mm) in diameter, driven, where practicable, into permanently damp earth and separated from lightning conductors as covered in Section 800-13 and at least 6 ft (1.83 m) from electrodes of other systems. Steam or hot water pipes or lightning-rod conductors shall not be employed as electrodes for protectors.

(c) **Electrode Connection.** Connections to grounding electrodes shall comply with Section 250-70. Connectors, clamps, fittings, or lugs used to attach grounding conductors and bonding jumpers to grounding electrodes or to each other that are to be concrete-encased or buried in the earth shall be suitable for its application.

(d) **Bonding of Electrodes.** A bonding jumper not smaller than No. 6 copper or equivalent shall be connected between the communications grounding electrode and power grounding electrode system at the building or structure served where separate electrodes are used. Bonding together of all separate electrodes shall be permitted.

Exception: At mobile homes as covered in Section 800-41.

FPN No. 1: See Section 250-60 for use of air terminals (lightning rods).

FPN No. 2: Bonding together of all separate electrodes will limit potential differences between them and between their associated wiring systems.

800-41. Primary Protector Grounding and Bonding at Mobile Homes.

(a) **Grounding.** Where there is no mobile home service equipment located in sight from and not more than 30 ft (9.14 m) from the exterior wall of the mobile home it serves, or there is no mobile home disconnecting means grounded in accordance with Section 250-32 and located within sight from and not more than 30 ft (9.14 m) from the exterior wall of the mobile home it serves, the primary protector ground shall be in accordance with Sections 800-40(b)(2) and (3).

(b) **Bonding.** The primary protector grounding terminal or grounding electrode shall be bonded to the metal frame or available grounding terminal of the mobile home with a copper grounding conductor not smaller than No. 12 under any of the following conditions:

(1) Where there is no mobile home service equipment or disconnecting means as in (a), or
(2) The mobile home is supplied by cord and plug.

E. Communications Wires and Cables Within Buildings

800-48. Raceways for Communications Wires and Cables. Where communications wire and cables are installed in a raceway, the raceway shall be of a type permitted in Chapter 3 and installed in accordance with Chapter 3.

Exception: Listed nonmetallic communications raceway identified as general purpose, riser, or plenum in accordance with Section 800-51 and installed in accordance with Sections 331-7 through 331-14, where the requirements applicable to electrical nonmeallic tubing shall apply.

800-49. Fire Resistance of Communications Wires and Cables. Communications wires and cables installed as wiring within a building shall be listed as being resistant to the spread of fire in accordance with Sections 800-50 and 800-51.

800-50. Listing, Marking, and Installation of Communications Wires and Cables. Communications wires and cables installed as wiring within buildings shall be listed as being suitable for the purpose and installed in accordance with Section 800-52. Communications cables and undercarpet communications wires shall be marked in accordance with Table 800-50. The cable voltage rating shall not be marked on the cable or on the under-carpet communications wire.

FPN: Voltage markings on cables may be misinterpreted to suggest that the cables may be suitable for Class 1, electric light, and power applications.

Exception No. 1: Voltage markings shall be permitted where the cable has multiple listings and voltage marking is required for one or more of the listings.

Exception No. 2: Listing and marking shall not be required where the cable enters the building from the outside and is continuously enclosed in a rigid metal conduit system or an intermediate metal conduit system and such conduit systems are grounded to an electrode in accordance with Section 800-40(b).

Exception No. 3: Listing and marking shall not be required where the length of the cable within the the building, measured from its point of entrance, does not exceed 50 ft

(15.2 m) and the cable enters the building from the outside and is terminated in an enclosure or on a listed primary protector.

> FPN No. 1: Splice cases or terminal boxes, both metallic and plastic types, are typically used as enclosures for splicing or terminating telephone cables.

> FPN No. 2: This exception limits the length of unlisted outside plant cable to 50 ft (15.2 m), while Section 800-30(b) requires that the primary protector shall be located as close as practicable to the point at which the cable enters the building. Therefore, in installations requiring a primary protector, the outside plant cable may not be permitted to extend 50 ft (15.2 m) into the building if it is practicable to place the primary protector closer than 50 ft (15.2 m) to the entrance point.

Exception No. 4: Multipurpose cables shall be considered as being suitable for the purpose and shall be permitted to substitute for communications cables as provided for in Section 800-53(f).

Table 800-50. Cable Markings

Cable Marking	Type	Reference
MPP	Multipurpose plenum cable	800-51(g) and 800-53(a)
CMP	Communications plenum cable	800-51(a) and 800-53(a)
MPR	Multipurpose riser cable	800-51(g) and 800-53(b)
CMR	Communications riser cable	800-51(b) and 800-53(b)
MPG	Multipurpose general-purpose cable	800-51(g) and 800-53(d)
CMG	Communications general-purpose cable	800-51(c) and 800-53(d)
MP	Multipurpose general-purpose cable	800-51(g) and 800-53(d)
CM	Communications general-purpose cable	800-51(d) and 800-53(d)
CMX	Communications cable, limited use	800-51(e) and 800-53(d), Exception Nos. 1, 2, 3, and 4
CMUC	Under-carpet communications wire and cable	800-51(f) and 800-53(d), Exception No. 5

> FPN No. 1: Cable types are listed in descending order of fire resistance rating, and multipurpose cables are listed above communications cables because multipurpose cables may substitute for communications cables.

> FPN No. 2: See the referenced sections for permitted uses.

800-51. Listing Requirements for Communications Wires and Cables and Communications Raceways. Communications wires and cables shall have a voltage rating of not less than 300 volts and shall be listed in accordance with (a) through (i), and communications raceways shall be listed in accordance with (j) through (l). Conductors in communications cables, other than in a coaxial cable, shall be copper.

> FPN: See Section 800-4 for listing requirement for equipment.

(a) Type CMP. Type CMP communications plenum cable shall be listed as being suitable for use in ducts, plenums, and other spaces used for environmental air and shall also be listed as having adequate fire-resistant and low smoke-producing characteristics.

> FPN: One method of defining low smoke-producing cables is by establishing an acceptable value of the smoke produced when tested in accordance with the *Standard Method of Test for Fire and Smoke Characteristics of Wires and Cables,* NFPA 262-1994, to a maximum peak optical density of 0.5 and a maximum average optical density of 0.15. Similarly, one method of defining fire-resistant cables is by establishing a maximum allowable flame travel distance of 5 ft (1.52 m) when tested in accordance with the same test.

(b) Type CMR. Type CMR communications riser cable shall be listed as being suitable for use in a vertical run in a shaft or from floor to floor and shall also be listed as having fire-resistant characteristics capable of preventing the carrying of fire from floor to floor.

> FPN: One method of defining fire-resistant characteristics capable of preventing the carrying of fire from floor to floor is that the cables pass the requirements of the *Standard Test for Flame Propagation Height of Electrical and Optical-Fiber Cable Installed Vertically in Shafts,* ANSI/UL 1666-1997.

(c) Type CMG. Type CMG general-purpose communications cable shall be listed as being suitable for general-purpose communications use, with the exception of risers and plenums, and shall also be listed as being resistant to the spread of fire.

> FPN: One method of defining resistant to the spread of fire is for the damage (char length) not to exceed 4 ft 11 in. (1.5 m) when performing the vertical flame test — cables in cable trays, as described in *Test Methods for Electrical Wires and Cables,* CSA C22.2 No. 0.3-M 1985.

(d) Type CM. Type CM communications cable shall be listed as being suitable for general-purpose communications use, with the exception of risers and plenums, and shall also be listed as being resistant to the spread of fire.

> FPN: One method of defining resistant to the spread of fire is that the cables do not spread fire to the top of the tray in the vertical-tray flame test in the *Reference Standard for Electrical Wires, Cables and Flexible Cords,* ANSI/UL 1581-1991. Another method of defining resistant to the spread of fire is for the damage (char length) not to exceed 4 ft 11 in. (1.5 m) when performing the vertical flame test — cables in cable trays, as described in *Test Methods for Electrical Wires and Cables,* CSA C22.2 No. 0.3-M-1985.

(e) Type CMX. Type CMX limited use communications cable shall be listed as being suitable for use in dwellings and for use in raceway and shall also be listed as being resistant to flame spread.

> FPN: One method of determining that cable is resistant to flame spread is by testing the cable to the VW-1 (vertical-wire) flame test in the *Reference Standard for Electrical Wires, Cables and Flexible Cords*, ANSI/UL 1581-1991.

(f) Type CMUC Under-Carpet Wire and Cable. Type CMUC under-carpet communications wire and cable shall be listed as being suitable for under-carpet use and shall also be listed as being resistant to flame spread.

> FPN: One method of determining that cable is resistant to flame spread is by testing the cable to the VW-1 (vertical-wire) flame test in the *Reference Standard for Electrical Wires, Cables and Flexible Cords*, ANSI/UL 1581-1991.

(g) Multipurpose (MP) Cables. Cables that meet the requirements for Types CMP, CMR, CMG, and CM and also satisfy the requirements of Section 760-71(b) for multiconductor cables and Section 760-71(h) for coaxial cables shall be permitted to be listed and marked as multipurpose cable Types MPP, MPR, MPG, and MP, respectively.

(h) Communications Wires. Communications wires, such as distributing frame wire and jumper wire, shall be listed as being resistant to the spread of fire.

> FPN: One method of defining resistant to the spread of fire is that the cables do not spread fire to the top of the tray in the vertical-tray flame test in the *Reference Standard for Electrical Wires, Cables and Flexible Cords*, ANSI/UL 1581-1991.
>
> Another method of defining resistant to the spread of fire is for the damage (char length) not to exceed 4 ft 11 in. (1.5 m) when performing the vertical flame test — cables in cable trays, as described in *Test Methods for Electrical Wires and Cables*, CSA C22.2 No. 0.3-M-1985.

(i) Hybrid Power and Communications Cable. Listed hybrid power and communications cable shall be permitted where the power cable is a listed Type NM or NM-B conforming to the provisions of Article 336, and the communications cable is a listed Type CM, and the jackets on the listed NM or NM-B and listed CM cables are rated for 600 volts minimum, and the hybrid cable is listed as being resistant to the spread of fire.

> FPN: One method of defining resistant to the spread of fire is that the cables do not spread fire to the top of the tray in the vertical-tray flame test in the *Reference Standard for Electrical Wires, Cables and Flexible Cords*, ANSI/UL 1581-1991.
>
> Another method of defining resistant to the spread of fire is for the damage (char length) not to exceed 4 ft 11 in. (1.5 m) when performing the vertical flame test — cables in cable trays, as described in *Test Methods for Electrical Wires and Cables*, CSA C22.2 No. 0.3-M-1985.

(j) Plenum Communications Raceways. Plenum communications raceways listed as plenum optical fiber raceways shall be permitted for use in ducts, plenums, and other spaces used for environmental air and shall also be listed as having adequate fire-resistant and low-smoke producing characteristics.

(k) Riser Communications Raceway. Riser communications raceways shall be listed as having adequate fire-resistant characteristics capable of preventing the carrying of fire from floor to floor.

(l) General-Purpose Communications Raceway. General-purpose communications raceways shall be listed as having adequate fire-resistant characteristics.

800-52. Installation of Communications Wires, Cables, and Equipment. Communications wires and cables from the protector to the equipment or, where no protector is required, communications wires and cables attached to the outside or inside of the building shall comply with (a) through (e).

(a) Separation from Other Conductors.

(1) In Raceways, Boxes, and Cables.

(a) *Other Power-Limited Circuits.* Communications cables shall be permitted in the same raceway or enclosure with cables of any of the following:

(1) Class 2 and Class 3 remote-control, signaling, and power-limited circuits in compliance with Article 725
(2) Power-limited fire alarm systems in compliance with Article 760
(3) Nonconductive and conductive optical fiber cables in compliance with Article 770
(4) Community antenna television and radio distribution systems in compliance with Article 820
(5) Low power network-powered broadband communications circuits in compliance with Article 830

(b) *Class 2 and Class 3 Circuits.* Class 1 circuits shall not be run in the same cable with communications circuits. Class 2 and Class 3 circuit conductors shall be permitted in the same cable with communications circuits, in which case the Class 2 and Class 3 circuits shall be classified as communications circuits and shall meet the requirements of this article. The cables shall be listed as communications cables or multipurpose cables.

Exception: Cables constructed of individually listed Class 2, Class 3, and communications cables under a common jacket shall not be required to be classified as communications cable. The fire-resistance rating of the composite cable shall be determined by the performance of the composite cable.

(c) *Electric Light, Power, Class 1, Nonpower-Limited Fire Alarm, and Medium Power Network-Powered Broadband Communications Circuits.*

1. In Raceways, Compartments, and Boxes. Communications conductors shall not be placed in any raceway, compartment, outlet box, junction box, or similar fitting with conductors of electric light, power, Class 1, nonpower-limited fire alarm or medium power network-powered broadband communications circuits.

Exception No. 1: Where all of the conductors of electric light, power, Class 1, nonpower-limited fire alarm, and medium power network-powered broadband communications circuits are separated from all of the conductors of communications circuits by a barrier.

Exception No. 2: Power conductors in outlet boxes, junction boxes, or similar fittings or compartments where such conductors are introduced solely for power supply to communications equipment. The power circuit conductors shall be routed within the enclosure to maintain a minimum of 0.25-in. (6.35-mm) separation from the communications circuit conductors.

(2) **Other Applications.** Communications wires and cables shall be separated at least 2 in. (50.8 mm) from conductors of any electric light, power, Class 1, nonpower-limited fire alarm, or medium power network-powered broadband communications circuits.

Exception No. 1: Where either (1) all of the conductors of the electric light, power, Class 1, nonpower-limited fire alarm, and medium power network-powered broadband communications circuits are in a raceway or in metal-sheathed, metal-clad, nonmetallic-sheathed, Type AC, or Type UF cables, or (2) all of the conductors of communications circuits are encased in raceway.

Exception No. 2: Where the communications wires and cables are permanently separated from the conductors of electric light, power, Class 1, nonpower-limited fire alarm, and medium power network-powered broadband communications circuits by a continuous and firmly fixed nonconductor, such as porcelain tubes or flexible tubing, in addition to the insulation on the wire.

(b) **Spread of Fire or Products of Combustion.** Installations in hollow spaces, vertical shafts, and ventilation or air-handling ducts shall be made so that the possible spread of fire or products of combustion will not be substantially increased. Openings around penetrations through fire resistance-rated walls, partitions, floors, or ceilings shall be firestopped using approved methods.

(c) **Equipment in Other Space Used for Environmental Air.** Section 300-22(c) shall apply.

(d) **Cable Trays.** Types MPP, MPR, MPG, and MP multipurpose cables and Types CMP, CMR, CMG, and CM communications cables shall be permitted to be installed in cable trays.

(e) **Support of Conductors.** Raceways shall be used for their intended purpose. Communications cables or wires shall not be strapped, taped, or attached by any means to the exterior of any conduit or raceway as a means of support.

Exception: Overhead (aerial) spans of communications cables or wires shall be permitted to be attached to the exterior of a raceway-type mast intended for the attachment and support of such conductors.

800-53. Applications of Listed Communications Wires and Cables, and Communications Raceways. Communications wires and cables shall comply with (a) through (f).

(a) **Plenum.** Cables installed in ducts, plenums, and other spaces used for environmental air shall be Type CMP. Also, listed plenum communications raceways shall be permitted to be installed in ducts and plenums as described in Section 300-22(b) and in other spaces used for environmental air as described in Section 300-22(c). Only Type CMP cable shall be permitted to be installed in these raceways.

Exception: Types CMP, CMR, CMG, CM, and CMX and communications wire installed in compliance with Section 300-22.

(b) **Riser.** Cables installed in vertical runs and penetrating more than one floor, or cables installed in vertical runs in a shaft, shall be Type CMR. Floor penetrations requiring Type CMR shall contain only cables suitable for riser or plenum use. Also, listed riser communications raceways shall be permitted to be installed in vertical riser runs in a shaft from floor to floor. Only Type CMR and CMP cables shall be permitted to be installed in these raceways.

FPN: See Section 800-52(b) for firestop requirements for floor penetrations.

Exception No. 1: Where the listed cables are encased in metal raceway or are located in a fireproof shaft having firestops at each floor.

Exception No. 2: Types CM and CMX cable in one- and two-family dwellings.

(c) Distributing Frames and Cross-Connect Arrays. Communications wires shall be used in distributing frames and cross-connect arrays.

Exception: Types CMP, CMR, CMG, and CM cables shall be permitted to be used.

(d) Other Wiring Within Buildings. Cables installed in building locations other than the locations covered in (a), (b), and (c) shall be Type CMG or Type CM. Only Type CMG, CM, CMR, or CMP cables shall be permitted to be installed in general purpose communications raceways.

Exception No. 1: Where listed communications wires and cables are enclosed in raceway of a type included in Chapter 3.

Exception No. 2: Type CMX communications cable in non-concealed spaces where the exposed length of cable does not exceed 10 ft (3.05 m).

Exception No. 3: Type CMX communications cables that are less than 0.25 in. (6.35 mm) in diameter and installed in one- or two-family dwellings.

Exception No. 4: Type CMX communications cables that are less than 0.25 in. (6.35 mm) in diameter and installed in nonconcealed spaces in multifamily dwellings.

Exception No. 5: Type CMUC under-carpet communications wires and cables installed under carpet.

(e) Hybrid Power and Communications Cable. Hybrid power and communications cable listed in accordance with Section 800-51(i) shall be permitted to be installed in one- and two-family dwellings.

(f) Cable Substitutions. Substitutions for communications cables listed in Table 800-53 shall be considered suitable for the purpose and shall be permitted.

Table 800-53. Cable Uses and Permitted Substitutions

Cable Type	Use	References	Permitted Substitutions
CMP	Communications plenum cable	800-53(a)	MPP
CMR	Communications riser cable	800-53(b)	MPP, CMP, MPR
CMG, CM	Communications general-purpose cable	800-53(d)	MPP, CMP, MPR, CMR, MPG, MP
CMX	Communications cable, limited use	800-53(d)	MPP, CMP, MPR, CMR, MPG, MP, CMG, CM

Note: See Figure 800-53, Cable substitution hierarchy.

Type CM—Communications cables
Type MP—Multipurpose cable

A → B Cable A shall be permitted to be used in place of cable B.

Figure 800-53   Cable substitution hierarchy.

Article 810 — Radio and Television Equipment

A. General

810-1. Scope. This article covers antenna systems for radio and television receiving equipment, amateur radio transmitting and receiving equipment, and certain features of transmitter safety. This article covers antennas such as multi-element, vertical rod, and dish, and also covers the wiring and cabling that connects them to equipment. This article does not cover equipment and antennas used for coupling carrier current to power line conductors.

810-2. Other Articles. Wiring from the source of power to and between devices connected to the interior wiring system shall comply with Chapters 1 through 4 other than as modified by Parts A and B of Article 640. Wiring for audio signal processing, amplification, and reproduction equipment shall comply with Article 640. Coaxial cables that connect antennas to equipment shall comply with Article 820.

810-3. Community Television Antenna. The antenna shall comply with this article. The distribution system shall comply with Article 820.

810-4. Radio Noise Suppressors. Radio interference eliminators, interference capacitors, or noise suppressors connected to power-supply leads shall be of a listed type. They shall not be exposed to physical damage.

810-5. Definitions. See Article 100.

B. Receiving Equipment — Antenna Systems

810-11. Material. Antennas and lead-in conductors shall be of hard-drawn copper, bronze, aluminum alloy, copper-clad steel, or other high-strength, corrosion-resistant material.

Exception: Soft-drawn or medium-drawn copper shall be permitted for lead-in conductors where the maximum span between points of support is less than 35 ft (10.67 m).

810-12. Supports. Outdoor antennas and lead-in conductors shall be securely supported. The antennas or lead-in conductors shall not be attached to the electric service mast. They shall not be attached to poles or similar structures carrying open electric light or power wires or trolley wires of over 250 volts between conductors. Insulators supporting the antenna conductors shall have sufficient mechanical strength to safely support the conductors. Lead-in conductors shall be securely attached to the antennas.

810-13. Avoidance of Contacts with Conductors of Other Systems. Outdoor antennas and lead-in conductors from an antenna to a building shall not cross over open conductors of electric light or power circuits and shall be kept well away from all such circuits so as to avoid the possibility of accidental contact. Where proximity to open electric light or power service conductors of less than 250 volts between conductors cannot be avoided, the installation shall be such as to provide a clearance of at least 2 ft (610 mm).

Where practicable, antenna conductors shall be installed so as not to cross under open electric light or power conductors.

810-14. Splices. Splices and joints in antenna spans shall be made mechanically secure with approved splicing devices or by such other means as will not appreciably weaken the conductors.

810-15. Grounding. Masts and metal structures supporting antennas shall be grounded in accordance with Section 810-21.

810-16. Size of Wire-Strung Antenna — Receiving Station.

(a) Size of Antenna Conductors. Outdoor antenna conductors for receiving stations shall be of a size not less than given in Table 810-16(a).

(b) Self-Supporting Antennas. Outdoor antennas, such as vertical rods, dishes, or dipole structures, shall be of corrosion-resistant materials and of strength suitable to withstand ice and wind loading conditions, and shall be located well away from overhead conductors of electric light and power circuits of over 150 volts to ground, so as to avoid the possibility of the antenna or structure falling into or making accidental contact with such circuits.

Table 810-16(a). Size of Receiving Station Outdoor Antenna Conductors

Material	Minimum Size of Conductors Where Maximum Open Span Length Is		
	Less than 35 ft	35 ft to 150 ft	Over 150 ft
Aluminum alloy, hard-drawn copper	19	14	12
Copper-clad steel, bronze, or other high-strength material	20	17	14

Note: For SI units, 1 ft = 0.3048 m.

810-17. Size of Lead-in — Receiving Station. Lead-in conductors from outside antennas for receiving stations shall, for various maximum open span lengths, be of such size as to have a tensile strength at least as great as that of the conductors for antennas as specified in Section 810-16. Where the lead-in consists of two or more conductors that are twisted together, are enclosed in the same covering, or are concentric, the conductor size shall, for various maximum open span lengths, be such that the tensile strength of the combination will be at least as great as that of the conductors for antennas as specified in Section 810-16.

810-18. Clearances — Receiving Stations.

(a) Outside of Buildings. Lead-in conductors attached to buildings shall be installed so that they cannot swing closer than 2 ft (610 mm) to the conductors of circuits of 250 volts or less between conductors, or 10 ft (3.05 m) to the conductors of circuits of over 250 volts between conductors, except that in the case of circuits not over 150 volts between conductors, where all conductors involved are supported so as to ensure permanent separation, the clearance shall be permitted to be reduced but shall not be less than 4 in. (102 mm). The clearance between lead-in conductors and any conductor forming a part of a lightning rod system shall not be less than 6 ft (1.83 m) unless the bonding referred to in Section 250-60 is accomplished. Underground conductors shall be separated at least 12 in. (305 mm) from conductors of any light or power circuits or Class 1 circuits.

Exception: Where the electric light or power conductors, Class 1 conductors, or lead-in conductors are installed in raceways or metal cable armor.

(b) Antennas and Lead-ins — Indoors. Indoor antennas and indoor lead-ins shall not be run nearer than 2 in. (50.8 mm) to conductors of other wiring systems in the premises.

Exception No. 1: Where such other conductors are in metal raceways or cable armor.

Exception No. 2: Where permanently separated from such other conductors by a continuous and firmly fixed nonconductor, such as porcelain tubes or flexible tubing.

(c) In Boxes or Other Enclosures. Indoor antennas and indoor lead-ins shall be permitted to occupy the same box or enclosure with conductors of other wiring systems where separated from such other conductors by an effective permanently installed barrier.

810-19. Electric Supply Circuits Used in Lieu of Antenna — Receiving Stations. Where an electric supply circuit is used in lieu of an antenna, the device by which the radio receiving set is connected to the supply circuit shall be listed.

810-20. Antenna Discharge Units — Receiving Stations.

(a) Where Required. Each conductor of a lead-in from an outdoor antenna shall be provided with a listed antenna discharge unit.

Exception: Where the lead-in conductors are enclosed in a continuous metallic shield that is either permanently and effectively grounded or is protected by an antenna discharge unit.

(b) Location. Antenna discharge units shall be located outside the building or inside the building between the point of entrance of the lead-in and the radio set or transformers, and as near as practicable to the entrance of the conductors to the building. The antenna discharge unit shall not be located near combustible material or in a hazardous (classified) location as defined in Article 500.

(c) Grounding. The antenna discharge unit shall be grounded in accordance with Section 810-21.

810-21. Grounding Conductors — Receiving Stations. Grounding conductors shall comply with (a) through (j).

(a) Material. The grounding conductor shall be of copper, aluminum, copper-clad steel, bronze, or similar corrosion-resistant material. Aluminum or copper-clad aluminum grounding conductors shall not be used where in direct contact with masonry or the earth or where subject to corrosive conditions. Where used outside, aluminum or copper-clad aluminum shall not be installed within 18 in. (457 mm) of the earth.

(b) Insulation. Insulation on grounding conductors shall not be required.

(c) Supports. The grounding conductors shall be securely fastened in place and shall be permitted to be directly attached to the surface wired over without the use of insulating supports.

Exception: Where proper support cannot be provided, the size of the grounding conductors shall be increased proportionately.

(d) Mechanical Protection. The grounding conductor shall be protected where exposed to physical damage, or the size of the grounding conductors shall be increased proportionately to compensate for the lack of protection. Where the grounding conductor is run in a metal raceway, both ends of the raceway shall be bonded to the grounding conductor or to the same terminal or electrode to which the grounding conductor is connected.

(e) Run in Straight Line. The grounding conductor for an antenna mast or antenna discharge unit shall be run in as straight a line as practicable from the mast or discharge unit to the grounding electrode.

(f) Electrode. The grounding conductor shall be connected as follows.

(1) To the nearest accessible location on the following:

(a) The building or structure grounding electrode system as covered in Section 250-50
(b) The grounded interior metal water piping system as covered in Section 250-104(a)
(c) The power service accessible means external to enclosures as covered in Section 250-92(b)
(d) The metallic power service raceway
(e) The service equipment enclosure, or
(f) The grounding electrode conductor or the grounding electrode conductor metal enclosures; or

(2) If the building or structure served has no grounding means, as described in (f)(1), to any one of the individual electrodes described in Section 250-50; or

(3) If the building or structure served has no grounding means, as described in (f)(1) or (f)(2), to an effectively grounded metal structure or to any of the individual electrodes described in Section 250-52.

(g) Inside or Outside Building. The grounding conductor shall be permitted to be run either inside or outside the building.

(h) Size. The grounding conductor shall not be smaller than No. 10 copper, No. 8 aluminum, or No. 17 copper-clad steel or bronze.

(i) Common Ground. A single grounding conductor shall be permitted for both protective and operating purposes.

(j) Bonding of Electrodes. A bonding jumper not smaller than No. 6 copper or equivalent shall be connected between the radio and television equipment grounding electrode and the power grounding electrode system at the building or structure served where separate electrodes are used.

C. Amateur Transmitting and Receiving Stations — Antenna Systems

810-51. Other Sections. In addition to complying with Part C, antenna systems for amateur transmitting and receiving stations shall also comply with Sections 810-11 through 810-15.

810-52. Size of Antenna. Antenna conductors for transmitting and receiving stations shall be of a size not less than given in Table 810-52.

Table 810-52. Size of Amateur Station Outdoor Antenna Conductors

	Minimum Size of Conductors Where Maximum Open Span Length Is	
Material	Less than 150 ft	Over 150 ft
Hard-drawn copper	14	10
Copper-clad steel, bronze, or other high-strength material	14	12

Note: For SI units, 1 ft = 0.3048 m.

810-53. Size of Lead-in Conductors. Lead-in conductors for transmitting stations shall, for various maximum span lengths, be of a size at least as great as that of conductors for antennas as specified in Section 810-52.

810-54. Clearance on Building. Antenna conductors for transmitting stations, attached to buildings, shall be firmly mounted at least 3 in. (76 mm) clear of the surface of the building on nonabsorbent insulating supports, such as treated pins or brackets equipped with insulators having not less than 3-in. (76-mm) creepage and airgap distances. Lead-in conductors attached to buildings shall also comply with these requirements.

Exception: Where the lead-in conductors are enclosed in a continuous metallic shield that is permanently and effectively grounded, they shall not be required to comply with these requirements. Where grounded, the metallic shield shall also be permitted to be used as a conductor.

810-55. Entrance to Building. Except where protected with a continuous metallic shield that is permanently and effectively grounded, lead-in conductors for transmitting stations shall enter buildings by one of the following methods:

(1) Through a rigid, noncombustible, nonabsorbent insulating tube or bushing
(2) Through an opening provided for the purpose in which the entrance conductors are firmly secured so as to provide a clearance of at least 2 in. (50.8 mm), or
(3) Through a drilled window pane

810-56. Protection Against Accidental Contact. Lead-in conductors to radio transmitters shall be located or installed so as to make accidental contact with them difficult.

810-57. Antenna Discharge Units — Transmitting Stations. Each conductor of a lead-in for outdoor antennas shall be provided with an antenna discharge unit or other suitable means that will drain static charges from the antenna system.

Exception No. 1: Where protected by a continuous metallic shield that is permanently and effectively grounded.

Exception No. 2: Where the antenna is permanently and effectively grounded.

810-58. Grounding Conductors — Amateur Transmitting and Receiving Stations. Grounding conductors shall comply with (a) through (c).

(a) Other Sections. All grounding conductors for amateur transmitting and receiving stations shall comply with Sections 810-21(a) through (j).

(b) Size of Protective Grounding Conductor. The protective grounding conductor for transmitting stations shall be as large as the lead-in, but not smaller than No. 10 copper, bronze, or copper-clad steel.

(c) Size of Operating Grounding Conductor. The operating grounding conductor for transmitting stations shall not be less than No. 14 copper or its equivalent.

D. Interior Installation — Transmitting Stations

810-70. Clearance from Other Conductors. All conductors inside the building shall be separated at least 4 in.

(102 mm) from the conductors of any electric light, power, or signaling circuit.

Exception No. 1: As provided in Article 640.

Exception No. 2: Where separated from other conductors by raceway or some firmly fixed nonconductor, such as porcelain tubes or flexible tubing.

810-71. General. Transmitters shall comply with (a) through (c).

(a) Enclosing. The transmitter shall be enclosed in a metal frame or grille, or separated from the operating space by a barrier or other equivalent means, all metallic parts of which are effectively connected to ground.

(b) Grounding of Controls. All external metal handles and controls accessible to the operating personnel shall be effectively grounded.

(c) Interlocks on Doors. All access doors shall be provided with interlocks that will disconnect all voltages of over 350 volts between conductors when any access door is opened.

Article 820 — Community Antenna Television and Radio Distribution Systems

A. General

820-1. Scope. This article covers coaxial cable distribution of radio frequency signals typically employed in community antenna television (CATV) systems.

820-2. Definitions. See Article 100. For the purposes of this article, the following additional definitions apply.

Exposed. An exposed cable is one that is in such a position that, in case of failure of supports and insulation, contact with another circuit may result.

FPN: See Article 100 for two other definitions of *Exposed.*

Point of Entrance. The point within a building at which the cable emerges from an external wall, from a concrete floor slab, or from a rigid metal conduit or an intermediate metal conduit grounded to an electrode in accordance with Section 820-40(b).

Premises. The land and buildings of a user located on the user side of utility-user network point of demarcation.

820-3. Locations and Other Articles. Circuits and equipment shall comply with (a) through (g).

(a) Spread of Fire or Products of Combustion. Section 300-21.

(b) Ducts, Plenums, and Other Air-Handling Spaces. Section 300-22, where installed in ducts or plenums or other spaces used for environmental air.

Exception: As permitted in Section 820-53(a).

(c) Installation and Use. Section 110-3(b) shall apply.

(d) Installations of Conductive and Nonconductive Optical Fiber Cables. Article 770.

(e) Communications Circuits. Article 800.

(f) Network-Powered Broadband Communications Systems. Article 830.

(g) Alternate Wiring Methods. The wiring methods of Article 830 shall be permitted to substitute for the wiring methods of Article 820.

FPN: Use of Article 830 wiring methods will facilitate the upgrading of Article 820 installations to network-powered broadband applications.

820-4. Energy Limitations. The coaxial cable shall be permitted to deliver low-energy power to equipment that is directly associated with the radio frequency distribution system if the voltage is not over 60 volts and if the current supply is from a transformer or other device that has energy-limiting characteristics.

820-5. Access to Electrical Equipment Behind Panels Designed to Allow Access. Access to equipment shall not be denied by an accumulation of cables that prevents removal of panels, including suspended ceiling panels.

820-6. Mechanical Execution of Work. Community antenna television and radio distribution systems shall be installed in a neat and workmanlike manner. Cables shall be supported by the building structure in such a manner that the cable will not be damaged by normal building use.

FPN: One way to determine accepted industry practice is to refer to nationally recognized standards such as *Commercial Building Telecommunications Cabling Standard*, ANSI/EIA/TIA 568-A-1995; *Commercial Building Standard for Telecommunications Pathways and Spaces*, ANSI/EIA/TIA 569-1990; and *Residential and Light Commercial Telecommunications Wiring Standard*, ANSI/EIA/TIA 570-1991.

B. Cables Outside and Entering Buildings

820-10. Outside Cables. Coaxial cables, prior to the point of grounding, as defined in Section 820-33, shall comply with (a) through (f).

(a) On Poles. Where practicable, conductors on poles shall be located below the electric light, power, Class 1, or non-power-limited fire alarm circuit conductors and shall not be attached to a crossarm that carries electric light or power conductors.

(b) Lead-in Clearance. Lead-in or aerial-drop cables from a pole or other support, including the point of initial attachment to a building or structure, shall be kept away from electric light, power, Class 1, or nonpower-limited fire alarm circuit conductors so as to avoid the possibility of accidental contact.

Exception: Where proximity to electric light, power, Class 1, or nonpower-limited fire alarm circuit service conductors cannot be avoided, the installation shall be such as to provide clearances of not less than 12 in. (305 mm) from light, power, Class 1, or nonpower-limited fire alarm circuit service drops.

(c) On Masts. Aerial cable shall be permitted to be attached to an above-the-roof raceway mast that does not enclose or support conductors of electric light or power circuits.

(d) Above Roofs. Cables shall have a vertical clearance of not less than 8 ft (2.44 m) from all points of roofs above which they pass.

Exception No. 1: Auxiliary buildings such as garages and the like.

Exception No. 2: A reduction in clearance above only the overhanging portion of the roof to not less than 18 in. (457 mm) shall be permitted if (1) not more than 4 ft (1.22 m) of communications service drop conductors pass above the roof overhang, and (2) they are terminated at a raceway mast or other approved support.

Exception No. 3: Where the roof has a slope of not less than 4 in. (102 mm) in 12 in. (305 mm), a reduction in clearance to not less than 3 ft (914 mm) shall be permitted.

(e) Between Buildings. Cables extending between buildings and also the supports or attachment fixtures shall be acceptable for the purpose and shall have sufficient strength to withstand the loads to which they may be subjected.

Exception: Where a cable does not have sufficient strength to be self-supporting, it shall be attached to a supporting messenger cable that, together with the attachment fixtures or supports, shall be acceptable for the purpose and shall have sufficient strength to withstand the loads to which they may be subjected.

(f) On Buildings. Where attached to buildings, cables shall be securely fastened in such a manner that they will be separated from other conductors as follows.

(1) Electric Light or Power. The coaxial cable shall have a separation of at least 4 in. (102 mm) from electric light, power, Class 1, or nonpower-limited fire alarm circuit conductors not in raceway or cable, or be permanently separated from conductors of the other system by a continuous and firmly fixed nonconductor in addition to the insulation on the wires.

(2) Other Communications Systems. Coaxial cable shall be installed so that there will be no unnecessary interference in the maintenance of the separate systems. In no case shall the conductors, cables, messenger strand, or equipment of one system cause abrasion to the conductors, cable, messenger strand, or equipment of any other system.

(3) Lightning Conductors. Where practicable, a separation of at least 6 ft (1.83 m) shall be maintained between any coaxial cable and lightning conductors.

820-11. Entering Buildings.

(a) Underground Systems. Underground coaxial cables in a duct, pedestal, handhole, or manhole that contains electric light or power conductors or Class 1 circuits shall be in a section permanently separated from such conductors by means of a suitable barrier.

(b) Direct-Buried Cables and Raceways. Direct-buried coaxial cable shall be separated at least 12 in. (305 mm) from conductors of any light or power or Class 1 circuit.

Exception No. 1: Where electric service conductors or coaxial cables are installed in raceways or have metal cable armor.

Exception No. 2: Where electric light or power branch-circuit or feeder conductors or Class 1 circuit conductors are installed in a raceway or in metal-sheathed, metal-clad, or Type UF or Type USE cables; or the coaxial cables have metal cable armor or are installed in a raceway.

C. Protection

820-33. Grounding of Outer Conductive Shield of a Coaxial Cable. Where coaxial cable is exposed to lightning or to accidental contact with lightning arrester conductors or power conductors operating at a voltage of over 300 volts to ground, the outer conductive shield of the coaxial cable shall be grounded at the building premises as close to the point of cable entry as practicable. For purposes of this section, the point at which the exposed cable enters shall be considered to be the point of emergence through an exterior wall, a concrete floor slab, or from a rigid or intermediate metal conduit grounded to an electrode in accordance with Section 820-40(b).

For purposes of this section, grounding located at mobile home service equipment located in sight from and not more than 30 ft (9.14 m) from the exterior wall of the mobile home it serves, or at a mobile home disconnecting means grounded in accordance with Section 250-32 and located in sight from and not more than 30 ft (9.14 m) from the exterior wall of the mobile home it serves, shall be considered to meet the requirements of this section.

FPN: Selecting a grounding location to achieve the shortest practicable grounding conductor will help limit potential differences between CATV and other metallic systems.

(a) Shield Grounding. Where the outer conductive shield of a coaxial cable is grounded, no other protective devices shall be required.

(b) Shield Protection Devices. Grounding of a coaxial drop cable shield by means of a protective device that does not interrupt the grounding system within the premises shall be permitted.

D. Grounding Methods

820-40. Cable Grounding. Where required by Section 820-33, the shield of the coaxial cable shall be grounded as specified in (a) through (d).

(a) Grounding Conductor.

(1) Insulation. The grounding conductor shall be insulated and shall be listed as suitable for the purpose.

(2) Material. The grounding conductor shall be copper or other corrosion-resistant conductive material, stranded or solid.

(3) Size. The grounding conductor shall not be smaller than No. 14. It shall have a current-carrying capacity approximately equal to that of the outer conductor of the coaxial cable.

(4) Run in Straight Line. The grounding conductor shall be run to the grounding electrode in as straight a line as practicable.

(5) Physical Protection. Where subject to physical damage, the grounding conductor shall be adequately protected. Where the grounding conductor is run in a metal raceway, both ends of the raceway shall be bonded to the grounding conductor or the same terminal or electrode to which the grounding conductor is connected.

(b) Electrode. The grounding conductor shall be connected as follows:

(1) To the nearest accessible location on the following:

(a) The building or structure grounding electrode system as covered in Section 250-50
(b) The grounded interior metal water piping system as covered in Section 250-104(a)

(c) The power service accessible means external to enclosures as covered in Section 250-92(b)
(d) The metallic power service raceway
(e) The service equipment enclosure
(f) The grounding electrode conductor or the grounding electrode conductor metal enclosure, or
(g) To the grounding conductor or to the grounding electrode of a building or structure disconnecting means that is grounded to an electrode as covered in Section 250-32; or

(2) If the building or structure served has no grounding means as described in (b)(1), to any one of the individual electrodes described in Section 250-50; or

(3) If the building or structure served has no grounding means as described in (b)(1) or (b)(2), to an effectively grounded metal structure or to any one of the individual electrodes described in Section 250-52.

(c) Electrode Connection. Connections to grounding electrodes shall comply with Section 250-70.

(d) Bonding of Electrodes. A bonding jumper not smaller than No. 6 copper or equivalent shall be connected between the antenna systems grounding electrode and the power grounding electrode system at the building or structure served where separate electrodes are used.

Exception: At mobile homes as covered in Section 820-42.

FPN No. 1: See Section 250-60 for use of air terminals (lightning rods).

FPN No. 2: Bonding together of all separate electrodes will limit potential differences between them and between their associated wiring systems.

820-41. Equipment Grounding. Unpowered equipment and enclosures or equipment powered by the coaxial cable shall be considered grounded where connected to the metallic cable shield.

820-42. Bonding and Grounding at Mobile Homes.

(a) Grounding. Where there is no mobile home service equipment located in sight from and not more than 30 ft (9.14 m) from the exterior wall of the mobile home it serves, or there is no mobile home disconnecting means grounded in accordance with Section 250-32 and located within sight from and not more than 30 ft (9.14 m) from the exterior wall of the mobile home it serves, the coaxial cable shield ground, or surge arrester ground, shall be in accordance with Sections 820-40(b)(2) and (3).

(b) Bonding. The coaxial cable shield grounding terminal, surge arrester grounding terminal, or grounding electrode shall be bonded to the metal frame or available grounding terminal of the mobile home with a copper grounding conductor not smaller than No. 12 under any of the following conditions:

(1) Where there is no mobile home service equipment or disconnecting means as in (a), or
(2) Where the mobile home is supplied by cord and plug

E. Cables Within Buildings

820-49. Fire Resistance of CATV Cables. Coaxial cables installed as wiring within buildings shall be listed as being resistant to the spread of fire in accordance with Sections 820-50 and 820-51.

820-50. Listing, Marking, and Installation of Coaxial Cables. Coaxial cables in a building shall be listed as being suitable for the purpose, and cables shall be marked in accordance with Table 820-50. The cable voltage rating shall not be marked on the cable.

FPN: Voltage markings on cables may be misinterpreted to suggest that the cables may be suitable for Class 1, electric light, and power applications.

Exception No. 1: Voltage markings shall be permitted where the cable has multiple listings and voltage marking is required for one or more of the listings.

Exception No. 2: Listing and marking shall not be required where the cable enters the building from the outside and is run in rigid metal conduit or intermediate metal conduit, and such conduits are grounded to an electrode in accordance with Section 820-40(b).

Exception No. 3: Listing and marking shall not be required where the length of the cable within the building, measured from its point of entrance, does not exceed 50 ft (15.2 m) and the cable enters the building from the outside and is terminated at a grounding block.

Table 820-50. Cable Markings

Cable Marking	Type	Reference
CATVP	CATV plenum cable	820-51(a) and 820-53(a)
CATVR	CATV riser cable	820-51(b) and 820-53(b)
CATV	CATV cable	820-51(c) and 820-53(c)
CATVX	CATV cable, limited use	820-51(d) and 820-53(c), Exception Nos. 1, 2, 3, and 4

FPN No. 1: Cable types are listed in descending order of fire-resistance rating.

FPN No. 2: See the referenced sections for listing requirements and permitted uses.

820-51. Additional Listing Requirements. Cables shall be listed in accordance with (a) through (d).

(a) Type CATVP. Type CATVP community antenna television plenum cable shall be listed as being suitable for use in ducts, plenums, and other spaces used for environmental air and shall also be listed as having adequate fire-resistant and low smoke-producing characteristics.

FPN: One method of defining low smoke-producing cables is by establishing an acceptable value of the smoke produced when tested in accordance with the *Standard Method for Test for Fire and Smoke Characteristics of Wires and Cables,* NFPA 262 -1994, to a maximum peak optical density of 0.5 and a maximum average optical density of 0.15. Similarly, one method of defining fire-resistant cables is by establishing maximum allowable flame travel distance of 5 ft (1.52 m) when tested in accordance with the same test.

(b) Type CATVR. Type CATVR community antenna television riser cable shall be listed as being suitable for use in a vertical run in a shaft or from floor to floor and shall also be listed as having fire-resistant characteristics capable of preventing the carrying of fire from floor to floor.

FPN: One method of defining fire-resistant characteristics capable of preventing the carrying of fire from floor to floor is that the cables pass the requirements of the *Standard Test for Flame Propagation Height of Electrical and Optical-Fiber Cable Installed Vertically in Shafts,* ANSI/UL 1666-1997.

(c) Type CATV. Type CATV community antenna television cable shall be listed as being suitable for general purpose CATV use, with the exception of risers and plenums, and shall also be listed as being resistant to the spread of fire.

FPN: One method of defining resistant to the spread of fire is that the cables do not spread fire to the top of the tray in the vertical-tray flame test in the *Reference Standard for Electrical Wires, Cables and Flexible Cords,* ANSI/UL 1581-1991.
   Another method of defining resistant to the spread of fire is for the damage (char length) not to exceed 4 ft 11 in. (1.5 m) when performing the vertical flame test — cables in cable trays, as described in *Test Methods for Electrical Wires and Cables,* CSA C22.2 No. 0.3-M-1985.

(d) Type CATVX. Type CATVX limited-use community antenna television cable shall be listed as being suitable for use in dwellings and for use in raceway and shall also be listed as being resistant to flame spread.

FPN: One method of determining that cable is resistant to flame spread is by testing the cable to the VW-1 (vertical-wire) flame test in the *Reference Standard for Electrical Wires, Cables and Flexible Cords,* ANSI/UL 1581-1991.

820-52. Installation of Cables and Equipment. Beyond the point of grounding, as defined in Section 820-33, the cable installation shall comply with (a) through (e).

(a) Separation from Other Conductors.

(1) In Raceways and Boxes.

(a) *Other Circuits.* Coaxial cables shall be permitted in the same raceway or enclosure with jacketed cables of any of the following:

(1) Class 2 and Class 3 remote-control, signaling, and power-limited circuits in compliance with Article 725
(2) Power-limited fire alarm systems in compliance with Article 760
(3) Communications circuits in compliance with Article 800
(4) Nonconductive and conductive optical fiber cables in compliance with Article 770
(5) Low power network-powered broadband communications circuits in compliance with Article 830

(b) *Electric Light, Power, Class 1, Nonpower-Limited Fire Alarm, and Medium Power Network-Powered Broadband Communications Circuits.* Coaxial cable shall not be placed in any raceway, compartment, outlet box, junction box, or other enclosures with conductors of electric light, power, Class 1, nonpower-limited fire alarm, or medium power network-powered broadband communications circuits.

Exception No. 1: Where all of the conductors of electric light, power, Class 1, nonpower-limited fire alarm, and medium power network-powered broadband communications circuits are separated from all of the coaxial cables by a barrier.

Exception No. 2: Power circuit conductors in outlet boxes, junction boxes, or similar fittings or compartments where such conductors are introduced solely for power supply to the coaxial cable system distribution equipment. The power circuit conductors shall be routed within the enclosure to maintain a minimum 0.25-in. (6.35-mm) separation from coaxial cables.

(2) Other Applications. Coaxial cable shall be separated at least 2 in. (50.8 mm) from conductors of any electric light, power, Class 1, nonpower-limited fire alarm, or medium power network-powered broadband communications circuits.

Exception No. 1: Where either (1) all of the conductors of electric light, power, Class 1, nonpower-limited fire alarm, and medium power network-powered broadband communications and circuits are in a raceway, or in metal-sheathed, metal-clad, nonmetallic-sheathed, Type AC, or Type UF cables, or (2) all of the coaxial cables are encased in raceway.

Exception No. 2: Where the coaxial cables are permanently separated from the conductors of electric light, power, Class

1, nonpower-limited fire alarm, and medium power network-powered broadband communications circuits by a continuous and firmly fixed nonconductor, such as porcelain tubes or flexible tubing, in addition to the insulation on the wire.

(b) Spread of Fire or Products of Combustion. Installations in hollow spaces, vertical shafts, and ventilation or air-handling ducts shall be made so that the possible spread of fire or products of combustion will not be substantially increased. Openings around penetrations through fire resistance-rated walls, partitions, floors, or ceilings shall be firestopped using approved methods.

(c) Equipment in Other Space Used for Environmental Air. Section 300-22(c) shall apply.

(d) Hybrid Power and Coaxial Cabling. The provisions of Section 780-6 shall apply for listed hybrid power and coaxial cabling in closed-loop and programmed power distribution.

(e) Support of Conductors. Raceways shall not be used as a means of support for coaxial cables.

820-53. Applications of Listed CATV Cables. CATV cables shall comply with (a) through (d).

(a) Plenum. Cables installed in ducts, plenums, and other spaces used for environmental air shall be Type CATVP.

Exception: Types CATVP, CATVR, CATV, and CATVX cables installed in compliance with Section 300-22.

(b) Riser. Cables installed in vertical runs and penetrating more than one floor, or cables installed in vertical runs in a shaft, shall be Type CATVR. Floor penetrations requiring Type CATVR shall contain only cables suitable for riser or plenum use.

Exception No. 1: Types CATV and CATVX cables encased in metal raceway or located in a fireproof shaft having firestops at each floor.

Exception No. 2: Types CATV and CATVX cables in one- and two-family dwellings.

FPN: See Section 820-52(b) for the firestop requirements for floor penetrations.

(c) Other Wiring Within Buildings. Cables installed in building locations other than the locations covered in (a) and (b) shall be Type CATV.

Exception No. 1: Type CATVX cable enclosed in raceway.

Exception No. 2: Type CATVX cable in nonconcealed spaces where the exposed length of cable does not exceed 10 ft (3.05 m).

Exception No. 3: Type CATVX cables that are less than 0.375 in. (9.52 mm) in diameter and installed in one- or two-family dwellings.

Exception No. 4: Type CATVX cables that are less than 0.375 in. (9.52 mm) in diameter and installed in nonconcealed spaces in multifamily dwellings.

(d) Cable Substitutions. The substitutions for community antenna television cables listed in Table 820-53 shall be considered suitable for the purpose and shall be permitted.

Table 820-53. Coaxial Cable Uses and Permitted Substitutions

Cable Type	Use	References	Permitted Substitutions
CATVP	Coaxial plenum cable	820-53(a)	CMP
CATVR	Coaxial riser cable	820-53(b)	CATVP, CMP, CMR
CATV	Coaxial general-purpose cable	820-53(c)	CATVP, CMP, CATVR, CMR, CMG, CM
CATVX	Coaxial cable, limited use	820-53(c)	CATVP, CMP, CATVR, CMR, CATV, CMG, CM

Note: See Figure 820-53, Cable substitution hierarchy.

FPN: The substitute cables in Table 820-53 are only coaxial-type cables.

Type CATV—Community antenna television cables
Type CM—Communications cables

A → B Coaxial cable A shall be permitted to be used in place of coaxial cable B.

Figure 820-53   Cable substitution hierarchy.

Article 830 — Network-Powered Broadband Communications Systems

A. General

830-1. Scope. This article covers network-powered broadband communications systems that provide any combination of voice, audio, video, data, and interactive services through a network interface unit.

FPN No. 1: A typical basic system configuration includes a cable supplying power and broadband signal to a network interface unit that converts the broadband signal to the component signals. Typical cables are coaxial cable with both broadband signal and power on the center conductor, composite metallic cable with a coaxial member for the broadband signal and a twisted pair for power, and composite optical fiber cable with a pair of conductors for power. Larger systems may also include network components such as amplifiers that require network power.

FPN No. 2: See Section 90-2(b)(4) for installations of broadband communications systems that are not covered.

830-2. Definitions. See Article 100. For purposes of this article, the following additional definitions apply.

Block. A square or portion of a city, town, or village enclosed by streets and including the alleys so enclosed, but not any street.

Exposed to Accidental Contact with Electrical Light or Power Conductors. The circuit is in such a position that, in case of failure of supports or insulation, contact with another circuit may result.

Fault Protection Device. An electronic device intended for the protection of personnel that functions under fault conditions, such as network-powered broadband communications cable short or open circuit, to limit the current or voltage, or both, for a low power network-powered broadband communications circuit and provide acceptable protection from electric shock.

Network Interface Unit (NIU). A device that converts a broadband signal into component voice, audio, video, data, and interactive services signals. The NIU provides isolation between the network power and the premises signal circuits. The NIU may also contain primary and secondary protectors.

Network-Powered Broadband Communications Circuit. The circuit extending from the communications utility's serving terminal or tap up to and including the NIU.

FPN: A typical single-family network-powered communications circuit consists of a communications drop or communications service cable and an NIU, and includes the communications utility's serving terminal or tap where it is not under the exclusive control of the communications utility.

Point of Entrance. The point within a building at which the cable emerges from an external wall, from a concrete floor slab, or from a rigid metal conduit or an intermediate metal conduit grounded to an electrode in accordance with Section 830-40(b).

Premises Wiring. The circuits located on the user side of the network interface unit.

830-3. Locations and Other Articles. Circuits and equipment shall comply with (a) through (d).

(a) Spread of Fire or Products of Combustion. Section 300-21.

(b) Ducts, Plenums, and Other Air-Handling Spaces. Section 300-22, where installed in ducts or plenums or other spaces used for environmental air.

Exception: As permitted in Section 830-55(a).

(c) Installation and Use. Section 110-3(b).

(d) Output Circuits. As appropriate for the services provided, the output circuits derived from the network interface unit shall comply with the requirements of the following:

(1) Installations of communications circuits — Article 800
(2) Installations of community antenna television and radio distribution circuits — Article 820

Exception: Section 830-30(b)(3) shall apply where protection is provided in the output of the NIU.

(3) Installations of optical fiber cables — Article 770
(4) Installations of Class 2 and Class 3 circuits — Article 725
(5) Installations of power-limited fire alarm circuits — Article 760

830-4. Power Limitations. Network-powered broadband communications systems shall be classified as having low or medium power sources as defined in Table 830-4.

830-5. Network-Powered Broadband Communications Equipment and Cables. Network-powered broadband communications equipment and cables shall be listed as suitable for the purpose.

Exception No. 1: This listing requirement shall not apply to community antenna television and radio distribution system coaxial cables that were installed prior to January 1, 2000, in accordance with Article 820 and are used for low power network-powered broadband communications circuits. See Section 830-9.

Exception No. 2: Substitute cables for network-powered broadband communications cables shall be permitted as shown in Table 830-58.

(a) Listing and Marking. Listing and marking of network-powered broadband communications cables shall comply with (1) or (2).

(1) Type BMU, Type BM, and Type BMR Cables. Network-powered broadband communications medium power underground cable, Type BMU; network-powered

Table 830-4. Limitations for Network-Powered Broadband Communications Systems

Network Power Source	Low	Medium
Circuit voltage, V_{max} (volts) (Note 1)	0–100	0–150
Power limitation, VA_{max} (volt-amperes) (Note 1)	250	250
Current limitation, I_{max} (amperes) (Note 1)	$1000/V_{max}$	$1000/V_{max}$
Maximum power rating (volt-amperes)	100	100
Maximum voltage rating (volts)	100	150
Maximum overcurrent protection (amperes) (Note 2)	$100/V_{max}$	NA

Notes:

1. V_{max}, I_{max}, and VA_{max} are determined with the current-limiting impedance in the circuit (not bypassed) as follows:

V_{max}—Maximum system voltage regardless of load with rated input applied.

I_{max}—Maximum system current under any noncapacitive load, including short circuit, and with overcurrent protection bypassed if used. I_{max} limits apply after 1 minute of operation.

VA_{max}—Maximum volt-ampere output after 1 minute of operation regardless of load and overcurrent protection bypassed if used.

2. Overcurrent protection is not required where the current-limiting device provides equivalent current limitation and the current-limiting device does not reset until power or the load is removed.

broadband communications medium power cable, Type BM; and network-powered broadband communications medium power riser cable, Type BMR, shall be factory-assembled cables consisting of a jacketed coaxial cable, a jacketed combination of coaxial cable and multiple individual conductors, or a jacketed combination of an optical fiber cable and multiple individual conductors. The insulation for the individual conductors shall be rated for 300 volts minimum. Cables intended for outdoor use shall be listed as suitable for the application. Cables shall be marked in accordance with Section 310-11. Type BMU cables shall be jacketed and listed as being suitable for outdoor underground use. Type BM cables shall be listed as being suitable for general purpose use, with the exception of risers and plenums, and shall also be listed as being resistant to the spread of fire. Type BMR cables shall be listed as being suitable for use in a vertical run in a shaft or from floor to floor and shall also be listed as having fire-resistant characteristics capable of preventing the carrying of fire from floor to floor.

FPN No. 1: One method of defining resistant to spread of fire is that the cables do not spread fire to the top of the tray in the vertical tray flame test in the *Reference Standard for Electrical Wires, Cables and Flexible Cords,* ANSI/UL 1581-1991. Another method of defining resistant to the spread of fire is for the damage (char length) not to exceed

4 ft. 11 in. (1.5 m) when performing the CSA vertical flame test for cables in cable trays, as described in *Test Methods for Electrical Wires and Cables,* CSA C22.2 No. 0.3-M-1985.

FPN No. 2: One method of defining fire-resistant characteristics capable of preventing the carrying of fire from floor to floor is that the cables pass the requirements of the *Standard Test for Flame Propagation Height of Electrical and Optical-Fiber Cable Installed Vertically in Shafts,* ANSI/UL 1666-1997.

(2) Type BLU, Type BLX, and Type BLP Cables. Network-powered broadband communications low power underground cable, Type BLU; limited use network-powered broadband communications low power cable, Type BLX; and network-powered broadband communications low power plenum cable, Type BLP, shall be factory assembled cables consisting of a jacketed coaxial cable, a jacketed combination of coaxial cable and multiple individual conductors, or a jacketed combination of an optical fiber cable and multiple individual conductors. The insulation for the individual conductors shall be rated for 300 volts minimum. Cables intended for outdoor use shall be listed as suitable for the application. Cables shall be marked in accordance with Section 310-11. Type BLU cables shall be jacketed and listed as being suitable for outdoor underground use. Type BLX limited-use cables shall be listed as being suitable for use outside, for use in dwellings, and for use in raceways, and shall also be listed as being flame retardant. Type BLP cables shall be listed as being suitable for use in ducts, plenums, and other spaces for environmental air and shall also be listed as having adequate fire-resistant and low smoke-producing characteristics.

FPN No. 1: One method of determining that cable is flame retardant is by testing the cable to VW-1 (vertical-wire) flame test in the *Reference Standard for Electrical Wires, Cables and Flexible Cords,* ANSI/UL 1581-1991.

FPN No. 2: One method of defining low smoke-producing cable is by establishing an acceptable value of the smoke produced when tested in accordance with *Standard Method of Test for Fire and Smoke Characteristics of Wires and Cables,* NFPA 262-1994, to a maximum peak optical density of 0.5 and a maximum average optical density of 0.15. Similarly, one method of defining fire-resistant cables is by establishing maximum allowable flame travel distance of 5 ft (1.52 m) when tested in accordance with the same test.

830-6. Access to Electrical Equipment Behind Panels Designed to Allow Access. Access to equipment shall not be denied by an accumulation of cables that prevents removal of panels, including suspended ceiling panels.

830-7. Mechanical Execution of Work. Network-powered broadband communications circuits and equipment shall be installed in a neat and workmanlike manner. Cables shall be supported by the building structure in such a manner that the cable will not be damaged by normal building use.

FPN: One way to determine accepted industry practice is to refer to nationally recognized standards such as *Commercial Building Telecommunications Cabling Standard,* ANSI/EIA/TIA 568-A-1995; *Commercial Building Standard for Telecommunications Pathways and Spaces,* ANSI/TIA/EIA 569-A-1997; and *Residential and Light Commercial Telecommunications Wiring Standard,* ANSI/EIA/TIA 570-1991, or other ANSI-approved installation standards.

830-8. Hazardous (Classified) Locations. Network-powered broadband communications circuits and equipment installed in a location that is classified in accordance with Article 500 shall comply with the applicable requirements of Chapter 5.

B. Cables Outside and Entering Buildings

830-9. Entrance Cables. Cables installed outdoors shall be listed as suitable for the application. In addition, network-powered broadband communications cables located outside and entering buildings shall comply with (a) and (b).

(a) Medium Power Circuits. Medium power network-powered broadband communications circuits located outside and entering buildings shall be installed using Type BMU, Type BM, or Type BMR network-powered broadband communications medium power cables.

(b) Low Power Circuits. Low power network-powered broadband communications circuits located outside and entering buildings shall be installed using Type BLU or Type BLX low power network-powered broadband communications cables. Cables shown in Table 830-58 shall be permitted to substitute.

Exception: Outdoor community antenna television and radio distribution system coaxial cables installed prior to January 1, 2000, and installed in accordance with Article 820, shall be permitted for low-power type, network-powered broadband communications circuits.

830-10. Aerial Cables. Aerial network-powered broadband communications cables shall comply with (a) through (g).

(a) On Poles. Where practicable, network-powered broadband communications cables on poles shall be located below the electric light, power, Class 1, or nonpower-limited fire alarm circuit conductors and shall not be attached to a crossarm that carries electric light or power conductors.

(b) Climbing Space. The climbing space through network-powered broadband communications cables shall comply with the requirements of Section 225-14(d).

(c) Lead-in Clearance. Lead-in or aerial-drop network-powered broadband communications cables from a pole or other support, including the point of initial attachment to a building or structure, shall be kept away from electric light, power, Class 1, or nonpower-limited fire alarm circuit conductors so as to avoid the possibility of accidental contact.

Exception: Where proximity to electric light, power, Class 1, or nonpower-limited fire alarm circuit service conductors cannot be avoided, the installation shall be such as to provide clearances of not less than 12 in. (305 mm) from light, power, Class 1, or nonpower-limited fire alarm circuit service drops.

(d) Clearance from Ground. Overhead spans of network-powered broadband communication cables shall conform to not less than the following:

9.5 ft (2.89 m) — above finished grade, sidewalks, or from any platform or projection from which they might be reached and accessible to pedestrians only

11.5 ft (3.50 m) — over residential property and driveways, and those commercial areas not subject to truck traffic

15.5 ft (4.72 m) — over public streets, alleys, roads, parking areas subject to truck traffic, driveways on other than residential property, and other land traversed by vehicles such as cultivated, grazing, forest, and orchard

FPN: These clearances have been specifically chosen to correlate with the *National Electrical Safety Code,* ANSI C2-1997, Table 232-1, which provides for clearances of wires, conductors, and cables above ground and roadways, rather than using the clearances referenced in Section 225-18. Because Article 800 and Article 820 have had no required clearances, the communications industry has used the clearances from the NESC for their installed cable plant.

(e) Over Pools. Clearance of network-powered broadband communications cable in any direction from the water level, edge of pool, base of diving platform, or anchored raft shall be not less than 22 ft (6.71 m). Clearance in any direction to the diving platform or tower shall be not less than 14 ft (4.27 m).

Exception No. 1: Where the pool is fully enclosed by a solid or screened permanent structure.

FPN: These clearances have been specifically chosen to correlate with Section 680-8 and to incorporate the more stringent clearance requirements of the *National Electrical Safety Code,* ANSI C2-1997, Table 234-3, which provides for clearances of wires, conductors, and cables over pools.

Exception No. 2: Where the cables are located more than 10 ft (3.05 m) horizontally from the following:

(1) The edge of the pool
(2) Diving structure or diving tower, or
(3) Observation stands, towers, or platforms

(f) Above Roofs. Network-powered broadband communications cables shall have a vertical clearance of not less than 8 ft (2.44 m) from all points of roofs above which they pass.

Exception No. 1: Auxiliary buildings such as garages and the like.

Exception No. 2: A reduction in clearance above only the overhanging portion of the roof to not less than 18 in. (457 mm) shall be permitted if (1) not more than 4 ft (1.22 m) of the broadband communications drop cables pass above the roof overhang, and (2) they are terminated at a through-the-roof raceway or support.

Exception No. 3: Where the roof has a slope of not less than 4 in. (102 mm) in 12 in. (305 mm), a reduction in clearance to not less than 3 ft (914 mm) shall be permitted.

(g) Final Spans. Final spans of network-powered broadband communications cables without an outer jacket shall be permitted to be attached to the building, but they shall be kept not less than 3 ft (914 mm) from windows that are designed to be opened, doors, porches, balconies, ladders, stairs, fire escapes, or similar locations.

Exception: Conductors run above the top level of a window shall be permitted to be less than the 3-ft (914-mm) requirement above.

Overhead network-powered broadband communications cables shall not be installed beneath openings through which materials may be moved, such as openings in farm and commercial buildings, and shall not be installed where they will obstruct entrance to these building openings.

(h) Between Buildings. Network-powered broadband communications cables extending between buildings and also the supports or attachment fixtures shall be acceptable for the purpose and shall have sufficient strength to withstand the loads to which they may be subjected.

Exception: Where a network-powered broadband communications cable does not have sufficient strength to be self-supporting, it shall be attached to a supporting messenger cable that, together with the attachment fixtures or supports, shall be acceptable for the purpose and shall have sufficient strength to withstand the loads to which they may be subjected.

(i) On Buildings. Where attached to buildings, network-powered broadband communications cables shall be securely fastened in such a manner that they will be separated from other conductors as follows.

(1) Electric Light or Power. The network-powered broadband communications cable shall have a separation of at least 4 in. (102 mm) from electric light, power, Class 1, or nonpower-limited fire alarm circuit conductors not in raceway or cable, or be permanently separated from conductors of the other system by a continuous and firmly fixed nonconductor in addition to the insulation on the wires.

(2) Other Communications Systems. Network-powered broadband communications cables shall be installed so that there will be no unnecessary interference in the maintenance of the separate systems. In no case shall the conductors, cables, messenger strand, or equipment of one system cause abrasion to the conductors, cables, messenger strand, or equipment of any other system.

(3) Lightning Conductors. Where practicable, a separation of at least 6 ft (1.83 m) shall be maintained between any network-powered broadband communications cable and lightning conductors.

(4) Protection from Damage. Network-powered broadband communications cables attached to buildings and located within 8 ft (2.44 m) of finished grade shall be protected by enclosures, raceways, or other approved means.

Exception: A low power network-powered broadband communications circuit that is equipped with a listed fault protection device, appropriate to the network-powered broadband communications cable used, and located on the network side of the network-powered broadband communications cable being protected.

830-11. Underground Circuits Entering Buildings.

(a) Underground Systems. Underground network-powered broadband communications cables in a duct, pedestal, handhole, or manhole that contains electric light, power conductors, nonpower-limited fire alarm circuit conductors or Class 1 circuits shall be in a section permanently separated from such conductors by means of a suitable barrier.

(b) Direct-Buried Cables and Raceways. Direct-buried network-powered broadband communications cables shall be separated at least 12 in. (305 mm) from conductors of any light, power, nonpower-limited fire alarm circuit conductors or Class 1 circuit.

Exception No. 1: Where electric service conductors or network-powered broadband communications cables are installed in raceways or have metal cable armor.

Exception No. 2: Where electric light or power branch-circuit or feeder conductors, nonpower-limited fire alarm circuit conductors, or Class 1 circuit conductors are installed in a raceway or in metal-sheathed, metal-clad, or Type UF or Type USE cables; or the network-powered broad-

band communications cables have metal cable armor or are installed in a raceway.

(c) Mechanical Protection. Direct-buried cable, conduit, or other raceways shall be installed to meet the minimum cover requirements of Table 830-11. In addition, direct-buried cables emerging from the ground shall be protected by enclosures, raceways, or other approved means extending from the minimum cover distance required by Table 830-11 below grade to a point at least 8 ft (2.44 m) above finished grade. In no case shall the protection be required to exceed 18 in. (457 mm) below finished grade. Type BMU and BLU direct-buried cables emerging from the ground shall be installed in rigid metal conduit, intermediate metal conduit, rigid nonmetallic conduit, or other approved means extending from the minimum cover distance required by Table 830-11 below grade to the point of entrance.

Exception: A low power network-powered broadband communications circuit that is equipped with a listed fault protection device, appropriate to the network-powered broadband communications cable used, and located on the network side of the network-powered broadband communications cable being protected.

(d) Pools. Underground cables shall not be permitted under the pool or within the area extending 5 ft (1.52 m) horizontally from the inside wall of the pool.

Exception: Where space limitations prevent cables from being routed 5 ft (1.52 m) or more from the pool, such wiring shall be permitted where installed in rigid metal conduit, intermediate metal conduit, or a nonmetallic raceway system. All metal conduit shall be corrosion resistant and suitable for the location. The minimum burial depth shall be as follows:

Wiring Method	Minimum Burial (in.)
Rigid metal conduit	6
Intermediate metal conduit	6
Nonmetallic raceways listed for direct burial	18
Other approved raceways*	18

Note: For SI units, 1 in. = 25.4 mm.

*Raceways approved for burial only where concrete encased shall require a concrete envelope not less than 2 in. (50.8 mm) thick.

C. Protection

830-30. Primary Electrical Protection.

(a) Application. Primary electrical protection shall be provided on all network-powered broadband communications conductors that are neither grounded nor interrupted and are

Table 830-11. Network-Powered Broadband Communications Systems Minimum Cover Requirements, Burial in Inches (*Cover* is the shortest distance measured between a point on the top surface of any direct-buried cable, conduit, or other raceway and the top surface of finished grade, concrete, or similar cover.)

Location of Wiring Method or Circuit	Direct Burial Cables (in.)	Rigid Metal Conduit or Intermediate Metal Conduit (in.)	Nonmetallic Raceways Listed for Direct Burial; Without Concrete Encasement or Other Approved Raceways (in.)
All locations not specified below	18	6	12
In trench below 2-in. thick concrete or equivalent	12	6	6
Under a building (in raceway only)	0	0	0
Under minimum of 4-in. thick concrete exterior slab with no vehicular traffic and the slab extending not less than 6 in. beyond the underground installation	12	4	4
One- and two-family dwelling driveways and outdoor parking areas and used only for dwelling related purposes	12	12	12

Notes:

1. For SI units, 1 in. = 25.4 mm.

2. Raceways approved for burial only where concrete encased shall require a concrete envelope not less than 2 in. (50.8 mm) thick.

3. Lesser depths shall be permitted where cables rise for terminations or splices or where access is otherwise required.

4. Where solid rock is encountered, all wiring shall be installed in metal or nonmetallic raceway permitted for direct burial. The raceways shall be covered by a minimum of 2 in. (50.8 mm) of concrete extending down to rock.

5. Low power network-powered broadband communications circuits using directly buried community antenna television and radio distribution system coaxial cables that were installed outside and entering buildings prior to January 1, 2000, in accordance with Article 820 shall be permitted where buried to a minimum depth of 12 in. (30.48 cm).

run partly or entirely in aerial cable not confined within a block. Also, primary electrical protection shall be provided on all aerial or underground network-powered broadband communications conductors that are neither grounded nor interrupted and are located within the block containing the building served so as to be exposed to lightning or accidental contact with electric light or power conductors operating at over 300 volts to ground.

Exception: Where electrical protection is provided on the derived circuit(s) (output side of the NIU) in accordance with Section 830-30(b)(3).

FPN No. 1: On network-powered broadband communications conductors not exposed to lightning or accidental contact with power conductors, providing primary electrical protection in accordance with this article will help protect against other hazards, such as ground potential rise caused by power fault currents, and above-normal voltages induced by fault currents on power circuits in proximity to the network-powered broadband communications conductors.

FPN No. 2: Network-powered broadband communications circuits are considered to have a lightning exposure unless one or more of the following conditions exist.

(1) Circuits in large metropolitan areas where buildings are close together and sufficiently high to intercept lightning.

(2) Interbuilding cable runs of 140 ft (42.7 m) or less, directly buried or in underground conduit, where a continuous metallic cable shield or a continuous metallic conduit containing the cable is bonded to each building grounding electrode system.

(3) Areas having an average of five or fewer thunderstorm days per year and earth resistivity of less than 100 ohmmeters. Such areas are found along the Pacific coast.

(1) Fuseless Primary Protectors. Fuseless-type primary protectors shall be permitted where power fault currents on all protected conductors in the cable are safely limited to a value no greater than the current-carrying capacity of the primary protector and of the primary protector grounding conductor.

(2) Fused Primary Protectors. Where the requirements listed under (1) are not met, fused-type primary protectors shall be used. Fused-type primary protectors shall consist of an arrester connected between each conductor to be protected and ground, a fuse in series with each conductor to be protected, and an appropriate mounting arrangement. Fused primary protector terminals shall be marked to indicate line, instrument, and ground, as applicable.

(b) Location. The location of the primary protector, where required, shall comply with (1), (2), or (3).

(1) A listed primary protector shall be applied on each network-powered broadband communications cable external to and on the network side of the network interface unit.

(2) The primary protection function shall be an integral part of and contained in the network interface unit. The network interface unit shall be listed for the purpose and shall have an external marking indicating that it contains primary electrical protection.

(3) The primary protector(s) shall be provided on the derived circuit(s) (output side of the NIU), and the combination of the NIU and the protector(s) shall be listed for the purpose.

A primary protector, whether provided integrally or external to the network interface unit, shall be located as close as practicable to the point of entrance.

For purposes of this section, a network interface unit and any externally provided primary protectors located at mobile home service equipment located in sight from and not more than 30 ft (9.14 m) from the exterior wall of the mobile home it serves, or at a mobile home disconnecting means grounded in accordance with Section 250-32 and located in sight from and not more than 30 ft (9.14 m) from the exterior wall of the mobile home it serves, shall be considered to meet the requirements of this section.

FPN: Selecting a network interface unit and primary protector location to achieve the shortest practicable primary protector grounding conductor will help limit potential differences between communications circuits and other metallic systems.

(c) Hazardous (Classified) Locations. The primary protector or equipment providing the primary protection function shall not be located in any hazardous (classified) location as defined in Article 500 or in the vicinity of easily ignitible material.

Exception: As permitted in Sections 501-14, 502-14, and 503-12.

830-33. Grounding or Interruption of Metallic Members of Network-Powered Broadband Communications Cables. The shields of network-powered broadband communications cables used for communications or powering shall be grounded at the building as close to the point of entrance as practicable. Metallic cable members not used for communications or powering shall be grounded or interrupted by an insulating joint or equivalent device as close to the point of entrance as practicable.

For purposes of this section, grounding or interruption of network-powered broadband communications cable metallic members installed at mobile home service equipment located in sight from and no more than 30 ft (9.14 m) from the exterior wall of the mobile home it serves, or at a mobile home disconnecting means grounded in accordance with Section 250-24 and located in sight from and not more than 30 ft (9.14 m) from the exterior wall of the mobile home it serves, shall be considered to meet the requirements of this section.

FPN: Selecting a grounding location to achieve the shortest practicable grounding conductor will help limit potential differences between the network-powered broadband communications circuits and other metallic systems.

D. Grounding Methods

830-40. Cable, Network Interface Unit, and Primary Protector Grounding. Network interface units containing protectors, NIUs with metallic enclosures, primary protectors, and the grounded metallic members of the network-powered broadband communications cable shall be grounded as specified in (a) through (d).

(a) Grounding Conductor.

(1) Insulation. The grounding conductor shall be insulated and shall be listed as suitable for the purpose.

(2) Material. The grounding conductor shall be copper or other corrosion-resistant conductive material, stranded or solid.

(3) Size. The grounding conductor shall not be smaller than No. 14, and shall have a current-carrying capacity approximately equal to that of the grounded metallic member(s) and protected conductor(s) of the network-powered broadband communications cable. The grounding conductor shall not be required to exceed No. 6.

(4) Run in Straight Line. The grounding conductor shall be run to the grounding electrode in as straight a line as practicable.

(5) Physical Protection. Where subject to physical damage, the grounding conductor shall be adequately protected. Where the grounding conductor is run in a metal raceway, both ends of the raceway shall be bonded to the grounding conductor or the same terminal or electrode to which the grounding conductor is connected.

(b) Electrode. The grounding conductor shall be connected as follows.

(1) To the nearest accessible location on the following:

(a) The building or structure grounding electrode system as covered in Section 250-50
(b) The grounded interior metal water piping system as covered in Section 250-104(a)
(c) The power service accessible means external to enclosures as covered in Section 250-92(b)
(d) The metallic power service raceway
(e) The service equipment enclosure
(f) The grounding electrode conductor or the grounding electrode conductor metal enclosure, or
(g) To the grounding conductor or to the grounding electrode of a building or structure disconnecting means that is grounded to an electrode as covered in Section 250-32

For purposes of this section, the mobile home service equipment or the mobile home disconnecting means, as described in Section 830-33, shall be considered accessible.

(2) If the building or structure served has no grounding means as described in (b)(1), to any one of the individual electrodes described in Section 250-50; or

(3) If the building or structure served has no grounding means, as described in (b)(1) or (b)(2), to an effectively grounded metal structure or to a ground rod or pipe not less than 5 ft (1.52 m) in length and ½ in. (12.7 mm) in diameter, driven, where practicable, into permanently damp earth and separated from lightning conductors as covered in Section 800-13 and at least 6 ft (1.83 m) from electrodes of other systems. Steam or hot water pipes or lightning rod conductors shall not be employed as electrodes for protectors, NIUs with integral protection, grounded metallic members, NIUs with metallic enclosures, and other equipment.

(c) Electrode Connection. Connections to grounding electrodes shall comply with Section 250-70. Connectors, clamps, fittings, or lugs used to attach grounding conductors and bonding jumpers to grounding electrodes or to each other that are to be concrete encased or buried in the earth shall be suitable for its application.

(d) Bonding of Electrodes. A bonding jumper not smaller than No. 6 copper or equivalent shall be connected between the network-powered broadband communications system grounding electrode and the power grounding electrode sys-

tem at the building or structure served where separate electrodes are used.

Exception: At mobile homes as covered in Section 830-42.

FPN No. 1: See Section 250-60 for use of lightning rods.

FPN No. 2: Bonding together of all separate electrodes will limit potential differences between them and between their associated wiring systems.

830-42. Bonding and Grounding at Mobile Homes.

(a) Grounding. Where there is no mobile home service equipment located in sight from and not more than 30 ft (9.14 m) from the exterior wall of the mobile home it serves, or there is no mobile home disconnecting means grounded in accordance with Section 250-32 and located within sight from and not more than 30 ft (9.14 m) from the exterior wall of the mobile home it serves, the network-powered broadband communications cable, network interface unit, and primary protector ground, shall be installed in accordance with Sections 830-40(b)(2) and (3).

(b) Bonding. The network-powered broadband communications cable grounding terminal, network interface unit grounding terminal, if present, and primary protector grounding terminal shall be bonded together with a copper bonding conductor not smaller than No. 12. The network-powered broadband communications cable grounding terminal, network interface unit grounding terminal, primary protector grounding terminal, or the grounding electrode shall be bonded to the metal frame or available grounding terminal of the mobile home with a copper bonding conductor not smaller than No. 12 under any of the following conditions:

(1) Where there is no mobile home service equipment or disconnecting means as in (a), or
(2) Where the mobile home is supplied by cord and plug

E. Network-Powered Broadband Communications Systems Wiring Methods Within Buildings

830-54. Medium Power Network-Powered Broadband Communications System Wiring Methods. Medium power network-powered broadband communications systems shall be installed within buildings using listed Type BM or Type BMR, network-powered broadband communications medium power cables.

(a) Ducts, Plenums, and Other Air-Handling Spaces. Section 300-22 shall apply.

(b) Riser. Cables installed in vertical runs and penetrating more than one floor, or cables installed in vertical runs in a shaft, shall be Type BMR. Floor penetrations requiring

Type BMR shall contain only cables suitable for riser or plenum use.

Exception No. 1: Type BM cables encased in metal raceway or located in a fireproof shaft that has firestops at each floor.

Exception No. 2: Type BM cables in one- and two-family dwellings.

(c) Other Wiring. Cables installed in locations other than the locations covered in (a) and (b) shall be Type BM.

Exception: Type BMU cable where the cable enters the building from the outside and is run in rigid metal conduit or intermediate metal conduit, and such conduits are grounded to an electrode in accordance with Section 830-40(b).

830-55. Low Power Network-Powered Broadband Communications System Wiring Methods. Low power network-powered broadband communications systems shall be installed within buildings using listed Type BLX or Type BLP network-powered broadband communications low power cables.

(a) Ducts, Plenums, and Other Air-Handling Spaces. Cables installed in ducts, plenums, and other spaces used for environmental air shall be Type BLP.

Exception: Type BLX cables installed in compliance with Section 300-22.

(b) Riser. Cables installed in vertical runs and penetrating more than one floor, or cables installed in vertical runs in a shaft, shall be Type BLP or Type BMR. Floor penetrations requiring Type BMR shall contain only cables suitable for riser or plenum use.

Exception No. 1: Type BLX cables encased in metal raceway or located in a fireproof shaft that has firestops at each floor.

Exception No. 2: Type BLX cable that is less than 0.375 in. (9.52 mm) in diameter in one- and two-family dwellings.

(c) Other Wiring. Cables installed in locations other than the locations covered in (a) and (b) shall be Type BLP or Type BM.

Exception No. 1: Type BLX cable enclosed in raceway.

Exception No. 2: Type BLU cables where the cable enters the building from the outside and is run in rigid metal conduit or intermediate metal conduit, and such conduits are grounded to an electrode in accordance with Section 830-40(b).

Exception No. 3: Type BLX cable that is less than 0.375 in. (9.52 mm) in diameter in one- and two-family dwellings.

Exception No. 4: Type BLX cable where the length of cable within the building does not exceed 50 ft (15.2 m) and the cable enters the building from outside and is terminated at a grounding block or a primary protection location.

> FPN: This exception limits the length of Type BLX cable to 50 ft (15.2 m), while Section 830-30(b) requires that the primary protector, or NIU with integral protection, be located as close as practicable to the point at which the cable enters the building. Therefore, in installations requiring a primary protector, or NIU with integral protection, Type BLX cable may not be permitted to extend 50 ft (15.2 m) into the building if it is practicable to place the primary protector closer than 50 ft (15.2 m) to the entrance point.

830-56. Protection Against Physical Damage. Section 300-4 shall apply.

830-57. Bends. Bends in network broadband cable shall be made so as not to damage the cable.

830-58. Installation of Network-Powered Broadband Communications Cables and Equipment. Cable and equipment installations within buildings shall comply with (a) through (e), as applicable.

(a) Separation of Conductors.

(1) In Raceways and Enclosures.

(a) *Low and Medium Power Network-Powered Broadband Communications Circuit Cables.* Low and medium power network-powered broadband communications cables shall be permitted in the same raceway or enclosure.

(b) *Low Power Network-Powered Broadband Communications Circuit Cables.* Low power network-powered broadband communications cables shall be permitted in the same raceway or enclosure with jacketed cables of any of the circuits shown below:

(1) Class 2 and Class 3 remote-control, signaling, and power-limited circuits in compliance with Article 725
(2) Power-limited fire alarm systems in compliance with Article 760
(3) Communications circuits in compliance with Article 800
(4) Nonconductive and conductive optical fiber cables in compliance with Article 770
(5) Community antenna television and radio distribution systems in compliance with Article 820

(c) *Medium Power Network-Powered Broadband Communications Circuit Cables.* Medium power network-powered broadband communications cables shall not be permitted in the same raceway or enclosure with conductors of any of the circuits shown below:

(1) Class 2 and Class 3 remote-control, signaling, and power-limited circuits in compliance with Article 725

(2) Power-limited fire alarm systems in compliance with Article 760

(3) Communications circuits in compliance with Article 800

(4) Conductive optical fiber cables in compliance with Article 770

(5) Community antenna television and radio distribution systems in compliance with Article 820

(d) *Electric Light, Power, Class 1, Nonpower-Limited Fire Alarm Circuits.* Network-powered broadband communications cable shall not be placed in any raceway, compartment, outlet box, junction box, or similar fittings with conductors of electric light, power, Class 1, or nonpower-limited fire alarm circuit cables.

Exception No. 1: Where all of the conductors of electric light, power, Class 1, nonpower-limited fire alarm circuits are separated from all of the network-powered broadband communications cables by a barrier.

Exception No. 2: Power circuit conductors in outlet boxes, junction boxes, or similar fittings or compartments where such conductors are introduced solely for power supply to the network-powered broadband communications system distribution equipment. The power circuit conductors shall be routed within the enclosure to maintain a minimum 0.25-in. (6.35-mm) separation from network-powered broadband communications cables.

(2) Other Applications. Network-powered broadband communications cable shall be separated at least 2 in. (50.8 mm) from conductors of any electric light, power, Class 1, and nonpower-limited fire alarm circuits.

Exception No. 1: Where either (1) all of the conductors of electric light, power, Class 1, and nonpower-limited fire alarm circuits are in a raceway, or in metal-sheathed, metal-clad, nonmetallic-sheathed, Type AC, or Type UF cables, or (2) all of the network-powered broadband communications cables are encased in raceway.

Exception No. 2: Where the network-powered broadband communications cables are permanently separated from the conductors of electric light, power, Class 1, and nonpower-limited fire alarm circuits by a continuous and firmly fixed nonconductor, such as porcelain tubes or flexible tubing, in addition to the insulation on the wire.

(b) Spread of Fire or Products of Combustion. Installations in hollow spaces, vertical shafts, and ventilation or air-handling ducts shall be made so that the possible spread of fire or products of combustion will not be substantially increased. Openings around penetrations through fire–resistance-rated walls, partitions, floors, or ceilings shall be firestopped using approved methods.

(c) Equipment in Other Space Used for Environmental Air. Section 300-22(c) shall apply.

(d) Support of Conductors. Raceways shall be used for their intended purpose. Network-powered broadband communications cables shall not be strapped, taped, or attached by any means to the exterior of any conduit or raceway as a means of support.

(e) Cable Substitutions. The substitutions for network-powered broadband cables listed in Table 830-58 shall be permitted. All cables in Table 830-58, other than network-powered broadband cables, shall be coaxial cables.

Table 830-58. Cable Substitutions

Cable Type	Permitted Cable Substitutions
BM	BMR
BLP	MPP, CMP, CL3P
BLX	MPP, CMP, CL3P, MPR, CMR, CL3R, MPG, MP, CMG, CM, CL3, CMX, CL3X, BMR, BM, BLP

FPN: The substitute multipurpose, communications, and Class 3 cables in Table 830-58 are only coaxial-type cables.

CHAPTER 9
Tables

Table 1. Percent of Cross Section of Conduit and Tubing for Conductors

Number of Conductors	All Conductor Types
1	53
2	31
Over 2	40

FPN No. 1: Table 1 is based on common conditions of proper cabling and alignment of conductors where the length of the pull and the number of bends are within reasonable limits. It should be recognized that, for certain conditions, a larger size conduit or a lesser conduit fill should be considered.

FPN No. 2: When pulling three conductors or cables into a raceway, if the ratio of the raceway (inside diameter) to the conductor or cable (outside diameter) is between 2.8 and 3.2, jamming can occur. While jamming can occur when pulling four or more conductors or cables into a raceway, the probability is very low.

Notes to Tables

(1) See Appendix C for the maximum number of conductors and fixture wires, all of the same size (total cross-sectional area including insulation) permitted in trade sizes of the applicable conduit or tubing.

(2) Table 1 applies only to complete conduit or tubing systems and is not intended to apply to sections of conduit or tubing used to protect exposed wiring from physical damage.

(3) Equipment grounding or bonding conductors, where installed, shall be included when calculating conduit or tubing fill. The actual dimensions of the equipment grounding or bonding conductor (insulated or bare) shall be used in the calculation.

(4) Where conduit or tubing nipples having a maximum length not to exceed 24 in. (610 mm) are installed between boxes, cabinets, and similar enclosures, the nipples shall be permitted to be filled to 60 percent of their total cross-sectional area, and Section 310-15(b)(2)(a) adjustment factors need not apply to this condition.

(5) For conductors not included in Chapter 9, such as multiconductor cables, the actual dimensions shall be used.

(6) For combinations of conductors of different sizes, use Tables 5 and 5A for dimensions of conductors and Table 4 for the applicable conduit or tubing dimensions.

(7) When calculating the maximum number of conductors permitted in a conduit or tubing, all of the same size (total cross-sectional area including insulation), the next higher whole number shall be used to determine the maximum number of conductors permitted when the calculation results in a decimal of 0.8 or larger.

(8) Where bare conductors are permitted by other sections of this *Code*, the dimensions for bare conductors in Table 8 shall be permitted.

(9) A multiconductor cable of two or more conductors shall be treated as a single conductor for calculating percentage conduit fill area. For cables that have elliptical cross sections, the cross-sectional area calculation shall be based on using the major diameter of the ellipse as a circle diameter.

Table 4. Dimensions and Percent Area of Conduit and Tubing (Areas of Conduit or Tubing for the Combinations of Wires Permitted in Table 1, Chapter 9)

Electrical Metallic Tubing						Electrical Nonmetallic Tubing					
Trade Size (in.)	Internal Diameter (in.)	Total Area 100% (in.^{2})	2 Wires 31% (in.^{2})	Over 2 Wires 40% (in.^{2})	1 Wire 53% (in.^{2})	Trade Size (in.)	Internal Diameter (in.)	Total Area 100% (in.^{2})	2 Wires 31% (in.^{2})	Over 2 Wires 40% (in.^{2})	1 Wire 53% (in.^{2})
½	0.622	0.304	0.094	0.122	0.161	½	0.560	0.246	0.076	0.099	0.131
¾	0.824	0.533	0.165	0.213	0.283	¾	0.760	0.454	0.141	0.181	0.240
1	1.049	0.864	0.268	0.346	0.458	1	1.000	0.785	0.243	0.314	0.416
1¼	1.380	1.496	0.464	0.598	0.793	1¼	1.340	1.410	0.437	0.564	0.747
1½	1.610	2.036	0.631	0.814	1.079	1½	1.570	1.936	0.600	0.774	1.026
2	2.067	3.356	1.040	1.342	1.778	2	2.020	3.205	0.994	1.282	1.699
2½	2.731	5.858	1.816	2.343	3.105	2½	—	—	—	—	—
3	3.356	8.846	2.742	3.538	4.688	3	—	—	—	—	—
3½	3.834	11.545	3.579	4.618	6.119	3½	—	—	—	—	—
4	4.334	14.753	4.573	5.901	7.819	4	—	—	—	—	—

Flexible Metal Conduit						Intermediate Metal Conduit					
Trade Size (in.)	Internal Diameter (in.)	Total Area 100% (in.^{2})	2 Wires 31% (in.^{2})	Over 2 Wires 40% (in.^{2})	1 Wire 53% (in.^{2})	Trade Size (in.)	Internal Diameter (in.)	Total Area 100% (in.^{2})	2 Wires 31% (in.^{2})	Over 2 Wires 40% (in.^{2})	1 Wire 53% (in.^{2})
⅜	0.384	0.116	0.036	0.046	0.061	⅜	—	—	—	—	—
½	0.635	0.317	0.098	0.127	0.168	½	0.660	0.342	0.106	0.137	0.181
¾	0.824	0.533	0.165	0.213	0.282	¾	0.864	0.586	0.182	0.235	0.311
1	1.020	0.817	0.253	0.327	0.433	1	1.105	0.959	0.297	0.384	0.508
1¼	1.275	1.277	0.396	0.511	0.677	1¼	1.448	1.646	0.510	0.658	0.872
1½	1.538	1.857	0.576	0.743	0.984	1½	1.683	2.223	0.689	0.889	1.178
2	2.040	3.269	1.013	1.307	1.732	2	2.150	3.629	1.125	1.452	1.923
2½	2.500	4.909	1.522	1.964	2.602	2½	2.557	5.135	1.592	2.054	2.722
3	3.000	7.069	2.191	2.827	3.746	3	3.176	7.922	2.456	3.169	4.199
3½	3.500	9.621	2.983	3.848	5.099	3½	3.671	10.584	3.281	4.234	5.610
4	4.000	12.566	3.896	5.027	6.660	4	4.166	13.631	4.226	5.452	7.224

Liquidtight Flexible Nonmetallic Conduit (Type LFNC-B*)						Liquidtight Flexible Nonmetallic Conduit (Type LFNC-A*)					
Trade Size (in.)	Internal Diameter (in.)	Total Area 100% (in.^{2})	2 Wires 31% (in.^{2})	Over 2 Wires 40% (in.^{2})	1 Wire 53% (in.^{2})	Trade Size (in.)	Internal Diameter (in.)	Total Area 100% (in.^{2})	2 Wires 31% (in.^{2})	Over 2 Wires 40% (in.^{2})	1 Wire 53% (in.^{2})
⅜	0.494	0.192	0.059	0.077	0.102	⅜	0.495	0.192	0.060	0.077	0.102
½	0.632	0.314	0.097	0.125	0.166	½	0.630	0.312	0.097	0.125	0.165
¾	0.830	0.541	0.168	0.216	0.287	¾	0.825	0.535	0.166	0.214	0.283
1	1.054	0.872	0.270	0.349	0.462	1	1.043	0.854	0.265	0.341	0.452
1¼	1.395	1.528	0.474	0.611	0.810	1¼	1.383	1.501	0.465	0.600	0.796
1½	1.588	1.979	0.614	0.792	1.049	1½	1.603	2.017	0.625	0.807	1.069
2	2.033	3.245	1.006	1.298	1.720	2	2.063	3.341	1.036	1.336	1.771

*Corresponds to Section 351-22(2).   *Corresponds to Section 351-22(1).

Table 4. (Continued)

Trade Size (in.)	Internal Diameter (in.)	Total Area 100% (in.²)	2 Wires 31% (in.²)	Over 2 Wires 40% (in.²)	1 Wire 53% (in.²)	Trade Size (in.)	Internal Diameter (in.)	Total Area 100% (in.²)	2 Wires 31% (in.²)	Over 2 Wires 40% (in.²)	1 Wire 53% (in.²)
	Liquidtight Flexible Metal Conduit						**Rigid Metal Conduit**				
⅜	0.494	0.192	0.059	0.077	0.102	⅜	—	—	—	—	—
½	0.632	0.314	0.097	0.125	0.166	½	0.632	0.314	0.097	0.125	0.166
¾	0.830	0.541	0.168	0.216	0.287	¾	0.836	0.549	0.170	0.220	0.291
1	1.054	0.872	0.270	0.349	0.462	1	1.063	0.888	0.275	0.355	0.470
1¼	1.395	1.528	0.474	0.611	0.810	1¼	1.394	1.526	0.473	0.610	0.809
1½	1.588	1.979	0.614	0.792	1.049	1½	1.624	2.071	0.642	0.829	1.098
2	2.033	3.245	1.006	1.298	1.720	2	2.083	3.408	1.056	1.363	1.806
2½	2.493	4.879	1.513	1.952	2.586	2½	2.489	4.866	1.508	1.946	2.579
3	3.085	7.475	2.317	2.990	3.962	3	3.090	7.499	2.325	3.000	3.975
3½	3.520	9.731	3.017	3.893	5.158	3½	3.570	10.010	3.103	4.004	5.305
4	4.020	12.692	3.935	5.077	6.727	4	4.050	12.883	3.994	5.153	6.828
5	—	—	—	—	—	5	5.073	20.213	6.266	8.085	10.713
6	—	—	—	—	—	6	6.093	29.158	9.039	11.663	15.454
	Rigid PVC Conduit, Schedule 80						**Rigid PVC Conduit, Schedule 40, and HDPE Conduit**				
½	0.526	0.217	0.067	0.087	0.115	½	0.602	0.285	0.088	0.114	0.151
¾	0.722	0.409	0.127	0.164	0.217	¾	0.804	0.508	0.157	0.203	0.269
1	0.936	0.688	0.213	0.275	0.365	1	1.029	0.832	0.258	0.333	0.441
1¼	1.255	1.237	0.383	0.495	0.656	1¼	1.360	1.453	0.450	0.581	0.770
1½	1.476	1.711	0.530	0.684	0.907	1½	1.590	1.986	0.616	0.794	1.052
2	1.913	2.874	0.891	1.150	1.1523	2	2.047	3.291	1.020	1.316	1.744
2½	2.290	4.119	1.277	1.647	2.183	2½	2.445	4.695	1.455	1.878	2.488
3	2.864	6.442	1.997	2.577	3.414	3	3.042	7.268	2.253	2.907	3.852
3½	3.326	8.688	2.693	3.475	4.605	3½	3.521	9.737	3.018	3.895	5.161
4	3.786	11.258	3.490	4.503	5.967	4	3.998	12.554	3.892	5.022	6.654
5	4.768	17.855	5.535	7.142	9.463	5	5.016	19.761	6.126	7.904	10.473
6	5.709	25.598	7.935	10.239	13.567	6	6.031	28.567	8.856	11.427	15.141
	Type A, Rigid PVC Conduit						**Type EB, PVC Conduit**				
½	0.700	0.385	0.119	0.154	0.204	½	—	—	—	—	—
¾	0.910	0.650	0.202	0.260	0.345	¾	—	—	—	—	—
1	1.175	1.084	0.336	0.434	0.575	1	—	—	—	—	—
1¼	1.500	1.767	0.548	0.707	0.937	1¼	—	—	—	—	—
1½	1.720	2.324	0.720	0.929	1.231	1½	—	—	—	—	—
2	2.155	3.647	1.131	1.459	1.933	2	2.221	3.874	1.201	1.550	2.053
2½	2.635	5.453	1.690	2.181	2.890	2½	—	—	—	—	—
3	3.230	8.194	2.540	3.278	4.343	3	3.330	8.709	2.700	3.484	4.616
3½	3.690	10.694	3.315	4.278	5.668	3½	3.804	11.365	3.523	4.546	6.024
4	4.180	13.723	4.254	5.489	7.273	4	4.289	14.448	4.479	5.779	7.657
5	—	—	—	—	—	5	5.316	22.195	6.881	8.878	11.764
6	—	—	—	—	—	6	6.336	31.530	9.774	12.612	16.711

Table 5. Dimensions of Insulated Conductors and Fixture Wires

Type: AF, FFH-2, RFH-1, RFH-2, RH, RHH*, RHW*, RHW-2*, RHH, RHW, RHW-2, SF-1, SF-2, SFF-1, SFF-2, TF, TFF, THHW, THW, THW-2, TW, XF, XFF

Type	Size (AWG or kcmil)	Approximate Diameter (in.)	Approximate Area (in.^{2})
RFH-2,	18	0.136	0.0145
FFH-2	16	0.148	0.0172
RH	14	0.163	0.0209
	12	0.182	0.0260
RHW-2, RHH, RHW	14	0.193	0.0293
	12	0.212	0.0353
RH, RHH, RHW, RHW-2	10	0.236	0.0437
	8	0.326	0.0835
	6	0.364	0.1041
	4	0.412	0.1333
	3	0.440	0.1521
	2	0.472	0.1750
	1	0.582	0.2660
	1/0	0.622	0.3039
	2/0	0.668	0.3505
	3/0	0.720	0.4072
	4/0	0.778	0.4754
	250	0.895	0.6291
	300	0.950	0.7088
	350	1.001	0.7870
	400	1.048	0.8626
	500	1.133	1.0082
	600	1.243	1.2135
	700	1.314	1.3561
	750	1.348	1.4272
	800	1.380	1.4957
	900	1.444	1.6377
	1000	1.502	1.7719
	1250	1.729	2.3479
	1500	1.852	2.6938
	1750	1.966	3.0357
	2000	2.072	3.3719
SF-2, SFF-2	18	0.121	0.0115
	16	0.133	0.0139
	14	0.148	0.0172
SF-1, SFF-1	18	0.091	0.0065
RFH-1, AF, XF, XFF	18	0.106	0.0080
AF, TF, TFF, XF, XFF	16	0.118	0.0109
AF, TW, XF, XFF	14	0.133	0.0139
TW	12	0.152	0.0181
	10	0.176	0.0243
	8	0.236	0.0437
RHH*, RHW*, RHW-2*, THHW, THW, THW-2	14	0.163	0.0209

Table 5. (Continued)

Type: AF, RHH*, RHW*, RHW-2*, THHN, THHW, THW, THW-2, TFN, TFFN, THWN, THWN-2, XF, XFF

Type	Size (AWG or kcmil)	Approximate Diameter (in.)	Approximate Area (in.^{2})
RHH*, RHW*, RHW-2*, THHW, THW, AF, XF, XFF,	12	0.182	0.0260
	10	0.206	0.0333
RHH*, RHW*, RHW-2*, THHW, THW, THW-2	8	0.266	0.0556
TW, THW, THHW, THW-2, RHH*, RHW*, RHW-2*	6	0.304	0.0726
	4	0.352	0.0973
	3	0.380	0.1134
	2	0.412	0.1333
	1	0.492	0.1901
	1/0	0.532	0.2223
	2/0	0.578	0.2624
	3/0	0.630	0.3117
	4/0	0.688	0.3718
	250	0.765	0.4596
	300	0.820	0.5281
	350	0.871	0.5958
	400	0.918	0.6619
	500	1.003	0.7901
	600	1.113	0.9729
	700	1.184	1.1010
	750	1.218	1.1652
	800	1.250	1.2272
	900	1.314	1.3561
	1000	1.372	1.4784
	1250	1.539	1.8602
	1500	1.662	2.1695
	1750	1.776	2.4773
	2000	1.882	2.7818
TFN, TFFN	18	0.084	0.0055
	16	0.096	0.0072
THHN, THWN, THWN-2	14	0.111	0.0097
	12	0.130	0.0133
	10	0.164	0.0211
	8	0.216	0.0366
	6	0.254	0.0507
	4	0.324	0.0824
	3	0.352	0.0973
	2	0.384	0.1158
	1	0.446	0.1562
	1/0	0.486	0.1855
	2/0	0.532	0.2223
	3/0	0.584	0.2679
	4/0	0.642	0.3237
	250	0.711	0.3970
	300	0.766	0.4608

Table 5. (Continued)

Type: FEP, FEPB, PAF, PAFF, PF, PFA, PFAH, PFF, PGF, PGFF, PTF, PTFF, TFE, THHN, THWN, THWN-2, Z, ZF, ZFF

Type	Size (AWG or kcmil)	Approximate Diameter (in.)	Approximate Area (in.^{2})
THHN, THWN, THWN-2	350	0.817	0.5242
	400	0.864	0.5863
	500	0.949	0.7073
	600	1.051	0.8676
	700	1.122	0.9887
	750	1.156	1.0496
	800	1.188	1.1085
	900	1.252	1.2311
	1000	1.310	1.3478
PF, PGFF, PGF, PFF, PTF, PAF, PTFF, PAFF	18	0.086	0.0058
	16	0.098	0.0075
PF, PGFF, PGF, PFF, PTF, PAF, PTFF, PAFF, TFE, FEP, PFA, FEPB, PFAH	14	0.113	0.010
TFE, FEP, PFA, FEPB, PFAH	12	0.132	0.0137
	10	0.156	0.0191
	8	0.206	0.0333
	6	0.244	0.0468
	4	0.292	0.0670
	3	0.320	0.0804
	2	0.352	0.0973
TFE, PFAH	1	0.422	0.1399
TFE, PFA, PFAH, Z	1/0	0.462	0.1676
	2/0	0.508	0.2027
	3/0	0.560	0.2463
	4/0	0.618	0.3000
ZF, ZFF	18	0.076	0.0045
	16	0.088	0.0061
Z, ZF, ZFF	14	0.103	0.0083
Z	12	0.122	0.0117
	10	0.156	0.0191
	8	0.196	0.0302
	6	0.234	0.0430
	4	0.282	0.0625
	3	0.330	0.0855
	2	0.362	0.1029
	1	0.402	0.1269

Table 5. (Continued)

Type: KF-1, KF-2, KFF-1, KFF-2, XHH, XHHW, XHHW-2, ZW

Type	Size (AWG or kcmil)	Approximate Diameter (in.)	Approximate Area (in.^{2})
XHHW, ZW, XHHW-2, XHH	14	0.133	0.0139
	12	0.152	0.0181
	10	0.176	0.0243
	8	0.236	0.0437
	6	0.274	0.0590
	4	0.322	0.0814
	3	0.350	0.0962
	2	0.382	0.1146
XHHW, XHHW-2, XHH	1	0.442	0.1534
	1/0	0.482	0.1825
	2/0	0.528	0.2190
	3/0	0.58	0.2642
	4/0	0.638	0.3197
	250	0.705	0.3904
	300	0.76	0.4536
	350	0.811	0.5166
	400	0.858	0.5782
	500	0.943	0.6984
	600	1.053	0.8709
	700	1.124	0.9923
	750	1.158	1.0532
	800	1.190	1.1122
	900	1.254	1.2351
	1000	1.312	1.3519
	1250	1.479	1.7180
	1500	1.602	2.0157
	1750	1.716	2.3127
	2000	1.822	2.6073
KF-2, KFF-2	18	0.063	0.0031
	16	0.075	0.0044
	14	0.090	0.0064
	12	0.109	0.0093
	10	0.133	0.0139
KF-1, KFF-1	18	0.057	0.0026
	16	0.069	0.0037
	14	0.084	0.0055
	12	0.103	0.0083
	10	0.127	0.0127

*Types RHH, RHW, and RHW-2 without outer covering.

Table 5A. Compact Aluminum Building Wire Nominal Dimensions* and Areas

Size (AWG or kcmil)	Bare Conductor			Types THW and THHW		Type THHN		Type XHHW		Size (AWG or kcmil)
	Number of Strands	Diameter (in.)		Approximate Diameter (in.)	Approximate Area (in.^{2})	Approximate Diameter (in.)	Approximate Area (in.^{2})	Approximate Diameter (in.)	Approximate Area (in.^{2})	
8	7	0.134		0.255	0.0510	—	—	0.224	0.0394	8
6	7	0.169		0.290	0.0660	0.240	0.0452	0.260	0.0530	6
4	7	0.213		0.335	0.0881	0.305	0.0730	0.305	0.0730	4
2	7	0.268		0.390	0.1194	0.360	0.1017	0.360	0.1017	2
1	19	0.299		0.465	0.1698	0.415	0.1352	0.415	0.1352	1
1/0	19	0.336		0.500	0.1963	0.450	0.1590	0.450	0.1590	1/0
2/0	19	0.376		0.545	0.2332	0.495	0.1924	0.490	0.1885	2/0
3/0	19	0.423		0.590	0.2733	0.540	0.2290	0.540	0.2290	3/0
4/0	19	0.475		0.645	0.3267	0.595	0.2780	0.590	0.2733	4/0
250	37	0.520		0.725	0.4128	0.670	0.3525	0.660	0.3421	250
300	37	0.570		0.775	0.4717	0.720	0.4071	0.715	0.4015	300
350	37	0.616		0.820	0.5281	0.770	0.4656	0.760	0.4536	350
400	37	0.659		0.865	0.5876	0.815	0.5216	0.800	0.5026	400
500	37	0.736		0.940	0.6939	0.885	0.6151	0.880	0.6082	500
600	61	0.813		1.050	0.8659	0.985	0.7620	0.980	0.7542	600
700	61	0.877		1.110	0.9676	1.050	0.8659	1.050	0.8659	700
750	61	0.908		1.150	1.0386	1.075	0.9076	1.090	0.9331	750
1000	61	1.060		1.285	1.2968	1.255	1.2370	1.230	1.1882	1000

*Dimensions are from industry sources.

Table 8. Conductor Properties

Size (AWG or kcmil)	Area (Circular Mils)	Conductors				Direct-Current Resistance at 75°C (167°F)		
		Stranding		Overall		Copper		Aluminum
		Quantity	Diameter (in.)	Diameter (in.)	Area (in.^{2})	Uncoated (ohm/1000 ft)	Coated (ohm/1000 ft)	(ohm/1000 ft)
18	1620	1	—	0.040	0.001	7.77	8.08	12.8
18	1620	7	0.015	0.046	0.002	7.95	8.45	13.1
16	2580	1	—	0.051	0.002	4.89	5.08	8.05
16	2580	7	0.019	0.058	0.003	4.99	5.29	8.21
14	4110	1	—	0.064	0.003	3.07	3.19	5.06
14	4110	7	0.024	0.073	0.004	3.14	3.26	5.17
12	6530	1	—	0.081	0.005	1.93	2.01	3.18
12	6530	7	0.030	0.092	0.006	1.98	2.05	3.25
10	10380	1	—	0.102	0.008	1.21	1.26	2.00
10	10380	7	0.038	0.116	0.011	1.24	1.29	2.04
8	16510	1	—	0.128	0.013	0.764	0.786	1.26
8	16510	7	0.049	0.146	0.017	0.778	0.809	1.28
6	26240	7	0.061	0.184	0.027	0.491	0.510	0.808
4	41740	7	0.077	0.232	0.042	0.308	0.321	0.508
3	52620	7	0.087	0.260	0.053	0.245	0.254	0.403
2	66360	7	0.097	0.292	0.067	0.194	0.201	0.319
1	83690	19	0.066	0.332	0.087	0.154	0.160	0.253
1/0	105600	19	0.074	0.372	0.109	0.122	0.127	0.201
2/0	133100	19	0.084	0.418	0.137	0.0967	0.101	0.159
3/0	167800	19	0.094	0.470	0.173	0.0766	0.0797	0.126
4/0	211600	19	0.106	0.528	0.219	0.0608	0.0626	0.100
250	—	37	0.082	0.575	0.260	0.0515	0.0535	0.0847
300	—	37	0.090	0.630	0.312	0.0429	0.0446	0.0707
350	—	37	0.097	0.681	0.364	0.0367	0.0382	0.0605
400	—	37	0.104	0.728	0.416	0.0321	0.0331	0.0529
500	—	37	0.116	0.813	0.519	0.0258	0.0265	0.0424
600	—	61	0.099	0.893	0.626	0.0214	0.0223	0.0353
700	—	61	0.107	0.964	0.730	0.0184	0.0189	0.0303
750	—	61	0.111	0.998	0.782	0.0171	0.0176	0.0282
800	—	61	0.114	1.030	0.834	0.0161	0.0166	0.0265
900	—	61	0.122	1.094	0.940	0.0143	0.0147	0.0235
1000	—	61	0.128	1.152	1.042	0.0129	0.0132	0.0212
1250	—	91	0.117	1.289	1.305	0.0103	0.0106	0.0169
1500	—	91	0.128	1.412	1.566	0.00858	0.00883	0.0141
1750	—	127	0.117	1.526	1.829	0.00735	0.00756	0.0121
2000	—	127	0.126	1.632	2.092	0.00643	0.00662	0.0106

Notes:

1. These resistance values are valid **only** for the parameters as given. Using conductors having coated strands, different stranding type, and, especially, other temperatures changes the resistance.

2. Formula for temperature change: $R_2 = R_1[1 + \alpha(T_2 - 75)]$ where: $\alpha_{cu} = 0.00323$, $\alpha_{AL} = 0.00330$

3. Conductors with compact and compressed stranding have about 9 percent and 3 percent, respectively, smaller bare conductor diameters than those shown. See Table 5A for actual compact cable dimensions.

4. The IACS conductivities used: bare copper = 100%, aluminum = 61%.

5. Class B stranding is listed as well as solid for some sizes. Its overall diameter and area is that of its circumscribing circle.

FPN: The construction information is per NEMA WC8-1992. The resistance is calculated per National Bureau of Standards Handbook 100, dated 1966, and Handbook 109, dated 1972.

Table 9. Alternating-Current Resistance and Reactance for 600-Volt Cables, 3-Phase, 60 Hz, 75°C (167°F) — Three Single Conductors in Conduit

| | Ohms to Neutral per 1000 Feet | | | | | | | | | | | | | | | |
| | X_L (Reactance) for All Wires | | Alternating-Current Resistance for Uncoated Copper Wires | | | Alternating-Current Resistance for Aluminum Wires | | | Effective Z at 0.85 PF for Uncoated Copper Wires | | | Effective Z at 0.85 PF for Aluminum Wires | | | |
Size (AWG or kcmil)	PVC, Aluminum Conduits	Steel Conduit	PVC Conduit	Aluminum Conduit	Steel Conduit	PVC Conduit	Aluminum Conduit	Steel Conduit	PVC Conduit	Aluminum Conduit	Steel Conduit	PVC Conduit	Aluminum Conduit	Steel Conduit	Size (AWG or kcmil)
14	0.058	0.073	3.1	3.1	3.1	—	—	—	2.7	2.7	2.7	—	—	—	14
12	0.054	0.068	2.0	2.0	2.0	3.2	3.2	3.2	1.7	1.7	1.7	2.8	2.8	2.8	12
10	0.050	0.063	1.2	1.2	1.2	2.0	2.0	2.0	1.1	1.1	1.1	1.8	1.8	1.8	10
8	0.052	0.065	0.78	0.78	0.78	1.3	1.3	1.3	0.69	0.69	0.70	1.1	1.1	1.1	8
6	0.051	0.064	0.49	0.49	0.49	0.81	0.81	0.81	0.44	0.45	0.45	0.71	0.72	0.72	6
4	0.048	0.060	0.31	0.31	0.31	0.51	0.51	0.51	0.29	0.29	0.30	0.46	0.46	0.46	4
3	0.047	0.059	0.25	0.25	0.25	0.40	0.41	0.40	0.23	0.24	0.24	0.37	0.37	0.37	3
2	0.045	0.057	0.19	0.20	0.20	0.32	0.32	0.32	0.19	0.19	0.20	0.30	0.30	0.30	2
1	0.046	0.057	0.15	0.16	0.16	0.25	0.26	0.25	0.16	0.16	0.16	0.24	0.24	0.25	1
1/0	0.044	0.055	0.12	0.13	0.12	0.20	0.21	0.20	0.13	0.13	0.13	0.19	0.20	0.20	1/0
2/0	0.043	0.054	0.10	0.10	0.10	0.16	0.16	0.16	0.11	0.11	0.11	0.16	0.16	0.16	2/0
3/0	0.042	0.052	0.077	0.082	0.079	0.13	0.13	0.13	0.088	0.092	0.094	0.13	0.13	0.14	3/0
4/0	0.041	0.051	0.062	0.067	0.063	0.10	0.11	0.10	0.074	0.078	0.080	0.11	0.11	0.11	4/0
250	0.041	0.052	0.052	0.057	0.054	0.085	0.090	0.086	0.066	0.070	0.073	0.094	0.098	0.10	250
300	0.041	0.051	0.044	0.049	0.045	0.071	0.076	0.072	0.059	0.063	0.065	0.082	0.086	0.088	300
350	0.040	0.050	0.038	0.043	0.039	0.061	0.066	0.063	0.053	0.058	0.060	0.073	0.077	0.080	350
400	0.040	0.049	0.033	0.038	0.035	0.054	0.059	0.055	0.049	0.053	0.056	0.066	0.071	0.073	400
500	0.039	0.048	0.027	0.032	0.029	0.043	0.048	0.045	0.043	0.048	0.050	0.057	0.061	0.064	500
600	0.039	0.048	0.023	0.028	0.025	0.036	0.041	0.038	0.040	0.044	0.047	0.051	0.055	0.058	600
750	0.038	0.048	0.019	0.024	0.021	0.029	0.034	0.031	0.036	0.040	0.043	0.045	0.049	0.052	750
1000	0.037	0.046	0.015	0.019	0.018	0.023	0.027	0.025	0.032	0.036	0.040	0.039	0.042	0.046	1000

Notes:

1. These values are based on the following constants: UL-type RHH wires with Class B stranding, in cradled configuration. Wire conductivities are 100 percent IACS copper and 61 percent IACS aluminum, and aluminum conduit is 45 percent IACS. Capacitive reactance is ignored, since it is negligible at these voltages. These resistance values are valid only at 75°C (167°F) and for the parameters as given, but are representative for 600-volt wire types operating at 60 Hz.

2. *Effective Z* is defined as $R \cos(\theta) + X \sin(\theta)$, where θ is the power factor angle of the circuit. Multiplying current by effective impedance gives a good approximation for line-to-neutral voltage drop. Effective impedance values shown in this table are valid only at 0.85 power factor.

For another circuit power factor (*PF*), effective impedance (*Ze*) can be calculated from R and X_L values given in this table as follows:

$$Ze = R \times PF + X_L \sin[\arccos(PF)].$$

Tables 11(a) and 11(b)

For listing purposes, Tables 11(a) and 11(b) provide the required power source limitations for Class 2 and Class 3 power sources. Table 11(a) applies for alternating-current sources, and Table 11(b) applies for direct-current sources.

The power for Class 2 and Class 3 circuits shall be either (1) inherently limited, requiring no overcurrent protection, or (2) not inherently limited, requiring a combination of power source and overcurrent protection. Power sources designed for interconnection shall be listed for the purpose.

As part of the listing, the Class 2 or Class 3 power source shall be durably marked where plainly visible to indicate the class of supply and its electrical rating. A Class 2 power source not suitable for wet location use shall be so marked.

Exception: Limited power circuits used by listed information technology equipment.

Overcurrent devices, where required, shall be located at the point where the conductor to be protected receives its supply and shall not be interchangeable with devices of higher ratings. The overcurrent device shall be permitted as an integral part of the power source.

Table 11(a). Class 2 and Class 3 Alternating-Current Power Source Limitations

Power Source		Inherently Limited Power Source (Overcurrent Protection Not Required)				Not Inherently Limited Power Source (Overcurrent Protection Required)			
		Class 2			Class 3	Class 2			Class 3
Source voltage V_{max} (volts) (see Note 1)		0 through 20*	Over 20 and through 30*	Over 30 and through 150	Over 30 and through 100	0 through 20*	Over 20 and through 30*	Over 30 and through 100	Over 100 and through 150
Power limitations VA_{max} (volt-amperes) (see Note 1)		—	—	—	—	250 (see Note 3)	250	250	N.A.
Current limitations I_{max} (amperes) (see Note 1)		8.0	8.0	0.005	$150/V_{max}$	$1000/V_{max}$	$1000/V_{max}$	$1000/V_{max}$	1.0
Maximum overcurrent protection (amperes)		—	—	—	—	5.0	$100/V_{max}$	$100/V_{max}$	1.0
Power source maximum nameplate rating	VA (volt-amperes)	$5.0 \times V_{max}$	100	$0.005 \times V_{max}$	100	$5.0 \times V_{max}$	100	100	100
	Current (amperes)	5.0	$100/V_{max}$	0.005	$100/V_{max}$	5.0	$100/V_{max}$	$100/V_{max}$	$100/V_{max}$

*Voltage ranges shown are for sinusoidal ac in indoor locations or where wet contact is not likely to occur. For nonsinusoidal or wet contact conditions, see Note 2.

Table 11(b). Class 2 and Class 3 Direct-Current Power Source Limitations

Power Source		Inherently Limited Power Source (Overcurrent Protection Not Required)					Not Inherently Limited Power Source (Overcurrent Protection Required)			
		Class 2				Class 3	Class 2			Class 3
Source voltage V_{max} (volts) (see Note 1)		0 through 20*	Over 20 and through 30*	Over 30 and through 60	Over 60 and through 150	Over 60 and through 100	0 through 20*	Over 20 and through 60*	Over 60 and through 100	Over 100 and through 150
Power limitations VA_{max} (volt-amperes) (see Note 1)		—	—	—	—	—	250 (see Note 3)	250	250	N.A.
Current limitations I_{max} (amperes) (see Note 1)		8.0	8.0	$150/V_{max}$	0.005	$150/V_{max}$	$1000/V_{max}$	$1000/V_{max}$	$1000/V_{max}$	1.0
Maximum overcurrent protection (amperes)		—	—	—	—	—	5.0	$100/V_{max}$	$100/V_{max}$	1.0
Power source maximum nameplate rating	VA (volt-amperes)	$5.0 \times V_{max}$	100	100	$0.005 \times V_{max}$	100	$5.0 \times V_{max}$	100	100	100
	Current (amperes)	5.0	$100/V_{max}$	$100/V_{max}$	0.005	$100/V_{max}$	5.0	$100/V_{max}$	$100/V_{max}$	$100/V_{max}$

*Voltage ranges shown are for continuous dc in indoor locations or where wet contact is not likely to occur. For interrupted dc or wet contact conditions, see Note 4.

Notes for Table 11(a) and 11(b)

1: V_{max}, I_{max}, and VA_{max} are determined with the current limiting impedance in the circuit (not bypassed) as follows.

V_{max}: Maximum output current under any noncapacitive load, including short circuit, and with overcurrent protection bypassed if used. Where a transformer limits the output current, I_{max} limits apply voltage regardless of load with rated input applied.

I_{max}: Maximum output after 1 minute of operation. Where a current-limiting impedance, listed for the purpose, or as part of a listed product, is used in combination with a nonpower-limited transformer or a stored energy source, e.g., storage battery, to limit the output current, I_{max} limits apply after 5 seconds.

VA_{max}: Maximum volt-ampere output after 1 minute of operation regardless of load and overcurrent protection bypassed if used.

2: For nonsinusoidal ac, V_{max} shall not be greater than 42.4 volts peak. Where wet contact (immersion not included) is likely to occur, Class 3 wiring methods shall be used or V_{max} shall not be greater than 15 volts for sinusoidal ac and 21.2 volts peak for nonsinusoidal ac.

3: If the power source is a transformer, VA_{max} is 350 or less when V_{max} is 15 or less.

4: For dc interrupted at a rate of 10 to 200 Hz, V_{max} shall not be greater than 24.8 volts peak. Where wet contact (immersion not included) is likely to occur, Class 3 wiring methods shall be used or V_{max} shall not be greater than 30 volts for continuous dc; 12.4 volts peak for dc that is interrupted at a rate of 10 to 200 Hz.

Tables 12(a) and 12(b)

For listing purposes, Tables 12(a) and 12(b) provide the required power source limitations for power-limited fire alarm sources. Table 12(a) applies for alternating-current sources, and Table 12(b) applies for direct-current sources.

The power for power-limited fire alarm circuits shall be either (1) inherently limited, requiring no overcurrent protection, or (2) not inherently limited, requiring the power to be limited by a combination of power source and overcurrent protection.

As part of the listing, the PLFA power source shall be durably marked where plainly visible to indicate that it is a power-limited fire alarm power source.

The overcurrent device, where required, shall be located at the point where the conductor to be protected receives its supply and shall not be interchangeable with devices of higher ratings. The overcurrent device shall be permitted as an integral part of the power source.

Table 12(a). PLFA Alternating-Current Power Source Limitations

Power Source		Inherently Limited Power Source (Overcurrent Protection Not Required)			Not Inherently Limited Power Source (Overcurrent Protection Required)		
Circuit voltage V_{max} (volts) (see Note 1)		0 through 20	Over 20 and through 30	Over 30 and through 100	0 through 20	Over 20 and through 100	Over 100 and through 150
Power limitations VA_{max} (volt-amperes) (see Note 1)		—	—	—	250 (see Note 2)	250	N.A.
Current limitations I_{max} (amperes) (see Note 1)		8.0	8.0	$150/V_{max}$	$1000/V_{max}$	$1000/V_{max}$	1.0
Maximum overcurrent protection (amperes)		—	—	—	5.0	$100/V_{max}$	1.0
Power source maximum nameplate ratings	VA (volt-amperes)	$5.0 \times V_{max}$	100	100	$5.0 \times V_{max}$	100	100
	Current (amperes)	5.0	$100/V_{max}$	$100/V_{max}$	5.0	$100/V_{max}$	$100/V_{max}$

Table 12(b). PLFA Direct-Current Power Source Limitations

Power Source		Inherently Limited Power Source (Overcurrent Protection Not Required)				Not Inherently Limited Power Source (Overcurrent Protection Required)		
Circuit voltage V_{max} (volts) (see Note 1)		0 through 20	Over 20 and through 30	Over 30 and through 100	Over 100 and through 250	0 through 20	Over 20 and through 100	Over 100 and through 150
Power limitations VA_{max} (volt-amperes) (see Note 1)		—	—	—	—	250 (see Note 2)	250	N.A.
Current limitations I_{max} (amperes) (see Note 1)		8.0	8.0	$150/V_{max}$	0.030	$1000/V_{max}$	$1000/V_{max}$	1.0
Maximum overcurrent protection (amperes)		—	—	—	—	5.0	$100/V_{max}$	1.0
Power source maximum nameplate ratings	VA (volt-amperes)	$5.0 \times V_{max}$	100	100	$0.030 \times V_{max}$	$5.0 \times V_{max}$	100	100
	Current (amperes)	5.0	$100/V_{max}$	$100/V_{max}$	0.030	5.0	$100/V_{max}$	$100/V_{max}$

Notes for Tables 12(a) and 12(b)

1. V_{max}, I_{max}, and VA_{max} are determined as follows.

   V_{max}: Maximum output voltage regardless of load with rated input applied.

   I_{max}: Maximum output current under any noncapacitive load, including short circuit, and with overcurrent protection bypassed if used. Where a transformer limits the output current, I_{max} limits apply after 1 minute of operation. Where a current-limiting impedance, listed for the purpose, is used in combination with a nonpower-limited transformer or a stored energy source, e.g., storage battery, to limit the output current, I_{max} limits apply after 5 seconds.

   VA_{max}: Maximum volt-ampere output after 1 minute of operation regardless of load and overcurrent protection bypassed if used. Current limiting impedance shall not be bypassed when determining I_{max} and VA_{max}.

2. If the power source is a transformer, VA_{max} is 350 or less when V_{max} is 15 or less.

APPENDIX A
Extract References

(See Section 90-3, Paragraph 4)

A-250-104(b)	*National Fuel Gas Code*, NFPA 54-1996	Paragraph 3.14(a)
A-500-5(a)(1)	*Recommended Practice for the Classification of Flammable Liquids, Gases, or Vapors and of Hazardous (Classified) Locations for Electrical Installations in Chemical Process Areas*, NFPA 497-1997	Section 1-3, Definitions, Combustible Material
A-500-5(a)(2)	*Recommended Practice for the Classification of Flammable Liquids, Gases, or Vapors and of Hazardous (Classified) Locations for Electrical Installations in Chemical Process Areas*, NFPA 497-1997	Section 1-3, Definitions, Combustible Material
A-500-5(a)(3)	*Recommended Practice for the Classification of Flammable Liquids, Gases, or Vapors and of Hazardous (Classified) Locations for Electrical Installations in Chemical Process Areas*, NFPA 497-1997	Section 1-3, Definitions, Combustible Material
A-500-5(a)(4)	*Recommended Practice for the Classification of Flammable Liquids, Gases, or Vapors and of Hazardous (Classified) Locations for Electrical Installations in Chemical Process Areas*, NFPA 497-1997	Section 1-3, Definitions, Combustible Material
A-500-5(b)(1)	*Recommended Practice for the Classification of Combustible Dusts and of Hazardous (Classified) Locations for Electrical Installations in Chemical Process Areas*, NFPA 499-1997	Section 1-3, Definitions, Combustible Dusts
A-500-5(b)(2)	*Recommended Practice for the Classification of Combustible and of Hazardous (Classified) Locations for Electrical Installations in Chemical Process Areas*, NFPA 499-1997	Section 1-3, Definitions, Combustible Dusts
A-505-7(a)	*Recommended Practice for the Classification of Flammable Liquids, Gases, or Vapors and of Hazardous (Classified) Locations in Chemical Process Areas*, NFPA 497-1997	Section 1-3, Definitions, Combustible Material
A-505-7(b)	*Recommended Practice for the Classification of Flammable Liquids, Gases, or Vapors and of Hazardous (Classified) Locations in Chemical Process Areas*, NFPA 497-1997	Section 1-3, Definitions, Combustible Material
A-505-7(c)	*Recommended Practice for the Classification of Flammable Liquids, Gases, or Vapors and of Hazardous (Classified) Locations in Chemical Process Areas*, NFPA 497-1997	Section 1-3, Definitions, Combustible Material
A-514-2	*Automotive and Marine Service Station Code*, NFPA 30A-1996	Sections 7-1 and 7-3
A-Table 514-2	*Automotive and Marine Service Station Code*, NFPA 30A-1996	Table 7
A-Fig. 514-2	*Automotive and Marine Service Station Code*, NFPA 30A-1996	Figure 7-1
A-514-5(b)	*Automotive and Marine Service Station Code*, NFPA 30A-1996	Paragraph 9-4.5
A-514-5(c)	*Automotive and Marine Service Station Code*, NFPA 30A-1996	Paragraph 9-5.3
A-515-2	*Flammable and Combustible Liquids Code*, NFPA 30-1996	Paragraphs 5-9.5.1, 5-9.5.3
A-Table 515-2	*Flammable and Combustible Liquids Code*, NFPA 30-1996	Table 5-9.5.3
A-Fig. 515-2	*Flammable and Combustible Liquids Code*, NFPA 30-1996	Figure 5-7.16
A-516-2(a)(1)	*Standard for Spray Application Using Flammable or Combustible Materials*, NFPA 33-1995	Section 1-6, Definitions, Spray Area
A-516-2(a)(2)	*Standard for Spray Application Using Flammable or Combustible Materials*, NFPA 33-1995	Section 1-6, Definitions, Spray Area
A-516-2(a)(3)	*Standard for Spray Application Using Flammable or Combustible Materials*, NFPA 33-1995	Section 1-6, Definitions, Spray Area
A-516-2(a)(4)	*Standard for Dipping and Coating Processes Using Flammable or Combustible Liquids*, NFPA 34-1995	Section 1-6, Definitions, Vapor Source, and Paragraph 4-2.2
A-516-2(a)(5)	*Standard for Dipping and Coating Processes Using Flammable or Combustible Liquids*, NFPA 34-1995	Paragraph 4-2.1
A-516-2(a)(6)	*Standard for Dipping and Coating Processes Using Flammable or Combustible Liquids*, NFPA 34-1995	Paragraph 4-3.1
A-516-2(b)(1)	*Standard for Spray Application Using Flammable or Combustible Materials*, NFPA 33-1995	Paragraph 4-3.1
A-516, Fig. 516-2(b)(1)	*Standard for Spray Application Using Flammable or Combustible Materials*, NFPA 33-1995	Figure 4-3.1
A-516-2(b)(2)	*Standard for Spray Application Using Flammable or Combustible Materials*, NFPA 33-1995	Paragraph 4-3.2
A-516, Fig. 516-2(b)(2)	*Standard for Spray Application Using Flammable or Combustible Materials*, NFPA 33-1995	Figures 4-3.2(a), 4-3.2(b)
A-516-2(b)(3)	*Standard for Spray Application Using Flammable or Combustible Materials*, NFPA 33-1995	Paragraph 4-3.3
A-516-2(b)(4)	*Standard for Spray Application Using Flammable or Combustible Materials*, NFPA 33-1995	Paragraph 4-3.4
A-516, Fig. 516-2(b)(4)	*Standard for Spray Application Using Flammable or Combustible Materials*, NFPA 33-1995	Figure 4-3.4
A-516-2(b)(5)	*Standard for Dipping and Coating Processes Using Flammable or Combustible Liquids*, NFPA 34-1995	Paragraph 4-2.3

A-516-2(b)(6)	*Standard for Dipping and Coating Processes Using Flammable or Combustible Liquids,* NFPA 34-1995	Paragraph 4-2.4
A-516, Fig. 516-2(b)(5)	*Standard for Dipping and Coating Processes Using Flammable or Combustible Liquids,* NFPA 34-1995	Figure 4-2(a)
A-516-2(c)	*Standard for Dipping and Coating Processes Using Flammable or Combustible Liquids,* NFPA 34-1995	Section 4-3.2
A-516-3(b)	*Standard for Spray Application Using Flammable or Combustible Materials,* NFPA 33-1995	Section 4-2
A-516-3(d)	*Standard for Spray Application Using Flammable or Combustible Materials,* NFPA 33-1995	Section 4-7
A-516-4	*Standard for Spray Application Using Flammable or Combustible Materials,* NFPA 33-1995	Chapter 9
A-516-5	*Standard for Spray Application Using Flammable or Combustible Materials,* NFPA 33-1995	Chapter 10
A-516-6	*Standard for Spray Application Using Flammable or Combustible Materials,* NFPA 33-1995	Chapter 13
A-517-30(b)(1)	*Standard for Health Care Facilities,* NFPA 99-1996	Paragraph 3-4.2.2.1
A-517-30(b)(2)	*Standard for Health Care Facilities,* NFPA 99-1996	Paragraph 3-4.2.2.1
A-517-30(b)(3)	*Standard for Health Care Facilities,* NFPA 99-1996	Paragraph 3-4.2.2.1
A-517-30(b)(4)	*Standard for Health Care Facilities,* NFPA 99-1996	Paragraph 12-3.3.2
A-517-31	*Standard for Health Care Facilities,* NFPA 99-1996	Paragraphs 3-4.2.2.2(a), 3-5.2.2.2
A-517-32	*Standard for Health Care Facilities,* NFPA 99-1996	Paragraph 3-4.2.2.2(b)
A-517-33	*Standard for Health Care Facilities,* NFPA 99-1996	Paragraph 3-4.2.2.2(c)
A-517-33(c)	*Standard for Health Care Facilities,* NFPA 99-1996	Paragraph 3-4.2.2.4(b)2.
A-517-34	*Standard for Health Care Facilities,* NFPA 99-1996	Paragraph 3-4.2.2.3(b)
A-517-34(a)	*Standard for Health Care Facilities,* NFPA 99-1996	Paragraph 3-4.2.2.3(c)
A-517-34(b)	*Standard for Health Care Facilities,* NFPA 99-1996	Paragraph 3-4.2.2.3(d)
A-517-35(a)	*Standard for Health Care Facilities,* NFPA 99-1996	Paragraph 3-4.1.1.2
A-517-40(a)	*Standard for Health Care Facilities,* NFPA 99-1996	Paragraphs 16-3.3.2, Exception; 17-3.3.2, Exception
A-517-41(a)	*Standard for Health Care Facilities,* NFPA 99-1996	Paragraph 3-5.2.2.1
A-517-41(b)	*Standard for Health Care Facilities,* NFPA 99-1996	Paragraph 3-5.2.2.1
A-517-42	*Standard for Health Care Facilities,* NFPA 99-1996	Paragraphs 3-5.2.2.2, 3-5.3.1
A-517-43	*Standard for Health Care Facilities,* NFPA 99-1996	Paragraph 3-5.2.2.3
A-517-44(a)	*Standard for Health Care Facilities,* NFPA 99-1996	Paragraphs 3-5.1, 3-4.1.1.2
A-517-44(b)	*Standard for Health Care Facilities,* NFPA 99-1996	Paragraphs 3-5.1, 3-4.1.1.3, 16-3.3.2.1, 17-3.3.2.1
A-517-45(b)	*Standard for Health Care Facilities,* NFPA 99-1996	Paragraphs 3-6.2.2.2, 13-3.4.1
A-517-45(c)(1)	*Standard for Health Care Facilities,* NFPA 99-1996	Paragraph 3-6.1
A-517-45(c)(2)	*Standard for Health Care Facilities,* NFPA 99-1996	Paragraph 3-6.3.1.1
A-517-45(c)(3)	*Standard for Health Care Facilities,* NFPA 99-1996	Paragraph 3-6.3.1.2
A-517-50(b)	*Standard for Health Care Facilities,* NFPA 99-1996	Paragraphs 3-6.2.2.2, 14-3.4.1, 15-3.4.1
A-517-50(c)(1)	*Standard for Health Care Facilities,* NFPA 99-1996	Paragraph 3-6.1
A-517-50(c)(2)	*Standard for Health Care Facilities,* NFPA 99-1996	Paragraph 3-6.3.1.1
A-517-50(c)(3)	*Standard for Health Care Facilities,* NFPA 99-1996	Paragraph 3-6.3.1.2
A-517-60(a)(1)	*Standard for Health Care Facilities,* NFPA 99-1996	ANNEX 2, Sections 2-1 and 2-2
A-517-61(a)(1)	*Standard for Health Care Facilities,* NFPA 99-1996	ANNEX 2, Paragraph 2-6.3.2
A-517-61(a)(3)	*Standard for Health Care Facilities,* NFPA 99-1996	ANNEX 2, Para graphs 2-2.1, 2-4.5, 2-4.6, 2-4.7
A-517-160(a)(2)	*Standard for Health Care Facilities,* NFPA 99-1996	Paragraph 3-3.2.2.1
A-517-160(a)(4)	*Standard for Health Care Facilities,* NFPA 99-1996	Paragraphs 12-4.1.2.6(d), 12-4.1.2.6(e)

A-695-3(a)	*Standard for the Installation of Centrifugal Fire Pumps,* NFPA 20-1996	Section 6-2 and Paragraphs 6-2.1, 6-2.2
A-695-3(b)	*Standard for the Installation of Centrifugal Fire Pumps,* NFPA 20-1996	Paragraphs 6-2.3.1, 6-2.4.1
A-695-4	*Standard for the Installation of Centrifugal Fire Pumps,* NFPA 20-1996	Paragraphs 6-2.4.3, 6-3.2.2
A-695-7	*Standard for the Installation of Centrifugal Fire Pumps,* NFPA 20-1996	Paragraph 6-3.1.2
A-695-10	*Standard for the Installation of Centrifugal Fire Pumps,* NFPA 20-1996	Paragraphs 6-4.1.1, 7-1.1.1, 7-7, 7-8.1, 7-9.1, 9-1.1.1.
A-695-14(a)	*Standard for the Installation of Centrifugal Fire Pumps,* NFPA 20-1996	Paragraph 7-5.2.5
A-695-14(b)	*Standard for the Installation of Centrifugal Fire Pumps,* NFPA 20-1996	Paragraph 7-4.5.6
A-695-14(c)	*Standard for the Installation of Centrifugal Fire Pumps,* NFPA 20-1996	Paragraph 7-8.1.3
A-695-14(d)	*Standard for the Installation of Centrifugal Fire Pumps,* NFPA 20-1996	Paragraph 9-3.5.1

APPENDIX B

This appendix is not a part of the requirements of this Code but is included for informational purposes only.

B-310-15(b)(1). Formula Application Information. This appendix provides application information for ampacities calculated under engineering supervision.

B-310-15(b)(2). Typical Applications Covered by Tables. Typical ampacities for conductors rated 0 through 2000 volts are shown in Tables B-310-1 through B-310-10. Underground electrical duct bank configurations, as detailed in Figures B-310-3, B-310-4, and B-310-5, are utilized for conductors rated 0 through 5000 volts. In Figures B-310-2 through B-310-5, where adjacent duct banks are used, a separation of 5 ft (1.52 m) between the centerlines of the closest ducts in each bank or 4 ft (1.22 m) between the extremities of the concrete envelopes is sufficient to prevent derating of the conductors due to mutual heating. These ampacities were calculated as detailed in the basic ampacity paper, *The Calculation of the Temperature Rise and Load Capability of Cable Systems,* by J. H. Neher and M. H. McGrath, AIEE Paper 57-660. For additional information concerning the application of these ampacities, see *Power Cable Ampacities,* IEEE/ICEA Standard S-135/P-46-426 and IEEE Standard 835-1994, *Standard Power Cable Ampacity Tables.*

Typical values of thermal resistivity (Rho) are as follows:

Average soil (90 percent of USA) = 90

Concrete = 55

Damp soil (coastal areas, high water table) = 60

Paper insulation = 550

Polyethylene (PE) = 450

Polyvinyl chloride PVC = 650

Rubber and rubber-like = 500

Very dry soil (rocky or sandy) = 120

Thermal resistivity, as used in this appendix, refers to the heat transfer capability through a substance by conduction. It is the reciprocal of thermal conductivity and is normally expressed in the units °C-cm/watt. For additional information on determining soil thermal resistivity (Rho), see *Guide for Soil Thermal Resistivity Measurements,* ANSI/IEEE Standard 442-1996.

B-310-15(b)(3). Criteria Modifications. Where values of load factor and Rho are known for a particular electrical duct bank installation and they are different from those shown in a specific table or figure, the ampacities shown in the table or figure can be modified by the application of factors derived from the use of Figure B-310-1.

Where two different ampacities apply to adjacent portions of a circuit, the higher ampacity can be used beyond the point of transition, a distance equal to 10 ft (3.05 m) or 10 percent of the circuit length figured at the higher ampacity, whichever is less.

Where the burial depth of direct burial or electrical duct bank circuits are modified from the values shown in a figure or table, ampacities can be modified as shown in (1) and (2) as follows.

(1) Where burial depths are increased in part(s) of an electrical duct run to avoid underground obstructions, no decrease in ampacity of the conductors is needed, provided the total length of parts of the duct run increased in depth to avoid obstructions is less than 25 percent of the total run length.

(2) Where burial depths are deeper than shown in a specific underground ampacity table or figure, an ampacity derating factor of 6 percent per increased foot (305 mm) of depth for all values of Rho can be utilized. No rating change is needed where the burial depth is decreased.

B-310-15(b)(4). Electrical Ducts. The term *electrical duct(s)* is defined in Section 310-60.

B-310-15(b)(5). Tables B-310-6 and B-310-7.

(1) To obtain the ampacity of cables installed in two electrical ducts in one horizontal row with 7.5-in. (191-mm) center-to-center spacing between electrical ducts, similar to Figure B-310-2, Detail 1, multiply the ampacity shown for one duct in Tables B-310-6 and B-310-7 by 0.88.

(2) To obtain the ampacity of cables installed in four electrical ducts in one horizontal row with 7.5-in. (191-mm) center-to-center spacing between electrical ducts, similar to Figure B-310-2, Detail 2, multiply the ampacity shown for three electrical ducts in Tables B-310-6 and B-310-7 by 0.94.

B-310-15(b)(6). Electrical Ducts Utilized in Figure B-310-2. If spacing between electrical ducts, as shown in Figure B-310-2, is less than specified in Figure B-310-2, where electrical ducts enter equipment enclosures from underground, the ampacity of conductors contained within such electrical ducts need not be reduced.

B-310-15(b)(7). Examples Showing Use of Figure B-310-1 for Electrical Duct Bank Ampacity Modifications. Figure B-310-1 is used for interpolation or extrapolation for values of Rho and load factor for cables installed in electrical ducts. The upper family of curves shows the variation in ampacity and Rho at unity load factor in terms of I_1, the ampacity for Rho = 60, and 50 percent load factor. Each curve is designated for a particular ratio I_2/I_1, where I_2 is the ampacity at Rho = 120 and 100 percent load factor.

The lower family of curves shows the relationship between Rho and load factor that will give substantially the same ampacity as the indicated value of Rho at 100 percent load factor.

As an example, to find the ampacity of a 500 kcmil copper cable circuit for six electrical ducts as shown in Table B-310-5: At the Rho = 60, LF = 50, I_1 = 583; for Rho = 120 and LF = 100, I_2 = 400. The ratio I_2/I_1 = 0.686. Locate Rho = 90 at the bottom of the chart and follow the 90 Rho line to the intersection with 100 percent load factor where the equivalent Rho = 90. Then follow the 90 Rho line to I_2/I_1 ratio of 0.686 where F = 0.74. The desired ampacity = 0.74 × 583 = 431, which agrees with the table for Rho = 90, LF = 100.

To determine the ampacity for the same circuit where Rho = 80 and LF = 75, using Figure B-310-1, the equivalent Rho = 43, F = 0.855, and the desired ampacity = 0.855 × 583 = 498 amperes. Values for using Figure B-310-1 are found in the electrical duct bank ampacity tables of this appendix.

Where the load factor is less than 100 percent and can be verified by measurement or calculation, the ampacity of electrical duct bank installations can be modified as shown. Different values of Rho can be accommodated in the same manner.

Table B-310-1. Ampacities of Two or Three Insulated Conductors, Rated 0 through 2000 Volts, Within an Overall Covering (Multiconductor Cable), in Raceway in Free Air Based on Ambient Air Temperature of 30°C (86°F)

Size	Temperature Rating of Conductor. See Table 310-13.						Size
	60°C (140°F)	75°C (167°F)	90°C (194°F)	60°C (140°F)	75°C (167°F)	90°C (194°F)	
	Types TW, UF	Types RH, RHW, THHW, THW, THWN, XHHW, ZW	Types THHN, THHW, THW-2, THWN-2, RHH, RWH-2, USE-2, XHHW, XHHW-2, ZW-2	Type TW	Types RH, RHW, THHW, THW, THWN, XHHW	Types THHN, THHW, THW-2, THWN-2, RHH, RWH-2, USE-2, XHHW, XHHW-2, ZW-2	
AWG or kcmil	COPPER			ALUMINUM OR COPPER-CLAD ALUMINUM			AWG or kcmil
14	16*	18*	21*	—	—	—	14
12	20*	24*	27*	16*	18*	21*	12
10	27*	33*	36*	21*	25*	28*	10
8	36	43	48	28	33	37	8
6	48	58	65	38	45	51	6
4	66	79	89	51	61	69	4
3	76	90	102	59	70	79	3
2	88	105	119	69	83	93	2
1	102	121	137	80	95	106	1
1/0	121	145	163	94	113	127	1/0
2/0	138	166	186	108	129	146	2/0
3/0	158	189	214	124	147	167	3/0
4/0	187	223	253	147	176	197	4/0
250	205	245	276	160	192	217	250
300	234	281	317	185	221	250	300
350	255	305	345	202	242	273	350
400	274	328	371	218	261	295	400
500	315	378	427	254	303	342	500
600	343	413	468	279	335	378	600
700	376	452	514	310	371	420	700
750	387	466	529	321	384	435	750
800	397	479	543	331	397	450	800
900	415	500	570	350	421	477	900
1000	448	542	617	382	460	521	1000
Ambient Temp. (°C)	For ambient temperatures other than 30°C (86°F), multiply the ampacities shown above by the appropriate factor shown below.						Ambient Temp. (°F)
21–25	1.08	1.05	1.04	1.08	1.05	1.04	70–77
26–30	1.00	1.00	1.00	1.00	1.00	1.00	79–86
31–35	0.91	0.94	0.96	0.91	0.94	0.96	88–95
36–40	0.82	0.88	0.91	0.82	0.88	0.91	97–104
41–45	0.71	0.82	0.87	0.71	0.82	0.87	106–113
46–50	0.58	0.75	0.82	0.58	0.75	0.82	115–122
51–55	0.41	0.67	0.76	0.41	0.67	0.76	124–131
56–60	—	0.58	0.71	—	0.58	0.71	133–140
61–70	—	0.33	0.58	—	0.33	0.58	142–158
71–80	—	—	0.41	—	—	0.41	160–176

*Unless otherwise specifically permitted elsewhere in this *Code,* the overcurrent protection for these conductor types shall not exceed 15 amperes for No. 14, 20 amperes for No. 12, and 30 amperes for No. 10 copper; or 15 amperes for No. 12 and 25 amperes for No. 10 aluminum and copper-clad aluminum.

Table B-310-3. Ampacities of Multiconductor Cables with Not More than Three Insulated Conductors, Rated 0 Through 2000 Volts, in Free Air Based on Ambient Air Temperature of 40°C (104°F) (For Types TC, MC, MI, UF, and USE Cables)

Size	Temperature Rating of Conductor. See Table 310-13.								Size
	60°C (140°F)	75°C (167°F)	85°C (185°F)	90°C (194°F)	60°C (140°F)	75°C (167°F)	85°C (185°F)	90°C (194°F)	
AWG or kcmil	COPPER				ALUMINUM OR COPPER-CLAD ALUMINUM				AWG or kcmil
18	—	—	—	11*	—	—	—	—	18
16	—	—	—	16*	—	—	—	—	16
14	18*	21*	24*	25*	—	—	—	—	14
12	21*	28*	30*	32*	18*	21*	24*	25*	12
10	28*	36*	41*	43*	21*	28*	30*	32*	10
8	39	50	56	59	30	39	44	46	8
6	52	68	75	79	41	53	59	61	6
4	69	89	100	104	54	70	78	81	4
3	81	104	116	121	63	81	91	95	3
2	92	118	132	138	72	92	103	108	2
1	107	138	154	161	84	108	120	126	1
1/0	124	160	178	186	97	125	139	145	1/0
2/0	143	184	206	215	111	144	160	168	2/0
3/0	165	213	238	249	129	166	185	194	3/0
4/0	190	245	274	287	149	192	214	224	4/0
250	212	274	305	320	166	214	239	250	250
300	237	306	341	357	186	240	268	280	300
350	261	337	377	394	205	265	296	309	350
400	281	363	406	425	222	287	317	334	400
500	321	416	465	487	255	330	368	385	500
600	354	459	513	538	284	368	410	429	600
700	387	502	562	589	306	405	462	473	700
750	404	523	586	615	328	424	473	495	750
800	415	539	604	633	339	439	490	513	800
900	438	570	639	670	362	469	514	548	900
1000	461	601	674	707	385	499	558	584	1000
Ambient Temp. (°C)	For ambient temperatures other than 40°C (104°F), multiply the ampacities shown above by the appropriate factor shown below.								Ambient Temp. (°F)
21–25	1.32	1.20	1.15	1.14	1.32	1.20	1.15	1.14	70–77
26–30	1.22	1.13	1.11	1.10	1.22	1.13	1.11	1.10	79–86
31–35	1.12	1.07	1.05	1.05	1.12	1.07	1.05	1.05	88–95
36–40	1.00	1.00	1.00	1.00	1.00	1.00	1.00	1.00	97–104
41–45	0.87	0.93	0.94	0.95	0.87	0.93	0.94	0.95	106–113
46–50	0.71	0.85	0.88	0.89	0.71	0.85	0.88	0.89	115–122
51–55	0.50	0.76	0.82	0.84	0.50	0.76	0.82	0.84	124–131
56–60	—	0.65	0.75	0.77	—	0.65	0.75	0.77	133–140
61–70	—	0.38	0.58	0.63	—	0.38	0.58	0.63	142–158
71–80	—	—	0.33	0.44	—	—	0.33	0.44	160–176

*Unless otherwise specifically permitted elsewhere in this *Code,* the overcurrent protection for these conductor types shall not exceed 15 amperes for No. 14, 20 amperes for No. 12, and 30 amperes for No. 10 copper; or 15 amperes for No. 12 and 25 amperes for No. 10 aluminum and copper-clad aluminum.

Table B-310-5. Ampacities of Single Insulated Conductors, Rated 0 through 2000 Volts, in Nonmagnetic Underground Electrical Ducts (One Conductor per Electrical Duct), Based on Ambient Earth Temperature of 20°C (68°F), Electrical Duct Arrangement per Figure B-310-2, Conductor Temperature 75°C (167°F)

Size (kcmil)	3 Electrical Ducts (Fig. B-310-2, Detail 2) Types RHW, THHW, THW, THWN, XHHW, USE			6 Electrical Ducts (Fig. B-310-2, Detail 3) Types RHW, THHW, THW, THWN, XHHW, USE			9 Electrical Ducts (Fig. B-310-2, Detail 4) Types RHW, THHW, THW, THWN, XHHW, USE			3 Electrical Ducts (Fig. B-310-2, Detail 2) Types RHW, THHW, THW, THWN, XHHW, USE			6 Electrical Ducts (Fig. B-310-2, Detail 3) Types RHW, THHW, THW, THWN, XHHW, USE			9 Electrical Ducts (Fig. B-310-2, Detail 4) Types RHW, THHW, THW, THWN, XHHW, USE			Size (kcmil)
	COPPER									ALUMINUM OR COPPER-CLAD ALUMINUM									
	RHO 60 LF 50	RHO 90 LF 100	RHO 120 LF 100	RHO 60 LF 50	RHO 90 LF 100	RHO 120 LF 100	RHO 60 LF 50	RHO 90 LF 100	RHO 120 LF 100	RHO 60 LF 50	RHO 90 LF 100	RHO 120 LF 100	RHO 60 LF 50	RHO 90 LF 100	RHO 120 LF 100	RHO 60 LF 50	RHO 90 LF 100	RHO 120 LF 100	
250	410	344	327	386	295	275	369	270	252	320	269	256	302	230	214	288	211	197	250
350	503	418	396	472	355	330	446	322	299	393	327	310	369	277	258	350	252	235	350
500	624	511	484	583	431	400	545	387	360	489	401	379	457	337	313	430	305	284	500
750	794	640	603	736	534	494	674	469	434	626	505	475	581	421	389	538	375	347	750
1000	936	745	700	864	617	570	776	533	493	744	593	557	687	491	453	629	432	399	1000
1250	1055	832	781	970	686	632	854	581	536	848	668	627	779	551	508	703	478	441	1250
1500	1160	907	849	1063	744	685	918	619	571	941	736	689	863	604	556	767	517	477	1500
1750	1250	970	907	1142	793	729	975	651	599	1026	796	745	937	651	598	823	550	507	1750
2000	1332	1027	959	1213	836	768	1030	683	628	1103	850	794	1005	693	636	877	581	535	2000

Ambient Temp. (°C)	For ambient temperatures other than 20°C (68°F), multiply the ampacities shown above by the appropriate factor shown below.																		Ambient Temp. (°F)
6–10	1.09			1.09			1.09			1.09			1.09			1.09			43–50
11–15	1.04			1.04			1.04			1.04			1.04			1.04			52–59
16–20	1.00			1.00			1.00			1.00			1.00			1.00			61–68
21–25	0.95			0.95			0.95			0.95			0.95			0.95			70–77
26–30	0.90			0.90			0.90			0.90			0.90			0.90			79–86

Table B-310-6. Ampacities of Three Insulated Conductors, Rated 0 through 2000 Volts, Within an Overall Covering (Three-Conductor Cable) in Underground Electrical Ducts (One Cable per Electrical Duct) Based on Ambient Earth Temperature of 20°C (68°F), Electrical Duct Arrangement per Figure B-310-2, Conductor Temperature 75°C (167°F)

Size (AWG or kcmil)	1 Electrical Duct Fig. B-310-2, Detail 1 Types RHW, THHW, THW, THWN, XHHW, USE			3 Electrical Ducts (Fig. B-310-2, Detail 2) Types RHW, THHW, THW, THWN, XHHW, USE			6 Electrical Ducts (Fig. B-310-2, Detail 3) Types RHW, THHW, THW, THWN, XHHW, USE			1 Electrical Duct Fig. B-310-2, Detail 1 Types RHW, THHW, THW, THWN, XHHW, USE			3 Electrical Ducts (Fig. B-310-2, Detail 2) Types RHW, THHW, THW, THWN, XHHW, USE			6 Electrical Ducts (Fig. B-310-2, Detail 3) Types RHW, THHW, THW, THWN, XHHW, USE			Size (AWG or kcmil)
	COPPER									ALUMINUM OR COPPER-CLAD ALUMINUM									
	RHO 60 LF 50	RHO 90 LF 100	RHO 120 LF 100	RHO 60 LF 50	RHO 90 LF 100	RHO 120 LF 100	RHO 60 LF 50	RHO 90 LF 100	RHO 120 LF 100	RHO 60 LF 50	RHO 90 LF 100	RHO 120 LF 100	RHO 60 LF 50	RHO 90 LF 100	RHO 120 LF 100	RHO 60 LF 50	RHO 90 LF 100	RHO 120 LF 100	
8	58	54	53	56	48	46	53	42	39	45	42	41	43	37	36	41	32	30	8
6	77	71	69	74	63	60	70	54	51	60	55	54	57	49	47	54	42	39	6
4	101	93	91	96	81	77	91	69	65	78	72	71	75	63	60	71	54	51	4
2	132	121	118	126	105	100	119	89	83	103	94	92	98	82	78	92	70	65	2
1	154	140	136	146	121	114	137	102	95	120	109	106	114	94	89	107	79	74	1
1/0	177	160	156	168	137	130	157	116	107	138	125	122	131	107	101	122	90	84	1/0
2/0	203	183	178	192	156	147	179	131	121	158	143	139	150	122	115	140	102	95	2/0
3/0	233	210	204	221	178	158	205	148	137	182	164	159	172	139	131	160	116	107	3/0
4/0	268	240	232	253	202	190	234	168	155	209	187	182	198	158	149	183	131	121	4/0
250	297	265	256	280	222	209	258	184	169	233	207	201	219	174	163	505	144	132	250
350	363	321	310	340	267	250	312	219	202	285	252	244	267	209	196	245	172	158	350
500	444	389	375	414	320	299	377	261	240	352	308	297	328	254	237	299	207	190	500
750	552	478	459	511	388	362	462	314	288	446	380	372	413	314	293	374	254	233	750
1000	628	539	518	579	435	405	522	351	321	521	447	430	480	361	336	433	291	266	1000

Ambient Temp. (°C)	For ambient temperatures other than 20°C (68°F), multiply the ampacities shown above by the appropriate factor shown below.																		Ambient Temp. (°F)
6–10	1.09			1.09			1.09			1.09			1.09			1.09			43–50
11–15	1.04			1.04			1.04			1.04			1.04			1.04			52–59
16–20	1.00			1.00			1.00			1.00			1.00			1.00			61–68
21–25	0.95			0.95			0.95			0.95			0.95			0.95			70–77
26–30	0.90			0.90			0.90			0.90			0.90			0.90			79–86

Table B-310-7. Ampacities of Three Single Insulated Conductors, Rated 0 Through 2000 Volts, in Underground Electrical Ducts (Three Conductors per Electrical Duct) Based on Ambient Earth Temperature of 20°C (68°F), Electrical Duct Arrangement per Figure B-310-2, Conductor Temperature 75°C (167°F)

Size (AWG or kcmil)	1 Electrical Duct (Fig. B-310-2, Detail 1) Types RHW, THHW, THW, THWN, XHHW, USE COPPER			3 Electrical Ducts (Fig. B-310-2, Detail 2) Types RHW, THHW, THW, THWN, XHHW, USE			6 Electrical Ducts (Fig. B-310-2, Detail 3) Types RHW, THHW, THW, THWN, XHHW, USE			1 Electrical Duct (Fig. B-310-2, Detail 1) Types RHW, THHW, THW, THWN, XHHW, USE ALUMINUM OR COPPER-CLAD ALUMINUM			3 Electrical Ducts (Fig. B-310-2, Detail 2) Types RHW, THHW, THW, THWN, XHHW, USE			6 Electrical Ducts (Fig. B-310-2, Detail 3) Types RHW, THHW, THW, THWN, XHHW, USE			Size (AWG or kcmil)
	RHO 60 LF 50	RHO 90 LF 100	RHO 120 LF 100	RHO 60 LF 50	RHO 90 LF 100	RHO 120 LF 100	RHO 60 LF 50	RHO 90 LF 100	RHO 120 LF 100	RHO 60 LF 50	RHO 90 LF 100	RHO 120 LF 100	RHO 60 LF 50	RHO 90 LF 100	RHO 120 LF 100	RHO 60 LF 50	RHO 90 LF 100	RHO 120 LF 100	
8	63	58	57	61	51	49	57	44	41	49	45	44	47	40	38	45	34	32	8
6	84	77	75	80	67	63	75	56	53	66	60	58	63	52	49	59	44	41	6
4	111	100	98	105	86	81	98	73	67	86	78	76	79	67	63	77	57	52	4
3	129	116	113	122	99	94	113	83	77	101	91	89	83	77	73	84	65	60	3
2	147	132	128	139	112	106	129	93	86	115	103	100	108	87	82	101	73	67	2
1	171	153	148	161	128	121	149	106	98	133	119	115	126	100	94	116	83	77	1
1/0	197	175	169	185	146	137	170	121	111	153	136	132	144	114	107	133	94	87	1/0
2/0	226	200	193	212	166	156	194	136	126	176	156	151	165	130	121	151	106	98	2/0
3/0	260	228	220	243	189	177	222	154	142	203	178	172	189	147	138	173	121	111	3/0
4/0	301	263	253	280	215	201	255	175	161	235	205	198	219	168	157	199	137	126	4/0
250	334	290	279	310	236	220	281	192	176	261	227	218	242	185	172	220	150	137	250
300	373	321	308	344	260	242	310	210	192	293	252	242	272	204	190	245	165	151	300
350	409	351	337	377	283	264	340	228	209	321	276	265	296	222	207	266	179	164	350
400	442	376	361	394	302	280	368	243	223	349	297	284	321	238	220	288	191	174	400
500	503	427	409	460	341	316	412	273	249	397	338	323	364	270	250	326	216	197	500
600	552	468	447	511	371	343	457	296	270	446	373	356	408	296	274	365	236	215	600
700	602	509	486	553	402	371	492	319	291	488	408	389	443	321	297	394	255	232	700
750	632	529	505	574	417	385	509	330	301	508	425	405	461	334	309	409	265	241	750
800	654	544	520	597	428	395	527	338	308	530	439	418	481	344	318	427	273	247	800
900	692	575	549	628	450	415	554	355	323	563	466	444	510	365	337	450	288	261	900
1000	730	605	576	659	472	435	581	372	338	597	494	471	538	385	355	475	304	276	1000

Ambient Temp. (°C)	For ambient temperatures other than 20°C (68°F), multiply the ampacities shown above by the appropriate factor shown below.																		Ambient Temp. (°F)
6–10	1.09			1.09			1.09			1.09			1.09			1.09			43–50
11–15	1.04			1.04			1.04			1.04			1.04			1.04			52–59
16–20	1.00			1.00			1.00			1.00			1.00			1.00			61–68
21–25	0.95			0.95			0.95			0.95			0.95			0.95			70–77
26–30	0.90			0.90			0.90			0.90			0.90			0.90			79–86

Table B-310-8. Ampacities of Two or Three Insulated Conductors, Rated 0 Through 2000 Volts, Cabled Within an Overall (Two- or Three-Conductor) Covering, Directly Buried in Earth, Based on Ambient Earth Temperature of 20°C (68°F), Arrangement per Figure B-310-2, 100 Percent Load Factor, Thermal Resistance (Rho) of 90

Size (AWG or kcmil)	1 Cable (Fig. B-310-2, Detail 5) 60°C (140°F) UF	75°C (167°F) RHW, THHW, THW, THWN, XHHW, USE	2 Cables (Fig. B-310-2, Detail 6) 60°C (140°F) UF	75°C (167°F) RHW, THHW, THW, THWN, XHHW, USE	1 Cable (Fig. B-310-2, Detail 5) 60°C (140°F) UF	75°C (167°F) RHW, THHW, THW, THWN, XHHW, USE	2 Cables (Fig. B-310-2, Detail 6) 60°C (140°F) UF	75°C (167°F) RHW, THHW, THW, THWN, XHHW, USE	Size (AWG or kcmil)
	COPPER				ALUMINUM OR COPPER-CLAD ALUMINUM				
8	64	75	60	70	51	59	47	55	8
6	85	100	81	95	68	75	60	70	6
4	107	125	100	117	83	97	78	91	4
2	137	161	128	150	107	126	110	117	2
1	155	182	145	170	121	142	113	132	1
1/0	177	208	165	193	138	162	129	151	1/0
2/0	201	236	188	220	157	184	146	171	2/0
3/0	229	269	213	250	179	210	166	195	3/0
4/0	259	304	241	282	203	238	188	220	4/0
250	—	333	—	308	—	261	—	241	250
350	—	401	—	370	—	315	—	290	350
500	—	481	—	442	—	381	—	350	500
750	—	585	—	535	—	473	—	433	750
1000	—	657	—	600	—	545	—	497	1000

Ambient Temp. (°C)	For ambient temperatures other than 20°C (68°F), multiply the ampacities shown above by the appropriate factor shown below.								Ambient Temp. (°F)
6–10	1.12	1.09	1.12	1.09	1.12	1.09	1.12	1.09	43–50
11–15	1.06	1.04	1.06	1.04	1.06	1.04	1.06	1.04	52–59
16–20	1.00	1.00	1.00	1.00	1.00	1.00	1.00	1.00	61–68
21–25	0.94	0.95	0.94	0.95	0.94	0.95	0.94	0.95	70–77
26–30	0.87	0.90	0.87	0.90	0.87	0.90	0.87	0.90	79–86

Note: For ampacities of Type UF cable in underground electrical ducts, multiply the ampacities shown in the table by 0.74.

Table B-310-9. Ampacities of Three Triplexed Single Insulated Conductors, Rated 0 Through 2000 Volts, Directly Buried in Earth Based on Ambient Earth Temperature of 20°C (68°F), Arrangement per Figure B-310-2, 100 Percent Load Factor, Thermal Resistance (Rho) of 90

Size (AWG or kcmil)	See Fig. B-310-2, Detail 7 60°C (140°F) UF	75°C (167°F) USE	See Fig. B-310-2, Detail 8 60°C (140°F) UF	75°C (167°F) USE	See Fig. B-310-2, Detail 7 60°C (140°F) UF	75°C (167°F) USE	See Fig. B-310-2, Detail 8 60°C (140°F) UF	75°C (167°F) USE	Size (AWG or kcmil)
	COPPER				ALUMINUM OR COPPER-CLAD ALUMINUM				
8	72	84	66	77	55	65	51	60	8
6	91	107	84	99	72	84	66	77	6
4	119	139	109	128	92	108	85	100	4
2	153	179	140	164	119	139	109	128	2
1	173	203	159	186	135	158	124	145	1
1/0	197	231	181	212	154	180	141	165	1/0
2/0	223	262	205	240	175	205	159	187	2/0
3/0	254	298	232	272	199	233	181	212	3/0
4/0	289	339	263	308	226	265	206	241	4/0
250	—	370	—	336	—	289	—	263	250
350	—	445	—	403	—	349	—	316	350
500	—	536	—	483	—	424	—	382	500
750	—	654	—	587	—	525	—	471	750
1000	—	744	—	665	—	608	—	544	1000

Ambient Temp. (°C)	For ambient temperatures other than 20°C (68°F), multiply the ampacities shown above by the appropriate factor shown below.								Ambient Temp. (°F)
6–10	1.12	1.09	1.12	1.09	1.12	1.09	1.12	1.09	43–50
11–15	1.06	1.04	1.06	1.04	1.06	1.04	1.06	1.04	52–59
16–20	1.00	1.00	1.00	1.00	1.00	1.00	1.00	1.00	61–68
21–25	0.94	0.95	0.94	0.95	0.94	0.95	0.94	0.95	70–77
26–30	0.87	0.90	0.87	0.90	0.87	0.90	0.87	0.90	79–86

Table B-310-10. Ampacities of Three Single Insulated Conductors, Rated 0 Through 2000 Volts, Directly Buried in Earth Based on Ambient Earth Temperature of 20°C (68°F), Arrangement per Figure B-310-2, 100 Percent Load Factor, Thermal Resistance (Rho) of 90

Size (AWG or kcmil)	See Fig. B-310-2, Detail 9		See Fig. B-310-2, Detail 10		See Fig. B-310-2, Detail 9		See Fig. B-310-2, Detail 10		Size (AWG or kcmil)
	60°C (140°F)	75°C (167°F)	60°C (140°F)	75°C (167°F)	60°C (140°F)	75°C (167°F)	60°C (140°F)	75°C (167°F)	
	TYPES				TYPES				
	UF	USE	UF	USE	UF	USE	UF	USE	
	COPPER				ALUMINUM OR COPPER-CLAD ALUMINUM				
8	84	98	78	92	66	77	61	72	8
6	107	126	101	118	84	98	78	92	6
4	139	163	130	152	108	127	101	118	4
2	178	209	165	194	139	163	129	151	2
1	201	236	187	219	157	184	146	171	1
1/0	230	270	212	249	179	210	165	194	1/0
2/0	261	306	241	283	204	239	188	220	2/0
3/0	297	348	274	321	232	272	213	250	3/0
4/0	336	394	309	362	262	307	241	283	4/0
250	—	429	—	394	—	335	—	308	250
350	—	516	—	474	—	403	—	370	350
500	—	626	—	572	—	490	—	448	500
750	—	767	—	700	—	605	—	552	750
1000	—	887	—	808	—	706	—	642	1000
1250	—	979	—	891	—	787	—	716	1250
1500	—	1063	—	965	—	862	—	783	1500
1750	—	1133	—	1027	—	930	—	843	1750
2000	—	1195	—	1082	—	990	—	897	2000

Ambient Temp. (°C)	For ambient temperatures other than 20°C (68°F), multiply the ampacities shown above by the appropriate factor shown below.								Ambient Temp. (°F)
6–10	1.12	1.09	1.12	1.09	1.12	1.09	1.12	1.09	43–50
11–15	1.06	1.04	1.06	1.04	1.06	1.04	1.06	1.04	52–59
16–20	1.00	1.00	1.00	1.00	1.00	1.00	1.00	1.00	61–68
21–25	0.94	0.95	0.94	0.95	0.94	0.95	0.94	0.95	70–77
26–30	0.87	0.90	0.87	0.90	0.87	0.90	0.87	0.90	79–86

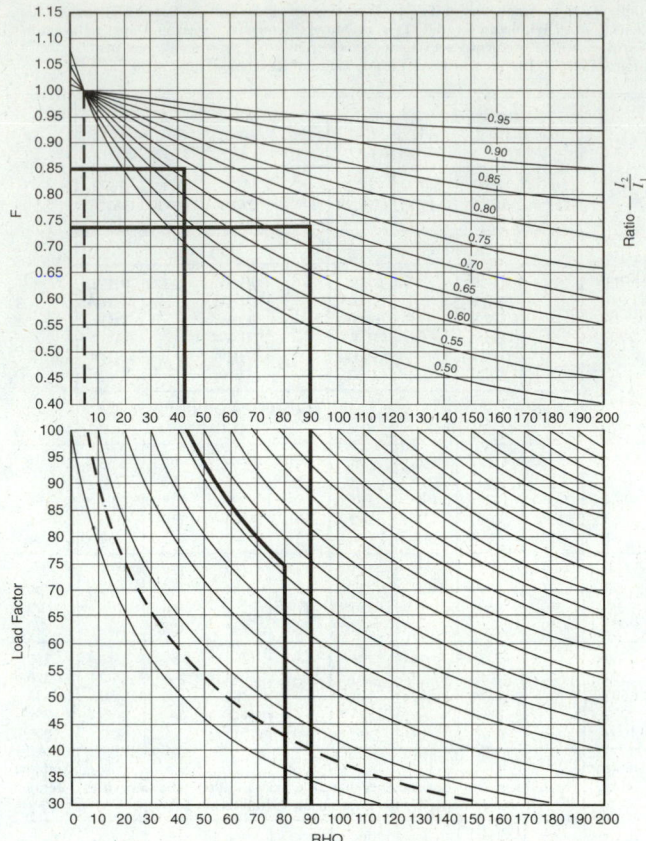

Figure B-310-1   Interpolation chart for cables in a duct bank I_1 = ampacity for Rho = 60, 50 LF; I_2 = ampacity for Rho = 120, 100 LF (load factor); desired ampacity = F × I_1.

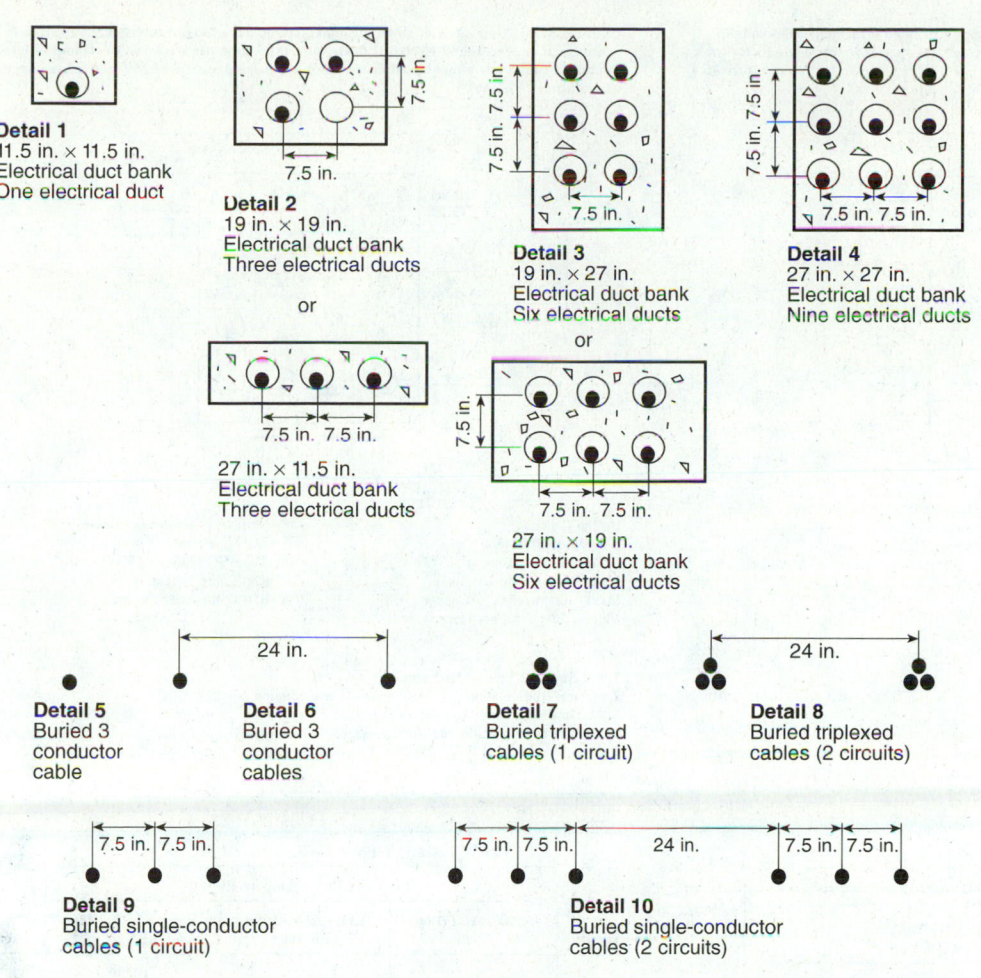

Detail 1
11.5 in. × 11.5 in.
Electrical duct bank
One electrical duct

Detail 2
19 in. × 19 in.
Electrical duct bank
Three electrical ducts

or

27 in. × 11.5 in.
Electrical duct bank
Three electrical ducts

Detail 3
19 in. × 27 in.
Electrical duct bank
Six electrical ducts

or

27 in. × 19 in.
Electrical duct bank
Six electrical ducts

Detail 4
27 in. × 27 in.
Electrical duct bank
Nine electrical ducts

Detail 5
Buried 3
conductor
cable

Detail 6
Buried 3
conductor
cables

Detail 7
Buried triplexed
cables (1 circuit)

Detail 8
Buried triplexed
cables (2 circuits)

Detail 9
Buried single-conductor
cables (1 circuit)

Detail 10
Buried single-conductor
cables (2 circuits)

Note 1: Minimum burial depths to top electrical ducts or cables shall be in accordance with Section 300-5. Maximum depth to the top of electrical duct banks shall be 30 in. and maximum depth to the top of direct buried cables shall be 36 in.

Note 2: For two and four electrical duct installations with electrical ducts installed in a single row, see Section B-310-15(b)(5).

Note 3: For SI units: 1 in. = 25.4 mm; 1 ft = 0.3048 m.

Legend

Backfill (earth or concrete)

◯ Electrical duct

● Cable or cables

Figure B-310-2   Cable installation dimensions for use with Tables B-310-5 through B-310-10.

Figure B-310-3 Ampacities of single insulated conductors rated 0 through 5000 volts in underground electrical ducts (three conductors per electrical duct) nine single-conductor cables per phase based on ambient earth temperature of 20°C (68°F), conductor temperature 75°C (167°F).

Figure B-310-4 Ampacities of single insulated conductors rated 0 through 5000 volts in nonmagnetic underground electrical ducts (one conductor per electrical duct) four single-conductor cables per phase based on ambient earth temperature of 20°C (68°F), conductor temperature 75°C (167°F).

Design Criteria
Neutral and Electrical Grounding Conductor (EGC) Duct = 6 in.
Phase Ducts = 3 to 5 in.
Conductor Material = Copper
Number of Cables per Duct = 3

Number of Cables per Phase = 9
Rho Concrete = Rho Earth 5
Rho PVC Duct = 650
Rho Cable Insulation = 500
Rho Cable Jacket = 650

For SI units: 1 in. = 25.4 mm.

Notes:
1. Neutral configuration per Section 300-5(i), Exception No. 2, for isolated phase installations in nonmagnetic ducts.
2. Phasing is A, B, C in rows or columns. Where magnetic electrical ducts are used, conductors are installed A, B, C per electrical duct with the neutral and all equipment grounding conductors in the same electrical duct. In this case, the 6-in. trade size neutral duct is eliminated.
3. Maximum harmonic loading on the neutral conductor cannot exceed 50 percent of the phase current for the ampacities shown in the table.
4. Metallic shields of Type MV-90 cable shall be grounded at one point only where using A, B, C phasing in rows or columns.

Design Criteria
Neutral and Electrical Grounding Conductor (EGC) Duct = 6 in.
Phase Ducts = 3 in. (min.)
Conductor Material = Copper
Number of Cables per Duct = 1

Number of Cables per Phase = 4
Rho Concrete = Rho Earth − 5
Rho PVC Duct = 650
Rho Cable Insulation = 500
Rho Cable Jacket = 650

Notes:
1. For SI units, 1 in. = 25.4 mm.
2. Neutral configuration per Section 300-5(i), Exception No. 2.
3. Maximum harmonic loading on the neutral conductor cannot exceed 50 percent of the phase current for the ampacities shown in the table.
4. Metallic shields of Type MV-90 cable shall be grounded at one point only.

Size	TYPES RHW, THHW, THW, THWN, XHHW, USE, OR MV-90*			Size
	Total per Phase Ampere Rating			
kcmil	RHO EARTH 60 LF 50	RHO EARTH 90 LF 100	RHO EARTH 120 LF 100	kcmil
250	2340 (260A/Cable)	1530 (170A/Cable)	1395 (155A/Cable)	250
350	2790 (310A/Cable)	1800 (200A/Cable)	1665 (185A/Cable)	350
500	3375 (375A/Cable)	2160 (240A/Cable)	1980 (220A/Cable)	500

Size	TYPES RHW, THHW, THW, THWN, XHHW, USE, OR MV-90*			Size
	Total Per Phase Ampere Rating			
kcmil	RHO EARTH 60 LF 50	RHO EARTH 90 LF 100	RHO EARTH 120 LF 100	kcmil
750	2520 (705A/Cable)	1860 (465A/Cable)	1680 (420A/Cable)	750
1000	3300 (825A/Cable)	2140 (535A/Cable)	1920 (480A/Cable)	1000
1250	3700 (925A/Cable)	2380 (595A/Cable)	2120 (530A/Cable)	1250
1500	4060 (1015A/Cable)	2580 (645A/Cable)	2300 (575A/Cable)	1500
1750	4360 1090A/Cable	2740 (685A/Cable)	2460 (615A/Cable)	1750

Ambient Temp. (°C)	For ambient temperatures other than 20°C (68°F), multiply the ampacities shown above by the appropriate factor shown below.					Ambient Temp. (°F)
6–10	1.09	1.09	1.09	1.09	1.09	43–50
11–15	1.04	1.04	1.04	1.04	1.04	52–59
16–20	1.00	1.00	1.00	1.00	1.00	61–68
21–25	0.95	0.95	0.95	0.95	0.95	70–77
26–30	0.90	0.90	0.90	0.90	0.90	79–86

*Limited to 75°C conductor temperature.

Ambient Temp. (°C)	For ambient temperatures other than 20°C (68°F), multiply the ampacities shown above by the appropriate factor shown below.					Ambient Temp. (°F)
6–10	1.09	1.09	1.09	1.09	1.09	43–50
11–15	1.04	1.04	1.04	1.04	1.04	52–59
16–20	1.00	1.00	1.00	1.00	1.00	61–68
21–25	0.95	0.95	0.95	0.95	0.95	70–77
26–30	0.90	0.90	0.90	0.90	0.90	79–86

*Limited to 75°C conductor temperature.

Figure B-310-5 Ampacities of single insulated conductors rated 0 through 5000 volts in nonmagnetic underground electrical ducts (one conductor per electrical duct) five single-conductor cables per phase based on ambient earth temperature of 20°C (68°F), conductor temperature 75°C (167°F).

Design Criteria
Neutral and Electrical Grounding Conductor (EGC) Duct = 6 in.
Phase Ducts = 3 in. (min.)
Conductor Material = Copper
Number of Cables per Duct = 1

Number of Cables per Phase = 5
Rho Concrete = Rho Earth − 5
Rho PVC Duct = 650
Rho Cable Insulation = 500
Rho Cable Jacket = 650

Notes:
1. For SI units, 1 in. = 25.4 mm.
2. Neutral configuration per Section 300-5(i), Exception No. 2.
3. Maximum harmonic loading on the neutral conductor cannot exceed 50 percent of the phase current for the ampacities shown in the table.
4. Metallic shields of Type MV-90 cable shall be grounded at one point only.

Size	TYPES RHW, THHW, THW, THWN, XHHW, USE, OR MY-90*			Size
	Total Per Phase Ampere Rating			
kcmil	RHO EARTH 60 LF 50	RHO EARTH 90 LF 100	RHO EARTH 120 LF 100	kcmil
2000	5575 (1115A/Cable)	3375 (675A/Cable)	3000 (600A/Cable)	2000

Ambient Temp. (°C)	For ambient temperatures other than 20°C (68°F), multiply the ampacities shown above by the appropriate factor shown below.					Ambient Temp. (°F)
6–10	1.09	1.09	1.09	1.09	1.09	43–50
11–15	1.04	1.04	1.04	1.04	1.04	52–59
16–20	1.00	1.00	1.00	1.00	1.00	61–68
21–25	0.95	0.95	0.95	0.95	0.95	70–77
26–30	0.90	0.90	0.90	0.90	0.90	79–86

*Limited to 75°C conductor temperature.

Table B-310-11. Adjustment Factors for More than Three Current-Carrying Conductors in a Raceway or Cable with Load Diversity

Number of Current-Carrying Conductors	Percent of Values in Tables as Adjusted for Ambient Temperature if Necessary
4–6	80
7–9	70
10–24	70*
25–42	60*
43–85	50*

*These factors include the effects of a load diversity of 50 percent.

FPN: The ampacity limit for the number of current-carrying conductors in 10 through 85 is based on the formula shown below. For greater than 85 conductors, special calculations are required that are beyond the scope of this table.

$$A_2 = \sqrt{\frac{0.5N}{E}} \times (A_1) \text{ or } A_1, \text{ whichever is less}$$

Where:

A_1 = ampacity from Tables 310-16; 310-18; B-310-1; B-310-6; and B-310-7 multiplied by the appropriate factor from Table B-310-11.

N = total number of conductors used to obtain multiplying factor from Table B-310-11.

E = desired number of current-carrying conductors in the raceway or cable.

A_2 = ampacity limit for the current-carrying conductors in the raceway or cable.

Example No. 1

Calculate the ampacity limit for 12 No. 14 THWN current-carrying conductors (75°C) in a raceway that contains 24 conductors.

$$A_2 = \sqrt{\frac{(0.5)(24)}{12}} \times 20(0.7)$$
$$= 14 \text{ amperes (i.e., 50 percent diversity)}$$

Example No. 2

Calculate the ampacity limit for 18 No. 14 THWN current-carrying conductors (75°C) in a raceway that contains 24 conductors.

$$A_2 = \sqrt{\frac{(0.5)(24)}{18}} \times 20(0.7) = 11.5 \text{ amperes}$$

APPENDIX C
Conduit and Tubing Fill Tables for Conductors and Fixture Wires of the Same Size

Table C1. Maximum Number of Conductors and Fixture Wires in Electrical Metallic Tubing *(Based on Table 1, Chapter 9)*

CONDUCTORS

Type	Conductor Size (AWG/kcmil)	Trade Size (in.) ½	¾	1	1¼	1½	2	2½	3	3½	4
RH	14	6	10	16	28	39	64	112	169	221	282
	12	4	8	13	23	31	51	90	136	177	227
RHH, RHW, RHW,	14	4	7	11	20	27	46	80	120	157	201
	12	3	6	9	17	23	38	66	100	131	167
RH, RHH, RHW, RHW-2	10	2	5	8	13	18	30	53	81	105	135
	8	1	2	4	7	9	16	28	42	55	70
	6	1	1	3	5	8	13	22	34	44	56
	4	1	1	2	4	6	10	17	26	34	44
	3	1	1	1	4	5	9	15	23	30	38
	2	1	1	1	3	4	7	13	20	26	33
	1	0	1	1	1	3	5	9	13	17	22
	1/0	0	1	1	1	2	4	7	11	15	19
	2/0	0	1	1	1	2	4	6	10	13	17
	3/0	0	0	1	1	1	3	5	8	11	14
	4/0	0	0	1	1	1	3	5	7	9	12
	250	0	0	0	1	1	1	3	5	7	9
	300	0	0	0	1	1	1	3	5	6	8
	350	0	0	0	1	1	1	3	4	6	7
	400	0	0	0	1	1	1	2	4	5	7
	500	0	0	0	0	1	1	2	3	4	6
	600	0	0	0	0	1	1	1	3	4	5
	700	0	0	0	0	0	1	1	2	3	4
	750	0	0	0	0	0	1	1	2	3	4
	800	0	0	0	0	0	1	1	2	3	4
	900	0	0	0	0	0	1	1	1	3	3
	1000	0	0	0	0	0	1	1	1	2	3
	1250	0	0	0	0	0	0	1	1	1	2
	1500	0	0	0	0	0	0	1	1	1	1
	1750	0	0	0	0	0	0	1	1	1	1
	2000	0	0	0	0	0	0	1	1	1	1
TW	14	8	15	25	43	58	96	168	254	332	424
	12	6	11	19	33	45	74	129	195	255	326
	10	5	8	14	24	33	55	96	145	190	243
	8	2	5	8	13	18	30	53	81	105	135
RHH*, RHW*, RHW-2*, THHW, THW THW-2,	14	6	10	16	28	39	64	112	169	221	282
RHH*, RHW*, RHW-2*, THHW, THW	12	4	8	13	23	31	51	90	136	177	227
	10	3	6	10	18	24	40	70	106	138	177
RHH*, RHW*, RHW-2*, THHW, THW, THW-2	8	1	4	6	10	14	24	42	63	83	106

*Types RHH, RHW, and RHW-2 without outer covering.

Table C1. Continued

CONDUCTORS

Type	Conductor Size (AWG/kcmil)	Trade Size (in.) ½	¾	1	1¼	1½	2	2½	3	3½	4
RHH*, RHW*, RHW-2*, TW, THW, THHW, THW-2	6	1	3	4	8	11	18	32	48	63	81
	4	1	1	3	6	8	13	24	36	47	60
	3	1	1	3	5	7	12	20	31	40	52
	2	1	1	2	4	6	10	17	26	34	44
	1	1	1	1	3	4	7	12	18	24	31
	1/0	0	1	1	2	3	6	10	16	20	26
	2/0	0	1	1	1	3	5	9	13	17	22
	3/0	0	1	1	1	2	4	7	11	15	19
	4/0	0	0	1	1	1	3	6	9	12	16
	250	0	0	1	1	1	3	5	7	10	13
	300	0	0	1	1	1	2	4	6	8	11
	350	0	0	0	1	1	1	4	6	7	10
	400	0	0	0	1	1	1	3	5	7	9
	500	0	0	0	1	1	1	3	4	6	7
	600	0	0	0	1	1	1	2	3	4	6
	700	0	0	0	0	1	1	1	3	4	5
	750	0	0	0	0	1	1	1	3	4	5
	800	0	0	0	0	1	1	1	3	3	5
	900	0	0	0	0	0	1	1	2	3	4
	1000	0	0	0	0	0	1	1	2	3	4
	1250	0	0	0	0	0	1	1	1	2	3
	1500	0	0	0	0	0	1	1	1	1	2
	1750	0	0	0	0	0	0	1	1	1	2
	2000	0	0	0	0	0	0	1	1	1	1
THHN, THWN, THWN-2	14	12	22	35	61	84	138	241	364	476	608
	12	9	16	26	45	61	101	176	266	347	443
	10	5	10	16	28	38	63	111	167	219	279
	8	3	6	9	16	22	36	64	96	126	161
	6	2	4	7	12	16	26	46	69	91	116
	4	1	2	4	7	10	16	28	43	56	71
	3	1	1	3	6	8	13	24	36	47	60
	2	1	1	3	5	7	11	20	30	40	51
	1	1	1	1	4	5	8	15	22	29	37
	1/0	1	1	1	3	4	7	12	19	25	32
	2/0	0	1	1	2	3	6	10	16	20	26
	3/0	0	1	1	1	3	5	8	13	17	22
	4/0	0	1	1	1	2	4	7	11	14	18
	250	0	0	1	1	1	3	6	9	11	15
	300	0	0	1	1	1	3	5	7	10	13
	350	0	0	1	1	1	2	4	6	9	11
	400	0	0	0	1	1	1	4	6	8	10
	500	0	0	0	1	1	1	3	5	6	8
	600	0	0	0	1	1	1	2	4	5	7
	700	0	0	0	1	1	1	2	3	4	6
	750	0	0	0	0	1	1	1	3	4	5
	800	0	0	0	0	1	1	1	3	4	5
	900	0	0	0	0	1	1	1	3	3	4
	1000	0	0	0	0	1	1	1	2	3	4
FEP, FEPB, PFA, PFAH, TFE	14	12	21	34	60	81	134	234	354	462	590
	12	9	15	25	43	59	98	171	258	337	430
	10	6	11	18	31	42	70	122	185	241	309
	8	3	6	10	18	24	40	70	106	138	177
	6	2	4	7	12	17	28	50	75	98	126
	4	1	3	5	9	12	20	35	53	69	88
	3	1	2	4	7	10	16	29	44	57	73
	2	1	1	3	6	8	13	24	36	47	60
PFA, PFAH, TFE	1	1	1	2	4	6	9	16	25	33	42
PFA, PFAH, TFE, Z	1/0	1	1	1	3	5	8	14	21	27	35
	2/0	0	1	1	3	4	6	11	17	22	29
	3/0	0	1	1	2	3	5	9	14	18	24
	4/0	0	1	1	1	2	4	8	11	15	19
Z	14	14	25	41	72	98	161	282	426	556	711
	12	10	18	29	51	69	114	200	302	394	504
	10	6	11	18	31	42	70	122	185	241	309
	8	4	7	11	20	27	44	77	117	153	195
	6	3	5	8	14	19	31	54	82	107	137
	4	1	3	5	9	13	21	37	56	74	94
	3	1	2	4	7	9	15	27	41	54	69
	2	1	1	3	6	8	13	22	34	45	57
	1	1	1	2	4	6	10	18	28	36	46

*Types RHH, RHW, and RHW-2 without outer covering.

Table C1. Continued

CONDUCTORS

Type	Conductor Size (AWG/ kcmil)	Trade Size (in.)									
		½	¾	1	1¼	1½	2	2½	3	3½	4
XHH,	14	8	15	25	43	58	96	168	254	332	424
XHHW,	12	6	11	19	33	45	74	129	195	255	326
XHHW-2,	10	5	8	14	24	33	55	96	145	190	243
ZW	8	2	5	8	13	18	30	53	81	105	135
	6	1	3	6	10	14	22	39	60	78	100
	4	1	2	4	7	10	16	28	43	56	72
	3	1	1	3	6	8	14	24	36	48	61
	2	1	1	3	5	7	11	20	31	40	51
XHH,	1	1	1	1	4	5	8	15	23	30	38
XHHW,	1/0	1	1	1	3	4	7	13	19	25	32
XHHW-2	2/0	0	1	1	2	3	6	10	16	21	27
	3/0	0	1	1	1	3	5	9	13	17	22
	4/0	0	1	1	1	2	4	7	11	14	18
	250	0	0	1	1	1	3	6	9	12	15
	300	0	0	1	1	1	3	5	8	10	13
	350	0	0	1	1	1	2	4	7	9	11
	400	0	0	0	1	1	1	4	6	8	10
	500	0	0	0	1	1	1	3	5	6	8
	600	0	0	0	1	1	1	2	4	5	6
	700	0	0	0	0	1	1	2	3	4	6
	750	0	0	0	0	1	1	1	3	4	5
	800	0	0	0	0	1	1	1	3	4	5
	900	0	0	0	0	1	1	1	3	3	4
	1000	0	0	0	0	0	1	1	2	3	4
	1250	0	0	0	0	0	1	1	1	2	3
	1500	0	0	0	0	0	1	1	1	1	3
	1750	0	0	0	0	0	0	1	1	1	2
	2000	0	0	0	0	0	0	1	1	1	1

FIXTURE WIRES

Type	Conductor Size (AWG/ kcmil)	Trade Size (in.)					
		½	¾	1	1¼	1½	2
FFH-2,	18	8	14	24	41	56	92
RFH-2,	16	7	12	20	34	47	78
RFHH-3							
SF-2, SFF-2	18	10	18	30	52	71	116
	16	8	15	25	43	58	96
	14	7	12	20	34	47	78
SF-1, SFF-1	18	18	33	53	92	125	206
AF, RFH-1, RFHH-2, TF, TFF, XF, XFF	18	14	24	39	68	92	152
AF, RFHH-2, TF, TFF, XF, XFF	16	11	19	31	55	74	123
AF, XF, XFF	14	8	15	25	43	58	96
TFN, TFFN	18	22	38	63	108	148	244
	16	17	29	48	83	113	186
PF, PFF, PGF, PGFF, PAF, PTF, PTFF, PAFF	18	21	36	59	103	140	231
	16	16	28	46	79	108	179
	14	12	21	34	60	81	134
ZF, ZFF, ZHF, HF, HFF	18	27	47	77	133	181	298
	16	20	35	56	98	133	220
	14	14	25	41	72	98	161
KF-2, KFF-2	18	39	69	111	193	262	433
	16	27	48	78	136	185	305
	14	19	33	54	93	127	209
	12	13	23	37	64	87	144
	10	8	15	25	43	58	96
KF-1, KFF-1	18	46	82	133	230	313	516
	16	33	57	93	161	220	362
	14	22	38	63	108	148	244
	12	14	25	41	72	98	161
	10	9	16	27	47	64	105
AX, XF, XFF	12	4	8	13	23	31	51
	10	3	6	10	18	24	40

Note: This table is for concentric stranded conductors only. For compact stranded conductors, Table C1(A) should be used.

Table C1(A). Maximum Number of Compact Conductors in Electrical Metallic Tubing (Based on Table 1, Chapter 9)

COMPACT CONDUCTORS

Type	Conductor Size (AWG/ kcmil)	Trade Size (in.)									
		½	¾	1	1¼	1½	2	2½	3	3½	4
THW,	8	2	4	6	11	16	26	46	69	90	115
THW-2,	6	1	3	5	9	12	20	35	53	70	89
THHW	4	1	2	4	6	9	15	26	40	52	67
	2	1	1	3	5	7	11	19	29	38	49
	1	1	1	1	3	4	8	13	21	27	34
	1/0	1	1	1	3	4	7	12	18	23	30
	2/0	0	1	1	2	3	5	10	15	20	25
	3/0	0	1	1	1	3	5	8	13	17	21
	4/0	0	1	1	1	2	4	7	11	14	18
	250	0	0	1	1	1	3	5	8	11	14
	300	0	0	1	1	1	3	5	7	9	12
	350	0	0	1	1	1	2	4	6	8	11
	400	0	0	0	1	1	1	4	6	8	10
	500	0	0	0	1	1	1	3	5	6	8
	600	0	0	0	1	1	1	2	4	5	7
	700	0	0	0	1	1	1	2	3	4	6
	750	0	0	0	0	1	1	1	3	4	5
	1000	0	0	0	0	1	1	1	2	3	4
THHN,	8	—	4	—	13	—	29	—	78	—	130
THWN,	6	2	4	7	13	18	29	52	78	102	130
THWN-2	4	1	3	4	8	11	18	32	48	63	81
	2	1	1	3	6	8	13	23	34	45	58
	1	1	1	2	4	6	10	17	26	34	43
	1/0	1	1	1	3	5	8	14	22	29	37
	2/0	1	1	1	3	4	7	12	18	24	30
	3/0	0	1	1	2	3	6	10	15	20	25
	4/0	0	1	1	1	3	5	8	12	16	21
	250	0	1	1	1	1	4	6	10	13	16
	300	0	0	1	1	1	3	5	8	11	14
	350	0	0	1	1	1	3	5	7	10	12
	400	0	0	1	1	1	2	4	6	9	11
	500	0	0	0	1	1	1	4	5	7	9
	600	0	0	0	1	1	1	3	4	6	7
	700	0	0	0	1	1	1	2	4	5	7
	750	0	0	0	1	1	1	2	4	5	6
	1000	0	0	0	0	1	1	1	3	3	4
XHHW,	8	3	5	8	15	20	34	59	90	117	149
XHHW-2	6	1	4	6	11	15	25	44	66	87	111
	4	1	3	4	8	11	18	32	48	63	81
	2	1	1	3	6	8	13	23	34	45	58
	1	1	1	2	4	6	10	17	26	34	43
	1/0	1	1	1	3	5	8	14	22	29	37
	2/0	1	1	1	3	4	7	12	18	24	31
	3/0	0	1	1	2	3	6	10	15	20	25
	4/0	0	1	1	1	3	5	8	13	17	21
	250	0	1	1	1	2	4	7	10	13	17
	300	0	0	1	1	1	3	6	9	11	14
	350	0	0	1	1	1	3	5	8	10	13
	400	0	0	1	1	1	2	4	7	9	11
	500	0	0	0	1	1	1	4	6	7	9
	600	0	0	0	1	1	1	3	4	6	8
	700	0	0	0	1	1	1	2	4	5	7
	750	0	0	0	1	1	1	2	3	5	6
	1000	0	0	0	0	1	1	1	3	4	5

Definition: Compact stranding is the result of a manufacturing process where the standard conductor is compressed to the extent that the interstices (voids between strand wires) are virtually eliminated.

Table C2. Maximum Number of Conductors and Fixture Wires in Electrical Nonmetallic Tubing (Based on Table 1, Chapter 9)

CONDUCTORS

Type	Conductor Size (AWG/ kcmil)	Trade Size (in.)					
		½	¾	1	1¼	1½	2
RH	14	4	8	15	27	37	61
	12	3	7	12	21	29	49
RHH, RHW, RHW-2	14	3	6	10	19	26	43
	12	2	5	9	16	22	36
RH, RHH, RHW, RHW-2	10	1	4	7	13	17	29
	8	1	1	3	6	9	15
	6	1	1	3	5	7	12
	4	1	1	2	4	6	9
	3	1	1	1	3	5	8
	2	0	1	1	3	4	7
	1	0	1	1	1	3	5
	1/0	0	0	1	1	2	4
	2/0	0	0	1	1	1	3
	3/0	0	0	1	1	1	3
	4/0	0	0	1	1	1	2
	250	0	0	0	1	1	1
	300	0	0	0	1	1	1
	350	0	0	0	1	1	1
	400	0	0	0	1	1	1
	500	0	0	0	0	1	1
	600	0	0	0	0	1	1
	700	0	0	0	0	0	1
	750	0	0	0	0	0	1
	800	0	0	0	0	0	1
	900	0	0	0	0	0	1
	1000	0	0	0	0	0	1
	1250	0	0	0	0	0	0
	1500	0	0	0	0	0	0
	1750	0	0	0	0	0	0
	2000	0	0	0	0	0	0
TW	14	7	13	22	40	55	92
	12	5	10	17	31	42	71
	10	4	7	13	23	32	52
	8	1	4	7	13	17	29
RHH*, RHW*, RHW-2*, THHW, THW, THW-2	14	4	8	15	27	37	61
RHH*, RHW*, RHW-2*, THHW, THW	12	3	7	12	21	29	49
	10	3	5	9	17	23	38
RHH*, RHW*, RHW-2*, THHW, THW, THW-2	8	1	3	5	10	14	23
RHH*, RHW*, RHW-2*, TW, THW, THHW, THW-2	6	1	2	4	7	10	17
	4	1	1	3	5	8	13
	3	1	1	2	5	7	11
	2	1	1	2	4	6	9
	1	0	1	1	3	4	6
	1/0	0	1	1	2	3	5
	2/0	0	1	1	1	3	5
	3/0	0	0	1	1	2	4
	4/0	0	0	1	1	1	3
	250	0	0	1	1	1	2
	300	0	0	0	1	1	2
	350	0	0	0	1	1	1
	400	0	0	0	1	1	1
	500	0	0	0	1	1	1
	600	0	0	0	0	1	1
	700	0	0	0	0	1	1
	750	0	0	0	0	1	1
	800	0	0	0	0	1	1
	900	0	0	0	0	0	1
	1000	0	0	0	0	0	1
	1250	0	0	0	0	0	1
	1500	0	0	0	0	0	0
	1750	0	0	0	0	0	0
	2000	0	0	0	0	0	0

*Types RHH, RHW, and RHW-2 without outer covering.

Table C2. Continued

CONDUCTORS

Type	Conductor Size (AWG/ kcmil)	Trade Size (in.)					
		½	¾	1	1¼	1½	2
THHN, THWN, THWN-2	14	10	18	32	58	80	132
	12	7	13	23	42	58	96
	10	4	8	15	26	36	60
	8	2	5	8	15	21	35
	6	1	3	6	11	15	25
	4	1	1	4	7	9	15
	3	1	1	3	5	8	13
	2	1	1	2	5	6	11
	1	1	1	1	3	5	8
	1/0	0	1	1	3	4	7
	2/0	0	1	1	2	3	5
	3/0	0	1	1	1	3	4
	4/0	0	0	1	1	2	4
	250	0	0	1	1	1	3
	300	0	0	1	1	1	2
	350	0	0	0	1	1	2
	400	0	0	0	1	1	1
	500	0	0	0	1	1	1
	600	0	0	0	1	1	1
	700	0	0	0	0	1	1
	750	0	0	0	0	1	1
	800	0	0	0	0	1	1
	900	0	0	0	0	1	1
	1000	0	0	0	0	0	1
FEP, FEPB, PFA, PFAH, TFE	14	10	18	31	56	77	128
	12	7	13	23	41	56	93
	10	5	9	16	29	40	67
	8	3	5	9	17	23	38
	6	1	4	6	12	16	27
	4	1	2	4	8	11	19
	3	1	1	4	7	9	16
	2	1	1	3	5	8	13
PFA, PFAH, TFE	1	1	1	1	4	5	9
PFA, PFAH, TFE, Z	1/0	0	1	1	3	4	7
	2/0	0	1	1	2	4	6
	3/0	0	1	1	1	3	5
	4/0	0	1	1	1	2	4
Z	14	12	22	38	68	93	154
	12	8	15	27	48	66	109
	10	5	9	16	29	40	67
	8	3	6	10	18	25	42
	6	1	4	7	13	18	30
	4	1	3	5	9	12	20
	3	1	1	3	6	9	15
	2	1	1	3	5	7	12
	1	1	1	2	4	6	10
XHH, XHHW, XHHW-2, ZW	14	7	13	22	40	55	92
	12	5	10	17	31	42	71
	10	4	7	13	23	32	52
	8	1	4	7	13	17	29
	6	1	3	5	9	13	21
	4	1	1	4	7	9	15
	3	1	1	3	6	8	13
	2	1	1	2	5	6	11
XHH, XHHW, XHHW-2	1	1	1	1	3	5	8
	1/0	0	1	1	3	4	7
	2/0	0	1	1	2	3	6
	3/0	0	1	1	1	3	5
	4/0	0	0	1	1	2	4
	250	0	0	1	1	1	3
	300	0	0	1	1	1	3
	350	0	0	1	1	1	2
	400	0	0	0	1	1	1
	500	0	0	0	1	1	1
	600	0	0	0	1	1	1
	700	0	0	0	0	1	1
	750	0	0	0	0	1	1
	800	0	0	0	0	1	1
	900	0	0	0	0	1	1
	1000	0	0	0	0	0	1
	1250	0	0	0	0	0	1
	1500	0	0	0	0	0	1
	1750	0	0	0	0	0	0
	2000	0	0	0	0	0	0

Table C2. Continued

FIXTURE WIRES

Type	Conductor Size (AWG/kcmil)	½	¾	1	1¼	1½	2
FFH-2, RFH-2, RFHH-3	18	6	12	21	39	53	88
	16	5	10	18	32	45	74
SF-2, SFF-2	18	8	15	27	49	67	111
	16	7	13	22	40	55	92
	14	5	10	18	32	45	74
SF-1, SFF-1	18	15	28	48	86	119	197
AF, RFH-1, RFHH-2, TF, TFF, XF, XFF	18	11	20	35	64	88	145
AF, RFHH-2, TF, TFF, XF, XFF	16	9	16	29	51	71	117
AF, XF, XFF	14	7	13	22	40	55	92
TFN, TFFN	18	18	33	57	102	141	233
	16	13	25	43	78	107	178
PF, PFF, PGF, PGFF, PAF, PTF, PTFF, PAFF	18	17	31	54	97	133	221
	16	13	24	42	75	103	171
	14	10	18	31	56	77	128
ZF, ZFF, ZHF, HF, HFF	18	22	40	70	125	172	285
	16	16	29	51	92	127	210
	14	12	22	38	68	93	154
KF-2, KFF-2	18	31	58	101	182	250	413
	16	22	41	71	128	176	291
	14	15	28	49	88	121	200
	12	10	19	33	60	83	138
	10	7	13	22	40	55	92
KF-1, KFF-1	18	38	69	121	217	298	493
	16	26	49	85	152	209	346
	14	18	33	57	102	141	233
	12	12	22	38	68	93	154
	10	7	14	24	44	61	101
AF, XF, XFF	12	3	7	12	21	29	49
	10	3	5	9	17	23	38

Note: This table is for concentric stranded conductors only. For compact stranded conductors, Table C2(A) should be used.

Table C2(A). Maximum Number of Compact Conductors in Electrical Nonmetallic Tubing *(Based on Table 1, Chapter 9)*

COMPACT CONDUCTORS

Type	Conductor Size (AWG/kcmil)	½	¾	1	1¼	1½	2
THW, THW-2, THHW	8	1	3	6	11	15	25
	6	1	2	4	8	11	19
	4	1	1	3	6	8	14
	2	1	1	2	4	6	10
	1	0	1	1	3	4	7
	1/0	0	1	1	3	4	6
	2/0	0	1	1	2	3	5
	3/0	0	1	1	1	3	4
	4/0	0	0	1	1	2	4
	250	0	0	1	1	1	3
	300	0	0	1	1	1	2
	350	0	0	0	1	1	2
	400	0	0	0	1	1	1
	500	0	0	0	1	1	1
	600	0	0	0	1	1	1
	700	0	0	0	0	1	1
	750	0	0	0	0	1	1
	1000	0	0	0	0	0	1
THHN, THWN, THWN-2	8	—	—	—	—	—	—
	6	1	4	7	12	17	28
	4	1	2	4	7	10	17
	2	1	1	3	5	7	12
	1	1	1	2	4	5	9
	1/0	1	1	1	3	5	8
	2/0	0	1	1	3	4	6
	3/0	0	1	1	2	3	5
	4/0	0	1	1	1	2	4
	250	0	0	1	1	1	3
	300	0	0	1	1	1	3
	350	0	0	1	1	1	2
	400	0	0	0	1	1	2
	500	0	0	0	1	1	1
	600	0	0	0	1	1	1
	700	0	0	0	1	1	1
	750	0	0	0	1	1	1
	1000	0	0	0	0	1	1
XHHW, XHHW-2	8	2	4	8	14	19	32
	6	1	3	6	10	14	24
	4	1	2	4	7	10	17
	2	1	1	3	5	7	12
	1	1	1	2	4	5	9
	1/0	1	1	1	3	5	8
	2/0	0	1	1	3	4	7
	3/0	0	1	1	2	3	5
	4/0	0	1	1	1	3	4
	250	0	0	1	1	1	3
	300	0	0	1	1	1	3
	350	0	0	1	1	1	3
	400	0	0	1	1	1	2
	500	0	0	0	1	1	1
	600	0	0	0	1	1	1
	700	0	0	0	1	1	1
	750	0	0	0	1	1	1
	1000	0	0	0	0	1	1

Definition: Compact stranding is the result of a manufacturing process where the standard conductor is compressed to the extent that the interstices (voids between strand wires) are virtually eliminated.

Table C3. Maximum Number of Conductors and Fixture Wires in Flexible Metal Conduit (*Based on Table 1, Chapter 9*)

CONDUCTORS

Type	Conductor Size (AWG/kcmil)	½	¾	1	1¼	1½	2	2½	3	3½	4
RH	14	6	10	15	24	35	62	94	135	184	240
	12	5	8	12	19	28	50	75	108	148	193
RHH, RHW, RHW-2	14	4	7	11	17	25	44	67	96	131	171
	12	3	6	9	14	21	37	55	80	109	142
RH, RHH, RHW, RHW-2	10	3	5	7	11	17	30	45	64	88	115
	8	1	2	4	6	9	15	23	34	46	60
	6	1	1	3	5	7	12	19	27	37	48
	4	1	1	2	4	5	10	14	21	29	37
	3	1	1	1	3	5	8	13	18	25	33
	2	1	1	1	3	4	7	11	16	22	28
	1	0	1	1	1	2	5	7	10	14	19
	1/0	0	1	1	1	2	4	6	9	12	16
	2/0	0	1	1	1	1	3	5	8	11	14
	3/0	0	0	1	1	1	3	5	7	9	12
	4/0	0	0	1	1	1	2	4	6	8	10
	250	0	0	0	1	1	1	3	4	6	8
	300	0	0	0	1	1	1	2	4	5	7
	350	0	0	0	1	1	1	2	3	5	6
	400	0	0	0	0	1	1	1	3	4	6
	500	0	0	0	0	1	1	1	3	4	5
	600	0	0	0	0	0	1	1	1	2	3
	700	0	0	0	0	0	1	1	1	3	3
	750	0	0	0	0	0	1	1	1	2	3
	800	0	0	0	0	0	1	1	1	2	3
	900	0	0	0	0	0	1	1	1	2	3
	1000	0	0	0	0	0	1	1	1	1	3
	1250	0	0	0	0	0	0	1	1	1	1
	1500	0	0	0	0	0	0	1	1	1	1
	1750	0	0	0	0	0	0	1	1	1	1
	2000	0	0	0	0	0	0	0	1	1	1
TW	14	9	15	23	36	53	94	141	203	277	361
	12	7	11	18	28	41	72	108	156	212	277
	10	5	8	13	21	30	54	81	116	158	207
	8	3	5	7	11	17	30	45	64	88	115
RHH*, RHW*, RHW-2*, THHW, THW, THW-2	14	6	10	15	24	35	62	94	135	184	240
RHH*, RHW*, RHW-2*, THHW, THW	12	5	8	12	19	28	50	75	108	148	193
	10	4	6	10	15	22	39	59	85	115	151
RHH*, RHW*, RHW-2*, THHW, THW, THW-2	8	1	4	6	9	13	23	35	51	69	90
RHH*, RHW*, RHW-2*, TW, THW, THHW, THW-2	6	1	3	4	7	10	18	27	39	53	69
	4	1	1	3	5	7	13	20	29	39	51
	3	1	1	3	4	6	11	17	25	34	44
	2	1	1	2	4	5	10	14	21	29	37
	1	1	1	1	2	4	7	10	15	20	26
	1/0	0	1	1	1	3	6	9	12	17	22
	2/0	0	1	1	1	3	5	7	10	14	19
	3/0	0	1	1	1	2	4	6	9	12	16
	4/0	0	0	1	1	1	3	5	7	10	13
	250	0	0	1	1	1	3	4	6	8	11
	300	0	0	1	1	1	2	3	5	7	9
	350	0	0	0	1	1	1	3	4	6	8
	400	0	0	0	1	1	1	3	4	6	7
	500	0	0	0	1	1	1	2	3	5	6
	600	0	0	0	0	1	1	1	3	4	5
	700	0	0	0	0	1	1	1	2	3	4
	750	0	0	0	0	1	1	1	2	3	4
	800	0	0	0	0	1	1	1	1	3	4
	900	0	0	0	0	0	1	1	1	3	3
	1000	0	0	0	0	0	1	1	1	2	3
	1250	0	0	0	0	0	1	1	1	1	2
	1500	0	0	0	0	0	0	1	1	1	1
	1750	0	0	0	0	0	0	1	1	1	1
	2000	0	0	0	0	0	0	1	1	1	1

*Types RHH, RHW, and RHW-2 without outer covering.

Table C3. Continued

CONDUCTORS

Type	Conductor Size (AWG/kcmil)	½	¾	1	1¼	1½	2	2½	3	3½	4
THHN, THWN, THWN-2	14	13	22	33	52	76	134	202	291	396	518
	12	9	16	24	38	56	98	147	212	289	378
	10	6	10	15	24	35	62	93	134	182	238
	8	3	6	9	14	20	35	53	77	105	137
	6	2	4	6	10	14	25	38	55	76	99
	4	1	2	4	6	9	16	24	34	46	61
	3	1	1	3	5	7	13	20	29	39	51
	2	1	1	3	4	6	11	17	24	33	43
	1	1	1	1	3	4	8	12	18	24	32
	1/0	1	1	1	2	4	7	10	15	20	27
	2/0	0	1	1	1	3	6	9	12	17	22
	3/0	0	1	1	1	2	5	7	10	14	18
	4/0	0	1	1	1	1	4	6	8	12	15
	250	0	0	1	1	1	3	5	7	9	12
	300	0	0	1	1	1	3	4	6	8	11
	350	0	0	1	1	1	2	3	5	7	9
	400	0	0	0	1	1	1	3	5	6	8
	500	0	0	0	1	1	1	2	4	5	7
	600	0	0	0	0	1	1	1	3	4	5
	700	0	0	0	0	1	1	1	3	4	5
	750	0	0	0	0	1	1	1	2	3	4
	800	0	0	0	0	1	1	1	2	3	4
	900	0	0	0	0	0	1	1	1	3	4
	1000	0	0	0	0	0	1	1	1	1	3
FEP, FEPB, PFA, PFAH, TFE	14	12	21	32	51	74	130	196	282	385	502
	12	9	15	24	37	54	95	143	206	281	367
	10	6	11	17	26	39	68	103	148	201	263
	8	4	6	10	15	22	39	59	85	115	151
	6	2	4	7	11	16	28	42	60	82	107
	4	1	3	5	7	11	19	29	42	57	75
	3	1	2	4	6	9	16	24	35	48	62
	2	1	1	3	5	7	13	20	29	39	51
PFA, PFAH, TFE	1	1	1	2	3	5	9	14	20	27	36
PFA, PFAH, TFE, Z	1/0	1	1	1	3	4	8	11	17	23	30
	2/0	1	1	1	2	3	6	9	14	19	24
	3/0	0	1	1	1	3	5	8	11	15	20
	4/0	0	1	1	1	2	4	6	9	13	16
Z	14	15	25	39	61	89	157	236	340	463	605
	12	11	18	28	43	63	111	168	241	329	429
	10	6	11	17	26	39	68	103	148	201	263
	8	4	7	11	17	24	43	65	93	127	166
	6	3	5	7	12	17	30	45	65	89	117
	4	1	3	5	8	12	21	31	45	61	80
	3	1	2	4	6	8	15	23	33	45	58
	2	1	1	3	5	7	12	19	27	37	49
	1	1	1	2	4	6	10	15	22	30	39
XHH, XHHW, XHHW-2, ZW	14	9	15	23	36	53	94	141	203	277	361
	12	7	11	18	28	41	72	108	156	212	277
	10	5	8	13	21	30	54	81	116	158	207
	8	3	5	7	11	17	30	45	64	88	115
	6	1	3	5	8	12	22	33	48	65	85
	4	1	2	4	6	9	16	24	34	47	61
	3	1	1	3	5	7	13	20	29	40	52
	2	1	1	3	4	6	11	17	24	33	44
XHH, XHHW, XHHW-2	1	1	1	1	3	5	8	13	18	25	32
	1/0	1	1	1	2	4	7	10	15	21	27
	2/0	0	1	1	2	3	6	9	13	17	23
	3/0	0	1	1	1	3	5	7	10	14	19
	4/0	0	1	1	1	2	4	6	9	12	15
	250	0	0	1	1	1	3	5	7	10	13
	300	0	0	1	1	1	3	4	6	8	11
	350	0	0	1	1	1	2	4	5	7	9
	400	0	0	0	1	1	1	3	5	7	8
	500	0	0	0	1	1	1	3	4	5	7
	600	0	0	0	0	1	1	1	3	4	5
	700	0	0	0	0	1	1	1	3	4	5
	750	0	0	0	0	1	1	1	2	3	4
	800	0	0	0	0	1	1	1	2	3	4
	900	0	0	0	0	0	1	1	1	3	4
	1000	0	0	0	0	0	1	1	1	3	3
	1250	0	0	0	0	0	1	1	1	1	3
	1500	0	0	0	0	0	0	1	1	1	2
	1750	0	0	0	0	0	0	1	1	1	1
	2000	0	0	0	0	0	0	1	1	1	1

Table C3. (Continued)

FIXTURE WIRES

Type	Conductor Size (AWG/kcmil)	½	¾	1	1¼	1½	2
FFH-2, RFH-2, RFHH-3	18	8	14	22	35	51	90
	16	7	12	19	29	43	76
SF-2, SFF-2	18	11	18	28	44	64	113
	16	9	15	23	36	53	94
	14	7	12	19	29	43	76
SF-1, SFF-1	18	19	32	50	78	114	201
AF, RFH-1, RFHH-2, TF, TFF, XF, XFF	18	14	24	37	58	84	148
AF, RFHH-2, TF, TFF, XF, XFF	16	11	19	30	47	68	120
AF, XF, XFF	14	9	15	23	36	53	94
TFN, TFFN	18	23	38	59	93	135	237
	16	17	29	45	71	103	181
PF, PFF, PGF, PGFF, PAF, PTF, PTFF, PAFF	18	22	36	56	88	128	225
	16	17	28	43	68	99	174
	14	12	21	32	51	74	130
ZF, ZFF, ZHF, HF, HFF	18	28	47	72	113	165	290
	16	20	35	53	83	121	214
	14	15	25	39	61	89	157
KF-2, KFF-2	18	41	68	105	164	239	421
	16	28	48	74	116	168	297
	14	19	33	51	80	116	204
	12	13	23	35	55	80	140
	10	9	15	23	36	53	94
KF-1, KFF-1	18	48	82	125	196	285	503
	16	34	57	88	138	200	353
	14	23	38	59	93	135	237
	12	15	25	39	61	89	157
	10	10	16	25	40	58	103
AF, XF, XFF	12	5	8	12	19	28	50
	10	4	6	10	15	22	39

Note: This table is for concentric stranded conductors only. For compact stranded conductors, Table C3(A) should be used.

Table C3(A). **Maximum Number of Compact Conductors in Flexible Metal Conduit** (*Based on Table 1, Chapter 9*)

COMPACT CONDUCTORS

Type	Conductor Size (AWG/kcmil)	½	¾	1	1¼	1½	2	2½	3	3½	4
THW, THHW, THW-2	8	2	4	6	10	14	25	38	55	75	98
	6	1	3	5	7	11	20	29	43	58	76
	4	1	2	3	5	8	15	22	32	43	57
	2	1	1	2	4	6	11	16	23	32	42
	1	1	1	1	3	4	7	11	16	22	29
	1/0	1	1	1	2	3	6	10	14	19	25
	2/0	0	1	1	1	3	5	8	12	16	21
	3/0	0	1	1	1	2	4	7	10	14	18
	4/0	0	1	1	1	1	4	6	8	11	15
	250	0	0	1	1	1	3	4	7	9	12
	300	0	0	1	1	1	2	4	6	8	10
	350	0	0	1	1	1	2	3	5	7	9
	400	0	0	0	1	1	1	3	5	6	8
	500	0	0	0	1	1	1	3	4	5	7
	600	0	0	0	0	1	1	1	3	4	6
	700	0	0	0	0	1	1	1	3	4	5
	750	0	0	0	0	1	1	1	2	3	5
	1000	0	0	0	0	0	1	1	1	3	4
THHN, THWN, THWN-2	8	—	—	—	—	—	—	—	—	—	—
	6	3	4	7	11	16	29	43	62	85	111
	4	1	3	4	7	10	18	27	38	52	69
	2	1	1	3	5	7	13	19	28	38	49
	1	1	1	2	3	5	9	14	21	28	37
	1/0	1	1	1	3	4	8	12	17	24	31
	2/0	1	1	1	2	4	6	10	14	20	26
	3/0	0	1	1	1	3	5	8	12	17	22
	4/0	0	1	1	1	2	4	7	10	14	18
	250	0	1	1	1	1	3	5	8	11	14
	300	0	0	1	1	1	3	5	7	9	12
	350	0	0	1	1	1	3	4	6	8	10
	400	0	0	1	1	1	2	3	5	7	9
	500	0	0	0	1	1	1	3	4	6	8
	600	0	0	0	1	1	1	2	3	5	6
	700	0	0	0	0	1	1	1	3	4	6
	750	0	0	0	0	1	1	1	3	4	5
	1000	0	0	0	0	0	1	1	1	3	4
XHHW, XHHW-2	8	3	5	8	13	19	33	50	71	97	127
	6	2	4	6	9	14	24	37	53	72	95
	4	1	3	4	7	10	18	27	38	52	69
	2	1	1	3	5	7	13	19	28	38	49
	1	1	1	2	3	5	9	14	21	28	37
	1/0	1	1	1	3	4	8	12	17	24	31
	2/0	1	1	1	2	4	7	10	15	20	26
	3/0	0	1	1	1	3	5	8	12	17	22
	4/0	0	1	1	1	2	4	7	10	14	18
	250	0	1	1	1	1	4	5	8	11	14
	300	0	0	1	1	1	3	5	7	9	12
	350	0	0	1	1	1	3	4	6	8	11
	400	0	0	1	1	1	2	4	5	7	10
	500	0	0	0	1	1	1	3	4	6	8
	600	0	0	0	1	1	1	2	3	5	6
	700	0	0	0	0	1	1	1	3	4	6
	750	0	0	0	0	1	1	1	3	4	5
	1000	0	0	0	0	1	1	1	2	3	4

Definition: Compact stranding is the result of a manufacturing process where the standard conductor is compressed to the extent that the interstices (voids between strand wires) are virtually eliminated.

Table C4. Maximum Number of Conductors and Fixture Wires in Intermediate Metal Conduit
(Based on Table 1, Chapter 9)

CONDUCTORS

Type	Conductor Size (AWG/ kcmil)	½	¾	1	1¼	1½	2	2½	3	3½	4
RH	14	6	11	18	31	42	69	98	151	202	261
	12	5	9	14	25	34	56	79	122	163	209
RHH, RHW, RHW-2	14	4	8	13	22	30	49	70	108	144	186
	12	4	6	11	18	25	41	58	89	120	154
RH, RHH, RHW, RHW-2	10	3	5	8	15	20	33	47	72	97	124
	8	1	3	4	8	10	17	24	38	50	65
	6	1	1	3	6	8	14	19	30	40	52
	4	1	1	3	5	6	11	15	23	31	41
	3	1	1	2	4	6	9	13	21	28	36
	2	1	1	1	3	5	8	11	18	24	31
	1	0	1	1	2	3	5	7	12	16	20
	1/0	0	1	1	1	3	4	6	10	14	18
	2/0	0	1	1	1	2	4	6	9	12	15
	3/0	0	0	1	1	1	3	5	7	10	13
	4/0	0	0	1	1	1	3	4	6	9	11
	250	0	0	1	1	1	1	3	5	6	8
	300	0	0	0	1	1	1	3	4	6	7
	350	0	0	0	1	1	1	2	4	5	7
	400	0	0	0	1	1	1	2	3	5	6
	500	0	0	0	1	1	1	1	3	4	5
	600	0	0	0	0	1	1	1	2	3	4
	700	0	0	0	0	1	1	1	2	3	4
	750	0	0	0	0	1	1	1	1	3	4
	800	0	0	0	0	0	1	1	1	3	3
	900	0	0	0	0	0	1	1	1	2	3
	1000	0	0	0	0	0	1	1	1	2	3
	1250	0	0	0	0	0	1	1	1	1	2
	1500	0	0	0	0	0	0	1	1	1	1
	1750	0	0	0	0	0	0	1	1	1	1
	2000	0	0	0	0	0	0	1	1	1	1
TW	14	10	17	27	47	64	104	147	228	304	392
	12	7	13	21	36	49	80	113	175	234	301
	10	5	9	15	27	36	59	84	130	174	224
	8	3	5	8	15	20	33	47	72	97	124
RHH*, RHW*, RHW-2*, THHW, THW, THW-2	14	6	11	18	31	42	69	98	151	202	261
RHH*, RHW*, RHW-2*, THHW, THW	12	5	9	14	25	34	56	79	122	163	209
	10	4	7	11	19	26	43	61	95	127	163
RHH*, RHW*, RHW-2*, THHW, THW, THW-2	8	2	4	7	12	16	26	37	57	76	98
RHH*, RHW*, RHW-2*, TW, THW, THHW, THW-2	6	1	3	5	9	12	20	28	43	58	75
	4	1	2	4	6	9	15	21	32	43	56
	3	1	1	3	6	8	13	18	28	37	48
	2	1	1	3	5	6	11	15	23	31	41
	1	1	1	1	3	4	7	11	16	22	28
	1/0	1	1	1	3	4	6	9	14	19	24
	2/0	1	1	1	2	3	5	8	12	16	20
	3/0	0	1	1	1	3	4	6	10	13	17
	4/0	0	1	1	1	2	4	5	8	11	14
	250	0	0	1	1	1	3	4	7	9	12
	300	0	0	1	1	1	2	4	6	8	10
	350	0	0	1	1	1	2	3	5	7	9
	400	0	0	0	1	1	1	3	4	6	8
	500	0	0	0	1	1	1	2	4	5	7
	600	0	0	0	1	1	1	1	3	4	5
	700	0	0	0	0	1	1	1	3	4	5
	750	0	0	0	0	1	1	1	2	3	4
	800	0	0	0	0	1	1	1	2	3	4
	900	0	0	0	0	1	1	1	2	3	4
	1000	0	0	0	0	0	1	1	1	3	3
	1250	0	0	0	0	0	1	1	1	1	3
	1500	0	0	0	0	0	1	1	1	1	2
	1750	0	0	0	0	0	0	1	1	1	1
	2000	0	0	0	0	0	0	1	1	1	1

*Types RHH, RHW, and RHW-2 without outer covering.

Table C4. Continued

CONDUCTORS

Type	Conductor Size (AWG/ kcmil)	½	¾	1	1¼	1½	2	2½	3	3½	4
THHN, THWN, THWN-2	14	14	24	39	68	91	149	211	326	436	562
	12	10	17	29	49	67	109	154	238	318	410
	10	6	11	18	31	42	68	97	150	200	258
	8	3	6	10	18	24	39	56	86	115	149
	6	2	4	7	13	17	28	40	62	83	107
	4	1	3	4	8	10	17	25	38	51	66
	3	1	2	4	6	9	15	21	32	43	56
	2	1	1	3	5	7	12	17	27	36	47
	1	1	1	2	4	5	9	13	20	27	35
	1/0	1	1	1	3	4	8	11	17	23	29
	2/0	1	1	1	3	4	6	9	14	19	24
	3/0	0	1	1	2	3	5	7	12	16	20
	4/0	0	1	1	1	2	4	6	9	13	17
	250	0	0	1	1	1	3	5	8	10	13
	300	0	0	1	1	1	3	4	7	9	12
	350	0	0	1	1	1	2	4	6	8	10
	400	0	0	1	1	1	2	3	5	7	9
	500	0	0	0	1	1	1	3	4	6	7
	600	0	0	0	1	1	1	2	3	5	6
	700	0	0	0	1	1	1	1	3	4	5
	750	0	0	0	1	1	1	1	3	4	5
	800	0	0	0	0	1	1	1	3	4	5
	900	0	0	0	0	1	1	1	2	3	4
	1000	0	0	0	0	1	1	1	2	3	4
FEP, FEPB, PFA, PFAH, TFE	14	13	23	38	66	89	145	205	317	423	545
	12	10	17	28	48	65	106	150	231	309	398
	10	7	12	20	34	46	76	107	166	221	285
	8	4	7	11	19	26	43	61	95	127	163
	6	3	5	8	14	19	31	44	67	90	116
	4	1	3	5	10	13	21	30	47	63	81
	3	1	3	4	8	11	18	25	39	52	68
	2	1	2	4	6	9	15	21	32	43	56
PFA, PFAH, TFE	1	1	1	2	4	6	10	14	22	30	39
PFA, PFAH, TFE, Z	1/0	1	1	1	4	5	8	12	19	25	32
	2/0	1	1	1	3	4	7	10	15	21	27
	3/0	0	1	1	2	3	6	8	13	17	22
	4/0	0	1	1	1	3	5	7	10	14	18
Z	14	16	28	46	79	107	175	247	381	510	657
	12	11	20	32	56	76	124	175	271	362	466
	10	7	12	20	34	46	76	107	166	221	285
	8	4	7	12	21	29	48	68	105	140	180
	6	3	5	9	15	20	33	47	73	98	127
	4	1	3	6	10	14	23	33	50	67	87
	3	1	2	4	7	10	17	24	37	49	63
	2	1	1	3	6	8	14	20	30	41	53
	1	1	1	3	5	7	11	16	25	33	43
XHH, XHHW, XHHW-2, ZW	14	10	17	27	47	64	104	147	228	304	392
	12	7	13	21	36	49	80	113	175	234	301
	10	5	9	15	27	36	59	84	130	174	224
	8	3	5	8	15	20	33	47	72	97	124
	6	1	4	6	11	15	24	35	53	71	92
	4	1	3	4	8	11	18	25	39	52	67
	3	1	2	4	7	9	15	21	33	44	56
	2	1	1	3	5	7	12	18	27	37	47
XHH, XHHW, XHHW-2	1	1	1	2	4	5	9	13	20	27	35
	1/0	1	1	1	3	5	8	11	17	23	30
	2/0	1	1	1	3	4	6	9	14	19	25
	3/0	0	1	1	2	3	5	7	12	16	20
	4/0	0	1	1	1	2	4	6	10	13	17
	250	0	0	1	1	1	3	5	8	11	14
	300	0	0	1	1	1	3	4	7	9	12
	350	0	0	1	1	1	3	4	6	8	10
	400	0	0	1	1	1	2	3	5	7	9
	500	0	0	0	1	1	1	3	4	6	8
	600	0	0	0	1	1	1	2	3	5	6
	700	0	0	0	1	1	1	1	3	4	5
	750	0	0	0	1	1	1	1	3	4	5
	800	0	0	0	0	1	1	1	3	4	5
	900	0	0	0	0	1	1	1	2	3	4
	1000	0	0	0	0	1	1	1	2	3	4
	1250	0	0	0	0	0	1	1	1	2	3
	1500	0	0	0	0	0	1	1	1	1	2
	1750	0	0	0	0	0	0	1	1	1	2
	2000	0	0	0	0	0	0	1	1	1	1

Table C4. Continued

FIXTURE WIRES

Type	Conductor Size (AWG/kcmil)	½	¾	1	1¼	1½	2
FHH-2, RFH-2, RFHH-3	18	9	16	26	45	61	100
	16	8	13	22	38	51	84
SF-2, SFF-2	18	12	20	33	57	77	126
	16	10	17	27	47	64	104
	14	8	13	22	38	51	84
SF-1, SFF-1	18	21	36	59	101	137	223
AF, RFH-1, RFHH-2, TF, TFF, XF, XFF	18	15	26	43	75	101	165
AF, RFH-2, TF, TFF, XF, XFF	16	12	21	35	60	81	133
AF, XF, XFF	14	10	17	27	47	64	104
TFN, TFFN	18	25	42	69	119	161	264
	16	19	32	53	91	123	201
PF, PFF, PGF, PGFF, PAF, PTF, PTFF, PAFF	18	23	40	66	113	153	250
	16	18	31	51	87	118	193
	14	13	23	38	66	89	145
ZF, ZFF, ZHF, HF, HFF	18	30	52	85	146	197	322
	16	22	38	63	108	145	238
	14	16	28	46	79	107	175
KF-2, KFF-2	18	44	75	123	212	287	468
	16	31	53	87	149	202	330
	14	21	36	60	103	139	227
	12	14	25	41	70	95	156
	10	10	17	27	47	64	104
KF-1, KFF-1	18	52	90	147	253	342	558
	16	37	63	103	178	240	392
	14	25	42	69	119	161	264
	12	16	28	46	79	107	175
	10	10	18	30	52	70	114
AX, XF, XFF	12	5	9	14	25	34	56
	10	4	7	11	19	26	43

Note: This table is for concentric stranded conductors only. For compact stranded conductors, Table C4(A) should be used.

Table C4(A). Maximum Number of Compact Conductors in Intermediate Metal Conduit *(Based on Table 1, Chapter 9)*

COMPACT CONDUCTORS

Type	Conductor Size (AWG/kcmil)	½	¾	1	1¼	1½	2	2½	3	3½	4
THW, THW-2, THHW	8	2	4	7	13	17	28	40	62	83	107
	6	1	3	6	10	13	22	31	48	64	82
	4	1	2	4	7	10	16	23	36	48	62
	2	1	1	3	5	7	12	17	26	35	45
	1	1	1	1	4	5	8	12	18	25	32
	1/0	1	1	1	3	4	7	10	16	21	27
	2/0	0	1	1	3	4	6	9	13	18	23
	3/0	0	1	1	2	3	5	7	11	15	20
	4/0	0	1	1	1	2	4	6	9	13	16
	250	0	0	1	1	1	3	5	7	10	13
	300	0	0	1	1	1	3	4	6	9	11
	350	0	0	1	1	1	2	4	6	8	10
	400	0	0	1	1	1	2	3	5	7	9
	500	0	0	0	1	1	1	3	4	6	8
	600	0	0	0	1	1	1	2	3	5	6
	700	0	0	0	1	1	1	1	3	4	5
	750	0	0	0	1	1	1	1	3	4	5
	1000	0	0	0	0	1	1	1	2	3	4
THHN, THWN, THWN-2	8	—	—	—	—	—	—	—	—	—	—
	6	3	5	8	14	19	32	45	70	93	120
	4	1	3	5	9	12	20	28	43	58	74
	2	1	1	3	6	8	14	20	31	41	53
	1	1	1	3	5	6	10	15	23	31	40
	1/0	1	1	2	4	5	9	13	20	26	34
	2/0	1	1	1	3	4	7	10	16	22	28
	3/0	0	1	1	3	4	6	9	14	18	24
	4/0	0	1	1	2	3	5	7	11	15	19
	250	0	1	1	1	2	4	6	9	12	15
	300	0	0	1	1	1	3	5	7	10	13
	350	0	0	1	1	1	3	4	7	9	11
	400	0	0	1	1	1	2	4	6	8	10
	500	0	0	1	1	1	2	3	5	7	9
	600	0	0	0	1	1	1	2	4	5	7
	700	0	0	0	1	1	1	2	3	5	6
	750	0	0	0	1	1	1	1	3	4	6
	1000	0	0	0	0	1	1	1	2	3	4
XHHW, XHHW-2	8	3	6	9	16	22	37	52	80	107	138
	6	2	4	7	12	16	27	38	59	80	103
	4	1	3	5	9	12	20	28	43	58	74
	2	1	1	3	6	8	14	20	31	41	53
	1	1	1	3	5	6	10	15	23	31	40
	1/0	1	1	2	4	5	9	13	20	26	34
	2/0	1	1	1	3	4	7	11	17	22	29
	3/0	0	1	1	3	4	6	9	14	18	24
	4/0	0	1	1	2	3	5	7	11	15	20
	250	0	1	1	1	2	4	6	9	12	16
	300	0	0	1	1	1	3	5	8	10	13
	350	0	0	1	1	1	3	4	7	9	12
	400	0	0	1	1	1	3	4	6	8	11
	500	0	0	1	1	1	2	3	5	7	9
	600	0	0	0	1	1	1	2	4	5	7
	700	0	0	0	1	1	1	2	3	5	6
	750	0	0	0	1	1	1	1	3	4	6
	1000	0	0	0	0	1	1	1	2	3	4

Definition: Compact stranding is the result of a manufacturing process where the standard conductor is compressed to the extent that interstices (voids between strand wires) are virtually eliminated.

Table C5. Maximum Number of Conductors and Fixture Wires in Liquidtight Flexible Nonmetallic Conduit (Type FNMC-B*) (Based on Table 1, Chapter 9)

Type	Conductor Size (AWG/kcmil)	3/8	1/2	3/4	1	1 1/4	1 1/2	2
RH	14	3	6	10	16	29	38	62
	12	3	5	8	13	23	30	50
RHH, RHW, RHW-2,	14	2	4	7	12	21	27	44
	12	1	3	6	10	17	22	36
RH, RHH, RHW, RHW-2	10	1	3	5	8	14	18	29
	8	1	1	2	4	7	9	15
	6	1	1	1	3	6	7	12
	4	0	1	1	2	4	6	9
	3	0	1	1	1	4	5	8
	2	0	1	1	1	3	4	7
	1	0	0	1	1	1	3	5
	1/0	0	0	1	1	1	2	4
	2/0	0	0	1	1	1	1	3
	3/0	0	0	0	1	1	1	3
	4/0	0	0	0	1	1	1	2
	250	0	0	0	0	1	1	1
	300	0	0	0	0	1	1	1
	350	0	0	0	0	1	1	1
	400	0	0	0	0	1	1	1
	500	0	0	0	0	1	1	1
	600	0	0	0	0	0	1	1
	700	0	0	0	0	0	0	1
	750	0	0	0	0	0	0	1
	800	0	0	0	0	0	0	1
	900	0	0	0	0	0	0	1
	1000	0	0	0	0	0	0	1
	1250	0	0	0	0	0	0	0
	1500	0	0	0	0	0	0	0
	1750	0	0	0	0	0	0	0
	2000	0	0	0	0	0	0	0
TW	14	5	9	15	25	44	57	93
	12	4	7	12	19	33	43	71
	10	3	5	9	14	25	32	53
	8	1	3	5	8	14	18	29
RHH†, RHW†, RHW-2†, THHW, THW, THW-2	14	3	6	10	16	29	38	62
RHH†, RHW†, RHW-2†, THHW, THW	12	3	5	8	13	23	30	50
	10	1	3	6	10	18	23	39
RHH†, RHW†, RHW-2†, THHW, THW, THW-2	8	1	1	4	6	11	14	23
RHH†, RHW†, RHW 2†, TW, THW, THHW, THW-2	6	1	1	3	5	8	11	18
	4	1	1	1	3	6	8	13
	3	1	1	1	3	5	7	11
	2	0	1	1	2	4	6	9
	1	0	1	1	1	3	4	7
	1/0	0	0	1	1	2	3	6
	2/0	0	0	1	1	2	3	5
	3/0	0	0	1	1	1	2	4
	4/0	0	0	0	1	1	1	3
	250	0	0	0	1	1	1	3
	300	0	0	0	1	1	1	2
	350	0	0	0	0	1	1	1
	400	0	0	0	0	1	1	1
	500	0	0	0	0	1	1	1
	600	0	0	0	0	1	1	1
	700	0	0	0	0	0	1	1
	750	0	0	0	0	0	1	1
	800	0	0	0	0	0	1	1
	900	0	0	0	0	0	0	1
	1000	0	0	0	0	0	0	1
	1250	0	0	0	0	0	0	1
	1500	0	0	0	0	0	0	0
	1750	0	0	0	0	0	0	0
	2000	0	0	0	0	0	0	0

*Corresponds to Section 351-22(2).

†Types RHH, RHW, and RHW-2 without outer covering.

Table C5. Continued

Type	Conductor Size (AWG/kcmil)	3/8	1/2	3/4	1	1 1/4	1 1/2	2
THHN, THWN, THWN-2	14	8	13	22	36	63	81	133
	12	5	9	16	26	46	59	97
	10	3	6	10	16	29	37	61
	8	1	3	6	9	16	21	35
	6	1	2	4	7	12	15	25
	4	1	1	2	4	7	9	15
	3	1	1	1	3	6	8	13
	2	1	1	1	3	5	7	11
	1	0	1	1	1	4	5	8
	1/0	0	1	1	1	3	4	7
	2/0	0	0	1	1	2	3	6
	3/0	0	0	1	1	1	3	5
	4/0	0	0	1	1	1	2	4
	250	0	0	0	1	1	1	3
	300	0	0	0	1	1	1	3
	350	0	0	0	1	1	1	2
	400	0	0	0	0	1	1	1
	500	0	0	0	0	1	1	1
	600	0	0	0	0	1	1	1
	700	0	0	0	0	1	1	1
	750	0	0	0	0	0	1	1
	800	0	0	0	0	0	1	1
	900	0	0	0	0	0	1	1
	1000	0	0	0	0	0	0	1
FEP, FEPB, PFA, PFAH, TFE	14	7	12	21	35	61	79	129
	12	5	9	15	25	44	57	94
	10	4	6	11	18	32	41	68
	8	1	3	6	10	18	23	39
	6	1	2	4	7	13	17	27
	4	1	1	3	5	9	12	19
	3	1	1	2	4	7	10	16
	2	1	1	1	3	6	8	13
PFA, PFAH, TFE	1	0	1	1	2	4	5	9
PFA, PFAH, TFE, Z	1/0	0	1	1	1	3	4	7
	2/0	0	1	1	1	3	4	6
	3/0	0	0	1	1	2	3	5
	4/0	0	0	1	1	1	2	4
Z	14	9	15	26	42	73	95	156
	12	6	10	18	30	52	67	111
	10	4	6	11	18	32	41	68
	8	2	4	7	11	20	26	43
	6	1	3	5	8	14	18	30
	4	1	1	3	5	9	12	20
	3	1	1	2	4	7	9	15
	2	0	1	1	3	6	7	12
	1	0	1	1	2	5	6	10
XHH, XHHW, XHHW-2, ZW	14	5	9	15	25	44	57	93
	12	4	7	12	19	33	43	71
	10	3	5	9	14	25	32	53
	8	1	3	5	8	14	18	29
	6	1	1	3	6	10	13	22
	4	1	1	2	4	7	9	16
	3	1	1	1	3	6	8	13
	2	1	1	1	3	5	7	11
XHH, XHHW, XHHW-2	1	0	1	1	1	4	5	8
	1/0	0	1	1	1	3	4	7
	2/0	0	0	1	1	2	3	6
	3/0	0	0	1	1	1	3	5
	4/0	0	0	1	1	1	2	4
	250	0	0	0	1	1	1	3
	300	0	0	0	1	1	1	3
	350	0	0	0	1	1	1	2
	400	0	0	0	0	1	1	1
	500	0	0	0	0	1	1	1
	600	0	0	0	0	1	1	1
	700	0	0	0	0	1	1	1
	750	0	0	0	0	0	1	1
	800	0	0	0	0	0	1	1
	900	0	0	0	0	0	1	1
	1000	0	0	0	0	0	0	1
	1250	0	0	0	0	0	0	1
	1500	0	0	0	0	0	0	1
	1750	0	0	0	0	0	0	0
	2000	0	0	0	0	0	0	0

Table C5. Continued

			FIXTURE WIRES					
Type	Conductor Size (AWG/kcmil)	3/8	1/2	3/4	1	1¼	1½	2
FFH-2, RFH-2	18	5	8	15	24	42	54	89
	16	4	7	12	20	35	46	75
SF-2, SFF-2	18	6	11	19	30	53	69	113
	16	5	9	15	25	44	57	93
	14	4	7	12	20	35	46	75
SF-1, SFF-1	18	11	19	33	53	94	122	199
AF, RFH-1, RFHH-2, TF, TFF, XF, XFF	18	8	14	24	39	69	90	147
AF, RFHH-2, TF, TFF, XF, XFF	16	7	11	20	32	56	72	119
AF, XF, XFF	14	5	9	15	25	44	57	93
TFN, TFFN	18	14	23	39	63	111	144	236
	16	10	17	30	48	85	110	180
PF, PFF, PGF, PGFF, PAF, PTF, PTFF, PAFF	18	13	21	37	60	105	136	223
	16	10	16	29	46	81	105	173
	14	7	12	21	35	61	79	129
HF, HFF, ZF, ZFF, ZHF	18	17	28	48	77	136	176	288
	16	12	20	35	57	100	129	212
	14	9	15	26	42	73	95	156
KF-2, KFF-2	18	24	40	70	112	197	255	418
	16	17	28	49	79	139	180	295
	14	12	19	34	54	95	123	202
	12	8	13	23	37	65	85	139
	10	5	9	15	25	44	57	93
KF-1, KFF-1	18	29	48	83	134	235	304	499
	16	20	34	58	94	165	214	350
	14	14	23	39	63	111	144	236
	12	9	15	26	42	73	95	156
	10	6	10	17	27	48	62	102
AF, XF, XFF	12	3	5	8	13	23	30	50
	10	1	3	6	10	18	23	39

Note: This table is for concentric stranded conductors only. For compact stranded conductors, Table C5(A) should be used.

Table C5(A). Maximum Number of Compact Conductors in Liquidtight Flexible Nonmetallic Conduit (Type FNMC-B*) *(Based on Table 1, Chapter 9)*

			COMPACT CONDUCTORS					
Type	Conductor Size (AWG/kcmil)	3/8	1/2	3/4	1	1¼	1½	2
THW, THW-2, THHW	8	1	2	4	7	12	15	25
	6	1	1	3	5	9	12	19
	4	1	1	2	4	7	9	14
	2	1	1	1	3	5	6	11
	1	0	1	1	1	3	4	7
	1/0	0	1	1	1	3	4	6
	2/0	0	0	1	1	2	3	5
	3/0	0	0	1	1	1	3	4
	4/0	0	0	1	1	1	2	4
	250	0	0	0	1	1	1	3
	300	0	0	0	1	1	1	2
	350	0	0	0	1	1	1	2
	400	0	0	0	0	1	1	1
	500	0	0	0	0	1	1	1
	600	0	0	0	0	1	1	1
	700	0	0	0	0	1	1	1
	750	0	0	0	0	0	1	1
	1000	0	0	0	0	0	1	1
THHN, THWN, THWN-2	8	—	—	—	—	—	—	—
	6	1	2	4	7	13	17	28
	4	1	1	3	4	8	11	17
	2	1	1	1	3	6	7	12
	1	0	1	1	2	4	6	9
	1/0	0	1	1	1	4	5	8
	2/0	0	1	1	1	3	4	6
	3/0	0	0	1	1	2	3	5
	4/0	0	0	1	1	1	3	4
	250	0	0	1	1	1	1	3
	300	0	0	0	1	1	1	3
	350	0	0	0	1	1	1	2
	400	0	0	0	1	1	1	2
	500	0	0	0	0	1	1	1
	600	0	0	0	0	1	1	1
	700	0	0	0	0	1	1	1
	750	0	0	0	0	1	1	1
	1000	0	0	0	0	0	1	1
XHHW, XHHW-2	8	1	3	5	9	15	20	33
	6	1	2	4	6	11	15	24
	4	1	1	3	4	8	11	17
	2	1	1	1	3	6	7	12
	1	0	1	1	2	4	6	9
	1/0	0	1	1	1	4	5	8
	2/0	0	1	1	1	3	4	7
	3/0	0	0	1	1	2	3	5
	4/0	0	0	1	1	1	3	4
	250	0	0	1	1	1	1	3
	300	0	0	0	1	1	1	3
	350	0	0	0	1	1	1	3
	400	0	0	0	1	1	1	2
	500	0	0	0	0	1	1	1
	600	0	0	0	0	1	1	1
	700	0	0	0	0	1	1	1
	750	0	0	0	0	1	1	1
	1000	0	0	0	0	0	1	1

*Corresponds to Section 351-22(2).

Definition: Compact stranding is the result of a manufacturing process where the standard conductor is compressed to the extent that the interstices (voids between strand wires) are virtually eliminated.

Table C6. Maximum Number of Conductors and Fixture Wires in Liquidtight Flexible Nonmetallic Conduit (Type FNMC-A*) (Based On Table 1, Chapter 9)

Type	Conductor Size (AWG/kcmil)	⅜	½	¾	1	1¼	1½	2
RH	14	3	6	10	16	28	38	64
	12	3	4	8	13	23	31	51
RHH, RHW, RHW-2	14	2	4	7	11	20	27	45
	12	1	3	6	9	17	23	38
RH, RHH, RHW, RHW-2	10	1	3	5	8	13	18	30
	8	1	1	2	4	7	9	16
	6	1	1	1	3	5	7	13
	4	0	1	1	2	4	6	10
	3	0	1	1	1	4	5	8
	2	0	1	1	1	3	4	7
	1	0	0	1	1	1	3	5
	1/0	0	0	1	1	1	2	4
	2/0	0	0	1	1	1	1	4
	3/0	0	0	0	1	1	1	3
	4/0	0	0	0	1	1	1	3
	250	0	0	0	0	1	1	1
	300	0	0	0	0	1	1	1
	350	0	0	0	0	1	1	1
	400	0	0	0	0	1	1	1
	500	0	0	0	0	0	1	1
	600	0	0	0	0	0	1	1
	700	0	0	0	0	0	0	1
	750	0	0	0	0	0	0	1
	800	0	0	0	0	0	0	1
	900	0	0	0	0	0	0	1
	1000	0	0	0	0	0	0	1
	1250	0	0	0	0	0	0	0
	1500	0	0	0	0	0	0	0
	1750	0	0	0	0	0	0	0
	2000	0	0	0	0	0	0	0
TW	14	5	9	15	24	43	58	96
	12	4	7	12	19	33	44	74
	10	3	5	9	14	24	33	55
	8	1	3	5	8	13	18	30
RHH†, RHW†, RHW-2†, THHW, THW, THW-2	14	3	6	10	16	28	38	64
RHH†, RHW†, RHW-2†, THHW, THW	12	3	4	8	13	23	31	51
	10	1	3	6	10	18	24	40
RHH†, RHW†, RHW-2†, THHW, THW, THW-2	8	1	1	4	6	10	14	24
RHH†, RHW†, RHW-2†, TW, THW, THHW, THW-2	6	1	1	3	4	8	11	18
	4	1	1	1	3	6	8	13
	3	1	1	1	3	5	7	11
	2	0	1	1	2	4	6	10
	1	0	1	1	1	3	4	7
	1/0	0	0	1	1	2	3	6
	2/0	0	0	1	1	1	3	5
	3/0	0	0	1	1	1	2	4
	4/0	0	0	0	1	1	1	3
	250	0	0	0	1	1	1	3
	300	0	0	0	1	1	1	2
	350	0	0	0	0	1	1	1
	400	0	0	0	0	1	1	1
	500	0	0	0	0	1	1	1
	600	0	0	0	0	1	1	1
	700	0	0	0	0	0	1	1
	750	0	0	0	0	0	1	1
	800	0	0	0	0	0	1	1
	900	0	0	0	0	0	0	1
	1000	0	0	0	0	0	0	1
	1250	0	0	0	0	0	0	1
	1500	0	0	0	0	0	0	1
	1750	0	0	0	0	0	0	0
	2000	0	0	0	0	0	0	0

*Corresponds to Section 351-22(1)

†Types RHH, RHW, and RHW-2 without outer covering.

Table C6. Continued

Type	Conductor Size (AWG/kcmil)	⅜	½	¾	1	1¼	1½	2
THHN, THWN, THWN-2	14	8	13	22	35	62	83	137
	12	5	9	16	25	45	60	100
	10	3	6	10	16	28	38	63
	8	1	3	6	9	16	22	36
	6	1	2	4	6	12	16	26
	4	1	1	2	4	7	9	16
	3	1	1	1	3	6	8	13
	2	1	1	1	3	5	7	11
	1	0	1	1	1	4	5	8
	1/0	0	1	1	1	3	4	7
	2/0	0	0	1	1	2	3	6
	3/0	0	0	1	1	1	3	5
	4/0	0	0	1	1	1	2	4
	250	0	0	0	1	1	1	3
	300	0	0	0	1	1	1	3
	350	0	0	0	1	1	1	2
	400	0	0	0	0	1	1	1
	500	0	0	0	0	1	1	1
	600	0	0	0	0	1	1	1
	700	0	0	0	0	1	1	1
	750	0	0	0	0	0	1	1
	800	0	0	0	0	0	1	1
	900	0	0	0	0	0	1	1
	1000	0	0	0	0	0	0	1
FEP, FEPB, PFA, PFAH, TFE	14	7	12	21	34	60	80	133
	12	5	9	15	25	44	59	97
	10	4	6	11	18	31	42	70
	8	1	3	6	10	18	24	40
	6	1	2	4	7	13	17	28
	4	1	1	3	5	9	12	20
	3	1	1	2	4	7	10	16
	2	1	1	1	3	6	8	13
PFA, PFAH, TFE	1	0	1	1	2	4	5	9
PFA, PFAH, TFE, Z	1/0	0	1	1	1	3	5	8
	2/0	0	1	1	1	3	4	6
	3/0	0	0	1	1	2	3	5
	4/0	0	0	1	1	1	2	4
Z	14	9	15	25	41	72	97	161
	12	6	10	18	29	51	69	114
	10	4	6	11	18	31	42	70
	8	2	4	7	11	20	26	44
	6	1	3	5	8	14	18	31
	4	1	1	3	5	9	13	21
	3	1	1	2	4	7	9	15
	2	1	1	1	3	6	8	13
	1	1	1	1	2	4	6	10
XHH, XHHW, XHHW-2, ZW	14	5	9	15	24	43	58	96
	12	4	7	12	19	33	44	74
	10	3	5	9	14	24	33	55
	8	1	3	5	8	13	18	30
	6	1	1	3	5	10	13	22
	4	1	1	2	4	7	10	16
	3	1	1	1	3	6	8	14
	2	1	1	1	3	5	7	11
XHH, XHHW, XHHW-2	1	0	1	1	1	4	5	8
	1/0	0	1	1	1	3	4	7
	2/0	0	0	1	1	2	3	6
	3/0	0	0	1	1	1	3	5
	4/0	0	0	1	1	1	2	4
	250	0	0	0	1	1	1	3
	300	0	0	0	1	1	1	3
	350	0	0	0	1	1	1	2
	400	0	0	0	0	1	1	1
	500	0	0	0	0	1	1	1
	600	0	0	0	0	1	1	1
	700	0	0	0	0	1	1	1
	750	0	0	0	0	0	1	1
	800	0	0	0	0	0	1	1
	900	0	0	0	0	0	1	1
	1000	0	0	0	0	0	0	1
	1250	0	0	0	0	0	0	1
	1500	0	0	0	0	0	0	1
	1750	0	0	0	0	0	0	0
	2000	0	0	0	0	0	0	0

Table C6. Continued

FIXTURE WIRES

Type	Conductor Size (AWG/kcmil)	3/8	1/2	3/4	1	1 1/4	1 1/2	2
FFH-2, RFH-2, RFHH-3	18	5	8	14	23	41	55	92
	16	4	7	12	20	35	47	77
SF-2, SFF-2	18	6	11	18	29	52	70	116
	16	5	9	15	24	43	58	96
	14	4	7	12	20	35	47	77
SF-1, SFF-1	18	12	19	33	52	92	124	205
AF, RFH-1, RFHH-2, TF, TFF, XF, XFF	18	8	14	24	39	68	91	152
AF, RFHH-2, TF, TFF, XF, XFF	16	7	11	19	31	55	74	122
AF, XF, XFF	14	5	9	15	24	43	58	96
TFN, TFFN	18	14	22	39	62	109	146	243
	16	10	17	29	47	83	112	185
PF, PFF, PGF, PGFF, PAF, PTF, PTFF, PAFF	18	13	21	37	59	103	139	230
	16	10	16	28	45	80	107	178
	14	7	12	21	34	60	80	133
HF, HFF, ZF, ZFF, ZHF	18	17	27	47	76	133	179	297
	16	12	20	35	56	98	132	219
	14	9	15	25	41	72	97	161
KF-2, KFF-2	18	25	40	69	110	193	260	431
	16	17	28	48	77	136	183	303
	14	12	19	33	53	94	126	209
	12	8	13	23	36	64	86	143
	10	5	9	15	24	43	58	96
KF-1, KFF-1	18	29	48	82	131	231	310	514
	16	21	33	57	92	162	218	361
	14	14	22	39	62	109	146	243
	12	9	15	25	41	72	97	161
	10	6	10	17	27	47	63	105
AF, XF, XFF	12	3	4	8	13	23	31	51
	10	1	3	6	10	18	24	40

Note: This table is for concentric stranded conductors only. For compact stranded conductors, Table C6(A) should be used.

Table C6(A). Maximum Number of Compact Conductors in Liquidtight Flexible Nonmetallic Conduit (Type FNMC-A*) *(Based on Table 1, Chapter 9)*

COMPACT CONDUCTORS

Type	Conductor Size (AWG/kcmil)	3/8	1/2	3/4	1	1 1/4	1 1/2	2
THW, THW-2, THHW	8	1	2	4	6	11	16	26
	6	1	1	3	5	9	12	20
	4	1	1	2	4	7	9	15
	2	1	1	1	3	5	6	11
	1	0	1	1	1	3	4	8
	1/0	0	1	1	1	3	4	7
	2/0	0	0	1	1	2	3	5
	3/0	0	0	1	1	1	3	5
	4/0	0	0	1	1	1	2	4
	250	0	0	0	1	1	1	3
	300	0	0	0	1	1	1	3
	350	0	0	0	1	1	1	2
	400	0	0	0	0	1	1	1
	500	0	0	0	0	1	1	1
	600	0	0	0	0	1	1	1
	700	0	0	0	0	1	1	1
	750	0	0	0	0	0	1	1
	1000	0	0	0	0	0	1	1
THHN, THWN, THWN-2	8	—	—	—	—	—	—	—
	6	1	2	4	7	13	18	29
	4	1	1	3	4	8	11	18
	2	1	1	1	3	6	8	13
	1	0	1	1	2	4	6	10
	1/0	0	1	1	1	3	5	8
	2/0	0	1	1	1	3	4	7
	3/0	0	0	1	1	2	3	6
	4/0	0	0	1	1	1	3	5
	250	0	0	1	1	1	1	3
	300	0	0	0	1	1	1	3
	350	0	0	0	1	1	1	3
	400	0	0	0	1	1	1	2
	500	0	0	0	0	1	1	1
	600	0	0	0	0	1	1	1
	700	0	0	0	0	1	1	1
	750	0	0	0	0	1	1	1
	1000	0	0	0	0	0	1	1
XHHW, XHHW-2	8	1	3	5	8	15	20	34
	6	1	2	4	6	11	15	25
	4	1	1	3	4	8	11	18
	2	1	1	1	3	6	8	13
	1	0	1	1	2	4	6	10
	1/0	0	1	1	1	3	5	8
	2/0	0	1	1	1	3	4	7
	3/0	0	0	1	1	2	3	6
	4/0	0	0	1	1	1	3	5
	250	0	0	1	1	1	2	4
	300	0	0	0	1	1	1	3
	350	0	0	0	1	1	1	3
	400	0	0	0	1	1	1	2
	500	0	0	0	0	1	1	1
	600	0	0	0	0	1	1	1
	700	0	0	0	0	1	1	1
	750	0	0	0	0	1	1	1
	1000	0	0	0	0	0	1	1

*Corresponds to Section 351-22(1).

Definition: Compact stranding is the result of a manufacturing process where the standard conductor is compressed to the extent that the interstices (voids between strand wires) are virtually eliminated.

Table C7. Maximum Number of Conductors and Fixture Wires in Liquidtight Flexible Metal Conduit *(Based on Table 1, Chapter 9)*

CONDUCTORS

Type	Conductor Size (AWG/kcmil)	½	¾	1	1¼	1½	2	2½	3	3½	4
RH	14	6	10	16	29	38	62	93	143	186	243
	12	5	8	13	23	30	50	75	115	149	195
RHH, RHW, RHW-2	14	4	7	12	21	27	44	66	102	133	173
	12	3	6	10	17	22	36	55	84	110	144
RH, RHH, RHW, RHW-2,	10	3	5	8	14	18	29	44	68	89	116
	8	1	2	4	7	9	15	23	36	46	61
	6	1	1	3	6	7	12	18	28	37	48
	4	1	1	2	4	6	9	14	22	29	38
	3	1	1	1	4	5	8	13	19	25	33
	2	1	1	1	3	4	7	11	17	22	29
	1	0	1	1	1	3	5	7	11	14	19
	1/0	0	1	1	1	2	4	6	10	13	16
	2/0	0	1	1	1	1	3	5	8	11	14
	3/0	0	0	1	1	1	3	4	7	9	12
	4/0	0	0	1	1	1	2	4	6	8	10
	250	0	0	0	1	1	1	3	4	6	8
	300	0	0	0	1	1	1	2	4	5	7
	350	0	0	0	1	1	1	2	3	5	6
	400	0	0	0	1	1	1	1	3	4	6
	500	0	0	0	0	1	1	1	3	4	5
	600	0	0	0	0	1	1	1	2	3	4
	700	0	0	0	0	0	1	1	1	3	3
	750	0	0	0	0	0	1	1	1	2	3
	800	0	0	0	0	0	1	1	1	2	3
	900	0	0	0	0	0	1	1	1	2	3
	1000	0	0	0	0	0	1	1	1	1	3
	1250	0	0	0	0	0	0	0	1	1	1
	1500	0	0	0	0	0	0	1	1	1	1
	1750	0	0	0	0	0	0	0	1	1	1
	2000	0	0	0	0	0	0	0	1	1	1
TW	14	9	15	25	44	57	93	140	215	280	365
	12	7	12	19	33	43	71	108	165	215	280
	10	5	9	14	25	32	53	80	123	160	209
	8	3	5	8	14	18	29	44	68	89	116
RHH*, RHW*, RHW-2*, THHW, THW, THW-2	14	6	10	16	29	38	62	93	143	186	243
RHH*, RHW*, RHW-2*, THHW, THW	12	5	8	13	23	30	50	75	115	149	195
	10	3	6	10	18	23	39	58	89	117	152
RHH*, RHW*, RHW-2*, THHW, THW, THW-2	8	1	4	6	11	14	23	35	53	70	91
RHH*, RHW*, RHW-2*, TW, THW, THHW, THW-2	6	1	3	5	8	11	18	27	41	53	70
	4	1	1	3	6	8	13	20	30	40	52
	3	1	1	3	5	7	11	17	26	34	44
	2	1	1	2	4	6	9	14	22	29	38
	1	1	1	1	3	4	7	10	15	20	26
	1/0	0	1	1	2	3	6	8	13	17	23
	2/0	0	1	1	2	3	5	7	11	15	19
	3/0	0	1	1	1	2	4	6	9	12	16
	4/0	0	0	1	1	1	3	5	8	10	13
	250	0	0	1	1	1	3	4	6	8	11
	300	0	0	1	1	1	2	3	5	7	9
	350	0	0	0	1	1	1	3	5	6	8
	400	0	0	0	1	1	1	3	4	6	7
	500	0	0	0	1	1	1	2	3	5	6
	600	0	0	0	1	1	1	1	3	4	5
	700	0	0	0	0	1	1	1	2	3	4
	750	0	0	0	0	1	1	1	2	3	4
	800	0	0	0	0	1	1	1	2	3	4
	900	0	0	0	0	0	1	1	1	3	3
	1000	0	0	0	0	0	1	1	1	2	3
	1250	0	0	0	0	0	1	1	1	1	2
	1500	0	0	0	0	0	0	1	1	1	2
	1750	0	0	0	0	0	0	1	1	1	1
	2000	0	0	0	0	0	0	1	1	1	1

*Types RHH, RHW, and RHW-2 without outer covering.

Table C7. Continued

CONDUCTORS

Type	Conductor Size (AWG/kcmil)	½	¾	1	1¼	1½	2	2½	3	3½	4
THHN, THWN, THWN-2	14	13	22	36	63	81	133	201	308	401	523
	12	9	16	26	46	59	97	146	225	292	381
	10	6	10	16	29	37	61	92	141	184	240
	8	3	6	9	16	21	35	53	81	106	138
	6	2	4	7	12	15	25	38	59	76	100
	4	1	2	4	7	9	15	23	36	47	61
	3	1	1	3	6	8	13	20	30	40	52
	2	1	1	3	5	7	11	17	26	33	44
	1	1	1	1	4	5	8	12	19	25	32
	1/0	1	1	1	3	4	7	10	16	21	27
	2/0	0	1	1	2	3	6	8	13	17	23
	3/0	0	1	1	1	3	5	7	11	14	19
	4/0	0	1	1	1	2	4	6	9	12	15
	250	0	0	1	1	1	3	5	7	10	12
	300	0	0	1	1	1	3	4	6	8	11
	350	0	0	1	1	1	2	3	5	7	9
	400	0	0	0	1	1	1	3	5	6	8
	500	0	0	0	1	1	1	2	4	5	7
	600	0	0	0	1	1	1	1	3	4	6
	700	0	0	0	1	1	1	1	3	4	5
	750	0	0	0	0	1	1	1	3	3	5
	800	0	0	0	0	1	1	1	2	3	4
	900	0	0	0	0	1	1	1	2	3	4
	1000	0	0	0	0	0	1	1	1	3	3
FEP, FEPB, PFA, PFAH, TFE	14	12	21	35	61	79	129	195	299	389	507
	12	9	15	25	44	57	94	142	218	284	370
	10	6	11	18	32	41	68	102	156	203	266
	8	3	6	10	18	23	39	58	89	117	152
	6	2	4	7	13	17	27	41	64	83	108
	4	1	3	5	9	12	19	29	44	58	75
	3	1	2	4	7	10	16	24	37	48	63
	2	1	1	3	6	8	13	20	30	40	52
PFA, PFAH, TFE	1	1	1	2	4	5	9	14	21	28	36
PFA, PFAH, TFE, Z	1/0	1	1	1	3	4	7	11	18	23	30
	2/0	1	1	1	3	4	6	9	14	19	25
	3/0	0	1	1	2	3	5	8	12	16	20
	4/0	0	1	1	1	2	4	6	10	13	17
Z	14	20	26	42	73	95	156	235	360	469	611
	12	14	18	30	52	67	111	167	255	332	434
	10	8	11	18	32	41	68	102	156	203	266
	8	5	7	11	20	26	43	64	99	129	168
	6	4	5	8	14	18	30	45	69	90	118
	4	2	3	5	9	12	20	31	48	62	81
	3	2	2	4	7	9	15	23	35	45	59
	2	1	1	3	6	7	12	19	29	38	49
	1	1	1	2	5	6	10	15	23	30	40
XHH, XHHW, XHHW-2, ZW	14	9	15	25	44	57	93	140	215	280	365
	12	7	12	19	33	43	71	108	165	215	280
	10	5	9	14	25	32	53	80	123	160	209
	8	3	5	8	14	18	29	44	68	89	116
	6	1	3	6	10	13	22	33	50	66	86
	4	1	2	4	7	9	16	24	36	48	62
	3	1	1	3	6	8	13	20	31	40	52
	2	1	1	3	5	7	11	17	26	34	44
XHH, XHHW, XHHW-2	1	1	1	1	4	5	8	12	19	25	33
	1/0	1	1	1	3	4	7	10	16	21	28
	2/0	0	1	1	2	3	6	9	13	17	23
	3/0	0	1	1	1	3	5	7	11	14	19
	4/0	0	1	1	1	2	4	6	9	12	16
	250	0	0	1	1	1	3	5	7	10	13
	300	0	0	1	1	1	3	4	6	8	11
	350	0	0	1	1	1	2	3	5	7	10
	400	0	0	1	1	1	1	3	5	6	8
	500	0	0	0	1	1	1	2	4	5	7
	600	0	0	0	1	1	1	1	3	4	6
	700	0	0	0	1	1	1	1	3	4	5
	750	0	0	0	0	1	1	1	3	3	5
	800	0	0	0	0	1	1	1	2	3	4
	900	0	0	0	0	0	1	1	1	3	4
	1000	0	0	0	0	0	1	1	1	3	3
	1250	0	0	0	0	0	1	1	1	1	3
	1500	0	0	0	0	0	1	1	1	1	2
	1750	0	0	0	0	0	0	1	1	1	2
	2000	0	0	0	0	0	0	1	1	1	2

Table C7. Continued

FIXTURE WIRES

Type	Conductor Size (AWG/kcmil)	½	¾	1	1¼	1½	2
FFH-2, RFH-2, RFHH-3	18	8	15	24	42	54	89
	16	7	12	20	35	46	75
SF-2, SFF-2	18	11	19	30	53	69	113
	16	9	15	25	44	57	93
	14	7	12	20	35	46	75
SF-1, SFF-1	18	19	33	53	94	122	199
AF, RFH-1, RFHH-2, TF, TFF, XF, XFF	18	14	24	39	69	90	147
AF, RFHH-2, TF, TFF, XF, XFF	16	11	20	32	56	72	119
AF, XF, XFF	14	9	15	25	44	57	93
TFN, TFFN	18	23	39	63	111	144	236
	16	17	30	48	85	110	180
PF, PFF, PGF, PGFF, PAF, PTF, PTFF, PAFF	18	21	37	60	105	136	223
	16	16	29	46	81	105	173
	14	12	21	35	61	79	129
HF, HFF, ZF, ZFF, ZHF	18	28	48	77	136	176	288
	16	20	35	57	100	129	212
	14	15	26	42	73	95	156
KF-2, KFF-2	18	40	70	112	197	255	418
	16	28	49	79	139	180	295
	14	19	34	54	95	123	202
	12	13	23	37	65	85	139
	10	9	15	25	44	57	93
KF-1, KFF-1	18	48	83	134	235	304	499
	16	34	58	94	165	214	350
	14	23	39	63	111	144	236
	12	15	26	42	73	95	156
	10	10	17	27	48	62	102
AF, XF, XFF	12	5	8	13	23	30	50
	10	3	6	10	18	23	39

Note: This table is for concentric stranded conductors only. For compact stranded conductors, Table C7(A) should be used.

Table C7(A). Maximum Number of Compact Conductors in Liquidtight Flexible Metal Conduit (Based on Table 1, Chapter 9)

COMPACT CONDUCTORS

Type	Conductor Size (AWG/kcmil)	⅜	½	¾	1	1¼	1½	2	2½	3	3½	4
THW, THW-2, THHW	8	1	2	4	7	12	15	25	38	58	76	99
	6	1	1	3	5	9	12	19	29	45	59	77
	4	1	1	2	4	7	9	14	22	34	44	57
	2	1	1	1	3	5	6	11	16	25	32	42
	1	0	1	1	1	3	4	7	11	17	23	30
	1/0	0	1	1	1	3	4	6	10	15	20	26
	2/0	0	0	1	1	2	3	5	8	13	16	21
	3/0	0	0	1	1	1	3	4	7	11	14	18
	4/0	0	0	1	1	1	2	4	6	9	12	15
	250	0	0	0	1	1	1	3	4	7	9	12
	300	0	0	0	1	1	1	2	4	6	8	10
	350	0	0	0	1	1	1	2	3	5	7	9
	400	0	0	0	0	1	1	1	3	5	6	8
	500	0	0	0	0	1	1	1	3	4	5	7
	600	0	0	0	0	1	1	1	1	3	4	6
	700	0	0	0	0	1	1	1	1	3	4	5
	750	0	0	0	0	0	1	1	1	3	3	5
	1000	0	0	0	0	0	1	1	1	1	3	4
THHN, THWN, THWN-2	8	—	—	—	—	—	—	—	—	—	—	—
	6	1	2	4	7	13	17	28	43	66	86	112
	4	1	1	3	4	8	11	17	26	41	53	69
	2	1	1	1	3	6	7	12	19	29	38	50
	1	0	1	1	2	4	6	9	14	22	28	37
	1/0	0	1	1	1	4	5	8	12	19	24	32
	2/0	0	1	1	1	3	4	6	10	15	20	26
	3/0	0	0	1	1	2	3	5	8	13	17	22
	4/0	0	0	1	1	1	3	4	7	10	14	18
	250	0	0	1	1	1	1	3	5	8	11	14
	300	0	0	0	1	1	1	3	4	7	9	12
	350	0	0	0	1	1	1	2	4	6	8	11
	400	0	0	0	1	1	1	2	3	5	7	9
	500	0	0	0	0	1	1	1	3	5	6	8
	600	0	0	0	0	1	1	1	2	4	5	6
	700	0	0	0	0	1	1	1	1	3	4	6
	750	0	0	0	0	1	1	1	1	3	4	5
	1000	0	0	0	0	0	1	1	1	2	3	4
XHHW, XHHW-2	8	1	3	5	9	15	20	33	49	76	98	129
	6	1	2	4	6	11	15	24	37	56	73	95
	4	1	1	3	4	8	11	17	26	41	53	69
	2	1	1	1	3	6	7	12	19	29	38	50
	1	0	1	1	2	4	6	9	14	22	28	37
	1/0	0	1	1	1	4	5	8	12	19	24	32
	2/0	0	1	1	1	3	4	7	10	16	20	27
	3/0	0	0	1	1	2	3	5	8	13	17	22
	4/0	0	0	1	1	1	3	4	7	11	14	18
	250	0	0	1	1	1	1	3	5	8	11	15
	300	0	0	0	1	1	1	3	5	7	9	12
	350	0	0	0	1	1	1	3	4	6	8	11
	400	0	0	0	1	1	1	2	4	6	7	10
	500	0	0	0	0	1	1	1	3	5	6	8
	600	0	0	0	0	1	1	1	2	4	5	6
	700	0	0	0	0	1	1	1	1	3	4	6
	750	0	0	0	0	1	1	1	1	3	4	5
	1000	0	0	0	0	0	1	1	1	2	3	4

Definition: Compact stranding is the result of a manufacturing process where the standard conductor is compressed to the extent that the interstices (voids between strand wires) are virtually eliminated.

Table C8. Maximum Number of Conductors and Fixture Wires in Rigid Metal Conduit (*Based on Table 1, Chapter 9*)

CONDUCTORS

Type	Conductor Size (AWG/kcmil)	½	¾	1	1¼	1½	2	2½	3	3½	4	5	6
RH	14	6	10	17	29	39	65	93	143	191	246	387	558
	12	5	8	13	23	32	52	75	115	154	198	311	448
RHH, RHW, RHW-2	14	4	7	12	21	28	46	66	102	136	176	276	398
	12	3	6	10	17	23	38	55	85	113	146	229	330
RH, RHH, RHW, RHW-2	10	3	5	8	14	19	31	44	68	91	118	185	267
	8	1	2	4	7	10	16	23	36	48	61	97	139
	6	1	1	3	6	8	13	18	29	38	49	77	112
	4	1	1	2	4	6	10	14	22	30	38	60	87
	3	1	1	2	4	5	9	12	19	26	34	53	76
	2	1	1	1	3	4	7	11	17	23	29	46	66
	1	0	1	1	1	3	5	7	11	15	19	30	44
	1/0	0	1	1	1	2	4	6	10	13	17	26	38
	2/0	0	1	1	1	2	4	5	8	11	14	23	33
	3/0	0	0	1	1	1	3	4	7	10	12	20	28
	4/0	0	0	1	1	1	3	4	6	8	11	17	24
	250	0	0	0	1	1	1	3	4	6	8	13	18
	300	0	0	0	1	1	1	2	4	5	7	11	16
	350	0	0	0	1	1	1	2	4	5	6	10	15
	400	0	0	0	1	1	1	1	3	4	6	9	13
	500	0	0	0	1	1	1	1	3	4	5	8	11
	600	0	0	0	0	1	1	1	2	3	4	6	9
	700	0	0	0	0	1	1	1	1	3	4	6	8
	750	0	0	0	0	0	1	1	1	3	3	5	8
	800	0	0	0	0	0	1	1	1	2	3	5	7
	900	0	0	0	0	0	1	1	1	2	3	5	7
	1000	0	0	0	0	0	1	1	1	1	3	4	6
	1250	0	0	0	0	0	0	1	1	1	1	3	5
	1500	0	0	0	0	0	0	1	1	1	1	3	4
	1750	0	0	0	0	0	0	1	1	1	1	2	4
	2000	0	0	0	0	0	0	0	1	1	1	2	3
TW	14	9	15	25	44	59	98	140	216	288	370	581	839
	12	7	12	19	33	45	75	107	165	221	284	446	644
	10	5	9	14	25	34	56	80	123	164	212	332	480
	8	3	5	8	14	19	31	44	68	91	118	185	267
RHH*, RHW*, RHW-2*, THHW, THW, THW-2	14	6	10	17	29	39	65	93	143	191	246	387	558
RHH*, RHW*, RHW-2*, THHW, THW	12	5	8	13	23	32	52	75	115	154	198	311	448
	10	3	6	10	18	25	41	58	90	120	154	242	350
RHH*, RHW*, RHW-2*, THHW, THW, THW-2	8	1	4	6	11	15	24	35	54	72	92	145	209
RHH*, RHW*, RHW-2, TW, THW, THHW, THW-2	6	1	3	5	8	11	18	27	41	55	71	111	160
	4	1	1	3	6	8	14	20	31	41	53	83	120
	3	1	1	3	5	7	12	17	26	35	45	71	103
	2	1	1	2	4	6	10	14	22	30	38	60	87
	1	1	1	1	3	4	7	10	15	21	27	42	61
	1/0	0	1	1	2	3	6	8	13	18	23	36	52
	2/0	0	1	1	2	3	5	7	11	15	19	31	44
	3/0	0	1	1	1	2	4	6	9	13	16	26	37
	4/0	0	0	1	1	1	3	5	8	10	14	21	31
	250	0	0	1	1	1	3	4	6	8	11	17	25
	300	0	0	1	1	1	2	3	5	7	9	15	22
	350	0	0	0	1	1	1	3	5	6	8	13	19
	400	0	0	0	1	1	1	3	4	6	7	12	17
	500	0	0	0	1	1	1	2	3	5	6	10	14
	600	0	0	0	1	1	1	1	3	4	5	8	12
	700	0	0	0	1	1	1	1	2	3	4	7	10
	750	0	0	0	0	1	1	1	2	3	4	7	10
	800	0	0	0	0	1	1	1	2	3	4	6	9
	900	0	0	0	0	1	1	1	1	3	4	6	8
	1000	0	0	0	0	0	1	1	1	2	3	5	8
	1250	0	0	0	0	0	1	1	1	1	2	4	6
	1500	0	0	0	0	0	1	1	1	1	2	3	5
	1750	0	0	0	0	0	0	1	1	1	1	3	4
	2000	0	0	0	0	0	0	1	1	1	1	3	4

*Types RHH, RHW, and RHW-2 without outer covering.

Table C8. Continued

CONDUCTORS

Type	Conductor Size (AWG/kcmil)	½	¾	1	1¼	1½	2	2½	3	3½	4	5	6
THHN, THWN, THWN-2	14	13	22	36	63	85	140	200	309	412	531	833	1202
	12	9	16	26	46	62	102	146	225	301	387	608	877
	10	6	10	17	29	39	64	92	142	189	244	383	552
	8	3	6	9	16	22	37	53	82	109	140	221	318
	6	2	4	7	12	16	27	38	59	79	101	159	230
	4	1	2	4	7	10	16	23	36	48	62	98	141
	3	1	1	3	6	8	14	20	31	41	53	83	120
	2	1	1	3	5	7	11	17	26	34	44	70	100
	1	1	1	1	4	5	8	12	19	25	33	51	74
	1/0	1	1	1	3	4	7	10	16	21	27	43	63
	2/0	0	1	1	2	3	6	8	13	18	23	36	52
	3/0	0	1	1	1	3	5	7	11	15	19	30	43
	4/0	0	1	1	1	2	4	6	9	12	16	25	36
	250	0	0	1	1	1	3	5	7	10	13	20	29
	300	0	0	1	1	1	3	4	6	8	11	17	25
	350	0	0	1	1	1	2	3	5	7	10	15	22
	400	0	0	1	1	1	2	3	5	7	8	13	20
	500	0	0	0	1	1	1	2	4	5	7	11	16
	600	0	0	0	1	1	1	1	3	4	6	9	13
	700	0	0	0	1	1	1	1	3	4	5	8	11
	750	0	0	0	0	1	1	1	3	4	5	7	11
	800	0	0	0	0	1	1	1	2	3	4	7	10
	900	0	0	0	0	1	1	1	1	3	4	6	9
	1000	0	0	0	0	1	1	1	1	3	4	6	8
FEP, FEPB, PFA, PFAH, TFE	14	12	22	35	61	83	136	194	300	400	515	808	1166
	12	9	16	26	44	60	99	142	219	292	376	590	851
	10	6	11	18	32	43	71	102	157	209	269	423	610
	8	3	6	10	18	25	41	58	90	120	154	242	350
	6	2	4	7	13	17	29	41	64	85	110	172	249
	4	1	3	5	9	12	20	29	44	59	77	120	174
	3	1	2	4	7	10	17	24	37	50	64	100	145
	2	1	1	3	6	8	14	20	31	41	53	83	120
PFA, PFAH, TFE	1	1	1	2	4	6	9	14	21	28	37	57	83
PFA, PFAH, TFE, Z	1/0	1	1	1	3	5	8	11	18	24	30	48	69
	2/0	1	1	1	3	4	6	9	14	19	25	40	57
	3/0	0	1	1	2	3	5	8	12	16	21	33	47
	4/0	0	1	1	1	2	4	6	10	13	17	27	39
Z	14	15	26	42	73	100	164	234	361	482	621	974	1405
	12	10	18	30	52	71	116	166	256	342	440	691	997
	10	6	11	18	32	43	71	102	157	209	269	423	610
	8	4	7	11	20	27	45	64	99	132	170	267	386
	6	3	5	8	14	19	31	45	69	93	120	188	271
	4	1	3	5	9	13	22	31	48	64	82	129	186
	3	1	2	4	7	9	16	22	35	47	60	94	136
	2	1	1	3	6	8	13	19	29	39	50	78	113
	1	1	1	2	5	6	10	15	23	31	40	63	92
XHH, XHHW, XHHW-2, ZW	14	9	15	25	44	59	98	140	216	288	370	581	839
	12	7	12	19	33	45	75	107	165	221	284	446	644
	10	5	9	14	25	34	56	80	123	164	212	332	480
	8	3	5	8	14	19	31	44	68	91	118	185	267
	6	1	3	6	10	14	23	33	51	68	87	137	197
	4	1	2	4	7	10	16	24	37	49	63	99	143
	3	1	1	3	6	8	14	20	31	41	53	84	121
	2	1	1	3	5	7	12	17	26	35	45	70	101
XHH, XHHW, XHHW-2	1	1	1	1	4	5	9	12	19	26	33	52	76
	1/0	1	1	1	3	4	7	10	16	22	28	44	64
	2/0	0	1	1	2	3	6	9	13	18	23	37	53
	3/0	0	1	1	1	3	5	7	11	15	19	30	44
	4/0	0	1	1	1	2	4	6	9	12	16	25	36
	250	0	0	1	1	1	3	5	7	10	13	20	30
	300	0	0	1	1	1	3	4	6	9	11	18	25
	350	0	0	1	1	1	2	3	6	7	10	15	22
	400	0	0	1	1	1	2	3	5	7	9	14	20
	500	0	0	0	1	1	1	2	4	5	7	11	16
	600	0	0	0	1	1	1	1	3	4	6	9	13
	700	0	0	0	1	1	1	1	3	4	5	8	11
	750	0	0	0	0	1	1	1	3	4	5	7	11
	800	0	0	0	0	1	1	1	2	3	4	7	10
	900	0	0	0	0	1	1	1	2	3	4	6	9
	1000	0	0	0	0	1	1	1	1	3	4	6	8
	1250	0	0	0	0	0	1	1	1	2	3	4	6
	1500	0	0	0	0	0	1	1	1	1	2	4	5
	1750	0	0	0	0	0	0	1	1	1	1	3	5
	2000	0	0	0	0	0	0	1	1	1	1	3	4

Table C8. Continued

FIXTURE WIRES

Type	Conductor Size (AWG/kcmil)	½	¾	1	1¼	1½	2
FFH-2, RFH-2, RFHH-3	18	8	15	24	42	57	94
	16	7	12	20	35	48	79
SF-2, SFF-2	18	11	19	31	53	72	118
	16	9	15	25	44	59	98
	14	7	12	20	35	48	79
SF-1, SFF-1	18	19	33	54	94	127	209
AF, RFH-1, RFHH-2, TF, TFF, XF, XFF	18	14	25	40	69	94	155
AF, RFHH-2, TF, TFF, XF, XFF	16	11	20	32	56	76	125
AF, XF, XFF	14	9	15	25	44	59	98
TFN, TFFN	18	23	40	64	111	150	248
	16	17	30	49	84	115	189
PF, PFF, PGF, PGFF, PAF, PTF, PTFF, PAFF	18	21	38	61	105	143	235
	16	16	29	47	81	110	181
	14	12	22	35	61	83	136
HF, HFF, ZF, ZFF, ZHF	18	28	48	79	135	184	303
	16	20	36	58	100	136	223
	14	15	26	42	73	100	164
KF-2, KFF-2	18	40	71	114	197	267	439
	16	28	50	80	138	188	310
	14	19	34	55	95	129	213
	12	13	23	38	65	89	146
	10	9	15	25	44	59	98
KF-1, KFF-1	18	48	84	136	235	318	524
	16	34	59	96	165	224	368
	14	23	40	64	111	150	248
	12	15	26	42	73	100	164
	10	10	17	28	48	65	107
AF, XF, XFF	12	5	8	13	23	32	52
	10	3	6	10	18	25	41

Note: This table is for concentric stranded conductors only. For compact stranded conductors, Table C8(A) should be used.

Table C8(A). Maximum Number of Compact Conductors in Rigid Metal Conduit (*Based on Table 1, Chapter 9*)

COMPACT CONDUCTORS

Type	Conductor Size (AWG/kcmil)	½	¾	1	1¼	1½	2	2½	3	3½	4	5	6
THW, THW-2, THHW	8	2	4	7	12	16	26	38	59	78	101	158	228
	6	1	3	5	9	12	20	29	45	60	78	122	176
	4	1	2	4	7	9	15	22	34	45	58	91	132
	2	1	1	3	5	7	11	16	25	33	43	67	97
	1	1	1	1	3	5	8	11	17	23	30	47	68
	1/0	1	1	1	3	4	7	10	15	20	26	41	59
	2/0	0	1	1	2	3	6	8	13	17	22	34	50
	3/0	0	1	1	1	3	5	7	11	14	19	29	42
	4/0	0	1	1	1	2	4	6	9	12	15	24	35
	250	0	0	1	1	1	3	4	7	9	12	19	28
	300	0	0	1	1	1	3	4	6	8	11	17	24
	350	0	0	1	1	1	2	3	5	7	9	15	22
	400	0	0	1	1	1	1	3	5	7	8	13	20
	500	0	0	0	1	1	1	3	4	5	7	11	17
	600	0	0	0	1	1	1	1	3	4	6	9	13
	700	0	0	0	1	1	1	1	3	4	5	8	12
	750	0	0	0	0	1	1	1	3	4	5	7	11
	1000	0	0	0	0	1	1	1	1	3	4	6	9
THHN, THWN, THWN-2	8	—	—	—	—	—	—	—	—	—	—	—	—
	6	2	5	8	13	18	30	43	66	88	114	179	258
	4	1	3	5	8	11	18	26	41	55	70	110	159
	2	1	1	3	6	8	13	19	29	39	50	79	114
	1	1	1	2	4	6	10	14	22	29	38	60	86
	1/0	1	1	1	4	5	8	12	19	25	32	51	73
	2/0	1	1	1	3	4	7	10	15	21	26	42	60
	3/0	0	1	1	2	3	6	8	13	17	22	35	51
	4/0	0	1	1	1	3	5	7	10	14	18	29	42
	250	0	1	1	1	2	4	5	8	11	14	23	33
	300	0	0	1	1	1	3	4	7	10	12	20	28
	350	0	0	1	1	1	3	4	6	8	11	17	25
	400	0	0	1	1	1	2	3	5	7	10	15	22
	500	0	0	0	1	1	1	3	5	6	8	13	19
	600	0	0	0	1	1	1	2	4	5	6	10	15
	700	0	0	0	1	1	1	1	3	4	6	9	13
	750	0	0	0	1	1	1	1	3	4	5	9	13
	1000	0	0	0	0	1	1	1	2	3	4	6	9
XHHW, XHHW-2	8	3	5	9	15	21	34	49	76	101	130	205	296
	6	2	4	6	11	15	25	36	56	75	97	152	220
	4	1	3	5	8	11	18	26	41	55	70	110	159
	2	1	1	3	6	8	13	19	29	39	50	79	114
	1	1	1	2	4	6	10	14	22	29	38	60	86
	1/0	1	1	1	4	5	8	12	19	25	32	51	73
	2/0	1	1	1	3	4	7	10	16	21	27	43	62
	3/0	0	1	1	2	3	6	8	13	17	22	35	51
	4/0	0	1	1	1	3	5	7	11	14	19	29	42
	250	0	1	1	1	2	4	5	8	11	15	23	34
	300	0	0	1	1	1	3	5	7	10	13	20	29
	350	0	0	1	1	1	3	4	6	9	11	18	25
	400	0	0	1	1	1	2	4	6	8	10	16	23
	500	0	0	0	1	1	1	3	5	6	8	13	19
	600	0	0	0	1	1	1	2	4	5	7	10	15
	700	0	0	0	1	1	1	1	3	4	6	9	13
	750	0	0	0	1	1	1	1	3	4	5	8	12
	1000	0	0	0	0	1	1	1	2	3	4	7	10

Definition: Compact stranding is the result of a manufacturing process where the standard conductor is compressed to the extent that the interstices (voids between strand wires) are virtually eliminated.

Table C9. Maximum Number of Conductors and Fixture Wires in Rigid PVC Conduit, Schedule 80 (Based on Table 1, Chapter 9)

CONDUCTORS

Type	Conductor Size (AWG/kcmil)	½	¾	1	1¼	1½	2	2½	3	3½	4	5	6
RH	14	4	8	13	23	32	55	79	123	166	215	341	490
	12	3	6	10	19	26	44	63	99	133	173	274	394
RHH, RHW, RHW-2	14	3	5	9	17	23	39	56	88	118	153	243	349
	12	2	4	7	14	19	32	46	73	98	127	202	290
RH, RHH, RHW, RHW-2	10	1	3	6	11	15	26	37	59	79	103	163	234
	8	1	1	3	6	8	13	19	31	41	54	85	122
	6	1	1	2	4	6	11	16	24	33	43	68	98
	4	1	1	1	3	5	8	12	19	26	33	53	77
	3	0	1	1	3	4	7	11	17	23	29	47	67
	2	0	1	1	3	4	6	9	14	20	25	41	58
	1	0	1	1	1	2	4	6	9	13	17	27	38
	1/0	0	0	1	1	1	3	5	8	11	15	23	33
	2/0	0	0	1	1	1	3	4	7	10	13	20	29
	3/0	0	0	1	1	1	3	4	6	8	11	17	25
	4/0	0	0	0	1	1	2	3	5	7	9	15	21
	250	0	0	0	1	1	1	2	4	5	7	11	16
	300	0	0	0	1	1	1	2	3	5	6	10	14
	350	0	0	0	1	1	1	1	3	4	5	9	13
	400	0	0	0	0	1	1	1	3	4	5	8	12
	500	0	0	0	0	1	1	1	2	3	4	7	10
	600	0	0	0	0	0	1	1	1	3	3	6	8
	700	0	0	0	0	0	1	1	1	2	3	5	7
	750	0	0	0	0	0	1	1	1	2	3	5	7
	800	0	0	0	0	0	1	1	1	2	3	4	7
	1000	0	0	0	0	0	1	1	1	1	2	4	5
	1250	0	0	0	0	0	0	1	1	1	1	3	4
	1500	0	0	0	0	0	0	1	1	1	1	2	4
	1750	0	0	0	0	0	0	0	1	1	1	2	3
	2000	0	0	0	0	0	0	0	1	1	1	1	3
TW	14	6	11	20	35	49	82	118	185	250	324	514	736
	12	5	9	15	27	38	63	91	142	192	248	394	565
	10	3	6	11	20	28	47	67	106	143	185	294	421
	8	1	3	6	11	15	26	37	59	79	103	163	234
RHH*, RHW*, RHW-2*, THHW, THW, THW-2	14	4	8	13	23	32	55	79	123	166	215	341	490
RHH*, RHW*, RHW-2*, THHW, THW	12	3	6	10	19	26	44	63	99	133	173	274	394
	10	2	5	8	15	20	34	49	77	104	135	214	307
RHH*, RHW*, RHW-2*, THHW, THW, THW-2	8	1	3	5	9	12	20	29	46	62	81	128	184
RHH*, RHW*, RHW-2*, TW, THW, THHW, THW-2	6	1	1	3	7	9	16	22	35	48	62	98	141
	4	1	1	3	5	7	12	17	26	35	46	73	105
	3	1	1	2	4	6	10	14	22	30	39	63	90
	2	1	1	1	3	5	8	12	19	26	33	53	77
	1	0	1	1	2	3	6	8	13	18	23	37	54
	1/0	0	1	1	1	3	5	7	11	15	20	32	46
	2/0	0	1	1	1	2	4	6	10	13	17	27	39
	3/0	0	0	1	1	1	3	5	8	11	14	23	33
	4/0	0	0	1	1	1	3	4	7	9	12	19	27
	250	0	0	0	1	1	2	3	5	7	9	15	22
	300	0	0	0	1	1	1	3	5	6	8	13	19
	350	0	0	0	1	1	1	2	4	6	7	12	17
	400	0	0	0	1	1	1	2	4	5	7	10	15
	500	0	0	0	1	1	1	1	3	4	5	9	13
	600	0	0	0	0	1	1	1	2	3	4	7	10
	700	0	0	0	0	1	1	1	2	3	4	6	9
	750	0	0	0	0	0	1	1	1	3	4	6	8
	800	0	0	0	0	0	1	1	1	3	3	6	8
	900	0	0	0	0	0	1	1	1	2	3	5	7
	1000	0	0	0	0	0	1	1	1	2	3	5	7
	1250	0	0	0	0	0	1	1	1	1	2	4	5
	1500	0	0	0	0	0	0	1	1	1	1	3	4
	1750	0	0	0	0	0	0	1	1	1	1	3	4
	2000	0	0	0	0	0	0	0	1	1	1	2	3

*Types RHH, RHW, and RHW-2 without outer covering.

Table C9. Continued

CONDUCTORS

Type	Conductor Size (AWG/kcmil)	½	¾	1	1¼	1½	2	2½	3	3½	4	5	6
THHN, THWN, THWN-2	14	9	17	28	51	70	118	170	265	358	464	736	1055
	12	6	12	20	37	51	86	124	193	261	338	537	770
	10	4	7	13	23	32	54	78	122	164	213	338	485
	8	2	4	7	13	18	31	45	70	95	123	195	279
	6	1	3	5	9	13	22	32	51	68	89	141	202
	4	1	1	3	6	8	14	20	31	42	54	86	124
	3	1	1	3	5	7	12	17	26	35	46	73	105
	2	1	1	2	4	6	10	14	22	30	39	61	88
	1	0	1	1	3	4	7	10	16	22	29	45	65
	1/0	0	1	1	2	3	6	9	14	18	24	38	55
	2/0	0	1	1	1	3	5	7	11	15	20	32	46
	3/0	0	1	1	1	2	4	6	9	13	17	26	38
	4/0	0	0	1	1	1	3	5	8	10	14	22	31
	250	0	0	1	1	1	3	4	6	8	11	18	25
	300	0	0	0	1	1	2	3	5	7	9	15	22
	350	0	0	0	1	1	1	3	5	6	8	13	19
	400	0	0	0	1	1	1	3	4	6	7	12	17
	500	0	0	0	1	1	1	2	3	5	6	10	14
	600	0	0	0	0	1	1	1	3	4	5	8	12
	700	0	0	0	0	1	1	1	2	3	4	7	10
	750	0	0	0	0	1	1	1	2	3	4	7	9
	800	0	0	0	0	1	1	1	2	3	4	6	9
	900	0	0	0	0	0	1	1	1	3	3	6	8
	1000	0	0	0	0	0	1	1	1	2	3	5	7
FEP, FEPB, PFA, PFA, PFAH, TFE	14	8	16	27	49	68	115	164	257	347	450	714	1024
	12	6	12	20	36	50	84	120	188	253	328	521	747
	10	4	8	14	26	36	60	86	135	182	235	374	536
	8	2	5	8	15	20	34	49	77	104	135	214	307
	6	1	3	6	10	14	24	35	55	74	96	152	218
	4	1	2	4	7	10	17	24	38	52	67	106	153
	3	1	1	3	6	8	14	20	32	43	56	89	127
	2	1	1	3	5	7	12	17	26	35	46	73	105
PFA, PFAH, TFE	1	1	1	1	3	5	8	11	18	25	32	51	73
PFA, PFAH, TFE, Z	1/0	0	1	1	3	4	7	10	15	20	27	42	61
	2/0	0	1	1	2	3	5	8	12	17	22	35	50
	3/0	0	1	1	1	2	4	6	10	14	18	29	41
	4/0	0	0	1	1	1	4	5	8	11	15	24	34
Z	14	10	19	33	59	82	138	198	310	418	542	860	1233
	12	7	14	23	42	58	98	141	220	297	385	610	875
	10	4	8	14	26	36	60	86	135	182	235	374	536
	8	3	5	9	16	22	38	54	85	115	149	236	339
	6	2	4	6	11	16	26	38	60	81	104	166	238
	4	1	2	4	8	11	18	26	41	55	72	114	164
	3	1	2	3	5	8	13	19	30	40	52	83	119
	2	1	1	2	5	6	11	16	25	33	43	69	99
	1	0	1	2	4	5	9	13	20	27	35	56	80
XHH, XHHW, XHHW-2, ZW	14	6	11	20	35	49	82	118	185	250	324	514	736
	12	5	9	15	27	38	63	91	142	192	248	394	565
	10	3	6	11	20	28	47	67	106	143	185	294	421
	8	1	3	6	11	15	26	37	59	79	103	163	234
	6	1	2	4	8	11	19	28	43	59	76	121	173
	4	1	1	3	6	8	14	20	31	42	55	87	125
	3	1	1	3	5	7	12	17	26	36	47	74	106
	2	1	1	2	4	6	10	14	22	30	39	62	89
XHH, XHHW, XHHW-2	1	0	1	1	3	4	7	10	16	22	29	46	66
	1/0	0	1	1	2	3	6	9	14	19	24	39	56
	2/0	0	1	1	1	3	5	7	11	16	20	32	46
	3/0	0	1	1	1	2	4	6	9	13	17	27	38
	4/0	0	0	1	1	1	3	5	8	11	14	22	32
	250	0	0	1	1	1	3	4	6	9	11	18	26
	300	0	0	1	1	1	2	3	5	7	10	15	22
	350	0	0	0	1	1	1	3	5	6	8	14	20
	400	0	0	0	1	1	1	3	4	6	7	12	17
	500	0	0	0	1	1	1	2	3	5	6	10	14
	600	0	0	0	0	1	1	1	3	4	5	8	11
	700	0	0	0	0	1	1	1	2	3	4	7	10
	750	0	0	0	0	1	1	1	2	3	4	6	9
	800	0	0	0	0	1	1	1	1	3	4	6	9
	900	0	0	0	0	0	1	1	—	3	3	5	8
	1000	0	0	0	0	0	1	1	1	2	3	5	7
	1250	0	0	0	0	0	1	1	1	1	2	4	6
	1500	0	0	0	0	0	0	1	1	1	1	3	5
	1750	0	0	0	0	0	0	1	1	1	1	3	4
	2000	0	0	0	0	0	0	1	1	1	1	2	4

Table C9. Continued

Table C9(A). Maximum Number of Compact Conductors in Rigid PVC Conduit, Schedule 80 *(Based on Table 1, Chapter 9)*

FIXTURE WIRES

Type	Conductor Size (AWG/kcmil)	½	¾	1	1¼	1½	2
FFH-2, RFH-2, RFHH-3	18	6	11	19	34	47	79
	16	5	9	16	28	39	67
SF-2, SFF-2	18	7	14	24	43	59	100
	16	6	11	20	35	49	82
	14	5	9	16	28	39	67
SF-1, SFF-1	18	13	25	42	76	105	177
AF, RFH-1, RFHH-2, TF, TFF, XF, XFF	18	10	18	31	56	77	130
AF, RFHH-2, TF, TFF, XF, XFF	16	8	15	25	45	62	105
AF, XF, XFF	14	6	11	20	35	49	82
TFN, TFFN	18	16	29	50	90	124	209
	16	12	22	38	68	95	159
PF, PFF, PGF, PGFF, PAF, PTF, PTFF, PAFF	18	15	28	47	85	118	198
	16	11	22	36	66	91	153
	14	8	16	27	49	68	115
HF, HFF, ZF, ZFF, ZHF	18	19	36	61	110	152	255
	16	14	27	45	81	112	188
	14	10	19	33	59	82	138
KF-2, KFF-2	18	28	53	88	159	220	371
	16	19	37	62	112	155	261
	14	13	25	43	77	107	179
	12	9	17	29	53	73	123
	10	6	11	20	35	49	82
KF-1, KFF-1	18	33	63	106	190	263	442
	16	23	44	74	133	185	310
	14	16	29	50	90	124	209
	12	10	19	33	59	82	138
	10	7	13	21	39	54	90
AF, XF, XFF	12	3	6	10	19	26	44
	10	2	5	8	15	20	34

Note: This table is for concentric stranded conductors only. For compact stranded conductors, Table C9(A) should be used.

COMPACT CONDUCTORS

Type	Conductor Size (AWG/kcmil)	½	¾	1	1¼	1½	2	2½	3	3½	4	5	6
THW, THW-2, THHW	8	1	3	5	9	13	22	32	50	68	88	140	200
	6	1	2	4	7	10	17	25	39	52	68	108	155
	4	1	1	3	5	7	13	18	29	39	51	81	116
	2	1	1	1	4	5	9	13	21	29	37	60	85
	1	0	1	1	3	4	6	9	15	20	26	42	60
	1/0	0	1	1	2	3	6	8	13	17	23	36	52
	2/0	0	1	1	1	3	5	7	11	15	19	30	44
	3/0	0	0	1	1	2	4	6	9	12	16	26	37
	4/0	0	0	1	1	1	3	5	8	10	13	22	31
	250	0	0	1	1	1	2	4	6	8	11	17	25
	300	0	0	0	1	1	2	3	5	7	9	15	21
	350	0	0	0	1	1	1	3	5	6	8	13	19
	400	0	0	0	1	1	1	3	4	6	7	12	17
	500	0	0	0	1	1	1	2	3	5	6	10	14
	600	0	0	0	0	1	1	1	3	4	5	8	12
	700	0	0	0	0	1	1	1	2	3	4	7	10
	750	0	0	0	0	1	1	1	2	3	4	7	10
	1000	0	0	0	0	0	1	1	1	2	3	5	8
THHN, THWN, THWN-2	8	—	—	—	—	—	—	—	—	—	—	—	—
	6	1	3	6	11	15	25	36	57	77	99	158	226
	4	1	1	3	6	9	15	22	35	47	61	98	140
	2	1	1	2	5	6	11	16	25	34	44	70	100
	1	1	1	1	3	5	8	12	19	25	33	53	75
	1/0	0	1	1	3	4	7	10	16	22	28	45	64
	2/0	0	1	1	2	3	6	8	13	18	23	37	53
	3/0	0	1	1	1	3	5	7	11	15	19	31	44
	4/0	0	0	1	1	2	4	6	9	12	16	25	37
	250	0	0	1	1	1	3	4	7	10	12	20	29
	300	0	0	1	1	1	3	4	6	8	11	17	25
	350	0	0	0	1	1	2	3	5	7	9	15	22
	400	0	0	0	1	1	1	3	5	6	8	13	19
	500	0	0	0	1	1	1	2	4	5	7	11	16
	600	0	0	0	1	1	1	1	3	4	6	9	13
	700	0	0	0	0	1	1	1	3	4	5	8	12
	750	0	0	0	0	1	1	1	3	4	5	8	11
	1000	0	0	0	0	0	1	1	1	3	3	5	8
XHHW, XHHW-2	8	1	4	7	12	17	29	42	65	88	114	181	260
	6	1	3	5	9	13	21	31	48	65	85	134	193
	4	1	1	3	6	9	15	22	35	47	61	98	140
	2	1	1	2	5	6	11	16	25	34	44	70	100
	1	1	1	1	3	5	8	12	19	25	33	53	75
	1/0	0	1	1	3	4	7	10	16	22	28	45	64
	2/0	0	1	1	2	3	6	8	13	18	24	38	54
	3/0	0	1	1	1	3	5	7	11	15	19	31	44
	4/0	0	0	1	1	2	4	6	9	12	16	26	37
	250	0	0	1	1	1	3	5	7	10	13	21	30
	300	0	0	1	1	1	3	4	6	8	11	17	25
	350	0	0	1	1	1	2	3	5	7	10	15	22
	400	0	0	0	1	1	1	3	5	7	9	14	20
	500	0	0	0	1	1	1	2	4	5	7	11	17
	600	0	0	0	1	1	1	1	3	4	6	9	13
	700	0	0	0	0	1	1	1	3	4	5	8	12
	750	0	0	0	0	1	1	1	2	3	5	7	11
	1000	0	0	0	0	0	1	1	1	3	3	6	8

Definition: Compact stranding is the result of a manufacturing process where the standard conductor is compressed to the extent that the interstices (voids between strand wires) are virtually eliminated.

Table C10. Maximum Number of Conductors and Fixture Wires in Rigid PVC Conduit, Schedule 40 and HDPE Conduit (Based on Table 1, Chapter 9)

CONDUCTORS

Type	Conductor Size (AWG/kcmil)	½	¾	1	1¼	1½	2	2½	3	3½	4	5	6
RH	14	5	9	16	28	38	63	90	139	186	240	378	546
	12	4	8	12	22	30	50	72	112	150	193	304	439
RHH, RHW, RHW-2	14	4	7	11	20	27	45	64	99	133	171	269	390
	12	3	5	9	16	22	37	53	82	110	142	224	323
RH, RHH, RHW, RHW-2	10	2	4	7	13	18	30	43	66	89	115	181	261
	8	1	2	4	7	9	15	22	35	46	60	94	137
	6	1	1	3	5	7	12	18	28	37	48	76	109
	4	1	1	2	4	6	10	14	22	29	37	59	85
	3	1	1	1	4	5	8	12	19	25	33	52	75
	2	1	1	1	3	4	7	10	16	22	28	45	65
	1	0	1	1	1	3	5	7	11	14	19	29	43
	1/0	0	1	1	1	2	4	6	9	13	16	26	37
	2/0	0	0	1	1	1	3	5	8	11	14	22	32
	3/0	0	0	1	1	1	3	4	7	9	12	19	28
	4/0	0	0	1	1	1	2	4	6	8	10	16	24
	250	0	0	0	1	1	1	3	4	6	8	12	18
	300	0	0	0	1	1	1	2	4	5	7	11	16
	350	0	0	0	1	1	1	2	3	5	6	10	14
	400	0	0	0	1	1	1	1	3	4	6	9	13
	500	0	0	0	0	1	1	1	3	4	5	8	11
	600	0	0	0	0	1	1	1	2	3	4	6	9
	700	0	0	0	0	0	1	1	1	3	3	6	8
	750	0	0	0	0	0	1	1	1	2	3	5	8
	800	0	0	0	0	0	1	1	1	2	3	5	7
	900	0	0	0	0	0	1	1	1	2	3	5	7
	1000	0	0	0	0	0	1	1	1	1	3	4	6
	1250	0	0	0	0	0	0	1	1	1	1	3	5
	1500	0	0	0	0	0	0	1	1	1	1	3	4
	1750	0	0	0	0	0	0	1	1	1	1	2	3
	2000	0	0	0	0	0	0	0	1	1	1	2	3
TW	14	8	14	24	42	57	94	135	209	280	361	568	822
	12	6	11	18	32	44	72	103	160	215	277	436	631
	10	4	8	13	24	32	54	77	119	160	206	325	470
	8	2	4	7	13	18	30	43	66	89	115	181	261
RHH*, RHW*, RHW-2*, THHW, THW, THW-2	14	5	9	16	28	38	63	90	139	186	240	378	546
RHH*, RHW*, RHW-2*, THHW, THW*	12	4	8	12	22	30	50	72	112	150	193	304	439
	10	3	6	10	17	24	39	56	87	117	150	237	343
RHH*, RHW*, RHW-2*, THHW, THW, THW-2	8	1	3	6	10	14	23	33	52	70	90	142	205
RHH*, RHW*, RHW-2*, TW, THW, THHW, THW-2	6	1	2	4	8	11	18	26	40	53	69	109	157
	4	1	1	3	6	8	13	19	30	40	51	81	117
	3	1	1	3	5	7	11	16	25	34	44	69	100
	2	1	1	2	4	6	10	14	22	29	37	59	85
	1	0	1	1	3	4	7	10	15	20	26	41	60
	1/0	0	1	1	2	3	6	8	13	17	22	35	51
	2/0	0	1	1	1	3	5	7	11	15	19	30	43
	3/0	0	1	1	1	2	4	6	9	12	16	25	36
	4/0	0	0	1	1	1	3	5	8	10	13	21	30
	250	0	0	1	1	1	3	4	6	8	11	17	25
	300	0	0	1	1	1	2	3	5	7	9	15	21
	350	0	0	0	1	1	1	3	5	6	8	13	19
	400	0	0	0	1	1	1	3	4	6	7	12	17
	500	0	0	0	1	1	1	2	3	5	6	10	14
	600	0	0	0	0	1	1	1	3	4	5	8	11
	700	0	0	0	0	1	1	1	2	3	4	7	10
	750	0	0	0	0	1	1	1	2	3	4	6	10
	800	0	0	0	0	1	1	1	2	3	4	6	9
	900	0	0	0	0	0	1	1	1	3	3	6	8
	1000	0	0	0	0	0	1	1	1	2	3	5	7
	1250	0	0	0	0	0	1	1	1	1	2	4	6
	1500	0	0	0	0	0	1	1	1	1	1	3	5
	1750	0	0	0	0	0	0	1	1	1	1	3	4
	2000	0	0	0	0	0	0	1	1	1	1	3	4

*Types RHH, RHW, and RHW-2 without outer covering.

Table C10. Continued

CONDUCTORS

Type	Conductor Size (AWG/kcmil)	½	¾	1	1¼	1½	2	2½	3	3½	4	5	6
THHN, THWN, THWN-2	14	11	21	34	60	82	135	193	299	401	517	815	1178
	12	8	15	25	43	59	99	141	218	293	377	594	859
	10	5	9	15	27	37	62	89	137	184	238	374	541
	8	3	5	9	16	21	36	51	79	106	137	216	312
	6	1	4	6	11	15	26	37	57	77	99	156	225
	4	1	2	4	7	9	16	22	35	47	61	96	138
	3	1	1	3	6	8	13	19	30	40	51	81	117
	2	1	1	3	5	7	11	16	25	33	43	68	98
	1	1	1	1	3	5	8	12	18	25	32	50	73
	1/0	1	1	1	3	4	7	10	15	21	27	42	61
	2/0	0	1	1	2	3	6	8	13	17	22	35	51
	3/0	0	1	1	1	3	5	7	11	14	18	29	42
	4/0	0	1	1	1	2	4	6	9	12	15	24	35
	250	0	0	1	1	1	3	4	7	10	12	20	28
	300	0	0	1	1	1	3	4	6	8	11	17	24
	350	0	0	1	1	1	2	3	5	7	9	15	21
	400	0	0	0	1	1	1	3	5	6	8	13	19
	500	0	0	0	1	1	1	2	4	5	7	11	16
	600	0	0	0	1	1	1	1	3	4	5	9	13
	700	0	0	0	0	1	1	1	3	4	5	8	11
	750	0	0	0	0	1	1	1	2	3	4	7	11
	800	0	0	0	0	1	1	1	2	3	4	7	10
	900	0	0	0	0	1	1	1	2	3	4	6	9
	1000	0	0	0	0	1	1	1	1	2	3	6	8
FEP, FEPB, PFA, PFAH, TFE	14	11	20	33	58	79	131	188	290	389	502	790	1142
	12	8	15	24	42	58	96	137	212	284	366	577	834
	10	6	10	17	30	41	69	98	152	204	263	414	598
	8	3	6	10	17	24	39	56	87	117	150	237	343
	6	2	4	7	12	17	28	40	62	83	107	169	244
	4	1	3	5	8	12	19	28	43	58	75	118	170
	3	1	2	4	7	10	16	23	36	48	62	98	142
	2	1	1	3	6	8	13	19	30	40	51	81	117
PFA, PFAH, TFE	1	1	1	2	4	5	9	13	20	28	36	56	81
PFA, PFAH, TFE, Z	1/0	1	1	1	3	4	8	11	17	23	30	47	68
	2/0	0	1	1	3	4	6	9	14	19	24	39	56
	3/0	0	1	1	2	3	5	7	12	16	20	32	46
	4/0	0	1	1	1	2	4	6	9	13	16	26	38
Z	14	13	24	40	70	95	158	226	350	469	605	952	1376
	12	9	17	28	49	68	112	160	248	333	429	675	976
	10	6	10	17	30	41	69	98	152	204	263	414	598
	8	3	6	11	19	26	43	62	96	129	166	261	378
	6	2	4	7	13	18	30	43	67	90	116	184	265
	4	1	3	5	9	12	21	30	46	62	80	126	183
	3	1	2	4	6	9	15	22	34	45	58	92	133
	2	1	1	3	5	7	12	18	28	38	49	77	111
	1	1	1	2	4	6	10	14	23	30	39	62	90
XHH, XHHW, XHHW-2, ZW	14	8	14	24	42	57	94	135	209	280	361	568	822
	12	6	11	18	32	44	72	103	160	215	277	436	631
	10	4	8	13	24	32	54	77	119	160	206	325	470
	8	2	4	7	13	18	30	43	66	89	115	181	261
	6	1	3	5	10	13	22	32	49	66	85	134	193
	4	1	2	4	7	9	16	23	35	48	61	97	140
	3	1	1	3	6	8	13	19	30	40	52	82	118
	2	1	1	3	5	7	11	16	25	34	44	69	99
XHH, XHHW, XHHW-2	1	1	1	1	3	5	8	12	19	25	32	51	74
	1/0	1	1	1	3	4	7	10	16	21	27	43	62
	2/0	0	1	1	2	3	6	8	13	17	23	36	52
	3/0	0	1	1	1	3	5	7	11	14	19	30	43
	4/0	0	1	1	1	2	4	6	9	12	15	24	35
	250	0	0	1	1	1	3	5	7	10	13	20	29
	300	0	0	1	1	1	3	4	6	8	11	17	25
	350	0	0	1	1	1	2	3	5	7	9	15	22
	400	0	0	0	1	1	1	3	5	6	8	13	19
	500	0	0	0	1	1	1	2	4	5	7	11	16
	600	0	0	0	1	1	1	1	3	4	5	9	13
	700	0	0	0	0	1	1	1	3	4	5	8	11
	750	0	0	0	0	1	1	1	2	3	4	7	11
	800	0	0	0	0	1	1	1	2	3	4	7	10
	900	0	0	0	0	1	1	1	2	3	4	6	9
	1000	0	0	0	0	0	1	1	1	3	3	6	8
	1250	0	0	0	0	0	1	1	1	1	3	4	6
	1500	0	0	0	0	0	1	1	1	1	2	4	5
	1750	0	0	0	0	0	0	1	1	1	1	3	5
	2000	0	0	0	0	0	0	1	1	1	1	3	4

Table C10. Continued

FIXTURE WIRES

Type	Conductor Size (AWG/kcmil)	Trade Size (in.)					
		½	¾	1	1¼	1½	2
FFH-2, RFH-2, RFHH-3	18	8	14	23	40	54	90
	16	6	12	19	33	46	76
SF-2, SFF-2	18	10	17	29	50	69	114
	16	8	14	24	42	57	94
	14	6	12	19	33	46	76
SF-1, SFF-1	18	17	31	51	89	122	202
AF, RFH-1, RFHH-2, TF, TFF, XF, XFF	18	13	23	38	66	90	149
AF, RFHH-2, TF, TFF, XF, XFF	16	10	18	30	53	73	120
AF, XF, XFF	14	8	14	24	42	57	94
TFN, TFFN	18	20	37	60	105	144	239
	16	16	28	46	80	110	183
PF, PFF, PGF, PGFF, PAF, PTF, PTFF, PAFF	18	19	35	57	100	137	227
	16	15	27	44	77	106	175
	14	11	20	33	58	79	131
HF, HFF, ZF, ZFF, ZHF	18	25	45	74	129	176	292
	16	18	33	54	95	130	216
	14	13	24	40	70	95	158
KF-2, KFF-2	18	36	65	107	187	256	424
	16	26	46	75	132	180	299
	14	17	31	52	90	124	205
	12	12	22	35	62	85	141
	10	8	14	24	42	57	94
KF-1, KFF-1,	18	43	78	128	223	305	506
	16	30	55	90	157	214	355
	14	20	37	60	105	144	239
	12	13	24	40	70	95	158
	10	9	16	26	45	62	103
AF, XF, XFF	12	4	8	12	22	30	50
	10	3	6	10	17	24	39

Note: This table is for concentric stranded conductors only. For compact stranded conductors, Table C10(A) should be used.

Table C10(A). Maximum Number of Compact Conductors in Rigid PVC Conduit, Schedule 40 and HDPE Conduit *(Based on Table 1, Chapter 9)*

COMPACT CONDUCTORS

Type	Conductor Size (AWG/kcmil)	Trade Size (in.)											
		½	¾	1	1¼	1½	2	2½	3	3½	4	5	6
THW, THW-2, THHW	8	1	4	6	11	15	26	37	57	76	98	155	224
	6	1	3	5	9	12	20	28	44	59	76	119	173
	4	1	1	3	6	9	15	21	33	44	57	89	129
	2	1	1	2	5	6	11	15	24	32	42	66	95
	1	1	1	1	3	4	7	11	17	23	29	46	67
	1/0	0	1	1	3	4	6	9	15	20	25	40	58
	2/0	0	1	1	2	3	5	8	12	16	21	34	49
	3/0	0	1	1	1	3	5	7	10	14	18	29	42
	4/0	0	1	1	1	2	4	5	9	12	15	24	35
	250	0	0	1	1	1	3	4	7	9	12	19	27
	300	0	0	1	1	1	2	4	6	8	10	16	24
	350	0	0	1	1	1	2	3	5	7	9	15	21
	400	0	0	0	1	1	1	3	5	6	8	13	19
	500	0	0	0	1	1	1	2	4	5	7	11	16
	600	0	0	0	1	1	1	1	3	4	5	9	13
	700	0	0	0	0	1	1	1	3	4	5	8	12
	750	0	0	0	0	1	1	1	2	3	5	7	11
	1000	0	0	0	0	1	1	1	1	3	4	6	9
THHN, THWN, THWN-2	8	—	—	—	—	—	—	—	—	—	—	—	—
	6	2	4	7	13	17	29	41	64	86	111	175	253
	4	1	2	4	8	11	18	25	40	53	68	108	156
	2	1	1	3	5	8	13	18	28	38	49	77	112
	1	1	1	2	4	6	9	14	21	29	37	58	84
	1/0	1	1	1	3	5	8	12	18	24	31	49	72
	2/0	0	1	1	3	4	7	9	15	20	26	41	59
	3/0	0	1	1	2	3	5	8	12	17	22	34	50
	4/0	0	1	1	1	3	4	6	10	14	18	28	41
	250	0	0	1	1	1	3	5	8	11	14	22	32
	300	0	0	1	1	1	3	4	7	9	12	19	28
	350	0	0	1	1	1	3	4	6	8	10	17	24
	400	0	0	1	1	1	2	3	5	7	9	15	22
	500	0	0	0	1	1	1	3	4	6	8	13	18
	600	0	0	0	1	1	1	2	4	5	6	10	15
	700	0	0	0	1	1	1	1	3	4	5	9	13
	750	0	0	0	1	1	1	1	3	4	5	8	12
	1000	0	0	0	0	1	1	1	2	3	4	6	9
XHHW, XHHW-2	8	3	5	8	14	20	33	47	73	99	127	200	290
	6	1	4	6	11	15	25	35	55	73	94	149	215
	4	1	2	4	8	11	18	25	40	53	68	108	156
	2	1	1	3	5	8	13	18	28	38	49	77	112
	1	1	1	2	4	6	9	14	21	29	37	58	84
	1/0	1	1	1	3	5	8	12	18	24	31	49	72
	2/0	1	1	1	3	4	7	10	15	20	26	42	60
	3/0	0	1	1	2	3	5	8	12	17	22	34	50
	4/0	0	1	1	1	3	5	7	10	14	18	29	42
	250	0	0	1	1	1	4	5	8	11	14	23	33
	300	0	0	1	1	1	3	4	7	9	12	19	28
	350	0	0	1	1	1	3	4	6	8	11	17	25
	400	0	0	1	1	1	2	3	5	7	10	15	22
	500	0	0	0	1	1	1	3	4	6	8	13	18
	600	0	0	0	1	1	1	2	4	5	6	10	15
	700	0	0	0	1	1	1	1	3	4	5	9	13
	750	0	0	0	1	1	1	1	3	4	5	8	12
	1000	0	0	0	0	1	1	1	2	3	4	6	9

Definition: Compact stranding is the result of a manufacturing process where the standard conductor is compressed to the extent that the interstices (voids between strand wires) are virtually eliminated.

Table C11. Maximum Number of Conductors and Fixture Wires in Type A, Rigid PVC Conduit (*Based on Table 1, Chapter 9*)

Type	Conductor Size (AWG/kcmil)	½	¾	1	1¼	1½	2	2½	3	3½	4
RH	14	7	12	20	34	44	70	104	157	204	262
	12	6	10	16	27	35	56	84	126	164	211
RHH, RHW, RHW-2	14	5	9	15	24	31	49	74	112	146	187
	12	4	7	12	20	26	41	61	93	121	155
RH, RHH, RHW, RHW-2	10	3	6	10	16	21	33	50	75	98	125
	8	1	3	5	8	11	17	26	39	51	65
	6	1	2	4	6	9	14	21	31	41	52
	4	1	1	3	5	7	11	16	24	32	41
	3	1	1	3	4	6	9	14	21	28	36
	2	1	1	2	4	5	8	12	18	24	31
	1	0	1	1	2	3	5	8	12	16	20
	1/0	0	1	1	2	3	5	7	10	14	18
	2/0	0	1	1	1	2	4	6	9	12	15
	3/0	0	1	1	1	1	3	5	8	10	13
	4/0	0	0	1	1	1	3	4	7	9	11
	250	0	0	1	1	1	1	3	5	7	8
	300	0	0	1	1	1	1	3	4	6	7
	350	0	0	0	1	1	1	2	4	5	7
	400	0	0	0	1	1	1	2	4	5	6
	500	0	0	0	1	1	1	1	3	4	5
	600	0	0	0	0	1	1	1	2	3	4
	700	0	0	0	0	1	1	1	2	3	4
	750	0	0	0	0	1	1	1	1	3	4
	800	0	0	0	0	1	1	1	1	3	3
	900	0	0	0	0	0	1	1	1	2	3
	1000	0	0	0	0	0	1	1	1	2	3
	1250	0	0	0	0	0	1	1	1	1	2
	1500	0	0	0	0	0	0	1	1	1	1
	1750	0	0	0	0	0	0	1	1	1	1
	2000	0	0	0	0	0	0	1	1	1	1
TW	14	11	18	31	51	67	105	157	235	307	395
	12	8	14	24	39	51	80	120	181	236	303
	10	6	10	18	29	38	60	89	135	176	226
	8	3	6	10	16	21	33	50	75	98	125
RHH*, RHW*, RHW-2*, THHW, THW, THW-2	14	7	12	20	34	44	70	104	157	204	262
RHH*, RHW*, RHW-2*, THHW, THW	12	6	10	16	27	35	56	84	126	164	211
	10	4	8	13	21	28	44	65	98	128	165
RHH*, RHW*, RHW-2*, THHW, THW, THW-2	8	2	4	8	12	16	26	39	59	77	98
RHH*, RHW*, RHW-2*, TW, THW, THHW, THW-2	6	1	3	6	9	13	20	30	45	59	75
	4	1	2	4	7	9	15	22	33	44	56
	3	1	1	4	6	8	13	19	29	37	48
	2	1	1	3	5	7	11	16	24	32	41
	1	1	1	1	3	5	7	11	17	22	29
	1/0	1	1	1	3	4	6	10	14	19	24
	2/0	0	1	1	2	3	5	8	12	16	21
	3/0	0	1	1	1	3	4	7	10	13	17
	4/0	0	1	1	1	2	4	6	9	11	14
	250	0	0	1	1	1	3	4	7	9	12
	300	0	0	1	1	1	2	4	6	8	10
	350	0	0	1	1	1	2	3	5	7	9
	400	0	0	1	1	1	1	3	5	6	8
	500	0	0	0	1	1	1	2	4	5	7
	600	0	0	0	1	1	1	1	3	4	5
	700	0	0	0	1	1	1	1	3	4	5
	750	0	0	0	1	1	1	1	3	3	4
	800	0	0	0	0	1	1	1	2	3	4
	900	0	0	0	0	1	1	1	2	3	4
	1000	0	0	0	0	1	1	1	1	3	3
	1250	0	0	0	0	0	1	1	1	1	3
	1500	0	0	0	0	0	1	1	1	1	2
	1750	0	0	0	0	0	0	1	1	1	1
	2000	0	0	0	0	0	0	1	1	1	1

*Types RHH, RHW, and RHW-2 without outer covering.

Table C11. Continued

Type	Conductor Size (AWG/kcmil)	½	¾	1	1¼	1½	2	2½	3	3½	4
THHN, THWN, THWN-2	14	16	27	44	73	96	150	225	338	441	566
	12	11	19	32	53	70	109	164	246	321	412
	10	7	12	20	33	44	69	103	155	202	260
	8	4	7	12	19	25	40	59	89	117	150
	6	3	5	8	14	18	28	43	64	84	108
	4	1	3	5	8	11	17	26	39	52	66
	3	1	2	4	7	9	15	22	33	44	56
	2	1	1	3	6	8	12	19	28	37	47
	1	1	1	2	4	6	9	14	21	27	35
	1/0	1	1	2	4	5	8	11	17	23	29
	2/0	1	1	1	3	4	6	10	14	19	24
	3/0	0	1	1	2	3	5	8	12	16	20
	4/0	0	1	1	1	3	4	6	10	13	17
	250	0	1	1	1	2	3	5	8	10	14
	300	0	0	1	1	1	3	4	7	9	12
	350	0	0	1	1	1	2	4	6	8	10
	400	0	0	1	1	1	2	3	5	7	9
	500	0	0	1	1	1	1	3	4	6	7
	600	0	0	0	1	1	1	2	3	5	6
	700	0	0	0	1	1	1	1	3	4	5
	750	0	0	0	1	1	1	1	3	4	5
	800	0	0	0	1	1	1	1	3	4	5
	900	0	0	0	1	1	1	1	2	3	4
	1000	0	0	0	0	1	1	1	2	3	4
FEP, FEPB, PFA, PFAH, TFE	14	15	26	43	70	93	146	218	327	427	549
	12	11	19	31	51	68	106	159	239	312	400
	10	8	13	22	37	48	76	114	171	224	287
	8	4	8	13	21	28	44	65	98	128	165
	6	3	5	9	15	20	31	46	70	91	117
	4	1	4	6	10	14	21	32	49	64	82
	3	1	3	5	8	11	18	27	40	53	68
	2	1	2	4	7	9	15	22	33	44	56
PFA, PFAH, TFE	1	1	1	3	5	6	10	15	23	30	39
PFA, PFAH, TFE, Z	1/0	1	1	2	4	5	8	13	19	25	32
	2/0	1	1	1	3	4	7	10	16	21	27
	3/0	1	1	1	3	3	6	9	13	17	22
	4/0	0	1	1	2	3	5	7	11	14	18
Z	14	18	31	52	85	112	175	263	395	515	661
	12	13	22	37	60	79	124	186	280	365	469
	10	8	13	22	37	48	76	114	171	224	287
	8	5	8	14	23	30	48	72	108	141	181
	6	3	6	10	16	21	34	50	76	99	127
	4	2	4	7	11	15	23	35	52	68	88
	3	1	3	5	8	11	17	25	38	50	64
	2	1	2	4	7	9	14	21	32	41	53
	1	1	1	3	5	7	11	17	26	33	43
XHH, XHHW, XHHW-2, ZW	14	11	18	31	51	67	105	157	235	307	395
	12	8	14	24	39	51	80	120	181	236	303
	10	6	10	18	29	38	60	89	135	176	226
	8	3	6	10	16	21	33	50	75	98	125
	6	2	4	7	12	15	24	37	55	72	93
	4	1	3	5	8	11	18	26	40	52	67
	3	1	2	4	7	9	15	22	34	44	57
	2	1	1	3	6	8	12	19	28	37	48
XHH, XHHW, XHHW-2,	1	1	1	3	4	6	9	14	21	28	35
	1/0	1	1	2	4	5	8	12	18	23	30
	2/0	1	1	1	3	4	6	10	15	19	25
	3/0	0	1	1	2	3	5	8	12	16	20
	4/0	0	1	1	1	3	4	7	10	13	17
	250	0	1	1	1	2	3	5	8	11	14
	300	0	0	1	1	1	3	5	7	9	12
	350	0	0	1	1	1	3	4	6	8	10
	400	0	0	1	1	1	2	3	5	7	9
	500	0	0	1	1	1	1	3	4	6	8
	600	0	0	0	1	1	1	2	3	5	6
	700	0	0	0	1	1	1	1	3	4	5
	750	0	0	0	1	1	1	1	3	4	5
	800	0	0	0	1	1	1	1	3	4	5
	900	0	0	0	0	1	1	1	2	3	4
	1000	0	0	0	0	1	1	1	2	3	4
	1250	0	0	0	0	0	1	1	1	2	3
	1500	0	0	0	0	0	1	1	1	1	2
	1750	0	0	0	0	0	1	1	1	1	2
	2000	0	0	0	0	0	0	1	1	1	1

Table C11. Continued

	FIXTURE WIRES					
Type	Conductor Size (AWG/kcmil)	½	¾	1	1¼	1½	2
FFH-2, RFH-2, RFHH-3	18	10	18	30	48	64	100
	16	9	15	25	41	54	85
SF-2, SFF-2	18	13	22	37	61	81	127
	16	11	18	31	51	67	105
	14	9	15	25	41	54	85
SF-1, SFF-1	18	23	40	66	108	143	224
AF, RFH-1, RFHH-2, TF, TFF, XF, XFF	18	17	29	49	80	105	165
AF, RFHH-2, TF, TFF, XF, XFF	16	14	24	39	65	85	134
AF, XF, XFF	14	11	18	31	51	67	105
TFN, TFFN	18	28	47	79	128	169	265
	16	21	36	60	98	129	202
PF, PFF, PGF, PGFF, PAF, PTF, PTFF, PAFF	18	26	45	74	122	160	251
	16	20	34	58	94	124	194
	14	15	26	43	70	93	146
HF, HFF, ZF, ZFF, ZHF	18	34	58	96	157	206	324
	16	25	42	71	116	152	239
	14	18	31	52	85	112	175
KF-2, KFF-2	18	49	84	140	228	300	470
	16	35	59	98	160	211	331
	14	24	40	67	110	145	228
	12	16	28	46	76	100	157
	10	11	18	31	51	67	105
KF-1, KFF-1	18	59	100	167	272	357	561
	16	41	70	117	191	251	394
	14	28	47	79	128	169	265
	12	18	31	52	85	112	175
	10	12	20	34	55	73	115
AF, XF, XFF	12	6	10	16	27	35	56
	10	4	8	13	21	28	44

Note: This table is for concentric stranded conductors only. For compact stranded conductors, Table C11(A) should be used.

Table C11(A). Maximum Number of Compact Conductors in Type A, Rigid PVC Conduit
(Based on Table 1, Chapter 9)

	COMPACT CONDUCTORS									
Type	Conductor Size (AWG/kcmil)	½	¾	1	1¼	1½	2	2½	3	3½	4
THW, THW-2, THHW	8	3	5	8	14	18	28	42	64	84	107
	6	2	4	6	10	14	22	33	49	65	83
	4	1	3	5	8	10	16	24	37	48	62
	2	1	1	3	6	7	12	18	27	36	46
	1	1	1	2	4	5	8	13	19	25	32
	1/0	1	1	1	3	4	7	11	16	21	28
	2/0	1	1	1	3	4	6	9	14	18	23
	3/0	0	1	1	2	3	5	8	12	15	20
	4/0	0	1	1	1	3	4	6	10	13	17
	250	0	1	1	1	1	3	5	8	10	13
	300	0	0	1	1	1	3	4	7	9	11
	350	0	0	1	1	1	2	4	6	8	10
	400	0	0	1	1	1	2	3	5	7	9
	500	0	0	1	1	1	1	3	4	6	8
	600	0	0	0	1	1	1	2	3	5	6
	700	0	0	0	1	1	1	1	3	4	5
	750	0	0	0	1	1	1	1	3	4	5
	1000	0	0	0	0	1	1	1	2	3	4
THHN, THWN, THWN-2	8	—	—	—	—	—	—	—	—	—	—
	6	3	5	9	15	20	32	48	72	94	121
	4	1	3	6	9	12	20	30	45	58	75
	2	1	2	4	7	9	14	21	32	42	54
	1	1	1	3	5	7	10	16	24	31	40
	1/0	1	1	2	4	6	9	13	20	27	34
	2/0	1	1	1	3	5	7	11	17	22	28
	3/0	1	1	1	3	4	6	9	14	18	24
	4/0	0	1	1	2	3	5	8	11	15	19
	250	0	1	1	1	2	4	6	9	12	15
	300	0	1	1	1	1	3	5	8	10	13
	350	0	0	1	1	1	3	4	7	9	11
	400	0	0	1	1	1	2	4	6	8	10
	500	0	0	1	1	1	2	3	5	7	9
	600	0	0	0	1	1	1	3	4	5	7
	700	0	0	0	1	1	1	2	3	5	6
	750	0	0	0	1	1	1	2	3	4	6
	1000	0	0	0	0	1	1	1	2	3	4
XHHW, XHHW-2	8	4	6	11	18	23	37	55	83	108	139
	6	3	5	8	13	17	27	41	62	80	103
	4	1	3	6	9	12	20	30	45	58	75
	2	1	2	4	7	9	14	21	32	42	54
	1	1	1	3	5	7	10	16	24	31	40
	1/0	1	1	2	4	6	9	13	20	27	34
	2/0	1	1	1	3	5	7	11	17	22	29
	3/0	1	1	1	3	4	6	9	14	18	24
	4/0	0	1	1	2	3	5	8	12	15	20
	250	0	1	1	1	2	4	6	9	12	16
	300	0	1	1	1	1	3	5	8	10	13
	350	0	0	1	1	1	3	5	7	9	12
	400	0	0	1	1	1	3	4	6	8	11
	500	0	0	1	1	1	2	3	5	7	9
	600	0	0	0	1	1	1	3	4	5	7
	700	0	0	0	1	1	1	2	3	5	6
	750	0	0	0	1	1	1	2	3	4	6
	1000	0	0	0	0	1	1	1	2	3	4

Definition: Compact stranding is the result of a manufacturing process where the standard conductor is compressed to the extent that the interstices (voids between strand wires) are virtually eliminated.

Table C12. Maximum Number of Conductors in Type EB, PVC Conduit (Based on Table 1, Chapter 9)

CONDUCTORS

Type	Conductor Size (AWG/kcmil)	2	3	3½	4	5	6
RH	14	74	166	217	276	424	603
	12	59	134	175	222	341	485
RHH, RHW, RHW-2	14	53	119	155	197	303	430
	12	44	98	128	163	251	357
RH, RHH, RHW, RHW-2	10	35	79	104	132	203	288
	8	18	41	54	69	106	151
	6	15	33	43	55	85	121
	4	11	26	34	43	66	94
	3	10	23	30	38	58	83
	2	9	20	26	33	50	72
	1	6	13	17	21	33	47
	1/0	5	11	15	19	29	41
	2/0	4	10	13	16	25	36
	3/0	4	8	11	14	22	31
	4/0	3	7	9	12	18	26
	250	2	5	7	9	14	20
	300	1	5	6	8	12	17
	350	1	4	5	7	11	16
	400	1	4	5	6	10	14
	500	1	3	4	5	9	12
	600	1	3	3	4	7	10
	700	1	2	3	4	6	9
	750	1	2	3	4	6	9
	800	1	2	3	4	6	8
	900	1	1	2	3	5	7
	1000	1	1	2	3	5	7
	1250	1	1	1	2	3	5
	1500	0	1	1	1	3	4
	1750	0	1	1	1	3	4
	2000	0	1	1	1	2	3
TW	14	111	250	327	415	638	907
	12	85	192	251	319	490	696
	10	63	143	187	238	365	519
	8	35	79	104	132	203	288
RHH*, RHW*, RWH-2*, THHW, THW, THW-2	14	74	166	217	276	424	603
RHH*, RHW*, RHW-2*, THHW, THW	12	59	134	175	222	341	485
	10	46	104	136	173	266	378
RHH*, RHW*, RHW-2*, THHW, THW THW-2	8	28	62	81	104	159	227
RHH*, RHW*, RHW-2*, TW, THW, THHW, THW-2	6	21	48	62	79	122	173
	4	16	36	46	59	91	129
	3	13	30	40	51	78	111
	2	11	26	34	43	66	94
	1	8	18	24	30	46	66
	1/0	7	15	20	26	40	56
	2/0	6	13	17	22	34	48
	3/0	5	11	14	18	28	40
	4/0	4	9	12	15	24	34
	250	3	7	10	12	19	27
	300	3	6	8	11	17	24
	350	2	6	7	9	15	21
	400	2	5	7	8	13	19
	500	1	4	5	7	11	16
	600	1	3	4	6	9	13
	700	1	3	4	5	8	11
	750	1	3	4	5	7	11
	800	1	3	3	4	7	10
	900	1	2	3	4	6	9
	1000	1	2	3	4	6	8
	1250	1	1	2	3	4	6
	1500	1	1	1	2	4	6
	1750	1	1	1	2	3	5
	2000	0	1	1	1	3	4

*Types RHH, RHW, and RHW-2 without outer covering.

Table C12. Continued

CONDUCTORS

Type	Conductor Size (AWG/kcmil)	2	3	3½	4	5	6
THHN, THWN, THWN-2	14	159	359	468	595	915	1300
	12	116	262	342	434	667	948
	10	73	165	215	274	420	597
	8	42	95	124	158	242	344
	6	30	68	89	114	175	248
	4	19	42	55	70	107	153
	3	16	36	46	59	91	129
	2	13	30	39	50	76	109
	1	10	22	29	37	57	80
	1/0	8	18	24	31	48	68
	2/0	7	15	20	26	40	56
	3/0	5	13	17	21	33	47
	4/0	4	10	14	18	27	39
	250	4	8	11	14	22	31
	300	3	7	10	12	19	27
	350	3	6	8	11	17	24
	400	2	6	7	10	15	21
	500	1	5	6	8	12	18
	600	1	4	5	6	10	14
	700	1	3	4	6	9	12
	750	1	3	4	5	8	12
	800	1	3	4	5	8	11
	900	1	3	3	4	7	10
	1000	1	2	3	4	6	9
FEP, FEPB, PFA, PFAH, TFE	14	155	348	454	578	888	1261
	12	113	254	332	422	648	920
	10	81	182	238	302	465	660
	8	46	104	136	173	266	378
	6	33	74	97	123	189	269
	4	23	52	68	86	132	188
	3	19	43	56	72	110	157
	2	16	36	46	59	91	129
PFA, PFAH, TFE	1	11	25	32	41	63	90
PFA, PFAH, TFE, Z	1/0	9	20	27	34	53	75
	2/0	7	17	22	28	43	62
	3/0	6	14	18	23	36	51
	4/0	5	11	15	19	29	42
Z	14	186	419	547	696	1069	1519
	12	132	297	388	494	759	1078
	10	81	182	238	302	465	660
	8	51	115	150	191	294	417
	6	36	81	105	134	206	293
	4	24	55	72	92	142	201
	3	18	40	53	67	104	147
	2	15	34	44	56	86	122
	1	12	27	36	45	70	99
XHH, XHHW, XHHW-2, ZW	14	111	250	327	415	638	907
	12	85	192	251	319	490	696
	10	63	143	187	238	365	519
	8	35	79	104	132	203	288
	6	26	59	77	98	150	213
	4	19	42	56	71	109	155
	3	16	36	47	60	92	131
	2	13	30	39	50	77	110
XHH, XHHW, XHHW-2	1	10	22	29	37	58	82
	1/0	8	19	25	31	48	69
	2/0	7	16	20	26	40	57
	3/0	6	13	17	22	33	47
	4/0	5	11	14	18	27	39
	250	4	9	11	15	22	32
	300	3	7	10	12	19	28
	350	3	6	9	11	17	24
	400	2	6	8	10	15	22
	500	1	5	6	8	12	18
	600	1	4	5	6	10	14
	700	1	3	4	6	9	12
	750	1	3	4	5	8	12
	800	1	3	4	5	8	11
	900	1	3	3	4	7	10
	1000	1	2	3	4	6	9
	1250	1	1	2	3	5	7
	1500	1	1	1	3	4	6
	1750	1	1	1	2	4	5
	2000	0	1	1	1	3	5

NOTE: This table is for concentric stranded conductors only. For compact stranded conductors, Table C12(A) should be used.

Table C12(A). Maximum Number of Compact Conductors in Type EB, PVC Conduit *(Based on Table 1, Chapter 9)*

Type	Conductor Size (AWG/kcmil)	Trade Size (in.)					
		2	3	3½	4	5	6
THW, THW-2, THHW	8	30	68	89	113	174	247
	6	23	52	69	87	134	191
	4	17	39	51	65	100	143
	2	13	29	38	48	74	105
	1	9	20	26	34	52	74
	1/0	8	17	23	29	45	64
	2/0	6	15	19	24	38	54
	3/0	5	12	16	21	32	46
	4/0	4	10	14	17	27	38
	250	3	8	11	14	21	30
	300	3	7	9	12	19	26
	350	3	6	8	11	17	24
	400	2	6	7	10	15	21
	500	1	5	6	8	12	18
	600	1	4	5	6	10	14
	700	1	3	4	6	9	13
	750	1	3	4	5	8	12
	1000	1	2	3	4	7	9
THHN, THWN, THWN-2	8	—	—	—	—	—	—
	6	34	77	100	128	196	279
	4	21	47	62	79	121	172
	2	15	34	44	57	87	124
	1	11	25	33	42	65	93
	1/0	9	22	28	36	56	79
	2/0	8	18	23	30	46	65
	3/0	6	15	20	25	38	55
	4/0	5	12	16	20	32	45
	250	4	10	13	16	25	35
	300	4	8	11	14	22	31
	350	3	7	9	12	19	27
	400	3	6	8	11	17	24
	500	2	5	7	9	14	20
	600	1	4	6	7	11	16
	700	1	4	5	6	10	14
	750	1	4	5	6	9	14
	1000	1	3	3	4	7	10
XHHW, XHHW-2	8	39	88	115	146	225	320
	6	29	65	85	109	167	238
	4	21	47	62	79	121	172
	2	15	34	44	57	87	124
	1	11	25	33	42	65	93
	1/0	9	22	28	36	56	79
	2/0	8	18	24	30	47	67
	3/0	6	15	20	25	38	55
	4/0	5	12	16	21	32	46
	250	4	10	13	17	26	37
	300	4	8	11	14	22	31
	350	3	7	10	12	19	28
	400	3	7	9	11	17	25
	500	2	5	7	9	14	20
	600	1	4	6	7	11	16
	700	1	4	5	6	10	14
	750	1	3	5	6	9	13
	1000	1	3	4	5	7	10

Definition: Compact stranding is the result of a manufacturing process where the standard conductor is compressed to the extent that the interstices (voids between strand wires) are virtually eliminated.

APPENDIX D
Examples

Selection of Conductors. In the following examples, the results are generally expressed in amperes (A). To select conductor sizes, refer to the 0 through 2000 volt (V) ampacity tables of Article 310 and Section 310-5 that pertains to such tables.

Voltage. For uniform application of Articles 210, 215, and 220, a nominal voltage of 120, 120/240, 240, and 208Y/120 V is used in computing the ampere load on the conductor.

Fractions of an Ampere. Except where the computations result in a major fraction of an ampere (0.5 or larger), such fractions are permitted to be dropped.

Power Factor. Calculations in the following examples are based, for convenience, on the assumption that all loads have the same power factor (PF).

Ranges. For the computation of the range loads in these examples, Column A of Table 220-19 has been used. For optional methods, see Columns B and C of Table 220-19. Except where the computations result in a major fraction of a kilowatt (0.5 or larger), such fractions are permitted to be dropped.

SI Units. For metric conversions, 1 ft^2 = 0.093 m^2 and 1 ft = 0.3048 m.

Example No. D1(a). One-Family Dwelling

The dwelling has a floor area of 1500 ft^2, exclusive of an unfinished cellar not adaptable for future use, unfinished attic, and open porches. Appliances are a 12-kW range and a 5.5-kW, 240-V dryer. Assume range and dryer kW ratings equivalent to kVA ratings in accordance with Sections 220-18 and 220-19.

Computed Load *[see Section 220-10]*

General Lighting Load: 1500 ft^2 at 3 VA per ft^2 = 4500 VA

Minimum Number of Branch Circuits Required
[see Section 210-11(e)(1)]

General Lighting Load: 4500 VA ÷ 120 V = 37.5 A
  This requires three 15-A, 2-wire or two 20-A, 2-wire circuits
Small Appliance Load: Two 2-wire, 20-A circuits
  [see Section 210-11(c)(1)]
Laundry Load: One 2-wire, 20-A circuit *[see Section 210-11(c)(2)]*
Bathroom Branch Circuit: One 2-wire, 20-A circuit (no additional load calculation is required for this circuit) *[see Section 210-11(c)(3)]*

Minimum Size Feeder Required *[see Section 220-10]*

General Lighting	4500 VA
Small Appliance	3000 VA
Laundry	1500 VA
Total	9000 VA
3000 VA at 100%	3000 VA
9000 VA − 3000 VA − 6000 VA at 35%	2100 VA
Net Load	5100 VA
Range (*see Table 220-19*)	8000 VA
Dryer (*see Table 220-18*)	5500 VA
Net Computed Load	18,600 VA

Net Computed Load for 120/240-V, 3-wire, single-phase service or feeder

$$18,600 \text{ VA} \div 240 \text{ V} = 77.5 \text{ A}$$

Net computed load exceeds 10 kVA. Section 230-42(b) would require service conductors to be 100 A.

Calculation for Neutral for Feeder and Service

Lighting and Small Appliance Net Load	5100 VA
Range: 8000 VA at 70% (*see Section 220-22*)	5600 VA
Dryer: 5500 VA at 70% (*see Section 220-22*)	3850 VA
Total	14,550 VA

Computed Load for Neutral

$$14,550 \text{ VA} \div 240 \text{ V} = 60.6 \text{ A}$$

Example No. D1(b). One-Family Dwelling

Assume same conditions as Example No. D1(a), plus addition of one 6-A, 230-V, room air-conditioning unit and one 12-A, 115-V, room air-conditioning unit,* one 8-A, 115-V rated waste disposer, and one 10-A, 120-V, rated dishwasher.* See Article 430 for general motors and Article 440, Part G, for air-conditioning equipment. Motors have nameplate ratings of 115 V and 230 V for use on 120-V and 240-V nominal voltage systems.

From Example No. D1(a), feeder current is 78 A (3-wire, 240 V).

*(For feeder neutral, use larger of the two appliances for unbalance.)

	Line A	Neutral	Line B
Amperes from Example No. D1(a)	78	61	78
One 230-V air conditioner	6	—	6
One 115-V air conditioner and 120-V dishwasher	12	10	10
One 115-V disposer	—	8	8
25% of largest motor (*see Section 430-24*)	3	2	2
Total amperes per line	99	81	104

• Therefore, the service would be rated 110 A

Example No. D2(a). Optional Calculation for One-Family Dwelling, Heating Larger than Air Conditioning *[see Section 220-30]*

The dwelling has a floor area of 1500 ft^2, exclusive of an unfinished cellar not adaptable for future use, unfinished attic, and open porches. It has a 12-kW range, a 2.5-kW water heater, a 1.2-kW dishwasher, 9 kW of electric space heating installed in five rooms, a 5-kW clothes dryer, and a 6-A, 230-V, room air-conditioning unit. Assume range, water heater, dishwasher, space heating, and clothes dryer kW ratings equivalent to kVA.

Air Conditioner kVA Calculation

$$6 \text{ A} \times 230 \text{ V} \div 1000 = 1.38 \text{ kVA}$$

This 1.38 kVA (*Section 220-30(c)(1)*) is less than 40% of 9 kVA of separately controlled electric heat (*Section 220-3(c)(6)*), so the 1.38 kVA need not be included in the service calculation.

General Load

1500 ft^2 at 3 VA	4500 VA
Two 20-A appliance outlet circuits at 1500 VA each	3000 VA
Laundry circuit	1500 VA
Range (at nameplate rating)	12,000 VA
Water heater	2500 VA
Dishwasher	1200 VA
Clothes dryer	5000 VA
Total	29,700 VA

Application of Demand Factor (*see Section 220-30(b)*)

First 10 kVA of general load at 100%	10,000 VA
Remainder of general load at 40% (19.7 kVA × 0.4)	7880 VA
Total of general load	17,880 VA
9 kVA of heat at 40% (9000 VA × 0.4) =	3600 VA
Total	21,480 VA

Calculated Load for Service Size

$$21,480 \text{ VA} \div 240 \text{ V} = 89.5 \text{ A}$$

The minimum service size would be 100 A.

Feeder Neutral Load, per Section 220-22

1500 ft² at 3 VA	4500 VA
Three 20-A circuits at 1500 VA	4500 VA
Total	9000 VA
3000 VA at 100%	3000 VA
9000 VA − 3000 VA = 6000 VA at 35%	2100 VA
Subtotal	5100 VA
Range: 8 kVA at 70%	5600 VA
Clothes dryer: 5 kVA at 70%	3500 VA
Dishwasher	1200 VA
Total	15,400 VA

Calculated Load for Neutral

$$15{,}400 \text{ VA} \div 240 \text{ V} = 64.2 \text{ A}$$

Example No. D2(b). Optional Calculation for One-Family Dwelling, Air Conditioning Larger than Heating [see Section 220-30]

The dwelling has a floor area of 1500 ft², exclusive of an unfinished cellar not adaptable for future use, unfinished attic, and open porches. It has two 20-A small appliance circuits, one 20-A laundry circuit, two 4-kW wall-mounted ovens, one 5.1-kW counter-mounted cooking unit, a 4.5-kW water heater, a 1.2-kW dishwasher, a 5-kW combination clothes washer and dryer, six 7-A, 230-V room air-conditioning units, and a 1.5-kW permanently installed bathroom space heater. Assume wall-mounted ovens, counter-mounted cooking unit, water heater, dishwasher, and combination clothes washer and dryer kW ratings equivalent to kVA.

Air Conditioning kVA Calculation

$$\text{Total amperes} = 6 \text{ units} \times 7 \text{ A} = 42 \text{ A}$$

$$42 \text{ A} \times 240 \text{ V} \div 1000 = 10.08 \text{ kVA (assume PF} = 1.0)$$

Load Included at 100%

Air Conditioning: Included below (Section 220-30(c)(1))
Space Heater: Omit (Section 220-30(c)(5))

General Load

1500 ft² at 3 VA	4500 VA
Two 20-A small appliance circuits at 1500 VA each	3000 VA
Laundry circuit	1500 VA
Two ovens	8000 VA
One cooking unit	5100 VA
Water heater	4500 VA
Dishwasher	1200 VA
Washer/dryer	5000 VA
Total general load	32,800 VA
First 10 kVA at 100%	10,000 VA
Remainder at 40% (22.8 kVA × 0.4 × 1000)	9120 VA
Subtotal general load	19,120 VA
Air conditioning	10,080 VA
Total	29,200 VA

Calculated Load for Service

$$29{,}200 \text{ VA} \div 240 \text{ V} = 122 \text{ A (service rating)}$$

Feeder Neutral Load, per Section 220-22

Assume that the two 4-kVA wall-mounted ovens are supplied by one branch circuit, the 5.1-kVA counter-mounted cooking unit by a separate circuit.

1500 ft² at 3 VA	4500 VA
Three 20-A circuits at 1500 VA	4500 VA
Subtotal	9000 VA
3000 VA at 100%	3000 VA
9000 VA − 3000 VA = 6000 VA at 35%	2100 VA
Subtotal	5100 VA

Two 4-kVA ovens plus one 5.1-kVA cooking unit = 13.1 kVA. Table 220-19 permits 55% demand factor or 13.1 kVA × 0.55 = 7200 VA feeder capacity.

Subtotal from above	5100 VA
Ovens and cooking unit: 7200 VA × 70% for neutral load	5040 VA
Clothes washer/dryer: 5 kVA × 70% for neutral load	3500 VA
Dishwasher	1200 VA
Total	14,840 VA

Calculated Load for Neutral

$$14{,}840 \text{ VA} \div 240 \text{ V} = 61.83 \text{ A (use 62 A)}$$

Example No. D2(c). Optional Calculation for One-Family Dwelling with Heat Pump (Single-Phase, 240/120-Volt Service)
(see Section 220-30)

The dwelling has a floor area of 2000 ft², exclusive of an unfinished cellar not adaptable for future use, unfinished attic, and open porches. It has a 12-kW range, a 4.5-kW water heater, a 1.2-kW dishwasher, a 5-kW clothes dryer, and a 2½-ton (24-A) heat pump with 15 kW of backup heat.

Heat Pump kVA Calculation

$$24 \text{ A} \times 240 \text{ V} \div 1000 = 5.76 \text{ kVA}$$

This 5.76 kVA is less than 15 kVA of the backup heat; therefore, the heat pump load need not be included in the service calculation (see Section 220-30(c)).

General Load

2000 ft² at 3 VA	6000 VA
Two 20-A appliance outlet circuits at 1500 VA each	3000 VA
Laundry circuit	1500 VA
Range (at nameplate rating)	12,000 VA
Water heater	4500 VA
Dishwasher	1200 VA
Clothes dryer	5000 VA
Subtotal general load	33,200 VA
First 10 kVA of general load at 100%	10,000 VA
Remainder of general load at 40% (23,200 VA × 0.4)	9280 VA
Total net general load	19,280 VA

Heat Pump and Supplementary Heat*

$$240 \text{ V} \times 24 \text{ A} = 5760 \text{ VA}$$

15-kW Electric Heat:

$$5760 \text{ VA} + 15{,}000 \text{ VA} = 20{,}760 \text{ VA} = 20.76 \text{ kVA}$$
$$20.76 \text{ kVA} \times 100\% = 20.76 \text{ kVA}$$

*If supplementary heat is not on at same time as heat pump, heat pump kVA need not be added to total.

Totals

Net general load	19,280 VA
Heat pump and supplementary heat	20,760 VA
Total	40,040 VA

Calculated Load for Service

$$40.04 \text{ kVA} \times 1000 \div 240 \text{ V} = 166.8 \text{ A}$$

This dwelling unit would be permitted to be served by a 175-A service.

Example No. D3. Store Building

A store 50 ft by 60 ft, or 3000 ft², has 30 ft of show window. There are a total of 80 duplex receptacles. The service is a 120/240 V, single phase 3-wire service. Actual connected lighting load is 8500 VA.

Computed Load *(see Section 220-10)*

Noncontinuous Loads:

Receptacle Load *(see Section 220-13)*	
80 receptacles at 180 VA	14,400 VA
10,000 VA at 100%	10,000 VA
14,400 VA − 10,000 VA = 4400 VA at 50%	2,200 VA
Subtotal	12,200 VA

Continuous Loads:

General Lighting*

3000 ft² at 3 VA per ft²	9000 VA
Show Window Lighting	
30 ft at 200 VA per ft	6000 VA
Outside Sign Circuit [see Section 600-5(b)(3)]	1200 VA
Subtotal	16,200 VA
Subtotal from noncontinuous	12,200 VA
Total noncontinuous + continuous loads =	28,400 VA

*In the example, 125% of the actual connected lighting load (8500 VA × 1.25 = 10,625 VA) is less than 125% of the load from Table 220-3(a), so the minimum lighting load from Table 220-3(a) is used in the calculation. Had the actual lighting load been greater than the value computed from Table 220-3(a), 125% of the actual connected lighting load would have been used.

Minimum Number of Branch Circuits Required

General Lighting: Branch circuits need only be installed to supply the actual connected load [see Section 210-11(d)].

$$8500 \text{ VA} \times 1.25 = 10,625 \text{ VA}$$

$$10,625 \text{ VA} \div 240 \text{ V} = 44 \text{ A for 3-wire, 120/240 V}$$

The lighting load would be permitted to be served by 2-wire or 3-wire, 15- or 20-A circuits with combined capacity equal to 44 A or greater for 3-wire circuits or 88 A or greater for 2-wire circuits. The feeder capacity as well as the number of branch-circuit positions available for lighting circuits in the panelboard must reflect the full calculated load of 9000 VA × 1.25 = 11,250 VA.

Show Window

$$6000 \text{ VA} \times 1.25 = 7500 \text{ VA}$$

$$7500 \text{ VA} \div 240 \text{ V} = 31 \text{ A for 3-wire, 120/240 V}$$

The show window lighting is permitted to be served by 2-wire or 3-wire circuits with a capacity equal to 31 A or greater for 3-wire circuits or 62 A or greater for 2-wire circuits.

Receptacles required by Section 210-62 are assumed to be included in the receptacle load above if these receptacles do not supply the show window lighting load.

Receptacles

Receptacle Load: 14,400 VA ÷ 240 V = 60 A for 3-wire, 120/240 V

The receptacle load would be permitted to be served by 2-wire or 3-wire circuits with a capacity equal to 60 A or greater for 3-wire circuits or 120 A or greater for 2-wire circuits.

Minimum Size Feeder (or Service) Overcurrent Protection
[see Sections 215-3 or 230-90(a)]

Subtotal noncontinuous loads	12,200 VA
Subtotal continuous load at 125%	
(16,200 VA × 1.25)	20,250 VA
Total	32,450 VA

$$32,450 \text{ VA} \div 240 \text{ V} = 135 \text{ A}$$

The next higher standard size is 150 A (see Section 240-6).

Minimum Size Feeders (or Service Conductors) Required
[see Sections 215-2, 215-3, and 230-42(a)]

For 120/240-V, 3-wire system,

$$32,450 \text{ VA} \div 240 \text{ V} = 135 \text{ A}$$

Service or feeder conductor is 1/0 Cu per Section 215-3 and Table 310-16 (with 75°C terminations).

Example No. D4(a). Multifamily Dwelling

A multifamily dwelling has 40 dwelling units.

Meters are in two banks of 20 each with individual feeders to each dwelling unit.

One-half of the dwelling units are equipped with electric ranges not exceeding 12 kW each. Assume range kW rating equivalent to kVA rating in accordance with Section 220-19. Other half of ranges are gas ranges.

Area of each dwelling unit is 840 ft².

Laundry facilities on premises are available to all tenants. Add no circuit to individual dwelling unit. Add 1500 VA for each laundry circuit to house load and add to the example as a "house load."

Computed Load for Each Dwelling Unit (see Article 220)

General Lighting: 840 ft² at 3 VA per ft² = 2520 VA
Special Appliance Load: Electric range (see Section 220-19) = 8000 VA

Minimum Number of Branch Circuits Required for Each Dwelling Unit (see Section 220-4)

General Lighting Load: 2520 VA ÷ 120 V = 21 A or two 15-A, 2-wire or two 20-A, 2-wire circuits
Small Appliance Load: Two 2-wire circuits of No. 12 wire [see Section 210-11(c)(1)]
Range Circuit: 8000 VA ÷ 240 V = 33 A. A circuit of two No. 8 conductors and one No. 10 conductor as permitted by Section 220-22 [see Section 210-19(c)]

Minimum Size Feeder Required for Each Dwelling Unit (see Section 215-2)

Computed Load (see Article 220):

General Lighting	2520 VA
Small Appliance (two 20-ampere circuits)	3000 VA
Subtotal Computed Load (without ranges)	5520 VA

Application of Demand Factor (see Table 220-11)

First 3000 VA at 100%	3000 VA
5520 VA – 3000 VA = 2520 VA at 35%	882 VA
Net Computed Load (without ranges)	3882 VA
Range Load	8000 VA
Net Computed Load (with ranges)	11,882 VA

Size of Each Feeder (see Section 215-2)

For 120/240-V, 3-wire system (without ranges)

$$\text{Net computed load of 3882 VA} \div 240 \text{ V} = 16.2 \text{ A}$$

For 120/240-V, 3-wire system (with ranges)

$$\text{Net computed load of 11,882 VA} \div 240 \text{ V} = 49.5 \text{ A}$$

Feeder Neutral

Lighting and Small Appliance	3882 VA
Range: 8000 VA at 70% (see Section 220-22)	
(only for apartments with electric range)	5600 VA
Net Computed Load (neutral)	9482 VA

Calculated Load for Neutral

$$9482 \text{ VA} \div 240 \text{ V} = 39.5 \text{ A}$$

Minimum Size Feeder Required from Service Equipment to Meter Bank (For 20 Dwelling Units — 10 with Ranges)

Total Computed Load:

Lighting and Small Appliance	
20 units × 5520 VA	110,400 VA

Application of Demand Factor

First 3000 VA at 100%	3000 VA
110,400 VA – 3000 VA = 107,400 VA at 35%	37,590 VA
Net Computed Load	40,590 VA
Range: 10 ranges (less than 12 kVA)	
(see Col. A, Table 220-19)	25,000 VA
Net Computed Load (with ranges)	65,590 VA

Net computed load for 120/240-V, 3-wire system,

$$65,590 \text{ VA} \div 240 \text{ V} = 273 \text{ A}$$

Feeder Neutral

Lighting and Small Appliance	40,590 VA
Range: 25,000 VA at 70% (see Section 220-22)	17,500 VA
Net Computed Load (neutral)	58,090 VA

Calculated Load for Neutral

$$58,090 \text{ VA} \div 240 \text{ V} = 242 \text{ A}$$

Further Demand Factor *(Section 220-22)*

200 A at 100%	200 A
242 A – 200 A = 42 A at 70%	29 A
Net Computed Load (neutral)	229 A

Minimum Size Main Feeders (or Service Conductors) Required (Less House Load) (For 40 Dwelling Units — 20 with Ranges)

Total Computed Load:
Lighting and Small Appliance Load

40 units × 5520 VA	220,800 VA

Application of Demand Factor (from Table 220-11)

First 3000 VA at 100%	3000 VA
Next 120,000 VA − 3000 VA =	
117,000 VA at 35%	40,950 VA
Remainder 220,800 VA − 120,000 VA =	
100,800 VA at 25%	25,200 VA
Net Computed Load	69,150 VA
Range: 20 ranges (less than 12 kVA)	
(see Col. A, Table 220-19)	35,000 VA
Net Computed Load	104,150 VA

For 120/240-V, 3-wire system

$$\text{Net computed load of } 104,150 \text{ VA} \div 240 \text{ V} = 434 \text{ A}$$

Feeder Neutral

Lighting and Small Appliance Load	69,150 VA
Range: 35,000 VA at 70%	
(see Section 220-22)	24,500 VA
Computed Load (neutral)	93,650 VA

$$93,650 \text{ VA} \div 240 \text{ V} = 390 \text{ A}$$

Further Demand Factor *(see Section 220-22)*

200 A at 100%	200 A
390 A – 200 A = 190 A at 70%	133 A
Net Computed Load (neutral)	333 A

[See Tables 310-16 through 310-21, and Section 310-15(b)(2) and (b)(4).]

Example No. D4(b). Optional Calculation for Multifamily Dwelling

A multifamily dwelling equipped with electric cooking and space heating or air conditioning has 40 dwelling units.

Meters are in two banks of 20 each with house metering and individual feeders to each dwelling unit.

Each dwelling unit is equipped with an electric range of 8-kW nameplate rating, four 1.5-kW separately controlled 240-V electric space heaters, and a 2.5-kW, 240-V electric water heater. Assume range, space heater, and water heater kW ratings equivalent to kVA.

A common laundry facility is available to all tenants *[see Section 210-52(f), Exception No. 1].*

Area of each dwelling unit is 840 ft^2.

Computed Load for Each Dwelling Unit *(see Article 220)*

General Lighting Load:

840 ft^2 at 3 VA per ft^2	2520 VA

Other loads:

Electric range	8000 VA
Electric heat: 6 kVA	
(or air conditioning if larger)	6000 VA
Electric water heater	2500 VA

Minimum Number of Branch Circuits Required for Each Dwelling Unit

General Lighting: 2520 VA ÷ 120 V = 21 A
Two 15-A, 2-wire, or two 20-A, 2-wire circuits
Small Appliance Load: Two 2-wire circuits of No. 12
[see Section 210-11(c)(1)]

Range Circuit: 8000 VA × 80% ÷ 240 V = 27 A
Three No. 10 conductors as permitted in Column C of Table 220-19
Space Heating: 6000 VA ÷ 240 V = 25 A
Number of circuits *(see Section 210-11)*

Minimum Size Feeder Required for Each Dwelling Unit
(see Section 215-2)

Computed Load *(see Article 220):*

General Lighting	2520 VA
Small Appliance (two 20-A circuits)	3000 VA
Subtotal Computed Load	5520 VA
(without range and space heating)	

Application of Demand Factor

First 3000 VA at 100%	3000 VA
5520 VA – 3000 VA = 2520 VA at 35%	882 VA
Net Computed Load	3882 VA
(without range and space heating)	
Range	6400 VA
Space Heating *(see Section 220-15)*	6000 VA
Water Heater	2500 VA
Net Computed Load	18,782 VA
(for individual dwelling unit)	

Size of Each Feeder

For 120/240-V, 3-wire system,

$$\text{Net computed load of } 18,782 \text{ VA} \div 240 \text{ V} = 78 \text{ A}$$

Feeder Neutral *(see Section 220-22)*

Lighting and Small Appliance	3882 VA
Range: 6400 VA at 70% *(see Section 220-22)*	4480 VA
Water and Space Heating (no neutral): 240 V	0 VA
Net Computed Load (neutral)	8362 VA

Calculated Load for Neutral

$$8362 \text{ VA} \div 240 \text{ V} = 35 \text{ A}$$

Minimum Size Feeder Required from Service Equipment to Meter Bank (for 20 Dwelling Units)

Total Computed Load:

Lighting and Small Appliance Load	
20 units × 5520 VA	110,400 VA
Water and Space Heating Load	
20 units × 8500 VA	170,000 VA
Range: 20 × 8000 VA	160,000 VA
Net Computed Load	440,400 VA
(20 dwelling units)	
Net Computed Load Using Optional Calculation	
(see Table 220-32)	
440,400 VA × 0.38	167,352 VA

$$167,352 \text{ VA} \div 240 \text{ V} = 697 \text{ A}$$

Minimum Size Main Feeder Required (Less House Load) (For 40 Dwelling Units)

Computed Load:

Lighting and Small Appliance Load	
40 units × 5520 VA	220,800 VA
Water and Space Heating Load	
40 units × 8500 VA	340,000 VA
Range: 40 ranges × 8000 VA	320,000 VA
Net Computed	880,800 VA
(40 dwelling units)	

Net Computed Load Using Optional Calculation *(see Table 220-32)*

$$880,800 \text{ VA} \times 0.28 = 246,624 \text{ VA}$$

$$246,624 \text{ VA} \div 240 \text{ V} = 1028 \text{ A}$$

Feeder Neutral Load for Feeder from Service Equipment to Meter Bank (For 20 Dwelling Units)

Lighting and Small Appliance Load

20 units × 5520 VA	110,400 VA
First 3000 VA at 100%	3000 VA
110,400 VA − 3000 VA = 107,400 VA at 35%	37,590 VA
Net Computed Load	40,590 VA
20 ranges: 35,000 VA at 70%	
(see Table 220-19 and Section 220-22)	24,500 VA
Total	65,090 VA

$$65,090 \text{ VA} \div 240 \text{ V} = 271 \text{ A}$$

Further Demand Factor (see Section 220-22)

First 200 A at 100%	200 A
Balance: 271 A − 200 A = 71 A at 70%	50 A
Total	250 amperes

Feeder Neutral Load for Main Feeder (Less House Load) (For 40 Dwelling Units)

Lighting and Small Appliance Load

40 units × 5520 VA	220,800 VA
First 3000 VA at 100%	3000 VA
Next 120,000 VA − 3000 VA = 117,000 VA at 35%	40,950 VA
Remainder 220,800 VA − 120,000 VA = 100,800 VA at 25%	25,200 VA
Net Computed Load	69,150 VA
40 ranges: 55,000 VA at 70%	38,500 VA
(see Table 220-19 and Section 220-22)	
Total	107,650 VA

$$107,650 \text{ VA} \div 240 \text{ V} = 449 \text{ A}$$

Further Demand Factor (see Section 220-22)

First 200 A at 100%	200 A
Balance: 449 A − 200 A = 249 A at 70%	174 A
Total	374 A

Example No. D5(a). Multifamily Dwelling Served at 208Y/120 Volts, Three Phase

All conditions and calculations are the same as for the multifamily dwelling of Example No. D4(a) served at 120/240 V, single phase except service to each dwelling unit would be two phase legs and neutral.

Minimum Number of Branch Circuits Required for Each Dwelling Unit (see Section 210-11)

Range Circuit:

$$8000 \text{ VA} \div 208 \text{ V} = 38 \text{ A}$$

A circuit of two No. 8 conductors and one No. 10 conductor as permitted by Section 210-19(c)

Minimum Size Feeder Required for Each Dwelling Unit (see Section 215-2)

For 120/208-V, 3-wire system (without ranges),
Net Computed Load of 3882 VA ÷ 2 legs ÷ 120 V/leg = 16.2 A
For 120/208-V, 3-wire system (with ranges),

Net Computed Load (range) of 8000 VA ÷ 208 V = 38.5 A
Total Load (range + lighting) = 38.5 A + 16.2 A = 54.7 A

Feeder Neutral: (range) of 8000 VA × 70% =
5600 VA ÷ 208 V = 26.9 A

Total Load: (range + lighting) = 26.9 A + 16.2 A = 43.1 A

Minimum Size Feeder Required from Service Equipment to Meter Bank (for 20 Dwelling Units — 10 with Ranges)

For 208Y/120-V, 3-phase, 4-wire system,
Ranges:
Maximum number between any two phase legs = 4
2 × 4 = 8.

Table 220-19 demand = 23,000 VA
Per phase demand = 23,000 VA ÷ 2 = 11,500 VA
Equivalent 3-phase load = 34,500 VA
Net Computed Load (total):

$$40,590 \text{ VA} + 34,500 \text{ VA} = 75,090 \text{ VA}$$
$$75,090 \text{ VA} \div (208 \text{ V})(1.732) = 208.4 \text{ A}$$

Feeder Neutral Size:

Net Computed Lighting and Appliance Load & Equivalent Range Load

$$40,590 \text{ VA} + (34,500 \text{ VA} \times 70\%) = 64,740 \text{ VA}$$

Net Computed Neutral Load:

$$64,740 \text{ VA} \div (208 \text{ V})(1.732) = 179.6 \text{ A}$$

Minimum Size Main Feeder Required (Less House Load) (For 40 Dwelling Units — 20 with Ranges)

For 208Y/120-V, 3-phase, 4-wire system,
Ranges:
Maximum number between any two phase legs = 7
2 × 7 = 14.
Table 220-19 demand = 29,000 VA
Per phase demand = 29,000 VA ÷ 2 = 14,500 VA
Equivalent 3-phase load = 43,500 VA
Net Computed Lighting and Appliance Load & Equivalent Range Load:

$$69,150 \text{ VA} + 43,500 \text{ VA} = 112,650 \text{ VA}$$
$$112,650 \text{ VA} \div (208 \text{ V})(1.732) = 312.7 \text{ A}$$

Main Feeder Neutral Size:

$$69,150 \text{ VA} + (43,500 \text{ VA} \times 70\%) = 99,600 \text{ VA}$$
$$99,600 \text{ VA} \div (208 \text{ V})(1.732) = 276.5 \text{ A}$$

Further Demand Factor (see Section 220-22)

200 A at 100%	200.0 A
276.5 A − 200 A = 76.5 A at 70%	53.6 A
Net Computed Load (neutral)	253.6 A

Example No. D5(b). Optional Calculation for Multifamily Dwelling Served at 208Y/120 Volts, Three Phase

All conditions and calculations are the same as for Optional Calculation for the multifamily dwelling of Example No. D4(b) served at 120/240 V, single phase except service to each dwelling unit would be two phase legs and neutral.

Minimum Number of Branch Circuits Required for Each Dwelling Unit (see Section 210-11)

Range Circuit:

$$8000 \text{ VA} \times 80\% \div 208 \text{ V} = 30.7 \text{ A}$$

A circuit of two No. 8 conductors and one No. 10 conductor as permitted by Section 210-19(c)
Space Heating: 6000 VA ÷ 208 V = 28.8 A
Two 20-ampere, 2-pole circuits using No. 12 conductors

Minimum Size Feeder Required for Each Dwelling Unit

120/208-V, 3-wire circuit,
Net computed load of 18,782 VA ÷ 208 V = 90.3 A
Net computed lighting load (line to neutral):

$$3882 \text{ VA} \div 2 \text{ legs} \div 120 \text{ V per leg} = 16.2 \text{ amperes}$$

Line to line = 14,900 VA ÷ 208 V = 71.6 A
Total Load = 16.2 A + 71.6 A = 87.8 A

Minimum Size Feeder Required from Service Equipment to Meter Bank (For 20 Dwelling Units)

Net Computed Load

$$167,352 \text{ VA} \div (208 \text{ V})(1.732) = 464.9 \text{ A}$$

Feeder Neutral Load:

$$65,080 \text{ VA} \div (208 \text{ V})(1.732) = 180.65 \text{ A}$$

Minimum Size Main Feeder Required (Less House Load) (For 40 Dwelling Units)

Net computed load:

$$246,624 \text{ VA} \div (208 \text{ V})(1.732) = 684.6 \text{ A}$$

Main Feeder Neutral Load:

$$107,650 \text{ VA} \div (208 \text{ V})(1.732) = 298.8 \text{ A}$$

Further Demand Factor *(see Section 220-22)*

200 A at 100%	200.0 A
298.8 A – 200 A = 98.8 A at 70%	69.2 A
Net Computed Load (neutral)	269.2 A

Example No. D6. Maximum Demand for Range Loads

Table 220-19, Column A applies to ranges not over 12 kW. The application of Note 1 to ranges over 12 kW (and not over 27 kW) and Note 2 to ranges over 8¾ kW (and not over 27 kW) is illustrated in the following two examples:

Ranges All the Same Rating *(see Table 220-19, Note 1)*

Assume 24 ranges, each rated 16 kW.

From Table 220-19 Column A, the maximum demand for 24 ranges of 12-kW rating is 39 kW.

16 kW exceeds 12 kW by 4.

5% × 4 = 20% (5% increase for each kW in excess of 12)

39 kW × 20% = 7.8 kW increase

39 kW + 7.8 kW = 46.8 kW (value to be used in selection of feeders)

Ranges of Unequal Rating *(see Table 220-19, Note 2)*

Assume 5 ranges, each rated 11 kW; 2 ranges, each rated 12 kW; 20 ranges, each rated 13.5 kW; 3 ranges, each rated 18 kW.

5 ranges × 12 kW = 60 kW (use 12 kW for range rated less than 12)	
2 ranges × 12 kW = 24 kW	
20 ranges × 13.5 kW = 270 kW	
3 ranges × 18 kW = 54 kW	
30 ranges total kW = 408 kW	

408 kW ÷ 30 ranges = 13.6 kW (average to be used for computation)

From Table 220-19 Column A, the demand for 30 ranges of 12-kW rating is 15 kW + (1 kW × 30 ranges) = 45 kW. 13.6 kW exceeds 12 kW by 1.6 kW (use 2 kW).

5% × 2 = 10% (5% increase for each kW in excess of 12 kW)

45 kW × 10% = 4.5 kW increase

45 kW + 4.5 kW = 49.5 kW (value to be used in selection of feeders)

Example No. D8. Motor Circuit Conductors, Overload Protection, and Short-Circuit and Ground-Fault Protection
(see Sections 240-6, 430-6, 430-22, 430-23, 430-24, 430-32, 430-34, 430-52, and 430-62, Tables 430-150 and 430-152)

Determine the minimum required conductor ampacity, the motor overload protection, the branch-circuit short-circuit and ground-fault protection, and the feeder protection, for three induction-type motors on a 480-V, 3-phase feeder, as follows:

(a) One 25-hp, 460-V, 3-phase, squirrel-cage motor, nameplate full-load current 32 A, Design B, Service Factor 1.15

(b) Two 30-hp, 460-V, 3-phase, wound-rotor motors, nameplate primary full-load current 38 A, nameplate secondary full-load current 65 A, 40°C rise.

Conductor Ampacity

The full-load current value used to determine the minimum required conductor ampacity is obtained from Table 430-150 *[see Section 430-6(a)]* for the squirrel-cage motor and the primary of the wound-rotor motors. To obtain the minimum required conductor ampacity, the full-load current is multiplied by 1.25 *[see Sections 430-22 and 430-23(a)]*.

For the 25-hp motor,

$$32 \text{ A} \times 1.25 = 40 \text{ A}$$

For the 30-hp motors,

$$38 \text{ A} \times 1.25 = 47.5 \text{ A}$$
$$65 \text{ A} \times 1.25 = 81.25 \text{ A}$$

Motor Overload Protection

Where protected by a separate overload device, the motors are required to have overload protection rated or set to trip at not more than 125% of the nameplate full-load current *[see Sections 430-6(a) and 430-32(a)(1)]*.

For the 25-hp motor,

$$32 \text{ A} \times 1.25 = 40.0 \text{ A}$$

For the 30-hp motors,

$$38 \text{ A} \times 1.25 = 47.5 \text{ A}$$

Where the separate overload device is an overload relay (not a fuse or circuit breaker), and the overload device selected at 125% is not sufficient to start the motor or carry the load, the trip setting is permitted to be increased in accordance with Section 430-34.

Branch-Circuit Short-Circuit and Ground-Fault Protection

The selection of the rating of the protective device depends on the type of protective device selected, in accordance with Section 430-52 and Table 430-152. The following is for the 25-hp motor.

Nontime-Delay Fuse: The fuse rating is 300% × 32 A = 96 A. The next larger standard fuse is 100 A *[see Sections 240-6 and 430-52(c)(1), Exception No. 1]*. If the motor will not start with a 100-A nontime-delay fuse, the fuse rating is permitted to be increased to 125 A because this rating does not exceed 400% *[see Section 430-52(c)(1), Exception No. 2a]*.

Time-Delay Fuse: The fuse rating is 175% × 32 A = 56 A. The next larger standard fuse is 60 A *[see Sections 240-6 and 430-52(c)(1), Exception No. 1]*. If the motor will not start with a 60-A time-delay fuse, the fuse rating is permitted to be increased to 70 A because this rating does not exceed 225% *[see Section 430-52(c)(1), Exception No. 2b]*.

Feeder Short-Circuit and Ground-Fault Protection

The rating of the feeder protective device is based on the sum of the largest branch-circuit protective device (example is 100 A) plus the sum of the full-load currents of the other motors, or 100 A + 40 A + 40 A = 180 A. The nearest standard fuse that does not exceed this value is 175 A *[see Sections 240-6 and 430-62(a)]*.

Example No. D9. Feeder Ampacity Determination for Generator Field Control
[see Sections 220-10, 430-24, 430-24 Exception No. 1, 620-13, 620-14, 620-61, and Tables 430-22(b) and 620-14]

Determine the conductor ampacity for a 460-V 3-phase, 60-Hz ac feeder supplying a group of six elevators. The 460-V ac drive motor nameplate rating of the largest MG set for one elevator is 40 hp and 52 A, and the remaining elevators each have a 30-hp, 40-A, ac drive motor rating for their MG sets. In addition to a motor controller, each elevator has a separate motion/operation controller rated 10 A continuous to operate microprocessors, relays, power supplies, and the elevator car door operator. The MG sets are rated continuous.

Conductor Ampacity

Conductor ampacity is determined as follows:

(a) Per Sections 620-13(d) and 620-61(b)(1), use Table 430-22(b) for intermittent duty (elevators). For intermittent duty using a continuous rated motor, the percentage of nameplate current rating to be used is 140%.

(b) For the 30-hp ac drive motor,

$$140\% \times 40 \text{ A} = 56 \text{ A}$$

For the 40-hp ac drive motor,

$$140\% \times 52 \text{ A} = 73 \text{ A}$$

(c) The total conductor ampacity is the sum of all the motor currents.

$$(1 \text{ motor} \times 73 \text{ A}) + (5 \text{ motors} \times 56 \text{ A}) = 353 \text{ A}$$

(d) Per Section 620-14 and Table 620-14, the conductor (feeder) ampacity would be permitted to be reduced by the use of a demand factor. Constant loads are not included *(see Section 620-14, FPN)*. For six elevators, the demand factor is 0.79.
Therefore, feeder diverse ampacity = 0.79 × 353 A = 279 A.

(e) Per Sections 430-24 and 215-3

Controller continuous current = 125% × 10 A = 12.5 A

(f) The total feeder ampacity is the sum of the diverse current and all the controller continuous current.

$$I_{total} = 279 \text{ A} + (6 \text{ elevators} \times 12.5 \text{ A}) = 354 \text{ A}$$

(g) This ampacity would be permitted to be used to select the wire size. See Figure D9.

elevator car door operator. Each transformer is rated 95 kVA with an efficiency of 90%.

Conductor Ampacity. Conductor ampacity is determined as follows:

(a) Calculate the nameplate rating of the transformer:

$$I = \frac{95 \text{ kVA} \times 1000}{\sqrt{3} \times 460 \text{ V} \times 0.90 \text{ eff.}} = 133 \text{ A}$$

(b) Per Section 620-13(d) for six elevators, the total conductor ampacity is the sum of all the currents.

$$6 \text{ elevators} \times 133 \text{ A} = 798 \text{ A}$$

(c) Per Section 620-14 and Table 620-14, the conductor (feeder) ampacity would be permitted to be reduced by the use of a demand factor. Constant loads are not included *(see Section 620-13, FPN No. 2)*. For six elevators, the demand factor is 0.79.
Therefore, feeder diverse ampacity = 0.79 × 798 A = 630 A

(d) Per Sections 430-24 and 215-3,

Controller continuous current = 125% × 10 A = 12.5 A

(e) The total feeder ampacity is the sum of the diverse current and all the controller constant current.

$$I_{total} = 630 \text{ A} + (6 \text{ elevators} \times 12.5 \text{ A}) = 705 \text{ A}$$

(f) This ampacity would be permitted to be used to select the wire size. See Figure D10.

Figure D9.

Example No. D10, Feeder Ampacity Determination for Adjustable Speed Drive Control
[see Sections 215-3, 430-24, 430-24(b), 620-13, 620-14, 620-61, and Tables 430-22(b) and 620-14]

Determine the conductor ampacity for a 460-V, 3-phase, 60-Hz ac feeder supplying a group of six identical elevators. The system is adjustable-speed SCR dc drive. The power transformers are external to the drive (motor controller) cabinet. Each elevator has a separate motion/operation controller connected to the load side of the main line disconnect switch rated 10 A continuous to operate microprocessors, relays, power supplies, and the

Figure D10.

Example No. D11. Mobile Home

A mobile home floor is 70 ft by 10 ft and has two small appliance circuits, a 1000-VA, 240-V heater, a 200-VA, 120-V exhaust fan, a 400-VA, 120-V dishwasher, and a 7000-VA electric range.

Lighting and Small Appliance Load

Lighting (70 ft × 10 ft × 3 VA per ft^2)	2100 VA
Small appliance (1500 VA × 2 circuits)	3000 VA
Laundry (1500 VA × 1 curcuit)	1500 VA
Subtotal	6600 VA
First 3000 VA at 100%	3000 VA
Remainder (6600 VA − 3000 VA = 3600 VA) × 35%	1260 VA
Total	4260 VA

4260 VA ÷ 240 V = 17.75 A per leg

Amperes per Leg

	Leg A	Leg B
Lighting and appliances	17.75	17.75
Heater (1000 VA ÷ 240 V)	4.20	4.20
Fan (200 VA × 125% ÷ 120 V)	2.08	—
Dishwasher (400 VA ÷ 120 V)	—	3.30
Range (7000 VA × 0.8 ÷ 240 V)	23.30	23.30
Total amperes per leg	47.33	48.55

Based on the higher current calculated for either leg, a minimum 50-A supply cord would be required.

For SI units, 1 ft^2 = 0.093 m^2 and 1 ft = 0.3048 m

Example No. D12. Park Trailer

A park trailer floor is 40 ft by 10 ft and has two small appliance circuits, a 1000-VA, 240-V heater, a 200-VA, 120-V exhaust fan, a 400-VA, 120-V dishwasher, and a 7000-VA electric range.

Lighting and Small Appliance Load

Lighting (40 ft × 10 ft × 3 VA per ft^2)	1200 VA
Small appliance (1500 VA × 2 circuits)	3000 VA
Laundry (1500 VA × 1 circuit)	1500 VA
Subtotal	5700 VA
First 3000 VA at 100%	3000 VA
Remainder (5700 VA − 3000 VA = 2700 VA) × 35%	945 VA
Total	3945 VA

3945 VA ÷ 240 V = 16.44 A per leg

Amperes per Leg

	Leg A	Leg B
Lighting and appliances	16.44	16.44
Heater (1000 VA ÷ 240 V)	4.20	4.20
Fan (200 VA × 125% ÷ 120 V)	2.08	—
Dishwasher (400 VA ÷ 120 V)	—	3.3
Range (7000 VA × 0.8 ÷ 240 V)	23.30	23.30
Total amperes per leg	46.02	47.24

Based on the higher current calculated for either leg, a minimum 50-A supply cord would be required.

For SI units, 1 ft^2 = 0.093 m^2 and 1 ft = 0.3048 m

APPENDIX E
Article 250 Cross Reference

This appendix is not a part of the recommendations of this NFPA document but is included for informational purposes only.

E1. Cross Reference from the 1999 NEC to the 1996 NEC

Topic	1999 NEC	1996 NEC	Topic	1999 NEC	1996 NEC
Part A. General			Underground Service Cable or Conduit	−84	−55
Scope	−1	−1	Other Conductor Enclosures and Raceways	−86	−33
General Requirements for Grounding and Bonding	−2	−51, −1 FPN 1 & 2	**Part E. Bonding**		
Application of Other Articles	−4	−2	General	−90	−70
Objectionable Current Over Grounding Conductors	−6	−21	Services	−92	−71
Connection of Grounding and Bonding Equipment	−8	−113	Method of Bonding at the Service	−94	−72
Protection of Ground Clamps and Fittings	−10	−117	Bonding Other Enclosures	−96	−75
Clean Surfaces	−12	−118	Bonding for Over 250 Volts	−97	−76
Part B. Circuit and System Grounding			Bonding Loosely Jointed Metal Raceways	−98	−77
Alternating-Current Circuits and Systems to Be Grounded	−20	−5	Bonding in Hazardous Classified) Locations	−100	−78
Alternating-Current Systems of 50 Volts to 1000 Volts Not Required to be Grounded	−21	−5	Equipment Bonding Jumpers	−102	−79
Circuits Not to Be Grounded	−22	−7	Bonding of Piping Systems and Exposed Structural Steel	−104	−80
Grounding Service-Supplied Alternating-Current Systems	−24	−23, −53(a)	Lightning Protection Systems	−106	−46
Conductor to Be Grounded — Alternating-Current Systems	−26	−25	**Part F. Equipment Grounding and Equipment Grounding Conductors**		
Main Bonding Jumpers	−28	−79, −53(b)	Equipment Fastened in Place or Connected by Permanent Wiring Methods (Fixed)	−110	−42
Grounding Separately Derived Alternating-Current Systems	−30	−26	Fastened in Place or Connected by Permanent Wiring Methods (Fixed) — Specific	−112	−43
Two or More Buildings or Structures Supplied from a Common Service	−32	−24	Equipment Connected by Cord and Plug	−114	−45
Portable and Vehicle-Mounted Generators	−34	−6	Nonelectric Equipment	−116	−44
High-Impedance Grounded Neutral Systems	−36	−27	Types of Equipment Grounding Conductors	−118	−91(b)
Part C. Grounding Electrode System and Grounding Electrode Conductor			Identification of Equipment Grounding Conductors	−119	−57(b)
Grounding Electrode System	−50	−81	Equipment Grounding Conductor Installation	−120	−92(c)
Made and Other Electrodes	−52	−83	Size of Equipment Grounding Conductors	−122	−95
Supplementary Grounding Electrodes	−54	−91(c)	Equipment Grounding Conductor Continuity	−124	−99
Resistance of Made Electrodes	−56	−84	Identification of Wiring Device Terminals	−126	−119
Common Grounding Electrode	−58	−54	**Part G. Methods of Equipment Grounding**		
Use of Air Terminals	−60	−86	Equipment Grounding Conductor Connections	−130	−50
Grounding Electrode Conductor Material	−62	−91(a)	Short Sections of Raceway	−132	−56
Grounding Electrode Conductor Installation	−64	−91(a), −92	Equipment Fastened in Place or Connected by Permanent Wiring Methods (Fixed) — Grounding	−134	−57
Size of Alternating-Current Grounding Electrode Conductor	−66	−94	Equipment Considered Effectively Grounded	−136	−58
Grounding Electrode Conductor Connection to Electrodes	−68	−112	Cord- and Plug-Connected Equipment	−138	−59
Methods of Grounding Conductor Connection to Electrodes	−70	−115	Frames of Ranges and Clothes Dryers	−140	−60
Part D. Enclosure, Raceway, and Service Cable Grounding					
Service Raceways and Enclosures	−80	−32			

Topic	1999 NEC	1996 NEC
Use of Grounded Circuit Conductor for Grounding Equipment	–142	–61
Multiple Circuit Connections	–144	–62
Connecting Receptacle Grounding Terminal to Box	–146	–74
Continuity and Attachment of Equipment Grounding Conductors to Boxes	–148	–114
Part H. Direct-Current Systems		
General	–160	(new)
Direct-Current Circuits and Systems to Be Grounded	–162	–3
Point of Connection for Direct-Current Systems	–164	–22
Size of Direct-Current Grounding Electrode Conductor	–166	–93
Direct-Current Bonding Jumper	–168	–79(d)
Ungrounded Direct-Current Separately Derived Systems	–169	(new)

Topic	1999 NEC	1996 NEC
Part J. Instruments, Meters, and Relays		
Instrument Transformer Circuits	–170	–121
Instrument Transformer Cases	–172	–122
Cases of Instruments, Meters, and Relays Operating at Less than 1000 Volts	–174	–123
Cases of Instruments, Meters, and Relays Operating Voltage 1 kV and Over	–176	–124
Instrument Grounding Conductor	–178	–125
Part K. Grounding of Systems and Circuits of 1 kV and Over (High Voltage)		
General	–180	–150
Derived Neutral Systems	–182	–151
Solidly Grounded Neutral Systems	–184	–152
Impedance Grounded Neutral Systems	–186	–153
Grounding of Systems Supplying Portable or Mobile Equipment	–188	–154
Grounding of Equipment	–190	–155

E2. Cross Reference from the 1996 NEC to the 1999 NEC

Topic	1996 NEC	1999 NEC
Part A. General		
Scope	–1	–1
Application of Other Articles	–2	–4
Part B. Circuit and System Grounding		
Direct-Current Systems	–3	–162
Alternating-Current Circuits and Systems to Be Grounded	–5	–20, –21
Portable and Vehicle-Mounted Generators	–6	–34
Circuits Not to Be Grounded	–7	–22
Part C. Location of System Grounding Connections		
Objectionable Current Over Grounding Conductors	–21	–6
Point of Connection for Direct-Current Systems	–22	–164
Grounding Service-Supplied Alternating-Current Systems	–23	–24
Two or More Buildings or Structures Supplied from a Common Service	–24	–32
Conductor to Be Grounded — Alternating-Current Systems	–25	–26
Grounding Separately Derived Alternating-Current Systems	–26	–30
High-Impedance Grounded Neutral System Connections	–27	–36
Part D. Enclosure and Raceway Grounding		
Service Raceways and Enclosures	–32	–80
Other Conductor Enclosures and Raceways	–33	–86
Part E. Equipment Grounding		
Equipment Fastened in Place or Connected by Permanent Wiring Methods (Fixed)	–42	–110

Topic	1996 NEC	1999 NEC
Fastened in Place or Connected by Permanent Wiring Methods (Fixed) Specific	–43	–112
Nonelectric Equipment	–44	–116
Equipment Connected by Cord and Plug	–45	–114
Spacing from Lightning Rods	–46	–106
Part F. Methods of Grounding		
Equipment Grounding Conductor Connections	–50	–130
Effective Grounding Path	–51	–2
Grounding Path to Grounding Electrode at Services	–53	–24, –28
Common Grounding Electrode	–54	–58
Underground Service Cable	–55	–84
Short Sections of Raceway	–56	–132
Equipment Fastened in Place or Connected by Permanent Wiring Methods (Fixed) — Grounding	–57	–119, –134
Equipment Considered Effectively Grounded	–58	–136
Cord- and Plug-Connected Equipment	–59	–138
Frames of Ranges and Clothes Dryers	–60	–140
Use of Grounded Circuit Conductor for Grounding Equipment	–61	–142
Multiple Circuit Connections	–62	–144
Part G. Bonding		
General	–70	–90
Service Equipment	–71	–92
Method of Bonding Service Equipment	–72	–94
Metal Armor or Tape of Service Cable	–73	(removed)

Topic	1996 NEC	1999 NEC
Connecting Receptacle Grounding Terminal to Box	−74	−146
Bonding Other Enclosures	−75	−96
Bonding for Over 250 Volts	−76	−97
Bonding Loosely Jointed Metal Raceways	−77	−98
Bonding in Hazardous (Classified) Locations	−78	−100
Main and Equipment Bonding Jumpers	−79	−28, −102, −168
Bonding of Piping Systems and Exposed Structural Steel	−80	−104

Part H. Grounding Electrode System

Topic	1996 NEC	1999 NEC
Grounding Electrode System	−81	−50
Made and Other Electrodes	−83	−52
Resistance of Made Electrodes	−84	−56
Use of Lightning Rods	−86	−60

Part J. Grounding Conductors

Topic	1996 NEC	1999 NEC
Material	−91	−54, −62, −64, −118
Installation	−92	−64, −120
Size of Direct-Current Grounding Electrode Conductor	−93	−166
Size of Alternating-Current Grounding Electrode Conductor	−94	−66
Size of Equipment Grounding Conductors	−95	−122
Outline Lighting	−97	(600-7)
Equipment Grounding Conductor Continuity	−99	−124

Part K. Grounding Conductor Connections

Topic	1996 NEC	1999 NEC
To Grounding Electrode	−112	−68
To Conductors and Equipment	−113	−8

Topic	1996 NEC	1999 NEC
Continuity and Attachment of Equipment Grounding Conductors to Boxes	−114	−148
Connection to Electrodes	−115	−70
Protection of Attachment	−117	−10
Clean Surfaces	−118	−12
Identification of Wiring Device Terminals	−119	−126

Part L. Instrument Transformers, Relays, Etc.

Topic	1996 NEC	1999 NEC
Instrument Transformer Circuits	−121	−170
Instrument Transformer Cases	−122	−172
Cases of Instruments, Meters, and Relays — Operating at Less than 1000 Volts	−123	−174
Cases of Instruments, Meters, and Relays — Operating Voltage 1 kV and Over	−124	−176
Instrument Grounding Conductor	−125	−178

Part M. Grounding of Systems and Circuits of 1 kV and Over (High Voltage)

Topic	1996 NEC	1999 NEC
General	−150	−180
Derived Neutral Systems	−151	−182
Solidly Grounded Neutral Systems	−152	−184
Impedance Grounded Neutral Systems	−153	−186
Grounding of Systems Supplying Portable or Mobile Equipment	−154	−188
Grounding of Equipment	−155	−190

Index

2002 NATIONAL ELECTRICAL CODE SCHEDULE
(2001 ANNUAL MEETING)

Date	Event
Nov. 5, 1999 (5:00 p.m. EST)	Receipt of Proposals
Jan. 10–22, 2000	Code-Making Panel Meetings
May 15–19, 2000	Correlating Committee Meeting
July 14, 2000	NEC-ROP to Mailing House
Oct. 27, 2000 (5:00 p.m. EST)	Closing Date for Comments
Dec. 4–16, 2000	Code-Making Panel Meetings (ROC)
March 12–16, 2001	Correlating Committee Meeting
April 16, 2001	NEC-ROC to Mailing House
May 20–24, 2001	NFPA Annual Meeting (San Diego, CA)
July 22, 2001	Standards Council Issuance

Anyone may submit proposals to amend the 1999 *Code*. A sample form for this purpose may be obtained from the Secretary of the Standards Council at NFPA headquarters, and a copy is included in this *Code*.

Method of Submitting a Proposal to Revise the *National Electrical Code*®

The following is based on the NFPA Regulations Governing Committee Projects, adopted by the Board of Directors on December 3, 1977 (last amended on November 16, 1997).

A proposal to revise the 1999 edition of the *National Electrical Code* must be submitted so that the proposal is received at NFPA headquarters by November 5, 1999, as indicated in the time schedule for the 2002 *National Electrical Code*. A proposal received after this date will be returned to the submitter. The proposal is to be sent to the Secretary of the Standards Council at NFPA Headquarters, 1 Batterymarch Park, P.O. Box 9101, Quincy, MA 02269-9101.

Each proposal must include the following:

1. Identification of the submitter (the person's name) and his or her affiliation (i.e., committee, organization, company), where appropriate
2. An indication that the proposal is for revision of the 1999 *National Electrical Code* and identification of the specific section number, table number (or equivalent identification) of the section, etc., to be revised
3. A statement of the problem and substantiation for proposal
4. The proposed text of the proposal, including the wording to be added, revised (and how revised), or deleted

Proposals that do not include all of the above information may not be acted on by the National Electrical Code Committee.

It is preferred that the forms available from NFPA for submittal of proposals be used. A separate proposal form should be used for revision of each section of the *Code*.

A proposal form appears on the following page.

The NFPA Codes and Standards Development Process

Since 1896, one of the primary purposes of the NFPA has been to develop and update the standards covering all areas of fire safety.

Calls for Proposals

The code adoption process takes place twice each year and begins with a call for proposals from the public to amend existing codes and standards or to develop the content of new fire safety documents.

Report on Proposals

Upon receipt of public proposals, the technical committee members meet to review, consider, and act on the proposals. The public proposals – together with the committee action on each proposal and committee-generated proposals – are published in the NFPA's Report on Proposals (ROP). The ROP is then subject to public review and comment.

Report on Comments

These public comments are considered and acted upon by the appropriate technical committees. All public comments – together with the committee action on each comment – are published as the Committee's supplementary report in the NFPA's Report on Comments (ROC).

The committee's report and supplementary report are then presented for adoption and open debate at either of NFPA's semi-annual meetings held throughout the United States and Canada.

Association Action

The Association meeting may, subject to review and issuance by the NFPA Standards Council, (a) adopt a report as published, (b) adopt a report as amended, contingent upon subsequent approval by the committee, (c) return a report to committee for further study, and (d) return a portion of a report to committee.

Standards Council Action

The Standards Council will make a judgement on whether or not to issue an NFPA document based upon the entire record before the Council, including the vote taken at the Association meeting on the technical committee's report.

Voting Procedures

Voting at an NFPA Annual or Fall Meeting is restricted to members of record for 180 days prior to the opening of the first general session of the meeting, except that individuals who join the Association at an Annual or Fall Meeting are entitled to vote at the next Fall or Annual Meeting.

"Members" are defined by Article 3.2 of the Bylaws as individuals, firms, corporations, trade or professional associations, institutes, fire departments, fire brigades, and other public or private agencies desiring to advance the purposes of the Association. Each member shall have one vote in the affairs of the Association. Under Article 4.5 of the Bylaws, the vote of such a member shall be cast by that member individually or by an employee designated in writing by the member of record who has registered for the meeting. Such a designated person shall not be eligible to represent more than one voting privilege on each issue, nor cast more than one vote on each issue.

Any member who wishes to designate an employee to cast that member's vote at an Association meeting in place of that member must provide that employee with written authorization to represent the member at the meeting. The authorization must be on company letterhead signed by the member of record, with the membership number indicated, and the authorization must be recorded with the President of NFPA or his designee before the start of the opening general session of the Meeting. That employee, irrespective of his or her own personal membership status, shall be privileged to cast only one vote on each issue before the Association.